BASIC ELECTRICAL ENGINEERING

By the Same Author :

- A Text Book of Electrical Machines
- Direct Current Machines
- Alternating Current Machines
- A Text Book of Power Plant Engineering
- Electrical Engineering Materials
- Elements of Electrical Engineering and Electronics
- Thermal Engineering
- A Text Book of Engineering Thermodynamics
- Thermal Science and Engineering
- Engineering Mechanics

BASIC ELECTRICAL ENGINEERING

As per Syllabus of
Chhattisgarh Swami Vivekanand Technical University Bhilai
BPUT, Odisha
&
Other Courses of B.E., B. Tech. (Electrical) and B.Sc. (Engg.)

For
First Year B.E. Students

By

R.K. RAJPUT
M.E. (Hons.), (Gold Medallist); Grad. (Mech. Engg. & Elect. Engg.)
M.I.E. (India); M.S.E.S.I.; M.I.S.T.E.; C.E. (India)
PATIALA (***Punjab***)

LAXMI PUBLICATIONS (P) LTD
An ISO 9001:2015 Company
BENGALURU • CHENNAI • GUWAHATI • HYDERABAD
KOCHI • KOLKATA • LUCKNOW • RANCHI
NEW DELHI

BASIC ELECTRICAL ENGINEERING

Printed and bound in India

Typeset at : Goswami Printers, Delhi.

First Edition : 2005, Reprint : 2008, 2012, 2015, 2018, 2019, 2021; Edition: 2026

EEE-0529-899-ELECTRICAL ENGINEERING-(CG)

ISBN : 978-81-3180-636-4

Price: ₹ 899.00

Branches

©	Bengaluru	080-26 75 69 30
©	Chennai	044-24 34 47 26
©	Guwahati	0361-254 36 69
©	Hyderabad	040-27 55 53 83
©	Kochi	0484-405 13 03
©	Kolkata	033-40 04 77 79
©	Lucknow	0522-430 36 13
©	Ranchi	0651-224 24 64

PUBLISHED IN INDIA BY

LAXMI PUBLICATIONS (P) LTD

An ISO 9001:2015 Company
113, GOLDEN HOUSE, GURUDWARA ROAD, DARYAGANJ,
NEW DELHI - 110002, INDIA

Corporate Office : B-33, Sector-5, Noida,
Gautambuddha Nagar (U.P.) - 201 301, INDIA
Telephone : +91-120-44 46 900
www.laxmipublications.com
info@laxmipublications.com

C—0000/026/06
Printed at: Ajit Printers, Delhi.

PREFACE TO THE FIRST EDITION

This book has been written specifically to meet exhaustively the requirements of the syllabus of the subject **"Basic Electrical Engineering"** of B.E. Examination of **Chhattisgarh Swami Vivekanand Technical University, Bhilai** and **BPUT, Orissa.**

The book includes comprehensive treatment of the subject-matter under wide range of topics mentioned in the syllabus, with a large number of the solved examples to support the text. Besides this, Highlights, Objective Type Questions, Theoretical Questions and Unsolved Examples have been added at the end of each chapter to make the book a comprehensive unit in all respects.

Any suggestions for the improvement of this book will be thankfully acknowledged and incorporated in the next edition.

—AUTHOR

SYLLABUS

Basic Electrical Engineering

1. **D.C. NETWORKS**

Kirchoff's laws, node voltage and mesh current methods, Delta-star and star-delta conversion ; Classification of Network Elements, Superposition principle, Thevenin's and Norton's theorems.

2. **SINGLE PHASE A.C. CIRCUITS**

Single phase EMF generation, average and effective values of sinusoids ; Solutions of R, L, C series circuits, the j operator, complex representation of impedances ; Phasor diagram, power factor, power in complex notation ; Solution of parallel and series-parallel circuits ; Resonance.

3. **THREE PHASE A.C. CIRCUITS**

Three phase EMF generation, delta and Y-connection, line and phase quantities ; Solution of three phase circuits, balanced supply voltage and balanced load ; Phasor diagram, measurement of power in three phase circuits ; Three phase four wire circuit ; Unbalanced circuits.

4. **MAGNETIC CIRCUITS**

Ampere's circuital law, B-H curve, solution of magnetic circuits ; Hysteresis and eddy current losses ; Relays, an application of magnetic force.

5. **TRANSFORMERS**

Construction, EMF equation, ratings ; Phasor diagram on no load and full load ; Equivalent circuits, regulation and efficiency calculations ; Open and short circuit tests ; Auto-transformed and three phase transformers.

6. **INDUCTION MOTORS**

The revolsing magnetic field, principle of operation, ratings ; Equivalent circuit ; Torquespeed characteristics ; Starters for squirrel cage and wound rotor type induction motors, Single phase induction motors.

7. **D.C. MACHINES**

Construction, EMF and torque equations ; Characteristics of D.C. generators and motors; Speed control of D.C. motor starters ; Armature reaction and commutation.

8. ELECTRICAL MEASURING INSTRUMENTS

D.C. PMMC instruments, shunts and multipliers, multi-meters ; Moving iron ammeters and voltmeters ; Dynamometer wattmeters ; A.C. wat-hour meters. Extension of instrument ranges.

9. POWER SUPPLY SYSTEMS

General structure of electrical power systems ; Power transmission and distribution via overhead lines and underground cables. Steam, hydro, gas and nuclear power generation.

CONTENTS

Chapters *Pages*

Chapters		Pages

Chapters *Pages*

Chapters *Pages*

1

D.C. Networks

1. Current electricity—Basic concepts—Atomic structure—Electric current—Electromotive force—Resistance—Resistors. 2. Definitions of important terms relating network. 3. Limitations of Ohm's law. 4. Kirchhoff's laws. 5. Applications of Kirchhoff's laws : Branch-current method—Maxwell's loop (or mesh) current method—Nodal voltage method. 6. Source Conversion. 7. Solving equations by derminants—Cramer's rule. 8. Superposition theorem. 9. Thevenin's theorem. 10. Norton's theorem. 11. Maximum power transfer theorem. 12. Delta star transformation. 13. Compensation theorem. 14. Reciprocity theorem. 15. Millman's theorem—Highlights—Objective Type Questions—Theoretical Questions—Unsolved Examples.

1. CURRENT ELECTRICITY—BASIC CONCEPTS

1.1. Atomic Structure

— An ***element*** is defined as a substance which cannot be decomposed into other substances. *The smallest particle of an element which takes part in chemical reaction is known as* **atom**.

— All matter is composed of *atoms* which are *infinitesimally small*.

— All atoms are made of ***electrons, protons*** and ***neutrons.*** Most solid materials are classed, from the stand point of electrical conductivity, as *conductors, semiconductors* or *insulators*. To be ***conductor,*** the substance must contain some *mobile electrons* that can move freely between atoms. These free electrons come only from the *valence* (outer) *orbit* of the atom. *Conductivity depends on the number of electrons in the valence orbit.*

"The energy level of an electron increases as its distance from the nucleus increases. Thus an electron in the second orbit possesses more energy than electron in the first orbit, electrons in the third orbit have higher energy than in the second orbit and so on. It follows, therefore, that electrons in the last orbit will possess very high energy. These high energy electrons are less bound to the nucleus and hence they are more mobile. It is the mobility of last orbit electrons that they acquire the property of combining with other atoms. Further due to this combining power of last orbit electrons of an atom they are called ***valence electrons"***.

- Atoms with *fewer than four valence electrons* are good *conductors*.
- Atoms with *more than four valence electrons* are *poor conductors*.
- Atoms *with four valence electrons* are *semiconductors*.

Important data of an atom :

(*i*) **Electron**

Mass of an electron $= 9.11 \times 10^{-31}$ kg

$\simeq \dfrac{1}{1840}$ mass of proton

Charge of electron $= -1.602 \times 10^{-19}$ coulomb

Diameter of an electron $\simeq 10^{-15}$ m

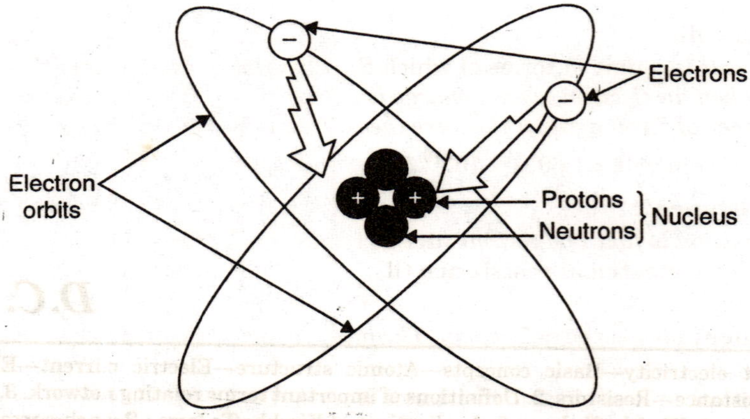

Fig. 1. Atomic structure : Electron, proton and neutron.

(ii) **Proton**

Mass of proton $= 1.67 \times 10^{-27}$ kg
Charge on proton $= + 1.602 \times 10^{19}$ coulomb.

(iii) **Neutron**

Mass of neutron $\simeq$ mass of proton $(= 1.67 \times 10^{-27}$ kg)
Charge of neutron = Nil
Diameter of nucleus ...... of the order of 10^{-14} m
Diameter of orbits $\simeq 10^4$ times the dia. of the molecule.

— Normally, the atoms are electrically neutral, that is, the number of electrons and protons are the same, cancelling each other's electrical force. Atoms "*stay together*" because *unlike charges attract each other*. The electrical force of the protons hold the electrons in their orbits. *Like electrical charges repel each other* so negatively charged electrons will not collide with each other.

Positive and negative ions :

When an electron is removed from a neutral atom, this atom becomes positively charged and is called *positive ion*. However if an electron is added to a neutral atom, it becomes negatively charged and is called a *negative ion*. Thus an atom becomes an ion by the gain or loss of an electron.

1.2. Electric Current

The controlled movement of electrons (or drift) through a substance is called **current.**

— Current is the *rate at which electrons move*. One *ampere* (unit of current) represents 6.28×10^{18} electrons passing a point each second (1 coulomb past a point in one second).

$$\left[\begin{array}{l}\text{Ampere} = \text{coulomb/second}\\ \text{One coulomb} = \text{charge of } 6.28 \times 10^{18} \text{ electrons}\end{array}\right]$$

1.3. Electromotive Force

- **Electromotive force** (e.m.f.) *is the force that causes a current of electricity to flow.*
- The **potential difference** (p.d.) *V*, *between two points in a circuit is the electrical pressure or voltage required to drive the current between them.*
- The **volt** is a unit of potential difference and electromotive force. It is defined as *the difference of potential across a resistance of 1 ohm carrying a current of 1 ampere.*

Electron volt :

Electron volt is a unit in terms of which the energies of atomic particles are expressed. *It is the work done when an electron, whose charge is e coulombs, is moved in an electric field through a potential difference of 1 volt against the force (newtons) acting on the charge.*

Thus 1 electron volt = 1.6021×10^{-19} J.

1.4. Resistance

The *opposition to flow of electrons* (due to bonds between protons and electrons, as well as to collisions) is called a **electrical resistance (R).**

Or

Resistance may also be defined as *"The property of the electric circuit which opposes the flow of current"*.

The practical unit of electric resistance is ohm (Ω). It (ohm) is defined as the *resistance in which a constant current of 1 ampere generates heat at the rate of 1 watt. One volt* applied across *1 ohm* will produce *1 ampere.*

1 Mega-ohm (MΩ) $= 10^6\ \Omega$
1 kilo-ohm (kΩ) $= 10^3\ \Omega$
1 milli-ohm (mΩ) $= 10^{-3}\ \Omega$
1 micro-ohm (μΩ) $= 10^{-6}\ \Omega$.

(*i*) **Laws of resistance :**

The resistance of a conductor, such as a wire, of uniform cross- section depends on the following factors :

(*i*) Length (l) : varies *directly* as its length, l
(*ii*) Cross-section (A) : varies *inversely* as the cross-section, A, of the conductor
(*iii*) Nature of the material (ρ)
(*iv*) Temperature of the conductor : It *almost varies directly* with the temperature

$$R = \rho \frac{l}{A} \quad \text{...(1)}$$

where ρ is known as specific resistance or resistivity.

Specific resistance or **resistivity** of a material may be defined as *"The resistance between the opposite faces of a metre cube of that material"*.

The unit of resistivity is ohm-metre (Ω-m).

(*ii*) **Conductance (G) :**

Conductance (G) is the reciprocal of resistance $\left(G = \frac{1}{R} = \frac{A}{\rho l} \right)$

(*iii*) **Conductivity (σ) :**

The *reciprocal of specific resistance* $\left(\sigma = \frac{1}{\rho} \right)$ of a material is called its *conductivity.*

The unit of conductivity $\left(\sigma = G \frac{l}{A} \right)$ is mho/metre.

(*iv*) **Temperature co-efficient of resistance :**

Temperature co-efficient of resistance at 0°C may be defined as follows :

"The change in resistance per ohm for change in temperature of 1°C from 0°C".

Over *large temperature ranges* the simple formula

$$R_t = R_0 (1 + \propto t) \quad \text{...(2)}$$

does not completely fit, but a formula of the type

$$R_t = R_0 (1 + \propto t + \beta t^2) \quad ...(3)$$

(where β is a smaller co-efficient)

applies.

Also $$\rho_t = \rho_0 (1 + \propto_0 . t)$$

where ρ_t and ρ_0 are the resistivities at $t°$ and 0°C respectively.

(*v*) **The Effect of temperature on resistance :**

The following points are worth noting :

- The resistance of *metal conductors 'increases'* (α, *i.e.*, temperature co-efficient of resistance being *positive*) with rise of temperature ; the rate of increase is very considerable for most pure metals, being as much as about $\frac{1}{150}$ of the total resistance for each centigrade rise in the case of iron ; the effect is smaller in case of alloys, and very small indeed for materials such as manganin and constantan which are therefore very suitable for making *standard resistances.*
- The resistance of semiconductors such as *carbon*, and all *electrolytes* 'decreases' as the temperature rises (α being *negative*).

(*vi*) **Ohm's Law :**

Ohm's law can be stated as follows :

"For *a fixed metal conductor, the temperature and other conditions remaining constant the current (I) through it is proportional to the potential difference (V) between its ends*".

In other words, $$\frac{V}{I} = \text{constant} \quad \text{or} \quad \frac{V}{I} = R \quad ...(4)$$

where R is the resistance of the conductor between the two points considered :

— The linear relationship ($I \propto V$) *does not apply to all non-metallic conductors.* For example, for silicone carbide, the relationship is given by :

$V = KI^x$ where K and x are constants and x is *less than unity*.

The following relations hold good :

(*i*) $P = VI = I^2R = \frac{V^2}{R}$

(*ii*) $I = \frac{P}{V} = \sqrt{\frac{P}{R}}$

(*iii*) $R = \frac{P}{I^2} = \frac{V^2}{P}$

(*iv*) $V = \frac{P}{I} = \sqrt{PR}$

where P = power in watts,
V = voltage in volts,
I = current in amperes, and
R = resistance in ohms.

Power is expressed in terms of kW (kilowatt = 1000 W) or MW (megawatt = 1000 kW or 10^6 W).

Electrical energy is expressed in terms of kWh (kilowatt hours)

$$1 \text{ kWh} = 1 \text{ kW} \times 1 \text{ hour} = 1000 \text{ watt-hours} (= 1000 \times 60 \times 60 \text{ watt-sec.})$$

Linear and non-linear resistors :

— A *linear resistor is one which obeys Ohm's law*. A circuit which contains only linear components is called a *linear circuit*.

— Such elements in which the *V/I* (volt-ampere) plots are not *straight lines but curves* are called *non-linear resistors* or *non-linear elements*.

Limitations of Ohm's law :

Ohm's law *does not apply under the following conditions* :

1. Electrolytes where enormous gases are produced on either electrode.
2. Non-linear resistors like vacuum radio valves, semiconductors, gas filled tubes etc.
3. Arc lamps.
4. Metals which get heated up due to flow of current.
5. Appliances like metal rectifiers, crystal detectors, etc. in which operation depends on the direction of current.

(*vii*) **Resistances in series**

Fig. 2 shows three resistances connected in series. Obviously *current flowing through each resistance will be same* but *voltage drop* across each of them will *vary* as per value of individual resistance.

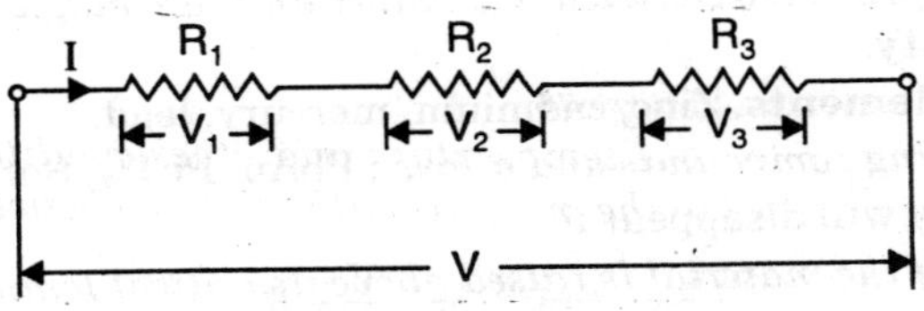

Fig. 2. Resistances in series.

Also the sum of all the voltage drops ($V_1 + V_2 + V_3$) is equal to the applied voltage (V).

i.e., $$V = V_1 + V_2 + V_3$$

$$IR = IR_1 + IR_2 + IR_3 \quad \text{[using Ohm's law : } V = IR\text{]}$$

i.e., $$R = R_1 + R_2 + R_3 \quad \text{...(5)}$$

where R is the *equivalent resistance* of series combination.

(*viii*) **Resistances in parallel**

Refer Fig. 3. In this case *voltage across each resistance will be same but current will be different* depending upon the value of the individual resistance.

i.e., $$I = I_1 + I_2 + I_3$$

$$\frac{V}{R} = \frac{V}{R_1} + \frac{V}{R_2} + \frac{V}{R_3}$$

$$\frac{1}{R} = \frac{1}{R_1} + \frac{1}{R_2} + \frac{1}{R_3} \quad \text{...(6)}$$

where R is the *equivalent resistance* of the parallel combination.

$$R = \frac{R_1R_2R_3}{R_1R_2 + R_2R_3 + R_3R_1} \quad \text{...[6 (a)]}$$

$$G = G_1 + G_2 + G_3 \quad \text{...(7)}$$

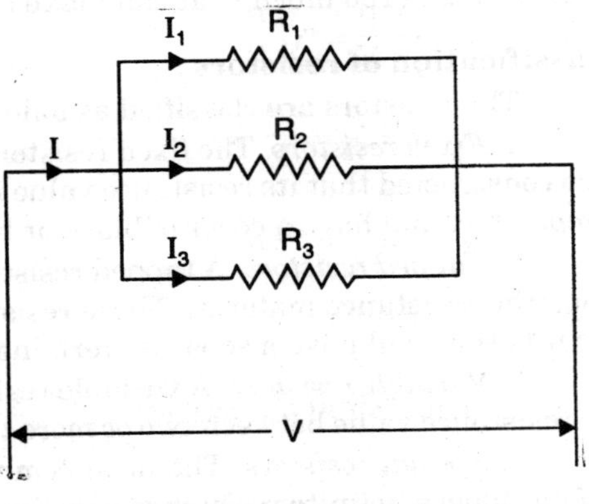

Fig. 3. Resistances in parallel.

Table 1. Resistivities and Temperature Co-efficients

S. No.	*Material*	*Resistivity in Ω-m at 20°C*	*Temperature coefficient at 20°C*
1.	Copper	1.59×10^{-8}	0.00428
2.	Aluminium	2.8×10^{-8}	0.0020
3.	Silver	1.52×10^{-8}	0.00377
4.	Platinum	11×10^{-8}	0.00340
5.	Iron	$(9 \text{ to } 15) \times 10^{-8}$	0.0070
6.	Mercury	19.9×10^{-8}	0.00089
7.	German silver (4 Cu ; 2 Ni ; 1 Zn)	20.2×10^{-8}	0.00027
8.	Constantan or Eureka	49×10^{-8}	– 0.00004 to + 0.00001
9.	Carbon	7000×10^{-8}	– 0.0005

(*ix*) Superconductivity

Equation $R_t = R[1 + \alpha(t - 20)]$ holds good for temperature below 20°C. But at *very low temperature, some metals acquire zero electrical resistance and zero magnetic induction ; the property known as* **superconductivity.**

Superconducting elements. Zinc, cadmium, mercury, lead.

Typical superconducting compounds and alloys : PbAu, $PbTl_2$, SnSb, CuS, NbN, NbB, ZrC.

The superconductivity will disappear if

(*i*) *The temperature of the material is raised above its critical temperature,*

Or

(*ii*) *a sufficiently strong magnetic field or current density is employed.*

1.5. Resistors

A resistor entails the following two main characteristics :

(*i*) Its *resistance (R)* in ohms. ...... The resistors are available from a fraction of an ohm to many mega ohms.

(*ii*) The *wattage rating.* ...... The power rating may be as high as several hundred watts or as low as $\frac{1}{10}$ watt. *Power rating indicates the maximum wattage the resistor can dissipate without excessive heat* (Too much heat can make the resistor burn open).

Classification of resistors :

The resistors are classified as follows :

1. *Fixed resistors.* The fixed resistor is the simplest type of resistor. Fixed means that the unit is so constructed that its resistance value is *constant* and *unchangeable.* These are made of a *carbon composition* and have a cover of black or brown hard plastics.

2. *Tapped resistors.* A tapped resistor is a resistor which has a tap, or connection, somewhere along the resistance material. These resistors are usually wire wound type. If they have more than one tap, they will have a separate terminal for each.

3. *Variable resistors.* A variable resistor has a movable contact that is used to adjust or select the resistance value between two or more terminals. A variable resistor is commonly called a *control.*

4. *Special resistors.* The most common type of special resistors is the fusible type. A *fusible resistor* have a definite resistance value and it protects the circuit much like a fuse. Another special

resistor is the *temperature compensating unit.* Such resistors are used to provide special control of circuit that must be extremely stable in their operation.

Schematic symbols for various resistors are shown in Fig. 4.

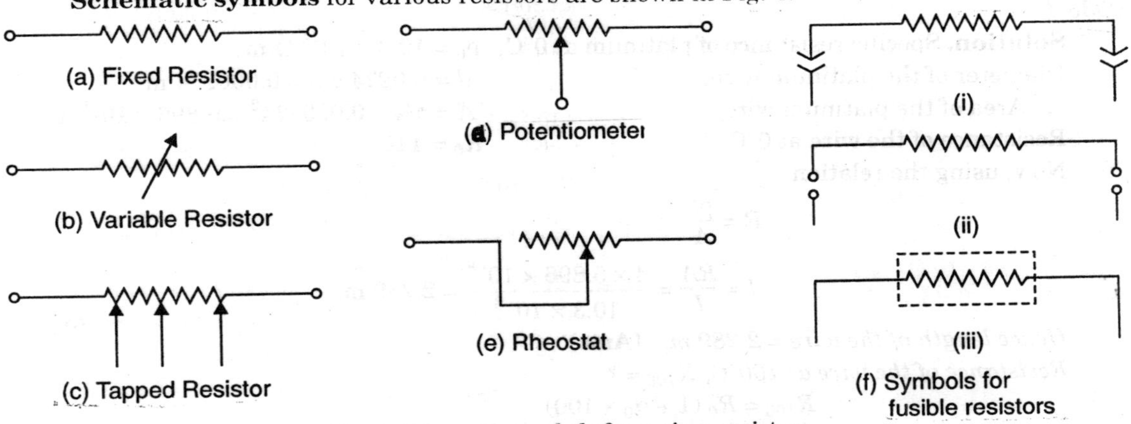

Fig. 4. Schematic symbols for various resistors.

The following types of resistors are used in electrical circuits :

(*i*) Carbon resistors.

(*ii*) Wire-wound resistors on ceramic or plastic forms (as in case of rheostats etc.)

(*iii*) Deposited carbon resistors on ceramic base.

(*iv*) Deposited metal resistors on ceramic base.

(*v*) Printed, painted or etched circuit resistors.

Example 1. *A semi-circular ring of copper has an inner radius of 12 cm, radial thickness 6 cm and axial thickness 8 cm. Find the resistance of the ring at 60°C between its two end faces.*

Assume : Specific resistance of copper at 20°C = 1.724×10^{-8} ohm-metre.

Temperature co-efficient of resistance of copper at 0°C = 0.0043/°C.

Solution. Refer Fig. 5.

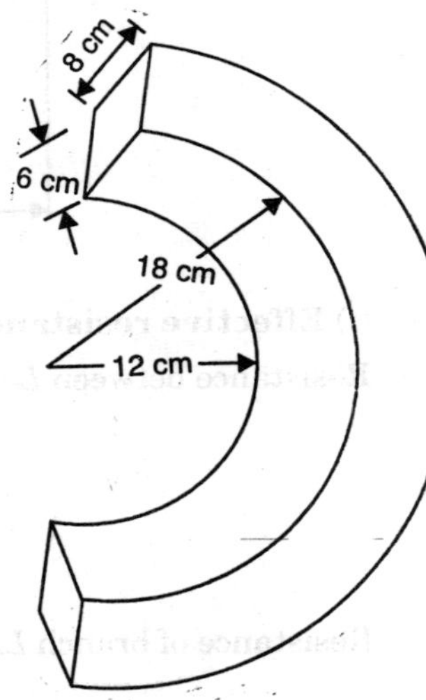

Fig. 5

Mean radius of the ring $= \dfrac{18 + 12}{2} = 15$ cm

Mean length between the faces = 15π cm = 0.47 m

Cross-section of the ring = $8 \times 6 = 48$ cm^2 = 48×10^{-4} m^2

Now $\alpha_0 = 0.0043/°C$

$$\alpha_{20} = \frac{\alpha_0}{1 + \alpha_0 (20 - 0)}$$

$$= \frac{0.0043}{1 + 0.0043 \times 20} = 0.00396$$

$$\rho_{60} = \rho_{20} [1 + \alpha_{20} (60 - 20)]$$

$$= 1.724 \times 10^{-8} [1 + 0.00395 \times 40]$$

$$= 1.997 \times 10^{-8} \ \Omega\text{-m}$$

Now $$R_{60} = \frac{\rho_{60} \times l}{A} = \frac{1.997 \times 10^{-8} \times 0.47}{48 \times 10^{-4}}$$

$$= \mathbf{1.955 \times 10^{-6} \ \Omega.} \quad \textbf{(Ans.)}$$

Example 2. *The specific resistance of platinum at 0°C is 10.3×10^{-8} Ω-m. How long must a wire of No. 32 S.W.G. (diameter = 0.0274 cm) platinum be to have a resistance of 4 Ω at 0°C.*

What will be resistance of a wire at 100°C, if the temperature co-efficient of platinum is 0.0038 ?

Solution. Specific resistance of platinum at 0°C, $\rho_0 = 10.3 \times 10^{-8}$ Ω-m

Diameter of the platinum wire, $d = 0.0274$ cm $= 0.000274$ m

∴ Area of the platinum wire, $A = \pi/4 \times 0.000274^2 = 5.896 \times 10^{-8}$ m²

Resistance of the wire at 0°C, $R_0 = 4\ \Omega$

Now, using the relation :

$$R = \frac{\rho l}{A}$$

$$\therefore \quad l = \frac{RA}{l} = \frac{4 \times 5.896 \times 10^{-8}}{10.3 \times 10^{-8}} = 2.289 \text{ m}$$

Hence length of the wire = 2.289 m. **(Ans.)**

Resistance of the wire at 100°C, R_{100} = ?

$$R_{100} = R_0 (1 + \alpha_0 \times 100)$$
$$= 4(1 + 0.0038 \times 100) = 5.52\ \Omega$$

Hence, $\mathbf{R_{100} = 5.52\ \Omega}$. **(Ans.)**

Example 3. *In the Fig. 6 is shown a combination of resistances. If the voltage across L and M is 75 volts find :*

(i) The effective resistance of the circuit.

(ii) Voltage drop across each resistance.

Solution. Refer Fig. 6.

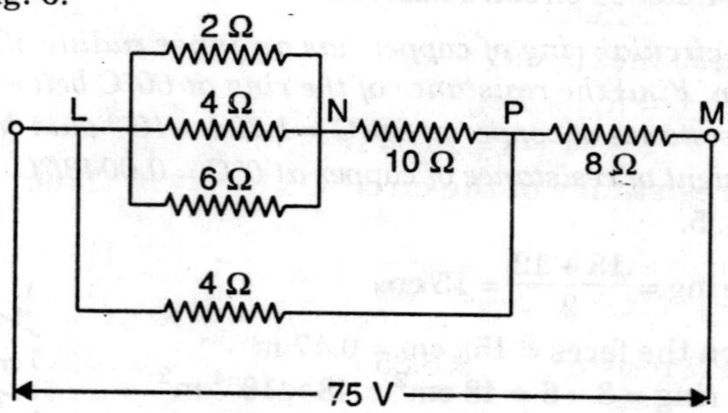

Fig. 6

(i) **Effective resistance of the circuit :**

Resistance between *L* and *N*,

$$\frac{1}{R_{LN}} = \frac{1}{2} + \frac{1}{4} + \frac{1}{6} = \frac{6+3+2}{12} = \frac{11}{12}\ \Omega$$

$$\therefore \quad R_{LN} = \frac{12}{11}\ \Omega \qquad \text{...(Fig. 6)}$$

Resistance of branch *LNP*,

$$R_{LNP} = \frac{12}{11} + 10 = \frac{122}{11} = 11.09\ \Omega \qquad \text{...(Fig. 7)}$$

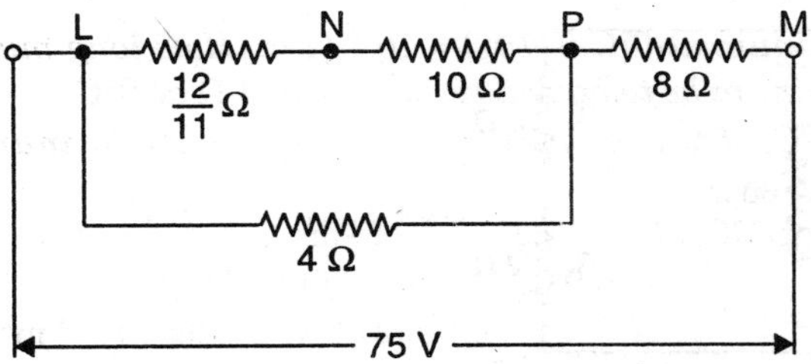

Fig. 7

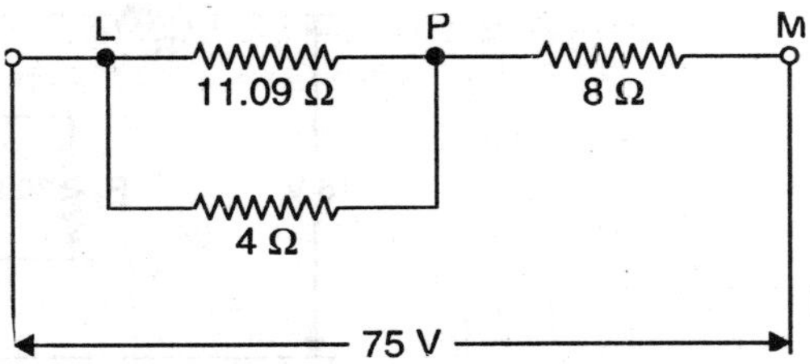

Fig. 8

Reactance between L and P

$$\frac{1}{R_{LP}} = \frac{1}{11.09} + \frac{1}{4}$$

[As there are two parallel paths between points L and P of resistances 11.09 Ω and 4 Ω]

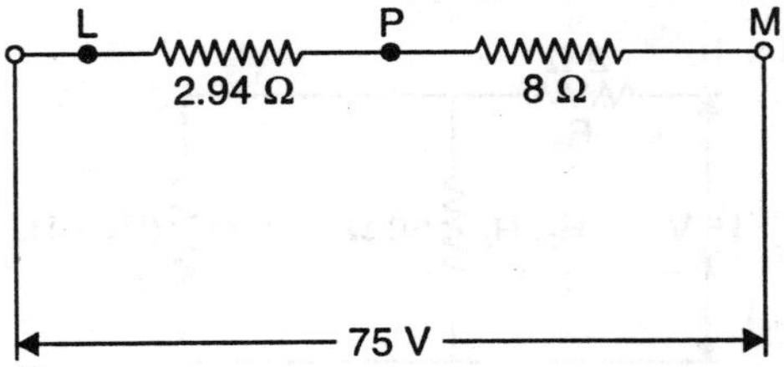

Fig. 9

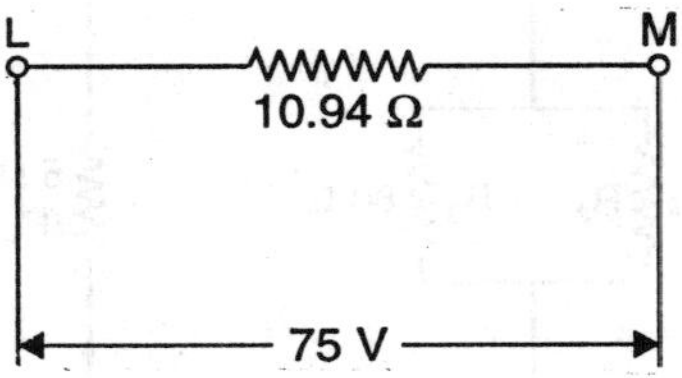

Fig. 10

$$= \frac{4 + 11.09}{44.36}$$

or $$R_{LP} = \frac{44.36}{15.09} = 2.94\ \Omega$$ ...(Fig. 8)

Effective resistance of the circuit,

$$R_{LM} = R_{LP} + R_{PM}$$
$$= 2.94 + 8 = 10.94\ \Omega$$

Hence effective resistance of the circuit = 10.94 Ω. (Ans.)

Total current through the circuit $= \dfrac{V}{R_{LM}} = \dfrac{75}{10.94} = 6.85$ A.

Current through 4 Ω resistor $= 6.85 \times \dfrac{11.09}{11.09 + 4} = 5.03$ A

Current in branch LNP $= 6.85 \times \dfrac{4}{11.09 + 4} = 1.82$ A

∴ **Voltage drop across 2, 4 and 6 Ω resistors**

$$= 1.82 \times R_{LN} = 1.82 \times \frac{12}{11} = \mathbf{1.98\ V.}\ \ \textbf{(Ans.)}$$

Voltage drop across 10 Ω resistor = 1.82 × 10 = **18.2 V. (Ans.)**

Voltage drop across 4 Ω resistor = 5.03 × 4 = **20.12 V. (Ans.)**

Voltage drop across 8 Ω resistor = 6.85 × 8 = **54.8 V. (Ans.)**

Example 4. *In the series-parallel circuit shown in the Fig. 11, find :*

(*i*) *The total resistance of the circuit.*

(*ii*) *The total current flowing through the circuit.*

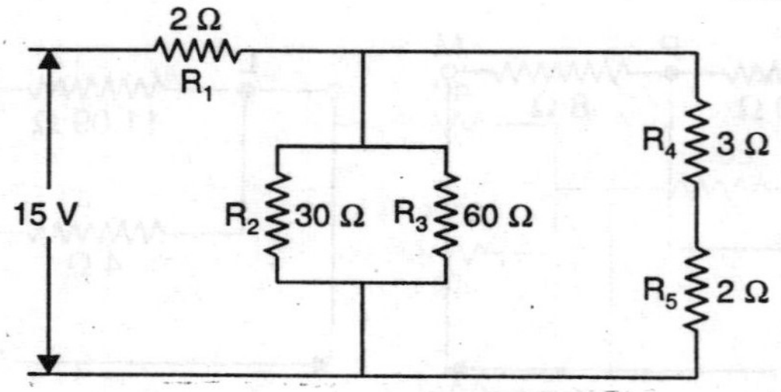

Fig. 11

Solution. (*i*) **Total resistance R_T :**

Resistors R_4 and R_5 are connected in series and have a net value of 5 Ω ($R_4 + R_5 = 3 + 2 = 5$) ...(Fig. 11)

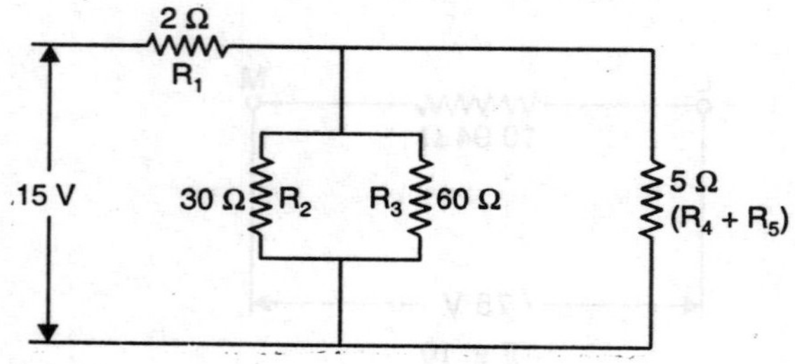

Fig. 12

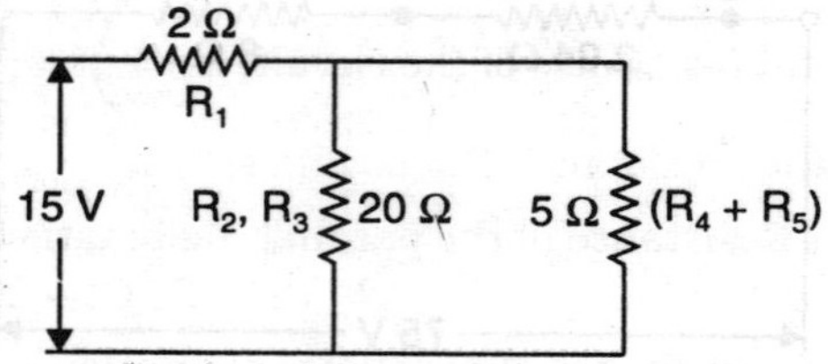

Fig. 13

The equivalent resistance of resistor R_2 and R_3 (connected in parallel)

$$= \frac{R_2 \times R_3}{R_2 + R_3} = \frac{30 \times 60}{30 + 60} = 20\ \Omega \qquad \text{...(Fig. 12)}$$

To simplify the circuit even more, we can combine the 20 Ω equivalent of R_2 and R_3, and 5 Ω equivalent of R_4 and R_5 (Fig. 13) :

$$R_{eq} = \frac{20 \times 5}{20 + 5} = \frac{100}{25} = 4\ \Omega$$

We now find the simple two-resistor series circuit shown in Fig. 14. This circuit includes a single resistance (4 Ω) which is equivalent of the series and parallel arrangement shown in Fig. 14 and R_2, R_3, R_4 and R_5. The total resistance of the circuit, R_T can now be found by addition :

$$R_T = R_1 + R_{eq} = 2 + 4 = 6\ \Omega. \quad \textbf{(Ans.)}$$

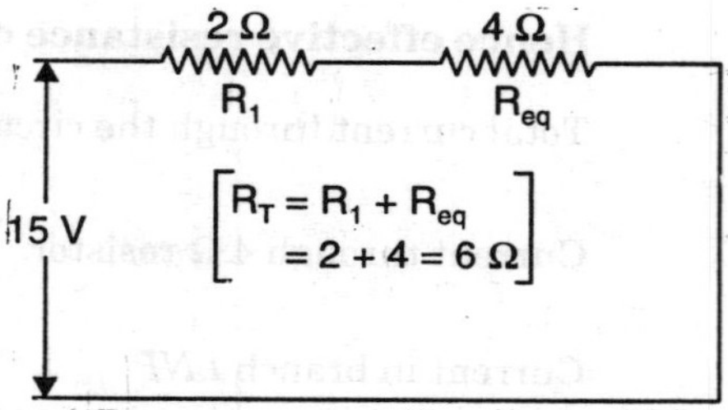

Fig. 14

(*ii*) **Total current, I_T :**

$$I_T = \frac{V}{R_T} = \frac{15}{6} = 2.5\ \text{A}$$

Hence total current = 2.5 A. (Ans.)

Example 5. *A resistance of 20 ohms is connected in series with a combination of two resistances arranged in parallel each of value 40 ohms. Determine the resistance R_3 which should be shunted across the parallel combination so that the total current drawn by the circuit is 1.5 A with applied voltage of 40 V.*

Solution. Refer Fig. 15.

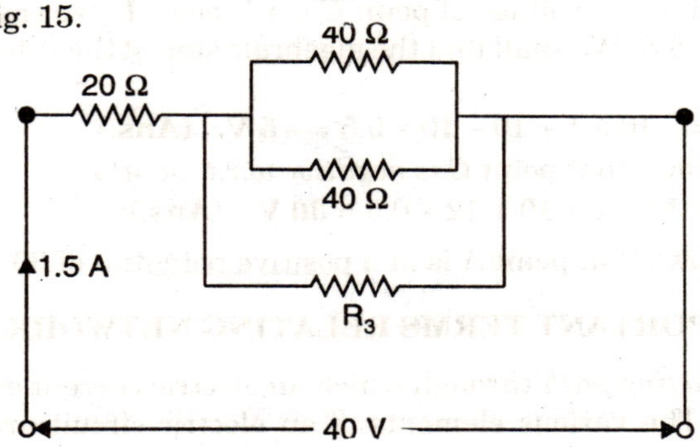

Fig. 15

Total resistance of the circuit, $R_T = \frac{V}{I} = \frac{40}{1.5} = 26.66$ ohms.

Series resistance R = 20 ohms

$\therefore$ Resistance of the parallel combination of resistances

$$R_T - R = 26.66 - 20 = 6.66 \text{ ohms}$$

Also resistance of parallel combination is given by :

$$\frac{1}{40} + \frac{1}{40} + \frac{1}{R_3} = \frac{1}{6.66} = 0.15$$

$\therefore$
$$\frac{1}{R_3} = 0.15 - \frac{1}{40} - \frac{1}{40} = 0.1$$

i.e., $\mathbf{R_3 = 10}$ **ohms. (Ans.)**

Example 6. *For the circuit shown in Fig. 16, find* V_{CE} *and* V_{AG}.

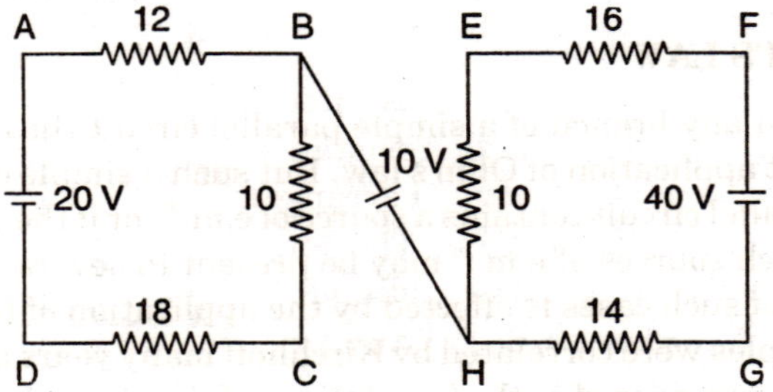

Fig. 16 **(Pune University)**

Solution. Refer Fig. 16.

- Let us consider the two battery circuits separately.

Current in 20 V battery circuit $ABCD = \frac{20}{(12 + 10 + 18)} = 0.5$ A

Current in 40 V battery circuit $EFGH = \frac{40}{10 + 16 + 14} = 1$ A

Voltage drops over different resistors can be found by using Ohm's law.

- In order to find V_{CE} *i.e.* voltage of point C w.r.t. point E, we shall start from E and go to C via points H and B. We shall find the algebraic sum of the voltage drops met on the way from point E to C.

$$V_{CE} = -10 \times 1 + 10 - 10 \times 0.5 = \mathbf{-5\ V.} \quad \textbf{(Ans.)}$$

The negative sign shows that point C *is negative w.r.t. point E*.

$$V_{AG} = 14 \times 1 + 10 + 12 \times 0.5 = \mathbf{30\ V.} \quad \textbf{(Ans.)}$$

The positive sign shows that point A is at a positive potential of 30 V w.r.t. point G.

2. DEFINITIONS OF IMPORTANT TERMS RELATING NETWORK

1. **Circuit.** A *conducting path* through which an electric current either flows or is intended to flow is called a *circuit*. The various elements of an electric circuit are called *parameters (e.g.* resistance, inductance and capacitance). These parameters may be *distributed* or *lumped*.

2. **Linear circuit.** The circuit whose parameters are *constant* (*i.e.* they do not change with voltage or current) is called a *linear circuit*.

3. **Non-linear circuit.** The circuit whose parameters *change* with voltage or current is called a *non-linear circuit*.

4. **Unilateral circuit.** A unilateral circuit is one whose properties or characteristics change with the direction of its operation (*e.g. diode rectifier*).

5. **Bilateral circuit.** It is that circuit whose properties or characteristics are same in either direction (*e.g. transmission line*).

6. **Electric network.** An electric network arises when a number of parameters or electric elements coexist or combine in any manner or arrangement.

7. **Active network.** An *active network* is one which contains one or more than one sources of *e.m.f.*

8. **Passive network.** A *passive network* is one which does not contain any source of e.m.f.

9. **Node.** A *node* is a junction in a circuit where two or more circuit elements are connected together.

10. **Branch.** The part of a network which lies between two junctions is called **branch**.

3. LIMITATIONS OF OHM'S LAW

In a series circuit or in any branch of a simple parallel circuit the calculation of the current is easily effected by the direct application of Ohm's law. But such a simple calculation is not possible if one of the branches of a parallel circuit contains a source of e.m.f., or if the current is to be calculated in a part of a network in which sources of e.m.f. may be present in several meshes or loops forming the network. The treatment of such cases is effected by the application of fundamental principles of electric circuits. These principles were correlated by Kirchhoff many years ago and enunciated in the form of *two laws*, which can be considered as the foundations of circuit analysis. Other, later, methods have been developed, which when applied to special cases considerably shorten the algebra and arithmetic computation compared with the original Kirchhoff's method.

4. KIRCHHOFF'S LAWS

For complex circuit computations, the following two laws first stated by Gutsav R. Kirchhoff (1824–87) are indispensable.

1. Kirchhoff's Point Law or Current Law (KCL). It states as follows :

The sum of the currents entering a junction is equal to the sum of the currents leaving the junction. Refer Fig. 17.

If the currents *towards* a junction are considered *positive* and those *away* from the same junction *negative*, then this law states that the *algebraic sum of all currents meeting at a common junction is zero.*

i.e., $\quad \Sigma$ Currents entering = Σ currents leaving

$$I_1 + I_3 = I_2 + I_4 + I_5 \quad \text{...[8 (a)]}$$

or $\quad I_1 + I_3 - I_2 - I_4 - I_5 = 0 \quad$ ...[8 (b)]

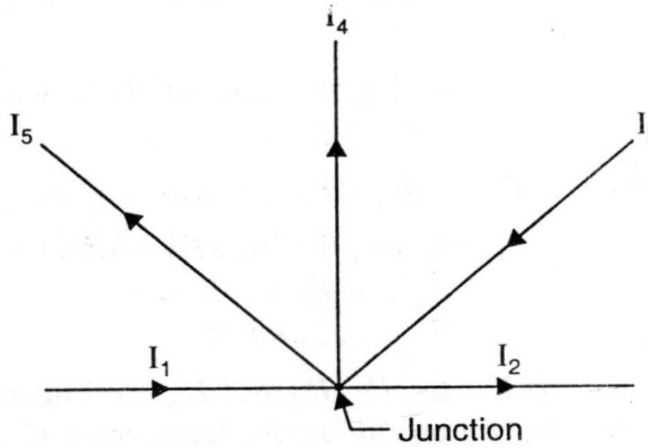

Fig. 17

2. Kirchhoff's Mesh Law or Voltage Law (KVL). It states as follows :

The sum of the e.m.fs (rises of potential) around any closed loop of a circuit equals the sum of the potential drops in that loop.

Considering a rise of potential as positive (+) and a drop of potential as negative (–), the *algebraic sum* of potential differences (voltages) around a closed loop of a circuit is zero :

$$\Sigma E - \Sigma IR \text{ drops} = 0 \text{ (around closed loop)}$$

i.e., $\quad \Sigma E = \Sigma IR \quad$ ...[9 (a)]

or $\quad \Sigma$ Potential rises = Σ potential drops $\quad$ ...[9 (b)]

To apply this law in practice, assume an arbitrary current direction for each branch current. The end of the resistor through which the current enters, is then positive, with respect to the other end. *If the solution for the current being solved turns out negative, then the direction of that current is opposite to the direction assumed.*

In tracing through any single circuit, whether it is by itself or a part of a network, the following **rules** must be applied :

1. A *voltage drop exists* when tracing through a resistance *with or in the same direction as the current*, or through a battery or generator against their voltage, that is from *positive (+) to negative (–)*. Refer Fig. 18.

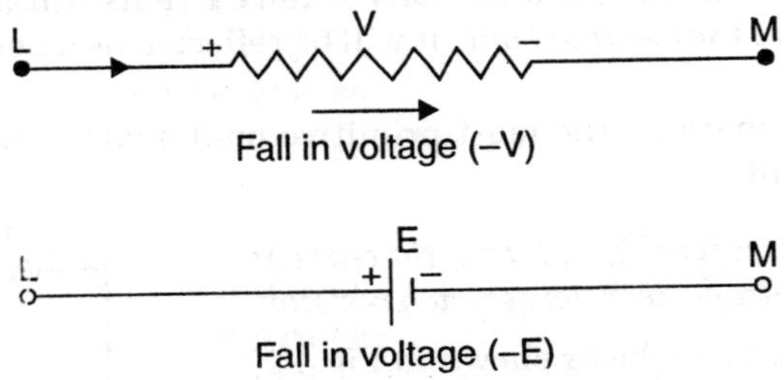

Fig. 18

2. A *voltage rise exists* when tracing through a resistance *against* or *in opposite direction to the current* or through a battery or a generator with their voltage that is from *negative (–) to positive (+)*. Refer Fig. 19.

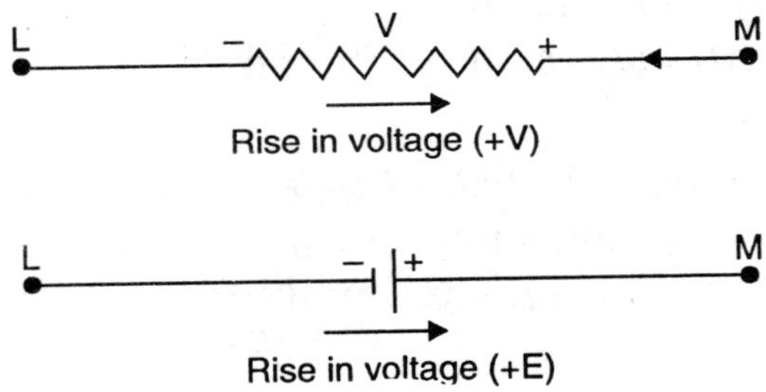

Fig. 19

Illustration. Consider a circuit shown in Fig. 20.

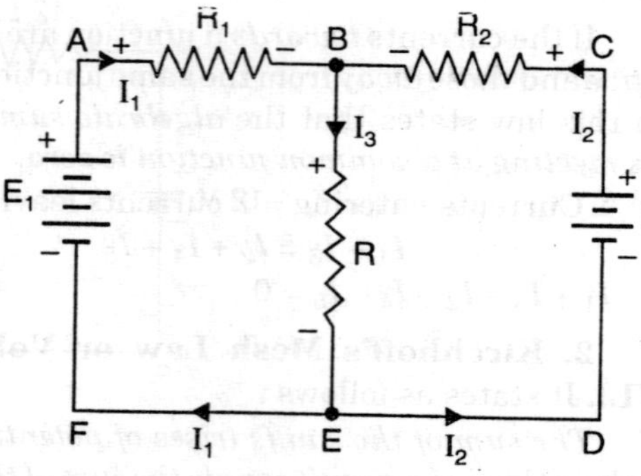

Fig. 20

Considering the loop **ABEFA**, we get

$$-I_1R_1 - I_3R_3 + E_1 = 0$$

or $\quad E_1 = I_1R_1 + I_3R_3 \quad$ (where $I_3 = I_1 + I_2$) ...(*i*)

Considering the loop **BCDEB**, we have

$$I_2R_2 - E_2 + I_3R_3 = 0 \qquad ...(ii)$$

or $\quad E_2 = I_2R_2 + I_3R_3$

If E_1, E_2, R_1, R_2 and R_3 are known, then I_1, I_2 and I_3 can be calculated from eqns. (*i*) and (*ii*).

5. APPLICATIONS OF KIRCHHOFF'S LAWS

Kirchhoff's laws may be employed in the following methods of solving networks :

1. Branch-current method
2. Maxwell's loop (or mesh) current method
3. Nodal voltage method.

5.1. Branch-Current Method

For a multi-loop circuit the following ***procedure*** is adopted for writing equations :

1. Assume currents in different branch of the network
2. Write down the smallest number of voltage drop loop equations so as to include all circuit elements ; these loop equations are independent.

If there are n nodes of three or more elements in a circuit, then write $(n - 1)$ equations as per current law.

3. Solve the above equations simultaneously.

The assumption made about the directions of the currents initially is arbitrary. In case the actual direction is *opposite to the assumed one*, it will be reflected as a negative value for that current in the answer.

The branch-current method (the most primitive one) involves more labour and is not used except for very simple circuits.

Example 7. *In the circuit of Fig. 21, find the current through each resistor and voltage drop across each resistor.*

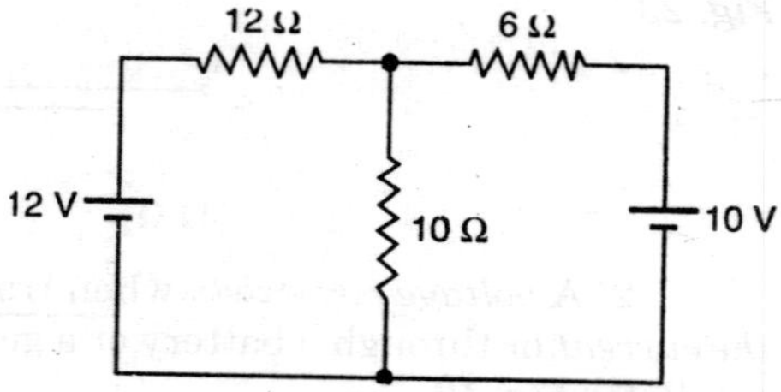

Fig. 21

Solution. Let the currents be as shown in Fig. 22.

Applying Kirchhoff's voltage law to the circuit **ABEFA**, we get

$$-12I_1 - 10(I_1 + I_2) + 12 = 0$$

$$-22I_1 - 10I_2 + 12 = 0$$

or $$+11I_1 + 5I_2 - 6 = 0 \qquad ...(i)$$

Circuit **BCDEB** gives,

$$6I_2 - 10 + 10(I_1 + I_2) = 0$$

$$10I_1 + 16I_2 - 10 = 0$$

$$5I_1 + 8I_2 - 5 = 0 \qquad ...(ii)$$

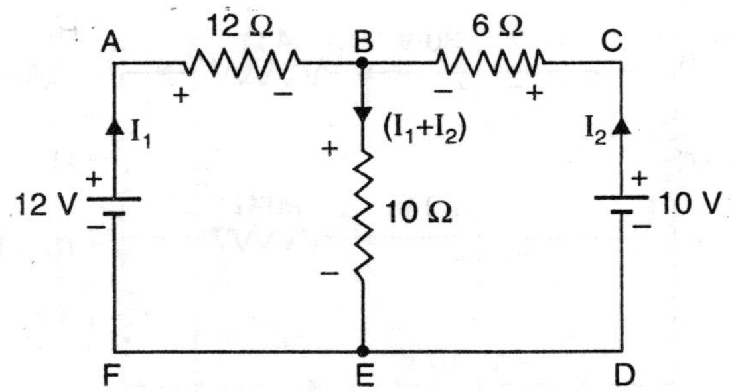

Fig. 22

Multiplying eqn. (*i*) by 5 and eqn. (*ii*) by 11 and subtracting, we get

$$55I_1 + 25I_2 - 30 = 0$$
$$55I_1 + 88I_2 - 55 = 0$$
$$- \quad - \quad +$$
$$-63I_2 + 25 = 0$$

i.e., $I_2 = 0.397$ A

Substituting this value in eqn. (*i*), we get

$$11I_1 + 5 \times 0.397 - 6 = 0$$

i.e., $I_1 = 0.365$ A

Hence, **Current through 12 Ω resistor, I_1 = 0.365 A. (Ans.)**

Current through 6 Ω resistor, I_2 = 0.397 A. (Ans.)

Current through 10 Ω resistor, $I_1 + I_2$ = 0.762 A. (Ans.)

The voltage drop across :

12 Ω resistor $= 0.365 \times 12 =$ **4.38 V. (Ans.)**

6 Ω resistor $= 0.397 \times 6 =$ **2.38 V. (Ans.)**

10 Ω resistor $= 0.762 \times 10 =$ **7.62 V. (Ans.)**

Example 8. *Find the magnitude and direction of currents in each of the batteries shown in Fig. 23.*

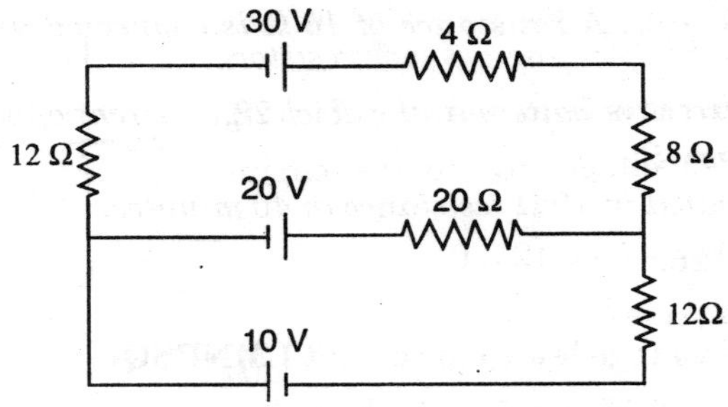

Fig. 23

Solution. Let the directions of currents I_1, I_2 and I_3 in the batteries be as shown in Fig. 24. Applying Kirchhoff's voltage law to the circuit **ABCFA**, we get

$$30 - 4I_1 - 8I_1 + 20I_2 - 20 - 12I_1 = 0$$

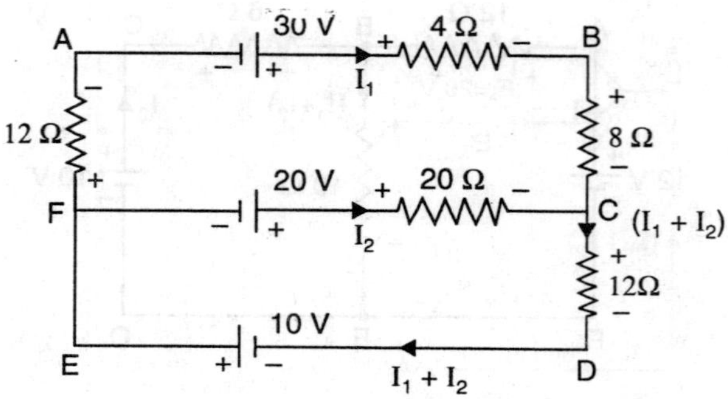

Fig. 24

$$-24I_1 + 20I_2 + 10 = 0$$

or $$12I_1 - 10I_2 - 5 = 0 \quad ...(i)$$

Circuit **ECDEF** gives,

$$20 - 20I_2 - 12(I_1 + I_2) + 10 = 0$$
$$20 - 20I_2 - 12I_1 - 12I_2 + 10 = 0$$
$$-12I_1 - 32I_2 + 30 = 0$$

or $$6I_1 + 16I_2 - 15 = 0 \quad ...(ii)$$

Multiplying eqn. (*ii*) by 2 and subtracting it from (*i*), we get

$$-42I_2 + 25 = 0$$

i.e., $$I_2 = 0.595 \text{ A}$$

Substituting this value of I_2 in eqn. (*i*), we get

$$12I_1 - 10 \times 0.595 - 5 = 0$$

or $$I_1 = 0.912 \text{ A}$$

Hence current through,

30 V battery, $I_1 = \mathbf{0.912\ A.}$ **(Ans).**

20 V battery, $I_2 = \mathbf{0.595\ A.}$ **(Ans.)**

10 V battery, $(I_1 + I_2) = \mathbf{1.507\ A.}$ **(Ans).**

Example 9. *The terminal resistances of batteries A and B are 2.5 Ω and 2 Ω respectively. The battery has an e.m.f. of 20 volts. A resistance of 10 Ω is connected across the battery terminals. Calculate :*

(*i*) *The discharge current of battery A, the discharge current of battery B being 1.75 A.*

(*ii*) *The e.m.f. of battery B.*

(*iii*) *The energy dissipated in 10 Ω resistance in 40 minutes.*

Solution. Refer Fig. 25.

(*i*) $\mathbf{I_1}$ **:**

Applying Kirchhoff's voltage law to the circuit **LMNPSQL**

$$20 - 2.5I_1 + 1.75 \times 2 - E_B = 0$$

i.e., $$E_B + 2.5I_1 - 23.5 = 0 \quad ...(i)$$

Circuit **LMNPTL** gives,

$$20 - 2.5I_1 - 10(I_1 + I_2) = 0$$

i.e., $$20 - 2.5I_1 - 10(I_1 + 1.75) = 0$$

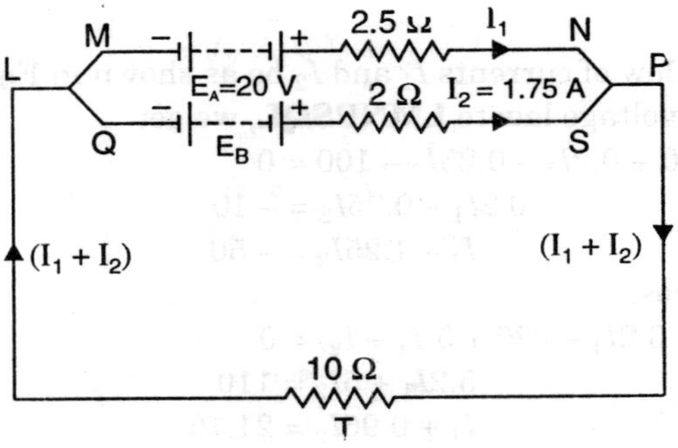

Fig. 25

or $$20 - 12.5I_1 - 17.5 = 0$$

or $$I_1 = 0.2 \text{ A.}$$

Hence discharge current of battery A = I_1 = 0.2 A. (Ans.)

(ii) **E_B :**

Substituting the value of I_1 in eqn. (*i*), we get

$$E_B + 2.5 \times 0.2 - 23.5 = 0$$

$\therefore$ $$E_B = 23 \text{ V}$$

Hence e.m.f. of battery B, E_B = 23 V. (Ans.)

(iii) **Energy dissipated in 10 Ω resistor :**

Energy dissipated $= I^2Rt$

$= (I_1 + I_2)^2 \times 10 \times (40 \times 60)$ joules

$= (0.2 + 1.75)^2 \times 10 \times (40 \times 60)$ joules

= 91260 joules. (Ans.)

Example 10. *A battery having an e.m.f. of 110 V and an internal resistance of 0.2 Ω is connected in parallel with another battery with e.m.f. of 100 V and a resistance of 0.25 Ω. The two in parallel are placed in series with a regulating resistance of 5 ohms and connected across 220 V mains. Calculate :*

(i) The magnitude and direction of the current in each battery.

(ii) The total current taken from the mains supply.

Solution. Refer Fig. 26.

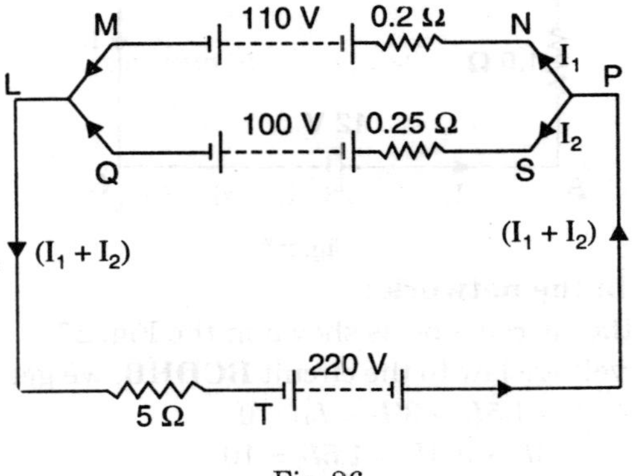

Fig. 26

(*i*) $\mathbf{I_1}$ **:,** $\mathbf{I_2}$ **:**

Let the directions of flow of currents I_1 and I_2 be as shown in Fig. 26.

Applying Kirchhoff's voltage law to **LMNPSQL**, we get

$$110 + 0.2I_1 - 0.25I_2 - 100 = 0$$

$$0.2I_1 - 0.25I_2 = -10$$

or $$I_1 - 1.25I_2 = -50 \quad ...(i)$$

Circuit **LMNPTL** gives,

$$110 + 0.2I_1 - 220 + 5(I_1 + I_2) = 0$$

$$5.2I_1 + 5I_2 = 110$$

$$I_1 + 0.96I_2 = 21.15 \quad ...(ii)$$

Subtracting (*ii*) from (*i*), we get

$$-2.21I_2 = -71.15$$

∴ $\mathbf{I_2 = 32.19\ A.}$ **(Ans.)**

and $\mathbf{I_1 = -9.75\ A.}$ **(Ans.)**

Since I_1 turns out to be negative, its actual direction of flow is *opposite* to that shown in Fig. 26. In other words it *is not a charging current but a discharging one*. However, I_2 is a *charging current*.

(*ii*) $\mathbf{(I_1 + I_2)}$ **:**

The total current taken from the mains supply,

$$I_1 + I_2 = -9.75 + 32.19 = \mathbf{22.44\ A.}\ \textbf{(Ans.)}$$

Example 11. *In the circuit shown in the Fig. 27 determine :*

(*i*) *All the currents in the network.*

(*ii*) *Voltages between the points.*

Solution. Refer Fig. 27.

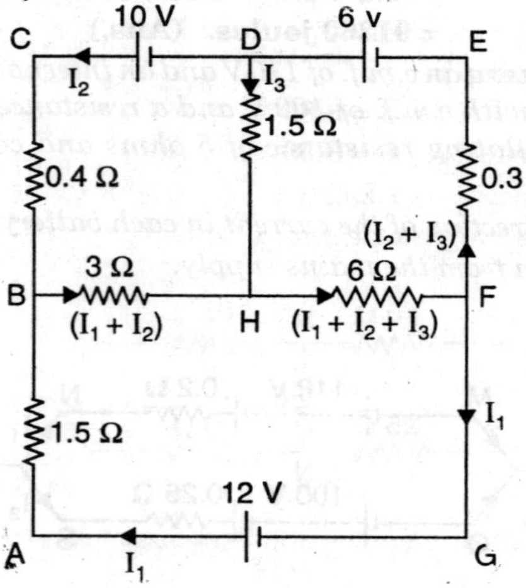

Fig. 27

(*i*) **All the currents in the network :**

Let the directions of the currents be as shown in the Fig. 27.

Applying Kirchhoff's voltage law to the circuit **BCDHB**, we get

$$0.4I_2 - 10 - 1.5I_3 + 3(I_1 + I_2) = 0$$

i.e., $$3I_1 + 3.4I_2 - 1.5I_3 = 10 \quad ...(i)$$

Circuit **HDEFH** gives,

$$1.5I_3 - 6 + 0.3(I_2 + I_3) + 6(I_1 + I_2 + I_3) = 0$$

$$6I_1 + 6.3I_2 + 7.8I_3 = 6 \quad ...(ii)$$

Circuit **ABHFGA** gives,

$$-1.5I_1 - 3(I_1 + I_2) - 6(I_1 + I_2 + I_3) + 12 = 0$$

$$-10.5I_1 - 9I_2 - 6I_3 + 12 = 0$$

i.e.,

$$10.5I_1 + 9I_2 + 6I_3 = 12 \quad ...(iii)$$

Multiplying eqn. (*ii*) by 2 and subtracting eqn. (*ii*) from eqn. (*i*), we get

$$0.5I_2 - 10.8I_3 = 14 \quad ...(iv)$$

Multiplying eqn. (*ii*) by 10.5 and eqn. (*iii*) by 6 and subtracting eqn. (*iii*) from eqn. (*ii*), we get

$$12.15I_2 + 45.9I_3 = -9 \quad ...(v)$$

Multiplying eqn. (*iv*) by 12.15 and eqn. (*v*) by 0.5 and subtracting eqn. (*v*) from eqn. (*iv*), we get

$$-154.17\,I_3 = 174.6$$

$\therefore$

$$I_3 = -1.132 \text{ A}$$

Substituting the value of I_3 in eqn. (*iv*), we get

$$0.5I_2 - 10.8 \times (-1.132) = 14$$

i.e.,

$$I_2 = 3.549 \text{ A}$$

Substituting the values of I_2 and I_3 in eqn. (*iii*), we get

$$10.5I_1 + 9 \times 3.549 + 6 \times (-1.132) = 12$$

i.e.,

$$I_1 = -1.252 \text{ A}$$

Hence the directions of I_1 and I_3 are actually *opposite to the assumed directions.*

The *current between B and H* = $I_1 + I_2 = -1.252 + 3.549$ = **2.297 A. (Ans.)**

The *current between H and F* = $I_1 + I_2 + I_3 = -1.252 + 3.549 - 1.132$ = **1.165 A. (Ans.)**

The current between H and F (and E) = $I_1 + I_3 = 3.549 - 1.132$ = **2.417 A. (Ans.)**

(*ii*) **Voltage between the points :**

Voltage across BH $= 2.297 \times 3$ = **6.891 V. (Ans.)**

Voltage across HF $= 1.165 \times 6$ = **6.99 V. (Ans.)**

Voltage across CE $= 12 - (-1.252 \times 1.5)$ = **13.878 V. (Ans.)**

Example 12. *Determine the magnitude and direction of flow of current in the branch MN for the circuit shown in Fig. 28.*

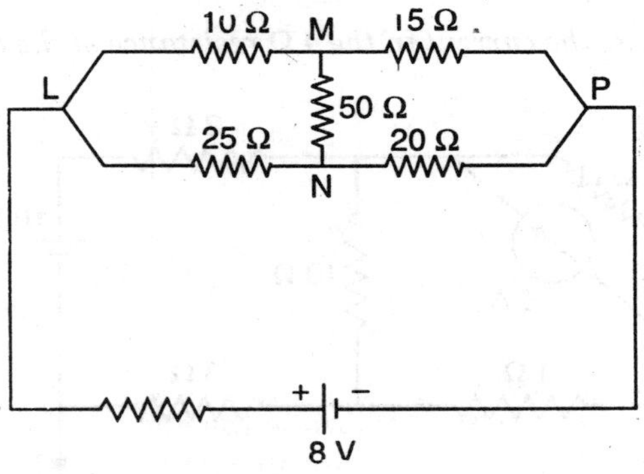

Fig. 28

Solution. Refer Fig. 29.

Let the directions and magnitudes of the currents flowing in the various circuits be as shown in Fig. 29.

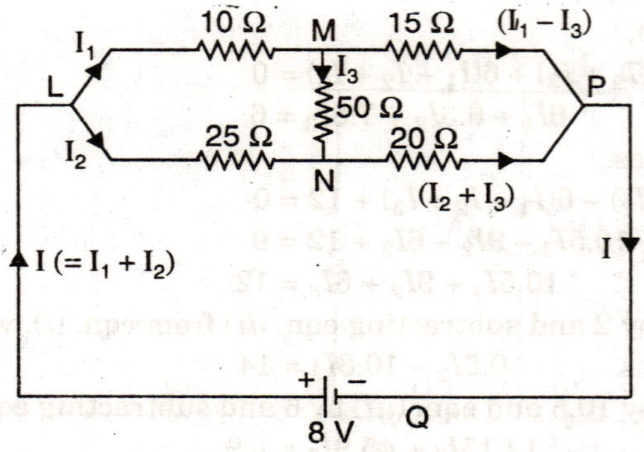

Fig. 29

Applying Kirchhoff's voltage law to the circuit **LMNL**, we get

$$-10I_1 - 50I_3 + 25I_2 = 0$$

or $$I_1 - 2.5I_2 + 5I_3 = 0 \quad ...(i)$$

Circuit **MPNM** gives,

$$-15(I_1 - I_3) + 20(I_2 + I_3) + 50I_3 = 0$$

$$-15I_1 + 20I_2 + 85I_3 = 0$$

or $$I_1 - 1.33I_2 - 5.66I_3 = 0 \quad ...(ii)$$

Circuit **LNPQL** gives,

$$-25I_2 - 20(I_2 + I_3) + 8 = 0$$

$$-45I_2 - 20I_3 + 8 = 0$$

or $$I_2 + 0.44I_3 = 0.177 \quad ...(iii)$$

Subtracting eqn. (*ii*) from eqn. (*i*), we get

$$-1.17I_2 + 10.66I_3 = 0$$

or $$I_2 - 9.11I_3 = 0 \quad ...(iv)$$

Subtracting eqn. (*iv*) from eqn. (*iii*), we get

$$9.55I_3 = 0.177$$

$$I_3 = 0.0185 \text{ A.}$$

Hence magnitude of current (I_3) flowing through the branch MN = 0.0185 A (from M to N). (Ans.)

Example 13. *Determine the current in the 4 Ω resistance of the circuit shown in Fig. 30.*

(Indore University)

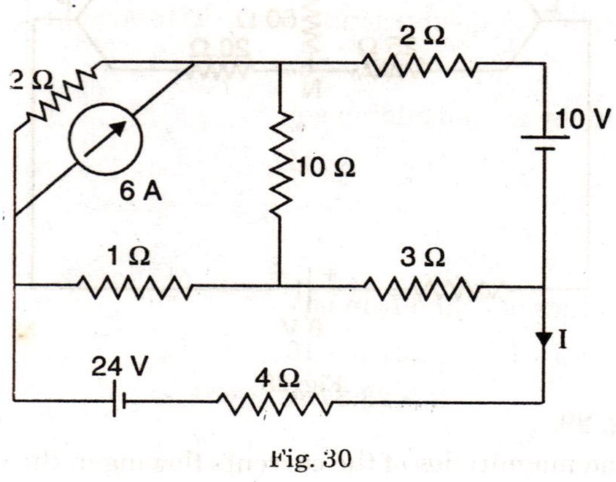

Fig. 30

Solution. Refer Fig. 31.

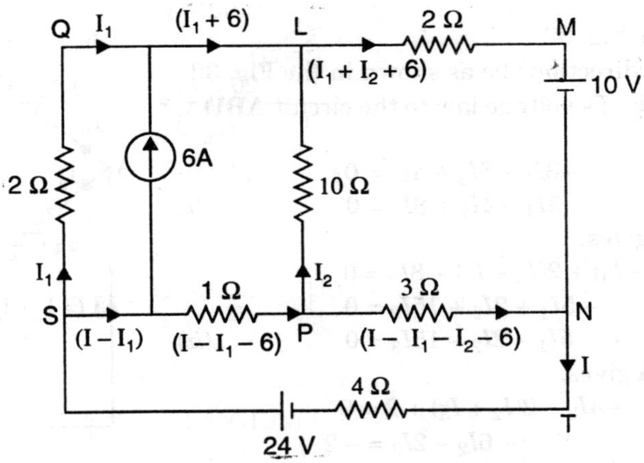

Fig. 31

Let the directions of various currents in different circuits be as shown in the Fig. 31.

Applying Kirchhoff's voltage law to the circuit **SQLPS**, we get

$$-2I_1 + 10I_2 + 1(I - I_1 - 6) = 0$$

$$I - 3I_1 + 10I_2 = 6 \quad ...(i)$$

Circuit **LMNPL** gives,

$$-2(I_1 + I_2 + 6) - 10 + 3(I - I_1 - I_2 - 6) - 10I_2 = 0$$

$$3I - 5I_1 - 15I_2 = 40 \quad ...(ii)$$

Circuit **SPNTS** gives,

$$-1(I - I_1 - 6) - 3(I - I_1 - I_2 - 6) - 4I + 24 = 0$$

$$-8I + 4I_1 + 3I_2 = -48$$

or $$8I - 4I_1 - 3I_2 = 48 \quad ...(iii)$$

Multiplying eqn. (*i*) by 3 and subtracting eqn. (*ii*) from eqn. (*i*), we get

$$-4I_1 + 45I_2 = -22$$

or $$I_1 - 11.25I_2 = 5.5 \quad ...(iv)$$

Multiplying eqn. (*i*) by 8 and subtracting eqn. (*iii*) from eqn. (*i*), we get

$$-20I_1 + 83I_2 = 0 \quad ...(v)$$

Multiplying eqn. (*iv*) by 20 and adding eqn. (*v*), we get

$$-142I_2 = 110$$

∴ $$I_2 = -0.774 \text{ A}$$

and $$I_1 = -3.212 \text{ A}$$

Substituting the values of I_1 and I_2 in eqn. (*i*), we get

$$I - 3 \times (-3.212) + 10 \times (-0.774) = 6$$

i.e. $$\mathbf{I = 23.37\ A.} \quad \textbf{(Ans.)}$$

Example 14. *Determine the current in each of the resistors of the network shown in the Fig. 32.*

Solution. Refer Fig. 32.

Let the current directions be as shown in the Fig. 32.

Applying Kirchhoff's voltage law to the circuit **ABDA**, we get

$$-3I_1 - 8I_3 + 4I_2 = 0$$

or $$3I_1 - 4I_2 + 8I_3 = 0 \qquad ...(i)$$

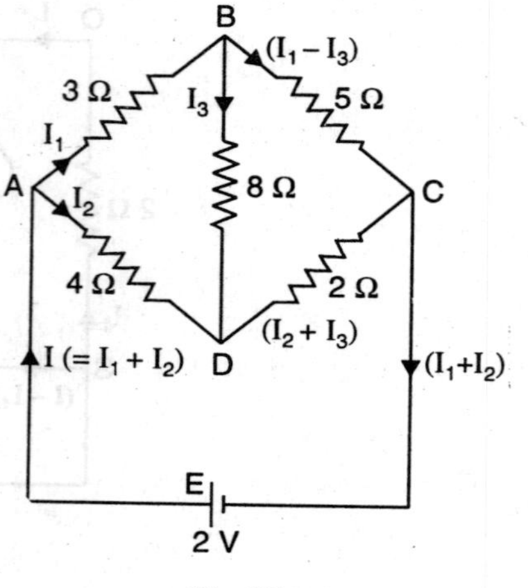

Fig. 32

Circuit **BCDB** gives,

$$-5(I_1 - I_3) + 2(I_2 + I_3) + 8I_3 = 0$$

$$-5I_1 + 2I_2 + 15I_3 = 0$$

$$5I_1 - 2I_2 - 15I_3 = 0 \qquad ...(ii)$$

Circuit **ADCEA** gives,

$$-4I_2 - 2(I_2 + I_3) + 2 = 0$$

$$-6I_2 - 2I_3 = -2$$

or $$3I_2 + I_3 = 1 \qquad ...(iii)$$

Multiplying eqn. (*i*) by 5 and eqn. (*ii*) by 3 and subtracting (*ii*) from (*i*), we get

$$15I_1 - 20I_2 + 40I_3 = 0$$

$$15I_1 - 6I_2 - 45I_3 = 0$$

$$\quad - \quad + \quad +$$

$$-14I_2 + 85I_3 = 0$$

or $$14I_2 - 85I_3 = 0 \qquad ...(iv)$$

Multiplying eqn. (*iii*) by 14 and eqn. (*iv*) by 3 and subtracting (*iv*) from (*iii*), we get

$$42I_2 + 14I_3 = 14$$

$$42I_2 - 255I_3 = 0$$

$$\quad - \quad +$$

$$269I_3 = 14$$

$\therefore$ $$I_3 = 0.052 \text{ A}$$

From eqn. (*iv*), $$I_2 = 0.316 \text{ A}$$

From eqn. (*i*), $$I_1 = 0.283 \text{ A}$$

Hence, Current through 3 Ω resistor = $\mathbf{I_1 = 0.283}$ **A. (Ans.)**

Current through 4 Ω resistor = $\mathbf{I_2 = 0.316}$ **A. (Ans.)**

Current through 8 Ω resistor = $\mathbf{I_3 = 0.052}$ **A. (Ans.)**

Current through 5 Ω resistor = $I_1 - I_3 =$ **0.231 A. (Ans.)**

Current through 2 Ω resistor = $I_2 + I_3 =$ **0.368 A. (Ans.)**

Example 15. *Determine the branch currents in the network of Fig. 33.*

Solution. Refer Fig. 33. Let the current directions be as shown.

Applying Kirchhoff's voltage law to the circuit **ABDA**, we get

$$5 - I_1 \times 1 - I_3 \times 1 + I_2 \times 1 = 0$$

or $$I_1 - I_2 + I_3 = 5 \qquad ...(i)$$

Circuit **BCDB** gives,

$$-(I_1 - I_3) \times 1 + 5 + (I_2 + I_3) \times 1 + I_3 \times 1 = 0$$

$$-I_1 + I_2 + 3I_3 = -5$$

or $$I_1 - I_2 - 3I_3 = 5 \qquad ...(ii)$$

Circuit **ADCEA** gives,

$$-I_2 \times 1 - (I_2 + I_3) \times 1 + 10 - (I_1 + I_2) \times 1 = 0$$

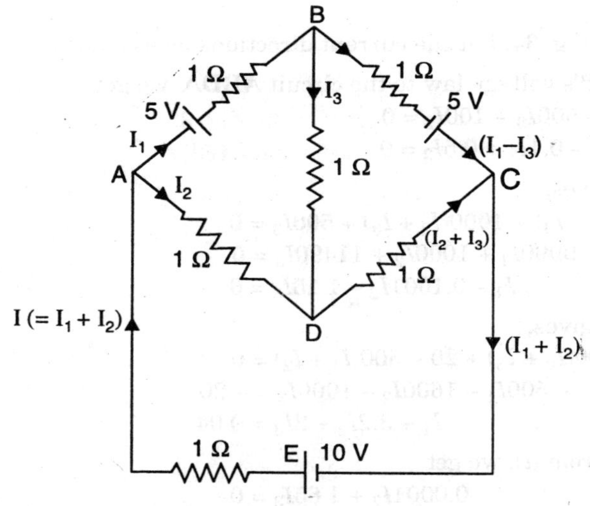

Fig. 33

$$-I_1 - 3I_2 - I_3 = -10$$

or $$I_1 + 3I_2 + I_3 = 10 \quad ...(iii)$$

Subtracting (*ii*) from (*i*), we get, $I_3 = 0$

From (*i*), we have, $I_1 - I_2 = 5$ ...(*iv*)

and from (*iii*), we have, $I_1 + 3I_2 = 10$ ...(*v*)

By solving (*iv*) and (*v*), we get, $I_2 = 1.25$ A

and $I_1 = 6.25$ A

Hence, Current in branch AB = **6.25 A. (Ans.)**

Current in branch BC = **6.25 A. (Ans.)**

Current in branch BD = **0. (Ans.)**

Current in branch AD = **1.25 A. (Ans.)**

Current in branch DC = **1.25 A. (Ans.)**

Current in branch CA = **7.5 A. (Ans.)**

Example 16. *Find the current in the galvanometer arm of the wheatstone bridge shown in Fig. 34.*

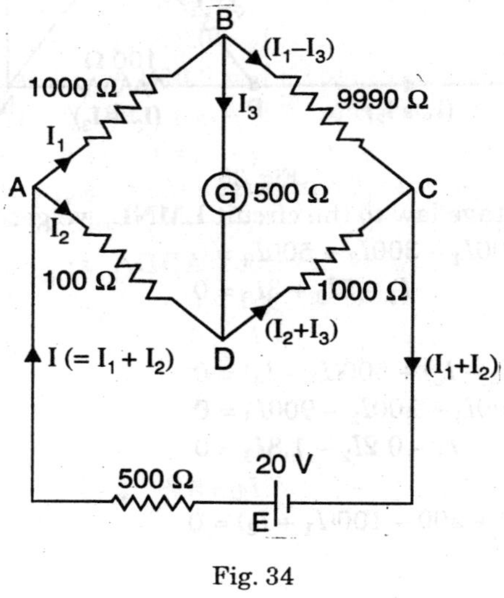

Fig. 34

Solution. Refer Fig. 34. Let the current directions be as shown.

Applying Kirchoff's voltage law to the circuit **ABDA**, we get

$$-1000I_1 - 500I_3 + 100I_2 = 0$$

$$I_1 - 0.1I_2 + 0.5I_3 = 0 \quad ...(i)$$

Circuit **BCDB** gives,

$$-9990(I_1 - I_3) + 1000(I_2 + I_3) + 500I_3 = 0$$

$$-9990I_1 + 1000I_2 + 11490I_3 = 0$$

$$I_1 - 0.1001I_2 - 1.15I_3 = 0 \quad ...(ii)$$

Circuit **ADCEA** gives,

$$-100I_2 - 1000(I_2 + I_3) + 20 - 500(I_1 + I_2) = 0$$

$$-500I_1 - 1600I_2 - 1000I_3 = -20$$

$$I_1 + 3.2I_2 + 2I_3 = 0.04 \quad ...(iii)$$

Subtracting (*ii*) from (*i*), we get

$$0.0001I_2 + 1.65I_3 = 0 \quad ...(iv)$$

Subtracting (*iii*) from (*ii*), we get

$$+3.3001I_2 + 3.15I_3 = +0.04 \quad ...(v)$$

Solving (*iv*) and (*v*), we get, $I_3 = 0.735 \times 10^{-6}$ A

Hence current in the galvanometer arm = 0.735 μA. (Ans.)

Example 17. *Determine the current supplied by the battery in the circuit shown in Fig. 35.*

(Bombay University)

Solution. Refer Fig. 35.

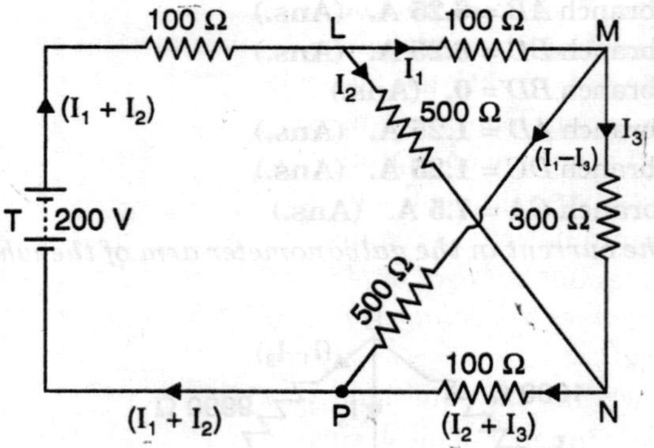

Fig. 35

Applying Kirchhoff's voltage law to the circuit **LMNL**, we get

$$-100I_1 - 300I_3 + 500I_2 = 0$$

$$I_1 - 5I_2 + 3I_3 = 0 \quad ...(i)$$

Circuit **MNPM** gives,

$$-300I_3 - 100(I_2 + I_3) + 500(I_1 - I_3) = 0$$

$$500I_1 - 100I_2 - 900I_3 = 0$$

$$I_1 - 0.2I_2 - 1.8I_3 = 0 \quad ...(ii)$$

Circuit **LMPTL** gives,

$$-100I_1 - 500(I_1 - I_3) + 200 - 100(I_1 + I_2) = 0$$

$$-700I_1 - 100I_2 + 500I_3 = -200$$
$$I_1 + 0.143I_2 - 0.714I_3 = 0.286 \quad ...(iii)$$

Multiplying (*i*) by 1.8 and (*ii*) by 3 and adding, we get

$$1.8I_1 - 9I_2 + 5.4I_3 = 0$$
$$3I_1 - 0.6I_2 - 5.4I_3 = 0$$
$$\overline{4.8I_1 - 9.6I_2 = 0}$$

or

$$I_1 - 2I_2 = 0 \quad ...(iv)$$

Multiplying (*ii*) by 0.714 and (*iii*) by 1.8 and subtracting, we get

$$0.714I_1 - 0.143I_2 - 1.285I_3 = 0$$
$$1.8I_1 + 0.257I_2 - 1.285I_3 = 0.515$$
$$- \quad - \quad + \quad -$$
$$\overline{-1.086I_1 - 0.4I_2 = -0.515}$$

or

$$I_1 + 0.368I_2 = 0.474 \quad ...(v)$$

Subtracting (*v*) from (*iv*), we get $2.368I_2 = 0.474$

i.e., $I_2 = 0.2$ A

and $I_1 = 0.4$ A

$\therefore$ **Current supplied by the battery** $= I_1 + I_2 = 0.2 + 0.46 =$ **0.6 A. (Ans.)**

5.2. Maxwell's Loop (or Mesh) Current Method

The method of *loop* or *mesh* currents is generally used in solving networks having some degree of complexity. Such a degree of complexity already begins for a network of three meshes. It might even be convenient at times to use the method of loop or mesh currents for solving a two-mesh circuit.

The *mesh-current method* is preferred to the general or branch-current method because the unknowns in the initial stage of solving a network are equal to the number of meshes, *i.e.*, the mesh currents. The necessity of writing the node-current equations, as done in the general or branch-current method where branch currents are used, is *obviated*. There are as many mesh-voltage equations as these are independent loop or mesh, currents. Hence, the M-mesh currents are obtained by solving the M-mesh voltages or loop equations for M unknowns. After solving for the mesh currents, only a matter of resolving the confluent mesh currents into the respective branch currents by very simple algebraic manipulations is required.

This method eliminates a great deal of tedious work involved in branch-current method and is best suited when energy sources are voltage sources rather than current sources. This method can be used only for *planar circuits*.

The **procedure** for writing the equations is as follows :

1. Assume the smallest number of mesh currents so that at least one mesh current links every element. As a matter of convenience, all mesh currents are assumed to have a *clockwise direction*.

The *number of mesh currents is equal to the number of meshes in the circuit*.

2. For each mesh write down the Kirchhoff's voltage law equation. Where more than one mesh current flows through an element, the algebraic sum of currents should be used. The algebraic sum of mesh currents may be sum or the difference of the currents flowing through the element depending on the direction of mesh currents.

3. Solve the above equations and from the mesh currents find the branch currents.

Fig. 36 shows two batteries E_1 and E_2 connected in a network consisting of three resistors. Let the loop currents for two meshes be I_1 and I_2 (both clockwise-assumed). It is obvious that current through R_3 (when considered as a part of first loop) is $(I_1 - I_2)$. However, when R_3 is considered part of the second loop, current through it is $(I_2 - I_1)$.

Applying Kirchhoff's voltage law to the two loops, we get

$$E_1 - I_1R_1 - R_3(I_1 - I_2) = 0$$

or
$$E_1 - I_1(R_1 + R_3) + I_2R_3 = 0 \quad \text{...Loop 1}$$

Similarly,
$$-I_2R_2 - E_2 - R_3(I_2 - I_1) = 0$$
$$-I_2R_2 - E_2 - I_2R_3 + I_1R_3 = 0$$

or
$$I_1R_3 - I_2(R_2 + R_3) - E_2 = 0 \quad \text{...Loop 2}$$

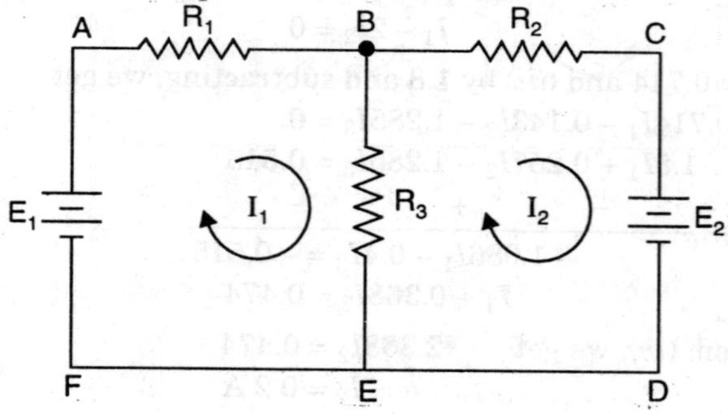

Fig. 36

The above two equations can be solved not only to find loop currents but branch currents as well.

Example 18. *Determine the currents through various resistors of the circuit shown in Fig. 37 using the concept of mesh currents.*

Solution. Refer Fig. 37.

Since there are two meshes, let the loop currents be as shown.

Applying Kirchhoff's law to loop 1, we get

$$24 - 4I_1 - 2(I_1 - I_2) = 0$$
$$-6I_1 + 2I_2 + 24 = 0$$

or
$$3I_1 - I_2 = 12 \quad \text{...}(i)$$

For loop 2, we have

$$-2(I_2 - I_1) - 6I_2 - 12 = 0$$
$$2I_1 - 8I_2 - 12 = 0$$
$$I_1 - 4I_2 = 6 \quad \text{...}(ii)$$

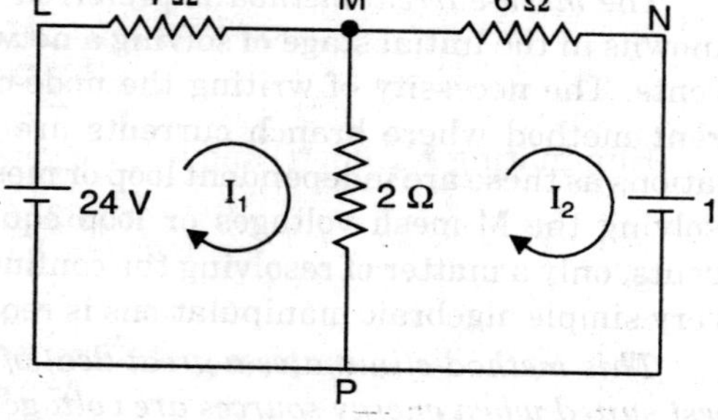

Fig. 37

Solving (*i*) and (*ii*), we get, $I_1 = \frac{42}{11}$ A

and $I_2 = -\frac{6}{11}$ A

Hence Current through 4 Ω resistor = $\frac{42}{11}$ **A (from L to M). (Ans.)**

Current through 6 Ω resistor = $\frac{6}{11}$ **A (from N to M). (Ans.)**

Current through 2 Ω resistor = $\frac{42}{11} - \left(-\frac{6}{11}\right) = \frac{48}{11}$ **A (from M to P). (Ans.)**

Example 19. *Determine the current supplied by each battery in the circuit shown in Fig. 38.* **(Aligarh University)**

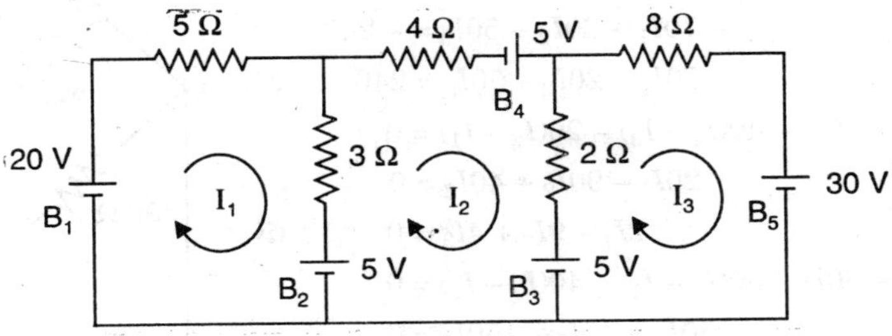

Fig. 38

Solution. Refer Fig. 38.

As there are three meshes, let the three loop currents be as shown.

Applying Kirchhoff's law to loop 1, we get

$$20 - 5I_1 - 3(I_1 - I_2) - 5 = 0$$

or $$8I_1 - 3I_2 = 15 \quad ...(i)$$

For loop 2, we have

$$-4I_2 + 5 - 2(I_2 - I_3) + 5 + 5 - 3(I_2 - I_1) = 0$$

$$3I_1 - 9I_2 + 2I_3 = -15 \quad ...(ii)$$

For loop 3, we have

$$-8I_3 - 30 - 5 - 2(I_3 - I_2) = 0$$

$$2I_2 - 10I_3 = 35 \quad ...(iii)$$

Eliminating I_1 from (*i*) and (*ii*), we get

$$63I_2 - 16I_3 = 165 \quad ...(iv)$$

Solving (*iii*) and (*iv*),we get

$$I_2 = 1.82 \text{ A and } I_3 = -3.15 \text{ A}$$

(– ve sign means direction of current is counter-clockwise)

Substituting the value of I_2 in (*i*), we get

$$I_1 = 2.56 \text{ A}$$

Current through battery B_1 (discharging current) = I_1 = **2.56 A. (Ans.)**

Current through battery B_2 (charging current) = $I_1 - I_2$ = 2.56 – 1.82 = **0.74 A. (Ans.)**

Current through battery B_3 (discharging current) = $I_2 + I_3$ = 1.82 + 3.15 = **4.97 A. (Ans.)**

Current through battery B_4 (discharging current) = I_2 = **1.82 A. (Ans.)**

Current through battery B_5 (discharging current) = I_3 = **3.15 A. (Ans.)**

Example 20. *Determine the currents through the different branches of the bridge circuit shown in Fig. 39.*

Solution. Refer Fig. 39.

The three mesh currents are assumed as shown.

The equations for the three meshes are :

For loop 1 : $240 - 20(I_1 - I_2) - 50(I_1 - I_3) = 0$

or $-70I_1 + 20I_2 + 50I_3 = -240$

or $70I_1 - 20I_2 - 50I_3 = 240$ ...(*i*)

For loop 2 : $-30I_2 - 40(I_2 - I_3) - 20(I_2 - I_1) = 0$

or $20I_1 - 90I_2 + 40I_3 = 0$

or $2I_1 - 9I_2 + 4I_3 = 0$ ...(*ii*)

For loop 3 : $-60I_3 - 50(I_3 - I_1) - 40(I_3 - I_2) = 0$

$50I_1 + 40I_2 - 150I_3 = 0$

$5I_1 + 4I_2 - 15I_3 = 0$ ...(*iii*)

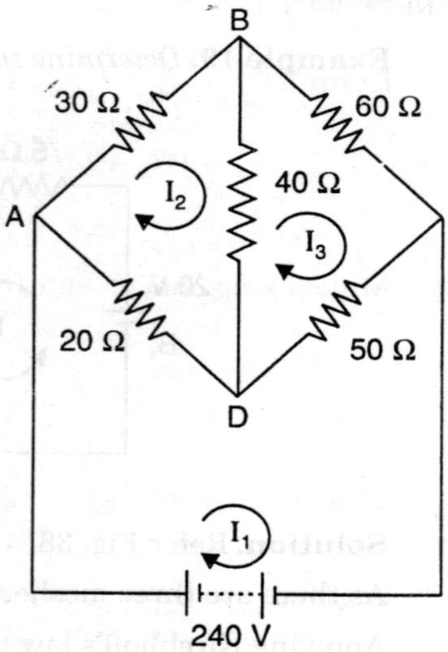

Fig. 39

Solving these equations, we get

$I_1 = 6.10$ A, $I_2 = 2.56$ A, $I_3 = 2.72$ A

Current through 30 Ω resistor = I_2 = **2.56 A (A to B). (Ans.)**

Current through 60 Ω resistor = I_3 = **2.72 A (B to C). (Ans.)**

Current through 20 Ω resistor = $I_1 - I_2 = 6.10 - 2.56$ = **3.54 A (A to D). (Ans.)**

Current through 50 Ω resistor = $I_1 - I_3 = 6.10 - 2.72$ = **3.38 A (D to C). (Ans.)**

Current through 40 Ω resistor = $I_3 - I_2 = 2.72 - 2.56$ = **0.16 A (D to B). (Ans.)**

Example 21. *Determine the current in the 8 Ω branch in the circuit shown in Fig. 40.* **(Nagpur University)**

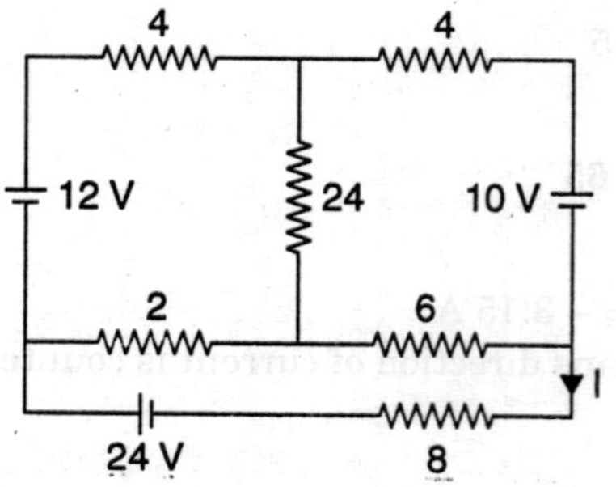

Fig. 40

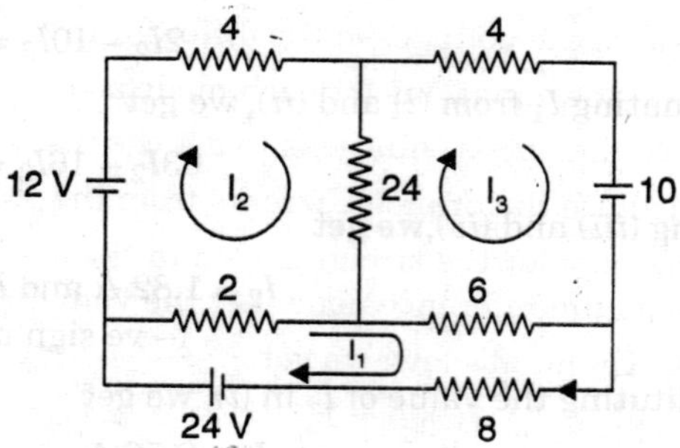

Fig. 41

Solution. The three loop currents are shown in Fig. 41 ; for these loops we have

Loop-1 : $-2(I_1 - I_2) - 6(I_1 - I_3) - 8I_1 + 24 = 0$

$-2I_1 + 2I_2 - 6I_1 + 6I_3 - 8I_1 + 24 = 0$

or $16I_1 - 2I_2 - 6I_3 = 24$

or $8I_1 - I_2 - 3I_3 = 12$ ...(*i*)

Loop-2 : $12 - 4I_2 - 24(I_2 - I_3) - 2(I_2 - I_1) = 0$

or $12 - 4I_2 - 24I_2 + 24I_3 - 2I_2 + 2I_1 = 0$

or $2I_1 - 30I_2 + 24I_3 = -12$

or $I_1 - 15I_2 + 12I_3 = -6$ ...(*ii*)

Loop-3 :

$$-24(I_3 - I_2) - 4I_3 - 10 - 6(I_3 - I_1) = 0$$

or $$-24I_3 + 24I_2 - 4I_3 - 10 - 6I_3 + 6I_1 = 0$$

or $$6I_1 + 24I_2 - 34I_3 = 10$$

or $$3I_1 + 12I_2 - 17I_3 = 5 \quad ...(iii)$$

Multiplying eqn. (*i*) and 15 and subtracting eqn. (*ii*) from (*i*), we get

$$\begin{array}{r} 120I_1 - 15I_2 - 45I_3 = 180 \\ I_1 - 15I_2 + 12I_3 = -6 \\ - \quad + \quad - \quad + \\ \hline 119I_1 \qquad - 57I_3 = 186 \end{array} \quad ...(iv)$$

Multiplying eqn. (*ii*) by 12 and eqn. (*iii*) by 15 and adding eqn. (*iii*) and eqn. (*ii*), we get

$$\begin{array}{r} 12I_1 - 180I_2 + 144I_3 = -72 \\ 45I_1 + 180I_2 - 255I_3 = 75 \\ \hline 57I_1 \qquad - 111I_3 = 3 \end{array} \quad ...(v)$$

Multiplying eqn. (*iv*) by 111 and eqn. (*v*) by 57 and subtracting eqn. (*v*) from eqn. (*iv*), we get

$$\begin{array}{r} 13209I_1 - 6327I_3 = 20646 \\ 3249I_1 - 6327I_3 = 171 \\ - \quad + \quad - \\ \hline 9960I_1 \qquad = 20475 \end{array}$$

$\therefore$ $\mathbf{I_1}$ (Current through 8 Ω resistance) $= \dfrac{20475}{9960} =$ **2.06 A. (Ans.)**

5.3. Nodal Voltage Method

Under this method the following ***procedure*** is adopted :

1. Assume the voltages of the different independent nodes.
2. Write the equations for each mode as per Kirchhoff's current law.
3. Solve the above equations to get the node voltages.
4. Calculate the branch currents from the values of node voltages.

Let us consider the circuit shown in the Fig. 42 *L* and *M* are the two independent nodes ; *M* can be taken as the reference node. Let the voltage of node *L* (with respect to *M*) be V_L.

Using Kirchhoff's law, we get

$$I_1 + I_2 = I_3 \quad ...(10)$$

Ohm's law gives $$I_1 = \frac{V_1}{R_1} = \frac{(E_1 - V_L)}{R_1}$$

$$I_2 = \frac{V_2}{R_2} = \frac{(E_2 - V_L)}{R_2} \quad ...(11)$$

$$I_3 = \frac{V_3}{R_3} = \frac{V_L}{R_3}$$

$$\frac{E_1 - V_L}{R_1} + \frac{E_2 - V_L}{R_2} = \frac{V_L}{R_3} \quad ...(12)$$

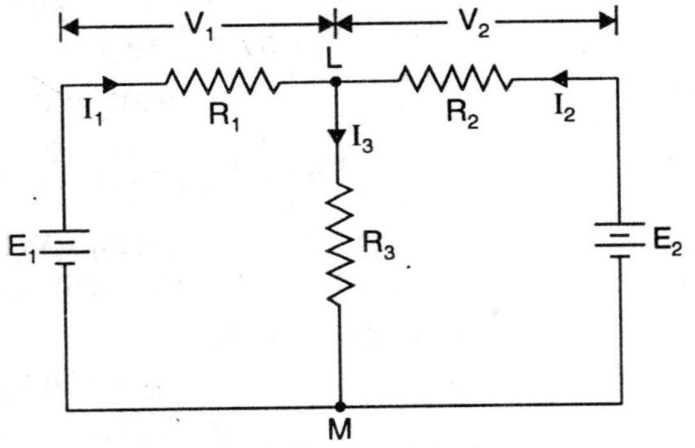

Fig. 42

Rearranging the terms, we get

$$V_L\left[\frac{1}{R_1} + \frac{1}{R_2} + \frac{1}{R_3}\right] - \frac{E_1}{R_1} - \frac{E_2}{R_2} = 0 \quad ...(13)$$

It may be noted that the above nodal equation contains the following terms :

(*i*) The node voltage multiplied by the sum of all conductances connected to that anode. This term is *positive*.

(*ii*) The node voltage at the other end of each branch (connected to this node) multiplied by the conductance of branch. These terms are *negative*.

— In this method of solving a network the *number of equations required for the solution is one less than the number of independent nodes in the network.*

— In general the nodal analysis *yields similar solutions.*

— The nodal method is very suitable for *computer work.*

Example 22. *For the circuit shown in Fig. 43 find the currents through the resistances* R_3 *and* R_4.

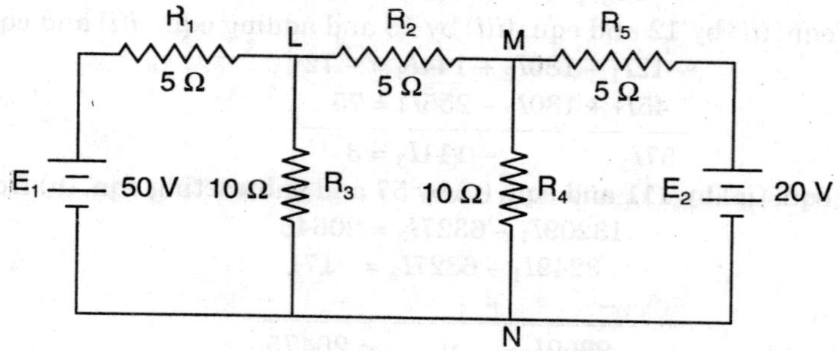

Fig. 43

Solution. Refer Fig. 43.

Let L, M and N = independent nodes, and

V_L and V_M = voltages of nodes L and M with respect to node N.

The nodal equations for the nodes L and M are :

$$V_L\left[\frac{1}{R_1}+\frac{1}{R_2}+\frac{1}{R_3}\right]-\frac{E_1}{R_1}-\frac{V_M}{R_2}=0 \quad ...(i)$$

$$V_M\left[\frac{1}{R_2}+\frac{1}{R_4}+\frac{1}{R_5}\right]-\frac{E_2}{R_5}-\frac{V_L}{R_2}=0 \quad ...(ii)$$

Substituting the values in (*i*) and (*ii*) and simplifying, we get

$$V_L\left(\frac{1}{5}+\frac{1}{5}+\frac{1}{10}\right)-\frac{50}{5}-\frac{V_M}{5}=0$$

or $$2.5V_L - V_M - 50 = 0 \quad ...(iii)$$

and $$V_M\left(\frac{1}{5}+\frac{1}{10}+\frac{1}{5}\right)-\frac{20}{5}-\frac{V_L}{5}=0$$

or $$2.5V_M - V_L - 20 = 0$$

or $$-V_L + 2.5V_M - 20 = 0 \quad ...(iv)$$

Solving (*iii*) and (*iv*), we get

$$V_L = 27.6 \text{ V}, \quad V_M = 19.05 \text{ V}$$

Current through $\mathbf{R_3} = \frac{V_L}{R_3} = \frac{27.6}{10} =$ **2.76 A. (Ans.)**

Current through $\mathbf{R_4} = \frac{V_M}{R_4} = \frac{19.05}{10} =$ **1.905 A. (Ans.)**

Example 23. *Using Node voltage method, find the current in 6 Ω resistance for the network shown in Fig. 44.*

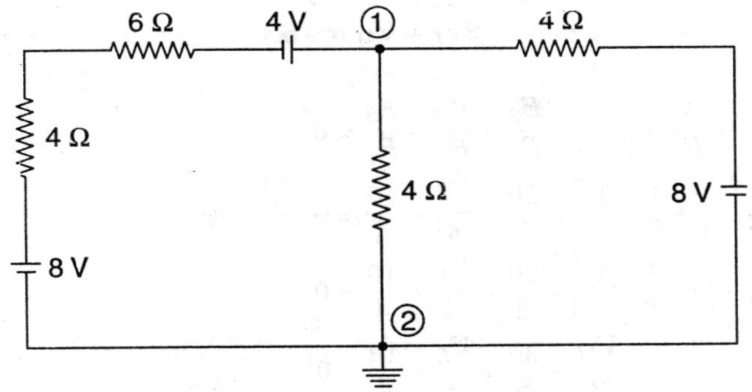

Fig. 44

Solution. Refer Fig. 44. Considering node 2 as the reference node and using node voltage method, we have

$$V_1\left[\frac{1}{(6+4)}+\frac{1}{4}+\frac{1}{4}\right]-\frac{8}{4}-\left(\frac{8+4}{10}\right)=0$$

(The reason for adding the two battery voltages of 4 V and 8 V is because they are connected in additive series).

or $$V_1\,(0.1 + 0.25 + 0.25) - 2 - 1.2 = 0$$

∴ $$V_1 = 5.33 \text{ V}$$

The current flowing through the 6 Ω resistance towards node 1 is

$$=\frac{12-5.33}{6+4}=\mathbf{0.667\ A.}\ \textbf{(Ans.)}$$

Example 24. *Find the branch currents in the circuit of Fig. 45 by nodal analysis.*

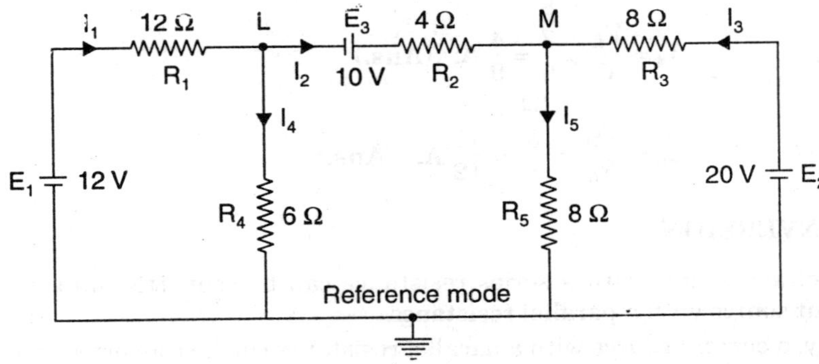

Fig. 45

Solution. Node L :

$$V_L\left[\frac{1}{R_1}+\frac{1}{R_2}+\frac{1}{R_4}\right]-\frac{E_1}{R_1}-\frac{V_M}{R_2}+\frac{E_3}{R_2}=0$$

Substituting the given data, we get

$$V_L\left[\frac{1}{12}+\frac{1}{4}+\frac{1}{6}\right]-\frac{12}{12}-\frac{V_M}{4}+\frac{10}{4}=0$$

$$V_L\left[\frac{1+3+2}{12}\right]-1-\frac{V_M}{4}+\frac{10}{4}=0$$

or $$0.5V_L-1-0.25V_M+2.5=0$$

or $$2V_L-V_M=-6 \quad \text{...}(i)$$

Node M :

$$V_M\left[\frac{1}{R_2}+\frac{1}{R_3}+\frac{1}{R_5}\right]-\frac{E_2}{R_3}-\frac{V_L}{R_2}-\frac{E_3}{R_2}=0$$

$$V_M\left[\frac{1}{4}+\frac{1}{8}+\frac{1}{8}\right]-\frac{20}{8}-\frac{V_L}{4}-\frac{10}{4}=0$$

or $$V_M\left[\frac{2+1+1}{8}\right]-\frac{20}{8}-\frac{V_L}{4}-\frac{10}{4}=0$$

or $$\frac{V_M}{2}-\frac{20}{8}-\frac{V_L}{4}-\frac{10}{4}=0$$

$$V_M-\frac{V_L}{2}=10 \quad \text{...}(ii)$$

From eqn. (*i*) and (*ii*), we get

$$V_L=\frac{8}{3}\,V;\ V_M=\frac{34}{3}\,V$$

$$\mathbf{I_1}=\frac{E_1-V_L}{R_1}=\frac{12-\frac{8}{3}}{12}=\frac{\mathbf{7}}{\mathbf{9}}\ \textbf{A. (Ans.)}$$

$$I_2=\frac{V_L+E_1-V_M}{R_2}=\frac{\frac{8}{3}+10-\frac{34}{3}}{4}=\frac{\mathbf{1}}{\mathbf{3}}\ \textbf{A. (Ans.)}$$

$$I_3=\frac{E_2-V_M}{R_3}=\frac{20-\frac{34}{3}}{8}=\frac{\mathbf{13}}{\mathbf{12}}\ \textbf{A. (Ans.)}$$

$$\mathbf{I_4}=\frac{V_L}{6}=\frac{\frac{8}{3}}{6}=\frac{\mathbf{4}}{\mathbf{9}}\ \textbf{A. (Ans.)}$$

$$\mathbf{I_5}=\frac{V_M}{R_5}=\frac{\frac{34}{3}}{8}=\frac{\mathbf{17}}{\mathbf{12}}\ \textbf{A. (Ans.)}$$

6. SOURCE CONVERSION

A given voltage source with a series resistance can be converted into (or replaced by) an equivalent current source with a parallel resistance.

Conversely, a current source with a parallel resistance can be converted into a voltage source with a series resistance.

- Suppose, we want to convert the voltage source of Fig. 46 (*i*) into an equivalent current source. To do so, we shall find the value of current, supplied by the source when a 'short' is put across terminals *L* and *M* as shown in Fig. 46 (*ii*) ; this current is $I=\frac{V}{R}$. A current source supplying this current *I* and having the same resistance *R* connected in *parallel* with it represents the equivalent source. It is shown in Fig. 46 (*iii*).

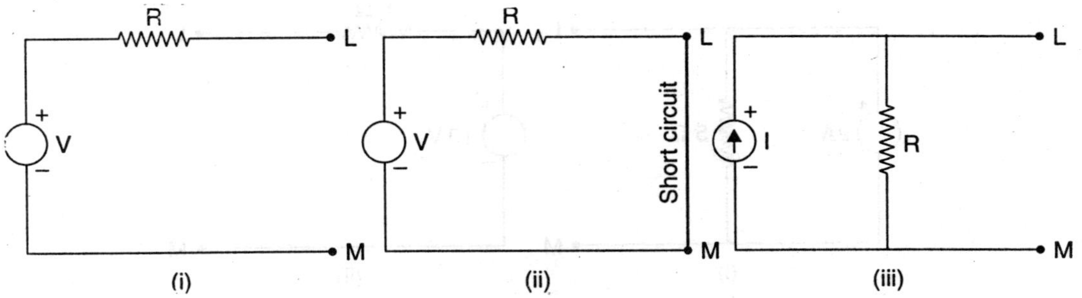

Fig. 46

- Similarly, a current source of I and a parallel resistance R can be converted into a voltage source of voltage $V \equiv IR$ and a resistance R in series with it.

It may be kept in mind that a voltage source-series resistance combination is equivalent to (or replaceable by) a constant source-parallel resistance combination if, and only if their *respective open-circuit voltages are equal, and respective short-circuit currents are equal.*

Example. Refer Fig. 46 (*i*). Voltage across terminals L and M when they are open (*i.e.* open-circuit voltage V_{OC}) is V itself because there is no drop across R. Short-circuit current across $LM = I = \frac{V}{R}$.

Refer Fig. 46 (*iii*). The open circuit voltage across LM = drop across $R = IR = V$.

If a short is placed across LM, whole of I passes through it because R is completely shorted out.

Example 25. *Convert the voltage source of Fig. 47 (i) into an equivalent current source.*

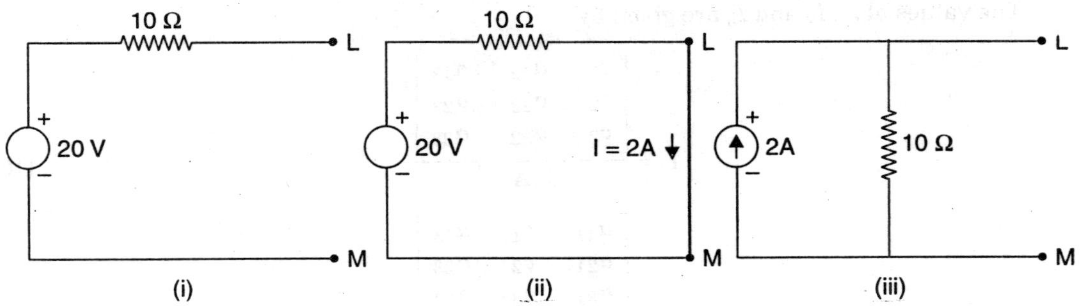

Fig. 47

Solution. As shown in Fig. 47 (*ii*), current obtained by putting a short across terminals L and M is $\frac{20}{10} = 2A$.

Hence the equivalent current source is as shown in Fig. 47 (*iii*).

Example 26. *Find the equivalent voltage source for the current source in Fig. 48 (i).*

Solution. The open-circuit voltage across termainals A and B in Fig. 48 (*i*) is V_{OC} = drop across $R = 2 \times 5 = 10$ V.

Hence, voltage source has a voltage of 10 V and the same resistance of 5 Ω connected in series [Fig. 48 (*ii*)].

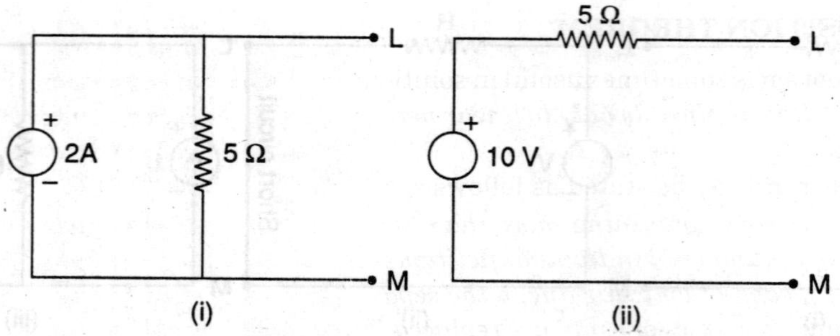

Fig. 48

7. SOLVING EQUATIONS BY DETERMINANTS—CRAMER'S RULE

If the number of equations is more than two, it is easier to get the solution by using determinants.

If I_1, I_2 and I_3 are the three unknowns in a system of three linear equations

$$a_{11}I_1 + a_{12}I_2 + a_{13}I_3 = C_1$$
$$a_{21}I_1 + a_{22}I_2 + a_{23}I_3 = C_2$$
$$a_{31}I_1 + a_{32}I_2 + a_{33}I_3 = C_3 \quad ...(14)$$

Then, the system can be written in matrix form as follows :

$$\begin{bmatrix} a_{11} & a_{12} & a_{13} \\ a_{21} & a_{22} & a_{23} \\ a_{31} & a_{32} & a_{33} \end{bmatrix} \begin{bmatrix} I_1 \\ I_2 \\ I_3 \end{bmatrix} = \begin{bmatrix} c_1 \\ c_2 \\ c_3 \end{bmatrix} \quad ...(15)$$

The values of I_1, I_2 and I_3 are given by

$$I_1 = \frac{\begin{bmatrix} c_1 & a_{12} & a_{13} \\ c_2 & a_{22} & a_{23} \\ c_3 & a_{32} & a_{33} \end{bmatrix}}{\Delta} \quad ...[16\ (a)]$$

$$I_2 = \frac{\begin{bmatrix} a_{11} & c_1 & a_{13} \\ a_{21} & c_2 & a_{23} \\ a_{31} & c_3 & a_{33} \end{bmatrix}}{\Delta} \quad ...[16\ (b)]$$

$$I_3 = \frac{\begin{bmatrix} a_{11} & a_{12} & c_1 \\ a_{21} & a_{22} & c_2 \\ a_{31} & a_{32} & c_3 \end{bmatrix}}{\Delta} \quad ...[16\ (c)]$$

where $\Delta = \begin{vmatrix} a_{11} & a_{12} & a_{13} \\ a_{21} & a_{22} & a_{23} \\ a_{31} & a_{32} & a_{33} \end{vmatrix}$

This is known as *Cramer's rule* and can be applied to any system of n linear equations provided Δ is not zero.

8. SUPERPOSITION THEOREM

This theorem is sometimes useful in solution of networks in which some branches may contain *sources of e.m.f. It is applicable only to linear networks where current is linearly related to voltage as per Ohm's law.*

This theorem may be stated as follows :

"In any network containing more than one source of e.m.f. the current in any branch is the algebraic sum of a number of individual fictitious currents (the number being equal to the number of sources of e.m.f.), each of which is due to the separate action of each source of e.m.f., taken in order, when the remaining sources of e.m.f. are replaced by conductors, the resistances of which are equal to the internal resistances of the respective sources".

The **procedure** of applying superposition theorem is as follows :

1. Replace all but one of the sources by their internal resistances. If the internal resistance of any source is small as compared to other resistances present in the network, the source is replaced by a short circuit.
2. Find the currents in different branches by using Ohm's law.
3. Repeat the process using each of the e.m.fs. as the sole e.m.f. each time.

The total current in any branch of the circuit is the algebraic sum of currents due to each source.

When finding total current in any branch, it is necessary to take into account the directions of the currents caused by each individual source, currents flowing in the *same direction* being *additive*, currents flowing in *opposite directions* being *subtractive*.

Explanation :

— In Fig. 49, I_1, I_2 and I represent the values of currents which are due to the simultaneous action of the two sources of e.m.f. in the network.

— In the Fig. 50 are shown the current values which would have been obtained if left-hand side battery had acted alone.

— Similarly Fig. 51 represents conditions obtained when right-hand side battery acts alone.

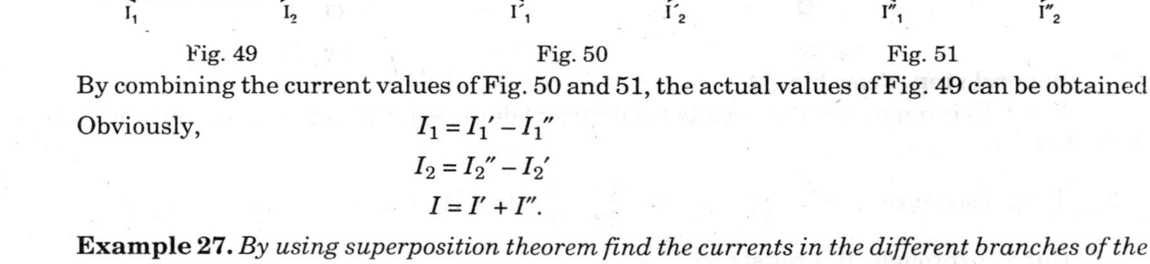

Fig. 49 Fig. 50 Fig. 51

By combining the current values of Fig. 50 and 51, the actual values of Fig. 49 can be obtained

Obviously,
$$I_1 = I_1' - I_1''$$
$$I_2 = I_2'' - I_2'$$
$$I = I' + I''.$$

Example 27. *By using superposition theorem find the currents in the different branches of the network shown in Fig. 52.*

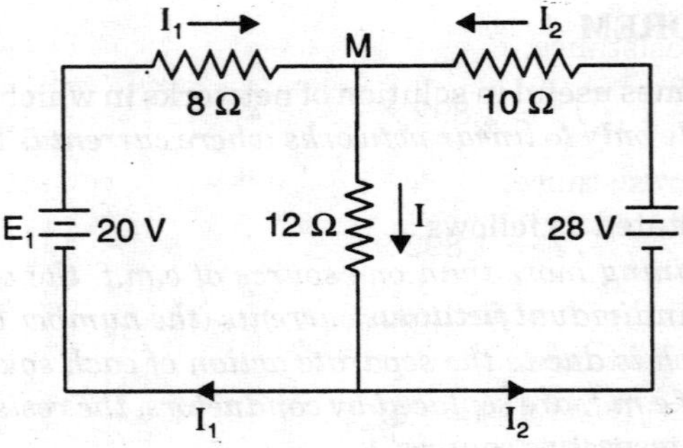

Fig. 52

Solution. $\mathbf{I_1}$ **: ,** $\mathbf{I_2}$ **: , I :**

First step. Refer Fig. 53.

Take e.m.f. E_1 only and replace e.m.f. E_2 by its zero internal resistance, the circuit is shown in Fig. 53.

Total resistance $= 8 + \dfrac{10 \times 12}{10 + 12} = 13.45\ \Omega$

Current through 8 Ω resistance,

$$I_1' = \frac{20}{13.45} = 1.487 \text{ A}$$

Current through 10 Ω resistance,

$$I_2' = 1.487 \times \frac{12}{12 + 10} = 0.81 \text{ A}$$

Current through 6 Ω resistance,

$$I' = 1.487 \times \frac{10}{12 + 10} = 0.675 \text{ A}$$

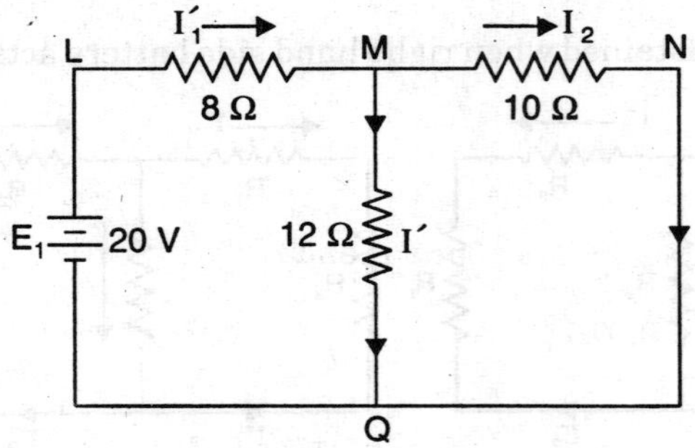

Fig. 53

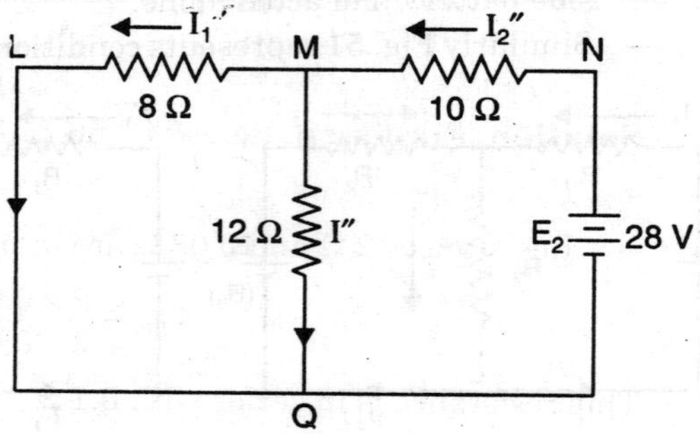

Fig. 54

Second step. Refer Fig. 54.

E.m.f. E_1 is removed/short circuited and current due to e.m.f. E_2 is found. The current is shown in the Fig. 54.

Total resistance $= 10 + \dfrac{12 \times 8}{12 + 8} = 14.8\ \Omega$

Current through 10 Ω resistance,

$$I_2'' = \frac{28}{14.8} = 1.892 \text{ A}$$

Current through 8 Ω resistance,

$$I_1'' = 1.892 \times \frac{12}{12+8} = 1.135 \text{ A}$$

Current through 12 Ω resistance,

$$I'' = 1.892 \times \frac{8}{12+8} = 0.757 \text{ A}$$

The total currents in different branches are :

Current through 8 Ω resistance,

$$\mathbf{I_1} = I_1' - I_1'' = 1.487 - 1.135 = \mathbf{0.352\ A\ (from\ L\ to\ M).}\quad \mathbf{(Ans.)}$$

Current through 10 Ω resistance,

$$\mathbf{I_2} = I_2'' - I_2' = 1.892 - 0.81 = \mathbf{1.082\ A\ (from\ N\ to\ M).}\quad \mathbf{(Ans.)}$$

Current through 12 Ω resistance,

$$\mathbf{I} = I' + I'' = 0.675 + 0.757 = \mathbf{1.432\ A\ (from\ M\ to\ Q).}\quad \mathbf{(Ans.)}$$

Example 28. *By using superposition theorem, find the current in 2 Ω resistance shown in Fig. 55. Internal resistances of the cells are negligible.*

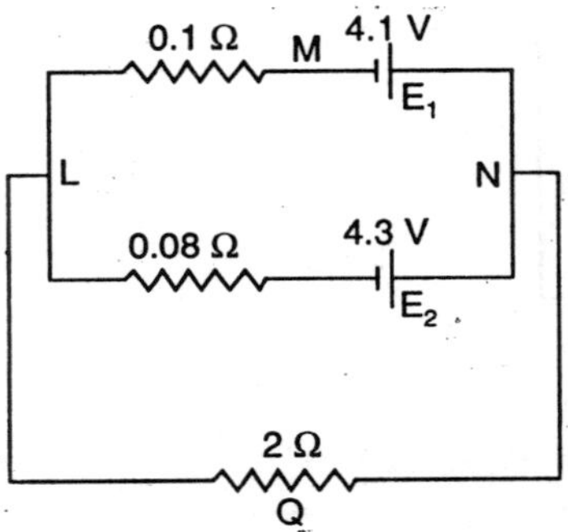

Fig. 55

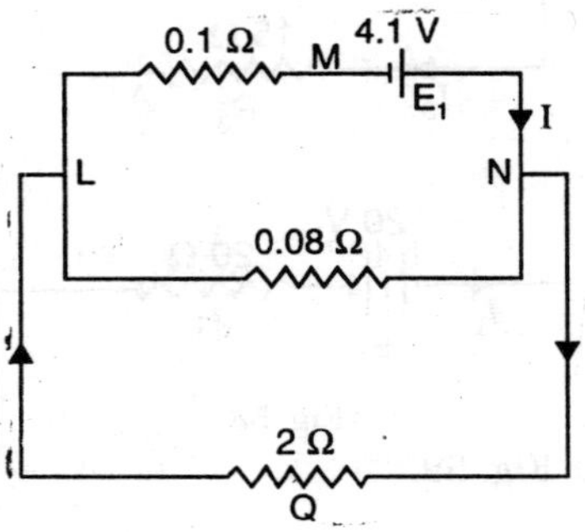

Fig. 56

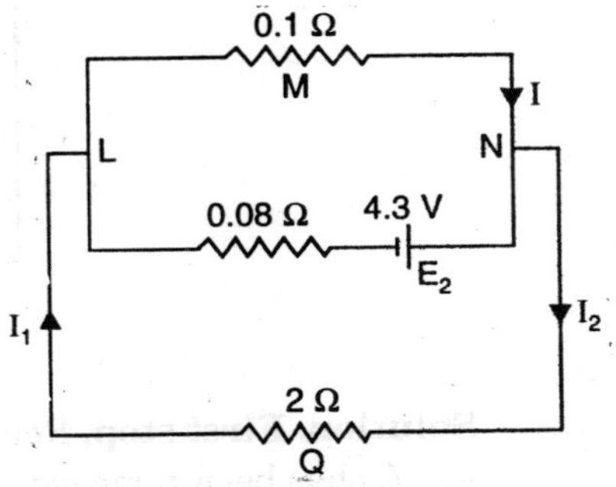

Fig. 57

Solution. First step. Refer Fig. 56.

— E_2 has been removed.

— Resistances 2 Ω and 0.08 Ω are in parallel across points L and N.

$$\therefore \qquad R_{LN} = \frac{2 \times 0.08}{2 + 0.08} = 0.076\ \Omega$$

This resistance is in series with 0.1 Ω.

Hence, total resistance offered to battery E_1

$$= 0.1 + 0.076 = 0.176\ \Omega$$

$$\therefore \text{ Current,} \qquad I = \frac{4.1}{0.176} = 23.3 \text{ A}$$

Current through 2 Ω resistance,

$$I_1 = 23.3 \times \frac{0.08}{0.08 + 2} = 0.896 \text{ A (from } N \text{ to } L)$$

Second step. Refer Fig. 57.

— E_1 has been removed.

— Combined resistance of paths NML and NQL

$$= \frac{2 \times 0.1}{2 + 0.1} = 0.095\ \Omega$$

Total resistance offered to E_2 $= 0.095 + 0.08 = 0.175\ \Omega$

$\therefore$ Current $$I = \frac{4.3}{0.175} = 24.57\ \text{A}$$

Again, $$I_2 = 24.57 \times \frac{0.1}{0.1 + 2} = 1.17\ \text{A}$$

Hence total current through the 2 Ω resistance when both batteries are present

$= I_1 + I_2 = 0.896 + 1.17 =$ **2.066 A. (Ans.)**

Example 29. *Using superposition theorem, find the currents in the circuit shown in the Fig. 58.*

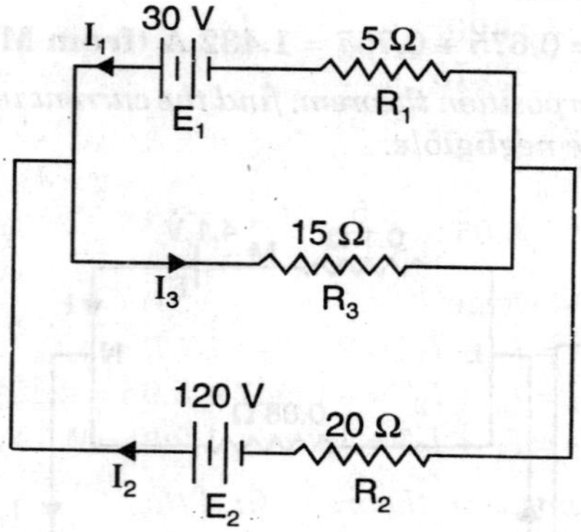

Fig. 58

Solution. First step. Refer Fig. 59.

— E_2 has been removed.

— Resistances R_2 and R_3 are in parallel,

$$\therefore \quad R_{\text{parallel}} = \frac{R_2 R_3}{R_2 + R_3} = \frac{20 \times 15}{20 + 15} = 8.57\ \Omega$$

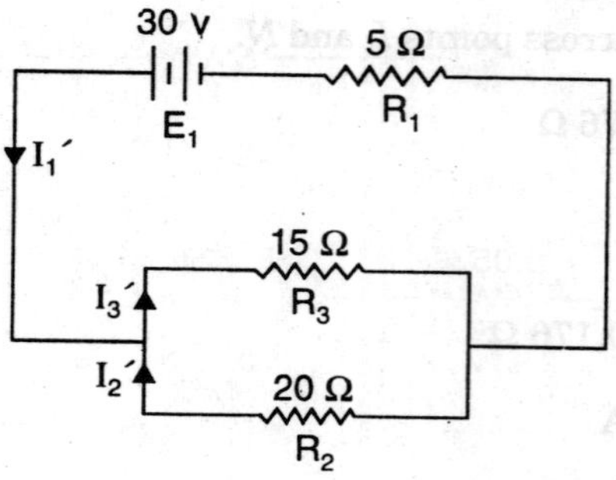

Fig. 59

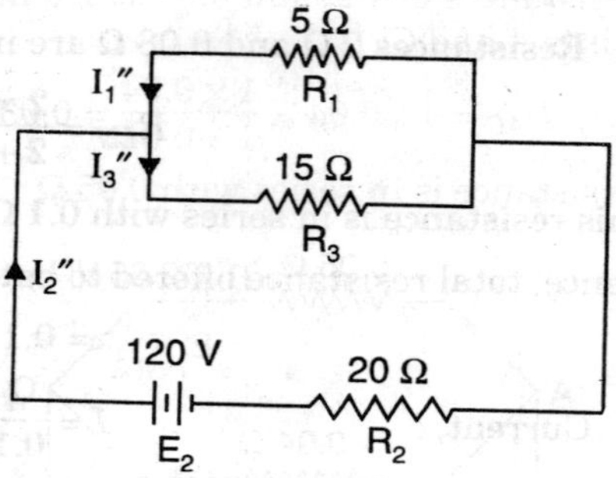

Fig. 60

This resistance is in series with R_1.

$\therefore$ Total resistance offered to battery E_1

$= 8.57 + 5 = 13.57\ \Omega$

Current $$I_1' = \frac{30}{13.57} = 2.21\ \text{A}.$$

Current $$I_2' = -2.21 \times \frac{15}{20+15} = -0.947 \text{ A}$$

Current $$I_3' = 2.21 \times \frac{20}{20+15} = 1.263 \text{ A.}$$

Second step. Refer Fig. 60.

— E_1 has been removed.

— Resistances R_1 and R_3 are in parallel.

$$\therefore \quad R_{parallel} = \frac{R_1 R_3}{R_1 + R_3} = \frac{5 \times 15}{5 + 15} = 3.75 \ \Omega$$

This resistance is in series with 20 Ω resistance.

∴ Total resistance offered to E_2

$$= 3.75 + 20 = 23.75 \ \Omega$$

Current $$I_2'' = \frac{120}{23.75} = 5.05 \text{ A}$$

Current $$I_1'' = -5.05 \times \frac{15}{5+15} = -3.79 \text{ A}$$

Current $$I_3'' = 5.05 \times \frac{5}{5+15} = 1.262 \text{ A}$$

Now superimposing the results, we get

$$I_1 = I_1' + I_1'' = 2.21 + (-3.79 \text{ A}) = \mathbf{-1.58 \text{ A.}} \quad \textbf{(Ans.)}$$
$$I_2 = I_2' + I_2'' = -0.947 + 5.05 = \mathbf{4.1 \text{ A.}} \quad \textbf{(Ans.)}$$
$$I_3 = I_3' + I_3'' = 1.263 + 1.262 = \mathbf{2.52 \text{ A.}} \quad \textbf{(Ans.)}$$

Example 30. *By using Superposition theorem, find the current in resistance R shown in Fig. 61 (i).*

$R_1 = 0.05 \ \Omega$, $R_2 = 0.04 \ \Omega$, $R = 1 \ \Omega$, $E_1 = 2.05$ V, $E_2 = 2.15$ V.

Internal resistances of the cells are negligible.

(Allahabad University)

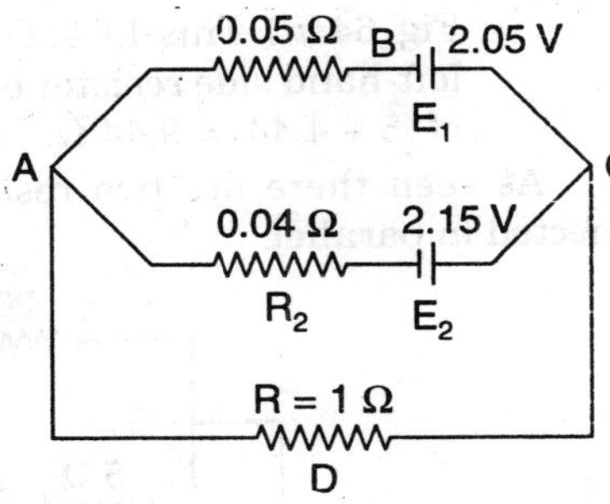

Fig. 61

Solution. Refer Fig. 62 (*i*). *E_2 has been removed :*

- Resistances of 1 Ω and 0.04 Ω are in parallel across points *A* and *C*.

$$R_{AC} = 1 \parallel 0.04 = \frac{1 \times 0.04}{1 + 0.04} = 0.038 \ \Omega$$

This resistance is in series with 0.05 Ω.

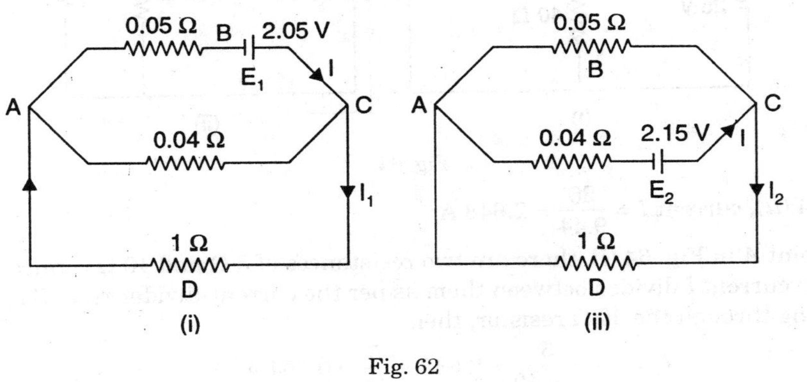

Fig. 62

Hence, total resistance offered to battery $E_1 = 0.05 + 0.038 = 0.088\ \Omega$

$$\text{Current } I = \frac{E_1}{0.088} = \frac{2.05}{0.088} = 23.3 \text{ A}$$

$\therefore$ Current through 1 Ω resistance $= \dfrac{0.04}{1.04} \times 23.3 = 0.896$ A from *C* to *A*.

Refer Fig. 62 (*ii*). *E_1 has been removed :*

- Combined resistance of paths *CBA* and *CDA*

$$= 1 \parallel 0.05 = \frac{1 \times 0.05}{1.05} = 0.048\ \Omega$$

Total resistance offered to $E_2 = 0.04 + 0.048 = 0.088\ \Omega$

$$\text{Current } I = \frac{2.15}{0.088} = 24.4 \text{ A}$$

Again, $$I_2 = \frac{24.4 \times 0.05}{1.05} = 1.16 \text{ A}$$

The current through 1 Ω resistance when both batteries are present.

$$= I_1 + I_2 = 0.896 + 1.16 = \mathbf{2.056\ A.}\ \ \textbf{(Ans.)}$$

Example 31. *Using Superposition theorem, find the current through the 40 Ω resistor of the circuit shown in Fig. 63.*

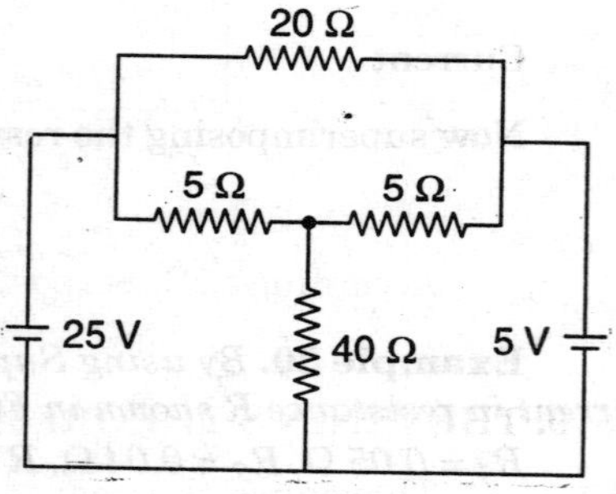

Fig. 63

Solution. Refer Fig. 64 (*i*). 5 V battery is replaced by a short-circuit and *25 V battery acts alone :*

- The right-hand side 5 Ω resistor becomes connected in parallel with 40 Ω resistor giving a combined resistance of $5 \parallel 40 = \dfrac{5 \times 40}{5 + 40} = 4.44\ \Omega$ as shown in Fig. 64 (*ii*). This 4.44 Ω resistance is in series with the left-hand side resistor of 5 Ω giving a total resistance of (5 + 4.44) = 9.44 Ω.

As seen there are two resistances of 20 Ω and 9.44 Ω connected in parallel.

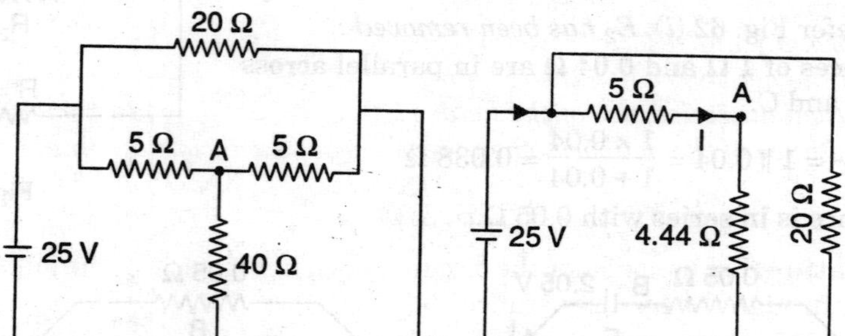

Fig. 64

In Fig. 64 (*ii*), current $I = \dfrac{25}{9.44} = 2.648$ A.

- At point *A* in Fig. 64 (*i*), there are two resistances of 5 Ω and 40 Ω connected in parallel, hence current *I* divides between them as per the current-divider rule. If I_1 is the current flowing through the 40 Ω resistor, then

$$I_1 = I \times \frac{5}{5 + 40} = 2.648 \times \frac{5}{45} = 0.294 \text{ A.}$$

Refer Fig. 65 (*i*). 25 V battery has been removed and *5 V battery acts alone* :

- As in previous case, there are two parallel branches of resistances 20 Ω and 9.44 Ω across the 5 V battery.

Current I through 9.44 Ω branch, $I = \frac{5}{9.44} = 0.53$ A

This current divides at point B between 5 Ω resistor and 40 Ω resistor. Current through 40 Ω resistor

$$I_2 = 0.53 \times \frac{5}{(40 + 5)} = 0.059 \text{ A}.$$

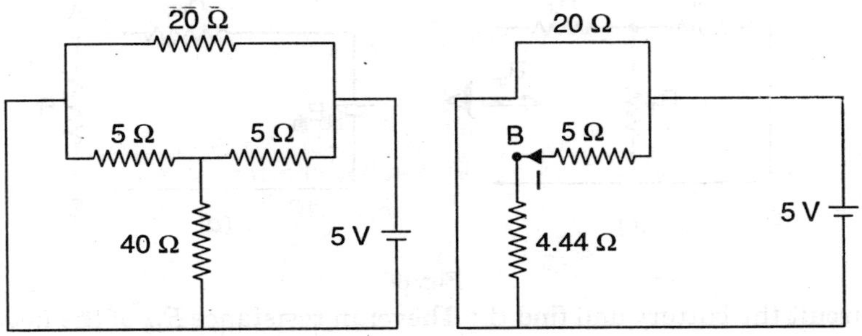

(i) (ii)

Fig. 65

According to Superposition theorem, total current through 40 Ω resistance
$= I_1 + I_2 = 0.294 + 0.059 =$ **0.353 A. (Ans.)**

9. THEVENIN'S THEOREM

Thevenin's theorem is quite useful when the current in one branch of a network is to be determined or when the current in an added branch is to be calculated.

"It states that for the purpose of determining the current in a resistor, R_L connected across two terminals of a network which contains sources of e.m.f. and resistors, the network can be replaced by a single source of e.m.f. and a series resistor, R_{th}. This e.m.f., E_{th} is equal to potential difference between the terminals of the network when the resistor, R, is removed : the resistance of series resistor, R_{th}, is equal to the equivalent resistance of the network with the resistor, R, removed (or as it is sometimes called, "the resistance of the network when viewed from the terminals under consideration")".

Hence $$I = \frac{E}{(R_L + R_{th})} \quad ...(17)$$

Explanation. Let us consider the circuit shown in Fig. 66 (*a*). The following steps are required to find current through the load resistance R_L.

1. Remove R_L from the circuit terminals A and B and redraw the circuit as shown in Fig. 66 (*b*). Obviously the terminals have been open circuited.

2. Calculate the open-circuit voltage ($V_{oc} = E_{th}$) which appears across terminals A and B, when they are open *i.e.* when R_L is removed. This voltage is E_{th} (Thevenin's voltage). A little thought will reveal that

$$E_{th} = \frac{ER_2}{R_1 + R_2} \quad ...(18)$$

$$\left[\because \ I = \frac{E}{R_1 + R_2} \text{ and } E_{th} = IR_2 = \frac{ER_2}{R_1 + R_2} \right]$$

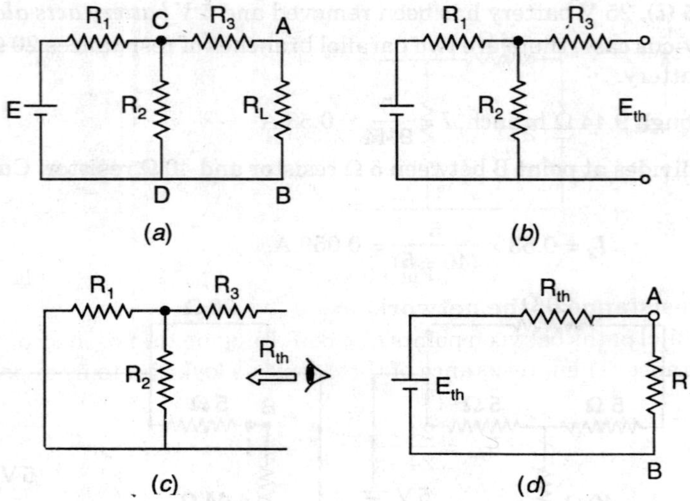

Fig. 66

3. Short circuit the battery and find the Thevenin resistance R_{th} of the network as seen from the terminals A and B [Fig. 66 (*c*)]

$$R_{th} = \frac{R_1 \times R_2}{R_1 + R_2} + R_3 \qquad ...(19)$$

4. Connect R_L back across the terminals A and B from where it was temporarily removed earlier [Fig. 66 (*d*)]. Current through R_L is given by

$$I = \frac{E_{th}}{(R_{th} + R_L)} \qquad ...(20)$$

Example 32. *With reference to the network shown in Fig. 67, by using Thevenin's theorem find the following :*

(i) The equivalent e.m.f. of the network when viewed from terminals L and M.

(ii) The equivalent resistance of the network when looked into from terminals L and M.

(iii) Current in the load resistance R_L of 30 Ω.

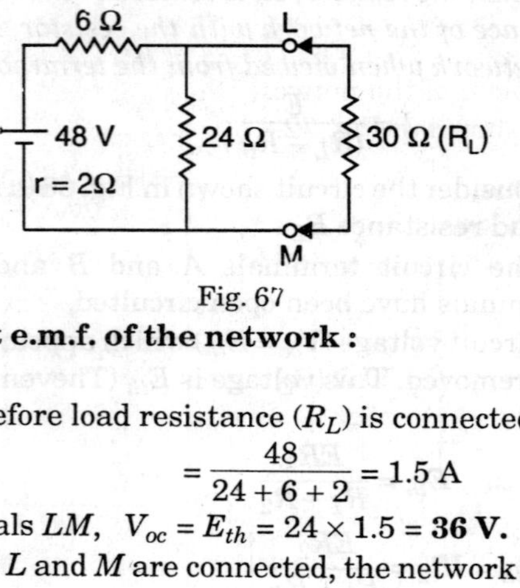

Fig. 67

Solution. (*i*) **Equivalent e.m.f. of the network :**

Refer Fig. 67.

Current in the network before load resistance (R_L) is connected

$$= \frac{48}{24 + 6 + 2} = 1.5 \text{ A}$$

∴ Voltage across terminals LM, $V_{oc} = E_{th} = 24 \times 1.5 = \mathbf{36\ V}$.

Hence, so far as terminals L and M are connected, the network has an **e.m.f. of 36 V** (and not 48 V). **(Ans.)**

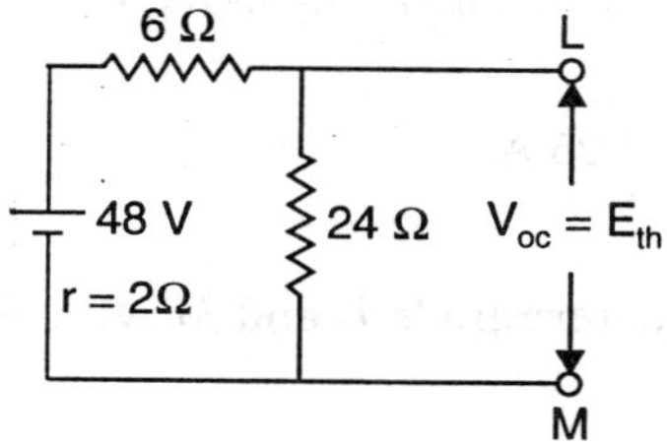

Fig. 68

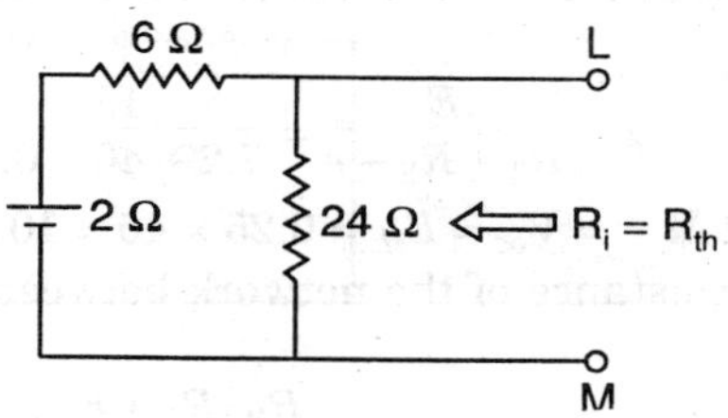

Fig. 69

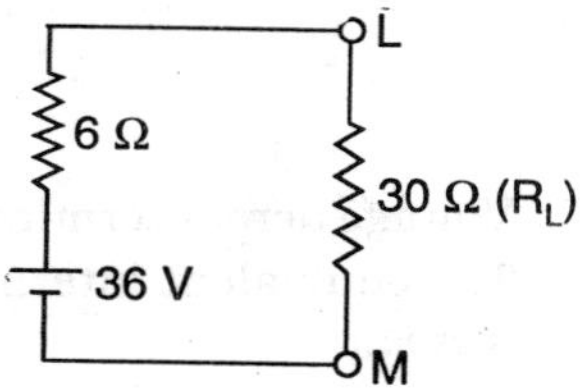

Fig. 70

(*ii*) **Equivalent resistance of the network :**

There are two parallel paths between points L and M. Imagine that battery of 48 V is removed but not its internal resistance. Then, resistance of the circuit as looked into from points L and M is (Fig. 69)

$$R_i = R_{th} = \frac{24 \times (6+2)}{24 + (6+2)} = \mathbf{6\ \Omega.} \quad \textbf{(Ans.)}$$

(*iii*) **Current in R_L, I :**

Refer Fig. 70.

$$I = \frac{E_{th}}{R_{th} + R_L} = \frac{36}{6+30} = \mathbf{1\ A.} \quad \textbf{(Ans.)}$$

Example 33. *Find the current through 50 ohms resistance in the circuit shown in Fig. 71. Use Thevenin's theorem.*

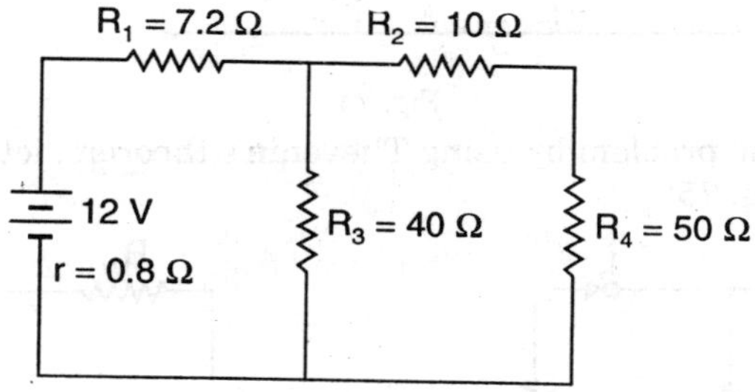

Fig. 71

Solution. To solve the problem of the network shown in Fig. 71 by Thevenin's theorem, let R_4 be assumed as disconnected as shown in Fig. 72.

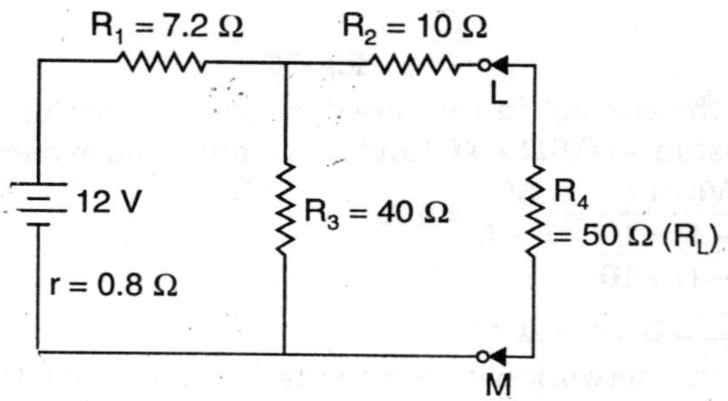

Fig. 72

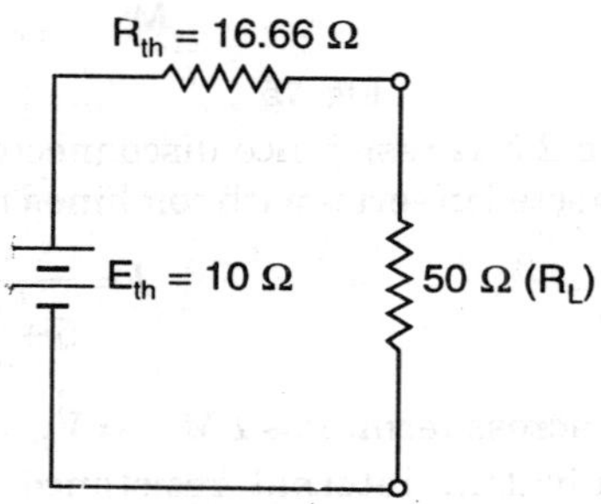

Fig. 73

With the resistance R_4 disconnected, the current in the closed circuit consisting of R_1, R_3 and r is,

$$I = \frac{E}{R_1 + R_3 + r} = \frac{12}{7.2 + 40 + 0.8} = 0.25 \text{ A.}$$

Voltage across terminals LM $= V_{oc} = E_{th} = 0.25 \times 40 = 10$ V

The equivalent internal resistance of the network between the terminals L and M with R_4 disconnected,

$$R_i = R_{th} = R_2 + \frac{R_3 (R_1 + r)}{R_3 + (R_1 + r)}$$

$$= 10 + \frac{40(7.2 + 0.8)}{40 + 7.2 + 0.8} = 10 + \frac{40 \times 8}{48} = 16.66\ \Omega.$$

$\therefore$ Current through 50 Ω (R_L) resistance (Refer Fig. 73),

$$I = \frac{E_{th}}{R_{th} + R_L} = \frac{10}{16.66 + 50} = \mathbf{0.15\ A.}\ \ \textbf{(Ans.)}$$

Example 34. *Using Thevenin's theorem find the current through the 2.5 Ω resistance in the circuit shown in the Fig. 74.*

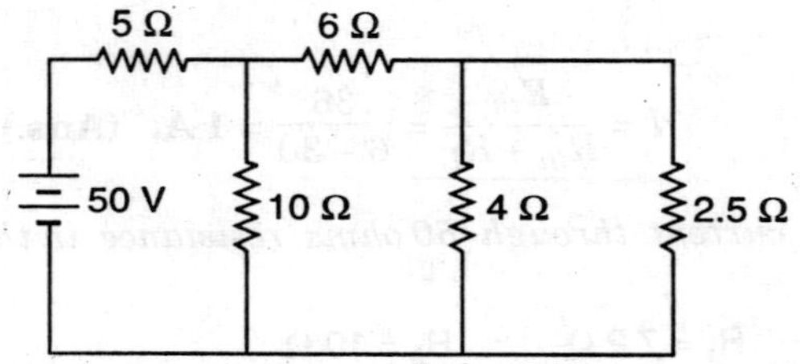

Fig. 74

Solution. To solve the problem by using Thevenin's theorem, let 2.5 Ω resistance be shown disconnected as shown in Fig. 75.

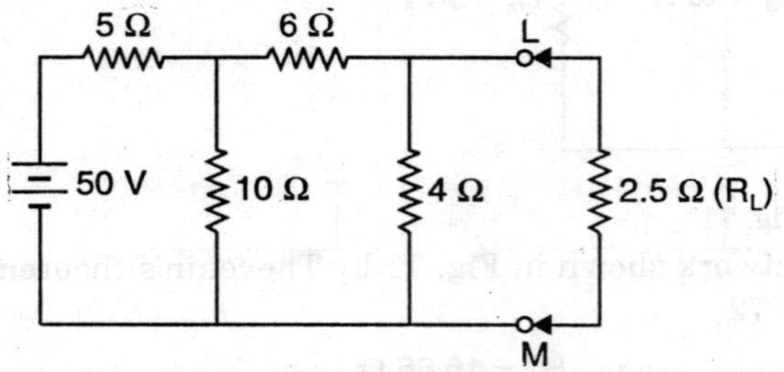

Fig. 75

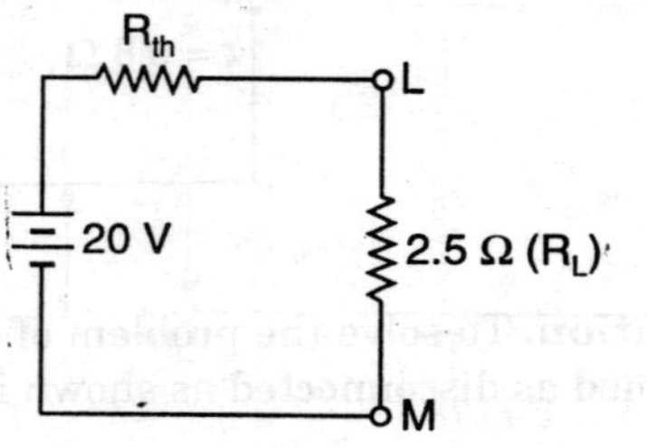

Fig. 76

With the 2.5 Ω resistance disconnected, the current in the closed circuit [Consisting of 5 Ω resistance connected in series with combined resistance of (6 Ω + 4 Ω) and 10 Ω (connected in parallel)],

$$I = \frac{50}{5 + \frac{(6+4) \times 10}{(6+4) + 10}} = \frac{50}{5 + 5} = 5 \text{ A}$$

Voltage across terminals LM $= V_{oc} = E_{th} = 5 \times 4 = 20$ V

The equivalent internal resistance of the network between the terminals L and M (with 2.5 Ω disconnected),

[Here R_{th} (or R_i) comprises of : 5 Ω and 10 Ω in parallel ; this combined resistance $\left(i.e. \frac{5 \times 10}{5+10} \Omega\right)$ is in series with 6 Ω. This total resistance, then, is in parallel with 4 Ω resistance.]

$$R_i = R_{th} = \frac{\left[\frac{5 \times 10}{5+10} + 6\right] \times 4}{\left[\left(\frac{5 \times 10}{5+10} + 6\right) + 4\right]} = \frac{9.33 \times 4}{13.33} = 2.8 \ \Omega$$

∴ Current through 2.5 Ω resistance (Refer Fig. 76),

$$I = \frac{E_{th}}{R_{th} + R_L} = \frac{20}{2.8 + 2.5} = \mathbf{3.77\ A.} \quad \textbf{(Ans.)}$$

Example 35. *Using Thevenin's theorem, calculate the potential difference across terminals L and M in Fig. 77.*

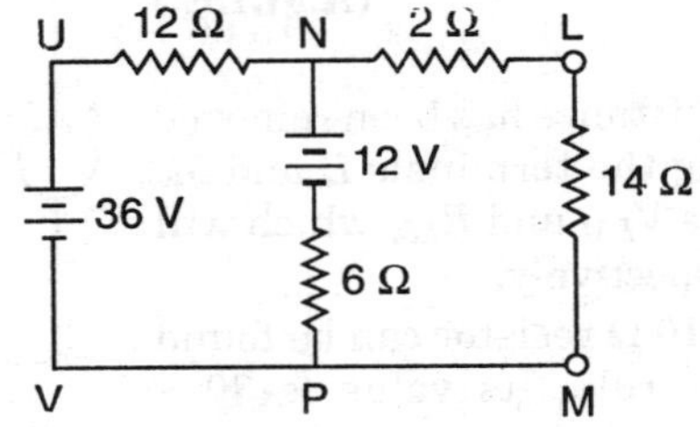

Fig. 77

Solution. First-step : To find V_{oc} :

Remove 14 Ω resistance thereby open-circuiting terminals *L* and *M* (see Fig. 78). Obviously there is no current in through 2 Ω resistor and hence no drop across it.

∴ $V_{LM} = V_{OC} = V_{NP}$

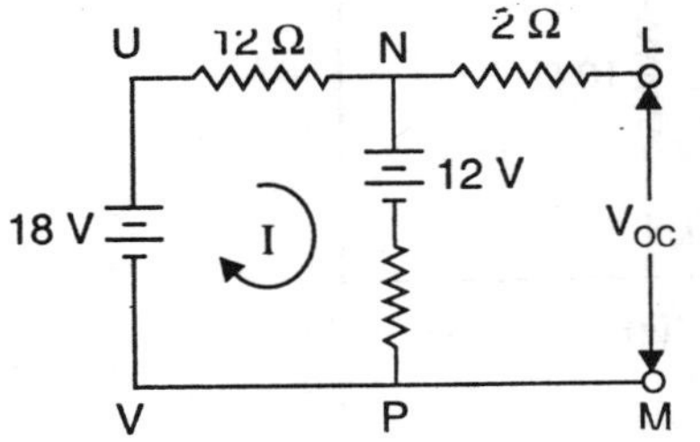

Fig. 78

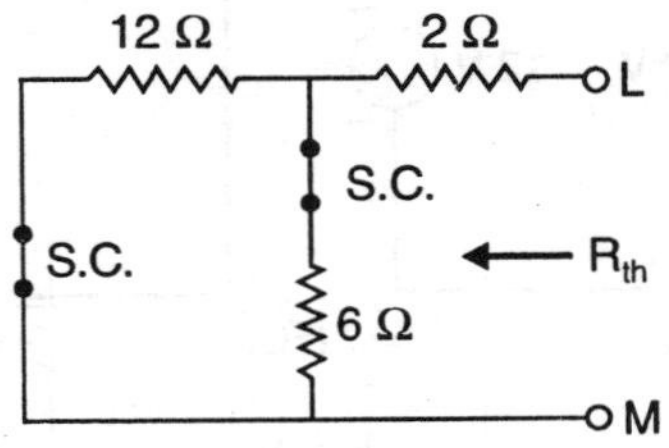

Fig. 79

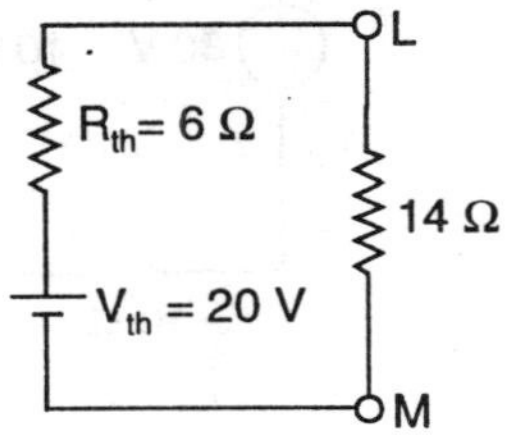

Fig. 80

As seen, current *I* flows due to the combined action of the two batteries. Net voltage in NPVU circuit = 36 – 12 = 24 V.

— Total resistance $= 12 + 6 = 18 \ \Omega$

Hence $I = \frac{24}{18} = \frac{4}{3}$ A

$$V_{NP} = 12 + \text{drop across } 6 \ \Omega \text{ resistor}$$
$$= 12 + 6 \times \frac{4}{3} = 20 \text{ V}$$

∴ $V_{OC} = V_{th} = 20$ V

Second step : To find R_i or R_{th}

As shown in Fig. 79, the two batteries have been replaced by short-circuits (S.C.) since their internal resistances are zero.

As seen, $R_i = R_{th} = 2 + 6 \parallel 12 = 2 + \dfrac{12 \times 6}{12 + 6} = 6\ \Omega$

The Thevenin's equivalent circuit is shown in Fig. 80 where the 14 Ω resistance has been reconnected across terminals L and M.

Third step : To find p.d. across L and M

The p.d. across L and M can be found with the help of *Proportional Voltage Formula*,

p.d. across L and $M = 20 \times \dfrac{14}{14+6} = $ **14 V (Ans.)**

[Alternatively : $I = \dfrac{20}{6+14} = 1$ A

∴ p.d. across 14 Ω resistance $= 1 \times 14 = 14$ V.]

Example 36. *Using Thevenin theorem calculate the current through the 8 Ω resistor of the circuit of Fig. 81.*

(A.M.I.E.)

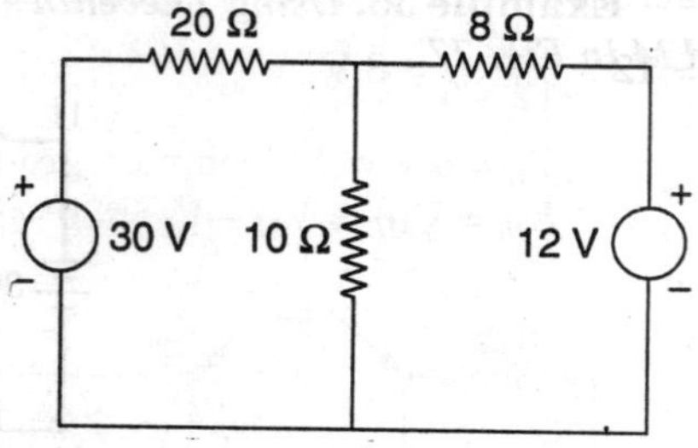

Fig. 81

Solution.

- *Refer Fig. 82 (i).* 8 Ω resistance has been removed thereby open-circuiting the terminals L and M. We shall now calculate V_{LM} and R_{LM} which will give us E_{th} and R_{th} respectively.

The potential drop across 10 Ω resistor can be found with the help of voltage divider rule. Its value is $30 \times \dfrac{10}{(10+20)} = 10$ V.

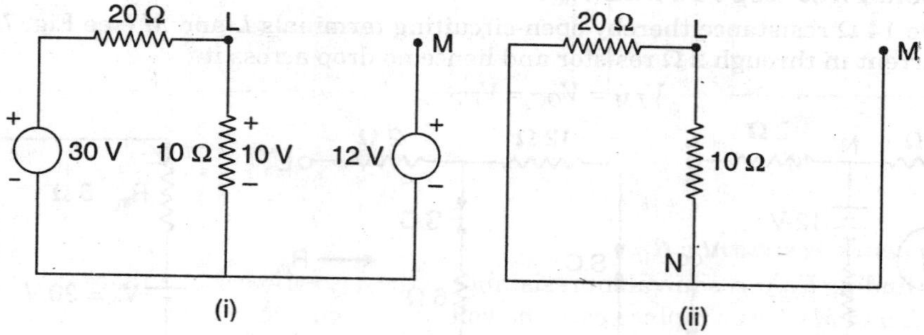

Fig. 82

In order to find V_{LM} we shall go from point M to point L in the clockwise direction and find the algebraic sum of the voltages met on the way.

$$V_{LM} = -12 + 10 = -2 \text{ V}$$

It means that point L is negative w.r.t. M, or point M is at a higher potential than point L by 2 volts.

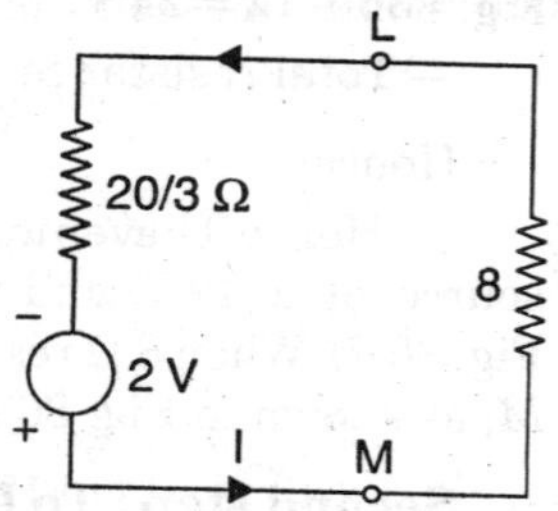

Fig. 83

- *Refer Fig. 82 (ii).* The two voltage sources have been short-circuited. The resistance of the network as viewed from points L and M is the same as viewed from points L and N

$$R_{LM} = R_{LN} = 1 \parallel 20 = \frac{10 \times 20}{10 + 20} = \frac{20}{3}\ \Omega$$

- Thevenin's equivalent source is shown in Fig. 83 in which 8 Ω resistor has been joined back across terminals L and M.

Polarity of the voltage source is worth noting.

$$I = \frac{2}{(20/3) + 8} = \textbf{0.136 A from M to L. (Ans.)}$$

Example 37. *Using Thevenin's theorem, calculate the current flowing through the 8 Ω resistor of Fig. 84.*

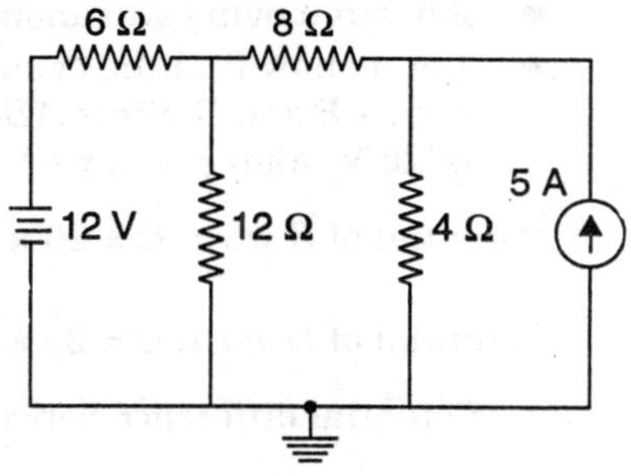

Fig. 84

Solution. ***Thevenin voltage E_{th} :***

Refer Fig. 85 (*i*). 8 Ω resistor is removed. Since full 5 A current passes through 4 Ω resistor, drop across it is 5 × 4 = 20 V. Hence V_M = 20 V (w.r.t. the common ground).

Two resistors 6 Ω and 12 Ω are connected in series across the 12 V battery. Hence drop across 12 Ω resistor $= 12 \times \frac{12}{12 + 6} = 8$ V.

V_L = 8 V w.r.t. common ground

$E_{th} = V_{ML} = V_M - V_L = 20 - 8 = 12$ V with M at a higher potential.

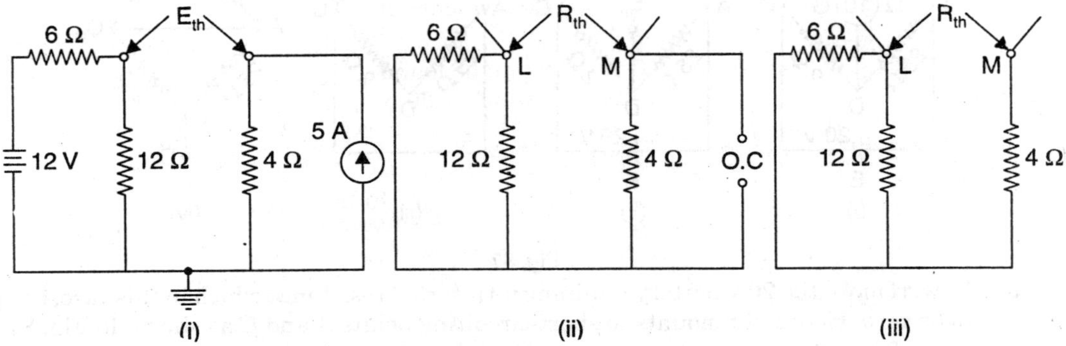

Fig. 85

Thevenin resistance R_{th} :

For finding R_{th} (*i.e.* equivalent resistance of the network as looked back into the open-circuited terminals L and M), let us replace both the voltage and current sources. Since voltage source has no internal resistance, it would be replaced by a short circuit *i.e.* zero resistance. However current source would be removed and replaced by an 'open' *i.e.* infinite resistance. In that case, the circuit becomes as shown in Fig. 85 (*ii*). As seen from Fig. 85 (*iii*),

$$R_{th} = 12 \,||\, 6 + 4 = \frac{12 \times 6}{12 + 6} + 4 = 8\ \Omega.$$

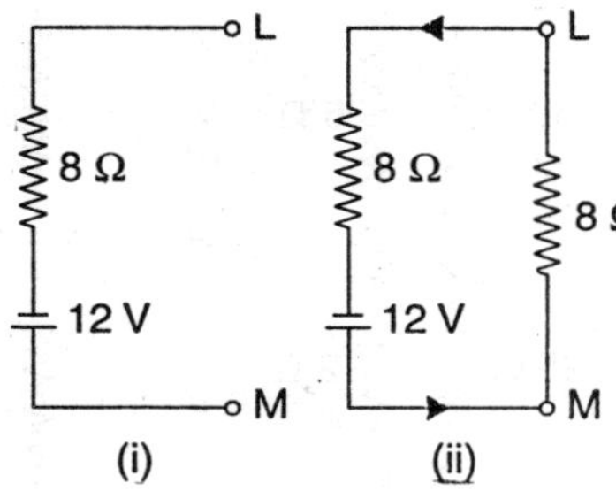

Fig. 86

Hence Thevenin's equivalent circuit consists of a voltage source of a 12 V and a series resistance of 8 Ω as shown in Fig. 86 (*i*). When 8 Ω resistor is connected across terminals L and M, as shown in Fig. 86 (*ii*).

$$I = \frac{12}{8 + 8} = \textbf{0.75 A from M to L. (Ans.)}$$

Example 38. *The four arms of a Wheatstone bridge have the following resistances :*

AB = 50 Ω, BC = 5 Ω, CD = 2 Ω, DA = 25 Ω. A galvanometer of 10 Ω resistance is connected across BD. If a potential difference of 20 V is maintained across AC, using Thevenin's theorem, calculate the current through the galvanometer.

Solution. Refer Fig. 87 (*i*).

- After removing galvanometer from Fig. 87 (*i*), we get the circuit of Fig. 87 (*ii*).
- Let us now find the open-circuit voltage V_{OC} (also called Thevenin voltage E_{th}) between points *B* and *D*. Since *ABC* (as well as *ADC*) is a potential divider on which a voltage drop of 20 V takes place, we have

Potential of *B* w.r.t. $C = 20 \times \frac{5}{50+5} = 1.818$ V.

Potential of *D* w.r.t. $C = 20 \times \frac{2}{25+2} = 1.481$ V.

∴ Potential difference between *B* and *D* is V_{OC} or $E_{th} = 1.818 - 1.481 = 0.337$ V.

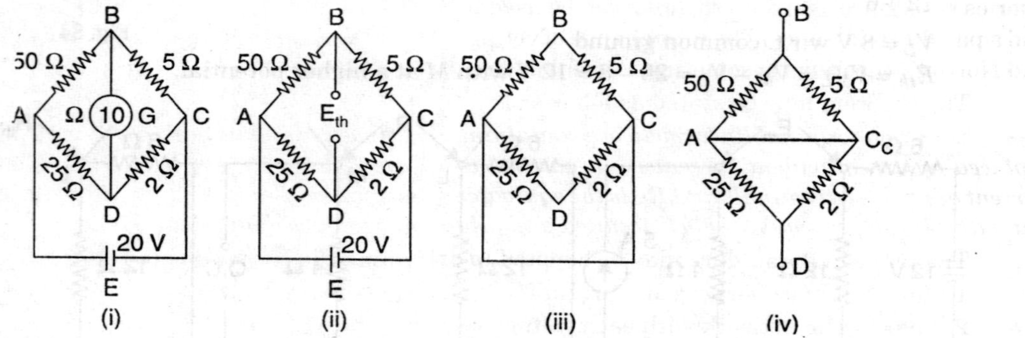

Fig. 87

- Now, remove the 20 V battery retaining its internal resistance which in this case, happens to be zero. Hence, it amounts to short-circuiting points *A* and *C* as shown in Fig. 87 (*iv*).
- Next, let us find the resistance of the whole network as viewed from points *B* and *D*. It may be seen that electrically speaking points *A* and *C* have become one as shown in

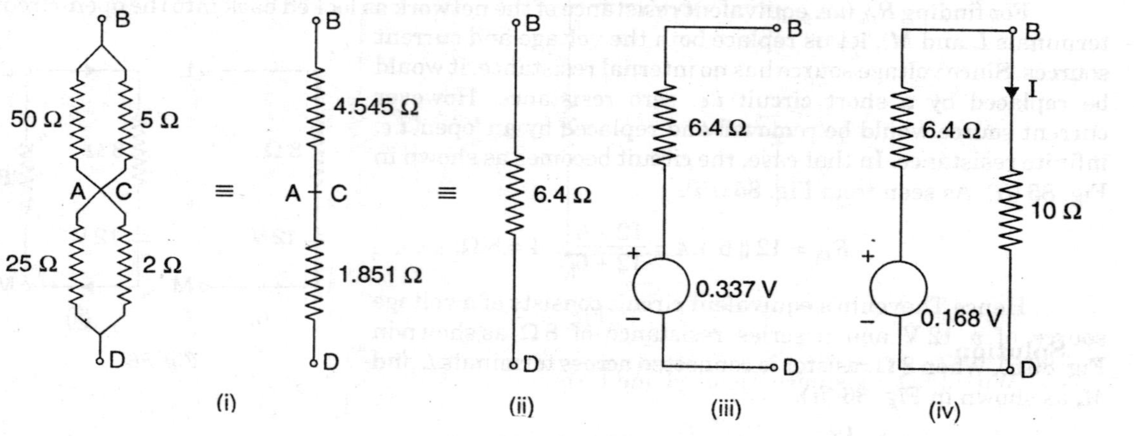

Fig. 88

Fig. 88 (*i*) Further it may be observed that *BA* is in parallel with *BC* and *AD* is in parallel with *CD*. Hence

$$R_{BD} = 5 \parallel 50 + 25 \parallel 2 = \frac{5 \times 50}{5 + 50} + \frac{25 \times 2}{25 + 2} = 6.4\ \Omega$$

- Now, so far points *B* and *D* are concerned, the network has a voltage source of 0.337 V and internal resistance $R_i = 6.4\ \Omega$. This Thevenin's source is shown in Fig. 88 (*iii*).
- Finally, let us connect the galvanometer, which was initially removed, to this Thevenin source and calculate the current flowing through it. As seen from Fig. 88 (*iv*), we have

current through the galvanometer, $I = \frac{0.337}{6.4 + 10} =$ **0.02 A. (Ans.)**

10. NORTON'S THEOREM

Whereas Thevenin's theorem was used to simplify a network to a constant-voltage source and a series resistance, Norton's theorem can be used to resolve a network into a constant-current source and a parallel resistance. The interchange of voltage sources and current sources by use of Thevenin's and Norton's theorems is sometimes useful in circuit analysis.

The theorem may be stated as follows :

"Any two-terminal linear network containing independent voltage and current sources may be replaced by an equivalent current I_N in parallel with a resistance R_N where I_N is the short circuit current at network terminals and R_N is the equivalent resistance of network as seen from the terminals but with all voltage sources short circuited and all current sources open circuited."

The following procedure may be adopted to determine the Norton's equivalent circuit :

1. Calculate the short circuit current (I_N) at the network terminals.
2. Redraw the network with each voltage source replaced by a short circuit in series with its internal resistance and each current source by an open circuit in parallel with its internal resistance.
3. Calculate the resistance (R_N) of the redrawn network as seen from the network terminals. (The resistance R_N is the same value as used in Thevenin's equivalent circuit).

Example 39. *By using Norton's theorem find the current in the 12 Ω resistance of the circuit shown in Fig. 89.*

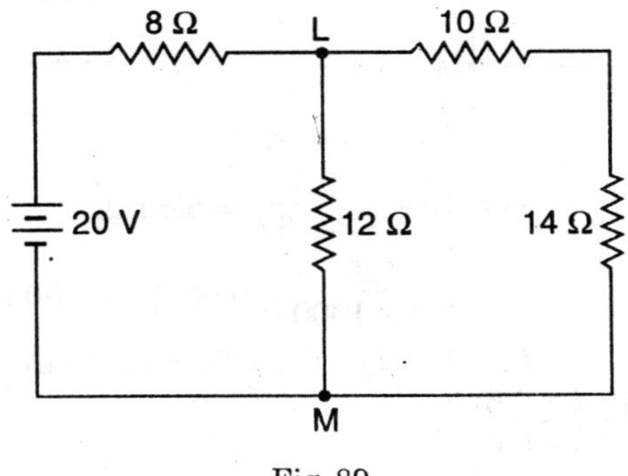

Fig. 89

Solution.

— With 12 Ω resistance removed and terminals *L-M* short circuited, short-circuit current,

$$I_N = \frac{20}{8} = 2.5\ \text{A}.$$

— With 20 V battery replaced by a short circuit, the resistance of the network as seen from terminals L and M is

$$R_N = \frac{8(10+14)}{8+(10+14)} = 6\ \Omega$$

— The Norton's equivalent circuit is shown in Fig. 90. The current through 12 Ω resistance is $I = 2.5 \times \frac{6}{6+12} = \mathbf{0.833\ A.}$ **(Ans.)**

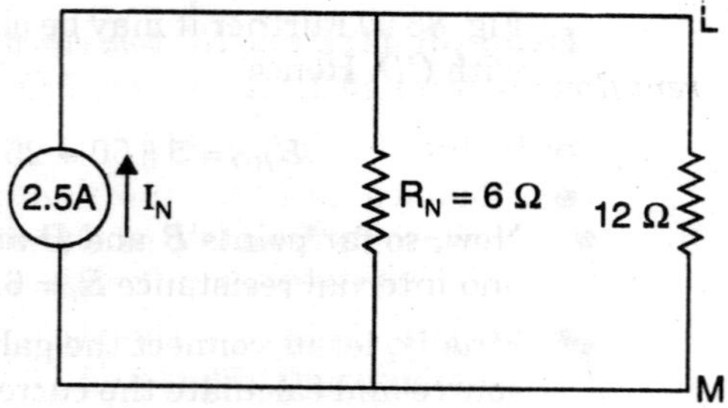

Fig. 90

Example 40. *For the network shown in Fig. 91 derive Norton's equivalent circuit and find the current through 24 k Ω resistance.*

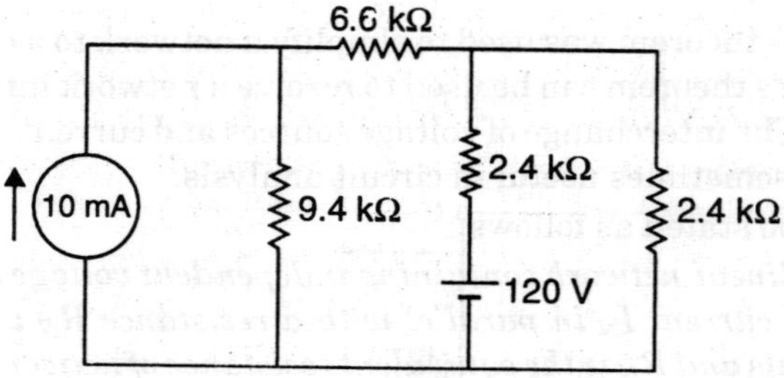

Fig. 91

Solution. With 24 kΩ resistance removed and terminals L-M short circuited, superposition theorem gives,

$$I_N = I_1 + I_2$$

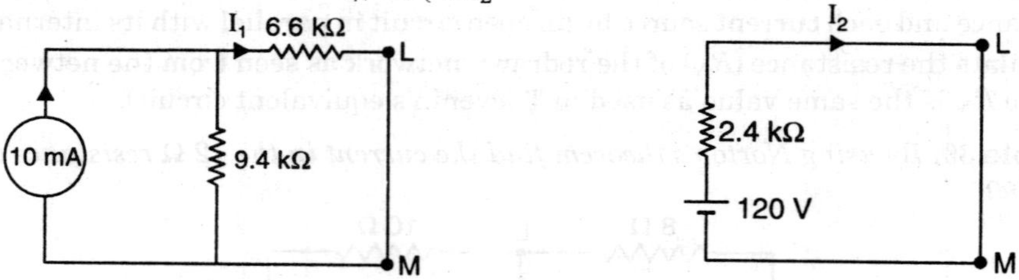

Fig. 92

From Fig. 92 (*a*), $I_1 = 10 \times \frac{9.4}{9.4+6.6} = 5.8$ mA

From Fig. 92 (*b*), $I_2 = \frac{120}{2.4 \times 1000} = 0.05$ A or 50 mA

$\therefore$ $I_N = I_1 + I_2 = 5.8 + 50 = 55.8$ mA

$$R_N = \frac{2.4 \times (9.4+6.6)}{2.4+(9.4+6.6)}$$

$$= 2.08\ \Omega$$

The Norton's equivalent circuit is shown in Fig. 93. Current through 12 kΩ resistance,

$$I = 55.8 \times \frac{2.08}{2.08+24} = \mathbf{4.45\ A.}\ \textbf{(Ans.)}$$

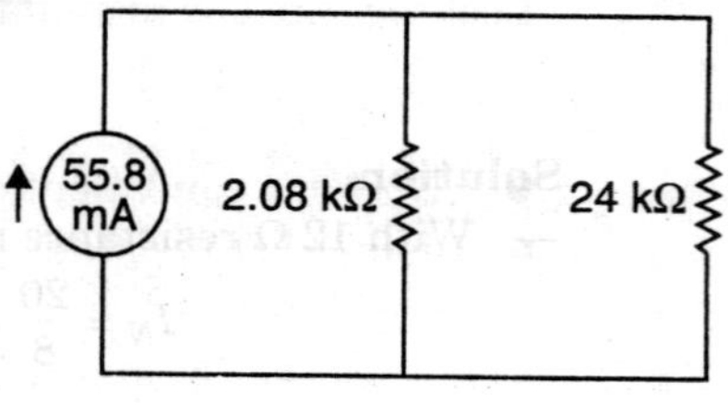

Fig. 93

Example 41. *Using Norton's theorem, calculate current flowing through 10 Ω resistor of Fig. 94.*

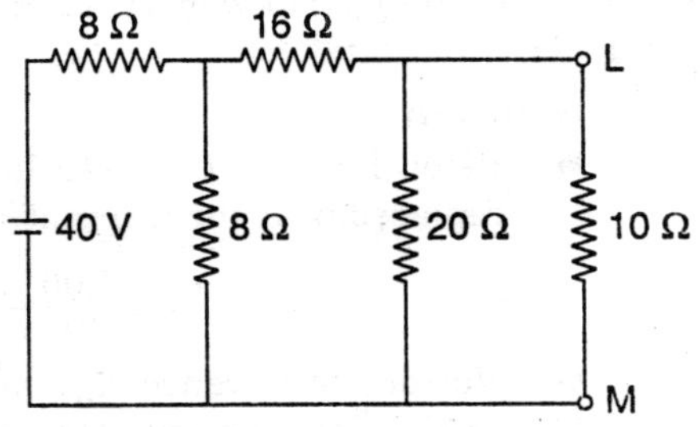

Fig. 94

Solution. Refer Fig. 94.

- Remove 10 Ω resistor and put a short across terminals L and M as shown in Fig. 95 (*i*). As seen 20 Ω resistor also becomes short-circuited.
- Let us find I_{SC}. The battery sees a parallel combination of 8 Ω and 16 Ω in series with a 8 Ω resistance. Total resistance seen by battery $= 8 + 8 \parallel 16 = 8 + \dfrac{8 \times 16}{8 + 16} = \dfrac{40}{3}\ \Omega.$

Hence, $I = \dfrac{40}{(40/3)} = 3$ A. This current divides at N of Fig. 95 (*i*).

Current going along path NLM gives I_{SC}. Its value is $\dfrac{3 \times 8}{24} = 1$ A.

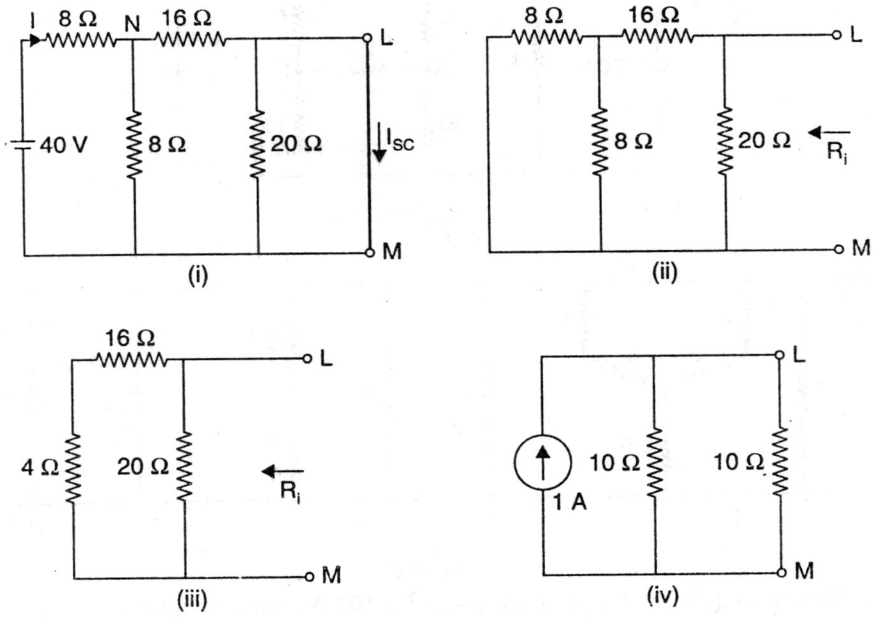

Fig. 95

- In Fig. 95 (*ii*) battery has been removed leaving behind its internal resistance which, in this case is zero.

Looking into the terminals L and M, resistance of the network [Fig. 95 (*iii*)],

$$R_i = 20 \parallel 20 = \frac{20 \times 20}{20 + 20} = 10\ \Omega$$

- Norton's equivalent circuit is shown in Fig. 95 (*iv*).
- Join 10 Ω resistance back across terminals L and M. The current flowing through this resistance (10 Ω)

$$= 1 \times \frac{10}{10 + 10} = \mathbf{0.5\ A.}\ \ \textbf{(Ans.)}$$

Example 42. *Using Norton's theorem calculate current through a load of 8 Ω in the circuit shown in Fig. 96.* **(A.M.I.E.)**

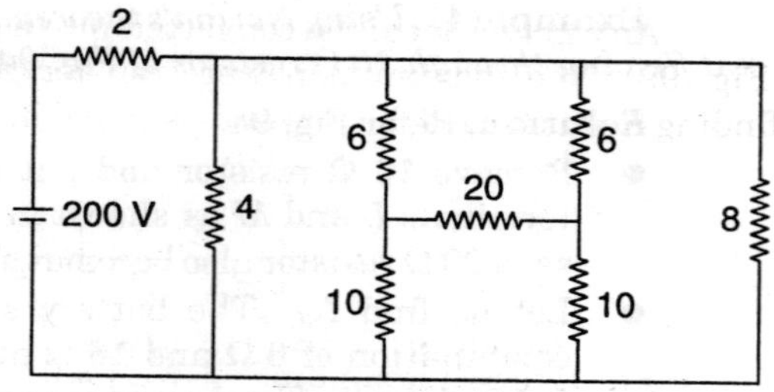

Fig. 96

Solution.

- Refer Fig. 97. It shows load impedance replaced by a short circuit.

$$I_{SC} = I_N = \frac{200}{2} = 100 \text{ A}.$$

- Norton's resistance R_N can be found by looking into the open terminals of Fig. 98.

For this purpose Δ *ABC* has been replaced by its equivalent star. As seen R_N is equal to $\frac{8}{7}$ Ω.

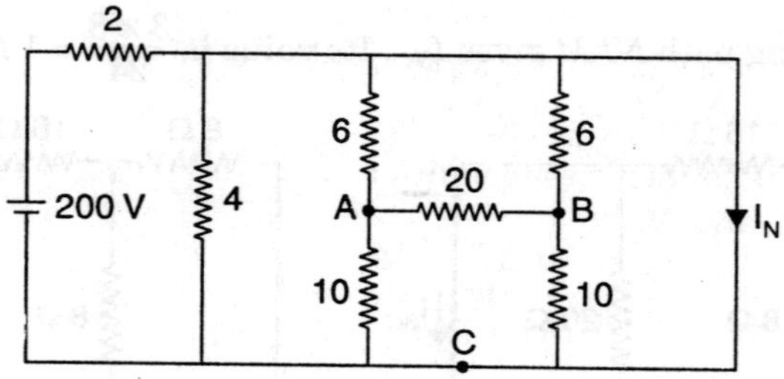

Fig. 97

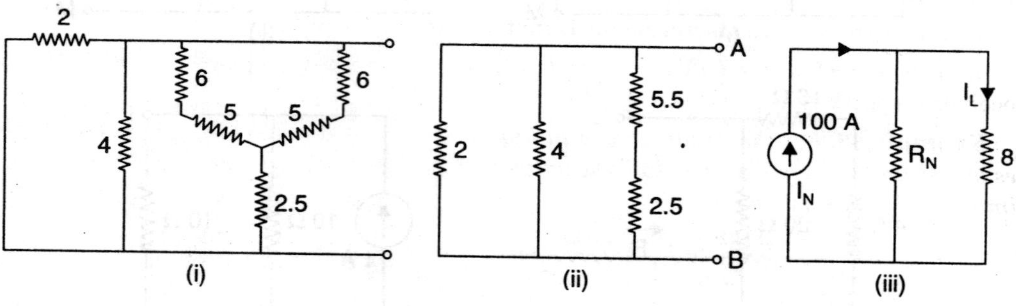

Fig. 98

Thus, Norton's equivalent circuit consists of a 100 A source having a parallel resistance of $\frac{8}{7}$ Ω as shown in Fig. 98 (*iii*).

$$\therefore \quad I_L = 100 \times \frac{(8/7)}{8 + (8/7)} = \mathbf{12.5\ A.} \quad \textbf{(Ans.)} \quad \text{(Using current divider rule).}$$

11. MAXIMUM POWER TRANSFER THEOREM

This theorem is particularly useful for analysing *communication networks*. It is stated as follows :

"Maximum power output is obtained from a network when the load resistance is equal to the output resistance of the network as seen from the terminals of the load".

Any network can be converted into a single voltage source by the use of Thevenin's theorem (Fig. 99). The maximum power transfer theorem aims at finding R_L such that the power dissipated in R_L is maximum.

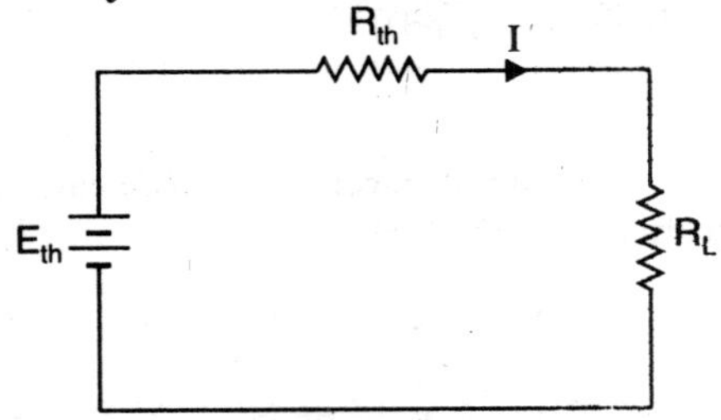

Fig. 99

$$P = I^2 R_L$$

$$= \left(\frac{E_{th}}{R_{th} + R_L} \right)^2 R_L \qquad ...(21)$$

For P to be maximum, $\frac{dP}{dR_L} = 0$

Differentiating eqn. (21), we have

$$\frac{dP}{dR_L} = \frac{E_{th}^2 \, [(R_{th} + R_L)^2 - 2R_L(R_{th} + R_L)]}{(R_{th} + R_L)^4}$$

$$\therefore \qquad \frac{E_{th}^2 \, [(R_{th} + R_L)^2 - 2R_L \, (R_{th} + R_L)]}{(R_{th} + R_L)^4} = 0$$

From which, $R_L = R_{th}$ ...(22)

It is worth noting that under these conditions the voltage across the load is half the open-circuit voltage at the terminals L and M.

$$\therefore \text{ Maximum power, } P_{max} = \left[\frac{E_{th}}{(R_L + R_L)} \right]^2 R_L = \frac{E_{th}^2}{4R_L} \qquad ...[22\,(a)]$$

The process of adjusting the load resistance for maximum power transfer is called *'load matching'*. This is done in the following typical cases :

(*i*) *Motor cars*—here starter motor is matched to the battery.

(*ii*) *Telephone lines and TV aerial leads*—these are matched to the telephone instrument and TV receiver respectively.

Example 43. *For the circuit shown in Fig. 100, find the current through R_L when it takes on values of 5 Ω and 25 Ω. Also, calculate the value of R_L for which the power dissipated in it would be maximum and find this power.*

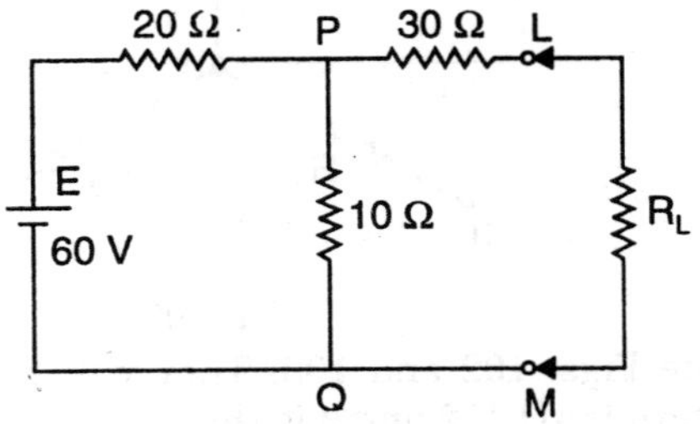

Fig. 100

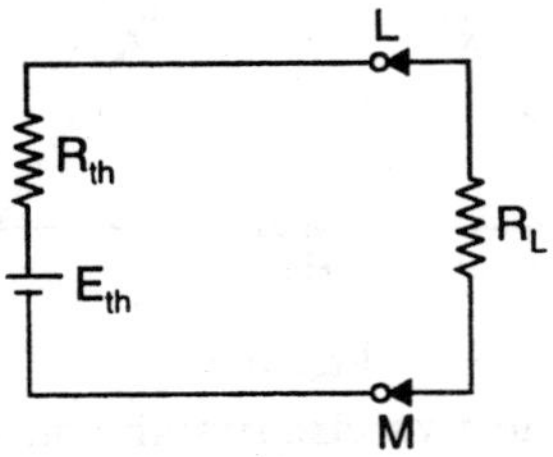

Fig. 101

Solution. The open circuit voltage V_{oc} (also called Thevenin's voltage E_{th}) which appears across terminals L and M is equal to the voltage drop across 10 Ω resistance.

Current flowing through the circuit EPQ $= \frac{60}{20 + 10} = 2$ A

$\therefore$ Voltage drop over 10 Ω resistance $= 2 \times 10 = 20$ V

Hence $V_{oc} = E_{th} = 20$ V

The resistance of the circuit as looked into the network from points L and M (when battery has been removed),

$$R_i = R_{th} = 30 + 10 \parallel 20 = 30 + \frac{10 \times 20}{10 + 20} = 36.67\ \Omega$$

The whole circuit up to LM can now be replaced by a single source of e.m.f. and single resistance as shown in Fig. 101.

(*i*) When $R_L = 5\ \Omega$, $$I = \frac{E_{th}}{R_{th} + R_L} = \frac{20}{36.67 + 5} = \mathbf{0.48\ A.}\quad \textbf{(Ans.)}$$

(*ii*) When $R_L = 25\ \Omega$, $$I = \frac{20}{36.67 + 25} = \mathbf{0.324\ A.}\quad \textbf{(Ans.)}$$

According to the maximum power transfer theorem, power drawn by R_L would be maximum when $R_L = R_i$ or when $R_L = 36.67\ \Omega$

$$\therefore\ \text{Maximum power drawn by } R_L = I^2 R_L = \left(\frac{E_{th}}{R_{th} + R_L}\right)^2 R_L$$

$$= \left(\frac{E_{th}}{R_L + R_L}\right)^2 R_L = \frac{E_{th}^2}{4R_L} \qquad (\because\ R_{th} = R_L)$$

$$= \frac{20^2}{4 \times 36.67} = \mathbf{2.72\ W.}\quad \textbf{(Ans.)}$$

12. DELTA STAR TRANSFORMATION

When networks having a large number of branches are to be solved by the use of Kirchhoff's law, a great difficulty is experienced in solving several simultaneous equations. Such complicated networks, however, can be simplified by successively replacing delta meshes by equivalent star systems and *vice versa*.

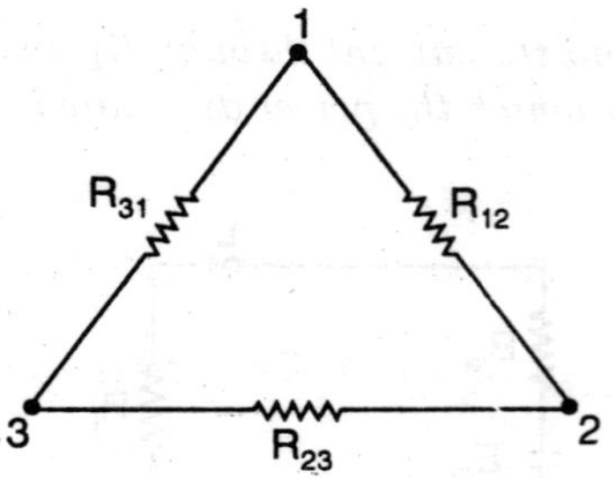

Fig. 102

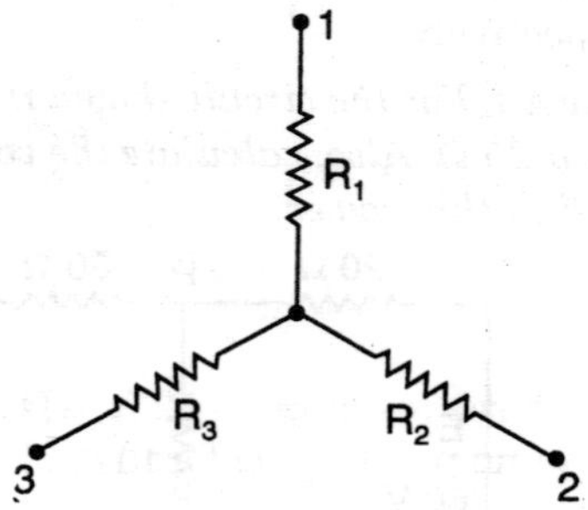

Fig. 103

Consider the two circuits shown in the Figs. 102 and 103. They will be *equivalent* if the resistance measured between any two of the terminals 1, 2 and 3 is the same in the two cases.

$$[R_{12}]_Y = [R_{12}]_\Delta \qquad ...(23)$$

or
$$R_1 + R_2 = \frac{R_{12}(R_{23} + R_{31})}{R_{12} + R_{23} + R_{31}} \qquad ...(24)$$

Similarly
$$R_2 + R_3 = \frac{R_{23}(R_{31} + R_{12})}{R_{12} + R_{23} + R_{31}} \qquad ...(25)$$

and
$$R_3 + R_1 = \frac{R_{31}(R_{12} + R_{23})}{R_{12} + R_{23} + R_{31}} \qquad ...(26)$$

Solving eqns. (24), (25) and (26) simultaneously, we get

$$R_1 = \frac{R_{12}R_{31}}{R_{12} + R_{23} + R_{31}} \quad ...(27)$$

$$R_2 = \frac{R_{23}R_{12}}{R_{12} + R_{23} + R_{31}} \quad ...(28)$$

$$R_3 = \frac{R_{31}R_{23}}{R_{12} + R_{23} + R_{31}} \quad ...(29)$$

From above it may be noted that *resistance of each arm of the star is given by the product of the resistances of the two delta sides that meet at its end divided by the sum of the three delta resistances.*

From eqns. (24) to (26), eqns. for star to delta conversion can also be obtained. These are as follows :

$$R_{12} = \frac{R_1R_2 + R_2R_3 + R_3R_1}{R_3} \quad ...(30)$$

$$R_{23} = \frac{R_1R_2 + R_2R_3 + R_3R_1}{R_1} \quad ...(31)$$

$$R_{31} = \frac{R_1R_2 + R_2R_3 + R_3R_1}{R_2} \quad ...(32)$$

In electronics, star and delta circuits are generally referred to as τ and π circuits respectively.

Example 44. *Fig. 104 shows a number of resistances connected in delta and star. Using star/delta conversion method complete the network resistance measured between (i) L and M (ii) M and N and (iii) N and L.*

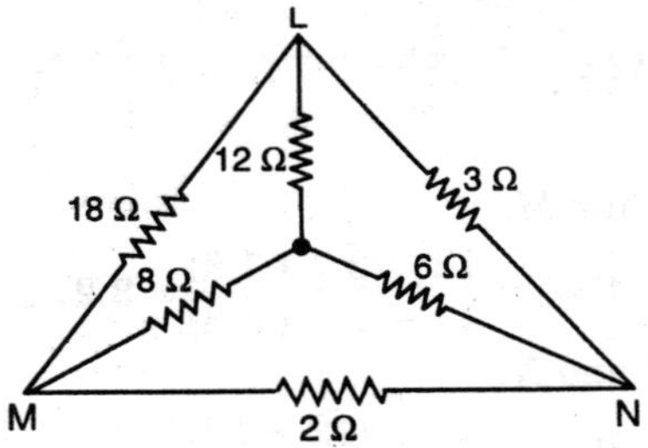

Fig. 104

Solution. Three resistances 12 Ω, 6 Ω and 8 Ω are star connected. Transform them into delta with ends at the same points as before.

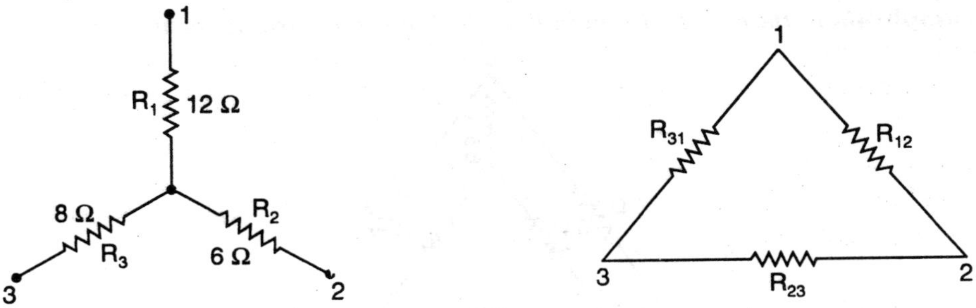

(*i*) (*ii*)

Fig. 105

Refer Fig. 105.
$$R_{12} = \frac{R_1R_2 + R_2R_3 + R_3R_1}{R_3} = \frac{12 \times 6 + 6 \times 8 + 8 \times 12}{8} = 27\ \Omega$$
$$R_{23} = \frac{R_1R_2 + R_2R_3 + R_3R_1}{R_1} = \frac{12 \times 6 + 6 \times 8 + 8 \times 12}{12} = 18\ \Omega$$
Similarly
$$R_{31} = \frac{12 \times 6 + 6 \times 8 + 8 \times 12}{6} = 36\ \Omega.$$

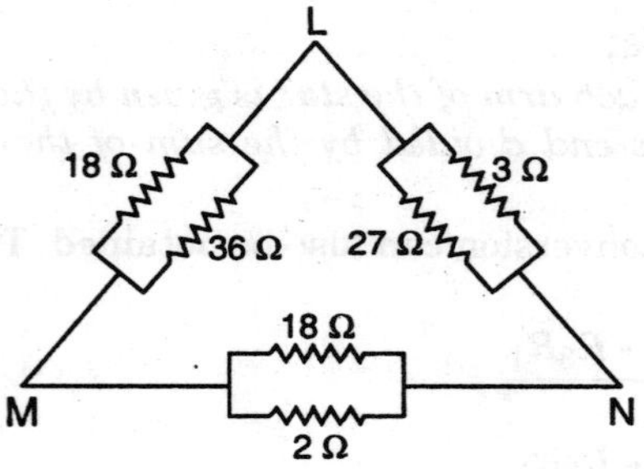

Fig. 106

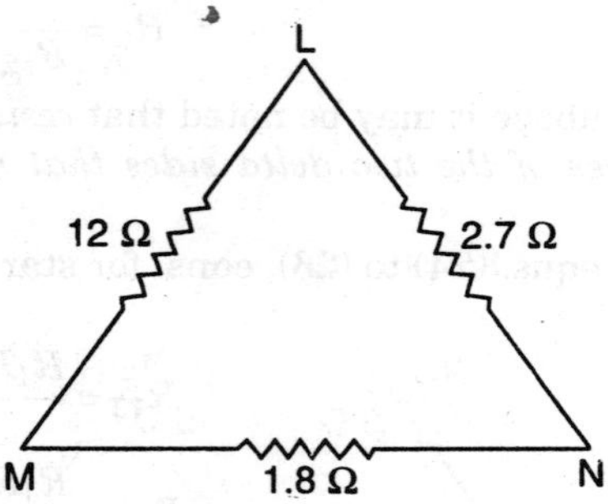

Fig. 107

Fig. 106 shows this transformed circuit connected to original delta connected resistances in the circuit 18 Ω, 3 Ω and 2 Ω.

Here 18 Ω and 36 Ω are in parallel ;
3 Ω and 27 Ω are in parallel, and
2 Ω and 18 Ω are in parallel.

These resistances are equivalent to :
$$\frac{18 \times 36}{18 + 36} = 12\ \Omega\ ;\ \frac{3 \times 27}{3 + 27} = 2.7\ \Omega \text{ and } \frac{2 \times 18}{2 + 18} = 1.8\ \Omega$$

This is shown in Fig. 107.

(*i*) Resistance between *L* and *M*,
$$\mathbf{R_{LM}} = \frac{12 \times (2.7 + 1.8)}{12 + (2.7 + 1.8)} = \mathbf{3.27\ \Omega}. \quad \textbf{(Ans.)}$$

(*ii*) Resistance between *M* and *N*,
$$\mathbf{R_{MN}} = \frac{1.8 \times (12 + 2.7)}{1.8 + 12 + 2.7} = \mathbf{1.6\ \Omega}. \quad \textbf{(Ans.)}$$

(*iii*) Resistance between *N* and *L*,
$$\mathbf{R_{NL}} = \frac{2.7 \times (12 + 1.8)}{2.7 + (12 + 1.8)} = \mathbf{2.25\ \Omega}. \quad \textbf{(Ans.)}$$

Example 45. *In the circuit shown in Fig. 108, find the resistance between M and N.*

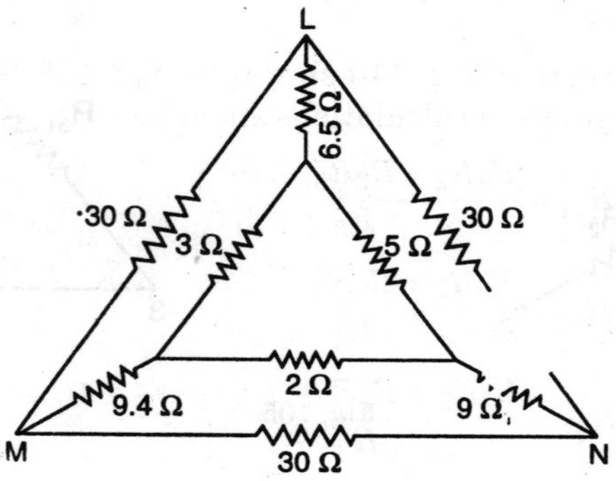

Fig. 108

Solution. Connecting the 1 2 3 delta [Fig. 109 (*i*)] to equivalent star [Fig. 109 (*ii*)]

$$R_1 = \frac{R_{12}R_{31}}{R_{12}+R_{23}+R_{31}} = \frac{5\times 3}{5+2+3} = 1.5\ \Omega$$

$$R_2 = \frac{R_{23}R_{12}}{R_{12}+R_{23}+R_{31}} = \frac{2\times 5}{5+2+3} = 1\ \Omega$$

and

$$R_3 = \frac{R_{31}R_{23}}{R_{12}+R_{23}+R_{31}} = \frac{3\times 2}{5+2+3} = 0.6\ \Omega$$

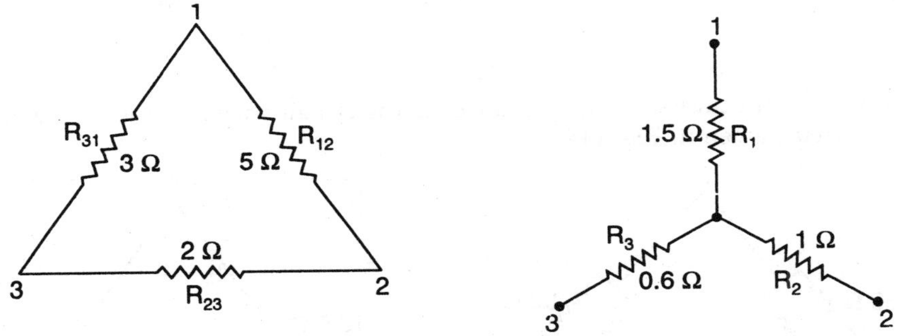

(*i*) (*ii*)

Fig. 109

Thus the original circuit reduces to that shown in Fig. 110 which further reduces to the circuit shown in Fig. 111.

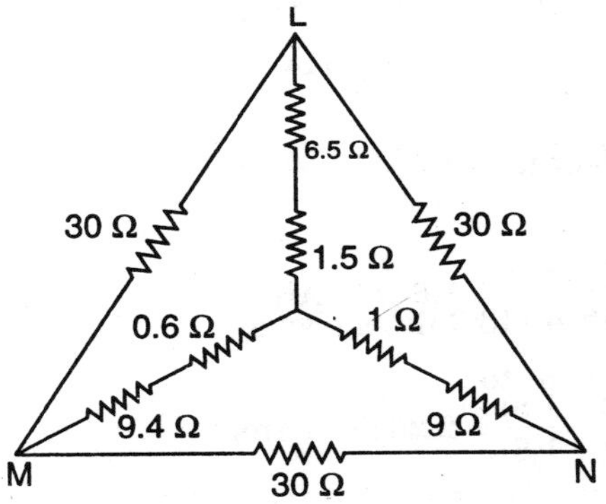

Fig. 110

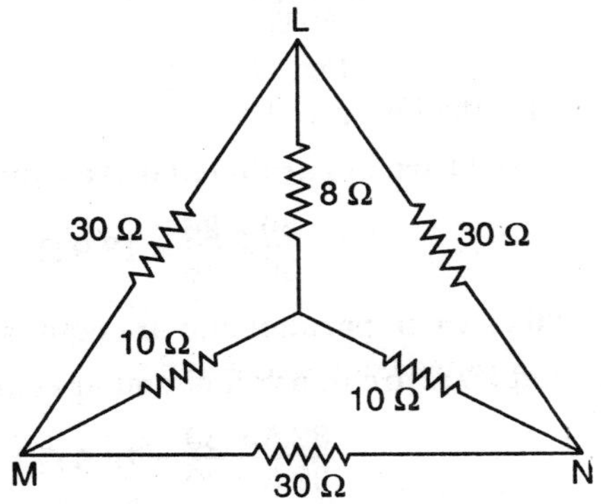

Fig. 111

Now, the inner star circuit of Fig. 110 shown as Fig. 111 is equivalent to the delta circuit shown in Fig. 112 (*ii*) as appears from calculations given below :

$$R_{12} = \frac{R_1R_2+R_2R_3+R_3R_1}{R_3} = \frac{8\times 10+10\times 10+10\times 8}{10} = 26\ \Omega$$

$$R_{23} = \frac{R_1R_2+R_2R_3+R_3R_1}{R_1} = \frac{8\times 10+10\times 10+10\times 8}{8} = 32.5\ \Omega$$

$$R_{31} = \frac{R_1R_2+R_2R_3+R_3R_1}{R_2} = \frac{8\times 10+10\times 10+10\times 8}{10} = 26\ \Omega.$$

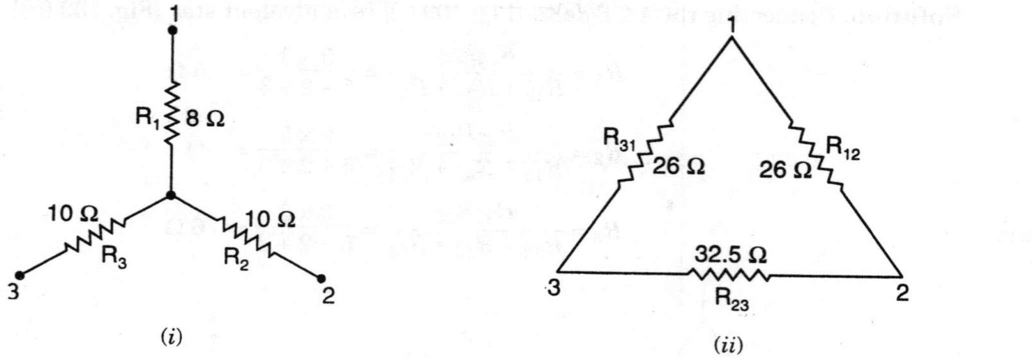

Fig. 112

∴ The given circuit thus ultimately reduces to the circuit shown in Fig. 113 which in turn is equivalent to the circuit given in Fig. 114.

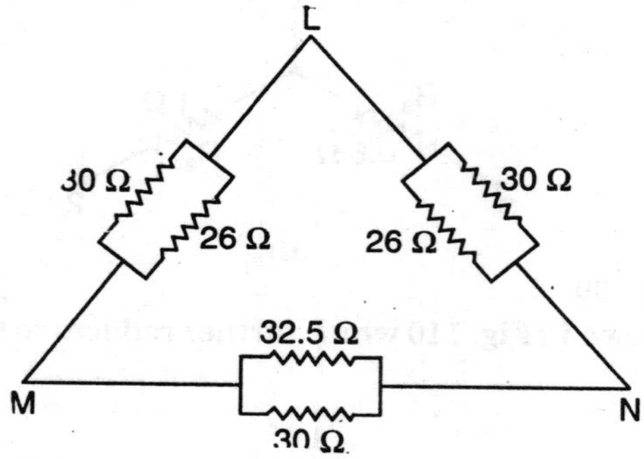

Fig. 113

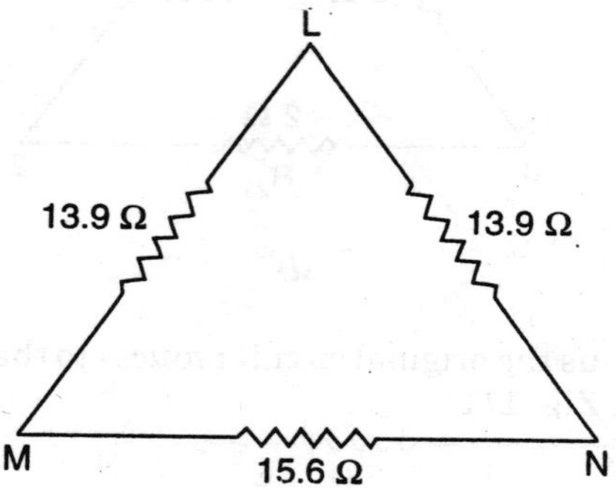

Fig. 114

In Fig. 113, the observe that :

30 Ω and 26 Ω are in parallel and are equivalent to :

$$\frac{30 \times 26}{30 + 26} = 13.9\ \Omega.$$

30 and 26 Ω are in parallel and are equivalent to : 13.9 Ω (as above)

32.5 Ω and 30 Ω are in parallel and are equivalent to :

$$\frac{32.5 \times 30}{32.5 + 30} = 15.6\ \Omega$$

Hence total resistance between M and N,

$$\mathbf{R_{MN}} = \frac{15.6 \times (13.9 + 13.9)}{15.6 + (13.9 + 13.9)}$$

$$= \frac{433.68}{43.4} = \mathbf{9.99\ \Omega.} \quad \textbf{(Ans.)}$$

Example 46. *Find the current I supplied by the battery for Fig. 115, using delta/star transformation.*

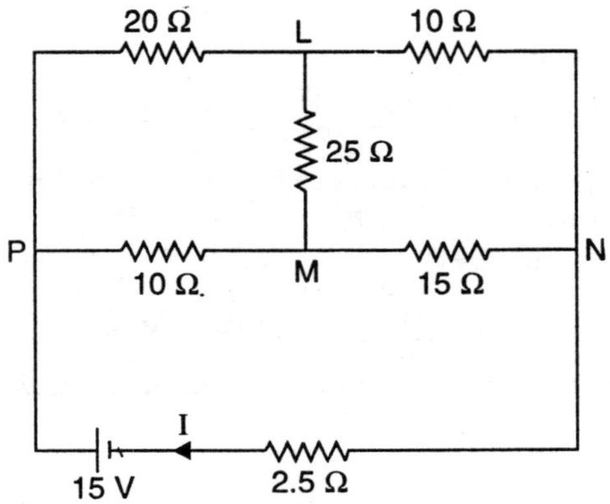

Fig. 115

Solution. Delta connected resistances 25 Ω, 10 Ω and 15 Ω are transformed to equivalent star as given below : (See Fig. 116)

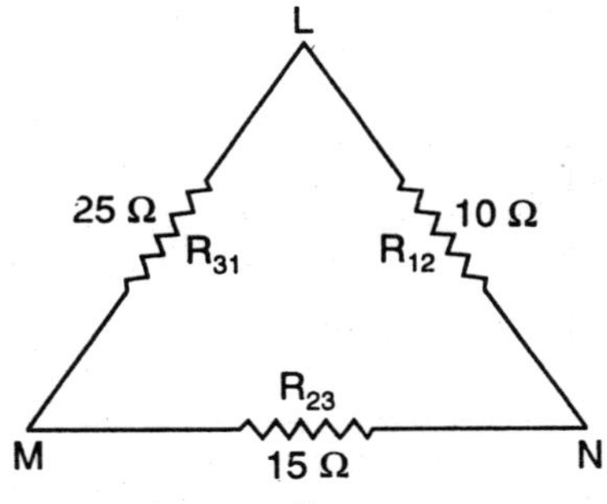

Fig. 116

$$R_1 = \frac{R_{12}R_{31}}{R_{12}+R_{23}+R_{31}} = \frac{10 \times 25}{10+15+25} = 5\ \Omega$$

$$R_2 = \frac{R_{23}R_{12}}{R_{12}+R_{23}+R_{31}} = \frac{15 \times 10}{10+15+25} = 3\ \Omega$$

$$R_3 = \frac{R_{31}R_{23}}{R_{12}+R_{23}+R_{31}} = \frac{25 \times 15}{10+15+25} = 7.5\ \Omega$$

The given circuit thus reduces to the circuit shown in Fig. 117.

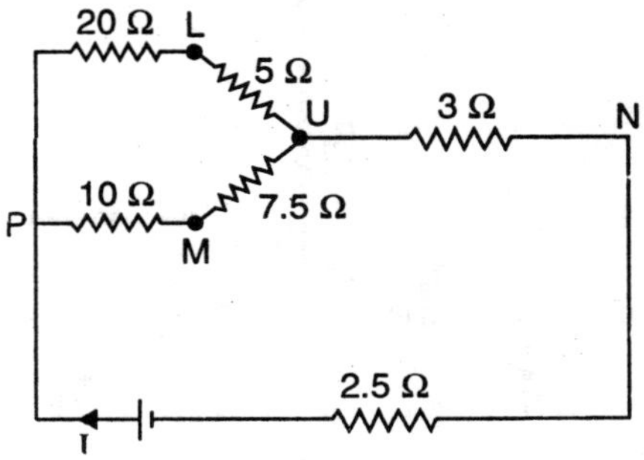

Fig. 117

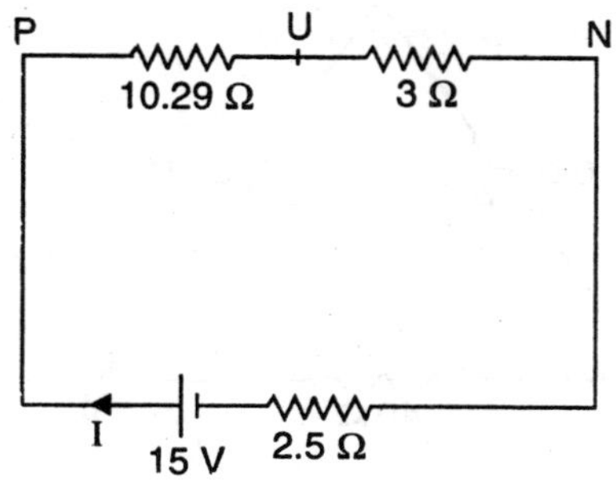

Fig. 118

The equivalent resistance of

$$(20 + 5)\ \Omega \parallel (10 + 7.5)\ \Omega = \frac{25 \times 17.5}{25 + 17.5} = 10.29\ \Omega$$

Thus the given circuit ultimately reduces to the circuit shown in Fig. 118.

Total resistance = 10.29 + 3 + 2.5 = 15.79 Ω

Hence current through the battery,

$$I = \frac{15}{15.79} = \mathbf{0.95\ A.}\ \ \textbf{(Ans.)}$$

Example 47. *Find the current in the galvanometer arm in the network shown in Fig. 119, using delta/star transformation.*

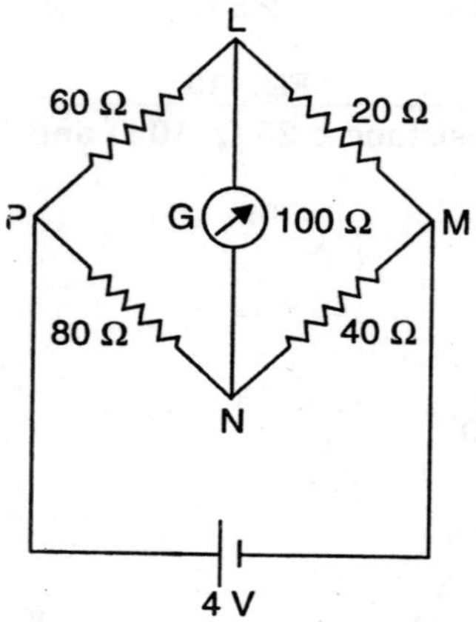

Fig. 119

Solution. Delta connected resistances 100 Ω, 20 Ω and 40 Ω are transformed to equivalent star as given below : (See Fig. 120)

$$R_1 = \frac{R_{12}R_{31}}{R_{12} + R_{23} + R_{31}} = \frac{20 \times 100}{20 + 40 + 100} = 12.5\ \Omega$$

$$R_2 = \frac{R_{23}R_{12}}{R_{12} + R_{23} + R_{31}} = \frac{40 \times 20}{20 + 40 + 100} = 5\ \Omega$$

$$R_3 = \frac{R_{31}R_{23}}{R_{12} + R_{23} + R_{31}} = \frac{100 \times 40}{20 + 40 + 100} = 25\ \Omega$$

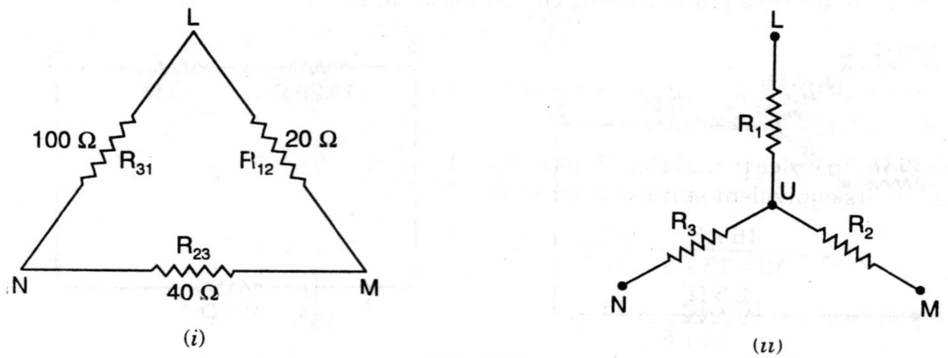

Fig. 120

The given circuit thus reduces to the circuit shown in Fig. 121.

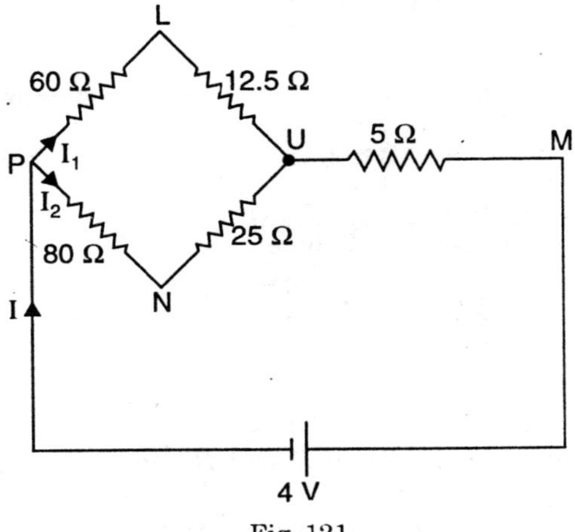

Fig. 121

The equivalent resistance, $R_{PU} = \frac{(60 + 12.5) \times (80 + 25)}{(60 + 12.5) + (80 + 25)} = 42.9\ \Omega$

∴ Total resistance of the circuit,

$$R_{PM} = 42.9 + 5 = 47.9\ \Omega$$

∴ The main current, $I = \frac{4}{47.9} = 0.0835$ A

Let I_1 = current in the arm *NLP*,

I_2 = current in the arm *LP*, and

I_3 = current in the galvanometer.

Then, $$I_1 = 0.0835 \times \left[\frac{(80 + 25)}{(80 + 25) + (60 + 12.5)}\right] = 0.0494 \text{ A}$$

and $$I_2 = 0.0835 \left[\frac{(60 + 12.5)}{(80 + 25) + (60 + 12.5)}\right] = 0.0341 \text{ A.}$$

∴ Potential difference between L and P

$= 0.0494 \times 60 = 2.964$ V

and between P and N $= 0.0341 \times 80 = 2.728$ V

and between L and N $= 2.964 - 2.728 = 0.236$ V

Hence the current flowing through the galvanometer,

$$I_3 = \frac{0.236}{100} = \mathbf{0.00236\ A.}\ \textbf{(Ans.)}$$

Example 48. *Find the input resistance of the circuit between the points A and B of Fig. 122.*

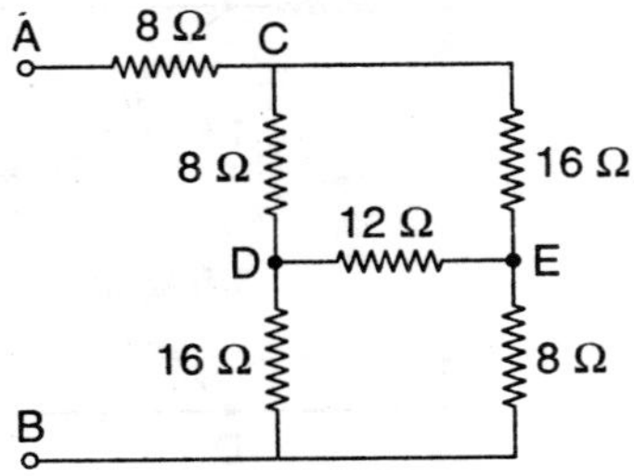

Fig. 122

Solution. In order to find R_{AB} let us convert the delta *CDE* of Fig. 122 into its equivalent star as shown in Fig. 123 (*i*).

$$R_{CS} = \frac{16 \times 8}{16 + 12 + 8} = \frac{32}{9}\ \Omega$$

$$R_{ES} = \frac{16 \times 12}{16 + 12 + 8} = \frac{48}{9}\ \Omega$$

$$R_{DS} = \frac{12 \times 8}{16 + 12 + 8} = \frac{24}{9}\ \Omega$$

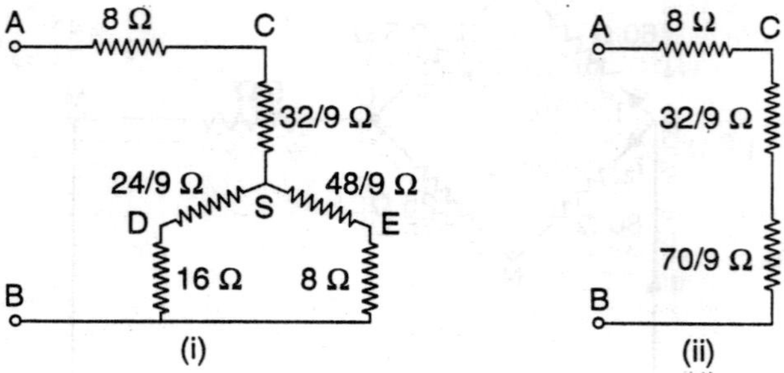

Fig. 123

The two parallel resistances between S and B can be reduced to a single resistance as follows :

$$\left(\frac{24}{9} + 16\right) \Big|\Big| \left(\frac{48}{9} + 8\right) = \frac{168}{9} \Big|\Big| \frac{120}{9}$$

$$= \frac{(168/9) \times (120/9)}{(168/9) + (120/9)} = \frac{70}{9}\ \Omega$$

As seen from Fig. 123 (*ii*),

$$\mathbf{R_{AB}} = 8 + \left(\frac{32}{9}\right) + \left(\frac{70}{9}\right) = \frac{\mathbf{174}}{\mathbf{9}}\ \Omega. \ \textbf{(Ans.)}$$

Example 49. *Using Delta-star conversion find resistance between terminals AB in Fig. 124.*

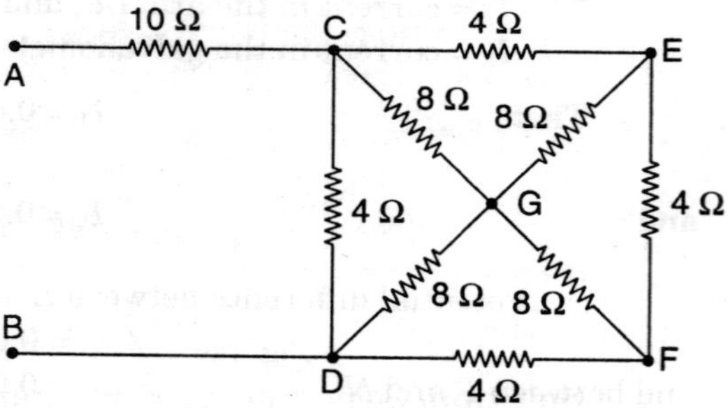

Fig. 124

Solution.

- Applying delta-star conversion to CGD and EGF of Fig. 124, we get Fig. 125 (*i*).
- Further simplification reduces the circuit to its equivalents as in Fig. 125 (*ii*) and later as in Fig. 125 (*iii*).
- Convert CHJ to equivalent star as in Fig. 125 (*iv*).
- Using series-parallel combinations, we have

$$\mathbf{R_{AB}} = (10 + 0.66) + [(0.752 + 1.6) \,||\, (2.64 + 5.6)]$$

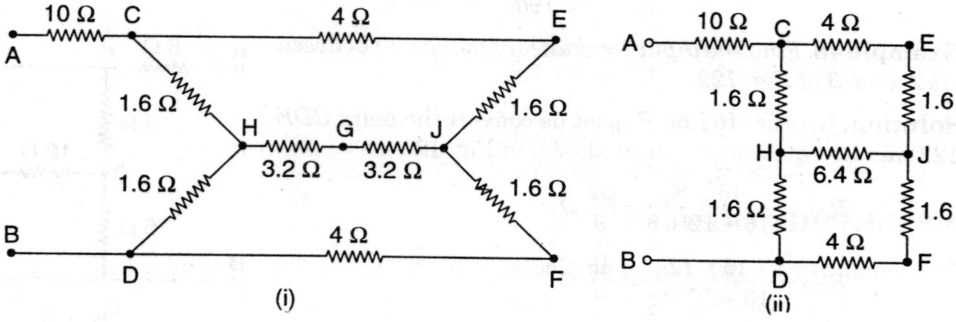

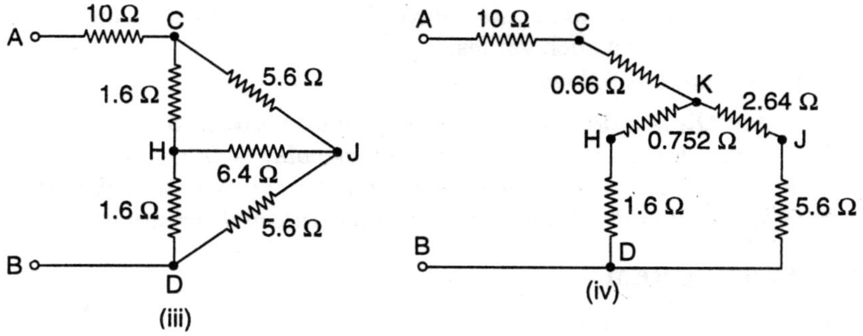

Fig. 125

$$= 10.66 + \frac{2.352 \times 8.24}{(2.352 + 8.24)} = \mathbf{12.49\ \Omega.}\ \textbf{(Ans.)}$$

Example 50. *Find the resistance at the A-B terminals in the electric circuit of Fig. 126, using Δ-Y transformations.* **(U.P.T.U. 2001)**

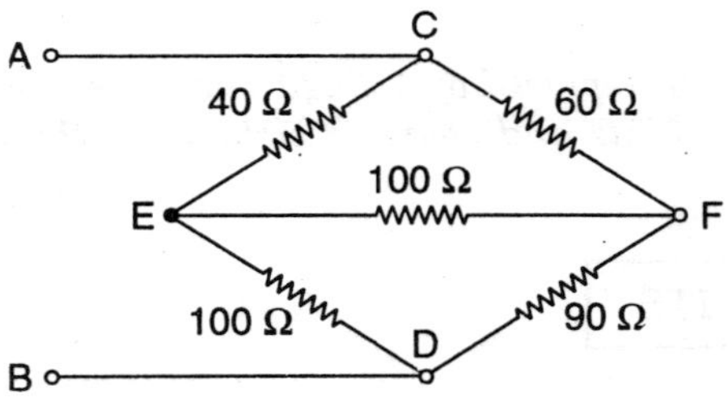

Fig. 126

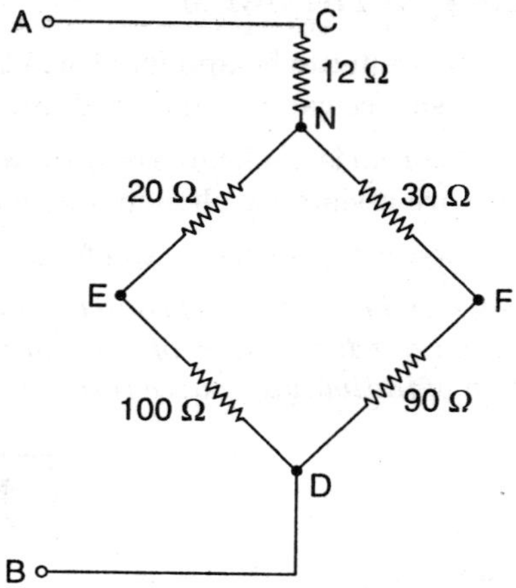

Fig. 127

Solution. Converting *CEF* into star connections, we get (Refer Fig. 127) :

$$CN = \frac{60 \times 40}{60 + 40 + 100} = 24\ \Omega$$

$$NE = \frac{40 \times 100}{60 + 40 + 100} = 20\ \Omega$$

$$NF = \frac{60 \times 100}{60 + 40 + 100} = 30\ \Omega$$

$$R_{ND} = \frac{(100 + 20) \times (90 + 30)}{120 + 120} = 60\ \Omega$$

$$\therefore \quad R_{AB} = 12 + 60 = \mathbf{72\ \Omega.}\ \textbf{(Ans.)}$$

13. COMPENSATION THEOREM

The compensation theorem is particularly useful for the following purposes :

(*i*) To calculate the sensitivity of a bridge network.

(*ii*) To analyse those networks where the values of the branch elements are varied and for studying the effect of tolerance on such values.

This theorem is stated as follows :

"If a change, say ΔR, is made in the resistance of any branch of a network when the current was originally I, then the change of current at any other point in the network may be calculated by assuming that an e.m.f. $- I\ \Delta R$ has been introduced into the changed branch while all other sources have their e.m.fs. suppressed and are represented by their internal resistances only".

14. RECIPROCITY THEOREM

The theorem is stated as follows :

"In any linear bilateral network, if a source of e.m.f. E in any branch produces a current I in any other branch, then the same e.m.f. acting in the second branch would produce the same current I in the first branch".

In other words, it simply means that E and I are mutually transferrable. The ratio E/I is known as the *transfer* resistance (or impedance in a.c. systems).

15. MILLMAN'S THEOREM

The theorem can be applied to a network having a combination of voltage and current sources since voltage source can be converted into a current source and *vice-versa*.

The theorem is applicable only to two sources connected directly in parallel. It is not applicable where there are resistance elements between the sources.

The theorem is stated in the following manner :

"Any number of current sources in parallel may be replaced by a single current source whose current is the algebraic sum of individual source currents and source resistance is the parallel combination of individual source resistances".

HIGHLIGHTS

1. Kirchhoff's laws :
   First law : Σ currents entering = Σ currents leaving
   Second law : Σ potential rises = Σ potential drops.
2. Applications of Kirchhoff's laws include the following :
   (*i*) Branch-current method
   (*ii*) Maxwell's loop (or mesh) current method
   (*iii*) Nodal voltage method.
3. The solutions of the networks involve the use of the following theorems :
   — Superposition theorem
   — Thevenin's theorem
   — Nortan's theorem
   — Maximum power transfer theorem
   — Delta star transformation
   — Compensation theorem
   — Reciprocity theorem
   — Millman's theorem.

OBJECTIVE TYPE QUESTIONS

Choose the Correct Answer

1. Kirchhoff's current law states that
   (*a*) net current flow at the junction is positive
   (*b*) algebraic sum of the currents meeting at the junction is zero
   (*c*) no current can leave the junction without some current entering it
   (*d*) total sum of currents meeting at the junction is zero.
2. According to Kirchhoff's voltage law, the algebraic sum of all *IR* drops and e.m.fs. in any closed loop of a network is always
   (*a*) negative (*b*) positive
   (*c*) determined by battery e.m.fs. (*d*) zero.
3. Kirchhoff's current law is applicable to only
   (*a*) junction in a network (*b*) closed loops in a network
   (*c*) electric circuits (*d*) electronic circuits.
4. Kirchhoff's voltage law is related to
   (*a*) junction voltages (*b*) battery e.m.fs. (*c*) *IR* drops (*d*) both (*a*) and (*b*)
   (*e*) none of the above.
5. Superposition theorem can be applied only to circuits having
   (*a*) resistive elements (*b*) passive elements
   (*c*) non-linear elements (*d*) linear bilateral elements.
6. The concept on which superposition theorem is based is
   (*a*) reciprocity (*b*) duality (*c*) non-linearity (*d*) linearity.
7. Thevenin resistance R_{th} is found
   (*a*) by removing voltage sources along with their internal resistances
   (*b*) by short-circuiting the given two terminals
   (*c*) between any two 'open' terminals
   (*d*) between same open terminals as for E_{th}.
8. An ideal voltage source should have
   (*a*) large value of e.m.f. (*b*) small value of e.m.f.
   (*c*) zero source resistance (*d*) infinite source resistance.
9. For a voltage source
   (*a*) terminal voltage is always lower than source e.m.f.
   (*b*) terminal voltage cannot be higher than source e.m.f.
   (*c*) the source e.m.f. and terminal voltage are equal.
10. To determine the polarity of the voltage drop across a resistor, it is necessary to know
   (*a*) value of current through the resistor (*b*) direction of current through the resistor
   (*c*) value of resistor (*d*) e.m.fs. in the circuit.

Answers

1. (*b*)	**2.** (*d*)	**3.** (*a*)	**4.** (*d*)	**5.** (*d*)	**6.** (*d*)	**7.** (*d*)
8. (*c*)	**9.** (*b*)	**10.** (*b*).				

THEORETICAL QUESTIONS

1. Define the following terms :
   Circuit, Electrical network, Active network, Node and Branch.
2. What are the limitations of Ohm's law ?
3. State and explain Kirchhoff's laws.
4. Discuss briefly application of Kirchhoff's laws.
5. Explain the nodal voltage method for solving networks. How are the nodal equations written ?
6. Explain Cramer's rule used for solving equations by determinants.
7. State and explain superposition theorem.
8. State Norton's theorem. List the steps for finding the current in a branch of a network with the help of this theorem.
9. State Thevenin's theorem.
10. State the maximum power transfer theorem and explain its importance.
11. State the compensation theorem and discuss its application.
12. State Millman's theorem.

UNSOLVED EXAMPLES

1. Determine the magnitude and direction of the current in each of the batteries *L*, *M* and *N* shown in the Fig. 128.

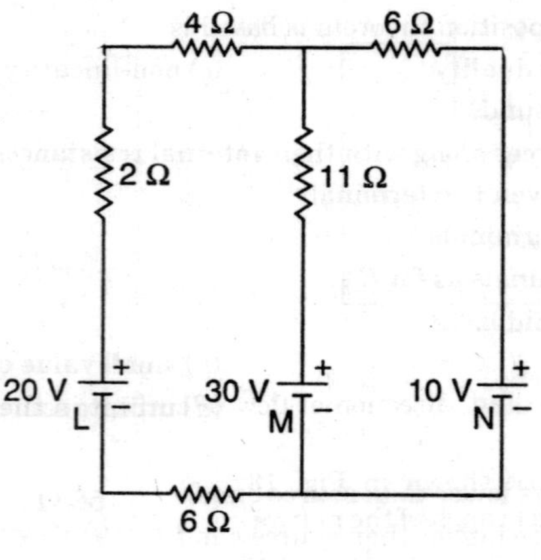

Fig. 128

2. Two batteries are connected in parallel. The e.m.f. and internal resistance of one are 110 V and 6 Ω respectively and the corresponding values for other are 130 V and 4 Ω respectively. A resistance of 20 Ω is connected across the parallel combination. Calculate :
   (*i*) The value and direction of the current in each battery.
   (*ii*) The terminal voltage. **[Ans.** (*i*) 0.1786 A, 5.2678 A (*ii*) 108.928 V]
   [**Hint.** Terminal voltage = $(I_1 + I_2)R$]
3. Two cells *A* and *B* are connected in parallel, unlike poles being joined together. The terminals of the cells are then joined by two resistors of 4 Ω and 2 Ω in parallel. The e.m.f. of *A* is 2 V, its internal resistance is 1 Ω ; the e.m.f. of *B* is 1 V, its internal resistance is 2 Ω. Find the current in each of the four branches of the circuit. **[Ans.** 4/3 A, 5/6 A, 1/6 A, 1/3 A]

4. What is the equivalent resistance of the network shown in Fig. 129 ? [**Ans.** 600 Ω] (*B.T.E. Delhi*)

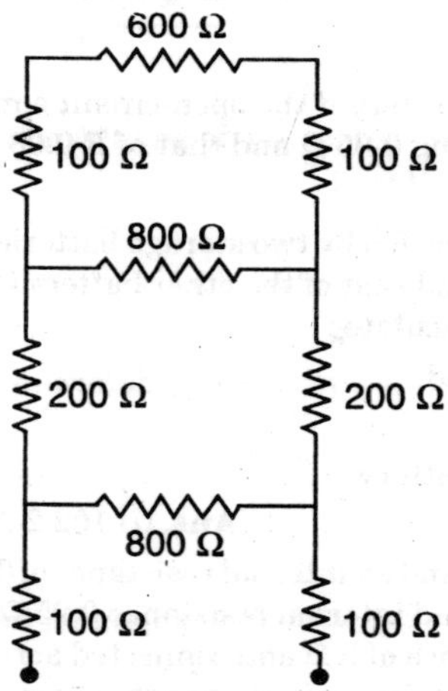

Fig. 129

5. Determine the currents in the three batteries (*L*, *M* and *N*) in the network shown in Fig. 130 and show their values and direction of flow on the diagram. Neglect the internal resistance of batteries.

[**Ans.** 2 A, 1 A, 0 A]

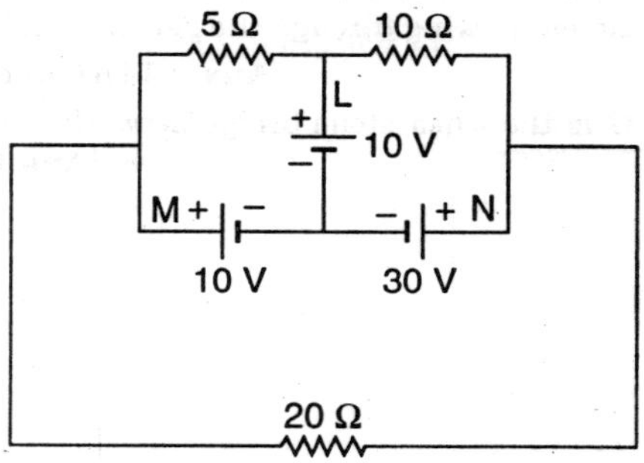

Fig. 130

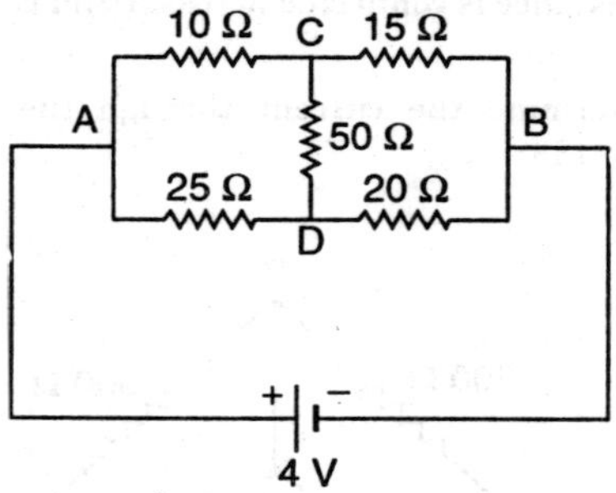

Fig. 131

6. Determine the magnitude and direction of flow of current in the branch *CD* for the circuit shown in the above Fig. 131. [**Ans.** 7/755 A from *C* to *D*]

7. For the lattice-type network shown in Fig. 132, calculate the current in each branch of the network. (*Bangladesh University*)

$$\left[\textbf{Ans. } I_{ab} = \frac{1}{50}\text{ A} ; I_{bed} = \frac{1}{100}\text{ A} ; I_{bc} = \frac{1}{100}\text{ A} ; \right.$$

$$\left. I_{ad} = \frac{1}{100}\text{ A} ; I_{dc} = \frac{1}{50}\text{ A} ; I_{cfa} = \frac{3}{100}\text{ A} \right]$$

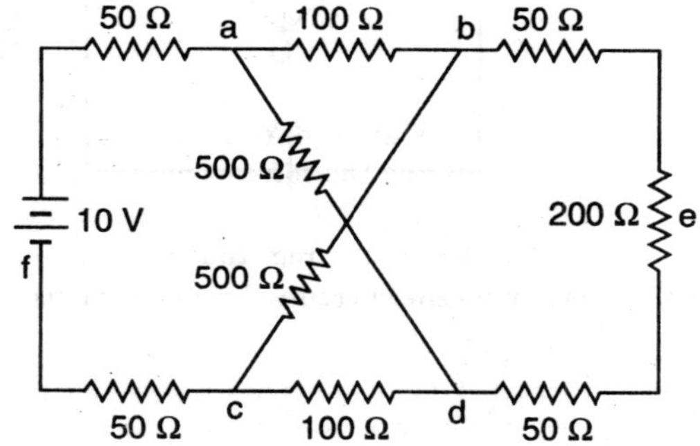

Fig. 132

8. Two storage batteries, A and B, are connected in parallel to supply a load the resistance of which is 1.2 Ω. Calculate :

(*i*) The current in this load.

(*ii*) The current supplied by each battery if the open-circuit e.m.f. of A is 12.5 V and that of B is 12.8 V, the internal resistance of A being 0.05 Ω and that of B 0.08 Ω. (*London University*)

[**Ans.** (*i*) 10.25 A (*ii*) 4 A (A), 6.25 A (B)]

9. A load having a resistance of 0.1 Ω is fed by two storage batteries connected in parallel. The open circuit e.m.f. of one battery (A) is 12.1 V and that of the other battery (B) is 11.8 V. The internal resistances are 0.03 Ω and 0.04 Ω respectively. Calculate :

(*i*) The current supplied to the load

(*ii*) The current in each battery

(*iii*) The terminal voltage of each battery. (*London University*)

[**Ans.** (*i*) 102.2 A, (*ii*) 62.7 A (A), 39.5 A (B), (*iii*) 10.22 V]

10. A battery having an e.m.f. of 110 V and an internal resistance of 0.2 Ω is connected in parallel with another battery having an e.m.f. of 100 V and internal resistance 0.25 Ω. The two batteries in parallel are placed in series with a regulating resistance of 5 Ω and connected across 200 V mains. Calculate :

(*i*) The magnitude and direction of the current in each battery.

(*ii*) The total current taken from the supply mains. (*Sumbhal University*)

[**Ans.** (*i*) 11.96 A (discharge) ; 30.43 A (charge) (*ii*) 18.47 A]

11. A wheatstone bridge $ABCD$ is arranged as follows : AB = 50 ohms, BC = 100 ohms and CD = 101 ohms. A galvanometer of 1000 ohms resistance is connected between B and D. A 2-volt battery having negligible resistance is connected across A and C. Estimate the current flowing through the galvanometer.

[**Ans.** 4.14 μA from D to B]

12. Determine the current through the galvanometer G in the wheatstone bridge network of the given Fig. 133. [**Ans.** 0.52 mA]

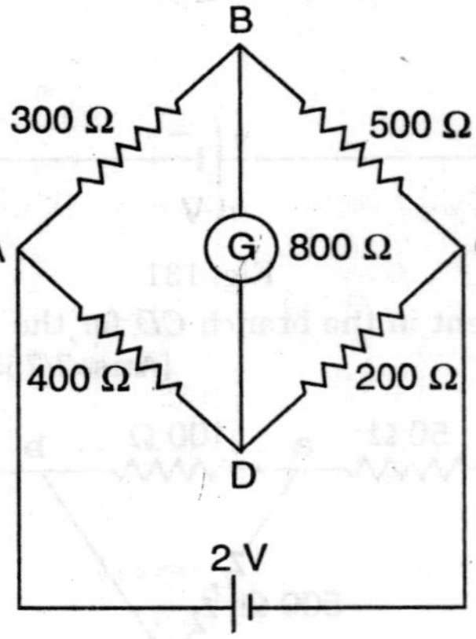

Fig. 133

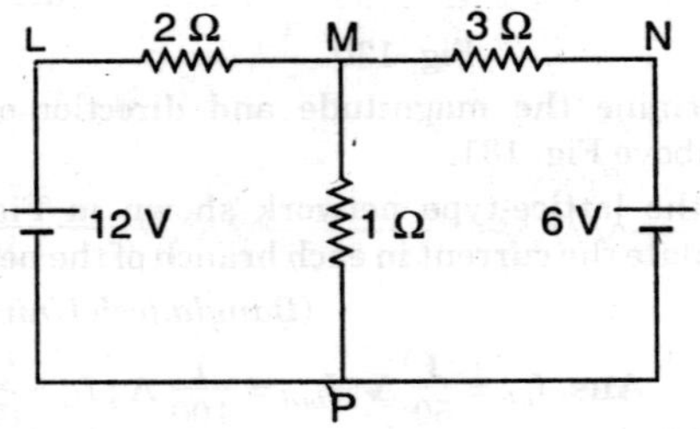

Fig. 134

13. Determine the currents through various resistors of the circuit shown in above Fig. 134.

$$\left[\textbf{Ans. } \frac{42}{11} \text{ A } (L \text{ to } M)\,;\ \frac{6}{11} \text{ A } (N \text{ to } M)\,;\ \frac{48}{11} \text{ A } (M \text{ to } P)\right]$$

14. Find I_1, I_2 and I_3 in the network shown in Fig. 135, using loop-current method. **[Ans.** 1 A, 2 A, 3 A]

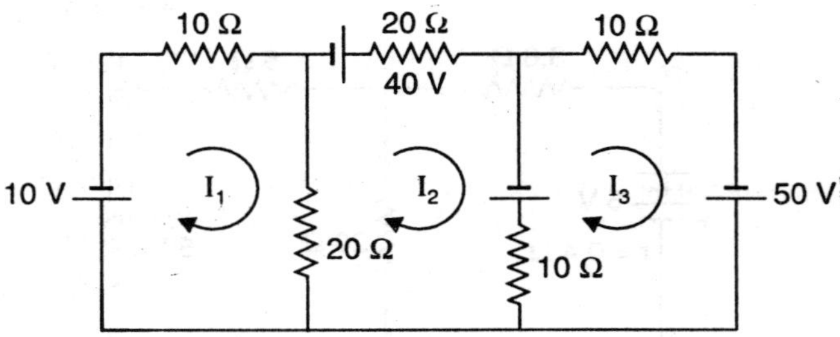

Fig. 135

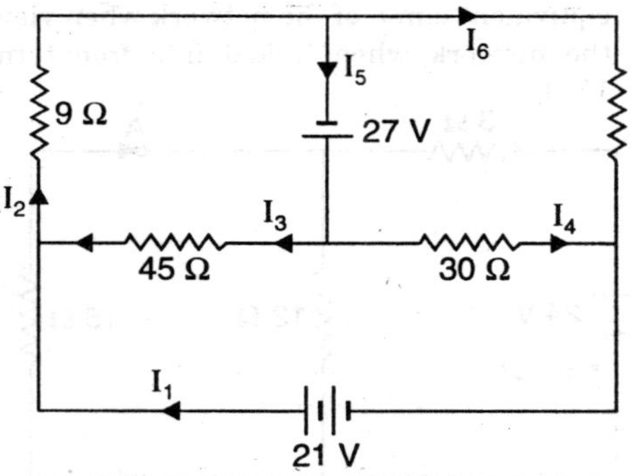

Fig. 136

15. Find the currents in the branches of network of Fig. 136 using nodal voltage method.
[Ans. $I_1 = 1.42$ A, $I_2 = 1.68$ A, $I_3 = 0.26$ A, $I_4 = 1.1$ A, $I_5 = 1.36$ A, $I_6 = 0.32$ A]

16. Using superposition theorem find the currents in the different branches of the network shown in Fig. 137. **[Ans.** $I_1 = 0.352$ A, $I_2 = 1.082$ A, $I = 1.432$ A]

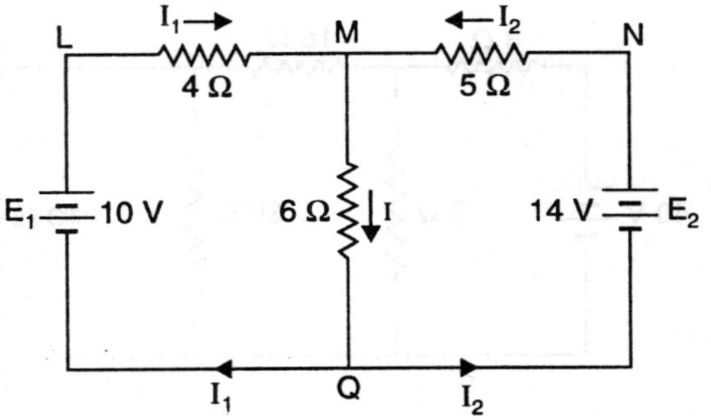

Fig. 137

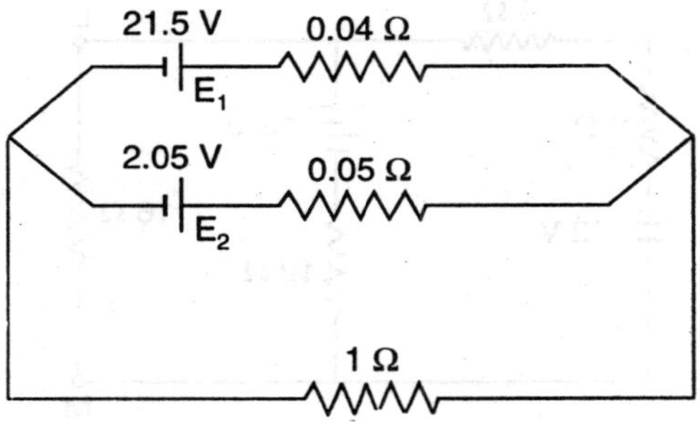

Fig. 138

17. By using superposition theorem, find the current in 1 Ω resistance in Fig. 138. Internal resistances of the cells are negligible. **[Ans.** 2.066 A]

18. Using superposition theorem find the current in 20 Ω resistor of the circuit shown in Fig. 139. **[Ans.** 0.443 A]

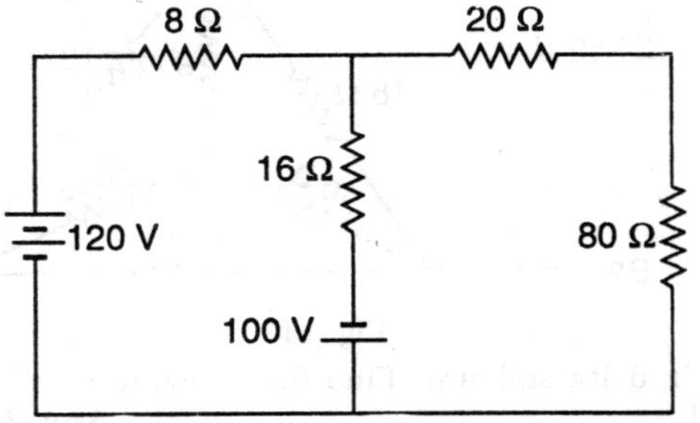

Fig. 139

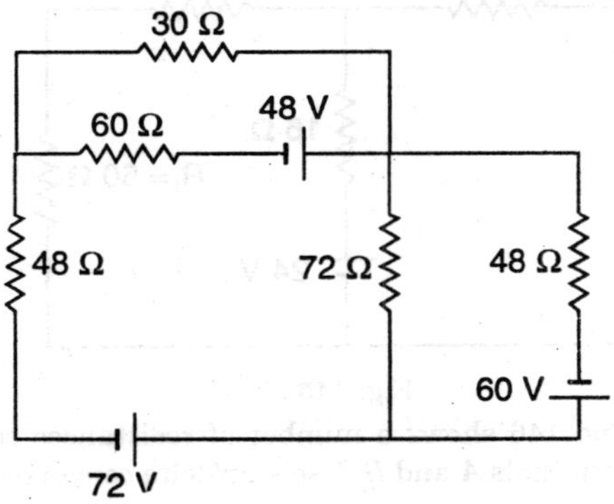

Fig. 140

19. Find the current in the 30 Ω resistor of the circuit shown in Fig. 140, using superposition theorem. **[Ans.** 0.662 A]

20. With reference to the network of Fig. 141, by applying Thevenin's theorem find the following : (*i*) The equivalent e.m.f. of the network when viewed from terminals *A* and *B* (*ii*) The equivalent resistance of the network when looked into from terminals *A* and *B*, and (*iii*) Current in the load resistance R_L of 15 Ω. [**Ans.** 1 A]

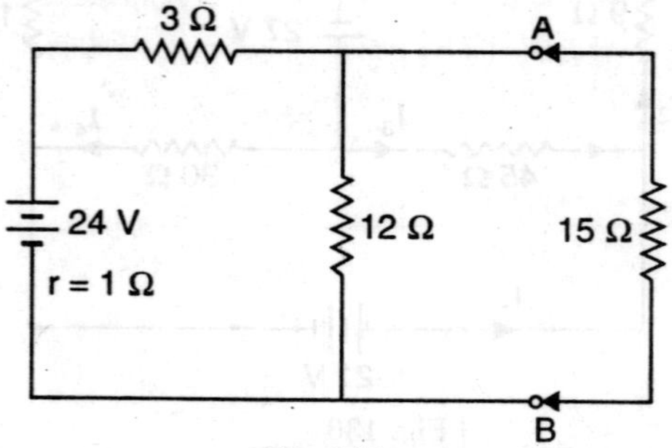

Fig. 141

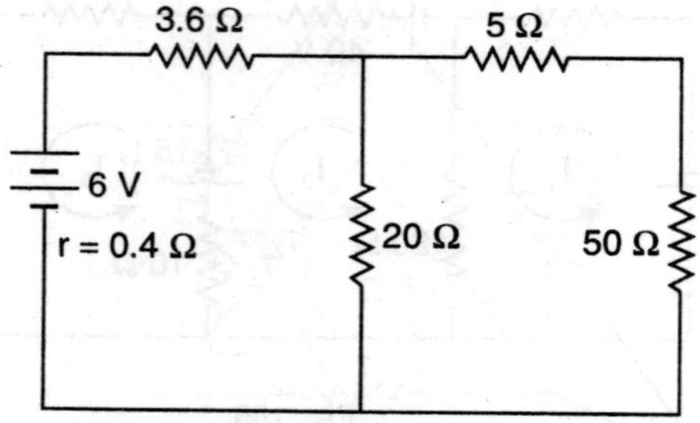

Fig. 142

21. Find the current through 50 Ω resistance in the circuit shown in Fig. 142. Use Thevenin's theorem. [**Ans.** 0.086 A]

22. Calculate the current in the 8 Ω resistor of Fig. 143 by using Thevenin's theorem. [**Ans.** 0.8 A]

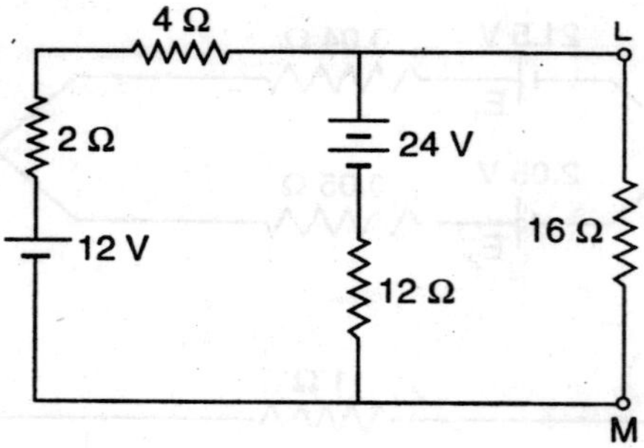

Fig. 143

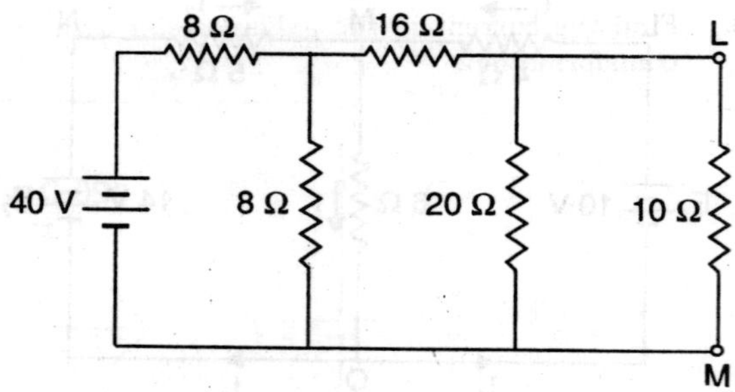

Fig. 144

23. Use Norton's theorem to calculate current flowing through 10 Ω resistor of Fig. 144.

24. Fig. 145 shows network with the value of load resistance $R_L = 50\ \Omega$. Develop Norton's equivalent circuit and determine the current and power delivered to R_L. [**Ans.** 0.114 A ; 0.65 W]

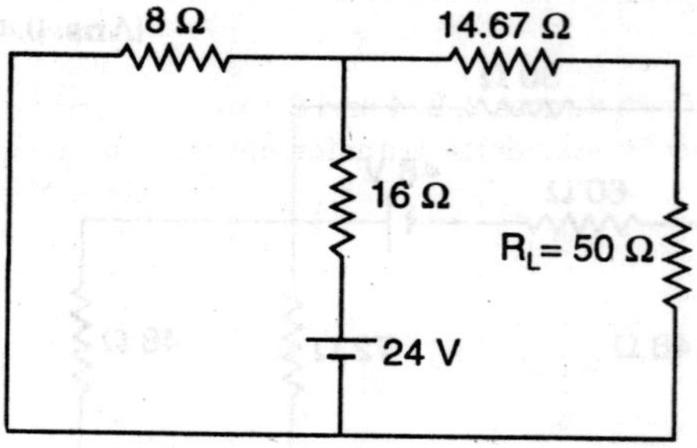

Fig. 145

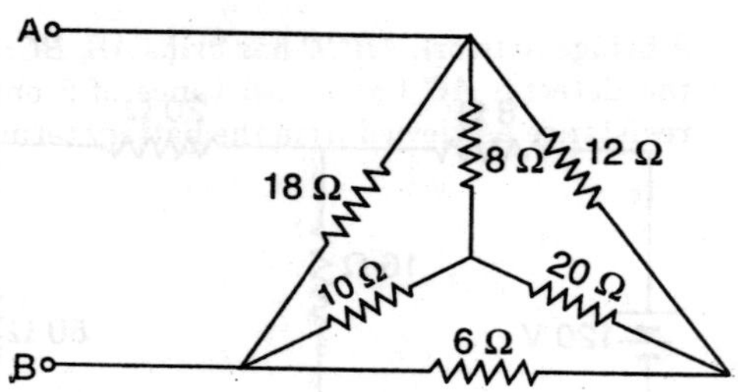

Fig. 146

25. Fig. 146 shows a number of resistances connected in delta and star. Find the resistance across the terminals *A* and *B*. Use star/delta conversion method. [**Ans.** 2.96 Ω]

26. In the given circuit, find the resistance between the points B and C (Fig. 147). **[Ans.** 5 Ω]

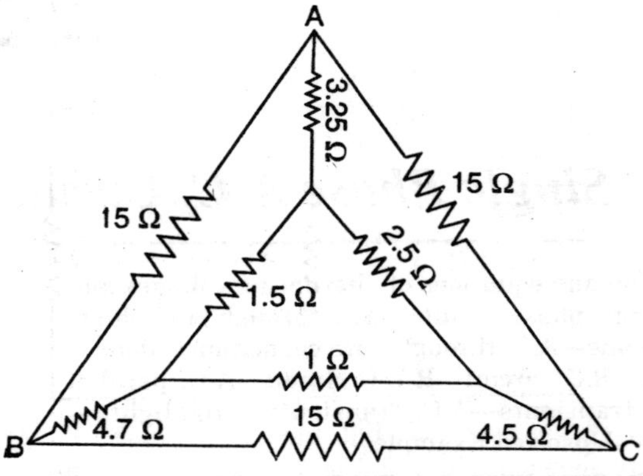

Fig. 147

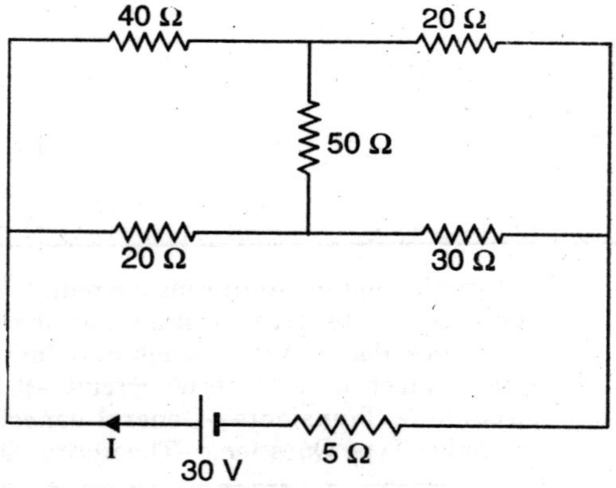

Fig. 148

27. By using delta/star transformation for Fig. 148, find the current I supplied by the battery. **[Ans.** 0.95 A]

28. Find the current in the galvanometer arm in the network shown in the Fig. 149, using delta/star transformation. **[Ans.** 0.236 A]

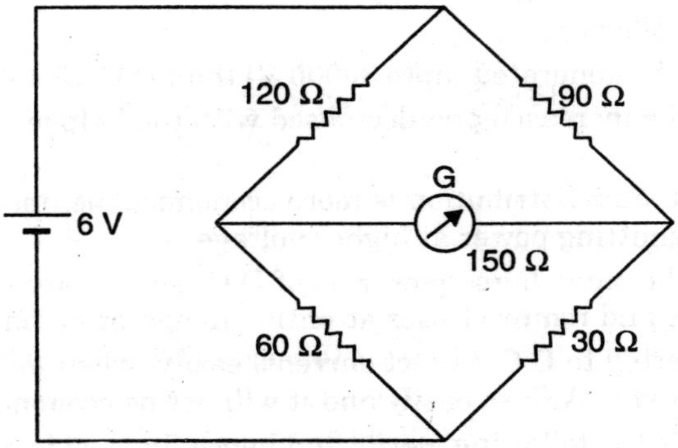

Fig. 149

29. A bridge network $ABCD$ has arms AB, BC, CD and DA of resistances 2, 2, 4 and 2 ohms respectively. If the detector AC has a resistance of 2 ohms, determine by star/delta transformation, the network resistance as viewed from the battery terminals BD. **[Ans.** 4.38 Ω]

2

Single Phase A.C. Circuits

1. Introduction to alternating current. 2. Generation and equations of alternating voltages and currents. 3. Alternating voltage and current. 4. Single phase circuits—A.C. through pure ohmic resistance alone—A.C. through pure inductance alone—A.C. through pure capacitance alone—Phasor algebra—A.C. series circuit—R-L circuit—R.C. circuit—R-L-C circuit—A.C. parallel circuits. 5. Transients—General aspects—D.C. transients—A.C. transients—Highlights—Objective Type Questions—Theoretical Questions—Unsolved Examples.

1. INTRODUCTION TO ALTERNATING CURRENT

A.C. means **alternating current**—*The current or voltage which alternates its direction and magnitude every time.* Now a days 95% of the total energy is produced, transmitted and distributed in A.C. supply.

The *reasons* are the following :

(*i*) More voltage can be generated (upto 33000 V) than D.C. (650 V only).

(*ii*) A.C. voltage can be increased and decreased with the help of a static machine called the '*transformer*'.

(*iii*) A.C. transmission and distribution is more economical as line material (say copper) can be saved by transmitting power at higher voltage.

(*iv*) A.C. motors for the same horse power as of D.C. motors are cheaper, lighter in weight, require less space and require lesser attention in operation and maintenance.

(*v*) A.C. can be converted to D.C. (direct current) easily, when and where required but D.C. cannot be converted to A.C. so easily and it will not be economical.

However, D.C. entails the following *merits* and hence finds wide applications.

(*i*) D.C. series motors are most suitable for traction purposes in tramway, railways, crains and lifts.

(*ii*) For electroplating, electrolytic and electrochemical processes (battery charging etc.), D.C. is required.

(*iii*) Arc lamps for search lights and cinema projectors work on D.C.

(*iv*) Arc welding is better than on A.C.

(*v*) Relay and operating time switches, etc., and circuit-breakers, D.C. works more efficiently.

(*vi*) In rolling mills, paper mills, colliery winding, etc., where fine speed control of speeds in both directions is required, D.C. motors are required.

2. GENERATION AND EQUATIONS OF ALTERNATING VOLTAGES AND CURRENTS

Generation of Alternating Voltages and Currents

Alternating voltages may be generated in the following two ways :

1. By *rotating a coil in a stationary magnetic field,* as shown in Fig. 1.
2. By *rotating a magnetic field within a stationary coil,* as shown in Fig. 2.

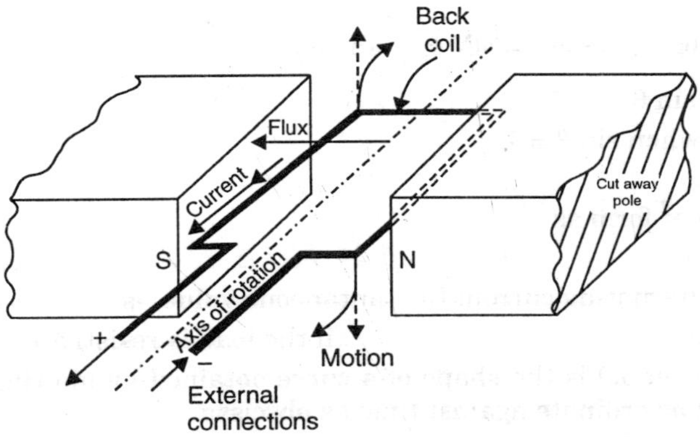

Fig. 1. Rotating a coil in a stationary magnetic field.

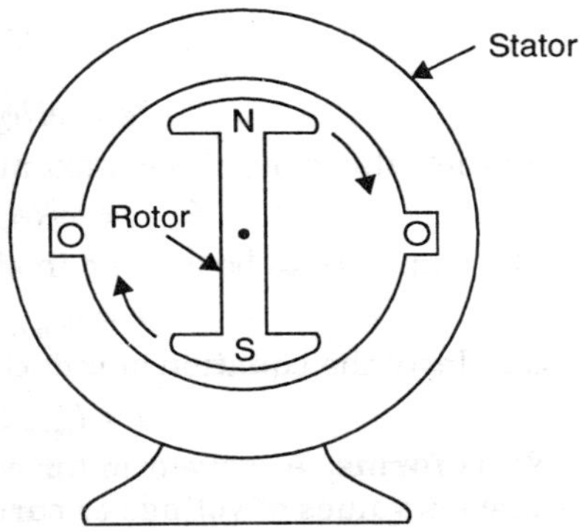

Fig. 2. Rotating a magnetic field within a stationary coil.

The value of the voltage generated in each case depends upon the following factors :

(*i*) The number of turns in the coils ;

(*ii*) The strength of the field ;

(*iii*) The speed at which the coil or magnetic field rotates.

- Out of the above two methods the *rotating-field method is mostly used in practice.*

Equations of Alternating Voltages and Currents

Fig. 3 shows a rectangular coil of N turns rotating clockwise with an angular velocity ω radians per second in a uniform magnetic field.

Since by Faraday's law, the voltage is proportional to the rate at which the conductor its across the magnetic field or to the rate of change of flux linkages, the shape of the wave of voltage applied to the external circuit will be determined by the *flux distribution in the air gap*. For a uniform field between the poles it is evident that *maximum* flux will link with the coil when its plane is in *vertical position i.e.*, perpendicular to the direction of flux between the poles. Also it is obvious that when the plane of coil is *horizontal no flux will link with the coil.*

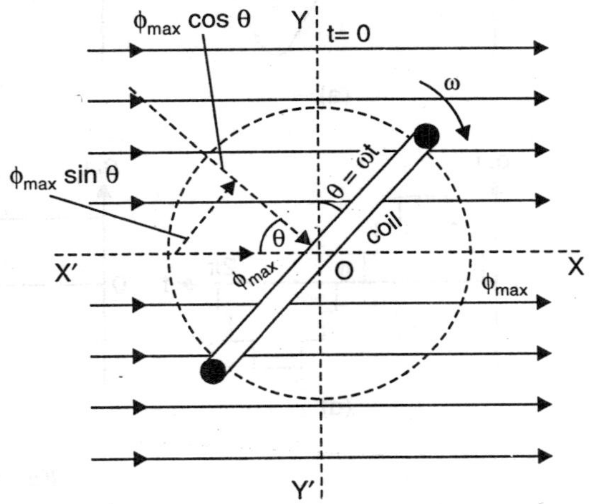

Fig. 3. A coil rotating in a magnet field.

If the position of the coil with reference to the vertical axis be denoted by θ the flux linking with the coil at any instant, as the coil rotates may be determined from the relation,

$$\phi = \phi_{max} \cos \theta$$
$$= \phi_{max} \cos \omega t \quad ...(i) \quad (\because \theta = \omega t)$$

where, ϕ_{max} = Maximum flux which can link with the coil, and

t = Time taken by the coil to move through an angle θ from vertical position.

Using Faraday's law to eqn. (*i*), in order to determine the voltage equation,

$$e = -N\frac{d\phi}{dt}$$ (where e is the *instantaneous* value of the induced e.m.f.)

$$= -N \frac{d}{dt}(\phi_{max} \cos \omega t) = \omega N \phi_{max} \sin \omega t$$

or $$e = \omega N \phi_{max} \sin \theta \qquad ...(ii)$$

As the value of e will be maximum when $\sin \theta = 1$,

$$\therefore \qquad E_{max} = \omega N \phi_{max}$$

The eqn. (*ii*) can be written in simpler form as

$$e = E_{max} \sin \theta \qquad ...(iii)$$

Similarly the equation of induced alternating current (instantaneous value) is

$$i = I_{max} \sin \theta \qquad \text{(if the load is resistive)} \qquad ...(iv)$$

Waveforms. A waveform (or wave-shape) is the shape of a curve obtained by plotting the instantaneous values of voltage or current as ordinate against time as abscissa.

Fig. 4 (*a, b, c, d, e*) shows irregular waveforms, but each cycle of current/voltage is an exactly replica of the previous one. Alternating e.m.fs and currents produced by machines usually both have positive and negative half waves, the same shape as shown. Fig. 4(*f*) represents a sine wave of A.C. This is the simplest possible waveform, and alternators are designed to give as nearly as possible a sine wave of e.m.f.

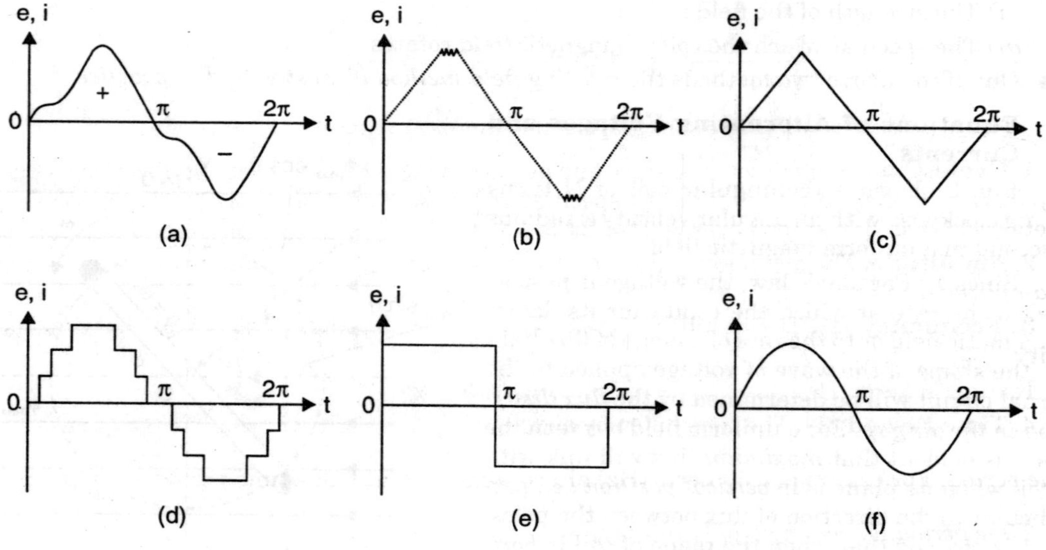

Fig. 4. Waveforms.

- In general, an *alternating current or voltage is one the circuit direction of which reverses at regularly recurring intervals.*
- The waves deviating from the standard sine wave are termed as *distorted waves.*
- *Complex waves* are those which depart from the ideal sinusoidal form. All alternating complex waves, which are periodic and have equal positive and negative half cycles can be shown to be made up of a number of pure sine waves, having different frequencies but all these frequencies are integral multiples of that of the lowest alternating wave, called the *fundamental* (or first harmonic). These waves of higher frequencies are called *harmonics.*

3. ALTERNATING VOLTAGE AND CURRENT

Modern alternators produce an e.m.f. which is for all practical purposes sinusoidal (*i.e.*, a sine curve), the equation between the e.m.f. and time being

$$e = E_{max} \sin \omega t \quad ...(1)$$

where, e = Instantaneous voltage ; E_{max} = Maximum voltage ;

ωt = Angle through which the armature has turned from neutral.

Taking the frequency as f hertz (cycles per second), the value of ω will be $2\pi f$, so that the equation reads

$$e = E_{max} \sin (2\pi f)t.$$

The graph of the voltage will be as shown in Fig. 5.

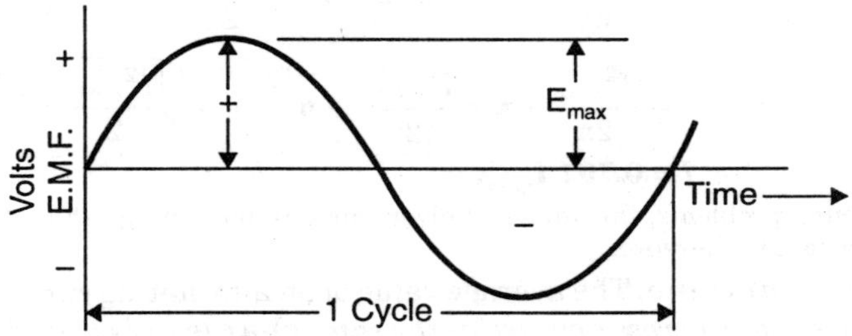

Fig. 5. The graph of the sinusoidal voltage.

1. Cycle. One complete set of positive and negative values of an alternating quantity is known as a *cycle*. A cycle may also sometimes be specified in terms of angular measure. In that case, one complete cycle is said to spread over 360° or 2π radians.

2. Amplitude. The maximum value, positive or negative, of an alternating quantity, is known as its *amplitude*.

3. Frequency (f). The number of cycles/second is called the frequency of the alternating quantity.

Its unit is *hertz* (Hz).

4. Time Period (T). The time taken by an alternating quantity to complete the cycle is called its *time period*. For example, a 50 hertz (Hz) alternating current has a time period of $\frac{1}{50}$ second.

Time period is reciprocal of frequency,

i.e.,
$$T = \frac{1}{f}\left(\text{or } f = \frac{1}{T}\right) \quad ...(2)$$

5. Root mean square (R.M.S.) value. The r.m.s. (or effective) value of an alternating current is given by that steady (D.C.) current *which when flowing through a given circuit for a given time produces the same* **heat** *as produced by the alternating current when flowing through the same circuit for the same time.*

R.M.S. value is the value which is taken for power purposes of any description. This value is obtained by finding the square root of the mean value of the squared ordinates for a cycle or half-cycle (See Fig. 5).

This is the value which is used for all power, lighting and heating purposes, as in these cases the power is proportional to the square of the voltage.

Refer Fig. 5.

The equation of sinusoidal alternating current is given as :

$$i = I_{max} \sin \theta$$

The mean of squares of the instantaneous values of current over half cycle is

$$I^2 = \int_0^{\pi} \frac{i^2 \, d\theta}{(\pi - 0)}$$

$$I^2 = \frac{1}{\pi} \int_0^{\pi} i^2 \, d\theta = \frac{1}{\pi} \int_0^{\pi} (I_{max} \sin \theta)^2 \, d\theta$$

$$= \frac{1}{\pi} \int_0^{\pi} I^2_{max} \sin^2 \theta \, d\theta = \frac{I^2_{max}}{\pi} \int_0^{\pi} \left(\frac{1 - \cos 2\theta}{2} \right) d\theta$$

$$= \frac{I^2_{max}}{2\pi} \int_0^{\pi} (1 - \cos 2\theta) \, d\theta = \frac{I^2_{max}}{2\pi} \left| \theta - \frac{\sin 2\theta}{2} \right|_0^{\pi}$$

$$= \frac{I^2_{max}}{2\pi} \times \pi = \frac{I^2_{max}}{2} \quad \text{or} \quad I = \sqrt{\frac{I^2_{max}}{2}} = \frac{I_{max}}{\sqrt{2}}$$

or $$\mathbf{I = 0.707\, I_{max}}. \qquad ...(3)$$

Note. While solving problems, the values of given current and voltage should always be taken as the r.m.s. values, unless indicated otherwise.

6. Average or mean value. The average value of an alternating current is expressed by *that steady current which transfers across any circuit the same* **charge** *as is transferred by that alternating current during the same time.*

The *mean value is only of use in connection with processes where the results depend on the current only, irrespective of the voltage, such as electroplating or battery charging.*

Refer Fig. 6.

The value of instantaneous current is given by

$$i = I_{max} \sin \theta$$

Refer Fig. 6. The value of instantaneous current is given by :

$$i = I_{max} \sin \theta \qquad [\theta = \omega t]$$

$$I_{av} = \frac{1}{(\pi - 0)} \int_0^{\pi} i d\theta$$

[Limits are taken from 0 to π, since only first half cycle is considered. For whole cycle, the average value of sine wave is zero.]

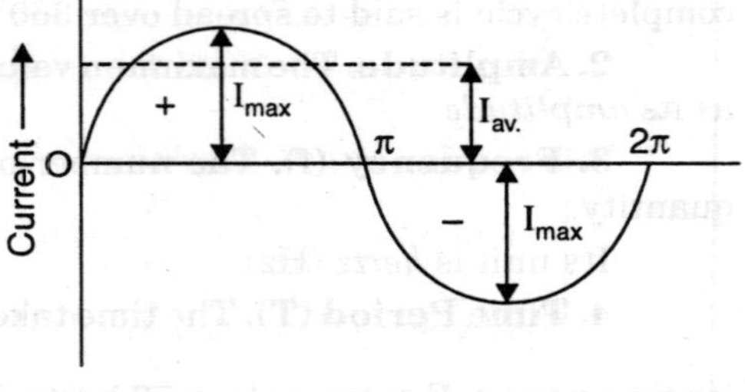

Fig. 6

$$= \frac{1}{\pi} \cdot \int_0^{\pi} I_{max} \cdot \sin \theta \, d\theta = \frac{1}{\pi} \cdot I_{max} \left| -\cos \theta \right|_0^{\pi}$$

$$= \frac{1}{\pi} \cdot I_{max} [1 - (-1)] = \frac{2}{\pi} \cdot I_{max}$$

or $$\mathbf{I_{av} = 0.637\, I_{max}}. \qquad ...(4)$$

Note. In case of unsymmetrical alternating current *viz.* half-wave rectified current the average value must always be taken over the whole cycle.

7. Form and Peak Factors

Form factor. *The ratio of r.m.s. (or effective) value to average value is the form factor (K_f) of the wave form.* It has use in voltage generation and instrument correction factors.

Peak factor. *The ratio of maximum value to the r.m.s. value is the peak factor (K_p) of the wave form.*

S. No.	*Wave form*	*Form factor* (K_f) $= \dfrac{\textit{r.m.s. value}}{\textit{average value}}$	*Peak factor* (K_p) $= \dfrac{\textit{max. value}}{\textit{r.m.s. value}}$
1.	**Sine wave :** 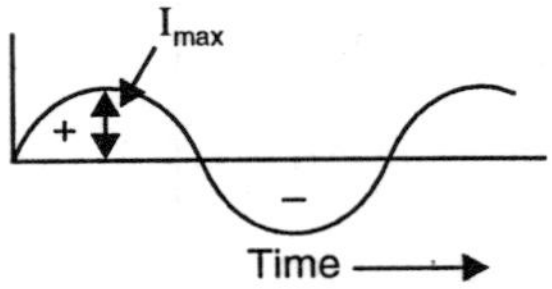 Fig. 7 R.M.S. value = $\dfrac{I_{max}}{\sqrt{2}} = 0.707\ I_{max}$ Average value = $\dfrac{2}{\pi} I_{max} = 0.637\ I_{max}$	$K_f = \dfrac{0.707\ I_{max}}{0.637\ I_{max}} = 1.11$	$K_p = \dfrac{I_{max}}{0.707\ I_{max}} = 1.41$
2.	**Half wave rectified sine wave :** 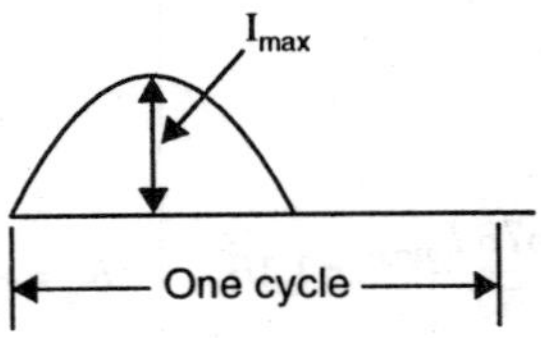 Fig. 8 R.M.S. value = $\dfrac{I_{max}}{2} = 0.5\ I_{max}$ Average Value = $\dfrac{1}{\pi} I_{max} = 0.318\ I_{max}$	$K_f = \dfrac{0.5\ I_{max}}{0.318\ I_{max}} = 1.57$	$K_p = \dfrac{I_{max}}{0.5\ I_{max}} = 2.0$
3.	**Full wave rectified sine wave :** 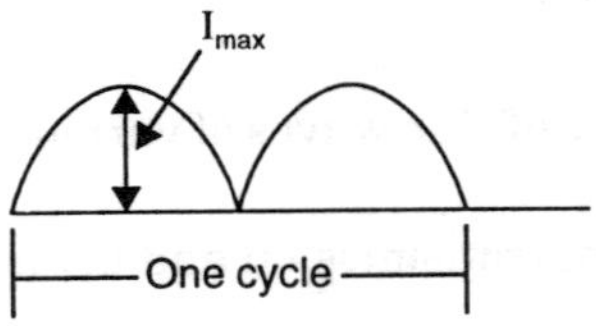 Fig. 9 R.M.S. value = $\dfrac{I_{max}}{\sqrt{2}} = 0.707\ I_{max}$ Average value = $\dfrac{2}{\pi} I_{max} = 0.637\ I_{max}$	$K_f = \dfrac{0.707\ I_{max}}{0.637\ I_{max}} = 1.11$	$K_p = \dfrac{I_{max}}{0.707\ I_{max}} = 1.41$

S. No.	*Wave form*	*Form factor* (K_f) $= \frac{r.m.s.\ value}{average\ value}$	*Peak factor* (K_p) $= \frac{max.\ value}{r.m.s.\ value}$
4.	**Rectangular wave :** 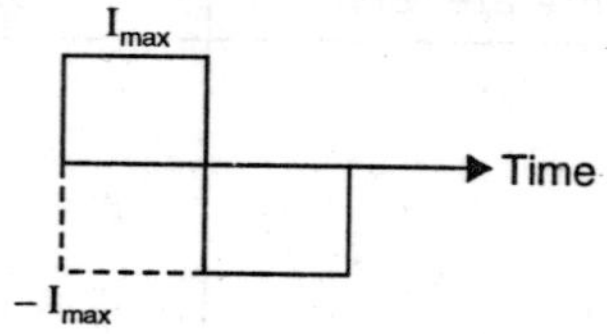 Fig. 10 R.M.S. value = I_{max} Average value = I_{max}	$K_f = 1$	$K_p = 1$
5.	**Triangular wave :** 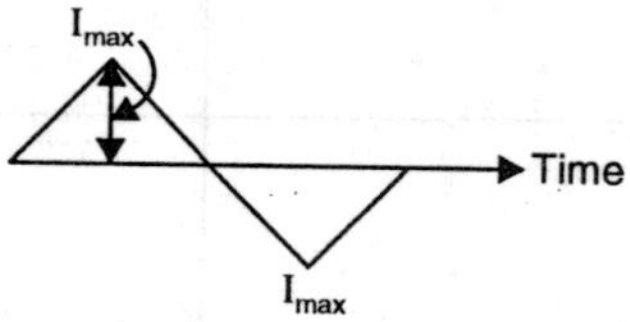 Fig. 11 R.M.S. = $\frac{I_{max}}{\sqrt{3}} = 0.578\ I_{max}$ Average value = $\frac{I_{max}}{2} = 0.5\ I_{max}$	$K_f = \frac{0.578\ I_{max}}{0.5\ I_{max}} = 1.16$	$K_p = \frac{I_{max}}{0.578\ I_{max}} = 1.73$

Reasons for using alternating current (or voltage) of sinusoidal form :

An alternating current (or voltage) of sinusoidal form is normally used because of the following reasons :

1. Mathematically, it is quite simple.
2. Its integrals and differentials both are sinusoidal.
3. It lends itself to vector representation.
4. A complex waveform can be analysed into a series of sine waves of various frequencies, and each such component can be dealt with separately.
5. This waveform is desirable for power generation, transmission and utilisation.

8. Phase and phase angle. The **'phase'** of an A.C. wave may be defined as *its position with respect to a reference axis or reference wave* and **'phase angle'** as *the angle of lead or lag with respect to the reference axis or with respect to another wave.*

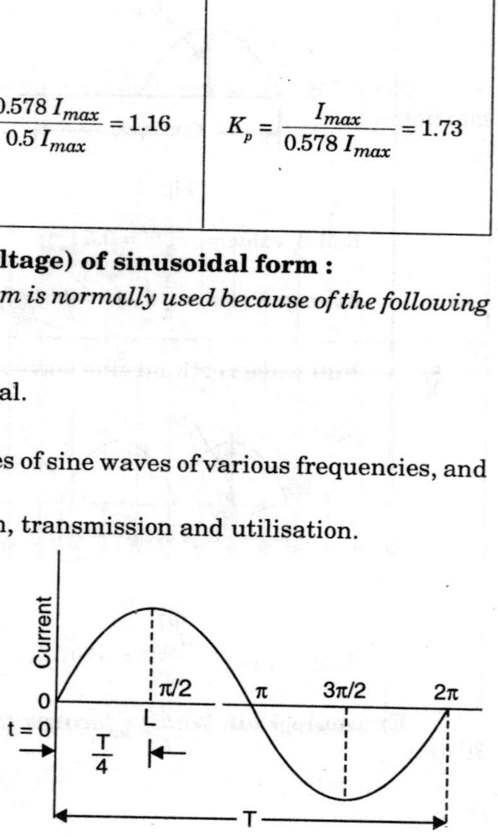

Fig. 12

Examples. The *phase* of current at point L is $\frac{T}{4}$ second where T is the time period or expressed in terms of angle θ, it is $\frac{\pi}{2}$ radian (Fig. 12). Similarly phase of the rotating coil at the instant shown in Fig. 13 is ωt which is therefore called its *phase angle*.

The e.m.fs. induced in both the coils (Fig. 13) will be of the same frequency and of sinusoidal shape, although the values of instantaneous e.m.f. induced will be different. However, the alternating e.m.fs. would reach their maximum and zero values at the *same time* as shown in Fig. 13 (*b*). Such alternating voltages or curve are said to *in phase* with each other.

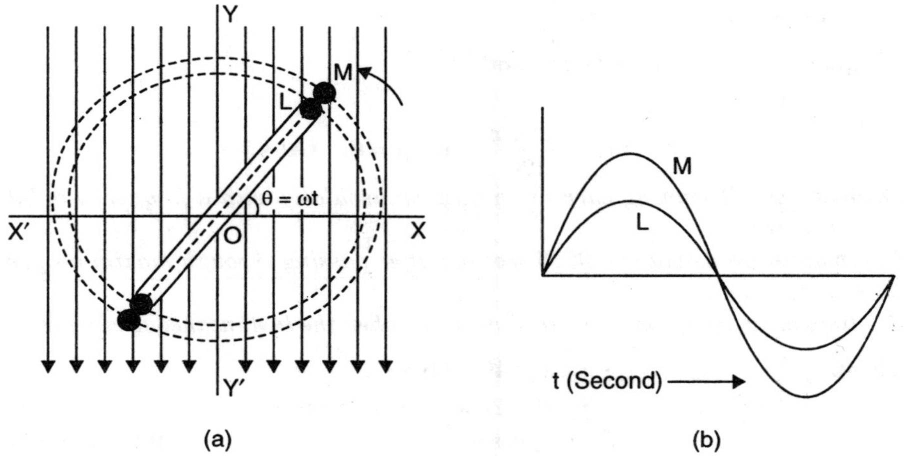

Fig. 13

Refer Fig. 14. *M lags* behind *L* by β and *N lags* behind *L* by (α + β) because they reach their maximum later.

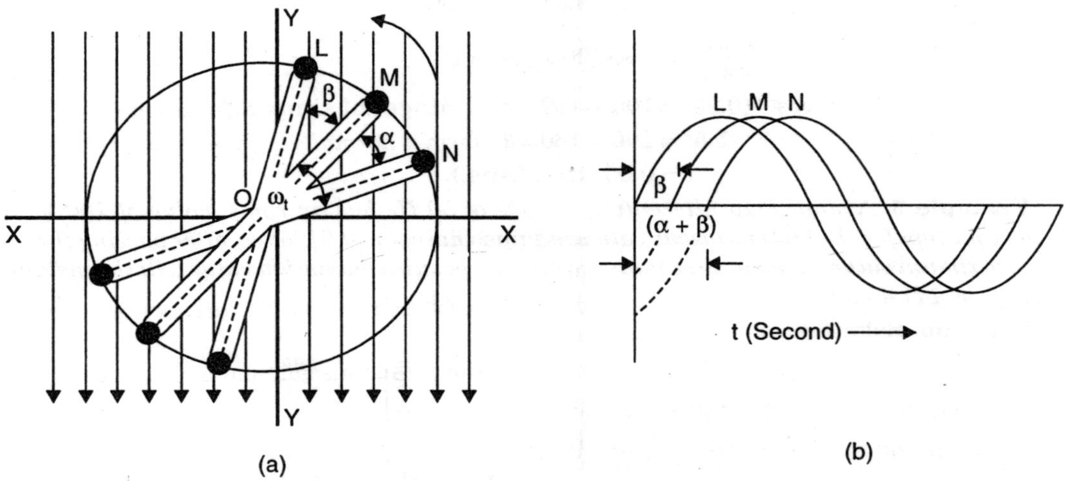

Fig. 14

Example 1. (*a*) *What is the equation of a 25 cycle current sine wave having r.m.s. value of 30 amps ?*

(*b*) *A 60 cycle engine-driven alternator has a speed of 1200 r.p.m. How many poles are there in the alternator ?*

Solution. We know that, $i = I_{max} \sin \omega t$

$= I_{max} \sin 2\pi f t$ $(\because \ \omega = 2\pi f)$

$= 30 \times \sqrt{2} \,.\, \sin (2\pi \times 25 \times t)$ $\left(\because \frac{\text{R.M.S. value}}{\text{Max. value}} = \frac{1}{\sqrt{2}}\right)$

$= \mathbf{42.42 \sin 157\, t.}$ **(Ans.)**

Using the relation, $f = \frac{Np}{120}$

where, f = frequency, N = speed in r.p.m., and

p = No. of poles

$\therefore$ $60 = \frac{1200\, p}{120}$ or $p = 6.$ **(Ans.)**

Example 2. *An alternating current varying sinusoidally with a frequency of 50 Hz has an r.m.s. value of 40 A. Find :*

(i) The instantaneous value 0.0025 seconds after passing through maximum positive value, and

(ii) The time measured from a maximum value when the instantaneous current is 14.14 A.

Solution. $I_{max} = \sqrt{2} \times 40 = 56.56$ A

$\omega = 2\pi f = 2\pi \times 50 = 100\pi$ radians

(i) $i = I_{max} \cos \omega t$ ... after +ve maximum value

$= 56.56 \cos 100\, \pi t$

$= 56.56 \cos (100\pi \times 0.0025)$ ...(t = 0.0025 s ...given)

$= 56.56 \cos 45°$ (Taking π = 180°)

= **40 A. (Ans.)**

(ii) $14.14 = 56.56 \cos (100 \times 180 \times t)$

or $\frac{14.14}{56.56} = \cos (100 \times 180 \times t)$

or $\cos^{-1}(0.25) = 100 \times 180 \times t$

$75.5° = 100 \times 180 \times t$

$\therefore$ t = **0.00419 s. (Ans.)**

Example 3. *A sinusoidal alternating voltage of 50 Hz has an r.m.s. value of 200 V. Write down the equation for the instantaneous value and find this value 0.0125 sec. after passing through a positive maximum value. At what time measured from a positive maximum value will the instantaneous voltage be 141.4 volts ?*

Solution. Refer Fig. 15.

$V_{max} = \sqrt{2} \times 200 = 282.2$ volts

$\omega = 2\pi f = 2\pi \times 50 = 100\, \pi$ rad/sec.

$\therefore$ Equation for the instantaneous voltage.

$V = V_{max} \sin \omega t$ (with reference to point O)

$= 282.2 \sin 100\, \pi t$ ...*(i)*

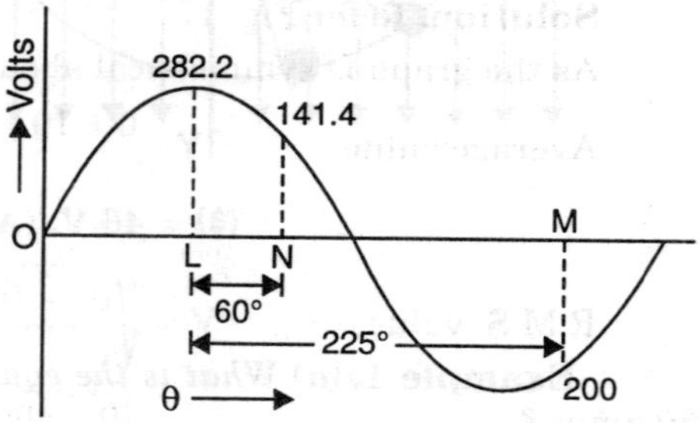

Fig. 15

Since the time (0.0125 sec.) is given from the point L (*i.e.*, from positive maximum value) the equation *(i)* when referred to point L can be written as

$v = 282.2 \sin (90° + 100\, \pi t)$

$= 282.2 \cos 100\, \pi t$

Hence instantaneous value of the voltage 0.0125 sec. after passing through + ve maximum value,

$$v = 282.2 \cos (100 \pi \times 0.0125) \quad \text{...(angle in radians)}$$
$$= 282.2 \cos (100 \times 180 \times 0.0125) \quad \text{...(angle in degrees)}$$
$$= 282.2 \cos 225° = 280.2 \times \left(-\frac{1}{\sqrt{2}}\right)$$
$$= \mathbf{-200\ V\ (point\ M).\ (Ans.)}$$

Also $v = 141.4$ V

$\therefore$ $141.4 = 282.2 \cos (100 \times 180 \times t)$

or $0.5 = \cos (100 \times 180 \times t)$

or $\cos^{-1} (0.5) = 100 \times 180 \times t$

or $60° = 100 \times 180 \times t$

or $t = \frac{1}{300}$ **sec. (point N). (Ans.)**

Example 4. (*a*) *What is the peak value of a sinusoidal alternating current of 4.78 r.m.s. amps ?*

(*b*) *What is the r.m.s. value of a rectangular wave with an amplitude of 9.87 volts ?*

(*c*) *What is the average value over half a cycle of a sinusoidal alternating current whose r.m.s. value is 31 A ?*

Solution. (*a*) Peak value, $I_{max} = \sqrt{2} \times 4.78 =$ **6.76 A. (Ans.)**

(*b*) Refer Fig. 16. If the first half-cycle is divided into n equal parts each of value V, then

$$\text{r.m.s. value} = \left(\frac{V^2 + V^2 + V^2 + \ldots\ldots}{n}\right)^{1/2}$$
$$= V = \mathbf{9.87\ volts.\ (Ans.)}$$

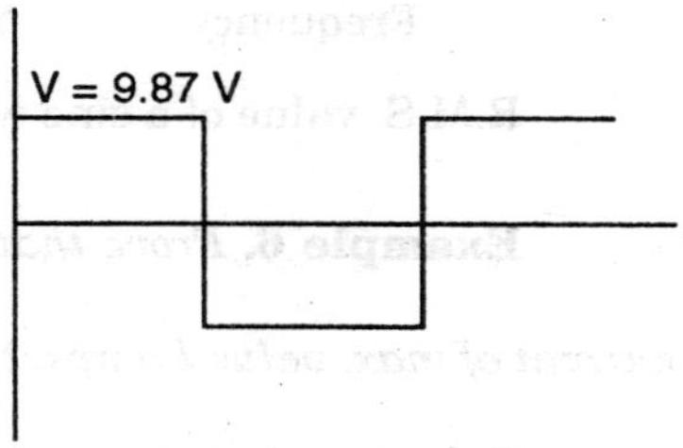

Fig. 16

(*c*) $I_{r.m.s.} = 31$ A

$$I_{av} = \frac{I_{r.m.s.}}{\text{form factor}}$$
$$= \frac{31}{1.11} = \mathbf{27.93\ A.\ (Ans.)}$$

Example 5. *The graph in Fig. 17 shows the variation of voltage with time. Use the graph to calculate the average and r.m.s. value of the voltage. What is the frequency of the voltage ? What would be the r.m.s. value of sine wave having the same peak value ?*

Solution. Refer Fig. 17.

As the graph is symmetrical about time axis, considering only the positive half cycle.

Average value, $$V_{av} = \frac{0 + 10 + 20 + 40 + 100 + 120 + 100 + 40 + 20 + 10}{10}$$
$$= \mathbf{46\ V.\ (Ans.)}$$

R.M.S. value, $$V = \sqrt{\frac{0^2 + 10^2 + 20^2 + 40^2 + 100^2 + 120^2 + 100^2 + 40^2 + 20^2 + 10^2}{10}}$$
$$= \sqrt{\frac{0 + 100 + 400 + 1600 + 10000 + 14400 + 10000 + 1600 + 400 + 100}{10}}$$

$$= \sqrt{\frac{38600}{10}} = \sqrt{3860} = \mathbf{62.1\ V.} \quad \textbf{(Ans.)}$$

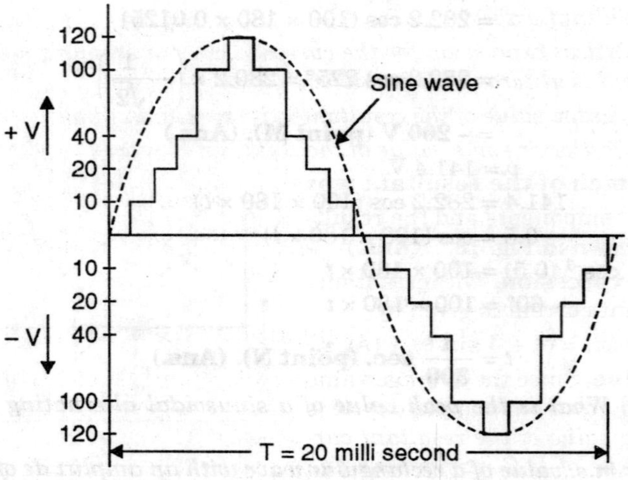

Fig. 17

Since the time period 'T' is 20 millisec.

$\therefore$ Frequency $\quad 'f' = \frac{1}{T} = \frac{1}{20 \times 10^{-3}} = \mathbf{50\ Hz.}$ **(Ans.)**

R.M.S. value of a sine wave of the same peak value

$$= 0.707 \times 120 = \mathbf{84.84\ V.} \quad \textbf{(Ans.)}$$

Example 6. *Prove that if a D.C. current of $I_{amps.}$ is superposed in a conductor by an A.C. current of max. value I amps, the r.m.s. value of the resultant is* $\sqrt{\frac{3}{2}}\, I$.

Solution. Let the *A.C. current* be $i = I \sin \theta$ where i is the instantaneous value of the A.C. current and I the *D.C. current.*

The r.m.s. value of $(I + i)$ over one complete cycle is,

$$= \sqrt{\frac{1}{2\pi} \int_0^{2\pi} (I + I \sin \theta)^2 \, d\theta}$$

$$= I \sqrt{\frac{1}{2\pi} \int_0^{2\pi} (I + 2 \sin \theta + \sin^2 \theta)}$$

$$= I \sqrt{\left\{ \frac{1}{2\pi} \int_0^{2\pi} 1 + 2 \sin \theta + \left(\frac{1 - \cos 2\theta}{2} \right) \right\}}$$

$$= I \sqrt{\frac{1}{2\pi} \left| \theta - 2 \cos \theta + \frac{\theta}{2} - \frac{\sin 2\theta}{4} \right|_0^{2\pi}}$$

$$= I \sqrt{\frac{1}{2\pi} (2\pi - 2 + \pi + 2)}$$

$$= I . \sqrt{\frac{3}{2}} . \quad \textbf{(Ans.)}$$

Example 7. *A resultant current wave is made up of two components : a 4A D.C. component and a 50 Hz A.C. component, which is of sinusoidal waveform and which has a maximum value of 4A.*

(*i*) *Sketch the resultant wave.*

(*ii*) *Write an analytical expression for the current wave, reckoning t = 0 at a point where the A.C. component is at zero value and where di / dt is positive.*

(*iii*) *What is the average value of the resultant current over a cycle ?*

(*iv*) *What is the effective or r.m.s. value of the resultant current ?*

Solution. (*i*) **Sketch of the resultant wave :**

The two current components and the resultant current wave are shown in Fig. 18. **(Ans.)**

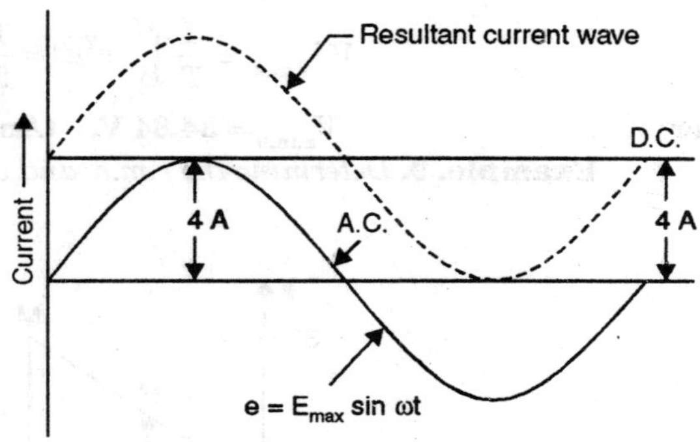

Fig. 18

(*ii*) **Analytical expression.** The instantaneous value of the resultant current is given by

$$i = (4 + 4 \sin \omega t) = \mathbf{(4 + 4 \sin \theta)}. \quad \textbf{(Ans.)}$$

(*iii*) **Average value.** Since the average value of the alternating current over one complete cycle is zero, hence the *average value of the resultant current is equal to the value of D.C. component i.e.,* **4A (Ans.)**

(*iv*) **Effective or r.m.s. value :**

Mean value of i^2 over complete cycle is

$$= \frac{1}{2\pi} \int_0^{2\pi} i^2 d\theta = \frac{1}{2\pi} \int_0^{2\pi} (4 + 4 \sin \theta)^2 \, d\theta$$

$$= \frac{1}{2\pi} \int_0^{2\pi} (16 + 32 \sin \theta + 16 \sin^2 \theta) \, d\theta$$

$$= \frac{1}{2\pi} \int_0^{2\pi} \left[16 + 32 \sin \theta + 16 \left(\frac{1 - \cos 2\theta}{2} \right) \right] d\theta$$

$$= \frac{1}{2\pi} \int_0^{2\pi} (24 + 32 \sin \theta - 8 \cos 2\theta) \, d\theta$$

$$= \frac{1}{2\pi} \left(24\,\theta - 32 \cos \theta - 8 \times \frac{\sin 2\,\theta}{2} \right)_0^{2\pi}$$

$$= \frac{1}{2\pi} [(48\pi - 32 \cos 2\pi - 4 \sin 4\pi) - (-32)] = \frac{48\pi}{2\pi} = 24 \text{ A}$$

∴ R.M.S. value, $I = \sqrt{24}$ = **4.9 A. (Ans.)**

Example 8. *Determine the average and effective values of the saw-tooth waveform shown in Fig. 19.* **(N.U.)**

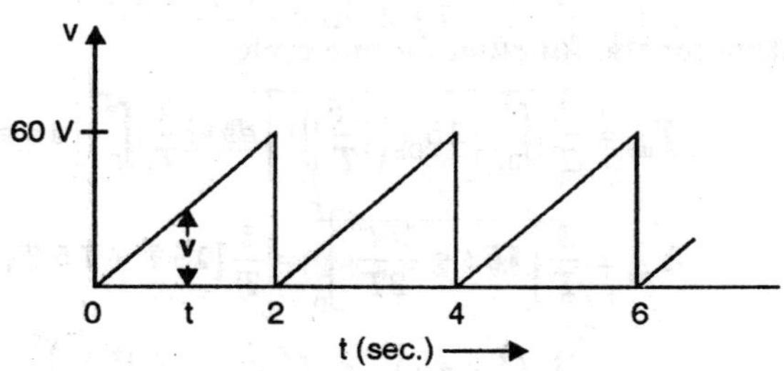

Fig. 19

Solution. Refer Fig. 19.

Since the voltage increases linearly, therefore,

$$V_{av} = \frac{60 + 0}{2} = \mathbf{30\ V.}\quad \textbf{(Ans.)}$$

The slope, for the internal $0 < t < 2$ is given by :

$$\text{Slope} = \frac{60}{2} = 30$$

$\therefore$ Instantaneous voltage, $v = 30t$ volts

The r.m.s. or effective value of the voltage,

$$V^2_{\text{r.m.s.}} = \frac{1}{T}\int_0^T v^2 dt = \frac{1}{2}\int_0^2 (30t)^2\, dt = 450 \int_0^2 t^2\, dt = 450 \left|\frac{t^3}{3}\right|_0^2 = 1200$$

or $$V_{\text{r.m.s}} = \mathbf{34.64\ V.}\quad \textbf{(Ans.)}$$

Example. 9. *Determine the r.m.s. and average values of the waveform shown in Fig. 20.* **[Indore University]**

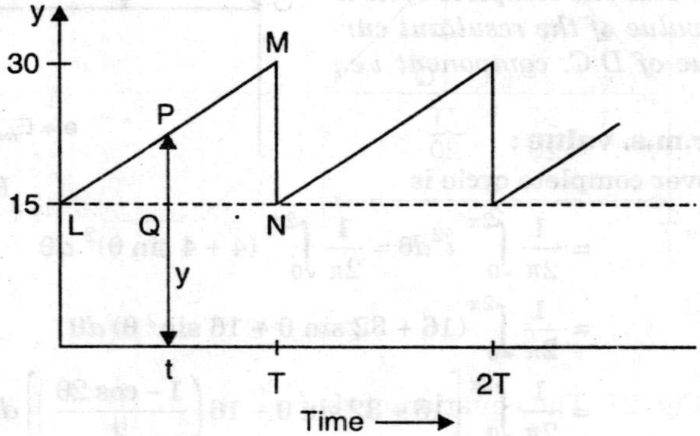

Fig. 20

Solution. Refer Fig. 20.

The slope the curve $LM = \frac{MN}{LN} = \frac{30 - 15}{T} = \frac{15}{T}$

Now consider the function y at any time 't'. We have,

$$\therefore \quad \frac{PQ}{LQ} = \frac{MN}{LN} = \frac{15}{T}$$

or $$\frac{y - 15}{t} = \frac{15}{T} \quad \text{or} \quad y = 15 + \left(\frac{15}{T}\right)t$$

This gives the equation for the function for one cycle.

$$Y_{av} = \frac{1}{T}\int_0^T \left[15 + \left(\frac{15}{T}\right)t\right] dt = \frac{1}{T}\int_0^T \left[15\, dt + \frac{15}{T}\,.\,t\,.\,dt\right]$$

or $$Y_{av} = \frac{1}{T}\left[15\,t + \frac{15t^2}{2T}\right]_0^T = \frac{1}{T}[15\,T + 7.5\,T] = \mathbf{22.5.}\quad \textbf{(Ans.)}$$

Mean square value $$= \frac{1}{T}\int_0^T y^2 dt = \frac{1}{T}\int_0^T \left\{15 + \left(\frac{15}{T}\right)t\right\}^2 dt$$

$$= \frac{1}{T} \int_0^T \left(225 + \frac{225t^2}{T^2} + \frac{450}{T} . t \right) dt$$

$$= \frac{1}{T} \left| 225\,t + \frac{225\,t^3}{3T^2} + \frac{450\,t^2}{2T} \right|_0^T$$

$$= \frac{1}{T} \mid 225\,T + 75\,T + 225\,T \mid = 525$$

∴ R.M.S. value $= \sqrt{525}$ = **22.9. (Ans.)**

Example 10. *Find the r.m.s. and average value of the trapezoidal current wave-form shown in the Fig. 21.* **[Bangalore University]**

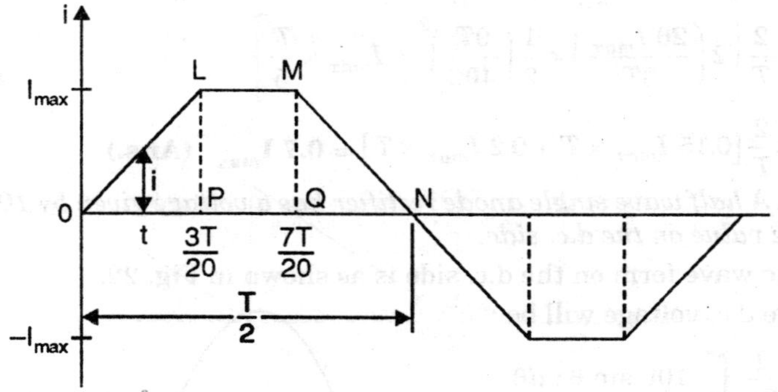

Fig. 21

Solution. Refer Fig. 21. The equation of the current for $0 < t < \frac{3T}{20}$ can be found from the relation,

$$\frac{i}{t} = \frac{I_{max}}{(3T/20)} \quad \text{or} \quad i = \frac{20\,I_{max}}{3T} . t$$

The equation of the current for $\frac{3T}{20} < t < \frac{7T}{20}$ is given by $i = I_{max}$. Remembering that ΔOLP is identical with MQN,

$$\textit{R.M.S. value of current, } I_{r.m.s.} = \sqrt{\frac{1}{(T/2)} \left[2 \int_0^{3T/20} i^2 dt + \int_{3T/20}^{7T/20} I_{max}^2 \, dt \right]}$$

$$= \sqrt{\frac{2}{T} \left[2 \left\{ \left(\frac{20\,I_{max}}{3T} \right)^2 \int_0^{3T/20} t^2 \, dt \right\} + I_{max}^2 \int_{3T/20}^{7T/20} dt \right]}$$

$$= \sqrt{\frac{2}{T} \left[2 \left\{ \left(\frac{20\,I_{max}}{3T} \right)^2 \left| \frac{t^3}{3} \right|_0^{3T/20} \right\} + I_{max}^2 \left| t \right|_{3T/20}^{7T/20} \right]}$$

$$= \sqrt{\frac{2}{T} \left[2 \times \frac{400\,I_{max}^2}{9T^2} \times \frac{27}{3 \times 8000} T^3 + I_{max}^2 \times \left(\frac{7T}{20} - \frac{3T}{20} \right) \right]}$$

$$= \sqrt{\frac{2}{T}\left(0.1\, I_{max}^2\, T + 0.2\, I_{max}^2\, T\right)} = \sqrt{0.6\, I_{max}^2} = \mathbf{0.775\ I_{max}}. \quad \textbf{(Ans.)}$$

Average value of current,

$$I_{av} = \frac{1}{T/2}\left[2\int_0^{3T/20} i\,dt + \int_{3T/20}^{7T/20} I_{max}\, dt\right] = \frac{2}{T}\left[2\int_0^{3T/20}\left(\frac{20\, I_{max}}{3T}\right) t\, dt + I_{max}\int_{3T/20}^{7T/20} dt\right]$$

$$= \frac{2}{T}\left[2\left(\frac{20\, I_{max}}{3T}\right)\left|\frac{t^2}{2}\right|_0^{3T/20} + I_{max}\left|t\right|_{3T/20}^{7T/20}\right]$$

$$= \frac{2}{T}\left[2\left(\frac{20\, I_{max}}{3T}\right)\times\frac{1}{2}\left(\frac{3T}{20}\right)^2 + I_{max}\left(\frac{7T}{20} - \frac{3T}{20}\right)\right]$$

$$= \frac{2}{T}\left[2\left(\frac{20\, I_{max}}{3T}\right)\times\frac{1}{2}\left(\frac{9T}{400}\right)^2 + I_{max}\times\frac{T}{5}\right]$$

$$I_{av} = \frac{2}{T}\left[0.15\, I_{max}\times T + 0.2\, I_{max}\times T\right] = \mathbf{0.7\ I_{max}}. \quad \textbf{(Ans.)}$$

Example 11. *A half wave single anode rectifier has a voltage given by 100 sin ωt applied to it. Estimate the average value on the d.c. side.*

Solution. The wave form on the d.c. side is as shown in Fig. 22.

Mean/Average d.c. voltage will be :

$$I_{av} = \frac{1}{2\pi}\int_0^{\pi} 100 \sin\theta \,.\, d\theta$$

$$= \frac{100}{2\pi}\left[-\cos\theta\right]_0^{\pi}$$

$$= \mathbf{31.83\ V.\ (Ans.)}$$

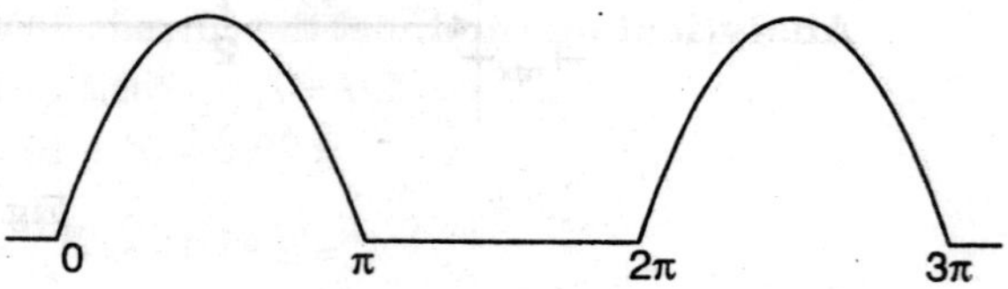

Fig. 22

Example 12. *Calculate from the first principles, the reading which would be indicated by a hot-wire ammeter in a circuit whose current waveform is given by 10 sin ωt + 3 sin 3ωt + 2 sin 5ωt.*

Solution. The expression for instantaneous current is :

$$i = 10 \sin \omega t + 3 \sin 3\omega t + 2 \sin 5\omega t$$

The hot-wire ammeter will read the "r.m.s. value" of the wave form.

Now $i^2 = (10 \sin \omega t + 3 \sin 3\omega t + 2 \sin 5\omega t)^2$

$$\therefore \quad \textit{R.M.S. value of the current} = \sqrt{\frac{1}{2\pi}\int_0^{2\pi} i^2 d(\omega t)}$$

$$= \left[\frac{1}{2\pi}\int_0^{2\pi} \left[(10 \sin \omega t + 3 \sin 3\omega t + 2 \sin 5\omega t)^2\, d(\omega t)\right]\right]^{1/2}$$

$$= \left[\frac{1}{2\pi}\int_0^{2\pi} \{10^2 \sin^2 \omega t + 3^2 \sin^2 3\omega t + 2^2 \sin^2 5\omega t + 2\times 10\times 3 \sin \omega t \sin 3\omega t + 2\times 10\times 2 \sin \omega t \sin 5\omega t + 2\times 3\times 2 \sin 3\omega t \,.\, \sin 5\omega t\}\, d(\omega t)\right]^{1/2}$$

$$= \left[\frac{1}{2\pi} \int_0^{2\pi} \left\{ \frac{10^2 (1 - \cos 2\omega t)}{2} + \frac{3^2}{2} (1 - \cos 6\omega t) + \frac{2^2}{2} (1 - \cos 10\omega t) + {}^* etc. \right\} d(\omega t) \right]^{1/2}$$

[*All the terms containing the product of two sines when integrated over the range 0 to 2π disappear. This is easily seen by splitting such terms into the difference of two cosines.]

$$= \left[\frac{1}{2\pi} \frac{(10^2 + 3^2 + 2^2)}{2} . 2\pi \right]^{1/2} = \left(\frac{10^2 + 3^2 + 2^2}{2} \right)^{1/2} = \left(\frac{113}{2} \right)^{1/2}$$

$= (56.5)^{1/2} =$ **7.52. (Ans.)**

Example 13. *Four branches A, B, C, D in an A.C. circuit meet at a junction point P. The currents in branches A, B, C flow towards P while the current in branch D flows away from P. The currents in branches A, B and C are*

$$i_A = 20 \sin 628t$$
$$i_B = 15 \sin (628t - \pi/6)$$
$$i_C = 25 \sin (628t + \pi/3)$$

Find an expression for the instantaneous value of current in branch D, and calculate (i) its frequency, and (ii) the heat (watts) that it would produce when flowing in a resistance of 5 ohms.

Solution.

Analytical method. Let the current flowing in the branch D be, $i_D = I_D \sin (628t + \phi)$

$$\Sigma H = I_{Amax} \cos \phi_1 + I_{Bmax} \cos \phi_2 + I_{Cmax} \cos \phi_3 + I_{Dmax} \cos \phi$$
$$= 20 \cos 0° + 15 \cos (-\pi/6) + 25 \cos \pi/3 + I_{Dmax} \cos \phi$$
$$= 20 + 15 \times \frac{\sqrt{3}}{2} + 25 \times \frac{1}{2} + I_{Dmax} \cos \phi$$
$$= 20 + 13 + 12.5 + I_{Dmax} \cos \phi = 45.5 + I_{Dmax} \cos \phi$$
$$\Sigma V = I_{Amax} \sin \phi_1 + I_{Bmax} \sin \phi_2 + I_{Cmax} \sin \phi_3 + I_{Dmax} \sin \phi$$
$$= 20 \sin 0° + 15 \sin (-\pi/6) + 25 \sin \pi/3 + I_{Dmax} \sin \phi$$
$$= 0 - 15 \times \frac{1}{2} + 25 \times \frac{\sqrt{3}}{2} + I_{Dmax} \sin \phi$$
$$= -7.5 + 21.65 + I_{Dmax} \sin \phi = 14.15 + I_{Dmax} \sin \phi$$

Since all the currents are meeting at point P,

$\therefore$ $\Sigma H = 0$

i.e., $45.5 + I_{Dmax} \cos \phi = 0$ or $I_{Dmax} \times \cos \phi = -45.5$ ...(*i*)

and $\Sigma V = 0$

i.e., $14.15 + I_{Dmcx} \sin \phi = 0$ or $I_{Dmax} \times \sin \phi = -14.15$ ...(*ii*)

From (*i*) and (*ii*), $I_{Dmax} = (-45.5)^2 + (-14.15)^2 = 47.6$ A.

$$\phi = \tan^{-1} \frac{-14.15}{-45.5} = 197° \text{ or } 3.44°$$

Hence the current in branch D follows the relation,

$$i_D = 47.6 \sin (678\, t + 3.44). \textbf{ (Ans.)}$$

(*i*) **Frequency** $= \frac{\omega}{2\pi} = \frac{628}{2\pi} = 100$ Hz. **(Ans.)**

(*ii*) **Heat produced** $= \left(\frac{I_{Dmax}}{\sqrt{2}} \right)^2 \times R = \left(\frac{47.6}{\sqrt{2}} \right)^2 \times 5 = 5620$ W. **(Ans.)**

4. SINGLE PHASE CIRCUITS

The study of circuits involves three basic types of units (*R, L, C i.e.*, resistance, reactance and capacitance respectively) and four possible series combination of them. The latter, in turn, may be arranged in many kinds of parallel, series-parallel, parallel-series or other complex circuits.

4.1. A.C. Through Pure Ohmic Resistance Alone

The circuit containing a pure resistance R is shown in Fig. 23 (a). Let the applied voltage be given by the equation,

$$v = V_{max} \sin \theta = V_{max} \sin \omega t \qquad ...(i)$$

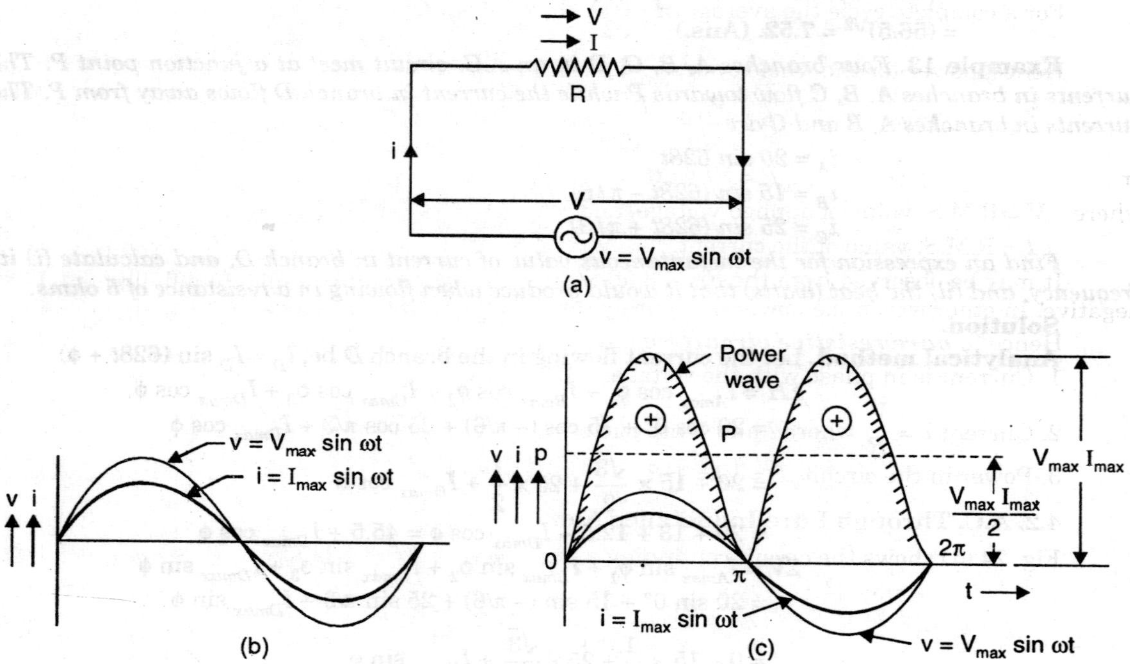

Fig. 23. A.C. through pure ohmic resistance alone.

Then the instantaneous value of current flowing through the resistance R will be,

$$i = \frac{v}{R} = \frac{V_{max} \sin \omega t}{R} \qquad ...(ii)$$

The value of current will be maximum

when $\qquad \sin \omega t = 1$ or ($\omega t = 90°$)

$$\therefore \qquad I_{max} = \frac{V_{max}}{R}$$

Substituting this value in eqn. (ii), we get

$$i = I_{max} \sin \omega t \qquad ...(iii)$$

Comparing (i) and (iii), we find that alternating *voltage and current are in phase with each other* as shown in Fig. 23 (b), also shown vectorially in Fig. 23 (c).

Power. Refer Fig. 23 (*c*)

Instantaneous power,

$$p = vi = V_{max} \sin \omega t \times I_{max} \sin \omega t = V_{max} I_{max} \sin^2 \omega t$$

$$= \frac{V_{max} I_{max}}{2} \times 2 \sin^2 \omega t = \frac{V_{max} I_{max}}{2} (1 - \cos 2 \omega t)$$

$$= \frac{V_{max}}{\sqrt{2}} \cdot \frac{I_{max}}{\sqrt{2}} - \frac{V_{max}}{\sqrt{2}} \cdot \frac{I_{max}}{\sqrt{2}} \cos 2\omega t$$

(Constant part) (Fluctuating part)

For a complete cycle the average of $\frac{V_{max}}{\sqrt{2}} \cdot \frac{I_{max}}{\sqrt{2}} \cos 2 \omega t$ *is zero.*

Hence, power for the whole cycle,

$$P = \frac{V_{max}}{\sqrt{2}} \cdot \frac{I_{max}}{\sqrt{2}} = V_{r.m.s} \cdot I_{r.m.s.}$$

or $$P = VI \text{ watt}$$

where V = R.M.S. value of applied voltage, and

I = R.M.S. value of the current.

It may be observed from the Fig. 23 (*c*) that no part of the power cycle at any time becomes negative. In other words the power in a purely resistive circuit *never becomes zero*.

Hence in ***pure resistive circuit*** we have :

1. Current is in phase with the voltage.
2. Current $I = \frac{V}{R}$ where I and V are r.m.s. values of current and voltage.
3. Power in the circuit, $P = VI = I^2R$.

4.2. A.C. Through Pure Inductance Alone

Fig. 24 (*a*) shows the circuit containing a pure inductance of L henry.

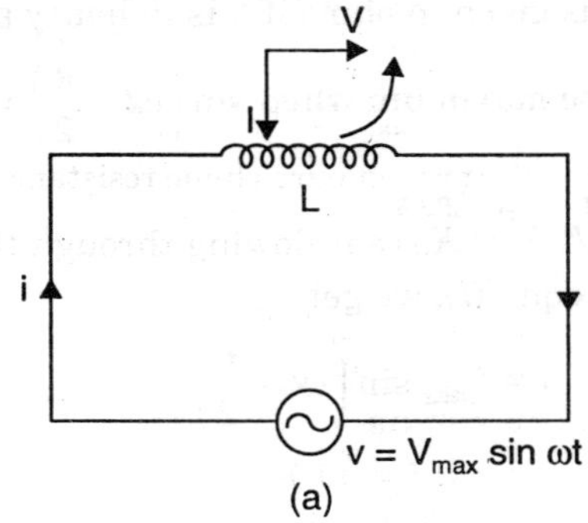

Fig. 24 (*a*)

Let the alternating voltage applied across the circuit be given by the equation,

$$v = V_{max} \sin \omega t \qquad ...(i)$$

Whenever an alternating voltage is applied to a purely inductive coil, a back e.m.f. is produced due to the self-inductance of the coil. This back e.m.f. opposes the rise or fall of the current through the coil. Since there is no ohmic drop in this case, therefore, the applied voltage has to overcome this induced e.m.f. only. Thus at every step,

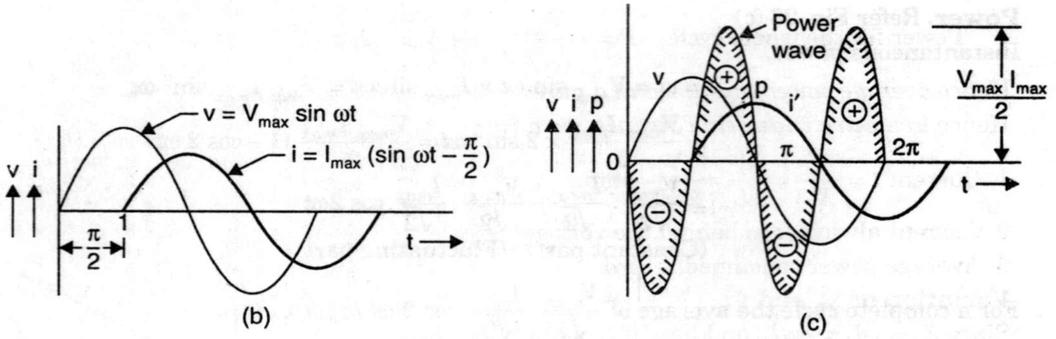

Fig. 24 (*b*), (*c*). A.C. through pure inductance alone. Resultant power zero.

$$v = L\frac{di}{dt}$$

or
$$V_{max} \sin \omega t = L\frac{di}{dt}$$

or
$$di = \frac{V_{max}}{L} \sin \omega t \, dt$$

Integrating both sides, we get

$$\int di = \int \frac{V_{max}}{L} \sin \omega t \, dt$$

or
$$i = \frac{V_{max}}{L}\left(-\frac{\cos \omega t}{\omega}\right) = \frac{V_{max}}{\omega L} \sin\left[\omega t - \frac{\pi}{2}\right]$$

or
$$i = \frac{V_{max}}{X_L} \sin\left[\omega t - \frac{\pi}{2}\right] \qquad ...(ii)$$

where $X_L = \omega L$ (opposition offered to the flow of alternating current by a pure inductances) and is called **Inductive reactance.** It is given in ohms if L is in henry and ω is in radian/second.

The value of current will be maximum when $\sin\left(\omega t - \frac{\pi}{2}\right) = 1$

$$\therefore \qquad I_{max} = \frac{V_{max}}{X_L}$$

Substituting this value in eqn. (*ii*), we get

$$i = I_{max} \sin\left(\omega t - \frac{\pi}{2}\right) \qquad ...(iii)$$

Power. Refer Fig. 24 (*c*)

Instantaneous power,
$$p = vi = V_{max} \sin \omega t \times I_{max} \sin\left(\omega t - \frac{\pi}{2}\right)$$

$$= -V_{max} I_{max} \sin \omega t \, . \cos \omega t$$

$$= -\frac{V_{max} I_{max}}{2} \times 2 \sin \omega t \cos \omega t$$

$$= -\frac{V_{max}}{\sqrt{2}} \cdot \frac{I_{max}}{\sqrt{2}} \cdot \sin 2\omega t$$

$\therefore$ Power for the whole cycle, $P = -\frac{V_{max}}{\sqrt{2}} \frac{I_{max}}{\sqrt{2}} \int_0^{2\pi} \sin 2\omega t = 0$

Hence *average power consumed in a pure inductive circuit is* **zero.**

Hence in a ***pure inductive circuit,*** we have :

1. Current $I = \frac{V}{X_L} = \frac{V}{\omega L} = \frac{V}{2\pi f L}$ amp.
2. Current always *lags* behind the voltage by *90°*.
3. Average power consumed is *zero.*

Variation of X_L and f :

Since $X_L = \omega L = 2\pi f L$, and here if L is constant, then

$$X_L \propto f$$

Fig. 25, shows the variation. As frequency is increased X_L increases and the current taken by the circuit decreases.

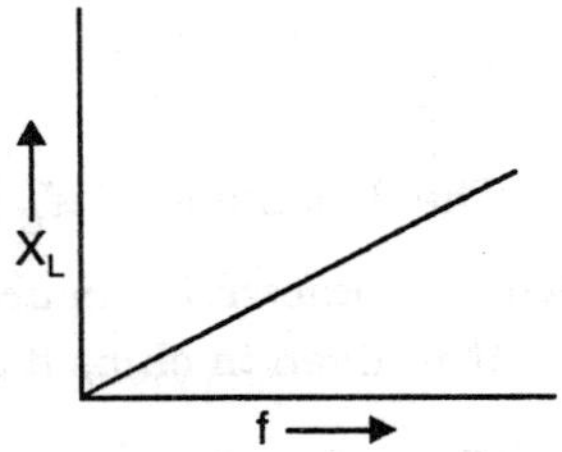

Fig. 25. Variation of X_L with f.

4.3. A.C. Through Pure Capacitance Alone

The circuit containing a pure capacitor of capacitance C farad is shown in Fig. 26 (*a*). Let the alternating voltage applied across the circuit be given by the equation,

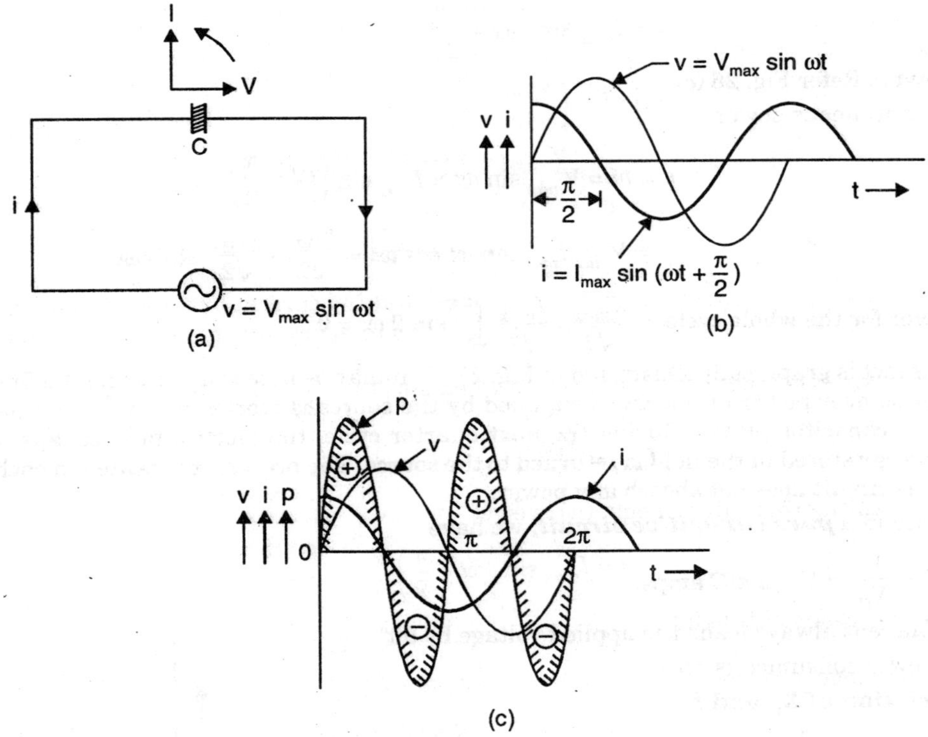

Fig. 26. A.C. through pure capacitance alone. Resultant power is zero.

$$v = V_{max} \sin \omega t \quad ...(i)$$

Charge on the capacitor at any instant,

$$q = C\,v$$

Current through the circuit,

$$i = \frac{dq}{dt} = \frac{d}{dt}(CV_{max} \sin \omega t) = \omega CV_{max} \cos \omega t$$

or

$$i = \frac{V_{max}}{1/\omega C} \sin\left(\omega t + \frac{\pi}{2}\right)$$

$\therefore$

$$i = \frac{V_{max}}{X_C} \sin\left(\omega t + \frac{\pi}{2}\right) \quad ...(ii)$$

The denominator $X_C = \frac{1}{\omega C}$ (opposition offered to the flow of alternating current by a pure capacitor) is known as *capacitive reactance.*

It is given in ohms if C is in farad and ω in radian/second.

The value of current will be maximum when $\sin\left(\omega t + \frac{\pi}{2}\right) = 1$

$\therefore$

$$I_{max} = \frac{V_{max}}{X_C}$$

Substituting this value in eqn. (*ii*), we get

$$i = I_{max} \sin\left(\omega t + \frac{\pi}{2}\right) \quad ...(iii)$$

Power. Refer Fig. 26 (*c*)

Instantaneous power,

$$p = vi = V_{max} \sin \omega t \times I_{max} \sin\left(\omega t + \frac{\pi}{2}\right)$$

$$= V_{max} I_{max} \sin \omega t \cos \omega t = \frac{V_{max}}{\sqrt{2}} \cdot \frac{I_{max}}{\sqrt{2}} \sin 2\omega t$$

Power for the whole cycle $= \frac{V_{max}}{\sqrt{2}} \cdot \frac{I_{max}}{\sqrt{2}} \int_0^{2\pi} \sin 2\,\omega t = 0$

This fact is graphically illustrated in Fig. 26 (*c*). It may be noted that, during the first quarter cycle, what so ever power or energy is supplied by the source is stored in the electric field set-up between the capacitor plates. During the next quarter cycle, the electric field collapses and the power or energy stored in the field is returned to the source. The process is repeated in each alternation and this circuit does not absorb any power.

Hence in a ***pure capacitive circuit,*** we have

1. $I = \frac{V}{X_C} = V \times 2\,\pi fC$ amps.
2. Current always leads the applied voltage by 90°.
3. Power consumed is *zero*.

Variation of X_C and f :

Since $X_C = \frac{1}{2\pi fC}$ and if C is kept constant, than

$$X_C \propto \frac{1}{f}$$

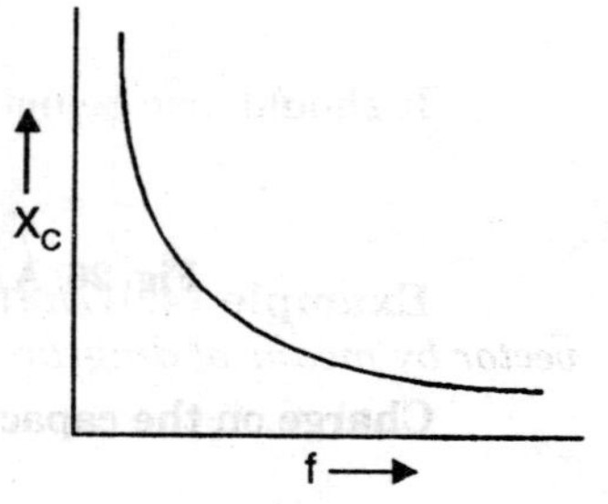

Fig. 27

Fig. 27. shows the variation. As the frequency increases X_C decreases, so the current increases.

4.4. Phasor Algebra

The following are the methods of representing vector quantities :

1. Symbolic notation
2. Trigonometrical form
3. Exponential form
4. Polar form.

A vector as shown in Fig. 28 may be described in the above forms as follows :

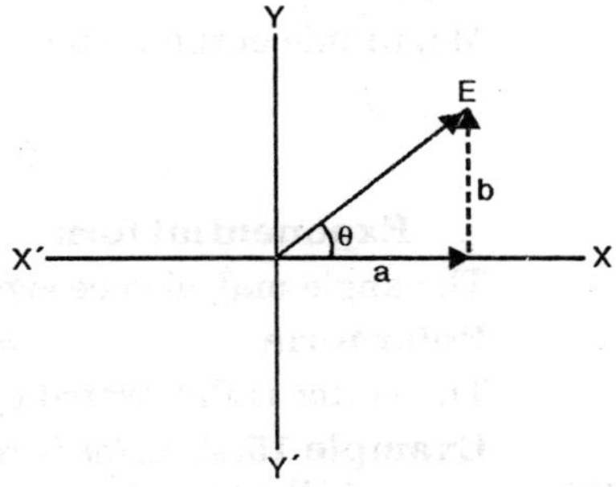

Fig. 28

1. *Symbolic notation* :

$$E = a + jb$$

2. *Trigonometrical form* :

$$E = \sqrt{a^2 + b^2}\ (\cos\theta + j\sin\theta)$$

$$= \sqrt{a^2 + b^2}\ (\cos\theta \pm j\sin\theta) \quad \text{......in general}$$

3. *Exponential form* : $E = \sqrt{a^2 \times b^2}\ e^{+j\theta}$

$$= \sqrt{a^2 + b^2}\ e^{\pm j\theta} \quad \text{......in general}$$

4. *Polar form* : $E = \sqrt{a^2 + b^2}\ \angle\theta$

$$= \sqrt{a^2 + b^2}\ \angle\pm\theta \quad \text{......in general}$$

Significance of operator j. The letter j used in the above expressions is a symbol of an operation. It is used to indicate the counter-clockwise rotation of a vector through 90°. It is assigned a value of $\sqrt{-1}$. The double operation of j on a vector rotates it counter-clockwise (*CCW*) through 180° and hence reverses its sense because, $j \times j = j^2 = \sqrt{(-1)^2} = -1$.

In general, each successive multiplication of j, rotates the phasor further by 90° as given below (Refer Fig. 29)

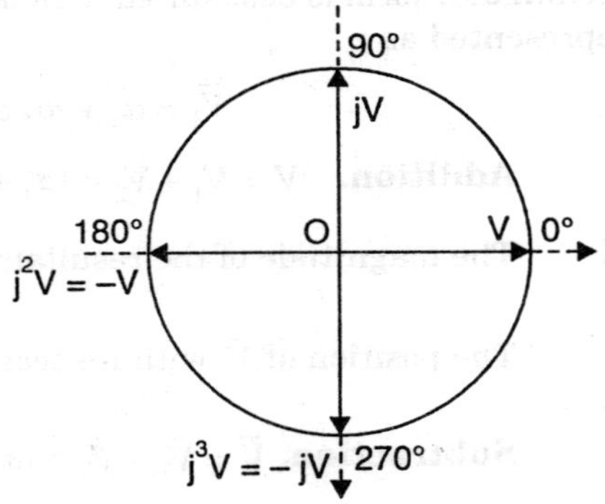

Fig. 29

$$j = \sqrt{-1}$$

...90° CCW rotation from *OX*-axis

$$j^2 = \left(\sqrt{-1}\right)^2 = -1$$

...180° CCW rotation from *OX*-axis

$$j^3 = \left(\sqrt{-1}\right)^3 = -\sqrt{-1} = -j$$

...270° CCW rotation from *OX*-axis

$$j^4 = \left(\sqrt{-1}\right)^4 = +1$$

...360° CCW rotation from *X*-axis

It should also be noted that,

$$\frac{1}{j} = \frac{j}{j^2} = \frac{j}{-1} = -j.$$

Example 14. *Write the equivalent potential and polar forms of vector 6 + j8. Also illustrate the vector by means of diagram.*

Solution. Refer Fig. 30.

Magnitude of the vector $= \sqrt{6^2 + 8^2} = 10, \ \tan\theta = \dfrac{8}{6}$

$\therefore \qquad \theta = \tan^{-1}\left(\dfrac{8}{6}\right) = 53.1°$

$\therefore$ **Exponential form = 10 $e^{j\,53.1°}$. (Ans.)**

The angle may also be *expressed in radians.*

Polar form **= 10 ∠ 53.1°. (Ans.)**

The vector is illustrated by means of diagram in Fig. 30.

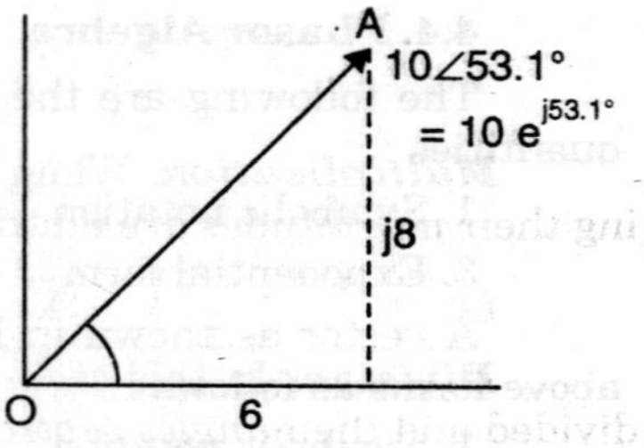

Fig. 30

Example 15. *A vector is represented by $30e^{-j2\pi/3}$. Write down the various equivalent forms of the vector and illustrate by means of a vector diagram, the magnitude and position of the above vector.*

Solution. Refer Fig. 31. Draw the vector in a direction making an angle of $\dfrac{2\pi}{3} = \dfrac{2 \times 180}{3}$ = 120° in the *clockwise* direction (since the angle is negative).

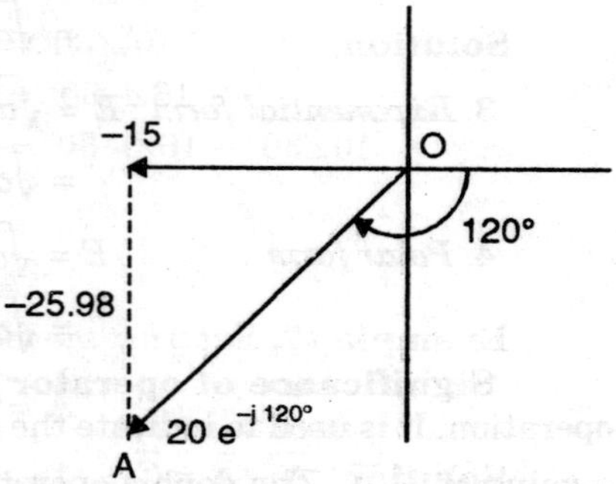

Fig. 31

(*i*) **Rectangular form :**

$a = 30 \cos(-120°) = -15$

$b = 30 \sin(-120°) = -25.98$

$\therefore$ Expression is **= (− 15 − j 25.98). (Ans.)**

(*ii*) **Polar form** **= 30 ∠ − 120°. (Ans.)**

Addition and subtraction of vector quantities :

For addition and subtraction of vector quantities rectangular form is best suited. Consider two voltage phasors represented as :

$$\overline{V_1} = a_1 + jb_1 \text{ and } \overline{V_2} = a_2 + jb_2$$

Addition. $\overline{V} = \overline{V_1} + \overline{V_2} = (a_1 + jb_1) + (a_2 + jb_2) = (a_1 + a_2) + j\,(b_1 + b_2)$

The magnitude of the resultant vector $\overline{V} = \sqrt{(a_1 + a_2)^2 + (b_1 + b_2)^2}$

The position of $\overline{V}$ with respect to X-axis is $\theta = \tan^{-1}\left(\dfrac{b_1 + b_2}{a_1 + a_2}\right)$

Subtraction. $\overline{V} = \overline{V_1} - \overline{V_2} = (a_1 + jb_1) - (a_2 + jb_2) = (a_1 - a_2) + j(b_1 - b_2)$

The magnitude of the resultant vector $\overline{V} = \sqrt{(a_1 - a_2)^2 + (b_1 - b_2)^2}$

The position of $\overline{V}$ with respect to X-axis is $\theta = \tan^{-1}\left(\dfrac{b_1 - b_2}{a_1 - a_2}\right)$

Multiplication and division of vector quantities :

It the vectors are represented in the polar exponential form, their multiplication and division becomes very easy and simple.

Consider two voltage phasors represented as

$$\overline{V_1} = a_1 + jb_1 = V_1\angle\theta_1, \text{ where } \theta_1 = \tan^{-1}\left(\frac{b_1}{a_1}\right)$$

$$\overline{V_2} = a_2 + jb_2 = V_2\angle\theta_2, \text{ where } \theta_2 = \tan^{-1}\left(\frac{b_2}{a_2}\right).$$

Multiplication. When the phasor quantities are represented in polar form, while multiplying their magnitudes are multiplied and their *angles added algebraically*,

i.e, $$\overline{V_1} \times \overline{V_2} = V_1\angle\theta_1 \times V_2\angle\theta_2 = V_1V_2\angle(\theta_1 + \theta_2)$$

Division. In this case, the magnitudes of phasor quantities (expressed in polar from) are divided and their *angles subtracted algebraically*

i.e. $$\frac{\overline{V_1}}{\overline{V_2}} = \frac{V_1\angle\theta_1}{V_2\angle\theta_2} = \frac{V_1}{V_2}\angle(\theta_1 - \theta_2).$$

Example 16. *Perform the following operation and express the final result in polar form :*

$$10\angle 30° + 16\angle -30°.$$

Solution. $10\angle 30° = 10(\cos 30° + j\sin 30°) = 8.66 + j5$

$$16\angle -30° = 16[\cos(-30°) + j\sin(-30°)] = 13.86 - j8$$

$$\therefore \quad 10\angle 30° + 16\angle -30° = (8.66 + j5) + (13.86 - j8) = 22.52 - j3$$

$$= (\sqrt{22.52^2 + 3^2})\tan^{-1}(-3/22.52)$$

$$= 22.72\tan^{-1}(-3/22.52) = \mathbf{22.72\angle -7.6°}. \quad \textbf{(Ans.)}$$

Example 17. *Subtract the following given vectors from one another.*

$$\overline{A} = 15 + j26 \text{ and } \overline{B} = -19.75 - j7.18.$$

Solution. $\overline{A} - \overline{B} = \overline{C} = (15 + j26) - (-19.75 - j7.18) = 34.75 + j33.18$

$$\therefore \quad \text{Magnitude of } \overline{C} = \sqrt{34.75^2 + 33.18^2} = \mathbf{48}$$

$$\text{Slope of } \overline{C} = \tan^{-1}(33.18/34.75) = 43.68°$$

$$\therefore \quad \overline{C} = \mathbf{48\angle 43.68°}. \quad \textbf{(Ans.)}$$

Example 18. *Perform the operation* $\frac{\overline{AB}}{\overline{C}}$ *and express the final result in polar form for the vectors given below :*

$$\overline{A} = 10 + j10 \ ; \ \overline{B} = 15\angle -120° \ ; \ \overline{C} = 5 + j0°.$$

Solution. Rearranging vectors $\overline{A}$ and in polar form, we have

$$\overline{A} = 10 + j10 = \sqrt{10^2 + 10^2}\ \tan^{-1}(10/10) = 14.14\angle 45°$$

$$\overline{C} = 5 + j0 = \sqrt{5^2 + 0^2}\ \tan^{-1}(0/5) = 5\angle 0°$$

$$\therefore \quad \frac{\overline{AB}}{\overline{C}} = \frac{14.14\angle 45° \times 15\angle -120°}{5\angle 0°} = \frac{14.14 \times 15}{5}\angle(45° - 120° - 0°)$$

$$= \mathbf{42.42\angle -75°}. \quad \textbf{(Ans.)}$$

Example 19. *The instantaneous values of two currents i_1 and i_2 are given as :*

$$i_1 = 5\sin\left(\omega t + \frac{\pi}{4}\right) \text{ and } i_2 = 2.5\cos\left(\omega t - \frac{\pi}{2}\right)$$

Find the r.m.s. value of $i_1 + i_2$ using complex number representation.

(Kerala University)

Solution. *Given :* $i_1 = 5\sin\left(\omega t + \frac{\pi}{4}\right)$,

$$i_2 = 2.5\cos\left(\omega t - \frac{\pi}{2}\right) = 2.5\sin\left[90° + \left(\omega t - \frac{\pi}{2}\right)\right] = 2.5\sin\omega t$$

$\therefore$ $I_{1(max)} = 5(\cos 45° + j\sin 45°) = (3.53 + j3.53)$

$I_{2(max)} = 2.5(\cos 0° + j\sin 0°) = (2.5 + j0)$

The maximum value of resultant current is

$$I_{max} = (3.53 + j3.53) + (2.5 + j0) = 6.03 + j3.53 = 6.987\angle 30.34°$$

$\therefore$ R.M.S. value $= \frac{6.987}{\sqrt{2}} = $ **4.94 A. (Ans.)**

Conjugate complex numbers

Two numbers are said to be conjugate if they *differ only in the algebraic sign of their quadrature components.* Accordingly, the numbers $(a + jb)$ and $(a - jb)$ are conjugate.

- The *sum* of two conjugate numbers gives in-phase or *active* component and their *difference* gives quadrature or *reactive* component.

i.e. $(a + jb) + (a - jb) = 2a$ (*i.e.,* active component), and

$(a + jb) - (a - jb) = j2b$ (*i.e.,* reactive component).

The resultant is the sum of two vertical components only.

- The resultant arising out of the *multiplication* of two conjugate numbers contains *no quadrature component*.

i.e., $(a + jb) \times (a - jb) = a^2 - j^2b^2 = a^2 + b^2$

The conjugate of a complex number is used to determine the *apparent power* of an A.C. circuit in complex form.

Power and roots of vectors/phasors

The *powers* and *roots* of vectors can be found conveniently in *polar form*. If the vector are not, in polar form, these should be converted into polar form before carrying out the algebraic operations, as mentioned below.

Powers. Consider a vector phasor quantity represented in polar form as $\bar{A} = A\angle\theta$,

Then $(\bar{A})^n = A^n\angle(n \times \theta)$

Example. Suppose it is required to find cube of the vector $4\angle 12°$

Then, $(4\angle 12°)^3 = (4)^3 \angle (3 \times 12°) = 64\angle 36°$

Roots. Consider a vector (phasor) quantity represented in polar form as $(\bar{A})^{1/n} = (A)^{1/n}\angle\theta/n$.

Example. Suppose it is required to find cube root of $125\angle 60°$

Then, $(125\angle 60°)^{1/3} = (125)^{1/3}\angle 60°/3 = 5\angle 20°$.

The 120° operator

In case of 3-phase work, where voltage vectors are displaced by 120° from one another (Fig. 32) it is convenient to use an operator, which rotates a vector/phasor through 120° toward or backwards without altering its length. This operator is 'a' and any operator which is multiplied by 'a' remains unaltered in magnitude but is rotated in CCW (counter-clockwise) direction by 120°.

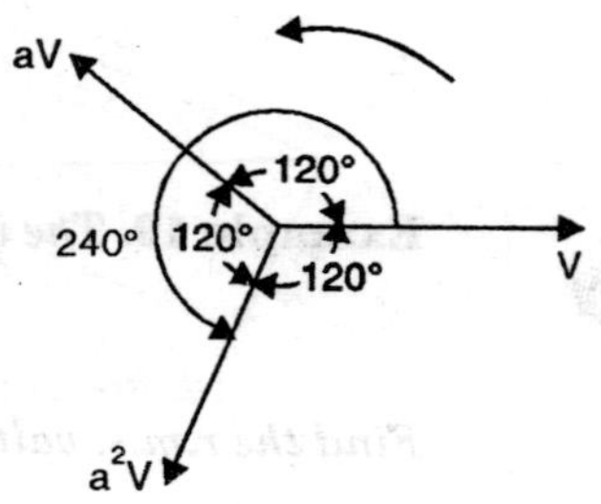

Fig. 32

$\therefore$ $a = 1\angle 120°$

In cartesian form, $a = \cos 120° + j\sin 120°$

$= -0.5 + j\,0.866$

Similarly, $a^2 = 1\angle 120° \times 1\angle 120° = 1\angle 240° = \cos 240° + j\sin 240° = -0.5 - j\,0.866$

Hence the operator 'a' will rotate the in CCW by 240° which is the same as rotating the vector in CW(clock-wise) direction by 120°.

$\therefore \qquad a^2 = 1\angle -120°$

Similarly $\qquad a^3 = 1\angle 360° = 1$ **(Numerically, a is equivalent to the cube root of unity.)**

4.5. A.C. Series Circuits

Under this heading we shall discuss *R-L, R-C* and *R-L-C* series circuits.

4.5.1. R-L circuit (Resistance and inductance in series)

Fig. 33 (a) shows a pure resistance R and a pure inductive coil of inductance L connected in series. Such a circuit is known as *R-L* circuit (usually met a cross in practice).

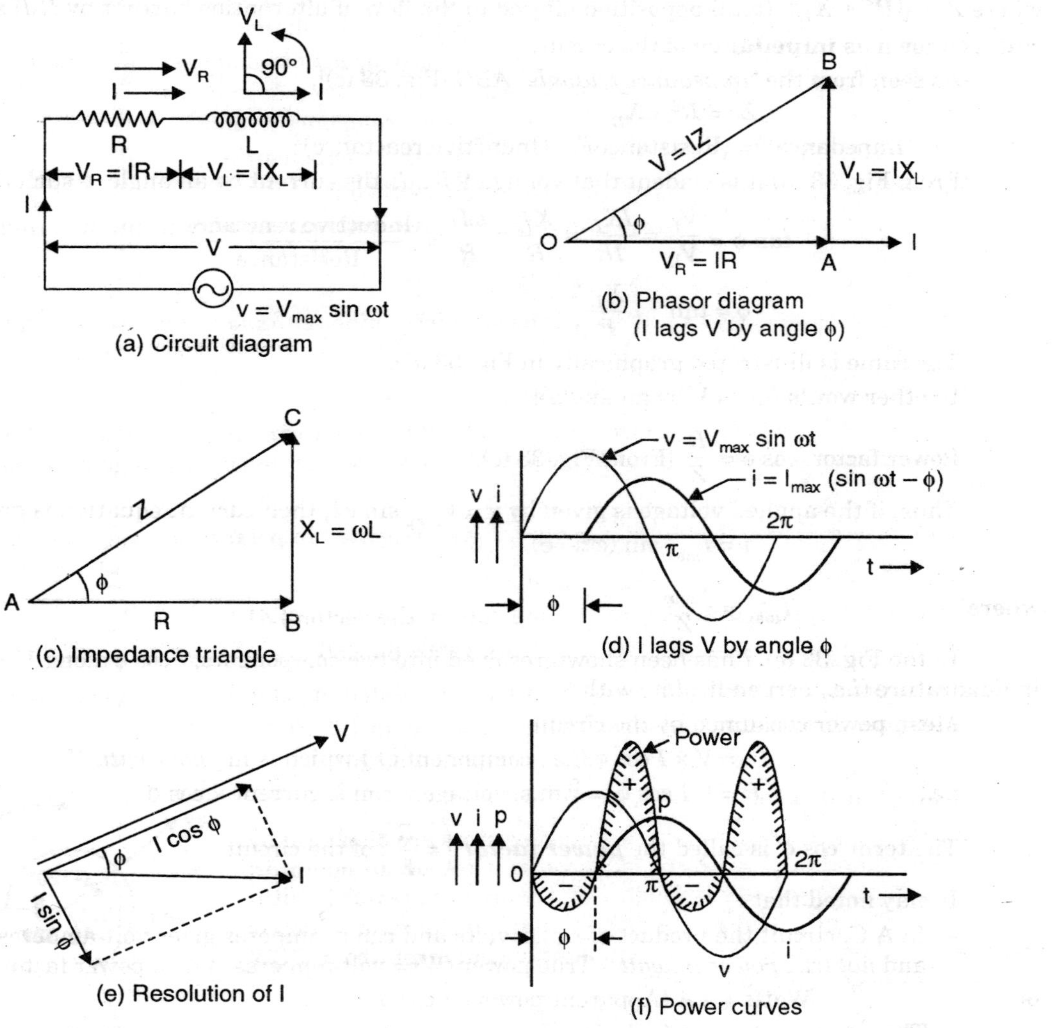

Fig. 33. *R-L* circuit (Resistance and inductance in series).

Let V = R.M.S. value of the applied voltage,

I = R.M.S. value of the resultant current,

$V_R = IR$ = Voltage drop across R(*in phase* with I), and

$V_L = IX_L$ = Voltage drop across L (coil), *ahead* of I by 90°.

The voltage drop V_R and V_L and shown in voltage triangle OAB in Fig. 33 (*b*), I being taken as the reference vector in the phasor diagram. Vector OA represents ohmic drop V_R and AB represents inductive drop V_L. Vector OB represents the applied voltage V which is the vector sum of the two (*i.e.*, V_R and V_L).

$$\therefore \quad V = \sqrt{V_R^2 + V_L^2} = \sqrt{(IR)^2 + (IX_L)^2} = I\sqrt{(R^2 + X_L^2)}$$

or
$$I = \frac{V}{\sqrt{R^2 + (X_L)^2}} = \frac{V}{Z}$$

where $Z = \sqrt{R^2 + X_L^2}$ (total opposition offered to the flow of alternating current by R-L series circuit) is known as **impedance** of the circuit.

As seen from the "*impedance triangle*" ABC [Fig. 33 (*c*)],

$$Z^2 = R^2 + X_L^2$$

i.e., (Impedance)² = (Resistance)² + (Inductive reactance)²

From Fig. 33 (*b*) it is evident that voltage V *leads* the current by an angle ϕ such that,

$$\tan\phi = \frac{V_L}{V_R} = \frac{IX_L}{IR} = \frac{X_L}{R} = \frac{\omega L}{R} = \frac{\text{Inductive reactance}}{\text{Resistance}}$$

$$\therefore \quad \phi = \tan^{-1}\left(\frac{X_L}{R}\right)$$

The same is illustrated graphically in Fig. 33 (*d*).

In other words I *lags* V by an angle ϕ.

Power factor, $\cos\phi = \frac{R}{Z}$ [From Fig. 33 (*c*)]

Thus, if the applied voltage is given by $v = V_{max}\sin\omega t$, then current equation is given as,

$$i = I_{max}\sin(\omega t - \phi),$$

where
$$I_{max} = \frac{V_{max}}{Z}$$

In the Fig. 33 (*e*), I has been shown resolved into two components, $I\cos\phi$ along V and $I\sin\phi$ in quadrature (*i.e.*, perpendicular) with V.

Mean power consumed by the circuit

$= V \times I\cos\phi$ (*i.e.*, component of I which is in *phase with* V

i.e., $P = VI\cos\phi$ (= r.m.s. voltage × r.m.s. current × cos ϕ)

The term 'cos ϕ' is called the ***power factor*** $\left(=\frac{R}{Z}\right)$ of the circuit

It may noted that :

— In A.C. circuit the product of r.m.s. volts and r.m.s. amperes gives volt-amperes (*i.e., VA*) and *not true power in watts*. True power (W) = volt-amperes (VA) × power factor

or Watts = VA (Apparent power) × cos ϕ

— The *power consumed is due to ohmic resistance only* since pure inductance consumes no power.

i.e. $P = VI\cos\phi = VI \times \frac{R}{Z} = \frac{V}{Z} IR = I \times IR = I^2R$, watts

$\left(\because \quad \cos\phi = R/Z \text{ and } \frac{V}{Z} = I\right)$

This shows that power is actually consumed in *resistance only* ; the *inductor does not consume any power*.

The power consumed in R-L circuit is shown graphically in Fig. 33 (*f*).

Thus in ***R-L circuit*** we have :

1. Impedance, $Z = \sqrt{R^2 + X_L^2}$ (where $X_L = \omega L = 2\pi \times fL$)

2. Current, $I = \frac{V}{Z}$

3. Power factor, $\cos\phi = \frac{R}{Z} \left(= \frac{\text{True power}}{\text{Apparent power}} = \frac{\text{W}}{\text{VA}} \right)$

[or angle of lag, $\phi = \cos^{-1}(R/Z)$]

4. Power consumed, $P = VI\cos\phi \left(= IZ \times I \times \frac{R}{Z} = I^2 R \right)$

Symbolic Notation :

$$Z = R + jX_L$$

The numerical value of impedance vector $= \sqrt{R^2 + X_L^2}$

The phase angle with the reference axis, $\phi = \tan^{-1}(X_L/R)$.

In *polar form :* $\bar{Z} = Z\angle\phi°$.

Apparent, Active (True or real) and Reactive Power :

Every circuit current has two components : (*i*) Active component and (*ii*) Reactive component.

"Active component" consumes power in the circuit while *"reactive component"* is responsible for the field which lags or leads the main current from the voltage.

In Fig. 34. active component is $I_{\text{active}} = I\cos\phi$, and reactive component is $I_{\text{reactive}} = I\sin\phi$

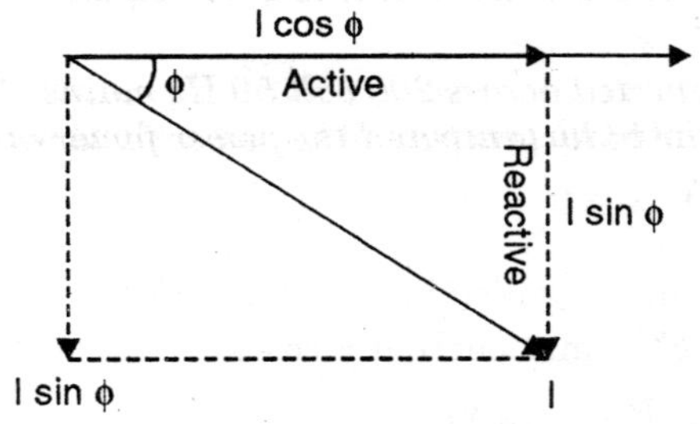

Fig. 34. Active and reactive components of circuit current I.

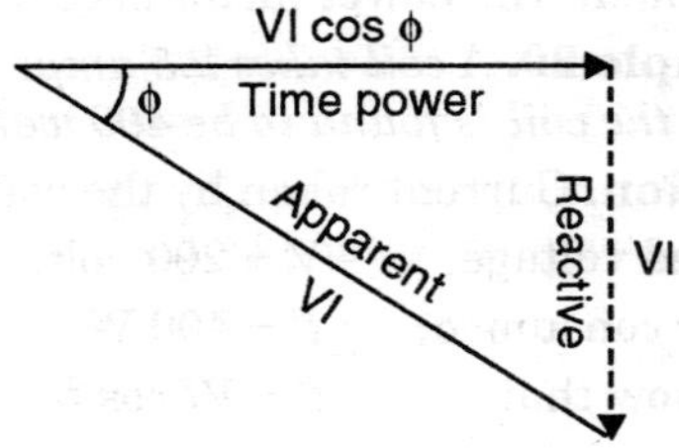

Fig. 35. Apparent, true and reactive power.

So, $I = \sqrt{(I_{\text{active}})^2 + (I_{\text{reactive}})^2}$

Refer Fig. 35.

(*i*) **Apparent power (S).** It *is given by the product of r.m.s. values of applied voltage and circuit current.*

$$\therefore \quad S = VI = (I \times Z) \,.\, I = I^2Z \text{ volt-amperes } (VA)$$

(*ii*) **Active or true or real power (P or W).** It *is the power which is actually dissipated in the circuit resistance.*

$$P = I^2R = VI \cos \phi \text{ watts}$$

(*iii*) **Reactive power (Q).** A pure inductor and a pure capacitor do not consume any power, since in a half cycle what so ever power is received from the source by these components the same is returned to the source. This power which flows back and forth (*i.e.,* in both directions in the circuit) or *reacts upon itself* is called *"reactive power."*

It may be noted that the *current in phase with the voltage produces active or true or real power while the current 90° out of phase with the voltage contributes to* reactive power.

In a R-L circuit, reactive power which is the power developed in the inductive reactance of the circuit, is given as :

$$Q = I^2 X_L = I^2Z \sin \phi = I \,.\, (IZ) \sin \phi$$
$$= VI \sin \phi \text{ volt-amperes-reactive } (VAR)$$

These three powers are shown in Fig. 35

Relation between *VA*, *W* and *VR*

$$W = VA \cos \phi \quad \text{...}(i)$$

$$VAR = VA \sin \phi \quad \text{...}(ii)$$

$$\therefore \quad VA = \frac{W}{\cos \phi} \quad \text{...[From } (i)]$$

$$\text{and,} \quad VA = \frac{VAR}{\sin \phi} \quad \text{...[From } (ii)]$$

$$\text{Power factor } (p.f.) = \frac{W}{VA} = \frac{\text{True power}}{\text{Apparent power}}$$

The larger bigger units of apparent, true and reactive power are *kVA* (or *MVA*), *kW*(or *MW*) and *kVAR* (or *MVAR*) respectively.

The power factor depends on the reactive power component. If it is made equal to the active power component, the power factor becomes unity.

Example 20. *A coil takes 2.5 amps. when connected across 200 volt 50 Hz mains. The power consumed by the coil is found to be 400 watts. Find the inductance and the power factor of the coil.*

Solution. Current taken by the coil, $I = 2.5$ A

Applied voltage, $V = 200$ volts

Power consumed, $P = 400$ W

We know that $P = VI \cos \phi$

$$\text{or} \quad 400 = 200 \times 2.5 \times \cos \phi \text{ or } \cos \phi = \frac{400}{200 \times 2.5} = 0.8$$

Hence power factor of coil is **0.8. (Ans.)**

$$\text{Impedance of the coil,} \quad Z = \frac{V}{I} = \frac{200}{2.5} = 80\ \Omega$$

$$\text{Also} \quad \frac{X_L}{Z} = \sin \phi$$

$$\therefore \quad X_L = Z \sin \phi$$

$$= 80 \sin \phi = 80 \sqrt{1 - \cos^2 \phi}$$

$$= 80 \sqrt{1 - 0.8^2} = 80 \times 0.6 = 48\ \Omega$$

But $X_L = 2\pi f L$

$\therefore$ $L = \dfrac{X_L}{2\pi f} = \dfrac{48}{2\pi \times 50} =$ **0.1529 H (henry). (Ans.)**

Example 21. *A 100 V, 80 W lamp is to be operated on 230 volts, 50 Hz A.C. supply. Calculate the inductance of the choke required to be connected in series with lamp for its operation. The lamp can be taken as equivalent to a non inductive resistance.*

Solution. Current through the lamp when connected across 100 V supply,

$$I = \frac{W}{V} = \frac{80}{100} = 0.8\ \text{A}$$

Resistance of the lamp, $R = \dfrac{V}{I} = \dfrac{100}{0.8} = 125\ \Omega$

If a choke of inductance L henry is connected in series with the lamp to operate it on 230 V, the current through the choke will also be 0.8 A.

The impedance of the circuit when choke is connected in series with the lamp,

$$Z = \frac{V}{I} = \frac{230}{0.8} = 287.5\ \Omega$$

Reactance of choke coil, $X_L = \sqrt{Z^2 - R^2} = \sqrt{287.5^2 - 125^2} = 258.5\ \Omega$

But $X_L = 2\pi f L$

or $L = \dfrac{X_L}{2\pi f} = \dfrac{258.5}{2\pi \times 50} = 0.825$ H

Hence inductance of choke coil, **L = 0.825 H. (Ans.)**

Example 22. *A coil has a resistance of 5 Ω and an inductance of 31.8 mH. Calculate the current taken by the coil and power factor when connected to 200 V, 50 Hz supply.*

Draw the vector diagram.

If a non-inductive resistance of 10 Ω is then connected in series with coil, calculate the new value of current and its power factor.

Solution. $R = 5\ \Omega$

$L = 31.8$ mH or 0.0318 H

$\therefore$ $X_L = 2\pi f L$

$= 2\pi \times 50 \times 0.0318 = 10\ \Omega$

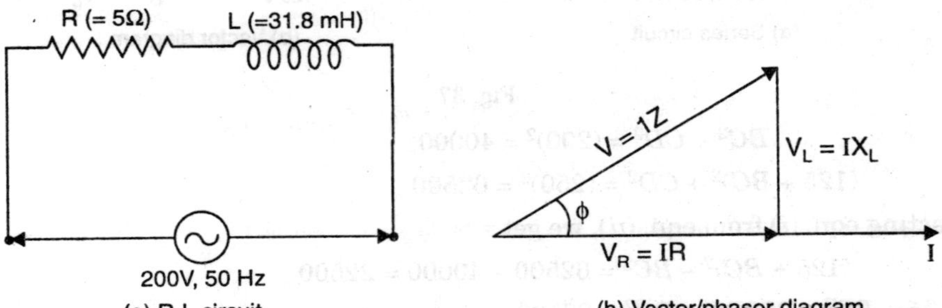

(a) R-L circuit (b) Vector/phasor diagram

Fig. 36

Impedance of the coil, $Z = \sqrt{R^2 + X_L^2}$

$= \sqrt{5^2 + 10^2} = 11.18\ \Omega$

$\therefore$ Current taken by the coil, $I = \frac{V}{Z} = \frac{200}{11.18}$ = **17.9 A. (Ans.)**

Power factor, $\cos\phi = \frac{R}{Z} = \frac{5}{11.18}$ = **0.4475. (Ans.)**

Fig. 36 (*b*) shows the vector diagram.

When non-inductive resistance of 10 Ω is connected in series with the coil :

Total resistance in the circuit, $R' = 5 + 10 = 15\ \Omega$

Reactance in the circuit, $X_L' = X_L = 10\ \Omega$

Impedance of the circuit, $Z' = \sqrt{R'^2 + X'_L{}^2} = \sqrt{15^2 + 10^2} = 18\ \Omega$

Current through the circuit, $I' = \frac{V}{Z'} = \frac{200}{18}$ = **11.11 A. (Ans.)**

Power factor of the circuit, $\cos\phi = \frac{R'}{Z'} = \frac{15}{18}$ = **0.833. (Ans.)**

Example 23. *A current of 5A flows through a non-inductive resistance in series with a choking coil when supplied at 250 V, 50 Hz. If the voltage across the resistance is 125 V and across the coil 200 V, calculate :*

(*i*) *Impedance, reactance and resistance of the coil,*

(*ii*) *The power absorbed by the coil,*

(*iii*) *The total power.*

Draw the vector diagram. **(Elect. Engg. Madras Univ.)**

Solution. Non-inductive resistance connected in series with coil $= \frac{125}{5} = 25\ \Omega$

Refer Fig. 37 (*b*).

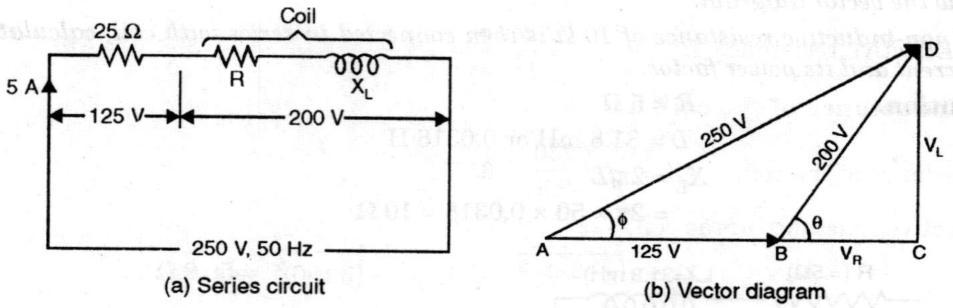

(a) Series circuit (b) Vector diagram

Fig. 37

$$BC^2 + CD^2 = (200)^2 = 40000 \quad ...(i)$$

$$(125 + BC)^2 + CD^2 = (250)^2 = 62500 \quad ...(ii)$$

Subtracting eqn. (*i*) from eqn. (*ii*), we get

$$(125 + BC)^2 - BC^2 = 62500 - 40000 = 22500$$

$$15625 + BC^2 + 250BC - BC^2 = 22500$$

$\therefore$ $BC = 27.5V$; $CD = \sqrt{(200)^2 - (27.5)^2} = 198.1$ V (*i*) **Coil impedance,** $Z = \frac{200}{5} = 40\ \Omega$. **(Ans.)**

$$V_R = IR = BC = 27.5 \quad \text{or} \quad 5R = 27.5$$

$$\mathbf{R} = \frac{27.5}{5} = \mathbf{5.5\ \Omega} \quad \textbf{(Ans.)}$$

Also, $V_L = IX_L = CD = 198.1$

$$\therefore \quad \mathbf{X_L} = \frac{198.1}{5} = \mathbf{39.62\ \Omega.} \quad \textbf{(Ans.)}$$

(*ii*) **Power absorbed by the coil,**

$$P = I^2R = 5^2 \times 5.5 = \mathbf{137.5\ W.} \quad \textbf{(Ans.)}$$

(*iii*) **The total power** $= VI \cos \phi = 250 \times 5 \times \frac{AC}{AD}$

$$= 250 \times 5 \times \frac{(125 + 27.5)}{250} = \mathbf{762.5\ W.} \quad \textbf{(Ans.)}$$

The vector diagram is shown in Fig. 37 (*b*). (Ans.)

Example 24. *An iron-cored coil has a D.C. resistance of 6 ohms. When it is connected to 230 V, 50 Hz mains, the current taken is 3.5 A at a power factor of 0.5. Determine :*

(*i*) *Effective resistance of the coil.*

(*ii*) *Inductance of the coil.*

(*iii*) *Resistance which represents the effect of the iron loss.*

Solution. *Given :* D.C. resistance (True resistance), R = 6 Ω ; supply voltage = 230 V, f = 50 Hz, I = 3.5 A ; *p.f.* = 0.5.

(*i*) **Effective resistance of the coil, R_e :**

Total power consumed by the iron-cored choke coil,

$$P = \text{Power loss in ohmic resistance + Iron loss in core} = I^2R + P_i$$

or $\frac{P}{I^2} = R + \frac{P_i}{I^2}$, where $\frac{P}{I^2}$ is known as *effective resistance* of the coil.

$$\therefore \quad \text{Effective resistance, } R_e = \frac{P}{I^2} = \frac{VI \cos \phi}{I^2} = \frac{230 \times 3.5 \times 0.5}{(3.5)^2} = \mathbf{32.86\ \Omega.} \quad \textbf{(Ans.)}$$

(*ii*) **Inductance of the coil, L :**

Impedance of the coil, $Z = \frac{V}{I} = \frac{230}{3.5} = 65.7\ \Omega$

Inductive reactance of the coil,

$$X_L = \sqrt{Z^2 - R_e^{\,2}} = \sqrt{(65.7)^2 - (32.86)^2} = 56.9\ \Omega$$

$$\therefore \quad L = \frac{X_L}{2\pi f} = \frac{56.9}{2\pi \times 50} = \mathbf{0.1811\ H.} \quad \textbf{(Ans.)}$$

(*iii*) **Resistance representing iron loss :**

Since $\frac{P}{I^2} = R + \frac{P_i}{I^2}$

Effective resistance, R_e = True resistance + Resistance representing iron loss

32.86 = 6 + Resistance representing iron loss

∴ Resistance representing iron loss = 32.86 – 6 = **26.86 Ω.** **(Ans.)**

Example 25. *Three coils connected in series across a 100 V, 50 Hz supply have the following parameters :*

$$R_1 = 18\ \Omega,\ L_1 = 0.012\ H\ ;\ R_2 = 12\ \Omega,\ L_2 = 0.036\ H\ ;\ R_3 = 3.6\ \Omega,\ L_3 = 0.072\ H$$

Determine the potential drop and phase angle for each coil.

Solution. Fig. 38. shows the circuit diagram.

Total resistance in the circuit, $R = R_1 + R_2 + R_3 = 18 + 12 + 3.6 = 33.6\ \Omega$

Total inductance in the circuit, $L = L_1 + L_2 + L_3 = 0.012 + 0.036 + 0.072 = 0.12$ H

Impedance of coil-1, $Z_1 = \sqrt{R_1^2 + (2\pi f L_1)^2} = \sqrt{(18)^2 + (2\pi \times 50 \times 0.012)^2} = 18.39\ \Omega$

Impedance of coil-2, $Z_2 = \sqrt{R_2^2 + (2\pi f_2 L_2)^2} = \sqrt{(12)^2 + (2\pi \times 50 \times 0.036)^2} = 16.49\ \Omega$

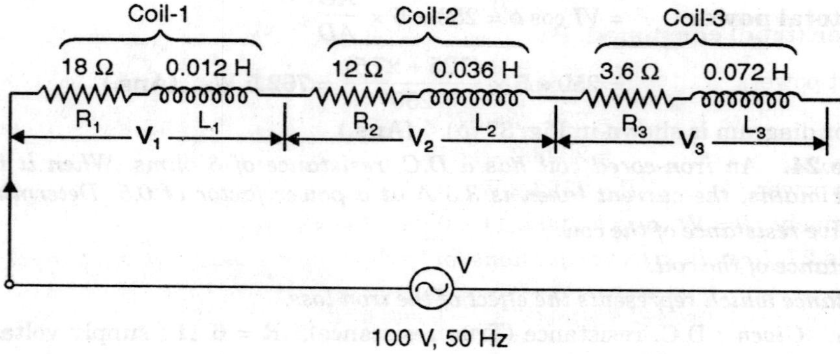

Fig. 38. Circuit diagram.

Impedance of coil–3, $Z_3 = \sqrt{R_3^2 + (2\pi f L_3)^2} = \sqrt{(3.6)^2 + (2\pi \times 50 \times 0.072)^2} = 22.90\ \Omega$

Impedance of the whole circuit, $Z = \sqrt{R^2 + (2\pi f L)^2} = \sqrt{(33.6)^2 + (2\pi \times 50 \times 0.12)^2} = 50.5\ \Omega$

Current through the circuit, $I = \dfrac{V}{Z} = \dfrac{100}{50.5} = 1.98\ A$

Potential drop across coil–1, $\mathbf{V_1} = IZ_1 = 1.98 \times 18.39 =$ **36.41 V (Ans.)**

Potential drop across coil–2, $\mathbf{V_2} = IZ_2 = 1.98 \times 16.49 =$ **32.65 V (Ans.)**

Potential drop across coil–3, $\mathbf{V_3} = IZ_3 = 1.98 \times 22.90 =$ **45.34 V (Ans.)**

Phase angle of coil–1, ϕ_1, $\cos^{-1}(R_1/Z_1) = \cos^{-1}(18/18.39) =$ **11.82°. (Ans.)**

Phase angle of coil–2, $\phi_2 = \cos^{-1}(R_2/Z_2) = \cos^{-1}(12/16.49) =$ **43.3°. (Ans.)**

Phase angle of coil–3, $\phi_3 = \cos^{-1}(R_3/Z_3) = \cos^{-1}(3.6/22.90) =$ **80.96°. (Ans.)**

Example 26. *An alternating voltage of (176 + j132) is applied to a circuit and the current in the circuit is given by (6.6 + j 8.8) A. Determine :*

(*i*) *Values of elements of the circuit.*

(*ii*) *Power factor of the circuit.*

(*iii*) *Power consumed.*

Solution. Given : Supply voltage, $\overline{V} = 176 + j132 = 220\ \angle 36.87°$

Circuit current, $\overline{I} = 6.6 + j\,8.8 = 11\ \angle 53.13°$

(*i*) **Values of elements of the circuit, R, C :**

Circuit impedance, $\bar{Z} = \frac{\bar{V}}{\bar{I}} = \frac{220 \angle 36.87°}{11 \angle 53.13°} = 20 \angle -16.26°$

$= 20 [\cos (-16.26°) + \sin (-16.26°)]$

$= 20 (0.96 - j\, 0.28) = (19.2 - j5.6)\ \Omega$

$\therefore \quad R = \mathbf{19.2\ \Omega}$ **(Ans.)**

$X_C = 5.6\ \Omega$ or $C = \frac{1}{2\pi f X_C} = \frac{1}{2\pi \times 50 \times 5.6}$ F = **568.4 μF.** **(Ans.)**

(*ii*) **Power factor of the circuit ; cos φ :**

$$\cos \phi = \frac{R}{Z} = \frac{19.2}{20} = \mathbf{0.96}$$ **(leading). (Ans.)**

(*iii*) **Power (true) consumed, P :**

Apparent power, $\bar{S} = \bar{V} \times \bar{I}$

$= 220 \angle 36.87° \times 11 \angle -53.13° = 2420 \angle -16.26°$

$= 2323.2 - j\, 677.6$

∴ True power, $P = \mathbf{2323.2\ W.}$ **(Ans.)**

(Alternatively : $P = VI \cos \phi = 220 \times 11 \times 0.96 = 2323.2$ W).

Example 27. *In a circuit, the equations for instantaneous voltage and current are given by,*

$$v = 141.4 \sin \left(\omega t - \frac{2\pi}{3}\right), \text{ volt and}$$

$$i = 7.07 \sin \left(\omega t - \frac{\pi}{2}\right), \text{ amp, where } \omega = 314 \text{ rad/sec.}$$

(*i*) *Sketch a neat phasor diagram for the circuit*

(*ii*) *Use polar notation to calculate impedance with phase angle.*

(*iii*) *Calculate average power.*

(*iv*) *Calculate the instantaneous power at the instant t = 0* **(Pune University)**

Solution. *Given :* $v = 141.4 \sin \left(\omega t - \frac{2\pi}{3}\right)$, and $i = 7.07 \sin \left(\omega t - \frac{\pi}{2}\right)$, where ω = 314 rad/s.

(*i*) **Phasor diagram :**

From the voltage equation, it is seen that the voltage lags behind the reference quantity by $\frac{2\pi}{3}$ rad or $2 \times \frac{180}{3} = 120°$. Similarly, current lags behind the reference quantity by $\frac{\pi}{2}$ rad or $\frac{180}{2}$ = 90°. Between themselves, voltage lags behind the current by (120° – 90°) = 30° as shown in Fig. 39 (*b*).

(*ii*) **Impedance with phase angle (polar notation)**

$$V = \frac{V_{max}}{\sqrt{2}} = \frac{141.4}{\sqrt{2}} = 100\ V;$$

$$I = \frac{I_{max}}{\sqrt{2}} = \frac{7.07}{\sqrt{2}} = 5A.$$

$\therefore \quad V = 100 \angle -120°$ and $I = 5 \angle -90°$

$\therefore \quad Z = \frac{V}{I} = \frac{100 \angle -120°}{5 \angle -90°} = \mathbf{20 \angle -30°\ \Omega}$ **(Ans.)**

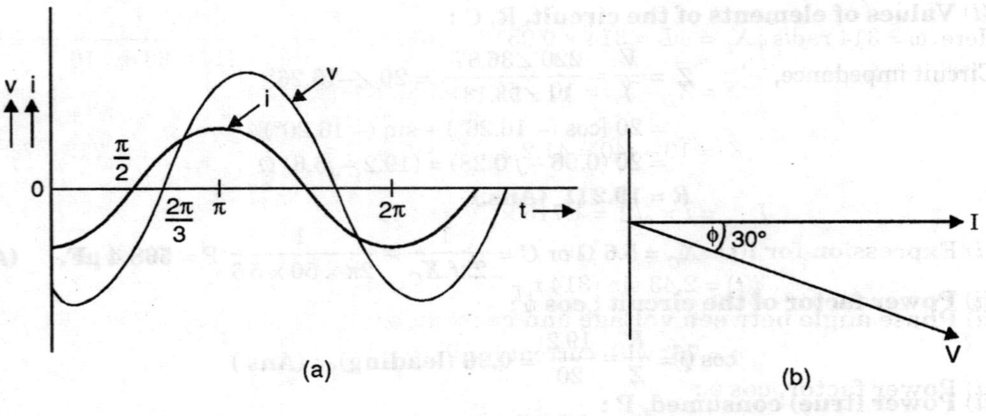

Fig. 39

(*iii*) **Average power :**

Average power = $VI \cos \phi$

$= 100 \times 5 \times \cos 30° =$ **433 W. (Ans.)**

(*iv*) **Instantaneous power at t = 0 :**

At $t = 0$; $v = 141.4 \sin (0 - 120°) = -122.45\ V$

$i = 7.07 \sin (0 - 90°) = -7.07\ A$

∴ Instantaneous power at $t = 0$,

$p = v_i = (-122.45) \times (-7.07) =$ **865.7 W. (Ans.)**

Example 28. *A voltage e(t) = 100 sin 314 t is applied to a series circuit consisting of 10 ohms resistance, 0.0318 henry inductance and a capacitor of 63.6 μF. Determine :*

(*i*) *Expression for i(t).*

(*ii*) *Phase angle between voltage and current.*

(*iii*) *Power factor.*

(*iv*) *Active power consumed.*

(*v*) *Peak value of pulsating energy.* **(Indore University)**

Soution. *Given :* $e(t) = 100 \sin 314\ t$, $R = 10\ \Omega$, $L = 0.0318$ H, $C = 63.6\ \mu F = 63.6 \times 10^{-6}\ F$.

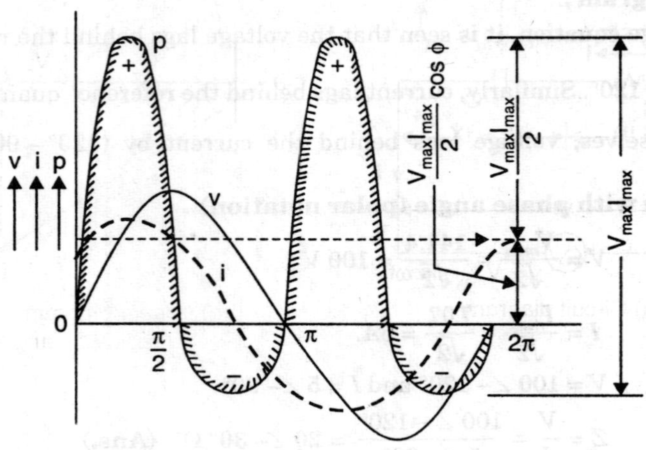

Fig. 40

Here $\omega = 314$ rad/s ; $X_L = \omega L = 314 \times 0.0318 = 10\ \Omega$; $X_C = \dfrac{1}{\omega C} = \dfrac{1}{314 \times 63.6 \times 10^{-6}} = 50\ \Omega$;

$$X = X_L - X_C = 10 - 50 = -40\ \Omega \text{ (capacitive)}$$

$$\bar{Z} = 10 - j40 = 41.2 \angle -76° \ ; \ \bar{I} = \frac{\bar{V}}{\bar{Z}} = \frac{(100/\sqrt{2}) \angle 0°}{41.2 \angle -76°} = 1.716 \angle 76°$$

$$I_{max} = I \times \sqrt{2} = 1.716 \times \sqrt{2} = 2.43 \text{ A}$$

(*i*) **Expression for i(t) :**

$$i(t) = 2.43 \sin(314\,t + 76°). \quad \textbf{(Ans.)}$$

(*ii*) **Phase angle between voltage and current, ϕ :**

ϕ = **76°** with current leading.

(*ii*) **Power factor, cos ϕ :**

$$\cos\phi = \cos 76° = \mathbf{0.24\ (lead).} \quad \textbf{(Ans.)}$$

(*iv*) **Active power consumed, P :**

$$P = VI\cos\phi = (100/\sqrt{2})(2.43/\sqrt{2}) \times 0.24 = \mathbf{29.16\ W.} \quad \textbf{(Ans.)}$$

(*iv*) **Peak value of pulsating energy :**

Refer. Fig. 40. The peak value of pulsating energy

$$= \frac{V_{max}\,I_{max}}{2} + \frac{V_{max}\,I_{max}}{2}\cos\phi$$

$$= \frac{V_{max}\,I_{max}}{2}(1 + \cos\phi) = \frac{100 \times 2.43}{2}(1 + 0.24) = \mathbf{150.66\ W.} \quad \textbf{(Ans.)}$$

4.5.2. R-C circuit (Resistance and capacitance in series)

Fig. 41. (*a*) shows a pure resistance *R* (ohms) and a pure capacitor of capacitance *C* (farads) connected in series. Such a circuit is known as *R-C* circuit.

Let, *V* = R.M.S. value of the applied voltage,

I = R.M.S. value of the resultant current,

$V_R = IR$ = Voltage drop across *R* (in phase with *I*) and

$V_C = IX_C$ = Voltage drop across *C*, lagging *I* by 90°.

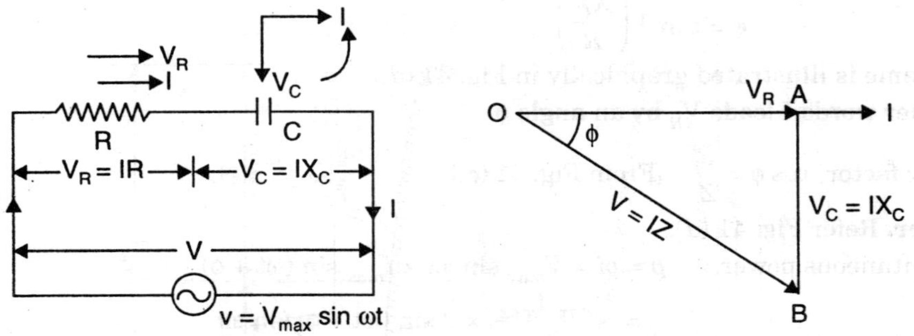

(a) Circuit diagram

(b) Phasor diagram (I leads V by angle ϕ)

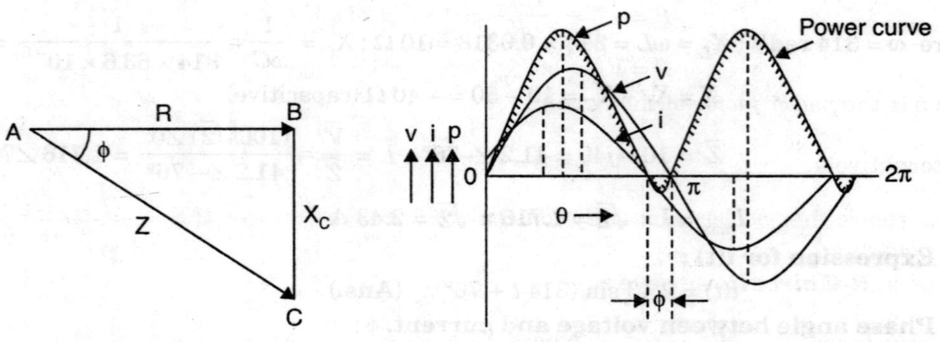

(c) Impedance triangle (d) Power curve

Fig. 41. *R-C* circuit (Resistance and capacitance in series).

Voltage drops V_R and V_L are shown in voltage triangle *OAB* in Fig. 41 (*b*) *I* being taken as the reference vector in the phasor diagram. Vector *OA* represents ohmic drop V_R and *AB* represents the capacitive drop V_C. Vector *OB* represents the applied voltage *V*, which is the vector sum of the V_R and V_C)

$$\therefore \quad V = \sqrt{V_R^2 + V_C^2} = \sqrt{(IR)^2 + (IX_C)^2} = I\sqrt{R^2 + X_C^2}$$

or
$$I = \frac{V}{\sqrt{R^2 + X_C^2}} = \frac{V}{Z}$$

where $Z = \sqrt{R^2 + X_C^2}$ (total opposition offered to the flow of alternating current by *R-C* series circuit) is known as the **impedance** of the circuit.

As seen from the *"impedance triangle" ABC* [Fig. 41 (*c*)],

$$Z^2 = R^2 + X_C^2$$

i.e., (Impedance)^2 = (Resistance)^2 + (Capacitive reactance)^2

From Fig. 41 (*b*) it is evident that *I* leads the voltage *V* by an angle φ such that,

$$\tan\phi = \frac{V_C}{V_R} = \frac{IX_C}{IR} = \frac{X_C}{R} = \frac{(1/\omega C)}{R} = \frac{\text{Capacitive reactance}}{\text{Resistance}}$$

$$\therefore \quad \phi = \tan^{-1}\left(\frac{X_C}{R}\right)$$

The same is illustrated graphically in Fig. 41 (*d*).

In other words I leads V_R by an angle φ.

Power factor, $\cos\phi = \dfrac{R}{Z}$ [From Fig. 41 (*c*)]

Power. Refer Fig. 41 (*d*),

Instantaneous power, $p = vi = V_{max} \sin \omega t \times I_{max} \sin (\omega t + \phi)$

$$= \frac{V_{max}\, I_{max}}{2} \times 2 \sin(\omega t + \phi) \sin \omega t$$

$$= \frac{V_{max}}{\sqrt{2}} \times \frac{I_{max}}{\sqrt{2}} \, [\cos\phi - \cos(2\,\omega t + \phi)]$$

Average power consumed in the circuit over a complete cycle,

$$P = \text{Average of } \frac{V_{max}}{\sqrt{2}} \cdot \frac{I_{max}}{\sqrt{2}} \cos\phi - \text{Average of } \frac{V_{max}}{\sqrt{2}} \cdot \frac{I_{max}}{\sqrt{2}} \cos(2\omega t + \phi)$$

or $$P = \frac{V_{max}}{\sqrt{2}} \cdot \frac{I_{max}}{\sqrt{2}} \cos \phi - \text{zero},$$

or $$P = V_{r.m.s.} \times I_{r.m.s.} \cos \phi = VI \cos \phi$$

where $\cos \phi$ is the *power factor* of the circuit :

Alternatively, $$P = VI \cos \phi = IZ \times I \times \frac{R}{Z} = I^2R$$

This shows that *power is actually consumed in resistance only* ; the *capacitor does not consume any power.*

Thus in ***R-C circuit***, we have :

1. Impedance, $Z = \sqrt{R^2 + X_C^2}$ $\left(\text{where } X_C = \frac{1}{\omega C} = \frac{1}{2\pi f, C \text{ being in farad}}\right)$

2. Current, $I = \frac{V}{Z}$

3. Power factor, $\cos \phi = \frac{R}{Z} \left(= \frac{\text{True power}}{\text{Apparent power}} = \frac{W}{VA}\right)$

[or angle of lead, $\phi = \cos^{-1} (R/Z)$]

4. Power consumed, $P = VI \cos \phi \; (= I^2R)$.

Example 29. *A capacitance of 20 μF and a resistance of 100 ohms are connected in series across 120 V, 60 Hz mains. Determine the average power expended in the circuit.*

Also draw the vector diagram.

Solution. $R = 100 \; \Omega$

$C = 20 \; \mu F = 20 \times 10^{-6}$ F (farad)

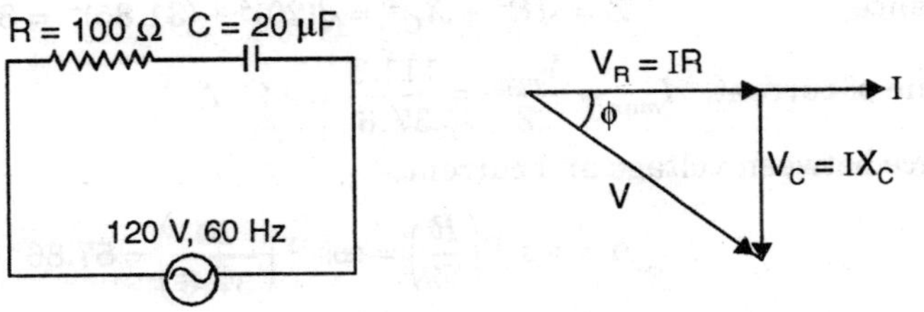

(*a*) R-C circuit (*b*) Vector/phasor diagram

Fig. 42

Capacitive reactance, $$X_C = \frac{1}{2\pi f C} = \frac{1}{2\pi \times 50 \times 20 \times 10^{-6}} = 159 \; \Omega$$

Impedance of the circuit, $$Z = \sqrt{R^2 + X_C^2} = \sqrt{100^2 + 150^2} = 188 \; \Omega$$

Current through the circuit $$I = \frac{V}{Z} = \frac{120}{188} = 0.638 \text{ A}$$

Power factor, $$\cos \phi = \frac{R}{Z} = \frac{100}{188} = 0.532$$

Average power expended in the circuit,

$$P_{av} = VI \cos \phi$$

$$= 120 \times 0.638 \times 0.532 = \mathbf{40.75 \; W.} \quad \textbf{(Ans.)}$$

Fig. 42 (*b*) shows the vector/phasor diagram.

Example 30. *A voltage $v = 100 \sin 314t - 50 \cos 314t$, is applied to a circuit having $R = 20\ \Omega$ in series with $C = 100\ \mu F$. Obtain expression for instantaneous current, r.m.s. value of current and the power in the circuit.* **(PTU, 1999)**

Solution. *Given* : $v = 100 \sin 314t - 50 \cos 314t$; $R = 20\ \Omega$; $C = 100\ \mu F$.

The R-C circuit and the phasor diagram for the given instantaneous voltage are shown in Figs. 43 and 44 respectively.

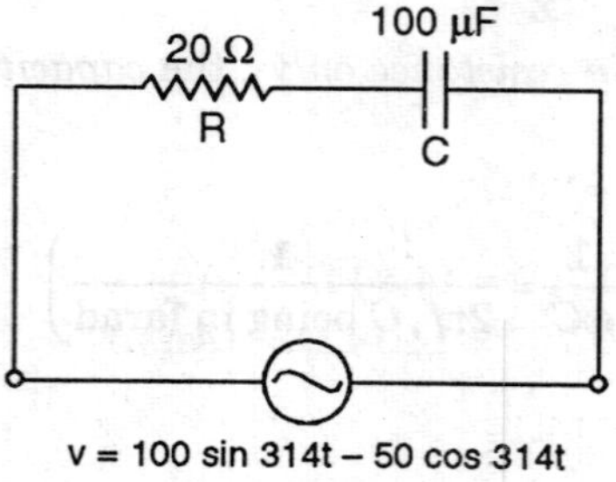

Fig. 43

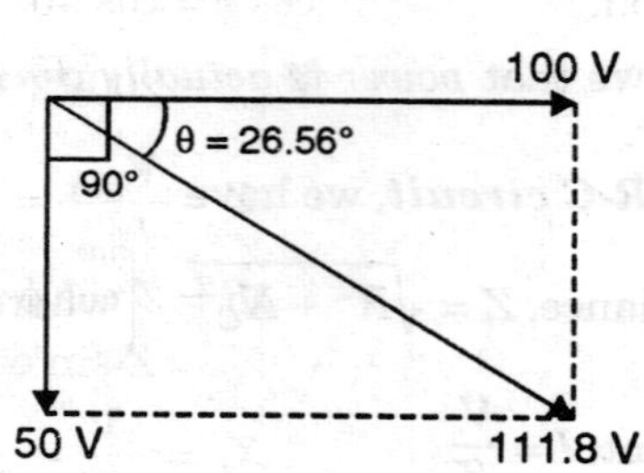

Fig. 44

Resultant voltage, $V_{max(R)} = \sqrt{(100)^2 + (50)^2} = 111.8$ V

Phase angle with the horizontal, $\theta = \tan^{-1}\left(-\dfrac{50}{100}\right) = -26.56°$

$\therefore$ $v = 111.8 \sin (314t - 26.56°)$

Now, $\omega t = 314\,t$ or $\omega = 314$

$\therefore$ Capacitive reactance, $X_C = \dfrac{1}{\omega C} = \dfrac{1}{314 \times 100 \times 10^{-6}} = 31.85\ \Omega$

Circuit impedance, $Z = \sqrt{R^2 + X_C^{\,2}} = \sqrt{(20)^2 + (31.85)^2} = 37.6\ \Omega$

Maximum value of current, $I_{max} = \dfrac{V_{max}}{Z} = \dfrac{111.8}{37.6} = 2.97$ A

Phase difference between voltage and current,

$$\phi = \cos^{-1}\left(\frac{R}{Z}\right) = \cos^{-1}\left(\frac{20}{37.6}\right) = 57.86° \text{ (leading)}$$

$\therefore$ *Instantaneous value of current*

$$i = 2.97 \sin (314t - 26.56° + 57.86°) = 2.97 \sin (314t + 31.3°)$$

or $$i = 2.97 [\sin 314t \,.\, \cos 31.3° + \cos 314t \,.\, \sin 31.3°]$$

or $$\mathbf{i = 2.54 \sin 314t + 1.54 \cos 314t.} \quad \textbf{(Ans.)}$$

RMS value of the current, $\mathbf{I} = \dfrac{I_{max}}{\sqrt{2}} = \dfrac{2.97}{\sqrt{2}} = \mathbf{2.1\ A.}$ **(Ans.)**

Power in the circuit, $P = VI \cos \phi$

$$= \frac{V_{max(R)}}{\sqrt{2}} \times I \cos \phi = \frac{111.8}{\sqrt{2}} \times 2.1 \times \cos (57.86°) = \mathbf{88.32\ W} \textbf{ (Ans.)}$$

Example 31. *A two element series circuit is connected across an A.C. source $e = 200\sqrt{2} \sin (\omega t + 20°)$ V. The current in the circuit then is found to be $i = 10\sqrt{2} \cos (314\,t - 25°)$ A. Determine parameters of the circuit.* **(Allahabad University)**

Solution. *Given* : $e = 200\sqrt{2} \sin (\omega t + 20°)$

$$i = 10\sqrt{2} \cos (314\,t - 25°)$$

Parameters of the circuit, cos ϕ, R, X_C and C :

The current i can be written as,

$$i = 10\sqrt{2}\ \sin(314\,t - 25° + 90°) = 10\sqrt{2}\ \sin(314\,t + 65°)$$

It is seen that applied voltage leads by 20° and current leads by 65° with regards to the reference quantity, their mutual difference is 65° – 20° = 45°.

Hence, p.f. $\cos\phi = \cos 45° = $ **0.707 (lead). (Ans.)**

Now, $V_{max} = 200\sqrt{2}$ and $I_{max} = 10\sqrt{2}$

$$\therefore \quad Z = \frac{V_{max}}{I_{max}} = \frac{200\sqrt{2}}{10\sqrt{2}} = 20\ \Omega$$

$\therefore$ $R = Z\cos\phi = 20 \times 0.707 =$ **14.14 Ω. (Ans.)**

$X_C = Z\sin\phi = 20 \times 0.707 =$ **14.14 Ω. (Ans.)**

Also,

$$X_C = \frac{1}{\omega C} = \frac{1}{2\pi f C} = 14.14$$

where

$$f = \frac{314}{2\pi} = 50\ \text{Hz}$$

$$\therefore \quad X_C = 14.14 = \frac{1}{2\pi \times 50 \times C}$$

or

$$C = \frac{1}{14.14 \times 2\pi \times 50}\ \text{F} = \mathbf{225.1\ \mu F.}\quad \textbf{(Ans.)}$$

Hence the *given circuit is an R-C circuit.*

4.5.3. R-L-C circuit (Resistance, inductance and capacitance in series)

Fig. 45 shows a *R-L-C* circuit.

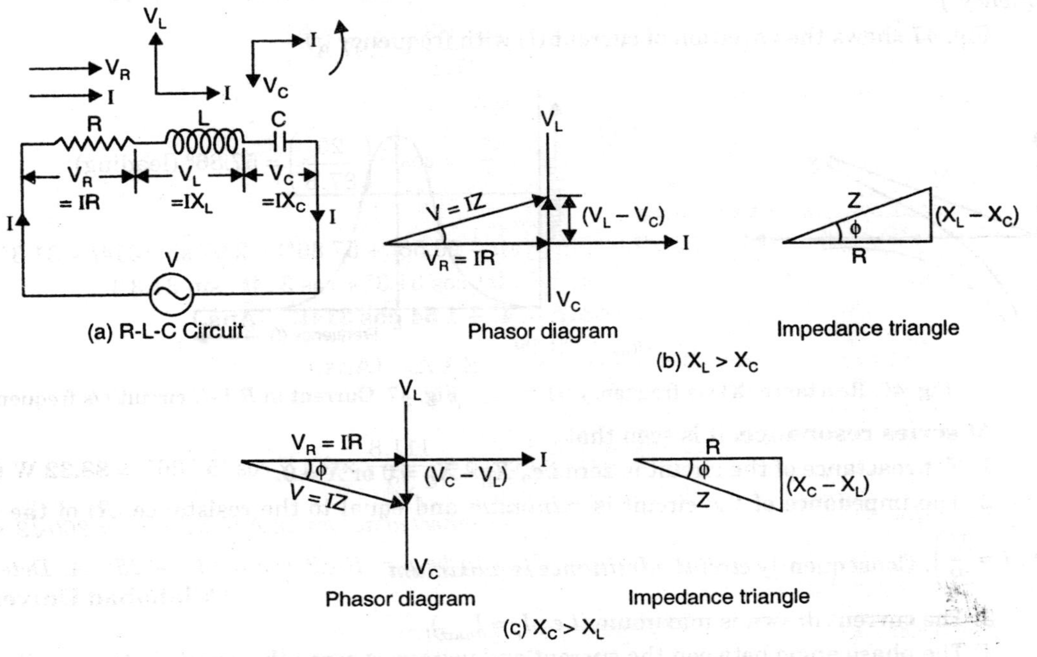

Fig. 45. Resistance, inductance and capacitance in series.

Important formulae :

1. Impedance, $Z = \sqrt{R^2 + (X_L - X_C)^2}$

$\left[\text{where } X_L = 2\pi fL, L \text{ in } henry \text{ and } X_C = \frac{1}{2\pi fC}, C \text{ in farad}\right]$

2. Current, $I = \frac{V}{Z}$

3. Power factor, $\cos\phi = \frac{R}{Z}$

$\left[\text{angle of lag (when } X_L > X_C) \text{ or lead (when } X_C > X_L), \phi = \cos^{-1}\frac{R}{Z}\right]$

4. Power consumed $= VI\cos\phi \; (= I^2R)$

Resonance in R-L-C circuits

Refer Fig. 45 (*a*).

The frequency of the voltage which gives the maximum value of the current in the circuit is called ***resonant frequency,*** and the circuit is said to be ***resonant.***

At resonance, $X_L = X_C$ (*i.e.,* $Z = R$)

i.e., $2\pi f_r L = \frac{1}{2\pi f_r C}$

$$f_r = \frac{1}{2\pi\sqrt{LC}} \quad ...(5)$$

where f_r = Resonance frequency in Hz ; L = Inductance in henry ; and C = Capacitance in farad.

Fig. 46 shows variation of X_L, X_C and X (total reactance = $X_L - X_C$) with variation of frequency f.

Fig. 47 shows the variation of current (I) with frequency (f).

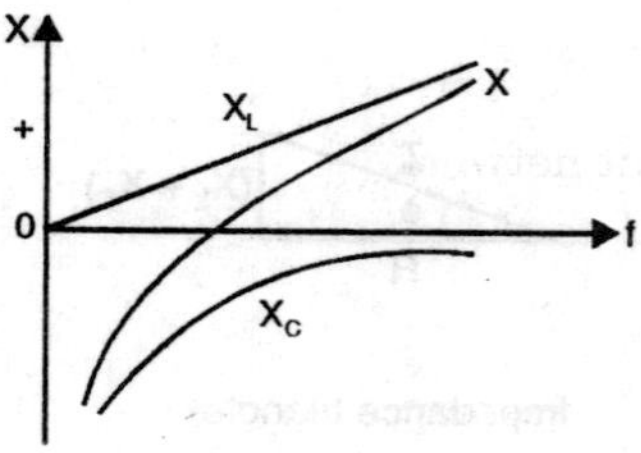

Fig. 46. Reactance (X) v/s frequency (*f*).

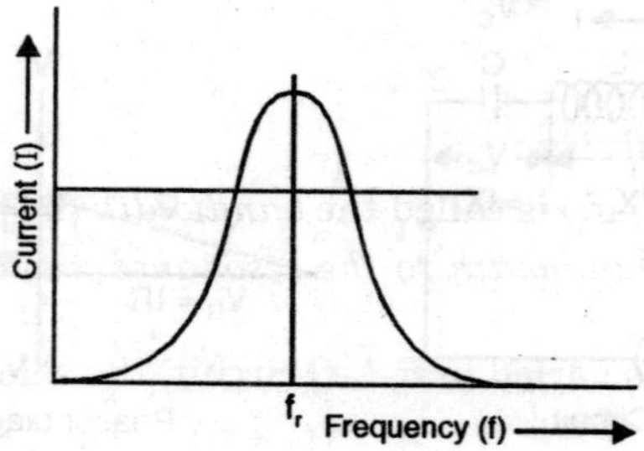

Fig. 47. Current in *R-L-C* circuit v/s frequency.

At **series resonance,** it is seen that :

1. Net reactance of the circuit is zero *i.e.,* $X_L - X_C = 0$ or $X = 0$.
2. The impedance of the circuit is *minimum* and equal to the resistance (R) of the circuit $\left(i.e., I = \frac{V}{R}\right)$. Consequently *circuit admittance is maximum.*
3. The current drawn is maximum (*i.e.,* $I_r = I_{max}$).
4. The phase angle between the current and voltage is zero ; the *power factor is unity.*

5. The resonant frequency is given by $f_r = \dfrac{1}{2\pi\sqrt{LC}}$; if the *frequency is below* the resonant frequency the net reactance in the circuit is *capacitive* and if the frequency is above the resonant frequency, the net reactance in the circuit is *inductive*.

6. Although $V_L = V_C$, yet V_{coil} is greater than V_C because of its resistance.

Half power frequencies, Bandwidth and Selectivity

Half power (cut-off) frequencies :

The half power frequencies *are those frequencies at which the power dissipation in the circuit is half of the power dissipation at resonant frequency f_r.* They are the corresponding frequencies f_1 and f_2 at the value of current $I = I_r/\sqrt{2}$; where I_r is the current at resonance in *R-L-C* series circuit (Refer Fig. 48).

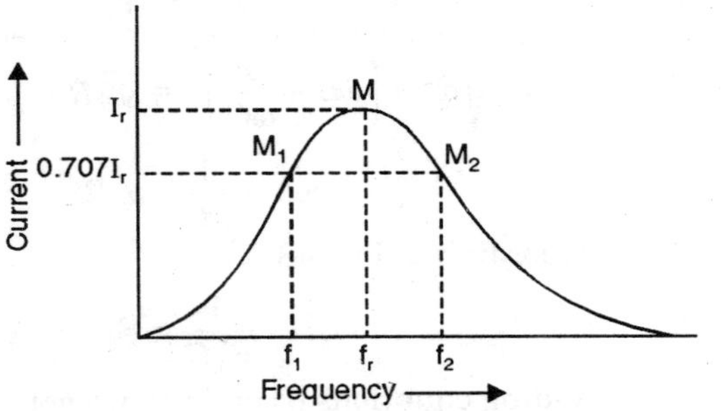

Fig. 48

Hence power, P_r drawn by the circuit at the *resonance* is

$$P_r = I_r^2 R \qquad ...(6)$$

Power in the circuit at $f_1 = \left(\dfrac{I_r}{\sqrt{2}}\right)^2 R = \dfrac{1}{2} I_r^2 R$ (= half the power at resonance)

Power in the circuit at $f_2 = \left(\dfrac{I_r}{\sqrt{2}}\right)^2 R = \dfrac{1}{2} I_r^2 R$ (= half the power at resonance)

Also,
$$f_1 = f_r - \frac{R}{4\pi L}$$
$$f_2 = f_r + \frac{R}{4\pi L}$$
$$f_2 - f_1 = \frac{R}{2\pi L}$$

Bandwidth and Selectivity :

The difference $(f_2 - f_1)$ is called the *bandwidth* (B_{hp}) of the resonant network.

The ratio of the bandwidth to the resonance frequency is defined as the ***selectivity*** of the circuit.

When frequency is varied in *R-L-C* circuit, the selectivity becomes

$$\frac{^*(f_2 - f_1)}{f_r} = \frac{1}{Q} \qquad ...(7)$$

where Q_r is the *quality factor* of the resonant circuit.

***Relation between bandwidth and quality factor in series resonant conditions :**

A series *R-L-C* circuit is considered. The resonant frequency and angular frequency are expressed by f_r and ω_r respectively. In the above circuit, the current (I) can be described as follows :

$$I = \frac{V}{\sqrt{R^2 + \left(\omega L - \dfrac{1}{\omega C}\right)^2}}$$

where *V, R, L* and *C* are the source voltage, resistance, inductance and capacitance of the circuit respectively.

The current, at a power, half of the maximum power developed at resonant frequency, is $\frac{1}{\sqrt{2}} I_r$, where I_r is the series resonant current *i.e.* $\frac{V}{R}$.

According to the definition of bandwidth,

$$\frac{1}{\sqrt{2}}\frac{V}{R} = \frac{V}{\sqrt{R^2 + \left(\omega L - \frac{1}{\omega C}\right)^2}}$$

or $$\sqrt{R^2 + \left(\omega L - \frac{1}{\omega C}\right)^2} = \sqrt{2}R \quad \text{or} \quad R^2 = \left(\omega L - \frac{1}{\omega C}\right)^2$$

or $$\omega L - \frac{1}{\omega C} = \pm R.$$

According to Fig. 48

$$\omega_2 L - \frac{1}{\omega_2 C} = R \quad ...(i) \qquad \omega_1 L - \frac{1}{\omega_1 C} = -R \quad ...(ii)$$

Adding equations (*i*) and (*ii*), we get

$$(\omega_1 + \omega_2)L - \frac{1}{C}\left(\frac{1}{\omega_1} + \frac{1}{\omega_2}\right) = 0$$

or $$(\omega_1 + \omega_2)L - \frac{1}{C}\left(\frac{\omega_1 + \omega_2}{\omega_1\omega_2}\right) = 0$$

Since $$\omega_1 + \omega_2 \neq 0,\ \omega_1\omega_2 = \frac{1}{LC}.$$

Again $\omega_r^2 = \frac{1}{LC}$, [$\omega_r$ is the angular frequency at resonant condition.]

Subtracting equations (*i*) and (*ii*), we have

$$L(\omega_2 - \omega_1) + \frac{1}{C}\left(\frac{\omega_2 - \omega_1}{\omega_2\omega_1}\right) = 2R$$

or $$(\omega_2 - \omega_1) + \frac{1}{LC}\frac{(\omega_2 - \omega_1)}{\omega_2\omega_1} = \frac{2R}{L}$$

or $$(\omega_2 - \omega_1) + \omega_1\omega_2 \cdot \frac{(\omega_2 - \omega_1)}{\omega_1\omega_2} = \frac{2R}{L} \qquad \left[\because \omega_1\omega_2 = \frac{1}{LC}\right]$$

or $$(\omega_2 - \omega_1) = \frac{R}{L}$$

or $$\frac{\omega_2 - \omega_1}{\omega_0} = \frac{R}{\omega_0 L}$$

[ω_0 is the angular frequency at resonant condition]

or $$\omega_2 - \omega_1 = \omega_0 \times \frac{1}{\frac{\omega_0 L}{R}}$$

or $$\omega_2 - \omega_1 = \frac{\omega_0}{Q} \qquad \left[\because Q = \frac{\omega_1 L}{R}\right]$$

or $$f_2 - f_1 = \frac{f_r}{Q} \qquad [\because \quad \omega = 2\pi f]$$

or *Bandwidth* at series resonant condition $= \dfrac{f_r}{Q}$.

$$\left[Q = \frac{\omega_r}{\text{bandwidth}} = \frac{\omega_r}{\omega_2 - \omega_1} = \frac{\omega_r}{R/L} = \frac{\omega_r L}{R} = \frac{L}{R\sqrt{LC}} = \frac{1}{R}\sqrt{\frac{L}{C}} \right]$$

Q-factor of a resonant series circuit :

The Q-factor of an R-L-C series circuit can be defined in the following different ways :

(*i*) ***Q-factor*** is *defined as the voltage magnification in the circuit at the time of resonance.*

Since at resonance current is maximum *i.e.,* $I_r = \dfrac{V}{R}$, the voltage across either coil or capacitor $= I_r X_{Lr}$ or $I_r X_{Cr}$ and supply voltage, $V = I_r R$.

$$\therefore \quad \text{Voltage magnification} = \frac{V_{Lr}}{V} = \frac{I_r X_{Lr}}{I_r R} = \frac{X_{Lr}}{R} = \frac{\omega_r L}{R} = \frac{\text{Reactance}}{\text{Resistance}}$$

or $$\frac{V_{Cr}}{V} = \frac{I_r \times C_r}{I_r R} = \frac{X_{Cr}}{R} = \frac{\text{Reactance}}{\text{Resistance}} = \frac{1}{\omega_r CR}$$

$$\therefore \quad Q\text{-factor} = \frac{\omega_r L}{R} = \frac{2\pi f_r L}{R} = \tan\theta \qquad ...[8(a)]$$

where θ is the circuit *power factor angle of the coil.*

(At resonance, circuit phase angle $\theta = 0$, and $Q = \tan\theta = 0$)

(*ii*) The ***Q-factor*** may also be *defined* as under :

$$Q\text{-factor} = 2\pi \frac{\text{maximum stored energy}}{\text{energy dissipated per cycle}} \qquad \text{...... in the circuit}$$

$$= 2\pi \frac{\frac{1}{2} L I_r^2}{I^2 R T_r} = 2\pi \frac{\frac{1}{2} L(\sqrt{2}I)^2}{I^2 R(1/f_r)} = \frac{I^2 \times 2\pi f_r L}{I^2 R} = \frac{\omega_r L}{R} \left(= \frac{1}{\omega_r CR} \right) \qquad \left(\because \quad T_r = \frac{1}{f_r} \right)$$

But resonant frequency, $f_r = \dfrac{1}{2\pi\sqrt{LC}}$ or $2\pi f_r = \dfrac{1}{\sqrt{LC}}$

Putting this value in eqn. 8(*a*), we get

$$Q\text{-factor} = \frac{2\pi f_r L}{R} = \frac{L}{R} \times \frac{1}{\sqrt{LC}} = \frac{1}{R}\sqrt{\frac{L}{C}} \qquad ...[8(b)]$$

In series resonance, higher quality factor *i.e.,* Q-factor means higher voltage magnification as well as higher selectivity of the tuning coil.

Example 32. *A resistance 12 Ω, an inductance of 0.15 H and a capacitance of 100 μF are connected in series across a 100 V, 50 Hz supply. Calculate :*

(*i*) *The current.*

(*ii*) *The phase difference between current and the supply voltage.*

(*iii*) *Power consumed.*

Draw the vector diagram of supply voltage and the line current.

Solution. *Given :* $R = 12\ \Omega$, $L = 0.15$ H or $X_L = 2\pi fL$

$$= 2\pi \times 50 \times 0.15 = 47.1\ \Omega$$

$$C = 100\ \mu F = 100 \times 10^{-6} F$$

or $$X_C = \frac{1}{2\pi fC} = \frac{1}{2\pi \times 50 \times 100 \times 10^{-6}} = 31.8\ \Omega$$

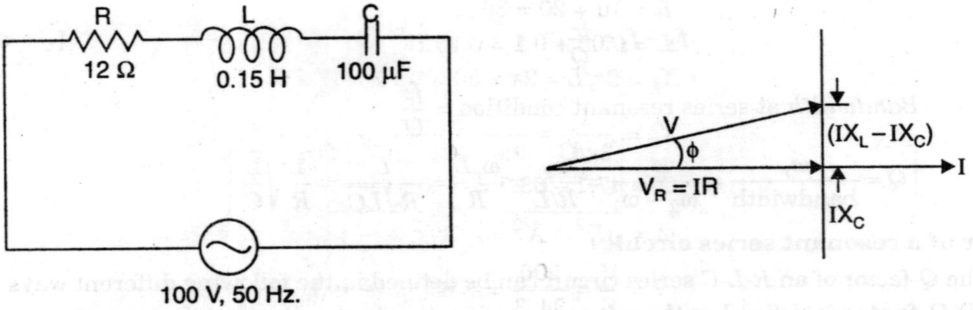

(*a*) R-L-C circuit (*b*) Vector/phasor diagram

Fig. 49

(*i*) **The current, I :** $Z = \sqrt{R^2 + (X_L - X_C)^2}$

$$= \sqrt{12^2 + (47.1 - 31.8)^2} = 19.43\ \Omega$$

Current, $I = \dfrac{V}{Z} = \dfrac{100}{19.43} = \mathbf{5.15\ A.\ (Ans.)}$

(*ii*) **Phase difference, ϕ :**

$$\phi = \cos^{-1} \frac{R}{Z} \left[\text{or } \tan^{-1} \frac{X_L - X_C}{R} \right]$$

$$= \cos^{-1} \frac{12}{19.43} \left[\text{or } \tan^{-1} \frac{15.3}{12.0} \right] = 52° \text{ (lag)}$$

Hence current lags **supply voltage by 52°. (Ans.)**

(*iii*) **Power consumed, P :**

$$P = VI \cos \phi$$

$$= 100 \times 5.15 \times \cos 52° = \mathbf{371.1\ W.}\quad \textbf{(Ans.)}$$

Fig. 49 (*a*), (*b*) show the circuit and vector/phasor diagrams respectively.

Example 33. *For the circuit shown in Fig. 50 find the values of (i) current I, (ii) V_1 and V_2 and (iii) p.f.*

Draw the vector diagram. **(Bangalore University)**

Solution. Refer Fig. 50

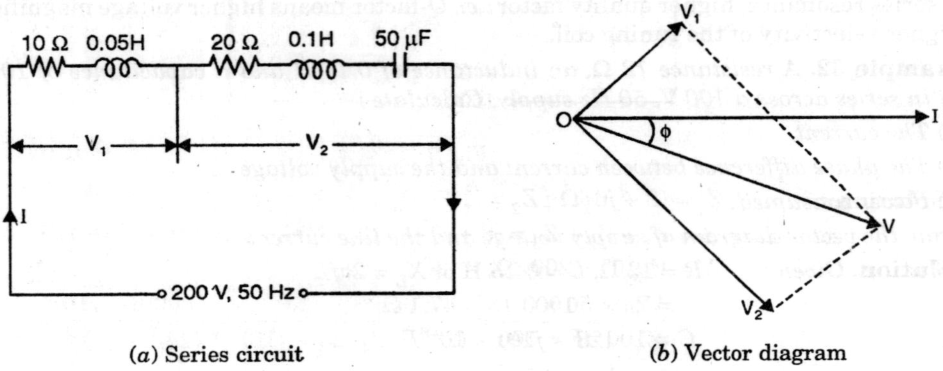

(*a*) Series circuit (*b*) Vector diagram

Fig. 50

$$R = 10 + 20 = 30\ \Omega$$
$$L = 0.05 + 0.1 = 0.15\ \text{H}$$
$$\therefore \quad X_L = 2\pi f L = 2\pi \times 50 \times 0.15 \doteq 47.1\ \Omega$$
$$X_C = \frac{1}{2\pi f C} = \frac{1}{2\pi \times 50 \times 10^{-6} \times 50} = 63.7\ \Omega$$
$$X = 47.1 - 63.7 = -16.6\ \Omega$$
$$Z = \sqrt{R^2 + X^2} = \sqrt{(30)^2 + (-16.6)^2} = 34.3\ \Omega$$

(*i*) $$I = \frac{V}{Z} = \frac{200}{34.3} = \mathbf{5.83\ A.\ (Ans.)}$$

(*ii*) $$X_{L1} = 2\pi \times 50 \times 0.05 = 15.7\ \Omega$$
$$Z_1 = \sqrt{10^2 + 15.7^2} = 18.6\ \Omega$$
$$\therefore \quad V_1 = IZ_1 = 5.83 \times 18.6 = \mathbf{108.4\ V.\ (Ans.)}$$
$$\phi_1 = \cos^{-1}(10/18.6) = 57.5° \text{ (lag)}$$
$$X_{L2} = 2\pi \times 50 \times 0.1 = 31.4\ \Omega$$
$$X = 31.4 - 63.7 = -32.3\ \Omega$$
$$Z_2 = \sqrt{20^2 + (-32.3)^2} = 38\ \Omega$$
$$\therefore \quad V_2 = IZ_2 = 5.83 \times 38 = \mathbf{221.5\ V.\ (Ans.)}$$
$$\phi_2 = \cos^{-1}(20/38) = 58.2° \text{ (lead)}$$

(*iii*) Combined p.f. $$= \cos\phi = \frac{R}{Z} = \frac{30}{34.3} = \mathbf{0.875\ (lead)}$$

Vector diagram is shown in Fig. 50 (*b*).

Example 34. *For the circuit shown in Fig. 51. Calculate :*

(*i*) *Current ;* (*ii*) *Voltage drops* V_1, V_2 *and* V_3 *;*

(*iii*) *Power absorbed by each importance ;* (*iv*) *Total power absorbed by the circuit.*

Take voltage vector along the reference axis.

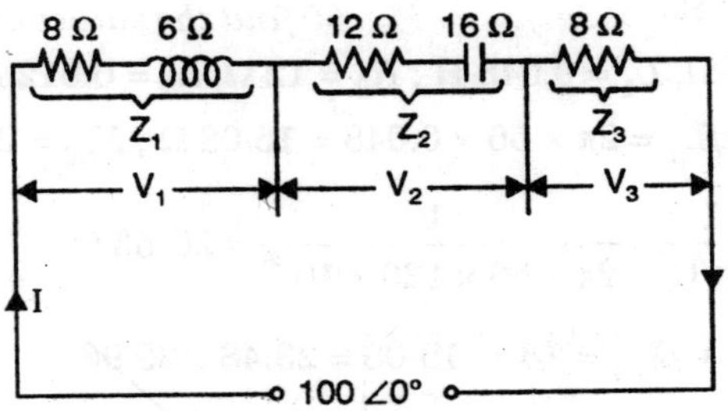

Fig. 51

(Pune University)

Solution. $Z_1 = (8 + j6)\ \Omega$; $Z_2 = (12 - j16)\ \Omega$; $Z_3 = (8 + j0)$

$$Z = Z_1 + Z_2 + Z_3 = (8 + j6) + (12 - j16) + (8 + j0) = (28 - j10)\ \Omega$$

Taking $V = V\angle 0° = 200\angle 0° = (200 + j0)$

$$\therefore \quad I = \frac{V}{Z} = \frac{100}{(28 - j10)} = \frac{200(28 + j10)}{(28 - j10)(28 + j10)} = \frac{100(28 + j10)}{(28)^2 + (10)^2}$$
$$= \frac{200(28 + j10)}{884} = 3.17 + j1.13$$

(*i*) Magnitude of current $= \sqrt{(3.17)^2 + (1.13)^2}$ = **3.36 A. (Ans.)**

(*ii*) $V_1 = IZ_1 = (3.17 + j1.13)(8 + j6)$

$= 3.17 \times 8 + 3.17 \times j6 + 8 \times j1.13 + j1.13 \times j6$

$= 25.36 + j19.02 + j9.04 - 6.78 =$ **18.58 + j28.06. (Ans.)**

$V_2 = IZ_2 = (3.17 + j1.13)(12 - j16)$

$= 38.04 - j50.72 + j13.56 + 18.08 =$ **56.12 – j37.16. (Ans.)**

$V_3 = IZ_3 = (3.17 + j1.13)(8 + j0) =$ **25.36 + j9.04. (Ans.)**

$[V = V_1 + V_2 + V_3 = (18.58 + j28.06) + (56.12 - j37.16) + (25.36 + j9.04) = 100 + j0$ (check)]

Example 35. *Fig. 52 shows a circuit connected to a 230 V, 50 Hz supply. Determine the following :*

(*i*) *Current drawn* (*ii*) *Voltages V_1 and V_2*

(*iii*) *Power factor.*

Draw also the phasor diagram.

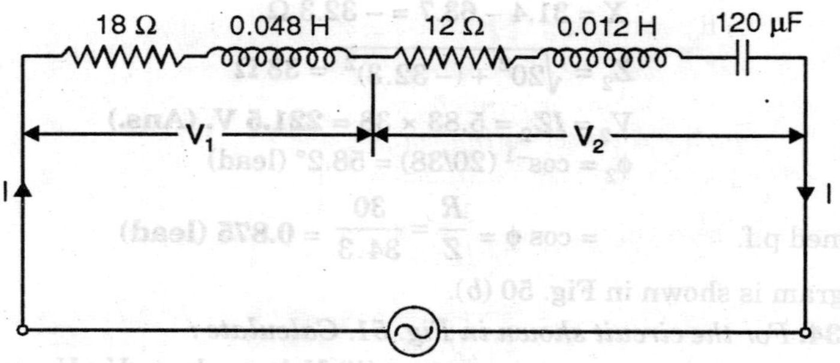

Fig. 52 **(Pune University)**

Solution. Refer Fig. 52.

Given : $R_1 = 18\ \Omega, L_1 = 0.048$ H ; $R_2 = 12\ \Omega, L_2 = 0.012\ H$; $C = 120\ \mu F = 120 \times 10^{-6}$ F.

$\therefore$ $X_{L1} = 2\pi f L_1 = 2\pi \times 50 \times 0.048 = 15.08\ \Omega$; $X_{L2} = 2\pi f L_2 = 2\pi \times 50 \times 0.012 = 3.77\ \Omega$

$$X_{C2} = \frac{1}{2\pi f C} = \frac{1}{2\pi \times 50 \times 120 \times 10^{-6}} = 26.53\ \Omega$$

Impedance, $\overline{Z}_1 = R_1 + jX_{L1} = 18 + j15.08 = 23.48 \angle 39.96°$

Impedance, $\overline{Z}_2 = R_2 + jX_{L2} - jX_C = 12 + j3.77 - j26.53 = 12 - j22.76 = 25.73 \angle -62.2°$

Total impedance, $\overline{Z} = \overline{Z}_1 + \overline{Z}_2 = (18 + j15.08) + (12 - j22.76) = 30 - j7.68 = 30.97 \angle -14.36°$

Current drawn, I :

Taking supply voltage as reference vector, $\overline{V} = V\angle 0° = 230\angle 0°$.

Current, $\overline{I} = \frac{\overline{V}}{\overline{Z}} = \frac{230\angle 0°}{30.97\angle -14.36°} =$ **7.43 ∠14.36°A. (Ans.)**

Voltages V_1 and V_2 :

Voltage, $\overline{V_1} = \overline{I}\overline{Z_1} = 7.43 \angle 14.36° \times 23.48 \angle 39.96°$

$= 7.43 \times 23.48 \angle(14.36° + 39.96°)$

$= \mathbf{174.46 \angle 54.32° \ V.}$ **(Ans.)**

Voltage $\overline{V_2} = \overline{I}\overline{Z_2} = 7.43 \angle 14.36° \times 25.73 \angle -62.2°$

$= 7.43 \times 25.73 \angle(14.36° - 62.2°)$

$= \mathbf{191.2 \angle -47.84°.}$ **(Ans.)**

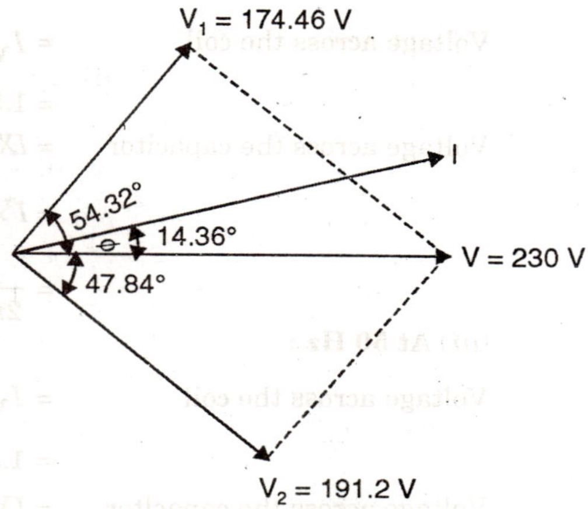

Fig. 53

Phase angle between supply voltage and current *i.e.*, V and I, $\phi = 14.36°$ (lead)

Power factor, cos φ :

$\cos \phi = \cos (14.36°)$

$= \mathbf{0.9687}$ **(leading). (Ans.)**

$[\overline{V} = \overline{V_1} + \overline{V_2} = 174.46 \angle 54.32° + 191.2 \angle -47.84° = 230 \angle 0°$ (check)]

Resonance, Q-Factor and Bandwidth

Example 36. *A circuit consisting of a coil having an inductance of 0.25 H and a resistance of 3 Ω is arranged in series with a capacitor of capacitance 20 μF. Calculate at what frequency resonance will take place and current flow if an alternating voltage of 40 V at the resonant frequency is applied to the circuit. Find also the voltage across the capacitor.*

Solution. Resonant frequency, $f_r = \dfrac{1}{2\pi\sqrt{LC}}$ [L in henry, C in farad]

$$= \frac{1}{2\pi\sqrt{0.25 \times 20 \times 10^{-6}}} = \mathbf{71.2\ Hz.\ (Ans.)}$$

At the resonant frequency, I_r (or I_{max}) $= \dfrac{V}{R} = \dfrac{40}{3} = \mathbf{13.33\ A.\ (Ans.)}$

Voltage across the capacitor, $V_c = IX_c$

$$= \frac{13.33}{2\pi fC} \text{ (where } C \text{ is in farad)}$$

$$= \frac{13.33}{2\pi \times 71.2 \times 20 \times 10^{-6}} = \mathbf{1489.8\ V.\ (Ans.)}$$

Example 37. *A coil of inductance 0.64 H and resistance 40 Ω is connected in series with a capacitor of capacitance 12 μF.*

Estimate :

(i) The frequency at which resonance will occur.

(ii) The voltage across the coil and capacitor, respectively and also the supply voltage when a current of 1.5 A at the resonant frequency is flowing.

(iii) The three voltages in (ii) with a current of 1.5 A flowing at a frequency of 50 Hz.

Solution. (*i*) $f_r = \dfrac{1}{2\pi\sqrt{LC}}$ [L is in henry, C is in farad]

$$= \frac{1}{2\pi\sqrt{0.64 \times 12 \times 10^{-6}}} = \mathbf{57.4\ Hz.\ (Ans.)}$$

(*ii*) At resonance the supply voltage = IR

$= 1.5 \times 40 = \mathbf{60\ V.\ (Ans.)}$

Voltage across the coil $= I\sqrt{R^2 + X_L^2}$

$= 1.5\sqrt{40^2 + (2\pi \times 57.4 \times 0.64)^2}$ = **351.4 V. (Ans.)**

Voltage across the capacitor $= IX_C$

$= I \times \dfrac{1}{2\pi fC}$ [C in μF]

$= \dfrac{1.5}{2\pi \times 57.4 \times 12 \times 10^{-6}}$ = **346.6 V. (Ans.)**

(*iii*) **At 50 Hz :**

Voltage across the coil $= I\sqrt{R^2 + X_L^2}$

$= 1.5\sqrt{40^2 + (2\pi \times 50 \times 0.64)^2}$ = **307.5 V. (Ans.)**

Voltage across the capacitor $= IX_C$

$= \dfrac{1.5}{2\pi \times 50 \times 12 \times 10^{-6}} \simeq$ **398 V. (Ans.)**

Voltage across the entire circuit,

$$V = I\sqrt{R^2 + (X_L - X_C)^2}$$

$$= 1.5\sqrt{(40)^2 + \left\{(2\pi \times 50 \times 0.64) - \left(\frac{1}{2\pi \times 50 \times 12 \times 10^{-6}}\right)\right\}^2}$$

$$= 1.5\sqrt{(40)^2 + (-64.2)^2} = \textbf{113.5 V. (Ans.)}$$

Example 38. *A coil having an inductance of 50 mH and resistance 10 Ω is connected in series with a 25 μF capacitor across a 200 V A.C. supply. Calculate :*

(*i*) *Resonance frequency of the circuit ;*

(*ii*) *Current flowing at resonance ;*

(*iii*) *Value of Q by using different data.* **(Bombay University)**

Solution. (*i*) $f_r = \dfrac{1}{2\pi\sqrt{LC}} = \dfrac{1}{2\pi\sqrt{50 \times 10^{-3} \times 25 \times 10^{-6}}}$ = **142.3 Hz. (Ans.)**

(*ii*) $I_{max} = \dfrac{V}{R} = \dfrac{200}{10}$ = **20 A. (Ans.)**

(*iii*) $Q = \dfrac{1}{R}\sqrt{\dfrac{L}{C}} = \dfrac{1}{10}\sqrt{\dfrac{50 \times 10^{-3}}{25 \times 10^{-6}}}$ = **4.47. (Ans.)**

Example 39. *A series R-L-C series circuit consists of R = 800 Ω, L = 80 mH and C = 8 pico-farad. The applied voltage across the circuit is 100 V. Determine :*

(*i*) *Resonant frequency of the circuit.*

(*ii*) *Q-factor of the circuit at the resonant frequency.*

(*iii*) *Bandwidth of the resonant circuit.*

(*iv*) *Frequencies at which the half power points occur.*

(*v*) *Bandwidth of the circuit.* **(Delhi University)**

Solution. *Given :* $R = 800\ \Omega$; $L = 80$ mH $= 0.08$ H ; $C = 8 \times 10^{-12}$ F ; $V = 100$ volts.

(*i*) **Resonant frequency of the circuit, f_r :**

$$f_r = \frac{1}{2\pi\sqrt{LC}} = \frac{1}{2\pi\sqrt{0.08 \times 8 \times 10^{-12}}} = \textbf{198.94 kHz.} \quad \textbf{(Ans.)}$$

(*ii*) **Q-factor at the resonant frequency :**

$$(Q\text{-factor})_{\text{resonant frequency}} = \frac{1}{R}\sqrt{\frac{L}{C}} = \frac{1}{800}\sqrt{\frac{0.08}{8\times10^{-12}}} = \mathbf{125.}\quad \textbf{(Ans.)}$$

(*iii*) **Bandwidth of the resonant circuit ; B_{hp} ;**

$$B_{hp} = \frac{R}{2\pi L} = \frac{800}{2\pi\times0.08} = \mathbf{1591.5\ Hz.}\quad \textbf{(Ans.)}$$

Also,
$$B_{hp} = \frac{f_r}{Q} = \frac{198.94\ \text{kHz}}{125} = 1.5915\ \text{kHz} = 1591.5\ \text{Hz} \qquad \text{...as above}$$

(*iv*) **Frequencies at which the half power points occur, ω_1, ω_2 :**

$$f_1 = f_r - \frac{R}{4\pi L} = 198.94\ (\text{kHz}) - \frac{800}{4\pi\times0.08}\times\frac{1}{1000}\ (\text{kHz}) = \mathbf{198.14\ kHz.}\quad \textbf{(Ans.)}$$

$$f_2 = f_r + \frac{R}{4\pi L} = 198.94 + \frac{800}{4\pi\times0.08}\times\frac{1}{1000} = \mathbf{199.74\ kHz.}\quad \textbf{(Ans.)}$$

(*v*) **Bandwidth of the resonant circuit :**

Bandwidth $= f_2 - f_1 = 199.74 - 198.14 =$ **1.6 kHz. (Ans.)**

Example 40. *A series R-L-C circuit consists of R = 20 Ω, L = 20 mH and C = 0.5 μF. If the circuit is connected to a 20 V variable frequency supply calculate the following :*

(*i*) *Resonant frequency f_r.*

(*ii*) *Resonance circuit Q-factor using L/C ratio.*

(*iii*) *Half-power bandwidth, using f_r and Q-factor.*

(*iv*) *Half-power bandwidth using the general formula for any bandwidth.*

(*v*) *Half-power bandwidth using the given component values.*

(*vi*) *Maximum power dissipated at f_r.* **(Delhi University)**

Solution. *Given :* R = 20 Ω ; L = 20 mH = 0.02 H ; C = 0.5 × 10⁻⁶ F ; V = 20 volts.

(*i*) **Resonant frequency,**
$$\mathbf{f_r} = \frac{1}{2\pi\sqrt{LC}} = \frac{1}{2\pi\sqrt{0.02\times0.5\times10^{-6}}} = \mathbf{1591\ Hz.}\quad \textbf{(Ans.)}$$

(*ii*) **Q-factor**
$$= \frac{1}{R}\sqrt{\frac{L}{C}} = \frac{1}{20}\sqrt{\frac{0.02}{0.5\times10^{-6}}} = \mathbf{10.}\quad \textbf{(Ans.)}$$

(*iii*) **Half-power bandwidth** $\left(\textit{using } \frac{L}{C} \textit{ ratio}\right)$, $\mathbf{B_{hp}} = \frac{f_r}{Q\text{-factor}} = \frac{1591}{10} = \mathbf{159.1\ Hz.}$ **(Ans.)**

(*iv*) **Half-power bandwidth** *(using the general formula),*

$$B_{hp} = \frac{f_r Q}{(Q)_{\text{resonance}}} = \frac{1591\times\tan 45^\circ}{10} = \mathbf{159.1\ Hz.}\quad \textbf{(Ans.)}$$

($\because$ At half power points, $Q = \tan\theta = \tan 45^\circ = 1$)

(*v*) **Half-power bandwidth** *(using component values),*

$$B_{hp} = \frac{R}{2\pi L} = \frac{20}{2\pi\times0.02} = \mathbf{1591.5.}\quad \textbf{(Ans.)}$$

(*vi*) **Maximum power dissipated at $\mathbf{f_r}$,**

$$P_r = I_r^2 R = \left(\frac{V}{R}\right)^2 R = \frac{V^2}{R} = \frac{(20)^2}{20} = \mathbf{20\ W.}\quad \textbf{(Ans.)}$$

4.6. A.C. Parallel Circuits

4.6.1. Introduction

Now-a-days, owing to multiple system of transmission and distribution, we come across parallel circuits (*i.e.*, impedances joined in parallel) more often. Practically all lighting and power circuits are constant voltage circuits with the loads connected in parallel. In a parallel A.C. circuit (like parallel D.C. circuit) the *voltage is the same across each branch.*

4.6.2. Methods for solving A.C. parallel circuits

The following *three* methods are available to solve such circuits :

1. Phasor or vector method
2. Admittance method
3. Vector algebra (symbolic method or *j*-method)

1. *Vector or phasor method :*

Consider a parallel circuit consisting of two branches of impedances $Z_1(R_1, L)$ and $Z_2(R_2, C)$ respectively, and connected in parallel across an alternating voltage V volts (r.m.s.), as shown in Fig. 54 (*a*). Since the two branches are connected in parallel therefore, the voltage across each branch is the same and equal to supply voltage V but currents through them will be different.

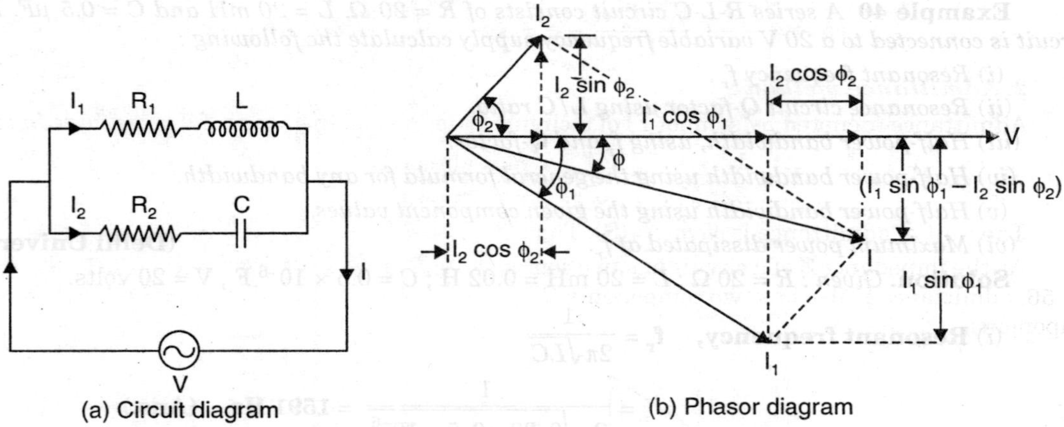

Fig. 54. Single-phase parallel circuit—Phasor method.

Branch-1 Impedance, $Z_1 = \sqrt{R_1^2 + X_L^2}$

Current, $I_1 = \dfrac{V}{Z_1}$;

Power factor, $\cos \phi_1 = \dfrac{R_1}{Z_1}$ or $\phi_1 = \cos^{-1}\left(\dfrac{R_1}{Z_1}\right)$;

Current, I_1 lags behind the applied voltage by ϕ_1.

Branch-2 Impedance, $Z_2 = \sqrt{R_2^2 + X_C^2}$;

Current, $I_2 = \dfrac{V}{Z_2}$;

Power factor, $\cos \phi_2 = \dfrac{R_2}{Z_2}$ or $\phi_2 = \cos^{-1}\left(\dfrac{R_2}{Z_2}\right)$

Current I_2 leads V by ϕ_2 [Fig. 54 (*b*)]

Resultant current I, which is phasor sum of I_1 and I_2, can be determined either by using parallelogram law of phasors, as shown in Fig. 54 (*b*) or by resolving branch currents I_1 and I_2 along X-axis and Y-axis and then determining the resultant of these components analytically.

Component of resultant current I along X-axis

= Sum of components of branch currents I_1 and I_2 along X-axis

or $$I \cos \phi = I_1 \cos \phi_2 + I_2 \cos \phi_2$$

Similarly, component of resultant current I along Y-axis

= Sum of components of branch currents I_1 and I_2 along Y-axis

or $$I \sin \phi = -I_1 \sin \phi_1 + I_2 \sin \phi_2$$

$$\therefore \quad I = \sqrt{(I_1 \cos \phi_1 + I_2 \cos \phi_2)^2 + (I_2 \sin \phi_2 - I_1 \sin \phi_1)^2} \quad ...(9)$$

and $$\tan \phi = \frac{I_2 \sin \phi_2 - I_1 \sin \phi_1}{I_1 \cos \phi_1 + I_2 \cos \phi_2} = \frac{Y\text{-component}}{X\text{-component}} \quad ...[9(a)]$$

If tan ϕ is +ve, then current I will lead the applied voltage V, if ϕ is –ve current I will lag behind the applied voltage V.

Power factor of the whole circuit is given by,

$$\cos \phi = \frac{I_1 \cos \phi_1 + I_2 \cos \phi_2}{I} = \frac{X\text{-component}}{I} \quad ...[9(b)]$$

2. *Admittance method :*

Admittance (denoted by symbol Y) of a circuit is defined as the *reciprocal of its impedance.*

$$\therefore \quad Y = \frac{1}{Z} = \frac{I}{V} \quad \text{or} \quad Y = \frac{\text{r.m.s. amperes}}{\text{r.m.s. volts}}$$

The unit of admittance is *siemens* (S). The old unit was mho (℧).

As the impedance Z of a circuit has two components R and X (See Fig. 55), similarly, shown in Fig. 56, admittance Y also has two components G (*conductance*-X-component) and B (*susceptance*-Y-component).

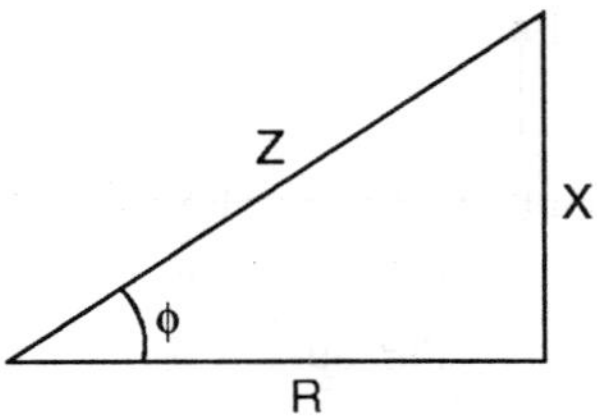

Fig. 55

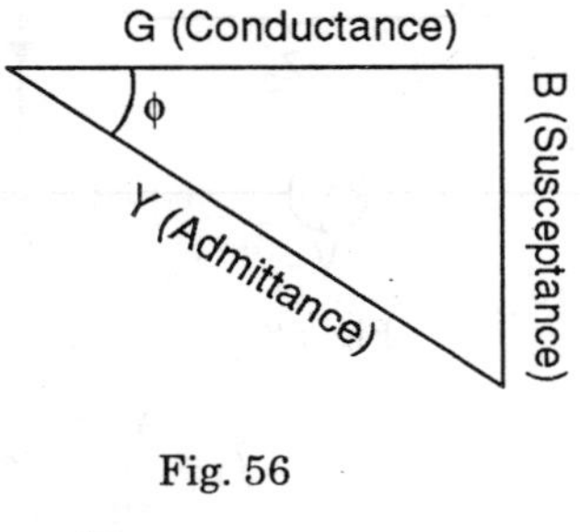

Fig. 56

Obviously, $$G = Y \cos \phi = \frac{1}{Z} \cdot \frac{R}{Z} = \frac{R}{Z^2} = \frac{R}{R^2 + X^2}$$

Similarly, $$B = Y \sin \phi = \frac{1}{Z} \cdot \frac{X}{Z} = \frac{X}{Z^2} = \frac{X}{R^2 + X^2}$$

$\therefore$ Admittance, $Y = \sqrt{G^2 + B^2}$ just as $Z = \sqrt{R^2 + X^2}$

The units of G, B and Y are in Siemens. Here, we shall consider capacitive susceptance as +ve and inductive capacitance as –ve.

Application of admittance method in solution of single-phase parallel circuits :

Refer Fig. 57. Determine conductance and susceptance of individual branches from the relations

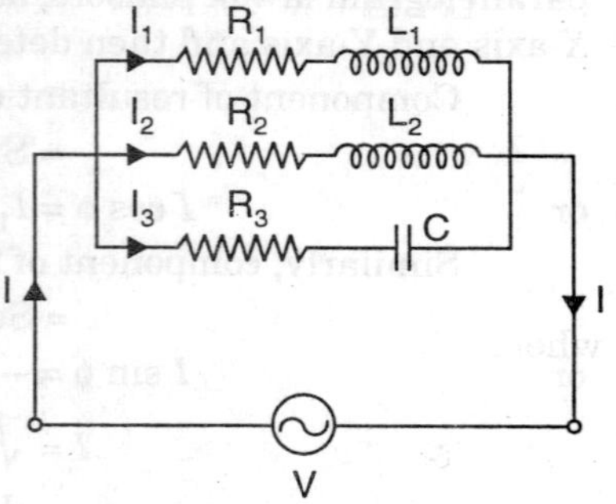

Fig. 57. Admittance method.

$$G = \frac{R}{Z^2} \text{ and } B = \frac{X}{Z^2}$$

Taking B as +ve if X is capacitive and as –ve if X is inductive. Let the conductances of the three branches of circuit shown in Fig. 57 be G_1, G_2 and G_3 respectively and susceptances be B_1, B_2 and B_3 respectively. Total conductance is found by merely adding the conductances of three branches. Similarly, total susceptance is found by algebraically adding the individual susceptances of different branches.

$\therefore$ Total conductance, $G = G_1 + G_2 + G_3$

and, total susceptance, $B = B_1 + B_2 + B_3$

$\therefore$ Total admittance $Y = \sqrt{G^2 + B^2}$ ...(10)

Total current, $I = VY$ ...(11)

Power factor, $\cos\phi = \frac{G}{Y}$ ...(12)

3. *Complex or Phasor algebra :*

Consider the parallel circuit shown in Fig. 58 in which two impedances $\overline{Z}_1$ and $\overline{Z}_2$, being in parallel, have the same potential difference across them

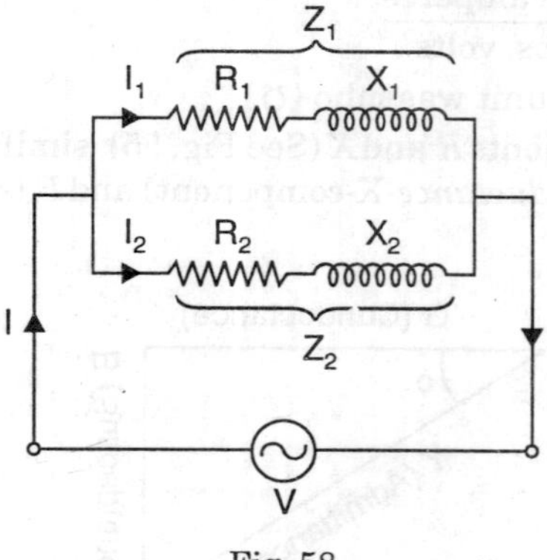

Fig. 58

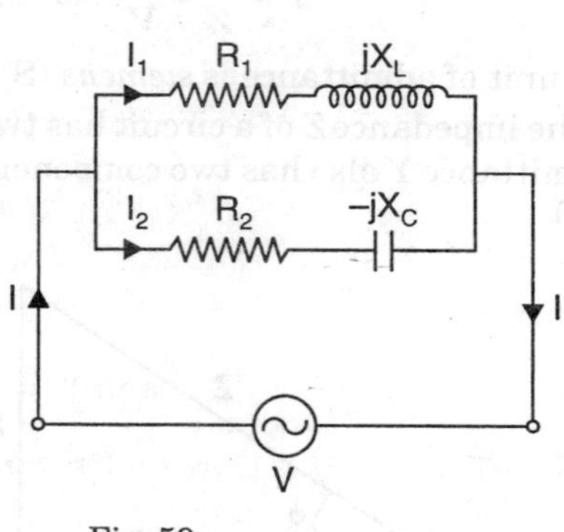

Fig. 59

Now, $$\overline{I}_1 = \frac{\overline{V}}{\overline{Z}_1} \text{ and } \overline{I}_2 = \frac{\overline{V}}{\overline{Z}_2}$$

and current $$\overline{I} = \overline{I}_1 + \overline{I}_2 = \frac{\overline{V}}{\overline{Z}_1} + \frac{\overline{V}}{\overline{Z}_2}$$

$$= \overline{V}\left(\frac{1}{\overline{Z}_1} + \frac{1}{\overline{Z}_2}\right) = \overline{V}(\overline{Y} + \overline{Y}_2) = \overline{V}\overline{Y} \quad ...(13)$$

where $\overline{Y}$ (= total admittance) = $\overline{Y}_1 + \overline{Y}_2$ ...(14)

It may be noted that the *admittances* are *added in parallel branches*, whereas *impedances are added for branches in series.*

It is most important to remember that *admittances and impedances being complex quantities must be added in complex form.*

Let us now consider the two parallel branches shown in Fig. 59, we have

$$\bar{Y}_1 = \frac{1}{\bar{Z}_1} = \frac{1}{R_1 + jX_L} = \frac{R_1 - jX_L}{(R_1 + jX_L)(R_1 - jX_L)}$$

$$= \frac{R_1 - jX_L}{R_1^2 + X_L^2} = \frac{R_1}{R_1^2 + X_L^2} - j\frac{X_L}{R_1^2 + X_L^2} = G_1 - jB_1$$

where $G_1 = \frac{R_1}{R^2 + X_L^2}$ ...Conductance of upper branch,

$B_1 = -\frac{X_L}{R_1^2 + X_L^2}$ ...Susceptance of upper branch.

Similarly, $$\bar{Y}_2 = \frac{1}{\bar{Z}_2} = \frac{1}{R_2 - jX_C} = \frac{(R_2 + jX_C)}{(R_2 - jX_C)(R_2 + jX_C)}$$

$$= \frac{R_2 + jX_C}{R_2^2 + X_C^2} = \frac{R_2}{R_2^2 + X_C^2} + j\frac{X_C}{R_2^2 + X_C^2} = G_2 + jB_2$$

Total admittance $\bar{Y} = \bar{Y}_1 + \bar{Y}_2 = (G - jB_1) + (G_2 + jB_2) = (G_1 + G_2) - j(B_1 - B_2) = G - jB$

$$\bar{Y} = \sqrt{(G_1 + G_2)^2 + (B_1 - B_2)^2} \quad ...(15)$$

and $$\phi = \tan^{-1}\left(\frac{B_1 - B_2}{G_1 + G_2}\right) \quad ...(16)$$

For admittance the *polar form* is :

$\bar{Y} = Y\angle\phi°$, where ϕ is as given above

$$Y = \sqrt{G^2 + B^2}\ \angle \tan^{-1}\left(\frac{B}{G}\right)$$

Total current $\bar{I} = \bar{V}\,\bar{Y}$; $I_1 = \bar{V}\,\bar{Y}_1$ and $I_2 = \bar{V}\,\bar{Y}_2$

If $\bar{V} = V\angle 0°$ and $\bar{Y} = Y\angle\phi$ then $\bar{I} = \bar{V}\,\bar{Y} = V\angle 0° \times Y\angle\phi = VY\angle\phi$

In general, *i.e.*, $\bar{V} = V\angle\alpha$ and $\bar{Y} = Y\angle\beta$, then

$$\bar{I} = \bar{V}\,\bar{Y} = V\angle\alpha \times Y\angle\beta = VY\angle(\alpha + \beta) \quad ...(17)$$

Thus, it is worth noting that when vector algebra is multiplied by admittance either in complex (rectangular) or polar form, the result is vector current in its proper phase relationship with respect to the voltage, irrespective of the axis to which the voltage may have been referred to.

Example 41. *A resistance of 60 Ω, an inductance of 0.18 H and a capacitance of 120 μF are connected in parallel across a 100 V, 50 Hz supply. Calculate :*

(i) Current in each path.

(ii) Resultant current.

(iii) Phase angle between the resultant current and the supply voltage.

(iv) Power factor of the circuit.

Solution. *Given* : $R = 60\ \Omega$; $L = 0.18$ H ; $C = 120\ \mu F = 120 \times 10^{-6}$ F, $V = 100$ volts, 50 Hz.

∴ Inductance reactance, $X_L = 2\pi fL = 2\pi \times 50 \times 0.18 = 56.55\ \Omega$, and

capacitance reactance, $$X_C = \frac{1}{2\pi fC} = \frac{1}{2\pi \times 50 \times 120 \times 10^{-6}} = 26.53\ \Omega$$

(*i*) **Current in each path :**

Current through reactance,

$$\mathbf{I_1} = \frac{V}{R} = \frac{100}{60} = \textbf{1.67 A in phase with voltage V.} \quad \textbf{(Ans.)}$$

Current through inductance,

$$\mathbf{I_2} = \frac{V}{X_L} = \frac{100}{56.55} = \textbf{1.77 A lagging behind voltage V by 90°.} \quad \textbf{(Ans.)}$$

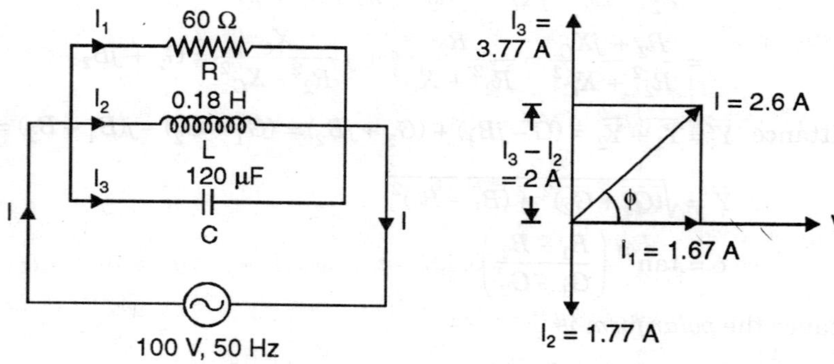

(a) Circuit diagram (b) Phasor diagram

Fig. 60

Current through capacitance,

$$\mathbf{I_3} = \frac{V}{X_C} = \frac{100}{26.53} = \textbf{3.77 A leading the voltage V by 90°.} \quad \textbf{(Ans.)}$$

The circuit and phasor diagrams are shown in Fig. 60 (*a*) and (*b*) respectively.

(*ii*) **Resultant current, I :**

Resultant current, $I = \sqrt{(I_1)^2 + (I_3 - I_2)^2} = \sqrt{(1.67)^2 + (3.77 - 1.77)^2}$ = **6 A. (Ans.)**

(*iii*) **Phase angle between the resultant current and the supply voltage, ϕ :**

$$\phi = \tan^{-1}\left(\frac{I_3 - I_2}{I_1}\right) = \tan^{-1}\left(\frac{2}{1.67}\right) = \textbf{50.14° (lead).} \quad \textbf{(Ans.)}$$

(*iv*) **Power factor of the circuit, cos ϕ :**

$$\cos\phi = \cos 50.14° = \textbf{0.641 (lead).} \quad \textbf{(Ans.)}$$

Example 42. *Determine the r.m.s. value of current in each branch and total current of the circuit shown in Fig. 61. Draw the phasor diagram.* **(PTU, Jan. 2000)**

Solution. Refer Fig. 61 (*a*).

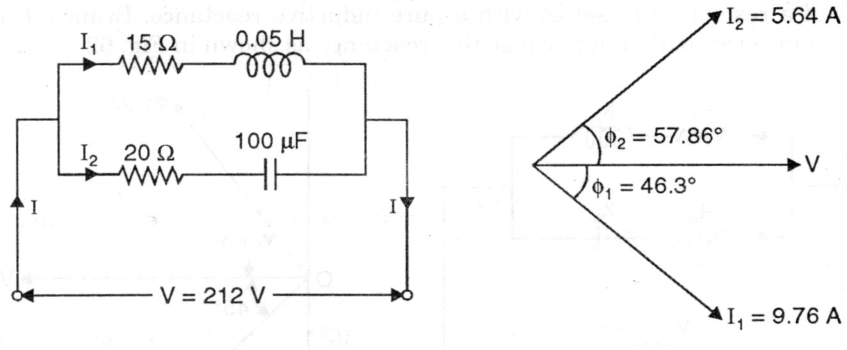

(*a*) Parallel circuit (*b*) Phasor diagram

Fig. 61

Branch No. 1 :

$$R_1 = 15\ \Omega,\ X_1 = 2\pi \times f \times L = 2\pi \times 50 \times 0.05 = 15.7\ \Omega$$

$$Z_1 = \sqrt{R_1^2 + X_1^2} = \sqrt{15^2 + 15.7^2} = 21.71\ \Omega$$

$$I_1 = \frac{V}{Z_1} = \frac{212}{21.71} = 9.76\ \text{A}$$

$$\cos\phi_1 = \frac{R_1}{Z_1} = \frac{15}{21.71} = 0.691\ [\text{or } \phi_1 = \cos^{-1}(0.691) = 46.3°\ (\textit{lagging})]$$

$$\sin\phi_1 = \sin(46.3°) = 0.723$$

Branch No. 2 :

$$R_2 = 20\ \Omega,\ X_2 = \frac{1}{2\pi fC} = \frac{1}{2\pi \times 50 \times 100 \times 10^{-6}} = 31.83\ \Omega$$

$$Z_2 = \sqrt{R_2^2 + X_2^2} = \sqrt{20^2 + 31.83^2} = 37.59\ \Omega$$

$$I_2 = \frac{V}{Z_2} = \frac{212}{37.59} = 5.64\ \text{A}$$

$$\cos\phi_2 = \frac{R_2}{Z_2} = \frac{20}{37.59} = 0.532 \qquad [\text{or } \phi_2 = \cos^{-1}(0.532) = 57.86°\ (\textit{leading})]$$

$$\sin\phi_2 = \sin(57.86°) = 0.847$$

The phasor diagram is shown in Fig. 60 (*b*)

X-components of I_1 and $I_2 = I_1 \cos\phi_1 + I_2 \cos\phi_2$

$$= 9.76 \times 0.691 + 5.64 \times 0.532 = 9.74\ \text{A}$$

Y-components of I_1 and $I_2 = -I_1 \sin\phi_1 + I_2 \sin\phi_2$

$$= -9.76 \times 0.723 + 5.64 \times 0.847 = -2.28\ \text{A}$$

Total current, $I = \sqrt{(9.74)^2 + (2.28)^2}$ = **10 A. (Ans.)**

Example 43. *The currents in each branch of a two-branched parallel circuit are given by the expressions $i_A = 7.07 \sin(314t - \pi/4)$ and $i_B = 21.2 \sin(314t + \pi/3)$. The supply voltage is given by the expression $v = 354 \sin 314t$. Derive a similar expression for the supply current and calculate the ohmic value of the components, assuming two pure components in each branch. State whether the reactive components are inductive or capacitive.* **(Elect. Engg. Calcutta University)**

Solution. From the given expressions of currents, we find that :

- i_A lags the voltage by $\pi/4$ radian or 45° and i_B leads it by $\pi/3$ radian or 60°. Hence branch A consists of a resistance in series with a pure inductive reactance. Branch B consists of a resistance in series with a pure capacitive reactance as shown in Fig. 62

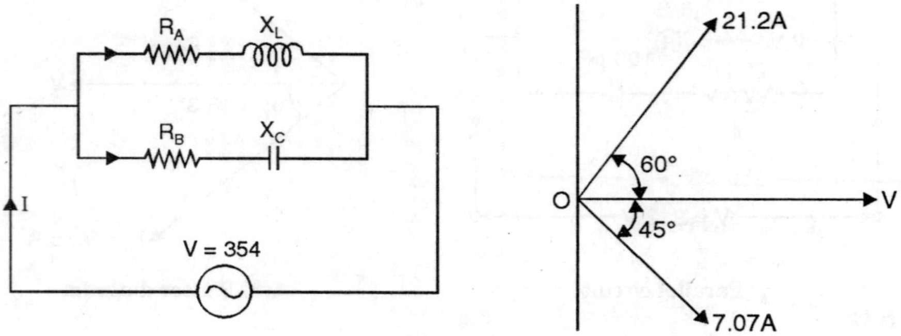

(*a*) Parallel circuit (*b*) Phasor diagram

Fig. 62

- Maximum value of current in branch A is 7.07 A and in branch B is 21.2 A.

The resultant current can be found as follows :

As seen from the vector diagram [Fig. 62 (*b*)],

X-component = 21.2 cos 60° + 7.07 cos 45° = 15.6 A

Y-component = 21.2 sin 60° – 7.07 sin 45° = 13.36 A

Maximum value of the resultant current

$$= \sqrt{(15.6)^2 + (13.36)^2} = 20.54 \text{ A}$$

$$\phi = \tan^{-1}(13.36/15.6) = 40.6° \text{ (lead)}$$

Hence the expression for the supply current is :

$$\mathbf{i = 20.54 \sin (314t + 40.6°) \text{ (Ans.)}}$$

$$Z_A = \frac{354}{7.07} = \mathbf{50\ \Omega}\ ;\ \cos \phi_A = \cos 45° = 0.707,\ \sin \phi_A = \sin 45° = 0.707$$

$$R_A = Z_A \cos \phi_A = 50 \times 0.707 = \mathbf{35.4\ \Omega}$$

$$X_L = Z_A \sin \phi_A = 50 \times 0.707 = \mathbf{35.4\ \Omega}.$$

$$Z_B = \frac{354}{21.2} = \mathbf{16.7\ \Omega.\ (Ans.)}$$

$$R_B = Z_B \cos 60° = 16.7 \times \cos 60° = \mathbf{8.35\ \Omega}$$

$$X_B = Z_B \sin 60° = 16.7 \times \sin 60° = \mathbf{14.46\ \Omega.\ (Ans.)}$$

Example 44. *Two impedances given by $Z_1 = (10 + j5)$ and $Z_2 = (8 + j6)$ are joined in parallel and connected across a voltage of $v = 200 + j0$. Calculate the circuit current, its phase and the branch currents. Draw the phasor diagram.* **(M.S. University Baroda)**

Solution. Refer Fig. 63 (*a*).

Branch A,
$$Y_1 = \frac{1}{Z_1} = \frac{1}{(10 + j5)}$$

$$= \frac{1}{(10 + j5)} \times \frac{10 - j5}{(10 - j5)} = \frac{10 - j5}{100 + 25} = \frac{10 - j5}{125}$$

$$= (0.08 - j0.04) \text{ siemens}$$

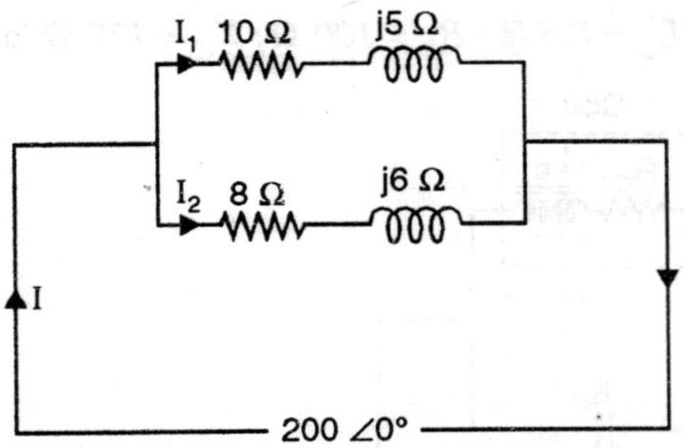

(a) Parallel circuit

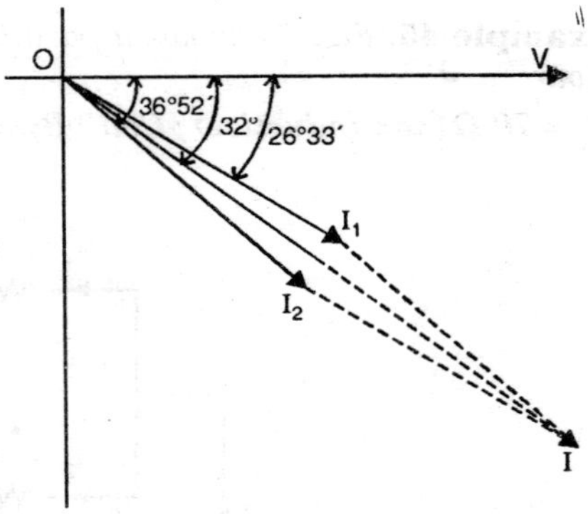

(b) Phasor diagram (not to scale)

Fig. 63

Branch B, $Y_2 = \frac{1}{Z_2} = \frac{1}{(8 + j6)}$

$$= \frac{1}{(8+j6)} \times \frac{(8-j6)}{(8-j6)} = \frac{8-j6}{64+36} = \frac{8-j6}{100}$$

$$= (0.08 - j0.06) \text{ siemens}$$

$$\therefore \quad Y = Y_1 + Y_2 = (0.08 - j0.04) + (0.08 - j0.06)$$

$$= (0.16 - j0.1) \text{ siemens}$$

Alternatively :

$$\frac{1}{Z} = \frac{1}{Z_1} + \frac{1}{Z_2} = \frac{Z_1 + Z_2}{Z_1 Z_2}$$

$$\therefore \quad Y = \frac{Z_1 + Z_2}{Z_1 Z_2} = \frac{(10 + j5) + (8 + j6)}{(10 + j5)(8 + j6)} = \frac{(18 + j11)}{50 + j100}$$

Rationalising the above expression, we get

$$Y = \frac{(18 + j11)(50 - j100)}{(50 + j100)(50 - j100)} = \frac{2000 - j1250}{12500}$$

$$= 0.16 - j0.1 \text{ (same as before)}$$

Now $V = 200\angle 0°$

$$\therefore \quad I = VY = (200 + j0)(0.16 - j0.1)$$

$$= 32 - j20 = \mathbf{37.74 \angle -32°} \text{ ...polar form. } \textbf{(Ans.)}$$

It lags behind the applied voltage by 32°.

Power factor $= \cos 32° = \mathbf{0.848}$. **(Ans.)**

$$I_1 = VY_1 = (200 + j0)(0.08 - j0.04)$$

$$= 16 - j8 = \mathbf{17.89 \angle -26° \, 33'}. \textbf{ (Ans.)}$$

It lags behind the applied voltage by 26° 33′.

$$I_2 = VY_2 = (200 + j0)(0.08 - j0.06)$$

$$= 16 - j12 = 20\angle -36° \, 52'$$

It lags behind the applied voltage by 36° 52′.

The phasor diagram is shown in Fig. 63 (b).

Example 45. *Fig. 64 shows a parallel circuit in which the values of the parameters are as given below :*

R_1 = 70 Ω (non-inductive) ; Coil : R_C = 30 Ω, L_C = 0.5 H ; R_2 = 100 Ω ; X_C = 157 Ω (at 50 Hz).

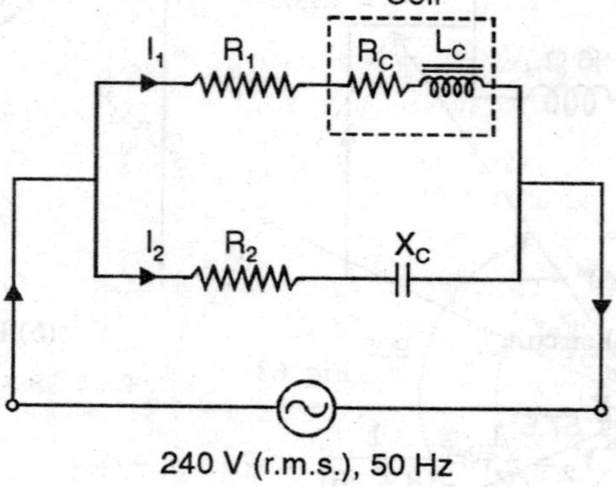

Fig. 64

(i) Determine the branch currents and the total current.

(ii) Draw the phasor diagram indicating the currents and voltages across coil and condenser.

(iii) If the A.C. source is replaced by an equivalent D.C. source, what current would be drawn by the circuit ?

(Pune University)

Solution. *Given :* R_1 = 70 Ω ; R_C = 30 Ω, L_C = 0.5 H ; R_2 = 100 Ω ; X_C = 157 Ω,

Applied voltage = 240 V (r.m.s.), 50 Hz

(i) The branch currents (I_1, I_2) and the total current (I) :

Resistance of the inductive branch $= R_1 + R_C = 70 + 30 = 100\ \Omega$

Reactance of the inductive branch $= 2\pi f L_C = 2\pi \times 50 \times 0.5 = 157\ \Omega$

Conductance of inductive branch,

$$G_1 = \frac{R_1 + R_C}{(R_1 + R_C)^2 + (2\pi f L_C)^2} = \frac{100}{(100)^2 + (157)^2} = 0.00288 \text{ S (siemens)}$$

Susceptance of inductive branch,

$$B_1 = \frac{2\pi f L_C}{(R_1 + R_C)^2 + (2\pi f L_C)^2}$$

$$= \frac{157}{(100)^2 + (157)^2} = -0.00453 \text{ S (Being inductive)}$$

Conductance of capacitive branch,

$$G_2 = \frac{R_2}{R_2^2 + X_C^2} = \frac{100}{(100)^2 + (157)^2} = 0.00288 \text{ S}$$

Susceptance of capacitive branch,

$$B_2 = \frac{X_C}{R_2^2 + X_C^2} = \frac{157}{(100)^2 + (157)^2} = 0.00453 \text{ S (Being capacitive)}$$

Total conductance of the circuit,

$$G = G_1 + G_2 = 0.00288 + 0.00288 = 0.00576 \text{ S}$$

Total susceptance of the circuit,

$$B = B_1 + B_2 = -0.00453 + 0.00453 = 0$$

Total admittance of the circuit,

$$Y = \sqrt{G^2 + B^2} = \sqrt{(0.00576)^2 + 0^2} = 0.00576 \text{ S}$$

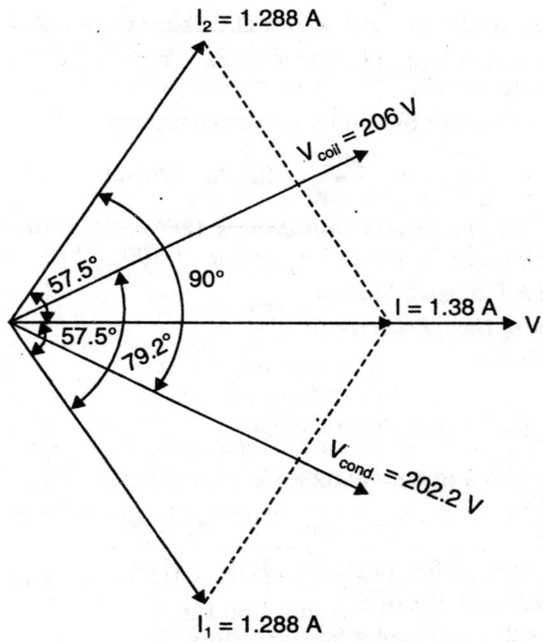

Fig. 65. Phasor diagram.

Current in inductive branch, $\mathbf{I_1} = V \times Y_1 = V\sqrt{G_1^2 + B_1^2}$

$$= 240\sqrt{(0.00288)^2 + (-0.00453)^2} = \mathbf{1.288\ A.} \quad \textbf{(Ans.)}$$

Phase angle, $\phi_1 = \tan^{-1}\left(\frac{B_1}{G_1}\right) = \tan^{-1}\left(\frac{-0.00453}{0.00288}\right) = \mathbf{-57.5°.} \quad \textbf{(Ans.)}$

Current in capacitive branch, $\mathbf{I_2} = V \times Y_2 = V\sqrt{G_2^2 + B_2^2}$

$$= 240\sqrt{(0.00288)^2 + (0.00453)^2} = \mathbf{1.288\ A.} \quad \textbf{(Ans.)}$$

Phase angle, $\phi_2 = \tan^{-1}\left(\frac{B_2}{G_2}\right) = \tan^{-1}\left(\frac{0.00453}{0.00288}\right) = \mathbf{57.5°.} \quad \textbf{(Ans.)}$

Total current, $\mathbf{I} = V \times Y = 240 \times 0.00576 = \mathbf{1.38\ A.} \quad \textbf{(Ans.)}$

Phase angle, $\phi = \tan^{-1}\left(\frac{B}{G}\right) = \tan^{-1}\left(\frac{0}{0.00576}\right) = 0°$

Voltage across condenser

$= I_2 X_C = 1.288 \times 157 =$ **202.2 V** *lagging behind I_2 by 90°.* **(Ans.)**

Voltage across coil $= I_1 \times \sqrt{(R_C)^2 + (2\pi f L_C)^2} = 1.288\sqrt{(30)^2 + (157)^2} = 206 \text{ V}$

Phase angle with current $I_1 = \tan^{-1}\left(\frac{157}{30}\right) = \mathbf{79.2°.} \quad \textbf{(Ans.)}$

(*ii*) **Phasor diagram :**

The phasor diagram indicating the currents and voltages across coil and condenser is shown in Fig. 65.

(*iii*) **Current drawn by the circuit when D.C. source is used :**

When energised by equivalent D.C. source (*i.e.,* 240 V D.C.), the capacitive branch will be open and current drawn by the circuit

$$= \text{Current drawn by inductive branch}$$

$$= \frac{V}{R_1 + R_C} = \frac{240}{70 + 30} = \mathbf{2.4\ A.}\quad \textbf{(Ans.)}$$

Example 46. *Fig. 66 shows two impedances (18 + j24) and (15 – j30) Ω connected in parallel ; across the combination is applied a voltage of 200∠53°8′. Determine :*

(i) kVA, kVAR and kW in each branch.

(ii) The power factor of the whole circuit.

Solution. Refer Fig. 66

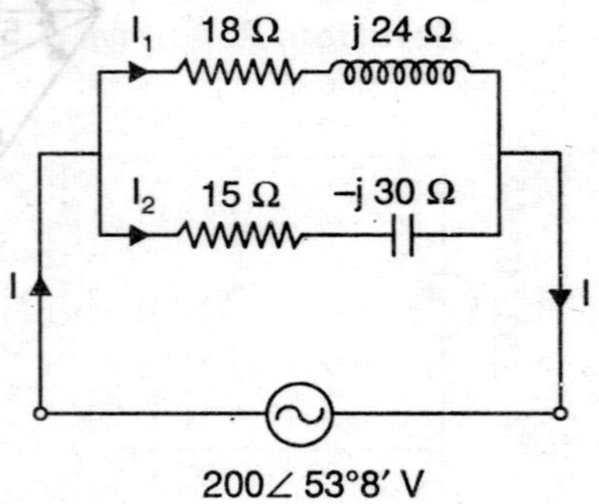

Fig. 66

$$\overline{Y}_1 = \frac{1}{18 + j24} = \frac{18 - j24}{(18 + j24)(18 - j24)}$$

$$= \frac{18 - j24}{900} = (0.02 - j0.0266)\ \text{S}$$

$$\overline{Y}_2 = \frac{1}{15 - j30} = \frac{15 + j30}{(15 + j30)(15 - j30)} = \frac{15 + j30}{1125}$$

$$= (0.0133 + j0.0266)\ \text{S}$$

Now, $V = 200\angle 53°8' = 200\ (\cos 53°8' + j \sin 53°8')$

$$= 200(0.6 + j0.8) = (120 + j160)\ \text{volts}$$

$$\overline{I}_1 = \overline{V}\,\overline{Y}_1 = (120 + j160)(0.02 - j0.0266)$$

$$= 2.4 + j3.2 - j3.2 + 4.26 = (6.66 + j0) \quad \text{...(along the reference axis)}$$

$\therefore \quad \overline{I}_2 = \overline{V}\,\overline{Y}_2 = (120 + j160)(0.0133 + j0.0266)$

$$= 1.6 + j3.2 + j2.13 - 4.26 = -\ 2.66 + j5.33\ \text{(leading)}$$

(*i*) **kVA, kVAR and kW in each branch :**

Power calculations can be calculated by the method of conjugates.

Branch 1 (*Inductive branch*)

The current conjugate of $(6.66 + j0)$ is $(6.66 - j0)$

$$\therefore \quad \overline{V}\,\overline{I}_1 = (120 + j160)(6.66 - j0) = 800 + j1066 \quad \therefore \quad \textbf{kW} = \frac{800}{1000} = \mathbf{0.8.}\quad \textbf{(Ans.)}$$

$$\therefore \quad \textbf{kVAR} = \frac{1066}{1000} = \mathbf{1.066.}\quad \textbf{(Ans.)}$$

The fact that it is positive merely shows that reactive *VA* (volt-amps) are due to lagging current

$$\textbf{kVA} = \sqrt{(kW)^2 + (kVAR)^2} = \sqrt{(0.8)^2 + (1.066)^2} = \mathbf{1.33.}\quad \textbf{(Ans.)}$$

Branch 2 (*Capacitive branch*) :

The current conjugate of $(-\ 2.66 + j5.33)$ is $(-\ 2.66 - j5.33)$

$$\therefore \quad \overline{V}\,\overline{I}_2 = (120 + j160)(-\ 2.66 - j5.33) = 533.6 - j1065.2$$

$$\therefore \quad \textbf{kW} = \frac{533.6}{1000} = \mathbf{0.5336.}\quad \textbf{(Ans.)} \quad ; \quad \textbf{kVAR} = \frac{-\ 1065.2}{1000} = \mathbf{-\ 1.065.}\quad \textbf{(Ans.)}$$

The negative sign merely indicates the reactive volt-amps are due to leading current.

$\therefore$ $$\textbf{kVA} = \sqrt{(0.5336)^2 + (-1.065)^2} = \textbf{1.191. (Ans.)}$$

(*ii*) **The power factor of the whole circuit, cos φ :**

$$\bar{Y} = \bar{Y}_1 + \bar{Y}_2 = (0.02 - j0.0266) + (0.0133 + j0.0266) = 0.0333 + j0$$

$$\bar{I} = \bar{V}\bar{Y} = (120 + j160)(0.0333 + j0) = 4 - j5.33 = 6.66 \angle 53°8'$$

or $$\bar{I} = \bar{I}_1 + \bar{I}_2 = (6.66 + j0) + (-2.66 + j5.33) = 4 - j5.33 \quad \text{(Same as above)}$$

Power factor of the circuit, cos φ = cos 0° = **1. (Ans.)**

(∵ Current is in phase with voltage)

Example 47. *For the circuit of Fig. 67, calculate the current supplied by the voltage source and the voltage across the current source.* **(PTU June 2000 ; May 2001)**

Solution. Refer Figs. 67 and 68.

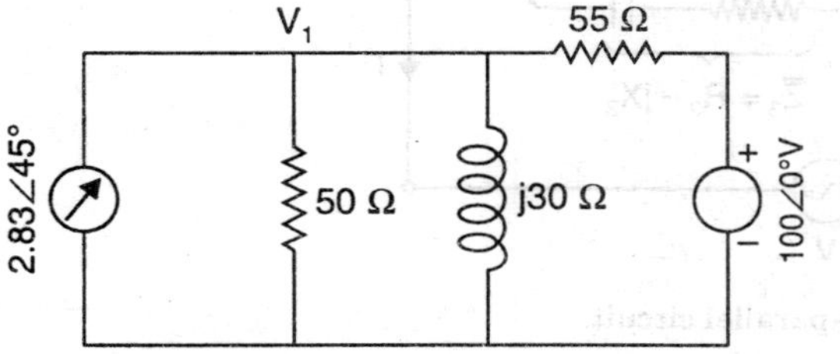

Fig. 67

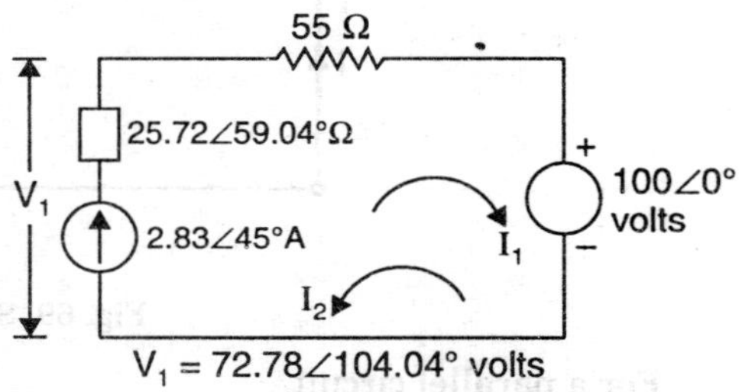

Fig. 68

Let the impedance of the current source be Z_i, then

$$\frac{1}{Z_i} = \frac{1}{50\angle 0°} + \frac{1}{30\angle 90°} = \frac{30\angle 90° + 50\angle 0°}{1500\angle 90°} = \frac{50 + j30}{1500\angle 90°} = \frac{58.31\angle 30.96°}{1500\angle 90°}$$

$$Z_i = \frac{1500\angle 90°}{58.31\angle 30.96°} = 25.72\angle 59.04° = (13.23 + j22.05)\ \Omega$$

Voltage across current source,

$$V_1 = IZ_i = 2.83\angle 45° \times 25.72\angle 59.04° = \textbf{72.78}\angle\textbf{104.04° volts. (Ans.)}$$

Current supplied by one source,

$$I_1 = \frac{V_1}{Z_i + R} = \frac{72.78\angle 104.04°}{13.25 + j22.05 + 55} = \frac{72.78\angle 104.04°}{68.25 + j22.05}$$

$$= \frac{72.78\angle 104.04°}{71.72\angle 17.91°} = 1.015\angle 86.13° = (0.0685 + j1.01)\ \text{A}$$

Current supplied by the other source,

$$I_2 \frac{V_2}{Z_i + R} = \frac{100\angle 0°}{71.72\angle 17.91°} = 1.394\angle(-17.91°) = 1.326 - j0.428$$

Current supplied by the voltage source,

$$I = I_2 - I_1 = 1.326 - j0.428 - 0.0685 - j1.01 = 1.2575 - j1.438$$

$$= \textbf{1.91}\angle\textbf{-48.83 amp. (Ans.)}$$

4.6.3. Series-Parallel Circuits

Series-parallel circuits may be solved by the following methods :

1. Admittance method.
2. Symbolic method.

1. *Admittance method :*

In series-parallel circuits, the parallel circuit is first reduced to an equivalent series circuit and then combined with the rest of the circuit as usual.

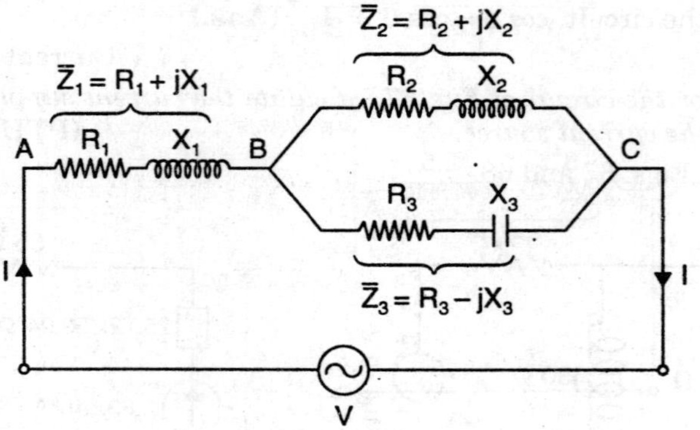

Fig. 69. Series-parallel circuit.

For a parallel circuit,

$$\text{Equivalent series resistance, } R_{eq} = Z \cos \phi = \frac{1}{Y} \cdot \frac{G}{Y} = \frac{G}{Y^2}$$

$$\text{Equivalent series reactance, } X_{eq} = Z \sin \phi = \frac{1}{Y} \cdot \frac{B}{Y} = \frac{B}{Y^2}$$

2. *Symbolic method :*

Refer series-parallel circuit shown in Fig. 69. First calculate the equivalent impedance of parallel branches and then add it to the series impedance to get the total impedance of the circuit. Then current flowing through the circuit is found as follows :

$$\bar{Y}_2 = \frac{1}{R_2 + jX_2} \quad \text{and} \quad \bar{Y}_3 = \frac{1}{R_3 - jX_3}$$

$$\therefore \quad \bar{Y}_{23} = \frac{1}{R_2 + jX_2} + \frac{1}{R_3 - jX_3}$$

$$\therefore \quad \bar{Z}_{23} = \frac{1}{\bar{Y}_{23}}; \ \bar{Z}_1 = R_1 + jX_1$$

$$\bar{Z} = \bar{Z}_{23} + \bar{Z}_1$$

$$\therefore \quad \bar{I} = \frac{\bar{V}}{\bar{Z}}.$$

Example 48. *For the circuit shown in Fig. 70 calculate :*

(i) Currents I_A, I_B and I_C.

(ii) Total power factor for the whole circuit.

Draw the complete phasor diagram.

Solution. Refer Fig. 70.

Given : $\bar{Z}_A = 2 + j1.5 = 2.5\angle 36.9°$; $\bar{Z}_B = 5 - j3.5 = 6.1\angle -35°$, $\bar{Z}_C = 3 + j2.5 = 3.9\angle 39.8°$

(*i*) **Currents I_A, I_B and I_C :**

$$\frac{1}{\bar{Z}_{AB}} = \frac{1}{\bar{Z}_A} + \frac{1}{\bar{Z}_B} \quad \text{or} \quad \bar{Z}_{AB} = \frac{\bar{Z}_A \bar{Z}_B}{\bar{Z}_A + \bar{Z}_B} = \frac{2.5\angle 36.9° \times 6.1\angle -35°}{(2 + j1.5) + (5 - j3.5)}$$

$$= \frac{15.25\angle 1.9°}{7 - j2} = \frac{15.25\angle 1.9°}{7.28\angle -16°} = 2.095\angle 17.9° \simeq 2 + j0.65$$

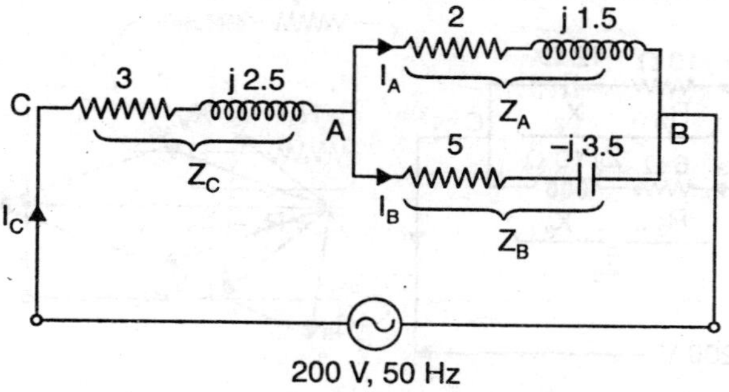

Fig. 70. Circuit diagram.

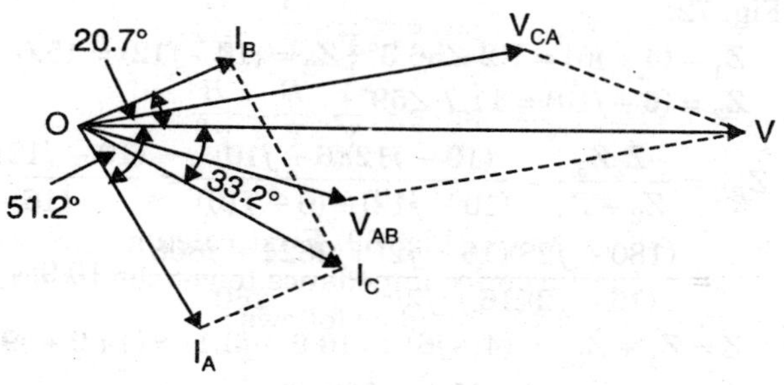

Fig. 71. Phasor diagram.

$$\bar{Z} = \bar{Z}_C + \bar{Z}_{AB} = (3 + j2.5) + (2 + j0.65) = 5 + j3.15 = 5.9\angle 32.2°$$

Let $\quad \bar{V} = 200\angle 0°$,

Then, $\quad \bar{\mathbf{I}}_C = \frac{\bar{V}}{\bar{Z}} = \frac{200\angle 0°}{5.9\angle 32.2°} = \mathbf{33.9\angle -32.2°}$. **(Ans.)**

$$\bar{\mathbf{I}}_A = \bar{I}_C \times \frac{\bar{Z}_B}{\bar{Z}_A + \bar{Z}_B} = 33.9\angle -32.2° \times \frac{6.1\angle -35°}{7.28\angle -16°} = \mathbf{28.4\angle -51.2°}. \quad \textbf{(Ans.)}$$

$$\bar{\mathbf{I}}_B = \bar{I}_C \times \frac{\bar{Z}_A}{\bar{Z}_A + \bar{Z}_B} = 33.9\angle -32.2° \times \frac{2.5\angle 36.9°}{7.28\angle -16°} = \mathbf{11.64\angle 20.7°}. \quad \textbf{(Ans.)}$$

(*ii*) **Total power factor in the whole circuit :**

The phase angle between V and total circuit current I_C is 32.2°.

∴ Power factor for the whole circuit ; cos φ = cos 32.2° = **0.846 (lag). (Ans.)**

For drawing the phasor diagram of Fig. 71, V_{CA} and V_B have to be calculated.

$$\bar{V}_{CA} = \bar{I}_C \bar{Z}_C = 33.9\angle -32.2° \times 3.9\angle 39.8° = 132.2\angle 7.6°$$

$$\bar{V}_{AB} = \bar{I}_C \bar{Z}_{AB} = 33.9\angle -32.2° \times 2.095\angle 17.9° = 71\angle -14.3°$$

Fig. 71 shows the complete phasor diagram.

Example 49. *Determine the current drawn by the following circuit (Fig. 72) when a voltage of 200 V is applied across the same. Draw the phasor diagram.*

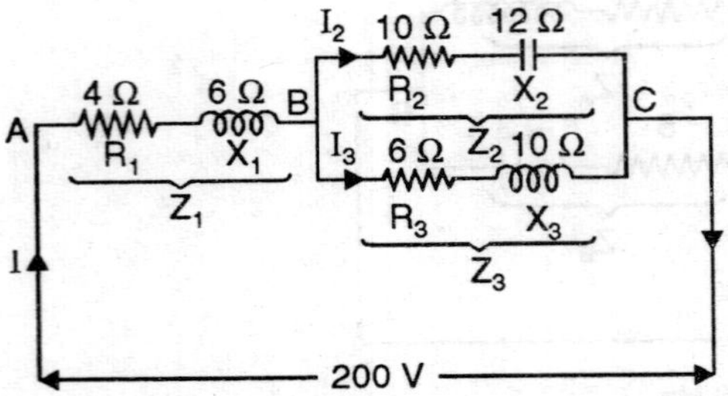

Fig. 72. Series parallel circuit

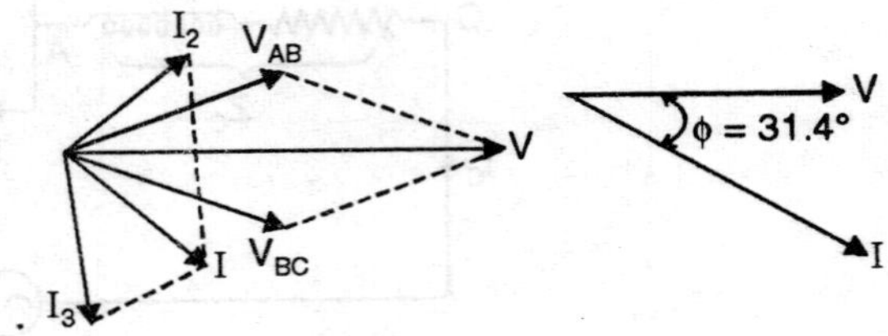

Fig. 73. Phasor diagram

(Elect. Engg. Bangalore University)

Solution. Refer Fig. 72.

$$Z_1 = (4 + j6) = 7.2\angle 56.3° \,;\; Z_2 = (10 - j12) = 15.6\angle -50.2°$$

$$Z_3 = (6 + j10) = 11.7\angle 59°$$

$$Z_{BC} = \frac{Z_2 Z_3}{Z_2 + Z_3} = \frac{(10 - j12)(6 + j10)}{(10 - j12) + (6 + j10)} = \frac{(10 - j12)(6 + j10)}{(16 - j2)} = \frac{(180 + j28)}{(16 - j2)}$$

$$= \frac{(180 + j28)(16 + j2)}{(16 - j2)(16 + j2)} = \frac{2824 + j808}{260} = 10.9 + j3.1 = 11.3\angle 15.9°$$

$$Z = Z_1 + Z_{BC} = (4 + j6) + (10.9 + j3.1) = (14.9 + j9.1) = 17.5\angle 31.4°$$

Assuming $V = 200\angle 0° \,;\; I = \dfrac{V}{Z} = \dfrac{200\angle 0°}{17.5\angle 31.4°} = 11.4\angle -31.4°$

For drawing the phasor diagram, let us find the following quantities :

$$V_{AB} = IZ_1 = 11.4\angle -31.4° \times 7.2\angle 56.3° = 82.08\angle 24.9°$$

$$V_{BC} = IZ_{BC} = 11.4\angle -31.4° \times 11.3\angle 15.9° = 128.8\angle -15.5°$$

$$I_2 = \frac{V_{BC}}{Z_L} = \frac{128.8\angle -15.5°}{15.6\angle -50.2°} = 8.26\angle 34.7°$$

$$I_3 = \frac{V_{BC}}{Z_3} = \frac{128.8\angle -15.5°}{11.7\angle 59°} = 11\angle -74.5°$$

Various currents and voltages are shown in their phase relationship in Fig. 73.

4.6.4. Resonance in parallel circuits

In case of a series circuit consisting of *R* (resistance), *L* (inductance) and *C* (capacitance), resonance takes place when $V_L = V_C$ *i.e.*, when $X_L = X_C$. In other words, resonance takes place when the power factor of the circuit approaches *unity*. The basic condition for resonance, *i.e.*, power factor of the entire circuit being unity, remains the same for parallel circuits also. Thus, *resonance in a parallel circuit will occur, when the power factor of the entire circuit becomes unity.*

A parallel circuit consisting of an inductive coil in parallel with a capacitor is shown in Fig. 74 (*a*). The phasor diagram of this circuit with applied voltage as the reference phasor is shown in Fig. 74 (*b*). The current drawn by the inductive branch lags the applied voltage by an angle ϕ_L. The current drawn by the capacitive branch leads the applied voltage by 90°.

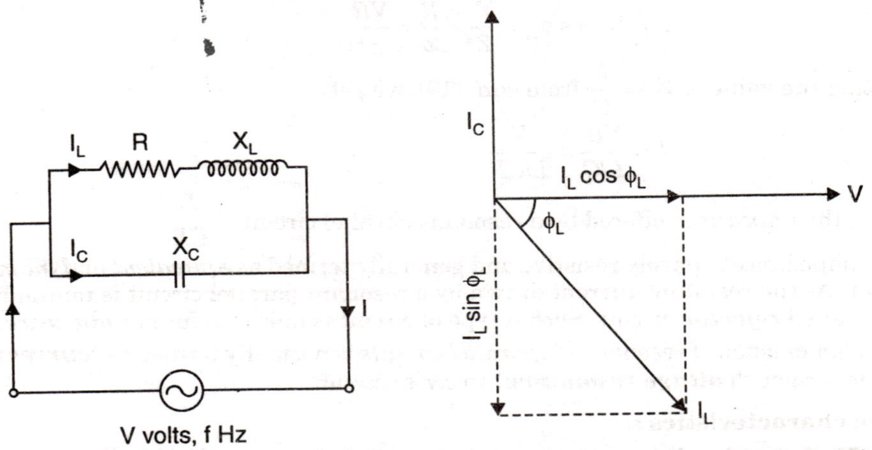

(a) Circuit diagram (Resonant circuit) (b) Phasor diagram (Resonant circuit)

Fig. 74. Resonance in a parallel circuit.

The power factor of the current becomes unity when the total current drawn by the entire circuit is in phase with the applied voltage. This will happen only when the current drawn by the capacitive branch I_C equals the active component of the current of the inductive branch I_L [Fig. 74 (*b*)].

Hence for *resonance* in parallel circuit.

$$I_C = I_L \sin \phi_L \qquad ...(18)$$

Now, $$I_L = \frac{V}{Z}\ ;\ \sin \phi_L = \frac{X_L}{Z} \text{ and } I_C = \frac{V}{X_C}$$

Hence, condition for resonance becomes

$$\frac{V}{X_C} = \frac{V}{Z} \times \frac{X_L}{Z} \quad \text{or} \quad Z^2 = X_L \times X_C$$

Now, $$X_L = \omega L,\ X_C = \frac{1}{\omega C}$$

$$\therefore \qquad Z^2 = \frac{\omega L}{\omega C} = \frac{L}{C} \qquad ...(19)$$

or $$R^2 + X_L{}^2 = R^2 + (2\pi f_r L)^2 = \frac{L}{C}$$

or $$(2\pi f_r L)^2 = \frac{L}{C} - R^2 \quad \text{or} \quad 2\pi f_r = \sqrt{\frac{1}{LC} - \frac{R^2}{L^2}}$$

or $$f_r = \frac{1}{2\pi}\sqrt{\frac{1}{LC} - \frac{R^2}{L^2}} \qquad ...(20)$$

This is the resonant frequency and is given in Hz if R is in ohm, L is in henry and C in farad.

If R is negligible, then

$$f_r = \frac{1}{2\pi\sqrt{LC}} \quad \text{...same as for series resonance.}$$

Current at resonance

Refer Fig. 74 (*b*). Since wattless component is zero, the circuit current is given as :

$$I = I_L \cos \phi_L = \frac{V}{Z} \times \frac{R}{Z} = \frac{VR}{Z^2}$$

Putting the value of $Z^2 = \frac{L}{C}$ from eqn. (19), we get

$$I = \frac{VR}{L/C} = \frac{V}{L/CR} \qquad \text{...(21)}$$

Thus, the *impedance* offered by a resonant parallel circuit $= \frac{L}{CR}$.

This impedance is purely resistive and generally termed as *equivalent* or *dynamic impedance* of the circuit. As the resultant current drawn by a resonant parallel circuit is minimum, the circuit is normally called ***rejector*** circuit. Such a type of circuit is quite useful in *radio work*.

The phenomenon of resonance in *parallel circuits* is normally termed as ***"current resonance"*** whereas it is termed ***"voltage resonance"*** in *series circuit*.

Resonance characteristics :

Fig. 75 shows the characteristics of the parallel circuit consisting of an inductance L and capacitance C in parallel plotted against frequency, the voltage applied being constant.

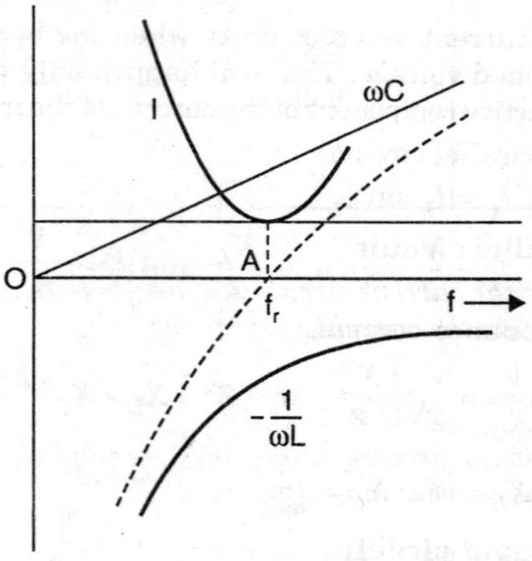

Fig. 75. Resonance characteristics.

Inductive susceptance, $\quad b = -\frac{1}{X_L} = -\frac{1}{\omega L} = -\frac{1}{2\pi f L}$.

Thus inductive susceptance is *inversely proportional* to the *frequency* and is represented by rectangular hyperbola in the fourth quadrant (because it is assumed –ve).

Capacitive susceptance, $\quad b = \frac{1}{X_C} \; \omega C = 2\pi f C$

Thus capacitive susceptance is *directly proportional to the frequency* and is represented by a straight line passing through the origin.

Net susceptance, B is the difference of the two susceptances and is represented by the dotted hyperbola. The net susceptance is zero at point A, hence *admittance is minimum* and is equal to G. Thus, at point A, *line current is minimum*. The frequency at which the total current becomes minimum is the resonance frequency f_r.

Evidently, below the resonant frequency, the inductive susceptance predominates, thus making the circuit current to be lagging, whereas beyond f_r capacitive susceptance predominates and the current leads the applied voltage. At resonant frequency f_r, the current is in phase with the applied voltage.

Hence at ***parallel resonance*** it is seen that :

1. The admittance of the circuit is *minimum* and is *equal to* the conductance of the circuit.
2. The *current drawn is minimum.*
3. The phase angle between the current and voltage is zero, the *power factor is unity.*
4. The resonant frequency is given by $f_r = \frac{1}{2\pi\sqrt{LC}}$ if the *resistance* in the inductance and capacitance branches is *negligible.*

4.6.5. Comparison of series and parallel resonant circuits

S. No.	*Aspects*	*Series circuit (R-L-C)*	*Parallel circuit (R-L and C)*
1.	*Impedance at resonance*	Minimum	Maximum
2.	*Current at resonance*	Maximum = $\frac{V}{R}$	Minimum = $V/(L/CR)$
3.	*Effective impedance*	R	L/CR
4.	*Power factor at resonance*	Unity	Unity
5.	*Resonant frequency*	$\frac{1}{2\pi\sqrt{LC}}$	$\frac{1}{2\pi}\sqrt{\left(\frac{1}{LC}-\frac{R^2}{L^2}\right)}$
6.	*It magnifies*	Voltage	Current
7.	*Magnification is*	$\frac{\omega L}{R}$	$\frac{\omega L}{R}$

4.6.6. Q-factor of a parallel circuit

It is defined as the *ratio of the current circulating between its two branches to the line current drawn from the supply or simply, as the current magnification.*

$$Q\text{-factor} = \frac{1}{R}\sqrt{\frac{L}{C}} \qquad ...(22)$$

- It may be noted that in *series circuits, Q-factor gives the voltage magnification, whereas in parallel circuits it gives the current magnification.*

Bandwidth of a parallel resonant circuit :

The bandwidth of a parallel circuit is defined in the same way as that for a series circuit. This circuit also has upper and lower half-power frequencies where power dissipated is half of that at resonant frequency.

The net susceptance B, at bandwidth frequencies, equals conductance. Hence,

At f_2 : $\quad B = B_{C2} - B_{L2}$

At f_1 : $\quad B = B_{L1} - B_{C1}$

Hence $Y = \sqrt{G^2 + B^2} = \sqrt{2}G$ and $\phi = \tan^{-1}\left(\frac{B}{G}\right) = \tan^{-1}(1) = 45°$

However, at off-resonant frequencies, $Y > G$ and $B_C \neq B_L$ and phase angle is greater than zero.

Example 50. *An inductive circuit of resistance 2 ohms and inductance 0.01 H is connected to a 250 V, 50 Hz supply.*

(i) What capacitance placed in parallel will produce resonance ?

(ii) Determine also the total current taken from the supply and the currents in the branch circuit. **(Kerala University)**

Solution. *Given :* $R = 2\ \Omega$; $L = 0.01$ H, Supply voltage = 250 V, 50 Hz.

(i) **Value of capacitance which will produce resonance, C :**

Now, $X_L = 2\pi f L = 2\pi \times 50 \times 0.01 = 3.14\ \Omega$

$$Z = \sqrt{R^2 + X_L{}^2} = \sqrt{2^2 + 3.14^2} = 3.72\ \Omega$$

We know that, $Z^2 = \dfrac{L}{C}$ or $C = \dfrac{L}{Z^2}$

$$\therefore \quad C = \frac{0.01}{(3.72)^2} = 722.6 \times 10^{-6}\ \text{F or } \mathbf{722.6\ \mu F.}\ \textbf{(Ans.)}$$

(ii) **Total current and currents in the branch circuits, I, I_L, I_C :**

$$\mathbf{I_{R\text{-}L}} = \frac{V}{Z} = \frac{250}{3.72} = \mathbf{67.2\ A.}\quad \textbf{(Ans.)}$$

$$\tan \phi_L = \frac{3.14}{2} = 1.57 \text{ or } \phi_L = \tan^{-1}(1.57) = \mathbf{57.5°}\quad \textbf{(Ans.)}$$

Hence, current in *R-L* branch lags the applied voltage by 57.5°.

$$\mathbf{I_C} = \frac{V}{X_C} = \frac{V}{1/\omega C} = \omega VC = 2\pi \times 50 \times 250 \times 722.6 \times 10^{-6} = \mathbf{56.75}\quad \textbf{(Ans.)}$$

This current leads the applied voltage by 90°.

Total current, $\mathbf{I} = I_{R\text{-}L} \cos\phi = 67.2 \cos 57.5° = \mathbf{36.1\ A.}$ **(Ans.)**

$$\left[\text{or } I = \frac{V}{L/CR} = \frac{VCR}{L} = \frac{250 \times 722.6 \times 10^{-6} \times 2}{0.01} = 36.1\ \text{A}\right]$$

5. TRANSIENTS

5.1. General Aspects

If a circuit is switched from one condition to another either by a change in the applied voltage or change in a circuit parameter, there *exists a transitional period during which the branch currents and voltage drops change from their former values to new ones. After transition period, the circuit becomes steady.*

The ***transient disturbances*** in the electrical circuits are *disturbances caused by sudden switching off and on or short circuit of the circuit and sudden change in the applied voltage.* The current developed in the circuit due to this disturbance is called the *"transient current"*. The *"resultant current"* in the circuit is the *steady state current with a transient current superimposed.* The *transient currents* are found to be *associated with the changes in stored energy in capacitors and inductors.* Hence in a *purely resistive circuit no transient current is developed since there is no stored energy in a resistor.*

Single energy and double energy transients :

- Single *energy transient* is the transient disturbance where only one form of energy, *either electromagnetic or electrostatic* is involved *e.g.*, transient disturbance in a circuit consisting of resistor and inductor *i.e.*, *R-L* circuit or a circuit consisting of resistor and capacitor *i.e.*, *R-C* circuit.
- *Double energy transient* is the transient disturbance where *both electromagnetic and electrostatic energies* are involved *e.g.*, transient disturbance in a circuit consisting of resistor, inductor and capacitor *i.e.*, *R-L-C* circuit.

5.2. D.C. Transients

(*i*) **R-L transients :**

In the *R-L* circuit shown in Fig. 76

$$i = \frac{V}{R}[1 - e^{-(R/L)t}] \quad ...(23)$$

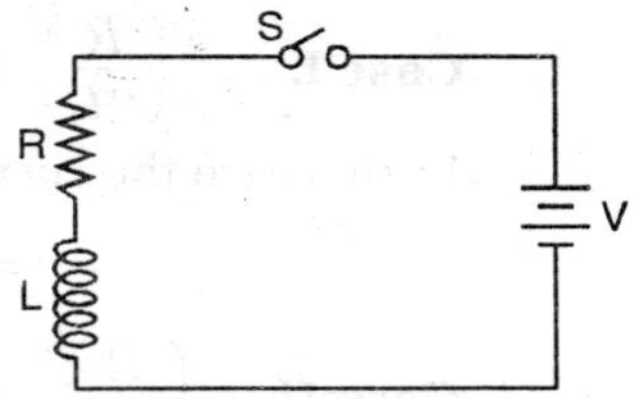

Fig. 76. *R-L* circuit.

The plot of *i* (exponential rise equation) versus time is shown in Fig. 77.

The time constant (λ) *for the above function is the time at which the exponent of* e *is unity.* Thus in this case time constant (λ) is L/R. At one time constant, the value of *i* will be

$$i = (1 - e^{-1}) = 1 - 0.368 = 0.632$$

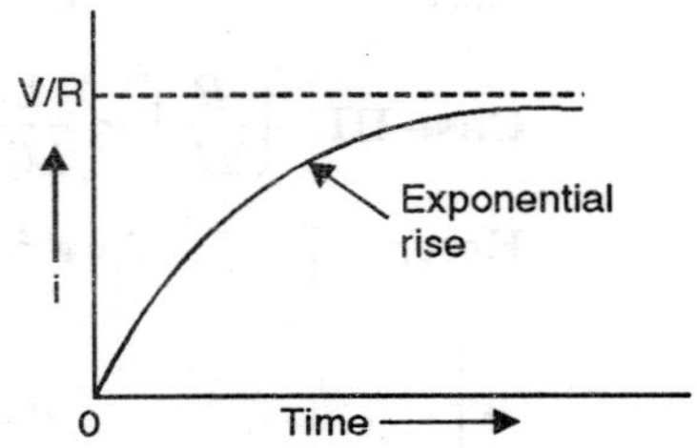

Fig. 77

At this time current will be *63.2% of its final value.*

The voltage across inductance,

$$v_L = L\frac{di}{dt} = Ve^{-(R/L)t} \quad ...(24)$$

and voltage across resistor,

$$v_R = V[1 - e^{-(R/L)t}] \quad ...(25)$$

The exponential rise of resistor voltage and exponential decay of inductor voltage are shown in Fig. 78.

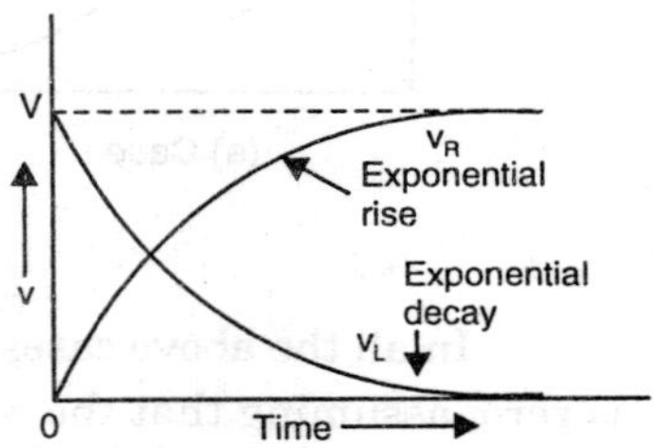

Fig. 78

Also, $v_R + v_L = V[1 - e^{-(R/L)t}] + Ve^{-(R/L)t} = V$

Power in the circuit elements is given by

$$P_R = \frac{V^2}{R}[1 - 2e^{-(R/L)t} + e^{-2(R/L)t}] \quad ...(26)$$

$$P_L = \frac{V^2}{R}[e^{-(R/L)t} - e^{-2(R/L)t}] \quad ...(27)$$

Total power,

$$P = P_R + P_L = \frac{V^2}{R}[1 - e^{-(R/L)t}] \quad ...(28)$$

(*ii*) **R-C transients :**

In the *R-C* circuit shown in Fig. 79,

$$i = \frac{V}{R}e^{-t/RC}$$

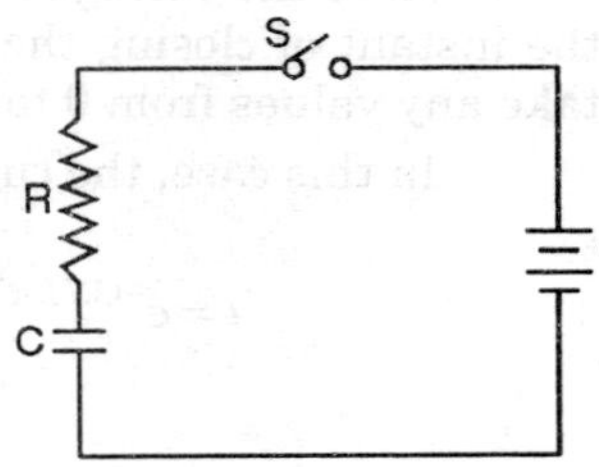

Fig. 79. *R-C* circuit.

Transients voltages across *R* and *C* are given by

$$v_R = Ve^{-t/RC} \quad ...(29)$$

$$v_C = V(1 - e^{-t/RC}) \quad ...(30)$$

Also, the power in circuit elements is given by

$$P_R = \frac{V^2}{R}e^{-2t/RC} \quad ...(31)$$

$$P_L = \frac{V^2}{R}(e^{-t/RC} - e^{-2t/RC}) \quad ...(32)$$

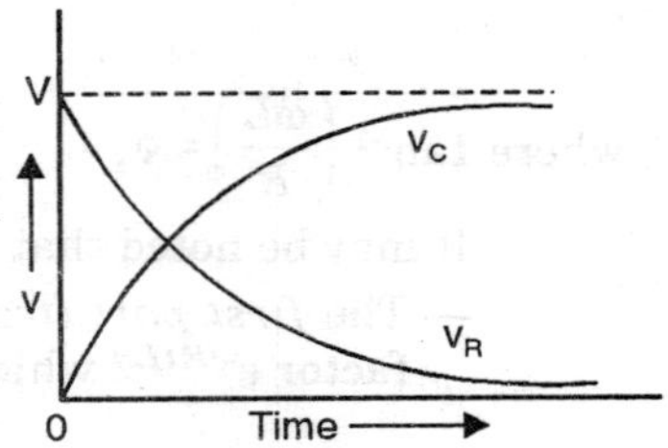

Fig. 80

(*iii*) **R-L-C transients :**

For *R-L-C* circuit shown in Fig. 81 the following integro-differential equation can be written as follows :

$$R_i + L\frac{di}{dt} + \frac{1}{C}\int idt = V \quad ...(33)$$

While solving for i, the following three cases are considered :

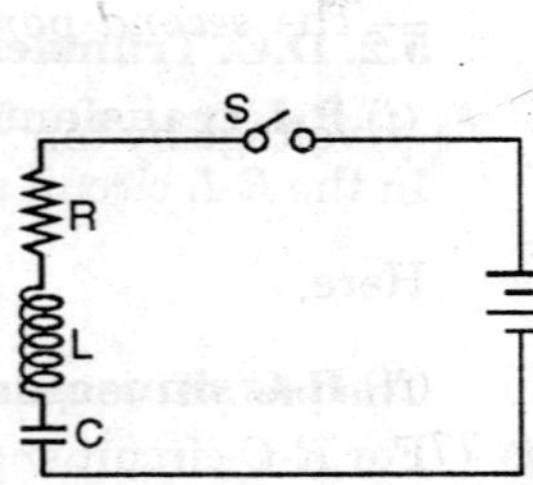

Fig. 81. *R-L-C* circuit.

Case I. $\left(\frac{R}{2L}\right)^2 > \frac{1}{LC}$

In this case the current is given by

$$i = C_1 e^{\alpha t}\ C_1(e^{\beta t} + C_2 e^{-\beta t}) \qquad ...(34)$$

Case II. $\left(\frac{R}{2L}\right)^2 = \frac{1}{LC}$

Here, $$i = e^{\alpha t}\ (C_1 + C_2 t) \qquad ...(35)$$

Case III. $\left(\frac{R}{2L}\right)^2 < \frac{1}{LC}$

Here, $$i = e^{\alpha t}\ (C_1 \cos \beta t + C_2 \sin \beta t) \qquad ...(36)$$

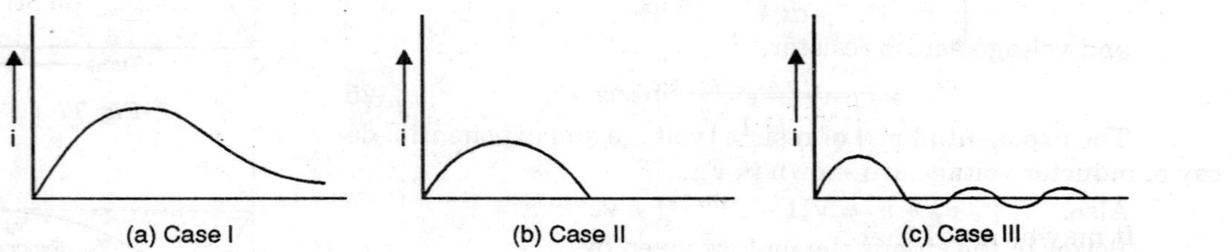

Fig. 82

In all the above cases the current contains the factor $e^{\alpha t}$ and since $\alpha = -R/2L$ the final value is zero, assuming that the complimentary function decays in a relatively short time. Fig. 82 shows the value of i for initial values zero and initial slope positive.

5.3. A.C. Transients

(i) **R-L sinusoidal transient :** Refer Fig. 83

Here the voltage function could be at any point in the period at the instant of closing the switch and therefore the phase angle ϕ can take any values from 0 to 2 π rad/sec.

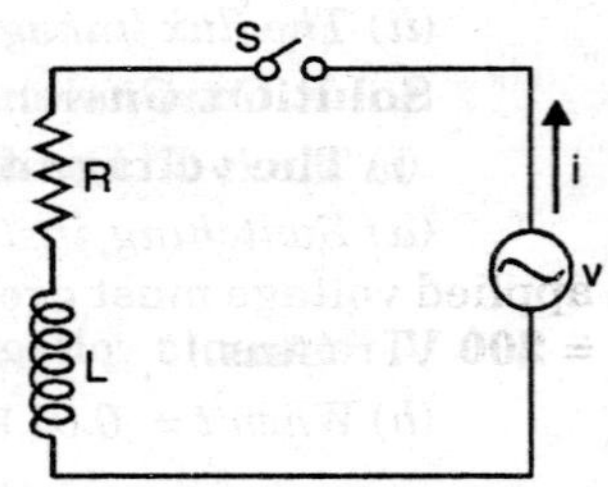

$v = V_{max} \sin (\omega t + \psi)$

Fig. 83. *R-L* circuit.

In this case, the current (i) is given by

$$i = e^{-(R/L)t}\left[\frac{-V_{max}}{\sqrt{R^2+\omega^2L^2}} \sin\left(\psi - \tan^{-1}\frac{\omega L}{R}\right)\right] + \frac{V_{max}}{\sqrt{R^2+\omega^2L^2}} \sin\left(\omega t + \psi - \tan^{-1}\frac{\omega L}{R}\right) \qquad ...(37)$$

where $\tan^{-1}\left(\frac{\omega L}{R}\right) = \phi$.

It may be noted that :

— The *first part (transient component of current, i_t,* of the above equation contains the factor $e^{-(R/L)t}$ which has a *value of zero in a relatively short time.*

— The *second part* of the above equation is the *steady current* (i_s) which lags the applied voltage by $\tan^{-1} \frac{\omega L}{R}$.

Here, $$\frac{V_{max}}{R^2 + \omega^2 L^2} = I_{max},\ \tan^{-1}\left(\frac{\omega L}{R}\right) = \phi$$

(*ii*) **R-C sinusoidal transient :**

For *R-C* circuit shown in Fig. 84 the basic equation is :

$$Ri + \frac{1}{C}\int v dt = V_{max} \sin(\omega t + \psi) \qquad ...(38)$$

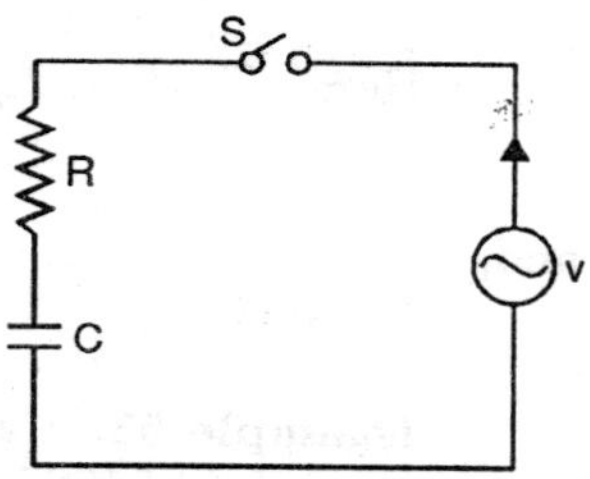

$v = V_{max} \sin(\omega t + \psi)$

Fig. 84. ***R-C*** **circuit.**

Here the current i is given by,

$$i = e^{-(t/RC)} \left[\frac{V_{max}}{R} \sin\psi - \frac{V_{max}}{\sqrt{R^2 + \left(\frac{1}{\omega C}\right)^2}} \sin\left(\psi + \tan^{-1}\frac{1}{\omega CR}\right) \right] + \frac{V_{max}}{\sqrt{R^2 + \left(\frac{1}{\omega C}\right)^2}} \sin\left(\omega t + \psi + \tan^{-1}\frac{1}{\omega CR}\right) \qquad ...(39)$$

It may be noted that :

—The *first part* of the above equation is *the transient with decay factor* $e^{-t/RC}$

—The *second part is the steady current which leads the applied voltage by* $\tan^{-1} \frac{1}{\omega CR}$.

Example 51. *A series circuit has R = 10 Ω and L = 0.1 H. A direct voltage of 200 V is suddenly applied to it. Calculate the following :*

(*i*) *The voltage drop across the inductance at the instance of switching on and at 0.01 second, and*

(*ii*) *The flux linkages at these instants.* **(Bombay University)**

Solution. *Given :* $R = 10\ \Omega$; $L = 0.1$ H ; $V = 200$ volts

(*i*) **The voltage drop across the inductance :**

(*a*) *Switching instant.* At the instant of switching on, $i = 0$, so that $iR = 0$ hence all the applied voltage must drop across the inductance only. Therefore, voltage drop across inductance = **200 V. (Ans.)**

(*b*) *When t = 0.01 second*

$$i = \frac{V}{R}[1 - e^{-(R/L)t}] \qquad ...[\text{Eqn. } 23]$$

$$= \frac{200}{10}[1 - e^{-(10/0.1) \times 0.01}] = 20(1 - e^{-1}) = 12.64 \text{ A}$$

and, $$iR = 12.64 \times 10 = 126.4 \text{ V}$$

∴ The voltage drop across the inductance = $\sqrt{(200)^2 - (126.4)^2}$ = **155 V. (Ans.)**

(*ii*) **The flux linkages :**

Now, $$L = \frac{N\phi}{i} \quad \text{or} \quad N\phi = Li$$

∴ Flux linkages ($N\phi$) = Li = 0.1 × 12.64 = **1.264 Wb-turns. (Ans.)**

Example 52. *A choke has a resistance of 50 Ω and inductance of 1.0 H. It is supplied with an A.C. voltage given by 141 sin 314t. Find the expression for transient component of the current flowing through the choke after the voltage is suddenly switched on.* **(Jadavpur University)**

Solution. *Given :* $R = 50\ \Omega$; $L = 1.0$ H ; $e = 141 \sin 314t$

Expression for transient component of the current :

The equation of the transient component of the current is given by :

$$i_t = I_{max} \sin \phi e^{-(R/L)t} \quad \text{...[Eqn. 37]}$$

Here $$I_{max} = \frac{V_{max}}{Z} = \frac{V_{max}}{\sqrt{R^2 + \omega^2 L^2}} = \frac{141}{\sqrt{(50)^2 + (314)^2(1.0)^2}} = 0.443 \text{ A}$$

and, $$\phi = \tan^{-1}\left(\frac{\omega L}{R}\right) = \tan^{-1}\left(\frac{314 \times 1.0}{50}\right) = 80.95°$$

∴ $$i_t = 0.443 \sin 80.95° \, e^{-(50/1.0)t} = \mathbf{0.437\, e^{-50t}}. \quad \textbf{(Ans.)}$$

Example 53. *A series circuit has R = 10 Ω and L = 0.1 H. A 50 Hz sinusoidal voltage of maximum value of 400 V is applied across this circuit. Find an expression for the value of current at any instant after the voltage is applied, assuming that the voltage is zero at the instant of application. Calculate its value 0.02 second after switching on.* **(Punjab University)**

Solution. *Given :* $R = 10\ \Omega$; $L = 0.1$ H ; $f = 50$ Hz ; $V_{max} = 400$ V ; $t = 0.02$ s.

The current consists of transient component and steady-state component. The equation of the resultant current is given by :

$$i = e^{-(R/L)t}\left[\frac{-V_{max}}{\sqrt{R^2 + \omega^2 L^2}} \sin\left(\psi - \tan^{-1}\frac{\omega L}{R}\right)\right]$$

(Transient current)

$$+ \frac{V_{max}}{\sqrt{R^2 + \omega^2 L^2}} \sin\left(\omega t + \psi - \tan^{-1}\frac{\omega L}{R}\right) \quad \text{...[Eqn. 37]}$$

(Steady-state current)

where $$\frac{V_{max}}{\sqrt{R^2 + \omega^2 L^2}} = I_{max}, \ \tan^{-1}\left(\frac{\omega L}{R}\right) = \phi$$

Here, $\psi = 0$ as per given condition, then

$$i = e^{-(R/L)t}\left[-\frac{V_{max}}{\sqrt{R^2 + \omega^2 L^2}} \sin\left(-\tan^{-1}\frac{\omega L}{R}\right)\right] + \frac{V_{max}}{\sqrt{R^2 + \omega^2 L^2}} \sin\left(\omega t - \tan^{-1}\frac{\omega L}{R}\right) \quad \text{...(i)}$$

Now, $$\frac{V_{max}}{R^2 + \omega^2 L^2} = I_{max} = \frac{400}{\sqrt{10^2 + (314)^2(0.1)^2}} = 12.14 \text{ A}$$

$$\tan^{-1}\frac{\omega L}{R} = \phi = \tan^{-1}\left(\frac{314 \times 0.1}{10}\right) = 72.3° = 1.262 \text{ rad.}$$

Substituting the value in eqn. (*i*), we get

$$i = e^{-(10/0.1) \times 0.02}\,[-12.14 \sin(-72.3°)] + 12.14 \sin(314 \times 0.02 - 1.262)$$

$$= e^{-2}\,[12.14 \sin(72.3°)] + 12.14 \sin \underset{\text{(rad)}}{(5.018)}$$

$$= 0.1353(12.14 \times 0.9527) + 12.14 \sin \underset{\text{(deg)}}{(287.5°)}$$

$$= 1.56 - 11.58 = -\mathbf{10.02\ A.} \quad \textbf{(Ans.)}$$

Example 54. *In a simple saw-tooth generator circuit with the thyratron switches on at 150 V and switches off at 10 V. If this circuit is supplied with 250 V D.C. source ; find the time period of saw-tooth wave. The resistance and capacitance have the values of 10 kΩ and 1 μF respectively.*

(Jadavpur University)

Solution. *Given* : R = 10 kΩ ; μ = 1 μF = 1 × 10^{-6} F ; Circuit voltage = 250 V ; Switching on voltage = 150 V ; Switching off voltage = 10 V.

Time period of saw-tooth wave : Refer Fig. 85

Let, V = Applied voltage

v_{C1} = Switching-off voltage of thyratron = 10 V

v_{C2} = Switching-on voltage of thyratron = 150 V

Now, $$v = V(1 - e^{-t/RC}) \qquad ...[\text{Eqn. 30}]$$

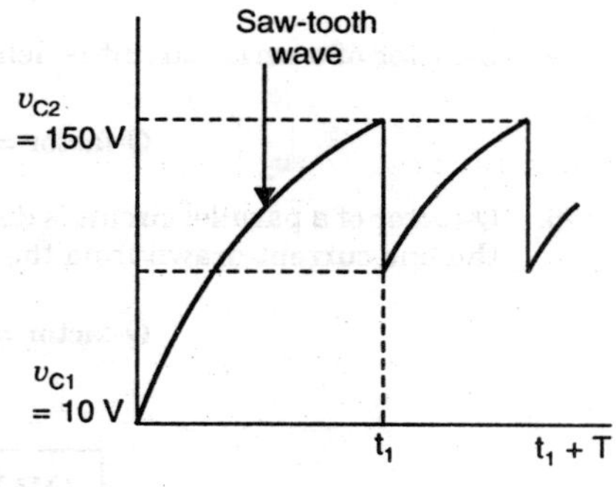

Fig. 85

$$\therefore \qquad v_{C1} = V(1 - e^{-t_1/RC}) \qquad ...(i)$$

$$v_{C2} = V[1 - e^{-(t_1 + T)/RC}] \qquad ...(ii)$$

where, T is the time period of the saw-tooth wave.

Eqns. (*i*) and (*ii*) can be written as :

$$V - v_{C1} = e^{-t_1/RC} \qquad ...(iii)$$

$$V - v_{C2} = e^{-(t_1 + T)/RC} \qquad ...(iv)$$

Dividing (*i*) by (*ii*), we get

$$\frac{V - v_{C1}}{V - v_{C2}} = \frac{e^{-t_1/RC}}{e^{-(t_1 + T)/RC}} = e^{[-t_1/RC - \{-(t_1 + T)/RC\}]} = e^{T/RC}$$

or $$T/RC = \ln\left(\frac{V - v_{C1}}{V - v_{C2}}\right) \quad \text{or} \quad T = \ln\left(\frac{V - v_{C1}}{V - v_{C2}}\right) \times RC$$

Here, v_{C1} = 10 V ; v_{C2} = 150 V

Substituting the values, we get

$$T = \ln\left(\frac{250 - 10}{250 - 150}\right) \times 10 \times 10^3 \times 1 \times 10^{-6} = 0.00875 \text{ s.} \quad \textbf{(Ans.)}$$

HIGHLIGHTS

1. Modern alternators produce an e.m.f. which is for all practical purposes sinusoidal (*i.e.* a sine curve), the equation between the e.m.f. and time being

$$e = E_{max} \sin \omega t$$

where e = instantaneous voltage ; E_{max} = maximum voltage;

ωt = angle through which the armature has turned from neutral.

2. The *r.m.s. value* of an alternating current is given by that steady (D.C.) current which when flowing through a given circuit produces the same *heat* as is produced by the alternating current when flowing through the same circuit for the same time.

$$I_{rms} = 0.707\, I_{max}.$$

3. The *average or mean value* of an alternating current is expressed by that steady current which transfers across any circuit the same **charge** as is transferred by that alternating current during the same time.

$$I_{av} = 0.637\, I_{max}$$

Form factor is the ratio of r.m.s. value to average value of the wave form.

Peak factor is the ratio of maximum value to the r.m.s. value of the wave form.

4. The frequency (series circuit) of the voltage which gives the maximum value of the current in the circuit is called *resonant frequency* (f_r) and the circuit is said to be *resonant*.

$$f_r = \frac{1}{2\pi\sqrt{LC}}.$$

5. *Q*-factor of a series circuit is defined as equal to the voltage magnification in the circuit at resonance.

$$Q\text{-factor} = \frac{1}{R}\sqrt{\frac{L}{C}}.$$

6. *Q*-factor of a parallel circuit is defined as the ratio of the current circulating between its two branches to the line current drawn from the supply or simply, as the current magnification.

$$Q\text{-factor} = \frac{1}{R}\sqrt{\frac{L}{C}}.$$

OBJECTIVE TYPE QUESTIONS

Choose the Correct Answer :

1. A sine wave has a frequency of 50 Hz. Its angular frequency is ...... radian/second.
(*a*) 100 π (*b*) 50 π
(*c*) 25 π (*d*) 5 π.
2. The reactance offered by a capacitor to alternating current of frequency 50 Hz is 20 Ω. If frequency is increased to 100 Hz, reactance becomes ...... ohms.
(*a*) 2.5 (*b*) 5
(*c*) 10 (*d*) 15.
3. The period of a wave is
(*a*) the same as frequency (*b*) time required to complete one cycle
(*c*) expressed in amperes (*d*) none of the above.
4. The form factor is the ratio of
(*a*) peak value to r.m.s. value (*b*) r.m.s. value to average value
(*c*) average value to r.m.s. value (*d*) none of the above.
5. The period of a sine wave is $\frac{1}{50}$ seconds. Its frequency is
(*a*) 20 Hz (*b*) 30 Hz
(*c*) 40 Hz (*d*) 50 Hz.
6. An A.C. current is given by $i = 200 \sin 100\pi t$. It will achieve a value of 100 A after ...... second.
(*a*) $\frac{1}{900}$ (*b*) $\frac{1}{800}$
(*c*) $\frac{1}{700}$ (*d*) $\frac{1}{600}$.
7. A heater is rated as 230 V, 10 kW, A.C. The value 230 V refers to
(*a*) average voltage (*b*) r.m.s. voltage
(*c*) peak voltage (*d*) none of the above.
8. If two sinusoids of the same frequency but of different amplitudes and phase angles are subtracted, the resultant is
(*a*) a sinusoid of the same frequency (*b*) a sinusoid of half the original frequency
(*c*) a sinusoid of double the frequency (*d*) not a sinusoid.

9. The peak value of a sine wave is 200 V. Its average value is
(*a*) 127.4 V (*b*) 141.4 V
(*c*) 282.8 V (*d*) 200 V.

10. If two sine waves of the same frequency have a phase difference of π radians, then
(*a*) both will reach their minimum values at the same instant
(*b*) both will reach their maximum values at the same instant
(*c*) when one wave reaches its maximum value, the other will reach its minimum value
(*d*) none of the above.

11. The r.m.s. value of a sine wave is 100 A. Its peak value is
(*a*) 70.7 A (*b*) 141.4 A
(*c*) 150 A (*d*) 282.8 A.

12. If two waves are expressed as $e_1 = E_{m_1} \sin(\omega t + \alpha_1)$ and $e_2 = E_{m_2} \sin(\omega t + \alpha_2)$, then
(*a*) e_1 is leading e_2 by $\angle(\alpha_2 - \alpha_1)$ (*b*) e_2 is leading e_1 by $\angle(\alpha_2 - \alpha_1)$
(*c*) e_2 is leading e_1 by $\angle(\alpha_1 - \alpha_2)$ (*d*) e_1 is in phase with e_2.

13. The voltage of domestic supply is 220 V. This figure represents
(*a*) mean value (*b*) r.m.s. value
(*c*) peak value (*d*) average value.

14. Two waves of the same frequency have opposite phase when the phase angle between them is•
(*a*) 360° (*b*) 180°
(*c*) 90° (*d*) 0°.

15. The power consumed in a circuit element will be least when the phase difference between the current and voltage is
(*a*) 180° (*b*) 90°
(*c*) 60° (*d*) 0°.

16. The r.m.s. value and mean value is the same in the case of
(*a*) triangular wave (*b*) sine wave
(*c*) square wave (*d*) half wave rectified sine wave.

17. For the same peak value which of the following wave will have the highest r.m.s. value ?
(*a*) square wave (*b*) half wave rectified sine wave
(*c*) triangular wave (*d*) sine wave.

18. For the same peak value, which of the following wave has the *least* mean value ?
(*a*) half wave rectified sine wave (*b*) triangular wave
(*c*) sine wave (*d*) square wave.

19. For a sine wave with peak value I_{max} the r.m.s. value is
(*a*) 0.5 I_{max} (*b*) 0.707 I_{max}
(*c*) 0.9 I_{max} (*d*) 1.414 I_{max}.

20. Form Factor is the ratio of
(*a*) average value/r.m.s. value (*b*) average value/peak value
(*c*) r.m.s. value/average value (*d*) r.m.s. value/peak value.

21. Form factor for a sine wave is
(*a*) 1.414 (*b*) 0.707
(*c*) 1.11 (*d*) 0.637.

22. For a sine wave with peak value E_{max} the average value is
(*a*) 0.636 E_{max} (*b*) 0.707 E_{max}
(*c*) 0.434 E_{max} (*d*) 1.414 E_{max}.

23. The current in a circuit is given by : $i = 100 \sin 314\, t$ amperes
The maximum value and frequency of current are
(*a*) $50\sqrt{2}$ A, 100 Hz (*b*) $100\sqrt{2}$ A, 100 Hz
(*c*) 100 A, 50 Hz (*d*) 70.7A, 50 Hz.

24. For a frequency of 200 Hz, the time period will be
(*a*) 0.05 s (*b*) 0.005 s
(*c*) 0.0005 s (*d*) 0.5 s.

25. The phase difference between voltage and current wave through a circuit element is given as 30°. The essential condition is that
(*a*) both waves must have same frequency
(*b*) both waves must have identical peak values
(*c*) both waves must have zero value at the same time
(*d*) none of the above.

26. An A.C. voltage of 50 Hz has a maximum value of 50 V. Its value after 1/600 second after the instant the current is zero, will be
(*a*) 5 V (*b*) 12.5 V
(*c*) 25 V (*d*) 43.3 V.

27. When two waves are in phase they have peak values at an interval of
(*a*) 180° (*b*) 120°
(*c*) 90° (*d*) none of the above.

28. For 200 V r.m.s. value triangular wave, the peak voltage will be
(*a*) 200 V (*b*) 222 V
(*c*) 282 V (*d*) 346 V.

29. A sine wave of voltage varies from zero to maximum of 200 V. How much is the voltage at the instant of 30° of the cycle ?
(*a*) 50 V (*b*) 82.8 V
(*c*) 100 V (*d*) 173.2 V.

30. How much r.m.s. current does a 300 W 200 V bulb take from the 200 V, 50 Hz power line ?
(*a*) 0.5 A (*b*) 1.5 A
(*c*) 2 A (*d*) 3 A.

31. Two sinusoidal currents are given by $i_1 = 100 \sin (\omega t + \pi/3)$, and $i_2 = 150 \sin (\omega t - \pi/4)$ The phase difference between them is ...... degrees.
(*a*) 15 (*b*) 50
(*c*) 60 (*d*) 105.

32. The r.m.s. value of a half-wave rectified current is 100 A. Its value for full-wave rectification would be ...... amperes.
(*a*) 141.4 (*b*) 200
(*c*) 200/π (*d*) 40/π.

33. From the two voltages equations $e_1 = E_{max} \sin 100\pi t$, and $e_2 = E_{max} \sin (100\, \pi t + \pi/6)$, it is obvious that
(*a*) 1 leads 2 by 30° (*b*) 2 lags behind 1
(*c*) 2 achieves its maximum value $\frac{1}{600}$ second before 1 does
(*d*) 1 achieves its zero value $\frac{1}{600}$ second before 2.

34. The r.m.s. value of a sinusoidal A.C. current is equal to its value at an angle of ...... degrees.
(*a*) 90 (*b*) 60
(*c*) 45 (*d*) 30.

35. Capacitive reactance is more when
(*a*) capacitance is less and frequency of supply is less
(*b*) capacitance is less and frequency of supply is more
(*c*) capacitance is more and frequency of supply is less
(*d*) capacitance is more and frequency of supply is more.

36. Time constant of a capacitive circuit increases with the
(*a*) increase of capacitance and decrease of resistance
(*b*) increase of capacitance and increase of resistance
(*c*) decrease of capacitance and decrease of resistance
(*d*) decrease of capacitance and increase of resistance.

37. In a series circuit on resonance, following will occur
(*a*) $V = V_R$ (*b*) $X_L = X_C$
(*c*) $Z = R$ (*d*) $V_L = V_C$
(*e*) all above.

38. In a series resonant circuit, the impedance of the circuit is
(*a*) minimum (*b*) maximum
(*c*) zero (*d*) none of the above.

39. Power factor of an electrical circuit is equal to
(*a*) R/Z
(*b*) cosine of phase angle difference between current and voltage
(*c*) kW/kVA (*d*) ratio of useful current to total current I_W/I
(*e*) all above.

40. The best place to install a capacitor is
(*a*) very near to inductive load (*b*) across the terminals of the inductive load
(*c*) far away from the inductive load (*d*) any where.

41. Poor power factor
(*a*) reduces load handling capability of electrical system
(*b*) results in more power losses in the electrical system
(*c*) overloads alternators, transformers and distribution lines
(*d*) results in more voltage drop in the line (*e*) results in all above.

42. Capacitors for power factor correction are rated in
(*a*) kW (*b*) kVA
(*c*) kV (*d*) kVAR.

43. In series resonant circuit, increasing inductance to its twice value and reducing capacitance to its half value
(*a*) will change the maximum value of current at resonance
(*b*) will change the resonance frequency
(*c*) will change the impedance at resonance frequency
(*d*) will increase the selectivity of the circuit.

44. Pure inductive circuit
(*a*) consumes some power on average
(*b*) does not take power at all from a line
(*c*) takes power from the line during some part of the cycle and then returns back to it during other part of the cycle
(*d*) none of the above.

45. Inductance affects the direct current flow
(*a*) only at the time of turning off (*b*) only at the time of turning on
(*c*) at the time of turning on and off (*d*) at all the time of operation.

46. Inductance of a coil varies
(*a*) directly as the cross-sectional area of magnetic core
(*b*) directly as square of number of turns (*c*) directly as the permeability of the core
(*d*) inversely as the length of the iron path (*e*) as (*a*) to (*d*).

47. All the rules and laws of D.C. circuit also apply to A.C. circuit containing
(*a*) capacitance only (*b*) inductance only
(*c*) resistance only (*d*) all above.

48. Time constant of an inductive circuit
(*a*) increases with increase of inductance and decrease of resistance
(*b*) increases with the increase of inductance and the increase of resistance
(*c*) increases with decrease of inductance and decrease of resistance
(*d*) increases with decrease of inductance and increase of resistance.

49. Power factor of an inductive circuit is usually improved by connecting capacitor to it in
(*a*) parallel (*b*) series
(*c*) either (*a*) or (*b*) (*d*) none of the above.

50. In a highly capacitive circuit the
(*a*) apparent power is equal to the actual power (*b*) reactive power is more than the apparent power
(*c*) reactive power is more than the actual power (*d*) actual power is more than its reactive power.

51. Power factor of the following circuit will be zero
(*a*) resistance (*b*) inductance
(*c*) capacitance (*d*) both (*b*) and (*c*).

52. Power factor of the following circuit will be unity
(*a*) inductance (*b*) capacitance
(*c*) resistance (*d*) both (*a*) and (*b*).

53. Power factor of the system is kept high
(*a*) to reduce line losses
(*b*) to maximise the utilization of the capacities of generators, lines and transformers
(*c*) to reduce voltage regulation of the line
(*d*) due to all above reasons.

54. The time constant of the capacitance circuit is defined as the time during which voltage
(*a*) falls to 36.8% of its final steady value (*b*) rises to 38.6% of its final steady value
(*c*) rises to 63.2% of its final steady value (*d*) none of the above.

55. In the *R-L-C* containing $R = 4.5\ \Omega$, $L = 0.06$ H, $C = 0.6\ \mu$F the power factor will be
(*a*) zero (*b*) lagging
(*c*) leading (*d*) unity.

ANSWERS

A. Choose the Correct Answer :

1. (*a*)	**2.** (*c*)	**3.** (*b*)	**4.** (*b*)	**5.** (*d*)	**6.** (*d*)
7. (*b*)	**8.** (*a*)	**9.** (*a*)	**10.** (*c*)	**11.** (*b*)	**12.** (*b*)
13. (*b*)	**14.** (*b*)	**15.** (*b*)	**16.** (*c*)	**17.** (*a*)	**18.** (*a*)
19. (*b*)	**20.** (*c*)	**21.** (*c*)	**22.** (*a*)	**23.** (*c*)	**24.** (*b*)
25. (*a*)	**26.** (*c*)	**27.** (*d*)	**28.** (*d*)	**29.** (*c*)	**30.** (*b*)
31. (*d*)	**32.** (*a*)	**33.** (*c*)	**34.** (*c*)	**35.** (*a*)	**36.** (*b*)
37. (*e*)	**38.** (*a*)	**39.** (*e*)	**40.** (*b*)	**41.** (*e*)	**42.** (*d*)
43. (*d*)	**44.** (*c*)	**45.** (*c*)	**46.** (*e*)	**47.** (*c*)	**48.** (*a*)
49. (*a*)	**50.** (*c*)	**51.** (*d*)	**52.** (*c*)	**53.** (*d*)	**54.** (*c*)
55. (*c*).					

THEORETICAL QUESTIONS

1. Define the following terms :
Circuit, Electrical network, Active network, Node and Branch.
2. What are the limitations of ohm's law ?
3. State and explain Kirchhoff's laws.
4. Discuss briefly application of Kirchhoff's laws.

5. Explain the nodal voltage method for solving networks. How are the nodal equations written ?
6. Explain Cramer's rule used for solving equations by determinants.
7. State and explain Superposition theorem.
8. State Norton's theorem. List the steps for finding the current in a branch of a network with the help of this theorem.
9. State Thevenin's theorem.
10. State the maximum power transfer theorem and explain its importance.
11. Define the following terms as applied to an alternating current :
Cycle, frequency, time period, amplitude.
12. What do you mean by the term "Phase difference" ?
13. Explain the following terms relating alternating current :
(*i*) R.M.S. value (*ii*) Average value
(*iii*) Form factor (*iv*) Peak factor.
14. Explain briefly the following as applied to A.C. series and parallel circuits :
(*i*) Resonance frequency (*ii*) Q-factor.
15. What do you mean by transient disturbances ?
16. Define single energy and double energy transients.

UNSOLVED EXAMPLES

1. An alternating current of frequency 60 Hz has a maximum value of 120 A. Write down the equation for its instantaneous value. Reckoning time from the instant the current is zero and is becoming positive, find :
(*i*) The instantaneous value after $\frac{1}{360}$ second ;
(*ii*) The time taken to reach 96 A for the first time. **[Ans.** 103.9 A, 0.00245 second]
2. An alternating current of frequency 50 Hz has a maximum value of 100 A. Calculate :
(*i*) Its value 1/600 second after the instant the current is zero and its value decreasing thereafter wards.
(*ii*) How many seconds after the instant the current is zero (increasing thereafter wards) will the current attain the value of 86.6 A ? **[Ans.** – 50 A, 1/300 s]
3. Calculate the r.m.s. value, the form factor of a periodic voltage having the following values for equal time intervals changing suddenly from one value to the next : 0, 5, 10, 20, 50, 60, 50, 20, 10, 5, 0, 5, 10 V etc. What would be the r.m.s. value of sine wave having the same peak value ?
[Ans. 31 V ; 23 V ; 1.35 ; (app.) ; 42.2 V]
4. A sinusoidally varying alternating current has an average value of 127.4 A. When its value is zero, then its rate of change is 62,800 A/s. Find the analytical expression for the sine wave.
[Ans. $i = 200 \sin 100\,\pi t$]
5. A coil of resistance 10 Ω and inductance 0.1 H is connected in series with a 150 μF capacitor across a 200 V, 50 Hz supply. Calculate (*i*) the inductive reactance, (*ii*) the capacitive reactance, (*iii*) the impedance (*iv*) the current, (*v*) the power factor, (*vi*) the voltage across the coil and the capacitor respectively.
[Ans. (*i*) 31.4 Ω, (*ii*) 21.2 Ω, (*iii*) 14.3 Ω, (*iv*) 14 A, (*v*) 0.7 lag (*vi*) 460 V, 297 V]
6. A circuit is made up of 10 Ω resistance, 12 mH inductance and 281.5 μF capacitance in series. The supply voltage is 100 V (constant). Calculate the value of the current when the supply frequency is (*i*) 50 Hz and (*ii*) 150 Hz. **[Ans.** 8 A leading ; 8 A lagging]
7. A coil having a resistance of 10 Ω and an inductance of 0.2 H is connected in series with a capacitor of 50.7 μF. The circuit is connected across a 100 V, 50 Hz A.C. supply. Calculate (*i*) the current flowing (*ii*) the voltage across the capacitor (*iii*) the voltage across the coil. Draw a vector diagram to scale.
[Ans. (*i*) 10 A (*ii*) 628 V (*iii*) 635 V]
8. A coil is in series with a 20 μF capacitor across a 230 V, 50 Hz supply. The current taken by the circuit is 8 A and the power consumed is 200 W. Calculate the inductance of the coil if the power factor of the circuit is (*i*) leading and (*ii*) lagging.

Sketch a vector diagram for each condition and calculate the coil power factor in each case.
[Ans. 0.415 H ; 0.597 H ; 0.0238 ; 0.0166]

9. An A.C. series circuit has a resistance of 10 Ω, an inductance of 0.2 H and a capacitance of 60 μF. Calculate :
(*i*) the resonant frequency (*ii*) the current and
(*iii*) the power at resonance.
Given that the applied voltage is 200 V. **[Ans.** 46 Hz ; 20 A ; 4 kW]

10. A circuit consists of an inductor which has a resistance of 10 Ω and an inductance of 0.3 H, in series with a capacitor of 30 μF capacitance. Calculate :
(*i*) The impedance of the circuit to currents of 40 Hz ;
(*ii*) The resonant frequency ;
(*iii*) The peak value of stored energy in joules when the applied voltage is 200 V at the resonant frequency.
[Ans. 58.31 Ω ; 53 Hz ; 120 J]

11. A resistor and a capacitor are connected in series with a variable inductor. When the circuit is connected to a 240 V, 50 Hz supply, the maximum current given by varying the inductance is 0.5 A. At this current, the voltage across the capacitor is 250 V. Calculate the values of the following :
(*i*) The resistance ; (*ii*) The capacitance ;
(*iii*) The inductance.
Neglect the resistance of the inductor. **[Ans.** 480 Ω, 6.36 μF ; 1.59 H]

12. A resistance, a capacitor and a variable inductance are connected in series across a 200 V, 50 Hz supply. The maximum current which can be obtained by varying the inductance is 314 mA and the voltage across the capacitor is then 300 V. Calculate the capacitance of the capacitor and the values of the inductance and resistance. **[Ans.** 3.33 μF, 3.04 H, 637 Ω]

13. A circuit consisting of a coil of resistance 12 Ω and inductance 0.15 H in series with a capacitor of 12 μF is connected to a variable frequency supply which has a constant voltage of 24 V.
Calculate : (*i*) The resonant frequency, (*ii*) The current in the circuit at resonance, (*iii*) The voltage across the capacitor and the coil at resonance. **[Ans.** (*i*) 153 Hz, (*ii*) 2 A, (*iii*) 224 V]

14. A resistance of 24 Ω, a capacitance of 150 μF and an inductance of 0.16 H are connected in series with each other. A supply at 240 V, 50 Hz is applied to the ends of the combination. Calculate (*i*) the current in the circuit (*ii*) the potential differences across each element of the circuit (*iii*) the frequency to which the supply would need to be changed so that the current would be at unity power-factor and find the current at this frequency. **[Ans.** (*i*) 6.37 A (*ii*) V_R = 152.8 V, V_C = 320 V, V_L = 123.3 V (*iii*) 32 Hz ; 10 A]

15. A coil-A of inductance 80 mH and resistance 120 Ω is connected to a 230 V, 50 Hz single-phase supply. In parallel with it is a 16 μF capacitor is series with a 40 Ω non-inductive resistor *B*. Determine
(*i*) The power factor of the combined circuit,
(*ii*) The total power taken from the supply. **[Ans.** (*i*) 0.945 lead (*ii*) 473 W]

16. A choking coil of inductance 0.08 H and resistance 12 ohm, is connected in parallel with a capacitor of 120 μF. The combination is connected to a supply at 240 V, 50 Hz. Determine the total current from the supply and its power factor. Illustrate your answers with a phasor diagram. **[Ans.** 3.94 A, 0.943 lag]

17. A choking coil having a resistance of 20 Ω and an inductance of 0.07 henry is connected with a capacitor of 60 μF capacitance which is in series with a resistor of 50 Ω. Calculate the total current and the phase angle when this arrangement is connected to 200 V, 50 Hz mains. **[Ans.** 7.15 A, 24° 39′ lag]

18. A coil of resistance of 15 Ω and inductance 0.05 H is connected in parallel with a non-inductive resistance of 20 Ω. Find (*i*) the current in each branch ; (*ii*) the total current (*iii*) the phase angle of whole arrangement for an applied voltage of 200 V at 50 Hz. **[Ans.** 9.22 A ; 10 A ; 22.1°]

19. A sinusoidal 50 Hz voltage of 200 V (r.m.s.) supplies the following three circuits which are in parallel : (*i*) a coil of inductance 0.03 H and resistance 3 Ω ; (*ii*) a capacitor of 400 μF in series with a resistance of 100 Ω ; (*iii*) a coil of inductance 0.02 H and resistance 7 Ω in series with a 300 μF capacitor. Find the total current supplied and draw a complete vector diagram. **[Ans.** 29.4 A]

20. In a series-parallel circuit, the two parallel branches *A* and *B* are in series with *C*. The impedances are $Z_A = (10 - j8)\ \Omega$, $Z_B = (9 - j6)\ \Omega$ and $Z_C = (100 + j0)$. Find the currents I_A and I_B and the phase difference between them. Draw the phasor diagram. **[Ans.** $I_A = 12.71 \angle -30°\ 58'$ $I_B = 15 \angle -35°\ 56'$; 4° 58′]

3

Three Phase A.C. Circuits

1. INTRODUCTION

- Generation, transmission and heavy-power utilisation of A.C. electric energy almost invariably involve a type of system or circuit called a *polyphase system* or *polyphase circuit*. In such a system, each voltage source consists of a group of voltages having relative magnitudes and phase angles. Thus, a *m-phase system will employ voltage sources which, conventionally, consist of m voltages substantially equal in magnitude and successively displaced by a phase angle of 360° / m.*
- A *3-phase system* will employ voltage sources which, conventionally, consist of *three voltages substantially equal in magnitude* and *displaced by phase angles of 120°*. Because it *possesses definite economic and operating advantages, the 3-phase system is by far the most common,* and consequently emphasis is placed on 3-phase *circuits*.

2. ADVANTAGES OF POLYPHASE SYSTEMS

The advantages of polyphase systems *over single-phase systems* are :

1. A polyphase transmission line requires less conductor material than a single-phase line for transmitting the same amount power at the same voltage.

2. For a given frame size a polyphase machine gives a higher output than a single-phase machine. For example, output of a 3-phase motor is 1.5 times the output of single-phase motor of same size.

3. Polyphase motors have a uniform torque where most of the single-phase motors have a pulsating torque.

4. Polyphase induction motors are self-starting and are more efficient. On the other hand single-phase induction motors are not self-starting and are less efficient.

5. Per unit of output, the polyphase machine is very much cheaper.

6. Power factor of a single-phase motor is lower than that of polyphase motor of the same rating.

7. Rotating field can be set up by passing polyphase current through stationary coils.

8. Parallel operation of polyphase alternators is simple as compared to that of single-phase alternators because of pulsating reaction in single-phase alternator.

It has been found that the above advantages are *best realised in the case of three-phase systems*. Consequently, the electric power is generated and transmitted in the form of three-phase system.

3. GENERATION OF THREE-PHASE VOLTAGES

- Let us consider an elementary 3-phase 2-pole generator as shown in Fig. 1. On the armature are three coils, ll', mm', and nn' whose axes are displaced 120° in space from each other.

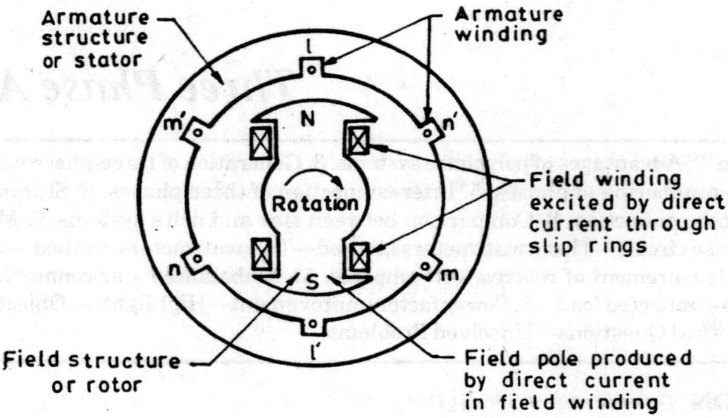

Fig. 1. Elementary 3-phase 2-pole generator.

- When the field is excited and rotated, voltages will be generated in the three phases in accordance with Faraday's law. If the field structure is so designed that the flux is distributed sinusoidally over the poles, the flux linking any phase will vary sinusoidally with time and sinusoidal voltages will be induced in three-phases. These three waves will be displaced 120 electrical degrees (Fig. 2) in time as a result of the phases being displaced

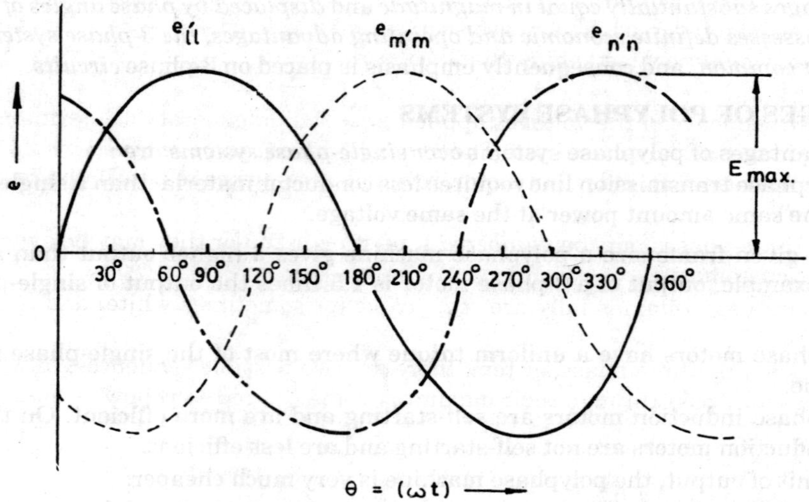

Fig. 2. Voltage waves generated in windings of Fig. 1.

120° in space. The corresponding phasor diagram is shown in Fig. 3. The equations of the instantaneous values of the three voltages (given by Fig. 2) are :

$$e_{l'l} = E_{max.} \sin \omega t$$
$$e_{m'm} = E_{max.} \sin (\omega t - 120°)$$
$$e_{n'n} = E_{max.} \sin (\omega t - 240°)$$

The sum of the above three e.m.fs. is always zero as shown below :

Resultant instantaneous e.m.f.

$= e_{l'l} + e_{m'm} + e_{n'n}$

$= E_{max.} \sin \omega t + E_{max.} \sin (\omega t - 120°) + E_{max.} \sin (\omega t - 240°)$

$= E_{max.} [\sin \omega t + (\sin \omega t \cos 120° - \cos \omega t \sin 120° + \sin \omega t \cos 240° - \cos \omega t \sin 240°)]$

$= E_{max.} [\sin \omega t + (- \sin \omega t \cos 60° - \cos \omega t \sin 60° - \sin \omega t \cos 60° + \cos \omega t \sin 60°)]$

$= E_{max.} (\sin \omega t - 2 \sin \omega t \cos 60°)$

$= E_{max.} (\sin \omega t - \sin \omega t) = 0.$

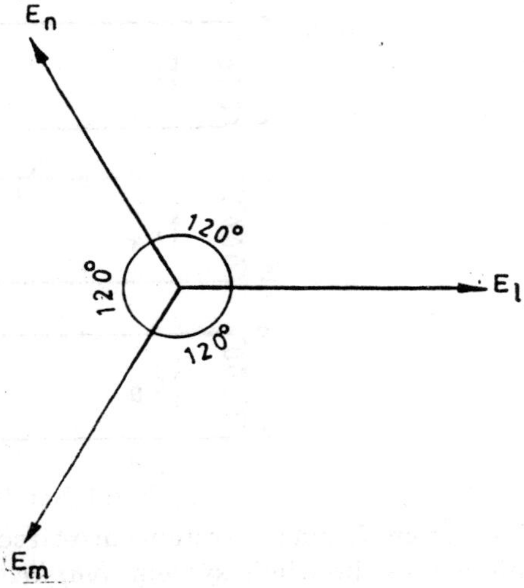

Fig. 3. Phasor diagram of generated voltages.

4. PHASE SEQUENCE AND NUMBERING OF PHASES

- *By phase sequence is meant the order in which the three phases attain their peak or maximum.*

In the generation of three-phase e.m.fs. in Fig. 2 clockwise rotation of the field system in Fig. 1 was assumed. This assumption made the e.m.f. of phase '**m**' lag behind that of '**l**' by 120° and in a similar way, made that of '**n**' lag behind that of '**m**' by 120° (or that of **l** by 240°). Hence, the order in which the e.m.fs. of phases **l, m** and **n** attain their maximum value is **lmn**. It is called the *phase order* or phase sequence **l** → **m** → **n**. If now the rotation of field structure of Fig. 1 is reversed *i.e.* made counter-clockwise, then the order in which three phases would attain their corresponding maximum voltages would also be reversed. The phase sequence would become **l** → **n** → **m**. This means that e.m.f. of phase '**n**' would now lag behind that of phase '**l**' by 120° instead of 240° as in the previous case.

The phase sequence of the voltages applied to a load, in general, is determined by the order in which the 3-phase lines are connected. The *phase sequence can be reversed* by *interchanging any pair of lines*. (In the case of an induction motor, reversal of sequence results in the reversed direction of motor rotation).

- The three-phases may be *numbered* **l, m, n** or 1, 2, 3 or they may be given three *colours* (as is customary).

The colours used commercially are *red, yellow* (or sometimes white) and *blue*. In this case sequence is RYB.

Evidently in any three-phase system, there are two possible sequences, in which three coils or phase voltages may pass through their maximum value *i.e.,* red → yellow → blue (RYB) or red → blue → yellow (RBY).

By convention :

RYB ...... taken as *positive*.

RBY ...... taken as *negative*.

5. INTER-CONNECTION OF THREE PHASES

Each coil of three phases has two terminals [one 'start' (*S*) and another 'finish' (*F*)] and if individual phase is connected to a separate load circuit, as shown in Fig. 4, we get a non-interlinked

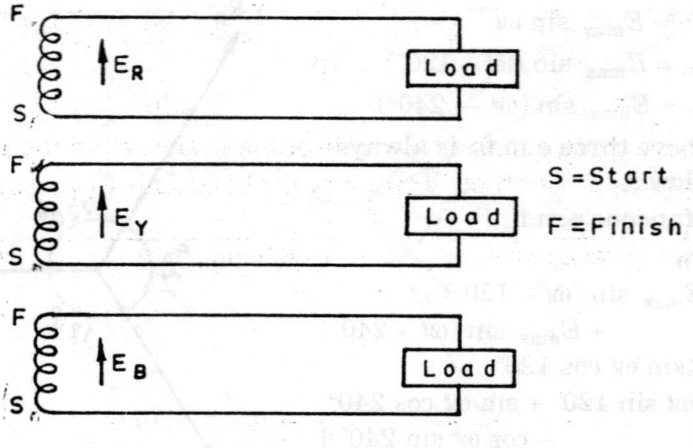

Fig. 4. Non-interlinked 3-phase system.

3-phase system. In such a system each circuit will require two conductors, therefore, 6 conductors in all. This makes the whole system *complicated* and *expensive*. Hence the *three phases are generally interconnected which results in substantial saving of copper*.

The general method of inter-connections are :

1. Star or Wye (*Y*) connection.
2. Mesh or delta (Δ) connection.

6. STAR OR WYE (Y) CONNECTION

- In this method of inter-connection the similar ends either the 'start' or 'finish' are joined together at point *N*. This common point *N* [Fig. 5 (*a*)] is called *star point* or *neutral point*. Ordinarily only three wires are carried to the external circuit giving 3-phase, 3-wire start connected system but sometimes a *fourth-wire*, known as *neutral wire* is carried to the neutral point of the external load circuit giving 3-phase, 4 wire star connected system.
- The *voltage between any line and the neutral point* (*i.e.* voltage across the phase winding) is called the **'phase voltage' (E_{ph})** ; while the *voltage available between any pair of terminals* (or outers) is called the **'line voltage' (E_L)**.

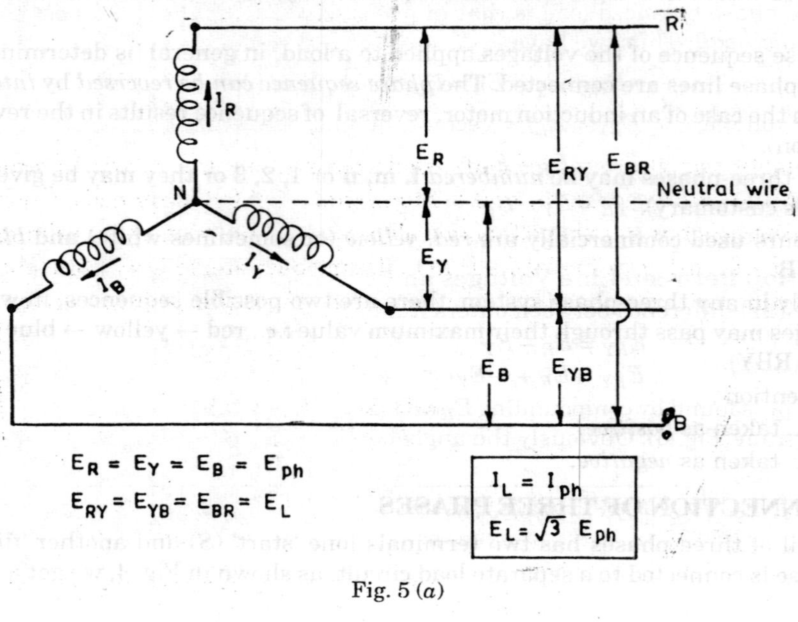

Fig. 5 (*a*)

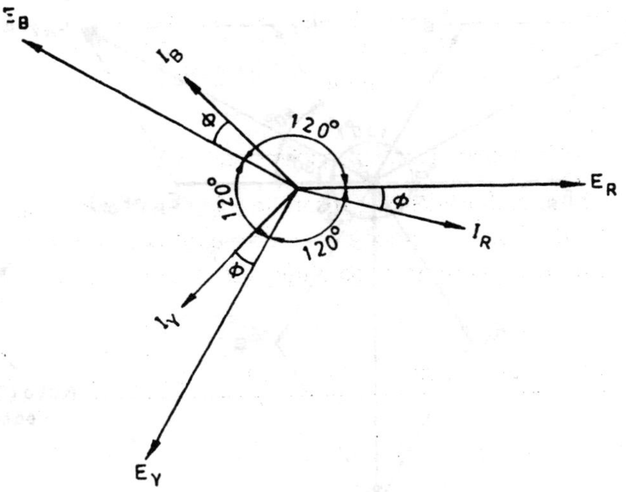

Fig. 5. (*b*) Star-connected three-phase net work.

- In star connection, as is evident in Fig. 5 (*a*) there are two-phase windings between each pair of terminals, but since their *similar* ends have been joined together, they are in *opposition.* Obviously, the instantaneous value of potential difference between any two terminals is the *arithmetic difference* of the two-phase e.m.fs. concerned. However, the r.m.s. value of this potential difference is given by the *vector difference* of the two-phase e.m.fs.

- Fig. 5 (*b*) shows the vector diagram for phase voltages and currents in a star connection where a *balanced system has been assumed.* [A balanced system is one in which (*i*) the voltages in all phases are equal in magnitude and differ in phase from one another by equal angles, in this case, the angle = 360/3 = 120°, (*ii*) the currents in the three phases are equal in magnitude and also differ in phase from one another by equal angles. A three-phase balanced load is that in which the loads connected across the three-phases are *identical*]. Thus, we have

$$E_R = E_Y = E_B = E_{ph} \text{ (phase e.m.f.)}$$

Line voltage, $E_{RY}(= E_L)$ = Vector difference of E_R and E_Y

$$= E_R - E_Y$$

Line voltage, $E_{YB} = E_Y - E_B$

Line voltage, $E_{BR} = E_B - E_R$.

(*a*) **Relation Between Line Voltages and Phase Voltages.** Refer Fig. 6.

The potential difference between outers R any Y is

$$E_{RY} = E_R - E_Y \qquad \text{[vector difference]}$$

or

$$E_{RY} = E_R + (-E_Y) \qquad \text{[vector sum]}$$

Hence, E_{RY} is found by compounding E_R and E_Y reversed and its value is given by the diagonal of the parallelogram (Fig. 6). Obviously the angle between E_R and E_Y reversed is 60° and the value of

$$E_{RY} \text{ (or } E_L) = \sqrt{E_R^2 + E_Y^2 + 2E_R E_Y \cos 60^\circ} = \sqrt{E_{ph}^2 + E_{ph}^2 + 2E_{ph} \times E_{ph} \times \frac{1}{2}} = \sqrt{3} E_{ph} \qquad ...(1)$$

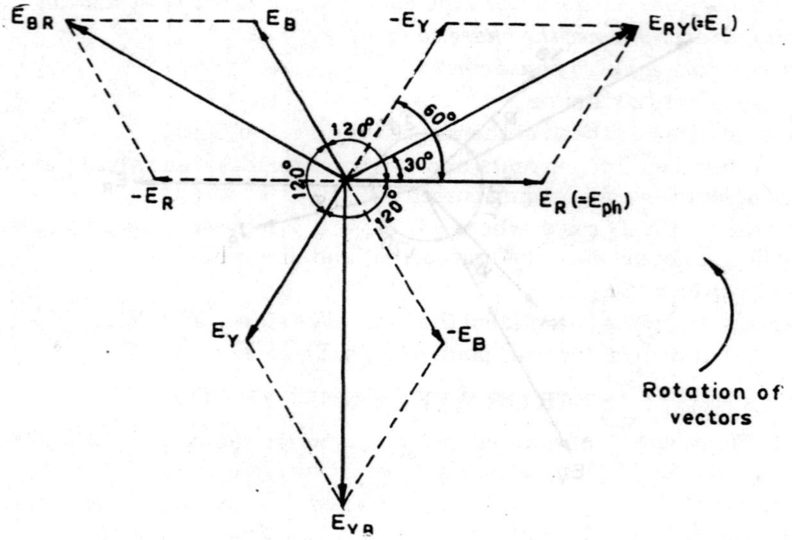

Fig. 6. Vector diagram for star connected net-work.

Similarly $E_{YB}(=E_L) = E_Y - E_B = \sqrt{3}E_{ph}$

and $E_{BR}(=E_L) = E_B - E_R = \sqrt{3}E_{ph}$

i.e., $E_{RY} = E_{YB} = E_{BR} = E_L = \sqrt{3}E_{ph}$

Hence, $E_L = \sqrt{3}E_{ph}$ ...(2)

(*i.e., Line voltage* = $\sqrt{3}$ *phase voltage*).

(*b*) **Relation between Line Currents and Phase Currents.** Since in star-connected system each line conductor is connected to separate phase, so the current flowing through the line and phase are same.

Current in outer (or line) $R = I_R$

Current in outer $Y = I_Y$

Current in outer $B = I_B$

Since $I_R = I_Y = I_B =$ say, I_{ph}—the phase current

∴ Line current, $I_L = I_{ph}$ ...(3)

(*c*) **Power.** If the phase current has a phase difference of ϕ with phase voltage,

Power per phase $= E_{ph}I_{ph}\cos\phi$

Total power (true), $P = 3 \times$ power per phase

$P = 3 \times E_{ph}I_{ph}\cos\phi$ ...[4 (*a*)]

Now $E_{ph} = \dfrac{E_L}{\sqrt{3}}$ and $I_{ph} = I_L$

Hence in terms of line values, the above expression becomes

$$P = 3 \times \frac{E_L}{\sqrt{3}} I_L \cos\phi$$

or

$$P = \sqrt{3}E_L I_L \cos\phi \qquad \text{...[4 (b)]}$$

(Apparent power $= \sqrt{3}E_L I_L$).

In a *balanced star-connected net work* the following points are worthnoting :

(*i*) *Line voltages are $\sqrt{3}$ times the phase voltages.*

(*ii*) *Line currents are equal to phase currents.*

(*iii*) Line voltages are 120° apart.

(*iv*) Line voltages are 30° ahead of the respective phase voltages.

(*v*) The angle between line currents and the corresponding line voltages is $(30° \pm \phi)$ + for lagging currents – ve for leading currents.

(*vi*) True power = $\sqrt{3}\, E_L I_L \cos\phi$, where ϕ is the angle between respective phase current and phase voltage, not between the line current and line voltage.

(*vii*) Apparent power = $\sqrt{3} E_L I_L$.

(*viii*) In balance system, the potential of the neutral or star point is zero.

$\therefore$ Potential at neutral (or star) point $= E_{NR} + E_{NY} + E_{NB} = 0$.

STAR OR WYE CONNECTION

Example 1. *Three equal impedances each having a resistance of 25 Ω and reactance of 40 Ω are connected in star to a 400 V, 3-phase, 50 Hz system. Calculate :*

(*i*) *The line current* (*ii*) *Power factor, and*

(*iii*) *Power consumed.*

Solution. Resistance per phase, $R_{ph} = 25\ \Omega$

Reactance per phase, $X_{ph} = 40\ \Omega$

Line voltage, $E_L = 400$ V

Line current, I_L :

Power factor, $\cos\phi$:

Power consumed, P :

Refer Fig. 7.

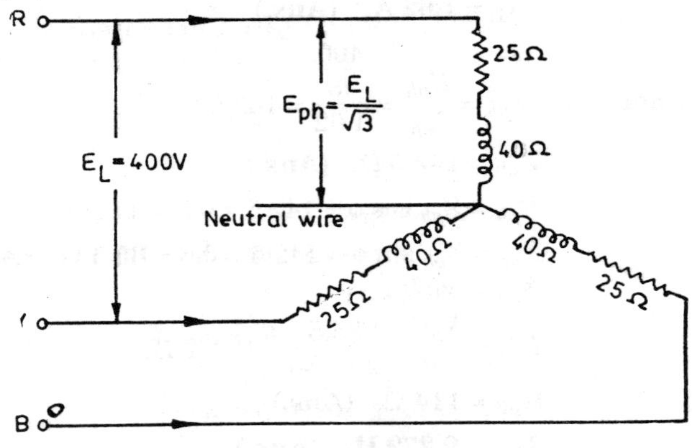

Fig. 7

Impedance per phase, $Z_{ph} = \sqrt{R_{ph}^2 + X_{ph}^2}$

$\therefore$ $Z_{ph} = \sqrt{25^2 + 40^2} = 47.17\ \Omega$

Phase voltage, $E_{ph} = \frac{E_L}{\sqrt{3}} = \frac{400}{\sqrt{3}} = 231$ V

Phase current, $I_{ph} = \frac{E_{ph}}{Z_{ph}} = \frac{231}{47.17} = 4.9$ A (app.)

(*i*) **Line current,** I_L = phase current, I_{ph}

$\therefore$ $\mathbf{I_L = 4.9\ A.}$ **(Ans.)**

(*ii*) **Power factor,** $\cos\phi = \frac{R_{ph}}{Z_{ph}} = \frac{95}{47.17} = \mathbf{0.53\ (lag).}$ **(Ans.)**

(*iii*) **Power consumed,** $P = \sqrt{3}E_L I_L \cos\phi = \sqrt{3} \times 400 \times 4.9 \times 0.53$

$= \mathbf{1800\ W\ (app.).}$ **(Ans.)**

[or $P = 3I_{ph}^2 R_{ph} = 3 \times 4.9^2 \times 25 = 1800$ W].

Example 2. *Three identical coils are connected in star to a 400 V (line voltage), 3-phase A.C. supply and each coil takes 300 W. If the power factor is 0.8 (lagging). Calculate :*

(*i*) *The line current,* (*ii*) *Impedance, and*

(*iii*) *Resistance and inductance of each coil.*

Solution. Line voltage, $E_L = 400$ V

Power taken by each coil, $P_{ph} = 300$ W

Power factor, $\cos\phi = 0.8$ (lagging)

$\mathbf{I_L\ ;\ Z\ ;\ R_{ph}\ ;\ L_{ph}}$ **:**

Phase voltage, $E_{ph} = \frac{E_L}{\sqrt{3}} = \frac{400}{\sqrt{3}}$ V

Also $P_{ph} = E_{ph} I_{ph} \cos\phi$

$300 = \frac{400}{\sqrt{3}} \times I_{ph} \times 0.8$

$\therefore$ $I_{ph} = \frac{300 \times \sqrt{3}}{400 \times 0.8} = 1.62$ A.

(*i*) **Line current,** I_L = phase current, I_{ph}

$\therefore$ $\mathbf{I_L = 1.62\ A.}$ **(Ans.)**

(*ii*) **Coil impedance,** $Z_{ph} = \frac{E_{ph}}{I_{ph}} = \frac{\frac{400}{\sqrt{3}}}{1.62} = 142.5\ \Omega$

$\therefore$ $\mathbf{Z_{ph} = 142.5\ \Omega.}$ **(Ans.)**

(*iii*) $R_{ph} = Z_{ph} \cos\phi = 142.5 \times 0.8 = 114\ \Omega$

Coil reactance, $X_{ph} = Z_{ph} \sin\phi = 142.5 \times 0.6 = \mathbf{85.5\ \Omega}$ **(Ans.)**

But $X_{ph} = 2\pi f L_{ph}$

$\therefore$ $L_{ph} = \frac{X_{ph}}{2\pi f} = \frac{85.5}{2\pi \times 50} = 0.272$ H.

Hence, $\mathbf{R_{ph} = 114\ \Omega.}$ **(Ans.)**

and $\mathbf{L_{ph} = 0.272\ H.}$ **(Ans.)**

Example 3. *In a 3-phase, 3-wire system with star-connected load the impedance of each phase is (3 + j4) Ω. If the line voltage is 230 V, calculate :*

(*i*) *The line current, and* (*ii*) *The power absorbed by each phase.*

Solution. Line voltage, $E_L = 230$ V

Resistance per phase, $R_{ph} = 3\ \Omega$

Reactance per phase, $X_{ph} = 4\ \Omega$

$\mathbf{I_L\ ;\ P_{ph}}$ **:**

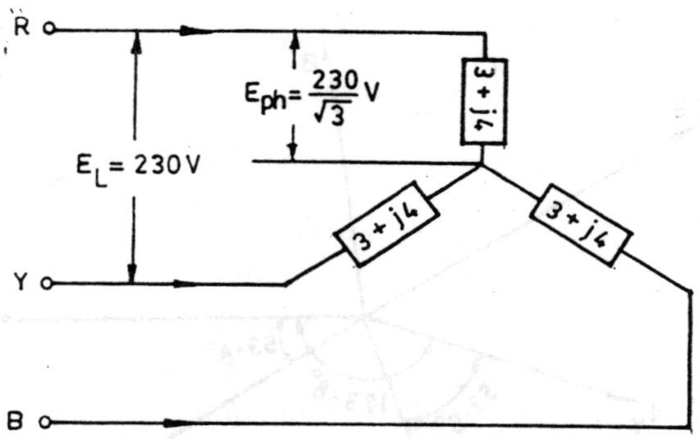

Fig. 8

Phase voltage, $E_{ph} = \frac{E_L}{\sqrt{3}} = \frac{230}{\sqrt{3}}$ V

Impedance per phase, $Z_{ph} = \sqrt{R_{ph}^2 + X_{ph}^2} = \sqrt{3^2 + 4^2} = 5\ \Omega$

Power factor, $\cos\phi = \frac{R}{Z} = \frac{3}{5} = 0.6$

Phase current, $I_{ph} = \frac{E_{ph}}{Z_{ph}} = \frac{230}{\sqrt{3} \times 5} = 26.56$ A

Line current, $\mathbf{I_L = I_{ph} = 26.56}$ **A. (Ans.)**

Power absorbed by each phase,

$$\mathbf{P_{ph}} = E_{ph} I_{ph} \cos\phi = \frac{230}{\sqrt{3}} \times 26.56 \times 0.6 = \mathbf{2116\ W\ (Ans.)}$$

$$[P_{ph} = I_{ph}^2 R_{ph} = 26.56^2 \times 3 = 2116 \text{ W}]$$

Solution by Symbolic Notation. In Fig. 9 E_R, E_Y and E_B are the phase voltages whereas I_R, I_Y and I_B are phase currents.

Taking E_R *as the reference vector,* we get

$$E_R = \frac{230}{\sqrt{3}} \angle 0° = 133 \angle 0° = 133 + j0 \text{ volt}$$

$$E_Y = 133 \angle -120° = 133\,(-0.5 - j0.866) = (-66.5 - j115) \text{ volts}$$

$$E_B = 133 \angle 120° = 133\,(-0.5 + j0.866) = (-66.5 + j115) \text{ volts}$$

$$Z = 3 + j4 = 5 \angle 53° 8'$$

$$I_R = \frac{E_R}{Z} = \frac{133 \angle 0°}{5 \angle 53° 8'} = 26.6 \angle -53° 8'$$

This current lags behind the reference voltage (E_R) by 53° 8′ (Fig. 9).

$$I_Y = \frac{E_Y}{Z} = \frac{133 \angle -120°}{5 \angle 53° 8'} = 26.6 \angle -173° 8'$$

It lags the reference vector *i.e.*, E_R by 173° 8′ which amounts to lagging behind its phase voltage E_Y by 53° 8′,

$$I_B = \frac{E_B}{Z} = \frac{133 \angle 120°}{5 \angle 53° 8'} = 26.6 \angle 66° 52'$$

This current leads E_R by 66° 52′ which is the same as *lagging* behind its phase voltage by 53° 8′.

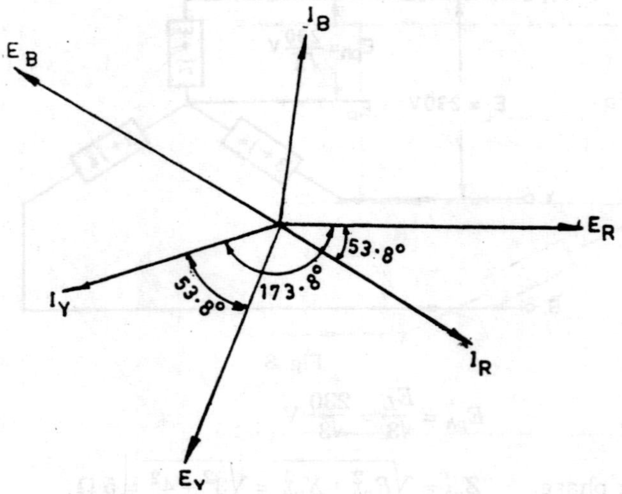

Fig. 9

Let us consider R-phase for calculation of power

$$E_R = (133 + j0)\ ;\ I_R = 26.6\ (0.6 - j0.8) = (15.96 - j21.28)$$

Using method of conjugates, we get

$$P_{VA} = (133 - j0)\ (15.96 - j21.28) = 2116 - j2830$$

∴ Real power absorbed/phase = 2116 W ...(As before)

Example 4. *A star-connected, 6000 V, 3-phase alternator is supplying 4000 kW at a power factor of 0.8. Calculate the active and reactive components of the current in each phase.*

Solution. Line voltage, $E_L = 6000$ V

Power supplied, $P = 4000$ kW

Power factor, $\cos \phi = 0.8$

Active and reactive components of current :

We know that, $P = \sqrt{3}\ E_L I_L \cos \phi$

$$4000 \times 1000 = \sqrt{3} \times 6000 \times I_L \times 0.8$$

i.e., $$I_L = \frac{4000 \times 1000}{\sqrt{3} \times 6000 \times 0.8} = 481 \text{ A}$$

∴ $$I_{ph} = I_L = 481 \text{ A}$$

Active component $= I_{ph} \cos \phi = 481 \times 0.8 =$ **384.8 A. (Ans.)**

Reactive component $= I_{ph} \sin \phi = 481 \times 0.6 =$ **288.6 A. (Ans.)**

Example 5. *In a 3-phase, 4-wire system, two phases have currents of 20 A and 12 A in lagging power factors of 0.8 and 0.6 respectively, while the third phase is open-circuited. Calculate the current in the neutral and draw the vector diagram.*

Solution. Refer Figs. 10 and 11.

$$\cos \phi_1 = 0.8$$

∴ $$\phi_1 = \cos^{-1} 0.8 = 36° \ 52' \text{ (lag)}$$

$$\cos \phi_2 = 0.6$$

∴ $$\phi_2 = \cos^{-1} 0.6 = 53° \ 8' \text{ (lag)}$$

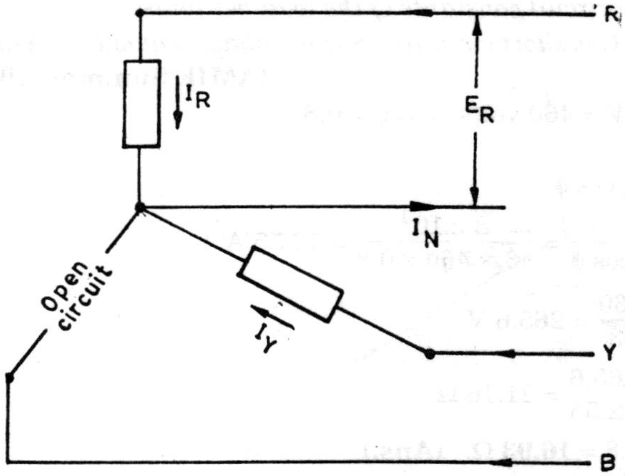

Fig. 10

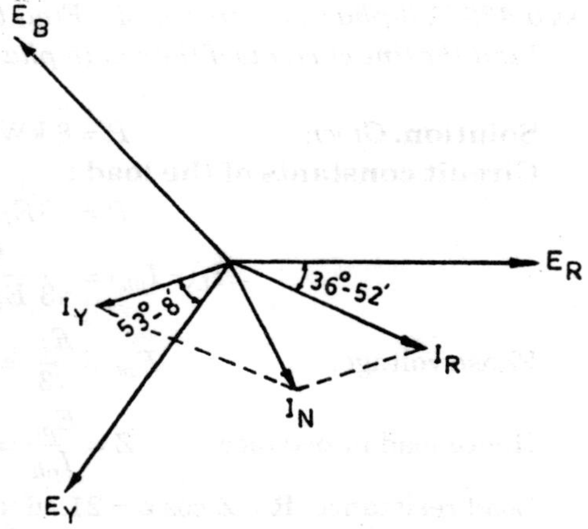

Fig. 11

Let E_R be the reference vector.

Then $\quad I_R = 20 \angle -36° \, 52' = 20\,(0.8 - j0.6) = (16 - j12)$

and $\quad I_Y = 12 \angle -173° \, 8' = 12(-1 - j0.12) = (-12 - j1.44)$

The current through the neutral

$$I_N = I_R + I_Y$$
$$= (16 - j12) + (-12 - j1.44) = 4 - j13.44 = 14 \angle -73° \, 24'$$

Hence, **current in the neutral = 14 ∠– 73° 24′. (Ans.)**

Example 6. *A balanced 3-phase star connected load of 100 kW takes a leading current of 80 A, when connected across a 3-phase, 1100 V, 50 Hz supply. Find the circuit constants of the load per phase.* **(AMIE Winter, 1997)**

Solution. *Given* : $P = 100$ kW ; $I_{ph}\ (= I_L) = 80$ A ; $E_L = 1100$ V ; $f = 50$ Hz

Circuit constants of the load per phase, R, C :

As the 3-ϕ load is balanced and star connected, line or phase current,

$$I_L = (I_{ph}) = \frac{P}{\sqrt{3}\, E_L \cos\phi} = \frac{100 \times 10^3}{\sqrt{3} \times 1100 \times \cos\phi}$$

or
$$\cos\phi = \frac{100 \times 10^3}{\sqrt{3} \times 1100 \times 80} = 0.656$$

Load impedance,
$$Z = \frac{E_{ph}}{I_{ph}} = \frac{(1100/\sqrt{3})}{80} = 7.94\ \Omega$$

$\therefore$
$$\mathbf{R} = Z \cos\phi = 7.94 \times 0.656 = \mathbf{5.2\ \Omega}. \quad \textbf{(Ans.)}$$

Now $X_C = \dfrac{1}{2\pi fC}$ as the current given is *leading* current.

$\therefore$
$$C = \frac{1}{2\pi f X_C}$$

But
$$X_C = Z \sin\phi = 7.94 \times 0.755 = 5.99\ \Omega$$

$\therefore$
$$\mathbf{C} = \frac{1}{2\pi \times 50 \times 5.99}\ \text{F}$$
$$= \frac{10^6}{2\pi \times 50 \times 5.99}\ \mu\text{F} = \mathbf{531.4\ \mu F}. \quad \textbf{(Ans.)}$$

Example 7. *Three identical star-connected coils take 8 kW at a power factor 0.8 when connected across a 460 V, 3-phase, 3-wire supply. Find the circuit constants of the load per phase.*

Find the line currents if the coil in phase C is short circuited. Supply phase sequence is A-B-C.

(AMIE Summer, 1997)

Solution. *Given* : $P = 8$ kW ; $V = 460$ volts ; $\cos \phi = 0.8$

Circuit constants of the load :

$$P = \sqrt{3} E_L I_L \cos \phi$$

$$I_L (= I_{ph}) = \frac{P}{\sqrt{3}\, E_L \cos \phi} = \frac{8 \times 10^3}{\sqrt{3} \times 460 \times 0.8} = 12.55 \text{ A}$$

Phase voltage, $E_{ph} = \frac{E_L}{\sqrt{3}} = \frac{460}{\sqrt{3}} = 265.6$ V

Hence load impedance, $Z = \frac{E_{ph}}{I_{ph}} = \frac{265.6}{12.55} = 21.16\ \Omega$

Load resistance, $\mathbf{R} = Z \cos \phi = 21.16 \times 0.8 = \mathbf{16.93\ \Omega}$. **(Ans.)**

Load reactance, $\mathbf{X} = Z \sin \phi = 21.16 \times 0.6 = \mathbf{12.7\ \Omega}$. **(Ans.)**

Line currents if the coil in phase C is short circuited :

The connection arrangement is shown in Fig. 12 when coil C is shorted.

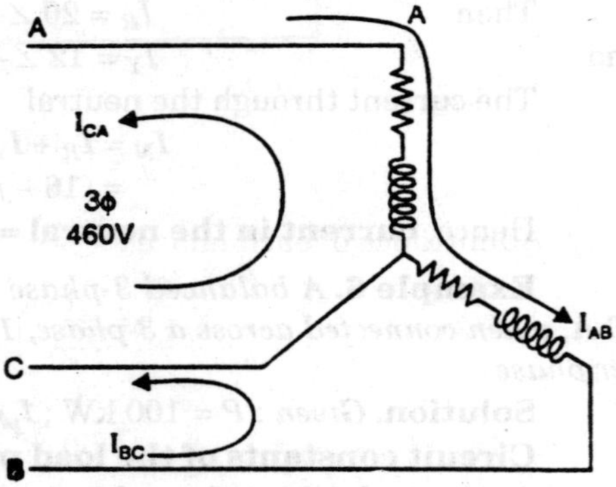

Fig. 12

Given that phase sequence is A-B-C.

Hence, $E_{AB} = 460 \angle 0°$

$E_{BC} = 460 \angle -120°$

$E_{CA} = 460 \angle +120°$

Also, $Z = 21.6 \angle 36.8°$

$$I_{AB} = \frac{E_{AB}}{2Z} = \frac{460 \angle 0°}{2 \times 21.16 \angle 36.8°} = 10.87 \angle -36.8°$$

$$I_{BC} = \frac{E_{BC}}{Z} = \frac{460 \angle -120°}{21.16 \angle 36.8°} = 21.74 \angle -156.8°$$

$$I_{CA} = \frac{E_{CA}}{Z} = \frac{460 \angle 120°}{21.16 \angle 36.8°} = 21.74 \angle 83.2°$$

Line current in line A $= I_{AB} - I_{CA}$

$= 10.87 \angle -36.8° - 21.74 \angle 83.2°$

$= 10.87(0.8 - j0.6) - 21.74(0.118 + j0.993)$

$= 8.70 - j6.52 - 2.56 - j21.59 = \mathbf{(6.14 - j28.11)A.}$ **(Ans.)**

Line current in line B $= I_{BC} - I_{AB}$

$= 21.74 \angle -156.8° - 10.87 \angle -36.8°$

$= 21.74(-0.92 - j0.394) - (8.70 - j6.52)$

$= -20.0 - j8.56 - 8.70 + j6.52 = \mathbf{(-28.70 - j2.04)A.}$ **(Ans.)**

Line current in line $C = I_{CA} - I_{BC}$

$= (-2.56 - j21.59) - (-20 - j8.56) = -2.56 - j21.59 + 20 + j8.56$

$= \mathbf{(17.44 - j13.03)A.}$ **(Ans.)**

Example 8. *In a star-connected load each phase consists of a resistance of 50 Ω in parallel with a capacitor of capacitance 16 μF. When it is connected to a 400 V, 3-phase, 5 Hz supply, calculate :*

(i) The line current, *(ii) The power factor,*

(iii) The power absorbed, and *(iv) The total kVA.*

Solution. Resistance per phase, $R = 50\ \Omega$

Capacitance, $C = 16\ \mu F = 16 \times 10^{-6}$ F

The circuit is shown in Fig. 13.

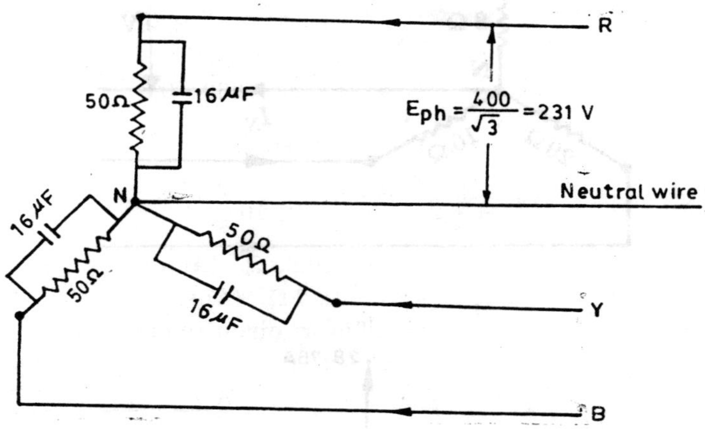

Fig. 13

(*i*) $$E_{ph} = \left(\frac{400}{\sqrt{3}}\right) \angle\ 0° = 231 \angle\ 0° = (231 + j0)$$

Admittance of each phase,

$$Y_{ph}\left(=\frac{1}{Z_{ph}}\right) = \frac{1}{R} + j\omega C$$

$$= \frac{1}{50} + j(2\pi \times 50 \times 16 \times 10^{-6}) = 0.02 + j0.005$$

$\therefore$ $$I_{ph} = E_{ph} \cdot Y_{ph} = 231\ (0.02 + j0.005) = 4.62 + j1.155 = 4.76 \angle\ 14°$$

For a star-connection, $I_{ph} = I_L$

$\therefore$ $$I_L = 4.76 \text{ A}$$

Hence, **line current = 4.76 A. (Ans.)**

(*ii*) **Power factor** $= \cos 14° =$ **0.97 (leading). (Ans.)**

(*iii*) Now $E_{ph} = (231 + j0)$; $I_{ph} = (4.62 + j1.155)$

$\therefore$ $$P_{VA} = (231 + j0)\ (4.62 - j1.155)$$

$$= 231 \times 4.62 - j1.155 \times 231 = 1067.22 - j266.8$$

$$= 1100 \angle -14° \quad \text{(per phase)}$$

$\therefore$ **Total power absorbed** $= 3 \times 1067.22 = 3201.66$ W = **3.201 kW. (Ans.)**

(*iv*) **Total volt-amperes** $= 3 \times 1100 = 3300$ VA = **3.3 kVA. (Ans.)**

Example 9. *A 3-phase, star-connected system with 230 V between each phase and neutral has resistances of 8, 10 and 20 Ω respectively in three phases, calculate :*

(*i*) *The current flowing in each phase,* (*ii*) *The neutral current, and*

(*iii*) *The total power absorbed.*

Solution. Refer Figs. 14 (*a*) and (*b*).

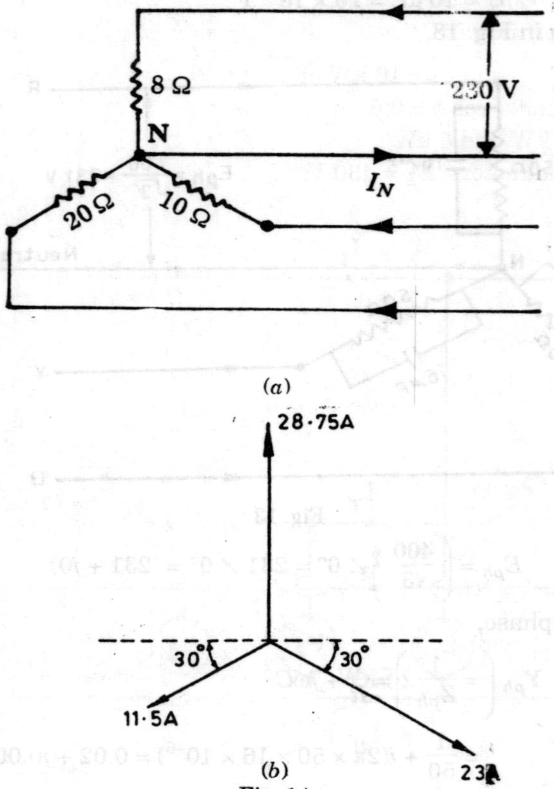

(*a*)

(*b*)

Fig. 14

Phase voltage, $E_{ph} = 230$ V

(*i*) **Current in 8 Ω resistor** $= \frac{230}{8} =$ **28.75 A. (Ans.)**

Current in 10 Ω resistor $= \frac{230}{10} =$ **23 A. (Ans.)**

Current in 12 Ω resistor $= \frac{230}{20} =$ **11.5 A. (Ans.)**

(*ii*) **The above currents are mutually displaced by 120°. The neutral current I_N is the vector sum of these three currents.**

I_N can be found by splitting up these three-phase currents into their *X*-components and *Y*-components and then by combining them together.

Σ*X*-components $= 23 \cos 30° - 11.5 \cos 30° = 11.5 \cos 30° = 9.96$ A

Σ*Y*-components $= 28.75 - 23 \sin 30° - 11.5 \sin 30° = 28.75 - 34.5 \sin 30° = 11.5$ A

∴ **Neutral current,** $I_N = \sqrt{(9.96)^2 + (11.5)^2} =$ **15.21 A. (Ans.)**

(*iii*) **Total power absorbed,**

$$P = 230\,(28.75 + 23 + 11.5) = \textbf{14547.5 W.} \quad \textbf{(Ans.)}$$

Example 10. *In a 3-phase, 4-wire system, there is a balanced 3-phase motor load taking 10 kW at a power factor of 0.8 lagging while lamps connected between phase conductors and the neutral are taking 2.5 kW, 2 kW and 5 kW respectively. Voltage between line conductors is 430 V. Calculate :*

(*i*) *The current in the neutral wire.* (*ii*) *The current in each conductor.*

Solution. Motor load = 10 kW or 10000 W

Power factor of motor load, $\cos \phi = 0.8$

Lamp loads : 2.5 kW, 2 kW and 5 kW

Voltage between line conductors, $E_L = 430$ V.

Refer Figs. 15 and 16.

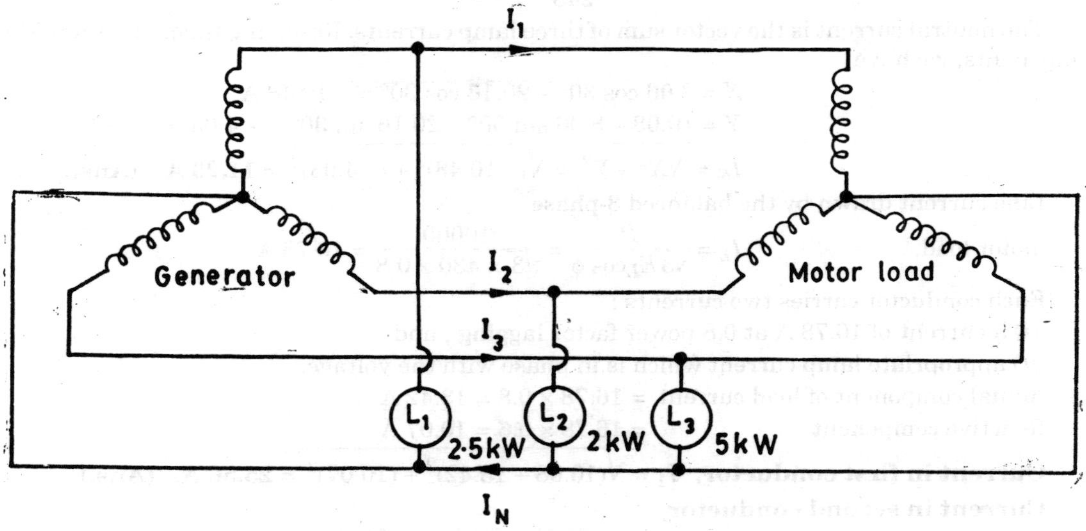

Fig. 15

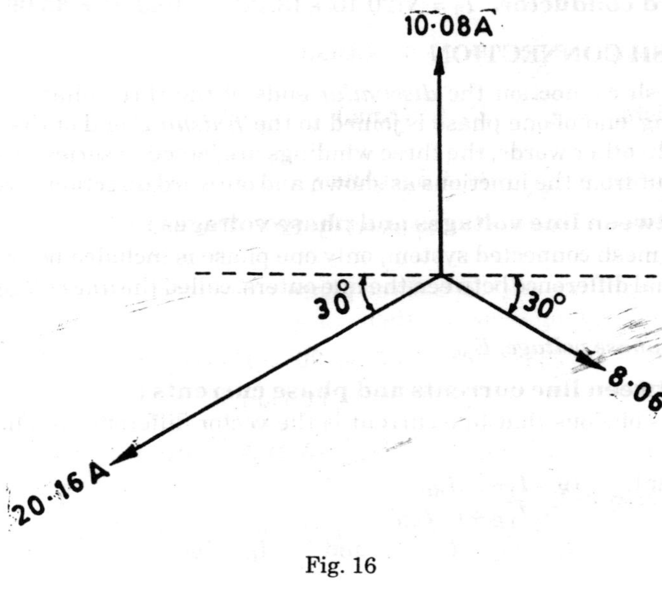

Fig. 16

Let us first find the current in the neutral wire due to lamp loads L_1 (2.5 kW), L_2 (2 kW) and L_3 (5 kW) respectively.

Phase voltage, $E_{ph} = \frac{E_L}{\sqrt{3}} = \frac{430}{\sqrt{3}} = 248$ V

Current through lamp L_1 $= \frac{2.5 \times 1000}{248} = 10.08$ A

Current through lamp L_2 $= \frac{2 \times 1000}{248} = 8.06$ A

Current through lamp L_3 $= \frac{5 \times 1000}{248} = 20.16$ A.

The neutral current is the vector sum of three lamp currents. Resolving them into their X and Y components, we have

$$X = 8.06 \cos 30° - 20.16 \cos 30° = -10.48 \text{ A}$$
$$Y = 10.08 - 8.06 \sin 30° - 20.16 \sin 30° = -4.03 \text{ A}$$

$\therefore$ $I_N = \sqrt{X^2 + Y^2} = \sqrt{(-10.48)^2 + (-4.03)^2} = $ **11.23 A. (Ans.)**

Line current drawn by the balanced 3-phase motor load, $I_L = \frac{P}{\sqrt{3} E_L \cos \phi} = \frac{10000}{\sqrt{3} \times 430 \times 0.8} = 16.78$ A

Each conductor carries two currents :

(*i*) a current of 16.78 A at 0.8 power factor lagging ; and

(*ii*) appropriate lamp current which is in phase with the voltage.

Actual component of load current $= 16.78 \times 0.8 = 13.42$ A

Reactive component $= 16.78 \times 0.6 = 10.07$ A

Current in first conductor, $I_1 = \sqrt{(10.08 + 13.42)^2 + (10.07)^2} = $ **25.56 A. (Ans.)**

Current in second conductor,

$$I_2 = \sqrt{(8.06 + 13.42)^2 + (10.07)^2} = \textbf{23.72 A (Ans.)}$$

Current in third conductor, $I_3 = \sqrt{(20.16 + 13.42)^2 + (10.07)^2} = $ **35.06 A. (Ans.)**

7. DELTA (Δ) OR MESH CONNECTION

In a delta or mesh connection the *dissimilar* ends of the three-phase windings are joined together *i.e.*, the *'starting'* end of one phase is joined to the *'finishing'* end of the other phase and so on as shown in Fig. 17. In other words, the three windings are joined in series to form a closed mesh. Three leads are taken out from the junctions as shown and outward directions are taken as positive.

(*a*) **Relation between line voltages and phase voltages :**

Since in delta or mesh connected system, only one phase is included between any pair of line outers, therefore potential difference between the line outers, called the *line voltage*, is equal to phase voltage.

i.e., *Line voltage*, E_L = *phase voltage*, E_{ph}.

(*b*) **Relation between line currents and phase currents :**

From Fig. 17 it is obvious that line current is the vector difference of phase currents of two phases concerned.

Thus, line current, $I_R = I_{YR} - I_{RB}$ (Vector difference)

$= I_{YR} + (-I_{RB})$ (Vector sum)

Similarly, $I_Y = I_{BY} - I_{YR}$ and $I_B = I_{RB} - I_{BY}$

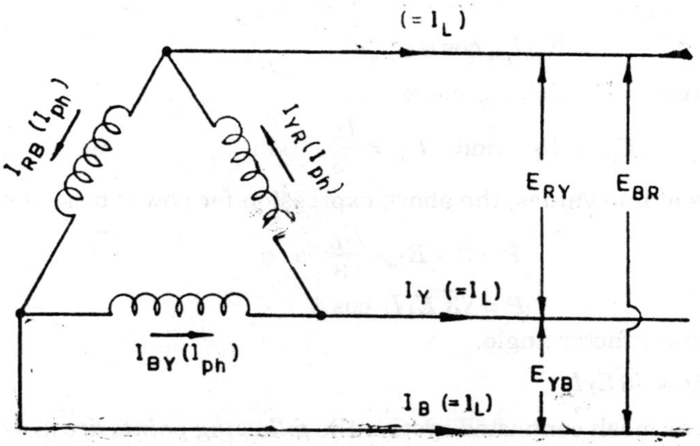

$E_{RY} = E_{YB} = E_{BR} = E_{ph} = E_L$

$I_R = I_Y = I_B = I_L \; (= \sqrt{3}\, I_{ph})$

Fig. 17. Delta or mesh connected diagram.

Refer Fig. 18. Since phase angle between phase current I_{YR} and $-I_{RB}$ is 60°,

$$\therefore \qquad I_R = \sqrt{I_{YR}^2 + I_{RB}^2 + 2I_{YR}I_{RB} \cos 60°}$$

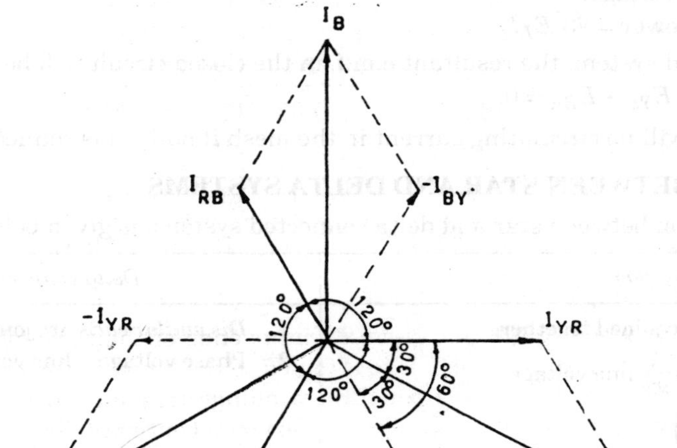

Fig. 18. Vector diagram for delta connected net work.

Assuming the delta connected system or network be balanced, the phase current in each winding is equal and let each be equal to I_{ph} (*i.e.*, $I_{YR} = I_{BY} = I_{RB} = I_{ph}$)

$$\therefore \qquad I_R \;(= I_L) = \sqrt{I_{ph}^2 + I_{ph}^2 + 2I_{ph}I_{ph} \cos 60°} = \sqrt{2I_{ph}^2 + 2I_{ph}^2 \times \tfrac{1}{2}} = \sqrt{3}\, I_{ph}$$

Similarly, $\qquad I_Y = I_B = \sqrt{3}\, I_{ph}$

Hence, $\qquad I_L = \sqrt{3}\; I_{ph} \qquad ...(5)$

(*i.e., line current* $= \sqrt{3}$ *phase current*)

(c) **Power :**

Power/phase $= E_{ph} I_{ph} \cos \phi$

Total power (true) $P = 3E_{ph} I_{ph} \cos \phi$

But $E_{ph} = E_L$ and $I_{ph} = \frac{I_L}{\sqrt{3}}$

Hence, in terms of line values, the above expression for power becomes

$$P = 3 \times E_L \times \frac{E_L}{\sqrt{3}} \cos \phi$$

or
$$P = \sqrt{3}\, E_L I_L \cos \phi \quad ...(6)$$

where ϕ = the phase power factor angle.

(*Apparent power* $= \sqrt{3} E_L I_L$)

In case of delta or mesh connected system the *following points are worthnoting* :

(*i*) *Line voltages are equal to phase voltages.*

(*ii*) Line currents are $\sqrt{3}$ times phase currents.

(*iii*) Line currents are 120° apart.

(*iv*) Line currents are 30° behind their respective phase currents.

(*v*) The angle between line currents and corresponding line voltages is $(30° \pm \phi)$ as in the star system.

(*vi*) True power $= \sqrt{3}\, E_L I_L \cos \phi$, where ϕ is the phase angle between respective phase current and phase voltage.

(*vii*) Apparent power $= \sqrt{3}\, E_L I_L$.

(*viii*) In balanced system, the resultant e.m.f. in the closed circuit will be zero.

i.e., $E_{RY} + E_{YB} + E_{BR} = 0$.

Hence, there will no circulating current in the mesh if no-load is connected to the lines.

8. COMPARISON BETWEEN STAR AND DELTA SYSTEMS

The comparison between star and delta connected systems is given below :

Star connected system	*Delta connected system*
1. *Similar ends* are joined together.	1. *Dissimilar ends* are joined.
2. Phase voltage $= \frac{1}{\sqrt{3}}$ line voltage $\left(i.e., E_{ph} = \frac{E_L}{\sqrt{3}}\right)$.	2. Phase voltage = line voltage (*i.e.*, $E_{ph} = E_L$).
3. Phase current = line current (*i.e.*, $I_{ph} = I_L$).	3. Phase current $= \frac{1}{\sqrt{3}} \times$ line current $\left(i.e., I_{ph} = \frac{I_L}{\sqrt{3}}\right)$.
4. Possible to carry neutral to the load.	4. Neutral wire not available.
5. Provides 3-phase 4-wire arrangement.	5. Provides 3-phase 3-wire arrangement.
6. Can be used for lighting as well as power load.	6. Can be used for power loads only.
7. Neutral wire of a star connected alternator can be connected to earth, so relays and protective devices can be provided in the star connected alternators for safety.	7. Not possible. Delta connected system is mostly used in transformer for running of small low voltage 3-phase motors and *best suited for rotary converters*.

DELTA OR MESH CONNECTION

Example 11. *Three identical coils connected in delta across 400 V, 50 Hz, 3-phase supply take a line current of 15 A at a power factor 0.8 lagging. Calculate :*

(*i*) *The phase current, and*

(*ii*) *The impedance, resistance and inductance of each winding.*

Solution. Line voltage, $E_L = 400$ V

Line current, $I_L = 15$ A

Power factor, $\cos\ \phi = 0.8$ lagging

$\mathbf{I_{ph}}$; $\mathbf{Z_{ph}}$; $\mathbf{R_{ph}}$; **L :**

Phase voltage, $E_{ph} = E_L = 400$ V

(*i*) **Phase current,** $I_{ph} = \frac{I_L}{\sqrt{3}} = \frac{15}{\sqrt{3}} = \mathbf{8.66\ A.}$ **(Ans.)**

(*ii*) **Impedance of each phase,** $\mathbf{Z_{ph}} = \frac{E_{ph}}{I_{ph}} = \frac{400}{8.66} = \mathbf{46.19\ \Omega.}$ **(Ans.)**

Resistance of each phase, $\mathbf{R_{ph}} = Z_{ph} \cos\phi = 46.19 \times 0.8 = \mathbf{36.95\ \Omega.}$ **(Ans.)**

Reactance of each phase, $X_{ph} = Z_{ph} \sin\phi = 46.19\sqrt{(1 - \cos^2\phi)}$

$= 46.19\sqrt{1 - (0.8)^2} = 27.71\Omega$

$\therefore$ **Inductance,** $\mathbf{L} = \frac{X_{ph}}{2\pi f} = \frac{27.71}{2\,\pi \times 50} = \mathbf{0.088\ H.}$ **(Ans.)**

Example 12. *A 220 V, 3-phase voltage is applied to a balanced delta-connected 3-phase load of phase impedance (6 + j8).*

(*i*) *Find the phasor current in each line.*

(*ii*) *What is the power consumed per phase ?*

(*iii*) *What is the phasor sum of the three line currents ? What does it have this value ?*

Solution. Refer Fig. 19.

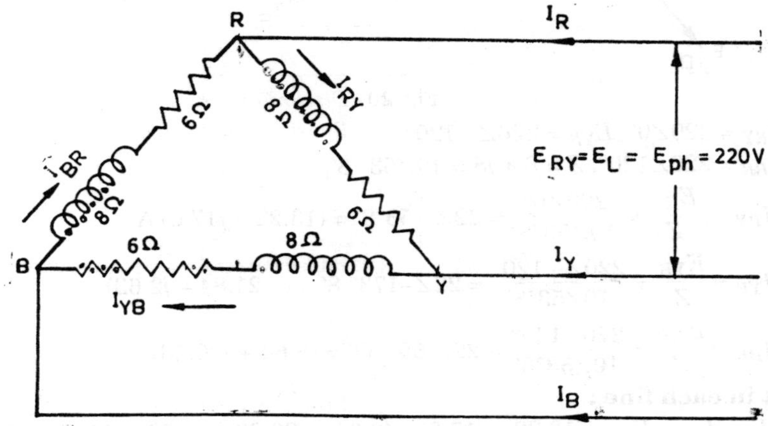

Fig. 19

Resistance per phase, $R_{ph} = 6\ \Omega$

Reactance per phase, $X_{ph} = 8\ \Omega$

$E_L = E_{ph} = 220$ V

Impedance per phase, $Z_{ph} = \sqrt{R_{ph}^2 + X_{ph}^2} = \sqrt{6^2 + 8^2} = 10\ \Omega$

(*i*) Phase current, $I_{ph} = \dfrac{E_{ph}}{Z_{ph}} = \dfrac{220}{10} = 22$ A

$\therefore$ **Line current,** $I_L = \sqrt{3} \times 22 =$ **38.1 A. (Ans.)**

(*ii*) **Power consumed per phase,**

$$P_{ph} = I_{ph}^2 \times R_{ph} = 22^2 \times 6 = \mathbf{2904\ W.}\ \textbf{(Ans.)}$$

(*iii*) Phasor sum would be zero because the three currents are equal in magnitudes and have a mutual phase difference of 120°.

Solution by Symbolic Notation. Let E_{RY} is taken as a reference vector (Fig. 20).

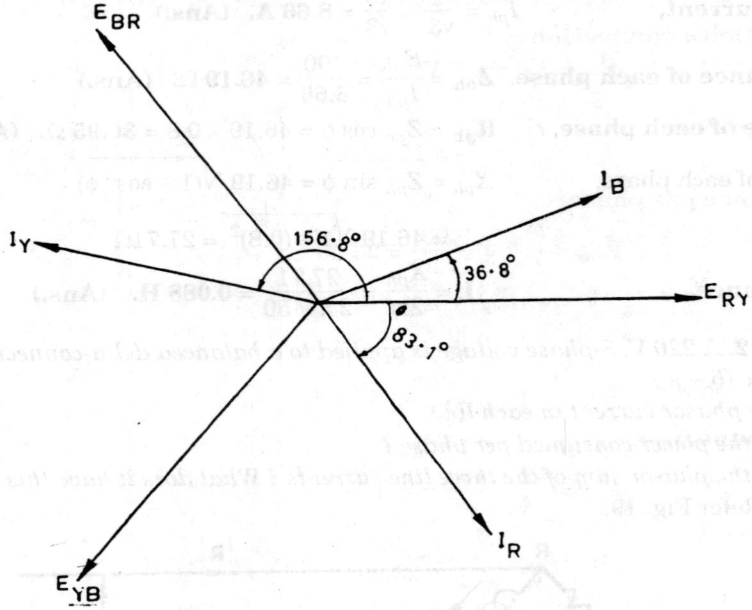

Fig. 20

$E_{RY} = 220\angle 0°$, $E_{YB} = 220\angle -120°$

$E_{BR} = 220\angle 120°$, $Z = 6 + j8 = 10\angle 53°\,8'$,

$$I_{RY} = \frac{E_{RY}}{Z} = \frac{220\angle 0°}{10\angle 53°8'} = 22\angle -53°8' = (13.22 - j17.6)\ \text{A}$$

$$I_{YB} = \frac{E_{YB}}{Z} = \frac{220\angle -120°}{10\angle 53°8'} = 22\angle -173°\,8' = (-21.84 - j2.63)$$

$$I_{BR} = \frac{E_{BR}}{Z} = \frac{220\angle 120°}{10\angle 53°8'} = 22\angle\, 66°\,52' = (8.64 + j20.23)$$

(*i*) **Current in each line :**

$I_R = I_{RY} - I_{BR} = (13.22 - j17.6) - (8.64 + j20.23) = 4.58 - j37.83 = \mathbf{38.1\angle 83.1°}$. **(Ans.)**

$I_Y = I_{YB} - I_{RY} = (-21.84 - j2.63) - (13.22 - j17.6)$

$= -21.84 - j2.63 - 13.22 + j17.60 = -35.06 + j14.97 = \mathbf{38.12\angle 156.8°}$. **(Ans.)**

$I_B = I_{BR} - I_{YB} = (8.64 + j20.23) - (-21.84 + j2.63) = 8.64 + j20.23 + 21.84 + j2.63$

$= 30.48 + j22.86 = \mathbf{38.1\angle 36.8°}$. **(Ans.)**

(*ii*) **Power consumed per phase :**

Using conjugate of voltage, we get for R-phase

$$P_{VA} = E_{RY} \cdot I_{RY} = (220 - j0)(13.22 - j17.6) = (2908.4 - j3872) \text{ volt ampere}$$

True power per phase = **2.908 kW. (Ans.)**

(*iii*) **Phase sum of the three line currents**

$$= I_R + I_Y + I_B$$
$$= (4.58 - j37.83) + (-35.06 + j14.96) + (30.48 + j22.86) = 0$$

Hence, **the phasor sum of three line currents drawn by a 'balanced load' is zero. (Ans.)**

Example 13. *A delta-connected balanced 3-phase load is supplied from a 3-phase, 400 V supply. The line current is 30 A and the power taken by the load is 12 kW. Find :*

(*i*) *Impedance in each branch ; and*

(*ii*) *The line current, power factor and power consumed if the same load is connected is star.*

Solution. Delta-connection :

$$E_{ph} = E_L = 400 \text{ V}$$
$$I_L = 30 \text{ A}$$

$$\therefore \quad I_{ph} = \frac{I_L}{\sqrt{3}} = \frac{30}{\sqrt{3}} = 17.32 \text{ A.}$$

(*i*) **Impedance per phase**

$$\mathbf{Z_{ph}} = \frac{E_{ph}}{I_{ph}} = \frac{400}{17.32} = \mathbf{23.09\ \Omega.\ \ (Ans.)}$$

Now
$$P = \sqrt{3} E_L I_L \cos \phi$$
$$12000 = \sqrt{3} \times 400 \times 30 \times \cos \phi.$$

or
$$\cos \phi \text{ (power factor)} = \frac{12000}{\sqrt{3} \times 400 \times 30} = 0.577$$

(*ii*) **Star-connection**

$$E_{ph} = \frac{E_L}{\sqrt{3}} = \frac{400}{\sqrt{3}} = 231 \text{ V}$$

$$\mathbf{I_L} = I_{ph} = \frac{E_{ph}}{Z_{ph}} = \frac{231}{23.09} = \mathbf{10\ A\ \ (Ans.)}$$

Power factor, $\cos \phi = 0.577$ (since impedance is same)

Power consumed $= \sqrt{3}\, E_L I_L \cos \phi = \sqrt{3} \times 400 \times 10 \times 0.577 =$ **3997.6 W (Ans.)**

Example 14. *Three 50 Ω non-inductive resistances are connected in (i) star, (ii) delta across a 400 V, 50 Hz., 3-phase mains. Calculate the power taken from the supply system in each case. In the event of one of the three resistances getting opened, what would be the value of the total power taken from the mains in each of the two cases.*

Solution. Star connection :

Phase voltage, $$E_{ph} = \frac{E_L}{\sqrt{3}} = \frac{400}{\sqrt{3}} = 231 \text{ V}$$

Phase current, $$I_{ph} = \frac{E_{ph}}{R_{ph}} = \frac{231}{50} = 4.62 \text{ A}$$

Power consumed, $$P = 3 I_{ph}^2 R_{ph} = 3 \times 4.62^2 \times 50 = \mathbf{3200\ W.\ \ (Ans.)}$$

$$[\text{or } P = \sqrt{3} E_L I_L \cos \phi = \sqrt{3} \times 400 \times 4.62 \times 1 = 3200 \text{ W}]$$

Delta connection :

Phase voltage, $E_{ph} = E_L = 400$ V

Phase current, $I_{ph} = \dfrac{E_{ph}}{R_{ph}} = \dfrac{400}{50} = 8$ A

Power consumed, $P = 3I_{ph}^2 R_{ph} = 3 \times 8^2 \times 50 =$ **9600 W. (Ans.)**

When one of the resistances is disconnected :

(*i*) **Star connection.** Refer Fig. 21.

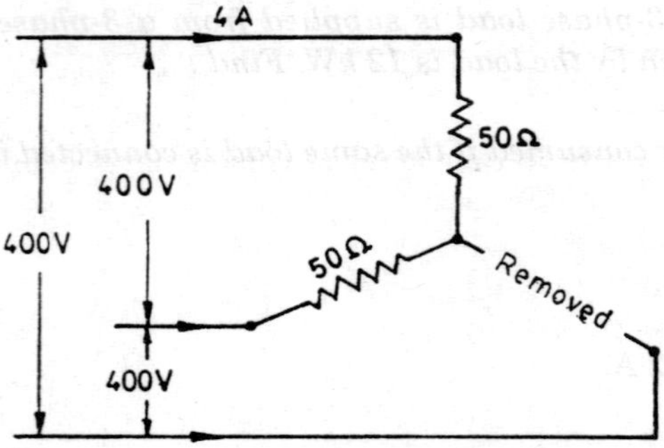

Fig. 21

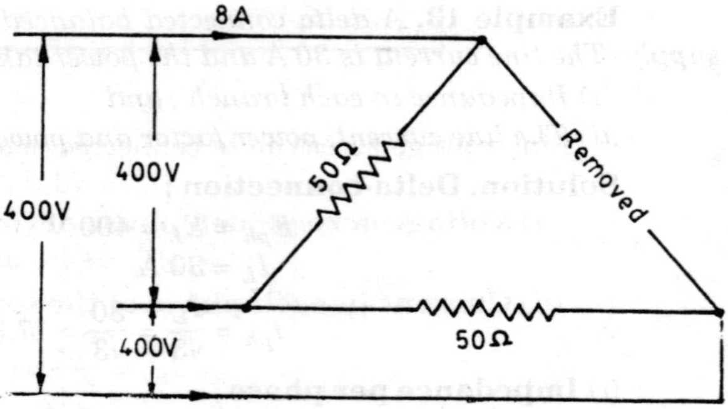

Fig. 22

When one of the resistances is disconnected, the circuit is no longer 3-phase but converted into single-phase circuit, having two resistances each of 50 ohm connected in series across supply of 400 V.

Hence line current, $I_L = \dfrac{E_L}{2R_{ph}} = \dfrac{400}{2 \times 50} = 4$ A

Power consumed, $P = 4^2 (50 + 50) =$ **1600 W. (Ans.)**

[or $P = VI \cos \phi = 400 \times 4 \times 1 = 1600$ W].

(*ii*) **Delta connection.** Refer Fig. 22.

Potential difference across each resistance, $E_L = 400$ V

Current in each resistance $= \dfrac{400}{50} = 8$ A

Power consumed in both resistances $= 2 \times 8^2 \times 50 =$ **6400 W. (Ans.)**

[or $P = 2 \times E_{ph} I_{ph} \cos \phi = 2 \times 400 \times 8 \times 1 = 6400$ W].

Example 15. *The secondary of a 3-phase star-connected transformer, which has a phase voltage of 230 V feeds a 3-phase delta connected load ; each phase of which has a resistance of 30 Ω and an inductive reactance of 40 Ω. Draw the circuit diagram of the system and calculate :*

(*i*) *The voltage across each phase of load,*

(*ii*) *The current in each phase of load,*

(*iii*) *The current in the transformer secondary windings, and*

(*iv*) *The total power taken from the supply and its power factor.*

Solution. Refer Fig. 23.

Resistance per phase, $R_{ph} = 20\ \Omega$

Reactance per phase, $X_{ph} = 40\ \Omega$

Phase voltage across transformer secondary,

$E_{ph} = 230$ V

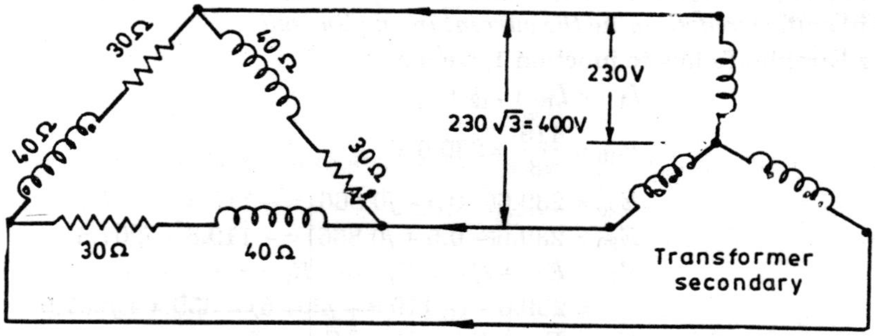

Fig. 23

Line voltage across delta connected load = line voltage across transformer secondary

$$= \sqrt{3} \times 230 = 400 \text{ V (app.)}$$

(*i*) **Voltage across each phase of the load,**

$$E_{ph} = E_L = \mathbf{400\ V.} \quad \textbf{(Ans.)}$$

(*ii*) **Current in each phase of the load,**

$$I_{ph} = \frac{E_{ph}}{Z_{ph}} = \frac{400}{\sqrt{R_{ph}^2 + X_{ph}^2}} = \frac{400}{\sqrt{30^2 + 40^2}} = \frac{400}{50} = \mathbf{8\ A.} \quad \textbf{(Ans.)}$$

(*iii*) **Current in the transformer secondary**

$$= \text{line current of load} = \sqrt{3} \times I_{ph} = \sqrt{3} \times 8 = \mathbf{13.86\ A.} \ \textbf{(Ans.)}$$

(*iv*) **Power factor,** $\cos \phi = \frac{R_{ph}}{Z_{ph}} = \frac{30}{50} = \mathbf{0.6.}$ **(Ans.)**

Total power consumed, $P = \sqrt{3} E_L I_L \cos \phi = \sqrt{3} \times 400 \times 13.86 \times 0.6 = \mathbf{5761.5\ W.}$ **(Ans.)**

Example 16. *Three non-inductive resistances, each of 50 ohms are connected in star to a 3-phase 415 volts supply. Three inductive coils of 100 ohms each connected in delta fashion are also connected to the same supply. Calculate :*

(*i*) *Line current* (*ii*) *Power factor of the system.*

(AMIE Winter, 1999)

Solution. The connection diagram and the vector diagram of the star and delta connected impedances are shown in Fig. 24 (*a*) and (*b*).

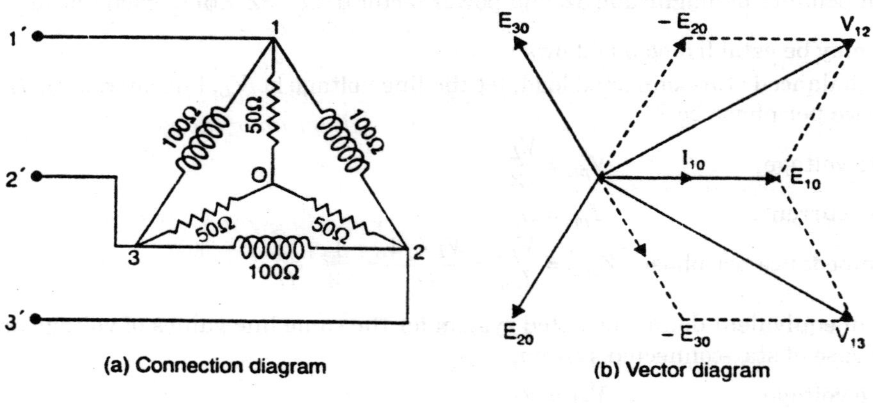

Fig. 24

The voltage E_{10} between the line 1 and neutral is taken along the X-axis. Since the load is **balanced**, it will *suffice to determine the current in one line only*.

Applying Kirchhoff's law to junction 1, we have

$$I_{11'} = I_{10} + I_{12} + I_{13}$$

Now $$E_{10} = \frac{415}{\sqrt{3}} = 239.6 \text{ V}$$

$$E_{20} = 239.6(-0.5 - j0.866) = -119.8 - j207.5$$
$$E_{30} = 239.6(-0.5 + j0.866) = -119.8 + j207.5$$

Now $$V_{12} = E_{10} + E_{02} = E_{10} + (-E_{20}) = E_{10} - E_{20}$$
$$= 239.6 - (-119.8 - j207.5) = 359.4 + j207.5$$
$$V_{13} = E_{10} + E_{03} = E_{10} - E_{30}$$
$$= 239.6 - (-119.8 + j207.5) = 359.4 - j207.5$$

(*i*) **Line current :**

Now, $$I_{10} = \frac{E_{10}}{50} = \frac{239.6}{50} = 4.792 + j0$$

$$I_{12} = \frac{V_{12}}{j100} = \frac{359.4 + j207.5}{j100} = 2.075 - j3.594$$

$$I_{13} = \frac{V_{13}}{j100} = \frac{359.4 - j207.5}{j100} = -2.075 - j3.594$$

$\therefore$ $$I_{11}' = I_{10} + I_{12} + I_{13}$$
$$= (4.792 + j0) + (2.075 - j3.594) + (-2.075 - j3.594)$$
$$= 4.792 - j7.188 = \mathbf{8.639 \angle -56.3°}.$$

Hence line current $= \mathbf{8.639\ A.}$ **(Ans.)**

Power factor $= \cos 56.3° = \mathbf{0.5548.}$ **(Ans.)**

☞**Example 17. (Balanced parallel loads)** *Three star-connected impedances $Z_1 = (20 + j37.7)\ \Omega$ per phase are in parallel with three delta-connected impedances $Z_2 = (30 - j159.3)\ \Omega$ per phase. The line voltage is 398 V. Find the line current, power and reactive volt-amperes taken by the combination.*

(AMIE Winter, 1998)

Solution. Any balanced star-connected system can be completely replaced by an equivalent delta-connected system or vice-versa because of their relationship between phase and line voltages and currents. For example, a balanced star-connected load having an impedance of magnitude Z with a power factor $\cos \phi$ (or $Z \angle\phi$) in each phase can be replaced by an equivalent delta-connected load having an impedance of magnitude $3Z$ and power factor ϕ (*i.e.*, $3Z \angle\phi$) in each phase.

This may be established as follows :

For a balanced star-connected load, let the line voltage be V_L, line current be I_L and impedance per phase be Z.

Phase voltage, $$V_{ph} = \frac{V_L}{\sqrt{3}}$$

Phase current, $$I_{ph} = I_L$$

$\therefore$ Impedance per phase, $$Z_{star} = \frac{V_{ph}}{I_{ph}} = \frac{V_L/\sqrt{3}}{I_L} = \frac{1}{\sqrt{3}} \cdot \frac{V_L}{I_L} \quad ...(i)$$

Now in equivalent delta-connected system for the same line values of voltage and current as in case of star-connected system,

Phase voltage, $$V_{ph} = V_L$$

Phase current, $I_{ph} = \frac{I_L}{\sqrt{3}}$

$\therefore$ Impedance per phase, $Z_{delta} = \frac{V_{ph}}{I_{ph}} = \frac{V_L}{I_L/\sqrt{3}} = \sqrt{3}\,\frac{V_L}{I_L}$ ...(*ii*)

Comparing (*i*) and (*ii*), we have

$$Z_{star} = \frac{1}{3} \times Z_{delta} \text{ or } Z_{delta} = 3Z_{star}$$

The equivalent delta load of the star-connected load is given as $3Z$, *i.e.*, $(60 + j113.1)\ \Omega$

Line voltage, $V_L = 398$ V

Phase voltage, $V_{ph} = 398$ V in case of delta-connections.

Taking phase voltage as reference phasor

$$V_{ph} = 398 \angle 0° \text{ V}$$

Phase current for impedance Z_1, $I_{ph1} = \frac{V_{ph}}{Z_1} = \frac{398 \angle 0°}{(60 + j113.1)}$

$$= \frac{398 \angle 0°}{128 \angle 62.05°} = 3.109 \angle -62.05° \text{ A or } (1.46 - j2.746) \text{ A}$$

Phase current to impedance Z_2, $I_{ph2} = \frac{398 \angle 0°}{(30 - j159.3)}$

$$= \frac{398 \angle 0°}{162.1 \angle -79.33°}$$

$$= 2.455 \angle 79.33° \text{ A or } (0.4545 + j2.412) \text{ A}$$

Total phase current, $I_{ph} = I_{ph1} + I_{ph2}$

$$= (1.46 - j2.746) + (0.4545 + j2.412)$$

$$= (1.914 - j0.334) \text{ A or } 1.943 \angle -9.9°$$

Total line current, $\mathbf{I_L} = \sqrt{3} I_{ph} = \sqrt{3} \times 1.943 = \mathbf{3.36\ A.}$ **(Ans.)**

Power, $\mathbf{P} = \sqrt{3} V_L I_L \cos \phi$

$$= \sqrt{3} \times 398 \times 3.36 \cos(-9.9°) = \mathbf{2281.7\ W.} \text{ (Ans.)}$$

Reactive power $= \sqrt{3}\, V_L I_L \sin \phi$

$$= \sqrt{3} \times 398 \times 3.36 \sin(-9.9°)$$

$$= -398.2\ VAR \text{ or } \mathbf{398.2\ VAR} \text{ (inductive). (Ans.)}$$

9. MEASUREMENT OF POWER IN 3-PHASE CIRCUIT

- The power in 3-phase load can be measured by using the following methods :
  1. Three wattmeters method.
  2. Two wattmeters method.
  3. One wattmeter method.
- A wattmeter consists of two coils : Refer Fig. 25.
  1. *Current coil*—possesses a *low resistance*.
  2. *Pressure or potential coil*—possesses a *high resistance*.

The '*current coil*' is connected in *series with the line carrying the current* and the '*pressure coil*' *is connected across the two points whose potential*

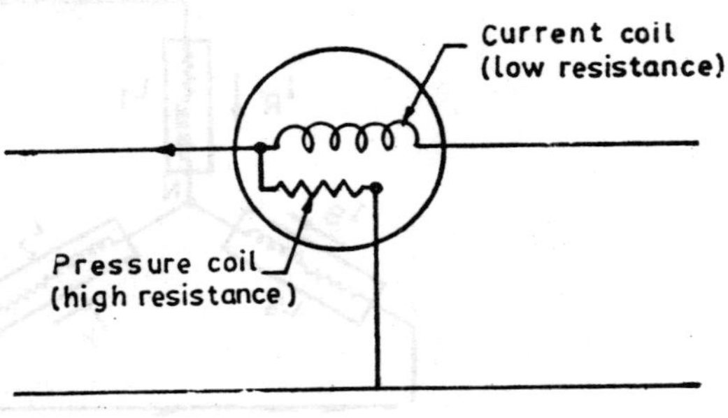

Fig. 25. Connections of a wattmeter.

difference is to be measured. A wattmeter shows a reading which is proportional to the *product* of the *current* through its current coil, the *potential difference* across its pressure coil and *cosine of the angle* between this voltage and current.

9.1. Three Wattmeters Method. Figs. 26 and 27 show the connection diagram for star-connected and delta-connected loads respectively. As indicated in the figures three wattmeters are connected in each of the three phases of the load whether star or delta connected. The current coil

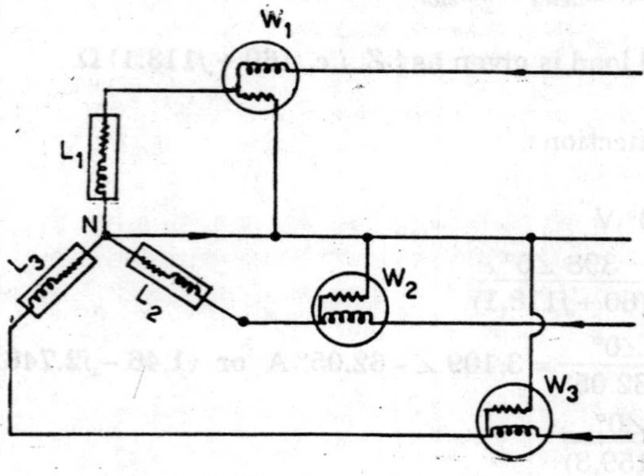

Fig. 26. Star-connected load.

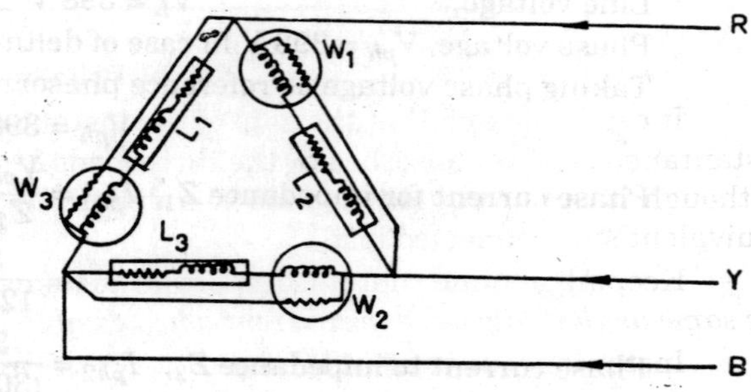

Fig. 27. Delta-connected load.

of each wattmeter carries the current of one phase only and the pressure coil measures the phase-voltage of the phase. Hence, *each wattmeter measures the power in a single phase. The total power in the load is given by the algebraic sum of the readings of the three wattmeters.*

While using this method following *difficulty* is met with :

- In case of star-connected load it is not always possible to get at neutral point which is required for connections (Fig. 26).
- In case of delta-connected load, under ordinary conditions it is not generally feasible to break into the phases of the load.

To measure power it is not necessary to use three wattmeters, two wattmeters can be used for the purpose as explained in the Article 9.2.

9.2. Two Wattmeters Method. Balanced or unbalanced load, Figs. 28 and 29 show connection diagrams for star-connected and delta-connected loads respectively. In this method the current coils of the two wattmeters are inserted in *any two* lines and the pressure (or potential) coil of each joined to the *third line.*

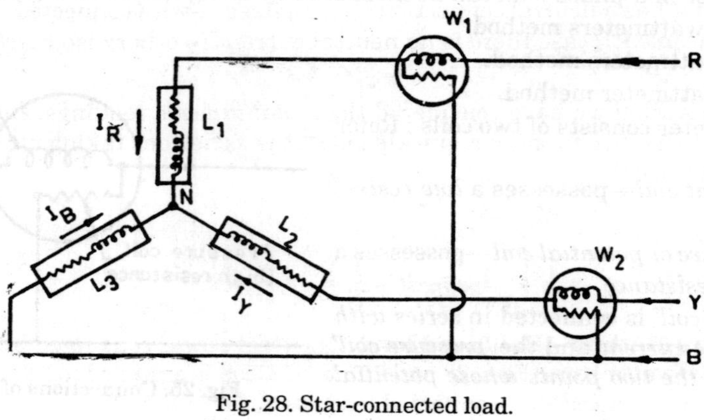

Fig. 28. Star-connected load.

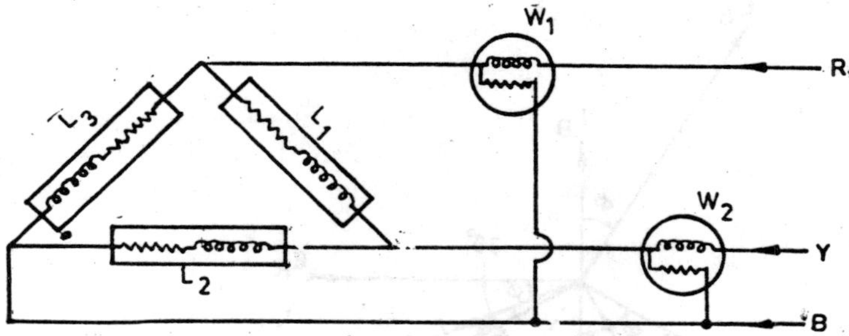

Fig. 29. Delta-connected load.

It can be proved that the sum of the instantaneous powers indicated by W_1 and W_2 gives the instantaneous power absorbed by the three loads L_1, L_2 and L_3. Let us consider a star-connected load (although it can be equally applied to a delta-connected load which can always be replaced by an equivalent star-connected load).

Keeping in mind that it is important to take *the direction of the voltage through the circuit as the same as that taken for the current when establishing the readings of the two wattmeters.*

Instantaneous current through $W_1 = i_R$

Instantaneous potential difference across $W_1 = e_{RB} = e_R - e_B$

Instantaneous power read by $W_1 = i_R(e_R - e_B)$

Instantaneous current through $W_2 = i_Y$

Instantaneous potential difference across $W_2 = e_{YB} = e_Y - e_B$

Instantaneous power read by $W_2 = i_Y(e_Y - e_B)$

$\therefore \quad W_1 + W_2 = i_R(e_R - e_B) + i_Y(e_Y - e_B) = i_R e_R + i_Y e_Y - e_B(i_R + i_Y)$

Now according to Kirchhoff's point law

$$i_R + i_Y + i_B = 0$$

$$\therefore \quad i_R + i_Y = -i_B$$

$$\therefore \quad W_1 + W_2 = i_R e_R + i_Y e_Y + i_B e_B = p_1 + p_2 + p_3$$

where p_1 = power absorbed by load L_1, p_2 = power absorbed by load L_2, and

p_3 = power absorbed by load L_3.

$$\therefore \quad W_1 + W_2 = \text{total power absorbed.}$$

This proof is true whether the load is *balanced* or *unbalanced*.

In case the load is star-connected, then it should have no neutral connection (*i.e.*, 3-phase, 3-wire connected) and if it has a neutral connection (*i.e.*, 3-phase, 4-wire connected) that it should be exactly balanced so that in each case there is no neutral current i_N otherwise Kirchhoff's point law will give $i_R + i_Y + i_B + i_N = 0$.

In the above derivation we have considered the *instantaneous* readings. In fact the moving system wattmeter, due to its inertia, cannot quickly follow the variations taking place in cycle, hence it indicates the *average power*.

$\therefore$ Total power $= W_1 + W_2$

$$= \frac{1}{T}\int_0^T i_R e_{RB}\, dt + \frac{1}{T}\int_0^T i_Y e_{YB}\, dt. \qquad ...(7)$$

Two Wattmeters Method—Balanced load. The total power consumed by a *balanced load* can be found by using two wattmeters (Figs. 28 and 29). When load is assumed inductive in Fig. 26, the vector diagram for such a balanced star-connected load is shown in Fig. 30.

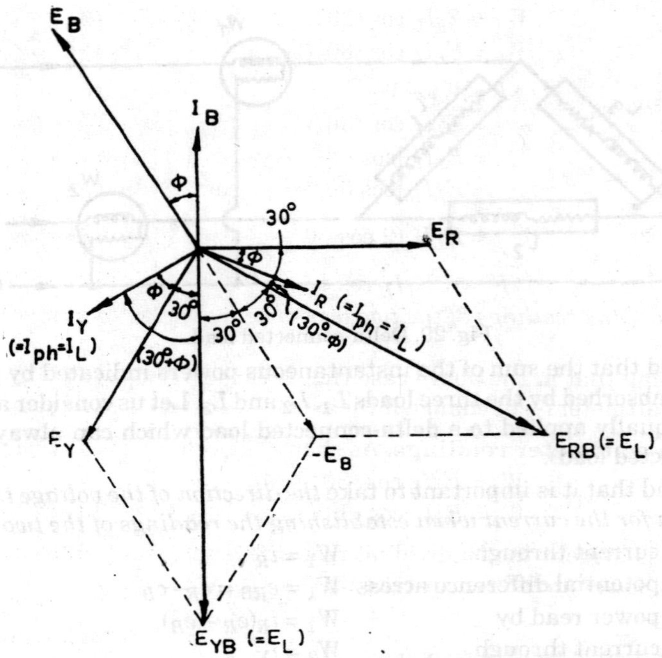

Fig. 30. Vector diagram—two wattmeters method.

Let us consider the problem in terms of r.m.s. values (instead of instantaneous values).

Let E_R, E_Y, E_B = r.m.s. values of the three phase voltages,

and I_R, I_Y, I_B = r.m.s. values of the currents.

Since these voltages and currents are assumed sinusoidal, they can be represented by vectors, the currents lagging behind their respective phase voltages by ϕ.

Refer Fig. 30.

Current through wattmeter $W_1 = I_R$

Potential difference across pressure coil of wattmeter

$$W_1 = E_{RB} = E_R - E_B \quad \text{(Vectorially)}$$

The value of E_{RB} is found by compounding E_R and E_B reversed as shown in Fig. 30. It may be observed that phase difference between E_{RB} and I_R = (30° – ϕ).

$\therefore$ Reading of wattmeter $W_1 = E_{RB} I_R \cos (30° - \phi)$ ...(*i*)

Similarly, current through wattmeter $W_2 = I_Y$

Potential difference across pressure coil of wattmeter

$$W_2 = E_{YB} = E_Y - E_B \quad \text{(Vectorially)}$$

The value of E_{YB} is found by compounding E_Y and E_B reversed as shown in Fig. 30. The phase difference between E_{YB} and I_Y = (30° + ϕ).

$\therefore$ Reading of wattmeter $W_2 = E_{YB} I_Y \cos (30° + \phi)$ ...(*ii*)

Since the load is balanced, $E_{RB} = E_{YB} = E_L$ (Line voltage)

and $I_R = I_Y = I_L$ (Line current)

$\therefore$ $$W_1 = E_L I_L \cos (30° - \phi)$$

and $$W_2 = E_L I_L \cos (30° + \phi)$$

$\therefore$ Total power,

$$P = W_1 + W_2$$
$$= E_L I_L \cos (30° - \phi) + E_L I_L \cos (30° + \phi)$$
$$= E_L I_L [\cos (30° - \phi) + \cos (30° + \phi)]$$
$$= E_L I_L [\cos 30° \cos \phi + \sin 30° \sin \phi + \cos 30° \cos \phi - \sin 30° \sin \phi]$$
$$= E_L I_L (2 \cos 30° \cos \phi) = E_L I_L \times 2 \times \frac{\sqrt{3}}{2} \cos \phi = \sqrt{3} E_L I_L \cos \phi$$

i.e., $$P = \sqrt{3} E_L I_L \cos \phi$$

Hence the *sum of the readings of the two wattmeters give the total power consumption in the 3-phase load.*

It is worth noting that in the above case the phase sequence of *RYB* has been assumed, the readings of the two wattmeters will change if the phase sequence is reversed.

Variations in wattmeter readings. As shown above that for a *lagging power factor*

$$W_1 = E_L I_L \cos (30° - \phi)$$

and $$W_2 = E_L I_L \cos (30° + \phi)$$

From above it is evident that individual readings of the wattmeters *not only depend on the load but also upon its power factor*. Let us take up the following cases :

(*i*) When $\phi = 0$

i.e., power factor is *unity* (*i.e.,* load is *resistive*)

Then $$W_1 = W_2 = E_L I_L \cos 30°$$

The reading of each wattmeter will be equal and opposite (*i.e.,* up-scale reading).

(*ii*) When $\phi = 60°$

i.e., power factor = 0.5 (lagging)

Then $$W_2 = E_L I_L \cos (30° + 60°) = 0$$

Hence, the power is measured by W_1 above.

(*iii*) When $90° > \phi > 60°$

i.e., $0.5 > \text{p.f.} > 0$

Then W_1 is still *positive* but reading of W_2 is *reversed.* For a leading p.f., conditions are just the opposite of this. In that case, W_1 will read *negative* because the phase angle between the current and voltage is *more than* 90°. For getting the total power, the reading of W_2 is to be *subtracted* from that of W_1. Under this condition, W_2 will read *'down scale' i.e., backwards*. Hence, *to obtain a reading on* W_2, *it is necessary to reverse either its pressure coil or current coil, usually the former.*

All readings taken after reversal of pressure coil are to be taken as negative.

(*iv*) When $\phi = 90°$

(*i.e.,* p.f. = 0 *i.e.,* pure inductive or capacitive load)

Then $$W_1 = E_L I_L \cos (30° - 90°) = E_L I_L \sin 30°$$

and $$W_2 = E_L I_L \cos (30° + 90°) = - E_L I_L \sin 30°$$

These two readings are equal in magnitude but opposite in sign

$\therefore$ $$W_1 + W_2 = 0.$$

So far we have considered *lagging angles* (taken as *positive*). Now let us discuss how the readings of wattmeters change when the power factor is *leading* one.

— For $\phi = +60°$ (lag) : $W_2 = 0$

— For $\phi = -60°$ (lead) : $W_1 = 0$

Thus we find that for angles of lead the readings of the two wattmeters are interchanged. Hence, when the power is *leading* :

$$W_1 = E_L I_L \cos(30° + \phi)$$
$$W_2 = E_L I_L \cos(30° - \phi).$$

Power Factor—When the load is 'balanced'. When load is balanced with a *lagging power factor* and the voltage and currents are sinusoidal :

$$W_1 + W_2 = E_L I_L \cos(30° - \phi) + E_L I_L \cos(30° + \phi) = \sqrt{3} E_L I_L \cos\phi \quad \text{...}(i)$$

Similarly,

$$W_1 - W_2 = E_L I_L \cos(30° - \phi) - E_L I_L \cos(30° + \phi)$$
$$= E_L I_L (2 \sin\phi \sin 30°) = E_L I_L \sin\phi \quad \text{...}(ii)$$

Dividing (*ii*) by (*i*), we get

$$\tan\phi = \frac{\sqrt{3}(W_1 - W_2)}{(W_1 + W_2)} \quad \text{...(8)}$$

For a *leading power*, this expression becomes

$$\tan\phi = -\frac{\sqrt{3}\,(W_1 - W_2)}{(W_1 + W_2)} \quad \text{...(9)}$$

After finding tan ϕ, hence ϕ, the value of power factor cos ϕ can be found (from trigonometrical tables).

One important point which must be kept in mind is that if W_2 reading has been taken after reversing the pressure coil *i.e.*, if W_2 is *negative*, then the eqn. (8) becomes

$$\tan\phi = \sqrt{3}\,\frac{W_1 - (-W_2)}{W_1 + (-W_2)}$$

or

$$\tan\phi = \sqrt{3}\,\frac{W_1 + W_2}{W_1 - W_2} \quad \text{...(10)}$$

The power factor may also be expressed in terms of ratio of the readings of the two wattmeters.

Let $\dfrac{\text{Smaller reading}}{\text{Larger reading}} = \dfrac{W_2}{W_1} = \alpha$

Then from eqn. (8) above, we have

$$\tan\phi = \frac{\sqrt{3}\left[1 - \left(\dfrac{W_2}{W_1}\right)\right]}{\left[1 + \left(\dfrac{W_2}{W_1}\right)\right]} = \frac{\sqrt{3}\,(1-\alpha)}{(1+\alpha)}$$

We know that, $\sec^2\phi = 1 + \tan^2\phi$

or

$$\frac{1}{\cos^2\phi} = 1 + \tan^2\phi \quad \text{or} \quad \cos^2\phi = \frac{1}{1+\tan^2\phi}$$

$$\cos\phi = \frac{1}{\sqrt{1+\tan^2\phi}} = \frac{1}{\sqrt{1 + \left[\dfrac{\sqrt{3}\,(1-\alpha)}{1+\alpha}\right]^2}}$$

or $$= \frac{1}{\sqrt{1 + 3\left(\frac{1-\alpha}{1+\alpha}\right)^2}} = \frac{1+\alpha}{\sqrt{(1+\alpha)^2 + 3(1-\alpha)^2}}$$

$$= \frac{1+\alpha}{\sqrt{1 + \alpha^2 + 2\alpha + 3(1 + \alpha^2 - 2\alpha)}} = \frac{1+\alpha}{\sqrt{4 + 4\alpha^2 - 4\alpha}}$$

$$= \frac{1+\alpha}{2\sqrt{1 - \alpha + \alpha^2}}$$

i.e., $$\cos\phi = \frac{1+\alpha}{2\sqrt{1-\alpha+\alpha^2}} \quad ...(11)$$

If a curve is plotted between α and $\cos\phi$, then the curve obtained will be as shown in Fig. 31, this curve is called **watt-ratio curve.**

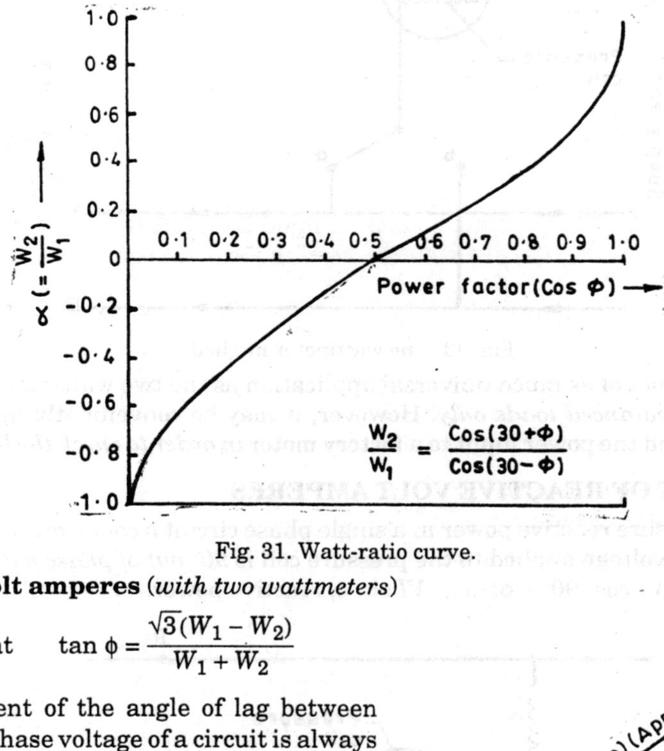

Fig. 31. Watt-ratio curve.

Reactive volt amperes (*with two wattmeters*)

We know that $$\tan\phi = \frac{\sqrt{3}(W_1 - W_2)}{W_1 + W_2}$$

As the tangent of the angle of lag between phase current and phase voltage of a circuit is always equal to the ratio of reactive power to the true power (Fig. 32). Hence, in case of a balanced load, the reactive power is given by $\sqrt{3}$ *times the difference of the readings of the two wattmeters* used to measure the power of a 3-phase circuit by two wattmeter method.

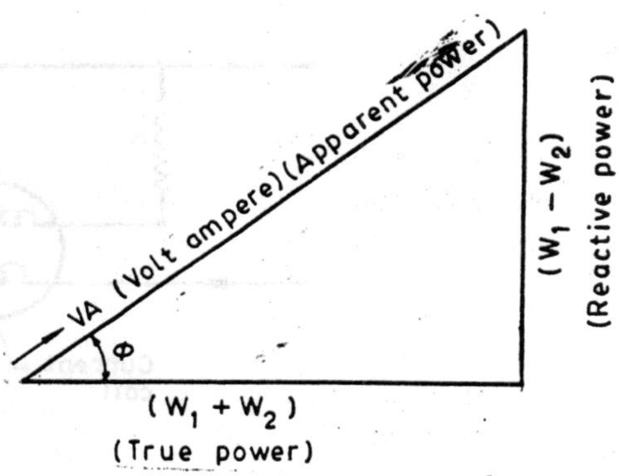

Fig. 32

Mathematical proof is as follows :

$$\sqrt{3}\,(W_1 - W_2) = \sqrt{3}\,[(E_L I_L \cos(30° - \phi) - E_L I_L \cos(30° + \phi)]$$
$$= \sqrt{3}\,E_L I_L\,[(\cos 30° \cos\phi + \sin 30° \sin\phi) - (\cos 30° \cos\phi - \sin 30° \sin\phi)]$$
$$= \sqrt{3}\,E_L I_L\,[\cos 30° \cos\phi + \sin 30° \sin\phi - \cos 30° \cos\phi + \sin 30° \sin\phi]$$
$$= \sqrt{3}\,E_L I_L\,(2 \sin 30° \sin\phi) = \sqrt{3}\,E_L I_L \sin\phi$$

i.e., $\sqrt{3}\,(W_1 - W_2) = \sqrt{3}\,E_L I_L \sin\phi.$

9.3. One Wattmeter Method. In this method the current coil is connected in any one line and the pressure coil is connected alternately between this and the other two lines as shown in Fig. 33. The two readings so obtained, for a balanced load, correspond to those obtained by normal two wattmeter method.

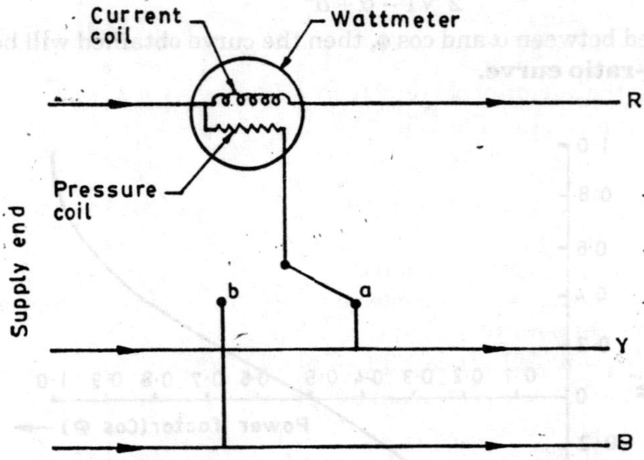

Fig. 33. One wattmeter method.

This method is not of as much universal application as the two wattmeter method because it is restricted to *fairly balanced loads only*. However, it may be conveniently applied, for instance, when it is desired to find the power input to a factory motor *in order to check the load upon the motor*.

10. MEASUREMENT OF REACTIVE VOLT AMPERES

In order to measure reactive power in a single phase circuit a *compensated wattmeter* is used. In this wattmeter the voltage applied to the pressure coil is 90° *out of phase with the actual voltage* and hence it will read $VI \cos(90° - \phi)$ *i.e.,* $VI \sin\phi$, reactive power.

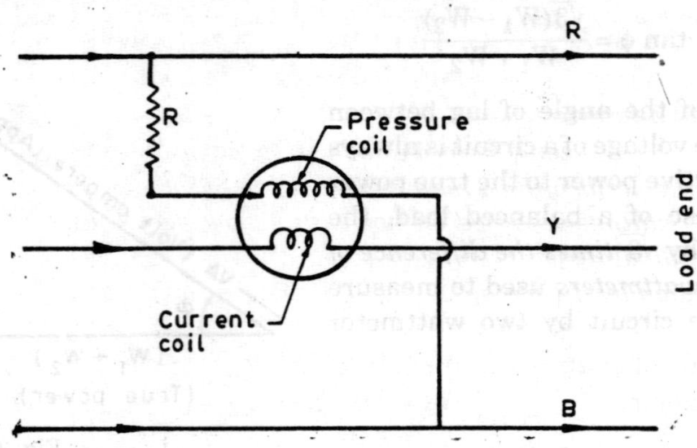

Fig. 34. Measurement of reactive volt-amperes.

In balanced three-phase circuit the reactive power can be determined by using one wattmeter. The necessary connections are shown in Fig. 34. The current coil is inserted in one line and the pressure coil is connected across the other two lines.

The current following through the current coil of the wattmeter

$$= I_Y = I_{ph} \text{ (say)}$$

The potential difference across the potential coil of wattmeter,

$E = E_R - E_B$

$= \sqrt{3}\, E_{ph}$ leading vector E_Y by 90°

$= \sqrt{3}\, E_{ph}$ leading the vector of current I_Y by (90° + φ)

∴ Reading of the wattmeter (Fig. 35)

$$= \sqrt{3}\, E_{ph} I_{ph} \cos(90° + \phi)$$

$$= -\sqrt{3}\, E_{ph} I_{ph} \sin\phi = -W_r$$

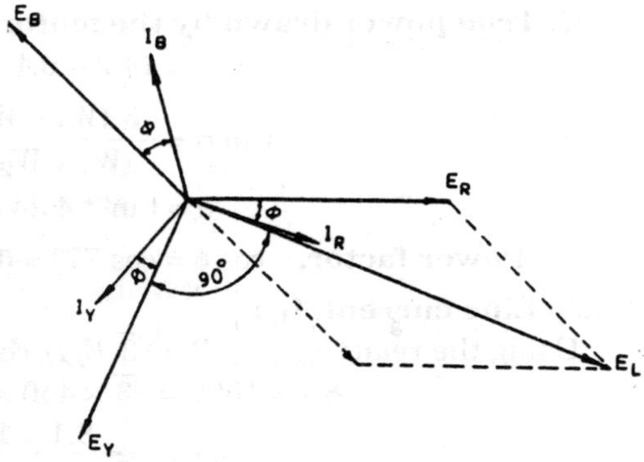

Fig. 35

Total reactive power of the circuit,

$$= 3\, E_{ph} I_{ph} \sin\phi = -\sqrt{3}\, W_r$$

Reactive power (as earlier started) can also be determined from two wattmeter readings connected for measurement of power.

$$W_1 - W_2 = E_L I_L \cos(30° - \phi) - E_L I_L \cos(30° + \phi)$$

or $$W_1 - W_2 = E_L I_L \times 2 \sin 30° \sin\phi$$

or $$E_L I_L \sin\phi = W_1 - W_2$$

Reactive power of load circuit.

$$W_r = \sqrt{3}\, E_L I_L \sin\phi = \sqrt{3}\,(W_1 - W_2).$$

POWER MEASUREMENT IN 3-PHASE CIRCUITS

Example 18. *The power input to a 3-phase induction motor is read by two wattmeters. The readings are 920 W and 300 W. Calculate the power factor of the motor.*

Solution. Reading of wattmeter, $W_1 = 920$ W

Reading of wattmeter, $W_2 = 300$ W

Power factor of the motor, cos φ :

Using the relation, $$\tan\phi = \frac{\sqrt{3}\,(W_1 - W_2)}{(W_1 + W_2)} = \frac{\sqrt{3}\,(920 - 300)}{(920 + 300)} = \frac{\sqrt{3} \times 620}{1220} = 0.88$$

∴ $$\phi = \tan^{-1} 0.88 = 41.35°$$

Power factor of the motor,

$$\cos\phi = \cos 41.35° = \mathbf{0.75\ (lag).}\ \ \textbf{(Ans.)}$$

Example 19. *While performing a load test on a 3-phase wound-rotor induction motor by two wattmeters method, the readings obtained on two wattmeters were + 14.2 kW and – 6.1 kW and the line voltage was 440 V. Calculate :*

(*i*) *True power drawn by the motor* (*ii*) *Power factor, and*

(*iii*) *Line current.*

Solution. Reading of wattmeter, $W_1 = 14.2$ kW

Reading of wattmeter, $W_2 = -6.1$ kW

Line voltage, $E_L = 440$ V

(*i*) **True power drawn by the motor**

$= 14.2 - 6.1 = \mathbf{8.1\ kW.}$ **(Ans.)**

(*ii*) $$\tan\phi = \frac{\sqrt{3}\,(W_1 - W_2)}{(W_1 + W_2)} = \frac{\sqrt{3}\,[14.2 - (-6.1)]}{[14.2 + (-6.1)]} = \frac{\sqrt{3} \times 20.3}{8.1} = 4.34$$

$\therefore$ $\phi = \tan^{-1} 4.34 = 77°$

$\therefore$ **Power factor,** $\cos\phi = \cos 77° = \mathbf{0.2249\ (lag).}$ **(Ans.)**

(*iii*) **Line current,** $\mathbf{I_L}$ **:**

Using the relation, $P = \sqrt{3}\, E_L I_L \cos\phi$

$$8.1 \times 1000 = \sqrt{3} \times 440 \times I_L \times 0.2249$$

$\therefore$ $$I_L = \frac{8.1 \times 1000}{\sqrt{3} \times 440 \times 0.2249} = 47.26 \text{ A}$$

Hence, **line current = 47.26 A. (Ans.)**

Example 20. *A 3-phase, 440 V motor load has a power factor of 0.6. Two wattmeters connected to measure the power show the input to be 25 kW. Find the reading on each instrument.*

Solution. Input power = 25 kW

Line voltage, $E_L = 440$ V

Power factor of the motor load, $\cos\phi = 0.6$

$\mathbf{W_1}$; $\mathbf{W_2}$:

Using the relation : $$\tan\phi = \frac{\sqrt{3}\,(W_1 - W_2)}{W_1 + W_2} \quad ...(i)$$

Now, $\cos\phi = 0.6$, $\therefore$ $\phi = 53.13°$

$\therefore$ $\tan\phi = \tan 53.13° = 1.333$

Also $W_1 + W_2 = 25$ kW (*given*) ...(*ii*)

Substituting these values in (*i*), we get

$$1.333 = \frac{\sqrt{3}\,(W_1 - W_2)}{25}$$

$\therefore$ $$W_1 - W_2 = \frac{1.333 \times 25}{\sqrt{3}} = 19.24 \text{ kW} \quad ...(iii)$$

From (*ii*) and (*iii*), we get $\mathbf{W_1 = 22.12\ kW}$

and $\mathbf{W_2 = 2.88\ kW.}$ **(Ans.)**

☞**Example 21.** *In a 3-phase circuit two wattmeters used to measure power indicate 1200 W and 600 W respectively. Find the power factor of the circuit :*

(*i*) *When both wattmeter readings are positive.*

(*ii*) *When the latter is obtained by reversing the current coil connections.*

(AMIE Summer, 1999)

Solution. (*i*) *When both wattmeter readings are positive :*

Reading of wattmeter, $W_1 = 1200$ W

Reading of wattmeter, $W_2 = 600$ W

We know that, $$\tan\phi = \frac{\sqrt{3}\,(W_1 - W_2)}{W_1 + W_2} = \frac{\sqrt{3}\,(1200 - 600)}{(1200 + 600)} = 0.577$$

$$\phi = \tan^{-1} 0.577 = 30°$$

Power factor, $\mathbf{\cos\phi = \cos 30° = 0.866\ (lag).}$ **(Ans.)**

(*ii*) *When the reading of wattmeter W_2 is obtained by reversing the coil connection* :

Reading of wattmeter, $W_1 = 1200$ W

Reading of wattmeter, $W_2 = -600$ W

We know that, $$\tan\phi = \frac{\sqrt{3}\,(W_1 - W_2)}{(W_1 + W_2)} = \frac{\sqrt{3}\,[1200 - (-600)]}{[1200 + (-600)]} = \frac{\sqrt{3} \times 1800}{600} = 5.196$$

or $$\phi = \tan^{-1} 5.196 = 79.1°$$

Hence, **power factor** $= \cos\phi = \cos 79.1° = 0.1889$. **(Ans.)**

Example 22. *In order to measure the power input and the power factor of an over-excited synchronous motor two wattmeters are used. If the meters indicate (– 3.5 kW) and (+ 8.0 kW) respectively. Calculate :*

(*i*) *Power factor of the motor.* (*ii*) *Power input to the motor.*

Solution. (*i*) *Since an over-excited synchronous motor runs with a leading power factor*, we should use the relation,

$$\tan\phi = -\frac{\sqrt{3}\,(W_1 - W_2)}{(W_1 + W_2)}$$

Moreover it is W_1 that gives negative reading and not W_2. Hence $W_1 = -3.5$ kW.

$$\therefore \quad \tan\phi = -\frac{\sqrt{3}\,(-3.5 - 8)}{(-3.5 + 8)} = -\frac{\sqrt{3} \times 11.5}{5.5} = 3.62$$

$$\therefore \quad \phi = \tan^{-1} 3.62 = 74.6° \text{ (lead)}$$

∴ **Power factor,** $\cos\phi = \cos 74.6° = 0.2655$ **(lead). (Ans.)**

(*ii*) **Power input** $= W_1 + W_2 = -3.5 + 8 =$ **4.5. W. (Ans.)**

Example 23. *Two wattmeters are used to measure power input to a 1.5 kV, 50 Hz, 3-phase motor running on full-load at an efficiency of 85 per cent. Their readings are 250 kW and 80 kW respectively. Calculate :*

(*i*) *Input,* (*ii*) *Power factor,*

(*iii*) *Line current, and* (*iv*) *Output.*

Solution. *Since the motor is running at full-load, its power factor must be greater than 0.5. Hence W_2 reading is positive*

$$\therefore \quad W_1 = +250 \text{ kW} \quad \text{and} \quad W_2 = +80 \text{ kW}$$

(*i*) **Input** $= W_1 + W_2 = 250 + 80 =$ **330 kW. (Ans.)**

(*ii*) $$\tan\phi = \frac{\sqrt{3}\,(W_1 - W_2)}{W_1 + W_2} = \frac{\sqrt{3}\,(250 - 80)}{(250 + 80)} = 0.892$$

$$\therefore \quad \phi = \tan^{-1} 0.892 = 41.74°$$

and **power factor** $= \cos\phi = \cos 41.74° =$ **0.746 (lag). (Ans.)**

(*iii*) **Power,** $$P = \sqrt{3}\, E_L I_L \cos\phi$$

$$\therefore \quad I_L = \frac{P}{\sqrt{3}\, E_L \cos\phi} = \frac{330 \times 1000}{\sqrt{3} \times 1.5 \times 1000 \times 0.746} = 170.27$$

Hence, **line current = 170.27 A. (Ans.)**

(*iv*) **Output** = input × efficiency = 330 × 0.85 = **280.5 kW. (Ans.)**

☞**Example 24.** *Two wattmeters are used to measure power input to a synchronous motor. Each of them indicates 60 kW. If the power factor be changed to 0.866 leading, determine the readings of the two wattmeters, the total input power remaining the same. Draw the vector diagram for the second condition of the load.*

Solution. Reading of wattmeter, $W_1 = 60$ kW

Reading of wattmeter, $W_2 = 60$ kW.

First case. In the first case, *both wattmeters read equal and positive. Hence, motor must be running at unity power factor.*

Second case. *Power factor is 0.866 leading.*

In this case :
$$W_1 = E_L I_L \cos (30° + \phi)$$
$$W_2 = E_L I_L \cos (30° - \phi)$$
$$\therefore \quad W_1 + W_2 = \sqrt{3}\, E_L I_L \cos \phi$$
$$W_1 - W_2 = -\, E_L I_L \sin \phi$$
$$\therefore \quad \tan \phi = -\frac{\sqrt{3}\,(W_1 - W_2)}{(W_1 + W_2)}$$

Since $\cos \phi = 0.866$ *(given)*
$$\therefore \quad \phi = \cos^{-1} 0.866 = 30°$$
and
$$\tan \phi = \tan 30° = \frac{1}{\sqrt{3}}$$
$$\therefore \quad \frac{1}{\sqrt{3}} = -\frac{\sqrt{3}\,(W_1 - W_2)}{120}$$
$$[\because\ W_1 + W_2 = 120 \text{ kW } (given)]$$
$$\therefore \quad W_1 - W_2 = -\frac{120}{3} = -40 \quad ...(i)$$
But
$$W_1 + W_2 = 120 \quad ...(ii)$$

From (*i*) and (*ii*), we get **W_1 = 40 kW, W_2 = 80 kW. (Ans.)**

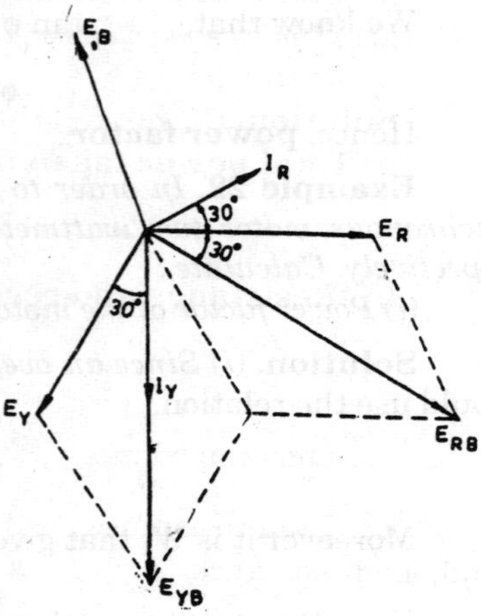

Fig. 36

For connection diagram, please refer to Fig. 29. The vector or phasor diagram is shown in Fig. 36.

Example 25. *Two wattmeters connected to read the total power in a 3-phase system supplying a balanced load read 10.5 kW and – 2.5 kW respectively. Calculate the total power and power factor.*

Also, explain the significance of (i) equal wattmeter readings and (ii) a zero reading on one wattmeter. **(AMIE Winter, 1998)**

Solution. *Given* : $W_1 = 10.5$ kW ; $W_2 = -\,2.5$ kW.

For two wattmeter method,

total power = $W_1 + W_2$ = 10.5 + (– 2.5) = 8.0 kW. (Ans.)

We know that,
$$\tan \phi = \frac{\sqrt{3}\,(W_1 - W_2)}{W_1 + W_2} = \frac{\sqrt{3}\,[10.5 - (-2.5)]}{[10.5 + (-2.5)]} = 2.8145$$
or
$$\phi = 70.44°$$

$\therefore$ **Power factor,** $\cos \phi = \cos 70.44° =$ **0.335. (Ans.)**

(*i*) For readings of the two wattmeters to be equal,
$$W_1 = W_2$$
or
$$\sqrt{3}\, E_L I_L \cos (30° - \phi) = \sqrt{3}\, E_L I_L \cos (30° + \phi)$$
or
$$\cos (30° - \phi) = \cos (30° + \phi)$$
$$\therefore \quad \phi = 0° \text{ or } \cos \phi = 1$$

i.e., for *unity power factor, the readings of two wattmeters are equal*. **(Ans.)**

(*ii*) As readings of wattmeters are
$$W_1 = \sqrt{3} E_L I_L \cos (30° - \phi), \text{ and}$$
$$W_2 = \sqrt{3}\, E_L I_L \cos (30° + \phi),$$
for *reading of one of the wattmeters to be zero ϕ must be 60°, which makes* **$W_2 = \sqrt{3}\, E_L I_L \cos 90° = 0$. (Ans.)**

Example 26. *A star connected 3-phase load consists of three similar impedances. When load is connected to a 3-phase 400 V, 50 Hz supply it takes 30A line current at a power factor of 0.8 lagging.*

(*i*) *Calculate the total power taken by the load.*

(*ii*) *Calculate the resistance and reactance of each phase of the load.*

(*iii*) *If power is measured by two wattmeters method, then calculate the reading of each wattmeter.* **(AMIE Winter, 1996)**

Solution. *Given* : $E_L = 400$ V ; $f = 50$ Hz ; $I_L = (I_{ph}) = 30$ A, $\cos \phi = 0.8$ lagging.

(*i*) **Total power taken by the load, P :**

$$P = \sqrt{3} E_L I_L \cos \phi$$
$$= \sqrt{3} \times 400 \times 30 \times 0.8 = 16627 \text{ W or } \mathbf{16.627 \text{ kW.}} \text{ (Ans.)}$$

(*ii*) **Resistance and reactance of each phase of the load, R, X :**

$$E_{ph} = \frac{E_L}{\sqrt{3}} = \frac{400}{\sqrt{3}} = 230.95 \text{ V}$$

$\therefore$ Load impedance, $Z = \frac{E_{ph}}{I_{ph}} = \frac{230.95}{30} = 7.7\ \Omega$

$\therefore$ Load resistance, $\mathbf{R} = Z \cos \phi = 7.7 \times 0.8 = \mathbf{6.16\ \Omega}$. **(Ans.)**

and, load reactance, $\mathbf{X} = Z \sin \phi = 7.7 \times 0.6 = \mathbf{4.62\ \Omega}$. **(Ans.)**

(*iii*) **Reading of each wattmeter, W_1, W_2 :**

Reading of wattmeter 1, $W_1 = E_L I_L \cos (30° + \phi)$

Reading of wattmeter 2, $W_2 = E_L I_L \cos (30° - \phi)$

where $\phi = \cos^{-1} 0.8 = 36.9°$

$\therefore$ $\mathbf{W_1} = 400 \times 30 \cos (30° + 36.9°) = 4708$ W or **4.708 kW. (Ans.)**

and $\mathbf{W_2} = 400 \times 30 \cos (30° - 36.9°) = 11913$ W or **11.913 kW. (Ans.)**

Example 27. *A 3-phase, 3-wire, 415 V system supplies a balanced load of 20 A at a power factor 0.8 lag. The current coil of wattmeter (1) is in phase R and of wattmeter 2 in phase B. Calculate :*

(*i*) *The reading on 1 when its voltage coil is across R and Y ;*

(*ii*) *The reading on 2 when its voltage coil is across B and Y ;*

(*iii*) *The reading on 1 when its voltage coil is across Y and B.* **(AMIE Winter, 1997)**

Solution. *Given* : $E_L = 415$ V, $I_L = 20$ A, $\cos \phi = 0.8$ (lag) or $\phi = \cos^{-1} 0.8 = 36.87°$ (lag)

(*i*) *The reading on 1 when its voltage coil is across R and Y and current coil is in R phase*

$$= E_L I_L \cos (30° - \phi) = 415 \times 20 \times \cos (30° - 36.87°)$$
$$= 8240 \text{ W or } \mathbf{8.24 \text{ kW.}} \text{ (Ans.)}$$

(*ii*) *Reading on wattmeter 2 when its voltage coil is across B and Y and current coil is in phase B*

$$= E_L I_L \cos (30° + \phi) = 415 \times 20 \cos (30° + 36.87°)$$
$$= 3260 \text{ W or } \mathbf{3.26 \text{ kW.}} \text{ (Ans.)}$$

(*iii*) *When the current coil is in one phase (phasor R) and the voltage coil is across other two phases (Y and R),* the w... $\frac{\text{total } kVAR}{\sqrt{3}}$ and

Total ... $(W_1 - W_2)$

$\therefore$ Wattmeter reading $= \frac{\text{Total } kVAR}{\sqrt{3}} = \frac{\sqrt{3}\,(W_1 - W_2)}{\sqrt{3}}$

$$W_1 - W_2 = 8240 - 3260 = \mathbf{4980 \text{ VAR.}} \text{ (Ans.)}$$

Example 28. *The ratio of the readings of two wattmeters connected to measure power in a balanced 3 – φ, 3 wire load is 5 : 3. The load is known to be inductive with a lagging power factor. Calculate the power factor of the load.* **(Punjab University)**

Solution. Let the ratio of wattmeter readings be *r i.e.*, $\frac{W_2}{W_1} = r$

Now,
$$\tan\phi = \sqrt{3}\left(\frac{W_1 - W_2}{W_1 + W_2}\right) = \sqrt{3}\left[\frac{1 - \frac{W_2}{W_1}}{1 + \frac{W_2}{W_1}}\right] = \sqrt{3}\left(\frac{1-r}{1+r}\right)$$

and power factor,
$$\cos\phi = \frac{1}{\sec\phi} = \frac{1}{\sqrt{1+\tan^2\phi}} = \frac{1}{\sqrt{1+3\left(\frac{1-r}{1+r}\right)^2}}$$

Substituting $r = \frac{W_2}{W_1} = \frac{3}{5} = 0.6$ in the above expression, we get

$$\cos\phi = \frac{1}{\sqrt{1+3\left(\frac{1-0.6}{1+0.6}\right)^2}} = \mathbf{0.918\ (lag).}\quad \textbf{(Ans.)}$$

Example 29. *A symmetrical star-connected three-phase load is made of three identical coils having an internal resistance of 12 Ω and inductance of 142 mH. The power taken by this load when connected to a balanced 3-phase, three wire supply of 400 V, 50 Hz is measured by using the two wattmeters method. The wattmeters used are similar and are of type of 400 V, 5 A, UPF, class 1 accuracy.*

(i) Determine expected readings of W_1 and W_2.

(ii) Compute the total power. **(GATE, 1995)**

Solution. Line voltage, $E_L = 400$ V

Phase voltage, $E_{ph} = \frac{400}{\sqrt{3}} = 231$ V

Coil resistance, $R_{ph} = 12\ \Omega$

Coil inductive reactance, $X_{ph} = 2\pi f L_{ph} = 2\pi \times 50 \times 142 \times 10^{-3} = 44.61\ \Omega$

Coil impedance, $Z_{ph} = \sqrt{R_{ph}^2 + X_{ph}^2} = \sqrt{12^2 + 44.61^2} = 46.2\ \Omega$

Coil phase angle, $\phi = \cos^{-1}\frac{R_{ph}}{Z_{ph}} = \cos^{-1}\left(\frac{12}{46.2}\right) = 74.94°$ (lagging)

Phase current, $I_{ph} = \frac{E_{ph}}{Z_{ph}} = \frac{231}{46.2} = 5$ A

(*i*) **Readings of wattmeters W_1, W_2 :**

Readings of wattmeter, $W_1 = E_L I_L \cos(30° - \phi)$

$= 400 \times 5.0 \cos(30° - 74.94°) =$ **1415.7 W. (Ans.)**

($\because\ I_L = I_{ph} = 5.0$ A)

Reading of wattmeter, $W_2 = E_L I_L \cos(30° + \phi)$

$= 400 \times 5.0 \times \cos(30° + 74.24°) = -$ **515.6 W. (Ans.)**

(*ii*) **Total power :**

Total power $= W_1 + W_2$

$= 1415.7 + (-515.6) =$ **900.1 W. (Ans.)**

Example 30. *Two wattmeters are being used to measure power of a balanced load of 30 A at power factor 0.8 lagging being supplied by a 3-phase, 3-wire, 440 V supply. Calculate :*

(*i*) *Power consumed.* (*ii*) *Reading of wattmeter No. 1*

(*iii*) *Reading of wattmeter No. 2.*

Solution. Line voltage, $E_L = 440$ V

Line current, $I_L = 30$ A

Power factor, $\cos \phi = 0.8$

Phase angle, $\phi = \cos^{-1} 0.8 = 36.87°$

(*i*) **Power consumed,** $P = \sqrt{3}\, E_L I_L \cos \phi = \sqrt{3} \times 440 \times 30 \times 0.8$

$= 18290$ W or **18.29 kW. (Ans).**

(*ii*) **Reading of wattmeter No. 1**

$= E_L I_L \cos (30° - \phi) = 440 \times 30 \cos (30° - 36.87°) =$ **13105 W. (Ans.)**

(*iii*) **Reading of wattmeter No. 2**

$= E_L I_L \cos (30° + \phi) = 440 \times 30 \times \cos (30° + 36.87°)$

= 5185 W (app.). (Ans.)

☞**Example 31.** *A balanced star-connected load, each having a resistance of 10 Ω and inductive reactance of 20 Ω is connected to a 400 V, 50 Hz supply. The phase sequence is RYB. Two wattmeters connected to read total power have their current coils connected in the red and blue lines respectively. Calculate the reading of each wattmeter.*

Draw the circuit and vector diagram.

Solution. Resistance per phase, $R_{ph} = 10\ \Omega$

Reactance per phase, $X_{ph} = 20\ \Omega$

Line voltage, $E_L = 400$ V

Impedance per phase, $Z_{ph} = \sqrt{R_{ph}^2 + X_{ph}^2} = \sqrt{10^2 + 20^2} = 22.36\ \Omega$

Phase voltage, $E_{ph} = \dfrac{E_L}{\sqrt{3}} = \dfrac{400}{\sqrt{3}} = 231$ V

$$I_{ph}\ (= I_L) = \frac{E_{ph}}{Z_{ph}} = \frac{231}{22.36} = 10.33 \text{ A}$$

Power factor, $\cos \phi = \dfrac{R_{ph}}{Z_{ph}} = \dfrac{10}{22.36} = 0.447$

$\therefore$ $\phi = 63.43°$

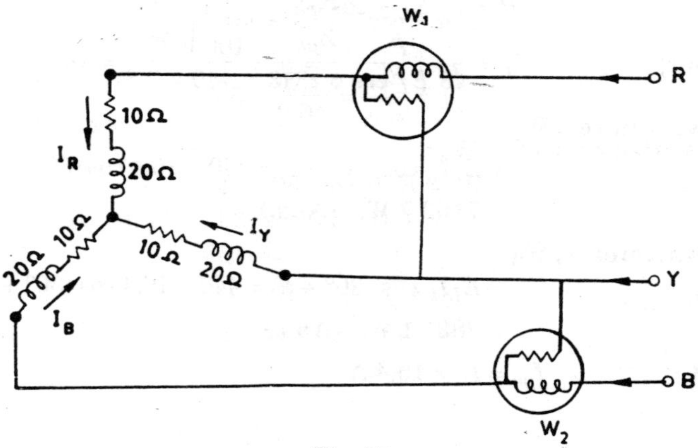

Fig. 37

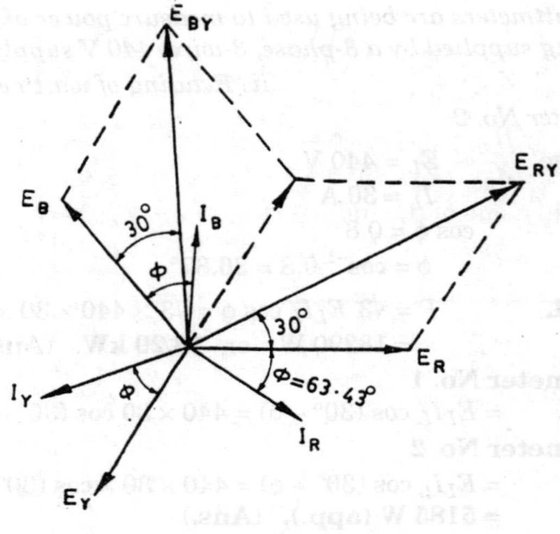

Fig. 38

The circuit and vector diagrams are shown in Figs. 37 and 38 respectively.

Reading of wattmeter, $W_1 = E_{RY} \times I_R \times \cos(30° + \phi) = E_L \times I_L \times \cos(30° + \phi)$

$= 400 \times 10.33 \cos(30° + 63.43°) = -$ **247 W (app.). (Ans.)**

Reading of wattmeter, $W_2 = E_{BY} \times I_B \times \cos(30° - \phi) = E_L \times I_L \times \cos(30° - 63.43°)$

$= 400 \times 10.33 \times \cos(30° - 63.43°) =$ **3448 W (app.). (Ans.)**

☞**Example 32.** *Three equal star-connected inductors take 10 kW at power factor 0.8 when connected to a 440 V, 3 φ, 3-wire supply. Give the connection diagram of two single-phase wattmeters to measure the power in the circuit. Allocate the reading of each wattmeter. Find the line currents if one of the inductors is short-circuited. Draw the vector diagrams of the currents and voltages under the condition.*

Solution. Power, $P = 10 \text{ kW} = 10000 \text{ W}$

Line voltage, $E_L = 440 \text{ V}$

Power factor, $\cos\phi = 0.8$ (lag)

∴ $\phi = 36.9°$ (lag)

The connection diagram of two single-phase wattmeters to measure the power in the circuit is shown in Figs. 39 and 40.

$$P = \sqrt{3}\, E_L I_L \cos\phi$$

∴ Line current, $I_L = \dfrac{P}{\sqrt{3}\, E_L \cos\phi} = \dfrac{10000}{\sqrt{3} \times 440 \times 0.8} = 16.4 \text{ A}.$

Reading of wattmeter, W_1

$= E_L I_L \cos(30° - \phi) = 440 \times 16.4 \cos(30° - 36.9°)$

= 7163.7 W. (Ans.)

Reading of wattmeter, W_2

$= E_L I_L \cos(30° + \phi) = 440 \times 16.4 \cos(30° + 36.9°)$

= 2831.1 W. (Ans.)

Phase current, $I_{ph} = I_L = 16.4 \text{ A}$

Phase impedance, $Z_{ph} = \dfrac{E_{ph}}{I_{ph}} = \dfrac{\frac{440}{\sqrt{3}}}{16.4} = 15.49\ \Omega$

Phase resistance, $R_{ph} = Z_{ph} \cos \phi = 15.49 \times 0.8 = 12.39\ \Omega$

Phase reactance, $X_{ph} = Z_{ph} \sin \phi = 15.49 \times 0.6 = 9.29\ \Omega$

The circuit diagram when one of the inductors is short-circuited is shown in Fig. 39.

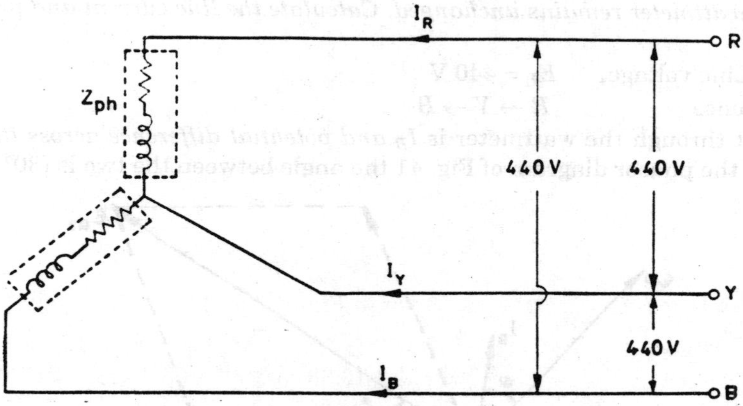

Fig. 39

Current in line R, $I_R = \dfrac{E_{RY}}{Z_{ph}} = \dfrac{440}{15.49} = 28.4$ A

Current in line B, $I_B = \dfrac{E_{BY}}{Z_{ph}} = \dfrac{440}{15.49} = 28.4$ A

The currents in line R and line B are equal in magnitude but differ in phase by 60°, as shown in Fig. 40.

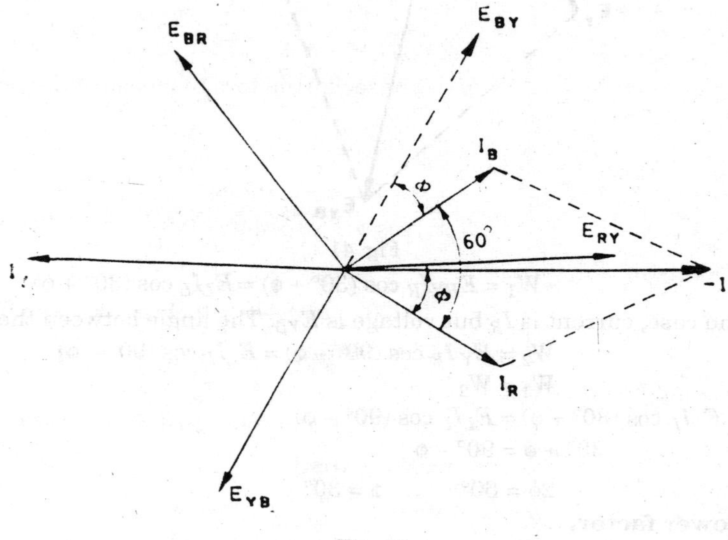

Fig. 40

Current in line Y, $I_Y = -(I_R + I_B)$

Magnitude of current flowing in line Y,

$$I_Y = 2I_R \cos 30° = 2 \times 28.4 \times \cos 30° = \mathbf{49.19\ A.}\ \ \textbf{(Ans.)}$$

Example 33. *A 3-phase, 3-wire balanced load with a lagging power factor is supplied at 440 V (between lines). A single phase wattmeter (scaled in kW) when connected with its current coil in the R-line and voltage coil between R and Y lines gives a reading of 8 kW. When the same terminals of the voltage coil are switched over to Y-line and B-line, the current coil connections remaining the same, the reading of the wattmeter remains unchanged. Calculate the line current and power factor of the load.*

Solution. Line voltage, $E_L = 440$ V

Phase sequence $R \to Y \to B$

The current through the wattmeter is I_R *and potential difference across its pressure coil is* E_{RY}. As seen from the phasor diagram of Fig. 41 the angle between the two is (30° + φ).

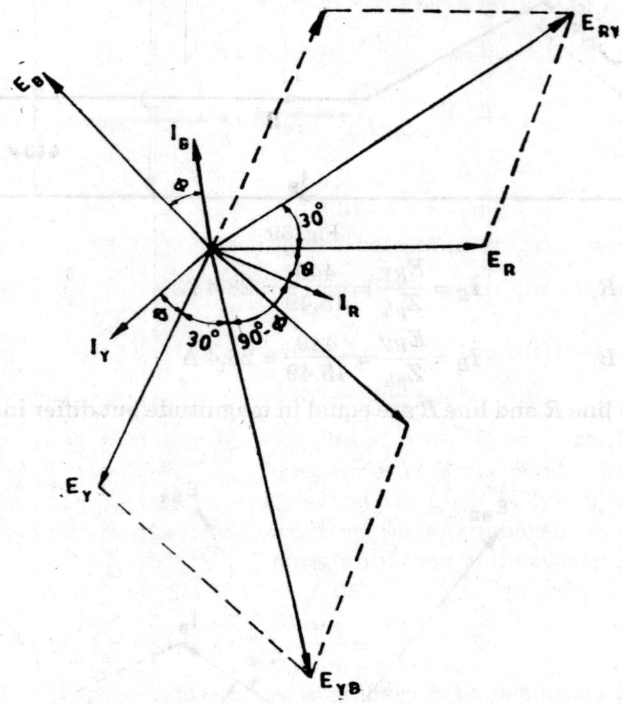

Fig. 41

$$\therefore \qquad W_1 = E_{RY} I_R \cos (30° + \phi) = E_L I_L \cos (30° + \phi) \qquad ...(i)$$

In the second case, current is I_R but voltage is E_{YB}. The angle between the two is (90° – φ).

$$\therefore \qquad W_2 = E_Y I_R \cos (90° - \phi) = E_L I_L \cos (90° - \phi) \qquad ...(ii)$$

Since $W_1 = W_2$ *(given)*

$$E_L I_L \cos (30° + \phi) = E_L I_L \cos (90° - \phi)$$

$$\therefore \qquad 30° + \phi = 90° - \phi$$

or $$2\phi = 60° \qquad \therefore \ \phi = 30°$$

∴ **Load power factor,**

$$\cos \phi = \cos 30° = \mathbf{0.866\ (lag).}\ \ \textbf{(Ans.)}$$

Now, $W_1 = W_2 = 8 \text{ kW} = 8000 \text{ W}$

From (*ii*), we get $8000 = E_L I_L \cos(90° - \phi) = 440 \times I_L \cos(90° - 30°)$

$\therefore$ $$I_L = \frac{8000}{440 \times \cos 60°} = 36.36 \text{ A}$$

Hence, **line current = 36.36 A. (Ans.)**

Example 34. *A wattmeter indicates 6 kW when its current coil is connected in red phase and its potential coil is connected between the neutral and red phase of a symmetrical 3-phase system supplying a balanced load of 40 A at 440 V. What will be reading of the instrument if the current coil connections remain unchanged, the potential coil be connected between blue and yellow phases ?*

Solution. When the wattmeter current coil is connected in red phase and its pressure coil is connected between red phase and neutral, the wattmeter reads per phase $E_{ph} I_{ph} \cos\phi$

or $$E_{ph} I_{ph} \cos\phi = 6 \text{ kW} = 6000 \text{ W}$$

But $$E_{ph} = \frac{440}{\sqrt{3}} \text{ V and } I_{ph} = 40 \text{ A} \quad \textit{(given)}$$

$\therefore$ Power factor, $$\cos\phi = \frac{6000}{E_{ph} I_{ph}} = \frac{6000}{\frac{440}{\sqrt{3}} \times 40} = 0.59$$

and $$\sin\phi = \sqrt{1 - (0.59)^2} = 0.807$$

When the pressure coil is connected between blue and yellow phases and current coil is connected in red phase, the instrument will read $\frac{\text{Total kVAR}}{\sqrt{3}}$ or $\sqrt{3}$ kVAR per phase.

$\therefore$ **Reading of instrument** $= \sqrt{3} E_{ph} I_{ph} \sin\phi = \sqrt{3} \times \frac{440}{\sqrt{3}} \times 40 \times 0.807$

$= 14203$ or **14.203 kVAR. (Ans.)**

Example 35. *A wattmeter reads 5.54 kW when its current coil is connected in R phase and its voltage coil is connected between the neutral and R phase of a symmetrical 3-phase system supplying a balanced load of 30 A at 400 V. What will be the reading on the instrument if the connections to the current coil remain unchanged and the voltage coil is connected between B and Y phases ? Take phase sequence RYB. Draw corresponding phasor diagram.* **(Panjab University)**

Solution. Phase current, $I_{ph} = I_L = 30 \text{ A}$

Phase voltage, $$E_{ph} = \frac{E_L}{\sqrt{3}} = \frac{400}{\sqrt{3}} = 231 \text{ V}$$

Power supplied per phase, P = reading of wattmeter
$= 5.54$ kW

Load power factor, $$\cos\phi = \frac{P}{E_{ph} \cdot I_{ph}}$$
$$= \frac{5.54 \times 10^3}{231 \times 30} = 0.8$$

or $$\phi = \cos^{-1} 0.8 = 36.87°$$

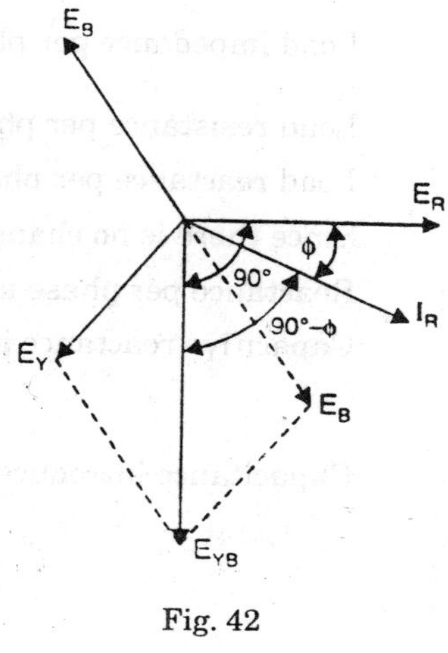

Fig. 42

In the *second case* when pressure coil is connected across phases B and Y, the phase angle between current in phase R and voltage across voltage coil is $(90° - \phi)$ as obvious from the phasor diagram, shown in Fig. 42.

$\therefore$ Wattmeter reading $= E_{YB}\, I_R \cos(90° - \phi)$
$= 400 \times 30 \times \cos(90° - 36.87°)$
$= 7200$ W or **7.2 kW. (Ans.)**

Example 36. *Two wattmeters A and B give readings as 5000 W and 1000 W respectively during the power measurement of 3 – φ, 3 wire balanced load system. (a) Calculate the power and power factor (i) when both meters read direct and (ii) when one of them reads in reverse (b) if the voltage of the circuit is 400 V, what is the value of the capacitance which must be introduced in each phase to cause the whole of the power to appear on A ? The frequency of supply is 50 Hz.*

(Nagpur University, 1998)

Solution. *Given* : $W_1 = 5000$ W ; $W_2 = 1000$ W ; $V = 400$ volts ; $f = 50$ Hz.

(*a*) **Power (P) and power factor (cos φ) :**

(*i*) *When both meters read direct :*

Total power, $P = W_1 + W_2 = 5000 + 1000 = 6000$ W or **6 kW. (Ans.)**

We know that, $\tan\phi = \dfrac{\sqrt{3}\,(W_1 - W_2)}{W_1 + W_2} = \dfrac{\sqrt{3}\,(5000 - 1000)}{(5000 + 1000)} = 1.1547$

or $\phi = \tan^{-1}(1.1547) = 49.1°$

$\therefore$ Power factor, $\cos\phi = \cos 49.1° =$ **0.655 (lagging). (Ans.)**

(*ii*) *When one of the meters reads in reverse :*

Here, $W_2 = -1000$ W

Total power, $P = W_1 + W_2 = 5000 + (-1000) = 4000$ W or **4 kW. (Ans.)**

Again, $\tan\phi = \dfrac{\sqrt{3}\,(W_1 - W_2)}{W_1 + W_2} = \dfrac{\sqrt{3}\,[5000 - (-1000)]}{(5000 - 1000)} = 2.598$

or $\phi = \tan^{-1} 2.598 = 68.95°$

$\therefore$ $\cos\phi = \cos 68.95° =$ **0.359 (lagging). (Ans.)**

(*b*) **Capacitance to be introduced in each phase, C :**

Reading of wattmeter, *B will be zero and whole of the power would appear on wattmeter A when power factor cos φ′ = 0.5 or φ′ = 60°*. It means power factor is to be improved from 0.359 to 0.5.

Load cement per phase, $I_{ph}\,(= I_L) = \dfrac{P}{\sqrt{3}\,E_L \cos\phi} = \dfrac{4000}{\sqrt{3} \times 400 \times 0.359} = 16.08$ A

Load impedance per phase, $Z_{ph} = \dfrac{E_{ph}}{I_{ph}} = \dfrac{400/\sqrt{3}}{16.08} = 14.36\ \Omega$

Load resistance per phase, $R_{ph} = Z_{ph} \cos\phi = 14.36 \times 0.359 = 5.156\ \Omega$

Load reactance per phase, $X_{ph} = Z_{ph} \sin\phi = 14.36 \sin\cos^{-1}(0.359) = 13.4\ \Omega$

Since there is no change in resistance, therefore,

Reactance per phase in the circuit $X'_{ph} = R_{ph} \,.\, \tan\phi' = 5.156 \tan 60° = 8.93\ \Omega$

Capacitive reactance introduced in each phase,

$$X_c = X_{ph} - X'_{ph} = 13.4 - 8.93 = 4.47\ \Omega$$

Capacitance introduced in each phase, $\mathbf{C} = \dfrac{1}{2\pi f X_L}$

$$= \frac{1}{2\pi \times 50 \times 4.47} \times 10^6\ \mu\text{F} = \mathbf{712\ \mu F.}\ \textbf{(Ans.)}$$

☞**Example 37.** *A 3-phase, 434 V, 50 Hz supply is connected to a 3-phase, star-connected induction motor and synchronous motor. Impedance of each phase of induction motor is 1.25 + j2.17 Ω. The three-phase synchronous motor is over-excited and it draws a current of 120 A at 0.87 leading power factor. Two wattmeters are connected in usual manner to measure power drawn by the two motors. Calculate :*

(*i*) *Reading on each wattmeter,* (*ii*) *Combined power factor.*

Solution. Line voltage, E_L = 434 V

Impedance of each phase of induction motor,

$$Z_1 = 1.25 + j2.17\ \Omega$$

Current drawn by the synchronous motor = 120 A at p.f. 0.87 leading

Reading on each wattmeter :

Combined power factor :

Let us assume that the synchronous motor is star-connected. *Since it is over-excited it has a leading power factor.* The connections and phasor diagrams are shown in Figs. 43 and 44.

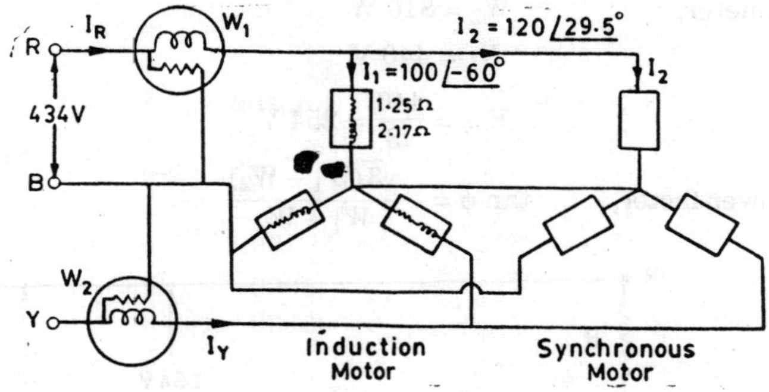

Fig. 43

Phase voltage in each case $= \dfrac{434}{\sqrt{3}} = 250$ V

$Z_1 = 1.25 + j2.17 = 2.5\angle 60°$

$I_1 = \dfrac{250}{2.5} = 100$ A

lagging the reference vector E_R by 60°

$\therefore \quad I_1 = 100\angle -60° = 50 - j86.6$

$I_2 = 120\angle 29.5° = 104.4 + j59.1$

$I_R = I_1 + I_2 = (50 - j86.6) + (104.4 + j59.1)$

$= (154.4 - j27.5) = 156.8\angle -10.1°$

(*i*) As shown in Fig. 44, I_R lags E_R by 10.1°
Similarly, I_Y lags E_Y by 10.1°.

As seen from Fig. 43, current through W_1 is I_R and voltage across it is $E_{RB} = E_R - E_B$. As seen, E_{RB} = 434 V lagging by 30°. Phase difference between E_{RB} and I_R is 30 – 10.1 = 19.9°.

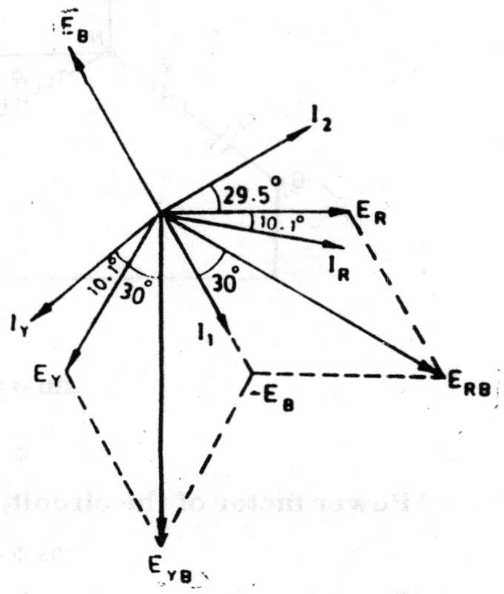

Fig. 44

Reading of W_1 = 434 × 156.8 × cos 19.9°

= **63988 W. (Ans.)**

Current I_Y is also (like I_R) the vector sum of the line currents draws by the two motors. It is equal to 156.8 A and lags behind its respective phase voltage E_Y by 10.1°. Current through W_2 is I_Y and voltage across it is $E_{YB} = E_Y - E_B$. As seen E_{YB} = 43 V.

Phase difference between E_{YB} and I_Y = 30° + 10.1° = 40.1° (lag)

$\therefore$ **Reading of wattmeter, W_2** = 434 × 156.8 × cos 40.1° = **52054 W. (Ans.)**

(*ii*) **Combined power factor** = cos 10.1° = **0.984 (lag). (Ans.)**

☞**Example 38.** *A 3-phase balanced supply with a line voltage of 440 V at a frequency of 50 Hz supplies a star-connected balance load. Each phase of the load consists of a resistance and a capacitor joined in series and the readings on two wattmeters connected to measure the total power supplied are 810 W and 2100 W, both positive. Calculate :*

(*i*) *Power factor of the circuit,* (*ii*) *Line current, and*

(*iii*) *Capacitance of each capacitor.*

Solution. Reading of wattmeter, W_1 = 2100 W

Reading of wattmeter, W_2 = 810 W

Line voltage, E_L = 440 V

Phase voltage, $E_{ph} = \dfrac{440}{\sqrt{3}} = 254$ V

For a leading power factor, $\tan \phi = -\dfrac{\sqrt{3}(W_1 - W_2)}{W_1 + W_2}$

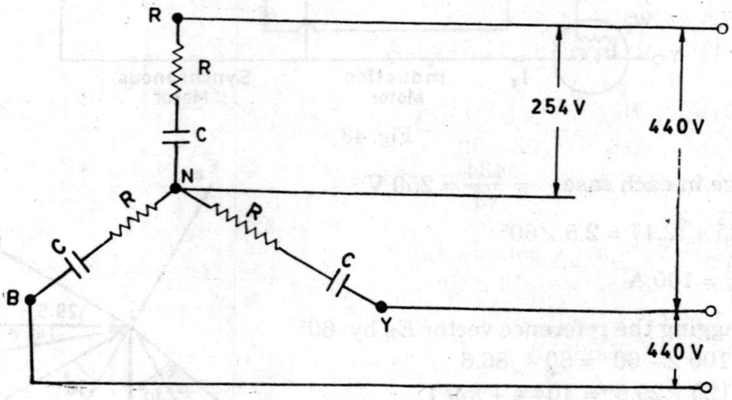

Fig. 45

$\therefore \quad \tan \phi = -\sqrt{3} \times \dfrac{(2100 - 810)}{(2100 + 810)} = -0.767$

$\therefore \quad \phi = -37.48°.$

(*i*) **Power factor of the circuit,**

cos φ = cos (– 37.48°) = **0.793 (leading). (Ans.)**

(*ii*) Power, $P = \sqrt{3}\, E_L I_L \cos \phi$

$(W_1 + W_2) = \sqrt{3} E_L I_L \cos \phi$ $\quad [\because P = W_1 + W_2]$

$(2100 + 810) = \sqrt{3} \times 440 \times I_B \times 0.793$

$\therefore \qquad I_L = \dfrac{2910}{\sqrt{3} \times 440 \times 0.793} = 4.81 \text{ A}$

Hence, **line current = 4.81 A. (Ans.)**

(*iii*) Impedance per phase, $Z_{ph} = \dfrac{E_{ph}}{I_{ph}} = \dfrac{254}{4.81}$ $\qquad [\because\ I_{ph} = I_L = 4.81 \text{ A}]$

$= 52.8\ \Omega$

$Z_{ph} = 52.8\ \angle -37.48° = 41.9 - j32.12$

$\therefore$ Capacitive reactance, $X_C = 32.12\ \Omega$ or $\dfrac{1}{2\pi fC} = 32.12$

$\therefore \qquad C = \dfrac{1}{2\pi f \times 32.12} \text{ F} = \dfrac{10^6}{2\pi \times 50 \times 32.12}\ \mu\text{F} = 99.1\ \mu\text{F}$

Hence, **capacitance of each capacitor = 99.1 μF. (Ans.)**

Example 39. *A 3-phase, 3-wire, CBA system supplies a Δ-connected load in which $Z_{AB} = 25\ \angle 90°$; $Z_{BC} = 15\ \angle 30°$; $Z_{CA} = 20\ \angle 0°$ ohms. Line voltage is 230 V. Find the wattmeter readings when the two wattmeters method is used with meters in lines A and B.* **(A.M.I.E. 1998)**

Solution. *Given* : $Z_{AB} = 25\ \angle 90°\ \Omega$; $Z_{BC} = 15\ \angle 30°\ \Omega$; $Z_{CA} = 20\ \angle 0°\ \Omega$; $E_L = 230$ V.

Wattmeter readings W_1, W_2 :

Let the voltage E_{AC} be the reference phasor, then

$E_{AC} = 230\ \angle 0°$; $E_{CB} = 230\ \angle -120°$ V and $E_{BA} = 230\ \angle 120°$

Phase current,

$$I_{CA} = \frac{E_{CA}}{Z_{CA}} = \frac{-E_{AC}}{Z_{CA}} = \frac{-230\ \angle 0°}{20\ \angle 0°}$$

$= -11.5\ \angle 0°$ A or $(-11.5 + j0)$ A

Phase current,

$$I_{AB} = \frac{E_{AB}}{Z_{AB}} = \frac{-E_{BA}}{Z_{AB}} = \frac{-230\ \angle 120°}{25\ \angle 90°}$$

$= -9.2\ \angle 30°$ A or $(-7.967 - j4.6)$A

Phase current,

$$I_{BC} = \frac{E_{BC}}{Z_{BC}} = \frac{-E_{CB}}{Z_{BC}} = \frac{-230\ \angle -120°}{15\ \angle 30°}$$

$= -15.333\ \angle -150°$

$= (13.279 + j7.666)$ A

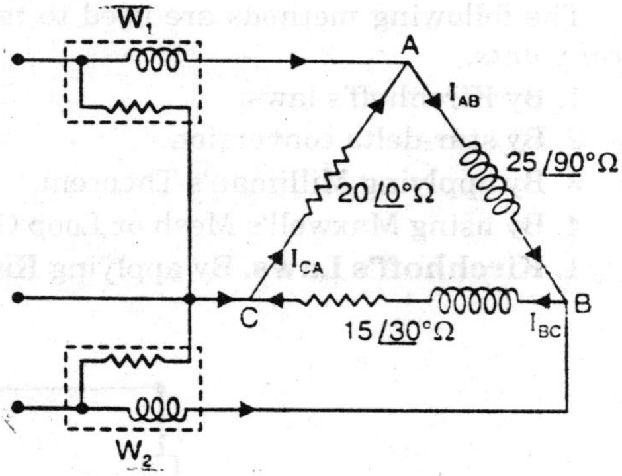

Fig. 46

Now line current, $I_A = I_{AB} - I_{CA} = (-7.967 - j4.6) - (-11.5 + j0)$

$= (3.533 - j4.6) = 5.8\ \angle -52.47°$

Line current, $I_B = I_{BC} - I_{AB} = (13.279 + j7.666) - (-7.967 - j4.6)$

$= (21.246 + j12.266) = 24.533\ \angle 30°$

Reading of wattmeter W_1 = Real part of $^*E_{AC} \times I_A$

= Real part of $230\ \angle 0° \times 5.8\ \angle -52.47°$

= **812.64 W. (Ans.)** (*means conjugate of)

Reading of wattmeter W_2 = Real part of $^*E_{BC} \times I_B$

$= -230\ \angle -120° \times 24.533\ \angle 30°$ = **2443.3 W. (Ans.)**

11. UNBALANCED STAR-CONNECTED LOAD

The following two cases will be considered :

1. Three-phase four-wire system.
2. Three-phase three-wire system.

11.1. Three Phase Four Wire System

- In this type of system a total of *four conductors* are used, three for three phases and one for neutral.
- This system is *very flexible* and is *used for distribution of electric power to domestic, commercial, industrial and other consumers.*
- The line voltage is $\sqrt{3}$ times the phase voltage. The domestic and other single-phase loads needing higher voltages are connected between lines. *The currents in three lines will not be equal and the phasor sum of line currents would return through neutral. The current in the neutral must be equal and opposite to the sum of line currents.*

11.2. Three-Phase Three-Wire System. In case a star-connected load is unbalanced and if there is *no neutral wire,* then its star point is isolated from the star-point of the generator. The potential of the load star-point is different from that of the generator star-point. The potential of the former is subject to variations according to the unbalance of the load and under certain conditions of loading, the potentials of the two star-points may differ considerably. Such an isolated load star-point or neutral point is called a *'floating'* neutral point *because its potential is always changing and not fixed.* All star-connected unbalanced load supplied from polyphase systems without a neutral wire, have floating neutral points. *Any unbalancing of the load causes variations not only of the potential of star-point but also of voltages across the different branches of the load.* Hence in that case, *phase voltage is not* $\frac{1}{\sqrt{3}}$ *of the line voltage.* In general, the phase voltages are not only unequal in magnitude but also subtend angles other than 120° with one another.

The following methods are used to tackle *unbalanced star-connected loads having isolated neutral points.*

1. By Kirchhoff's laws.
2. By star-delta conversion.
3. By applying Milliman's Theorem.
4. By using Maxwell's Mesh or Loop Current Method.

1. **Kirchhoff's Laws.** By applying Kirchhoff's laws, we get (Refer Fig. 47)

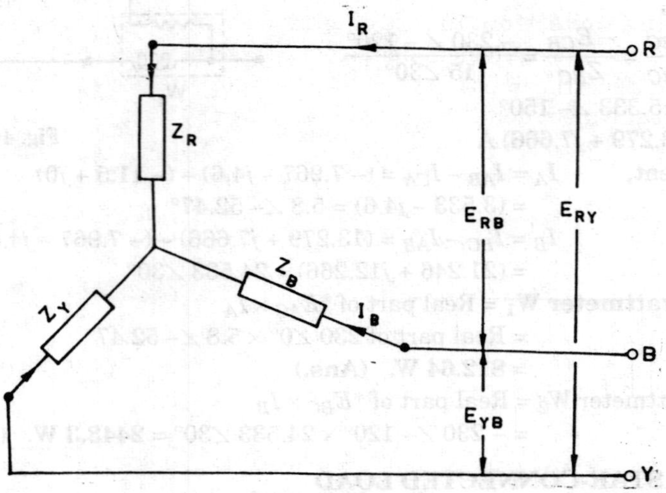

Fig. 47

$$E_{RY} = I_R Z_R - I_Y Z_Y \quad \ldots(i)$$

$$E_{YB} = I_Y Z_Y - I_B Z_B \quad \ldots(ii)$$

$$E_{BR} = I_B Z_B - I_R Z_R \quad ...(iii)$$

and $$I_R + I_Y + I_B = 0 \quad ....(iv)$$

Adding (*i*) and (*iii*), we get

$$E_{RY} + E_{BR} = I_R Z_R - I_Y Z_Y + I_B Z_B - I_R Z_R$$
$$= I_B Z_B - I_Y Z_Y \quad ...(v)$$

or $$I_B = \frac{E_{RY} + E_{BR} + I_Y Z_Y}{Z_B}$$

Substituting $I_Y + I_B = -I_R$ from (*iv*) in (*iii*), we get

$$E_{BR} = I_B Z_B + (I_Y + I_B)\, Z_R = I_B(Z_B + Z_R) + I_Y Z_R \quad ...(vi)$$

Substituting the value of I_B from (*iv*) in (*vi*), we get

$$E_{BR} = \left(\frac{E_{RY} + E_{BR} + I_Y Z_Y}{Z_B}\right)(Z_B + Z_R) + I_Y Z_R$$

or $$E_{BR}Z_B = E_{RY}Z_B + E_{BR}Z_B + I_Y Z_Y Z_B + E_{RY}Z_R + E_{BR}Z_R + I_Y Z_Y Z_R + I_Y Z_R Z_B$$

or $$I_Y(Z_R Z_Y + Z_Y Z_B + Z_B Z_R) = -Z_R(E_{RY} + E_{BR}) - E_{RY}Z_B$$

or $$I_Y(Z_R Z_Y + Z_Y Z_B + Z_B Z_R) = Z_R E_{YB} - Z_B E_{RY} \qquad (\because\ E_{RY} + E_{YB} + E_{BR} = 0)$$

or $$I_Y = \frac{E_{YB} Z_R}{Z_R Z_Y + Z_Y Z_B + Z_B Z_R} - \frac{E_{RY} Z_B}{Z_R Z_Y + Z_Y Z_B + Z_B Z_R}$$

i.e., $$I_Y = \frac{E_{YB}}{Z_Y + Z_B + \dfrac{Z_Y Z_B}{Z_R}} - \frac{E_{RY}}{Z_R + Z_Y + \dfrac{Z_R Z_Y}{Z_B}}$$

Similarly $$I_B = \frac{E_{BR}}{Z_B + Z_R + \dfrac{Z_B Z_R}{Z_Y}} - \frac{E_{YB}}{Z_Y + Z_B + \dfrac{Z_Y Z_B}{Z_R}}$$

and $$I_R = \frac{E_{RY}}{Z_R + Z_Y + \dfrac{Z_R Z_Y}{Z_B}} - \frac{E_{BR}}{Z_B + Z_R + \dfrac{Z_B Z_R}{Z_Y}}.$$

2. **Star-delta Conversion.** In case of unbalanced star-connected load it is somewhat more convenient to convert it into equivalent delta-connected load and solve it.

If Z_R, Z_Y and Z_B be the impedances of the given star-connected load and Z_{RY}, Z_{YB}, Z_{BR} be the impedances of the equivalent delta-connected load (Fig. 48), then

$$Z_{RY} = Z_R + Z_Y + \frac{Z_R Z_Y}{Z_B} = \frac{Z_R Z_B + Z_{YB} + Z_{RY}}{Z_B}$$

$$Z_{YB} = Z_Y + Z_B + \frac{Z_Y Z_B}{Z_R} = \frac{Z_Y Z_R + Z_B Z_R + Z_Y Z_B}{Z_R}$$

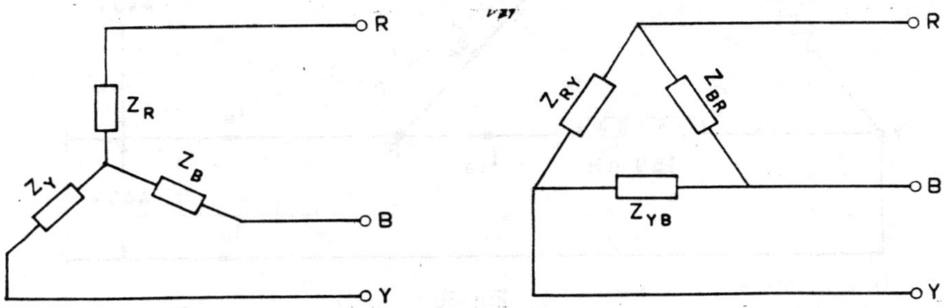

Fig. 48

$$Z_{BR} = Z_B + Z_R + \frac{Z_B Z_R}{Z_Y} = \frac{Z_B Z_Y + Z_R Z_Y + Z_B Z_R}{Z_Y}.$$

Now the phase currents and line currents can be determined. (Refer worked examples).

3. **Milliman's Theorem**. Refer Fig. 49. A number of linear bilateral admittances Y_1, Y_2, Y_3, ..., Y_n are connected to a common point (or mode) O'. The voltages of the free ends of these admittances with respect to another common point O are V_{1O}, V_{2O}, V_{3O}, ..., V_{nO}. Then according to this theorem, the voltage of O' with respect to A is given by

$$V_{O'O} = \frac{V_{1O}\,Y_1 + V_{2O}Y_2 + V_{3O}\,Y_3 + ...\,V_{nO}Y_n}{Y_1 + Y_2 + Y_2 + ... + Y_n}$$

$$V_{O'O} = \frac{\sum_{k=1}^{n} V_{kO}\,Y_k}{\sum_{k=1}^{n} Y_k}.$$

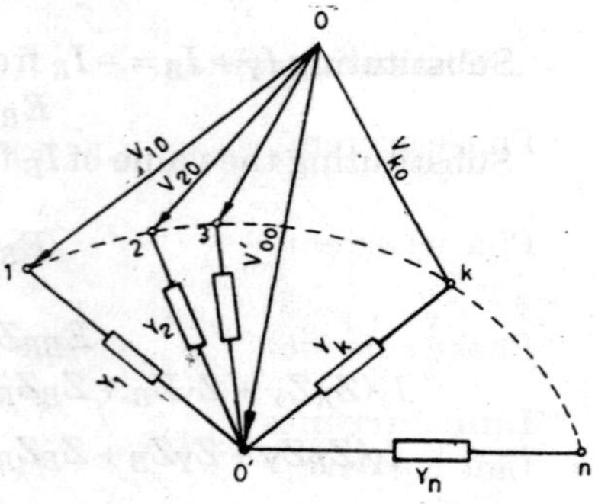

Fig. 49

12. UNBALANCED DELTA CONNECTED LOAD

When the load impedance in the three phases are not equal in magnitude or phase or both, the load is said to be *unbalanced*. If three unequal loads are connected to form a delta and connected across a 3-phase supply the currents in the three loads will not be equal in magnitude and/or phase. The three-phase currents and the line currents will also be unbalanced.

Example 40. *A 440 V, 50 Hz, 3-phase supply has delta-connected load having 50 Ω between R and Y, 159 mH between Y and B and 15.9 μF between B and R. Find* :

(*i*) *The line current for the sequence RYB.*

(*ii*) *The value of star-connected balanced resistors for the same power.*

Solution. Phase voltage (E_{ph}) = line voltage (E_L) = 440 V

The potential difference across three-phases are given by

$$E_{RY} = 440(1 + j0) = 440\angle 0° \text{ V}$$
$$E_{YB} = 440(-0.5 - j0.866) \text{ or } 440 \angle -120° \text{ V}$$
$$E_{BR} = 440(-0.5 + j0.866) \text{ or } 440 \angle 120° \text{ V}$$

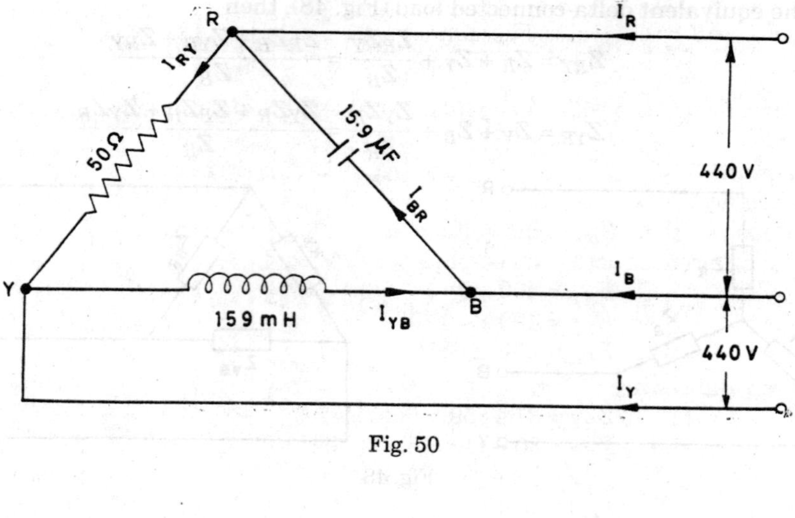

Fig. 50

Impedance, $Z_{RY} = (50 + j0) = 50 \angle 0° \ \Omega$

Impedance, $Z_{YB} = (0 + j2\pi fL) = (0 + j2\pi \times 50 \times 159 \times 10^{-3})$

$= (0 + j50)$ or $50 \angle 90° \ \Omega$

Impedance, $Z_{BR} = \left(0 - \dfrac{j}{2\pi fC}\right)$

$= \left(0 - \dfrac{j}{2\pi \times 50 \times 15.9 \times 10^{-6}}\right) = (0 - j200) = 200\angle -90°$

(*i*) **Phase currents :**

Phase current, $I_{RY} = \dfrac{E_{RY}}{Z_{RY}} = \dfrac{440\angle 0°}{50\angle 0°} = 8.8(1 + j0)$ A

Phase current, $I_{YB} = \dfrac{E_{YB}}{Z_{YB}} = \dfrac{440 \angle -120°}{50 \angle 90°} = 8.8 \angle -210°$

Phase current $I_{BR} = \dfrac{E_{BR}}{Z_{BR}} = \dfrac{440 \angle 120°}{200\angle -90°} = 2.2 \angle 210° = 2.2(-0.866 - j0.5)$ A.

Line currents :

Line current, $I_R = I_{RY} - I_{BR} = 8.8(1 + j0) - 2.2(-0.866 - j0.5)$

$= 8.8 + j0 + 1.905 + j1.1 = 10.7 + j1.1 = \mathbf{10.75 \angle 5.8°}$ **A. (Ans.)**

Line current, $I_Y = I_{YB} - I_{RY} = 8.8(-0.866 + j0.5) - 8.8(1 + j0)$

$= -7.62 + j4.4 - 8.8 - j0$

$= -16.42 + j4.4 = \mathbf{16.99 \angle 164.9°}$ **A. (Ans.)**

Line current, $I_B = I_{BR} - I_{YB} = 2.2(-0.866 - j0.5) - 8.8(-0.866 + j0.5)$

$= -1.905 - j1.1 + 7.62 - j4.4 = 5.715 - j5.5$

$= \mathbf{7.93 \angle -43.9°}$ **A. (Ans.)**

(*ii*) **Power supplied :**

$$P = I_{RY}^2 \times R = 8.8^2 \times 50 = 3872 \text{ W}$$

Let the resistance per phase be R_{ph}, which when connected in star across 440 V, 3-phase supply take power of 3872 W.

Line current, $I_L = I_{ph} = \dfrac{E_L/\sqrt{3}}{R_{ph}} = \dfrac{440/\sqrt{3}}{R_{ph}}$

Power supplied, $P = \sqrt{3}\, E_L I_L$

$\therefore \quad 3872 = \sqrt{3} \times 440 \times \dfrac{440/\sqrt{3}}{R_{ph}}$

$\therefore \quad \mathbf{R_{ph}} = \dfrac{\sqrt{3} \times 440 \times 440/\sqrt{3}}{3872} = \dfrac{440 \times 440}{3872} = \mathbf{50\ \Omega}$. **(Ans.)**

☞**Example 41.** *Three loads (31 + j59), (30 – j40) and (80 + j60) Ω are connected in delta across 400 V, 3-phase supply. Find :*

(*i*) *Phase currents,* (*ii*) *Line currents, and*

(*iii*) *Power consumed.*

Solution. Phase voltage, E_{ph} = line voltage, E_L = 400 V

The potential differences across three-phases are given by :

$$E_{RY} = 400 \angle 0°$$
$$E_{YB} = 400 \angle -120°$$
$$E_{BR} = 400 \angle 120°$$

The phase impedances are :

$$Z_{RY} = 31 + j59 = 66.6 \angle 62.3°$$
$$Z_{YB} = 30 - j40 = 50 \angle -53.1°$$
$$Z_{BR} = 80 + j60 = 100 \angle 36.9°$$

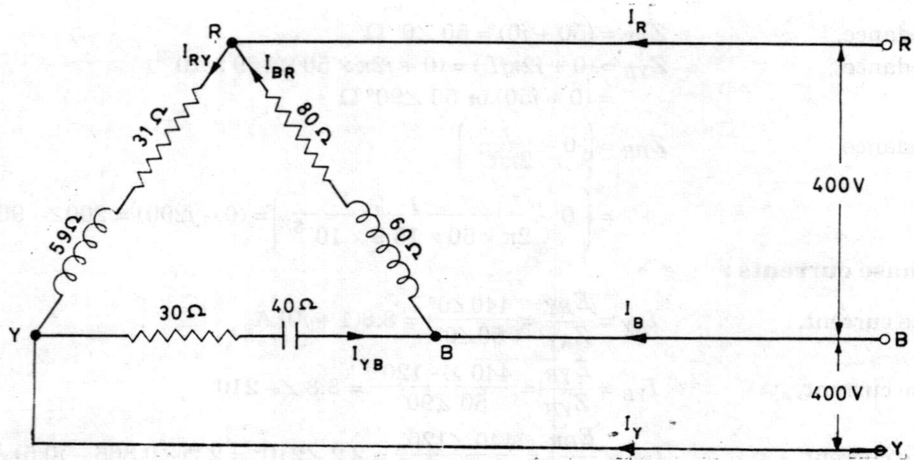

Fig. 51

Phase currents :

$$\mathbf{I_{RY}} = \frac{E_{RY}}{Z_{RY}} = \frac{400\ \angle 0^\circ}{66.6\angle 62.3^\circ} = \mathbf{6\ \angle -62.3^\circ\ A.}\ \ \textbf{(Ans.)}$$

$$\mathbf{I_{YB}} = \frac{E_{YB}}{Z_{YB}} = \frac{400\ \angle -120^\circ}{50\angle -53.1^\circ} = \mathbf{8\ \angle -66.9^\circ\ A.}\ \ \textbf{(Ans.)}$$

$$\mathbf{I_{BR}} = \frac{E_{BR}}{Z_{BR}} = \frac{400\ \angle 120^\circ}{100\angle 36.9^\circ} = \mathbf{4\ \angle 83.1\ A.}\ \ \textbf{(Ans.)}$$

Line currents :

$$\begin{aligned}\mathbf{I_R} &= I_{RY} - I_{BR} = 6\ \angle -62.3^\circ - 4\ \angle 83.1^\circ\\ &= 6(0.465 - j0.885) - 4(0.12 + j0.993)\\ &= 2.79 - j5.31 - 0.48 - j3.97 = 2.31 - j\,9.28 = \mathbf{9.56\ \angle -76^\circ\ A.}\ \ \textbf{(Ans.)}\\ \mathbf{I_Y} &= I_{YB} - I_{RY} = 8\ \angle -66.9^\circ - 6\ \angle -62.3^\circ\\ &= 8(0.392 - j0.92) - 6(0.465 - j0.885)\\ &= (3.136 - j7.36) - (2.79 - j5.31)\\ &= 0.346 - j2.05 = \mathbf{2.07\ \angle -80.4^\circ\ A.}\ \ \textbf{(Ans.)}\\ \mathbf{I_B} &= I_{BR} - I_{YB} = 4\ \angle 83.1^\circ - 8\ \angle -66.9^\circ\\ &= 4(0.12 + j0.993) - 8(0.392 - j0.92)\\ &= 0.48 + j3.97 - 3.136 + j7.36\\ &= -2.656 + j11.33 = \mathbf{11.637\ \angle 103.2^\circ\ A.}\ \ \textbf{(Ans.)}\end{aligned}$$

Total power is the sum of power in different phases

$$\begin{aligned}P &= E_{RY} \times I_{RY} + E_{YB} \times I_{YB} + E_{BR} \times I_{BR}\\ &= 400\ \angle 0^\circ \times 6\ \angle -62.3^\circ + 400\ \angle -120^\circ \times 8\ \angle -66.9^\circ\\ &\qquad + 400\ \angle 120^\circ \times 4\angle 83.1^\circ\\ &= 400 \times 6\cos(-62.3^\circ) + 400 \times 8\cos(53.1^\circ) + 400 \times 4\cos(-36.9^\circ)\\ &= 1115.6 + 1921.3 + 1279.5 = 4316.4\ \text{W}\end{aligned}$$

[OR

$$\begin{aligned}P &= I_{RY}^2 R_{RY} + I^2{}_{YB} \times R_{YB} + I^2{}_{BR} \times R_{BR}\\ &= 6^2 \times 31 + 8^2 \times 30 + 4^2 \times 80 = 1116 + 1920 + 1280 = 4316\ \text{W}\end{aligned}$$]

Hence, **total power = 4316.4 W. (Ans.)**

Example 42. *For the unbalanced delta-connected load of Fig. 52, find the phase currents, line currents and total power consumed by the load, when phase sequence is :*

(*i*) *RYB* (*ii*) *RBY.*

(AMIE Advanced Electrical M/Cs Winter, 1998)

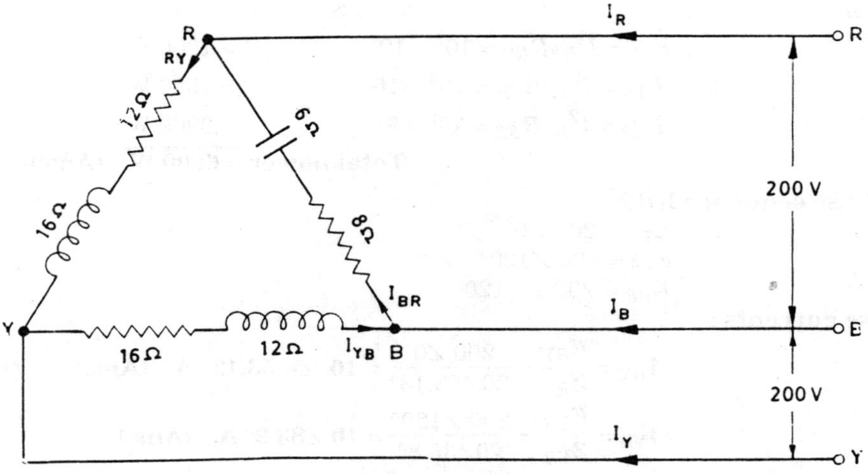

Fig. 52

Solution. (*i*) **Phase sequence RYB :**

The potential differences across three phases are :

$E_{RY} = 200 \angle 0°$

$E_{YB} = 200 \angle -120°$

$E_{BR} = 200 \angle 120°$

Impedance, $Z_{RY} = 12 + j16 = 20 \angle 53.13°$

Impedance, $Z_{YB} = 16 + j12 = 20 \angle 36.8°$

Impedance, $Z_{BR} = 8 - j6 = 10 \angle -36.8°$

Phase currents :

Phase current, $$\mathbf{I_{RY}} = \frac{E_{RY}}{Z_{RY}} = \frac{200 \angle 0°}{20 \angle 53.13°} = \mathbf{10 \angle -53.13°\ A.\ \ (Ans.)}$$

Similarly, $$\mathbf{I_{YB}} = \frac{E_{YB}}{Z_{YB}} = \frac{200 \angle -120°}{20 \angle 36.8°} = \mathbf{10 \angle -156.8°\ A.\ \ (Ans.)}$$

$$\mathbf{I_{BR}} = \frac{E_{RB}}{Z_{BR}} = \frac{200 \angle 120°}{10 \angle -36.8°} = \mathbf{20 \angle 156.8°\ A.\ \ (Ans.)}$$

Line currents :

Line current, $\mathbf{I_R} = I_{RY} - I_{BR} = 10 \angle -53.13° - 12 \angle 156.8°$

$= 10(0.60 - j0.8) - 20\,(-0.72 + j0.394) = 6.0 - j8 + 18.4 - j7.88$

$= 24.4 - j15.88 = \mathbf{29.11 \angle -33°\ A.\ \ (Ans.)}$

Similarly, $\mathbf{I_Y} = I_{YB} - I_{RY} = 10 \angle -156.8° - 10 \angle -53.13°$

$= 10(-0.919 - j0.394) - 10(0.6 - j0.8) = -9.19 - j3.94 - 6 + j8$

$= -15.19 + j4.06 = \mathbf{15.72 \angle 165°\ A.\ \ (Ans.)}$

$\mathbf{I_B} = I_{BR} - I_{YB} = 20 \angle 156.8° - 10 \angle -156.8°$

$= 20\,(0.919 + j0.394) - 10(-0.919 - j0.394)$

$= -18.38 + j7.88 + 9.19 + j3.94 = -9.19 + j11.82 = 14.97 \angle 127.9°$

[Check $\Sigma I = (24.4 - j15.88) + (-15.19 + j4.06) + (-9.19 + j11.82) = 0$]

Power :

$$P_{RY} = I^2_{RY}R_{RY} = 10^2 \times 12 = 1200 \text{ W}$$
$$P_{YB} = I^2_{YB}R_{YB} = 10^2 \times 16 = 1600 \text{ W}$$
$$P_{BR} = I^2_{BR}R_{BR} = 20^2 \times 8 = 3200 \text{ W}$$

Total power = 6000 W. (Ans.)

(*ii*) **Phase sequence RBY :**

Here,
$$E_{RY} = 200 \angle 0°$$
$$E_{YB} = 200 \angle 120°$$
$$E_{BR} = 200 \angle -120°$$

Phase currents :

$$\mathbf{I_{RY}} = \frac{E_{RY}}{Z_{RY}} = \frac{200 \angle 0°}{20 \angle 53.13°} = \mathbf{10 \angle -53.13° \text{ A. (Ans.)}}$$

$$\mathbf{I_{YB}} = \frac{E_{YB}}{Z_{YB}} = \frac{200 \angle 120°}{20 \angle 36.8°} = \mathbf{10 \angle 83.2° \text{ A. (Ans.)}}$$

$$\mathbf{I_{BR}} = \frac{E_{BR}}{Z_{BR}} = \frac{200 \angle -120°}{10 \angle -36.8°} = \mathbf{20 \angle -83.2° \text{ A. (Ans.)}}$$

Line currents :

$$\mathbf{I_R} = I_{RY} - I_{BR} = 10 \angle -53.13° - 20 \angle -83.2°$$
$$= 10(0.6 - j0.8) - 20(0.118 - j0.993) = 6 - j8 - 2.36 + j19.86$$
$$= 3.64 + j11.86 = \mathbf{12.4 \angle 72.9° \text{ A. (Ans.)}}$$

$$\mathbf{I_Y} = I_{YB} - I_{RY} = 10 \angle 83.2° - 10 \angle -53.13°$$
$$= 10(0.118 + j\,0.993) - 10(0.6 - j0.8)$$
$$= 1.18 + j9.93 - 6 + j8 = -4.82 + j17.93 = \mathbf{18.56 \angle 105° \text{ A. (Ans.)}}$$

$$\mathbf{I_B} = I_{BR} - I_{YB} = 20 \angle -83.2° - 10 \angle 83.2°$$
$$= 20(0.118 - j0.993) - 10(0.118 + j0.993)$$
$$= 2.36 - j19.86 - 1.18 - j9.93 = 1.18 - j29.79 = \mathbf{29.81 \angle -87.7°. \text{ (Ans.)}}$$

Power :

$$P_{RY} = 10^2 \times 12 = 1200 \text{ W}$$
$$P_{YB} = 10^2 \times 16 = 1600 \text{ W}$$
$$P_{BR} = 20^2 \times 8 = 3200 \text{ W}$$

Total power **= 6000 W. (Ans.)**

Example 43. *A 3-phase, 400 V, 50 Hz system has the following load connected in delta :*

Between R and Y lines : a non-reactive resistor of 50 Ω.

Between Y and B lines : a coil having a resistance of 10 Ω and reactance of 30 Ω.

Between B and R lines : a loss free capacitor of 15 μ F.

Calculate : (i) The phase currents (ii) The line currents.

Assume phase sequence to be RYB.

Solution. $E_{ph} = E_L = 400$ V

The potential difference across the three phases are :

$$E_{RY} = 400 \angle 0° \text{ V}$$
$$E_{YB} = 400 \angle -120° \text{ V}$$
$$E_{BR} = 400 \angle 120° \text{ V}$$

Impedances

$$Z_{RY} = 50 + j0 = 50 \angle 0° \ \Omega$$
$$Z_{YB} = 10 + j30 = 31.62 \angle 71.56° \ \Omega$$

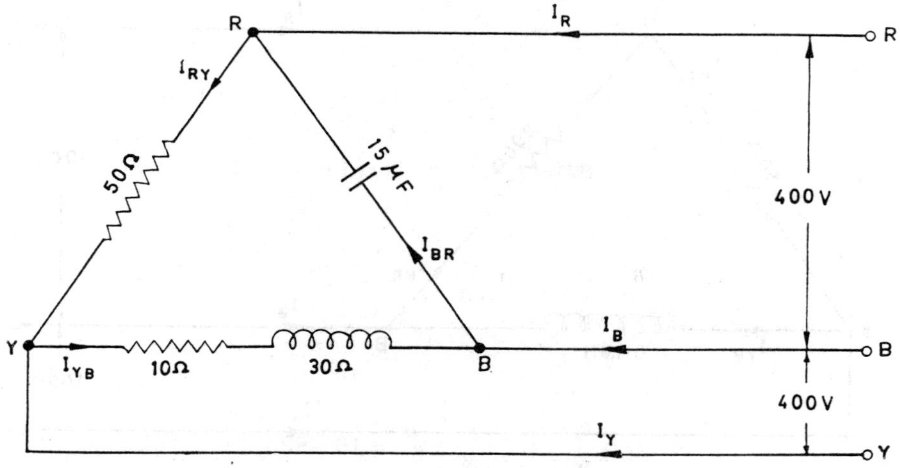

Fig. 53

$$Z_{BR} = \left(0 - \frac{j}{2\pi fC}\right) = \left(0 - \frac{j}{2\pi \times 50 \times 15 \times 10^{-6}}\right)$$
$$= (0 - j212) = 212 \angle -90° \ \Omega.$$

Phase currents :

$$\mathbf{I_{RY}} = \frac{E_{RY}}{Z_{RY}} = \frac{400 \angle 0°}{50\angle 0°} = \mathbf{8 \angle 0° \ A.} \quad \textbf{(Ans.)}$$

$$\mathbf{I_{YB}} = \frac{E_{YB}}{Z_{YB}} = \frac{400 \angle -120°}{31.62\angle 71.56°} = \mathbf{12.65 \angle -191.56° \ A.} \quad \textbf{(Ans.)}$$

$$\mathbf{I_{BR}} = \frac{E_{BR}}{Z_{BR}} = \frac{400 \angle 120°}{212\angle -90°} = \mathbf{1.886 \angle 210° \ A.} \quad \textbf{(Ans.)}$$

Line currents :

$$\mathbf{I_R} = I_{RY} - I_{BR} = 8 \angle 0° - 1.886 \angle 210°$$
$$= 8 - 1.886\,(-0.866 - j0.5) = 8 + 1.633 + j0.943$$
$$= 9.633 + j0.943 = \mathbf{9.68 \angle 5.6° \ A.} \quad \textbf{(Ans.)}$$
$$\mathbf{I_Y} = I_{YB} - I_{RY} = 12.65 \angle -191.56° - 8 \angle 0°$$
$$= 12.65(-0.979 + j0.2) - 8 = -12.38 + j2.53 - 8$$
$$= -20.38 + j2.53 = \mathbf{20.53 \angle 7.07° \ A.} \quad \textbf{(Ans.)}$$
$$\mathbf{I_B} = I_{BR} - I_{YB} = 1.886 \angle 210° - 12.65 \angle -191.56°$$
$$= 1.886\,(-0.866 - j0.5) - 12.65(-0.979 + j0.2)$$
$$= -1.633 - j\,0.943 + 12.38 - j2.53$$
$$= 10.747 - j3.473 = \mathbf{11.29 \angle 17.4°}. \quad \textbf{(Ans.)}$$

☞**Example 44.** *A balanced 3-phase system supplies an unbalanced 3-phase delta-connected load made up of resistors 50 Ω and 100 Ω and a reactor having an inductance of 0.15 H with negligible resistance. Line voltage = 100 V at 50 Hz. Calculate :*

(i) The total power in the system.

(ii) The total volt amperes reactive.

Solution. The delta-connected load and its phasor diagram are shown in Figs. 54 and 55 respectively.

Reactance, $X_{YB} = 2\pi fL = 2\pi \times 50 \times 0.15 = 47.12 \ \Omega$

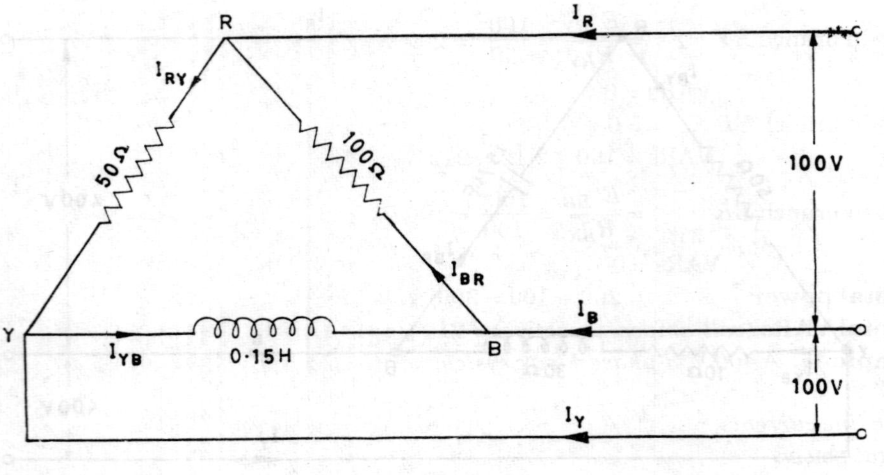

Fig. 54

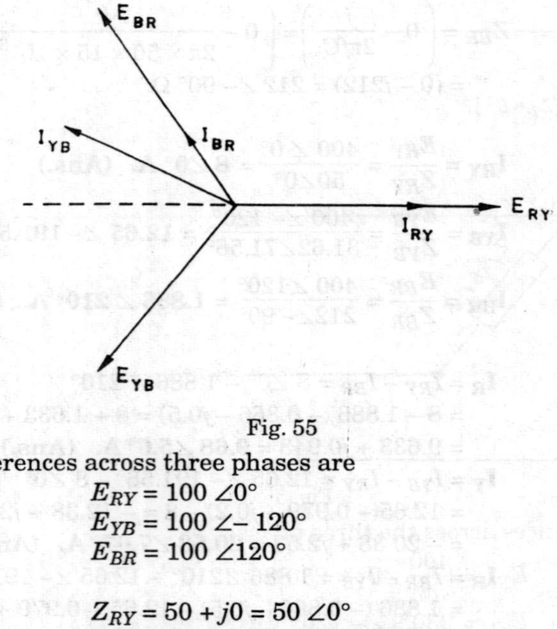

Fig. 55

The potential differences across three phases are

$$E_{RY} = 100 \angle 0°$$
$$E_{YB} = 100 \angle -120°$$
$$E_{BR} = 100 \angle 120°$$

Impedances :

$$Z_{RY} = 50 + j0 = 50 \angle 0°$$
$$Z_{YB} = 0 + j47.12 = 47.12 \angle 90°$$
$$Z_{BR} = 100 + j0 = 100 \angle 0°.$$

Phase currents :

$$I_{RY} = \frac{E_{RY}}{Z_{RY}} = \frac{100 \angle 0°}{50 \angle 0°} = 2 \angle 0° = 2 + j0$$

$$I_{YB} = \frac{E_{YB}}{Z_{YB}} = \frac{100 \angle -120°}{47.12 \angle 90°} = 2.12 \angle -210°$$

$$= 2.12\,(-0.866 + j0.5) = -1.836 + j1.06$$

$$I_{BR} = \frac{E_{BR}}{Z_{BR}} = \frac{100 \angle 120°}{100 \angle 0°} = 1 \angle 120° = -0.5 + j0.866$$

Watts in branch RY $= \dfrac{E^2_{RY}}{R_{RY}} = \dfrac{100^2}{50} = 200$ W

VARs = 0

Watts in branch YB = 0

VARs = 100 × 2.12 = 212 (lag)

Watts in branch BR $= \dfrac{E^2_{BR}}{R_{BR}} = \dfrac{100^2}{100} = 100$ W

VARs = 0

(*i*) **Total power** = 200 + 100 = **300 W. (Ans.)**

(*ii*) **Total VARs** (*volt-amperes reactive*) = **212 (lag). (Ans.)**

Example 45. *A 400 V, 3-phase 4 wire system has the loads 6 – j8, 6 + j0 and 4 + j3 Ω connected in star. Find :*

(*i*) *The line currents,* (*ii*) *Neutral current, and*

(*iii*) *Total power.*

Solution. The connections are shown in Fig. 56.

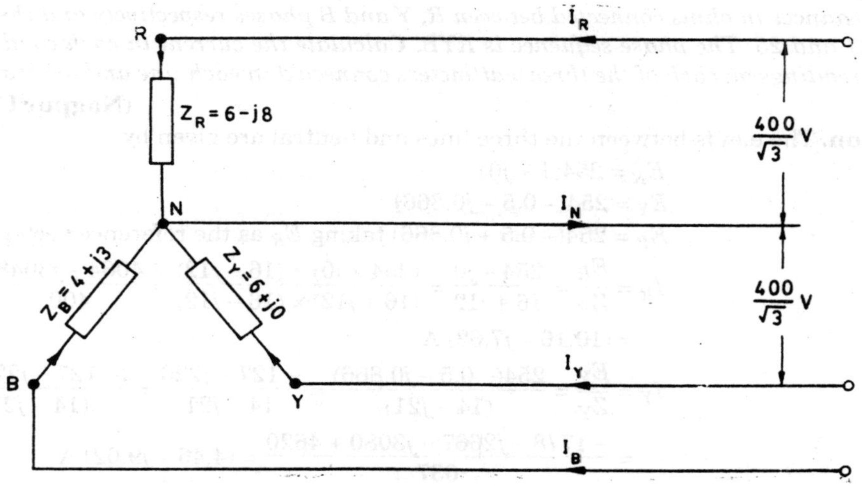

Fig. 56

The potential differences across the three phases are :

$$E_{RN} = \frac{400}{\sqrt{3}} \angle 0° = 230.9 \angle 0° \text{ V}$$

$$E_{YN} = 230.9 \angle -120° \text{ V}$$

$$E_{BN} = 230.9 \angle 120° \text{ V}$$

Impedances :

$$Z_R = 6 - j8 = 10 \angle -53.13°$$

$$Z_Y = 6 + j0 = 6 \angle 0°$$

$$Z_B = 4 + j3 = 5 \angle 36.9°$$

(*i*) **Line currents** :

$$\mathbf{I_R} = I_{RN} = \frac{E_R}{Z_R} = \frac{230.9 \angle 0°}{10 \angle -53.13°} = \mathbf{23.09 \angle 53.13° \text{ A.}} \quad \textbf{(Ans).}$$

$$\mathbf{I_Y} = I_{YN} = \frac{E_Y}{Z_Y} = \frac{230.9 \angle -120°}{6 \angle 0°} = \mathbf{38.48 \angle -120° \text{ A.}} \quad \textbf{(Ans.)}$$

$$\mathbf{I_B} = I_{BN} = \frac{E_B}{Z_B} = \frac{230.9 \angle 120°}{5 \angle 36.9°} = \mathbf{46.18 \angle 83.1° \ A.} \ \textbf{(Ans.)}$$

(*ii*) **The current in the neutral wire :**

$$\mathbf{I_N} = -(I_R + I_Y + I_B)$$
$$= -[23.09 \angle 53.13° + 38.48 \angle -120° + 46.18 \angle 83.1°]$$
$$= -[23.09(0.6 + j0.8) + 36.48(-0.5 - j0.866) + 46.18(0.12 + j0.99)]$$
$$= -[13.85 + j18.47 - 18.24 - j31.59 + 5.54 + j45.72]$$
$$= -(1.15 + j32.6) = -1.15 - j32.6 = \mathbf{32.62 \angle -92° \ A.} \ \textbf{(Ans.)}$$

Power :

Power taken by phase R,

$$P_B = I_R^2 \times R_R = 23.09^2 \times 6 = 3199 \text{ W (app.)}.$$

Similarly, $$P_Y = I_Y^2 \times R_Y = 38.48^2 \times 6 = 8884 \text{ W (app.)}.$$

and $$P_B = I_B^2 \times R_B = 46.18^2 \times 4 = 8530 \text{ W (app.)}.$$

Total power taken $= P_R + P_Y + P_B = 3199 + 8884 + 8530 = \mathbf{20613 \ W.}$ **(Ans.)**

Example 46. *A 440/254 volt, 3 phase, 4-core supplies an unbalanced load represented by the following impedances in ohms connected between R, Y and B phases respectively and the neutral 16 + j12 ; 14 – j21 and 25. The phase sequence is RYB. Calculate the current in each conductor of the cable and the readings on each of the three wattmeters connected in each line and neutral.*

(Nagpur University)

Solution. The e.m.fs between the three lines and neutral are given by :

$$E_R = 254(1 + j0)$$
$$E_Y = 254(-0.5 - j0.866)$$
$$E_B = 254(-0.5 + j0.866) \text{ taking } E_R \text{ as the reference vector}$$

$$I_R = \frac{E_R}{Z_R} = \frac{254 + j0}{16 + j12} = \frac{(254 + j0) \times (16 - j12)}{(16 + j12) \times (16 - j12)} = \frac{4064 - j3048}{400}$$
$$= (10.16 - j7.62) \text{ A}$$

$$I_Y = \frac{E_Y}{Z_Y} = \frac{254(-0.5 - j0.866)}{(14 - j21)} = \frac{-127 - j220}{14 - j21} = \frac{(-127 - j220) \times (14 + j21)}{(14 - j21)(14 + j21)}$$
$$= \frac{-1778 - j2667 - j3080 + 4620}{637} = (4.46 - j9.02) \text{ A}$$

$$I_B = \frac{E_B}{Z_B} = \frac{254(-0.5 + j0.866)}{25} = \frac{-127 + j220}{25} = (-5.08 + j8.8) \text{ A}$$

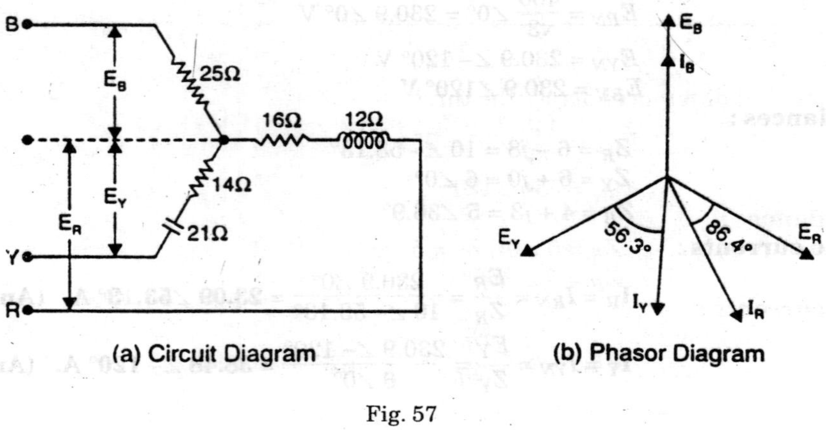

(a) Circuit Diagram (b) Phasor Diagram

Fig. 57

Now current in neutral wire,

$$\mathbf{I_N} = -(I_R + I_Y + I_B)$$
$$= -[(10.16 - j7.62) + (4.46 - j9.02) + (-5.08 + j8.8)]$$
$$= -9.54 + j7.84$$

Evaluating these currents, we get

$$\mathbf{I_R} = \sqrt{10.16^2 + (-7.62)^2} = \mathbf{12.7\ A.}\ \textbf{(Ans.)}$$
$$\mathbf{I_Y} = \sqrt{4.46^2 + (-9.02)^2} = \mathbf{10.06\ A.}\ \textbf{(Ans.)}$$
$$\mathbf{I_B} = \sqrt{(-5.08)^2 + (8.8)^2} = \mathbf{10.16\ A.}\ \textbf{(Ans.)}$$
$$\mathbf{I_N} = \sqrt{(-9.54)^2 + (7.84)^2} = \mathbf{12.35\ A.}\ \textbf{(Ans.)}$$

Power in R-phase $= I_R^2 \times R_R = (12.7)^2 \times 16 = \mathbf{2580.6\ W.}$ **(Ans.)**

Power in Y-phase $= I_Y^2 \times R_Y = (10.06)^2 \times 14 = \mathbf{1416.8\ W.}$ **(Ans.)**

Power in B-phase $= I_B^2 \times R_B = (10.16)^2 \times 25 = \mathbf{2580.6\ W.}$ **(Ans.)**

☞**Example 47.** *A 3-phase, 4-wire system having a 440 V line to line has the following loads connected between the respective lines and neutral :*

$$Z_R = 5\ \angle 0°\ \Omega,\ Z_Y = 5\ \angle 37°\ \Omega,\ Z_B = 5\ \angle -53°\ \Omega.$$

Calculate the current in neutral wire and power taken by each load when phase sequence is :
(i) RYB *(ii) RBY.*

Solution. (*i*) **Phase sequence RYB** (Figs. 58 and 59).

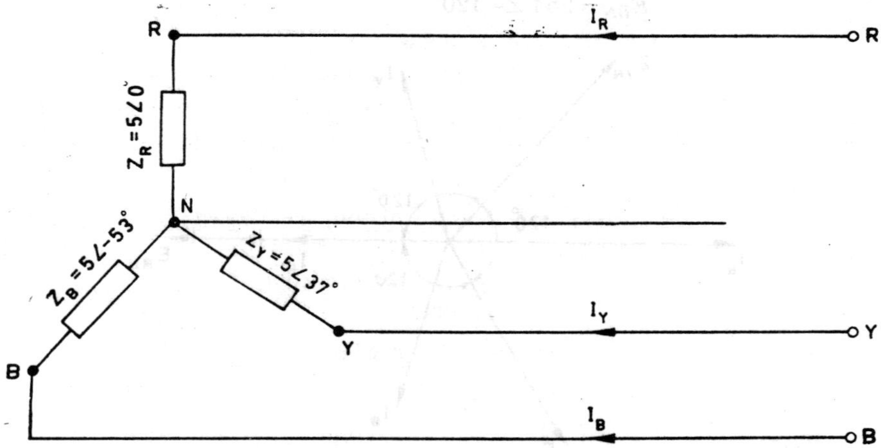

Fig. 58

The potential differences across the three phases are :

$$E_{RN} = \frac{440}{\sqrt{3}}\angle 0° = 254\ \angle 0°,\quad E_{YN} = 254\ \angle -120°$$
$$E_{BN} = 254\ \angle 120°$$

Impedances :

$$Z_R = 5\ \angle 0°,\ Z_Y = 5\ \angle 37°$$
$$Z_B = 5\ \angle -53°$$

Line currents :

$$I_R = I_{RN} = \frac{E_{RN}}{Z_R} = \frac{254\ \angle 0°}{5\ \angle 0°} = 50.8\ \angle 0°$$

$$I_Y = I_{YN} = \frac{E_{YN}}{Z_Y} = \frac{254 \angle -120°}{5 \angle 37°} = 50.8 \angle -157°$$

$$I_B = I_{BN} = \frac{E_{BN}}{Z_B} = \frac{254 \angle 120°}{5 \angle -53°} = 50.8 \angle 173°$$

Current in the neutral :

$$\mathbf{I_N} = -(I_R + I_Y + I_B)$$
$$= -(50.8 \angle 0° + 50.8 \angle -157° + 50.8 \angle 173°)$$
$$= -[50.8 + 50.8(-0.92 - j0.39) + 50.8(-0.99 + j0.122)]$$
$$= -[50.8 - 46.74 - j19.81 - 50.3 + j6.2]]$$
$$= -(-46.24 - j13.61) = 46.24 + j13.61$$
$$= \mathbf{48.2 \angle 16.4°}. \text{ (Ans.)}$$

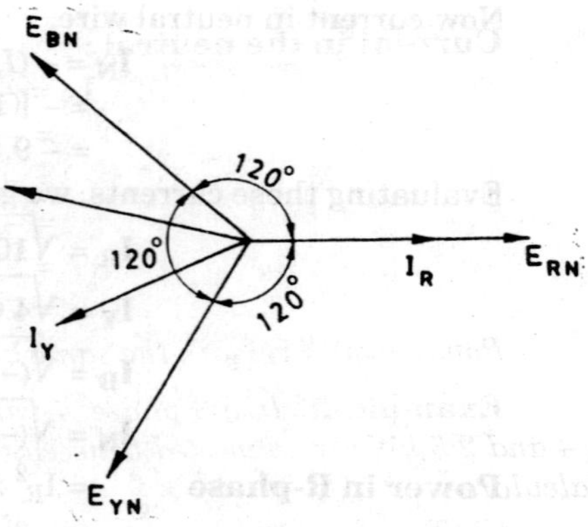

Fig. 59

Now $R_R = 5\ \Omega$, $R_Y = 5 \cos 37° = 4\ \Omega$,
$R_B = 5 \cos(-53°) = 3\ \Omega$

Power taken by phase R,

$$P_R = 50.8^2 \times 5 = \mathbf{12903\ W.}\ \textbf{(Ans.)}$$

Similarly, $$P_Y = 50.8^2 \times 4 = \mathbf{10322\ W.}\ \textbf{(Ans.)}$$

and $$P_B = 50.8^2 \times 3 = \mathbf{7742\ W.}\ \textbf{(Ans.)}$$

Phase Sequence RBY (Fig. 60) :

$$E_{RN} = 254 \angle 0°, E_{YN} = 254 \angle 120°$$
$$E_{BN} = 254 \angle -120°$$

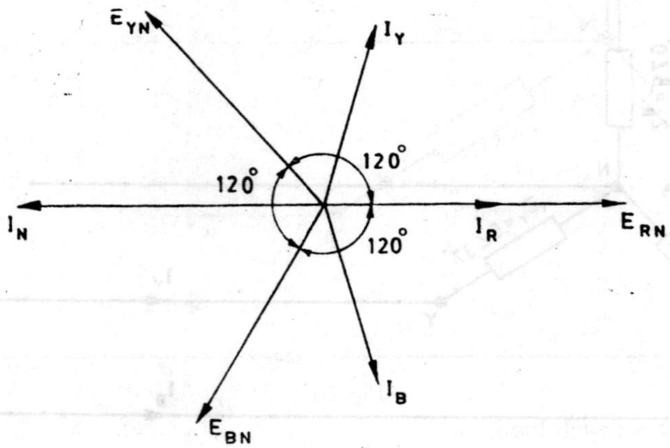

Fig. 60

Line currents :

$$I_R = I_{RN} = \frac{E_{RN}}{Z_R} = \frac{254 \angle 0°}{5 \angle 0°} = 50.8 \angle 0°$$

$$I_Y = I_{YN} = \frac{E_{YN}}{Z_Y} = \frac{254 \angle 120°}{5 \angle 37°} = 50.8 \angle 83°$$

$$I_B = I_{BN} = \frac{E_{BN}}{Z_B} = \frac{254 \angle -120°}{5 \angle -53°} = 50.8 \angle -67°$$

Current in the neutral :

$$\mathbf{I_N} = -(I_R + I_Y + I_B)$$
$$= -(50.8 \angle 0° + 50.8 \angle 83° + 50.8 \angle -67°)$$
$$= -[50.8 + 50.8(0.122 + j0.992) + 50.8(0.39 - j0.92)]$$
$$= -(50.8 + 6.197 + j50.4 + 19.81 - j46.74)$$
$$= -(76.81 + j3.66) = -76.81 - j3.66$$
$$= \mathbf{76.89 \angle -177.3° \ A.} \quad \textbf{(Ans.)}$$

Power would remain the same because magnitude of branch currents is unaltered.

Example 48. *In a 3-phase, 4-wire system, the line voltage is 400 V and non-inductive loads of 5, 4 and 2.5 kW are connected between the three lines conductors and the neutral, as shown in Fig. 61. Calculate :*

(*i*) *The current in lines ; and*

(*ii*) *The current in the neutral conductor.*

Solution. Refer Fig. 61.

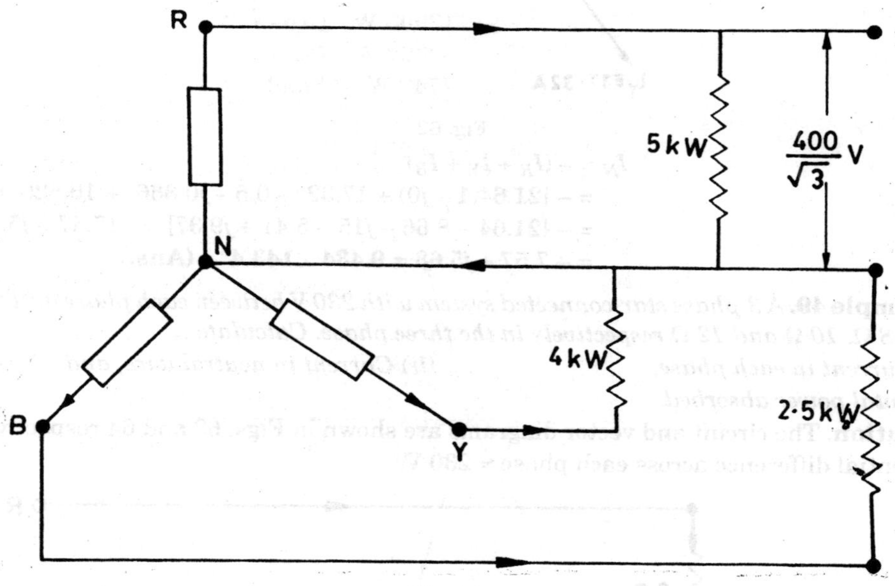

Fig. 61

Phase voltage across each load,

$$E_{ph} = \frac{E_L}{\sqrt{3}} = \frac{400}{\sqrt{3}} = 231 \text{ V}$$

(*i*) Current in phase *R*,

$$I_R = \text{(line current)} = \frac{\text{kW} \times 1000}{E_{ph} \cos \phi} = \frac{5 \times 1000}{231 \times 1} = \mathbf{21.64 \ A.} \quad \textbf{(Ans.)}$$

Current in phase *Y*, $I_Y = \dfrac{4 \times 1000}{231 \times 1} = \mathbf{17.32 \ A.}$ **(Ans.)**

Current in phase *B*, $I_B = \dfrac{2.5 \times 1000}{231 \times 1} = \mathbf{10.82 \ A.}$ **(Ans.)**

These currents are mutually 120° out of phase as the circuit is purely resistive. Taking I_R as the reference vector, line currents may be given as

$$I_R = 21.64 \angle 0° = 21.64\,(1 + j0)$$
$$I_Y = 17.32 \angle -120° = 17.32\,(-0.5 - j0.866)$$
$$I_B = 10.82 \angle 120° = 10.82\,(-0.5 + j0.866)$$

(*ii*) **Current in the neutral wire** (Refer Fig. 62) :

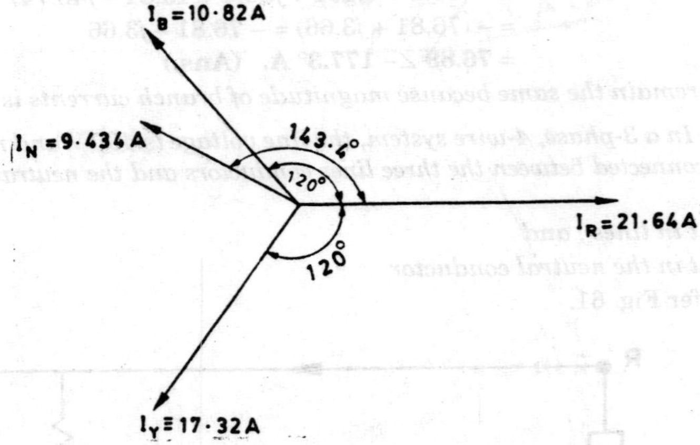

Fig. 62

$$I_N = -(I_R + I_Y + I_B)$$
$$= -[21.64(1 - j0) + 17.32\,(-0.5 - j0.866) + 10.82(-0.5 + j0.866)]$$
$$= -[21.64 - 8.66 - j15 - 5.41 + j9.37] = -(7.57 - j5.63)$$
$$= -7.57 + j5.63 = \mathbf{9.434 \angle 143.4°}.\ \textbf{(Ans.)}$$

Example 49. *A 3-phase star-connected system with 230 V between each phase and neutral has resistances 8 Ω, 10 Ω and 12 Ω respectively in the three phase. Calculate :*

(*i*) *Current in each phase,* (*ii*) *Current in neutral wire, and*

(*iii*) *Total power absorbed.*

Solution. The circuit and vector diagrams are shown in Figs. 63 and 64 respectively.

Potential difference across each phase = 230 V.

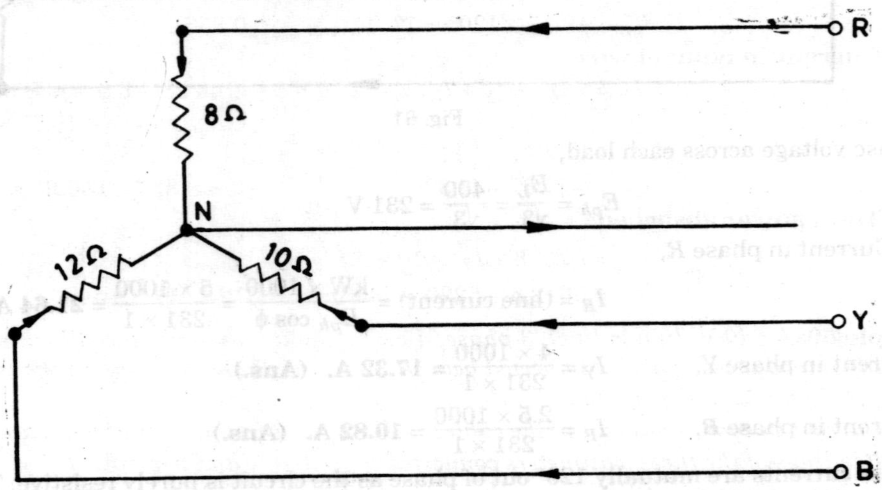

Fig. 63

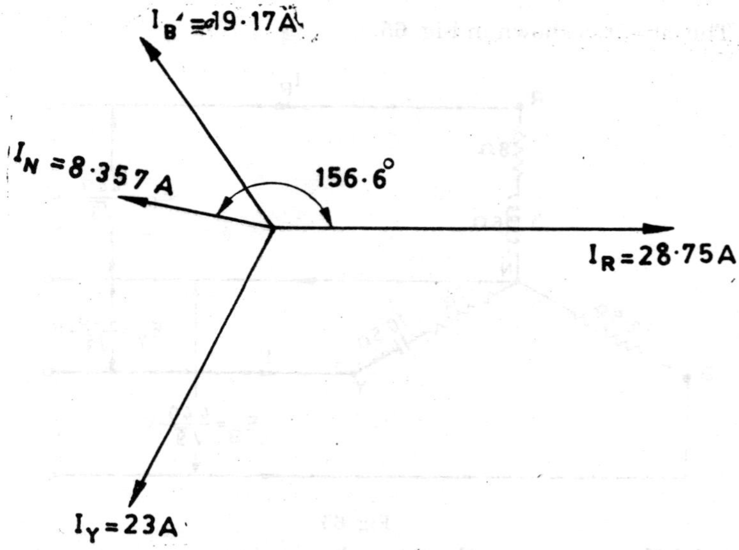

Fig. 64

(*i*) **Phase currents :**

Current in 8 Ω resistance $= \frac{230}{8} = $ **28.75 A. (Ans.)**

Current in 10 Ω resistance $= \frac{230}{10} = $ **23 A. (Ans.)**

Current in 12 Ω resistance $= \frac{230}{12} = $ **19.17 A. (Ans.)**

Since the circuit is purely resistive, so currents will be in phase with their respective voltages and will be mutually 120° apart. Taking I_R as the reference vector (Fig. 64) the line currents are given as :

$$I_R = 28.75 \angle 0° = 28.75(1 + j0)$$
$$I_Y = 23 \angle -120° = 23(-0.5 - j0.866)$$
$$I_B = 19.17 \angle 120° = 19.17(-0.5 + j0.866).$$

(*ii*) **Current in neutral wire,**

$$I_N = -(I_R + I_Y + I_B) = -[28.75(1 + j0) + 23(-0.5 - j0.866) + 19.17(-0.5 + j0.866)$$
$$= -[28.75 - 11.5 - j19.92 - 9.58 + j16.6]$$
$$= -[7.67 - j3.32] = -7.67 + j3.32 = \mathbf{8.357 \angle 156.6°\ A.}\ \textbf{(Ans.)}$$

(*iii*) **Total power absorbed**

$$= I_R^2 \times 8 + I_Y^2 \times 10 + I_B^2 \times 12 = 28.75^2 \times 8 + 23^2 \times 10 + 19.17^2 \times 12$$
$$= 6612.5 + 5290 + 4409.8 = \mathbf{16312.3\ W.}\ \textbf{(Ans.)}$$

☞**Example 50.** *A 440 V (line-to-line), 3-phase, 4-core supplies an unbalanced load represented by the following impedances in ohms connected between R, Y and B phases respectively and the neutral 8 + j6, 7 – j10.5 and 12.5. Calculate :*

(i) Current in each conductor of the cable, (ii) Current in the neutral wire, and

(iii) Readings of the three wattmeters connected in each line and neutral.

The phase sequence is RYB.

Solution. The circuit is shown in Fig. 65.

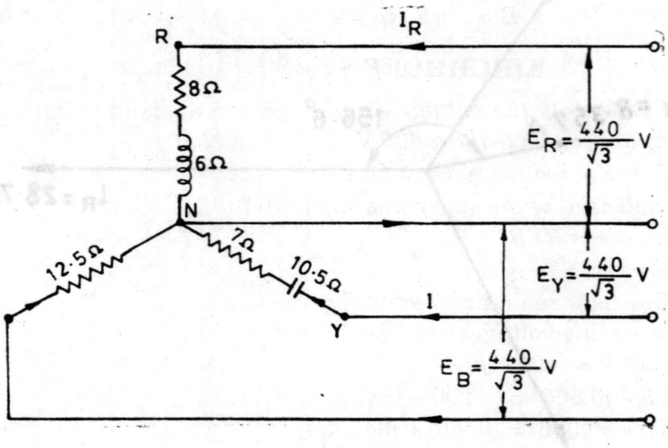

Fig. 65

The potential differences across the three phases are :

$$E_{RN} = \frac{440}{\sqrt{3}} = 254 \angle 0°$$

$$E_{YN} = 254 \angle -120°$$

$$E_{BN} = 254 \angle 120°.$$

Impedances :

$$Z_R = 8 + j6 = 10 \angle 36.9°$$

$$Z_Y = 7 - j10.5 = 12.62 \angle -56.3°$$

$$Z_B = 12.5 = 12.5 \angle 0°.$$

(*i*) **Phase currents :**

$$\mathbf{I_R} = I_{RN} = \frac{E_{RN}}{E_R} = \frac{254 \angle 0°}{10 \angle 36.9°}$$

$$= \mathbf{25.4 \angle -36.9° \text{ A. (Ans.)}}$$

$$\mathbf{I_Y} = I_{YN} = \frac{E_{YN}}{Z_Y} = \frac{254 \angle -120°}{12.62 \angle -56.3°}$$

$$= \mathbf{20.12 \angle -63.7° \text{ A. (Ans.)}}$$

$$\mathbf{I_B} = I_{BN} = \frac{E_{BN}}{Z_B} = \frac{254 \angle 120°}{12.5 \angle 0°}$$

$$= \mathbf{20.32 \angle 120° \text{ A. (Ans.)}}$$

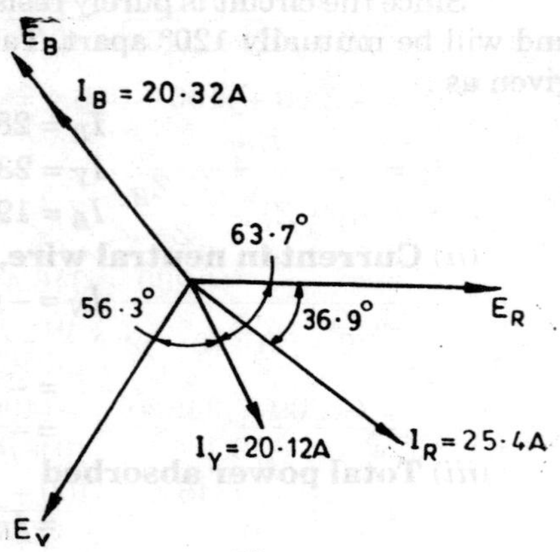

Fig. 66

(*ii*) **Current in the neutral wire :**

$$I_N = -(I_R + I_Y + I_B)$$

$$= -[25.4 \angle -36.9° + 20.12 \angle -63.7° + 20.32 \angle 120°]$$

$$= -[25.4(+0.8 - j0.6) + 20.12(0.44 - j0.896) + 20.32(-0.5 + j0.866)]$$

$$= -(20.32 - j15.24 + 8.85 - j18.03 - 10.16 + j17.6)$$

$$= -(19.01 - j15.67) = -19.01 + j15.67 = \mathbf{24.64 \angle 39.5° \text{ A. (Ans.)}}$$

(*iii*) **Readings of wattmeters :**

Power in phase, $R = I_R^2 \times R_R = 25.4^2 \times 8 =$ **5161.3 W. (Ans.)**

Power in phase, $Y = I^2_Y \times R_Y = 20.12^2 \times 7 =$ **2833.7 W. (Ans.)**

Power in phase, $B = I^2_B \times R_B = 20.32^2 \times 12.5 =$ **5161.3 W. (Ans.)**

KIRCHHOFF'S LAWS (Article 11)

Example 51. *Calculate the voltage drop across each of the following three star-connected impedances when they are connected to a 400 V, 3-phase system :*

$$Z_R = j40\ \Omega,\ Z_Y = j2\ \Omega,\ Z_B = -j2\ \Omega.$$

Also find the potential of the star-point above ground.

The phase sequence is RYB.

Solution. Refer Fig. 67.

Considering E_{RY} as a reference vector. The vector expressions for the three line voltages are :

$E_{RY} = 400(1 + j0)$ V $= (400 + j0)$ V

$E_{YB} = 400(-0.5 - j0.866) = (-200 - j346.4)$ V

$E_{BR} = 400(-0.5 + j0.866) = (-200 + j346.4)$ V

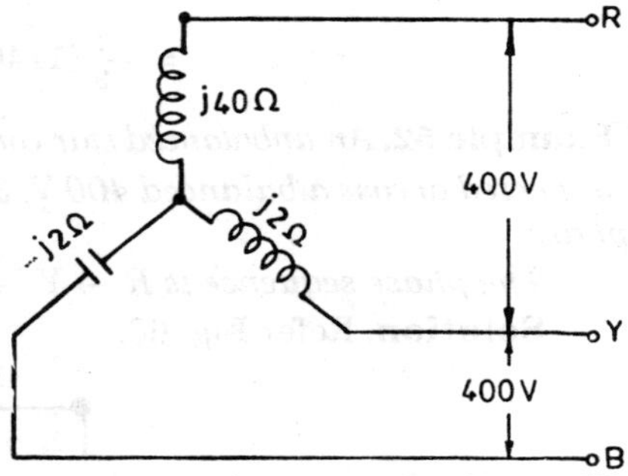

Fig. 67

Impedances :

$Z_R = (0 + j40)$, $Z_Y = (0 + j2)$

$Z_B = (0 - j2)$

Now for finding I_R, I_Y and I_B, using the relations :

$$I_R = \frac{E_{RY}}{Z_R + Z_Y + \dfrac{Z_R Z_Y}{Z_B}} - \frac{E_{BR}}{Z_B + Z_R + \dfrac{Z_B Z_R}{Z_Y}}$$

$$= \frac{400 + j0}{(0 + j40) + (0 + j2) + \dfrac{(0 + j40)(0 + j2)}{(0 - j2)}} - \frac{(-200 + j346.4)}{(0 - j2) + (0 + j40) + \dfrac{(0 - j2)(0 + j40)}{(0 + j2)}}$$

$$= \frac{(400 + j0)}{(0 + j42) - (j40)} - \frac{(-200 + j346.4)}{(0 + j38) - j40} = -\frac{400}{j2} - \frac{(-200 + j346.4)}{-j2}$$

$$= -j200 + j100 + 346.4 = (173.2 - j100) \text{ A or } 200 \angle -30° \text{ A}$$

$$I_Y = \frac{E_{YB}}{Z_Y + Z_B + \dfrac{Z_Y Z_B}{Z_R}} - \frac{E_{RY}}{Z_R + Z_Y + \dfrac{Z_R Z_Y}{Z_B}}$$

$$= \frac{(-200 - j346.4)}{(0 + j2) + (0 - j2) + \dfrac{(0 + j2)(0 - j2)}{(0 + j40)}} - \frac{(400 + j0)}{(0 + j40) + (0 + j2) + \dfrac{(0 + j40)(0 + j2)}{(0 - j2)}}$$

$$= \frac{(-200 - j346.4)}{-j0.1} - \frac{(400 + j0)}{(0 + j42) - j40}$$

$$= \frac{-200 - j346.4}{-j0.1} - \frac{(400 + j0)}{j2} = -j2000 + 3464 + j200 = 3464 - j1800$$

$$= 3904 \angle -27.45° \text{ A}$$

$$I_B = \frac{E_{BR}}{Z_B + Z_R + \dfrac{Z_B Z_R}{Z_Y}} - \frac{E_{YB}}{Z_Y + Z_B + \dfrac{Z_Y Z_B}{Z_R}}$$

$$= \frac{(-200 + j346.4)}{(0 - j2) + (0 + j40) + \dfrac{(0 - j2)(0 + j40)}{(0 + j2)}} - \frac{(-200 - j346.4)}{(0 + j2) + (0 - j2) + \dfrac{(0 + j2)(0 - j2)}{(0 + j40)}}$$

$$= \frac{(-200 + j346.4)}{j38 - j40} - \frac{(-200 - j346.4)}{-j0.1} = \frac{(-200 + j346.4)}{-j2} - \frac{(-200 - j346.4)}{-j0.1}$$

$$= -j100 - 173.2 + j2000 - 3464 = -3637.2 + j1900 = 4104 \angle 152.4° \text{ A}.$$

Voltage drops across impedances :

Voltage drop, $E_R = I_R Z_R = (173.2 - j100) \times j40 =$ **(4000 + j6928) V. (Ans.)**

Voltage drop, $E_Y = I_Y Z_Y = (3464 - j1800)(j2) =$ **(3600 + j6928) V. (Ans.)**

Voltage drop, $E_B = I_B Z_B = (-3637.2 + j1900)(-j2) =$ **(3800 + j7274.4) V. (Ans.)**

Potential of star point above ground :

$$= -\frac{1}{3}(E_R + E_Y + E_B)$$

$$= -\frac{1}{3}[(4000 + j6928) + (3600 + j6928) + (3800 + j7274.4)]$$

$$= -\frac{1}{3}(11400 + j17656) = -(3800 + j7043.5) = -\mathbf{8003\ V.}\ \textbf{(Ans.)}$$

☞**Example 52.** *An unbalanced star-connected load shown in Fig. 68, $Z_R = j40\ \Omega$, $Z_Y = -j40\ \Omega$, $Z_B = 40\ \Omega$ is connected across a balanced 400 V, 3-phase supply. Find the reading of the wattmeter connected in Y-phase.*

The phase sequence is $R \to Y \to B$.

Solution. Refer Fig. 68.

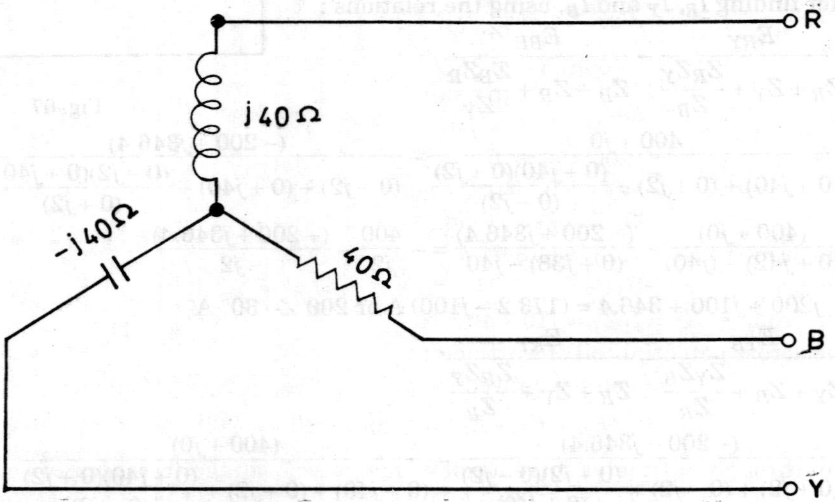

Fig. 68

Taking E_{RY} as reference vector

$$E_{RY} = 400 \angle 0° = 400(1 + j0)$$
$$E_{YB} = 400 \angle -120° = 400(-0.5 - j0.866)$$
$$E_{BR} = 400 \angle 120° = 400(-0.5 + j0.866)$$

Impedances are

$$Z_R = (0 + j40), \quad Z_Y = (0 - j40), \quad Z_B = (40 + j0)$$

Now,
$$I_R = \frac{E_{RY}}{Z_R + Z_Y + \dfrac{Z_R Z_Y}{Z_B}} - \frac{E_{BR}}{Z_B + Z_R + \dfrac{Z_B Z_R}{Z_Y}}$$

$$= \frac{(400 + j0)}{j40 - j40 + \frac{j40 \times (-j40)}{40}} - \frac{400(-0.5 + j0.866)}{40 + j40 + \frac{40 \times j40}{-j40}}$$

$$= \frac{400}{40} - \frac{-200 + j346.4}{40 + j40 - 40}$$

$$= 10 - \frac{-200 + j346.4}{j40} = 10 - \frac{+j200 - 346.4}{40}$$

$$= 10 - j5 + 8.66 = (1.34 - j5) \text{ A}$$

$$I_Y = \frac{E_{YB}}{Z_Y + Z_B + \frac{Z_Y Z_B}{Z_R}} - \frac{E_{RY}}{Z_R + Z_Y + \frac{Z_R Z_Y}{Z_B}}$$

$$= \frac{400(-0.5 - j0.866)}{-j40 + 40 + \frac{(-j40) \times 40}{j40}} - \frac{400 + j0}{j40 - j40 + \frac{j40(-j40)}{40}}$$

$$= \frac{-200 - j346.4}{-j40 + 40 - 40} - \frac{400 + j0}{40}$$

$$= -j5 + 8.66 - 10 = (-1.34 - j5) \text{ A}$$

$$I_B = \frac{E_{BR}}{Z_B + Z_R + \frac{Z_B Z_R}{Z_Y}} - \frac{E_{YB}}{Z_Y + Z_B + \frac{Z_Y Z_B}{Z_R}}$$

$$= \frac{400(0.5 + j0.866)}{40 + j40 + \frac{40 \times j40}{-j40}} - \frac{400(-0.5 - j0.866)}{-j40 + 40 + \frac{(-j40) \times 40}{j40}}$$

$$= \frac{200 + j346.4}{40 + j40 - 40} - \frac{-200 - j346.4}{-j40 + 40 - 40} = \frac{200 + j346.4}{j40} - \frac{-200 - j346.4}{-j40}$$

$$= 8.66 + j5 + j5 - 8.66 = j10 \text{ A}$$

Voltage drop, $E_R = I_R Z_R = (1.34 - j5) \times j40 = (200 + j53.6)$ V

Voltage drop, $E_Y = I_Y Z_Y = (-1.34 - j5) \times (-j40) = (-200 + j53.6)$ V

Voltage drop, $E_B = I_B Z_B = j10 \times 40 = (0 + j400)$ V

Voltage across potential coil of wattmeter

= voltage between supply neutral and load neutral

$$= \frac{1}{3}(E_R + E_Y + E_B) = \frac{1}{3}(200 + j53.6 - 200 + j53.6 + j400) = j169 \text{ V}$$

Current through wattmeter current coil × conjugate of voltage across wattmeter potential coil

$$= I_Y \times (-j169) = (-1.34 - j5) \times (-j169) = (-845 + j226.5)$$

∴ **Wattmeter reading = – 845 W. (Ans.)**

STAR-DELTA CONVERSION

☞**Example 53.** *An unbalanced star-connected load has branch impedances of $Z_R = 10 \angle 30° \ \Omega$, $Z_Y = 20 \angle 45° \ \Omega$ and $Z_B = 20 \angle 60° \ \Omega$ and is connected across a balanced 3-phase, 3-wire supply of 440 V. Find :*

(i) Line currents, and *(ii) Voltage across each impedance.*

Use star-delta conversion method.

Solution. Taking E_{RY} as reference vector

Then $E_{RY} = 440 \angle 0°$, $E_{YB} = 440 \angle -120°$

$E_{BR} = 440 \angle 120°$

Refer Fig. 69.

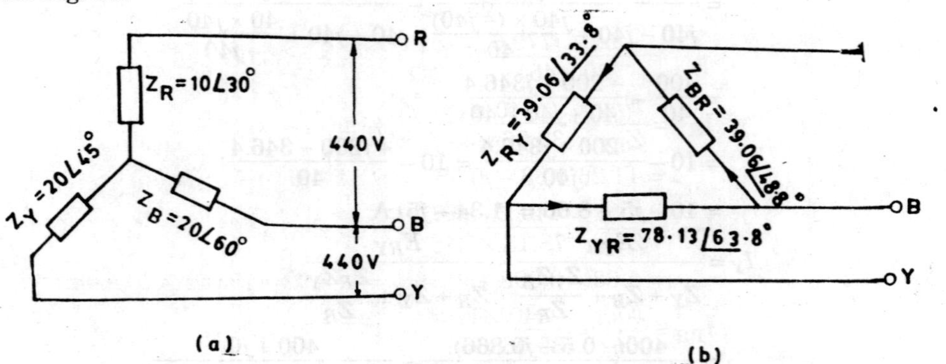

Fig. 69

Also $Z_R = 10 \angle 30°$ *(given)*

$Z_Y = 20 \angle 45°$ *(given)*

$Z_B = 20 \angle 60°$ *(given)*

Impedances of equivalent delta-connected load (Refer Fig. 69) :

Now
$$Z_{RY} = Z_R + Z_Y + \frac{Z_R R_Y}{Z_B} = \frac{Z_R Z_B + Z_Y Z_B + Z_R Z_Y}{Z_B}$$

$$= \frac{(10 \angle 30°)(20 \angle 60°) + (20 \angle 45°)(20 \angle 60°) + (10 \angle 30°)(20 \angle 45°)}{20 \angle 60°}$$

$$= \frac{200 \angle 90° + 400 \angle 105° + 200 \angle 75°}{20 \angle 60°}$$

$$= \frac{200(0 + j1) + 400(-0.26 + j0.966) + 200(0.26 + j0.966)}{20 \angle 60°}$$

$$= \frac{j200 - 104 + j386.4 + 52 + j193.2}{20 \angle 60°}$$

$$= \frac{-52 + j779.6}{20 \angle 60°} = \frac{781.3 \angle 93.8°}{20 \angle 60°} = 39.06 \angle 33.8°\ \Omega$$

$$Z_{YB} = Z_Y + Z_B + \frac{Z_Y Z_B}{Z_R} = \frac{Z_Y Z_R + Z_B Z_R + Z_Y Z_B}{Z_R}$$

$$= \frac{(20 \angle 45°)(10 \angle 30°) + (20 \angle 60°)(10 \angle 30°) + (20 \angle 45°)(20 \angle 60°)}{10 \angle 30°}$$

$$= \frac{200 \angle 75° + 200 \angle 90° + 400 \angle 105°}{10 \angle 30°}$$

$$= \frac{200(0.62 + j0.966) + j200 + 400(-0.26 + j0.966)}{10 \angle 30°}$$

$$= \frac{52 + j193.2 + j200 - 104 + j386.4}{10 \angle 30°}$$

$$= \frac{-52 + j779.6}{10 \angle 30°} = \frac{781.3 \angle 93.8°}{10 \angle 30°} = 78.13 \angle 63.8°\ \Omega$$

$$Z_{BR} = Z_B + Z_R + \frac{Z_B Z_R}{Z_Y} = \frac{Z_B Z_Y + Z_R Z_Y + Z_B Z_R}{Z_Y}$$

$$= \frac{(20 \angle 60°)(20 \angle 45°) + (10 \angle 30°)(20 \angle 45°) + (20 \angle 60°)(10 \angle 30°)}{20 \angle 45°}$$

$$= \frac{400\ \angle 105^\circ + 200\ \angle 75^\circ + 200\angle 90^\circ}{20\ \angle 45^\circ}$$

$$= \frac{-\ 52 + j779.6}{20\ \angle 45^\circ} = \frac{781.3\ \angle 93.8}{20\ \angle 45^\circ} = 39.06\ \angle 48.8^\circ\ \Omega$$

$$\mathbf{I_{RY}} = \frac{E_{RY}}{Z_{RY}} = \frac{440\ \angle 0^\circ}{39.06\ \angle 33.8^\circ} = 11.26\ \angle -\ 33.8^\circ\ \text{A}$$

$$= 11.26(0.83 - j0.56) = \mathbf{(9.34 - j6.3)\ A.\ \ (Ans.)}$$

$$\mathbf{I_{YB}} = \frac{E_{YB}}{Z_{YB}} = \frac{440\ \angle -\ 120^\circ}{78.13\ \angle 63.8^\circ} = 5.63\ \angle -\ 183.8^\circ$$

$$= 5.63(-\ 0.998 + j0.066) = \mathbf{(-\ 5.62 + j0.37)\ A.\ \ (Ans.)}$$

$$\mathbf{I_{BR}} = \frac{E_{BR}}{Z_{BR}} = \frac{440\ \angle 120^\circ}{39.06\ \angle 48.8^\circ} = 11.26\ \angle 71.2^\circ$$

$$= 11.26(0.322 + j0.947) = \mathbf{(3.625 + j10.66)\ A.\ \ (Ans.)}$$

(*i*) **Line currents :**

$$\mathbf{I_R} = I_{RY} - I_{BR} = (9.34 - j6.3) - (3.625 + j10.66)$$

$$= 5.715 - j16.96 = \mathbf{17.89\ \angle -\ 71.4^\circ\ A.\ \ (Ans.)}$$

$$\mathbf{I_Y} = I_{YB} - I_{RY} = (-\ 5.62 + j0.37) - (9.34 - j6.3)$$

$$= -\ 14.96 + j6.67 = \mathbf{16.4\ \angle 156^\circ\ A.\ \ (Ans.)}$$

$$\mathbf{I_B} = I_{BR} - I_{YB} = (3.625 + j10.66) - (-\ 5.62 + j0.37)$$

$$= 9.245 + j10.29 = \mathbf{13.833\ \angle 48^\circ\ A.\ \ (Ans.)}$$

These are the currents in the phases of the star-connected unbalanced load. Let us find voltage drop across each star-connected branch impedance.

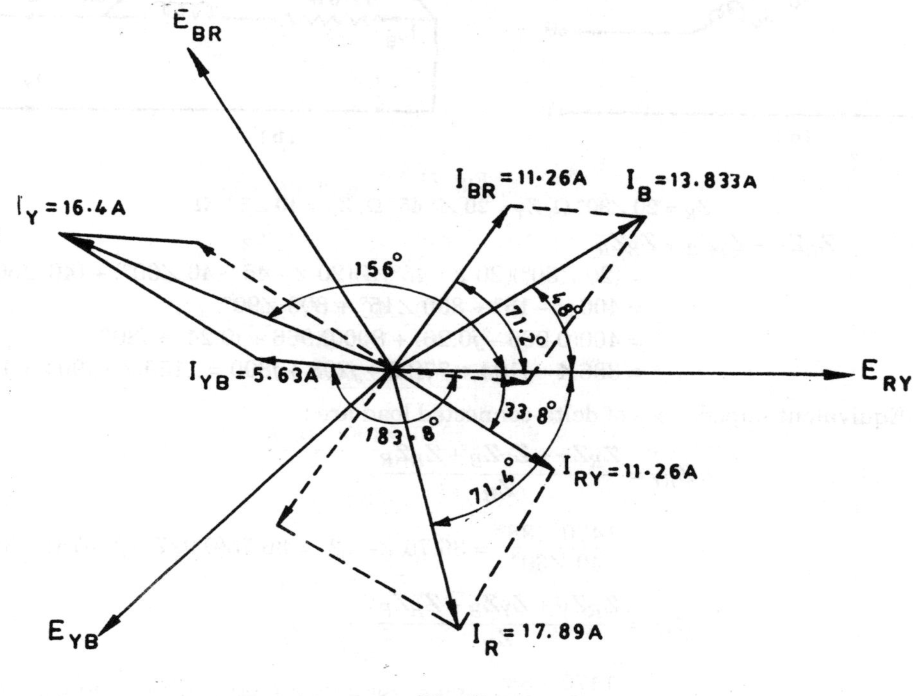

Fig. 70

(*ii*) **Voltage across impedances :**

Voltage across $Z_R = I_R \times Z_R$

$= 17.89 \angle -71.4° \times 10 \angle 30° = \mathbf{178.9 \angle -41.4° \ V.}$ **(Ans.)**

Voltage across $Z_Y = I_Y \times Z_Y = 16.4 \angle 156° \times 20 \angle 45°$

$= \mathbf{328 \angle 210° \ V.}$ **(Ans.)**

Voltage across $Z_B = I_B \times Z_B = 13.833 \angle 48° \times 20 \angle 60°$

$= \mathbf{276.67 \angle 108° \ V.}$ **(Ans.)**

The phasor diagram is shown in Fig. 70.

☞**Example 54.** *The branch impedances of an unbalanced star-connected load are :* $Z_R = 20 \angle 30° \ \Omega$, $Z_Y = 20 \angle -45° \ \Omega$, $Z_B = 40 \angle 60° \ \Omega$. *It is connected across a balanced 3-phase, 3-wire supply of 200 V. Find :*

(*i*) *Line currents, and* (*ii*) *Voltage across each impedance.*

Use star-delta conversion method.

Solution. The unbalanced star-connected load and its equivalent delta-connected load are shown in Fig. 71.

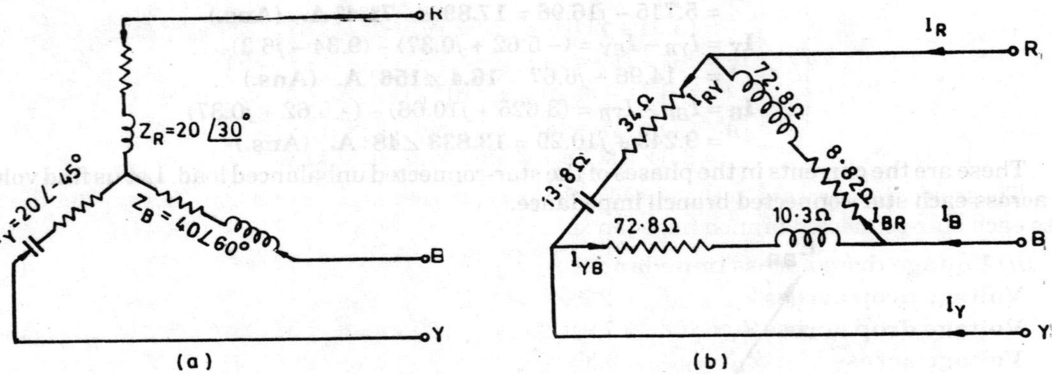

Fig. 71

$Z_R = 20 \angle 30° \ \Omega, \ Z_Y = 20 \angle -45° \ \Omega, \ Z_B = 40 \angle 60° \ \Omega$

Now, $Z_R Z_Y + Z_Y Z_B + Z_B Z_R$

$= (20 \angle 30°)(20 \angle -45°) + (20 \angle -45°)(40 \angle 60°) + (40 \angle 60°)(20 \angle 30°)$

$= 400 \angle -15° + 800 \angle 15° + 800 \angle 90°$

$= 400(0.966 - j0.26) + 800(0.966 + j0.24) + j800$

$= 386.4 - j104 + 772.8 + j208 + j800 = 1159.2 + j904 = 1470 \angle 38°$

$\therefore$ Equivalent impedances of delta-connected load are :

$$Z_{RY} = \frac{Z_R Z_Y + Z_Y Z_B + Z_B Z_R}{Z_B}$$

$$= \frac{1470 \angle 38°}{40 \angle 60°} = 36.75 \angle -22° = 36.75(0.927 - j0.375) = 34 - j13.8$$

$$Z_{YB} = \frac{Z_R Z_Y + Z_Y Z_B + Z_B Z_R}{Z_R}$$

$$= \frac{1470 \angle 38°}{20 \angle 30°} = 73.5 \angle 8° = 73.5(0.99 + j0.14) = 72.8 + j10.3$$

$$Z_{BR} = \frac{Z_R Z_Y + Z_Y Z_B + Z_B Z_R}{Z_Y} = \frac{1470 \angle 38^\circ}{20 \angle -45^\circ}$$

$$= 73.5 \angle 83^\circ = 73.5(0.12 + j0.99) = 8.82 + j72.8$$

Taking E_{RY} as the difference vector and assuming sequence as $R \to Y \to B$,

$$E_{RY} = 200 \angle 0^\circ, E_{YB} = 200 \angle -120^\circ, E_{BR} = 200 \angle 120^\circ$$

$$I_{RY} = \frac{E_{RY}}{Z_{RY}} = \frac{200 \angle 0^\circ}{36.75 \angle -22^\circ} = 5.44 \angle 22^\circ$$

$$= 5.44(0.927 + j0.375) = 5.04 + j2.04$$

$$I_{YB} = \frac{E_{YB}}{Z_{YB}} = \frac{200 \angle -120^\circ}{73.5 \angle 8^\circ} = 2.72 \angle -128^\circ$$

$$= 2.72(-0.616 - j0.79) = -1.67 - j2.15$$

$$I_{BR} = \frac{E_{BR}}{Z_{BR}} = \frac{200 \angle 120^\circ}{73.5 \angle 83^\circ} = 2.72 \angle 37^\circ = 2.72(0.8 + j0.6) = 2.17 + j1.63$$

(*i*) **Line Currents :**

$$\mathbf{I_R} = I_{RY} - I_{BR}$$
$$= (5.04 + j2.04) - (2.17 + j1.63) = 2.87 + j0.41 = \mathbf{2.9 \angle 8.1^\circ\ A.}\ \textbf{(Ans.)}$$
$$\mathbf{I_Y} = I_{YB} - I_{RY} = (-1.67 - j2.15) - (5.04 + j2.04)$$
$$= -6.71 - j4.19 = \mathbf{7.91 \angle -148^\circ\ A.}\ \textbf{(Ans.)}$$
$$\mathbf{I_B} = I_{BR} - I_{YB} = (2.17 + j1.63) - (-1.67 - j2.15)$$
$$= 3.84 + j3.78 = \mathbf{5.39 \angle 44.5^\circ\ A.}\ \textbf{(Ans.)}$$

These are the currents in the phases of star-connected unbalanced load. Let us find the drop across each star-connected branch impedance.

(*ii*) **Voltage drop across impedances** :

Voltage drop across $\mathbf{Z_R}$ $= I_R Z_R = 2.9 \angle 8.1^\circ \times 20 \angle 30^\circ = \mathbf{58 \angle 38.1^\circ\ V.}$ **(Ans.)**

Voltage drop across $\mathbf{Z_Y}$ $= I_Y Z_Y = 7.91 \angle -148^\circ \times 20 \angle -45^\circ = \mathbf{158.2 \angle -193^\circ\ V.}$ **(Ans.)**

Voltage across $\mathbf{Z_B} = I_B Z_B = 5.39 \angle 44.5^\circ \times 40 \angle 60^\circ = \mathbf{215.6 \angle 104.5^\circ\ V.}$ **(Ans.)**

Example 55. *A star-connected load consisting of non-inductive r esistors of 30, 12 and 20 ohms connected to the R, Y and B lines respectively is fed by a 300 V (line), 3-phase supply. The phase sequence is RYB. Calculate the voltage across each resistor.*

Solution. Refer Fig. 72.

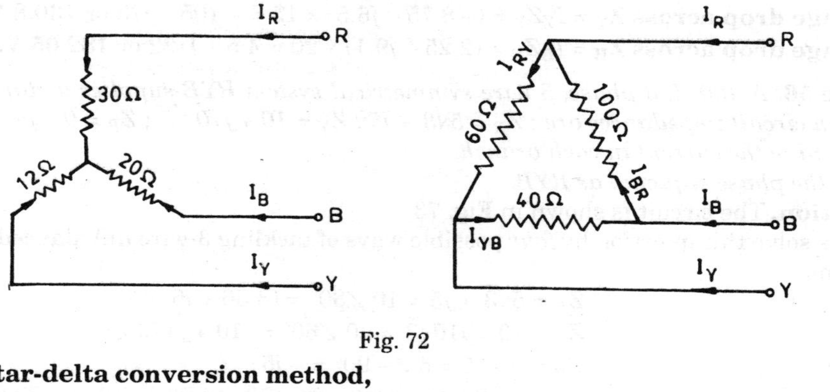

Fig. 72

Using star-delta conversion method,

Impedances are :

$$Z_R = 30\ \Omega, \quad Z_Y = 12\ \Omega, \quad Z_B = 20\ \Omega$$

Now $Z_RZ_Y + Z_YZ_B + Z_{BR} = 30 \times 12 + 12 \times 20 + 20 \times 30 = 360 + 240 + 600 = 1200\ \Omega$

Equivalent impedances of delta-connected load are :

$$Z_{RY} = \frac{Z_RZ_Y + Z_YZ_B + Z_{BR}}{Z_B} = \frac{1200}{20} = 60\ \Omega$$

$$Z_{YB} = \frac{Z_RZ_Y + Z_YZ_B + Z_BZ_R}{Z_R} = \frac{1200}{30} = 40\ \Omega$$

$$Z_{BR} = \frac{Z_RZ_Y + Z_YZ_B + Z_BZ_R}{Z_Y} = \frac{1200}{12} = 100\ \Omega$$

Phase currents are :

$$I_{RY} = \frac{E_{RY}}{Z_{RY}} = \frac{300}{60} = 5\text{ A}$$

$$I_{YB} = \frac{E_{YB}}{Z_{YB}} = \frac{300}{40} = 7.5\text{ A}$$

$$I_{BR} = \frac{E_{BR}}{Z_{BR}} = \frac{300}{100} = 3\text{ A}$$

Each current is in phase with its own voltage because the load is purely resistive.

To find line currents for delta-connection using vector algebra and choosing E_{RY} as the reference axis, we get

$$I_{RY} = 5 + j0$$
$$I_{YB} = 7.5(-0.5 - j0.866) = -3.75 - j6.5$$
$$I_{BR} = 3(-0.5 + j0.866) = -1.5 + j2.6$$

Line currents for delta-connection are :

$I_R = I_{RY} - I_{BR}$

$= (5 + j0) - (-1.5 + j2.6) = (6.5 - j2.6)$ **or 7 A in magnitude**

$I_Y = I_{YB} - I_{RY}$

$= (-3.75 - j6.5) - (5 + j0) = -8.75 - j6.5$ **or 10.9 A in magnitude**

$I_B = I_{BR} - I_{YB} = (-1.5 + j2.6) - (-3.75 - j6.5)$

$= 2.25 - j9.1$ **or 9.37 A in magnitude**

These line currents for delta-connection are the phase currents for star-connection.

Voltage drop across each limb of star-connected load is :

Voltage drop across $\mathbf{Z_R}$ $= I_RZ_R = (6.5 - j2.6) \times 30 = 195 - j78$ or **210 V. (Ans.)**

Voltage drop across $\mathbf{Z_Y}$ $= I_YZ_Y = (-8.75 - j6.5) \times 12 = -105 - j78$ or **130.8 V. (Ans.)**

Voltage drop across $\mathbf{Z_B}$ $= I_BZ_B = (2.25 - j9.1) \times 20 = 4.5 - j182$ or **182.05 V. (Ans.)**

☞**Example 56.** *A 400 V, 3-phase, 3-wire symmetrical system RYB supplies a star-connected load whose branch circuit impedances are :* $Z_R = 5\sqrt{3} + j5$; $Z_Y = 10 + j10\sqrt{3}$; $Z_B = 0 - j5$.

Determine the current in each branch.

Take the phase sequence as RYB.

Solution. The circuit is shown in Fig. 73.

Let us solve this question by *four* possible ways of tackling 3-wire unbalanced star-connected load problems.

Now
$$Z_R = 5\sqrt{3} + j5 = 10\angle 30° = (8.66 + j5)$$
$$Z_Y = 10 + j10\sqrt{3} = 20\angle 60° = (10 + j17.32)$$
$$Z_B = 0 - j5 = 5\angle -90° = -j5$$

Also, let
$$E_{RY} = 400\angle 0° = (400 + j0)$$
$$E_{YB} = 400\angle -120° = (-200 - j346)$$
$$E_{BR} = 400\angle 120° = (-200 + j346)$$

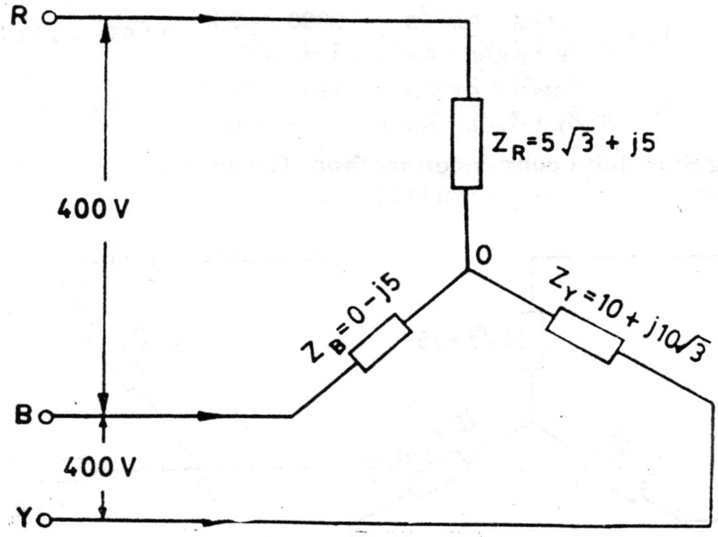

Fig. 73

1. By using Kirchhoff's law :

We know that

$$I_R = \frac{E_{RY}}{Z_R + Z_Y + \frac{Z_R Z_Y}{Z_B}} - \frac{E_{BR}}{Z_B + Z_R + \frac{Z_B Z_R}{Z_Y}}$$

$$= \frac{E_{RY} Z_B}{Z_R Z_B + Z_Y Z_B + Z_R Z_Y} - \frac{E_{BR} Z_Y}{Z_{BY} + Z_{RY} + Z_{BR}} \quad ...(i)$$

$$I_Y = \frac{E_{YB} Z_R}{Z_R Z_Y + Z_Y Z_B + Z_B Z_R} - \frac{E_{RY} Z_B}{Z_{RY} + Z_{YB} + Z_{BR}} \quad ...(ii)$$

and

$$I_B = \frac{E_{BR} Z_Y}{Z_R Z_Y + Z_Y Z_B + Z_B Z_R} - \frac{E_{YB} \times Z_R}{Z_{RY} + Z_{YB} + Z_{BR}} \quad ...(iii)$$

Now $Z_R Z_Y + Z_Y Z_B + Z_B Z_R$

$= (10 \angle 30°)(20° \angle 60°) + (20 \angle 60°)(5 \angle -90°) + (5 \angle -90°)(10 \angle 30°)$

$= 200 \angle 90° + 100 \angle -30° + 50 \angle -60°$

$= j200 + (86.6 - j50) + (25 - j43.3)$

$= 111.6 + j106.7 = 154.4 \angle 43.7°$

$E_{RY} Z_B - E_{BR} Z_Y = 400 \times 5 \angle -90° - 400 \angle 120° \times 20 \angle 60°$

$= 2000 \angle -90° - 8000 \angle 180° = -j2000 - (-8000)$

$= 8000 - j2000 = 8246 \angle -14°$

$E_{YB} Z_R - E_{RY} Z_B = 400 \angle -120° \times 10 \angle 30° - 400 \angle 0° \times 5 \angle -90°$

$= 4000 \angle -90° - 2000 \angle -90°$

$= -j4000 + j2000 = -j2000 = 2000 \angle -90°$

$E_{BR} Z_Y - E_{YB} Z_R = 400 \angle 120° \times 20 \angle 60° - 400 \angle -120° \times 10 \angle 30°$

$= 8000 \angle 180° - 4000 \angle -90° = -8000 + j4000 = 8944 \angle 153.4°$

Currents in each branch [from (*i*), (*ii*) and (*iii*)]

$$\mathbf{I_R} = \frac{E_{RY} Z_B - E_{BR} Z_Y}{Z_R Z_Y + Z_Y Z_B + Z_B Z_R} = \frac{8246 \angle -14°}{154.4 \angle 43.7°} = \mathbf{53.4 \angle -57.7° \text{ A.}} \quad \textbf{(Ans.)}$$

$$\mathbf{I_Y} = \frac{E_{YB}Z_R - E_{RY}Z_B}{Z_RZ_Y + Z_YZ_B + Z_BZ_R} = \frac{2000\ \angle -90^\circ}{154.4\ \angle 43.7} = \mathbf{12.95\ \angle -133.7^\circ\ A.\ \ (Ans.)}$$

$$\mathbf{I_B} = \frac{E_{BR}Z_Y - Z_{YB}Z_R}{Z_RZ_Y + Z_YZ_B + Z_BZ_R} = \frac{8944\ \angle 153.4^\circ}{154.4\angle 43.7} = \mathbf{57.93\ \angle 109.7^\circ\ A.\ \ (Ans.)}$$

2. By using Star-delta conversion method. The unbalanced star-connected load and its equivalent delta-connected load are shown in Fig. 74.

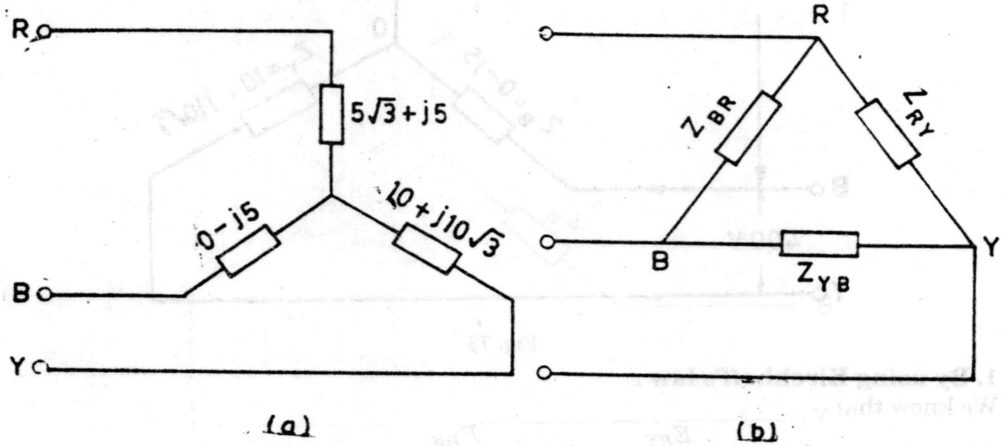

Fig. 74

Now, $\quad Z_RZ_Y + Z_YZ_B + Z_BZ_R = 154.4\ \angle 43.7^\circ$ [Already calculated]

Equivalent delta-connected impedances are :

$$Z_{RY} = \frac{Z_RZ_Y + Z_YZ_B + Z_BZ_R}{Z_B} = \frac{154.4\ \angle 43.7^\circ}{5\ \angle -90^\circ} = 30.88\ \angle 133.7^\circ$$

$$Z_{YB} = \frac{Z_RZ_Y + Z_YZ_B + Z_BZ_R}{Z_R} = \frac{154.4\ \angle 43.7^\circ}{10\ \angle 30^\circ} = 15.44\ \angle 13.7^\circ$$

$$Z_{BR} = \frac{Z_RZ_Y + Z_YZ_B + Z_BZ_R}{Z_Y} = \frac{154.4\ \angle 43.7^\circ}{20\ \angle 60^\circ} = 7.72\ \angle -16.3^\circ$$

Phase currents are :

$$I_{RY} = \frac{E_{RY}}{Z_{RY}} = \frac{400\ \angle 0^\circ}{30.88\ \angle 133.7} = 12.95\ \angle -133.7^\circ = (-8.95 - j9.36)$$

$$I_{YB} = \frac{E_{YB}}{Z_{YB}} = \frac{400\ \angle -120^\circ}{15.44\ \angle 13.7} = 25.9\ \angle -133.7^\circ = (-17.89 - j18.72)$$

$$I_{BR} = \frac{E_{BR}}{Z_{BR}} = \frac{400\ \angle 120^\circ}{7.72\ \angle -16.3^\circ} = 51.8\ \angle 136.3^\circ = (-37.45 + j35.78)$$

Line currents for delta-connection are :

$$\mathbf{I_R} = I_{RY} - I_{BR}$$
$$= (-8.95 - j9.36) - (-37.45 + j35.78)$$
$$= 28.5 - j45.14 = \mathbf{53.38\ \angle 57.7^\circ\ A.\ \ (Ans.)}$$

$$\mathbf{I_Y} = I_{YB} - I_{RY} = (-17.89 - j18.72) - (-8.95 - j9.36)$$
$$= -8.94 - j9.36 = 12.94\ \angle -133.7^\circ$$

$$\mathbf{I_B} = I_{BR} - I_{YB}$$
$$= (-37.45 + j35.78) - (-17.89 - j18.72) = -19.56 + j54.5 = 57.9 \angle 109.7°$$
$$\Sigma I = I_R + I_Y + I_B = 0 \text{ ...... as a check.}$$

3. By using Maxwell's Loop Current Method :

Fig. 75 shows the loop or mesh currents.

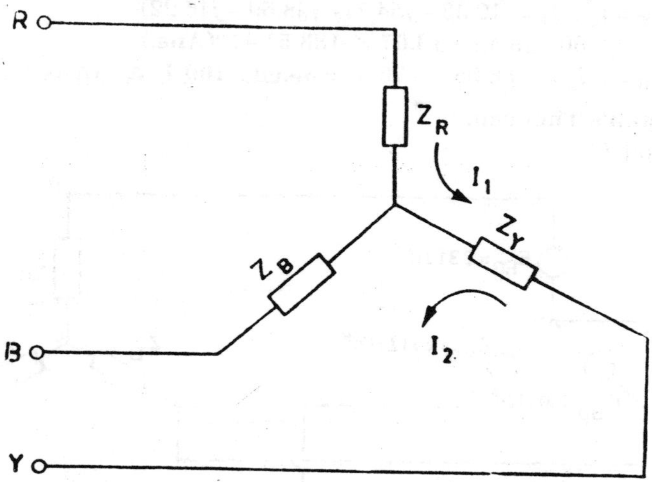

Fig. 75

It may be noted that $I_R = I_1$

$I_Y = I_2 - I_1$

and $I_B = -I_2$.

Considering drops across R and Y arms, we get

$$I_1 Z_R + Z_Y(I_1 - I_2) = E_{RY}$$

or
$$I_1(Z_R + Z_Y) - I_2 Z_Y = E_{RY} \quad ...(i)$$

Similarly, considering the arms Y and B, we have

$$(I_2 - I_1)\, Z_Y + I_2 Z_B = E_{YB}$$
$$-I_1 Z_Y + I_2(Z_Y + Z_B) = E_{YB} \quad ...(ii)$$

Solving for I_1 and I_2, we get

$$I_1 = \frac{E_{RY}(Z_Y + Z_B) + E_{YB} Z_Y}{(Z_R + Z_Y)(Z_Y + Z_B) - Z_Y^2}$$

$$I_2 = \frac{E_{YB}(Z_R + Z_Y) + E_{RY} Z_Y}{(Z_R + Z_Y)(Z_Y + Z_B) - Z_Y^2}$$

$$I_1 = \frac{400(10 + j12.32) + 400 \angle -120 \times 20 \angle 60°}{(18.66 + j22.32)(10 + j12.32) - 400 \angle 120°}$$

$$= \frac{4000 + j4928 + 4000 - j6928}{(186.6 + j229.9 + j223.2 - 275) - (-200 + j346.4)}$$

$$= \frac{8000 - j2000}{111.6 + j106.7} = \frac{8246.2 \angle -14°}{154 \angle 43.7°}$$

$$= 53.5 \angle -57.7° = (28.59 - j45.22)$$

$$I_2 = \frac{(-200 - j346)(18.66 + j22.32) + 400(10 + j17.32)}{111.6 + j106.7}$$

$$= \frac{-373.2 - j4464 - j6456.4 + 7723 + 4000 + j6928}{111.6 + j106.7}$$

$$= \frac{8000 - j4000}{111.6 + j106.7} = \frac{8944 \angle -26.6^\circ}{154 \angle 43.7^\circ} = 58.08 \angle -70.3^\circ = 19.58 - j54.7$$

$\therefore$ $\mathbf{I_R} = I_1 = \mathbf{53.5 \angle -57.7^\circ}$ **A. (Ans.)**

$\mathbf{I_Y} = I_2 - I_1 = (19.58 - j54.7) - (28.59 - j45.22)$

$= -90 - j9.48 = \mathbf{13.07 \angle -133.5^\circ}$ **A. (Ans.)**

$\mathbf{I_B} = -I_2 = -58.08 \angle -70.3^\circ = \mathbf{58.08 \angle 109.7^\circ}$ **A. (Ans.)**

4. Using Milliman's Theorem :

Refer Figs. 76 and 77.

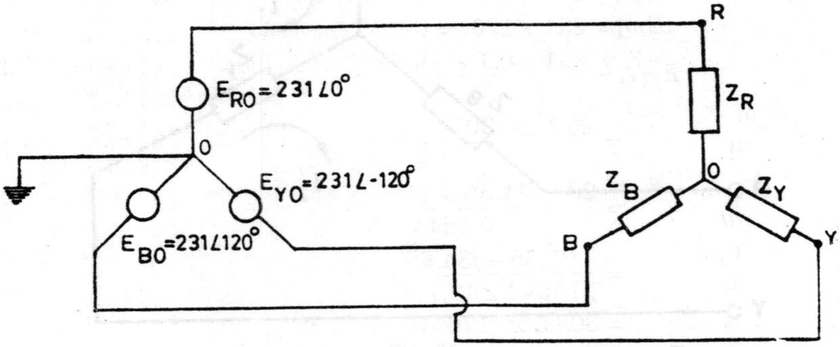

Fig. 76

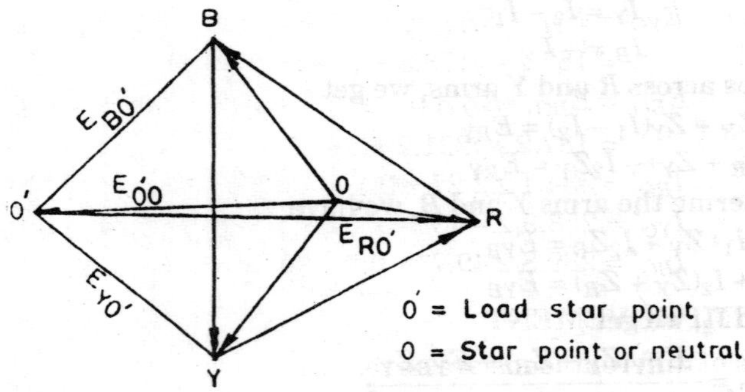

Fig. 77

According to Milliman's theorem, the voltage of the *load* star point *O′* with respect to the star-point or neutral *O* of the generator or supply (normally zero potential) is given by

$$E_{O'O} = \frac{E_{RO}Y_R + E_{YO}Y_Y + E_{BO}Y_B}{Y_R + Y_Y + Y_B}$$

where E_{RO}, E_{YO} and E_{BO} are the phase voltages of the generator or 3-phase supply.

Voltage across each phase of the load is

$$E_{RO'} = E_{RO} - E_{O'O}$$

$$E_{YO'} = E_{YO} - E_{O'O}$$

$$E_{BO'} = E_{BO} - E_{O'O}$$

Now, $I_{RO'} = (E_{RO} - E_{O'O})\,Y_R,\ I_{YO'} = (E_{YO} - E_{O'O})Y_Y$

$E_{BO'} = (E_{BO} - E_{O'O})Y_B$

Here,
$$Y_R = \frac{1}{10 \angle 30^\circ} = 0.1 \angle -30^\circ = (0.0866 - j0.05)$$
$$Y_Y = \frac{1}{20 \angle 60^\circ} = 0.05 \angle -60^\circ = (0.025 - j0.0433)$$
$$Y_B = \frac{1}{5 \angle -90^\circ} = 0.2 \angle 90^\circ = (0 + j0.2)$$
$\therefore$
$$Y_R + Y_Y + Y_B = (0.0866 - j0.05) + (0.025 - j0.0433) + (0 + j0.2)$$
$$= 0.1116 + j0.1067 = 0.1544 \angle 43.7^\circ$$
Let
$$E_{RO} = \frac{400}{\sqrt{3}} \angle 0^\circ = (231 + j0)$$
$$E_{YO} = 231 \angle -120^\circ = (-115.5 - j200)$$
$$E_{BO} = 231 \angle 120^\circ = (-115.5 + j200)$$
$$E_{O'O} = \frac{231 \times 0.1 \angle -30^\circ + 231 \angle -120^\circ \times 0.05 \angle -60^\circ + 231 \angle 120^\circ \times 0.2 \angle 90^\circ}{0.1544 \angle 43.7^\circ}$$
$$= \frac{23.1 \angle -30^\circ + 11.55 \angle -180^\circ + 46.2 \angle 210^\circ}{0.1544 \angle 43.7^\circ}$$
$$= \frac{20 - j11.55 - 11.55 - 40 - j23.1}{0.1544 \angle 43.7^\circ}$$
$$= \frac{-31.55 - j34.65}{0.1544 \angle 43.7^\circ} = \frac{46.86 \angle -132.3^\circ}{0.1544 \angle 43.7^\circ}$$
$$= 303.5 \angle -176^\circ = (-302.7 - j21.2)$$
$$E_{RO'} = E_{RO} - E_{O'O} = 231 - (-302.7 - j21.2)$$
$$= 533.7 + j21.2 = 534.1 \angle 2.27^\circ$$
$$E_{YO'} = E_{YO} - E_{O'O} = -115.5 - j200 + 302.7 + j21.2$$
$$= 187.2 - j178.8 = 258.9 \angle -43.7^\circ$$
$$E_{BO'} = E_{BO} - E_{O'O} = -115.5 + j200 + 302.7 + j21.2$$
$$= 187.2 + j221.2 = 289.8 \angle 49.7^\circ$$
$\therefore$
$$I_{RO'} = 534.1 \angle 2.27 \times 0.1 \angle -30^\circ = 53.41 \angle -27.7^\circ$$
$$I_{YO'} = 258.9 \angle -43.7^\circ \times 0.05 \angle -60^\circ = 12.94 \angle -103.7^\circ$$
$$I_{BO'} = 289.8 \angle 49.7^\circ \times 0.2 \angle 90^\circ = 57.96 \angle 139.7^\circ.$$

13. POWER FACTOR IMPROVEMENT

The power factors of different loads are as follows :

Type of load	**Power factor (range)**
Heating and lighting	0.95 to unity
Motor loads	0.5 to 0.9
Single-phase motors	as low as 0.4
Electric welding units	0.2 to 0.3

The power factor is given by :

$$\cos \phi = \frac{kW}{kVA} \quad \text{or} \quad kVA = \frac{kW}{\cos \phi}$$

In case of *single-phase supply*,

$$kVA = \frac{VI}{1000} \quad \text{or} \quad I = \frac{1000\ kVA}{V}$$

$\therefore$
$$I \propto kVA$$

In case of *three phase supply*,

$$\text{kVA} = \frac{\sqrt{3}\, E_L I_L}{1000} \quad \text{or} \quad I_L = \frac{1000\ \text{kVA}}{\sqrt{3}\, E_L}$$

$$\therefore \qquad I_L \propto \text{kVA}$$

In each case, the *kVA is directly proportional to current.*

The higher current due to poor power factor affects the system and lead to the following undesirable results :

1. Rating of alternators and transformers are proportional to their output current hence inversely proportional to power factor, therefore large *generators and transformers are required to deliver same load but at a low power factor.*

2. When a load having a low lagging power is switched on, there is a large voltage drop in the supply voltage because of the increased voltage drop in the supply lines and transformers. This drop in voltage adversely affects the starting torque of motors and necessitates *expensive voltage stabilizing equipment* for keeping the consumer's voltage fluctuations within the statutory limits.

3. The cross-sectional area of the bus-bar, and the contact surface of switch gear is required to be *enlarged* for the same power to be delivered but at a low power factor.

4. To transmit same power at low power factor, more current will have to be carried by the transmission line or the distributor or cable. If the current density in the line is to be kept constant *the size of the conductor will have to be increased.* Thus *more copper is required to deliver the same load but at a low power factor.*

5. Copper losses are proportional to the square of the current hence inversely proportional to the square of the power factor *i.e.* more copper losses at low power factor and hence *poor efficiency*.

Thus we find that the capital cost of transformers, alternators, distributors, transmission lines etc. *increase with the low power factor.*

Causes of Low Power Factor :

(*i*) All A.C. motors (except overexcited synchronous motors and certain types of commutators motors) and transformers operate at lagging power factor.

(*ii*) Due to typical characteristic of the arc, arc lamps operate at low power factor.

(*iii*) When there is increase in supply voltage, which usually occurs during low load periods (such as lunch hours, night hours etc.) the magnetizing current of inductive reactances increases and power factor of the electrical plant as a whole decreases.

(*iv*) Arc and induction furnaces etc. operate at a very low lagging power factor.

(*v*) Due to improper maintenance and repairs of motors the power factor at which motors operate fall.

Methods of Improving Power Factor. The power factor may improve by using the following methods :

1. Use of high power factor motors.
2. Use of induction motors with phase advancers.
3. Use of static capacitors.
4. Use of capacitance boosters.
5. Use of synchronous condensers.

POWER FACTOR IMPROVEMENT

Example 57. *A 40-MVA, 11 kV 3-phase alternator supplies full-load at a lagging power factor of 0.6. Find the percentage increase in earning capacity if the power factor is increased to 0.9.*

Solution. Rating of the alternator = 40 MVA
Supply voltage = 11 kV
Initial power factor = 0.6 (lagging)
Final power factor after increase = 0.9 (lagging)

Percentage increase in earning capacity :

The earning capacity is proportional to the power (in MW or kW) supplied by the alternator.
MW supplied at 0.6 lagging $= 40 \times 0.6 = 24$
MW supplied at 0.9 lagging $= 40 \times 0.9 = 36$
Increase in MW $= 36 - 24 = 12$

The increase in earning capacity is proportional to 12.

$\therefore$ **Percentage increase in earning capacity** $= \frac{12}{24} \times 100 =$ **50%. (Ans.)**

Example 58. *A 3-phase, 40 kW, 440 V, 50 Hz induction motor operates on full-load with an efficiency of 90 per cent and at a power factor of 0.8 lagging. Calculate the total kVA rating of capacitors required to raise the full-load power factor to 0.9 lagging. What will be capacitance per phase if the capacitances are :*

(*i*) *Delta connected* (*ii*) *Star-connected.*

Solution. Motor output = 40 kW
Supply voltage = 440 V
Efficiency of motor = 90%

Motor power input, $P = \frac{\text{output}}{\text{efficiency}} = \frac{40}{0.9} = 44.44$ kW

Power factor 0.8 (lagging)

$$\cos \phi_1 = 0.8$$
$$\therefore \quad \phi_1 = \cos^{-1} 0.8 = 36.9°$$
$$\tan \phi_1 = 0.75$$
$$\text{Motor kVAR}_1 = P \tan \phi_1 = 44.44 \times 0.75 = 33.33$$

Power factor 0.9 (lagging)

Motor power input = 44.44 kW [As before]

[Power is same as before since the capacitors are *loss free i.e.*, they *do not absorb any power*]

$$\cos \phi_2 = 0.9$$
$$\therefore \quad \phi_2 = 25.8°$$
$$\tan \phi_2 = 0.484$$
$$\therefore \quad \text{Motor kVAR}_2 = P \tan \phi_2 = 44.44 \times 0.484 = 21.5.$$

The difference in values of kVAR is *due to the capacitors which supply leading kVAR to partially neutralize the lagging kVAR of the machine.*

$\therefore$ **Leading kVAR supplied by capacitors is**

$= \text{kVAR}_1 - \text{kVAR}_2 = 33.33 - 21.5$
$= 11.83$ ...... MQ or NT.

Since capacitors are loss-free, their kVAR is same as kVA

$\therefore$ Total kVA rating of capacitors = **11.83 (Ans.)**

$$\text{kVAR/capacitor} = \frac{11.83}{3} = 3.943$$

$\therefore$ VAR/capacitor = 3943.

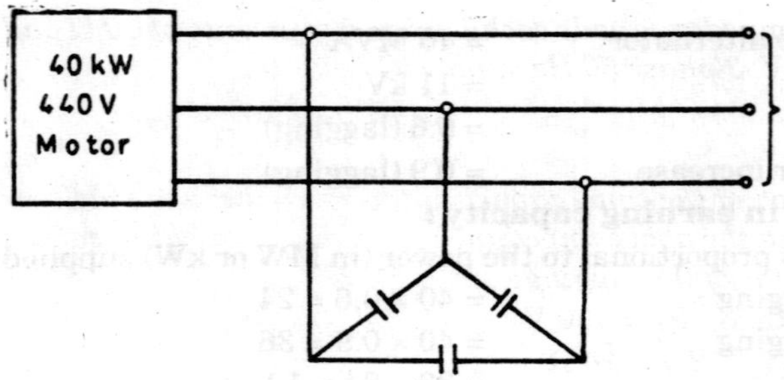

Fig. 78

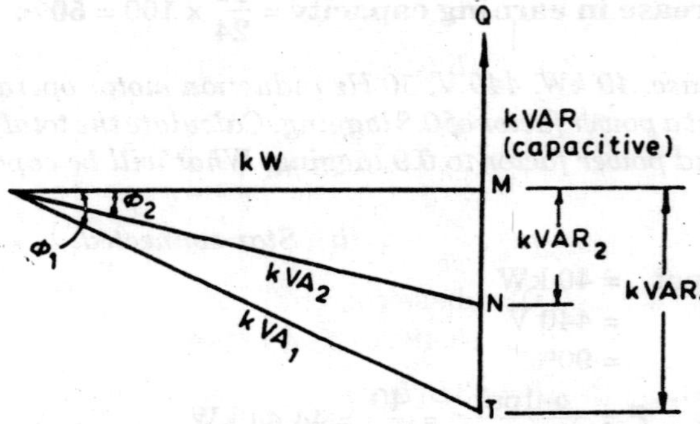

Fig. 79

(*i*) **Delta connection.** In delta-connection, voltage across each capacitor is 440 V.

Current drawn by each capacitor, $I_C = \dfrac{3943}{440} = 8.96$ A

Now,
$$I_C = \frac{V}{X_C} = \frac{V}{\dfrac{1}{2\pi fC}} = 2\pi fCV$$

∴
$$C = \frac{I_C}{2\pi fV} = \frac{6.96}{2\pi \times 50 \times 440} = 50.35 \times 10^{-6} \text{ F} = 50.35 \text{ μF}$$

Hence, **capacitance of each capacitor = 50.35 μF. (Ans.)**

(*ii*) **Star-connection.** In star-connection, voltage across each capacitor

$$= \frac{440}{\sqrt{3}} = 254 \text{ V}$$

Current drawn by each capacitor,

$$I_C = \frac{3943}{254} = 15.52 \text{ A}$$

Now,
$$I_C = \frac{V}{X_C} = \frac{V}{\dfrac{1}{2\pi fC}} \; 2\pi fCV$$

∴
$$C = \frac{I_C}{2\pi fV} = \frac{15.52}{2\pi \times 50 \times 440} = 112.3 \times 10^{-6} \text{ F} = 112.3 \text{ μF}$$

Hence, **capacitance of each capacitor = 112.3 μF. (Ans.)**

Example 59. *Three impedance coils each having a resistance of 10 Ω and a reactance of 7.5 Ω are connected in star to a 400 V, 3-phase 50 Hz supply. Calculate :*

(*i*) *Power factor* (*ii*) *Line current, and*

(*iii*) *Power supplied.*

If three capacitors, each of the same capacitance, are connected in delta to the same supply so as to from parallel circuit with the above impedance coils, calculate the capacitance of each capacitor to obtain resultant power factor of 0.9 lagging.

Solution. Supply voltage, $E_L = 400$ V [Line-to-line]

Phase resistance, $R_{ph} = 10\ \Omega$

Phase reactance, $X_{ph} = 7.5\ \Omega$

Phase voltage, $E_{ph} = \frac{400}{\sqrt{3}} = 231$ V

Phase impedance, $Z_{ph} = \sqrt{R_{ph}^2 + X_{ph}^2} = \sqrt{10^2 + (7.5)^2} = 12.5\ \Omega$

(*i*) **Power factor,** $\cos\phi_1 = \frac{R_{ph}}{Z_{ph}} = \frac{10}{12.5} = \mathbf{0.8.}$ **(Ans.)**

Phase current, $I_{ph} = \frac{E_{ph}}{Z_{ph}} = \frac{231}{12.5} = \mathbf{18.48\ A.}$ **(Ans.)**

(*ii*) Line current, (I_L) = phase current, (I_{ph})

∴ $I_L = \mathbf{18.48\ A.}$ **(Ans.)**

(*iii*) **Power supplied,** $P = \sqrt{3}E_LI_L\cos\phi = \sqrt{3} \times 400 \times 18.48 \times 0.8 = \mathbf{10243\ W.}$ **(Ans.)**

Motor $VAR_1 = \sqrt{3} \times E_LI_L\sin\phi_1 = \sqrt{3} \times 400 \times 18.48 \times 0.6 = \mathbf{7682.}$ **(Ans.)**

When capacitors are connected :

Power factor, $\cos\phi_2 = 0.9$

∴ $\phi_2 = 25.8°$

and $\tan\phi_2 = 0.4834$

Since capacitors themselves do not absorb any power, power remains the same *i.e.*, 10243 W even when capacitors are connected. The only thing that changes is the VAR.

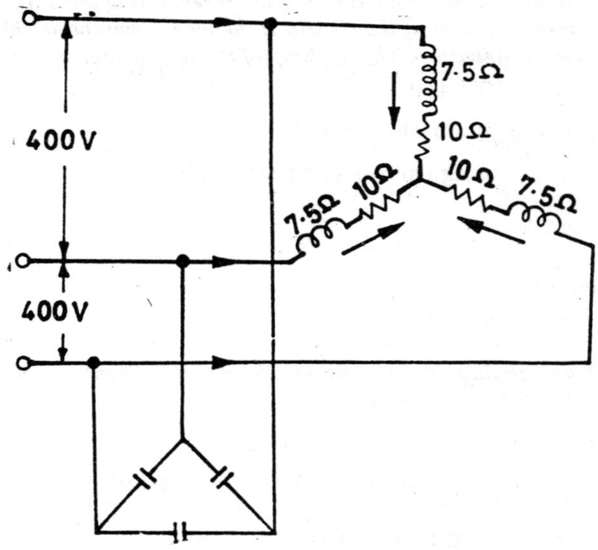

Fig. 80

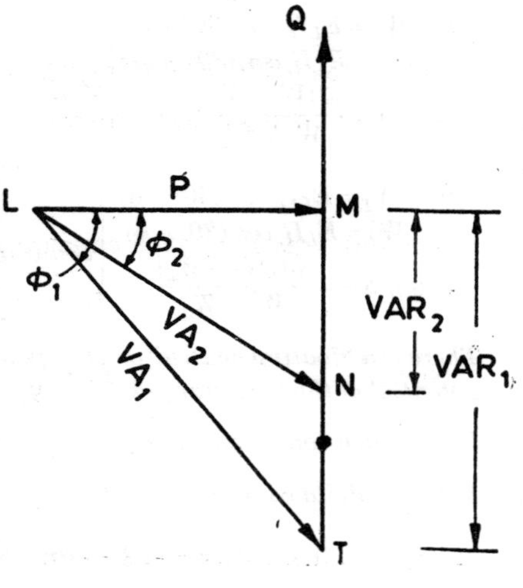

Fig. 81

Now $\quad VAR_2 = P \tan \phi_2 = 10243 \times 0.4834 = 4951$

Leading VAR supplied by three capacitors is

$$= VAR_1 - VAR_2 = 7682 - 4951 = 2731 = MQ \text{ or } NT \text{ (Fig. 81)}$$

$$VAR/capacitor = \frac{2731}{3} = 910.3$$

For delta connection voltage across each capacitor is 400 V

$$\therefore \quad I_C = \frac{910.3}{400} = 2.275 \text{ A}$$

Also
$$I_C = \frac{V}{\frac{1}{2\pi fC}} = V \times 2\pi fC$$

$$\therefore \quad C = \frac{I_C}{V \times 2\pi f} = \frac{2.275}{400 \times 2 \times 3.142 \times 50} = 18.10 \times 10^{-6} \text{ F or } 18.10 \ \mu\text{F}$$

Hence, **capacitance of each capacitor = 18.10 μF. (Ans.)**

HIGHLIGHTS

1. *In a star-connected system*

$$E_{ph} = \frac{E_L}{\sqrt{3}}$$
$$I_{ph} = I_L$$
$$P = \sqrt{3} E_L I_L \cos \phi$$

2. *In a delta-connected system*

$$E_{ph} = E_L$$
$$I_{ph} = \frac{I_L}{\sqrt{3}}$$
$$P = \sqrt{3} E_L I_L \cos \phi.$$

3. *Two-wattmeters method is generally used to measure 3-phase power. In this method, the current coils of the two wattmeters are connected in any two lines and their potential coils to the remaining third line. The sum of the two wattmeters readings gives the total power in the circuit. If the load is balanced, then its power factor can also be calculated from these two readings. The readings of the two wattmeters are :*

(*i*) $W_1 = E_L I_L \cos (30° - \phi)$, $W_2 = E_L I_L \cos (30° + \phi)$, $\tan \phi = \frac{\sqrt{3}(W_1 - W_2)}{W_1 + W_2}$ — *Lagging power factor*

(*ii*) $W_1 = E_L I_L \cos (30° + \phi)$, $W_2 = E_L I_L \cos (30° - \phi)$, $\tan \phi = -\frac{\sqrt{3}(W_1 - W_2)}{W_1 + W_2}$ — *Leading power factor*

4. *There is a 'floating neutral' or 'star' point in an unbalanced star-connected load without neutral wire. The potential of the load star point is different from that of the generator star point.*

Under such conditions, the phase voltage of load is **not** $\frac{1}{\sqrt{3}}$ *of the line voltage.*

5. *In a star-delta connection*

$$Z_{RY} = Z_R + Z_Y + \frac{Z_R Z_Y}{Z_B} = \frac{Z_R Z_B + Z_Y Z_B + Z_R Z_Y}{Z_B}$$

$$Z_{YB} = Z_Y + Z_B + \frac{Z_Y Z_B}{Z_R} = \frac{Z_Y Z_R + Z_B Z_R + Z_Y Z_B}{Z_R}$$

$$Z_{BR} = Z_B + Z_R + \frac{Z_B Z_R}{Z_Y} = \frac{Z_B Z_Y + Z_R Z_Y + Z_B Z_R}{Z_Y}.$$

6. *In a single-phase as well as in a three-phase system, the kVA is directly proportional to the current I. The disadvantage of a lower power factor is that the current required for a given power is very high, which fact leads to many undesirable results. The power factor may be improved by the following :*
   — *Static capacitors*
   — *Phase advancers*
   — *Synchronous capacitors*
   — *Capacitor boosters*
   — *High power factor motors.*

OBJECTIVE TYPE QUESTIONS

A. Choose the Correct Answer :

1. In two wattmeters method of power measurement, one of the wattmeters will show negative reading when the load power factor angle is strictly
   (*a*) less than 30° (*b*) less than 60°
   (*c*) greater than 30° (*d*) greater than 60°. **[U.P.S.C., 1996]**
2. Three identical impedances are connected in delta to a 3-phase supply of 400 V. The line-current is 34.65 A and the total power taken is 14.4 kW. The resistance of the load in each phase (in ohm) is
   (*a*) 20 (*b*) 16
   (*c*) 12 (*d*) 10. **[U.P.S.C., 1993]**
3. The power in a 3-phase system is given by $\sqrt{3} V_L I_L \cos \phi$ where ϕ is the phase angle between
   (*a*) line-voltage and line current (*b*) phase voltage and phase current
   (*c*) line voltage and phase current (*d*) phase voltage and line current.
4. Which of the following statements associated with 3-phase delta connected circuits is *true* ?
   (*a*) Line voltage is equal to phase voltage (*b*) Line current is equal to phase current
   (*c*) Line voltage is $\sqrt{3}$ time of phase voltage (*d*) Line currents are 60° apart.
5. In a 3-phase supply, floating neutral is undesirable because it may result in ...... across the load.
   (*a*) unequal line voltages (*b*) high voltage
   (*c*) low voltage.
6. Phase reversal in a 4-wire unbalanced load supplied from a balanced 3-phase supply causes change in
   (*a*) the power consumed (*b*) magnitude of phase currents
   (*c*) only the magnitude of the neutral current
   (*d*) magnitude as well as phase angle of the neutral current.
7. A 3-phase star connected symmetrical load consumes *P* watts of power from a balanced supply. If the same load is connected in delta to the same supply, the power consumption will be
   (*a*) *P* (*b*) $\sqrt{3}P$
   (*c*) 3*P* (*d*) not determined from the given data.
   [U.P.S.C., 1995]
8. Three unequal impedances are connected in star to a 3-phase system. The sum of three line currents will be
   (*a*) equal to the each line current (*b*) zero
   (*c*) none of these.

9. Three equal impedances are first connected in delta across a $3-\phi$ balanced supply. If the same impedances are connected in star across the same supply

(*a*) phase currents will be one-third (*b*) line currents will be one-third
(*c*) power consumed will be one-third (*d*) none of the above.

10. Which of the following is a four wire system ?

(*a*) Delta (*b*) Star
(*c*) Both delta and star (*d*) Neither delta nor star.

11. In a 3-phase balanced star-connected load, neutral current is equal to

(*a*) zero (*b*) I_P
(*c*) I_L (*d*) unpredictable.

12. Which of the following equations is valid for a 3-phase 4-wire balanced star-connected load ?

(*a*) $I_R + I_Y + I_B = I_N = 0$ (*b*) $I_R + I_Y - I_B = I_N$
(*c*) $I_R - I_Y + I_B = I_N$ (*d*) $\dfrac{V_B + V_R + V_Y}{Z} = I_N$.

13. Three unequal impedances are connected in delta to a 3-phase, 3-wire system.

(*a*) The voltages across the three phases will be different
(*b*) Both of the phase currents and line currents will be unbalanced
(*c*) Phase currents will be unbalanced but the line currents will be balanced
(*d*) None of the above.

14. The relationship between the line and phase voltages of a delta-connected circuit is given by

(*a*) $V_L = V_P$ (*b*) $V_L = \sqrt{3}V_P$
(*c*) $V_L = \dfrac{V_P}{\sqrt{2}}$ (*d*) $V_L = \dfrac{2}{\pi} V_P$. **[B.T.E. M.P. 1996]**

15. In case of a delta connected load, if one resistor is removed, the power will become

(*a*) zero (*b*) one-third
(*c*) two-third (*d*) none of the above.

16. Readings of 1154 and 577 watts are obtained when two wattmeters method was used on a balanced load. The delta connected load impedance for a system of 100 V will be

(*a*) $15 \angle\pm 30°$ (*b*) $15 \angle+ 30°$
(*c*) $15 \angle- 30°$ (*d*) $15 \angle+ 90°$. **[U.P.S.C., 1993]**

17. Which of the following statements is true about two-wattmeters method for power measurement in 3-phase circuit ?

(*a*) Power can be measured using two wattmeters method only for star-connected 3-phase circuits
(*b*) When two wattmeters show identical readings, the power factor is 0.5
(*c*) When power factor is unity, one of the wattmeter reads zero
(*d*) When the readings of the two wattmeters are equal but of opposite sign, the power factor is zero. **[U.P.S.C., 1992]**

18. The minimum number of wattmeters to measure power in a 3-phase unbalanced star-connected load is

(*a*) one (*b*) two
(*c*) three (*d*) four.

19. While measuring power in a three phase load by two-wattmeters method, the readings of two wattmeters will be equal and opposite when

(*a*) *pf* is unity (*b*) load is balanced
(*c*) phase angle is between 60° and 90° (*d*) the load is purely inductive. **[U.P.S.C., 1995]**

20. In the measurement of 3-phase power by two-wattmeters method, if the two wattmeter readings are equal the power factor of the circuit is

(*a*) 0.8 lagging (*b*) 0.8 leading
(*c*) zero (*d*) unity. **[A.M.I.E., 1998]**

21. Which of the following are the necessary conditions for an entire 3-ϕ system to be balanced ?

1. The line voltages are equal in magnitude
2. The phase differences between successive line voltages are equal
3. The impedances in each of the phases are identical.

Select the correct answer using the codes given below :

Codes :

(*a*) 1, 2 and 3 (*b*) 1 and 3

(*c*) 1 and 2 (*d*) 2 and 3. **[U.P.S.C., 1996]**

22. The power delivered to a 3-phase load can be measured by the use of 2 wattmeters only when the

(*a*) load is balanced (*b*) load is unbalanced

(*c*) 3-phase load is connected to the source through 3-wires

(*d*) 3-phase load is connected to the source through 4-wires. **[U.P.S.C., 1995]**

23. The ratio of the readings of two wattmeters connected to measure power in a balanced 3-phase load is 5 : 3 and the load is inductive. The power factor of the load is

(*a*) 0.917 lead (*b*) 0.917 lag

(*c*) 0.6 lead (*d*) 0.6 lag. **[U.P.S.C., 1994]**

24. In two phase system phase voltages differ by

(*a*) 60° (*b*) 90°

(*c*) 120° (*d*) 180°.

25. W_1 and W_2 are the readings of two wattmeters used to measure power of a 3-phase balanced load. The reactive power drawn by the load is

(*a*) $W_1 + W_2$ (*b*) $W_1 - W_2$

(*c*) $\sqrt{3}(W_1 + W_2)$ (*d*) $\sqrt{3}(W_1 - W_2)$. **[A.M.I.E., 1997]**

26. In a two wattmeters method of measuring power in a 3-phase system one of the wattmeters reads negative implying

(*a*) wattmeter connection is faulty (*b*) load is unbalanced

(*c*) power flow is in the reverse direction (*d*) power factor is less than 0.5. **[A.M.I.E., 1998]**

27. v_{RN}, v_{YN} and v_{BN} are the instantaneous line to neutral voltages and i_R, i_Y and i_B are instantaneous line currents in a balanced three phase circuit, the computation $v_{RN}(i_Y - i_B) - (v_{YN} - v_{BN})i_R$ will yield a quantity proportional to the

(*a*) active power (*b*) power factor

(*c*) reactive power (*d*) complex power. **[GATE, 1993]**

28. If the *R*-phase of a 3-phase star connected alternator becomes reversely connected by mistake, it will affect

(*a*) E_{RY} and E_{YB} (*b*) E_{YB} and E_{BR}

(*c*) E_{RY} and E_{BR} (*d*) E_{RY}, E_{YB} and E_{BR}.

29. When phase sequence at the 3-phase load is reversed

(*a*) phase currents are changed in magnitude

(*b*) phase currents change in phase angle but not in magnitude

(*c*) total power consumed is changed

(*d*) phase powers are changed.

30. Lamp load is being supplied from a 3-phase, 4-wire, 230/400 V ac supply system. If a 3-phase motor is now switched on across the same supply, then neutral current will

(*a*) increase (*b*) decrease

(*c*) remain unchanged (*d*) unpredictable.

31. Lamps of 40 W, 100 W and 200 W ratings are connected in different phases of a 3-phase, 4-wire supply. If the neutral wire breaks, then the lamp likely to fuse first will be

(*a*) 200 W (*b*) 100 W

(*c*) 40 W (*d*) all lamps.

32. The instantaneous values of currents in phases *R* and *Y* of a 3-phase system are 25 A each. For a phase sequence of *BRY*, the instantaneous value of current in phase *B* is
(*a*) 25 A (*b*) 50 A
(*c*) 12.5 A (*d*) 43.3 A.

33. The power measurement in balanced 3-ϕ circuit can be done by
(*a*) one wattmeter method only (*b*) two wattmeters method only
(*c*) three wattmeters method only (*d*) any one of the above. **[B.T.E. M.P., 1998]**

ANSWERS

1. (*d*)	**2.** (*c*)	**3.** (*b*)	**4.** (*a*)	**5.** (*a*)	**6.** (*d*)	**7.** (*c*)
8. (*b*)	**9.** (*c*)	**10.** (*b*)	**11.** (*a*)	**12.** (*a*)	**13.** (*b*)	**14.** (*a*)
15. (*c*)	**16.** (*b*)	**17.** (*d*)	**18.** (*b*)	**19.** (*d*)	**20.** (*d*)	**21.** (*a*)
22. (*c*)	**23.** (*b*)	**24.** (*b*)	**25.** (*d*)	**26.** (*d*)	**27.** (*c*)	**28.** (*c*)
29. (*b*)	**30.** (*c*)	**31.** (*c*)	**32.** (*b*)	**33.** (*d*).		

B. Say 'Yes' or 'No' :

1. Polyphase system means the system which consists of numerous windings or circuits.
2. Electrical displacement = $\dfrac{\text{180 electrical degrees}}{\text{number of phases}}$.
3. A load circuit is said to be balanced when the loads (impedances) connected in various phases are same in magnitude as well as in phase.
4. Rotating magnetic field can be set up by passing polyphase current through stationary coils.
5. Polyphase system is less capable and reliable than single phase system.
6. Parallel operation of polyphase alternators is simple as compared to that of single phase alternators because of pulsating reaction in single-phase alternators.
7. Double script notation is usually preferred in A.C. electrical circuits.
8. Phase sequence is the order or sequence in which the currents or voltages in different phases attain their maximum values one after the other.
9. If the phase sequence is reversed by interchanging any two terminals of the 3-phase supply, the motor would rotate in the same direction.
10. Star-connected system is obtained by joining together similar ends either the start or the finish, the other ends are joined to the line wires.
11. The common point at which similar (start or finish) ends, are connected is called the neutral or star point.
12. In a balanced star-connected system line voltage = phase voltage.
13. In a balanced delta-connected system line current is equal to phase current.
14. Delta-connected system is best suited for rotary converters.
15. Three wattmeters method is useful for measuring power in 3-phase, 3-wire load circuits.
16. Power in a 3-phase circuit = $\sqrt{3}\, E_L I_L \cos \phi$.
17. In the case of balanced three-phase circuit the reactive power can be determined by use of one wattmeter.
18. A load having a low lagging power factor gives a poor voltage regulation.
19. Static capacitors are installed to improve the power factor of a group of A.C. motors.
20. Static capacitors are practically loss free.

ANSWERS

1. Yes	**2.** No	**3.** Yes	**4.** Yes	**5.** No	**6.** Yes	**7.** Yes
8. Yes	**9.** No	**10.** Yes	**11.** Yes	**12.** No	**13.** No	**14.** Yes
15. No	**16.** Yes	**17.** Yes	**18.** Yes	**19.** Yes	**20.** Yes.	

THEORETICAL QUESTIONS

1. State the advantages of A.C. polyphase supply system over single-phase system.
2. Why is the number of phases in a polyphase system always three rather than any other number ?
3. Explain clearly what is meant by 'phase sequence' of 3-phase voltages.
4. What are the two systems in which three-phases can be connected ? What are the advantages and disadvantages of each system ?
5. What are the advantages of inter-connecting the 3-phases ?
6. Derive an expression for power in a 3-phase star-connected system in terms of (*i*) phase values and (*ii*) line values of voltages and currents.
7. Derive the numerical relationship between line and phase currents for a balanced 3-phase delta-connected load.
8. Deduce an expression for power in a 3-phase balanced load circuit. Show that it is the same irrespective of the load being connected in star or delta.
9. Compare the star and delta connections in a 3-phase system.
10. Enumerate various methods for 3-phase power measurement, and describe in detail two wattmeters method for 3-phase power measurement.
11. Prove that the power in a balanced 3-phase circuit can be deduced from the readings of two wattmeters. Draw the relevant connection diagram and vector diagram.
12. Discuss two-wattmeters method for power measurement in 3-phase system and obtain a relation for the power factor.
13. Derive an expression for star-connected impedances equivalent to three delta-connected impedances.
14. What does the following signify in case of measurement of balanced 3-phase power by two-wattmeters method :

    (*i*) When the reading of one wattmeter is zero.

    (*ii*) When the readings of two wattmeters are equal.

    (*iii*) When the readings of two wattmeters are equal and opposite.

    In each case give reasons to your answer.
15. Explain what will be the consequences of disconnecting the neutral in a 3-phase, 4-wire system.
16. What are the disadvantages of having a low lagging power factor in a system.
17. Enumerate the various equipments which can be employed to improve the power factor.

UNSOLVED EXAMPLES

Star/Delta Connections

1. Three equal impedances each having a resistance of 20 Ω and reactance of 15 Ω are connected in star to a 400 V, 3-phase, 50 Hz system. Calculate :

    (*i*) The line current, (*ii*) The power factor, and

    (*iii*) The power consumed. **[Ans.** 9.24 A ; 0.8 (lag) ; 5120 W]
2. Three equal impedances are star-connected to a 3-phase, 50 Hz supply. If the resistance and reactance of each branch are 25 Ω and 38 Ω respectively, calculate :

    (*i*) The line current, and (*ii*) The power consumed. **[Ans.** 5.28 A ; 2086 W]
3. Three resistances of 20 Ω each are connected in star across the 400 V, 3-phase A.C. supply. Calculate :

    (*i*) The line and phase currents, (*ii*) Phase voltages, and

    (*iii*) Total power taken. **[Ans.** (*i*) 11.55 A ; (*ii*) 231 V ; (*iii*) 8.0 kW]

4. A star-connected, 3-phase load consists of three similar impedances. When the load is connected to a 3-phase, 500 V, 50 Hz supply, the line current is 28.85 A and the power factor is 0.8 lagging, calculate :
(*i*) The total power taken by the load, and (*ii*) The resistance of each phase of the load.
[**Ans.** 20 kW ; 8 Ω]

5. Three non-inductive resistances each of 50 Ω are connected in star across 400 V, 3-phase A.C. supply. Calculate the current through each. Calculate the current if they were connected in delta across the same supply. [**Ans.** 4.62 A ; 8 A]

6. Three identical coils are connected in star to a 200 V (line voltage), 3-phase, A.C. supply and each coil takes 200 W. The power factor is 0.8 (lagging), calculate :
(*i*) The line current, (*ii*) Impedance, and
(*iii*) Resistance and inductance of each coil. [**Ans.** 2.165 A ; 53.334 Ω, 42.662 Ω, 0.102 H]

7. In a three-phase, 3-wire system with star-connected load the impedance of each phase is $(6 + j8)$ Ω. If the line voltage is 230 V, calculate :
(*i*) The line current, and (*ii*) The power absorbed by each phase.
[**Ans.** 13.3 A ; 1067 W]

8. A balanced star-connected load of $(8 + j6)$ Ω per phase is connected to 3-phase, 230 V supply. Find :
(*i*) Line current, (*ii*) Power factor,
(*iii*) Power, (*iv*) Reactive volt-amperes, and
(*v*) Total volt-amperes. [**Ans.** 13.28 A ; 0.8 (lag) ; 4.232 W ; 3174 ; 5290]

9. A star-connected, 5000 V, 3-phase alternator is supplying 3,000 kW at power factor of 0.8. Calculate the active and reactive components of the current in each phase. [**Ans.** 346.2 A ; 260 A]

10. In a 3-phase, 4-wire system, two phases have currents of 10 A and 6 A in lagging power factors of 0.8 and 0.6 respectively, while the third phase is open-circuited. Calculate the current in the neutral and sketch the vector diagram. [**Ans.** $7 \angle -73° 24'$]

11. In a star-connected load each phase consists of a resistance of 100 Ω in parallel with a capacitor of capacitance 31.8 μF. When it is connected to a 416 V, 3-phase, 50 Hz supply, calculate :
(*i*) The line current, (*ii*) The power factor,
(*iii*) The power absorbed, and (*iv*) The total kVA.
[**Ans.** 3.39 A ; 0.707 (leading) ; 1.728 kW ; 2.443 kVA]

12. Three identical coils connected in delta across 400 V, 50 Hz, 3- phase supply takes a line current of 17.32 A at power factor 0.5 lagging. Determine :
(*i*) The current in each phase, and
(*ii*) Resistance, inductance and impedance of each phase winding. [**Ans.** 10 A ; 20 Ω ; 0.11 H, 40 Ω]

13. A 220 V, 3-phase voltage is applied to a balanced delta-connected 3-phase load of phase impedance $(15 + j20)$ Ω. Find :
(*i*) The phasor current in each line. (*ii*) What is the power consumed per phase ?
(*iii*) What is the phasor sum of the three line currents ? Why does it have this value ?
[**Ans.** 15.24 A ; 1161.6 W ; zero]

14. Calculate (*i*) line current and (*ii*) the total power absorbed when three coils each having a resistance of 10 Ω and reactance of 7 Ω are connected (*a*) in star and (*b*) in delta across a 400 V, 3-phase supply.
[**Ans.** 18.93 A, 10750 W ; 56.7 A, 32250 W]

15. A delta-connected balanced 3-phase load is supplied from a 3-phase 400 V supply. The line current is 20 A and power taken by the load is 10000 W. Find :
(*i*) Impedance in each branch ; and
(*ii*) The line current, power factor and power consumed if the same load is connected in star.
[**Ans.** 34.6 Ω, 20/3 A, 0.7217, 3330 W]

16. A balanced 3-phase load consists of resistances of 4 Ω each. Determine the total power when the resistances connected to 400 V supply are :

(*i*) Star connected (*ii*) Delta connected. [**Ans.** 40 kW ; 120 kW]

17. Three non-inductive resistances, each of 100 Ω are connected in star across 400 V supply. Calculate the current through each. What would be the current through each if they are connected in delta across the same supply. [**Ans.** 2.31 A ; 4 A]

18. Three 100 Ω non-inductive resistances are connected in (*i*) star (*ii*) delta across a 400 V, 50 Hz, 3-phase mains. Calculate the power taken from the supply system in each case. In the event of one of the three resistances opened, what would be the value of the total power taken from the mains in each of the two cases. [**Ans.** 1600 W ; 4800 W]

19. A 3-phase delta connected load ; each phase of which has an inductive reactance of 40 Ω and a resistance of 25 Ω, is fed from the secondary of a 3-phase star-connected transformer, which has phase voltage of 240 V. Draw the circuit diagram of the system and calculate :

(*i*) The current in each phase of the load, (*ii*) The voltage across each phase of the load,
(*iii*) The current in the transformer secondary windings, and
(*iv*) The total power taken from the supply and its power factor. [**Ans.** 8.8 A ; 415.7 V ; 15.3 A ; 5820 W]

20. Three similar resistors are connected in star across 400 V, 3-phase supply. The line current is 5 A. Calculate the value of each resistor. To what value line voltage be changed to obtain the same current with the resistors connected in delta ? [**Ans.** 46.2 Ω ; 133.3 V]

21. A 3-phase, star-connected system with 230 V between each phase and neutral has a resistance of 4, 5 and 6 Ω respectively in three phases. Calculate :

(*i*) The current flowing in each phase, (*ii*) The neutral current, and
(*iii*) The total power absorbed. [**Ans.** 57.5 A, 46 A, 38.3 A ; 16.71 A ; 32610 W]

22. In a 3-phase, 4 wire system there is a balanced 3-phase motor load taking 20 kW at a power factor of 0.8 lagging while lamps connected between phase conductors and the neutral are taking 5, 4 and 10 kW respectively. Voltage between line conductors is 430 V. Calculate the current in each conductor and in the neutral wire of the feeder supplying these loads. [**Ans.** 51.2 A ; 47.5 A ; 80 A ; 22.4 A]

Power Measurement in 3-phase Circuits

23. The power input to a 3-phase induction motor is read by two wattmeters. The readings are 860 W and 240 W. What is the power factor of the motor ? [**Ans.** 0.7155 (lag)]

24. While performing a load test on a 3-phase wound rotor induction motor by two wattmeters method, the readings obtained on two wattmeters were + 12.5 kW and – 4.8 kW and the line voltage was 440 V. Calculate :

(*i*) Power drawn by the motor, (*ii*) Power factor, and
(*iii*) Line current. [**Ans.** 7.7 kW ; 0.2334 ; 43.3 A]

25. The input power to a 3-phase delta-connected motor was measured by two wattmeters method. The readings were 20.8 kW and – 6.8 kW and the line voltage was 400 V. Calculate :

(*i*) Input power, (*ii*) Power factor, and
(*iii*) Line current. [**Ans.** 14 kW ; 0.281 ; 72 A]

26. A 3-phase, 500 V motor load has a power factor of 0.4. Two wattmeters connected to measure the power show the input to be 30 kW. Find the reading on each instrument. [**Ans.** 35 kW ; – 5 kW]

27. The power in a 3-phase circuit is measured by two wattmeters. If the total power is 100 kW and power factor is 0.66 leading, what will be the reading of each wattmeter ? For what power factor will one of the wattmeters read zero ? [**Ans.** 17.15 kW ; 82.85 kW ; p.f. = 0.5]

28. Two wattmeters used to measure the power input in a 3-phase circuit indicate 1000 W and 500 W respectively. Find the power factor of the circuit

(*i*) when both wattmeter readings are positive ; and
(*ii*) when the latter is obtained by reversing the current coil connections.

[**Ans.** 0.866 (lag) ; 0.1889 (lag)]

29. Two wattmeters connected to read the total power in a 3-phase system supplying a balanced load read 10.5 kW and – 2.5 kW respectively. Calculate the total power and the power factor. Draw suitable phasor diagram, explain the significance of
(*i*) equal wattmeter readings ; and (*ii*) a zero reading on one wattmeter.
[Ans. 8 kW ; 0.3348 (lag) (*i*) unity p.f., (*ii*) 0.5 p.f.]

30. The power input to a 1000 V, 3-phase induction motor running on full-load is measured by two wattmeters which indicate 300 kW and 100 kW respectively. Determine :
(*i*) Input, (*ii*) Line current,
(*iii*) Power factor, and (*iv*) Output if the efficiency of the motor is 92%.
[Ans. 400 kW ; 303.5 A ; 0.756 (lag) ; 368 kW]

31. Two wattmeters are used for measuring the power input and the power factor of an over-excited synchronous motor. If the readings of the meters are – 2.0 kW and + 7.0 kW respectively, calculate the input and power factor of the motor. [Ans. 5 kW ; 0.3057 (leading)]

32. The power input to a 2-kV, 50 Hz, 3-phase motor running on full-load at an efficiency of 90 per cent is measured by two wattmeters which indicate 300 kW and 100 kW respectively. Calculate :
(*i*) Input, (*ii*) Power factor,
(*iii*) Line current, and (*iv*) Output. [Ans. 400 kW ; 0.756 (lag) ; 153 A ; 360 kW]

33. In a balanced three-phase system power is measured by two-wattmeters method and the ratio of the two wattmeter readings is 2 : 1. Find the power factor of the system. [Ans. 0.866 (log)]

34. The power input of a synchronous motor is measured by two wattmeters both of which indicate 50 kW. If the power factor of the motor be changed to 0.866 leading, determine the readings of the two wattmeters, the total input power remaining the same. Draw the vector diagram for the second condition of load. [Ans. 33.33 kW ; 66.67 kW]

35. Each phase of a 3-phase delta-connected load consists of an impedance $Z = 20 \angle 60° \ \Omega$. The line voltage is 440 V at 50 Hz. Calculate the power consumed by each phase impedance and the total power. What will be the readings of the two wattmeters connected ? [Ans. 4840 W ; 14520 W ; 14520 W. zero]

36. A 3-phase 3-wire, 415 V supplies a balanced load of 20 A at power factor 0.8 lagging. Two wattmeters are used to measure power. Calculate :
(*i*) Power, (*ii*) Reading of wattmeter No 1, and
(*iii*) Reading of wattmeter No. 2. [Ans. 11.5 kW ; 8240 W ; 3260 W]

37. A balanced load is supplied from a 3-phase, 400 V, 3-wire system whose power is measured by two wattmeters. If the total power supplied is 26 kW at 0.75 power factor lagging, find the readings of each of the two wattmeters. [Ans. 19.62 kW ; 6.38 kW]

38. A 3-phase 4-wire, star-connected system supplies only non-inductive load. The current in line *R* is 8 A, the current in line *Y* is 10 A and the current in line *B* is 6 A. The voltage from each line to neutral is 120 V. Find :
(*i*) Wattage shown by each of the three wattmeters, and
(*ii*) Power taken by the lighting load. [Ans. 960 W, 1200 W, 720 W ; 2880 W]

39. Three identical coils, each having a resistance of 20 Ω and reactance of 20 Ω are connected in (*i*) star ; and (*ii*) delta across 440 V, 3-phase supply. Calculate for each method of connection the line current and reading on each of the two wattmeters connected to measure power.
[Ans. 8.98 A ; 3817.5 W, 1022.5 W ; 26.95 A ; 12452.5 W ; 3067.5 W]

40. A balanced star-connected load, each having a resistance of 10 Ω and inductive reactance of 30 Ω is connected to a 400 V, 50 Hz supply. The phase sequence is RYB. Two wattmeters connected to read total power have their current coils connected in the red and blue lines respectively. Calculate the reading of each wattmeter. Draw the circuit and vector diagrams. [Ans. W_1 = – 585 W ; W_2 = 2185 W]

41. Three equal star-connected inductors take 8 kW at power factor 0.8 when connected to a 460 V, 3ϕ, 3-wire supply. Give the connection diagram of 2 single-phase wattmeters to measure the power in the circuit. Allocate the reading of each wattmeter. Find the line currents if one of the inductor is short-circuited. Draw the vector diagrams of the currents and voltages under the condition.

[**Ans.** 5732 W, 2268 W ; 21.74 A, 21.74 A]

42. A 3-phase, 3-wire balanced load with a lagging power factor is supplied at 400 V (between lines). A single-phase wattmeter (scaled in kW) when connected with its current coil in the *R*-line and the voltage coil between *R* and *Y* lines gives a reading of 6 kW. When the same terminals of the same voltage coil are switched over to *Y*-line and *B*-line, the current-coil connections remaining the same, the reading of the wattmeter remains unchanged. Calculate the line current and power factor of the load. Phase sequence is $R \rightarrow Y \rightarrow B$. [**Ans.** 30 A ; 0.866 (lag)]

43. A wattmeter reads 5.54 kW when its current coil is connected in red phase and its potential coil is connected between the neutral and red phase of a symmetrical 3-phase system supplying a balanced load of 30 A at 400 V. What will be the reading of the instrument if the current coil connections remains unchanged but the potential coil be connected between blue and yellow phase ? What does this figure represent ? [**Ans.** 7 kVAR]

44. A 3-phase, 434 V, 50 Hz supply is connected to a 3-phase star-connected induction motor and synchronous motor. Impedance of each phase of induction motor is $1.25 + j2.17\ \Omega$. The 3-phase synchronous motor is over-excited and it draws a current of 120 A at 0.87 leading power factor. Two wattmeters are connected in usual manner to measure power drawn by the two motors. Calculate :

(*i*) Reading on each wattmeter, and (*ii*) Combined power factor.

[**Ans.** 63970 W, 52050 W ; 0.9845 (lag)]

45. A star-connected balanced load is supplied from a 3-phase balanced supply with a line voltage of 416 V at a frequency of 50 Hz. Each phase of load consists of a resistance and a capacitor joined in series and the readings on two wattmeters connected to measure the total power supplied are 782 W and 1980 W, both positive. Calculate :

(*i*) The power factor of the circuit, (*ii*) The line current, and

(*iii*) The capacitance of each capacitor. [**Ans.** 0.8 (leading) ; 4.8 A ; 106 μF]

Unbalanced Load Circuits

46. A 400 V, 50 Hz, 3-phase supply has delta-connected load having 100 Ω between *R* and *Y*, 318 mH between *Y* and *B* and 31.8 μF between *B* and *R*. Find the line currents for the phase sequence *RYB*. Calculate the value of star-connected balanced resistor for the same power.

[**Ans.** 7.73 ∠15° A ; 7.73 ∠165° A ; 4 ∠270° A ; 100 Ω]

47. Three loads $(31 + j59)$, $(30 - j40)$ and $(80 + j60)\ \Omega$ are connected in delta across a 400 V, 3-phase supply. Find the phase currents, line currents and power.

[**Ans.** 3 ∠– 62.3° A ; 4 ∠– 66.9° A ; ∠83.1° A ; 4.78 ∠– 76° A ; 1.038 ∠– 80.35° A ; 5.819 ∠– 256.8° ; 1078.9 W]

48. A 3-phase, 400 V, 50 Hz system has the following load connected in delta : between *R* and *Y* lines a non-reactive resistor of 100 Ω, between *Y* and *B* lines a coil having a reactance of 66 Ω and a resistance of 20 Ω, between *B* and *R* lines a loss free capacitor of 30 μF. Calculate :

(*i*) The phase currents, and (*ii*) The line currents.

Assume the phase sequence to be *RYB*.

[**Ans.** 4 ∠0° A ; 6.324 ∠– 191.56° A ; 3.774 ∠210° A ; 7.5 ∠14.6° A ; 10.275 ∠173° A ; 4.3 ∠– 47.13° A]

49. A balanced 3-phase system supplies an unbalanced 3-phase *delta-connected* load made up of two resistors 100 Ω and 200 Ω and a reactor having an inductance of 0.3 H with negligible resistance. $E_L = 100$ V at 50 Hz. Calculate :

(*i*) The total power in the system, and (*ii*) The total volt-amperes reactive.

[**Ans.** 150 W ; 106 (lag)]

50. A 400 volt 3-phase, 4-wire system has the loads $12 - j16$, $12 + j0$ and $8 + j6\ \Omega$ connected in star. Find the line currents, neutral current and total power.

[**Ans.** 11.55 ∠53.1° A ; 19.25 ∠– 120° A ; 23.09 ∠83.1° A ; I_N = 15.47 ∠– 90.3° A ; 10310.11 W]

51. A 3-phase 3-wire system having a 254 V line-to-neutral has the following loads connected between the respective lines and neutral : $Z_R = 10\ \angle 0°\ \Omega$, $Z_Y = 10\ \angle 37°\ \Omega$, $Z_B = 10\ \angle -53°\ \Omega$. Calculate the current in the neutral wire and power taken by each load when phase sequence is

(*i*) *RYB* (*ii*) *RBY*.

[**Ans.** (*i*) 24.46 ∠16° 15′, 6452 W ; 5162 W ; 3871 (*ii*) 38.5 ∠– 177.3°; Power same as in (*i*)]

52. A 3-phase star connected system with 230 V between each phase and neutral has resistances of 4 Ω, 5 Ω and 6 Ω respectively in the three phases. Calculate :

(*i*) Current flowing in each phase, (*ii*) Neutral current, and

(*iii*) Total power absorbed. [**Ans.** 57.5 A, 46 A, 38.33 A ; 16.71 A ; 32620 W]

53. A 440/254 V, 3-phase, 4-core supplies an unbalanced load represented by the following impedances in ohms between *R*, *Y* and *B* phases respectively and the neutral $16 + j12$, $14 - j21$ and 25 Ω. The phase sequence is *RYB*. Calculate :

(*i*) The current in each conductor of the cable, and

(*ii*) Readings of three wattmeters connected in each line and neutral.

[**Ans.** 12.7 A, 10.06 A, 10.16 A ; 2580 W, 1413 W ; 2580 W]

54. Calculate the voltage across each of the following star-connected impedances when they are connected to a 400 V, 3-phase system, $Z_R = j20\ \Omega$, $Z_Y = j1\ \Omega$, $Z_B = -j1\ \Omega$.

Also find the potential of the star point above ground.

The phase sequence is *RYB*.

[**Ans.** $(4000 + j6928)$ V ; $(3600 + j6928)$ V ; $(3800 + j7274.4)$ V ; – 8000 V]

55. An unbalanced star-connected load comprising of the impedances $Z_R = (10 + j0)$, $Z_Y = (3 + j4)$, $Z_B = (0 - j10)$ is connected across a 3-phase 200 V circuit with balanced voltages. Find :

(*i*) Three line currents, and (*ii*) Voltage across each impedance.

The phase sequence is *RYB*.

[**Ans.** $(-20.2 + j4.54)$; $(7.24 + j5.48)$; $(12.78 - j10.02)$; $(-200.2 + j45.4)$; $(-0.2 + j45.4)$; $(-100.2 - j127.8)$]

56. An unbalanced star-connected load is supplied from a symmetrical 3-phase, 440 V, 3-wire system. The branch impedances of the load are $Z_1 = 5\ \angle 30°\ \Omega$, $Z_2 = 10\ \angle 45°\ \Omega$ and $Z_3 = 10\ \angle 60°\ \Omega$. Assuming the positive phase sequence of *RYB* determine the line currents and voltage across each impedance. Draw the phasor diagram.

[**Ans.** 35.75 ∠– 71.3° A ; 32.77 ∠156.1° A ; 27.67 ∠48° A ; 178.7 ∠– 41.3° V, 327.7 ∠201.1° V, 276.7 ∠108° V]

57. An unbalanced star-connected load has branch impedances of $Z_1 = 10\ \angle 30°\Omega$, $Z_2 = 30\ \angle -45°\ \Omega$, $Z_3 = 20\ \angle 60°\ \Omega$ and is connected across a balanced 3-phase, 3-wire supply of 200 V. Find the line currents and the voltage across each impedance using star-delta conversion method.

[**Ans.** 5.39 ∠8° 32′ A, 15.79 ∠– 148° 6′ A, 10.73 ∠44° 16′ A, 53.9 ∠38° 32′ V, 157.9 ∠– 193.6°, 214.6 ∠104° 16′]

58. A 300 V (line) 3-phase supply feeds a star-connected load consisting of non-inductive resistors of 15, 6 and 10 Ω connected to the *R*, *Y* and *B* lines respectively. The phase sequence is *RYB*. Calculate the voltage across each resistor. [**Ans.** $(195 - j78)$ V, $(-105 - j78)$ V, $(4.5 + j182)$ V]

59. A 400 V, 3-phase, 3-wire symmetrical system *RYB* supplies a star-connected load whose branch circuit impedances are : $Z_R = 10\sqrt{3} + j10$; $Z_Y = 20 + j20\sqrt{3}$; $Z_B = 0 - j10$.

Determine the current in each branch.

Take the phase sequences as $R \rightarrow Y \rightarrow B$. [**Ans.** 25.55 ∠– 57° 45′ ; 6.48 ∠– 133.7° ; 29 ∠109° 45′]

Power Factor Improvement

60. A 50-MVA, 11 kV 3-phase alternator supplies full-load at a lagging power factor of 0.7. What would be percentage increase in earning capacity if the power factor is increased to 0.95. [**Ans.** 35.7 per cent]

61. A 3-phase, 37.3 kW, 440 V, 50 Hz induction motor operates on full-load with an efficiency of 89% and a power factor of 0.85 lagging. Calculate the total kVA rating of capacitors required to raise the full-load power factor to 0.95 lagging. What will be the capacitance per phase if the capacitors are :

(*i*) Delta-connected, and (*ii*) Star-connected.

[**Ans.** 12.19 kVA ; 66.8 μF ; 200.4 μF]

62. Three impedance coils, each having a resistance of 20 Ω and a reactance of 15 Ω are connected in star to a 400 V, 3-phase, 50 Hz supply. Calculate :

(*i*) The line current, (*ii*) Power supplied, and (*iii*) The power factor.

If three capacitors, each of the same capacitance are connected in delta to the same supply as to form parallel circuit with the above impedance coils, calculate the capacitance of each capacitor to obtain a resultant power factor of 0.95 lagging. [**Ans.** 9.24 A, 5120 W, 0.8 (lag) ; 14.32 μF]

63. A balanced 3-phase load, each element of which consist of a 5 Ω resistance in series with an inductance of reactance 3 Ω, is connected in star across mains whose line potential difference is 400 V and frequency 50 Hz. What is the power consumed by the system ?

Three equal capacitors are connected across the mains in the mesh. What should their capacitance be to cause the total line current to be in phase with the line voltage ? [**Ans.** 23.55 kW ; 93.8 μF/phase]

4

Electromagnetism

1. MAGNETIC FIELD

- Substances that are magnetic have a north-seeking pole and south-seeking pole, corresponding to the earth's magnetic poles. In effect, *each magnet has a magnetic field around* it just as the earth does.
- The magnetic field is *strongest at the ends of the magnet*. In the *centre* of the magnet the *strength is negligible*. A magnetic field can be seen by placing a magnet under a piece of paper and then sprinkling iron filings onto the paper. The iron filings line up to show the magnetic field and its concentration, as shown in Fig. 1 (*a*).

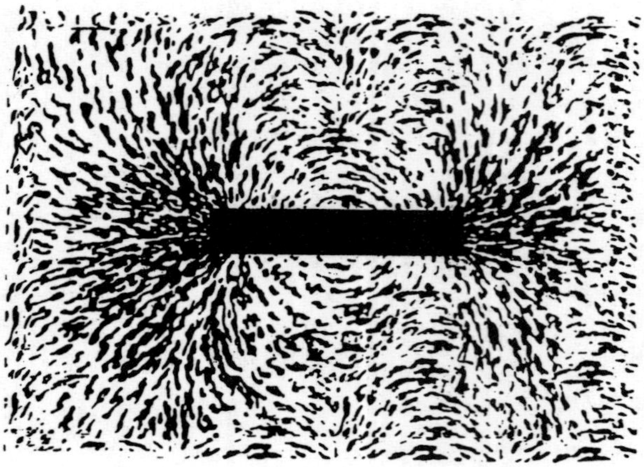

(*a*) Illustrating a magnetic field by the use of iron filings.

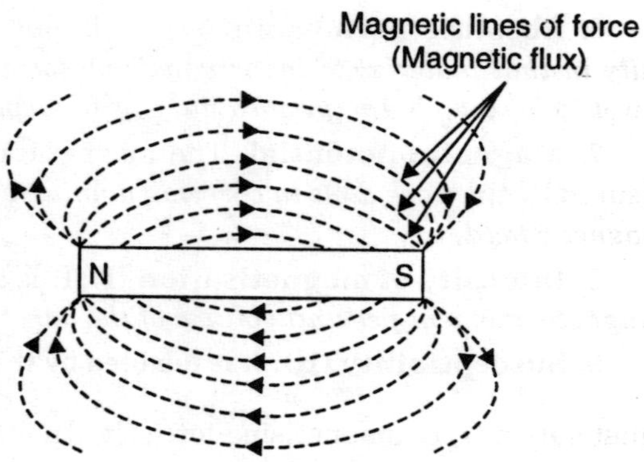

(*b*)

Fig. 1. Magnetic field around a magnet.

- The unlike poles of two magnets attract each other while the like poles repel each other (Fig. 2).

The following points are worth noting about magnets and magnetic lines of force.

1. Each magnet has two poles called north pole (*N*) and south pole (*S*).
2. Unlike poles attract, like poles repel.
3. A force field exists, around a magnet which is concentrated at the poles of the magnet. The strength of the magnetic field is indicated by the density or concentration of lines of force.
4. Lines of force repel each other.
5. Lines of force cannot intersect.
6. Lines of force form closed loops.

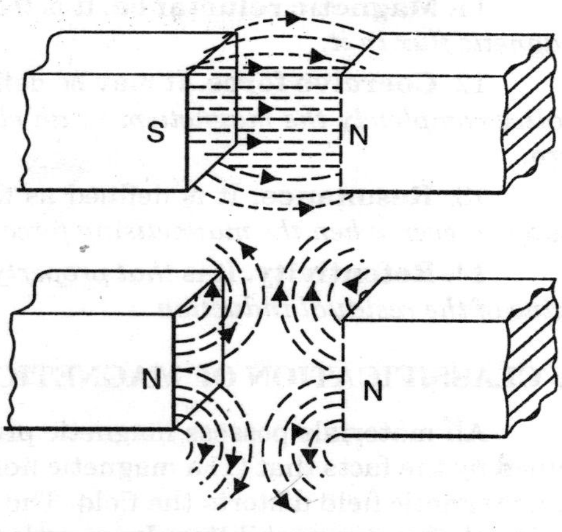

Fig. 2. Unlike poles attract each other ; like poles repel each other.

2. TERMS CONNECTED WITH MAGNETIC MATERIALS

1. **Magnetic force.** It is the *force exerted by one magnet on another either to attract it or repel it.*

2. **Unit pole strength.** It is defined as the *strength of that pole which when placed in vacuum at a distance of one metre from a similar and equal pole, repels it with a force of one newton.*

3. **Magnetic flux density (B).** It is usually represented by letter *B* and is defined as *the flux or lines of force passing per unit area through any substance through a plane at right angles to the direction* of magnetic flux ; it is measured in Wb/m^2 (or *T i.e.*, Tesla).

4. **Magnetic field strength.** It may be defined in the following two ways :

(*i*) Field strength at any point within a magnetic field is *the number of lines of force passing through a unit area round the point considered and held perpendicular to the lines.*

(*ii*) Field strength at any point within a magnetic field *is the force exerted by a unit North pole at that point.*

5. **Relative permeability (μ_r).** It is *the ratio of flux density (B) produced in that material to the flux density produced in vacuum by the same magnetising force (H). It is denoted by* μ_r.

6. **Absolute permeability (μ).** Absolute permeability of the material is the *ratio of flux density in that material to the magnetising force producing that flux density and is denoted by* μ ; $\mu = \mu_0\mu_r$ *where* μ_0 *is the permeability of free space having a value of* $4\pi \times 10^{-7}$ *H/m.*

7. **Magnetic potential.** The magnetic potential at any point within a magnetic field is measured by the *work done in carrying a unit North pole from infinity to that point against the force of magnetic field.*

8. **Intensity of magnetisation (I).** It is defined as *the pole strength per unit area of the bar or magnetic moment per unit volume of the bar.* It is denoted by letter I.

9. **Susceptibility (K).** It is denoted by the letter *K*, and is defined as the ratio of intensity of magnetisation (*I*) to magnetising force (*H*). In other words $K = \dfrac{I}{H}$.

10. **Magnetomotive force.** It is *that force which drives or tends to drive the flux through a magnetic circuit.* In short it is written as m.m.f. It is the product of number of turns (*N*) and current (*I*) in amperes in these turns *i.e.*, m.m.f. = *NI*.

11. **Magnetic reluctance.** It is *that property of the material which opposes the production of magnetic flux in it.*

12. **Coercive force.** It may be defined as the *demagnetising force which is necessary to neutralise completely the magnetism in an electromagnet after the value of magnetising force becomes zero.*

13. **Resonance.** It is defined as the *magnetic flux density which still persists in magnetic material even when the magnetising force is completely removed.* It is expressed in Wb/m^2.

14. **Retentivity.** It is *that property of magnetic material which is measured by its maximum value of the residual induction.*

3. CLASSIFICATION OF MAGNETIC MATERIALS

All materials possess magnetic properties to a greater or lesser degree and these are determined by the facts that (1) a magnetic field exert forces and torques on the bodies, (2) a body placed in a magnetic field distorts the field. The magnetic properties of the materials are characterised by their relative permeabilities. In accordance with the value of *relative permeability* the materials may be classified in the following three ways :

1. **Ferromagnetic materials.** The relative permeabilities of these materials are *much greater than unity* and are dependent on the field strengths. They attract the lines of force strongly [Fig. 3 (*a*)]. The principal ferromagnetic elements are *iron, cobalt, nickel*. Gadolinium however also comes under this classification. These have high susceptibility.

2. **Paramagnetic materials.** These have relative permeability *slightly greater than unity* and are magnetised slightly. They attract the lines of forces weakly [Fig. 3 (*b*)]. *Aluminium, platinum* and *oxygen* belong to this category.

3. **Diamagnetic materials.** The relative permeability of these materials is *slightly less than unity*. They repel the lines of force slightly [Fig. 3 (*c*)]. The examples are *Bismuth, silver, copper* and *hydrogen.*

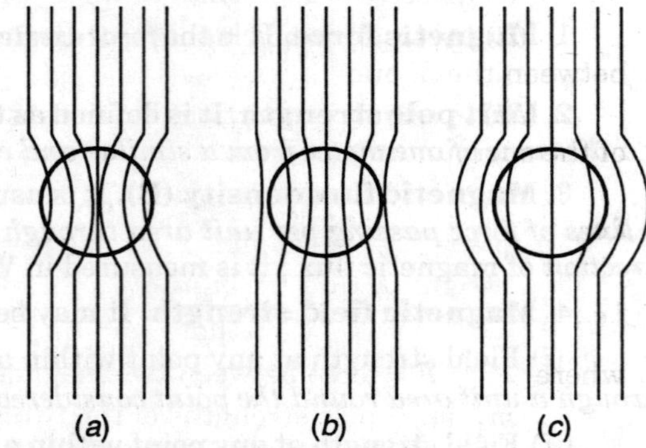

Fig. 3

4. MAGNETICALLY SOFT MATERIALS

The magnetically soft materials (suitable for making electromagnets) are characterised as follows :

1. They have high permeability.
2. The magnetic energy stored is not high.
3. They have negligible coercive force (due to which these are **not** suitable for making permanent magnets).
4. They have low resonance.

The different magnetically soft materials are enumerated below :

(*i*) Pure or ingot iron (*ii*) Cast iron
(*iii*) Carbon steels (*iv*) Silicon steels
(*v*) Manganese and nickel steels (*vi*) Permalloy
(*vii*) Mumetal (*viii*) Peminvar.

5. MAGNETICALLY HARD MATERIALS

Magnetically hard materials *are used for making* **permanent magnets** and have the following characteristics :

1. They possess high value of *BH* product.
2. High retentivity.
3. High coercivity.
4. Strong magnetic reluctance.
5. Hysteresis loop is more rectangular in shape.

Important magnetically hard materials are enumerated below :

(*i*) Tungsten steel (*ii*) Cobalt steel
(*iii*) Chromium steel (*iv*) Alnico
(*v*) Curife (*vi*) Hypernic.

6. LAWS OF MAGNETIC FORCE

Coulomb, through his experiments, found that the force between two magnetic poles placed in a medium is

(*i*) directly proportional to their pole strengths,

(*ii*) inversely proportional to the square of the distance between them, and

(*iii*) inversely proportional to the absolute permeability of the surrounding medium

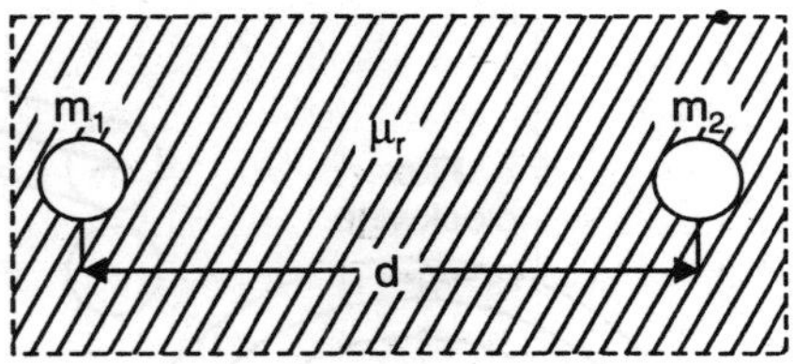

Fig. 4

i.e., $$F \propto \frac{m_1 m_2}{\mu d^2}$$

i.e., $$F = k\,\frac{m_1 m_2}{\mu d^2}$$

where F = force between two magnetic poles,
m_1, m_2 = pole strengths of the two poles respectively,
d = distance between the poles,
μ = absolute permeability, and
k = constant.

In the S.I. system, the value of the constant, $k = \frac{1}{4\pi}$

$$\therefore \quad F = \frac{m_1 m_2}{4\pi\mu d^2}\ \text{N} \qquad ...(1)$$

or $$F = \frac{m_1 m_2}{4\pi\mu_0\mu_r d^2} \quad \text{...... in a medium} \qquad ...[1\,(a)]$$

$$= \frac{m_1 m_2}{4\pi\mu_0 d^2} \quad \text{...... in air} \qquad ...[1\,(b)]$$

7. MAGNETIC FIELD DUE TO A CURRENT CARRYING CONDUCTOR

- When an electric current flows through a wire, a magnetic field is built up around the wire itself. This can be seen using a cardboard, iron filings, and a current-carrying wire. When the wire passed through the cardboard and the current flowing, iron filings are sprinkled onto the card board. They can be seen arranging themselves in a magnetic field (Fig. 5).

The magnetic lines of force are referred to as flux. Just as in a natural magnet, the field is strongest near the wire and diminishes as the distance from the wire increases.

Flux around a wire does have direction. Flux direction is determined by the direction of electron flow within the wire. As shown in Fig. 6, the *North pole of the compass needle indicates the direction of flux or magnetic field around the wire.* The **dot** *in the centre of the wire on the left indicated the point of the current-direction arrow coming* **toward the observer** ; the × at the right represents the tail of the current arrow **pointing away from the observer.** If the direction of electron flow within the wire is reversed, the compass needles will reverse themselves, indicating a change in flux direction.

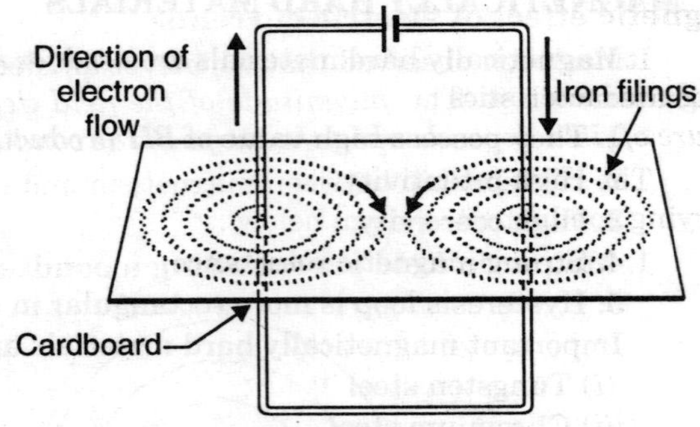

Fig. 5. Magnetic field around a wire that is carrying electric current.

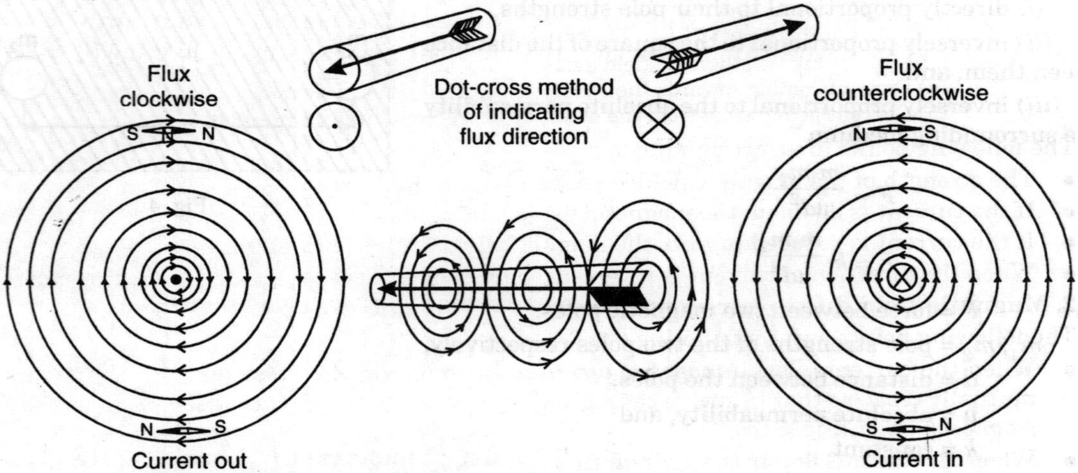

Fig. 6. Compasses indicate the direction of flux around a wire.

Right hand rule (or right hand screw rule)

The direction of the magnetic field can be found by using *right hand rule or the right hand screw rule*. The right hand rule states as follows :

"Grasp the wire in the right hand, with the thumb pointing in the direction of the current. The fingers will curl around the wire in the direction of the magnetic field".

Fig. 7, illustrates the rule.

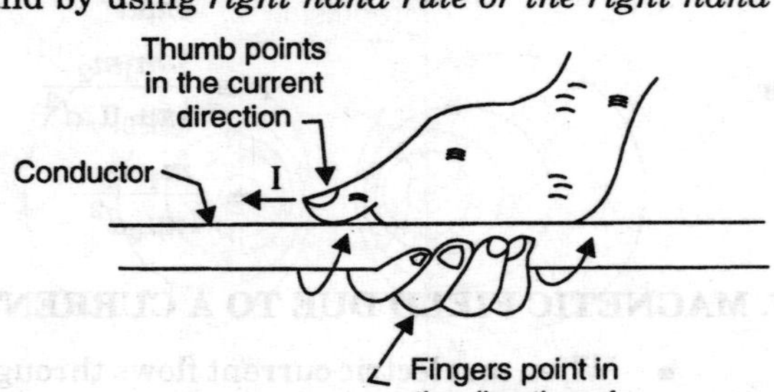

Fig. 7. Right hand rule (or right hand screw rule).

The right hand screw rule can be explained as follows :

As a wood screw is turned clockwise it moves (or progresses) into the wood. The *horizontal direction of the screw is analogous to the direction of current in a conductor. The circular motion of the screw shows the direction of the magnetic flux around the conductor.*

Magnetic effect of electric currents

It has been observed that whenever an electric current flows, a magnetic field is invariably associated with it. The *magnitude of the field depends on the magnitude of the current, and the nature of field depends upon the shape of the conductor*.

The relationship between magnetism and the electric current for various shapes of current carrying conductors is given below :

1. **Magnetic field surrounding a conductor :**

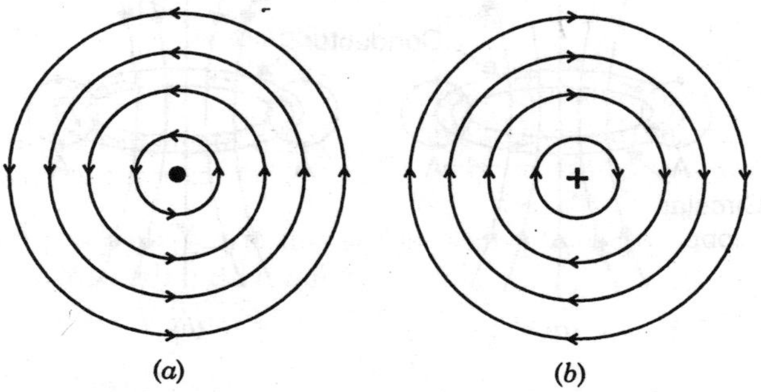

(*a*) (*b*)

Fig. 8. Magnetic field surrounding a conductor.
(*a*) Current coming towards the observer (or coming out of the page)
(*b*) Current going away from the observer (or current entering the page).

The following points are worth noting :

- **The strength of the magnetic field is proportional to the current.**
- **If the current is doubled, the magnetic field is doubled.**
- **If the current is reduced to zero, the magnetic field is also reduced to zero.**
- **When the direction of current is reversed, the magnetic field also reverses its direction.**

2. **Magnetic field due to currents in two parallel conductors :**

The following points are worth noting :

- **When the currents flow in *opposite direction* in the parallel conductors, the conductors mutually *repel*, so that in such a case, the current carrying conductors tend to *move away from each other* [Fig. 9 (*a*)].**
- **When the currents flow in the *same direction* in parallel conductors, the conductors mutually *attract*, so that in such a case, the current carrying conductors tend to move *towards each other* [Fig. 9 (*b*)].**

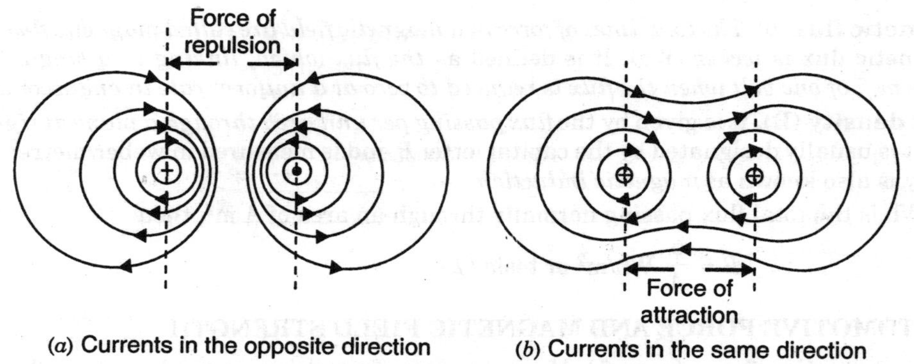

(*a*) Currents in the opposite direction (*b*) Currents in the same direction

Fig. 9. Force between parallel current carrying conductors.

3. **Magnetic field due to a circular loop :**

When a single conductor is looped to form a single turn of coil, the magnetic fields are set up as small loops around the coil shown as *AA* in Fig. 10 (*a*). The magnetic lines enter the inside of the coil on one face and leave it on the other. The *direction of the field in the centre of coil will be perpendicular to the plane of the loop* as shown in Fig. 10.

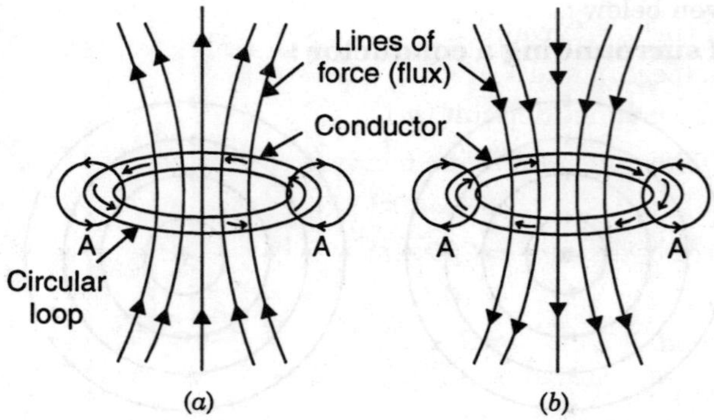

Fig. 10. Magnetic field due to circular loop.

4. **Magnetic field due to solenoid :**

A *solenoid is a coil of wire wound on a long straight core*. When a current passes through a solenoid a magnetic field is produced. The lines are approximately parallel to the axis of the solenoid until they reach the ends where they begin to diverge (magnetic field pattern long similar to bar magnet).

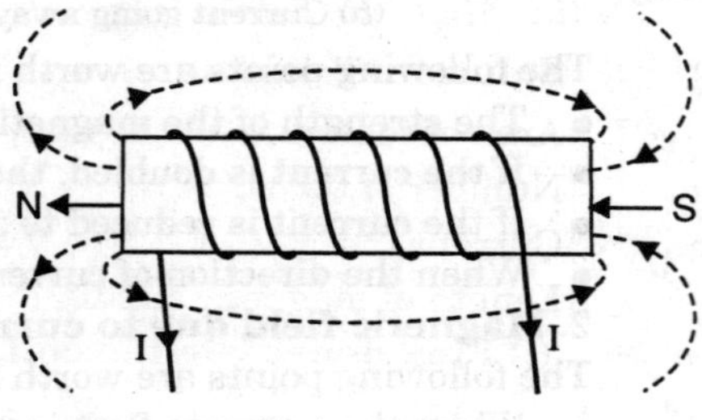

Fig. 11. Solenoid.

The solenoid produces strong magnetic fields for such applications as *relay coils, transformer coils and circuit breakers*.

The right hand rule for determining the direction of flux from a solenoid states as follows :

"Hold the solenoid in right hand such that the fingers point in the direction of current flow in the coils. The thumb will point towards N pole of the field".

8. MAGNETIC FLUX AND FLUX DENSITY

Magnetic flux (ϕ). *The total lines of force in a magnetic field are called magnetic flux*. The S.I. unit of magnetic flux is *weber (Wb)*. It is defined as *the flux which, linking in a single turn coil, induces an e.m.f. of one volt when the flux is reduced to zero at a uniform rate in one second.*

Flux density (B). It is given by the *flux passing per unit area through a plane at right angles to the flux*. It is usually designated by the capital letter *B* and is measured in weber/metre^2 (Wb/m^2). Flux density is also known as *magnetic induction*.

If ϕ Wb is the total flux passing normally through an area of A m^2, then

$$B = \frac{\phi}{A} \text{ Wb/m}^2 \text{ or tesla } (T) \qquad ...(2)$$

9. MAGNETOMOTIVE FORCE AND MAGNETIC FIELD STRENGTH

Magnetomotive force (m.m.f.). Magnetomotive force drives or tends to drive flux through a magnetic circuit and corresponds to electromotive force (e.m.f.) in an electric circuit. The m.m.f. of a coil is the *product of current in the coil and the number of turns of the coil* and has the unit of ampere turns (*AT*).

Magnetic field strength (H). The magnetic field strength (*H*) is defined as the *m.m.f. per metre length of magnetic circuit i.e.,*

$$H = \frac{NI}{l} \text{ AT/m}$$

where, N = number of turns of a coil,
I = current (amperes), and
l = length of the magnetic circuit (m).

Example 1. *The following data relate to an electromagnet :*

Total flux $= 8 \times 10^{-4}$ *Wb*
Cross-sectional area of the core $= 200$ *mm*^{2}
Number of turns $= 100$
Magnitude of current $= 2$ *A*
Length of the magnetic circuit $= 400$ *mm*

Calculate the following :
(*i*) *Flux density in the coil,*
(*ii*) *Magnetomotive force, and*
(*iii*) *Magnetic field strength.*

Solution. Total flux, $\phi = 8 \times 10^{-4}$ Wb
Area of the coil, $A = 200 \text{ mm}^2 = 2 \times 10^{-4} \text{ m}^2$
Number of turns, $N = 100$
Current, $I = 2$ A
Length of magnetic circuit, $l = 400$ mm $= 0.4$ m.

(*i*) **Flux density in the coil, B :**

$$B = \frac{\phi}{A} = \frac{8 \times 10^{-4}}{2 \times 10^{-4}} = \textbf{4 Wb/m}^2 \textbf{ or T. (Ans.)}$$

(*ii*) **Magnetomotive force, m.m.f. :**

$$\text{m.m.f.} = NI = 100 \times 2 = \textbf{200 AT. (Ans.)}$$

(*iii*) **Magnetic field strength, H :**

$$H = \frac{NI}{l} = \frac{200}{0.4} = \textbf{500 AT/m. (Ans.)}$$

10. FORCE ON A CURRENT-CARRYING CONDUCTOR LYING IN A MAGNETIC FIELD

It has been found that *whenever a current-carrying conductor is placed in a magnetic field, it experiences a force which acts in a direction perpendicular both to the direction of the current and the field.*

— Fig. 12 (*a*) shows the field set up by the poles.
— Fig. 12 (*b*) shows the conductor field due to flow of current in the conductor.
— Fig. 12 (*c*) shows the resultant field produced when the current-carrying conductor of [Fig. 12 (*b*) (*i*)] is inserted in the pole-field, *with the axis of the conductor at right angles to the direction of the flux.*

On the *upper side* of the conductor in Fig. 12 (*c*) the magnetizing forces of the field and of the current in the conductor are *additive*, while on the *lower side* these are subtractive. This explains why the resultant field is strengthened above and weakened below the conductor. Thus the conductor has a force on it which tends to move it *downwards*.

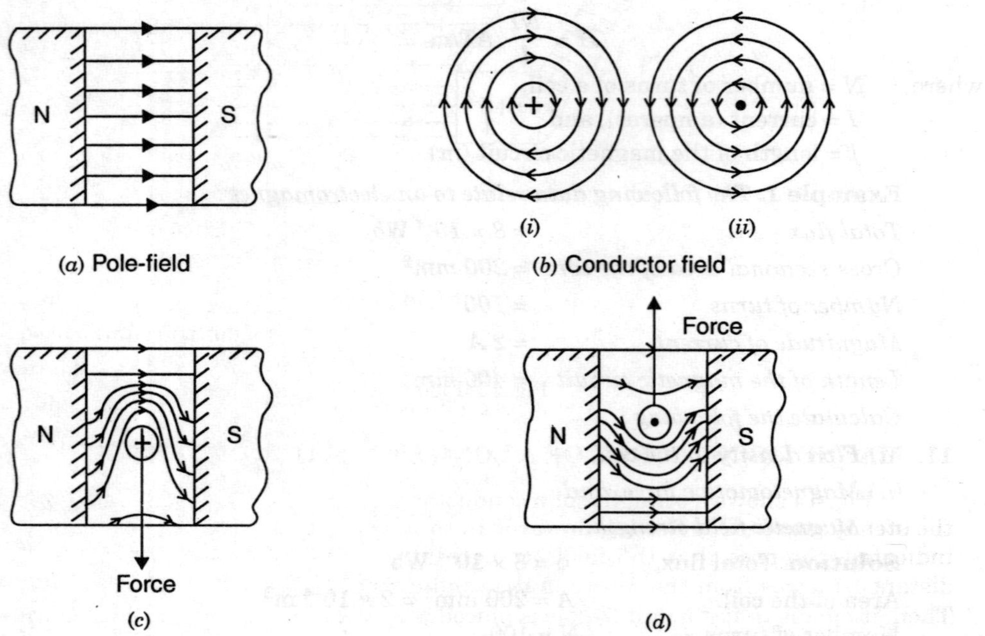

Fig. 12. Force on a current-carrying conductor lying in a magnetic field.

— Fig. 12 (*d*) shows that when the current in the conductor is reversed, the direction of the force is also reversed.

The force developed in the conductor is given by the relation :

$$F = BIl \text{ newtons} \quad ...(3)$$

$$(= \mu_0 \mu_r HIl \text{ newtons}) \quad ...[3\,(a)]$$

where F = force developed in the conductor

B = flux density, T (Wb/m^2)

I = current in the conductor, A

l = exposed length of the conductor, m

μ_0 = absolute permeability

μ_r = relative permeability

H = magnetising force.

The *direction of this force may be easily found by Fleming's left hand rule* which states as follows :

"Hold your left hand with index finger, middle finger and thumb at right angles. If the index finger points in the direction of the flux from north to south and middle finger points in the direction of the imposed voltage and its resulting conventional current flow, then the thumb will point in the direction of the force that is developed".

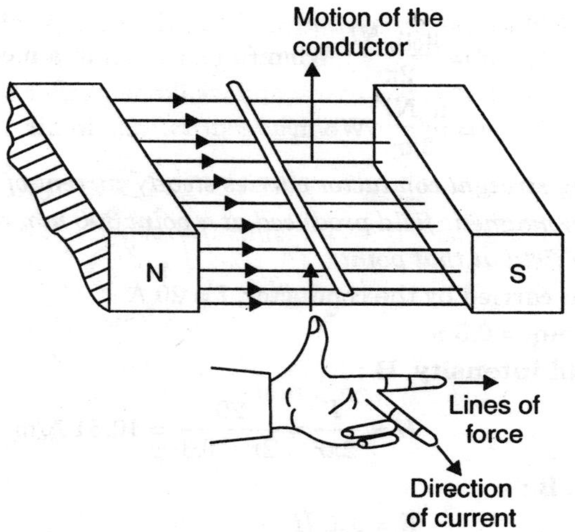

Fig. 13. Fleming's left hand rule.

11. MAGNETISING FORCE OF A LONG STRAIGHT CONDUCTOR

Fig. 14 shows a straight solid conductor in which the steady current I is flowing upwards in the direction indicated. Suppose that the field strength at point A distant r metres from the centre of the conductor is H. Then, it means that if a unit N-pole is placed at A, it will experience a force of H newtons. The direction of this force would be *tangential* to the circular line of force passing through A. If this unit N-pole is moved once round the conductor against this force, then work done is

$$= \text{force} \times \text{distance}$$
$$= H \times 2\pi r \text{ joules.}$$

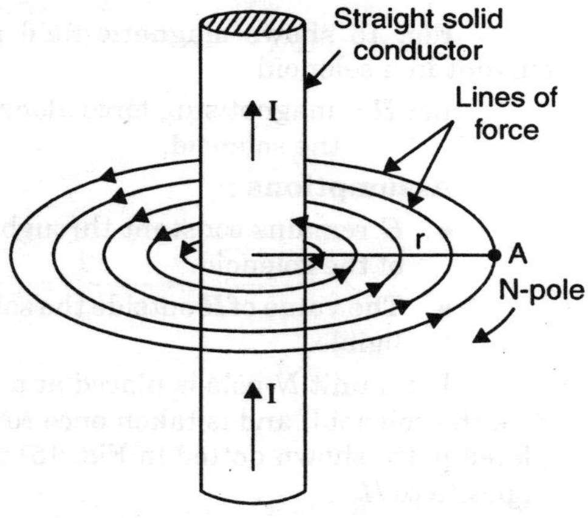

Fig. 14. Magnetic field produced by current in a straight solid conductor.

According to *work law* (It states that the *net work in joules done on or by a unit N-pole in moving around any single completed path in a magnetic field is numerically equal to the ampere-turns linked with its path*), this must be equal to the ampere-turns included within the path.

Ampere-turns linked $= NI$

$= 1 \times I = I$

$\therefore \quad H \times 2\pi r = I$

or $$H = \frac{1}{2\pi r} \text{ AT/m} \quad ...(4)$$

Also $$B = \mu H = \frac{\mu I}{2\pi r} \text{ Wb/m}^2 \quad ...(5)$$

(where $\mu = \mu_0 \mu_r$)

For N conductors, $$H = \frac{NI}{2\pi r} \text{ AT/m} \quad ...(6)$$

and $$B = \frac{\mu_0 \mu_r NI}{2\pi r} \text{ Wb/m}^2 \quad \text{...... in a medium} \qquad ...(7)$$

$$= \frac{\mu_0 NI}{2\pi r} \text{ Wb/m}^2 \quad \text{...... in air} \qquad ...[7(a)]$$

Example 2. *A long straight conductor carries steady current of 20 A. Find :*

(i) The intensity of magnetic field produced at a point 300 mm from the axis of the conductor.

(ii) Flux density of field at that point.

Solution. Current carried by the conductor, $I = 20$ A

Distance, $r = 300$ mm $= 0.3$ m

(*i*) **Magnetic field intensity, H :**

$$H = \frac{I}{2\pi r} = \frac{20}{2\pi \times 0.3} = 10.61 \text{ A/m. } \textbf{(Ans.)}$$

(*ii*) **Flux density, B :**

$$B = \mu_0 \mu_r H$$
$$= 4\pi \times 10^{-7} \times 1 \times 10.61$$
$$= \mathbf{133.3 \times 10^{-7}\ T\ (or\ Wb/m^2).\ (Ans.)}$$

12. MAGNETISING FORCE OF A LONG SOLENOID

Fig. 15 shows magnetic field produced by current in a solenoid.

Let H = magnetising force along the axis of the solenoid.

Assumptions :

- H remains constant through the length l of the solenoid.
- The value of H outside the solenoid is negligible.

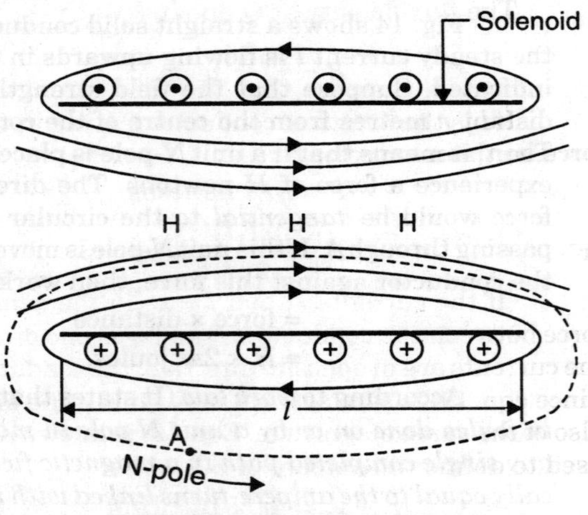

Fig. 15. Magnetic field produced by current in a solenoid.

Let a unit N-pole is placed at a point A (outside the solenoid) and is taken once round the completed path (shown dotted in Fig. 15) in a direction opposite to H.

Now work done is one round $= H \times l$ joules

The ampere-turns linked with this path $= NI$

[where N = number of turns of solenoid, and
I = current (A) passing through the solenoid.]

As per work law, $H \times l = N \times I$

or $$H = \frac{NI}{l} \text{ AT/m}$$

Also $$B = \frac{\mu_0 \mu_r NI}{l} \quad \text{......in a medium} \qquad ...(8)$$

$$= \frac{\mu_0 NI}{l} \quad \text{...... in air} \qquad ...[8(a)]$$

Example 3. *A solenoid of 300 turns is wound on a continuous ring of magnetic material of relative permeability 1000. If the flux density in the magnetic material of the core in the solenoid is 1.3 T, what is the current in the solenoid ? Mean diameter of the solenoid is 100 mm.*

Solution. Number of turns of the solenoid, $N = 300$

Relative permeability, $\mu_r = 1000$

Flux density, $B = 1.3$ T

Mean diameter of the solenoid, $d = 100$ mm $= 0.1$ m

$\therefore$ Mean circumference of the ring, $l = 2\pi r = 2\pi \times \frac{0.1}{2} = 0.314$ m

Current in the solenoid, I :

$$B = \frac{\mu_0 \mu_r N I}{l}$$

$$1.3 = \frac{4\pi \times 10^{-7} \times 1000 \times 300 \times I}{0.314}$$

$$\therefore \quad I = \frac{1.3 \times 0.314}{4\pi \times 10^{-7} \times 1000 \times 300} = \textbf{1.08 A. (Ans.)}$$

13. FORCE BETWEEN TWO PARALLEL CONDUCTORS—AMPERE'S LAW

Let P and Q be the two conductors, d metres apart, carrying currents I_1 and I_2 amperes as shown in Fig. 16.

The flux density at Q due to P,

$$B = \frac{\mu_0 I_1}{2\pi d} \text{ tesla}$$

If l is the length of conductor Q lying in this flux density, then force on this conductor,

$$F = B I_2 l \text{ newtons}$$

or $$F = \frac{\mu_0 I_1 I_2 l}{2\pi d} \text{ newtons} \quad ...(9)$$

(If the currents I_1 and I_2 are in the *same* direction, then the force between the conductors P and Q is one of *attraction* ; whereas if the currents are in opposite direction, the conductors *repel* each other). Since eqn. (9) is symmetrical in I_1 and I_2, the force on conductor P is also of the same magnitude. Eqn. (9) is known as *Ampere's law* and is used to define the ampere in S.I. units.

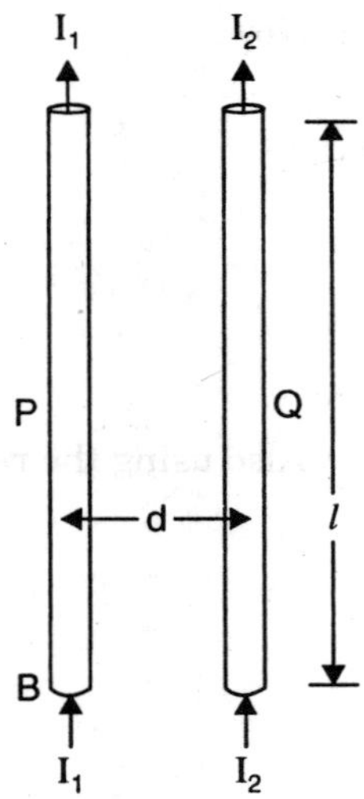

Fig. 16. Force between two parallel conductors.

$$\mu_0 = 4\pi \times 10^{-7} \text{ H/m}$$

If, $l = d = 1$ m,

and, $I_1 = I_2 = 1$ A,

then $F = 2 \times 10^{-7}$ N

Hence, one ampere current is defined as follows :

"An ampere is that current which when flowing in each of the two infinitely long parallel conductors situated in vacuum and separated 1 metre between centres, produces on each conductor a force of 2×10^{-7} N per metre length."

Example 4. *Two long parallel conductors carrying current, in the same direction, of 2500 A and 5000 A are placed with their axis 100 mm apart. Calculate the force between them in N/m.*

Solution. Current, $I_1 = 2500$ A

Currents, $I_2 = 5000$ A

Distance between the conductors, $d = 100$ mm $= 0.1$ m.

Force between the conductors per metre, F :

$$F = \frac{\mu_0 I_1 I_2 l}{2\pi d}$$

or
$$\frac{F}{l} = \frac{\mu_0 I_1 I_2}{2\pi d} = \frac{4\pi \times 10^{-7} \times 2500 \times 5000}{2\pi \times 0.1}$$

$$= \mathbf{25\ N/m.\ (Ans.)}$$

Example 5. *Two long straight parallel wires, standing in air 2 m apart, carry currents I_1 and I_2 in the same direction. The magnetic intensity at a point midway between the wires is 7.95 AT/m. If the force on each wire per unit length is 2.4×10^{-4} N, evaluate I_1 and I_2.* **(Madras University)**

Solution. Distance between the conductors, $d = 2$ m

Magnetic intensity midway between the wires = 7.95 AT/m

Force on each wire per unit length = 2.4×10^{-4} N

I_1, I_2 :

We know that when the two currents flow in the same direction, net field strength midway between the two conductors is the difference of the two field strengths.

Now,
$$H_1 = \frac{I_1}{2\pi r} = \frac{I_1}{2\pi \times \frac{2}{2}} = \frac{I_1}{2\pi} \qquad \left(\because\ r = \frac{d}{2} = \frac{2}{2} = 1\text{ m}\right)$$

and
$$H_2 = \frac{I_2}{2\pi r} = \frac{I_2}{2\pi \times \frac{2}{2}} = \frac{I_2}{2\pi}$$

$\therefore$
$$H_1 - H_2 = \frac{I_1}{2\pi} - \frac{I_2}{2\pi} = 7.95$$

or
$$I_1 - I_2 = 2\pi \times 7.95 = 50 \qquad ...(i)$$

Also using the relation,

$$F = \frac{\mu_0 I_1 I_2 l}{2\pi d}$$

or
$$\frac{F}{l} = \frac{\mu_0 I_1 I_2}{2\pi d}$$

or
$$2.4 \times 10^{-4} = \frac{4\pi \times 10^{-7} \times I_1 I_2}{2\pi \times 2} \qquad \left[\because\ \frac{F}{l} = 2.4 \times 10^{-4}\text{ N ...... given}\right]$$

or
$$I_1 I_2 = \frac{2.4 \times 10^{-4} \times 2\pi \times 2}{4\pi \times 10^{-7}} = 2400 \qquad ...(ii)$$

Substituting the value of I_1 from (*i*) in (*ii*), we get

$$(50 + I_2) \times I_2 = 2400$$

or
$$I_2^2 + 50 I_2 - 2400 = 0$$

or
$$I_2 = \frac{-50 \pm \sqrt{2500 + 9600}}{2}$$

$$= \frac{-50 \pm 110}{2} = 30\text{ A} \qquad \text{(taking + ve value only)}$$

$\therefore$
$$I_1 = 80\text{ A.}$$

Hence $\mathbf{I_1 = 80\ A}$ and $\mathbf{I_2 = 30\ A.}$ **(Ans.)**

14. TERMS CONNECTED WITH MAGNETIC CIRCUIT

A *magnetic circuit* is defined as *the route or path which is followed by a magnetic flux.*

Some important terms connected with magnetic circuits are discussed below :

1. **Permeability (μ).** Permeability of any material is a measure of ease with which the atoms can be arranged. The S.I. unit for permeability is henry/metre (H/m).

Mathematically, $\mu = \mu_0 \mu_r$ H/m ...(10)

where μ_0 = permeability of free space

$= 4\pi \times 10^{-7}$ H/m (in S.I. units)

μ_r = relative permeability (= 1 for air and non-magnetic materials)

Relative permeability (μ_r) is simply a numeric which expresses the degree to which the material is a better conductor of magnetic flux as compared to free space.

> **Note.** μ_r for *air* (and non-magnetic materials) = 1
> μ_r for diamagnetic materials = slightly less than one
> μ_r for paramagnetic materials = slightly higher than one
> μ_r for ferromagnetic materials = in the hundreds or thousands
> (μ_r for ferromagnetic materials changes with the magnetic field strength)

2. **Magnetomotive force (m.m.f.).** Magnetomotive force *drives or tends to drive flux through a magnetic circuit* and corresponds to electromotive force (e.m.f.) in a circuit. M.m.f. is equal to the work done in joules in carrying a unit magnetic pole once through the entire magnetic circuit.

M.m.f. is measured in **ampere-turns (AT).** Ampere-turns are found by *multiplying number of turns (N) of a magnetic circuit and the current in amperes (I)* in those turns.

Mathematically, $AT = NI$.

3. **Reluctance.** Just as the current in an electric circuit is limited by resistance, the flux in the magnetic circuit is limited by the *reluctance* (symbol S) of the magnetic circuit. Thus *reluctance is a measure of opposition offered by a magnetic circuit to the setting up of the flux.*

The reluctance of a magnetic circuit is given by,

$$S = \frac{l}{\mu A} = \frac{l}{\mu_0 \mu_r A} \qquad ...(11)$$

where, l = length of the magnetic circuit,

A = cross-sectional area of the magnetic circuit,

μ_0 = absolute permeability, and

μ_r = relative permeability.

Reluctance of a magnetic circuit is the **ratio of m.m.f. and flux.**

i.e.,
$$\text{Reluctance} = \frac{\text{m.m.f.}}{\text{flux}}$$

or
$$S = \frac{NI}{\phi} \qquad ...(12)$$

In other words the reluctance of a magnetic circuit is the number of ampere-turns required per weber of magnetic flux in the circuit. The *unit of reluctance* is AT/Wb. Since 1 AT/Wb = 1/henry, the unit of reluctance is "*reciprocal henry*".

4. **Permeance.** The *reciprocal of reluctance* is known as *permeance.*

i.e.,
$$\text{Permeance} = \frac{1}{\text{reluctance}} = \frac{1}{S}$$

It is measured in Wb/AT or *henry*.

5. **Reluctivity.** It is the *specific reluctance* and corresponds to resistivity which is specific resistance.

Relation between flux density (B) and magnetic field strength (H) :

Comparing eqns. (12) and (13), we get

$$\frac{l}{A\mu_0\mu_r} = \frac{NI}{\phi}$$

or $$\frac{\phi}{A\mu_0\mu_r} = \frac{NI}{l} \quad \text{(By rearranging)}$$

But $$\frac{\phi}{A} = B,$$

and $$\frac{NI}{l} = H$$

$\therefore$ $$\frac{B}{\mu_0\mu_r} = H$$

or $$B = \mu_0\mu_r H \quad ...(13)$$

or $$B = \mu H$$

[where μ (permeability) = $\mu_0\mu_r$]

Thus, *permeability is the ratio of flux density to magnetic field strength.*

15. COMPARISON OF ELECTRIC AND MAGNETIC CIRCUITS

The analogy between electric and magnetic circuits is given in the Table 1.

Table 1. Analogy between Electric and Magnetic Circuits (Similarities)

Aspect	*Electric circuit*	*Magnetic circuit*
	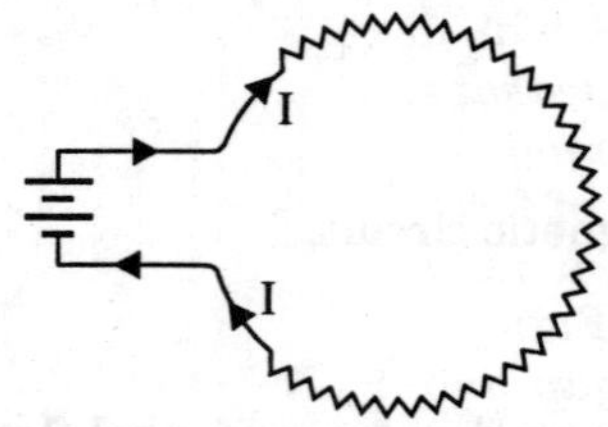 Fig. 17	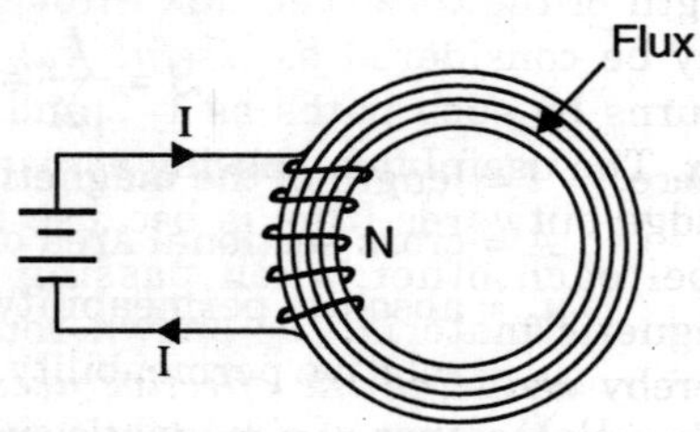 Fig. 18
1. *Equivalent circuit*	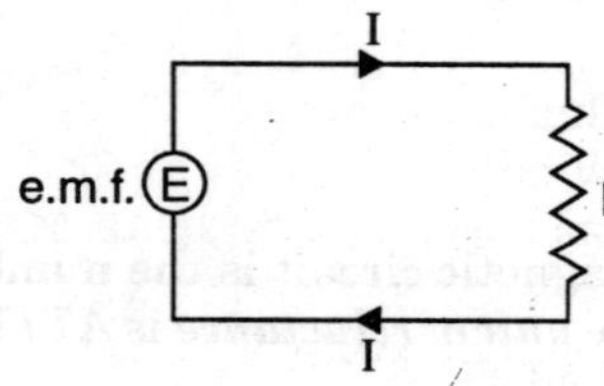	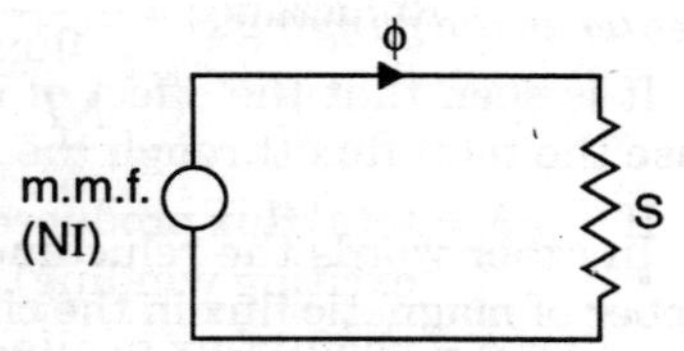
2. *Exciting force*	Battery voltage (E)	Ampere-turns (AT)
3. *Response*	Current (I)	Flux (ϕ)
4. *Ohm's law*	$I = \dfrac{E}{R}$	$\phi = \dfrac{\text{m.m.f.}}{\text{reluctance}} = \dfrac{NI}{S}$

5. *By dimensions*	$R = \rho \dfrac{l}{a}$ (Conductance = 1/R)	$S = \dfrac{l}{A\mu_0\mu_r}$ (Permeance = 1/S)
6. *Proportionality*	ρ	$\dfrac{1}{\mu} = \left(= \dfrac{l}{\mu_0\mu_r} \right)$
7. *Field intensity*	Electric field intensity = $\dfrac{E}{l}$ V/m	Magnetic field intensity = $\dfrac{NI}{l}$ AT/m
8. *Density*	Current density (A/m²)	Flux density (Wb/m²)

Differences

1. The *resistivity* of conductors is more or less *constant* but the *permeability* of the ferromagnetic materials *varies* greatly with magnetic field strength.

2. Strictly speaking, flux does not actually 'flow' in the sense in which an electric current flows.

3. In an electric circuit energy must be supplied to *maintain* the flow of electricity in the circuit, whereas (in a magnetic circuit) the magnetic flux once, it is set up, does not require any further supply of energy.

(**Example.** Once the flux produced by a current in the solenoid has attained its steady value, the energy subsequently absorbed by that solenoid is all dissipated as heat due to the resistance of the winding).

4. In ferromagnetic materials if the magnetic field strength is increased the flux density increases till a state of saturation is reached after which there is no increase in flux density for further increase in field strength. There is no such phenomenon in electric circuits.

16. MAGNETIC LEAKAGE AND FRINGING

Fig. 19 shows a metal ring symmetrically situated relative to the air gap in the iron ring. Let the magnetising winding be concentrated over a short length of the core. The flux through the *metal ring* may be considered as *useful flux* and that which returns by such paths as 1, 2 and 3 is the *leakage flux*. The useful flux passing across the gap tends to buldge outwards (this is because the lines of force repel each other when passing through a non-magnetic material) as shown roughly in Fig. 19, thereby *reducing the effective area of the gap and reducing the flux density in the gap*. This effect is referred to as **fringing** and the *longer the air gap, the greater is the fringing*.

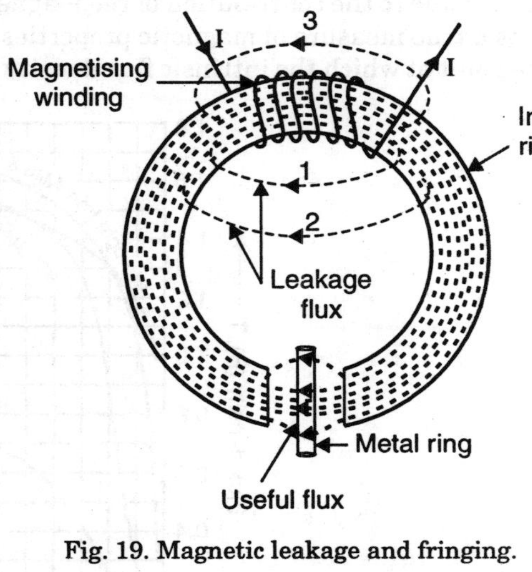

Fig. 19. Magnetic leakage and fringing.

It is seen that the effect of leakage flux is to increase the total flux through the exciting winding.

If ϕ_t = total flux produced (produced in the exciting winding), and

ϕ = useful flux available in the air-gap,

Then, leakage coefficient (or factor) $\lambda = \dfrac{\phi_t}{\phi}$

The value of λ for electrical machines is usually about 1.15 to 1.25.

[**Note.** Magnetic leakage can be minimised by placing the exciting coils or windings as close as possible to the air gap or to the points in the magnetic circuit where the flux is to be utilised.]

17. MAGNETISATION (B-H) CURVES

Consider a toroidal solenoid wound on a non-magnetic core, such as shown in Fig. 20. If the flux density is measured on the centre line of the toroid, the relationship between B and H is given by the straight line OA in Fig. 21. If now the space within an unmagnetised ferromagnetic material, the well known magnetisation curve $OBCDE$, is obtained. The magnetisation has many names being referred to as :

B-H curve, the magnetic saturation curve, the virgin curve or simply the saturation curve.

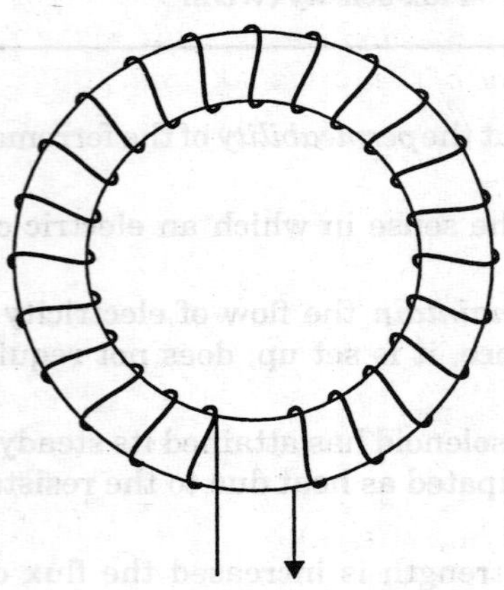

Fig. 20. Toroid.

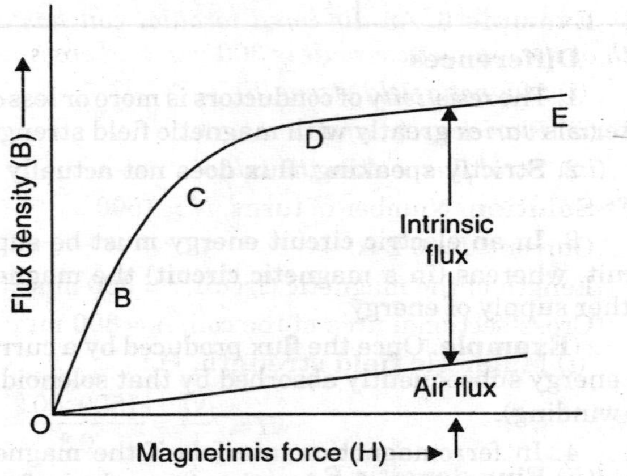

Fig. 21. Magnetising curve for a ferro-magnetic material.

The difference in flux between the saturation curve and the air line, OA at any magnetising force, is due to the contribution of the magnetic material. This flux is known as the *intrinsic flux* and gives a true measure of magnetic properties than does the total flux, especially at very high m.m.fs. The point at which the intrinsic flux density curve becomes horizontal gives the intrinsic saturation.

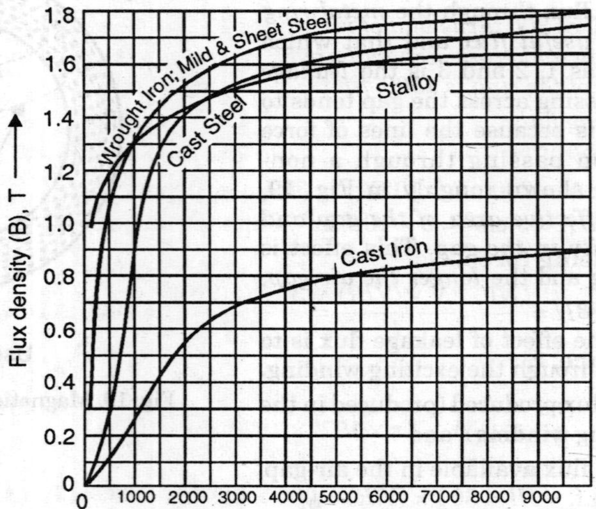

Fig. 22. *B-H* curves.

The magnetisation curves can be determined by the following methods provided the materials are in the form of a ring :

(*i*) By means of a ballistic galvanometer

(*ii*) By means of a fluxmeter.

The graphs in Fig. 22 show the relationship between the flux density (B) and the magnetic field strength obtained for different qualities of iron. The data for mild steel, wrought iron and sheet steel are so similar that they can be represented by a common graph. *Stalloy* is an alloy of iron and silicon commonly used in the construction of transformers and A.C. machines.

Example 6. *An air cored toroidal coil has 1500 turns and carries a current of 0.2 A. The length of the magnetic circuit is 300 mm and cross-sectional area of the coil is 800 mm². Calculate :*

(*i*) *Magnetic field strength,*

(*ii*) *Flux density, and*

(*iii*) *Total flux within the coil.*

Solution. Number of turns, $N = 1500$

Current, $I = 0.2$ A

Length of the magnetic circuit, $l = 300$ mm $= 0.3$ m

Cross-sectional area of the coil, $A = 800 \text{ mm}^2 = 800 \times 10^{-6} \text{ m}^2$

(*i*) **Magnetic field strength, H :**

$$H = \frac{NI}{l} = \frac{1500 \times 0.2}{0.3} = \mathbf{1000\ AT/m.} \quad \textbf{(Ans.)}$$

(*ii*) **Flux density, B :**

$$B = \mu_0 \mu_r H$$
$$= \mu_0 H \qquad (\because \text{ For air cored coil, } \mu_r = 1)$$
$$= 4\pi \times 10^{-7} \times 1000 = \mathbf{1.25 \times 10^{-3}\ T.} \quad \textbf{(Ans.)}$$

(*iii*) **Total flux within the coil, ϕ :**

$$\phi = B \times A$$
$$= 1.256 \times 10^{-3} \times 800 \times 10^{-6} = \mathbf{1.005 \times 10^{-6}\ Wb.} \quad \textbf{(Ans.)}$$

Example 7. *A coil of 300 turns and of resistance 10 Ω is wound uniformly over a steel ring of mean circumference 30 cm and cross-sectional area 9 cm². It is connected to a supply at 20 V D.C. If the relative permeability of the ring is 1500, find :*

(*i*) *The magnetising force,* (*ii*) *The reluctance,*

(*iii*) *The m.m.f., and* (*iv*) *The flux.* **(VTU, 1999)**

Solution. *Given* : $N = 300$; $R = 10\ \Omega$; $l = 30$ cm or 0.3 m ;

$A = 9 \text{ cm}^2 = 9 \times 10^{-4} \text{ m}^2$; Supply voltage, = 20 V ; $\mu_r = 1500$.

(*i*) **The magnetising force, H :**

$$H = \frac{NI}{l} = \frac{N \times (V/R)}{l} = \frac{30 \times (20/10)}{0.3} = \mathbf{2000\ AT/m.} \quad \textbf{(Ans.)}$$

(*ii*) **The reluctance, S :**

$$S = \frac{l}{\mu_0 \mu_r A} = \frac{0.3}{4\pi \times 10^{-7} \times 1500 \times 9 \times 10^{-4}} = \mathbf{176.84 \times 10^3\ AT/Wb.} \quad \textbf{(Ans.)}$$

(*iii*) **The m.m.f. :**

$$\text{m.m.f.} = NI = 300 \times (V/R) = 300 \times (20/10) = \mathbf{600\ AT.} \quad \textbf{(Ans.)}$$

(*iv*) **The flux, ϕ :**

$$\phi = \frac{\text{m.m.f.}}{\text{reluctance}} = \frac{600}{176.84 \times 10^3} = 3.39 \times 10^{-3} \text{ Wb} = \mathbf{3.39\ mWb.} \quad \textbf{(Ans.)}$$

Example 8. *A magnetic circuit consists of an iron ring of mean circumference 80 cm with cross-sectional area of 12 cm² throughout. A current of 1 A in the magnetising coil of 200 turns produces a total flux of 1.2 m Wb in the iron. Calculate :*

(*i*) *The flux density in the iron,*

(*ii*) *The absolute and relative permeability of iron,*

(*iii*) *The reluctance of the circuit.* **(P.U., 2000)**

Solution. *Given* : l = 80 cm = 0.8 m ; a = 12 cm² = 12 × 10⁻⁴ m² ;

I = 1 A ; N = 200 ; ϕ = 1.2 m Wb = 0.0012 Wb.

(*i*) **The flux density in the iron, B :**

$$B = \frac{\phi}{A} = \frac{0.0012}{12 \times 10^{-4}} = \mathbf{1\ Wb/m^2.\ (Ans.)}$$

(*ii*) **The absolute and relative permeability of iron, μ, μ_r :**

The magnetising force, $H = \dfrac{NI}{l} = \dfrac{200 \times 1}{0.8} = 250$ AT/m

Absolute permeability, $\mu = \mu_0\mu_r = \dfrac{B}{H}$ or $\mu = \dfrac{1}{250} = \mathbf{4 \times 10^{-3}}$ **units. (Ans.)**

and $\mu_r = \dfrac{\mu}{\mu_0} = \dfrac{4 \times 10^{-3}}{4\pi \times 10^{-7}} = \mathbf{3183.\ (Ans.)}$

(*iii*) **The reluctance of the circuit, S :**

$$S = \frac{l}{\mu_0\mu_r A} = \frac{0.8}{4\pi \times 10^{-7} \times 3183 \times 12 \times 10^{-4}}$$

$$= \mathbf{166.67 \times 10^3\ AT/Wb.\ (Ans.)}$$

Example 9. *An iron ring 200 mm mean diameter is made of 30 mm round iron of permeability 900, has an air gap 10 mm wide. It has 800 turns. If the current flowing through this winding is 6.8 amperes, determine :*

(*i*) *The magnetomotive force,*

(*ii*) *The total reluctance of the circuit,*

(*iii*) *The flux in the ring, and*

(*iv*) *The flux density in the ring.*

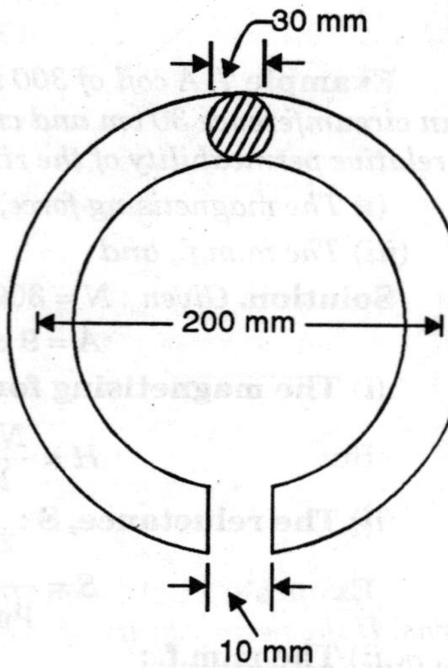

Fig. 23

Solution. Length of air gap, l_g = 10 mm = 0.01 m

Mean dia. of iron ring = 200 mm = 0.2 m

∴ Mean length of iron path,

$l_i = (\pi \times 0.2 - 0.01)$

= 0.618 m

Area of cross-section, $A = \dfrac{\pi}{4} \times 0.03^2$

= 7.068 × 10⁻⁴ m²

Number of turns, N = 800

Permeability of iron, μ_r = 900

Current through the winding, I = 68 A.

(*i*) **Magnetomotive force, m.m.f. :**

m.m.f. = NI ampere-turns

= 800 × 6.8 = **5440 AT. (Ans.)**

(*ii*) **Total reluctance of the circuit, S :**

$$S = \frac{l_i}{\mu_0\mu_r A} + \frac{l_9}{\mu_0 A} \qquad (\because \ \mu_r \text{ for air} = 1)$$

$$= \frac{0.618}{4\pi \times 10^{-7} \times 900 \times 7.068 \times 10^{-4}} + \frac{0.01}{4\pi \times 10^{-7} \times 7.068 \times 10^{-4}}$$

$$= \frac{1}{4\pi \times 10^{-7} \times 7.068 \times 10^{-4}} \left(\frac{0.618}{900} + 0.01 \right)$$

$$= \frac{1}{88.82 \times 10^{-11}} \times 0.0107 = \mathbf{0.12 \times 10^8\ AT/Wb.\ (Ans.)}$$

(*iii*) **Flux in the ring, ϕ :**

$$\phi = \frac{\text{m.m.f.}}{\text{reluctance}} = \frac{5440}{0.12 \times 10^8} = \mathbf{4.533 \times 10^{-4}\ Wb.\ (Ans.)}$$

(*iv*) **Flux density in the ring, B :**

$$B = \frac{\phi}{A} = \frac{4.533 \times 10^{-4}}{7.068 \times 10^{-4}} = \mathbf{0.64\ T.\ (Ans.)}$$

Example 10. *An iron of 15 cm diameter and 10 cm² cross-section with a saw cut 2 mm wide is wound with 200 turns of wire. The flux density is 1 Wb / m². The relative permeability of iron is 600. Find the exciting current. Neglect leakage.* **(K.U.)**

Solution. *Given :* $l_i = \pi \times 15 - \frac{2}{10} = 46.92$ cm = 0.4692 m

l_g = 2 mm = 0.002 m ; N = 200 ; B = 1 Wb/m², μ_r = 600.

Exciting current, I :

Total AT (ampere-turns) required

= AT required for iron ring (AT_i) + AT required for air gap (AT_g)

i.e.,

$$AT = AT_i + AT_g$$

$$= H_i l_i + H_g l_g$$

$$= \frac{B_i}{\mu_0\mu_r} l_i + \frac{B_g}{\mu_0} l_g$$

$$= \frac{1}{4\pi \times 10^{-7} \times 600} \times 0.4692 + \frac{1}{4\pi \times 10^{-7}} \times 0.002$$

$$(\because \ B_i = B_g = 1 \text{ Wb/m}^2)$$

$$= \frac{1}{4\pi \times 10^{-7}} \left(\frac{0.4692}{600} + 0.002 \right) = 2213.84$$

But total $AT = NI$

$$\therefore \qquad I = \frac{2213.84}{200} = \mathbf{11.07.\ (Ans.)}$$

Example 11. *An iron ring of mean length 1 metre has an air gap of 2 mm and a winding 400 turns. If the permeability of the iron is 400 when a current of 2.5 amperes flow through the coil, find the flux density.* **(Indore University)**

Solution. Mean length of iron ring (or iron path), l_i = 1 m (neglecting air gap length)

Length of air gap, l_g = 2 mm = 0.002 m

Number of turns, N = 400

Permeability of iron, $\mu_r = 400$

Current through the coil, $I = 2.5$ A

Flux density, B

Total AT required = AT required for iron ring (AT_i) + AT required for air gap (AT_g)

i.e.,
$$AT = AT_i + AT_g$$
$$= H_i l_i + H_g l_g$$
$$= \frac{B}{\mu_0 \mu_r} l_i + \frac{B}{\mu_0} l_g$$

i.e.,
$$AT = \frac{B}{\mu_0 \times 400} \times 1 + \frac{B}{\mu_0} \times 0.002$$
$$= \frac{B}{\mu_0}\left(\frac{1}{400} + 0.002\right) = 0.0045 \frac{B}{\mu_0}$$

No. of AT supplied = $NI = 400 \times 2.5 = 1000$

$$\therefore \quad 0.0045 \frac{B}{\mu_0} = 1000$$

and
$$B = \frac{1000\,\mu_0}{0.0045} = \frac{1000 \times 4\pi \times 10^{-7}}{0.0045} = \mathbf{0.279\ T.\ (Ans.)}$$

Example 12. *An iron ring of mean circumference of 900 mm and cross-sectional area of 250 mm² has flux of 500 μWb. Calculate the m.m.f. acting on the iron ring. An air gap of 2 mm is now cut in the ring. Find the flux in the ring if the m.m.f. remains same. A relative permeability of 1200 may be assumed.* **(Roorkee University)**

Solution. *Given :* $l_i = 900 - 2 = 898 = 0.898$ m ; $A = 250$ mm² $= 250 \times 10^{-6}$ m²

$\phi = 500\ \mu$ Wb $= 500 \times 10^{-6}$ Wb ; $l_g = 2$ mm $= 0.002$ m ; $\mu_r = 1200$.

(*i*) **Before the air gap is cut :**

$$\phi = \frac{\text{m.m.f.}}{\text{reluctance}} = \frac{NI}{l/\mu_0 \mu_r A}$$

$$\therefore \quad \textbf{M.m.f.,}\ NI\ (i.e., AT) = \frac{\phi l}{\mu_0 \mu_r A} = \frac{500 \times 10^{-6} \times 0.9}{4\pi \times 10^{-7} \times 1200 \times 250 \times 10^{-6}} = \mathbf{1193.7\ AT.\ (Ans.)}$$

(*ii*) **After the air gap is cut :**

$$\text{Total } AT = AT_i + AT_g$$

or
$$1193.7 = \frac{B_i}{\mu_0 \mu_r} l_i + \frac{B_g}{\mu_0} l_g$$
$$= \frac{B}{\mu_0}\left(\frac{l_i}{\mu_r} + l_g\right) \qquad (\because\ B_i = B_g = B)$$
$$= \frac{\phi}{\mu_0 A}\left(\frac{l_i}{\mu_r} + l_g\right)$$

or
$$1193.7 = \frac{\phi}{4\pi \times 10^{-7} \times 250 \times 10^{-6}}\left(\frac{0.898}{1200} + 0.002\right)$$
$$= \frac{\phi}{3141.6 \times 10^{-13}} \times 0.00275$$

$$\therefore \qquad \phi = \frac{1193.7 \times 3141.6 \times 10^{-13}}{0.00275} = 1.36 \times 10^{-4} \text{ or } 136 \times 10^{-6} \text{ Wb}$$

$$= \mathbf{136\ \mu Wb.\ (Ans.)}$$

Example 13. *A ring has a diameter of 24 cm and a cross-sectional area of 1000 mm². The ring is made up of semicircular sections of cast iron and cast steel with each joint having a reluctance equal to an air gap of 0.2 mm. Find the ampere-turns required to produce a flux of 8 × 10⁻⁴ Wb. The relative permeability of caststeel and cast iron are 900 and 170 respectively. Neglect fringing and leakage effects.* **(Gujrat University)**

Solution. *Given :* $\phi = 8 \times 10^{-4}$ Wb ; $A = 1000$ mm² $= 1000 \times 10^{-6} = 0.001$ m³ ;

Total length of air gap, $l_g = 2 \times 0.2 = 0.4$ mm $= 0.0004$ m ; $(\mu_r)_{c.i.} = 900$; $(\mu_r)_{c.s.} = 170$.

$$ATT = (AT)_{c.i.} + (AT)_g + (AT)_{c.s.}$$

$$= \frac{B}{\mu_0(\mu_r)_{c.i.}} l_{c.i.} + \frac{B}{\mu_0} . l_g + \frac{B}{\mu_0(\mu_r)_{c.s.}} l_{c.s.}$$

$$(\because \ B_{c.i.} = B_g = B_{i.s.} = B)$$

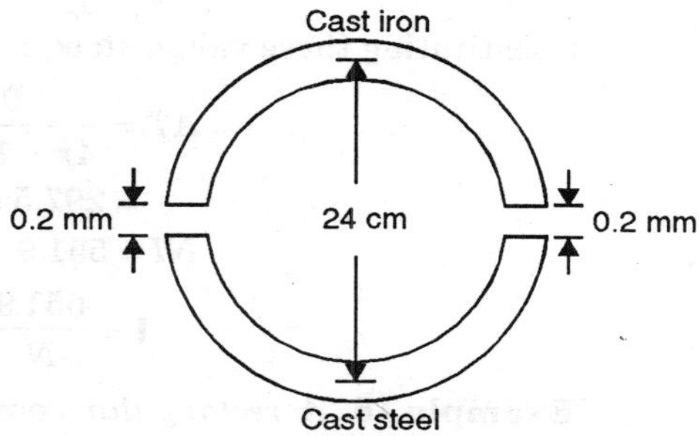

Fig. 24

or

$$AT = \frac{B}{\mu_0}\left[\frac{l_{c.i.}}{(\mu_r)_{c.i.}} + l_g + \frac{l_{c.s.}}{(\mu_r)_{c.s.}}\right] \qquad ...(i)$$

Here, $$B = \frac{\phi}{A} = \frac{8 \times 10^{-4}}{0.001} = 0.8 \text{ Wb/m}^2$$

$$l_{c.i.} = l_{c.s.} = \frac{\pi D}{2} = \frac{\pi \times 0.24}{2} = 0.377 \text{ m}$$

Substituting the values in eqn. (*i*), we get

$$AT = \frac{0.8}{4\pi \times 10^{-7}}\left[\frac{0.377}{900} + 0.0004 + \frac{0.377}{170}\right] = \mathbf{1933.\ (Ans.)}$$

Example 14. *A magnetic ring has a mean circumference of 1.8 m and is of 0.012 m² in cross-section, and is wound with 210 turns. A saw cut of 4.8 mm width is made in the ring. Calculate the magnetising current required to produce a flux of 0.8 milliweber in the air gap.*

Assume leakage factor as 1.25 and permeability of iron as 400.

Solution. Mean circumference of the ring = 1.8 m

Length of air gap, $l_g = 4.8$ mm $= 0.0048$ m

∴ Length of iron path $= 1.8 - l_g = 1.8 - 0.0048 = 1.795$ m

Area of cross-section, $A = 0.012$ m²

Flux in the air gap, $\phi_g = 0.8$ m Wb

Number of turns, $N = 210$

Permeability of iron, $\mu_r = 400$

Leakage factor, $\lambda = 1.25$

Magnetising current I :

Total ampere-turns required (*AT*) = *AT* required for iron ring + *AT* required for air gap

i.e.,

$$AT = AT_i + AT_g$$

$$= H_i l_i + H_g l_g$$

$$= \frac{B_i}{\mu_0 \mu_r} l_i + \frac{B_g}{\mu_0} . l_g \qquad ...(i)$$

Now, leakage factor $\lambda = \dfrac{\phi_i}{\phi_g} = 1.25$ (given)

[where ϕ_i = flux in the iron path, and ϕ_g = flux in the air gap]

$$\therefore \quad \frac{\phi_i}{0.8 \times 10^{-3}} = 1.25$$

$$\phi_i = 0.8 \times 10^{-3} \times 1.25 = 1 \times 10^{-3} \text{ Wb}$$

Hence,
$$B_i = \frac{\phi_i}{A} = \frac{1 \times 10^{-3}}{0.012} = 0.0833 \text{ T}$$

and
$$B_g = \frac{\phi_g}{A} = \frac{0.8 \times 10^{-3}}{0.012} = 0.0666 \text{ T}$$

Substituting these values in eqn. (*i*), we get

$$AT = \frac{0.0833}{4\pi \times 10^{-7} \times 400} \times 1.795 + \frac{0.0666}{4\pi \times 10^{-7}} \times 0.0048$$

$$= 297.5 + 254.4 = 551.9$$

i.e.,
$$NI = 551.9$$

$$\therefore \quad I = \frac{551.9}{N} = \frac{551.9}{210} = \mathbf{2.63\ A.\ (Ans.)}$$

Example 15. *A rectangular core shown in Fig. 25 has a mean length of magnetic path of 120 cm, cross-section of (3 cm × 3 cm), relative permeability of 1400 and an air gap of 4 mm cut in the core. The three coils carried by the core have number of turns $N_1 = 350$, $N_2 = 500$ and $N_3 = 500$; and the respective currents are 1.5 A, 3.5 A and 2.5 A. The directions of the currents are as shown. Find the flux in the air gap.* **(Pune University)**

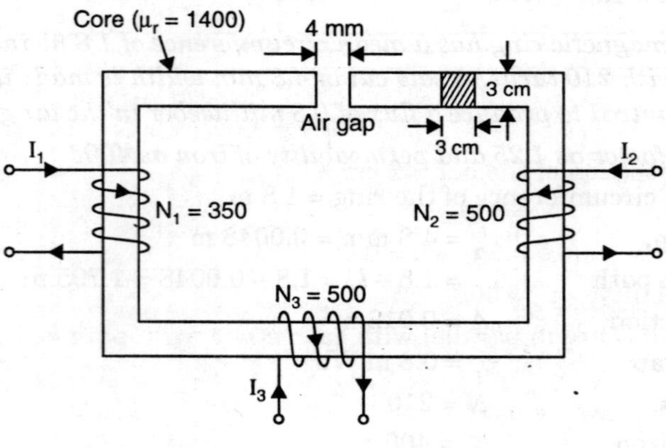

Fig. 25

Solution. *Given* : $l_i = 1.2 - 0.004 = 1.196$ m ; $\mu_r = 1400$; $l_g = 4$ mm $= 0.004$ m ;

$$A = 3 \times 3 = 9 \text{ cm}^2 = 9 \times 10^{-4} \text{ m}^2$$

$$N_1 = 350 \ ; N_2 = 500 \ ; N_3 = 500 \ ; I_1 = 1.5 \text{ A} \ ; I_2 = 3.5 \text{ A} \ ; I_3 = 2.5 \text{ A}.$$

Flux in the air gap, ϕ_g :

By applying the right-hand thumb rule, it is found that fluxes produced by the currents I_1 and I_2 are directed in the *clockwise* direction through the iron core whereas that produced by current I_3 is directed in the *anti-clockwise* direction through the core.

$\therefore$ Total m.m.f. $= N_1I_1 + N_2I_2 - N_3I_3$

$= 350 \times 1.5 + 500 \times 3.5 - 500 \times 2.5 = 1025\ AT$

Reluctance of the air gap $= \dfrac{l_g}{\mu_0 A} = \dfrac{0.004}{4\pi \times 10^{-7} \times 9 \times 10^{-4}} = 3.536 \times 10^6\ AT/\text{Wb}$

Reluctance of the iron path $= \dfrac{l_i}{\mu_0 \mu_r A} = \dfrac{1.196}{4\pi \times 10^{-7} \times 1400 \times 9 \times 10^{-4}} = 0.755 \times 10^6\ AT/\text{Wb}$

Total reluctance $= (3.536 + 0.755) \times 10^6 = 4.291 \times 10^6\ AT/\text{Wb}$

The flux in the air gap is the same as in the iron core.

Air gap flux, $\phi_g = \dfrac{\text{m.m.f.}}{\text{reluctance}} = \dfrac{1025}{4.291 \times 10^6} = 238.9 \times 10^{-6} =$ **238.9 μWb. (Ans.)**

Example 16. *A cast steel magnetic structure made for a bar of section 2 cm × 2 cm is shown in Fig. 26. Determine the current that the 600 turn magnetising coil on the left limb should carry so that a flux of 2 mWb is produced in the right limb. Take $\mu_r = 600$ and neglect leakage.*

(Allahabad University)

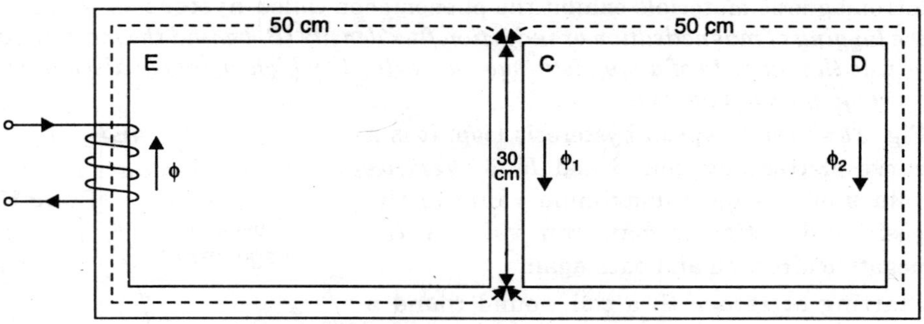

Fig. 26

Solution. Cross-section of the bar = 2 cm × 2 cm

Number of turns, $N = 600$

Flux, $\phi = 2\text{ mWb} = 0.002\text{ Wb}$

Relative permeability, $\mu_r = 600$

Since paths C and D are in parallel with each other w.r.t. path E, the m.m.f. (NI) across the two is the same.

i.e., $\phi_1 S_1 = \phi_2 S_2$

$$\phi_1 \times \frac{30}{\mu A} = 0.002 \times \frac{50}{\mu A} \qquad \left(\because\ \phi = \frac{NI}{S}\right)$$

$\therefore$ $\phi_1 = 0.00333$ Wb

$\therefore$ $\phi = \phi_1 + \phi_2 = 0.00333 + 0.002 = 0.00533$ Wb.

Total AT required $= AT$ required for path E

$+ AT$ required for either of the two paths C or D

$$= \frac{Bl}{\mu_0 \mu_r} + \frac{B_2 l_2}{\mu_0 \mu_r} = \frac{1}{\mu_0 \mu_r A}\ (\phi l + \phi_2 l_2) \qquad \left(\because\ B = \frac{\phi}{A}\right)$$

$$= \frac{1}{4\pi \times 10^{-7} \times 600 \times (2 \times 2 \times 10^{-4})} (0.00533 \times 0.5 + 0.002 \times 0.5)$$

$$= 3.316 \times 10^6 \times 3.665 \times 10^{-3} = 12153$$

$\therefore$ Current required, $I = \frac{12153}{600} = $ **20.26 A. (Ans.)**

18. LOSSES IN MAGNETIC MATERIALS

When magnetic materials are subjected to an alternating flux, two types of losses occur which are :

1. Hysteresis loss
2. Eddy current loss.

The above losses comprise the *total core loss*.

18.1. Hysteresis Loss

- Below *curie temperature* (it is the rising temperature at which the given material ceases to be ferromagnetic, or the falling temperature at which it becomes ferromagnetic) all ferromagnetic materials exhibit the phenomenon called **hysteresis** which is defined as *the lagging of magnetisation or induction flux density (B) behind the magnetising force (H) or it is that quality of a magnetic substance due to which energy is dissipated in it on the reversal of its magnetism.*
- Fig. 27 shows a typical hysteresis loop. It is a curve plotted between B and H for various values of H from a maximum value in the positive direction to maximum value in the negative direction and back again.

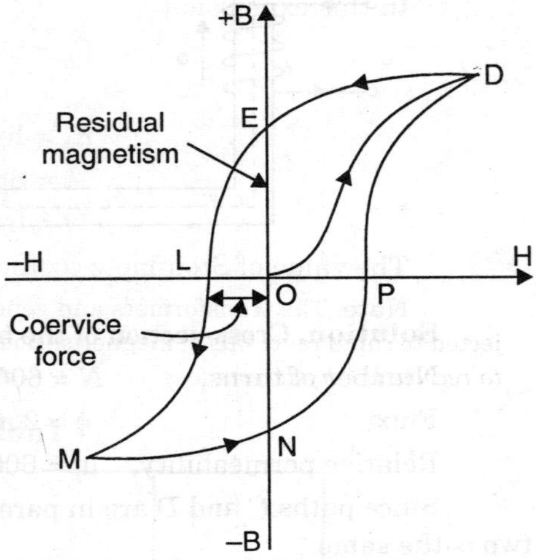

Fig. 27. Hysteresis loop for a ferromagnetic material.

— Starting at zero with a coil wound round a toroid of unmagnetised iron, the magnetisation curve follows curve OD. If the m.m.f. is gradually reduced, the flux curve follows the line DE. As the m.m.f. is reversed the flux falls into the point L and thence to M. As the m.m.f. is returned to zero, the flux traces out the path MN. Then, as m.m.f. is again increased, curve NPD is followed. *The area within the closed loop is a measure of energy lost during the cycle.* This energy is, in effect, a frictional loss and shows up as heat in the material. *The distance OE is a measure of the residual flux* left in the closed magnetic circuit when the current is zero. It is known as B_r, the *residual induction*. B_r describes circuits in which there are no air gaps, *e.g.* the iron toroid, should not be confused with *remanance*, a more general term which refers to magnetic induction remaining in the magnetic circuit (usually in the air gap when one is present) after magnetising force has been removed.

— The distance OL is known as $-H_c$, the *coercive force*, and is *the value of the demagnetising m.m.f. required to bring the residual or remanent magnetic induction to zero when such a loop is being traced out.*

- If a ferromagnetic substance is subjected to an alternating m.m.f., the first hysteresis loops traced out do not necessarily fall upon each other. When successive loops retrace preceding ones, the material is said to be in a cyclically magnetised condition. *For electro-magnet core materials, values of B_r and $-H_c$ are determined from a hysteresis loop taken when material is cyclically magnetised. Permanent magnet values, however are taken from the first hysteresis loop, since permanent magnets need be magnetised only once.*
- The *hysteresis loop equals the work which is necessary to reverse the direction of magnetisation.* The actual shape and area of loop depend on the internal structure and composition of the ferromagnetic substance.

— Thus *work done* (W) = *(area of B-H loop)* joules/m³/cycle.

It may be noted that while calculating the actual area, scales of B and H should be taken into consideration.

For example, if scales are :

1 cm = x AT/m ... for H

1 cm = y Wb/m² ... for B

then $W = x \cdot y$ (area of B-H loop) joules/m³/cycle.

— *Steinmetz* developed an empirical relationship to express this loss in following terms.

$$P_h = K_h f B_m^{\ k} \qquad \ldots(14)$$

In this expression

P_h = hysteresis loss in watt per m³ or per kg,

B_{max} = maximum flux density, Wb/m²,

K_h = hysteresis coefficient,

k = Steinmetz coefficient, and

f = frequency of magnetization, Hz.

The value of Steinmetz coefficient k is approximately 2 *for all modern magnetic materials.*

Note. The transformers and generators cores and armatures of the electric motors etc. which are subjected to rapid reversals of magnetisation should be made of such substances which have *low coefficient in order to reduce the hysteresis loss.*

Table 2. Hysteresis Coefficients

S. No.	*Material*	*Hysteresis coefficient K_h (joules / m³) × 10²*
1.	Cast iron	27.63 to 40.2
2.	Sheet iron	10.05
3.	Cast steel	7.54 to 30.14
4.	Hard cast steel	63 to 70.34
5.	Silicon steel (4.8 % Si)	1.91
6.	Hard tungsten steel	145.7
7.	Good dynamo sheet steel	5.02
8.	Mild steel castings	7.54 to 22.61
9.	Nickel	32.66 to 100.5
10.	Permalloy	0.25

18.2. Eddy Current Losses

- The term "eddy currents" is applied to *those electric currents which circulate within a mass of conducting material when the latter is situated in a varying magnetic field.* The conducting material may be considered as consisting of large number of closed conducting paths, each of which behaves like a short circuited winding. The varying magnetic field induces eddy e.m.fs. in these closed elemental paths giving rise to eddy currents. These eddy currents produce loss in power resulting in heating of materials. This loss is of considerable importance as it effects the efficiency and heating of electrical machines.
- *The eddy currents produce a magnetic field of their own which opposes the main magnetic field. As the effect of eddy currents is not uniform over the cross-section of the material, this results in a flux distribution which is not uniform, the flux density in the outer portions being greater than that at the centre. Thus there is a reduction in effective cross-section. The effect of eddy currents upon flux distribution is chiefly of importance in transformers where the material otherwise would be worked at a uniform flux density.*
- The magnetic materials used for varying magnetic fields are laminated (made up of thin sheets insulated from each other) so as to reduce eddy currents and associated losses, as by laminating, the *area of path of eddy currents is reduced giving rise to a large value of resistance.*

Eddy Current Loss in Thin Sheets :

Fig. 28 shows a thin plate of thickness t and width b, thickness being *considerably smaller than the width*. Let us suppose that this sheet carries a flux, $B = B_{max} \sin \omega t$, the field running parallel to the axis of the sheet. Eddy currents would flow in the sheet in the elemental paths as shown in Fig. 28.

Let us consider 1 metre axial length and also 1 metre width of sheet. Take an elemental path of thickness dx at a distance x from the axis. Flux enclosed by 1 metre length of this path,

$$\phi = B \text{ (flux density)} \times 2x \times 1 \text{ (area)} = 2xB$$
$$= 2xB \sin \omega t$$

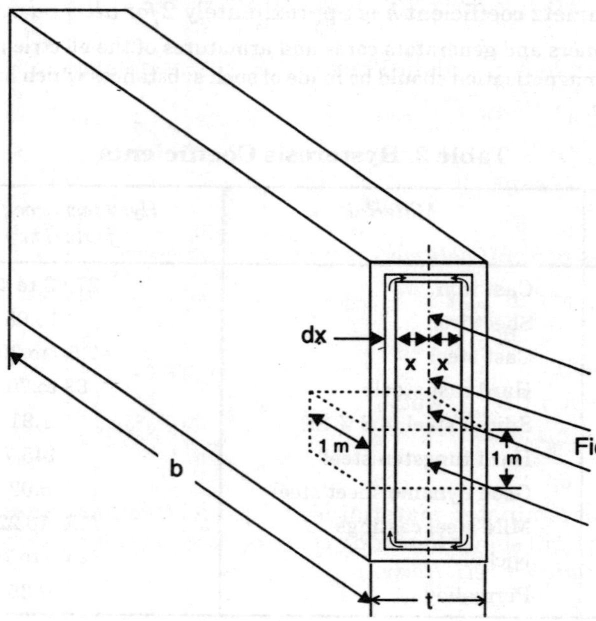

Fig. 28. Eddy currents in lamination.

Instantaneous value of eddy e.m.f. induced in the elemental path,

$$e_x = \frac{d\phi}{dt} = \frac{d}{dt}\,(2x\,B_{max}\sin\omega t)$$
$$= 2x\,B_{max}\cos\omega t$$
$$= 2x\,\omega B_{max}\cos\omega t.$$

R.m.s. value of eddy e.m.f. in the elemental path,

$$E_{r.m.s.} = \frac{1}{\sqrt{2}} \times E_{max}$$
$$= \frac{1}{\sqrt{2}} \times 2x\,\omega\,B_{max} \qquad [\because \quad E_{max} = 2x\omega\,B_{max}\ \textit{i.e.}, \text{ when } \cos\omega t = 1]$$
$$= \sqrt{2}x\,\omega\,B_{max} = \sqrt{2}x \times 2\pi f\,B_{max}$$
$$= 2\sqrt{2}x\,\pi f\,B_{max}$$

If ρ is the resistivity of material, the resistance of each elemental path

$$R_{xe} = \frac{2\times\rho}{1\times dx} = \frac{2\rho}{dx}$$

(Neglecting portions of path parallel to edges of plates).

∴ Eddy current in elemental path,

$$I_{xe} = \frac{E_{xe}}{R_{xe}} = \frac{2\sqrt{2}x\,\pi f\,B_{max}}{2\rho/dx}$$
$$= \frac{\sqrt{2}\,\pi f\,B_{max} x dx}{\rho}$$

Eddy current loss in this path with 1 metre length and 1 metre width is

$$P_{xe} = I_{xe}^{\ 2}\,R_{xe}$$
$$= \frac{2\pi^2 f^2 B^2{}_{max}(xdx)^2}{\rho^2} \times \frac{2\rho}{dx} = \frac{4\pi^2 f^2 B^2{}_{max}}{\rho}\,.\,x^2\,dx$$

∴ Total eddy current loss in a plate 1 metre wide, 1 metre long and t metre thick is

$$= \int_{x=0}^{x=t/2} \frac{4\pi^2 f^2 B^2{}_{max}}{\rho}\,.\,x^2\,dx$$
$$= \frac{\pi^2 f^2 B^2{}_{max}}{6\rho}\,t^3 \text{ watt} \qquad \text{...(15)}$$

∴ Eddy current loss *per unit volume*

$$= \frac{\pi^2 f^2 B^2{}_{max} t^2}{6\rho} \qquad \text{...(16)}$$

(the volume being $t\ m^3$)

$$= K_e f^2\,B^2{}_{max} \text{ watt} \qquad \text{...(17)}$$

where $$K_e = \frac{\pi^2 t^2}{6\rho} \qquad \text{...(18)}$$

— The thickness t of laminated material is of considerable importance as regards to eddy current loss. Eqn. (16) shows *that eddy current losses are reduced by using thinner plates and a material of higher resistivity*.

— Eqn. (16) is derived by *assuming that magnetic field is uniform throughout and parallel to the axis and also the sheets are very thin*. Therefore, as such, it *should not be applied to cases where frequency is abnormal, the sheets thick or the variation of flux density is not sinusoidal with time*.

The *eddy current loss can be* **reduced** in the following ways :

1. It has been found that this loss can be minimised by building up the required cross-section for the flux path by stacking thin pieces known as *"laminations"* as shown in Fig. 29. Since the e.m.fs. set up in the material by the varying flux are usually of small magnitude, the natural oxide on the surface of the sheet iron or steel from which the laminations are punched will effectively insulate the laminations from one another, and thus limit each eddy current path to a single lamination.

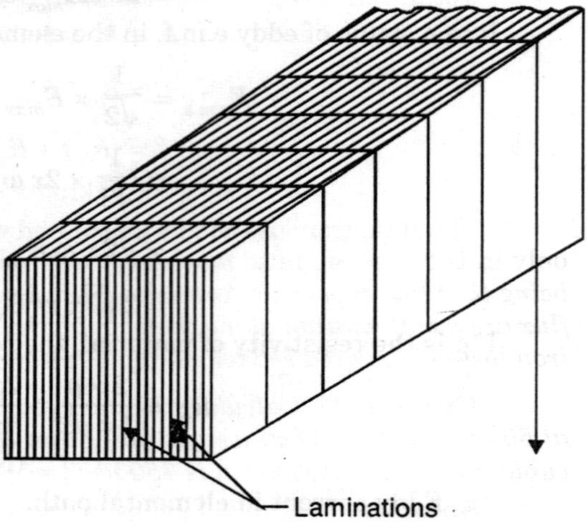

Fig. 29

2. This loss may also be reduced by grinding the ferromagnetic material to a powder and mixing it with a binder that effectively insulates the particles from one another. This mixture is then formed under pressure into the desired shape and heat treated. Magnetic cores for use in communication equipment are frequently made by this process.

- It may be noted that *iron losses* (hysteresis and eddy current losses) *if allowed to take place unchecked, not only reduce the efficiency of electrical equipment but also raise the temperature of the core*. Hence these losses should be kept *as small as is economically possible*.

Other Factors Affecting Core Loss

1. *Rivets and bolts*. Laminations must be held together. A common method is to use rivets or bolts, but they *tend to short circuit lamination insulation*. Rivets and bolts very often form part of a closed magnetic loop which, under A.C. operation, may *cause a serious loss of electrical energy*.

2. *Burrs*. As dies become worn, the burrs that they form on punched edges of lamination stock become larger and larger. Thus burrs are not always removed by deburring operation, and they *cut through sheet insulation forming contacts between laminations and allow excessive eddy currents to flow*. These are commonly referred to as *intersheet eddy currents*. Machining operations that are performed on stacked laminations, such as the grinding of induction motor stators or broaching of solenoid faces, often smear the lamination together, forming good connections between sheets. Many schemes, such as treating the faces with acid or alkali, have been tried in an attempt to eat away the metal particles after laminations have been assembled into apparatus, but this attempted cure often causes other troubles, such as rusting. Burning of the burrs with a flame has also been attempted, with varying success.

3. *Pressure*. Mechanical strains set up in magnetic materials either from *cold working during rolling or punching* or *distortion during final assembly operations*, affect magnetic properties in several ways. The hysteresis and eddy current losses as well as the magnetising current are increased. *The increased hysteresis loss due to punching depends upon the material, its thickness, and the configuration of the grains*. The loss may be as high as 300 per cent of the loss in the unpunched sheet. The strains produced by punching operations increase the loss primarily in the narrow section next to the punched edge. The per cent increase in hysteresis loss is much lower for high flux densities than for low ones.

The *effect of tension on silicon steel is to increase the permeability until high flux densities are reached. The effect of compression, however, is much more marked than that of tension and reduces the permeability considerably.*

Total iron losses

Total iron loss is the sum of hysteresis and eddy current loss and is given by the relation :

$$P_i = P_h + P_e$$

$$= K_h f B_{max}^k + K_e f^2 B_{max}^2 \text{ watts per m}^3 \text{ or per kg} \quad ...(19)$$

$$= (K_h f + K_e f^2) B_{max}^2 \text{ watt per m}^3 \text{ or per kg} \quad ...(20)$$

taking $k = 2$.

The total iron loss can be calculated with the aid of above expression but it gives good results only in the case of static *machinery and not in the case of dynamo electric machinery. The reason being that this expression is applicable when the field is simply alternating, but when the reversal of flux are due to rotational magnetic field, the losses do not follow the same laws. Therefore in practice, iron loss curves are used which represent the loss in watt per kg as a function of flux density.*

Example 17. *Calculate the energy lost per hour in a specimen of iron subjected to magnetisation at 50 c/s. The specimen weighs 50 kg and the hysteresis loop is equivalent in area to 250 joules per cubic metre. Density of iron is 7500 kg/m³.*

Solution. Loss = 250 joules/m³/cycle.

Volume of 1 kg of iron $= \dfrac{1}{\text{Density of iron}} = \dfrac{1}{7500}$

∴ Loss per kg $= 250 \times \dfrac{1}{7500}$ joules/cycle

No. of reversal/sec = 50

Weight of specimen = 50 kg

∴ Hysteresis loss/sec. $= 50 \times 250 \times \dfrac{1}{7500} \times 50 = \dfrac{250}{3}$ joules sec.

and Loss per hour $= \dfrac{250}{3} \times 60 \times 60 = \mathbf{3 \times 10^5}$ **joules. (Ans.)**

Example 18. *The hysteresis loop of a specimen of iron having 10 kg mass is equivalent in area to 250 joules/m³ of iron. Find the loss of energy per hour at the rate of 50 c/s. Assume the density of iron as 7500 kg/m³.*

Solution. Energy loss/m³/cycle = 250 joules

Density of iron = 7500 kg/m³

∴ Volume of iron/kg $= \dfrac{1}{7500}$ m³

Mass of iron specimen = 10 kg

∴ Total volume of iron $= 10 \times \dfrac{1}{7500}$ m³

No. of cycle per second = 50

Total No. of cycles per hour = 50 × 60 × 60

∴ Total loss $= \dfrac{10}{7500} \times 250 \times 50 \times 60 \times 60 =$ **60000 joules. (Ans.)**

Example 19. *The hysteresis loop for a certain magnetic material is drawn to the following scale : 1 cm = 4 AT / cm and 1 cm = 0.05 Wb / m², The loop area is 90 cm² and density of given material is 7800 kg / m³. Calculate the hysteresis loss in watts / kg at 50 c / s.*

Solution. Hysteresis loss = xy × (area of B-H loop) joules/m³/cycle

Also 1 cm = 4 AT/cm = 400 AT/m

and 1 cm = 0.05 Wb/m²

i.e., $x = 400$

and $y = 0.05$

∴ loss = 400 × 0.05 × 90 = 1800 joules/m³/cycle

Volume of 1 kg of material $= \frac{1}{7800}$ m³

∴ Loss/cycle $= 1800 \times \frac{1}{7800}$ joules

Number of reversals/sec. = 50

∴ Loss/kg $= 1800 \times \frac{1}{7800} \times 50$ joules/sec. or watts.

= 11.54 watts. (Ans.)

Example 20. *In a certain transformer, the hysteresis loss is 300 W when the maximum flux density is 0.9 Wb / m² and the frequency 50 c / s. What would be the hysteresis loss if the maximum flux density were increased to 1.1 Wb / m² and the frequency reduced to 40 c / s. Assume the hysteresis loss over this range to be proportional to* $B_{max}^{1.7}$.

Solution. We know that,

Hysteresis loss, $W_h = \eta B_{max}^{1.7} f V$ joules/sec.

$$\therefore \quad P_h \propto B_{max}^{1.7} f$$

In the *first case,* $P_h = 300$ W ; $B_{max} = 0.9$ Wb/m² ; $f = 50$ c/s.

In the *second case,* $P_h = ?$; $B_{max} = 1.1$ Wb/m² ; $f = 40$ c/s.

or
$$\frac{300}{P_h} = \frac{(0.9)^{1.7} \times 50}{(1.1)^{1.7} \times 40}$$

or
$$P_h = 300 \times \left(\frac{1.1}{0.9}\right)^{1.7} \times \frac{4}{5} = \textbf{337.7 watts. (Ans.)}$$

Example 21. *Calculate the energy loss per hour by hysteresis in 50 kg iron core which is subjected to a sinusoidal flux alternating at 50 c / s. The hysteresis loop for the core has an area of 150 cm², when plotted to a scale of 1 cm = 0.008 Wb / m², 1 cm = 20 AT / cm.*

Solution. Here 1 cm = 20 AT/cm = 2000 AT/m

1 cm = 0.008 Wb/m²

Area of hysteresis loop = 150 cm²

∴ Loss = xy × (area of B/H *loop*) joules/m³/cycle

= 2000 × 0.008 × 150 = 2400 joules/m³/cycle

Taking density of iron 7500 kg/m³.

Volume of core $= \frac{50}{7500}$ m³ $= \frac{1}{150}$ m²

Loss for total volume $= 2400 \times \frac{1}{150} = 16$ joules/cycle

No. of reversals per second $= 50$

$\therefore$ Loss/sec $= 16 \times 50 = 800$ joules

or Loss/hour $= 800 \times 3600 =$ **2880000 joules. (Ans.)**

Example 22. *Calculate the loss per kg in a specimen of alloy steel for a maximum flux density of 1.1 Wb/m² and a frequency of 50 Hz, using 0.5 mm thick sheets. The resistivity of alloy steel is 30 × 10⁻⁸ Ω m. The density is 7800 kg/m³. Hysteresis loss in each cycle is 380 W-s/m³.*

Solution. Maximum flux density, $B_{max} = 1.1$ Wb/m²

Frequency $= 50$ Hz

Thickness of sheet $= 0.5$ mm

Resistivity of alloy steel, $\rho = 30 \times 10^{-8}\ \Omega$ m

Density $= 7800$ kg/m³

Hysteresis loss/cycle $= 380$ W-s/m³

Total iron loss :

Eddy current loss is given by,

$$P_e = \frac{\pi^2 f^2 B_{max}^2 t^2}{6\rho} \text{ watt per m}^3$$

$$= \frac{\pi^2 f^2 B_{max}^2 t^2}{6\rho \times 7800} \text{ watt/kg}$$

$$= \frac{\pi^2 \times 50^2 \times 1.1^2 \times (0.5/1000)^2}{6 \times 30 \times 10^{-8} \times 7800} = 0.5316 \text{ W/kg}$$

Hysteresis loss, $P_h = \frac{380 \times 50}{7800} = 2.4359$ W/kg at 50 Hz.

$\therefore$ Total iron loss $= 0.5316 + 2.4359 =$ **2.9675 W/kg. (Ans.)**

Example 23. *A specimen of cold rolled grain oriented 0.3 mm thick stampings has a resistivity of 4.8 × 10⁻⁷ Ω m. The published hysteresis loop is essentially rectangular in form, with a coercive force, of 12 AT/m for all values of peak flux densities upto 1.6 Wb/m². The manufacturer quotes the iron loss in material as 1.2 W/kg, with a sinusoidal flux density 1.0 Wb/m² (peak) at 100 Hz. Calculate the loss in the material from its properties and compare it with the quoted value. The density is 7650 kg/m³.*

Solution. Thickness of stamping = 0.3 mm

Resistivity, $\rho = 4.8 \times 10^{-7}\ \Omega$ m

Density $= 7650$ kg/m³

The material is subjected to a peak flux density of 1 Wb/m² and therefore the value of coercive force is 12 AT/m for all densities. The hysteresis loop is thus a rectangle with sides 2 × 12 = 24 AT/m and 2 × 1 = 2 Wb/m².

Loss per cycle = area of hysteresis loop = 24 × 2 = 48 W-s/m³

Hysteresis loss/kg at 100 Hz $= \frac{48 \times 100}{7650} = 0.627$ W/kg

Eddy current loss, $P_e = \frac{\pi^2 f^2 B_{max}^2 t^2}{6\rho} \text{ W/m}^3 = \frac{\pi^2 f^2 B_{max}^2 t^2}{6\rho \times 7650} \text{ W/kg}$

$$= \frac{\pi^2 \times (100)^2 \times (1.0)^2 \times (0.3/1000)^2}{6 \times 4.8 \times 10^{-7} \times 7650} = 0.403 \text{ W/kg}$$

Total iron loss, $P_i = P_h + P_e = 0.627 + 0.403$
$= \mathbf{1.03\ W/kg.}$ **(Ans.)**

The calculated loss *is smaller* than the quoted. **(Ans.)**

19. FARADAY'S LAWS OF ELECTROMAGNETIC INDUCTION

"The phenomenon whereby an e.m.f. and hence current is induced in any conductor which is cut across or is cut by a magnetic flux is known as ***electromagnetic induction"*****.**

Faraday's First Law. It states as follows :

"Whenever the magnetic flux linked with a circuit changes, an e.m.f. is always induced in it".

Or

"Whenever a conductor cuts magnetic flux, an e.m.f. is induced in that conductor".

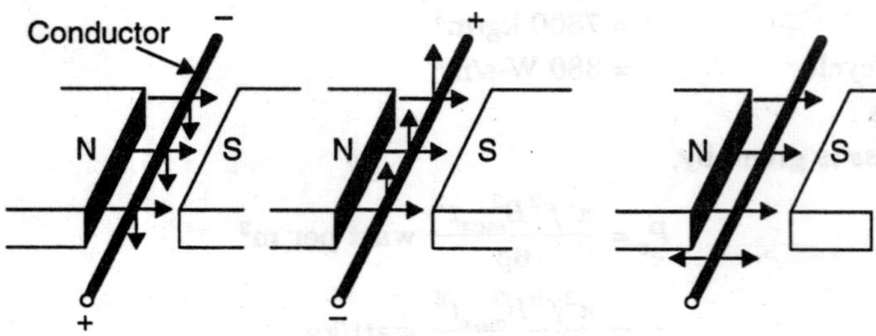

(*a*) Voltage induced across a wire moving downward.
(*b*) Voltage induced across a wire moving upward.
(*c*) No voltage is induced in a wire moving parallel to the field.

Fig. 30. When a conductor is moved across a magnetic field a voltage is induced in the conductor.

Faraday's Second Law. It states as follows :

"The magnitude of induced e.m.f. is equal to the rate of change of flux-linkages".

Mathematically, $$e = -N \frac{d\phi}{dt} \text{ volts} \quad ...(21)$$

where e = induced e.m.f.,

$\frac{d\phi}{dt}$ = rate of change of flux, and

N = number of turns of the coil.

[Usually, a *minus sign* is given to the right-hand side expression to signify the fact that the *induced e.m.f. sets up current in such a direction that magnetic effect produced by it opposes the very cause producing it.*]

Direction of induced e.m.f. and current :

The direction of the induced current may be found easily by applying either Fleming's right-hand Rule (Fig. 31) or Lenz's law. *Fleming's rule is used where induced e.m.f. is due to flux-cutting (i.e., dynamically induced e.m.f.) and Lenz's law when it is due to change by flux-linkages (i.e., statically induced e.m.f.).*

— **Lenz's law.** Fig. 32 shows induction of an e.m.f. in a simple circuit. The direction of the induced e.m.f. is determined by Lenz's law, *which states that the current produced by the induced e.m.f. opposes the change in flux.*

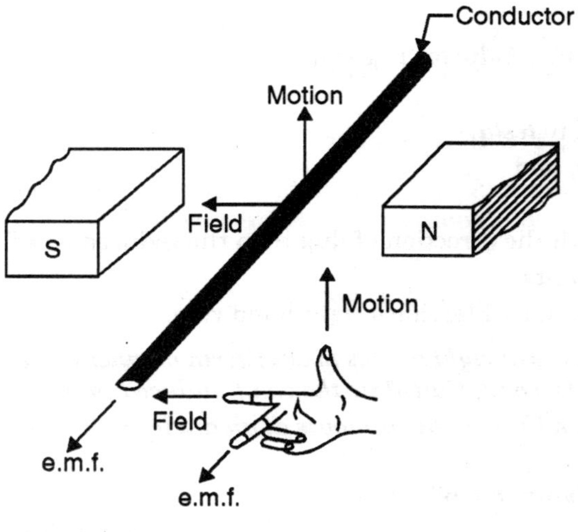

Fig. 31. Fleming's right-hand rule.

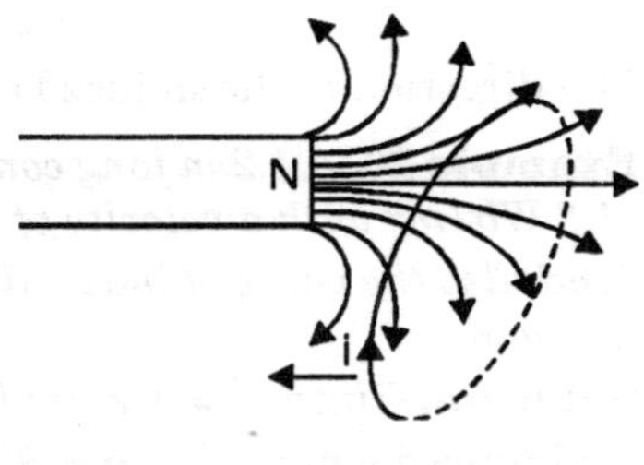

Fig. 32. Induction of e.m.f. in a simple circuit.

Lenz's law may also be stated as follows :

"In all cases of electromagnetic induction, an induced voltage will cause a current to flow in a closed circuit in such a direction that the magnetic field which is caused by that current will oppose the change that produced the current".

Example 24. *A coil has 480 turns and flux in it charges from 0.24 m Wb to 1.2 m Wb in 0.24 second. Calculate the voltage induced in the coil.*

Solution. *Given* : $N = 480$; $d\phi = 0.002 - 0.00024 = 0.00096$ Wb ; $dt = 0.24$ s

Induced e.m.f., e : $e = N\dfrac{d\phi}{dt} = \dfrac{480 \times 0.00096}{0.24} = \mathbf{1.9\ V.}$ **(Ans.)**

Example 25. *The field coils of a six-poles D.C. generator each having 600 turns, are connected in series. On the excitation of the field, there is a magnetic flux of 0.024 Wb / pole. If the field circuit is opened in 0.24 second and residual magnetism is 0.0024 Wb / pole, calculate the average voltage which is induced across field terminals. In which direction is this voltage directed relative to the direction of the current.*

Solution. Total number of turns $= 6 \times 600 = 3600$

Total initial flux $= 6 \times 0.024 = 0.144$ Wb

Total residual flux $= 6 \times 0.0024 = 0.0144$ Wb

Change in flux, $d\phi = 0.144 - 0.0144 = 0.1296$ Wb

Time of opening the circuit, $dt = 0.024$ second

$\therefore$ Induced e.m.f., $\mathbf{e} = N\dfrac{d\phi}{dt} = 3600 \times \dfrac{0.1296}{0.024} = \mathbf{19440\ V.}$ **(Ans.)**

The direction of this induced e.m.f. is the same as the initial direction of exciting current. **(Ans.)**

20. INDUCED E.M.F.

Induced e.m.f. may be of the following two types :

1. Dynamically induced e.m.f.
2. Statically induced e.m.f.

20.1. Dynamically Induced e.m.f. :

Refer Fig. 30. The e.m.f. induced (e) in the conductor is given by :

$$e = Blv \text{ volt} \qquad \ldots(22)$$

where B = flux density of the magnetic field in *tesla*,
l = length of the conductor in *metres*, and
v = velocity of the conductor in *m/s*.

If the conductor moves at an angle θ with the direction of flux then the induced e.m.f.

$$e = Blv \sin\theta \text{ volts} \qquad \ldots(23)$$

The direction of the induced e.m.f. is given by Fleming's right hand rule.

Example 26. *A 1.2 m long conductor moves at right angles to a uniform magnetic field of flux density 1.8 Wb / m² with a velocity of 60 metres / second. Calculate the e.m.f. induced in it.*

Find also the value of induced e.m.f. when the conductor moves at an angle of 30° to the direction of the field.

Solution. *Given* : l = 1.2 m ; B = 1.8 Wb/m² ; v = 60 m/s

(*i*) Induced e.m.f., $\mathbf{e} = Blv$ .... [See Fig. 33 (*a*)]

$= 1.8 \times 1.2 \times 60 =$ **129.6 V. (Ans.)**

(*ii*) Induced e.m.f., $\mathbf{e} = Blv \sin\theta$ .... [See Fig. 33 (*b*)]

$= 1.8 \times 1.2 \times 60 \times \sin 30° =$ **64.8 V. (Ans.)**

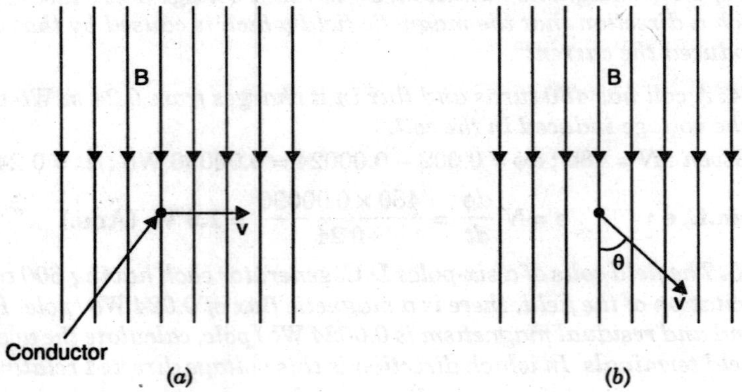

Fig. 33

Example 27. *In a four-pole dynamo, the flux / pole is 18 m Wb. If the armature is driven at 600 r.p.m., what is the average e.m.f. induced in one of the armature conductors ?*

Solution. Flux/pole = 18 m Wb ;

∴ Flux cut in one revolution $= 18 \times 4 = 72 \text{ m Wb} = 72 \times 10^{-3}$ Wb

Speed of the armature $= 600 \text{ r.p.m.} = \dfrac{600}{60} = 10$ r.p.s.

∴ Time taken for one revolution $= \dfrac{1}{10} = 0.1$ s

Average e.m.f. induced, $\mathbf{e} = N\dfrac{d\phi}{dt}$

$N = 1$; $d\phi = 72 \times 10^{-3}$ Wb ; $dt = 0.1$ s

∴ $\mathbf{e} = 1 \times \dfrac{72 \times 10^{-3}}{0.1} =$ **0.72 V. (Ans.)**

20.2. Statically Induced e.m.f. :

The e.m.f. induced by *variation of flux is termed as "statically induced e.m.f."*.

Statically induced e.m.f. can be further subdivided as follows :

(*i*) Self-induced e.m.f.

(*ii*) Mutually induced e.m.f.

20.2.1. Self-induced e.m.f. Self-induced e.m.f. is the *e.m.f. induced in a coil due to the change of its own flux linked with it. If the current through the coil (Fig. 34) is changed* then the flux linked with its own turns will also change which will produce in it, what its called *self-induced e.m.f.* $\left(e = -N\dfrac{d\phi}{dt}\right)$. The *direction of this e.m.f.* is given by *Lenz's law* (and would be such as *to oppose any change of flux which is, in fact, the very cause of its production*).

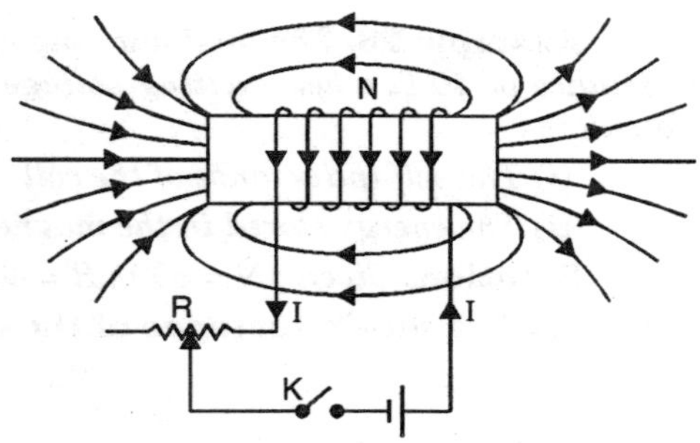

Fig. 34. Self-induced e.m.f. and self-induction.

Self-inductance

The *property of the coil due to which it opposes any increase or decrease of current or flux through it, is known as self-inductance*. It is measured in terms of self-induction L (in henry).

Self-induction is sometimes analogously called *electromagnetic or electrical inertia*.

Co-efficient of self-induction (L) may be found by the following relations :

1. $$L = \frac{N\phi}{i} \text{ henry (\textit{Weber-turns per ampere in the coil})} \quad ...(24)$$

2. $$L = \frac{\mu_0 \mu_r A N^2}{l} \text{ henry (H)} \quad ...(25)$$

3. $$L = \frac{e_L}{\dfrac{di}{dt}} \text{ henry (H)} \quad ...(26)$$

where N = number of turns of the solenoid,

A = area of cross-section,

e_L = induced e.m.f., and

$\dfrac{di}{dt}$ = rate of change of current.

Energy in electromagnetic field

If the current through an inductor is made to increase from 0 to i in a time 't', an e.m.f. e is induced in the coil and the total energy W absorbed by the inductor in time interval 0 to t is given as :

$$W = \int_0^t ei\, dt = \int_0^t \left(L\frac{di}{dt}\right) i\, dt$$

or

$$W = \int_0^i Li\, di = \frac{1}{2} L\, i^2 \text{ joules} \quad ...(27)$$

Thus the inductor has absorbed an energy which is proportional to inductance and the square of current. This energy is stored in the magnetic field. When the current changes from i to zero, this energy is returned to the source. Eqn. (27) can be written as :

$$L = \frac{2W}{i^2} \text{ henrys} \qquad ...(28)$$

Eqn. (28) is *an energy description of inductance.*

Example 28. *The field winding of a D.C. electromagnets is wound with 800 turns and has resistance of 40 Ω when exciting voltage is 230 V, the magnetic flux turbine the coil is 0.004 Wb. Calculate :*

(*i*) *The self inductance of the coil*

(*ii*) *The energy stored in the magnetic field.*

Solution. *Given* : $N = 800$; $R = 40\ \Omega$; $V = 230$ V ; $\phi = 0.004$ Wb,

(*i*) **The self-inductance of the coil, L :**

$$L = \frac{N\phi}{i} \qquad ...(\text{Eqn. } 24)$$

$$= \frac{N\phi}{(V/R)} = \frac{800 \times 0.004}{(230/40)} = \mathbf{0.556\ H.\ (Ans.)}$$

(*ii*) **The energy stored in the magnetic field, W :**

$$W = \frac{1}{2} L\, i^2 \text{ joules} \qquad ...(\text{Eqn. } 27)$$

$$= \frac{1}{2} \times 0.556 \times \left(\frac{230}{40}\right)^2 = \mathbf{9.19\ J.\ (Ans.)}$$

Example 29. *An iron ring 30 cm mean diameter is made of square iron of 2 cm × 2 cm cross-section and is uniformly wound with 400 turns of wire of 2 mm² cross-section. Calculate the value of the self-inductance of the coil. Assume* $\mu_r = 800$. **(Gwalior University)**

Solution. *Given* : $A = 2 \times 2 = 4\ \text{cm}^2 = 4 \times 10^{-4}\ \text{m}^2$; $N = 400$, $l = \pi d = \pi \times 0.3 = 0.3\ \pi m$;

μ_r (relative permeability) = 800

Self-inductance of will, L :

$$L = \frac{\mu_0 \mu_r A N^2}{l} \text{ henry} \qquad [\text{Eqn. } (25)]$$

$$= \frac{4\pi \times 10^{-7} \times 800 \times 4 \times 10^{-4} \times 400^2}{0.3\pi} = \mathbf{0.0683\ H.\ (Ans.)}$$

Example 30. *An air cored solenoid has a length of 60 cm and a diameter of 4 cm. Calculate its inductance if it has 1000 turns and also find the energy stored if the current rises from zero to 6 A.*

(VTU August, 2000)

Solution. *Given* : $l = 60$ cm = 0.6 m ; $d = 4$ cm = 0.04 m ; $N = 1000$; $i = 5 - 0 = 5$ A

Inductance, L :

$$L = \frac{\mu_0 \mu_r A N^2}{l} \qquad [\text{Eqn. } (25)]$$

$$= \frac{4\pi \times 10^{-7} \times 1 \times \left(\frac{\pi}{4} \times 0.04^2\right)^2 \times (1000)^2}{0.6}$$

($\because$ $\mu_r = 1$, since solenoid is air-cored)

$= 2.632 \times 10^{-3}$ H or 2.632 mH

Energy stored :

Energy stored $= \frac{1}{2} Li^2$

$= \frac{1}{2} \times 2.632 \times 10^{-3} \times 6^2 =$ **0.0474 joules. (Ans.)**

Example 31. *If a coil of 200 turns is linked with a flux of 0.01 Wb when carrying current of 10 A, calculate the inductance of the coil. If this current is uniformly reversed in 0.01 second, calculate the induced electromotive force.*

Solution. *Given* : $N = 200$; $\phi = 0.01$ Wb, $i = 10$ A ; $dt = 0.01$ s

$$L = \frac{N\phi}{i} = \frac{200 \times 0.01}{10} = 0.2 \text{ H}$$

Induced electromotive force, e_L :

$$e_L = L \frac{di}{dt} \quad \text{[Eqn. (26)]}$$

$$= 0.2 \times \frac{[10 - (-10)]}{0.01} = \mathbf{400 \text{ V.}} \text{ (Ans.)}$$

Example 32. *A coil of 1500 turns carrying a current of 5 A produces a flux of 2.5 m Wb. If the current is reversed in 0.2 second, find the average vlaue of e.m.f. induced in the coil. Also find the self-inductance of the coil.* **(VTU August, 2001)**

Solution. *Given : N* = 1500 ; $i = 5$ A ; $\phi = 2.5$ m Wb ; $dt = 0.2$ s.

We know that, $\mathbf{L} = \frac{N\phi}{i} = \frac{1500 \times (2.5 \times 10^{-3})}{5} = 0.75 \text{ H}$

∴ E.m.f. induced, $\mathbf{e_L} = L \frac{di}{dt} = 0.75 \times \frac{[5 - (-5)]}{0.2} =$ **37.5 V. (Ans.)**

Example 33. *A direct current of one ampere is passed through a coil of 6000 turns ; it produces a flux of 0.1 m Wb. Assuming that whole of this flux threads all the turns, what is the inductance of the coil ? What would be the voltage developed across the coil if the current were interrupted in 10^{-3} second ? What is the energy stored in the coil.*

Solution. *Given* : $i = 1$ A ; $N = 6000$; $\phi = 0.1$ m Wb $= 0.1 \times 10^{-3}$ Wb ; $dt = 10^{-3}$ s.

$$L = \frac{N\phi}{i} = \frac{6000 \times (0.1 \times 10^{-3})}{1} = \mathbf{0.6 \text{ H.}} \text{ (Ans.)}$$

Induced e.m.f. $e_L = L \cdot \frac{di}{dt} = \frac{0.6 \times 1}{10^{-3}} = 600 \text{ V}$

Energy stored in the coil $= \frac{1}{2} Li^2 = \frac{1}{2} \times 0.6 \times 1^2 =$ **0.3 J. (Ans.)**

20.2.2. Mutually induced e.m.f. Refer Fig. 35. *Production of e.m.f. in coil B due to change in current A is called mutually induced e.m.f.*

- Consider two coils *A* and *B* lying close to each other (Fig. 35). Coil *A* is joined to a *battery*, a *switch* (*K*) and a *variable resistance* (*R*) whereas coil *B* is connected to a *sensitive voltmeter V*.
- When current through *A* is established by closing the switch, its *magnetic field is set up and partly links with or threads through the coil B.* As the current through *A* is changed, the flux linked with *B* is also changed. Hence mutually induced *e.m.f.* is produced in *B* where *magnitude is given by Faraday's law and direction by Lenz's law.*

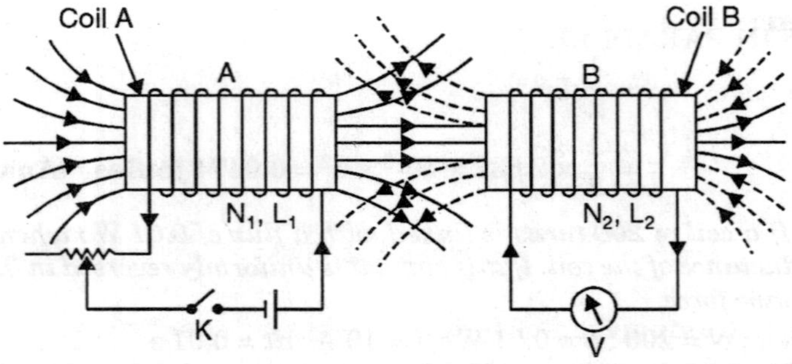

Fig. 35. Mutually induced e.m.f. and mutual inductance.

- Now, if battery is connected to B and the voltmeter across A, then the situation is reversed and now a change of current in B will produce mutually-induced *e.m.f. A*.

— From the examples considered above, it is evident that there is no movement of any conductor, the flux variations being brought about by variations in current strength only. Such an e.m.f. induced in one coil by the influence of the other coil is called (statically but) mutually induced e.m.f.

Mutual inductance

Mutual inductance may be defined *as the ability of one coil (or circuit) to produce e.m.f. in a nearly coil by induction when the current in the first coil changes.* This action being reciprocal, the second coil can also induce an e.m.f. in the first coil when current in the second coil changes. This ability of reciprocal induction is measured in terms of the coefficient of mutual induction M.

Co-efficient of mutual induction (M) may be found by the following relations :

1. $$M = \frac{N_2\phi_1}{i_1} \text{ henry (H)} \quad ...(29)$$

2. $$M = \frac{\mu_0\mu_r A N_1 N_2}{l} \text{ henry (H)} \quad ...(30)$$

3. $$M = -\frac{e_M}{\frac{di_1}{dt}} \quad ...(31)$$

Co-efficient of coupling (k) :

It is defined as the *ratio of mutual inductance between the coils and the square root of product of self-inductance of each coil.*

In other words, $$k = \frac{M}{\sqrt{L_1L_2}} \quad ...(32)$$

21. INDUCTANCES IN SERIES

In general, we have

$$L = L_1 + L_2 + 2M \text{ ...... if } \textit{m.m.fs. are additive} \quad ...(33)$$
$$L = L_1 + L_2 - 2M \text{ ...... if } \textit{m.m.fs. are subtractive} \quad ...(34)$$

22. INDUCTANCES IN PARALLEL

In general, we have

$$L = \frac{L_1L_2 - M^2}{L_1 + L_2 - 2M} \text{ when mutual field } \textit{assists} \text{ the separate fields} \quad ...(35)$$

$$L = \frac{L_1L_2 - M^2}{L_1 + L_2 + 2M} \text{ when two fields } \textit{oppose} \text{ each other} \quad ...(36)$$

Comparison of Inductance and Capacitance

Inductance	*Capacitance*
1. Conducts current (I)	1. Stores charge (Q)
2. $L = L_1 + L_2 + L_3 + ...$ in *series*	2. $\frac{1}{C} = \frac{1}{C_1} + \frac{1}{C_2} + \frac{1}{C_3} + ...$ in *series*
3. $\frac{1}{L} = \frac{1}{L_1} + \frac{1}{L_2} + \frac{1}{L_3} + ...$ in *parallel*	$C = C_1 + C_2 + C_3 + ...$ in *parallel*
4. Requires wire conductor.	4. Requires dielectric as insulator.

Example 34. *If a coil of 150 turns is linked with a flux of 0.01 Wb when carrying a current of 10 A ; calculate the inductance of the coil. If this current is uniformly reversed in 0.1 second, calculate the induced e.m.f. If a second coil of 100 turns is uniformly wound over the first coil, find the mutual inductance between the coils.* **(Pune University)**

Solution. *Given :* $N_1 = 150$; $\phi_1 = 0.01$ Wb ; $i_1 = 10$ A ; $di = 10 - (-10) = 20$ A ;
$dt = 0.1$ s ; $N_2 = 100$

Induced e.m.f. L_1 :

$$\mathbf{L_1} = \frac{N_1\phi_1}{i_1} = \frac{150 \times 0.01}{10} = \mathbf{0.15\ H.\ (Ans.)}$$

Mutual inductance, M :

$$\mathbf{M} = \frac{N_2\phi_1}{i_1} = \frac{100 \times 0.01}{10} = \mathbf{0.1\ H.\ (Ans.)}$$

Example 35. *Two coils having 30 and 600 turns respectively are wound side-by-side on a closed iron circuit of cross-section 100 sq. cm. and mean length 200 cm.*

(*i*) *Estimate the mutual inductance between the coils if the relative permeability of the iron is 2000.*

(*ii*) *If a current of zero ampere grows to 20 A in a time of 0.02 second in the first coil, find the e.m.f. induced in the second coil.* **(JNT University Warrangal)**

Solution. *Given :* $N_1 = 30$; $N_2 = 600$; $A = 100 \times 10^{-4}$ m^2 ; $l = 200$ cm = 2 m
$\mu_r = 2000$; $di = 20 - 0 = 20$ A ; $dt = 0.02$ s.

(*i*) **Mutual inductance, M :**

$$M = \frac{\mu_0\mu_r AN_1N_2}{l} \text{ H} \quad ...(37)$$

$$= \frac{4\pi \times 10^{-7} \times 2000 \times 100 \times 10^{-4} \times 30 \times 600}{2} = 0.226 \text{ H}$$

(*ii*) **Mutually induced e.m.f., e_M :**

$$e_M = M \cdot \frac{di_1}{dt} \qquad \text{...(38)}$$

$$= 0.226 \times \frac{20}{0.02} = \mathbf{226\ V.\ (Ans.)}$$

Example 36. *Two coils A and B are wound on the same ions core. There are 600 turns on A and 3600 turns on B. The current of 4 amperes through coil A produces a flux of 500 × 10⁻⁶ Wb in the core. If this current is reversed in 0.02 second, calculate average e.m.f. induced in coils A and B.*

Solution. *Given* : $N_1 = 600$; $N_2 = 3600$; $i_1 = 4$ A ; $\phi_1 = 500 \times 10^{-6}$ Wb ; $dt = 0.02$ s.

E.m.f. induced in coils A and B, e_1, e_2 :

Self-inductance of coil A, $L_1 = \dfrac{N_1\phi_1}{i_1} = \dfrac{600 \times 500 \times 10^{-6}}{4} = 0.075$ H

Mutual inductance between the coils, $M = \dfrac{N_2\phi_1}{i_1} = \dfrac{3600 \times 500 \times 10^{-6}}{4} = 0.45$ H

(Assuming flux created in coil *A* links completely with the coil *B*, *i.e.*, $\phi_2 = \phi_1$).

Average e.m.f. induced in coil *A*, $\mathbf{e_1} = L_1 \dfrac{di_1}{dt} = 0.075 \times \dfrac{[4-(-4)]}{0.02} = \mathbf{30\ V.\ (Ans.)}$

Average e.m.f. induced in coil *B*, $\mathbf{e_2} = M \dfrac{di_1}{dt} = 0.45 \times \dfrac{[4-(-4)]}{0.02} = \mathbf{1.80\ V.\ (Ans.)}$

Example 37. *The coils A of 11450 turns and B of 14500 turns lie in parallel planes so that 65 per cent of flux produced in A links coil B. It is found that a current of 6 A in A produces a flux of 0.7 m Wb while the same current in B produces 0.9 mWb. Determine :*

(*i*) *Mutual inductance* ; (*ii*) *Co-efficient of coupling.*

Solution. *Given* : $N_1 = 11450$; $N_2 = 14500$; $\phi_2 = 0.65\ \phi_1$; $i_1 = 6$ A ;

$\phi_1 = 0.7$ m Wb $= 0.7 \times 10^{-3}$ Wb ; $\phi_2 = 0.9$ m Wb $= 0.9 \times 10^{-3}$ Wb.

(*i*) **Mutual inductance M :**

$$\mathbf{M} = \frac{N_2\phi_2}{i_1} = \frac{14500 \times 0.65 \times 0.7 \times 10^{-3}}{6} = \mathbf{1.1\ H.\ (Ans.)}$$

(*ii*) **Co-efficient coupling *k* :**

$$L_1 = \frac{N_1\phi_1}{i_1} = \frac{11450 \times 0.7 \times 10^{-3}}{6} = 1.336 \text{ H}$$

$$L_2 = \frac{N_2\phi_2}{i_1} = \frac{14500 \times 0.9 \times 10^{-3}}{6} = 2.175 \text{ H}$$

$$\therefore \quad \mathbf{k} = \frac{M}{\sqrt{L_1 L_2}} \qquad \text{...[Eqn. (32)]}$$

$$= \frac{1.1}{\sqrt{1.336 \times 2.175}} = \mathbf{0.645.\ (Ans.)}$$

Example 38. *The combined inductance of two coils connected in series is 0.6 H or 0.1 H depending on the relative directions of the currents in the coils. If one of the coils when isolated has a self inductance of 0.2 H. Calculate :*

(*i*) *Mutual inductance* ; (*ii*) *Coupling coefficient.* **(Indore University)**

Solution. (*i*) $L = L_1 + L_2 + 2\,M$

or $0.6 = L_1 + L_2 + 2\,M$ ...(*i*)

and $0.1 = L_1 + L_2 - 2\,M$ ...(*ii*)

From (*i*) and (*ii*), we get **M = 0.125 H. (Ans.)**

(*ii*) Let $L_1 = 0.2$ H, then substituting this value in (*i*) above, we get $L_2 = 0.15$ H

$$\therefore \text{ Coupling coefficients, } \quad k = \frac{M}{\sqrt{L_1 L_2}} = \frac{0.125}{\sqrt{0.2 \times 0.15}} = \mathbf{0.72.\ (Ans.)}$$

HIGHLIGHTS

1. Laws of magnetic force :

$$F = \frac{m_1 m_2}{4\pi\mu_0\mu_r d^2}.$$

2. Flux density (*B*).

$$B = \frac{\phi}{A}.$$

3. Magnetic field strength (*H*).

$$H = \frac{NI}{l} \text{ AT/m.}$$

4. Force on a current carrying conductor lying in a magnetic field (*F*) :

$$F = BIl \text{ newtons}$$
$$(= \mu_0\mu_r HIl \text{ newtons}).$$

5. Magnetising force of a long straight conductor

$$H = \frac{NI}{2\pi r} \text{ AT/m and } B = \frac{\mu_0\mu_r NI}{2\pi r}.$$

6. Magnetising force of a long solenoid

$$H = \frac{NI}{l} \text{ AT/m and } B = \frac{\mu_0\mu_r NI}{l}.$$

7. Force between two parallel conductors

$$F = \frac{\mu_0 I_1 I_2 l}{2\pi d} \text{ newtons.}$$

8. **Permeability** is the ratio of flux density to magnetic field strength.
9. **Hysteresis** is defined as the lagging of magnetisation or induction flux density (*B*) behind the magnetising force (*H*).
10. *"The phenomenon whereby an e.m.f. and hence current is induced in any conductor which is cut across or is cut by a magnetic flux is known as* **electromagnetic induction.**
11. **Faraday's first law.** It states that : *"Whenever the magnetic flux linked with a circuit changes, an e.m.f. is always induced in it".*

Or

"Whenever a conductor cuts magnetic flux, an e.m.f. is induced in that conductor".

12. **Lenz's law.** It states that : *"In all cases of electromagnetic induction, an induced voltage will cause a current to flow in a closed circuit in such a direction that magnetic field which is caused by that current will oppose the change that produced the current".*
13. The e.m.f. induced by variation of flux is termed as "statically induced e.m.f.." It is further subdivided as :

(*i*) Self-induced e.m.f. ; (*ii*) Mutually induced e.m.f.

14. *The property of the coil due to which it opposes any increase or decrease of the current or flux through it, is known as* **self-inductance.**
15. *Mutual inductance* is the ability of one coil (or circuit) to produce e.m.f. in a near by coil by induction when the current in the first coil changes.
16. *Co-efficient of coupling* is defined as the ratio of mutual inductance between the coils and the square root of product of self-inductance of each coil.

OBJECTIVE TYPE QUESTIONS

A. Choose the correct answer :

1. An air gap is usually inserted in magnetic circuits to
(*a*) increase m.m.f (*b*) increase the flux (*c*) prevent saturation (*d*) none of the above.
2. The relative permeability of a ferromagnetic material is
(*a*) less than one (*b*) more than one
(*c*) more than 10 (*d*) more than 100 or 1000.
3. The unit of magnetic flux is
(*a*) henry (*b*) weber (*c*) ampere turn/metre (*d*) ampere/metre.
4. Permeability in a magnetic circuit corresponds to .... in an electric circuit.
(*a*) resistance (*b*) resistivity (*c*) conductivity (*d*) conductance.
5. Point out the *wrong* statement. Magnetic leakage is undesirable in electric machines because it
(*a*) lowers their power efficiency (*b*) increases their cost of manufacture
(*c*) leads to their increased weight (*d*) produces fringing.
6. Relative permeability of vacuum is
(*a*) 1 (*b*) 1 H/m (*c*) $1/4\pi$ (*d*) $4\pi \times 10^{-7}$ H/m.
7. The magnetising force (*H*) and magnetic flux density (*B*) are connected by the relation
(*a*) $B = \mu_r H/\mu_0$ (*b*) $B = \mu H$ (*c*) $B = H/\mu_0\mu_r$ (*d*) $B = \mu_0 H/\mu_r$.
8. Permanent magnets are normally made of
(*a*) alnico alloys (*b*) aluminium (*c*) cast iron (*d*) wrought iron.
9. Energy stored by a coil is doubled when its current is increased by ...... percent.
(*a*) 25 (*b*) 50 (*c*) 41.4 (*d*) 100.
10. Those magnetic materials are best suited for making armature and transformer cores which have ...... permeability and ..... hysteresis loss.
(*a*) high, high (*b*) low, high (*c*) high, low (*d*) low, low.
11. The rate of rise of current through an inductive coil is maximum
(*a*) at 63.2% of its maximum steady value (*b*) at the start of the current flow
(*c*) after one time constant (*d*) near the final maximum value of current.
12. When both the inductance and resistance of a coil are doubled the value of
(*a*) time constant remains unchanged (*b*) initial rate of rise of current is doubled
(*c*) final steady current is doubled (*d*) time constant is halved.
13. The initial rate of rise of current through a coil of inductance 10 H when suddenly connected to a D.C. supply of 200 V is ...... A/s
(*a*) 50 (*b*) 20 (*c*) 0.05 (*d*) 500.
14. A material for good magnetic memory should have
(*a*) low hysteresis loss (*b*) high permeability (*c*) low retentivity (*d*) high retentivity.
15. Conductivity is analogous to
(*a*) retentivity (*b*) resistivity (*c*) permeability (*d*) inductance.
16. In a magnetic material hysteresis loss takes place primarily due to
(*a*) rapid reversals of its magnetisation (*b*) flux density lagging behind magnetising force
(*c*) molecular friction (*d*) its high retentivity.

17. Those materials are well suited for making permanent magnets which have ...... retentivity and ...... coercivity
(*a*) low, high (*b*) high, high (*c*) high, low (*d*) low, low.

18. If the area of hysteresis loop of a material is large, the hysteresis loss in this material will be
(*a*) zero (*b*) small (*c*) large (*d*) none of the above.

19. Hard steel is suitable for making permanent magnets because
(*a*) it has good residual magnetism (*b*) its hysteresis loop has large area
(*c*) its mechanical strength is high (*d*) its mechanical strength is low.

20. Silicon steel is used in electrical machines because it has
(*a*) low coercivity (*b*) low retentivity
(*c*) low hysteresis loss (*d*) high coercivity.

21. Conductance is analogous to
(*a*) permeance (*b*) reluctance (*c*) flux (*d*) inductance.

22. The property of a material which opposes the creation of magnetic flux in it is known as
(*a*) reluctivity (*b*) magnetomotive force (*c*) permeance (*d*) reluctance.

23. The unit of retentivity is
(*a*) weber (*b*) weber/sq. m (*c*) ampere turn/metre (*d*) ampere turn.

24. Reciprocal of reluctance is
(*a*) reluctivity (*b*) permeance (*c*) permeability (*d*) susceptibility.

25. While comparing magnetic and electric circuits, the flux of magnetic circuit is compared with which parameter of electrical circuit ?
(*a*) E.m.f. (*b*) Current (*c*) Current density (*d*) Conductivity.

26. The unit of reluctance is
(*a*) metre/henry (*b*) henry/metre (*c*) henry (*d*) 1/henry.

27. A ferrite core has less eddy current loss than an iron core because
(*a*) ferrites have high resistance (*b*) ferrites are magnetic
(*c*) ferrites have low permeability (*d*) ferrites have high hysteresis.

28. A ferromagnetic core subjected to cycles of magnetisation will exhibit hysteresis when the cycle is
(*a*) rotating (*b*) alternating (*c*) pulsating (*d*) any of the above.

29. In order to minimise loss due to hysteresis, the magnetic material should have
(*a*) high resistivity (*b*) low hysteresis co-efficient
(*c*) large B-H loop area (*d*) high retentivity.

30. Hysteresis loss least depends on
(*a*) volume of material (*b*) frequency
(*c*) steinmetz co-efficient of material (*d*) ambient temperature.

31. Laminated cores, in electrical machines, are used to reduce
(*a*) copper loss (*b*) eddy current loss (*c*) hysteresis loss (*d*) all of the above.

32. The area of hysteresis loss is a measure of
(*a*) permittivity (*b*) permeance (*c*) energy loss per cycle (*d*) magnetic flux.

33. Permeance is to reluctance as conductance is to
(*a*) ampere turns (*b*) inductance (*c*) capacitance (*d*) resistance.

34. According to steinmetz hysteresis law, hysteresis loss in a material is proportional to
(*a*) $B^{3.6}$ (*b*) $B^{1.6}$ (*c*) $B^{1.2}$ (*d*) $B^{2.6}$.

35. The transformer cores operating at microwave frequency range, are usually made of
(*a*) carbon (*b*) copper (*c*) silicon steels (*d*) ferrites.

36. The property of coil by which a counter e.m.f. is induced in it when the current through the coil changes is known as
(*a*) self-inductance (*b*) mutual inductance (*c*) series aiding inductance (*d*) capacitance.

37. As per Faraday's laws of electromagnetic induction, an e.m.f. is induced in a conductor whenever it

(*a*) lies perpendicular to the magnetic flux (*b*) lies in a magnetic field

(*c*) cuts magnetic flux

(*d*) moves parallel to the direction of the magnetic field.

38. Which of the following circuit element stores energy in the electromagnetic field ?

(*a*) Inductance (*b*) Condenser (*c*) Variable resistor (*d*) Resistance.

39. The inductance of a coil will increase under all the following conditions *except*

(*a*) when more length for the same number of turns is provided

(*b*) when the number of turns of the coil increase

(*c*) when more area for each turn is provided

(*d*) when permeability of the core increases.

40. Higher the self-inductance of a coil,

(*a*) lesser its weber-turns (*b*) lower the e.m.f. induced

(*c*) greater the flux produced by it

(*d*) longer the delay in establishing steady current through it.

41. In an iron cored coil the iron core is removed so that the coil becomes an air cored coil. The inductance of the coil will

(*a*) increase (*b*) decrease

(*d*) remain the same (*d*) initially increase and then decrease.

42. An open coil has

(*a*) zero resistance and inductance (*b*) infinite resistance and zero inductance

(*c*) infinite resistance and normal inductance (*d*) zero resistance and high inductance.

43. Both the number of turns and the core length of an inductive coil are doubled. Its self-inductance will be

(*a*) unaffected (*b*) doubled (*c*) halved (*d*) quadrupled.

44. If current in a conductor increases then according to Lenz's law self-induced voltage will

(*a*) air the increasing current (*b*) tend to decrease the amount of current

(*c*) produce current opposite to the increasing current

(*d*) aid the applied voltage.

45. The direction of induced e.m.f. can be found by

(*a*) Laplace's law (*b*) Lenz's law

(*c*) Fleming's right hand rule (*d*) Kirchhoff's voltage law.

46. Air-core coils are practically free from

(*a*) hysteresis losses (*b*) eddy current losses (*c*) both (*a*) and (*b*) (*d*) none of the above.

47. The magnitude of the induced e.m.f. in a conductor depends on the

(*a*) flux density of the magnetic field (*b*) amount of flux cut

(*c*) amount of flux linkages (*d*) rate of change of flux-linkages.

48. Mutually inductance between two magnetically-coupled coils depends on

(*a*) permeability of the core (*b*) the number of their turns

(*c*) cross-sectional area of their common core (*d*) all of the above.

49. A laminated iron core has reduced eddy-current losses because

(*a*) more wire can be used with less D.C. resistance in coil

(*b*) the laminations are insulated from each other

(*c*) the magnetic flux is concentrated in the air gap of the core

(*d*) the laminations are stacked vertically.

50. The co-efficient of self-inductance for a coil is given as

(*a*) $\dfrac{NI}{\phi}$ (*b*) $\dfrac{N\phi}{I}$ (*c*) $\dfrac{NI^2}{\phi}$ (*d*) $\dfrac{N\phi}{I^2}$.

51. The law that the induced e.m.f. and current always oppose the cause producing them is due to
(a) Faraday (b) Lenz (c) Newton (b) Coulomb.

52. Which of the following is not a unit of inductance ?
(a) Henry (b) Coulomb/volt ampere
(c) Volt second per ampere (d) All of the above.

53. In case of an inductance, current is proportional to
(a) voltage across the inductance (b) magnetic field
(c) both (a) and (b) (d) neither (a) nor (b).

54. Which of the following circuit elements will oppose the change in circuit current ?
(a) Capacitance (b) Inductance (c) Resistance (d) All of the above.

55. For a purely inductive circuit which of the following is *true* ?
(a) Apparent power is zero (b) Relative power is zero
(c) Actual power of the circuit is zero
(d) Any capacitance even if present in the circuit will not be charged.

56. Which of the following is unit of inductance ?
(a) Ohm (b) Henry (c) Ampere turns (d) Webers/metre.

57. An e.m.f. of 16 volts is induced in a coil of inductance 4 H. The rate of change of current must be
(a) 64 A/s (b) 32 A/s (c) 16 A/s (d) 4 A/s.

58. The core of a coil has a length of 200 mm. The inductance of coil is 6 mH. If the core length is doubled, all other quantities, remaining the same, the inductance will be
(a) 3 mH (b) 12 mH (c) 24 mH (d) 48 mH.

59. The self-inductances of two coils are 8 mH and 18 mH. If the co-efficients of coupling is 0.5, the mutual inductance of the coils is
(a) 4 mH (b) 5 mH (c) 6 mH (d) 12 mH.

60. Two coils have inductances of 8 mH and 18 mH and a co-efficient of coupling of 0.5. If the two coils are connected in series aiding, the total inductance will be
(a) 32 mH (b) 38 mH (c) 40 mH (d) 48 mH.

61. A 200 turns coil has an inductance of 12 mH. If the number of turns is increased to 400 turns, all other quantities (area, length etc.) remaining the same, the inductance will be
(a) 6 mH (b) 14 mH (c) 24 mH (d) 48 mH.

62. Two coils have self-inductances of 10 H and 2 H, the mutual inductance being zero. If the two coils are connected in series, the total inductance will be
(a) 6 H (b) 8 H (c) 12 H (d) 24 H.

63. In case all the flux from the current in coil 1 links with coil 2, the co-efficient of coupling will be
(a) 2.0 (b) 1.0 (c) 0.5 (d) zero.

64. The energy in joules stored in the magnetic field of 0.15 H inductance with a 180 mA current will be
(a) 2.43 (b) 2.43×10^{-3} (c) 2.43×10^{-6} (d) 2.43×10^{-9}.

65. A coil with negligible resistance has 50 V across it with 10 mA. The inductive reactance is
(a) 50 ohms (b) 500 ohms (c) 1000 ohms (d) 5000 ohms.

ANSWERS

1. (c)	**2.** (d)	**3.** (b)	**4.** (c)	**5.** (a)	**6.** (a)	**7.** (b)
8. (a)	**9.** (c)	**10.** (c)	**11.** (b)	**12.** (a)	**13.** (b)	**14.** (d)
15. (c)	**16.** (d)	**17.** (b)	**18.** (c)	**19.** (a)	**20.** (c)	**21.** (a)
22. (d)	**23.** (b)	**24.** (b)	**25.** (b)	**26.** (d)	**27.** (d)	**28.** (d)
29. (b)	**30.** (d)	**31.** (b)	**32.** (c)	**33.** (d)	**34.** (b)	**35.** (d)
36. (a)	**37.** (c)	**38.** (a)	**39.** (a)	**40.** (d)	**41.** (b)	**42.** (b)
43. (b)	**44.** (c)	**45.** (b)	**46.** (c)	**47.** (d)	**48.** (d)	**49.** (b)

50. (*b*)	51. (*b*)	52. (*b*)	53. (*b*)	54. (*b*)	55. (*c*)	56. (*b*)
57. (*d*)	58. (*a*)	59. (*c*)	60. (*b*)	61. (*d*)	62. (*c*)	63. (*b*)
64. (*b*)	65. (*d*).					

THEORETICAL QUESTIONS

1. Define the following terms :
   (*i*) Magnetic flux density (*ii*) Relative permeability
   (*iii*) Absolute permeability (*iv*) Magnetomotive force.
2. State the differences between magnetically soft and magnetically hard materials.
3. Sketch the shape of magnetic field around a current-carrying conductor. How can the direction of the field be determined ?
4. What do you understand by the ferromagnetic, paramagnetic and diamagnetic materials ?
5. Sketch the magnetic field that occurs around two adjacent current carrying conductors (*i*) for currents in opposite direction (*ii*) for currents in the same direction.
6. Derive an expression for magnetising force of a long straight conductor.
7. Derive an expression for magnetising force of a long solenoid.
8. Derive an expression for force between two parallel conductors.
9. Define the following terms as applied to a magnetic circuit :
   Permeability ; Magnetomotive force (m.m.f.) ; Reluctance ; Permeance ; Reluctivity.
10. Give the analogy between electric and magnetic circuits (similarities).
11. What do you mean by magnetic leakage and fringing ?
12. Draw and explain a magnetising curve for a ferromagnetic material.
13. Explain briefly the following two types of losses which occur when magnetic materials are subjected to an alternating flux :
    (*i*) Hysteresis loss
    (*ii*) Eddy current loss.
14. What do you understand by "electromagnetic induction".
15. Explain briefly Faraday's first and second laws of electromagnetic induction.
16. State and explain Fleming's right hand rule.
17. What is "Lenz's law" ?
18. What is difference between the dynamically and statically induced e.m.fs. ?
19. Define the term "Self-inductance".
20. Derive an expression for energy stored in an electromagnetic field.
21. Explain briefly "Mutually induced e.m.f.".
22. What is "Mutual inductance" ?
23. What do you understand by "Co-efficient of coupling" ?

UNSOLVED EXAMPLES

MAGNETISM

1. The total flux of an electromagnet is 8×10^{-4} Wb. (*i*) If the cross-sectional area of the core is 100 mm^2, find the flux density in the core. (*ii*) The electromagnet has 50 turns and a current of 2 A flows through the coil. If the length of the magnetic circuit is 200 mm, find the m.m.f. and magnetic field strength.
   [**Ans.** (*i*) 8 T, (*ii*) 100 AT, 500 AT/m]
2. A long straight conductor carries a steady current of 10 A. Find (*i*) the intensity of magnetic field produced at a point 150 mm from the axis of the conductor and (*ii*) flux density of the field at that point.
   [**Ans.** 10.61 A/m, 133.3×10^{-7} T]
3. A solenoid of 400 turns is wound on a continuous ring of magnetic material of relative permeability 1200. If the flux density of the magnetic material of the core in the solenoid is 1.3 T, what is the current in the solenoid ? Mean diameter of the solenoid is 200 mm.
   [**Ans.** 1.35 A]

4. Two wires carrying current, in the same direction, of 5000 A and 10000 A are placed with their axis 50 mm apart. Calculate the force between them in N/m. [**Ans.** 200 N/m]
5. Two long parallel linear conductors carry 50 A. If the conductors are separated by 25 mm find the force per metre length on the conductor if the current flow is (*i*) in the opposite direction (*ii*) in the same direction. [**Ans.** (*i*) 0.02 N (repulsive) (*ii*) 0.02 (alternative)]
6. An air cored toroidal coil has 3000 turns and carries a current of 0.1 A. The length of the magnetic circuit is 150 mm and the cross-sectional area of the coil is 400 mm^2. Calculate : (*i*) Magnetic field strength, (*ii*) flux density and (*iii*) total flux within the coil.
[**Ans.** (*i*) 2000 AT/m, (*ii*) 2.51×10^{-3} T, (*iii*) 1.004×10^{-6} Wb]
7. A cast steel ring of 200 mm mean diameter and 300 mm^2 cross-section has a winding of 200 turns. If the relative permeability of steel is 1000, what current will be required produce a flux of 3.75×10^{-4} Wb.
[**Ans.** 3.17 A]
8. An iron ring of mean length 500 mm has an air gap of 1 mm and a winding of 200 turns. If the permeability of iron is 400 when a current of 1.25 A flows through the coil, find the flux density. [**Ans.** 0.134 T]
9. An iron ring 100 mm mean diameter is made of 15 mm round iron of permeability 900, has an air gap 5 mm wide. It has 400 turns. If the current through this winding is 3.4 A, determine : (*i*) magnetomotive force, (*ii*) the total reluctance of the circuit, (*iii*) the flux in the ring and (*iv*) the flux density in the ring.
[**Ans.** (*i*) 1360 AT, (*ii*) 0.2394×10^8 AT/Wb, (*iii*) 5680×10^{-8} Wb, (*iv*) 0.321 T]
10. A magnetic ring has a mean circumference of 1.5 m and is of 0.01 m^2 in cross-section, and is wound with 175 turns. A saw cut of 4 mm width is made in the ring. Calculate the magnetising current required to produce a flux of 0.8 m Wb in the air gap.
Assume leakage factor as 1.25 and permeability of iron as 400. [**Ans.** 3.16 A]

ELECTROMAGNETISM

11. A flux of 4×10^{-4} Wb is created by a current of 10 A flowing through a 100 turn coil. Calculate the inductance of coil corresponding to a complete reversal of the current in 0.01 second. Find the magnitude of induced e.m.f.
[**Ans.** 4×10^{-3} H ; 8 V]
12. The field winding of a D.C. electromagnet is wound with 960 turns and has resistance of 50 Ω when exciting voltage is 230 V, the magnetic flux linking with the coil is 0.005 Wb. Calculate : (*i*) The self inductance of the coil ; (*ii*) The energy stored in the magnetic field. [**Ans.** (*i*) 1.0435 H ; (*ii*) 11.04 J]
13. A solenoid has 1,200 turns and carries a current of 2 A. The iron core has a length of 0.4 m and cross-section of 80 cm^2, the relative permeability is 1,000. Calculate the self-induced e.m.f. in the solenoid, if the current is switched off in 0.01 second. [**Ans.** 7238 V]
14. A coil of 300 turns, wound on a core of non-magnetic material has an inductance of 10 mH. Calculate :
(*i*) The flux produced by a current of 5 A ;
(*ii*) The average value of e.m.f. induced when a current of 5 A is reversed in 8 milli seconds.
[**Ans.** (*i*) 0.1667 m Wb, (*ii*) 12.5 V]
15. Two coils A and B are wound on the same iron core. There are 500 turns on A and 3,000 turns on B. A current of 5 amperes through coil A produces a flux of 600×10^6 Wb in the core. If this current is reversed in 0.02 s, calculate the average e.m.f. induced in coils A and B. [**Ans.** 30 V ; 180 V]
16. Two coils, *A* of 12500 turns and *B* of 16000 turns, lie in parallel planes so that 60% of the flux produced in *A* links coil *B*. It is found that a current of 5 A in *A* produces a flux of 0.6×10^{-3} Wb while the same current in *B* produces 0.8×10^{-3} Wb. Determine :
(*i*) Mutual inductance ;
(*ii*) Coefficient of coupling. [**Ans.** (*i*) 1.15 H ; (*ii*) 0.586]

5

Transformers

1. GENERAL ASPECTS

Although the transformer is not classified as an electric machine, the principles of its operation are fundamental for the induction motor and synchronous machines. Since A.C. electric machines are normally built for low frequencies only the low-frequency power transformer will be considered in this text.

Function. *The function of a transformer, as the name implies, is to transform alternating current energy from one voltage into another voltage*. The transformer has *no rotating parts*, hence it is often called a *static transformer*.

When energy is transformed into a higher voltage the transformer is called a *step-up transformer* but when the case is otherwise it is called a *step-down transformer*. Most power transformers operate at constant voltage, *i.e.* if the power varies the current varies while the voltage remains fairly constant.

Applications. A transformer performs many important functions in prominent areas of electrical engineering.

- In *electrical power engineering* the transformer makes it possible to convert electric power from a generated voltage of about 11 kV (as determined by generator design limitations) to higher values of 132 kV, 220 kV, 400 kV, 500 kV and 765 kV thus permitting transmission of huge amounts of power along long distances to appropriate distribution points at tremendous savings in the cost of transmission lines as well as in power losses.
- At *distribution points* transformers are used to reduce these high voltages to a safe level of 400/230 volts for use in homes, offices etc.
- In *electric communication circuits* transformers are used for a variety of purposes *e.g.*, as an impedance transformation device to allow maximum transfer of power from the input circuit to the output device.

- In *radio and television circuits* input transformers, interstage transformers and output transformers are widely used.
- Transformers are also used in *telephone circuits, instrumentation circuits* and *control circuits*.

2. BASIC DEFINITIONS

- A *transformer* is a *static electromagnetic device designed for the transformation of the (primary) alternating current system into another (secondary) one of the same frequency with other characteristics, in particulars, other voltage and current*.
- As a rule a transformer consists of a core assembled of sheet transformer steel and two or several windings coupled *electromagnetically*, and in the case of *autotransformer, also electrically*.
- A transformer with two windings is called *double-wound transformer* ; a transformer with three or more windings is termed a *triple wound* or *multi-winding* one.
- According to the kind of current, transformers are distinguished as single-phase, three-phase and poly-phase ones. A *poly-phase transformer winding is a group of all phase windings of the same voltage*, connected to each other in a definite way.
- **Primary and secondary windings.** The transformer winding to which the energy of the alternating current is delivered is called the *primary winding* ; the other winding from which energy is received is called the *secondary winding*.
- In accordance with the names of the windings, all quantities pertaining to the primary winding as, for example, power, current, resistance etc., are also primary, and those pertaining to the secondary winding secondary.
- **h.v. and l.v. windings.** The winding connected to the circuit with the higher voltage is called the *high-voltage winding* (h.v.), the winding connected to the circuit with the lower voltage is called the *low-voltage winding*. (l.v.). If the secondary voltage is *less* than the primary one, the transformer is called a *step-down transformer* and if *more-a step-up transformer.*
- *A tapped transformer* is one whose windings are fitted with special taps for changing its voltage or current ratio.
- **Oil and dry transformers.** To avoid the deteremental effect of the air on the winding insulation and improve the cooling conditions of the transformer its core together with the windings assembled on it is immersed in a tank filled with transformer oil. Such transformers are called **oil transformers**. Transformers not immersed in oil are called **dry transformers.**

3. WORKING PRINCIPLE OF A TRANSFORMER

A transformer operates on the principle of *mutual inductance,* between two (and sometimes more) inductively coupled coils. It consists of two windings in close proximity as shown in Fig. 1. *The two windings are coupled by magnetic induction.* (There is no conductive connection between the windings). One of the windings called *primary* is energised by a sinusoidal voltage. The second winding, called *secondary* feeds the load. The alternating current in the primary winding sets up an alternating flux (ϕ) in the core. The secondary winding is linked by most of this flux and e.m.fs are induced in the two windings. The e.m.f. induced in the secondary winding drives a current through the load connected to the winding. Energy is transferred from the primary circuit to the secondary circuit through the medium of the magnetic field.

In brief, a transformer is a *device* that :

(*i*) *transfers electric power from one circuit to another* ;

(*ii*) *it does so without change of frequency ; and*

(*iii*) *it accomplishes this by electromagnetic induction (or mutual inductance).*

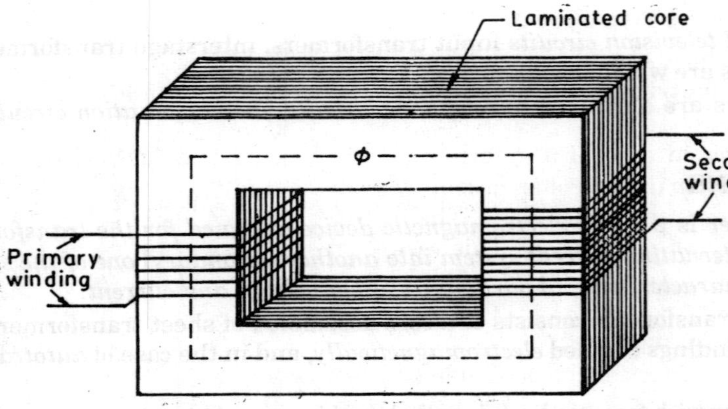

Fig. 1. Two winding transformer.

4. TRANSFORMER RATINGS

The rated quantities of a transformer, its power, voltage, frequency, etc. are given in Manufacturer's name plate, which should always be arranged so as to be accessible. But the term 'rated' can also be applied to quantities not indicated on the name plate, but relating to the rated duty, as for example, the rated efficiency, rated temperature conditions of the cooling medium, etc. :

- *The rated duty* of a transformer is determined by the quantities given in the name plate.
- The *rated power* of the transformer is the power at the secondary terminals, indicated in the name plate and expressed in kVA.
- *The rated primary voltage* is the voltage indicated in the transformer name plate ; if the primary is provided with taps, the rated tapped voltage is specially noted.
- *The rated secondary voltage* is the voltage across the transformer *secondary terminals at no-load* and with the rated voltage across the primary terminals ; if the secondary winding has taps, then their rated voltage is specially indicated.
- *The rated currents of the transformer*, primary and secondary, are the currents indicated in the name plate of the transformer and calculated by using the corresponding rated values of power and voltage.

5. KINDS OF TRANSFORMERS

The following kinds of transformers are the most important ones :

1. **Power transformers.** *For the transmission and distribution of electric power.*

2. **Auto-transformers.** *For converting voltages within relatively small limits to connect power systems of different voltages, to start A.C. motors etc.*

3. **Transformer for feed installations with static convertors.** (Mercury arc rectifiers, ignitrons, semi-conductor valves, etc.) *When converting A.C. into D.C.* (rectifying) *and convering D.C. into A.C. (inverting).*

4. **Testing transformers.** *For conducting tests at high and ultra-high voltages.*

5. **Power transformers for special applications.** Furnace, welding etc.

6. **Radio-transformers.** *It is used in radio engineering etc.*

Note. *Distribution transformers* should be designed to have maximum efficiency at a load much *lower than full-load* (*about 50 per cent*).

Power transformers should be designed to have maximum efficiency *at or near full-load.*

6. TRANSFORMER CONSTRUCTION

All transformers have the following essential elements :

1. Two or more **electrical windings** insulated from each other and from the core (except in auto-transformers).

2. A **core,** which in case of a single-phase distribution transformers usually comprises *cold-rolled silicon-steel strip* instead of an assembly of punched silicon-steel laminations such as are used in the larger power-transformer cores. The *flux path in the assembled core is parallel to the directions of steel's grain or 'orientation'.* This results in a *reduction in core losses* for a given flux density and frequency, or *it permits the use of higher core densities* and *reduced size of transformers for given core losses.*

Other necessary parts are :

- A *suitable container* for the assembled core and windings.
- A *suitable medium* for insulating the core and its windings from each other and from the container.
- Suitable *bushings* for insulating and bringing the terminals of the windings out of the case.

The two basic types of transformer construction are :

1. *The core type.*
2. *The shell type.*

The above two types differ in their relative arrangements of copper conductors and the iron cores. In the *'core type'*, the *copper virtually surrounds the iron core*, while *in the 'shell type', the iron surrounds the copper winding.*

6.1. Core Type Transformer. The completed magnetic circuit of the core-type transformer is in the shape of the hollow rectangle, exactly as shown in Fig. 2 in which I_0 is the no-load current and ϕ is the flux produced by it. N_1 and N_2 are the number of turns on the primary and secondary sides respectively.

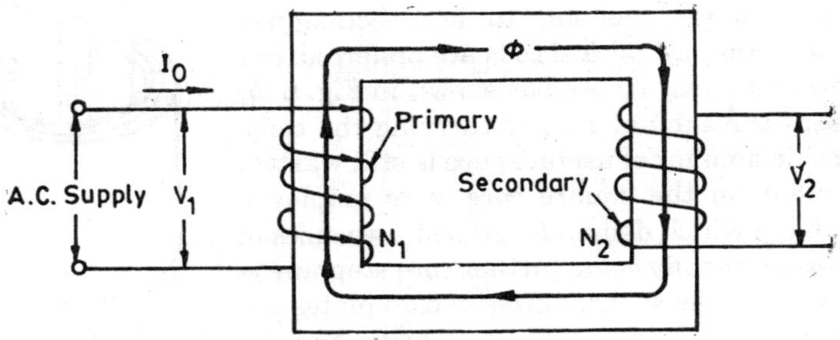

Fig. 2. Magnetic circuit of a core-type transformer.

The core is made up of *silicon-steel laminations* which are, either rectangular or L-shaped. With the coils wound on two legs the appearance is that of Fig. 3. If the two coils shown were the respective high and low-side coils as in Fig. 3, the *leakage reactance would be much too great.* In order to provide maximum *linkage* between windings, the group on *each leg is made up of both high-tension and low-tension coils.* This may be seen in Fig. 4, where a cross-sectional cut is taken across the legs of the core. By placing the high-voltage winding around the low-voltage winding, only one layer of high-voltage insulation is required, that between the two coils. *If the high-voltage coils were adjacent to the core, an additional high-voltage insulation layer would be necessary between the two coils and the iron core.*

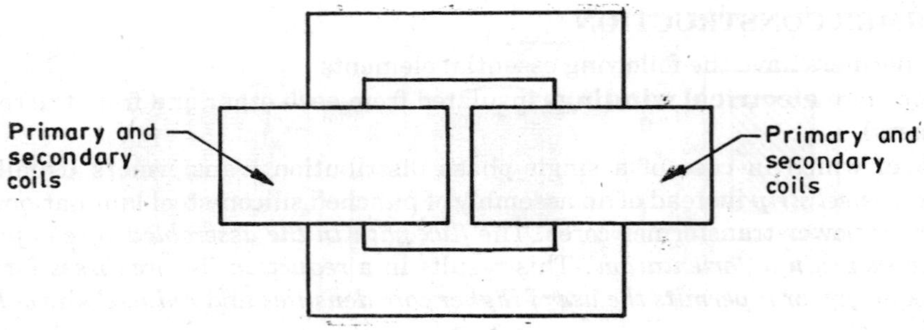

Fig. 3. Core-type transformer.

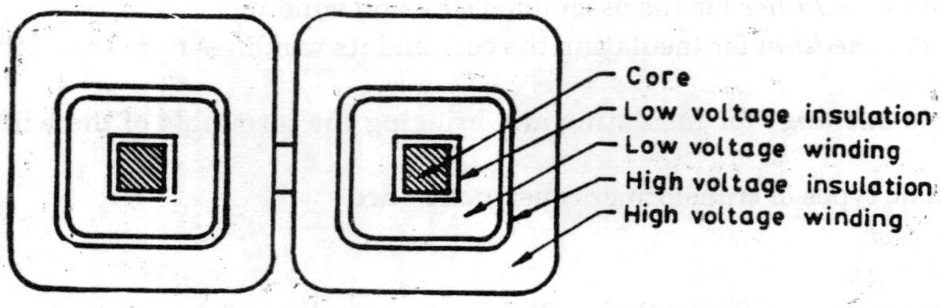

Fig. 4. Cross-section of a core-type transformer.

Fig. 5 shows the coils and laminations of a core-type transformer with a cruciform core and circular coils.

- Fig. 6 shows *the different types of cores used in core type transformers.*

Rectangular cores [Fig. 6 (*a*)] with rectangular cylindrical coils can be used for small size core-type transformers. For large size transformers it becomes wasteful to use rectangular cylindrical coils and so circular cylindrical coils are preferred. For such purposes, '*square cores*' may be used as shown in Fig. 6 (*b*) where circles represent the tubular former carrying the coils. Evidently a considerable amount of useful space is still wasted. A common improvement on the square core is to employ a '*cruciform core*' [Fig. 6 (*c*)] which demands, atleast, two sizes of core strips. For *very large transformers*, further core stepping is done as in Fig. 6 (*d*) where atleast three sizes of core plates are necessary. *Core stepping not only gives high space factor but also results in reduced length of the mean turn and the consequent I^2R loss. Three stepped core is the most commonly* used although more steps may be used for very large transformers as shown in Fig. [6 (*e*)].

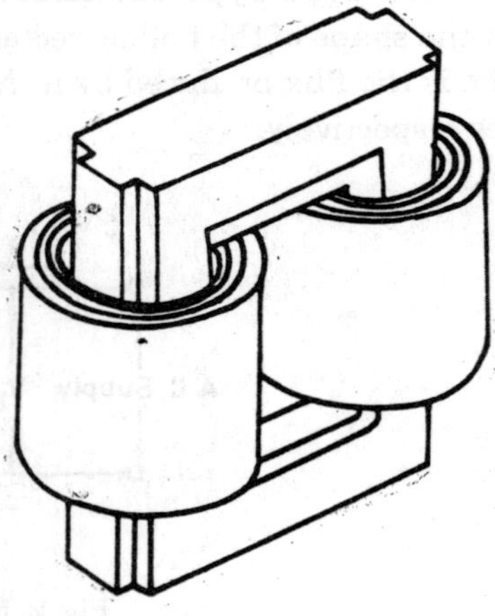

Fig. 5. Coils and laminations of a core-type transformer.

6.2. Shell Type Transformer. In the shell-type construction the iron almost entirely surrounds the copper (Fig. 7). The core is made up of E-shaped or F-shaped laminations which are stacked to give a rectangular figure eight. All the windings are placed on the centre leg, and in order to reduce leakage, each high-side coil is adjacent to a low-side coil. The coils actually occupy the entire space of both windows, are flat or pencake in shape, and are usually constructed of strip copper. Again, *to reduce the amount the high-voltage insulation required, the low-voltage coils are placed adjacent to the iron core.*

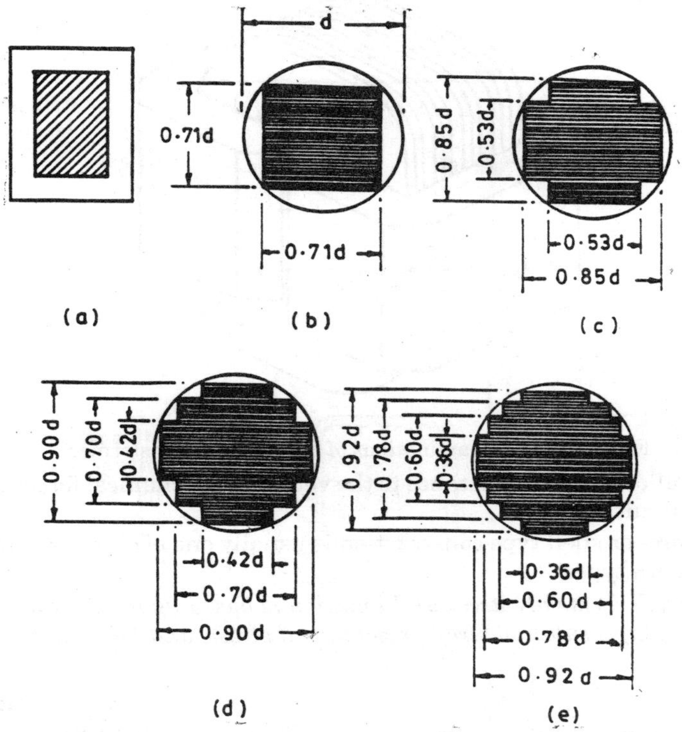

Fig. 6. Various types of cores.

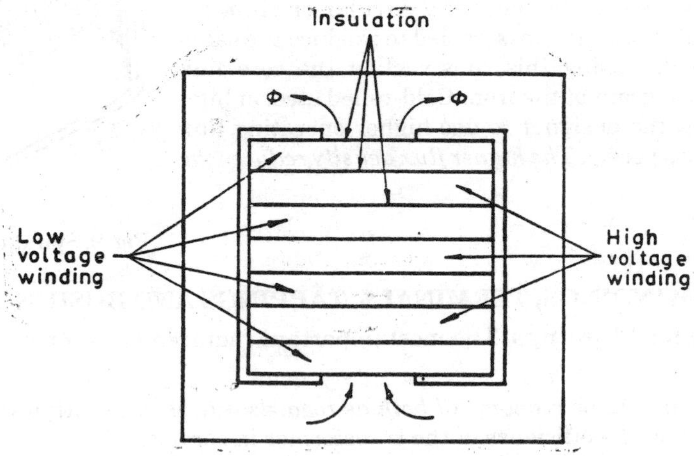

Fig. 7. Shell-type transformer.

Fig. 8 shows the coils and laminations of a typical shell-type transformer.

Choice of Core- or Shell Type Construction. In general, the core-type has a longer mean length of core and a shorter mean length of coil turn. The core type also has a smaller cross-section of iron and so will need a greater number of turns of wire, since, in general, not as high a flux may be reached in the core. However, *core type is better adopted for some high-voltage service since there is more room for insulation.* The *shell type has better provision for mechanically supporting and*

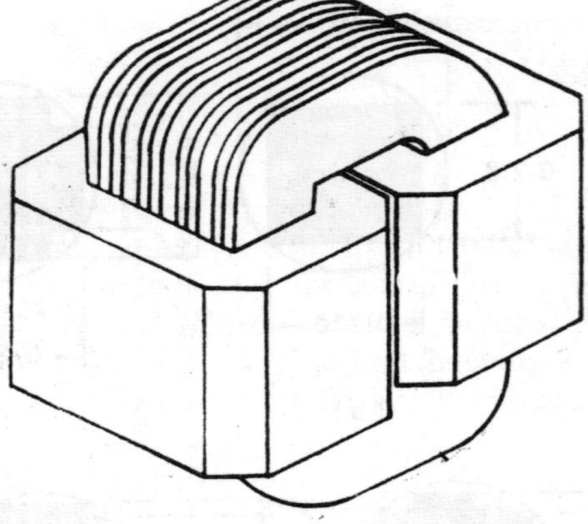

Fig. 8. Coils and laminations of a shell-type transformer.

bracing the coils. This allows better resistance to the very high mechanical forces that develop during a high-current short circuit.

The choice of core- or shell type construction is usually one of *cost*, for similar characteristics can be obtained with both types.

Both core and shell forms are used, and selection is based upon many factors such as voltage rating, kVA rating, weight, insulation stress, mechanical stress, and heat distribution.

6.3. Spiral Core Transformer. The typical spiral core is shown in Fig. 9. The core is assembled either of a *continuous* strip of transformer steel wound in the form of a circular or elliptical cylinder or of a group of short strips assembled to produce the same elliptical-shaped core. By using this construction the core flux always follows along the grain of the iron. Cold-rolled steel of high silicon content enables the designer to use higher operating flux densities with lower loss per kg. *The higher flux density reduces the weight per kVA.*

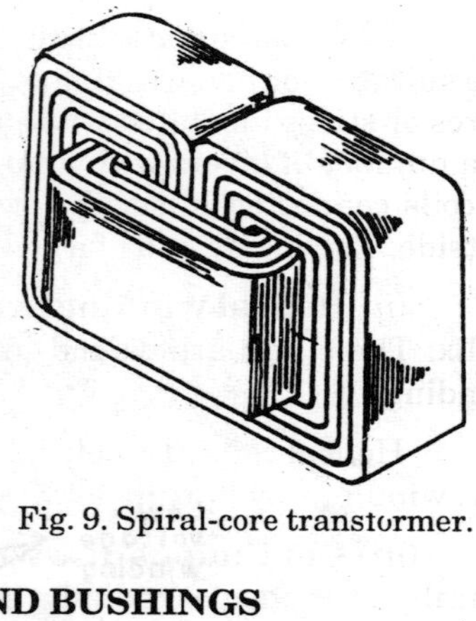

Fig. 9. Spiral-core transtormer.

7. TRANSFORMER WINDINGS, TERMINALS, TAPPINGS AND BUSHINGS

7.1. Transformer Windings. The most important requirements of transformer windings are :

1. The winding should be *economical both as regards initial cost*, with a view to the market availability of copper, and the efficiency of the transformer in service.

2. The heating conditions of the windings should *meet standard requirements*, since departure from these requirements towards allowing higher temperature will drastically shorten the service life of the transformer.

3. The winding should be *mechanically stable* in respect to the forces appearing when sudden short circuit of the transformer occur.

4. The winding should have the *necessary electrical strength in respect to over voltages*.

The different *types of windings* are *classified* and briefly discussed below :

1. Concentric windings :

(*i*) Cross-over (*ii*) Helical

(*iii*) Disc.

2. Sandwich windings

1. Concentric windings. Refer Fig. 10. These windings are *used for core type transformers.* Each limb is wound with a group of coils consisting of both primary and secondary turns which may be concentric cylinders. The l.v. winding is placed next to the core and h.v. winding on the outside. But the two windings can be sub-divided, *and interlaced with high tension and low tension section alternately to reduce leakage reactance.* These windings can be further divided as follows :

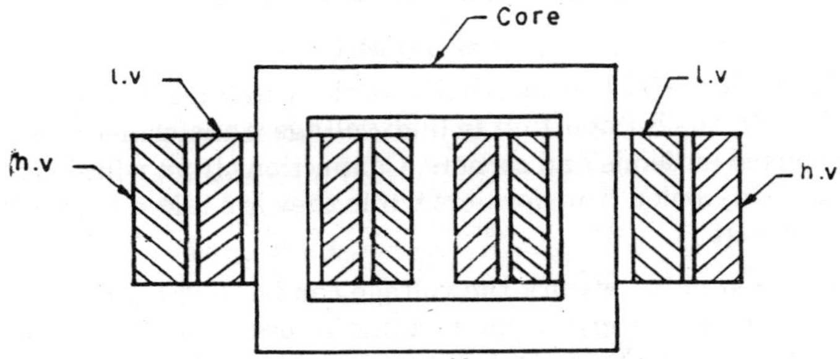

Fig. 10. Concentric coils.

(*i*) **Cross-over windings.** Cross-over windings are used for currents up to 20 A and so they are suitable for h.v. winding of small transformers. The conductors are either cotton covered round wires or strips insulated with paper. Cross-over coils are wound over formers and each coil consists of a number of layers with a number of turns per layer. The complete winding consists of a number of coils connected in series. Two ends of each coil are brought out, one from inside and one from outside. The inside end of a coil is connected to the outside end of the adjacent coil.

(*ii*) **Helical winding.** A helical winding consists of rectangular strips wound in the form of a helix. The strips are wound in parallel radially and each turn occupies the total radial depth of winding.

Helical coils are well suited for *l.v. windings of large transformers*. They can also be used for h.v. windings by putting extra insulation between layers in addition to insulation of conductors.

(*iii*) **Continuous disc winding.** This type of winding consists of a number of flat strips wound spirally from inside (radially) outwards. The conductor is used in such lengths as are sufficient for complete winding or section of winding between tappings. The conductor can either be a single strip or a number of strips in parallel, wound on the flat. This gives a robust construction for each disc. The discs are wound on insulating cylinders spaced from it by strips along the length of cylinder. The discs are separated from each other with press board sectors attached to the vertical strips. The vertical and horizontal spacers provide ducts for free circulation of oil which is in contact with every turn.

2. Sandwich coils. Sandwich coils (Fig. 11) are employed in transformers of *shell type.* Both high and low voltage windings are split into a number of sections. Each high voltage section lies between the low voltage sections.

The *advantage of sandwich coils is that their leakage can be easily controlled and so any desired value of leakage reactance can be had by the division of windings.*

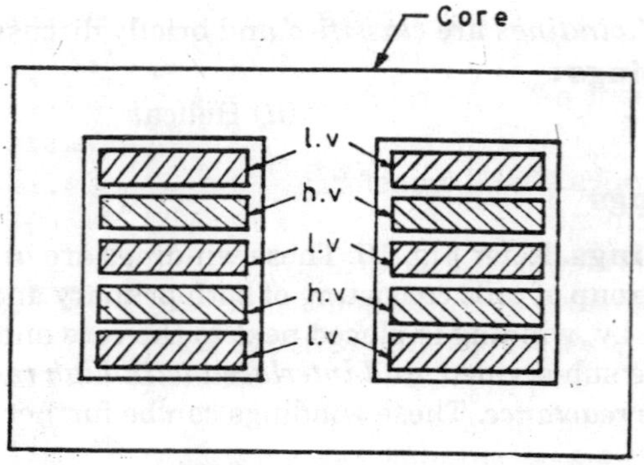

Fig. 11. Sandwich coils.

7.2. Terminals and Leads. The connection to the windings are of insulated copper rods or bars. The shape and size of leads is important in high voltage transformers owing to dielectric stress and corona which are caused at bends and corners. Connections from windings are directly taken to the busbars in the case of air-cooled transformers while they are taken to insulated bushings in the case of oil-cooled transformers.

7.3. Tappings. In a supply network the voltage can be controlled by changing the transformation ratio. This can be done by tapping the winding in order to alter the number of turns. The change in number of turns may be effected when the transformer is out of circuit (known as off load tap changing) or when on load (known as on load tap changing). *The tappings are provided on the high voltage winding because a fine voltage variation is obtained owing to large number of turns. It is difficult to obtain voltage variation within close percentage limits in low voltage winding as there are few turns and voltage per turn is a large percentage of the total voltage.*

In transformers, the tappings can be provided at :

(*i*) phase ends ; and

(*ii*) neutral point or in the middle of the windings.

- The advantage of providing tappings at *phase ends* is that the number of bushing insulators is reduced, this is important where the cover space is limited. Some transformers have reinforced insulation at the phase ends. It is essential that in such cases either the tapping should not be provided at end turns or the reinforcement should be carried beyond the lower tap.
- When the tappings are made at the *neutral point* the insulation between various parts is small. *This arrangement is economical especially in the case of high voltage transformers.*

7.4. Bushings. The bushings are *employed for insulating and bringing out terminals of the winding from the container to the external circuit.* For low-voltage transformers this is accomplished by employing *bushings of porcelain around the conductor at the point of entry.* For high voltages it is necessary to employ bushings of larger sizes. In modern transformers the problem is met by using *large porcelain or composition bushings for voltages as high as 33 kV, above that oil filled* or *condenser type bushings are used.*

8. TRANSFORMER COOLING

8.1. Cooling Methods. The transformers get heated due to iron and copper losses occurring in them. It is necessary to dissipate this heat so that the temperature of the windings is kept below the value at which the insulation begins to deteriorate. The *cooling of transformers is more difficult than that of rotating machines because the rotating machines create a turbulent air flow which assists in removing the heat generated due to losses.* Luckily the losses in transformers are comparatively small. Nevertheless the elaborate cooling arrangements have been devised to deal with the whole range of sizes.

As far as cooling methods are concerned, the transformers are of following two types :

1. Dry type. 2. Oil immersed type.

Dry Type Transformers. Small transformers upto *25 kVA size* are of the *dry type* and have the following cooling arrangements :

(*i*) **Air natural.** In this method the natural circulation of surrounding air is utilized to carry away the heat generated by losses. A sheet metal enclosure protects the winding from mechanical injury.

(*ii*) **Air blast.** Here the transformer is cooled by a continuous blast of cool air forced through the core and windings (Fig. 12). The blast is produced by a fan. The air supply must be filtered to prevent accumulation of dust in ventilating ducts.

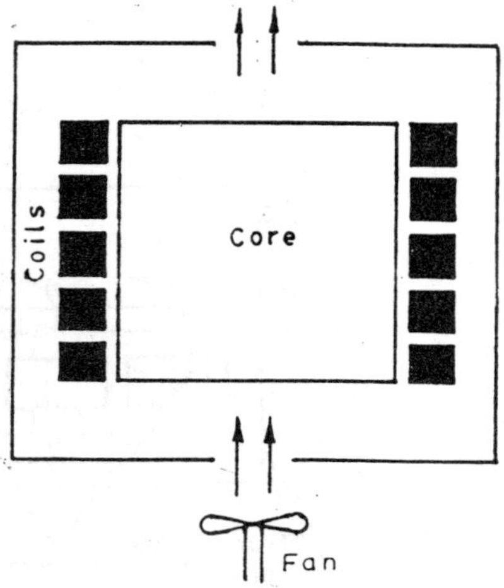

Fig. 12

Oil Immersed Transformers. In general most transformers are of oil immersed types. The *oil provides better insulation than air and it is a better conductor of heat than air*. Mineral oil is used for this purpose.

Oil immersed transformers are classified as follows :

(*i*) **Oil immersed self-cooled transformers.** The transformer is immersed in oil and heat generated in cores and windings is passed to oil by conduction. Oil in contact with heated parts rises and its place is taken by cool oil from the bottom. The natural oil transfers its heat to the tank walls from where heat is taken away by the ambient air. The oil gets cooler and falls to the bottom from where it is dissiptated into the surroundings. The tank surface is the best dissipator of heat but a plain tank will have to be excessively large, if used without any auxiliary means for high rating transformers. As both space and oil are costly, these auxiliary means should not increase the cubic capacity of the tank. The heat dissipating capacity can be increased by providing (*i*) corrugations, (*ii*) fins, (*iii*) tubes (Fig. 13) and (*iv*) radiator tanks.

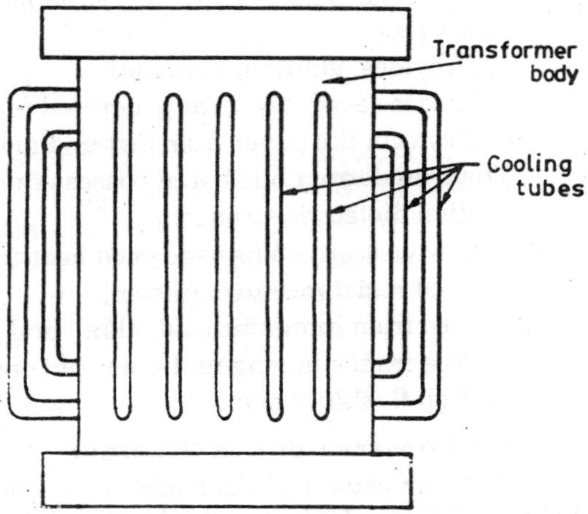

Fig. 13. Transformer with cooling tubes.

The *advantages of 'oil natural' cooling is that it does not clog the ducts and the windings are free from effects of moisture.*

(*ii*) **Oil immersed forced air-cooled transformers.** In this type of cooling, air is directed over the outer surfaces of the tank of the transformer immersed in oil.

(*iii*) **Oil immersed water-cooled transformers.** Heat is extracted from the oil by means of a stream of water pumped through a metallic coil immersed in the oil just below the top of the tank. The heated water is in turn cooled in a spray pond or a cooling tower.

(*iv*) **Oil immersed forced oil cooled transformers.** In such transformers heat is extracted from the oil by pumping the oil itself upward through the winding and then back by way of external radiators which may themselves be cooled by fans. The *extra cost of oil pumping equipment must of* *course be economically justified but it has incidentally the advantage of reducing the temperature difference between the top and bottom of enclosing tank.*

Fig. 14 shows the cooling of transformers having capacities from 10000 kVA and higher. In such cases air blast cooling of radiator is used.

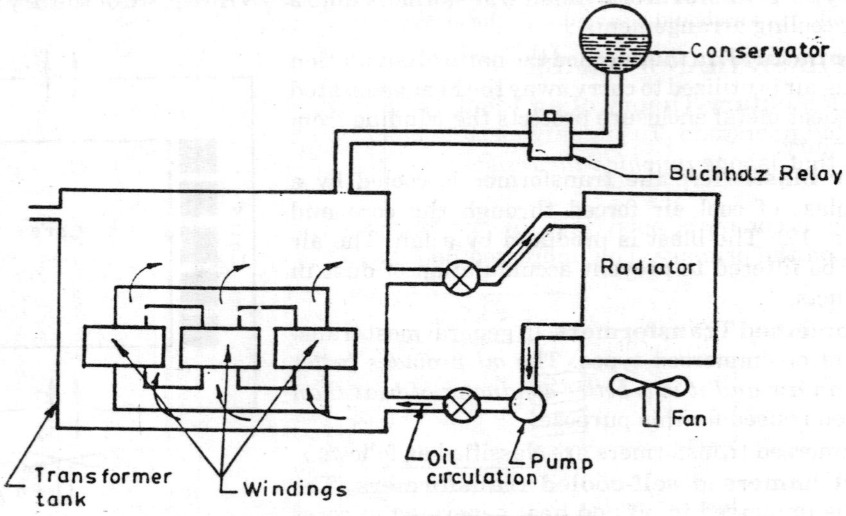

Fig. 14. Air blast cooling of radiator.

8.2. Transformer Oil. It is a *mineral oil obtained by refining crude petroleum*. It serves the following *purposes* :

(*i*) Provides additional insulation.
(*ii*) Carries away the heat generated in the core and coils.
(*iii*) Protects the paper from dirt and moisture.

The transformer oil should possess the following *properties* :

1. High dielectric strength.
2. Low viscosity to provide good heat transfer.
3. Good resistance to emulsion.
4. Free from inorganic acid, alkali and corrosive sulphur.
5. Free from sludging under normal operating conditions.
6. High flash/fire point.

8.3. Conservator and Breather

Conservator. The oil should not be allowed to come in contact with atmospheric air as it may take up moisture which may spoil its insulating properties. Also air may cause acidity and sludging of oil. To prevent this, many transformers are provided with conservators. The *function of a conservator (Fig. 14) is to take up contraction and expansion of oil without allowing it to come in contact with outside air*. The conservator consists of an air tight metal-drum fixed above the level of the top of the tank and connected with it by a pipe. The main tank is completely filled with oil when cold. The *conservator is partially filled with oil*. So the oil surface in contact with air is greatly reduced. The sludge thus formed remains in the conservator itself and does not go to the main tank.

Breather. When the temperature changes, the oil expands or contracts and there is a displacement of air. When the transformer cools, the oil level goes down, and air is drawn in. This is known as *breathing*. The air, coming in, is passed through an apparatus called *breather* for the purpose of extracting moisture. The *breather consists of a small vessel which contains a drying agent like silica get crystal impregnated with cobalt crystal*.

Note. *Sludging* means the slow formation of solid hydrocarbons due to heating and oxidation. The sludge deposit itself on the windings and cooling ducts producing overheating. This makes transformer still hotter producing more sludge. This process may continue till the transformer becomes unusable due to overheating. So the contact of oil with air should be avoided as the air contains oxygen.

9. SINGLE PHASE TRANSFORMER

9.1. Elementary Theory of an Ideal Transformer. The basic theory of a transformer is not difficult to understand. To simplify matters as much as possible, let us first consider *an ideal transformer*, that is, *one in which the resistance of the windings is negligible and the core has no losses*.

Let the secondary be open (Fig. 15), and let a sine wave of potential difference v_1 (Fig. 16) be impressed upon the primary. The impressed potential difference causes an alternating current to

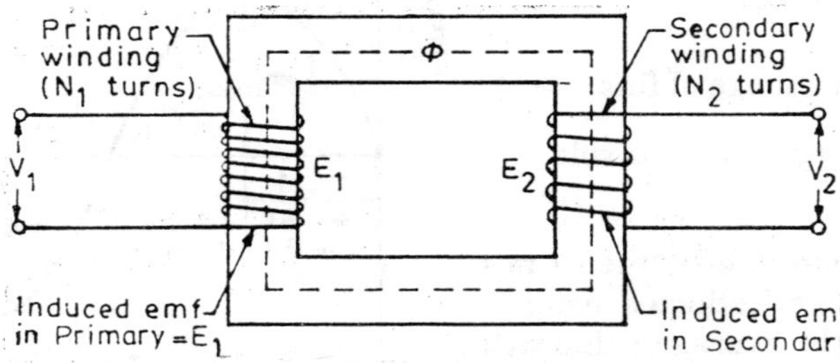

Fig. 15. Elementary diagram of an ideal transformer with an open secondary winding.

flow in the primary winding. Since the primary resistance is negligible and there are no losses in the core, the effective resistance is zero and the circuit is purely reactive. Hence the current wave i_m lags the impressed voltage wave v_1 by 90 time degrees, as shown in Fig. 17. The reactance of circuit is very high and the magnetizing current is very small. This current in the N_1 turns of the primary magnetizes the core and produces a flux ϕ that is at all times proportional to the current (if the permeability of the circuit is assumed to be constant), and therefore in time phase with the current. The flux, by its rate of change, induces in the primary winding E_1 which at every instant of time is equal in value and opposite in direction to V_1. It is called *counter e.m.f.* of the primary. The value which the primary current attains must be such that the flux which it produces in the core is of sufficient value to induce in the primary the required counter e.m.f.

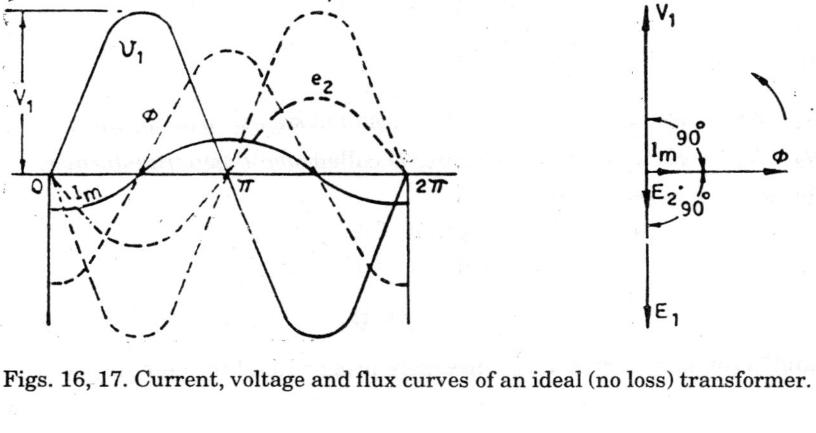

Figs. 16, 17. Current, voltage and flux curves of an ideal (no loss) transformer.

Since the flux also threads (or links) the secondary winding a voltage e_2 is induced in the secondary. This voltage is likewise proportional to the rate of change of flux and so is in time phase with e_1, but it may have any value depending upon the number of turns N_2 in the secondary.

9.2. E.M.F. Equation of a Transformer

Let N_1 = number of turns in primary,

N_2 = number of turns in secondary,

ϕ_m = maximum flux in the core, Wb.

$= B_m \times A$, [where B_m is the maximum flux density in the core and A is the core area], and

f = frequency of a.c. input, Hz.

Refer Fig. 18. Since the flux increases from its zero value to maximum value ϕ_m in one quarter of the cycle *i.e.* in $\frac{T}{4}$ or $\frac{1}{4f}$ second (T being time-period of the cycle),

∴ Average rate of change of flux

$$= \frac{\phi_m}{\frac{1}{4f}} = 4f\phi_{max} \text{ Wb/s or volt}$$

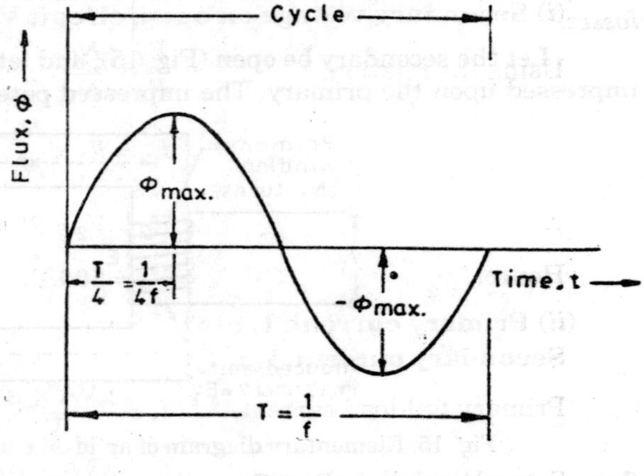

Fig. 18

If flux ϕ varies *sinusoidally,* then r.m.s. (root mean square) value of inducted e.m.f. is obtained by multiplying the average value with form factor.

$$\text{But, form factor} = \frac{\text{r.m.s. value}}{\text{average value}} = 1.11$$

∴ r.m.s. value of e.m.f./turn

$$= 1.11 \times 4f\phi_{max} = 4.44 f\phi_{max} \text{ volt}$$

Now, r.m.s. value of induced e.m.f. in the whole of primary winding,

$$E_1 = 4.44 f\phi_{max} N_1 \quad ...(1)$$

Similarly r.m.s. value of induced e.m.f. in secondary is,

$$E_2 = 4.44 f\phi_{max} N_2 \quad ...(2)$$

In an ideal transformer on no-load $V_1 = E_1$ and $V_2 = E_2$ (Fig. 15).

9.3. Voltage Transformation Ratio (K). The *transformation ratio is defined as the ratio of the secondary voltage to primary voltage.* It is denoted by the letter K.

From eqns. (1) and (2), $$\frac{E_2}{E_1} = \frac{N_2}{N_1} = K \quad ...(3)$$

- If $N_2 > N_1$ *i.e.* $K > 1$, then transformer is called *step-up transformer.*
- If $N_2 < N_1$ *i.e.* $K < 1$, then transformer is called *step-down transformer.*

Again for an *ideal transformer*

$$\text{Input } (VA) = \text{Output } (VA)$$

$$V_1 I_1 = V_2 I_2 \text{ or } E_1 I_1 = E_2 I_2$$

or $$\frac{I_2}{I_1} = \frac{E_1}{E_2} = \frac{N_1}{N_2} = \frac{1}{K} \quad ...(4)$$

i.e., Primary and secondary currents are inversely proportional to their respective turns.

Example 1. *A 40 kVA, single phase transformer has 400 turns on the primary and 100 turns on the secondary. The primary is connected to 2000 V, 50 Hz supply. Determine :*

(i) The secondary voltage on open circuit.

(ii) The current flowing through the two windings on full-load.

(iii) The maximum value of flux.

Solution. Rating $= 40$ kVA

Primary turns, $N_1 = 400$

Secondary turns, $N_2 = 100$

Primary induced voltage, $E_1 = V_1 = 2000$ V

(*i*) **Secondary voltage on open, circuit $\mathbf{V_2}$:**

Using the relation, $\dfrac{E_2}{E_1} = \dfrac{N_2}{N_1}$

$$E_2 = E_1 \times \frac{N_2}{N_1}$$

$\therefore$ $E_2 = V_2 = 2000 \times \dfrac{100}{400} = 500$ V

Hence, $\mathbf{V_2 = 500\ V.}$ **(Ans.)**

(*ii*) **Primary current, $\mathbf{I_1}$:**

Secondary current, $\mathbf{I_2}$:

Primary full-load current, $I_1 = \dfrac{\text{kVA} \times 100}{V_1} = \dfrac{40 \times 1000}{2000} = \mathbf{20\ A.}$ **(Ans.)**

Secondary full-current, $I_2 = \dfrac{\text{kVA} \times 1000}{V_2} = \dfrac{40 \times 1000}{500} = \mathbf{80\ A}$ **(Ans.)**

(*iii*) **Maximum value of flux, ϕ_{max} :**

Using e.m.f. equation, $E_1 = 4.44 f \phi_{max} N_1$

$\therefore$ $2000 = 4.44 \times 50 \times \phi_{max} \times 400$

$\therefore$ $\phi_{max} = \dfrac{2000}{4.44 \times 50 \times 400} = 0.0225$ Wb.

Hence, $\phi_{max} = \mathbf{0.0225\ Wb.}$ **(Ans.)**

Example 2. *The no-load ratio required in a single-phase 50 Hz transformer is 6600/600 V. If the maximum value of flux in the core is to be about 0.08 Wb, find the number of turns in each winding.*

Solution. Primary, $E_1 = V_1 = 6600$ V

Secondary, $E_2 = V_2 = 600$ V

Maximum value of flux $\phi_{max} = 0.08$ Wb.

Primary turns, $\mathbf{N_1}$:

Secondary turns, $\mathbf{N_2}$:

Using the relation, $E_1 = 4.44 f \phi_{max} N_1$

$$6600 = 4.44 \times 50 \times 0.08 \times N_1$$

$\therefore$ $N_1 = \dfrac{6600}{4.44 \times 50 \times 0.08} \simeq 372$

Hence, $\mathbf{N_1 = 372.}$ **(Ans.)**

Also $$\frac{E_2}{E_1} = \frac{N_2}{N_1}$$

∴ $$N_2 = \frac{E_2 N_1}{E_1} = \frac{600 \times 372}{6600} \simeq 34$$

Hence, $\mathbf{N_2 = 34.}$ **(Ans.)**

Example 3. *A single-phase transformer is connected to a 230 V, 50 Hz supply. The net cross-sectional area of the core is 60 cm². The number of turns in the primary is 500 and in the secondary 100. Determine :*

(i) Transformation ratio. *(ii) E.m.f. induced in secondary winding.*

(iii) Maximum value of flux density in the core.

Solution. Primary turns, $N_1 = 500$

Secondary turns, $N_2 = 100$

Primary, $E_1 = V_1 = 230$ V

Core area, $a = 60 \text{ cm}^2 = 60 \times 10^{-4} \text{ m}^2$

(*i*) **Transformation ratio, K :**

$$K = \frac{N_2}{N_1} = \frac{100}{500} = 0.2$$

Hence, **K = 0.2. (Ans.)**

(*ii*) **Maximum value of flux density, $\mathbf{B_{max}}$:**

Using the e.m.f. equation, $E_1 = 4.44 f \phi_{max} N_1$

∴ $$230 = 4.44 \times 50 \times \phi_{max} \times 500$$

or $$\phi_{max} = \frac{230}{4.44 \times 50 \times 500} = 0.00207 \text{ Wb}$$

Now, $$B_{max} = \frac{\phi_{max}}{A} = \frac{0.00207}{60 \times 10^{-4}} = 0.345 \text{ T}$$

[where T stands for tesla (Wb/m²)]

Hence, $\mathbf{B_{max} = 0.345}$ **T. (Ans.)**

(*iii*) **E.m.f. induced in the secondary winding, $\mathbf{E_2}$:**

Using the relation, $$\frac{E_2}{E_1} = \frac{N_2}{N_1}$$

$$\frac{E_2}{230} = \frac{100}{500}$$

∴ $\mathbf{E_2 = 46}$ **V. (Ans.)**

Example 4. *3300/300 V single-phase 300 kVA transformer has 1100 primary turns. Find :*

(i) Transformation ratio. *(ii) Secondary turns.*

(iii) Voltage/turn.

(iv) Secondary current when it supplies a load of 200 kW at 0.8 power factor lagging.

Solution. Primary, $E_1 = 3300$ V

$N_1 = 1100$

Secondary, $E_2 = 300$ V

Rating of the transformer = 300 kVA

Output = 200 kW

(*i*) **Transformation ratio, K** = ?

$$K = \frac{N_2}{N_1} = \frac{E_2}{E_1} = \frac{300}{3300} = \frac{1}{11}$$

Hence, $\mathbf{K = \frac{1}{11}}$. **(Ans.)**

(*ii*) **Secondary turns, N_2 = ?**

Using the relation, $\frac{E_2}{E_1} = \frac{N_2}{N_1}$

$\therefore$ $N_2 = \frac{E_2 N_1}{E_1} = \frac{300 \times 1100}{3300} = 100$

Hence, $\mathbf{N_2 = 100.}$ **(Ans.)**

(*iii*) **Voltage/turn** = ?

Voltage/turn $\frac{E_1}{N_1} = \frac{3300}{1100} = \mathbf{3\ V.}$ **(Ans.)**

(*iv*) **Secondary current,** $I_2 = \frac{\text{Output}}{V_2 \cos\phi} = \frac{200 \times 1000}{300 \times 0.8} = 833.33\ \text{A}$

Hence, $\mathbf{I_2 = 833.33\ A.}$ **(Ans.)**

Example 5. *The voltage per turn of a single-phase transformer is 1.1 V. When the primary winding is connected to a 220 V, 50 Hz A.C. supply, the secondary voltage is found to be 550 V. Find :*

(*i*) *Primary and secondary turns.*

(*ii*) *Core area if the maximum flux density is 1.1 T.*

Solution. Voltage per turn = 1.1 V

Primary, $E_1 = 220$ V

Secondary, $E_2 = 550$ V

Max. flux density, $B_{max} = 1.1$ T

(*i*) **Primary turns,** $\mathbf{N_1} = \frac{E_1}{1.1} = \frac{220}{1.1} = \mathbf{200.}$ **(Ans.)**

Secondary turns, $\mathbf{N_2} = \frac{E_2}{1.1} = \frac{550}{1.1} = \mathbf{500.}$ **(Ans.)**

(*ii*) **Core area A :**

Using the relation, $E_1 = 4.44 f \phi_{max} N_1$

$$220 = 4.44 \times 50 \times \phi_{max} \times 200$$

$\therefore$ $\phi_{max} = \frac{220}{4.44 \times 50 \times 200} = 0.004955\ \text{Wb}$

Core area, $A = \frac{\phi_{max}}{B_{max}} = \frac{0.004955}{1.1} = 0.004504\ \text{m}^2 = \mathbf{45.04\ cm^2.}$ **(Ans.)**

Example 6. *The core of 1000 kVA, 11000/550 V, 50 Hz, single-phase transformer has a cross-section of 20 cm × 20 cm. If the maximum core density is not to exceed 1.3 tesla, calculate :*

(*i*) *The number of h.v. and l.v. turns per phase.*

(*ii*) *The e.m.f. per turns.*

Assume a stacking factor of 0.9.

Solution. *Given :* $A = 20\ \text{cm} \times 20\ \text{cm} = 400 \times 10^{-4}\ \text{m}^2$; $B_{max} = 1.3\ \text{Wb/m}^2, f = 50$ Hz ;
$E_1 = 11000$ V ; $E_2 = 550$ V ; Stacking factor = 0.9

(*i*) **Number of turns N_1 (h.v.) ; N_2 (l.v.) :**

We know that, Flux = flux density × area of cross-section × stacking factor

$$\therefore \quad \phi_{max} = 1.3 \times 400 \times 10^{-4} \times 0.9 = 0.0468 \text{ Wb}$$

$$E_1 = 4.44 f\phi_{max} N_1$$

or $$11000 = 4.44 \times 50 \times 0.0468 \times N_1$$

$$\therefore \quad \mathbf{N_1} = \frac{11000}{4.44 \times 50 \times 0.0468} = 1058.75 \simeq \mathbf{1059.} \quad \textbf{(Ans.)}$$

and $$\mathbf{N_2} = \frac{E_2}{E_1} \times N_1 = \frac{550}{11000} \times 1059 = 52.95 \simeq \mathbf{53.} \quad \textbf{(Ans.)}$$

(*ii*) **The e.m.f. per turn :**

$$\text{E.M.F. per turn} = \frac{E_1}{N_1} = \frac{11000}{1059} = \mathbf{10.387 \text{ V.}} \quad \textbf{(Ans.)}$$

9.4. Transformer with Losses but no Magnetic Leakage. We shall consider the following two cases :

1. When transformer is on no-load.
2. When transformer is on load.

9.4.1. Transformer on no-load. A transformer is said to be on *no-load* if its secondary side is open and primary is connected to a sinusoidal alternating voltage V_1. The alternating applied voltage will cause *flow of alternating current in the primary winding which will create alternating flux*. This primary input current (I_0) under no-load conditions supply :

(*i*) *Iron losses in the core* (*i.e.* hysteresis loss and eddy current loss).

(*ii*) A very *small amount of copper losses in primary* (there being no copper loss in the secondary as it is open).

Thus I_0 is not at 90° behind V_1, but lags it by an angle $\phi_0 < 90°$.

No-load power input, $\quad P_0 = V_1 I_0 \cos \phi_0$ ...(5)

where $\cos \phi_0$ = primary power factor under no-load conditions.

As is evident from Fig. 19, primary current I_0 has the following two components :

(*i*) *Active* or *working or iron loss component* I_w. This component is in *phase with* V_1 and mainly supplies the iron loss plus small quantity of primary copper loss.

$$I_w = I_0 \cos \phi_0 \quad ...(i)$$

(*ii*) *Magnetising component* I_m. This component is *in quadrature with* V_1 and its function is to *sustain the alternating flux in the core*. It is *wattless*.

$$I_m = I_0 \sin \phi_0 \quad ...(ii)$$

Also $$I_0 = \sqrt{I_w^2 + I_m^2} \quad ...(iii)$$

The following points are *worthnoting* :

- The no-load primary current I_w is *very small as compared to the full-load primary current.*
- As I_0 is very small, the no-load primary copper loss is negligibly small which means that *no-load primary input is practically equal to the iron loss in the transformer.*

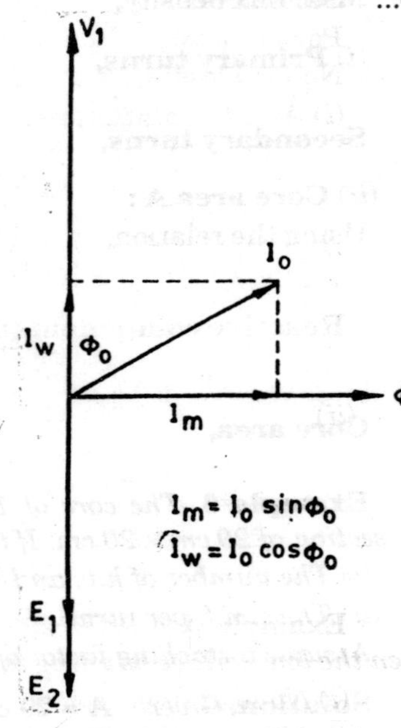

Fig. 19. No-load vector diagram.

- Since, it is primarily the core loss which is responsible for shift in the current vector, angle ϕ_0 is known as **hysteresis angle of advance.**

☞**Example 7.** *A 3300/300 V single-phase transformer gives 0.6 A and 60 W as ammeter and wattmeter readings when supply is given to the low voltage winding and high voltage winding is kept open, find :*

(i) Power factor of no-load current. *(ii) Magnetising component.*

(iii) Iron loss component.

Solution. No-load losses, $P_0 = 60$ W

No-load current, $I_0 = 0.6$ A

(*i*) $P_0 = VI_0 \cos \phi_0$

$\therefore$ $60 = 300 \times 0.6 \times \cos \phi_0$

or $$\cos \phi_0 = \frac{60}{300 \times 0.6} = 0.33 \text{ (lagging)}$$

Hence, **power factor of no-load current = 0.33. (Ans.)**

(*ii*) Magnetising component, $I_m = I_0 \sin \phi_0 = 0.6\sqrt{1 - \cos^2 \phi_0} = 0.6\sqrt{1 - (0.33)^2} = 0.566$ A

$\therefore$ I_m **= 0.566 A. (Ans.)**

(*iii*) Iron loss component, $I_w = I_0 \cos \phi_0 = 0.6 \times 0.33 =$ **0.198 A. (Ans.)**

[or $I_w = \sqrt{I_0^2 - I_m^2} = \sqrt{(0.6)^2 - (0.566)^2} = 0.198$ A].

☞**Example 8.** *Find (i) active and reactive components of no-load current ; and (ii) no-load current of a 440/220 V single-phase transformer if the power input on no- load to the high voltage winding is 80 W and power factor of no- load current is 0.3 lagging.*

Solution. Primary, $E_1 = 440$ V

Secondary, $E_2 = 220$ V

Power factor, $\cos \phi_0 = 0.3$ (lagging)

No-load losses, $P_0 = 80$ W

(*i*) **Active component** (or *wattful component*),

$$I_w = (I_0 \cos \phi_0) = \frac{P_0}{V_1} = \frac{80}{440} = \mathbf{0.182\ A.}\ \textbf{(Ans.)}$$

$$\cos \phi_0 = 0.3 \ ; \ \phi_0 = \cos^{-1}(0.3) = 72.54°$$

$\therefore$ $\tan \phi_0 = 3.18$

Reactive component (or *magnetising components*)

$$I_m = I_w \tan \phi_0 = 0.182 \times 3.18 = \mathbf{0.578\ A.}\ \textbf{(Ans.)}$$

(*ii*) $I_w = I_0 \cos \phi_0$

$$\therefore \quad I_0 = \frac{I_w}{\cos \phi_0} = \frac{0.182}{0.3} = \mathbf{0.606.}\ \textbf{(Ans.)}$$

or $[I_0 = \sqrt{I_w^2 + I_m^2} = \sqrt{(0.182)^2 + (0.578)^2} = 0.606$ A].

Example 9. *A 3300/220 V, 30 kVA, single-phase transformer takes a no-load current of 1.5 A when the low voltage winding is kept open. The iron loss component is equal to 0.4 A find :*

(i) No-load input power.

(ii) Magnetising component and power factor of no-load current.

Solution. Rating of transformer = 30 kVA

Primary, $E_1 = 3300$ V

Secondary, $E_2 = 220$ V

No-load current, $I_0 = 1.5$ A

Iron loss component, $I_m = 0.4$ A

(*i*) **No-load input power,**

$$P_0 = V_1 I_0 \cos \phi_0 = V_1 I_w \qquad (\because \; I_w = I_0 \cos \phi_0)$$
$$= 3300 \times 0.4 = \mathbf{1320\ W.\ (Ans.)}$$

(*ii*) **Magnetising component,**

$$I_m = \sqrt{I_0^2 - I_w^2} = \sqrt{1.5^2 - 0.4^2} = \mathbf{1.44\ A.\ (Ans.)}$$

No-load power factor,

$$\cos \phi_0 = \frac{I_w}{I_0} = \frac{0.4}{1.5} = \mathbf{0.267.\ (Ans.)}$$

9.4.2. Transformer on load. *The transformer is said to be loaded when the secondary circuit of a transformer is completed through an impedance or load.* The magnitude and phase of secondary current I_2 with respect to secondary terminal voltage will depend upon the characteristic of load, *i.e.* current I_2 will be in phase, lag behind and lead the terminal voltage V_2 respectively when the load is purely resistive, inductive and capacitive.

The secondary current I_2 sets up its own ampere-turns (= N_2I_2) and creates its own flux ϕ_2 opposing the main flux ϕ_0 created by no-load current I_0. The opposing secondary flux ϕ_2 weakens the primary flux ϕ_0 momentarily hence primary counter or back e.m.f. E_1 tends to be reduced. V_1 gains the upper hand over E_1 momentarily and hence causes more current to flow in primary. Let this additional primary current be I_2'. It is known as *load component of primary current*. The additional primary m.m.f. N_1I_2' sets up its own flux ϕ_2' which is in opposition to ϕ_2 (but is in the same direction as ϕ_0) and is equal to it in magnitude. Hence they *cancel* each other. Thus we find that the magnetic effects of secondary current I_2 are immediately neutralised by the additional primary current I_2' which is brought into existence exactly at the same instant as I_2.

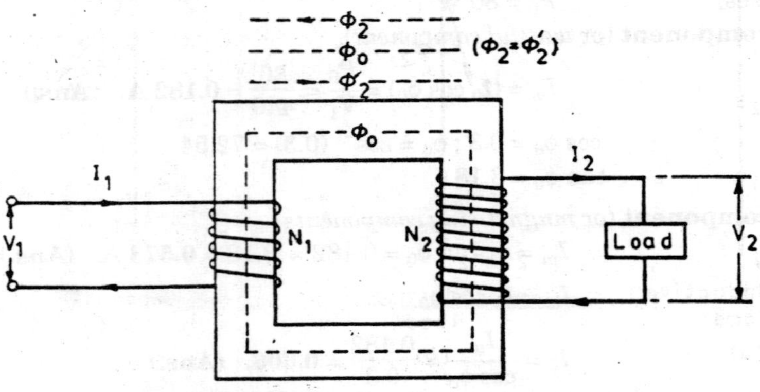

Fig. 20. An ideal transformer on load.

From above discussion it can be concluded that :

- Whatever be the load conditions, the *net flux passing through the core is approximately the same as at no-load.*
- Since the core flux remains constant at all loads, the core *loss almost remains constant under different loading conditions.*

Since $\phi_2 = \phi_2'$

$\therefore$ $N_2 I_2 = N_1 I_2'$

i.e., $$I_2' = \frac{N_2}{N_1} \times I_2 = KI_2 \qquad \left(\because \frac{N_2}{N_1} = K \right)$$

The total primary current is the vector sum of I_0 and I_2' ; the current I_2' is in *antiphase* with I_2 and K times in magnitude.

The components of primary current can be shown as below :

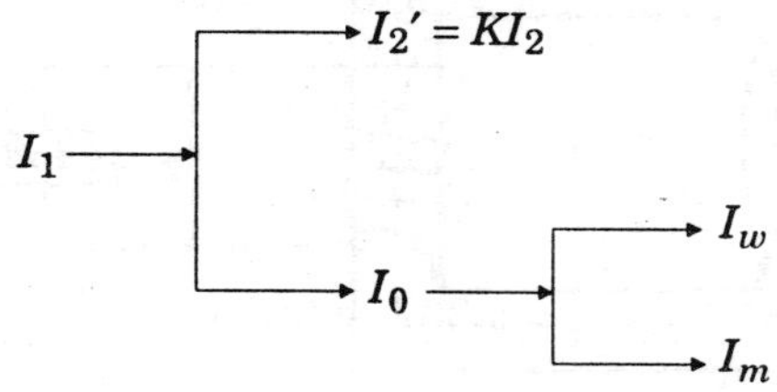

The vector diagrams for transformer on non-inductive, inductive and capacitive loads are shown in Fig. 21 (*a, b, c*).

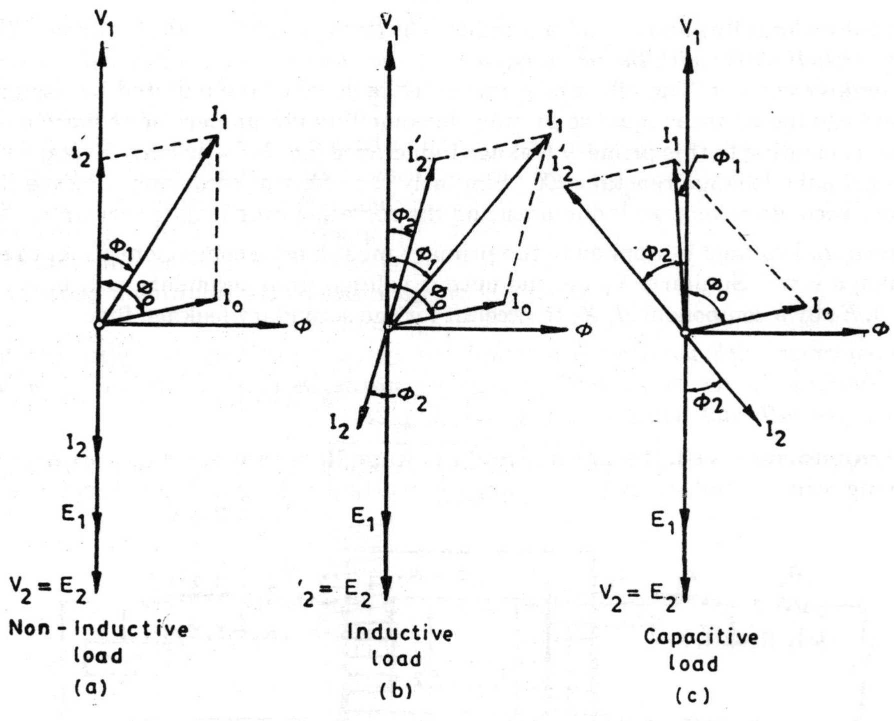

Fig. 21

9.5. Resistance and Magnetic Leakage. In an ideal transformer it is assumed that windings have got no resistance and there is no leakage flux. But in actual practice it is not possible to have an ideal transformer.

Resistance. In an actual transformer the primary as well secondary winding possess some resistance due to which some voltage drop takes place in them.

Magnetic leakage. Magnetic flux cannot be confined into a desired path. The greater portion of the flux (*i.e.*, the mutual flux) remains confined to the core and links both the windings but a small portion, called the *leakage flux,* completes its path through the air surrounding the coils. As shown in Fig. 22 each of the winding is associated with a leakage flux. Since the path of the leakage flux is largely in air the leakage flux and the voltage induced by it vary *linearly with the current. The primary leakage flux varies linearly with the primary current and the secondary leakage flux varies linearly with the secondary current.* The effect of primary leakage may be simulated by assigning to the primary a leakage inductance (equal to primary leakage flux per ampere of primary current). The reactance corresponding to this primary leakage inductance (*i.e.* $2\pi f \times$ primary leakage inductance) is known as primary leakage reactance X_1. Similarly the effect of secondary leakage flux may be simulated by a secondary leakage inductance and the corresponding leakage reactance X_2.

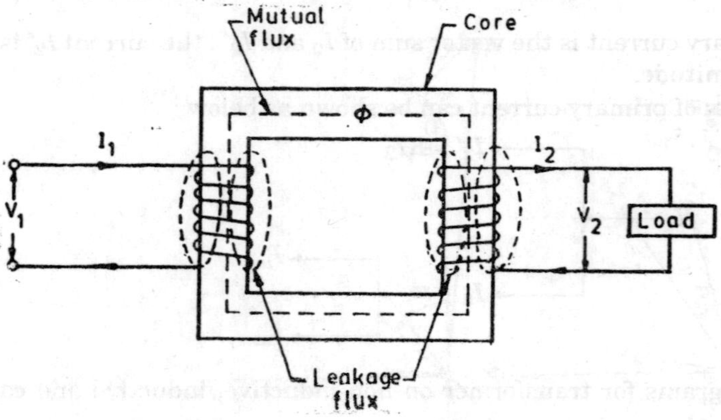

Fig. 22. Magnetic flux in a transformer.

The terminal voltage V_1 applied to the primary must have a component $jI_1 X_1$ to balance the primary leakage e.m.f. Similarly V_2 *i.e.* the output voltage from secondary will be less than the induced e.m.f. E_2 by a component $jI_2 X_2$ to account for the secondary leakage flux.

A transformer with winding resistance and magnetic leakage is equivalent to an ideal transformer (having no resistance and leakage resistance) having resistive and inductive coils connected in series with each winding as shown in Fig. 23.

9.6. Transformer with Resistance and Leakage Reactance. Let us consider transformer (Fig. 23) having primary and secondary windings of resistances R_1 and R_2 and reactances X_1 and X_2 respectively.

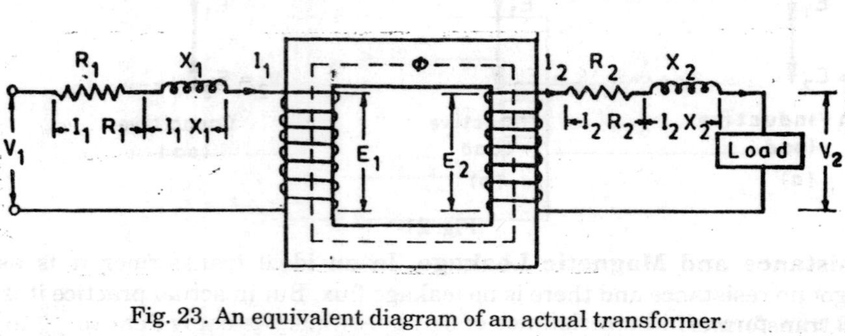

Fig. 23. An equivalent diagram of an actual transformer.

Impedance of primary winding, $Z_1 = R_1 + jX_1$
Impedance of secondary winding, $Z_2 = R_2 + jX_2$

The *vector diagrams* for different kinds of loads for such a transformer are shown in Fig. 24 (*a, b, c*). *In these diagrams, vectors for resistive drops are drawn parallel to current vectors* whereas *reactive drops are drawn perpendicular to the current vectors*. The angle ϕ_1 between V_1 and I_1 gives the power factor angle of the transformer.

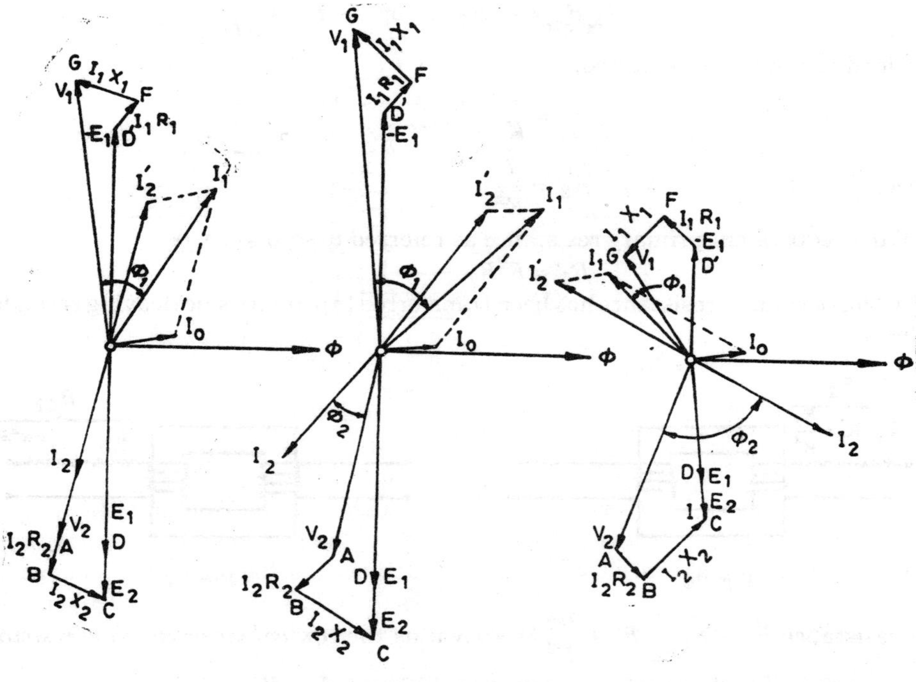

Fig. 24. (*a*) Unity p.f., (*b*) lagging p.f. and (*c*) leading p.f.

9.7. Equivalent Resistance and Reactance

Equivalent Resistance. In Fig. 25 is shown a transformer whose primary and secondary windings have resistances of R_1 and R_2 respectively (shown external to the windings). The resistances of the two windings can be transferred to any one of the two windings. If both the resistances are concentrated in one winding the calculations become simple since then we can work in one winding only.

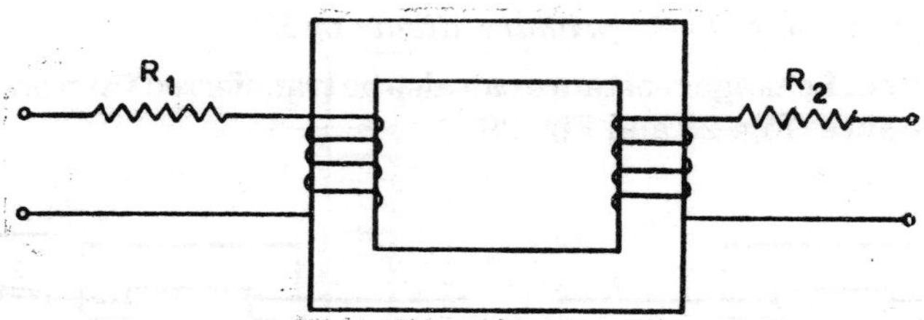

Fig. 25

We shall now prove that a resistance of R_2 in secondary is equivalent to $\dfrac{R_2}{K^2}$ in primary and its value will be denoted by R_2', — *the equivalent secondary resistance referred to primary.*

In the secondary, copper loss is $I_2^2R_2$. This loss is supplied by primary which takes a current of I_1. Hence, if R_2' is the equivalent resistance referred to primary which would have caused the same loss as R_2 in secondary, then

$$I_1^2R_2' = I_2^2R_2 \quad \text{or} \quad R_2' = \left(\frac{I_2}{I_1}\right)^2 R_2$$

If no-load current I_0 is neglected,

then,
$$\frac{I_2}{I_1} = \frac{1}{K}$$

Hence,
$$R_2' = \frac{R_2}{K^2}$$

Similarly, equivalent primary resistance as referred to secondary is

$$R_1' = K^2R_1$$

In Fig. 26, secondary resistance has been transferred to primary side (leaving secondary circuit resistanceless).

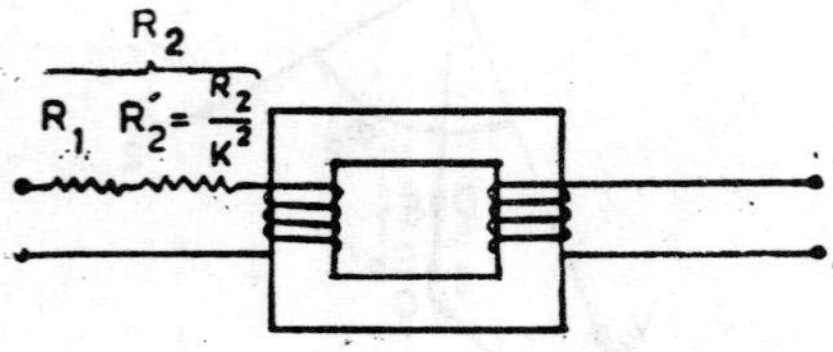

Fig. 26

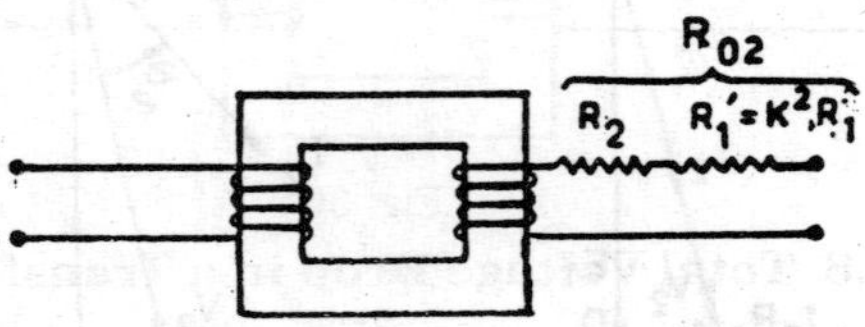

Fig. 27

The resistance $R_1 + R_2' = R_1 + \dfrac{R_2}{K^2}$ is known as the *equivalent or effective resistance of the transformer as referred to the primary* and may be designated as R_{01}.

Thus
$$R_{01} = R_1 + R_2' = R_1 + \frac{R_2}{K^2} \qquad \text{...(6)}$$

Similarly, the equivalent resistance of the transformer as referred to the secondary (Fig. 27) is

$$R_{02} = R_2 + R_1' = R_2 + K^2R_1 \qquad \text{...(7)}$$

The following points are worth remembering :

(*i*) When shifting resistance *to the secondary, multiply* it by K^2.

(*ii*) When shifting resistance *to the primary, divide* by K^2.

Leakage reactance. Leakage reactance can also be transferred from one winding to the other in the same way as resistance (Fig. 28 and Fig. 29).

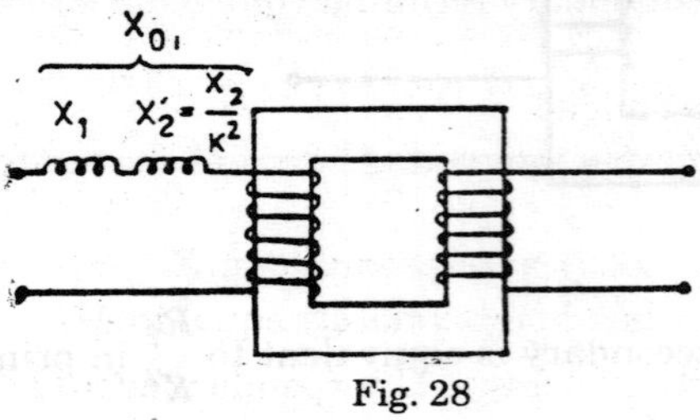

Fig. 28

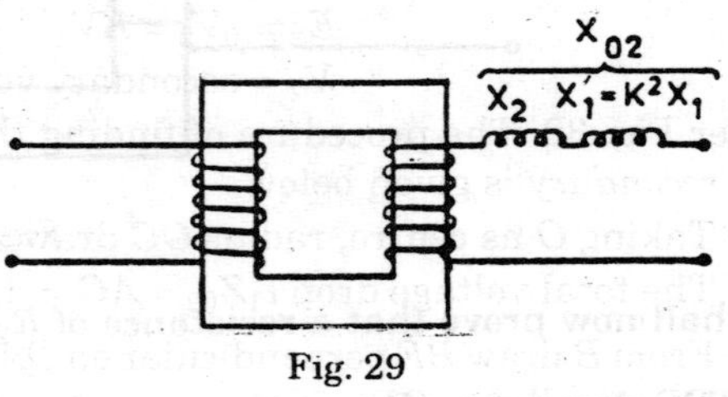

Fig. 29

$$X_2' = \frac{X_2}{K^2} \quad \text{and} \quad X_1' = K^2 X_1$$

and
$$X_{01} = X_1 + X_2' = X_1 + \frac{X_2}{K^2} \qquad ...(8)$$

$$X_{02} = X_2 + X_1' = X_2 + K^2 X_1 \qquad ...(9)$$

Total impedance. It is obvious that total impedance of the transformer as referred to primary is given by

$$Z_{01} = \sqrt{R_{01}^2 + X_{01}^2} \qquad \text{(Fig. 30)} \quad ...(10)$$

and
$$Z_{02} = \sqrt{R_0^2 + X_{02}^2} \qquad \text{(Fig. 31)} \quad ...(11)$$

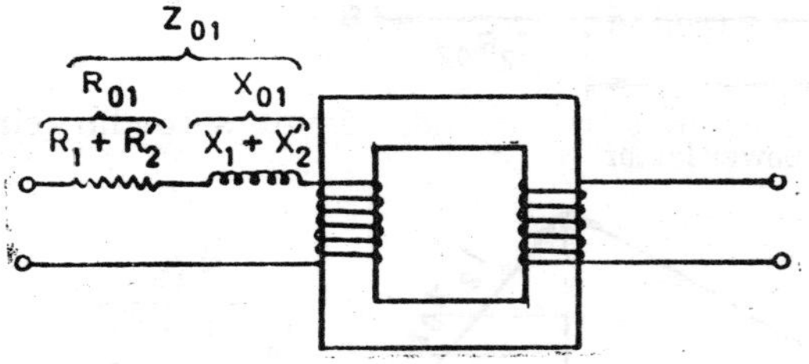

Fig. 30

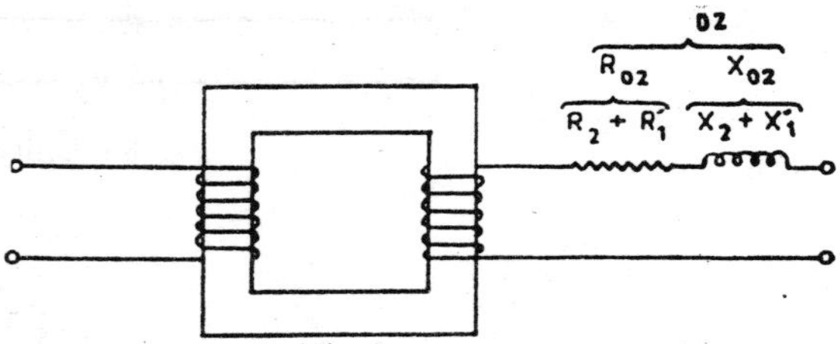

Fig. 31

9.8. Total Voltage Drop in a Transformer

(*i*) **Approximate voltage drop.** When there is no-load on the transformer, then,

$$V_1 = E_1 \text{ (approximately).}$$

and
$$E_2 = KE_1 = KV_1$$

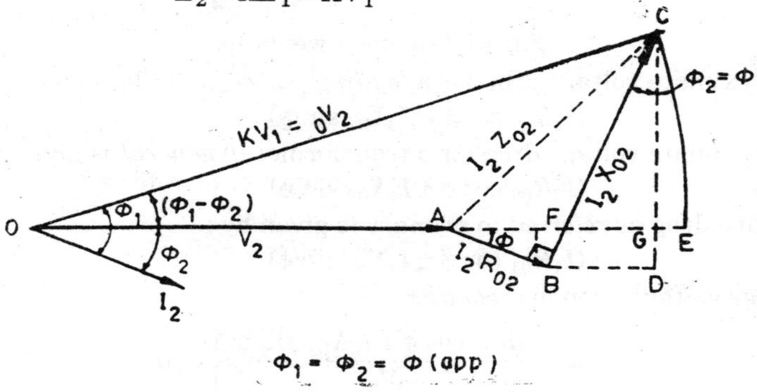

Fig. 32. Lagging power factor.

Also $E_2 = {}_0V_2$, where ${}_0V_2$ is secondary terminal voltage on no-load

or $E_2 = {}_0V_2 = KV_1$

V_2 = secondary voltage on load.

Refer Fig. 32. The procedure of finding the approximate voltage drop of the transformer as *referred to secondary* is given below :

- Taking O as centre, radius OC draw an arc cutting OA produced at E.
  The total voltage drop $I_1Z_{02} = AC = AE$ which is approximately equal to AG.
- From B draw BF perpendicular on OA produced. Draw CG perpendicular to OE and draw BD parallel to OE.

Approximate voltage drop $= AG = AF + FG = AF + BD$ $[\because \; FG = BD]$

$= I_2R_{02} \cos \phi + I_0X_{02} \sin \phi$

This is the value of approximate voltage drop for a *lagging power factor*.

Figs. 33 and 34 refer to *unity* and *leading power factor* respectively.

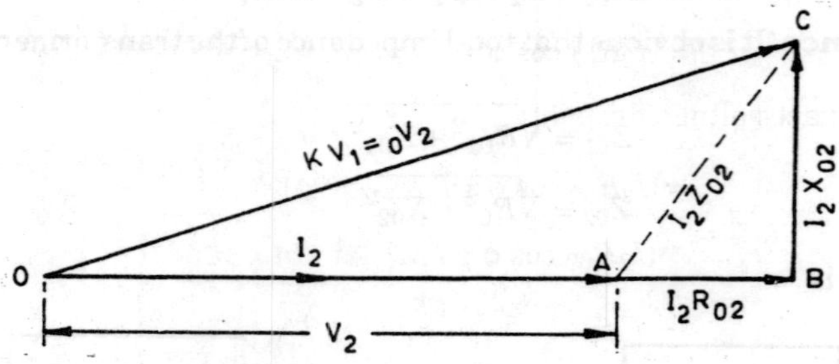

Fig. 33. Unity power factor.

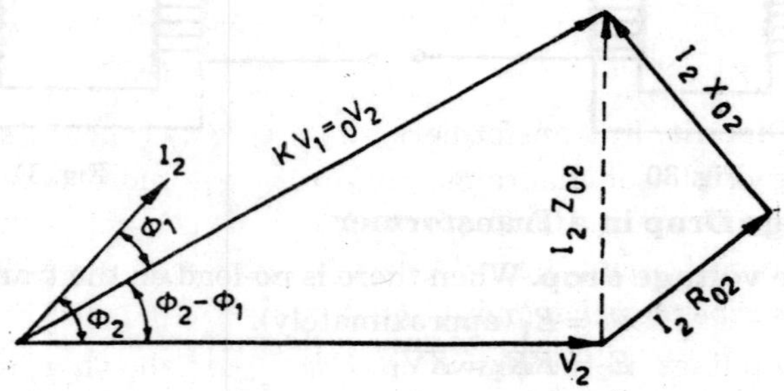

Fig. 34. Leading power factor.

- The approximate voltage drop for a *leading power factor* becomes :

$$(I_2R_{02} \cos \phi - I_2X_{02} \sin \phi)$$

- The approximate voltage drop for a transformer *in general* is given by :

$$(I_2R_{02} \cos \phi \pm I_1X_{02} \sin \phi) \quad ...(12)$$

- The voltage drop as *referred to primary* is given by :

$$(I_1R_{01} \cos \phi \pm I_1X_{02} \sin \phi)$$

- *Percentage* voltage drop in *secondary*

$$= \left(\frac{I_2R_{02} \cos \phi \pm I_2X_{02} \sin \phi}{{}_0V_2} \right) \times 100 \quad ...(13)$$

(*ii*) **Exact voltage drop.** When we refer to Fig. 32 we find that exact drop is AE and not AG. If quantity GE is added to AG the exact value of voltage drop will be obtained.

From right angled triangle OCG, we get

$$CG^2 = OC^2 - OG^2 = (OC + OG)(OC - OG)$$

$$= (OC + OG)(OE - OG) = GE \times 2OC$$

$$\therefore \quad GE = \frac{CG^2}{2OC}$$

Now $\quad CG = I_2X_{02} \cos \phi - I_2R_{02} \sin \phi$

$$\therefore \quad GE = \frac{(I_2X_{02} \cos \phi - I_2R_{02} \sin \phi)^2}{2\,{}_0V_2}$$

∴ The exact voltage drop for a *lagging* power factor is

$$= AG + GE$$

$$= (I_2R_{02} \cos \phi + I_2X_{02} \sin \phi) + \frac{(I_2X_{02} \cos \phi - I_2R_{02} \sin \phi)^2}{2{}_0V_2} \quad ...(14)$$

For a *leading* power, the expression becomes

$$= (I_2R_{02} \cos \phi - I_2X_{02} \sin \phi) + \frac{(I_2X_{02} \cos \phi + I_2R_{02} \sin \phi)^2}{2{}_0V_2} \quad ...(15)$$

In *general*, the exact voltage drop is

$$= (I_2R_{02} \cos \phi \pm I_2X_{02} \sin \phi) + \frac{(I_2X_{02} \cos \phi \mp I_2R_{02} \sin \phi)^2}{2{}_0V_2}$$

Percentage drop is

$$= \frac{(I_2R_{02} \cos \phi \pm I_2X_{02} \sin \phi) \times 100}{{}_0V_2} + \frac{(I_2X_{02} \cos \phi \mp I_2R_{02} \sin \phi)^2 \times 100}{2{}_0V_2{}^2} \quad ...(16)$$

[Upper signs ... for *lagging* power factor
Lower signs ... for *leading* power factor]

9.9. Equivalent Circuit. In transformers, the problems concerning voltages and currents can be solved by the use of phasor diagrams. However, it is more convenient to represent the transformer by an equivalent circuit. If an equivalent circuit is available the computations can be done by the direct application of circuit theory. An *equivalent circuit is merely a circuit interpretation of the equations which describe the behaviour of the device.*

The transformer windings, in the equivalent circuit, are shown as ideal. The resistance and leakage reactance of the primary and secondary are shown separately in the primary and secondary circuits. The effect of magnetising current is represented by X_0 connected in parallel across the winding. The effect of core loss is represented by a non-inductive resistance R_0 as shown in Fig. 35.

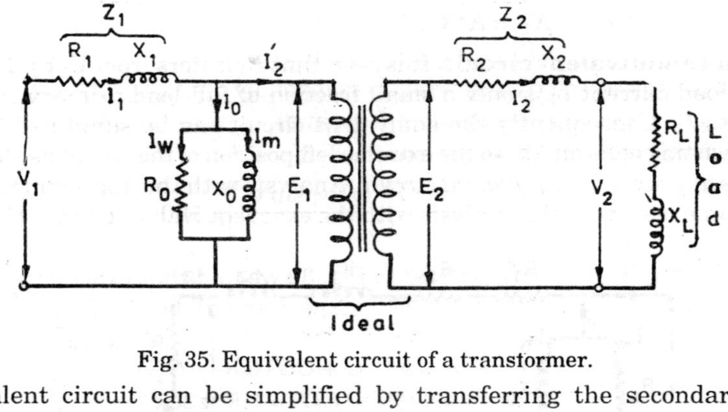

Fig. 35. Equivalent circuit of a transformer.

The equivalent circuit can be simplified by transferring the secondary resistances and reactances to the primary side in such a way that the ratio of E_2 to E_1 is not affected in magnitude or phase. Let R_2' be the resistance which must be placed in the primary circuit to produce the same drop as produced by R_2 in the secondary. Then R_2' causes a voltage drop in primary equal to $I_2'R_2'$.

The ratio of $I_2'R_2'$ and I_2R_2 must be the same as the turn ratio $\frac{N_1}{N_2}$.

Thus $$\frac{I_2' R_2'}{I_2 R_2} = \frac{N_1}{N_2} = \frac{1}{K} \quad \text{or} \quad R_2' = R_2 \times \frac{I_2}{I_2'} \times \frac{1}{K}$$

But $$\frac{I_2}{I_2'} = \frac{N_1}{N_2} = \frac{1}{K}$$

$\therefore$ $$R_2' = \frac{R_2}{K^2}$$

Similarly $$X_2' = \frac{X_2}{K^2}.$$

The load resistance and reactance can be transferred to the primary side in the same way. When all the secondary impedances have been transferred to the primary side, the winding need not be shown in the equivalent circuit. The exact equivalent circuit of the transformer with all impedances transferred to the primary side is shown in Fig. 36.

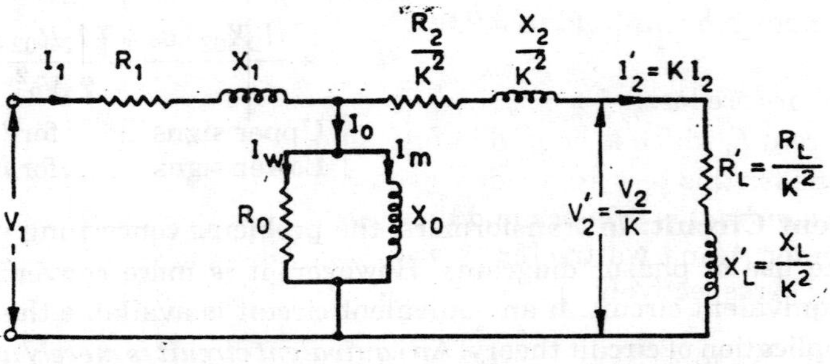

Fig. 36. Equivalent circuit with secondary impedances transferred to primary.

Just as the secondary impedances have been transferred to the primary it is possible to transfer the primary winding resistance and reactance to the secondary side. The primary resistance R_1 and leakage reactance X_1 when transferred to secondary are denoted by R_1' and X_1' and are given by

$$R_1' = K^2 R_1$$
$$X_1' = K^2 X_1.$$

Approximate equivalent circuit. It is seen that E_1 differs from V_1 by a very small amount. Moreover, the no-load current I_0 is only a small fraction of full-load primary current so that I_2' is practically equal to I_1. Consequently the equivalent circuit can be simplified by transferring the parallel branch consisting of R_0 and X_0 to the extreme left position of the circuit as shown in Fig. 37. This circuit is known as *approximate equivalent circuit*. Analysis with the approximate equivalent circuit gives almost the same results as the analysis with the exact equivalent circuit. *However, the analysis*

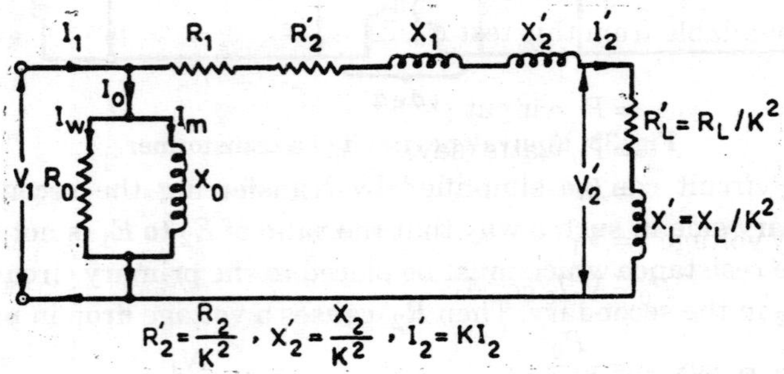

Fig. 37. Approximate equivalent circuit of transformer.

with the approximate equivalent circuit is simple because the resistances R_1 and R_2' and leakage reactances X_1 and X_2' can be combined.

9.10. Transformer Tests. The performance of a transformer can be calculated on the basis of its equivalent circuit which contains the following four main parameters :

(*i*) Equivalent resistance R_{01} as referred to primary (or secondary R_{02}).

(*ii*) Equivalent leakage resistance X_{01} as referred to primary (or secondary X_{02}).

(*iii*) Core loss conductance G_0 (or resistance R_0).

(*iv*) Magnetising susceptance B_0 (or reactance X_0).

These parameters or constants can be determined by the following *two tests* :

1. Open-circuit or no-load test.
2. Short-circuit or impedance test.

The above two tests are *convenient to perform and very economical because they furnish the required information without actually loading the transformer.*

9.10.1. Open-circuit or no-load test (O.C. Test). An open-circuit or no-load test is conducted to find :

(*i*) No-load loss or core loss.

(*ii*) No-load current I_0 which is helpful in finding R_0 and X_0.

The connections for this test are made as shown in Fig. 38. One winding of the transformer (*usually high voltage winding*) is *left open* and the other is connected to its supply of *normal voltage and frequency*. Ammeter *A* and wattmeter *W* are connected to measure *no-load current* (I_0) *and no-load input power* (P_0) respectively.

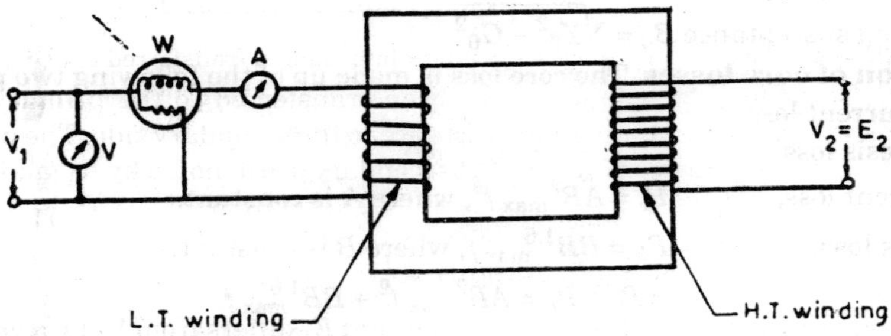

Fig. 38. Circuit diagram for open-circuit test.

As the primary no-load current I_0 (as measured by ammeter) is small (usually 3 to 10% of rated load current) copper loss is negligibly small in primary (L.T. winding) and nil in secondary winding (it being open). Hence the *wattmeter reading represents practically the core-loss under no-load conditions* (and *this loss is same for all loads*).

From the data available from this test R_0, X_0, $\cos \phi_0$ (no-load power factor), I_w and I_m can be calculated as follows :

Now, Iron loss $= P_i$ = input power on no-load

$= P_0$ watts (say)

No-load current $= I_0$

Applied primary voltage $= V_1$

Also $$P_0 = V_1 I_0 \cos \phi_0$$ (where $\cos \phi_0$ = no-load power factor)

$$\therefore \quad \cos \phi_0 = \frac{P_0}{V_1 I_0} \quad \ldots(17)$$

or $$\phi_0 = \cos^{-1} \frac{P_0}{V_1 I_0}$$

No-load current *wattful component*,

$$I_w = I_0 \cos \phi_0 = \frac{P_0}{V_1} \quad ...(18)$$

No-load current *magnetising component*,

$$I_m = \sqrt{I_0^2 - I_w^2} \quad ...(19)$$

The no-load resistance, $R_0 = \frac{V_1}{I_w} = \frac{V_1^2}{P_0}$ ...(20)

The no-load reactance, $X_0 = \frac{V_1}{I_m} = \frac{V_1}{\sqrt{I_0^2 - I_w^2}}$ ...(21)

The no-load vector diagram is shown in Fig. 19.

Since the current is practically all-exciting current when a transformer is on no-load (*i.e.* $I_0 = I_m$) and the voltage drop in primary leakage impedance is small, hence the exciting admittance Y_0 of the transformer is given by

$$I_0 = V_1 Y_0 \quad \text{or} \quad Y_0 = \frac{V_1}{I_0} \quad ...(22)$$

The exciting conductance $G_0 = \frac{P_0}{V_1^2}$ ...(23)

The exciting susceptance $B_0 = \sqrt{Y_0^2 - G_0^2}$ ...(24)

Separation of core losses. The core loss is made up of the following two parts :

(*i*) Eddy current loss.

(*ii*) Hysteresis loss.

Eddy current loss, $P_e = AB^2_{max} f^2$, where A is constant.

Hysteresis loss, $P_h = BB^{1.6}_{max} f$, where B is constant.

Total loss $= P_e + P_h = AB^2_{max} f^2 + BB^{1.6}_{max} f$.

The values of constants A and B can be found out by conducting two experiments using two different frequencies but the same maximum flux density ; thereafter eddy current and hysteresis loss can be found separately.

9.10.2. Short-circuit or impedance test (S.C. Test). This test is conducted to determine the following :

(*i*) Full-load copper loss.

(*ii*) Equivalent resistance and reactance *referred to metering side.*

In this test (Fig. 39) the terminals of the secondary winding (*usually low voltage winding*) are short-circuited by a thick conductor or through an ammeter which may serve the additional purpose of indicating rated load current. A low voltage, usually 5 to 10% of normal primary voltage, at correct frequency is applied to the primary and *is continuously increased till full-load currents flow in the primary as well as secondary windings* (as indicated by the respective ammeters).

Since applied voltage is very low so flux linking with the core is very small and therefore, iron losses are so small that these can be neglected, the reading of the wattmeter gives total copper losses at full-load.

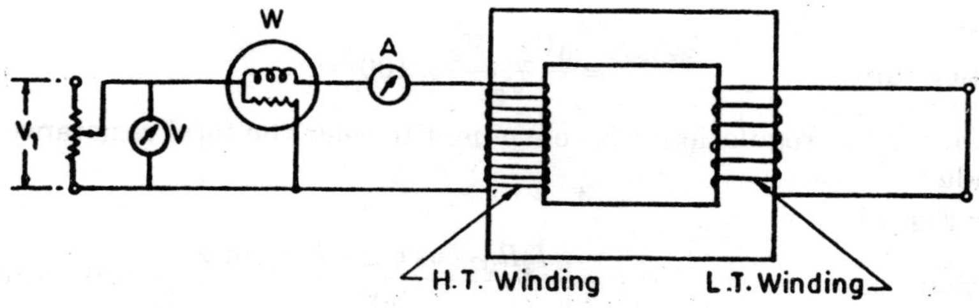

Fig. 39. Short-circuit test.

The equivalent circuit of the transformer under short-circuit condition is shown in Fig. 40.

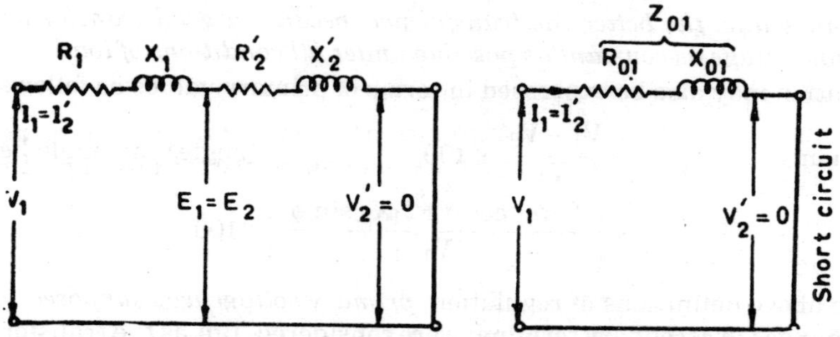

Fig. 40. Equivalent circuit of transformer under short-circuit condition.

Let V_{SC} = voltage required to circulate rated load currents

I_1 = reading of the ammeter on the primary side

Z_{01} = total impedance as referred to primary side

R_{01} = total resistance as referred to primary side

X_{01} = total reactance as referred to primary side.

Then, equivalent impedance as referred to primary side,

$$Z_{01} = \frac{V_{SC}}{I_1} \quad ...(25)$$

Also
$$P = I_1^2 R_{01}$$

$\therefore$
$$R_{01} = \frac{P}{I_1^2} \quad ...(26)$$

and
$$X_{01} = \sqrt{Z_{01}^2 - R_{01}^2} \quad ...(27)$$

9.11. Regulation of a Transformer

- Due to the resistances of the windings and leakage reactances voltage drop takes place in a transformer. Accordingly the output voltage under load conditions is different from the output voltage under no-load conditions. **Voltage regulation** is defined as :

"The change in secondary voltage when rated load at a specified power is removed".

It is specified as a percentage of the rated secondary voltage.

Thus, if $_0V_2$ = secondary terminal voltage at *no-load*

$= E_2 = KE_1 = KV_1$ because at no-load the impedance drop is negligible

V_2 = secondary terminal voltage on *full-load*

Then, % regulation (down)
$$= \frac{_0V_2 - V_2}{_0V_2} \times 100 \quad ...(28)$$

% regulation (up) $$= \frac{{}_0V_2 - V_2}{V_2} \times 100 \qquad ...(29)$$

For calculating the regulation it is convenient to refer the total resistance and reactance to the secondary side.

- Refer Fig. 32.

% regulation $$= \left(\frac{I_2R_{02} \cos \phi \pm I_2X_{02} \sin \phi}{{}_0V_2} \right) \times 100 \text{ (app.)} \qquad ...(30)$$

[+ sign for *lagging* power factor
− sign for *leading* power factor]

The less this value, the better the transformer, because a good transformer should keep its secondary terminal voltage as constant as possible under all conditions of load.

The regulation may also be expressed in terms of primary values as follows :

% regulation $$= \frac{V_1 - V_2'}{V_1} \times 100$$ (neglecting angle between V_1 and V_2')

$$= \left(\frac{I_1R_{01} \cos \phi \pm I_1X_0 \sin \phi}{V_1} \right) \times 100 \qquad ...(31)$$

- In the above definitions of regulation, *primary voltage was supposed to be kept constant* and changes in secondary terminal were considered. But as the transformer is loaded, the secondary terminal voltage falls (for lagging power factor) ; hence to keep the circuit voltage constant, the primary voltage must be increased. The rise *in primary voltage required to maintain rated output voltage from no-load to full-load at a given power factor expressed as percentage of rated primary voltage gives the regulation of the transformer.*

If the primary voltage has to be raised from its rated value V_1 to V_1', then

% regulation $$= \frac{V_1' - V}{V_1} \times 100 \qquad ...[31\ (a)]$$

9.12. Percentage Resistance and Reactance

- Invariably the equivalent resistance and reactance of a transformer are expressed in percent.

Percentage resistance is the *resistance drop in volts at rated current and frequency as a percentage of the rated voltage i.e.,*

$$\text{Percent } R = \frac{IR}{V} \times 100 \qquad ...(i)$$

Percentage reactance is the *reactance drop in volts at rated current and frequency expressed as a percentage of the rated voltage i.e.,*

$$\text{Percent } X = \frac{IX}{V} \times 100 \qquad ...(ii)$$

and $$\text{Percent } Z = \frac{IZ}{V} \times 100 \qquad ...(iii)$$

The important **advantage** of expressing resistance and reactance of a transformer in percentage is that the *percentage resistance and percentage reactance have same values whether determined referred to primary or secondary whereas when expressed in ohms they have different values when referred to primary and secondary.*

Further, *percent reactances and impedances are convenient for computing the current that will flow in a transformer when the secondary is accidentally shorted.* When the secondary is shorted,

$I_1 = \dfrac{E_1}{Z_0}$ or $E_1 = I_1 Z_0$ *i.e.*, when secondary is shorted, $I_1 Z_0$ is 100 percent of E_1. Therefore, the short-circuit current is given by the equation,

$$\text{Short circuit current ; } I_{SC} = \frac{100}{\text{Percent impedance}} \times \text{full load rated current} \quad ...(iv)$$

Unless the transformer is quite small, equivalent resistance is negligible and percent reactance may be substituted for percent impedance in the above equation (*iv*).

- **Another method of designating the resistances and reactances** is by the **per unit values**. The per unit values are equal to the percentage values divided by 100.

$$\therefore \quad \text{Per unit } R = \frac{IR}{V} \quad ...(v)$$

and

$$\text{Per unit } X = \frac{IX}{V} \quad ...(vi)$$

9.13. Transformer Losses. The losses in a transformer are classified as follows :

1. Iron losses (or core losses).
2. Copper losses.

1. **Iron or core losses.** It includes *hysteresis loss* and *eddy current loss.*

(*i*) **Hysteresis loss.** Since the flux in a transformer core is alternating, power is required for the continuous reversals of the elementary magnets of which the iron is composed. This loss is known as *hysteresis loss.*

$$\text{Hysteresis loss} = K_h f B^{1.6}_{max} \quad ...(32)$$

where f is the frequency in Hz, B_{max} is the maximum flux density in core and K_h is a constant.

(*ii*) **Eddy current loss.** This is due to the flow of eddy currents in the core. Thin laminations, insulated from each other, reduce the eddy current loss to small proportion.

$$\text{Eddy current loss} = K_e f^2 B^2_{max} \quad ...(33)$$

where K_e is a constant.

Iron or core loss is found from open circuit test. The input of the transformer when on no-load measures the core loss.

2. **Copper losses.** These losses are due to the ohmic resistance of the transformer windings.

$$\text{Total copper loss} = I_1^2 R_1 + I_2^2 R_2 = I_1^2 R_{01} = I_2^2 R_{02}$$

These losses, as is evident, are proportional to square of the current (or kVA)^2.

The value of copper losses is found from the *short-circuit test.*

9.14. Transformer Efficiency. The efficiency of a transformer at a particular load and power factor is defined as *the ratio of power output to power input.*

$$\therefore \text{ Efficiency} = \frac{\text{output}}{\text{input}}$$

$$= \frac{\text{output}}{\text{output + losses}} = \frac{\text{output}}{\text{output + cu loss + iron loss}} \quad ...(34)$$

or

$$\text{Efficiency} = \frac{\text{input} - \text{losses}}{\text{input}} = 1 - \frac{\text{losses}}{\text{input}} \quad ...(35)$$

It may be noted that efficiency is based on power output in watts and not in volt-amperes, although losses are proportional to volt-amperes. Hence at any volt-ampere load, the efficiency depends on power factor, being *maximum at unity power factor.*

Efficiency can be calculated by determining core losses from open-circuit test and copper losses from short-circuit test.

Condition for maximum efficiency :

Iron losses, P_i = hysteresis loss + eddy current loss

$$= P_h + P_e$$

Copper losses, $P_c = I_1^2 R_{01}$ or $I_2^2 R_{02}$

Considering primary side :

Input to primary $= V_1 I_1 \cos \phi_1$

Efficiency,
$$\eta = \frac{V_1 I_1 \cos \phi_1 - \text{losses}}{V_1 I_1 \cos \phi_1} = \frac{V_1 I_1 \cos \phi_1 - I_1^2 R_{01} - P_i}{V_1 I_1 \cos \phi_1}$$

$$= 1 - \frac{I_1 R_{01}}{V_1 \cos \phi_1} - \frac{P_i}{V_1 I_1 \cos \phi_1}$$

Differentiating both sides w.r.t. I_1, we get

$$\frac{d\eta}{dI_1} = 0 - \frac{R_{01}}{V_1 \cos \phi_1} + \frac{P_i}{V_1 I_1^2 \cos \phi_1}$$

For η to be maximum,

$$\frac{d\eta}{dI_1} = 0.$$ Hence the above equation reduces to

$$\frac{R_{01}}{V_1 \cos \phi_1} = \frac{P_i}{V_1 I_1^2 \cos \phi_1}$$

or $$P_i = I_1^2 R_{01} \text{ or } I_2^2 R_{02} \quad ...(36)$$

or **Copper losses = Iron losses**

The output current corresponding to maximum efficiency is

$$I_2 = \sqrt{\frac{P_i}{R_{02}}} \quad ...(37)$$

By proper design it is possible to make the maximum efficiency occur at any desired load.

Variation of efficiency with power factor. We know that transformers efficiency,

$$\eta = \frac{\text{output}}{\text{input}} = \frac{\text{input} - \text{losses}}{\text{input}}$$

$$= 1 - \frac{\text{losses}}{\text{input}} = 1 - \frac{\text{losses}}{(V_2 I_2 \cos \phi + \text{losses})}$$

Let, $$\frac{\text{losses}}{V_2 I_2} = \beta$$

$\therefore$ $$\eta = 1 - \left(\frac{\text{losses}/V_2 I_2}{\cos \phi + \text{losses}/V_2 I_2} \right) \quad \text{or} \quad = 1 - \frac{\beta}{(\cos \phi + \beta)}$$

or $$\eta = 1 - \frac{(\beta/\cos \phi)}{1 + (\beta/\cos \phi)} \quad ...(38)$$

The variations of efficiency with power factor at different loadings on a typical transformer are shown in Fig. 41.

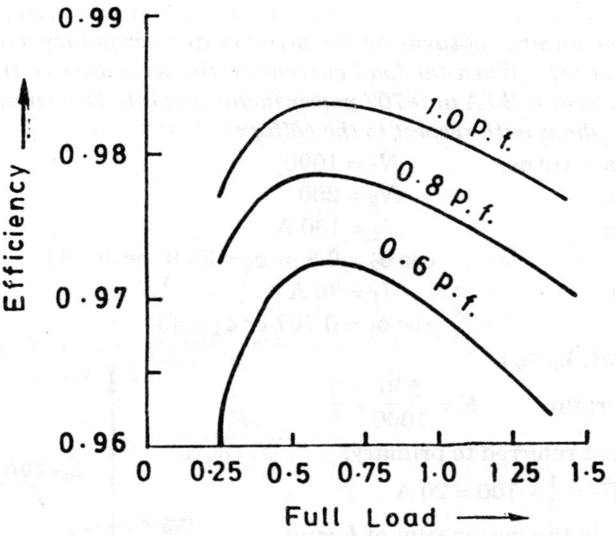

Fig. 41. Variations of efficiency with power factor at different loadings.

Transformer on Load

Example 10. *A 230 V / 115 V single-phase transformer takes a no-load current of 2 A at a power factor of 0.2 lagging with low voltage winding kept open. If the low voltage winding is now loaded to take a current of 15 A at 0.8 power factor lagging find the current taken by high voltage winding.*

Solution. Primary, $E_1 = V_1 = 230$ V

Secondary, $E_2 = V_2 = 115$ V

No-load current $I_0 = 2$ A

No-load power factor, $\cos \phi_0 = 0.2$ or $\phi_0 = 78.46°$ or $78° 29'$

Load power factor, $\cos \phi_2 = 0.8$ or $\phi_2 = 36.9°$ or $36° 54'$

Current taken by h.v. winding, I_1 :

Now, transformation ratio,

$$K = \frac{V_2}{V_1} = \frac{E_2}{E_1} = \frac{115}{230} = \frac{1}{2}$$

∴ Secondary current referred to primary,

$$I_2' = KI_2 = \frac{1}{2} \times 15 = 7.5 \text{ A}$$

Phase angle, between I_0 and I_2'

$$= 78.46° - 36.9° = 41.56° \text{ or } 41° 34'$$

Using parallelogram law of forces (Fig. 42), we get

$$I_1 = \sqrt{2^2 + 7.5^2 + 2 \times 2 \times 7.5 \times \cos 41.56°}$$

$$= 9.09 \text{ A}$$

Hence, **current taken by h.v. winding,**

I_1 = 9.09 A. (Ans.)

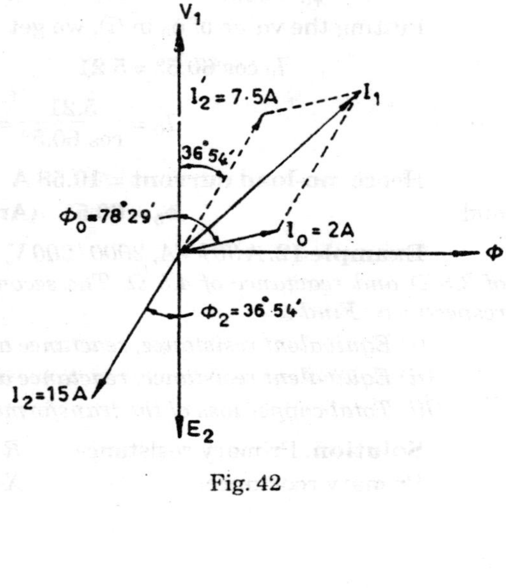

Fig. 42

Example 11. *The number of turns on the primary and secondary windings of a transformer are 1000 and 200 respectively. When the load current on the secondary is 100 A at 0.8 power factor lagging, the primary current is 30 A at 0.707 power factor lagging. Determine the no-load current of the transformer and its phase with respect to the voltage.*

Solution. Primary turns, $N_1 = 1000$

Secondary turns, $N_2 = 200$

Secondary current, $I_2 = 100$ A

Power factor, $\cos \phi_2 = 0.8$ or $\phi_2 = 36.9°$ or $36° 54'$

Primary current, $I_1 = 30$ A

Power factor, $\cos \phi_1 = 0.707$ or $\phi_1 = 45°$

No-load current, I_0, ϕ_0 :

Transformation ratio, $K = \frac{200}{1000} = \frac{1}{5}$

Secondary current referred to primary,

$$I_2' = KI_2 = \frac{1}{5} \times 100 = 20 \text{ A}$$

Refer Fig. 43. I_1 is the vector sum of I_0 and I_2'. Let I_0 lag behind V_1 by an angle ϕ_0.

Resolving currents into their X-and Y-components, we get

$$I_0 \cos \phi_0 + 20 \cos 36.9° = 30 \cos 45°$$

$$\therefore \quad I_0 \cos \phi_0 = 30 \cos 45° - 20 \cos 36.9°$$

$$= 21.21 - 16 = 5.21 \text{ A} \quad ...(i)$$

$$I_0 \sin \phi_0 + 20 \sin 36.9° = 30 \sin 45°$$

$$\therefore \quad I_0 \sin \phi_0 = 30 \sin 45° - 20 \sin 36.9°$$

$$= 21.21 - 12 = 9.21 \text{ A} \quad ...(ii)$$

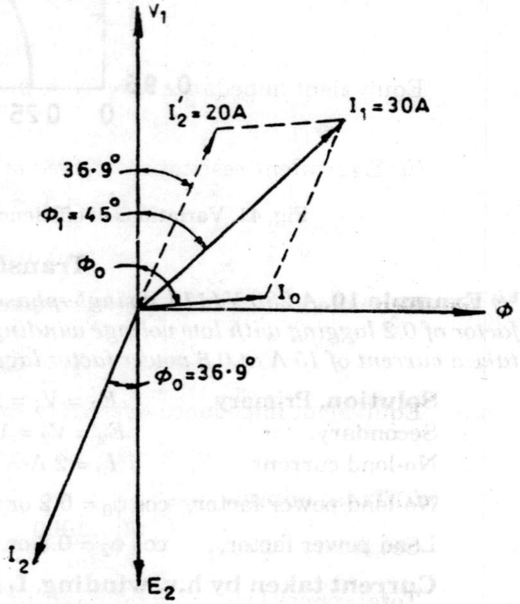

Fig. 43

From (*i*) and (*ii*), we get

$$\tan \phi_0 = \frac{9.21}{5.21} = 1.767$$

$$\therefore \quad \phi_0 = 60.5°$$

Putting the value of ϕ_0 in (*i*), we get

$$I_0 \cos 60.5° = 5.21$$

$$\therefore \quad I_0 = \frac{5.21}{\cos 60.5°} = 10.58 \text{ A}$$

Hence, **no-load current = 10.58 A**

and **$\phi_0 = 60.5°$. (Ans.)**

Example 12. *A 30 kVA, 2000/200 V, single-phase, 50 Hz transformer has a primary resistance of 3.5 Ω and reactance of 4.5 Ω. The secondary resistance and reactance are 0.015 Ω and 0.02 Ω respectively. Find :*

(*i*) *Equivalent resistance, reactance and impedance referred to primary.*

(*ii*) *Equivalent resistance, reactance and impedance referred to secondary.*

(*iii*) *Total copper loss of the transformer.*

Solution. Primary resistance, $R_1 = 3.5\ \Omega$

Primary reactance, $X_1 = 4.5\ \Omega$

Secondary resistance, $R_2 = 0.015\ \Omega$

Secondary reactance, $X_2 = 0.02\ \Omega$

Transformation ratio, $K = \frac{200}{2000} = \frac{1}{10}$

(*i*) Equivalent resistance *referred to primary*,

$$R_{01} = R_1 + R_2' = R_1 + \frac{R_2}{K^2} = 3.5 + \frac{0.015}{\left(\frac{1}{10}\right)^2} = \mathbf{5.0\ \Omega.}\ \textbf{(Ans.)}$$

Equivalent reactance *referred to primary*,

$$X_{01} = X_1 + X_2' = X_1 + \frac{X_2}{K^2} = 4.5 + \frac{0.02}{\left(\frac{1}{10}\right)^2} = \mathbf{6.5\ \Omega.}\ \textbf{(Ans.)}$$

Equivalent impedance *referred to primary*,

$$Z_{01} = \sqrt{(R_{01})^2 + (X_{01})^2} = \sqrt{(5)^2 + (6.5)^2} = \mathbf{8.2\ \Omega.}\ \textbf{(Ans.)}$$

(*ii*) Equivalent resistance *referred to secondary*,

$$R_{02} = R_2 + R_1' = R_2 + K^2R_1$$

$$= 0.015 + \left(\frac{1}{10}\right)^2 \times 3.5 = \mathbf{0.05\ \Omega.}\ \textbf{(Ans.)}$$

Equivalent reactance *referred to secondary*,

$$X_{02} = X_2 + X_1' = X_2 + K^2 X_1 = 0.02 + \left(\frac{1}{10}\right)^2 \times 4.5 = \mathbf{0.065\ \Omega.}\ \textbf{(Ans.)}$$

Equivalent impedance *referred to secondary*,

$$Z_{02} = \sqrt{(R_{02})^2 + (X_{02})^2} = \sqrt{(0.05)^2 + (0.065)^2} = \mathbf{0.082\ \Omega.}\ \textbf{(Ans.)}$$

(*iii*) **Total copper loss :**

Secondary current $= \frac{30 \times 1000}{200} = 150$ A

Total copper loss $= I_2^2 R_{02} = 150^2 \times 0.05 = \mathbf{1125\ W.}$ **(Ans.)**

☞ **Example 13.** *A single-phase transformer has the following data :*

Turn ratio 20 : 1 ; $R_1 = 20\ \Omega$, $X_1 = 80\ \Omega$; $R_2 = 0.04\ \Omega$; $X_2 = 0.2\ \Omega$. No-load current = 1.2 A leading the flux by 30°.

The secondary delivers 180 A at a terminal voltage of 400 V and at a power factor of 0.8 lagging. Determine by the aid of a vector diagram :

(*i*) *The primary applied voltage.* (*ii*) *The primary power factor.*

(*iii*) *The efficiency.*

Solution. Refer Fig. 44 :

(*i*) **Primary applied voltage, V_1 :**

Taking V_2 as the reference vector

$\therefore$

$$V_2 = 400 \angle 0° = 400 + j0$$

$$I_2 = 180(0.8 - j0.6) = 144 - j108$$

$$Z_2 = (0.04 + j0.2)$$

$$E_2 = V_2 + I_2Z_2 = (400 + j0) + (144 - j108) \times (0.04 + j0.2)$$

$$= 400 + (5.76 + j28.8 - j4.32 + 21.6) = (427.36 + j24.48) = 428.1 \angle 3.28°$$

Obviously,

$$\beta = 3.28°$$

$$E_1 = \frac{E_2}{K} = 20E_2 = 20(427.36 + j24.48) = 8547 + j490$$

$\therefore \quad -E_1 = -8547 - j490 = 8561 \angle 183.28°$

Secondary current referred to primary,

$$I_2' = -KI_2 = \frac{(-144 + j\,108)}{20} = -7.2 + j5.4$$

As seen from Fig. 44, I_0 leads V_2 by an angle $3.28° + 90° + 30° = 123.28°$

$$\therefore \quad I_0 = 1.2 \angle 123.28° = 1.2(\cos 123.28° + j \sin 123.28°)$$
$$= 1.2(-0.548 + j0.836) = -0.657 + j1.003.$$

Primary current,

$$I_1 = -I_2' + I_0 = (-7.2 + j5.4) + (-0.657 + j1.003)$$
$$= -7.857 + j6.403 = 10.14 \angle 140.8°$$
$$V_1 = -E_1 + I_1Z_1 = (-8547 - j490) + (-7.857 + j6.403)(20 + j80)$$
$$= -8547 - j490 + (-157.14 - j628.56 + j128.06 - 512.24)$$
$$= -9216.38 - j990.5 = \mathbf{9269 \angle 186.13°}. \quad \textbf{(Ans.)}$$

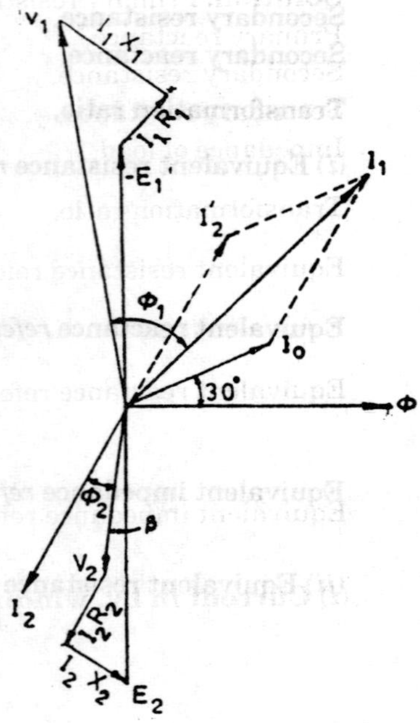

Fig. 44

(*ii*) **Primary power factor, cos ϕ_1 :**

Phase angle between V_1 and I_1,

$$\phi_1 = 186.13° - 140.8° = 45.33°$$

$\therefore$ Primary power factor $= \cos 45.33° = \mathbf{0.703}$ **(lag). (Ans.)**

(*iii*) **Efficiency :**

No-load primary input power

$$= V_1I_1 \cos \phi_0 = 9269 \times 1.2 \times \cos 60° = 5561.4 \text{ W}$$

$$R_{02} = R_2 + K^2R_1 = 0.04 + \left(\frac{1}{20}\right)^2 \times 20 = 0.09\ \Omega$$

Total copper losses as referred to secondary

$$= I_2^2R_{02} = (180)^2 \times 0.09 = 2916 \text{ W}$$

Output $= V_2I_2 \cos \phi_2 = 400 \times 180 \times 0.8 = 57600$ W

Total losses $= 5561.4 + 2916 = 8477.4$ W

Input $=$ Output $+$ Losses $= 57600 + 8477.4 = 66077.4$ W

$\therefore$ Efficiency, $\quad \eta = \frac{\text{output}}{\text{input}} = \frac{57600}{66077.4} = 0.872$ or **87.2%. (Ans.)**

Example 14. *The high voltage and low voltage windings of a 2200/220 V single-phase 50 Hz transformer has resistances of 4.8 Ω and 0.04 Ω and reactances 2 Ω and 0.018 Ω respectively. The low voltage winding is connected to a load having an impedance of (6 + j4) Ω. Determine* :

(*i*) *Current in l.v. winding,* (*ii*) *Current in h.v. winding,*

(*iii*) *Load voltage, and* (*iv*) *Power consumed by the load.*

Solution. Primary resistance, $R_1 = 4.8\ \Omega$

Primary reactance, $X_1 = 2\ \Omega$

Secondary resistance, $R_2 = 0.04\ \Omega$.

Secondary reactance, $X_2 = 0.018\ \Omega$

Impedance of load, $Z_L = (6 + j4)$

Transformation ratio, $K = \frac{E_2}{E_1} = \frac{220}{2200} = \frac{1}{10}$

Equivalent resistance referred to l.v. side,

$$R_{02} = R_2 + K^2R_1 = 0.04 + \left(\frac{1}{10}\right)^2 \times 4.8 = 0.088\ \Omega$$

Equivalent reactance referred to l.v. side,

$$X_{02} = X_2 + K^2X_1 = 0.018 + \left(\frac{1}{10}\right)^2 \times 2 = 0.038\ \Omega$$

Equivalent impedance referred to l.v. side,

$$Z_{02} = R_{02} + jX_{02} = (0.088 + j0.038)\ \Omega$$

(*i*) **Current in l.v. winding,**

$$I_2 = \frac{220}{Z_{02} + Z_L} = \frac{220}{(0.088 + j0.038) + (6 + j4)}$$

$$= \frac{220}{6.088 + j4.038} = \frac{220}{\sqrt{(6.088)^2 + (4.038)^2}} = \mathbf{30.11\ A.}\ \ \textbf{(Ans.)}$$

(*ii*) **Current in h.v. winding,**

$$I_1 \simeq I_2' = KI_2 = \frac{1}{10} \times 30.11 = \mathbf{3.011\ A.}\ \ \textbf{(Ans.)}$$

(*iii*) **Load voltage,** $= I_2Z_L = 30.11(6 + j4) = 30.11 \times \sqrt{(6)^2 + (4)^2} = \mathbf{217.12\ V.}$ **(Ans.)**

(*iv*) **Power consumed by the load,**

$$= I^2R_L = 30.11^2 \times 6 = \mathbf{5439.6\ W.}\ \ \textbf{(Ans.)}$$

☞**Example 15.** *A single-phase transformer has $Z_1 = 1.4 + j5.2\ \Omega$ and $Z_2 = 0.0117 + j0.0465\ \Omega$. The input voltage is 6600 V and the turn ratio is 10.6 : 1. The secondary feeds a load which draws 300 A at 0.8 power factor lagging. Find the secondary terminal voltage and the kW output. Neglect no-load current I_0.*

Solution. Impedance $Z_1 = 1.4 + j5.2\ \Omega$

Impedance $Z_2 = 0.0117 + j0.0465\ \Omega$

Input voltage $V_1 = 6600$ V

Turn ratio, $K = 10.6 : 1$

Secondary load current, $I_2 = 300$ A

Power factor, $\cos \phi_2 = 0.8$

Secondary terminal voltage, V_2 :

kW output :

Let the secondary terminal voltage $= V_2 \angle 0°$

$$I_2 = 300 \angle -36.87°\ \text{A}$$

$$E_2 = V_2 \angle 0° + I_2Z_2$$

$$= V_2\angle 0° + (300 \angle -36.87°)(0.0117 + j0.0465) = V_2 \angle 0° + 14.4 \angle 39°$$

$$E_1 = 10.6E_2 = 10.6V_2 \angle 0° + 152.64 \angle 39°$$

$$I_1 = \frac{300}{10.6} \angle -36.87° = 28.3 \angle -36.87°$$

$$V_1 = E_1 + I_1 Z_1$$
$$= 10.6V_2 \angle 0° + 152.64 \angle 39° + (28.3 \angle -36.87°)(1.4 + j5.2)$$
$$= 10.6V_2 + 152.64 \angle 39° + 152.4 \angle 38°$$
$$= 10.6V_2 + 118.62 + j96.06 + 120 + j93.83$$
$$= 10.6V_2 + 238.62 + j189.89$$

Since $\quad | V_1 | = 6600$ V

$\therefore \quad (10.6V_2 + 238.62)^2 + (189.89)^2 = 6600^2$

or $\quad V_2 \simeq$ **600 V**

Hence, **secondary terminal voltage = 600 V. (Ans.)**

Output :

Output $\quad = V_2 I_2 \cos \phi_2 = 600 \times 300 \times 0.8 = 144000$ W or **144 kW. (Ans.)**

Example 16. *The full-load copper loss on h.v. side of 100 kVA, 11000/317 V 1-phase transformer is 0.62 kW and on the l.v. side is 0.48 kW.*

(i) Calculate R_1, R_2 and R_2' in ohms :

(ii) The total reactance is 4 percent, find X_1, X_2 and X_2' in ohms if the reactance is divided in the same proportion as resistance.

Solution. *Given* : Rated kVA = 100 ; V_1 = 11000 V ; V_2 = 317 V ; loss on *h.v.* side = 0.62 kW ; loss of *l.v.* side = 0.48 kW ; total reactance = 4%.

(i) **R_1, R_2 and R_2' :**

h.v. side full-load current, $\quad I_1 = \dfrac{\text{rated kVA} \times 1000}{V_1} = \dfrac{100 \times 1000}{11000} = 9.09$ A

l.v. side full-load current, $\quad I_2 = \dfrac{\text{rated kVA} \times 1000}{V_2} = \dfrac{100 \times 1000}{317} = 315.46$ A

Primary winding resistance, $\quad \mathbf{R_1} = \dfrac{\text{full-load copper loss in primary winding}}{{I_1}^2}$

$$= \frac{0.62 \times 1000}{(9.09)^2} = \mathbf{7.5\ \Omega.\ (Ans.)}$$

Secondary winding resistance, $\quad \mathbf{R_2} = \dfrac{\text{full-load copper loss in secondary winding}}{{I_2}^2}$

$$= \frac{0.48 \times 1000}{(315.46)^2} = \mathbf{0.00482\ \Omega.\ (Ans.)}$$

Transformation ratio, $\quad K = \dfrac{V_2}{V_1} = \dfrac{317}{11000} = 0.02882$

Secondary resistance referred to primary,

$$R_2' = \frac{R_2}{K^2} = \frac{0.00482}{(0.02882)^2} = \mathbf{5.8\ \Omega.\ (Ans.)}$$

(ii) **X_1, X_2 and X_2' :**

Percentage reactance referred to *h.v.* side $= \dfrac{I_1 X_{01}}{V_1} \times 100 = 4$

or transformer total reactance referred to primary,

$$X_{01} = \frac{4 \times V_1}{100 I_1} = \frac{4 \times 11000}{100 \times 9.09} = 48.4\ \Omega.$$

Now, $$\frac{R_1}{R_2'} = \frac{X_1}{X_2'} \quad \text{...(Given)}$$

or $$\frac{R_1 + R_2'}{R_2'} = \frac{X_1 + X_2'}{X_2'}$$

or $$\frac{7.5 + 5.8}{5.8} = \frac{48.4}{X_2'} \quad (\because X_1 + X_2' = X_{01})$$

or $$\mathbf{X_2'} = \frac{48.4 \times 5.8}{(7.5 + 5.8)} = \mathbf{21.1\ \Omega.}\ \textbf{(Ans.)}$$

Primary winding reactance, $\mathbf{X_1} = X_{01} - X_2' = 48.4 - 21.1 = \mathbf{27.3\ \Omega.}$ **(Ans.)**

Secondary winding reactance, $\mathbf{X_2} = K^2 X_2' = \left(\frac{377}{11000}\right)^2 \times 21.1 = \mathbf{0.0175\ \Omega.}$ **(Ans.)**

Hysteresis and Eddy Current Losses

Example 17. *A 230 V, 2.5 kVA single-phase transformer has an iron loss of 100 W at 40-Hz and 70 W at 30-Hz. Find the hysteresis and eddy current losses at 50-Hz.*

Solution. Iron loss at 40-Hz = 110 W

Iron loss at 30-Hz = 75 W

Hysteresis and eddy current loss at 50-Hz :

We know that hysteresis loss,

$$P_h \propto f = Af$$

Eddy current loss, $$P_e \propto f^2 = Bf^2$$

Iron loss, $$P_i = P_h + P_e = Af + Bf^2$$

At 40-Hz :

$$100 = 40A + (40)^2 B$$

i.e., $$40A + 1600B = 100$$

or $$A + 40B = 2.5 \quad \text{...}(i)$$

At 30-Hz :

$$70 = 30A + (30)^2 B$$

i.e., $$30A + 900B = 70$$

or $$3A + 90B = 7 \quad \text{...}(ii)$$

Solving (*i*) and (*ii*), we get $A = \frac{11}{6}$, $B = \frac{1}{60}$.

Hysteresis loss at 50-Hz

$$= Af = \frac{11}{6} \times 50 = \mathbf{91.67\ W.}\ \textbf{(Ans.)}$$

Eddy current loss at 50-Hz

$$= Bf^2 = \frac{1}{60} \times 50^2 = \mathbf{41.67\ W.}\ \textbf{(Ans.)}$$

Example 18. *When a transformer is supplied at 400 V, 50-Hz the hysteresis loss is found to be 310 W and eddy current loss is found to be 260 W. Determine the hysteresis loss and eddy current loss when the transformer is supplied at 800 V, 100-Hz.*

Solution. Hysteresis loss at 400 V, 50 Hz = 310 W

Eddy current loss at 400 V, 50-Hz = 260 W

We know that,

Induced e.m.f., $E = 4.44 f\phi_{max} N = 4.44 (B_{max} A) f N$

i.e. $E \propto B_{max} f$

or $B_{max} \propto \frac{E}{f} = a \frac{E}{f}$ [a = constant]

In the **Ist case,** $B_{max_1} = a \cdot \frac{E_1}{f_1} = \frac{400}{50} a = 8a.$

In the **IInd case,** $B_{max_2} = a \frac{E_2}{f_2} = \frac{800}{100} a = 8a$

Hence, $B_{max_1} = B_{max_2}$

[$\because$ Core area A, and number of turns N remain the same]

Hence, $P_h = Af$ and $P_e = Bf^2$

$P_{h1} = 310$ W at f_1 = 50-Hz

$\therefore$ $A = \frac{310}{50} = 6.2$

Hysteresis loss at 100-Hz

$= Af_2 = 6.2 \times 100 =$ **620 W. (Ans.)**

$P_{e1} = 260$ W at f_1 = 50-Hz

$\therefore$ $B = \frac{260}{50^2} = 0.104$

Eddy current loss at 100-Hz

$= Bf_2^2 = 0.104 \times 100^2 =$ **1040 W. (Ans.)**

Transformer Tests (O.C. and S.C.) and Equivalent Circuit

☞**Example 19.** *A 50-Hz, single-phase transformer has a turn ratio of 5. The resistances are 0.8 Ω, 0.02 Ω and reactances are 4 Ω and 0.12 Ω for high-voltage and low-voltage windings respectively. Find :*

(i) The voltage to be applied to the h.v. side to obtain full-load current of 180 A in the l.v. winding on short-circuit.

(ii) The power factor on short-circuit.

Draw the equivalent circuit and vector diagram.

Solution. Turn ratio, $\frac{N_1}{N_2} = 5, K = \frac{1}{5}$

Primary resistance, $R_1 = 0.8\ \Omega$

Primary reactance, $X_1 = 4\ \Omega$

Secondary resistance, $R_2 = 0.02\ \Omega$

Secondary reactance, $X_2 = 0.12\ \Omega$

Total resistance referred to primary,

$$R_{01} = R_1 + R_2' = R_1 + \frac{R_2}{K^2} = 0.8 + 0.02 \times (5)^2 = 0.8 + 0.5 = 1.3\ \Omega$$

Total reactance referred to secondary,

$$X_{01} = X_1 + X_2' = X_1 + \frac{X_2}{K^2} = 4 + 0.12(5)^2 = 4 + 3 = 7\ \Omega$$

Total impedance referred to primary,

$$Z_{01} = \sqrt{R_{01}^2 + X_{01}^2} = \sqrt{(1.3)^2 + (7)^2} = 7.12 \text{ ohm}$$

Also $$I_1 = KI_2 = \frac{1}{5} \times 180 = 36 \text{ A}$$

(*i*) **Voltage to be applied** on h.v. side to obtain full-load current I_1,

$$V_{SC} = I_1 Z_{01} = 36 \times 7.12 = \mathbf{256.3\ V.}\quad \textbf{(Ans.)}$$

(*ii*) The **power factor** on short-circuit,

$$\cos \phi_0 = \frac{R_{01}}{Z_{01}} = \frac{1.3}{7.12} = \mathbf{0.1825.}\quad \textbf{(Ans.)}$$

Equivalent circuit and vector diagram are shown in Figs. 45, 46 and 47.

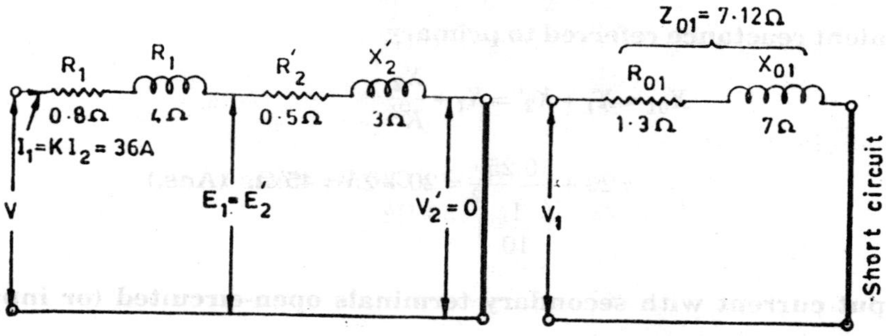

Fig. 45

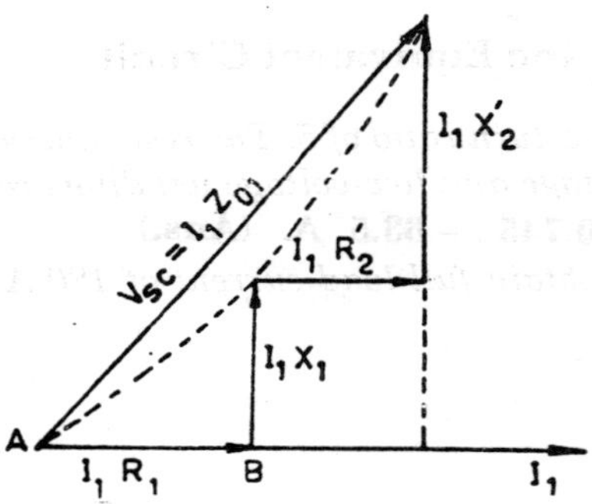

Fig. 46

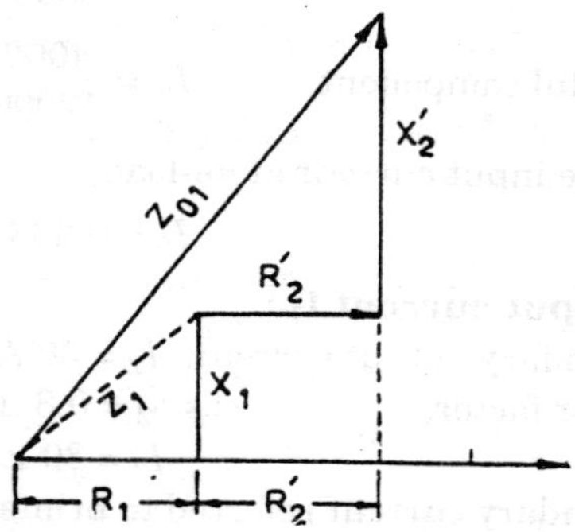

Fig. 47

Example 20. *A 12 kVA 4000/400 V transformer has primary and secondary winding resistances of 13 Ω and 0.15 Ω and leakage reactances of 20 Ω and 0.25 Ω respectively. The magnetising reactance is 6000 Ω and the resistance representing core loss is 12000 Ω. Determine :*

(*i*) *Equivalent resistance and reactance as referred to primary.*

(*ii*) *Input current with secondary terminals open circuited.*

(*iii*) *Input current when the secondary load current is 30 A at 0.8 power factor lagging.*

Solution. Transformation ratio, $K = \frac{400}{4000} = \frac{1}{10}$

Primary resistance, $R_1 = 13\ \Omega$
Primary reactance, $X_1 = 20\ \Omega$
Secondary resistance, $R_2 = 0.15\ \Omega$
Secondary reactance, $X_2 = 0.25\ \Omega$
Magnetising reactance, $X_0 = 6000\ \Omega$
Resistance representing the core loss, $R_0 = 12000\ \Omega$

(*i*) **Equivalent resistance referred to primary,**

$$R_{01} = R_1 + R_2' = R_1 + \frac{R_2}{K^2}$$

$$= 13 + \frac{0.15}{\left(\frac{1}{10}\right)^2} = 13 + 15 = \mathbf{28\ \Omega.}\ \textbf{(Ans.)}$$

Equivalent reactance referred to primary,

$$X_{01} = X_1 + X_2' = X_1 + \frac{X_2}{K^2}$$

$$= 20 + \frac{0.25}{\left(\frac{1}{10}\right)^2} = 20 + 25 = \mathbf{45\ \Omega.}\ \textbf{(Ans.)}$$

(*ii*) **Input current with secondary terminals open-circuited (or input current at no-load) :**

Magnetising component, $I_m = \frac{4000}{6000} = 0.667$ A

Wattful component, $I_w = \frac{4000}{12000} = 0.333$ A

Hence input current at no-load,

$$I_0 = 0.333 - j0.667 = \mathbf{0.745 \angle -63.5°\ A.}\ \textbf{(Ans.)}$$

(*iii*) **Input current $\mathbf{I_1}$:**

Secondary output current, $I_2 = 30$ A (*given*)
Power factor, $\cos \phi_2 = 0.8$ (lagging)
i.e., $I_2 = 30 \angle -36.9°$

Secondary current referred to primary,

$$I_2' = KI_2 = \frac{30}{10} \angle -36.9° = 3 \angle -36.9° = 2.4 - j1.8$$

∴ Input current, $I_1 = I_2' + I_0 = (2.4 - j1.8) + (0.333 - j0.667)$

$$= 2.733 - j2.467 = \mathbf{3.68 \angle -42.07°\ A.}\ \textbf{(Ans.)}$$

☞**Example 21.** *Obtain the approximate equivalent circuit of a given 200/2000 V single-phase 30 kVA transformer having the following test results :*

O.C. test : 200 V, 6.2 A, 360 W on l.v. side
S.C. test : 75 V, 18 A, 600 W on h.v. side.

Solution. O.C. test (l.v. side)

Primary voltage, $V_1 = 200$ V

No-load current, $I_0 = 6.2$ A

No-load loss, $P_0 = 360$ W

Now, $P_0 = V_1 I_0 \cos \phi_0$

$$360 = 200 \times 6.2 \times \cos \phi_0$$

$$\therefore \quad \cos \phi_0 = \frac{360}{200 \times 6.2} = 0.29$$

$$\sin \phi_0 = 0.957$$

Wattful component of no-load current,

$$I_w = I_0 \cos \phi_0 = 6.2 \times 0.29 = 1.8 \text{ A.}$$

Resistance representing the core loss,

$$R_0 = \frac{V_1}{I_w} = \frac{200}{1.8} = \mathbf{111.11\ \Omega.\ \ (Ans.)}$$

Magnetising component of no-load current,

$$I_m = I_0 \sin \phi_0 = 6.2 \times 0.95 = 5.93 \text{ A}$$

Magnetising reactance, $X_0 = \frac{V_1}{I_m} = \frac{200}{5.93} = \mathbf{33.7\ \Omega.\ \ (Ans.)}$

S.C. test (h.v. side) :

Short-circuit voltage, $V_{SC} = 75$ V

Short-circuit current, $I_{SC} = 18$ A

Losses, $P_{SC} = 600$ W

Impedance of transformer referred to h.v. side,

$$Z_{02} = \frac{V_{SC}}{I_{SC}} = \frac{75}{18} = 4.167\ \Omega$$

$$P_{SC} = I_{SC}^2 \times R_{02}$$

$$600 = 18^2 \times R_{02}$$

$$\therefore \quad R_{02} = \frac{600}{18^2} = 1.85\ \Omega$$

Transformation ratio, $K = \frac{2000}{200} = 10$

Referred to 200 V side :

$$Z_{01} = \frac{Z_{02}}{K^2} = \frac{4.167}{(10)^2} = 0.04167\ \Omega$$

$$R_{01} = \frac{R_{02}}{K^2} = \frac{1.85}{(10)^2} = 0.0185\ \Omega$$

$$\therefore \quad X_{01} = \sqrt{Z_{01}^2 - R_{01}^2} = \sqrt{(0.04167)^2 - (0.0185)^2} = \mathbf{0.0373\ \Omega.\ \ (Ans.)}$$

Approximate equivalent circuit is shown in Fig. 48.

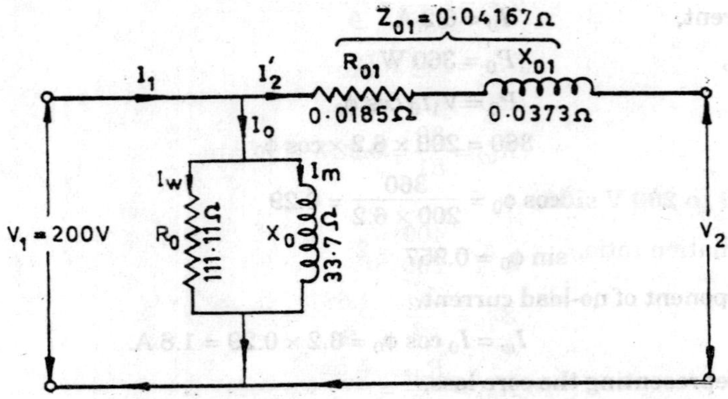

Fig. 48

Example 22. *The following readings were obtained on O.C. and S.C. tests on a 200/400 V, 50 Hz, single-phase transformer.*

O.C. test (l.v. side) : 200 V, 0.6 A, 60 W

S.C. test (h.v. side) : 15 V, 9 A, 80 W

Calculate the secondary voltage when delivering 4 kW at 0.8 power factor lagging, the primary voltage being 200 V.

Solution. O.C. test—l.v. side (*Instruments in the primary side* and secondary open)

Primary voltage, $V_1 = 200$ V

No-load current, $I_0 = 0.6$ A

No-load loss, $P_0 = 60$ W

Now, $P_0 = V_1 I_0 \cos\phi_0$

$60 = 200 \times 0.6 \times \cos\phi_0$

$$\therefore \quad \cos\phi_0 = \frac{60}{200 \times 0.6} = 0.5$$

$$\sin\phi_0 = 0.866$$

Wattful component of no-load current

$$I_w = I_0 \cos\phi_0 = 0.6 \times 0.5 = 0.3 \text{ A}$$

Magnetising component of no-load current,

$$I_m = I_0 \sin\phi_0 = 0.6 \times 0.866 = 0.52 \text{ A}$$

$\therefore$ Resistance representing the core loss,

$$R_0 = \frac{V_1}{I_w} = \frac{200}{0.3} = \mathbf{666.67\ \Omega.} \quad \textbf{(Ans.)}$$

Magnetising reactance, $X_0 = \frac{V_1}{I_m} = \frac{200}{0.52} = \mathbf{384.6\ \Omega.}$ **(Ans.)**

S.C. test—h.v. side (*Instruments in the secondary side* and primary short-circuited)

Short circuit voltage, $V_{SC} = 15$ V

Short circuit current, $I_{SC} = 9$ A

Losses, $P_{SC} = 80$ W

Impedance of transformer referred to secondary,

$$Z_{02} = \frac{V_{SC}}{I_{SC}} = \frac{15}{9} = 1.667\ \Omega$$

Also, $$P_{SC} = I_{SC}^2 \times R_{02}$$

$$80 = 9^2 \times R_{02}$$

$\therefore$ $$R_{02} = \frac{80}{81} = \mathbf{0.987\ \Omega.}\ \ \textbf{(Ans.)}$$

Referred to 200 V side :

Transformation ratio, $$K = \frac{400}{200} = 2$$

$$Z_{01} = \frac{Z_{02}}{K^2} = \frac{1.667}{2^2} = 0.417\ \Omega$$

$$R_{01} = \frac{R_{02}}{2^2} = \frac{0.987}{4} = 0.247\ \Omega$$

$\therefore$ $$X_{01} = \sqrt{Z_{01}^2 - R_{01}^2} = \sqrt{(0.417)^2 - (0.247)^2} = 0.336\ \Omega$$

Output, $$\text{kVA} = \frac{4}{0.8} = 5\ \text{kVA}$$

Output current, $$I_2 = \frac{5 \times 1000}{400} = \mathbf{12.5\ A.}\ \ \textbf{(Ans.)}$$

This value of I_2 is *approximate* because V_2 (which is to be calculated as yet) has been taken equal to 400 V (which in fact is equal to E_2 or ${}_0V_2$)

Now, $$Z_{02} = 1.667\ \Omega$$

$$R_{02} = 0.987\ \Omega$$

$\therefore$ $$X_{02} = \sqrt{Z_{02}^2 - R_{02}^2} = \sqrt{1.667^2 - 0.987^2} = 1.343\ \Omega$$

Total drop as referred to secondary

$$= I_2 (R_{02} \cos \phi_2 + X_{02} \sin \phi_2)$$

$$= 12.5(0.987 \times 0.8 + 1.343 \times 0.6) = 19.94\ \text{V}$$

$\therefore$ $$V_2 = 400 - 19.94 \simeq 380\ \text{V}$$

Hence, **secondary voltage = 380 V. (Ans.)**

Approximate equivalent circuit is shown in Fig. 49.

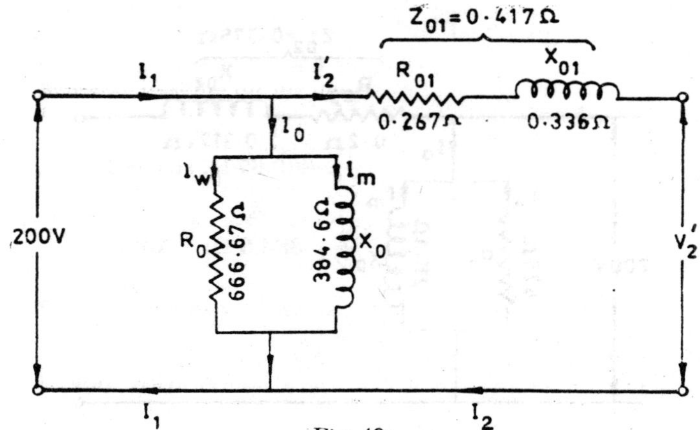

Fig. 49

Example 23. *A 4 kVA, 400/200 V, 50-Hz single-phase transformer has the following test data :*

O.C. test (l.v. side) 200 V, 1 A, 64 W
S.C. test (h.v. side) 15 V, 10 A, 80 W
Determine :
(i) Equivalent circuit referred to l.v. side, and
(ii) Secondary load voltage on full-load at 0.8 power factor lagging.

Solution. (*i*) **O.C. test—l.v. side :**

Voltage, $V_0 = 200$ V
No-load current, $I_0 = 1$ A
No-load loss, $P_0 = 64$ W
Now, $P_0 = V_0 I_0 \cos \phi_0$
$64 = 200 \times 1 \times \cos \phi_0$
$\therefore$ $\cos \phi_0 = 0.32$
$\sin \phi_0 = 0.9474$

Wattful component of no-load current,

$$I_w = I_0 \cos \phi_0 = 1 \times 0.32 = 0.32 \text{ A}$$

Magnetising component of no-load current,

$$I_m = I_0 \sin \phi_0 = 1 \times 0.9474 = 0.9474 \text{ A}$$

$\therefore$ Resistance representing the core loss,

$$R_0 = \frac{V_0}{I_w} = \frac{200}{0.32} = 625 \ \Omega. \ \textbf{(Ans.)}$$

Magnetising reactance, $X_0 = \frac{V_0}{I_m} = \frac{200}{0.9474} = 211.1 \ \Omega.$ **(Ans.)**

S.C. test—h.v. side :

Short-circuit voltage, $V_{SC} = 15$ V
Short-circuit current, $I_{SC} = 10$ A
Losses, $P_{SC} = 80$ W

Impedance of the circuit referred to h.v. side,

$$Z_{01} = \frac{V_{SC}}{I_{SC}} = \frac{15}{10} = 1.5 \ \Omega. \ \textbf{(Ans.)}$$

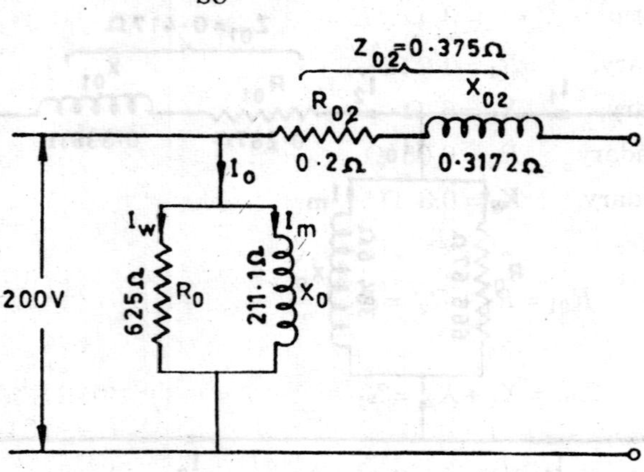

Fig. 50

Also $P_{SC} = I_{SC}^2 \times R_{01}$

$\therefore$ $80 = 10^2 \times R_{01}$

$\therefore$ $R_{01} = \frac{80}{100} = \mathbf{0.8\ \Omega}$. **(Ans.)**

Referred to l.v. side :

Transformation ratio, $K = \frac{200}{400} = \frac{1}{2}$

$$Z_{02} = K^2 Z_{01} = (\tfrac{1}{2})^2 \times 1.5 = \mathbf{0.375\ \Omega}. \quad \textbf{(Ans.)}$$

$$R_{02} = K^2 R_{01} = (\tfrac{1}{2})^2 \times 0.8 = \mathbf{0.2\ \Omega}. \quad \textbf{(Ans.)}$$

$$\therefore \quad X_{02} = \sqrt{Z_{02}^2 - R_{02}^2} = \sqrt{(0.375)^2 - (0.2)^2} = \mathbf{0.3172\ \Omega}. \quad \textbf{(Ans.)}$$

The approximate equivalent circuit is shown in Fig. 52.

(*ii*) **Secondary load voltage, V_2 :**

Secondary full-load current, $I_2 = \frac{4 \times 1000}{200} = 20$ A

$\cos \phi_2 = 0.8$

$\sin \phi_2 = 0.6$

$\therefore$ $I_2 = 20(0.8 - j0.6) = (16 - j12)$

$Z_{02} = R_{02} + jX_{02} = 0.2 + j0.3172$

Secondary load voltage, $V_2 = 200 - I_2 Z_{02} = 200 - (16 - j12)(0.2 + j0.3712)$

$= 200 - (3.2 + j5.075 - j2.4 + 3.806)$

$= 200 - (7 + j2.675) = 193 - j2.675 \simeq \mathbf{193\ V}$. **(Ans.)**

Example 24. *A single-phase step-down transformer has a turn ratio of 3. The resistance and reactance of the primary winding are 1.2 Ω and 6 Ω and those of the secondary winding are 0.05 Ω and 0.03 Ω respectively. If the h.v. winding is supplied at 230 V, 50-Hz with l.v. winding short circuited, find :*

(*i*) *Current in the l.v. winding,* (*ii*) *Copper loss in the transformer, and*
(*iii*) *Power factor.*

Solution. Turn ratio $= 3$, *i.e.*, $K = \frac{1}{3}$

Resistance of primary, $R_1 = 1.2\ \Omega$

Reactance of primary, $X_1 = 6\ \Omega$

Resistance of secondary, $R_2 = 0.05\ \Omega$

Reactance of secondary, $X_2 = 0.3\ \Omega$

Referred to *h.v. side*

$$R_{01} = R_1 + R_2' = R_1 + \frac{R_2}{K^2} = 1.2 + 0.05 \times (3)^2 = 1.65\ \Omega$$

$$X_{01} = X_1 + X_2' = X_1 + \frac{X_2}{K^2} = 6 + 0.3 \times (3)^2 = 8.7\ \Omega$$

$$Z_{01} = \sqrt{R_{01}^2 + X_{01}^2} = \sqrt{(1.65)^2 + (8.7)^2} = 8.855\ \Omega$$

Current in h.v. winding when l.v. winding is short-circuited,

$$I_{SC} = \frac{V_{SC}}{Z_{01}} = \frac{230}{8.855} = 25.97 \text{ A}$$

Neglecting I_0, $\quad I_1 = I_2' = 25.97$ A

(*i*) **Current in the l.v. winding,**

$$I_2 = \frac{I_2'}{K} = \frac{25.97}{\left(\frac{1}{3}\right)} = \mathbf{77.91\ A.\ (Ans.)}$$

(*ii*) **Total copper loss** $= I_1^2 R_{01} = (25.97)^2 \times 1.65 = \mathbf{1112.8\ W.}$ **(Ans.)**

(*iii*) **Primary power factor :**

Power input on short-circuit,

$$P_{SC} = V_{SC} I_{SC} \,.\, \cos \phi_{SC}$$

$$1112.8 = 230 \times 25.97 \times \cos \phi_{SC}$$

$$\therefore \quad \cos \phi_{SC} = \frac{1112.8}{230 \times 25.97} = \mathbf{0.1863.\ (Ans.)}$$

☞**Example 25.** *A single-phase, 3 kVA, 230/115 V, 50-Hz transformer has the following constants :*

Resistance : Primary 0.3 Ω, secondary 0.09 Ω.

Reactance : Primary 0.4 Ω, secondary 0.1 Ω.

Resistance of equivalent exciting circuit referred to primary, R_0 = 600 Ω.

Reactance of equivalent exciting circuit-referred to primary, X_0 = 200 Ω.

What would be the readings of the instruments when the transformer is connected for (i) O.C. test, (ii) S.C. test. In both tests supply is given to h.v. side.

Solution. Primary resistance, $\quad R_1 = 0.3\ \Omega$

Primary reactance, $\quad X_1 = 0.4\ \Omega$

Secondary resistance, $\quad R_2 = 0.09\ \Omega$

Secondary reactance, $\quad X_2 = 0.1\ \Omega$

$R_0 = 600\ \Omega$

$X_0 = 200\ \Omega$

(*i*) **O.C. test :**

Wattful component of no-load current,

$$I_w = \frac{V_1}{R_0} = \frac{230}{600} = 0.383 \text{ A}$$

Magnetising component of no-load current,

$$I_w = \frac{V_1}{X_0} = \frac{230}{200} = 1.15 \text{ A}$$

$\therefore$ No-load current, $\quad I_0 = \sqrt{I_w^2 + I_m^2} = \sqrt{(0.383)^2 + (1.15)^2} = 1.212$ A

Input on no-load to h.v. winding

$$= V_1 I_w = 230 \times 0.383 = 88.09 \text{ W}$$

Hence, the readings of the instruments are :

Voltmeter reading **= 230 V. (Ans.)**

Ammeter reading **= 1.212 A. (Ans.)**

Wattmeter reading **= 88.09 W. (Ans.)**

(*ii*) **S.C. test :**

Transformation ratio, $K = \frac{115}{230} = \frac{1}{2}$

Referred to *h.v. side*, $R_{01} = R_1 + R_2' = R_1 + \frac{R_2}{K^2} = 0.3 + \frac{0.09}{(\frac{1}{2})^2} = 0.66\ \Omega$

$$X_{01} = X_1 + X_2' = X_1 + \frac{X_2}{K^2} = 0.4 + \frac{0.1}{(\frac{1}{2})^2} = 0.8\Omega$$

$$\therefore \quad Z_{01} = \sqrt{R_{01}^2 + X_{01}^2} = \sqrt{(0.66)^2 + (0.8)^2} = 1.037\ \Omega$$

Full-load current in the h.v. winding

$$I_1 = \frac{3 \times 1000}{230} = 13.04\ \text{A}$$

Voltage to be applied to h.v. winding on short-circuiting l.v. winding to pass full-load current of 13.04 A through h.v. winding.

$$V_{SC} = I_1 Z_{01} = 13.04 \times 1.037 = 13.52\ \text{V}$$

Power supplied to h.v. winding on short-circuit

$$= I_1^2 . R_{01} = (13.04)^2 \times 0.66 = 112.2\ \text{W}$$

Hence, the readings of the instruments are :

Voltmeter reading = 13.52 V. (Ans.)
Ammeter reading = 13.04 A. (Ans.)
Wattmeter reading = 112.2 W. (Ans.)

Regulation and Efficiency of a Transformer

Example 26. *A 4kVA 220/440 V, 50 Hz, single-phase transformer gave the following test figures. No load test performed on 220 V side keeping 440 V side open : 220 V, 0.7 A, 60 W. Short circuit test performed short circuiting the 440 V side through an ammeter : 9V, 6A, 21.6W calculate.*

(*a*) *The magnetising current and the component corresponding to iron loss at normal voltage and frequency.*

(*b*) *The efficiency at full-load at unity power factor and the corresponding secondary terminal voltage.*

(*c*) *Draw the phasor diagram corresponding to the full load operation.* **(PTU, May, 2000)**

Solution. *Open circuit test* on 220 V (l.v.) side : $V_1 = 220$ V ; $P_0 = 60$ W ; $I_0 = 0.7$ A

Short circuit test on 440 V (h.v.) side : $V_{SC} = 9$V ; $I_{SC} = 6$A ; $P_{SC} = 21.6$ W

(*a*) $P_0 = V_1 I_0 \cos \phi_0$ or $60 = 220 \times 0.7 \times \cos \phi_0$

or $\cos \phi_0 = 0.3896$, and $\sin \phi_0 = 0.921$

$\therefore$ Iron loss component of current, $I_w = I_0 \cos \phi_0 = 0.7 \times 0.3896 =$ **0.273 A. (Ans.)**

Magnetising component of current, $I_m = I_0 \sin \phi_0 = 0.7 \times 0.921 =$ **0.6447 A. (Ans.)**

(*b*) Full-load secondary current, $I_2 = \frac{4 \times 1000}{440} = 9.1$ A

Full-load primary current, $I_1 = \frac{4 \times 1000}{220} = 18.2$ A

Full-load copper losses, $P_c = \left(\frac{I_1}{I_{sc}}\right)^2 \times P_{SC} = \left(\frac{18.2}{6}\right)^2 \times 21.6 = 198.7$ W

$\therefore$ **Full-load efficiency at unity power factor,**

$$\eta = \frac{4 \times 1000 \times 1}{4 \times 1000 \times 1 + 60 + 198.7} \times 100 = \mathbf{93.92\%.} \ \textbf{(Ans.)}$$

From short circuit test, we have

$$Z_{01} = \frac{V_{SC}}{I_{sc}} = \frac{9}{6} = 1.5\ \Omega$$

$$R_{01} = \frac{P_{SC}}{I_{sc}^2} = \frac{21.6}{6^2} = 0.6\ \Omega$$

$\therefore$ $$X_{01} = \sqrt{Z_{01}^2 - R_{01}^2} = \sqrt{1.5^2 - 0.6^2} = 1.375\ \Omega$$

Now, $$R_{02} = K^2 R_{01} = (2)^2 \times 0.6 = 2.4\ \Omega \qquad \left(\because K = \frac{440}{220} = 2\right)$$

$$X_{02} = K^2 X_{01} = 2^2 \times 1.375 = 5.5\ \Omega$$

Since $\cos \phi_2 = 1 \qquad \therefore \sin \phi_2 = \sin^{-1} (\cos^{-1} 1) = 0$

$$_0V_2^2 = (V_2 + I_2 R_{02})^2 + (I_2 X_{02})^2$$

$$(440)^2 = (V_2 + 9.1 \times 2.4)^2 + (9.1 \times 5.5)^2$$

or $$(V_2 + 21.84)^2 = (440)^2 - (9.1 \times 5.5)^2 = 191095$$

or $$\mathbf{V_2} = (\sqrt{191095} - 21.84) = \mathbf{415.3\ V.}\ \textbf{(Ans.)}$$

(*c*) **The phasor diagram of transformer corresponding to full-load operation is shown in Fig. 51.**

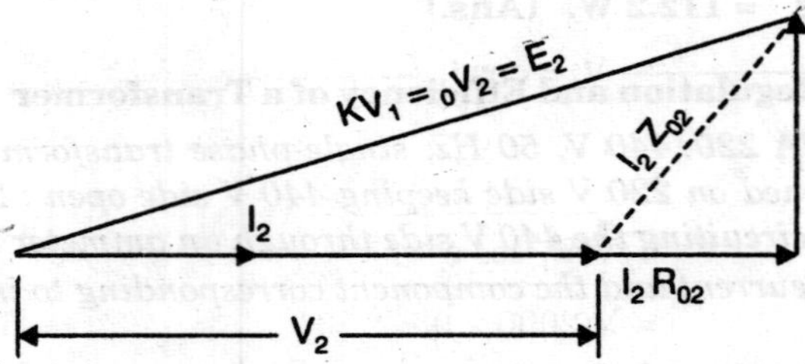

Fig. 51

Example 27. ***The parameters of the equivalent circuit of a 100 kVA, 2000/200 volt single-phase transformer are as follows :***

Primary resistance	*= 0.2 ohms ;*
Secondary resistance	*= 2 milli ohms.*
Primary leakage reactance	*= 0.45 ohms ;*
Secondary leakage reactance	*= 4.5 milli ohms.*
Core loss resistance	*= 10 kilo ohms ;*
Magnetizing reactance	*= 1.55 kilo ohms.*

Using the circuit referred to primary, determine the :

(*i*) *Voltage regulation.*

(*ii*) *Efficiency of the transformer operating at rated load with 0.8 lagging power factor.*

(PTU, June and Dec. 2000)

Solution. *Given* : Rating = 100 kVA ; $R_1 = 0.2\ \Omega$; $R_2 = 0.002\ \Omega$;

$X_1 = 0.45\ \Omega$; $X_2 = 0.0045\ \Omega$; $R_0 = 10000\ \Omega$; $X_0 = 1550\ \Omega$.

(*i*) **Voltage regulation :**

Transformation ratio, $K = \frac{200}{2000} = 0.1$

$$R_{01} = R_1 + R'_2 = R_1 + \frac{R_2}{K^2} = 0.2 + \frac{0.002}{0.1^2} = 0.4\ \Omega$$

$$X_{01} = X_1 + X'_2 = X_1 + \frac{X_2}{K^2} = 0.45 + \frac{0.0045}{0.1^2} = 0.9\ \Omega$$

Primary full-load current, $I_1 = \frac{100 \times 1000}{2000} = 50$ A,

$$\cos \phi = 0.8,\ \sin \phi = \sin [\cos^{-1} (0.8)] = 0.6$$

Refer Fig. 52.

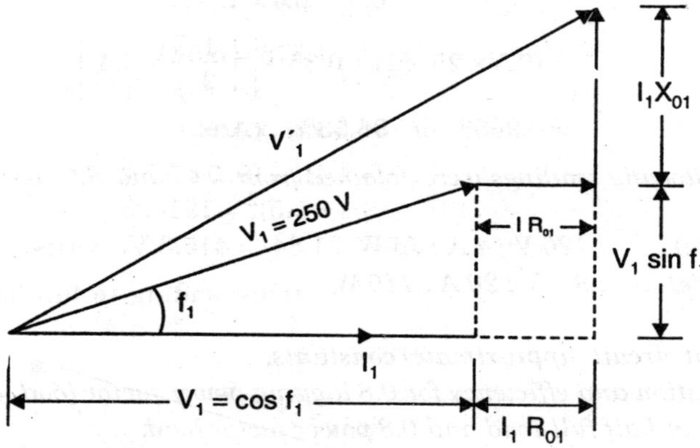

Fig. 52

Applied voltage, $V'_1 = \sqrt{(V_1 \cos \phi + I_1 R_{01})^2 + (V_1 \sin \phi + I_1 X_{01})^2}$

$$= \sqrt{(2000 \times 0.8 + 50 \times 0.4)^2 + (2000 \times 0.6 + 50 \times 0.9)^2} = 2043 \text{ V}$$

$\therefore$ % voltage regulation $= \frac{V'_1 - V_1}{V_1} \times 100 = \frac{2043 - 2000}{2000} \times 100 =$ **2.15%. (Ans.)**

(*ii*) **Efficiency of the transformer, η :**

Working component of no-load current, $I_w = \frac{V}{R_0} = \frac{2000}{10000} = 0.2$ A

Iron loss = $V_1 I_0 \cos \phi_0 = V_1 I_w = 2000 \times 0.2 = 400$ W

Full-load copper losses, $P_c = I_1^2 R_{01} = (50)^2 \times 0.4 = 1000$ W

Efficiency of the transformer at 0.8 p.f.,

$$\% \eta = \frac{\text{kVA} \times 1000 \times \cos \phi}{(\text{kVA} \times 1000 \times \cos \phi) + P_i + P_c} \times 100$$

$$= \frac{(100 \times 1000 \times 0.8)}{(100 \times 1000 \times 0.8 + 400 + 1000)} \times 100$$

= 98.28%. (Ans.)

Example 28. *In a 25 kVA, 2000/200 V transformer, the constant and variable losses are 350 W and 400 W respectively. Calculate the efficiency on u.p.f. at*

(*i*) *Full load, and* (*ii*) *Half full load.* **[PTU, 1999]**

Solution. Rating of transformer = 25 kVA

Constant or iron losses, P_i = 350 W or 0.35 kW

Variable or copper losses, P_c = 400 W or 0.4 kW

Power factor, $\cos\phi = 1$

Efficiency, η :

(*i*) **At full load :**

$$\eta = \frac{\text{output}}{\text{output} + \text{losses}} = \frac{kVA \cos\phi}{kVA \cos\phi + P_i + P_c}$$

$$= \frac{25 \times 1}{25 \times 1 + 0.35 + 0.4} = 0.9708 \text{ or } \mathbf{97.08\%}. \text{ (Ans.)}$$

(*ii*) **At half full load :**

$$\eta = \frac{(0.5 \times 25) \times 1}{(0.5 \times 25 \times 1) + 0.35 + \left[\left(\frac{1}{2}\right)^2 \times 0.4\right]}$$

$$= 0.9653 \text{ or } \mathbf{96.53\%}. \text{ (Ans.)}$$

☞**Example 29.** *The following readings were obtained from O.C. and S.C. tests on 8 kVA 400/120 V, 50-Hz transformer.*

O.C. test (l.v. side) : 120 V ; 4 A ; 75 W.

S.C. test (h.v. side) : 9.5 V ; 20 A ; 110 W.

Calculate :

(*i*) *The equivalent circuit (approximate) constants,*

(*ii*) *Voltage regulation and efficiency for 0.8 lagging power factor load, and*

(*iii*) *The efficiency at half full-load and 0.8 power factor load.*

Solution. (*i*) Transformation ratio, $K = \frac{120}{400} = \frac{3}{10}$

O.C. test :

It is seen from the O.C. test, that *primary open*, the secondary draws a no-load current of 4 A.

Since $K = \frac{3}{10}$, the corresponding no-load current $I_0 = 4 \times \frac{3}{10} = 1.2$ A.

Also $P_0 = V_1 I_0 \cos\phi_0$

∴ $75 = 400 \times 1.2 \times \cos\phi_0$

i.e., $\cos\phi_0 = \frac{75}{400 \times 1.2} = 0.156$

and $\sin\phi_0 = 0.987$

Now, wattful component of no-load current, I_0,

$$I_w = I_0 \cos\phi_0 = 1.2 \times 0.156 = 0.187 \text{ A}$$

and magnetising component of no-load current I_0,

$$I_m = I_0 \sin\phi_0 = 1.2 \times 0.987 = 1.184 \text{ A}$$

∴ Resistance representing the core loss,

$$R_0 = \frac{V_1}{I_w} = \frac{400}{0.187} = \mathbf{2139\ \Omega}. \text{ (Ans.)}$$

Magnetising reactance, $X_0 = \frac{V_1}{I_m} = \frac{400}{1.184} = \mathbf{337.8\ \Omega}$. **(Ans.)**

S.C. test :

During S.C. test the *instruments have been placed in primary.*

Here, $V_{SC} = 9.5$ V

$I_{SC} = 20$ A

$P_{SC} = 110$ W

Now, $Z_{01} = \frac{V_{SC}}{I_{SC}} = \frac{9.5}{20} = \mathbf{0.475\ \Omega}$. **(Ans.)**

$R_{01} = \frac{P_{SC}}{I_{SC}^2} = \frac{110}{(20)^2} = \mathbf{0.25\ \Omega}$. **(Ans.)**

∴ $X_{01} = \sqrt{Z_{01}^2 - R_{01}^2} = \sqrt{(0.475)^2 - (0.25)^2} = \mathbf{0.404\ \Omega}$. **(Ans.)**

The equivalent circuit is shown in Fig. 53.

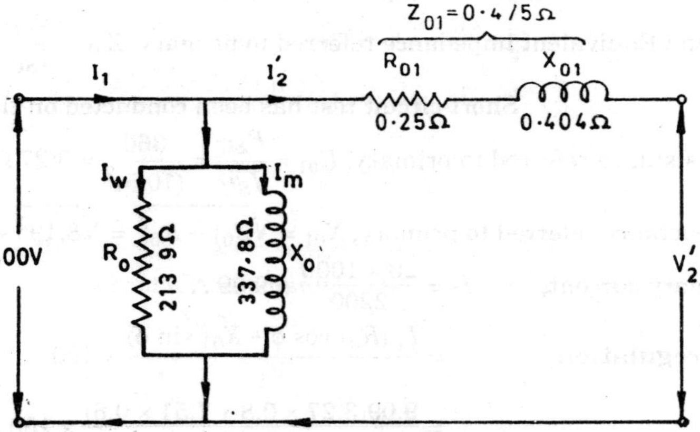

Fig. 53

(*ii*) Total *approximate voltage drop* as referred to primary is

$= I_1 (R_{01} \cos \phi + X_{01} \sin \phi)$

Now, full-load, $I_1 = \frac{8 \times 1000}{400} = 20$ A

∴ Voltage drop $= 20 (0.25 \times 0.8 + 0.404 \times 0.6) = 8.85$ V

% **Voltage regulation** $= \frac{8.85}{400} \times 100 = \mathbf{2.21\%}$. **(Ans.)**

Iron loss, $P_i = 75$ W

Copper loss, $P_c = 110$ W

Full-load losses $= P_i + P_c = 75 + 110 = 185$ W

∴ Full-load output $= 8 \times 1000 \times 0.8 = 6400$ W

∴ **Full-load efficiency,** $\eta = \frac{\text{output}}{\text{output + losses}} = \frac{6400}{6400 + 185} = 0.972$ or **97.2%. (Ans.)**

(*iii*) **Half full-load :**

Iron loss, $P_i = 75$ W

Copper loss, $P_c = (\frac{1}{2})^2 \times 110 = 27.5$ W

Total losses $= P_i + P_c = 75 + 27.5 = 102.5$ W

Output at half full-load $= \frac{1}{2} \times 8 \times 1000 \times 0.8 = 3200$ W

$\therefore$ **Efficiency** $= \dfrac{\text{output}}{\text{output + losses}} = \dfrac{3200}{3200 + 102.5} = 0.969$ or **96.9%. (Ans.)**

Example 30. *Consider a 20 kVA, 2200/220 V, 50 Hz transformer. The OC/SC test results are as follows :*

O.C. Test : 220 V, 4.2 A, 148 W (l.v. side)

S.C. Test : 86 V, 10.5 A, 360 W (h.v. side)

(i) Determine the regulation at 0.8 p.f. lagging at full load.

(ii) What is the p.f. on short-circuit ? **(Nagpur University, 1998)**

Solution. *Given* :

S.C. Test :

$$V_{SC} = 86 \text{ V}, I_{SC} = 10.5 \text{ A}, P_C = 360 \text{ W}$$

(*i*) **Regulation :** Equivalent impedance referred to primary, $Z_{01} = \dfrac{V_{SC}}{I_{SC}} = \dfrac{86}{10.5} = 8.19\ \Omega$

($\because$ Short-circuit test has been conducted on the h.v. (primary) side)

Equivalent resistance referred to primary, $R_{01} = \dfrac{P_{SC}}{I_{SC}^2} = \dfrac{360}{(10.5)^2} = 3.27\ \Omega$

Equivalent reactance referred to primary, $X_{01} = \sqrt{Z_{01}^2 - R_{01}^2} = \sqrt{8.19^2 - 3.27^2} = 7.51\ \Omega$

Full load primary current, $I_1 = \dfrac{20 \times 1000}{2200} = 9.09$ A

Percentage regulation

$$= \frac{I_1 (R_{01} \cos \phi + X_{01} \sin \phi)}{V_1} \times 100$$

$$= \frac{9.09(3.27 \times 0.8 + 7.51 \times 0.6)}{2200} \times 100 = \mathbf{2.94\%}. \text{ (Ans.)}$$

(*ii*) **P.f. on short-circuit, cos ϕ_{SC}** :

$$\cos \phi_{SC} = \frac{R_{01}}{Z_{01}} = \frac{3.27}{8.19} = \mathbf{0.399 \text{ (lag)}}. \text{ (Ans.)}$$

Example 31. *High voltage side short circuit test data for 20 kVA, 2300/230 V transformer are :*

Power = 250 watts ; Current = 8.7 A ; Voltage = 50 V.

Calculate equivalent impedance, resistance, reactance referred to h.v. side. Find the transformer regulation at 0.7 lagging power factor.

Solution. *Given* : **h.v. side :** $P_{SC} = 250$ W ; $I_{SC} = 8.7$ A ; $V_{SC} = 50$ V ; $\cos \phi = 0.7$.

From S.C. test with measurements performed on h.v. side the various parameters referred to *h.v.* side are :

$$R_{01} = \frac{P_{SC}}{I_{SC}^2} = \frac{250}{8.7^2} = 3.303\ \Omega$$

$$Z_{01} = \frac{V_{SC}}{I_{SC}} = \frac{50}{8.7} = 5.747\ \Omega$$

$$\therefore \quad X_{01} = \sqrt{Z_{01}^2 - R_{01}^2} = \sqrt{5.747^2 - 3.303^2} = 4.703\ \Omega$$

We know that, % Regulation $= \dfrac{I_1R_{01}\cos\phi + I_1X_{01}\sin\phi}{V_1} \times 100$

where $I_1 = \dfrac{20 \times 10^3}{2300} = 8.696$ A (rated current) ; and

$\cos\phi = 0.7$ (given) and hence $\sin\phi = \sin(\cos^{-1} 0.7) = 0.714$.

$\therefore$ **% Regulation** $= \dfrac{8.696 \times 3.303 \times 0.7 + 8.696 \times 4.703 \times 0.714}{2300} \times 100 =$ **2.14%. (Ans.)**

Example 32. *A 10 kVA, 2300/230 V, 50 Hz, distribution transformer has h.v. winding resistance 3.96 Ω and leakage resistance of 15.8 Ω. The l.v. winding has corresponding value of 0.0396 Ω and 0.158 Ω respectively. The transformer has a core loss of 58 W under normal operating conditions. Find :*

(i) Load terminal voltage when transformer delivers rated current at 0.8 p.f. lagging to a load on l.v. side, with h.v. side voltage held at rated value and compute efficiency at this load.

(ii) The h.v. side voltage necessary to maintain rated voltage at load terminals when the transformer is loaded as above. Is efficiency same as in (i) ? **(AMIE Summer, 1999)**

Solution. *Given* : $R_1 = 3.96\ \Omega$; $X_1 = 15.8\ \Omega$; $R_2 = 0.0396\ \Omega$; $X_2 = 0.158\ \Omega$; $P_i = 58$ W ; $p.f. = 0.8$ lagging

(*i*) Load terminal voltage :

Rated secondary current, $I_2 = \dfrac{10 \times 10^3}{230} = 43.48$ A

$$R_{02} = R_2 + K^2R_1 = 0.0396 + \left(\frac{230}{2300}\right)^2 \times 3.96 = 0.0792\ \Omega$$

$$X_{02} = X_2 + K^2X_1 = 0.158 + \left(\frac{230}{2300}\right)^2 \times 15.8 = 0.316\ \Omega$$

$\therefore$ Voltage drop on load side $= I_2(R_{02}\cos\phi + X_{02}\sin\phi)$

$= 43.48\,(0.0792 \times 0.8 + 0.316 \times 0.6) = 11$ V

$\therefore$ **Load terminal voltage** $= 230 - 11 =$ **219 V. (Ans.)**

Copper loss $(P_C) = I_2^2R_{02} = 43.48^2 \times 0.0792 = 149.73$ W

$$\therefore \quad \eta = \frac{\text{Output}}{\text{Output} + \text{Core loss } (P_i) + \text{Copper loss } (P_C)}$$

$= \dfrac{10 \times 10^3 \times 0.8}{(10 \times 10^3 \times 0.8) + 58 + 149.73} = 0.9747$ or **97.47%. (Ans.)**

(*ii*) To maintain rated voltage at load terminals of 230 V, the O.C. voltage required at secondary terminals would be 230 + 11 = 241 V. Hence primary induced voltage required

$= 241 \times \dfrac{2300}{230} =$ **2410 V. (Ans.)**

Example 33. *The following test results were obtained for a 1000/100 V, 100 kVA single-phase transformer :*

O.C. test :

Primary volts = 1000, secondary volts = 100, watts in primary = 950.

S.C. test :

Primary volts for full-load current = 20, watts in primary = 1000.

Determine the regulation and efficiency of the transformer at full-load and at 0.8 power factor lagging.

Solution. Transformation ratio, $K = \frac{100}{1000} = \frac{1}{10}$

O.C. test :

Primary volts, $V_1 = 1000$ V

Iron loss, $P_i = 950$ W $= 0.95$ kW

S.C. test :

Full-load copper loss, $P_c = 1000$ W $= 1$ kW

In S.C. test 20 V in primary are equivalent to $20 \times \frac{1}{10} = 2$ V in secondary.

Now, full-load secondary current,

$$I_2 = \frac{100 \times 1000}{100} = 1000 \text{ A}$$

Also $$Z_{02} = \frac{2}{1000} = 0.002\ \Omega$$

and $$R_{02} = \frac{1000}{(1000)^2} = 0.001\ \Omega$$

$\therefore$ $$X_{02} = \sqrt{Z_{02}^2 - R_{02}^2} = \sqrt{0.002^2 - 0.001^2} = 0.00173\ \Omega$$

Approximate voltage drop at 0.8 power factor lagging

$$= I_2(R_{02}\cos\phi + X_{02}\sin\phi)$$
$$= 1000(0.001 \times 0.8 + 0.00173 \times 0.6) = 1.838 \text{ V}$$

$\therefore$ **% regulation down** $= \frac{1.838}{100} \times 100 = $ **1.838%. (Ans.)**

Full-load output $= 100 \times 0.8 = 80$ kW

Total losses $= P_i + P_c = 0.95 + 1 = 1.95$ kW

$\therefore$ **Efficiency,** $\eta = \frac{\text{output}}{\text{output + losses}} = \frac{80}{80 + 1.95} = 0.976$ or **97.6%. (Ans.)**

Example 34. *The primary and secondary winding resistances of a 30 kVA, 6600/250 V single-phase transformer are 8 Ω and 0.015 Ω respectively. The equivalent leakage reactance as referred to the primary winding is 30 Ω. Find the full-load regulation for load power factors of :*

(i) unity *(ii) 0.8 lagging* *(iii) 0.8 leading.*

Solution. Transformer ratio, $K = \frac{250}{6600} = \frac{1}{26.4}$

Primary resistance, $R_1 = 8\ \Omega$

Secondary resistance, $R_2 = 0.015\ \Omega$

Equivalent leakage reactance as referred to primary,

$$X_{01} = 30\ \Omega$$

Now, $$R_{02} = R_2 + K^2R_1 = 0.015 + \left(\frac{1}{26.4}\right)^2 \times 8 = 0.02648\ \Omega$$

$$X_{02} = K^2X_{01} = \left(\frac{1}{26.4}\right)^2 \times 30 = 0.04304\ \Omega$$

Secondary full-load current, $I_2 = \frac{30 \times 1000}{250} = 120$ A.

(*i*) **Unity power factor,** $\cos \phi_2 = 1$:

Voltage drop $= I_2R_{02} \cos \phi_2 + I_2X_{02} \sin \phi_2 = 120 \times 0.02648 \times 1 + 0 = 3.177$ V

% **regulation** $= \frac{3.177}{250} \times 100 =$ **1.27%. (Ans.)**

(*ii*) **0.8 power factor (lagging)** :

Voltage drop $= I_2R_{02} \cos \phi_2 + I_2X_{02} \sin \phi_2$

$= 120 \times 0.02648 \times 0.8 + 120 \times 0.04304 \times 0.6 = 5.64$ V

% **regulation** $= \frac{564}{250} \times 100 =$ **2.256%. (Ans.)**

(*iii*) **0.8 power factor (leading) :**

Voltage drop $= I_2R_{02} \cos \phi_2 - I_2X_{02} \sin \phi_2$

$= 120 \times 0.02648 \times 0.8 - 120 \times 0.04304 \times 0.6 = -0.557$ V

% **regulation** $= -\frac{0.557}{250} \times 100 =$ **– 0.223%. (Ans.)**

Example 35. *Short-circuit test is conducted on a 5 kVA, 400 V/100 V single phase transformer with 100 V winding shorted. The input voltage at full load current is 40 V. The wattmeter, on the input reads 250 W. Find the power factor for which regulation at full load is zero.* **(GATE, 1994)**

Solution. *Given* : Rating of transformer : 5 kVA, 400 V/100 V

Input power, on short-circuit, $P_{SC} = 250$ W

The input voltage at full load current, $V_{SC} = 40$ V

Power factor for which regulation at full load is zero :

Input current I_{SC} = full load current, $I_1 = \frac{5 \times 1000}{400} = 12.5$ A

Equivalent resistance as referred to primary, $R_{01} = \frac{P_{SC}}{I_{SC}^2} = \frac{250}{(12.5)^2} = 1.6\ \Omega$

Equivalent impedance referred to primary, $Z_{01} = \frac{V_{SC}}{I_{SC}} = \frac{40}{12.5} = 3.2\ \Omega$

Equivalent reactance referred to primary, $X_{01} = \sqrt{Z_{01}^2 - R_{01}^2} = \sqrt{3.2^2 - 1.6^2} = 2.77\ \Omega$

Regulation will be zero when phase angle,

$$\phi = \tan^{-1}\frac{R_{02}}{X_{02}} = \tan^{-1}\frac{R_{01}}{X_{01}} = \tan^{-1}\left(\frac{1.6}{2.77}\right) = \mathbf{30^\circ}$$

or **power factor** $= \cos \phi = \cos (30°) =$ **0.866. (Ans.)**

Example 36. *A 50 MVA, 76.2 V/33 kV, 1-phase, 50 Hz, two-winding transformer with tap changer has percentage impedance of 0.5 + j7.0. What tapping must be used to maintain rated voltage at the secondary on*

(*i*) *full load at 0.8 lagging power factor, and*

(*ii*) *40 MVA load at 0.6 lagging power factor.*

Assume that the tap changer is provided on the h.v. side.

Solution. *Given* : Rated MVA = 50 ; $f = 50$ Hz ; $E_1 = 76.2$ V ; $E_2 = 33$ kV ;

percentage impedance $= 0.5 + j7.0$.

(*i*) Percentage regulation $= \dfrac{L_2R_{02}\cos\phi + I_2X_{02}\sin\phi}{E_2} \times 100$

$= \dfrac{I_2R_{02}}{E_2} \times 100 \cos\phi + \dfrac{I_2 X_0}{E_2} \times 100 \sin\phi$

$= \% R \cos\phi + \% X \sin\phi$

$= 0.5 \times 0.8 + 7 \times 0.6 = 4.6\%$

i.e., voltage to be raised for maintaining rated voltage = 4.6%

So tap setting required on h.v. side = – 4.6% or **4.6% down. (Ans.)**

(*ii*) At load of 40 MVA *i.e.* 80% of full load or 0.8 times full load,

voltage regulation $= 0.8\,(\% R \cos\phi + \% X \sin\phi)$

$= 0.8(0.5 \times 0.6 + 7 \times 0.8) = 4.72\%$

Tap setting required on h.v. side = **4.72% down. (Ans.)**

Note. *If the tap changer is on the primary side, tap setting will be down and if it is to be provided on secondary side, tap setting will be up for raising the secondary voltage on load.*

Example 37. *The percentage resistance and reactance of a transformer are 2% and 4% respectively. Find the approximate regulation on full-load at :*

(*i*) *Unity power factor,* (*ii*) *0.8 power factor lagging, and*

(*iii*) *0.8 power factor leading.*

Solution. Percentage resistance = 2%

Percentage reactance = 4%

Approximate % regulation = **% resistance × cos φ ± % reactance × sin φ**

(*i*) **Unity power factor :**

Approximate % regulation = % resistance cos φ (∵ $\sin\phi = 0$)

$= 2 \times 1 =$ **2%. (Ans.)**

(*ii*) **0.8 power factor lagging :**

Approximate % regulation = % resistance × cos φ + % reactance × sin φ

$= 2 \times 0.8 + 4 \times 0.6 = 1.6 + 2.4 =$ **4%. (Ans.)**

(*iii*) **0.8 power factor leading :**

Approximate % regulation = % resistance × cos φ – % reactance × sin φ

$= 2 \times 0.8 - 4 \times 0.6 = 1.6 - 2.4 =$ **– 0.8%. (Ans.)**

Example 38. *A single-phase 80 kVA, 2000/200 V, 50-Hz transformer has impedance drop of 8% and resistance drop of 4% :*

(*i*) *Find the regulation at full-load 0.8 power factor lagging.*

(*ii*) *At what power factor is the regulation zero.*

Solution. Impedance drop = 8%

Resistance drop = 4%

i.e., $\dfrac{I_2Z_{02}}{V_2} = \dfrac{8}{100}$ or $I_2Z_{02} = \dfrac{200 \times 8}{100} = 16$ V

and $\dfrac{I_2R_{02}}{V_2} = \dfrac{4}{100}$ or $I_2R_{02} = \dfrac{200 \times 4}{100} = 8$ V

∴ $I_2X_{02} = \sqrt{(I_2Z_{02})^2 - (I_2R_{02})^2} = \sqrt{16^2 - 8^2} = 13.86$ V.

(*i*) % regulation $= \dfrac{I_2R_{02}\cos\phi_2 + I_2X_{02}\sin\phi_2}{V_2}$

$$= \frac{8 \times 0.8 + 13.86 \times 0.6}{200} = \frac{6.4 \times 8.316}{200} = \mathbf{7.36\%.} \text{ (Ans.)}$$

(*ii*) *Regulation can be zero only when the power factor is leading.* Then

$$\frac{I_2R_{02}\cos\phi_2 - I_2X_{02}\sin\phi_2}{V_2} = 0$$

or $$I_2R_{02}\cos\phi_2 - I_2X_{02}\sin\phi_2 = 0$$

or $$I_2R_{02}\cos\phi_2 = I_2X_{02}\sin\phi_2$$

or $$\tan\phi_2 = \frac{I_2R_{02}}{I_2X_{02}} = \frac{8}{13.86} = 0.577 \quad \text{or} \quad \phi_2 = 30°$$

$\therefore$ **Power factor** = $\cos\phi_2$ = **0.866 leading. (Ans.)**

Example 39. *The high voltage of a single-phase 200 kVA 4400/220 V transformer takes a current of 35 A and power of 1250 W at 80 V when the low voltage winding is short-circuited. Determine :*

(*i*) *The voltage to be applied to the high voltage winding on full-load at 0.8 power factor lagging if the full-load secondary terminal voltage is to be kept at 220 V.*

(*ii*) *% regulation.*

Solution. (*i*) Transformation ratio, $K = \dfrac{220}{4400} = \dfrac{1}{20}$

Voltage applied to h.v. winding in S.C. test

$$V_{SC} = 80 \text{ V}$$

and $$I_{SC} = 35 \text{ A}$$

$$P_{SC} = 1250 \text{ W}$$

Referred to *h.v. side,*

$$Z_{01} = \frac{V_{SC}}{I_{SC}} = \frac{80}{35} = 2.28 \ \Omega$$

$$R_{01} = \frac{P_{SC}}{I_{SC}^2} = \frac{1250}{35^2} = 1.02 \ \Omega$$

$$X_{01} = \sqrt{Z_{01}^2 - R_{01}^2} = \sqrt{2.28^2 - 1.02^2} = 2.04 \ \Omega$$

Full-load current in h.v. winding,

$$I_1 = \frac{200 \times 1000}{4400} = 45.45 \text{ A}$$

Voltage drop in resistance $= I_1R_{01} = 45.45 \times 1.02 = 46.36$ V

Voltage drop in reactance $= I_1X_{01} = 45.45 \times 2.04 = 92.72$ V

Full-load secondary voltage referred to primary,

$$V_2' = \frac{V_2}{K} = \frac{220}{\frac{1}{20}} = 4400 \text{ V}$$

To get full-load secondary voltage equal to 220 V, supply voltage V_1 must be equal to the vector sum of 4400 V, I_1R_{01} and IX_{01}.

From Fig. 54,

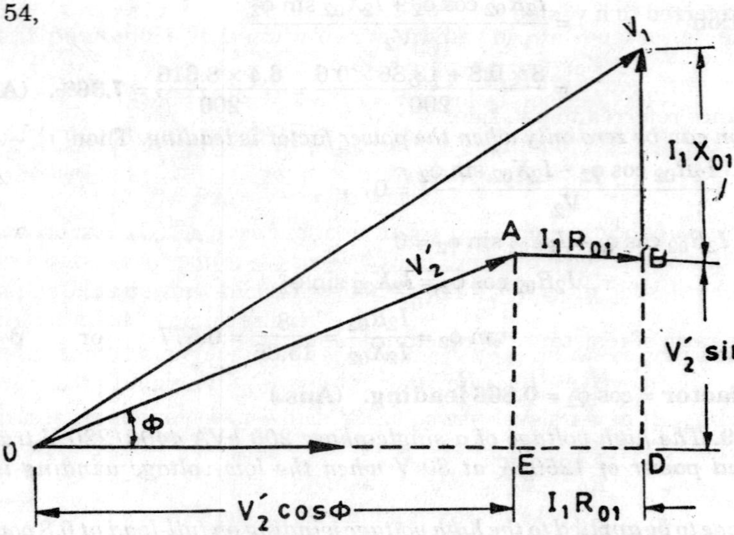

Fig. 54

$$OC^2 = OD^2 + CD^2 = (OE + ED)^2 + (BD + BC)^2$$

$$V_1^2 = (V_2' \cos\phi + I_1R_{01})^2 + (V_2' \sin\phi + I_1X_{01})^2$$

$$= (4400 \times 0.8 + 46.36)^2 + (4400 \times 0.6 + 92.72)^2$$

or $$V_1 = 4493 \text{ V}$$

Voltage to be applied to h.v. winding to get a full-load secondary voltage of 220 V = 4493 V. (Ans.)

(*ii*) % **regulation** $= \dfrac{V_1 - V_2'}{V_1} \times 100 = \dfrac{4493 - 4400}{4493} \times 100 =$ **2.07%. (Ans.)**

Example 40. *The high voltage and low voltage windings of a 50 kVA, 4400/220 V, 50-Hz transformer have resistances of 2.2 Ω and 0.005 Ω respectively. The full-load current is obtained when 160 V at 50-Hz is applied to h.v. winding with l.v. winding short-circuited. Find :*

(*i*) *The equivalent resistance and reactance of the transformer referred to h.v. side, and*

(*ii*) *Reactance of each winding.*

Assume that the ratio of resistance to reactance is the same for each winding and the full-load efficiency of transformer is 0.98.

Solution. (*i*) Transformation ratio, $K = \dfrac{220}{4400} = \dfrac{1}{20}$

Primary resistance, $R_1 = 22\ \Omega$

Secondary resistance, $R_2 = 0.005\ \Omega$

Full-load output $= 50 \text{ kVA} = 50 \times 10^3 \text{ VA}$

Full-load input $= \dfrac{\text{output}}{\text{efficiency}} = \dfrac{50 \times 10^3}{0.98}$ VA

Full-load input current, $I_1 = \dfrac{\text{input in VA}}{\text{supply voltage}} = \dfrac{50 \times 10^3}{0.98 \times 4400} = 11.6 \text{ A}$

Impedance referred to h.v. side

$$Z_{01} = \frac{V_{SC}}{I_{SC}} = \frac{160}{11.6} = 13.8\ \Omega$$

Total resistance referred to h.v. side,

$$R_{01} = R_1 + R_2' = R_1 + \frac{R_2}{K^2} = 2.2 + \frac{0.005}{\left(\frac{1}{20}\right)^2} = 2.2 + 2 = \mathbf{4.2\ \Omega.}\ \ \textbf{(Ans.)}$$

$$X_{01} = \sqrt{Z_{01}^2 - R_{01}^2} = \sqrt{(13.8)^2 - (4.2)^2} = \mathbf{13.15\ \Omega.}\ \ \textbf{(Ans.)}$$

(*ii*) Since the ratio of resistance to reactance remains same (given),

$$\therefore \qquad \frac{R_1}{X_1} = \frac{R_2}{X_2} = \frac{R_2'}{X_2'}$$

i.e.,
$$\frac{R_1}{R_2'} = \frac{X_1}{X_2'}$$

or
$$\frac{R_1 + R_2'}{R_2'} = \frac{X_1 + X_2'}{X_2'} \quad \text{or} \quad \frac{R_{01}}{R_2'} = \frac{X_{01}}{X_2'}$$

or
$$X_2' = \frac{R_2' X_{01}}{R_{01}} = \frac{2 \times 13.15}{4.2} = 6.26\ \Omega$$

Reactance of primary winding,

$$X_1 = X_{02} - X_2' = 13.15 - 6.26 = \mathbf{6.85\ \Omega.}\ \ \textbf{(Ans.)}$$

Reactance of secondary winding,

$$X_2 = K^2 X_2' = \left(\frac{1}{20}\right)^2 \times 6.26 = \mathbf{0.01565\ \Omega.}\ \ \textbf{(Ans.)}$$

☞**Example 41.** *The following test results were obtained in a 250/500 V transformer :*

O.C. test (l.v. side) : 250 V, 1 A, 80 W.

S.C. test (l.v. winding short-circuited) : 20 V, 12 A, 100 W.

Determine :

(i) The circuit constants.

(ii) The applied voltage and efficiency when the output is 10 A at 500 V and 0.8 power factor lagging.

Solution. Transformation ratio, $K = \frac{500}{250} = 2$

(*i*) **Circuit constants** :

O.C. test *(l.v. side)*

Voltage,	$V_1 = 250$ V
No-load current,	$I_0 = 1$ A
No-load loss,	$P_0 = 80$ W
Also,	$P_0 = V_1 I_0 \cos \phi_0$
	$80 = 250 \times 1 \times \cos \phi_0$

$$\cos \phi_0 = \frac{80}{250} = 0.32$$

Wattful component of no-load current I_0,

$$I_w = I_0 \cos \phi_0 = 1 \times 0.32 = 0.32 \text{ A}$$

Magnetising component of no-load current, I_0,

$$I_m = \sqrt{I_0^2 - I_w^2} = \sqrt{1^2 - 0.32^2} = \mathbf{0.95 \text{ A}}$$

Now, resistance representing the core loss,

$$R_0 = \frac{V_1}{I_w} = \frac{250}{0.32} = \mathbf{781.25\ \Omega.\ \ (Ans.)}$$

Magnetising reactance, $$X_0 = \frac{V_1}{I_m} = \frac{250}{0.95} = \mathbf{263.16\ \Omega.\ \ (Ans.)}$$

S.C. test *(l.v. winding short-circuited)*

Short-circuit voltage, $V_{SC} = 20$ V

Short-circuit current, $I_{SC} = 12$ A

Losses, $P_{SC} = 100$ W

As the primary is short-circuited all *values refer to secondary winding.*

$$\therefore \quad R_{02} = \frac{P_{SC}}{I_{SC}^2} = \frac{100}{(12)^2} = 0.694\ \Omega$$

$$Z_{02} = \frac{V_{SC}}{I_{SC}} = \frac{20}{12} = 1.677\ \Omega$$

and $$X_{02} = \sqrt{Z_{02}^2 - R_{02}^2} = \sqrt{(1.667)^2 - (0.694)^2} = 1.516\ \Omega$$

As R_0 and X_0 refer to primary, let us transfer these values to primary as follows :

$$R_{01} = \frac{R_{02}}{K^2} = \frac{0.694}{(2)^2} = \mathbf{0.174\ \Omega}$$

$$X_{01} = \frac{X_{02}}{K^2} = \frac{1.516}{(2)^2} = \mathbf{0.38\ \Omega}$$

$$Z_{01} = \frac{Z_{02}}{K^2} = \frac{1.669}{(2)^2} = \mathbf{0.417\ \Omega}$$

The equivalent circuit is shown in Fig. 55.

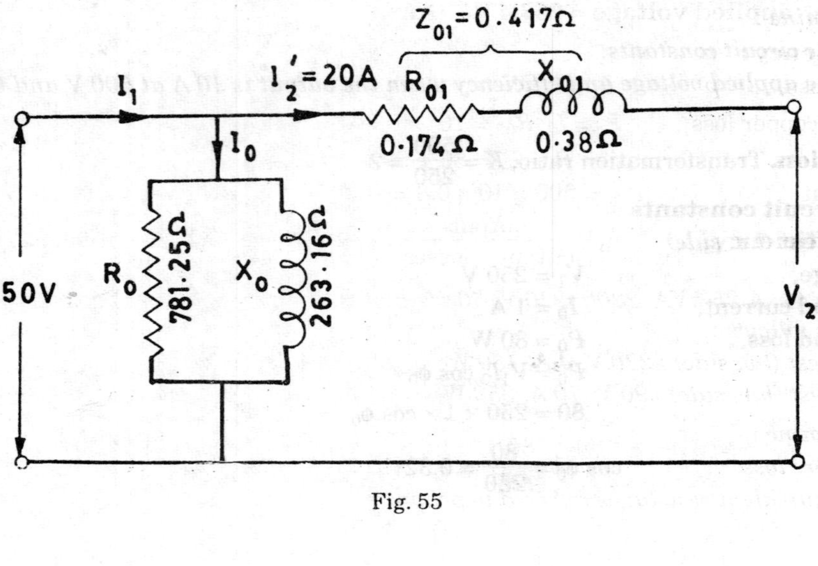

Fig. 55

(*ii*) **Applied voltage, V_1' :**

The applied voltage V_1' is the vector sum of V_1 and I_1Z_{01} (Fig. 56).

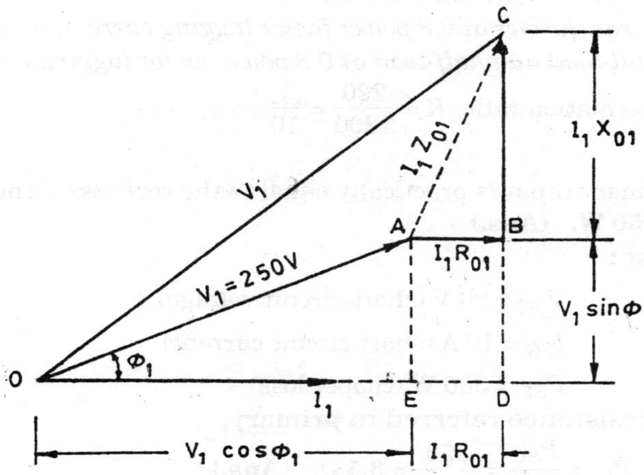

Fig. 56

Output current, $I_2 = 10$ A

$\therefore$ $I_1 = KI_2 = 2 \times 10 = 20$ A

Now, $I_1R_{01} = 20 \times 0.174 = 3.48$ V

$I_1X_{01} = 20 \times 0.38 = 7.6$ V

Refer Fig. 56. Neglecting the angle between V_1 and V_1', we have

$$OC^2 = OD^2 + DC^2$$

$$OC = V_1' = \sqrt{(OE + ED)^2 + (DB + BC)^2}$$

$$= \sqrt{(V_1 \cos \phi_1 + I_1R_{01})^2 + (V_1 \sin \phi_1 + I_1X_{01})^2}$$

$$= \sqrt{(250 \times 0.8 + 3.48)^2 + (250 \times 0.6 + 7.6)^2} = 257.4 \text{ V}$$

Hence, **applied voltage = 257.4 V. (Ans.)**

Efficiency :

Iron loss, $P_i = 80$ W

Total copper loss, $P_c = I_2^2R_{02} = 10^2 \times 0.694 = 69.4$ W

Total loss, $= P_i + P_c = 80 + 69.4 = 149.4$ W

Output $= 500 \times 10 \times 0.8 = 4000$ W

$\therefore$ **Efficiency,** $\eta = \dfrac{\text{output}}{\text{output + losses}} = \dfrac{4000}{4000 + 149.4}$ = **0.964** or **96.4%. (Ans.)**

☞**Example 42.** *A 25-kVA, 2200/220 V, 50-Hz distribution transformer is tested for efficiency and regulation as follows :*

O.C. test (l.v. side) : 220 V, 4 A, 150 W.

S.C. test (h.v. side) : 90 V, 10 A, 350 W.

Determine :

(i) Core loss,

(ii) Equivalent resistance referred to primary,

(*iii*) *Equivalent reactance referred to secondary,*
(*iv*) *Equivalent reactance referred to primary,*
(*v*) *Equivalent reactance referred to secondary,*
(*vi*) *Regulation of transformer at 0.8 power factor lagging current, and*
(*vii*) *Efficiency at full-load and half-load at 0.8 power factor lagging current.*

Solution. Transformation ratio, $K = \frac{220}{2200} = \frac{1}{10}$

(*i*) **Core loss :**

Since no-load primary input is practically equal to the core loss, hence, **core loss as found from no-load test, is 150 W. (Ans.)**

(*ii*) **From S.C. test :**

$$V_{SC} = 90 \text{ V (short-circuit voltage)}$$
$$I_{SC} = 10 \text{ A (short-circuit current)}$$
$$P_{SC} = 350 \text{ W (copper loss)}$$

$\therefore$ **Equivalent resistance referred to primary,**

$$R_{01} = \frac{P_{SC}}{I_{SC}^2} = \frac{350}{(10)^2} = \mathbf{3.5\ \Omega.\ (Ans.)}$$

(*iii*) **Equivalent resistance referred to secondary,**

$$R_{02} = K^2 R_{01} = \left(\frac{1}{10}\right)^2 \times 3.5 = \mathbf{0.035\ \Omega.\ (Ans.)}$$

(*iv*) Also
$$Z_{01} = \frac{V_{SC}}{I_{SC}} = \frac{90}{10} = 9\ \Omega$$

$\therefore$ **Equivalent reactance referred to primary,**

$$X_{01} = \sqrt{Z_{01}^2 - R_{01}^2} = \sqrt{9^2 - 3.5^2} = \mathbf{8.29\ \Omega.\ (Ans.)}$$

(*v*) **Equivalent reactance referred to secondary,**

$$X_{02} = K^2 X_{01} = \left(\frac{1}{10}\right)^2 \times 8.29 = \mathbf{0.0829\ \Omega.\ (Ans.)}$$

(*vi*) **% regulation :**

Let us find the rise in voltage necessary to maintain the output terminal voltage constant from no-load to full-load.

$$\text{Rated primary current} = \frac{25 \times 1000}{2200} = 11.36 \text{ A}$$

Now, using the relation,

$$V_1' = \sqrt{(V_1 \cos\phi + I_1 R_0)^2 + (V_1 \sin\phi + I_1 X_{01})^2}$$
$$= \sqrt{(2200 \times 0.8 + 11.36 \times 3.5)^2 + (2200 \times 0.6 + 11.36 \times 8.29)^2}$$
$$= 2289 \text{ V (app.)}$$

$\therefore$ **% regulation**
$$= \frac{V_1' - V_1}{V_1} \times 100 \qquad \text{(Art. 9.11)}$$
$$= \frac{2289 - 2200}{2200} \times 100 = \mathbf{4.045\%.\ (Ans.)}$$

We can get the same result by working in the secondary :

Rated secondary current,

$$I_2 = \frac{I_1}{K} = \frac{11.36}{1/10} = 113.6 \text{ A}$$

$$_0V_2 = \sqrt{(V_2 \cos \phi + I_2R_{02})^2 + (V_2 \sin \phi + I_2X_{02})^2}$$

$$= \sqrt{(220 \times 0.8 + 113.6 \times 0.035)^2 + (220 \times 0.6 + 113.6 \times 0.0829)^2}$$

$$= 228.9 \text{ V}$$

$\therefore$ % regulation $= \frac{_0V_2 - V_2}{V_2} \times 100 = \frac{228.9 - 220}{220} \times 100 =$ **4.045%. (Ans.)**

(*vii*) **Efficiency :**

Core loss, $P_i = 150$ W

Copper loss at full-load,

$$P_c = I_1^2R_{01} = 11.36^2 \times 3.5 = 451.7 \text{ W}$$

Copper loss at half-load,

$$P_c = \left(\frac{11.36}{2}\right)^2 \times 3.5 = 112.9 \text{ W}$$

$\therefore$ Efficiency at *full-load*

$$= \frac{\text{output}}{\text{output} + P_i + P_c} = \frac{25 \times 1000 \times 0.8}{(25 \times 1000 \times 0.8) + 150 + 451.7}$$

= **0.9707** or **97.07%. (Ans.)**

Efficiency at half *full-load*

$$= \frac{(25 \times 1000/2) \times 0.8}{(25 \times 1000/2) \times 0.8 + 150 + 112.9}$$ = **0.9744** or **97.44%. (Ans.)**

☞**Example 43.** *Two similar 100 kVA, single-phase transformers gave the following test readings when tested by Sumpner's test.*

Supply power = 2.4 kW.

Power supplied to secondary circuit in passing full-load current through it = 3.2 kW.

Find the efficiency and regulation of each transformer at unity power factor.

Solution. Core loss (or iron loss) of the transformers

= 2.4 kW

Iron loss of each transformer,

$$P_i = \frac{2.4}{2} = 1.2 \text{ kW}$$

Full-load copper loss of both transformers

= 3.2 kW

Full-load copper loss of each transformer

$$P_c = \frac{3.2}{2} = 1.6 \text{ kW}$$

Total full-load losses of each transformer

$$= P_i + P_c = 1.2 + 1.6 = 2.8 \text{ kW}$$

Output at full-load at unity power factor

$$= 100 \times 1 = 100 \text{ kW}$$

$\therefore$ **Full-load efficiency,** $\eta = \dfrac{\text{output}}{\text{output + losses}} = \dfrac{100}{100 + 2.8}$ = **0.9727** or **97.27%. (Ans.)**

Example 44. *A 10 kVA, 2500/250 V, single phase transformer gave the following test results :*

Open circuit test : 250 V, 0.8 A, 50 W

Short circuit test : 60 V, 3 A, 45 W

(i) Calculate the efficiency of half full-load at 0.8 p.f.

(ii) Calculate the load kVA at which maximum efficiency occurs and also the maximum efficiency at 0.8 p.f.

(iii) Compute the voltage regulation at 0.8 p.f. leading. **(AMIE Winter, 1998)**

Solution. Full rated current $= \dfrac{10 \times 1000}{2500} = 4$ A

Hence reading of wattmeter corresponding to full-load current of 4 A

$$= 45 \times \left(\frac{4}{3}\right)^2 = 80 \text{ W}$$

$\therefore$ Full-load copper losses = 80 W, and

Full-load iron losses = 50 W

(*i*) **Efficiency at half-load at 0.8 p.f. :**

$$\eta = \frac{\text{half-load output}}{\text{half-load output + losses}}$$

$$= \frac{(10 \times 1000 \times 0.5) \times 0.8}{[(10 \times 1000 \times 0.5) \times 0.8] + 50 + 80 \times (\frac{1}{2})^2}$$

$$= \frac{4000}{4000 + 50 + 20} = 0.983 \text{ or } \mathbf{98.3\%.}$$ **(Ans.)**

(*ii*) **Load kVA for maximum efficiency, and its value :**

For maximum efficiency :

Copper loss = Iron loss = 50 W

$\therefore$ Current at which maximum efficiency occurs $= \dfrac{50 \times 4}{80} = 2.5$ A

$\therefore$ **Load kVA** $= 10 \times \dfrac{2.5}{4}$ = **6.25 kVA. (Ans.)**

η_{max} (at 0.8 p.f.) $= \dfrac{(6.25 \times 1000 \times 0.8)}{(6.25 \times 1000 \times 0.8) + 50 + 50} = 0.98$ or **98%. (Ans.)**

(*iii*) **Voltage regulation at 0.8 p.f. leading :**

From short circuit test :

$$V_{SC} = 60 \text{ V} ; I_{SC} = 3 \text{ A} ; P_{SC} = 45 \text{ W}$$

Now,
$$R_{01} = \frac{P_{SC}}{I_{SC}^2} = \frac{45}{3^2} = 5\ \Omega$$

$$Z_{01} = \frac{V_{SC}}{I_{SC}} = \frac{60}{3} = 20\ \Omega$$

$\therefore$
$$X_{01} = \sqrt{Z_{01}^2 - R_{01}^2} = \sqrt{20^2 - 5^2} = 19.36\ \Omega$$

∴ Voltage drop at leading p.f. of 0.8

$$= I_1R_{01}\cos\phi - I_1X_{01}\sin\phi$$

$$= 4 \times 5 \times 0.8 - 4 \times 19.36 \times 0.6 = -30.46 \text{ V}$$

Hence % **voltage regulation** $= \dfrac{-30.46}{2500} \times 100 = \mathbf{-1.218\%}$. **(Ans.)**

Example 45. *A 50 kVA single-phase transformer has a full-load primary current of 250 A and total resistance referred to primary is 0.006 ohm. If the iron loss amounts to 200 W, find the efficiency on full-load and half-load at (i) unity power factor and (ii) 0.8 power factor.*

Solution. Full-load primary current,

$$I_1 = 250 \text{ A}$$

Total resistance referred to primary,

$$R_{01} = 0.006\ \Omega$$

Iron loss, $P_i = 200$ W

Full-load copper loss, $P_c = I_1^2R_{01} = (250)^2 \times 0.006 = 375$ W

Total full-load loss $= P_i + P_c = 200 + 375 = 575 \text{ W} = 0.575$ kW

Half full-load copper loss $= (\frac{1}{2})^2 P_c = (\frac{1}{2})^2 \times 375 = 93.75$ W

Total loss on half-load $= 200 + 93.75 = 293.75 \text{ W} = 0.294$ kW (say)

(*i*) **At unity power factor :**

Full-load output, $= 50 \times 1 = 50$ kW

Efficiency at full-load $= \dfrac{\text{output}}{\text{output + losses}} = \dfrac{50}{50 + 0.575} = \mathbf{0.9886}$ or **98.86%. (Ans.)**

Half-full-load output $= 50 \times \frac{1}{2} \times 1 = 25$ kW

Efficiency at half full-load at unity power factor

$$= \frac{25}{25 + 0.294} = \mathbf{0.9984} \text{ or } \mathbf{99.84\%}. \textbf{ (Ans.)}$$

(*ii*) **At 0.8 power factor :**

Full-load output $= 50 \times 0.8 = 40$ kW

Efficiency on full-load at 0.8 power factor

$$= \frac{40}{40 + 0.575} = \mathbf{0.9858} \text{ or } \mathbf{98.58\%}. \textbf{ (Ans.)}$$

Half full-load output $= 50 \times \frac{1}{2} \times 0.8 = 20$ kW

Efficiency on half full-load at 0.8 power factor

$$= \frac{20}{20 + 0.294} = \mathbf{0.9855} \text{ or } \mathbf{98.55\%}. \textbf{ (Ans.)}$$

☞**Example 46.** *A 40 kVA, single-phase transformer has an iron loss of 300 W and full-load copper loss of 600 W,*

(*i*) *Find the load at which maximum efficiency occurs and the value of maximum efficiency at unity power factor.*

(*ii*) *If the maximum efficiency occurs at 80% of full-load, find the new core loss and full-load copper loss assuming that total full-load loss is a constant.*

Solution. Rating of transformer = 40 kVA

Iron loss, $P_i = 300$ W

Full-load copper loss, $P_c = 600$ W

(*i*) Let the maximum efficiency occurs at x times full-load, then

$$x^2P_c = P_i$$

or $$x = \sqrt{\frac{P_i}{P_c}} = \sqrt{\frac{300}{600}} = 0.707 = 70.7\%$$

Hence, **efficiency occurs at 70.7% of full-load. (Ans.)**

Maximum efficiency :

Output at unity power factor

$$= 0.707 \times 40 \times 1 = 28.28 \text{ kW}$$

Total losses $= P_i + P_c = 2P_i$ $\left[\text{Because when efficiency is maximum, } P_i = P_c\right]$

$$= 2 \times 300 = 600 \text{ W}$$

$$= 0.6 \text{ kW}$$

$\therefore$ **Maximum efficiency**

$$= \frac{\text{output}}{\text{output + losses}} = \frac{28.28}{28.28 + 0.6} = \mathbf{0.9792} \text{ or } \mathbf{97.92\%}. \textbf{ (Ans.)}$$

(*ii*) **New core loss, P_i' :**

New copper loss, P_c' :

Maximum efficiency occurs at 80% of full-load

Now, $$P_i' + P_c' = P_i + P_c \quad \textit{(given)}$$

$$= 300 + 600 = 900 \text{ W}$$

Also $$0.8 = \sqrt{\frac{P_i'}{P_c'}} \quad \text{or} \quad \frac{P_i'}{P_c'} = [0.8]^2 = 0.64$$

or $$\frac{P_i'}{P_i' + P_c'} = \frac{0.64}{1 + 0.64} = 0.3902 \quad \text{or} \quad \frac{P_i'}{900} = 0.3902$$

$\therefore$ $$P_i' = 351.2 \text{ W}$$

and $$P_c' = 900 - P_i' = 900 - 351.2 = 548.8 \text{ W}$$

Hence, **new iron loss = 351.2 W. (Ans.)**

new copper loss = 548.8 W. (Ans.)

Example 47. *The primary and secondary resistances of a 1100/220 V transformer are 0.3 Ω and 0.02 Ω respectively. If iron loss amounts to 260 W determine the secondary current at which maximum efficiency occurs and find the maximum efficiency at 0.8 power factor.*

Solution. Transformation ratio, $K = \dfrac{220}{1100} = \dfrac{1}{5}$

Primary resistance, $R_1 = 0.3\ \Omega$

Secondary resistance, $R_2 = 0.02\ \Omega$

Iron loss, $P_i = 260$ W

Total resistance referred to secondary,

$$R_{02} = R_2 + K^2R_1 = 0.02 + \left(\frac{1}{5}\right)^2 \times 0.3 = \mathbf{0.032\ \Omega} \quad \textbf{(Ans).}$$

Let I_2 be the secondary current at maximum efficiency.

At maximum efficiency,

Copper loss = Iron loss

$$I_2^2R_{02} = 260$$

$\therefore$ $$I_2^2 = \frac{260}{R_{02}} = \frac{260}{0.032} = 8125 \text{ or } I_2 = 90.14 \text{ A}$$

Hence, **secondary current at maximum efficiency**

= 90.14 A. (Ans.)

Maximum efficiency at 0.8 power factor :

Output at maximum efficiency at 0.8 power factor

$= V_2 I_2 \cos \phi_2 = 220 \times 90.14 \times 0.8 = 15865$ W

Losses $= P_i + P_c = 2P_i$ $(\because \; P_i = P_c)$

$= 2 \times 260 = 520$ W

$\therefore$ **Maximum efficiency** $= \dfrac{\text{output}}{\text{output + losses}} = \dfrac{15865}{15865 + 520} =$ **0.9683** or **96.83%. (Ans.)**

9.15. All-day Efficiency. *All-day efficiency* is the *ratio of energy (kWh) delivered in a 24 hour period divided by the energy (kWh) input in the same length of time.*

$$\therefore \qquad \eta_{\text{all-day}} = \frac{\text{output in kWh}}{\text{input in kWh}} \text{ (for 24 hours)} \qquad ...(39)$$

Transformers used on residence-lighting circuits (and distribution circuits generally) are either idle or only lightly loaded during much of 24-hour period. However, they must at all times be connected to the line and ready to serve, so that the core losses are being supplied continually. *It is therefore very important that such transformers be designed for minimum core loss.* The copper losses are relatively less important, since they depend on the load. Because they are lightly loaded much of the time, *distribution transformers are designed for relatively large full-load copper loss and have their maximum power efficiencies at light loads.* This design results in improved all-day efficiency for these transformers. *Power transformers*, on the other hand are loaded more or less continuously and *are designed for full-load copper losses equal to about twice the no-load losses.*

To calculate all-day efficiency, it is necessary to know how the load on the transformer varies from hour to hour. The quotient obtained by dividing the energy output by the energy output plus energy losses over a 24-hour period yields the efficiency expressed as a decimal fraction.

The use of a *load factor* facilitates practical calculations.

Example 48. *A 15 kVA, 2000/200 V transformer has an iron loss of 250 W and full-load copper loss 350 W. During the day it is loaded as follows :*

No. of hours	*Load*	*Power factor*
9	$\frac{1}{4}$ *load*	*0.6*
7	*full-load*	*0.8*
6	$\frac{3}{4}$ *load*	*1.0*
2	*no-load*	—

Calculate the all-day efficiency.

Solution. Rating of transformer = 15 kVA

Iron loss, $P_i = 250$ W $= 0.25$ kW

Full-load copper loss, $P_c = 350$ W $= 0.35$ kW

Iron loss/day $= 0.25 \times 24 = 6$ kWh

Copper loss at $\frac{1}{4}$ *load* $= \left(\dfrac{1}{4}\right)^2 \times P_c = \dfrac{1}{16} \times 0.35 = 0.0218$ kW

Copper loss for 9 hours at $\frac{1}{4}$ load

$= 9 \times 0.0218 = 0.196$ kWh

Copper loss at *full-load* $= P_c = 0.35$ kW

Copper loss for 7 hours on full-load

$$= 7 \times 0.35 = 2.45 \text{ kWh}$$

Copper loss at $\frac{3}{4}$ *load* $= \left(\frac{3}{4}\right)^2 \times P_c = \frac{9}{16} \times 0.35 = 0.197 \text{ kW}$

Copper loss for 6 hours at $\frac{3}{4}$ load $= 0.197 \times 6 = 1.18$ kWh

Copper loss/day $= 0.196 + 2.45 + 1.18 = 3.826$ kWh

Total loss/day = iron loss/day + copper loss/day $= 6 + 3.826 = 9.826$ kWh

Total output/day $= \frac{1}{4} \times 15 \times 0.6 \times 9 + 15 \times 0.8 \times 7 + \frac{3}{4} \times 15 \times 1.0 \times 6$

$$= 20.25 + 84 + 67.5 = 171.75 \text{ kWh}$$

All-day efficiency $= \dfrac{\text{output}}{\text{output} + \text{losses}} = \dfrac{171.75}{171.75 + 9.826} = $ **0.9459** or **94.59%. (Ans.)**

9.16. Transformer Noise. The **"hum"** caused by energized power transformer, under *no-load conditions*, originates in the *core where the laminations tend to vibrate by magnetic forces*. The noise is transmitted through the oil to the tank side and thence to the surroundings.

The following are the *main factors which produce noise in transformers* :

1. *Magnetostriction* (occurrence of dimensional changes both parallel to, and perpendicular to the direction of magnetisation).
2. The mechanical vibrations caused by the laminations, depending upon the tightness of clamping, size, gauge, associated structural parts, etc.
3. The mechanical vibration of trank walls.
4. The damping.

The noise emission may be reduced by the following methods/means :

1. Prevention of vibration of core-plate by the use of a lower flux density and giving attention to constructional feature (such as clamping bolts, proportions and dimensions of the *'steps'* in plate width, tightness of clamping and uniformity of plates).
2. Using cushions, padding, or oil barriers to sound insulate the transformer from tank.
3. Designing suitably the tank and stiffeners to check tank wall vibration.
4. Sound insulating the tank from the ground or surrounding air.

However, the *noise problem cannot be solved completely.*

9.17. Auto-transformer. *A transformer in which part of the winding is common to both the primary and secondary circuits is known as an* **auto-transformer.** *The primary is electrically connected to the secondary, as well as magnetically coupled to it.*

Refer Fig. 57. *LM* is primary winding having N_1 turns and *MS* is secondary winding having N_2 turns. If no-load current and iron losses are neglected.

$$\frac{V_2}{V_1} = \frac{N_2}{N_1} = K.$$

The current in the section *MS* is vector difference of I_2 and I_1 ; but since the two currents are practically in phase opposition, the resultant is $(I_2 - I_1)$ where $I_2 > I_1$.

Saving of copper (in comparison to conventional two winding transformer) :

The volume and hence weight of copper is proportional to the length and area of cross-section of the conductors. But the *length of conductor is proportional to the number of turns and cross-section depends on current*. Hence the weight of copper is proportional to the *product of number of turns and current to be carried.*

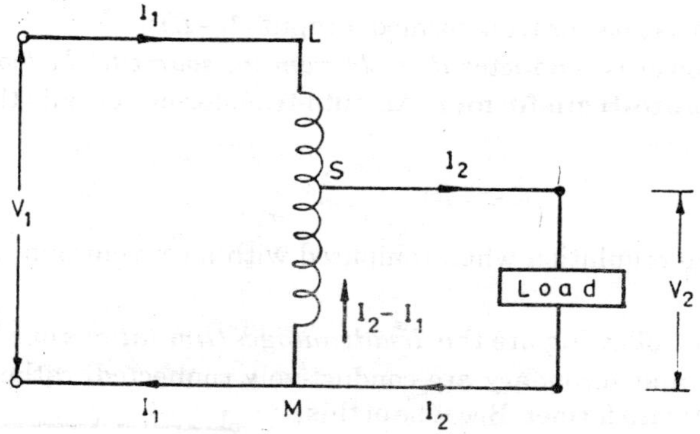

Fig. 56A. Auto-transformer.

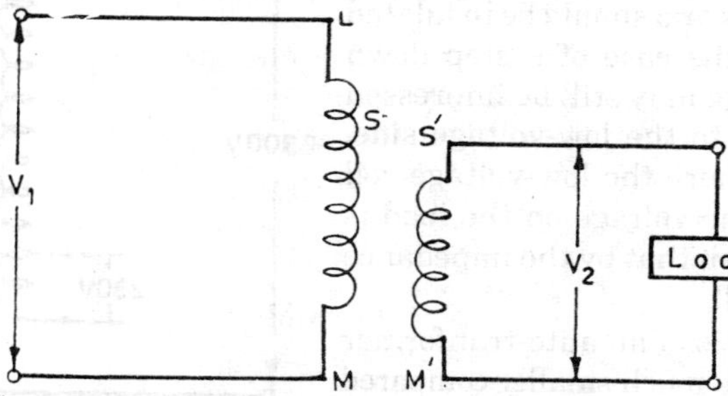

Fig. 57. Conventional two winding transformer.

Weight of copper in conventional two winding transformer

$$\propto (N_1I_1 + N_2I_2).$$

Weight of copper in auto-transformer

= weight of copper in section LS + weight of copper is section MS.

But weight of copper in section $LS \propto (N_1 - N_2)I_1$

and weight of copper in section $MS \propto N_2\,(I_2 - I_1)$

∴ Weight of copper in auto-transformer

$$\propto (N_1 - N_2)\,I_1 + N_2\,(I_2 - I_1)$$

$$\therefore \frac{\text{Weight of copper in auto-transformer } (W_a)}{\text{Weight of copper in ordinary transformer } (W_0)}$$

$$= \frac{(N_1 - N_2)\,I_1 + N_2\,(I_2 - I_1)}{N_1I_1 + N_2I_2} = \frac{(N_1 - 2N_2)\,I_1 + N_2I_2}{N_1I_1 + N_2I_2}$$

$$= \frac{\dfrac{N_1}{N_2} - 2 + \dfrac{I_2}{I_1}}{\dfrac{N_1}{N_2} + \dfrac{I_2}{I_1}} = \frac{\dfrac{1}{K} - 2 + \dfrac{1}{K}}{\dfrac{1}{K} + \dfrac{1}{K}}$$

$$= 1 - K \qquad \ldots(40) \qquad \left[\because \frac{N_2}{N_1} = K,\ \frac{I_2}{I_1} = \frac{1}{K}\right]$$

∴ Saving in copper $= W_0 - W_a = W_0 - (1 - K)\,W_0 = KW_0$

∴ Saving in copper $= K \times$ weight of copper in ordinary tranformer

It can be proved that power transformed = input $(1 - K)$

The rest of the power is conducted *directly from the source to the load.*

Advantages of auto-transformer. An auto-transformer entails the following *advantages* :

(*i*) Higher efficiency

(*ii*) Small size

(*iii*) Lower cost

(*iv*) Better voltage regulation when compared with a conventional two-winding transformer of the same rating.

Disadvantages. Following are the *disadvantages / limitations* of auto-transformers :

(*i*) The primary and secondary are conductively connected, rather than isolated as in the conventional (ordinary) transformer. Because of this, *both sides are subject to any stresses set up by disturbances on either side.* The low-voltage side is subject to high-voltage stress and should be insulated for the higher voltage. In the case of a step down transformer, the high voltage may still be impressed upon equipment connected to the low-voltage side. This is shown in Fig. 58 where the low-voltage coil has accidentally opened. The voltage on the load is nearly 2300 V, being less than that by the impedance volt drop in the primary coil.

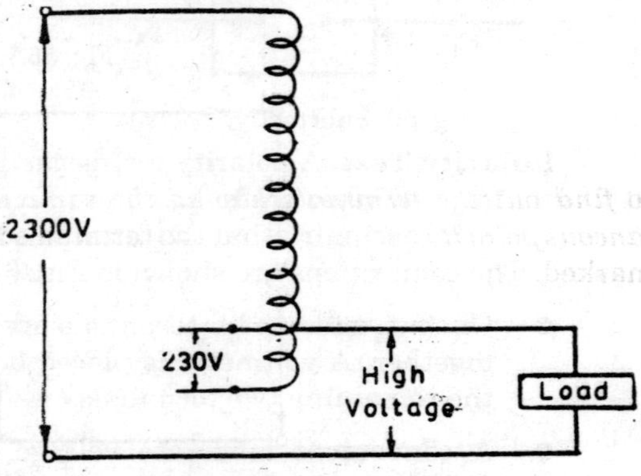

Fig. 58. Open circuit on auto-transformer.

(*ii*) As the voltage ratio of an auto-tranformer increases, the common coil is much smaller compared with the entire winding. This means that the economy gained is only a small part of the transformer and therefore this advantage is minimized.

Thus, because of the above disadvantages, lack of isolation and decreased economy, *auto-transformers are rarely used in ratios greater than 4 : 1, except for low-power devices on low- voltage systems.*

Uses. The auto-transformers find the following *applications* :

1. To tie together transmission or distribution circuits of slightly different voltages (*e.g.* 11000 V system with a 13200 V system).

2. To obtain partial line voltages for starting induction and synchronous motors with squirrel-cage windings.

3. To give a small boost to a distribution cable to correct for the voltage drop.

4. As furnace transformers for getting a convenient supply to suit the furnace winding from a 230 V supply.

5. As regulating transformers.

6. To obtain a neutral in a 3-wire A.C distribution system in the same way as a balancer set is used in a 3-wire D.C. distribution system.

7. A continuously variable auto-transformer finds useful applications in electrical testing laboratory.

9.18. Polarity of Transformers. In a transformer each of the primary terminals becomes alternately positive and negative with respect to the other and the same is true about the secondary terminals. If the transformer is to be used alone, the polarity is not important but if the transformer is to be used in parallel with another transformer the *instantaneous polarity is important* because the terminals having identical instantaneous polarity have to be connected together.

Refer Figs. 59 and 60. The terminals of the high voltage (h.v.) winding (*i.e.* H_1 and H_2) and the terminals of the low voltage (l.v.) winding (*i.e.* X_1 and X_2) are so marked that when the instantaneous voltage is directed from H_1 to H_2 in the high voltage winding, it is directed from X_1 to X_2 in the low voltage winding. In other words if the high voltage terminal H_1 is positive with respect to H_2 at any instant when the low voltage terminal X_1 will be positive with respect to X_2 at the same instant. It follows that when the terminals are arranged as in Fig. 59 the polarity is subtractive whereas the additive polarity is represented as in Fig. 60.

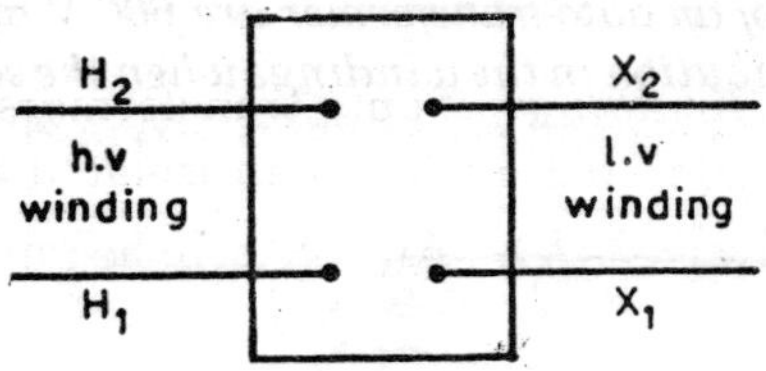

Fig. 59. Subtractive polarity.

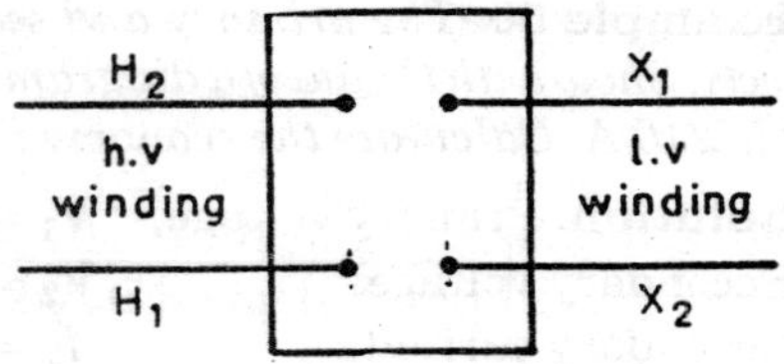

Fig. 60. Additive polarity.

Polarity Test. A polarity test is carried out to *find out the terminals having the same instantaneous polarity* assuming that the terminals are not marked. The connections are shown in Fig. 61.

- One h.t. and one l.t. terminals are joined together. A voltmeter is placed between the remaining two terminals.
- A convenient moderate voltage is impressed on the h.t. winding.

If the voltage V′, is 'greater' than the applied voltage V, then the transformer has 'additive' polarity.

If V′ is 'less' than V, the transformer has 'subtractive' polarity.

The terminals are then marked accordingly.

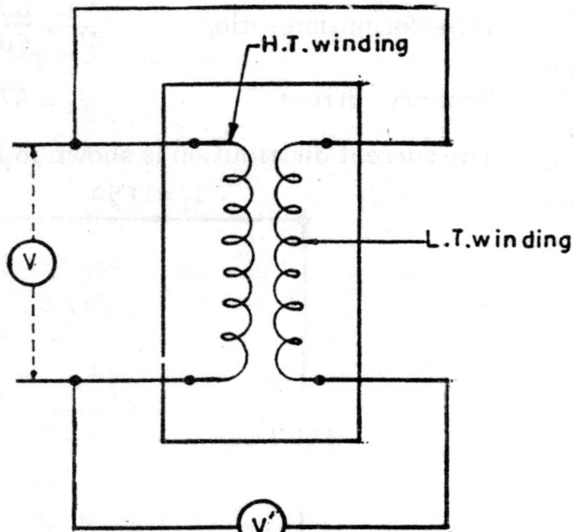

Fig. 61. Polarity test.

Example 49. *A load of 6 kW is supplied by an anto-transformer at 120 V and at unity power factor. If the primary voltage is 240 V, determine :*

(*i*) *Transformation ratio,* (*ii*) *Secondary current,*

(*iii*) *Primary current,*

(*iv*) *Number of turns across secondary if the total number of turns is 280.*

(*v*) *Power transformed, and*

(*vi*) *Power conducted directly from supply mains to load.*

Solution. Load suppied = 6 kW

Output voltage, $V_2 = 120$ V

Primary voltage, $V_1 = 240$ V

Total number of turns, $N_1 = 280$

(*i*) **Transformation ratio,** $K = \frac{V_2}{V_1} = \frac{120}{240} = \frac{1}{2}$. **(Ans.)**

(*ii*) **Secondary current,** $I_2 = \frac{6 \times 1000}{V_2 \cos \phi} = \frac{6000}{120 \times 1} = 50$ **A. (Ans.)**

(*iii*) **Primary current,** $I_1 = KI_2 = \frac{1}{2} \times 50 = \mathbf{25\ A.}$ **(Ans.)**

(*iv*) **Turns across secondary,** $N_2 = kN_1 = \frac{1}{2} \times 280 = \mathbf{140.}$ **(Ans.)**

(*v*) **Power transformed** $= \text{Load} \times (1 - K) = 6 \times (1 - \frac{1}{2}) = \mathbf{3\ kW.}$ **(Ans.)**

(*vi*) **Power conducted directly from supply mains**

$= 6 - 3 = \mathbf{3\ kW.}$ **(Ans.)**

Example 50. *The primary and secondary voltages of an auto-transformer are 600 V and 500 V respectively. Show with the aid of a diagram the current distribution in the windings when the secondary current is 210 A. Calculate the economy in copper.*

Solution. Primary voltage, $V_1 = 600$ V

Secondary voltage, $V_2 = 500$ V

Secondary current, $I_2 = 210$ A

Economy in copper :

Transformation ratio, $\frac{V_2}{V_1} = \frac{500}{600} = \frac{5}{6}$

Primary current, $I_1 = KI_2 = \frac{5}{6} \times 210 = 175$ A

The current distribution is shown in Fig. 62.

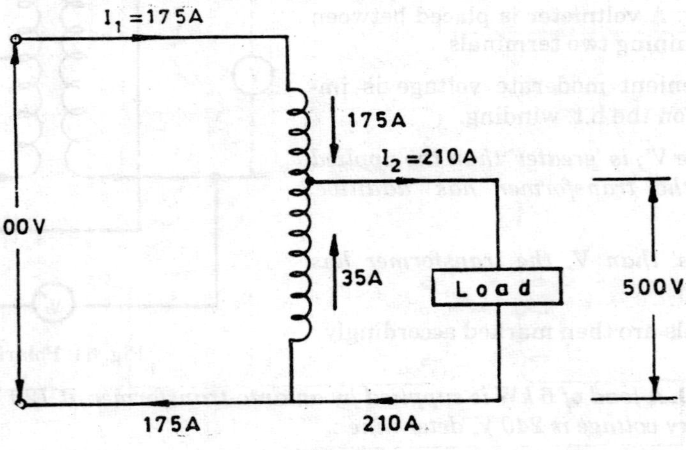

Fig. 62

We know that,

$$\frac{\text{Weight of copper on auto-transformer}}{\text{Weight of copper on ordinary transformer}} = 1 - K$$

Economy in copper

$$= \frac{\text{wt. of Cu on ordinary transformer} - \text{wt. of Cu on auto-transformer}}{\text{wt. of Cu on ordinary transformer}}$$

$$= 1 - \frac{\text{wt. of Cu on auto-transformer}}{\text{wt. of Cu on ordinary transformer}}$$

$= 1 - (1 - K) = K = \frac{5}{6}$ or **83.33%. (Ans.)**

Example 51. *A 4-kVA single-phase 50-Hz transformer has a full-load efficiency of 95.5% and iron loss of 45 W. The transformer is now connected as an auto-transformer to 220 V supply. If it delivers 4 kW load at unity power factor to a 110 V circuit, calculate the efficiency of the operation and the current drawn by the high-voltage side.*

Solution. Fig. 63 shows the connections for a 2-winding transformer. In Fig. 64 the same unit has been connected as an auto-transformer to a 220 V supply. Since the two-windings are connected in series, hence voltage across each is 110 V.

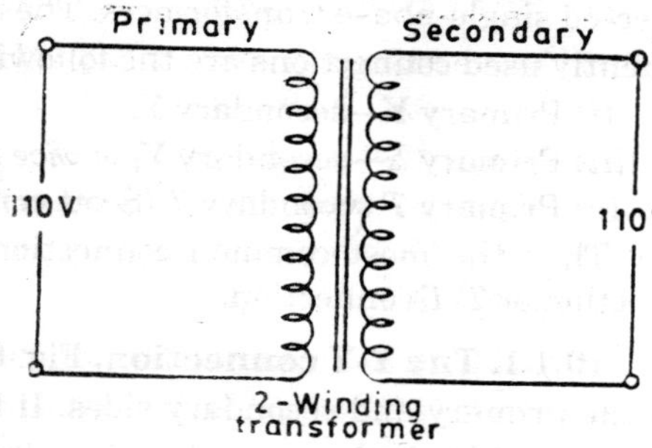

Fig. 63. Two-winding transformer.

In both connections the iron loss would remain the same. Since the auto-transformer windings will each carry but *half the current* as the conventional two-winding transformer, the copper losses will be $\frac{1}{4}$*th of previous value.*

Two-winding transformer :

Efficiency = 95.5% or 0.955

$$\therefore \quad 0.955 = \frac{\text{output}}{\text{output + losses}} = \frac{4000}{4000 + 45 + \text{copper loss}}$$

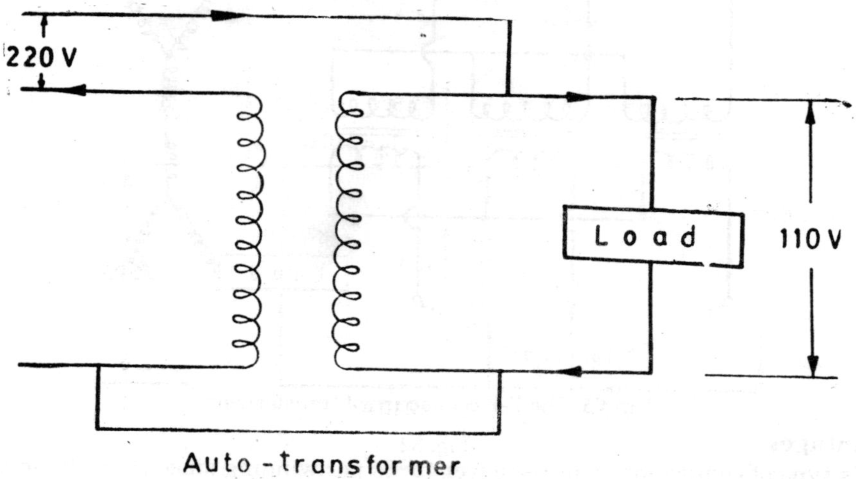

Fig. 64. Auto-transformer.

$$\therefore \text{ Copper loss } = \frac{4000}{0.955} - (4000 + 45) = 143.48 \text{ W}$$

Auto-transformer :

$$\text{Copper loss } = \frac{143.48}{4} = 35.87 \text{ W}$$

Iron loss = 45 W

$$\therefore \textbf{ Efficiency } = \frac{4000}{4000 + 35.87 + 45} = 0.9802 \text{ or } \mathbf{98.02\%}. \quad \textbf{(Ans.)}$$

10. THREE PHASE TRANSFORMER

10.1. Three-phase Transformer Connections. Virtually all power distribution is by poly-phase system of voltages. Three-phase transformations may be made with the use of properly connected single-phase transformers. These connections are in extensive commercial use. The most frequently used connections are the following :

(*i*) Primary *Y*—secondary *Y*. (*ii*) Primary Δ—secondary Δ.
(*iii*) Primary Δ—secondary *Y*, *or vice versa*. (*iv*) Primary and secondary open Δ.
(*v*) Primary *T* secondary *T* (Scott connection).

Thus the most common connections are *Y-Y*, Δ-Δ, *Y*-Δ, Δ-*Y*, open delta or *V-V* and Scott connection or *T-T* connection.

10.1.1. The Y-Y connection. Fig. 65 shows a bank of three transformers connected in *Y* on both the primary and secondary sides. If the ratio of transformation of each transformer is *K*, the same ratio will exist between the *line* voltages on the primary and secondary sides. This *connection will give satisfactory service only if the three-phase load is balanced* ; when the load is unbalanced, the electrical neutral will shift from its exact centre to a point that will make the line-to-neutral voltages unequal.

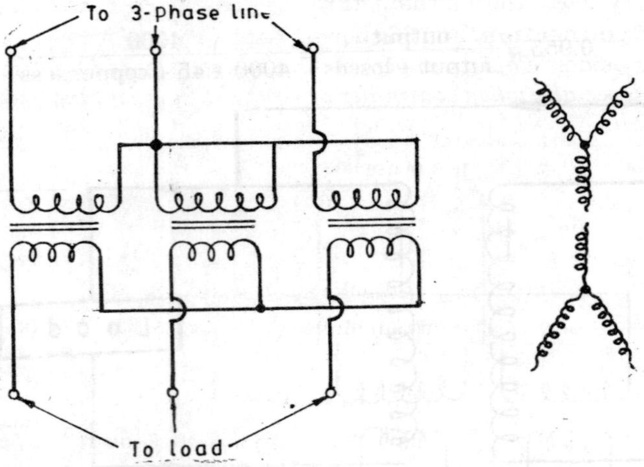

Fig. 65. The *Y-Y* connection of transformers.

Advantages :

1. This type of connection requires fewer turns per winding since the voltage across each is $\frac{1}{\sqrt{3}}$ times the line voltage ; hence it is *cheaper*.

2. The phase or winding current being equal to the line current, the cross-section of the winding wire is large, therefore the winding is stronger to bear stresses imposed upon it during heavy load or short circuit.

3. There is less dielectric stress on the insulating materials owing to lesser voltage *i.e.* $\frac{1}{\sqrt{3}}$ of line voltage.

The above advantages are enumerated with the understanding that, other things being equal, its rival is the delta-delta connection.

Disadvantages :

1. In case the load on the secondary side is unbalanced, as in the case of distributing network, the potential of the star-point will assume any value if the star-point is not earthed. This may impose full-line voltage on secondary windings. The shifting of the neutral point must be prevented by connecting the primary star-point to the star-point of the alternator winding.

2. In spite of grounding the star-point, if there is a third harmonic in the form of the alternator voltage, the third harmonic will appear in the voltages of the secondary side. This will cause triple frequency currents in the three-phase circuits. These currents when they flow in the neutral wire are additive and do not cancel out. Hence they will *cause interference to telephone lines located along the same route.*

3. The magnetising current in a transformer has third harmonic components. These currents will find a return path via the connection between the primary star-point of transformer and the neutral point of the alternator. However, if this connection is missing, these components will distort the flux wave which will produce a voltage having a third harmonic in each of the transformer, both on the primary and secondary sides. And, as before, if the star-point on the secondary is earthed, or grounded, triple harmonic currents will appear in the secondary circuit, and they will flow through the neutral wire causing interference to telephone lines in the vicinity.

4. If the star-points of both the primary and the secondary sides are not earthed, the regulation of the phases will be very poor if the load happen to be unbalanced as the case of distribution network.

10.1.2. The Δ-Δ connection. Fig. 66 shows a bank of transformers connected in Δ on both the primary and secondary sides. *This arrangement is generally used in systems in which the voltages are not very high and especially when continuity of service must be maintained even though one of the transformers should fail.*

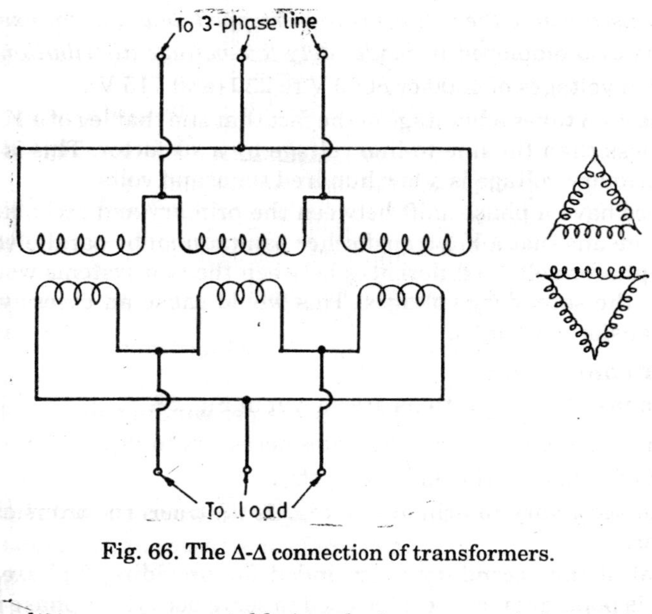

Fig. 66. The Δ-Δ connection of transformers.

Advantages :

1. The system voltages are more stable in relation to an unbalanced load.

2. If one transformer fails it may be switched out of the line and operation continued at a reduced power level. This is known as open-delta or *V-V* operation.

3. There is no distortion of flux, because the third harmonic component of magnetising current can flow in the delta connected primary windings without flowing in the line wires.

4. No difficulty is experienced due to unbalancing of loads on secondary side.

Disadvantages :

1. In comparison to *Y-Y* connections it requires more insulation.

2. The absence of star-point may be disadvantageous. If one line gets earthed due to fault, maximum voltage between windings and core will be full line voltage.

10.1.3. The Y-Δ connection. The *Y*-Δ connection is shown in Fig. 67.

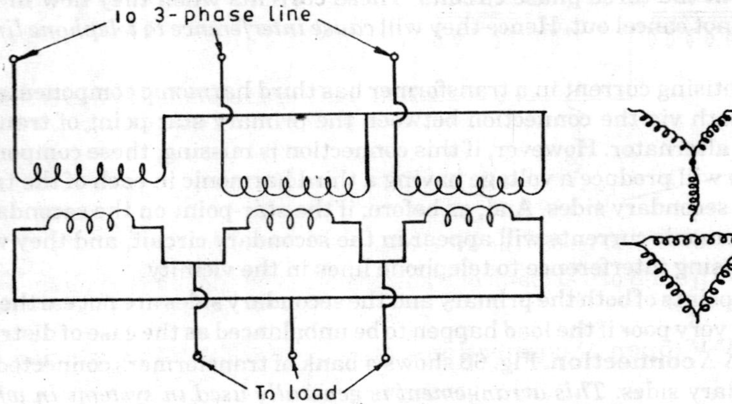

Fig. 67. The *Y*-Δ connection.

It is principally used where the voltage is to be stepped down, as for example, at the end of a transmission line. It is also employed in *moderately low-voltage distribution circuits for stepping down* from transmission voltages of 4000 or 8000 V to 230 (and 115 V).

- The *Y* connection takes advantage of the fact that one that leg of a *Y*, or the line-to-neutral voltage, is less than the line-to-line voltage by a $\sqrt{3}$ factor. This is especially important when the primary voltage is a few hundred thousand volts.
- The *Y*-Δ does have a phase shift between the primary and secondary voltages. This 30° phase shift means that a *Y*-Δ transformer bank *cannot* be paralleled with either a *Y-Y* or a Δ-Δ. The phasor voltage differences between the two systems would be around sin 30° = 0.5 times the secondary voltages. This would cause an excessive circulating current between transformer banks.

10.1.4. The Δ-Y connection

- The three-phase Δ-*Y* connections are shown in Fig. 68.

This type of connection is employed where it is necessary to *step up the voltage*, as for example, at the beginning of a high-tension transmission system.

- The ratio of secondary to primary voltage is $\sqrt{3}$ times the transformation ratio of each transformer.
- The neutral of the secondary is grounded for providing 3-phase 4-wire service. This connection is popular since it can be used to serve both the 3-phase power equipment and single-phase lighting circuit.
- This connection is not open to the objection of a floating neutral and voltage distortion because the existence of a Δ-connection allows a path for the third-harmonic currents. It would be observed that the primary and secondary line voltages and line currents are out of phase with each other by 30°. Because of this 30° shift, it is impossible to parallel such

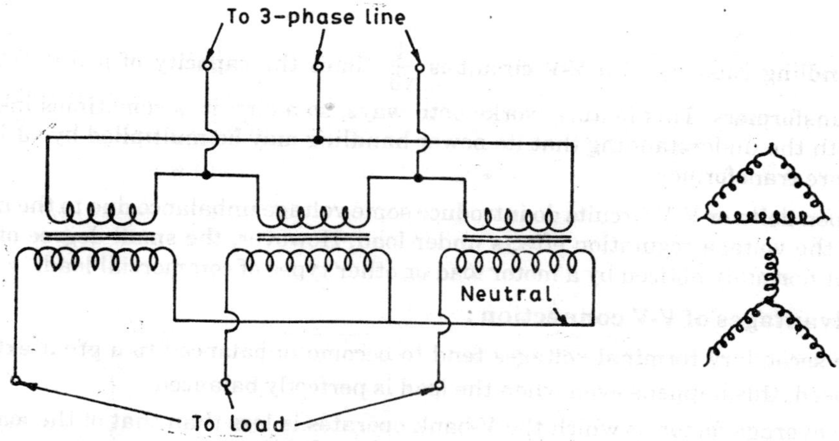

Fig. 68. The Δ-*Y* connection of transformers.

a bank with a Δ-Δ or *Y-Y* bank of transformers even though the voltage ratio are correctly adjusted.

10.1.5. The V-V (open-Δ) connections. Fig. 69 shows the *V-V* connection.

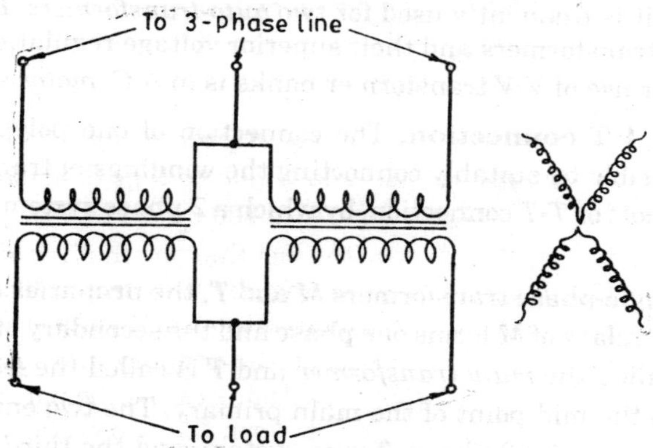

Fig. 69. The *V-V* (open Δ) connection of transformers.

If one of the transformers of a Δ-Δ bank is removed and a three-phase source is connected to the primaries (as shown in Fig. 66), three equal three-phase voltages will be measured at the secondary terminals at no-load. This method of transforming three-phase power, using two transformers, is called the open-delta, or *V-V* connection.

This type of connection is used in the following cases :

(*i*) When the three-phase load is comparatively small so that the installing does not warrant a three transformer bank.

(*ii*) When one of the transformers in a Δ-Δ bank fails, so that the service may be continued until the faulty transformer is repaired or a good one is substituted.

(*iii*) When it is anticipated that the future load will increase to warrant the chosing of the open Δ at some time later.

V-V connection has a number of *features* that are *advantageous* :

- Upon failure of the primary or secondary of one transformer of a complete Δ-Δ transformer circuit, the system reverts to a *V-V* circuit, so this is an *automatic standby*. The power-handling capacity of a *V-V* circuit is $\frac{1}{\sqrt{3}}$ times the capacity of a full Δ-Δ of the same transformers. This feature works both ways, so a circuit is sometimes installed as *V-V* with the understanding that its power handling may be multiplied by $\sqrt{3}$ by adding one more transformer.

- Open delta or *V-V* circuits do introduce some voltage unbalance due to the non-symmetry of the voltage regulation effects under load. However, the small degree of unbalance is not normally noticed by a motor load or other types of commercial load.

Disadvantages of V-V connection :

1. The secondary terminal voltages tend to become unbalanced to a great extent when the load is increased, this happens even when the load is perfectly balanced.

2. The average factor at which the *V*-bank operates is less than that of the load. This power factor is actually *86.6 per cent of the balanced load factor*. Another important point to note is that, except, for a balanced unity power factor load, the two transformers in the *V-V* bank operate at different power factors.

Uses of V-V connection :

(*i*) The *V-V* circuit is frequently used for two *auto-transformers*. Here advantage is taken of power handling of auto-transformers and their superior voltage regulation and efficiency.

(*ii*) Another major use of *V-V* transformer banks is in *A.C. motor starting*.

10.1.6. Scott or T-T connection. The connection of one polyphase system into another polyphase system is possible by suitably connecting the windings of transformers. One of the early types that was used is Scott or *T-T* connection, by which a 2-phase system is available from a 3-phase system or *vice-versa*.

In Fig. 70 two single-phase transformers *M* and *T*, the primaries of which are connected to a 3-phase supply. The secondary of *M* forms one phase and the secondary of *T* the other phase of a true 2-phase system. *M* is called the *main transformer* and *T* is called the *teaser*. One end of the teaser primary is connected to the mid-point of the main primary. The two ends of the main primary are connected to two lines wires of a 3-phase, 3-wire system, and the third line wire is connected to a tapping *X* on the teaser primary.

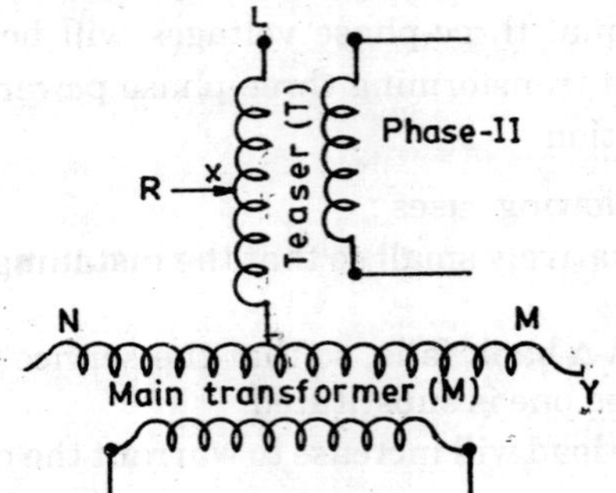

Fig. 70. Scott or *T-T* connection.

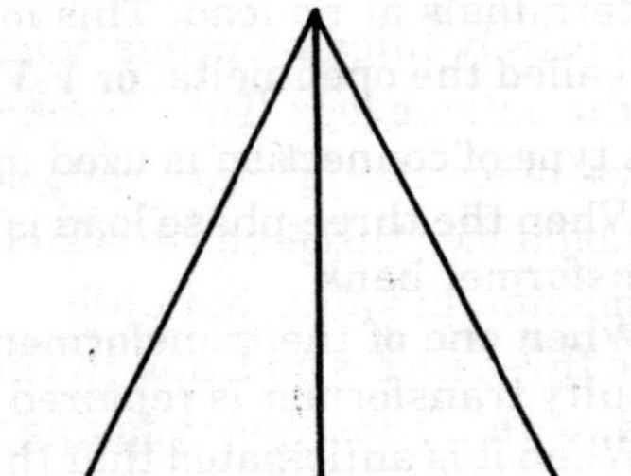

Fig. 71

If the supply voltages are assumed symmetrical, the triangle of voltages is *equilateral* as shown in Fig. 71. The vertical line $LS = \frac{\sqrt{3}}{2} NM$, so that it gives the relationship between the number of turns of two primaries, as

$$\frac{\text{Number of turns for the main transformer } M}{\text{Number of turns for the teaser } T} = \frac{100}{86.6}$$

Hence, if two identical single-phase transformers are to be used for Scott-connection, the primary one must have a tapping point brought out from its *mid-point of the primary,* and the secondary transformer must have a tapping X brought out from *86.6% of its primary turns.*

As the two-phase side is asymmetrical, there cannot be perfect balanced conditions on the three-phase side. However, neglecting the impedance of the windings, it can be shown that if load on the 2-phase side is balanced the 3-phase side is also balanced.

10.2. Three-phase Transformer Construction. The windings of three single-phase transformers can be wound on a common core. The *advantages* are :

1. One 3-phase transformer is *cheaper* than three single-phase transformers.
2. It has *slightly better efficiency and regulation.*
3. A 3-phase transformer takes *less floor space.*

On the other hand, from the point of view of standby, or same capacity, it is economical to have 3 single-phase transformers *plus* one spare rather than two 3-phase transformers one of which is a spare. However, in large central stations 3-phase transformers are often advantageous.

Disadvantages :

1. Three-phase transformers are much more difficult and costly to repair than are single-phase units.

2. When failure does occur and it becomes necessary to substitute a replacement unit to maintain service, the cost of spare is much greater than it would be were a single-phase transformer to be used as a replacement in a three-transformer bank.

3. There is a difficulty in transporting a heavier three-phase transformer compared with the moving of each of the three single-phase transformers.

Two general kinds of three-phase transformers are recognized, similar to single-phase transformers,depending upon the relative arrangements of windings and cores. These are the *core type* and the *shell type.*

Three-phase core type transformer. Fig. 72 shows three core-type transformers placed together so that they have a common path for the return magnetic circuit. Although the windows should be entirely filled by primary and secondary coils on each of the legs, only primary coils are shown on the outside legs. This simplifies the diagram, while it in no way changes the actual theory that follows, since the primary coils set up the flux. If the three transformers are identical in all respects, a balanced three-phase system of voltages will produce three fluxes in the cores which have the same maximum value, but differ in time phase by 120°. In the common leg of the three cores of Fig. 73, the three fluxes add, and the net flux is therefore always zero. The common leg may then be eliminated, with a subsequent saving in core material and size of transformer. A single polyphase transformer would be of impractical construction if it were the same as Fig. 72 with the centre leg omitted. Instead, the core-type polyphase transformer is manufactured so that it looks like that shown in Fig. 73. Actually, what we have done is, that axes of three coils have been moved into one plane. This causes the magnetic reluctance of coil M to differ somewhat from that of L and N. This

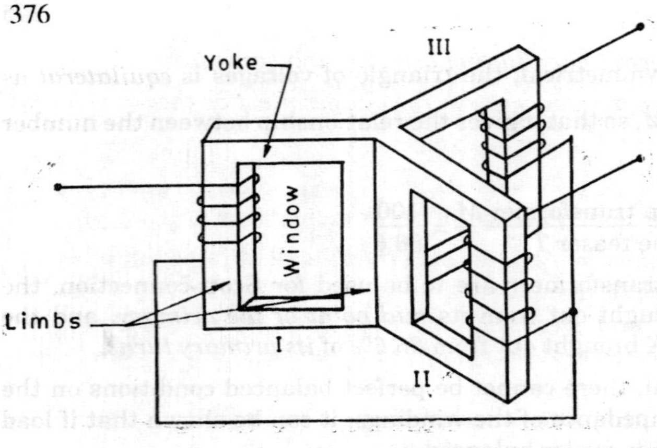

Fig. 72. Core-type transformers for polyphase transformation.

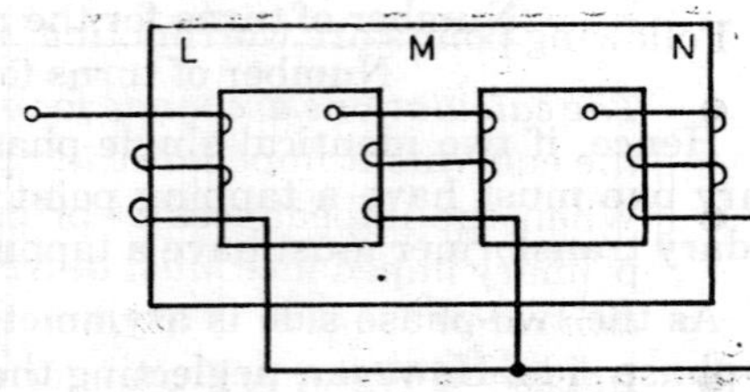

Fig. 73. Core-type three-phase transformer.

produces a slight unbalance in the three magnetizing currents, but the effect is not serious, especially under load.

Three-phase shell type transformer :

- **In Fig. 74 are shown three shell-type transformers stacked one above the other, with only the primary coils shown.**

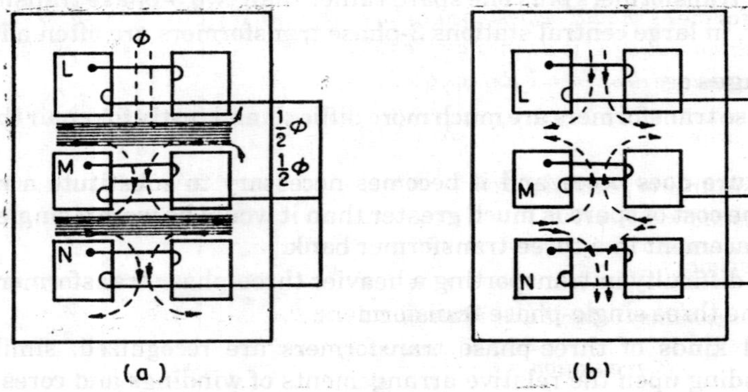

Fig. 74. Shell-type three-phase transformer (*a*) coil *M* wound in the same direction, (*b*) coil *M* wound in opposite direction.

- **In Fig. 74 (*a*) the three coils are wound in the same direction. The flux in the common core area between adjacent phases such as the shaded portions is thus equal to the *difference* of two of the phase fluxes. Since the fluxes are 120° apart in time, this mutual flux is equal to $\sqrt{3} \times (\frac{1}{2}\phi)$, or 0.866 of the flux in the centre leg. If the same flux density is maintained throughout, less iron is required in the common legs. Now, however if the centre coil, phase *M* is reversed, the flux in the common core is equal to the *sum* of the fluxes of two adjacent phases. This is shown in Fig. 74 (*b*). As in any three-phase system, the sum of two fluxes is equal to either flux alone, which in this case is 0.5 ϕ. This represents a further saving in iron, and for this reason shell-type three-phase transformers are *usually wound with centre coil opposing the two outside ones*.**

10.3. Parallel Operation of 3-phase Transformers. **The conditions for paralleling 3-phase transformers are same as that required for parallel operation of single-phase transformers with the following *additions* :**

(*i*) *The voltage ratio must refer to the terminal voltage of primary and secondary.*

(*ii*) *The phase displacement between primary and secondary voltages must be the same for all transformers which are to be paralleled.*

(*iii*) *The phase sequence must be the same.*

Following points are worthnoting while dealing with 3-phase transformers :

- *The calculations are made for one phase only*. The value of equivalent impedance used is the equivalent impedance per phase *referred to secondary*.
- When the impedances of primary and secondary windings are given separately then primary impedance *must be transferred to secondary by multiplying it with (transformation ratio)^2*.
- In case of Y/Δ or Δ/Y transformers the voltage ratios as given in the questions, *refer to terminal voltages and are quite difference from turn ratio.*

HIGHLIGHTS

1. *The function of a transformer is to transform alternating current energy from one voltage into another voltage. It operates on the principle of mutual inductance (between two or more inductively coupled coils).*
2. **Distribution transformers** *should be designed to have maximum efficiency at a load much lower than full-load (about 50%).*

   Power transformers *should be designed to have maximum efficiency at or near full-load.*
3. *The transformation ratio (K) is defined as the ratio of the secondary voltage to primary voltage.*
4. **Approximate** *voltage drop* $= I_2R_{02} \cos \phi \pm I_2 X_{02} \sin \phi$

   Exact *voltage drop* $= (I_2R_{02} \cos \phi \pm I_2X_{02}) + \dfrac{(I_2X_{02} \cos \phi \mp I_2R_{02} \sin \phi)^2}{2\ {}_0V_2}$
5. *Transfer of resistance or reactance from*

   Primary to secondary ...... $\times K^2$

   Secondary to primary ...... $\times \dfrac{1}{K^2}$
6. *The change in secondary voltage when rated load at a specified power is removed.*

$$\% \text{ regulation} = \frac{I_2R_{02} \cos \phi \pm I_2X_{02} \sin \phi}{{}_0V_2} \times 100$$

$$\eta_{all\text{-}day} = \frac{\text{output in kWh}}{\text{input in kWh}} \text{ (for 24 hours).}$$

7. *A transformer in which part of the winding is common to both the primary and secondary circuit is known as an* **auto-transformer.**

OBJECTIVE TYPE QUESTIONS

(A) Choose the Correct Answer :

1. Which of the following does not change in a transformer ?

   (*a*) Current (*b*) Voltage

   (*c*) Frequency (*d*) All of the above.
2. In a transformer the energy is conveyed from primary to secondary

   (*a*) through cooling coil (*b*) through air

   (*c*) by the flux (*d*) none of the above.

3. **A transformer core is laminated to**
(*a*) reduce hysteresis loss (*b*) reduce eddy current losses
(*c*) reduce copper losses (*d*) reduce all above losses.

4. **Which loss is not common between a transformer and rotating machines ?**
(*a*) Eddy current loss (*b*) Copper loss
(*c*) Windage loss (*d*) Hysteresis loss.

5. **The degree of mechanical vibrations produced by the laminations of a transformer depends on**
(*a*) tightness of clamping (*b*) gauge of laminations
(*c*) size of laminations (*d*) all of the above.

6. **The no-load current drawn by transformer is usually what percent of the full-load current ?**
(*a*) 0.2 to 0.5 per cent (*b*) 2 to 5 per cent
(*c*) 12 to 15 per cent (*d*) 20 to 30 per cent.

7. **In case there are burrs on the edges of the laminations of the transformer, it is likely to result in**
(*a*) vibrations (*b*) noise
(*c*) higher eddy current loss (*d*) higher hysteresis loss.

8. **The path of a magnetic flux in a transformer should have**
(*a*) high resistance (*b*) high reluctance
(*c*) low resistance (*d*) low reluctance.

9. **No-load test on a transformer is carried out to determine**
(*a*) copper loss (*b*) magnetising current
(*c*) magnetising current and loss (*d*) efficiency of the transformer.

10. **The dielectric strength of transformer oil is expected to be**
(*a*) 1 kV (*b*) 33 kV
(*c*) 100 kV (*d*) 330 kV.

11. **Sumpner's test is conducted on transformers to determine**
(*a*) temperature (*b*) stray losses
(*c*) all-day efficiency (*d*) none of the above.

12. **The permissible flux density in case of cold rolled grain oriented steel is around**
(*a*) 1.7 Wb/m^2 (*b*) 2.7 Wb/m^2
(*c*) 3.7 Wb/m^2 (*d*) 4.7 Wb/m^2.

13. **During the short-circuit test on a small transformer the frequency is increased from 50 Hz to 200 Hz. The copper losses will increase by a factor of**
(*a*) 16 (*b*) 4
(*c*) 1 (*d*) $\frac{1}{4}$.

14. **The efficiency of a transformer will be maximum when**
(*a*) copper losses = hysteresis losses (*b*) hysteresis losses = eddy current losses
(*c*) eddy current losses = copper losses (*d*) copper losses = iron losses.

15. **No-load current in a transformer**
(*a*) lags behind the voltage by about 75° (*b*) leads the voltage by about 75°
(*c*) lags behind the voltage by about 15° (*d*) leads the voltage by about 15°.

16. **The purpose of providing an iron core in a transformer is to**
(*a*) provide support to windings (*b*) reduce hysteresis loss
(*c*) decrease the reluctance of the magnetic path
(*d*) reduce eddy current losses.

17. **Which of the following is not a part of transformer installation ?**
(*a*) Conservator (*b*) Breather
(*c*) Buchholz relay (*d*) Exciter.

18. While conducting short-circuit test on a transformer the following side is short-circuited
(*a*) high voltage side (*b*) low voltage side
(*c*) primary side (*d*) secondary side.

19. In a transformer the toppings are generally provided on
(*a*) primary side (*b*) secondary side
(*c*) low voltage side (*d*) high voltage side.

20. The use of higher flux density in the transformer design
(*a*) reduces weight per kVA (*b*) reduces iron losses
(*c*) reduces copper losses (*d*) increases part load efficiency.

21. The chemical used in breather for transformer should have the quality of
(*a*) ionizing air (*b*) absorbing moisture
(*c*) cleansing the transformer oil (*d*) cooling the transformer oil.

22. The chemical used in breather is
(*a*) asbestos fibre (*b*) silica sand
(*c*) sodium chloride (*d*) silica gel.

23. If a pump motor is run on 2/3rd of its supply voltage, it will
(*a*) continue to deliver same power (*b*) burn
(*c*) stall (*d*) continue to run at lower speed.

24. An ideal transformer has infinite values of primary and secondary inductances. The statement is
(*a*) true (*b*) false.

25. The transformer ratings are usually expressed in terms of
(*a*) volts (*b*) amperes
(*c*) kW (*d*) kVA.

26. The noise resulting from vibrations of laminations set by magnetic force, is termed as
(*a*) magnetostriction (*b*) boo
(*c*) hum (*d*) zoom.

27. Hysteresis loss in a transformer varies as (B_{max} = maximum flux density)
(*a*) B_{max} (*b*) $B_{max}^{1.6}$
(*c*) B_{max}^{2} (*d*) $B_{max}^{2.4}$.

28. Material used for construction of transformer core is usually
(*a*) wood (*b*) copper
(*c*) aluminium (*d*) silicon steel.

29. The thickness of laminations used in a transformer is usually
(*a*) 0.4 mm to 0.5 mm (*b*) 4 mm to 5 mm
(*c*) 14 mm to 15 mm (*d*) 25 mm to 40 mm.

30. The function of conservator in a transformer is
(*a*) to protect against internal fault (*b*) to reduce copper as well as core losses
(*c*) to cool the transformer oil
(*d*) to take care of the expansion and contraction of transformer oil due to variation of temperature of surroundings.

31. The highest voltage for transmitting electrical power in India is
(*a*) 33 kV (*b*) 66 kV
(*c*) 132 kV (*d*) 400 kV.

32. In a transformer the resistance between its primary and secondary is
(*a*) zero (*b*) 1 Ω
(*c*) 1000 Ω (*d*) infinite.

33. A transformer oil must be free from
(*a*) sludge (*b*) odour
(*c*) gases (*d*) moisture.

34. A Buchholz relay can be installed on
(*a*) auto-transformers (*b*) air-cooled transformers
(*c*) welding transformers (*d*) oil cooled transformers.

35. Gas is usually not liberated due to dissociation of transformer oil unless the oil temperature exceeds
(*a*) 50°C (*b*) 80°C
(*c*) 100°C (*d*) 150°C.

36. The main reason for generation of harmonics in a transformer could be
(*a*) fluctuating load (*b*) poor insulation
(*c*) mechanical vibrations (*d*) saturation of core.

37. Distribution transformers are generally designed for maximum efficiency around
(*a*) 90% load (*b*) zero load
(*c*) 25% load (*d*) 50% load.

38. Which of the following properties is not necessarily desirable in the material for transformer core ?
(*a*) Mechanical strength (*b*) Low hysteresis loss
(*c*) High thermal conductivity (*d*) High permeability.

39. Helical coils can be used on
(*a*) low voltage side of high kVA transformers (*b*) high frequency transformers
(*c*) high voltage side of small capacity transformers
(*d*) high voltage side of high kVA rating transformers.

40. High frequency transformers sometimes make use of ferrite cores because it has
(*a*) high specific gravity (*b*) high resistance
(*c*) high hysteresis (*d*) low permeability.

41. Cross over windings are used on
(*a*) low voltage side of high kVA rating transformers
(*b*) current transformers
(*c*) high voltage side of high kVA rating transformers
(*d*) high voltage side of low kVA rating transformers.

42. While conducting short-circuit test on a transformer the following side is short-circuited.
(*a*) h.v. side (*b*) l.v. side
(*c*) primary side (*d*) secondary side.

43. A substance is placed in strong magnetic field. Which of the following will have maximum influence on the capacity of the substance to get magnetised ?
(*a*) Permeability (*b*) Susceptibility
(*c*) Permittivity (*d*) Resistivity.

44. Harmonics in transformer result in
(*a*) increased core losses (*b*) increased I^2R losses
(*c*) magnetic interference with communication circuits
(*d*) all of the above.

45. Tertiary winding is provided in transformers having
(*a*) mesh/star winding (*b*) mesh/mesh winding
(*c*) star/star winding (*d*) any of the above.

46. The core used in high frequency transformer is usually
(*a*) copper core (*b*) cost iron core
(*c*) air core (*d*) mild steel core.

47. The full-load copper loss of a transformer is 1600 W. At half-load the copper loss will be
(*a*) 6400 W (*b*) 1600 W
(*c*) 800 W (*d*) 400 W.

48. The value of flux involved in the e.m.f. equation of a transformer is
(*a*) average value (*b*) r.m.s. value
(*c*) maximum value (*d*) instantaneous value.

49. Silicon steel used in laminations mainly reduces

(*a*) hysteresis loss (*b*) eddy current losses

(*c*) copper losses (*d*) all of the above.

50. Which winding of the transformer has less cross-sectional area ?

(*a*) Primary winding (*b*) Secondary winding

(*c*) Low voltage winding (*d*) High voltage winding.

51. Power transformers are generally designed to have maximum efficiency around

(*a*) no-load (*b*) half-load

(*c*) near full-load (*d*) 10% overload.

52. Which of the following is the main advantage of an auto-transformer over a two winding transformer ?

(*a*) Hysteresis losses are reduced (*b*) Saving in winding material

(*c*) Copper losses are negligible (*d*) Eddy losses are totally eliminated.

53. During short-circuit test iron losses are negligible because

(*a*) the current on secondary side is negligible (*b*) the voltage on secondary side does not vary

(*c*) the voltage applied on primary side is low (*d*) full-load current is not supplied to the transformer.

54. Two transformers are connected in parallel. These transformers do not have equal percentage impedance. This is likely to result in

(*a*) short-circuiting of the secondaries

(*b*) power factor of one of the transformers is leading while that of the other lagging

(*c*) transformers having higher copper losses will have negligible core losses

(*d*) loading of the transformers not in proportion to their kVA ratings.

55. The changes in volume of transformer cooling oil due to variation of atmospheric temperature during day and night is taken care of by which part of transformer ?

(*a*) Conservator (*b*) Breather

(*c*) Bushings (*d*) Buchholz relay.

56. The transformer laminations are insulated from each other by

(*a*) mica strip (*b*) thin coat of varnish

(*c*) paper (*d*) any of the above.

57. Which type of winding is used in 3-phase shell-type transformer ?

(*a*) Circular type (*b*) Sandwich type

(*c*) Cylindrical type (*d*) Rectangular type.

58. During open circuit test of a transformer

(*a*) primary is supplied rated voltage (*b*) primary is supplied full-load current

(*c*) primary is supplied current at reduced voltage

(*d*) primary is supplied rated kVA.

59. Open circuit test on transformers is conducted to determine

(*a*) hysteresis losses (*b*) copper losses

(*c*) core losses (*d*) eddy current losses.

60. Short circuit test on transformers is conducted to determine

(*a*) hysteresis losses (*b*) copper losses

(*c*) core losses (*d*) eddy current losses.

61. For the parallel operation of single-phase transformers it is necessary that they should have

(*a*) same efficiency (*b*) same polarity

(*c*) same kVA rating (*d*) same number of turns on the secondary side.

62. The transformer oil should have ...... voltatility and ...... viscosity

(*a*) low ...... low (*b*) high ...... high

(*c*) low ...... high (*d*) high ...... low.

63. The function of breather in a transformer is
(*a*) to provide oxygen inside the tank (*b*) to cool the coils during reduced load
(*c*) to filter the transformer cooling oil
(*d*) to arrest flow of moisture when outside air enters the transformer.

64. The secondary winding of which of the following transformers is always kept closed ?
(*a*) Step-up transformer (*b*) Step-down transformer
(*c*) Potential transformer (*d*) Current transformer.

65. The size of a transformer core will depend on
(*a*) frequency (*b*) area of the core
(*c*) flux density of the core material (*d*) (*a*) and (*b*) both.

66. Natural air cooling is generally restricted for transformers up to
(*a*) 1.5 MVA (*b*) 5 MVA
(*c*) 15 MVA (*d*) 50 MVA.

67. A shell-type transformer has
(*a*) high eddy current losses (*b*) reduced magnetic leakage
(*c*) negligible hysteresis losses (*d*) none of the above.

68. A transformer can have regulation closer to zero
(*a*) on full-load (*b*) on overload
(*c*) on leading power factor (*d*) on zero power factor.

69. A transformer transforms
(*a*) voltage (*b*) current
(*c*) current and voltage (*d*) power.

70. Which of the following is not the standard voltage for power supply in India
(*a*) 11 kV (*b*) 33 kV
(*c*) 66 kV (*d*) 122 kV.

ANSWERS

1. (*c*)	**2.** (*c*)	**3.** (*b*)	**4.** (*a*)	**5.** (*d*)	**6.** (*b*)	**7.** (*c*)
8. (*d*)	**9.** (*c*)	**10.** (*b*)	**11.** (*a*)	**12.** (*a*)	**13.** (*a*)	**14.** (*d*)
15. (*a*)	**16.** (*c*)	**17.** (*d*)	**18.** (*b*)	**19.** (*c*)	**20.** (*a*)	**21.** (*b*)
22. (*d*)	**23.** (*c*)	**24.** (*b*)	**25.** (*d*)	**26.** (*c*)	**27** (*b*)	**28.** (*d*)
29. (*a*)	**30.** (*d*)	**31.** (*d*)	**32.** (*d*)	**33.** (*d*)	**34.** (*d*)	**35.** (*d*)
36. (*d*)	**37.** (*d*)	**38.** (*c*)	**39.** (*a*)	**40.** (*b*)	**41.** (*d*)	**42.** (*b*)
43. (*b*)	**44.** (*d*)	**45.** (*c*)	**46.** (*c*)	**47.** (*d*)	**48.** (*c*)	**49.** (*a*)
50. (*d*)	**51.** (*c*)	**52.** (*b*)	**53.** (*c*)	**54.** (*d*)	**55.** (*a*)	**56.** (*b*)
57. (*b*)	**58.** (*a*)	**59.** (*c*)	**60.** (*b*)	**61.** (*b*)	**62.** (*a*)	**63.** (*d*)
64. (*d*)	**65.** (*d*)	**66.** (*a*)	**67.** (*b*)	**68.** (*c*)	**69.** (*a*)	**70.** (*d*)

(B) Say 'Yes' or 'No'

1. When a transformer raises the voltage it is called the step-up transformer.
2. A transformer must not be connected to D.C. source.
3. The ratio of primary voltage to secondary voltage is known as transformation ratio.
4. An ideal transformer is one in which the resistance of the windings is negligible and the core has no losses.
5. Primary and secondary currents are directly proportional to their respective turns.
6. The function of the magnetising component of no-load current is to sustain the alternating flux in the core.
7. The no-load primary input is practically equal to the iron loss in the transformer.
8. A transformer is said to be loaded when the secondary circuit of a transformer is completed through an impedance load.

9. Magnetic flux can be confined into a designed path.
10. When shifting resistance to the secondary, divide it by K^2
11. An open-circuit test is conducted to find no-load or core loss.
12. Short-circuit test is conducted to find full-load copper loss.
13. The change in voltage when rated load at a specified power is removed is termed as voltage regulation.
14. Iron or core losses include copper loss and eddy current loss.
15. Iron or core losses are found from short-circuit test.
16. The efficiency of a transformer at a particular load and power factor is defined as the ratio of power output to power input.
17. Copper losses = iron losses is the condition for maximum efficiency of a transformer.
18. $\eta_{\text{all-day}} = \dfrac{\text{output in kWh}}{\text{input in kWh}}$ (for 24 hours)
19. A transformer in which part of the winding is common to both the primary and secondary circuits is known as auto-transformer.
20. The Δ-Δ connection is generally used in systems in which the voltages are not very high and especially when continuity of service must be maintained even though one of the transformers should fail.
21. The Y-Δ connection is principally used where the voltage is to be stepped up.
22. The Δ-Y connection is employed where it is necessary to step-up the voltage.
23. The V-V circuit is frequently used for two auto-transformers.
24. It is practicable to connect instruments and meters directly to the lines in high voltage circuits.
25. A potential transformer is a step down transformer used along with a low range voltmeter for measuring a high voltage.
26. The current transformer ratio is not equal to the ratio of secondary to primary turns, mainly because of the effect of the magnetising current.
27. A constant-current transformer is used to supply power to street lights which are connected in series.
28. In case of an induction regulator primary winding is stationary.

ANSWERS

1. Yes	**2.** Yes	**3.** No	**4.** Yes	**5.** No	**6.** Yes	**7.** Yes
8. Yes	**9.** No	**10.** No	**11.** Yes	**12.** Yes	**13.** Yes	**14.** No
15. No	**16.** Yes	**17.** Yes	**18.** Yes	**19.** Yes	**20.** Yes	**21.** No
22. Yes	**23.** Yes	**24.** No	**25.** Yes	**26.** Yes	**27.** Yes	**28.** No.

THEORETICAL QUESTIONS

1. What is a transformer ? How does it transfer electric energy from one circuit to another ?
2. Explain the principle of operation of a transformer.
3. What is meant by *transformer action* ? Under what conditions will it take place ?
4. If an alternating current is impressed on one coil, what will be the frequency of the induced voltage in another coil with which it is coupled ?
5. Enumerate the various kinds of transformers.
6. Explain the need for stepping up and stepping down voltages in a power system. How does a transformer accomplish ?
7. Why are transformer windings divided into several coils ?
8. What properties should a good transformer oil possess ?
9. What purposes are served by placing transformers in oil-filled tanks ?
10. Why are the tanks of some large transformers corrugated ?
11. Write a short note on 'transformer cooling'.
12. Enumerate and explain briefly different types of windings.

13. Derive an expression for induced e.m.f. in a transformer in terms of frequency, the maximum value of flux and the number of turns on the windings.
14. Derive an expression for the e.m.f. of an ideal transformer winding.
15. Explain the function of the active and reactive components of the no-load current of a static transformer.
16. Why are some transformers constructed with primary and secondary windings divided into two halves ?
17. Distinguish between *power* and *distribution* transformers.
18. Draw and explain the no-load phasor diagram for a single-phase transformer.
19. Draw the vector diagram of a power transformer under full-load condition.
20. What is meant by *equivalent resistance* of a transformer ? How may it be calculated in primary terms and secondary terms ?
21. What is meant by the *equivalent reactance* of a transformer ? How may it be calculated in primary terms and secondary terms ?
22. How can the equivalent impedance of a transformer be determined ?
23. Outline carefully the procedure for performing the short-circuit test.
24. What useful information is obtained from the short-circuit test ?
25. Outline carefully the procedure for performing the open-circuit test.
26. What useful information is obtained from open-circuit test ?
27. What are the two components of the core loss ?
28. How is the hysteresis loss affected by a change in flux density ?
29. Develop the equivalent circuit of a single phase transformer.
30. Draw the equivalent circuit of a transformer and show how the constants of the primary and secondary windings may be combined to give a simplified equivalent circuit with the values of constants given in terms of the secondary winding.
31. Describe the method of calculating the regulation and efficiency of a single-phase transformer by open circuit and short circuit tests.
32. Distinguish between the 'efficiency' and the 'regulation' of a transformer. Show how power factor affects both of them.
33. Explain with circuit diagrams, the open-circuit and short-circuit tests to be carried out in the laboratory for the determination of the parameters of a single-phase transformer. Derive the condition for maximum efficiency.
34. What is Sumpner's test ? Draw a circuit diagram to conduct this test and explain its principle.
35. Define all-day efficiency.
36. What is an auto-transformer ? What advantages are possessed by auto-transformers ?
37. Derive an expression for saving of copper when an auto-transformer is used.
38. State the applications of auto-transformers. Why is this transformer not used as a distribution transformer ? Prove that for the same capacity and voltage ratio, an auto-transformer requires less copper than an ordinary transformer.
39. What are the sources of heat in a transformer ? Describe briefly the various methods used for cooling of transformers.
40. To what does the '*polarity*' of a transformer refer ?

UNSOLVED EXAMPLES

E.m.f. Equation-Turn Ratio

1. The no-load ratio required in a single-phase 50-Hz transformer is 6600/300 V. If the maximum value of flux in the core is to be about 0.09 weber, find the number of turns in each winding.

[Ans. $N_1 = 330$; $N_2 = 15$]

2. A 20 kVA, single-phase transformer has 200 turns on the primary and 40 turns on the secondary. The primary is connected to 1000 V, 50-Hz supply. Determine :
   (*i*) the secondary voltage on open circuit ;
   (*ii*) the currents flowing through the two windings on full-load ; and
   (*iii*) the maximum value of flux. [**Ans.** (*i*) 200 V ; (*ii*) 20 A, 100 A ; (*iii*) 0.0225 Wb]
3. A single-phase transformer is connected to a 230 V, 50-Hz supply. The net cross-sectional area of the core is 50 cm^2. The number of turns in the primary is 460 and in the secondary 80. Determine :
   (*i*) Transformation ratio ; (*ii*) Peak value of flux density in the core ; and
   (*iii*) E.m.f. in the secondary winding. [**Ans.** 0.1739 ; 0.4504 T ; 40 V]
4. A 6600/440 V single-phase 600 kVA transformer has 1200 primary turns. Find :
   (*i*) Transformation ratio ; (*ii*) Secondary turns ;
   (*iii*) Voltage/turn ; and
   (*iv*) Secondary current when it supplies a load of 400 kW at 0.8 power factor lagging.
   $\left[\textbf{Ans. } \frac{1}{15} ; 80 ; 5.5 \text{ V} ; 1136 \text{ A} \right]$
5. Find the primary and secondary turns of a 3300/300 V, single-phase, 50-Hz, 30 kVA transformer if the flux in the core is to be about 0.06 Wb. Also determine the primary and secondary currents if the losses are to be neglected. [**Ans.** 264 ; 24 ; 100 A ; 9.09 A]
6. The voltage/turn of a single-phase transformer is 1.1 V, when the primary winding is connected to a 220 V, 50-Hz A.C. supply, the secondary voltage is found to be 660 V. Find :
   (*i*) Primary and secondary turns ; and
   (*ii*) Core area if the maximum flux density is 1.2 T. [**Ans.** 200 ; 600 ; 41.29 cm^2]

Transformer on No-load

7. A 2000/200 V single-phase transformer gives 0.5 A and 40 W as ammeter and wattmeter readings when supply is given to the low voltage winding and high voltage winding is kept open. Find :
   (*i*) The magnetising component, (*ii*) The iron loss component, and
   (*iii*) The power factor of no-load current. [**Ans.** 0.4582 A ; 0.2 A ; 0.4 lagging]
8. Find (*i*) active component and reactive components of no-load current and ; (*ii*) no-load current of a 230 V/115 V single-phase transformer if the power input on no-load to the high voltage winding is 70 W and power factor of no-load current is 0.25 lagging. [**Ans.** 0.3044 A ; 1.179 A ; 1.2176 A]
9. A single-phase transformer has 500 turns on the primary and 40 turns on the secondary winding. The mean length of the magnetic path in the core is 150 cm and the joints are equivalent to an air-gap of 0.1 mm. When a potential difference of 3000 V is applied to the primary, maximum flux density is 1.2 T. Calculate :
   (*i*) The cross-sectional area of the core, (*ii*) No-load secondary voltage,
   (*iii*) The no-load current drawn by the primary, and
   (*iv*) Power factor on no-load.
   Given that AT/cm for a flux density of 1.2 T in the iron to be 5, the corresponding iron loss to be 2 W/kg at 50 Hz and density of iron as 7.8 g/cm^3. [**Ans.** 225 cm^2 ; 240 V ; 1.208 A ; 0.1457]

Transformer on Load

10. A 230 V/115 V single phase transformer takes a no-load current of 1.7 A at a power factor of 0.18 lagging with low voltage winding kept open. If the low voltage winding is now loaded to take a current of 13 A at 0.8 power factor lagging find the current taken by high voltage winding. [**Ans.** 7.834 A]
11. A transformer has a primary winding of 800 turns and a secondary winding of 200 turns. When the load current on the secondary is 80 A at 0.8 power factor lagging, the primary current is 25 A at 0.707 power factor lagging. Determine graphically or otherwise the no-load current of the transformer and its phase with respect to the voltage. [**Ans.** I_0 = 5.93 A ; ϕ_0 = 73.3°]

12. A 8 : 1 step down, single-phase transformer takes a no-load current of 0.6 A at a power factor of 0.8 lagging with l.v. winding kept open. If the secondary is connected to a load taking a current of 80 A at (*i*) 0.8 power factor lagging ; and (*ii*) 0.8 power factor leading find the primary current and power factor.
[**Ans.** 10.49 A, 0.7996 lagging ; 9.817 A, 0.8333 leading]

13. A 20 kVA, 2000/200 V, single-phase, 50-Hz transformer has a primary resistance of 2.5 Ω and reactance of 4.8 Ω. The secondary resistance and reactance are 0.01 Ω and 0.018 Ω respectively. Find :
(*i*) Equivalent resistance referred to primary,
(*ii*) Equivalent impedance referred to primary,
(*iii*) Equivalent resistance, reactance and impedance referred to secondary, and
(*iv*) Total copper loss of the transformer.
[**Ans.** (*i*) 3.5 Ω ; (*ii*) 7.47 Ω ; (*iii*) 0.035, 0.066 Ω, 0.0747 Ω ; (*iv*) 350 W]

14. A 50 kVA, 4400/220 V transformer has R_1 = 3.45 Ω, R_2 = 0.009 Ω. The values of reactances are X_1 = 5.2 Ω and X_2 = 0.015 Ω. Calculate for the transformer :
(*i*) Equivalent resistance as referred to primary,
(*ii*) Equivalent resistance as referred to secondary,
(*iii*) Equivalent reactance as referred to both primary and secondary,
(*iv*) Equivalent impedance as referred to both primary and secondary, and
(*v*) Total copper loss, first using individual resistances of the two windings and secondly, using equivalent resistances as referred to each side.
[**Ans.** (*i*) 7.05 Ω ; (*ii*) 0.0176 Ω ; (*iii*) 11.2 Ω ; (*iv*) 13.23 Ω ; 0.0331 Ω ; (*iv*) 909 W]

15. A single-phase transformer has the following data :
Turn ratio = 19.5 : 1, R_1 = 25 Ω , X_1 = 100 Ω, R_2 = 0.06 Ω, X_2 = 0.25 Ω. No-load current = 1.25 A leading the flux by 30°.

The secondary delivers 200 A at a terminal voltage of 500 V and a power factor of 0.8 lagging. Determine by the aid of a vector diagram the primary applied voltage, the primary power factor and the efficiency.
[**Ans.** 12540 ∠186.7°, 0.698 (lag), 86.74%]

16. The high voltage and low voltage windings of a 1100/110 V single-phase 50 Hz transformer has resistances of 2.4 Ω and 0.02 Ω and reactances 1 Ω and 0.009 Ω respectively. The low voltage winding is connected to a load having an impedance of (3 + *j*2) Ω. Determine :
(*i*) Current in l.v. winding,
(*ii*) Current in h.v. winding,
(*iii*) The load voltage, and
(*iv*) Power consumed by the load.
[**Ans.** (*i*) 30.11 A ; (*ii*) 3.011 A ; 108.5 V (*iv*) 2720.7 W]

Equivalent Circuit and O.C. and S.C. Tests

17. The parameters of a 2300/230 V, 50 Hz transformer are given below :

R_1 = 0.286 Ω R_2' = 0.319 Ω
X_1 = 0.73 Ω X_2' = 0.73 Ω
R_0 = 250 Ω X_0 = 1250 Ω

The secondary load impedance Z_L = 0.387 + *j*0.29. Solve the exact equivalent circuit with normal voltage across the primary.
[**Ans.** η = 78.8%, % Regulation = 2.7%]

18. A 230 V, 3 kVA single-phase transformer has an iron loss of 100 W at 40-Hz and 70 W at 30-Hz. Find the hysteresis and eddy current losses at 50-Hz.
[**Ans.** 91.67 W, 41.67 W]

19. When a transformer is supplied at 400 V, 50-Hz the hysteresis loss is found to be 300 W and eddy current loss is found to be 250 W. Determine the hysteresis loss and eddy current loss when the transformer is supplied at 800 V, 100 Hz.
[**Ans.** 600 W ; 1000 W]

20. When the primary of a transformer rated at 2200 V, 50-Hz is supplied at 2200 V, 50-Hz the wattmeter gives a reading of 1200 W on no-load. When it is supplied at 1100 V, 25-Hz, the wattmeter gives a reading of 400 W on no-load. If the wattmeter is connected in the input circuit find the hysteresis loss and eddy current loss at normal voltage and frequency.
[**Ans.** 400 V ; 800 W]

21. A 4400 V, 50-Hz transformer has a hysteresis loss of 1000 W, eddy current loss of 1500 W and full-load copper loss of 3500 W. If the transformer is supplied at 6600 V, 75-Hz, what will be the losses ? Assume that the full-load current remains the same. [**Ans.** P_h = 1500 W, P_e = 3375 W]

22. A 50 Hz, single-phase transformer has a turn ratio of 6. The resistances are 0.9 Ω, 0.03 Ω and reactances are 5 Ω and 0.13 Ω for high-voltage and low voltage windings respectively. Find :

(*i*) The voltage to be applied to the h.v. side to obtain full-load current of 200 A in the l.v. winding on short-circuit.

(*ii*) The power factor on short-circuit.

Draw the equivalent circuit and show therein all the values. [**Ans.** 329.3 V, 0.2]

23. A 200/2000 V transformer is fed from a 200 V supply. The total winding resistance and leakage reactance as referred to low voltage side are 0.16 Ω and 7.0 Ω respectively. The representing core loss is 400 Ω and the magnetising reactance is 231 Ω. A load of impedance 596 + j444 Ω is connected across the secondary terminals. Calculate :

(*i*) Input current ; (*ii*) The secondary terminal voltage ;

(*iii*) Primary power factor. [**Ans.** (*i*) 25.96 ∠– 40.78° A ; (*ii*) 1859 V ; (*iii*) 0.757 lagging]

24. Determine the approximate equivalent circuit of a given 200/2000 V single-phase 40 kVA transformer having the following test results :

O.C. test : 200 V, 6.4 A, 384 W .... on l.v. side.

S.C. test : 78 V, 20 A, 620 W ...... on h.v. side. [**Ans.** R_{01} = 0.0155 Ω, X_{01} = 0.0358 Ω, R_0 = 104.2 Ω, X_0 = 32.75 Ω]

25. Determine the equivalent circuit of a 200/400 V, 50-Hz, single-phase transformer from the following test data :

O.C. test (l.v. side) : 200 V, 0.7 A, 70 W

S.C. test (h.v. side) : 15 V, 10 A, 85 W

Calculate the secondary voltage when delivering 5 kW at 0.8 power factor, lagging, the primary voltage being 200 V. [**Ans.** R_0 = 571.4 Ω, X_0 = 330 Ω, R_{01} = 0.21 Ω, X_{01} = 0.31 Ω, V_2 = 377.8 V]

26. A single-phase, 10 kVA, 500/250 V, 50-Hz transformer has the following constants :

Resistance : Primary 0.2 Ω, Secondary 0.5 Ω

Reactance : Primary 0.4 Ω, Secondary 0.1 Ω

Resistance or equivalent exciting circuit referred to primary, R_0 = 1500 Ω, reactance of the equivalent exciting circuit referred to primary, X_0 = 750 Ω.

What would be the readings of the instruments when the transformer is connected for the open-circuit and short-circuit tests ? [**Ans.** O.C. test : 500 V, 0.745 A, 167 W ; S.C. test : 46.8 V, 20 A, 880 W]

27. Find the approximate equivalent circuit of a single-phase 400/4400 V transformer having the following test readings :

O.C. test (l.v. side) : 400 V, 5.2 A, 600 W

S.C. test (h.v. side) : 155 V. 50 A, 1850 W [**Ans.** R_0 = 266.7 Ω, X_0 = 80.34 Ω, Z_{01} = 0.02563 Ω, R_{01} = 0.006115 Ω, X_{01} = 0.02488 Ω]

28. The following readings were taken in open-circuit and short-circuit tests on a single-phase 20 kVA, 2000/200 V transformer :

O.C. test (l.v. side) : 200 V, 2 A, 120 W

S.C. test (h.v. side) : 30 V, 10 A, 140 W

Determine :

(*i*) Equivalent circuit referred to l.v. side

(*ii*) Secondary terminal voltage on full-load at 0.8 power factor leading.

[**Ans.** (*i*) R_0 = 333.3 Ω, X_0 = 104.8 Ω, R_{02} = 0.014 Ω, X_{02} = 0.02654 Ω ; (*ii*) 200.47 V]

Regulation and Efficiency of a Transformer

29. The corrected instrument readings obtained from open and short-circuit tests on 10 kVA, 450/120 V, 50-Hz transformer are :

O.C. Test :

$V_1 = 120$ V, $I_1 = 4.2$ A, $W_1 = 80$ W ; V_1, W_1 and I_1 were read on the low-voltage side.

S.C. Test :

$V_1 = 9.65$ V ; $I_1 = 22.2$ A ; $W_1 = 120$ W with low-voltage winding short-circuited.

Calculate :

(*i*) The equivalent circuit (approximate) constants.
(*ii*) Efficiency and voltage regulation for 0.8 lagging power.
(*iii*) The efficiency at half full-load and 0.8 lagging power factor load.

[**Ans.** (*i*) $R_0 = 25.30\ \Omega$, $X_0 = 409\ \Omega$; (*ii*) 97.57%, 2.04% ; (*iii*) 97.34%]

30. The primary and secondary winding resistance of a 40 kVA, 6600/250 V single-phase transformer are 10 Ω and 0.02 Ω respectively. The equivalent leakage reactance as referred to the primary winding is 35 Ω. Find the full-load regulation for load power factors of (*i*) unity ; (*ii*) 0.8 lagging ; and (*iii*) 0.8 leading.

[**Ans.** (*i*) 2.202% ; (*ii*) 3.69% ; (*iii*) ; – 0.166 %]

31. The percentage resistance and reactance of a transformer are 2.5% and 4% respectively. Find the approximate regulation on full-load at

(*i*) unity power factor, (*ii*) 0.8 power factor lagging, and
(*iii*) 0.8 power factor leading. [**Ans.** (*i*) 2.5% ; (*ii*) 4.4% ; (*iii*) – 0.4%]

32. A single-phase 100 kVA, 2000/200 V, 50 Hz transformer has impedance drop of 10% and resistance drop of 5%.

(*i*) Find the regulation at full-load, 0.8 power factor lagging.
(*ii*) At what power factor is the regulation zero ? [**Ans.** 9.196% ; 0.866 leading]

33. The high voltage of a single-phase 200 kVA 4400/220 V transformer takes a current of 30 A and power of 1200 W at 75 V when the low voltage winding is short-circuited. Determine :

(*i*) The voltage to be applied to the high voltage winding on full-load 0.8 power factor lagging if the secondary terminal voltage is to be kept at 220 V, and
(*ii*) % regulation. [**Ans.** (*i*) 4506 V ; (*ii*) 2.355%]

34. A 20 kVA, 2200/220 V 50-Hz distribution transformer is tested for efficiency and regulation as follows ;

O.C. test (l.v. side) : 220 V, 4.2 A, 148 W.
S.C. test (h.v. side) : 86 V, 10.5 A, 360 W.

Determine :

(*i*) Core loss
(*ii*) Equivalent resistance referred to primary.
(*iii*) Equivalent resistance referred to secondary.
(*iv*) Equivalent reactance referred to primary.
(*v*) Equivalent reactance referred to secondary.
(*vi*) Regulation of transformer at 0.8 power factor lagging current.
(*vii*) Efficiency at full-load and half full-load at 0.8 power factor lagging current.

[**Ans.** 148 W, 3.26 Ω, 0.0326 Ω, 7.51 Ω, 0.0751 Ω, 2.95%, 97.4%, 97.3%]

35. A 100 kVA single-phase transformer has a full-load primary current of 400 A and total resistance referred to primary is 0.006 Ω. If the iron loss amounts to 500 W, find the efficiency on full-load and half-load at

(*i*) unity power factor, and (*ii*) 0.8 power factor.

[**Ans.** 98.58%, 98.56%, 98.22%, 98.19%]

Maximum Efficiency

36. A 50 kVA, single phase transformer has an iron loss of 400 W and full-load copper loss of 800 W.

(*i*) Find the load at which maximum efficiency occurs and the value of maximum efficiency at unity power factor.

(*ii*) If the maximum efficiency occurs at 80% of full-load, find the new core loss and full-load copper loss assuming that total full-load loss is constant.

[**Ans.** (*i*) 70.7% of full-load, 97.79% ; (*ii*) P_i = 468.36 W, P_c = 731.64 W]

37. A 1100/220 V transformer has a primary resistance of 0.25 Ω and secondary resistance of 0.03 Ω. If iron loss amounts to 250 W, determine the secondary current at which maximum efficiency occurs and find the maximum efficiency at 0.8 power factor. [**Ans.** 111.8 A ; 97.52%]

38. A single-phase 250 kVA transformer has an efficiency of 96% on full-load at 0.8 power factor and on half full-load at 0.8 power factor. Find :

(*i*) Iron loss, and (*ii*) full-load copper loss.

[**Ans.** P_i = 2.767 kW, P_c = 5.533 kW]

39. The efficiency of a 300 kVA, single-phase transformer is 97.8% when delivering full-load at 0.8 power factor lagging and 98.4% when delivering half full-load at unity power factor. Determine the efficiency at 80% of full-load at 0.8 power factor lagging. [**Ans.** 97.95%]

40. A 12 kVA, 400/200 V single-phase 50-Hz transformer has maximum efficiency of 95% at 85% of full-load at unity power factor. Determine the efficiency at full-load at 0.8 power factor lagging. [**Ans.** 93.76%]

41. A 50 kVA, 2000/250 V single-phase transformer has resistances of 1.1 Ω and 0.015 Ω and reactances of 4.4 Ω and 0.06 Ω for the high voltage and low voltage windings respectively. It has a maximum efficiency of 96.05% at 80% of full-load at unity p.f. The magnetising current for the h.v. side at 2000 V is 1.25 A. Find the readings of suitable instruments for open circuit test and short-circuit tests, supply being given to h.v. side in both cases.

[**Ans.** O.C. test : 2000 V, 1.316 A, 824 W
S.C. test : 212.3 V, 25 A, 1287 W]

42. A 100 kVA, 3300/300 V single-phase transformer has an efficiency of 97.5% both on full-load at unity p.f. and on half full-load at unity power factor. The power factor of the no-load current is 0.3 lagging and the regulation on full-load at 0.8 p.f. lagging is 3 per cent. Draw the equivalent circuit referred to the l.v. side.

[**Ans.** R_{01} = 105.3 Ω, X_0 = 33.11 Ω,
R_{02} = 0.01538 Ω, X_{02} = 0.02451 Ω]

43. The maximum efficiency of a single-phase 240 kVA, 2000/250 V transformer occurs at 70% of full-load and is equal to 98% at 0.8 p.f. lagging. Determine the efficiency and regulation on full-load at 0.8 p.f. lagging if the impedance of the transformer is 8 per cent. [**Ans.** 97.81%, 5.699%]

44. A 5 kVA, 230/115 V transformer takes 1.2 A and 75 W when 230 V is applied to h.v. winding and l.v. winding is kept open. It takes 21.75 A and 150 W when the l.v. winding is short-circuited and 17.4 V is applied to the h.v. winding. Find :

(*i*) The no-load current as a fraction of full-load input current at 0.8 p.f. lagging ;

(*ii*) Percentage regulation on full-load at 0.8 p.f. lagging ; and

(*iii*) The load at which maximum efficiency occurs and the maximum efficiency at unity p.f.

[**Ans.** 5.162% of full-load input current ; 6.566% ; 70.7% of full-load, 95.94%]

45. The efficiency of a 20 kVA, 2000/200 V transformer is 96.8% at full-load at unity p.f. and 96% at 60% of a load at 0.8 p.f. Find the regulation at full-load at :

(*i*) 0.8 p.f. lagging ; and (*ii*) 0.8 p.f. leading if the impedance is 7 per cent.

[**Ans.** (*i*) 5.469%, (*ii*) – 2.385%]

46. A 100 kVA, 2000/200 V single-phase transformer takes a current of 50 A and 2400 W at 100 V when the low voltage winding is short circuited. Determine the load-voltage and percentage regulation when delivering full-load current at 0.8 p.f. leading, the supply voltage being 2000 V.

[**Ans.** 200.142 V ; – 0.071%]

47. A single-phase, 25 kVA transformer has an iron loss of 240 W and full-load copper loss of 600 W.

(*i*) Find the load at which maximum efficiency occurs and maximum efficiency at 0.8 p.f.

(*ii*) If the maximum efficiency occurs at 70% of full-load, find the core loss and copper loss assuming the total loss to be the same as in the previous case.

[**Ans.** (*i*) 63.25% of full-load, 96.36% ;
(*ii*) 276.3 W, 563.7 W]

All-Day Efficiency

48. **A 20 kVA, 2000/200 V transformer has an iron loss of 300 W and full-load copper loss of 400 W. During the day it is loaded as follows :**

No. of hours	*Load*	*Power factor*
8	$\frac{1}{4}$ load	0.5
6	Full-load	0.8
6	$\frac{3}{4}$ load	unity
4	No-load	—

Find the all-day efficiency. **[Ans.** 94.84%**]**

49. **A 30 kVA transformer has got a maximum efficiency of 97% at 80% of load at unity p.f. During the day it is loaded as follows :**

No. of hours	*Load*	*Power factor*
10	4 kW	0.6 lag
9	20 kW	0.8 lag
5	24 kW	0.85 lag

Find the all-day efficiency. **[Ans.** 95.68%**]**

50. **The all-day efficiency of 200 kVA transformer is 96% when it is loaded as follows :**

No. of hours	*Load*	*Power factor*
12	120 kW	0.8
8	150 kW	unity
4	No-load	—

If maximum efficiency of this transformer occurs at 80% of full-load, find iron loss and full-load copper loss. **[Ans.** P_i = 2.645 kW, P_c = 4.133 kW**]**

51. **An auto transformer supplies a load of 5 kW at 125 V and at unity power factor. If the primary voltage is 250 V, determine :**

(*i*) Transformer ratio, **(*ii*) Secondary current,**
(*iii*) Primary current,
(*iv*) Number of turns across secondary if the total number of turns is 250,
(*v*) Power transformed, and
(*vi*) Power conducted directly from the supply mains to load. **[Ans.** $\frac{1}{2}$, 40 A, 20 A, 125, 9.5 kW, 2.5 kW**]**

52. **The primary and secondary voltages of an auto-transformer are 500 V and 400 V respectively. Show with the aid of diagram the current distribution in the windings when the secondary current is 100 A and calculate the economy in copper.** **[Ans.** 80 per cent**]**

53. **A 200/250 V auto-transformer draws power from a 200 V line and supplies a 5 kW load with a power factor of 0.8 lagging. A second load of 2 kW is supplied at unity power factor from 100 V winding. Neglecting losses, calculate the current drawn by the transformer from the 200 V line and its power factor.** **[Ans.** 42.64 A, 0.898**]**

54. **A 5 kVA, single-phase 50-Hz transformer has full-load efficiency of 95 per cent and an iron loss of 50 W. The transformer is now connected as an auto-transformer to a 220 V supply. If it delivers a 5 kW load at unity power factor to a 110 V circuit, calculate the efficiency of the operation and the current drawn by the high-voltage side.** **[Ans.** 79.27%, 23.2 A**]**

55. **A 480/120 V, 5 kVA two winding transformer is to be used as an auto-transformer to supply power at 480 V from 600 V source. Draw the connection diagram and determine the kVA capacity as an auto-transformer.** **[Ans.** 25 kVA**]**

6

Polyphase Induction Motors

1. GENERAL ASPECTS

Introduction. An induction motor is simply an *electric transformer* whose magnetic circuit is separated by an air gap into two relatively movable portions, one carrying the primary and the other the secondary winding. Alternating current supplied to the primary winding from an electric power system induces an opposing current in the secondary winding, when *latter is short-circuited* or *closed through an external impedance*. Relative motion between the primary and secondary structures is produced by the *electromagnetic forces corresponding to the power thus transferred across the air gap by induction*.

The *essential feature which distinguishes the induction machine from other types of electric motors is that the secondary currents are created solely by induction, as in a transformer instead of being supplied by a D.C. exciter or other external power source, as in synchronous and D.C. machines.*

Advantages. Three-phase induction motor is the *most commonly used motor* in industrial applications because of the *advantages* listed below :

1. Simple design
2. Rugged construction
3. Reliable operation
4. Low initial cost
5. Easy operation and simple maintenance
6. High efficiency
7. Simple control gear for starting and speed control.

Applications :

Induction motors are available with torque characteristics suitable for a *wide variety of loads* :

(*i*) The standard motor has a starting torque of about 120 to 150 per cent of full-load torque. Such motors are suitable for most applications.

(*ii*) For starting loads such as small refrigerating machines or plunger pumps operating against full pressure or belt conveyors, *high torque motors with a starting torque of twice normal full-load torque*, or more, are used.

(*iii*) For driving machines that use large flywheels to carry peak loads, such as punch presses and shears, a high-torque motor with a slip at full-load up to 10 per cent is available. *The high slip permits enough change in speed to make possible the proper functioning of the flywheel.*

(*iv*) By the use of a wound-rotor with suitable controller and external resistances connected in series with the rotor winding, it is possible to obtain any value of starting torque up to the maximum breakdown torque. *Such motors are well adapted as constant-speed drives for loads that have large friction loads to overcome at starting.*

2. CLASSIFICATION OF A.C. MOTORS

Different A.C. motors may be classified as follows :

1. According to the 'type of current' :

(*i*) Single-phase (*ii*) Three-phase.

2. According to 'speed' :

(*i*) Constant speed (*ii*) Variable speed

(*iii*) Adjustable speed.

3. According to 'principle of operation' :

(A) **Synchronous motors**

(*i*) Plain (*ii*) Super.

(B) **Asynchronous motors**

(*a*) Induction motors :

(*i*) Squirrel cage

—single

—double

(*ii*) Slip-ring (external resistance)

(*b*) Commutator motors :

(*i*) Series

—single-phase

—universal

(*ii*) Compensated

—conductively

—inductively.

4. According to 'structural features' :

(*i*) Open (*ii*) Enclosed

(*iii*) Semi-enclosed (*iv*) Ventilated

(*v*) Pipe-ventilated (*vi*) Riveted frame eye

(*vii*) Splash proof (*viii*) Totally enclosed fan-cooled

(*ix*) Explosion proof (*x*) Water proof.

3. CONSTRUCTIONAL DETAILS

3.1. The Stator

- The stator frame consists of a symmetrical and substantial casting, having feet cast integral with it. The stator core, consisting of high grade, low loss electrical sheet-steel stampings, is assembled in the frame under hydraulic pressure. The thickness of stampings/laminations is usually from 0.35 to 0.6 mm. The stator laminations are punched in one piece for small induction motor (Fig. 1). In induction machines of large size the stator core is assembled from a large number of *segmental laminations.*

The slots are *sometimes of the 'open type'* (*i.e.*, having parallel walls) for the accommodation of former wound coils. But the usual practice is to have practically *'enclosed slots' in order to reduce the effective* length of air-gap.

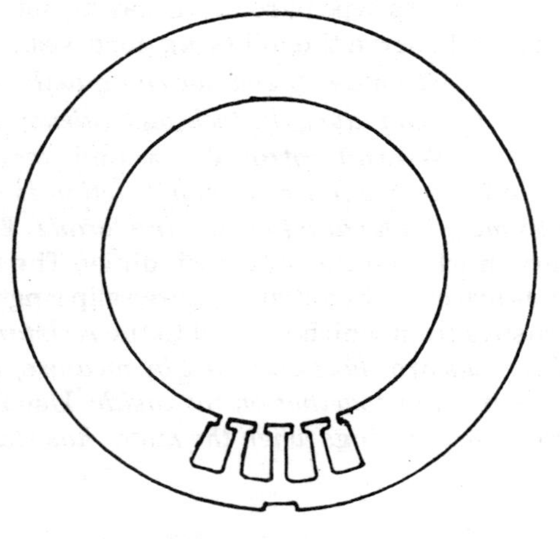

Fig. 1. Stator stamping.

- The *stator windings* are given the utmost care to make them mechanically and electrically sound, so as to ensure long life and high efficiency. After the winding is in position it is thoroughly dried out whilst still hot and is completely immersed in a high grade synthetic resin varnish. It is then *acid, alkali, moisture* and *oil proof.*

  For *small motors* working at ordinary voltages, *single layer mush winding* is used. For *medium size machines double layer lap winding with diamond shaped coils* is used. *Single layer concentric windings are used for large motors working at high voltages.*

- *Frames* of electrical machines house the stator core. Frames of small and medium sizes of induction motors have hollow cylindrical form and that of larger motors have the shape of a circular box. *In small induction motors,* having a frame diameter of up to about 150 cm, *the frame also supports the end shields.* The frame should be *strong and rigid* as rigidity is very important in the case of induction motors of large dimensions. *This is because the length of the air gap is very small and if the frame is not rigid, it would create an irregular air gap around the machine resulting in production of unbalanced magnetic pull.* Frames for small machines are made as a *single unit* and are usually cast. The frames of *medium and large sized* machines are *fabricated from rolled steel plates.*

3.2. The Rotor. The rotors are of two types :

1. Squirrel-cage ;
2. Wound rotor.

1. **Squirrel-cage.** The squirrel-cage rotor is made up of stampings (Fig. 2) which are keyed directly to the shaft. The slots are partially closed and the winding consists of embedded copper bars to which the short-circuited rings are brazed. The squirrel cage rotor is so robust that it is almost indestructible.

The great majority of present day induction motors are manufactured with squirrel-cage rotors, a *common practice being to employ winding of cast aluminium.* In this construction the assembled rotor laminations are placed in a mould after which molten aluminium is forced in, under pressure, to form bars, end rings and cooling fans as extension of end rings. This is known as *die cast rotor* and has become very popular as there are no joints and thus there is no possibility of high contact resistance.

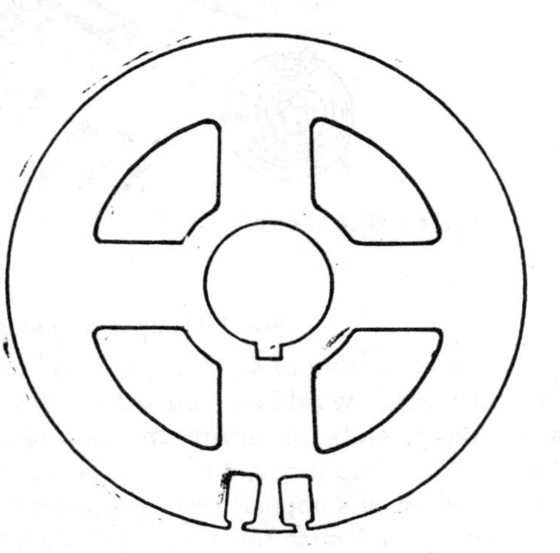

Fig. 2. Rotor stamping.

In this type of rotor, it may be noted that *slots are not made parallel to the shaft but they are 'skewed'* to serve the following purposes :

(*i*) *To make the motor run quietly by reducing the magnetic hum.*

(*ii*) *To reduce the locking tendency of the rotor.*

2. **Wound rotor.** The wound rotor has also slotted stampings and the *windings are former wound.* The *wound rotor construction is employed for induction motors requiring speed control or extremely high values of starting torque.* The wound rotor has completely insulated copper windings very much like the stator windings. The windings can be connected in star or delta and the three ends are brought out at the three slip rings. The current is collected from these slip rings with carbon brushes from which it is led to the *resistances for starting purposes.* When the motor is *running, the slip rings are short-circuited by means of a collar, which is pushed along the shaft and connects all the slip rings together on the inside.* Usually the *brushes* are *provided with a device for lifting them from the slip rings when the motor has started up, thus reducing the wear and the frictional losses.*

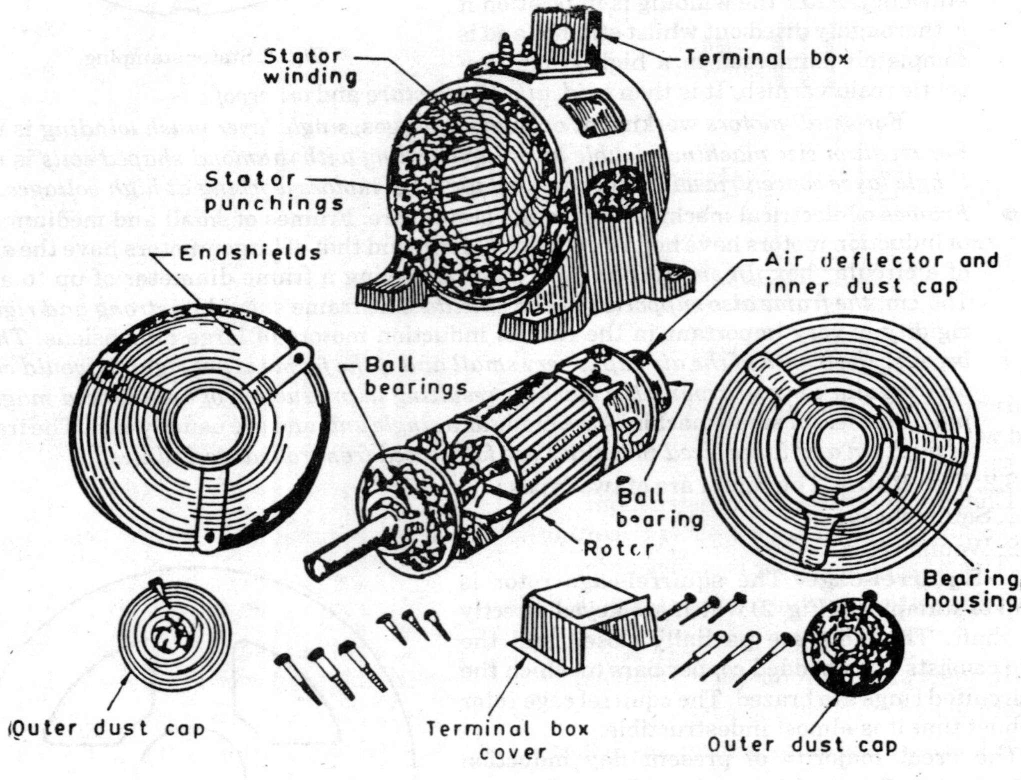

Fig. 3. Component parts of a small squirrel-cage induction motor.

The *number of slots in the rotor should never be equal to the number of slots in the stator.* If they are, there would be a *variation of reluctance of the magnetic path* from maximum, when teeth are opposite slots, to minimum when teeth are opposite teeth. The resulting flux pulsations would have a high frequency, since the periodic time would be the interval period for a tooth to occupy similar positions opposite two successive teeth. This will *not only cause extra iron loss but the rotor will tend to lock with the stator* if at the time of starting teeth are opposite teeth. The best plan is to make the number of the stator and the rotor teeth *prime to each other.*

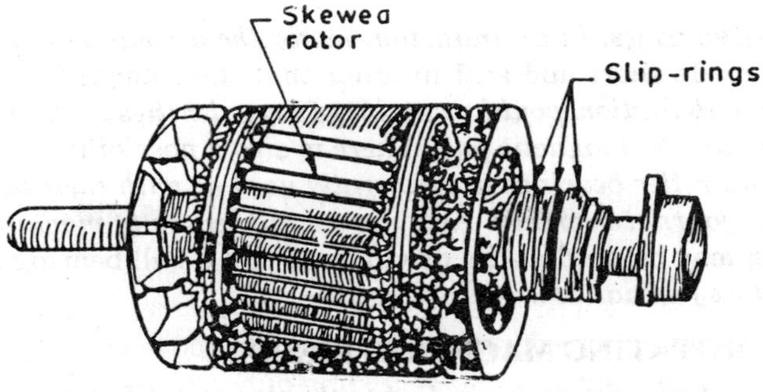

Fig. 4. Induction motor with phase-wound rotor, showing the three slip rings on the rotor shaft.

Figs. 5 and 6 show squirrel-cage and phase-wound induction motors respectively.

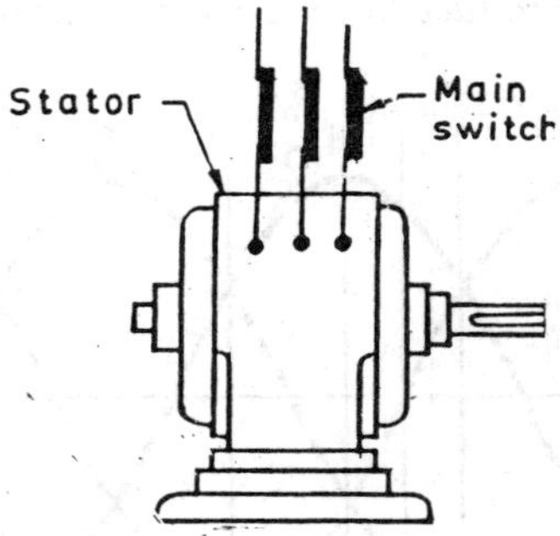

Fig. 5. Squirrel-cage motor.

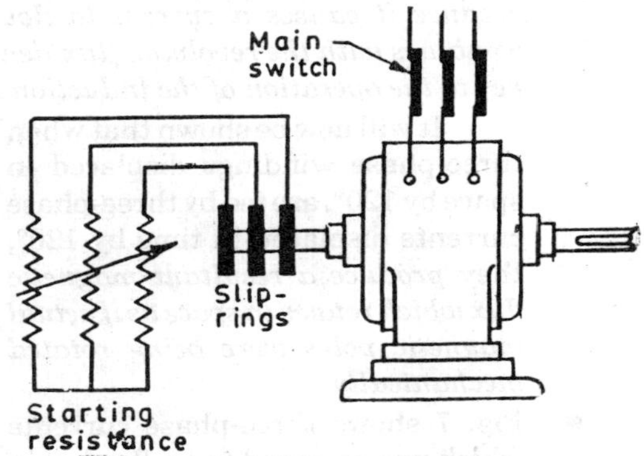

Fig. 6. Phase-wound motor connected to a three-phase star-connected starting resistance.

Advantages of a squirrel-cage motor over a phase-wound induction motor. As compared with a wound rotor a squirrel-cage induction motor entails the following *advantages* :

1. Slightly higher efficiency.

2. Cheaper and rugged in construction.

3. No slip rings, brush gear, short-circuiting devices, rotor terminals for starting rheostats are required. The star-delta starter is sufficient for starting.

4. It has better space factor for rotor slots, a shorter overhang and consequently a small copper loss.

5. It has a smaller rotor overhang leakage which gives a better power factor and a greater pull out torque and overload capacity.

6. It has bare end rings, a large space for fans and thus the cooling conditions are better.

The major *'disadvantage' of squirrel-cage motor* is that *it is not possible to insert resistance in the rotor circuit for the purpose of increasing the starting torque. The cage rotor has a smaller starting torque and large starting current as compared with wound rotor*.

3.3. Slip Rings. The slip rings for wound rotor machines are made of *either brass or phosphor bronze.* They are shrunk on to a cast iron sleeve with moulded silica insulation. This assembly is passed on to the rotor shaft. The slip rings are rotated either between the core and the bearing or on the shaft extension. In the latter case the shaft is made hollow to allow the three connections from rotor to slip rings to pass through bearings.

3.4. Shaft and Bearings. *In an induction motor the air gap is made as small as possible.* Therefore the shaft is made short and stiff in order that the rotor may not have any significant deflection, *as even a small deflection would create large irregularities in the air gap which would lead to production of an unbalanced magnetic pull. There is also a possibility of rotor and stator fouling with each other. Ball and roller bearings are generally used as with their use, accurate centering is much simpler than with journal bearings.* Also the *overall length of machine is reduced. For small motors,* a roller bearing may be used at the driving end and a ball bearing at the non-driving end. *For large and heavy rotors journal bearings are used.*

4. PRODUCTION OF ROTATING MAGNETIC FIELD

- *When a three-phase voltage is applied to the three-phase stator winding of the induction motor a rotating magnetic field is produced, which by transformer action induces a 'working' e.m.f. in the rotor winding. The rotor-induced e.m.f. is called a working e.m.f. because it causes a current to flow through the armature winding conductors. This combines with the revolving flux-density wave to produce torque. Thus revolving field is a key to the operation of the induction motor.*

  It will now be shown that when three-phase windings displaced in space by 120°, are fed by three-phase currents displaced in time by 120°, they *produce a resultant magnetic flux which rotates in space as if actual magnetic poles were being rotated mechanically.*

- Fig. 7 shows three-phase currents which are assumed to be flowing in phases '*l*', '*m*' and '*n*' respectively. These currents are time-displaced by the 120 electrical degrees.

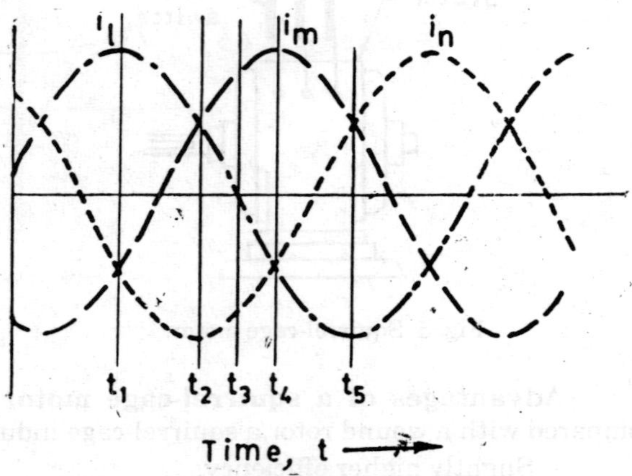

Fig. 7. Balanced three-phase alternating currents.

- Fig. 8 shows the stator structure and the three-phase winding. Each phase (normally distributed over 60 electrical degrees) for convenience is represented by a single coil. Thus coil *l-l'* represents the entire phase '*l*' winding having its flux axis directed along the vertical. This means that whenever phase '*l*' carries current it produces a flux field directed along the vertical-up or down. The right-hand flux rule readily verifies this statement. Similarly, the flux axis of phase '*m*' is 120 electrical degrees displaced from phase '*l*' ; and that of phase '*n*' is 120 electrical degrees displaced from phase '*m*'. The unprimed letters refer to the beginning terminals of each phase.

- **Magnitude and direction of the resultant flux corresponding to time instant t_1.** (Fig. 7). At this instant the current into phase '*l*' is at its *positive maximum* value while currents in phases '*m*' and '*n*' are at *one-half their maximum values.* In Fig. 8 it is arbitrarily assumed that when current in a given phase is positive, it flows into the paper with respect to the unprimed conductors. Thus since, at time t_1, i_l is positive, a cross is used for conductor '*l*' [Fig. 8 (*i*)]. Of course a dot is used for *l'* because it refers to the return connection. Then by the right-hand rule it follows that phase '*l*' produces a flux contribution directed downward along the vertical. Moreover, the magnitude of this contribution is the maximum value because the current is at maximum. Hence, $\phi_l = \phi_{max}$, where ϕ_{max} is the maximum flux per pole of phase '*l*'. It is important to understand that phase '*l*' really

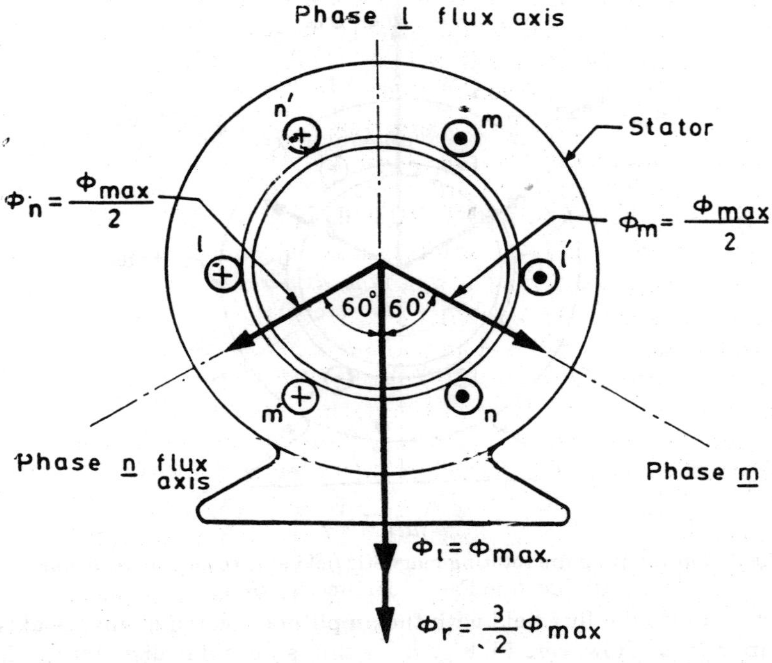
Phase l flux axis
Stator
$\Phi_n = \frac{\Phi_{max}}{2}$
$\Phi_m = \frac{\Phi_{max}}{2}$
n'
m
l
l'
m'
n
60°
60°
Phase n flux axis
Phase m flux axis
$\Phi_l = \Phi_{max}$
$\Phi_r = \frac{3}{2}\Phi_{max}$

(*i*)

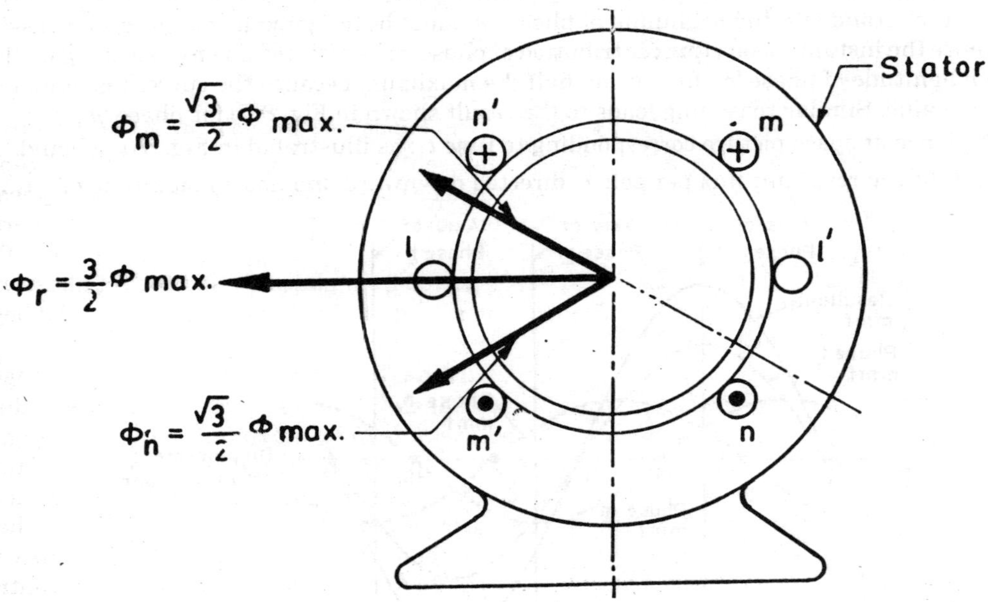
Stator
$\Phi_m = \frac{\sqrt{3}}{2}\Phi$ max.
$\Phi_r = \frac{3}{2}\Phi$ max.
$\Phi_n = \frac{\sqrt{3}}{2}\Phi$ max.
n'
m
l
l'
m'
n

(*ii*)

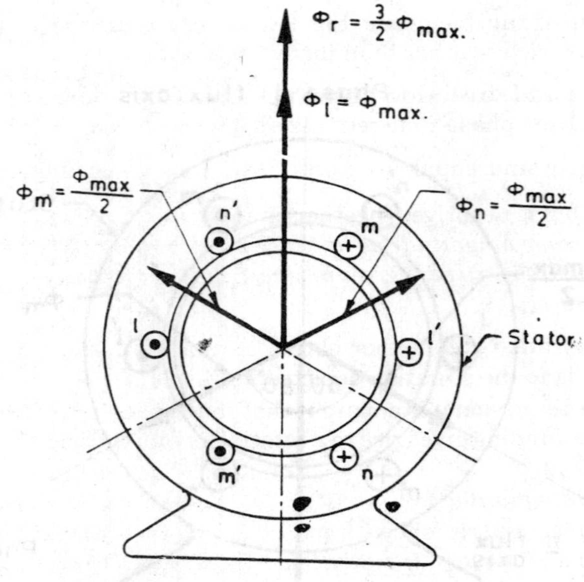

(iii)

Fig. 8. Representing the rotating magnetic field at three instants of time : (i) time t_1 in Fig. 7 ; (ii) time t_3 ; (iii) t_5.

produces a sinusoidal flux field with the amplitude located along the axis of phase '*l*' as shown in Fig. 9. However, in Fig. 8 (*i*) this sinusoidal distribution is conveniently represented by the vector ϕ_l.

For determining the magnitude and direction of the field contribution of phase '*m*' at time t_1, we first find that the current in phase '*m*' is negative with respect to that in phase '*l*'. Hence the conductor that stands for the beginning of phase '*m*' must be assigned a dot while '*m*' is assigned a cross. Hence the instantaneous flux contribution of phase '*m*' is directed downward along its flux axis and the magnitude of phase '*m*' flux is one-half the maximum because the current is at one-half its maximum value. Similar reasoning leads to the result shown in Fig. 8 (*i*) for phase '*n*'.

A glance at space picture corresponding to time t_1, as illustrated in Fig. 8 (*i*), should make it apparent that the resultant flux per pole is directed downward and has a magnitude of $\frac{1}{2}$ times the

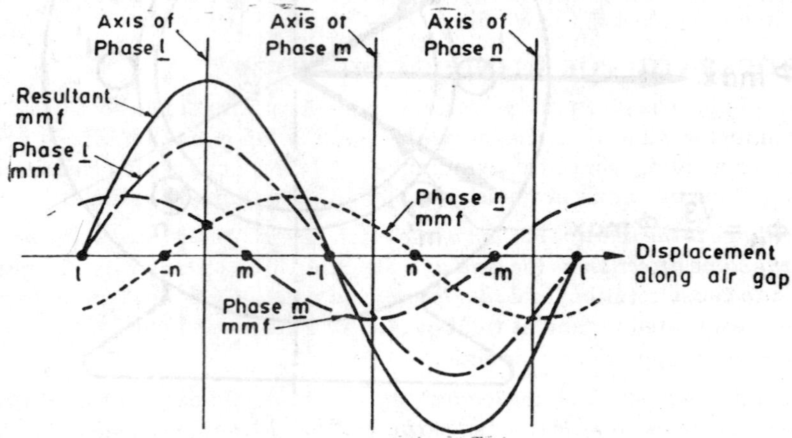

Fig. 9. Component and resultant field distributions corresponding to t_1 (Fig. 7).

maximum flux per pole of any one phase. Fig. 9 depicts the same results as Fig. 8 (*i*) but does so in terms of sinusoidal flux waves rather than flux vectors.

- **Magnitude and direction of the resultant flux corresponding to time instant t_3** [Fig. 8 (*ii*)]. Here phase '*l*' current is zero, yields no flux contribution. The current in phase '*m*' is positive and equal to $\frac{\sqrt{3}}{2}$ its maximum value. Phase '*n*' has the same current magnitude but is negative. Together phases '*m*' and '*n*' *combine to produce a resultant flux having the same magnitude as at time t_1*. See Fig. 8 (*ii*). It is important to note, too, that an *elapse of 90 electrical degrees in time results in rotation of the magnetic flux field of 90 electrical degrees.*
- **At time instant t_5.** A further elapse of time equivalent to an additional 90 electrical degrees leads to the situation depicted in Fig. 8 (*iii*).

From the above discussion it is apparent that the application of three-phase currents through a balanced three-phase winding gives rise to a rotating magnetic field that exhibits the following *two characteristics.*

1. It is of *constant amplitude.*
2. It is of *constant speed.* It follows from the fact that the resultant flux traverses through 2π electrical radians in space for every 2π electrical radians of variation in time for the phases currents. Hence for a 2-pole machines where electrical and mechanical degrees are identical, each cycle of variation of current produces one complete revolution of the flux field. Hence *this is a fixed relationship which is dependent upon the frequency of the currents* and number of poles for which the three-phase winding is designated. In the case where the winding is designed for *four* poles it requires *two cycles* of variation of the current to produce one revolution of the flux field. Therefore it follows that for a *p*-pole machine the relationship is :

$$N_s = \frac{120f}{p} \qquad ...(1)$$

or

$$f = \frac{pN_s}{120} \qquad ...[1\,(a)]$$

N_s (r.p.m.) is called the *synchronous speed* and all synchronous machines run at their respective synchronous speeds.

It may be noted that the speed of rotation of the field as described by eq. (1) is always given relative to the phase windings carrying the time-varying currents. Accordingly, if a situation arises where the *winding is itself revolving*, then the speed of rotation of the field relative to inertial force is different from it with respect to the winding.

5. THEORY OF OPERATION OF AN INDUCTION MOTOR

When a three-phase is given to the stator winding a rotating field is set-up. This field sweeps past the rotor (conductors) and by *virtue of relative motion,* an e.m.f. is induced in the conductors which form the rotor winding. Since this winding is in the form of a closed circuit, a current flows, the direction of which is, by Lenz's law, such as to oppose the change causing it.

Now, *the change is the relative motion of the rotating field and the rotor, so that, to oppose this, the rotor runs in the same direction as the field and* attempts to catch up with it. It is clear that torque must be produced to cause rotation, and *this torque is due to the fact that currents flow in the rotor conductors* which are situated in, and at right angles to, a magnetic field.

Fig. 10 shows the induction motor action.

- When the motor shaft is *not loaded*, the machine has only to rotate itself against the mechanical losses and the rotor speed is *very close to the synchronous speed.* However, the

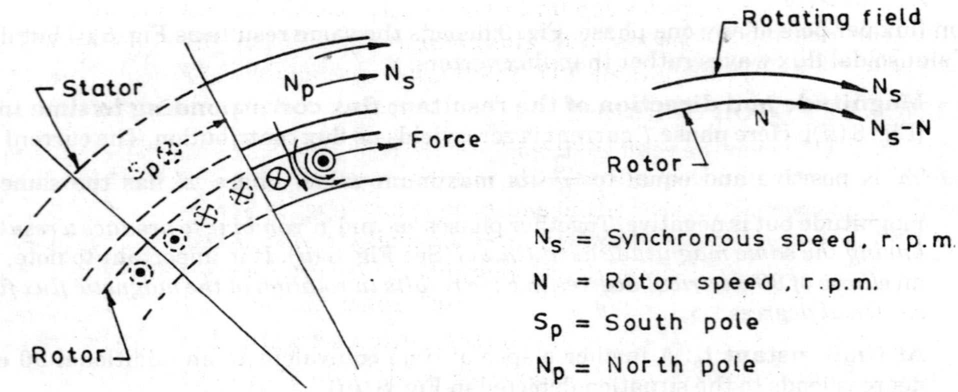

Fig. 10. Induction motor action.

rotor speed cannot become equal to the synchronous speed because if it does so, *the e.m.f. induced in the rotor winding would become zero and there will be no torque.* Hence the speed remains slightly less than the synchronous speed. If the motor shaft is *loaded,* the rotor will slow down and the relative speed of the rotor with respect to the stator rotating field will increase. The e.m.f. induced in the rotor winding will increase and will produce more rotor current which will increase the electromagnetic torque produced by the motor. Conditions of equilibrium are attained when the rotor speed has adjusted to a new value so that the electromagnetic torque is sufficient to balance the mechanical or load torque applied to the shaft. The speed of the motor *when running under full load conditions is somewhat less than the no-load speed.*

6. SLIP

- As earlier stated, *the rotor speed must always remain less than the synchronous speed. The difference between the synchronous speed and the rotor speed is known as* **'slip'.** It is usually expressed as a fraction of the synchronous speed. Thus slip s is

$$s = \frac{N_s - N}{N_s} \quad \text{...(2)}$$

or $$N = N_s(1 - s) \quad \text{...[2 (a)]}$$

where N_s = synchronous speed (r.p.m.)
N = motor speed (r.p.m.).

In practice the value of slip is very small. At no-load, slip is around 1% or so and at full-load it is around 3%. For large efficient machines the slip at full-load may be around 1% only. The induction motor, is therefore, a motor with substantially constant speed and fills the same role as *D.C. shunt motor*.

- When the rotor is stationary (standstill) its speed is zero and $s = 1$. The rotor cannot run at synchronous speed because then there will be no rotor e.m.f. and no rotor current and torque. If the rotor is to run at synchronous speed an external torque is necessary. *If the rotor is driven such that $N > N_s$, the slip becomes negative, the rotor torque opposes the external driving torque and the* **machine acts as induction generator.**

- The induction motor derives its name from the fact that the *current in the rotor circuit is induced from the stator.* There is no external connection to the rotor except for some special purposes.

If the rotor reactance at standstill is X_2 its value at slip 's' becomes sX_2. *This is very desirable,* for at no-load the reactance becomes almost negligible and the rotor impedance is now all resistance.

Further if the rotor resistance is small the rotor current is large, so that motor works with a large torque which brings the speed near to synchronous speed, *i.e.* the slip is reduced.

7. FREQUENCY OF ROTOR CURRENT

At standstill (*i.e.* when the rotor is stationary), the *frequency of the rotor current is the same as the supply frequency (f)*. But when the rotor starts revolving, then the frequency depends upon the relative speed or slip-speed. If f_r is the frequency of the rotor current, then

$$N_s - N = \frac{120 f_r}{p} \quad \ldots(i)$$

Also

$$N_s = \frac{120 f}{p} \quad \ldots(ii)$$

Dividing (*i*) by (*ii*), we get

$$\frac{N_s - N}{N_s} = \frac{f_r}{f} \quad \text{or} \quad s = \frac{f_r}{f}$$

or

$$f_r = sf \quad \ldots(3)$$

8. ROTOR E.M.F. AND ROTOR CURRENT

Rotor e.m.f :

When the rotor is stationary, an induction motor is equivalent to a 3-phase transformer with secondary short-circuited. Therefore, the induced e.m.f. per phase E_2 in the rotor at the instant of starting is given as :

$$E_2 = E_1 \times \frac{N_2}{N_1} \quad \ldots(4)$$

where E_1 = applied voltage per phase to primary *i.e.*, stator winding,
N_1 = number of stator turns, and
N_2 = number of rotor turns.

When the rotor starts gaining speed, the relative speed of the rotor with respect to stator flux *i.e.* slip, is decreased. Hence induced e.m.f. in the rotor, which is *directly proportional to the relative speed i.e. slip*, is also decreased and is given by sE_2. Hence *for slip 's', the induced e.m.f. in the rotor is s times the induced e.m.f. in the rotor at standstill.*

Rotor current :

Let R_2 = rotor resistance/phase,
L_2 = rotor inductance/phase, and
E_2 = induced e.m.f. of rotor/phase at standstill.

At standstill :

Induced e.m.f. of rotor/phase $= E_2$
Rotor winding resistance/phase $= R_2$
Rotor winding reactance/phase, $X_2 = 2\pi f L_2$ where f is the supply frequency
Rotor impedance/phase, $Z_2 = \sqrt{R_2^2 + X_2^2}$

$\therefore$ Rotor current/phase $= \dfrac{E_2}{Z_2} = \dfrac{E_2}{\sqrt{R_2^2 + X_2^2}}$.

At slip 's' :

Induced e.m.f. of rotor/phase $= sE_2$
Rotor winding resistance $= R_2$
Rotor winding reactance $= 2\pi f_r L_2 = 2\pi s f L_2 = s(2\pi f L_2) = sX_2$

Rotor winding impedance/phase $= \sqrt{R_2^2 + (sX_2)^2}$

$\therefore$ Rotor current/phase,
$$I_2 = \frac{sE_2}{\sqrt{R_2^2 + (sX_2)^2}}$$
$$= \frac{sE_2}{\sqrt{R_2^2 + s^2X_2^2}} = \frac{E_2}{\sqrt{(R_2/s)^2 + X_2^2}} \qquad ...(5)$$

The rotor currrent I_2 lags the rotor voltage E_2 by rotor power factor angle ϕ_2 given by
$$\phi_2 = \tan^{-1}\left(\frac{sX_2}{R}\right)$$

Power factor of rotor current,
$$\cos\phi_2 = \frac{R_2}{\sqrt{R_2^2 + s^2X_2^2}} = \frac{R_2/s}{\sqrt{(R_2/s)^2 + X_2^2}}. \qquad ...(6)$$

9. TORQUE AND POWER

The torque of an induction motor (being due to interaction of a rotor and stator fields),
$$T \propto \phi I_2 \cos\phi_2$$
where ϕ = flux of rotating stator,
I_2 = rotor current/phase, and
$\cos\phi_2$ = rotor power factor.

Since rotor e.m.f./phase at standstill, $E_2 \propto \phi$
$$\therefore \quad T \propto E_2 I_2 \cos\phi_2$$
or $T = kE_2I_2\cos\phi_2$ where k is any constant ...(7)

Substituting the value of I_2 and $\cos\phi_2$ from eqns. (5) and (6) in eqn. (7), we get
$$T = kE_2 \frac{sE_2}{\sqrt{R^2 + s^2X_2^2}} \times \frac{R_2}{\sqrt{R^2 + s^2X_2^2}}$$
i.e.,
$$T = \frac{ksR_2E_2^2}{R_2^2 + s^2X_2^2} \qquad ...(8)$$

9.1. Starting Torque. At start slip 's' = 1. Therefore, expression for starting torque may be obtained by putting $s = 1$ in eqn. 8.

Starting torque,
$$T_{st} = \frac{kR_2E_2^2}{R_2^2 + X_2^2}. \qquad ...(9)$$

9.2. Condition for Maximum Torque. The value of torque when motor is running is given by
$$T = \frac{ksR_2E_2^2}{R_2^2 + s^2X_2^2}$$

Torque will be maximum when,
$$\frac{sR_2}{R_2^2 + s^2X_2^2} \quad \text{or} \quad \frac{R_2}{\dfrac{R_2^2}{s} + sX_2^2}$$

or
$$\frac{R_2}{\left(\dfrac{R_2}{\sqrt{s}} - X_2\sqrt{s}\right)^2 + 2R_2X_2} \text{ is maximum, } viz., \; \frac{R_2}{\sqrt{s}} - X_2\sqrt{s} = 0$$

or $$s\ (= s_{mT}) = \frac{R_2}{X_2} \qquad ...(10)$$

(where s_{mT} = slip corresponding to maximum torque)

$\therefore$ Maximum torque, $$T_{max} = \frac{kE_2^2}{2X_2} \qquad ...(11)$$

From the above expression, the following conclusions can be drawn :

- Maximum torque is *independent* of rotor circuit resistance.
- Maximum torque varies *inversely as standstill reactance* of the rotor. Therefore to have maximum torque, standstill reactance (*i.e.* inductance) of the rotor should be kept as small as possible.
- The slip at which the maximum torque occurs depends upon the resistance of the rotor.

The condition for getting maximum torque at starting can be obtained by putting $s = 1$ in eqn. (10).

Thus, starting torque will be maximum if

$$\frac{R_2}{X_2} = s = 1 \quad \text{or} \quad R_2 = X_2.$$

9.3. Starting Torque of a Squirrel-Cage Motor. *The squirrel-cage rotor resistance is fixed* and *small* as compared to its reactance which is very large especially *at start* (because at standstill the frequency of rotor current is equal to that of supply frequency). Hence, the starting current I_2 of the rotor, though very large in magnitude, *lags by a very large angle* behind E_2 ; consequently the *starting torque per ampere is very poor.* It is roughly 1.5 times the full-load torque although the starting current is 5 to 7 times the full-load current. Thus such motors are *not suitable* for applications where these have to be *started against heavy loads.*

9.4. Starting Torque of a Slip Ring Motor. In a slip ring motor the torque is increased by *improving its power factor by adding external resistance in the rotor circuit from the star- connected rheostat* ; as the motor gains speed the rheostat resistance is gradually cut out. This additional resistance, however, increases the rotor impedance and so reduces the rotor current. At first, the effect of improved power factor *predominates the current-decreasing effect* of impedance, hence starting torque is increased. But after a certain point, the effect of increased impedance *predominates the effect of improved power factor and so the torque starts decreasing.*

9.5. Power. Eqn. (5) can be represented by a simple series circuit as shown in Fig. 11.

It is seen from this circuit that **per phase** power input (gross) to rotor,

$$P_g = E_2 I_2 \cos \phi_2$$

where $$\cos \phi_2 = \frac{R_2/s}{\sqrt{(R_2/s)^2 + X_2^2}}$$

$$\therefore \quad P_g = \frac{E_2}{\sqrt{(R_2/s)^2 + X_2^2}} \cdot I_2 \frac{R_2}{s} = I_2^2 \frac{R_2}{s} \qquad ...(12)$$

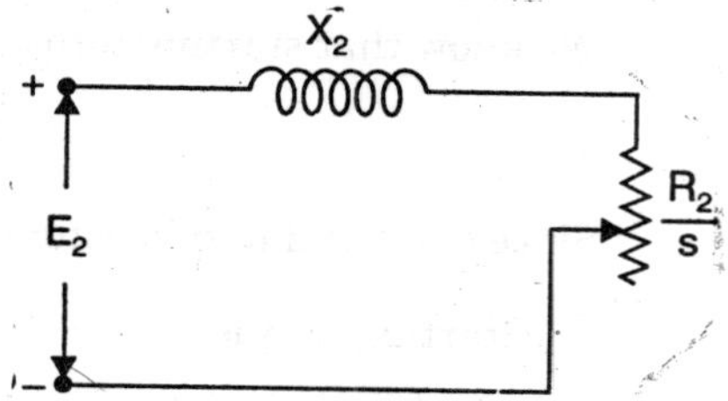

Fig. 11. Rotor equivalent circuit of an induction motor.

An examination of Fig. 11 also shows that per phase power input to rotor is equal to $I_2^2 \frac{R_2}{s}$ as the reactance is X_2 consumes no power.

P_g is the power transferred from stator to rotor across the air gap. In view of this, P_g is called the **air-gap power**. The expression for P_g may be written as

$$P_g = I_2^2 \frac{R_2}{s} = I_2^2 R_2 + I_2^2 R_2 \left(\frac{1-s}{s}\right)$$

= Rotor ohmic loss + internal mechanical power developed in rotor ($P_{mech.}$)

$$= sP_g + (1-s)P_g \quad ...(13\ a)$$

$$\therefore \quad P_{mech.} = (1-s)P_g = I_2^2 R_2 \left(\frac{1-s}{s}\right) \quad ...(13\ b)$$

$$\text{Rotor ohmic loss} = \left(\frac{s}{1-s}\right) P_{mech.} = sP_g \quad ...(13\ c)$$

Eqn. (12) reveals that ohmic loss

$$= I_2^2 R_2 = sP_g = s(\text{power input to rotor}) \quad ...(14)$$

Internal (or gross) torque developed per phase is given by

$$T_g = \frac{\text{internal mechanical power developed in rotor}}{\text{rotor speed in mechanical rad. per sec.}}$$

$$= \frac{P_{mech.}}{\omega_r} = \frac{(1-s)P_g}{(1-s)\omega_s} = \frac{P_g}{\omega_s} \quad ...(15)$$

Here $\omega_s \left(= \frac{2\pi N_s}{60}\right)$ is the *synchronous speed in mechanical radians per second.*

$$\text{Also} \quad T = \frac{P_g}{\omega_s} = \frac{1}{\omega_s} \times \frac{I_2^2 R_2}{s} = \frac{\text{Rotor ohmic loss}}{(\omega_s)\ \text{slip}} = \frac{1}{2\pi(N_s/60)} I_2^2 \frac{R_2}{s} \quad ...(16)$$

The power available at the shaft can be obtained from P_g as follows :

Output or shaft power, $P_{sh} = P_m$ – mechanical losses (friction, and windage losses)

or $\quad P_{sh} = P_g$ – rotor ohmic loss – friction and windage losses.

Output or shaft torque,

$$T_{sh} = \frac{P_{sh}}{\text{rotor speed}} = \frac{P_{sh}}{(1-s)\omega_s}$$

If stator input is known, then air-gap power P_g is given by

$$P_g = \text{stator power input} - \text{stator } I^2R \text{ loss} - \text{stator core loss.}$$

10. EFFECT OF CHANGE IN SUPPLY VOLTAGE ON STARTING TORQUE

We know that starting torque,

$$T_{st} = \frac{kR_2E_2^2}{R_2^2 + X_2^2}$$

Since e.m.f. induced in rotor (at standstill) $E_2 \propto \phi \propto V$

$$\therefore \quad \text{Starting torque,} \quad T_{st} = \frac{k'R_2V_2^2}{R_2^2 + X_2^2}.$$

where k' is another constant

or $\quad T_{st} \propto V^2$

i.e., Starting torque is proportional to the square of the applied voltage.

11. EFFECT OF CHANGE IN SUPPLY VOLTAGE ON TORQUE AND SLIP

The torque acting on the rotor when the motor is running with slip 's' is given by

$$T = \frac{ksR_2E_2^2}{R_2^2 + s^2X_2^2}$$

Since e.m.f. induced in rotor (at standstill), $E_2 \propto \phi \propto V$

$$\therefore \qquad T = \frac{k' s R_2 V^2}{R_2^2 + s^2 X_2^2}, \text{ where } k' \text{ is another constant}$$

Since the slip '*s*' at full-load is *very low*, therefore $s^2 X_2^2$ can be *neglected* in comparison with R_2^2.

$$\therefore \qquad T = \frac{k' s R_2 V^2}{R_2^2} = \frac{k' s V^2}{R_2} \quad \text{or} \quad T \propto sV^2$$

When the supply voltage is changed, it changes the torque under running condition also. With the decrease in supply voltage torque under running condition decreases, therefore, in order to *maintain the same torque, slip increases* or *speed decreases.*

12. FULL-LOAD TORQUE AND MAXIMUM TORQUE

Let, s_f = full-load slip of the motor, and

s_{mT} = slip corresponding to maximum torque $= \dfrac{R_2}{X_2}$

We know that,

Full-load torque, $$T_f = \frac{k s_f R_2 E_2^2}{R_2^2 + s_f^2 X^2}$$ [From Art. 9]

Maximum torque, $$T_m = \frac{k E_2^2}{2X_2}$$ [From Art. 9.2]

$$\therefore \qquad \frac{T_f}{T_m} = \frac{k s_f R_2 E_2^2 / (R_2^2 + s_f^2 X_2^2)}{k E_2^2 / (2X_2)} = \frac{2 s_f R_2 X_2}{R_2^2 + s_f^2 X_2^2}$$

$$= \frac{2 s_f R_2 X_2 / (X_2^2)}{(R_2^2 + s_f^2 X_2^2)/(X_2^2)}$$ [Dividing numerator and denominator by X_2^2]

$$= \frac{2 s_f (R_2/X_2)}{(R_2/X_2)^2 + s_f^2} = \frac{2 s_f s_{mT}}{s_{mT}^2 + s_f^2}$$

i.e., $$\frac{T_f}{T_m} = \frac{2 s_f s_{mT}}{s_f^2 + s_{mT}^2} \qquad \text{...(17)}$$

$$= \frac{2}{\dfrac{s_f}{s_{mT}} + \dfrac{s_{mT}}{s_f}} \qquad \text{...(17 a)}$$

[Dividing numerator and denominator by $s_f s_{mT}$]

13. STARTING TORQUE AND MAXIMUM TORQUE

$$\frac{T_{st}}{T_m} = \frac{k R_2 E_2^2 / (R_2^2 + X_2^2)}{k E_2^2 / (2X_2)} = \frac{2 R_2 X_2}{R_2^2 + X_2^2} = \frac{2(R_2/X_2)}{(R_2/X_2)^2 + 1}$$

(Dividing numerator and denominator by X_2^2)

or $$\frac{T_{st}}{T_m} = \frac{2 s_{mT}}{s^2_{mT} + 1} \qquad \text{...(18)}$$

14. TORQUE-SLIP AND TORQUE-SPEED CURVES

The expression for torque is as follows :

$$T = \frac{ksR_2E_2^2}{R_2^2 + s^2X_2^2}$$

From the above expression, it is evident, that

- Torque is zero when slip $s = 0$ (*i.e.*, speed is synchronous).
- When slip 's' *is very low* the value of the term sX_2 is very small and is negligible in comparison with R_2, therefore torque T *is approximately proportional* to slip 's' if rotor resistance R_2 is constant. This means that at speeds near to synchronous speed the *torque-speed* and *torque-slip* curves are approximately straight lines (Figs. 12 and 13).
- When the slip 's' increases (*i.e.* as the speed decreases with increase in load) torque increases and reaches its maximum value when $s = \frac{R_2}{X_2}$: The *maximum torque is also known as* **'pull-out'** or **'break-down' torque.**

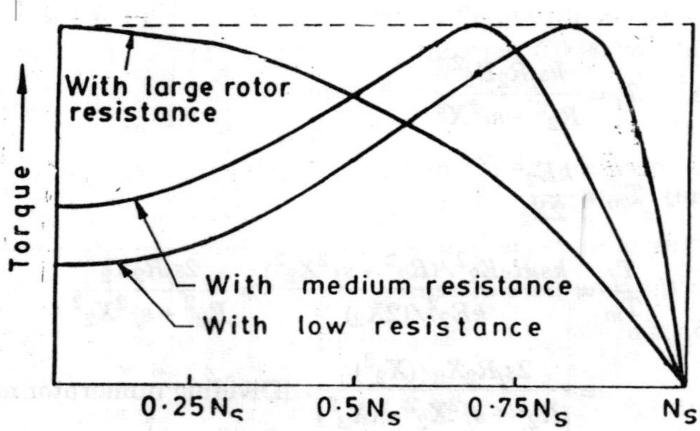

Fig. 12. Torque-speed curves.

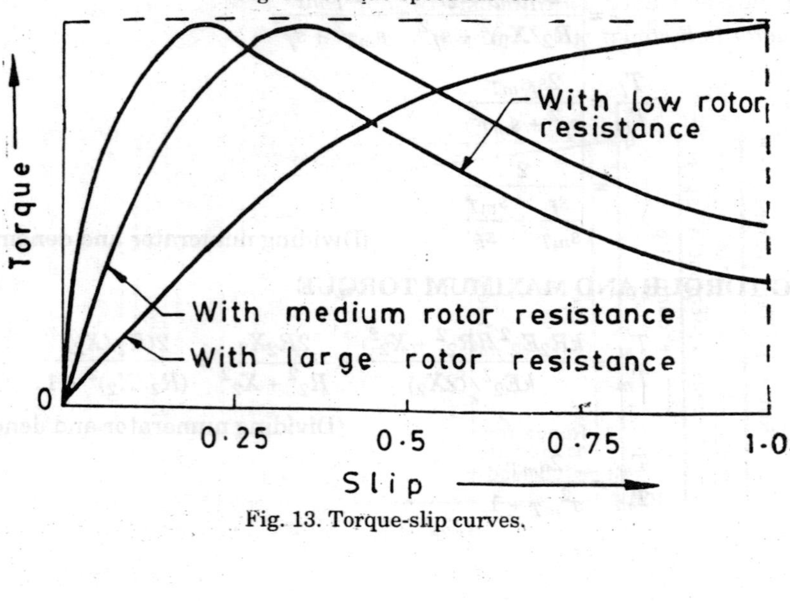

Fig. 13. Torque-slip curves.

- When the slip is further increased the torque decreases. The result is that *motor slows down* and *eventually stops. The motor operates for the value of slip between zero and that corresponding to maximum torque.*

With higher slip, R_2 becomes negligible as compared to sX_2 and torque varies as follows :

$$T \propto \frac{s}{s^2 X_2^2} \propto \frac{1}{s} \text{ if standstill reactance is constant.}$$

This means that *speed-torque* or *slip-torque* curves are rectangular hyperbola with the speed or slip beyond that corresponding to maximum torque. Figs. 12 and 13 show the *torque-speed* and *torque-slip* curves for different values of rotor resistance. It is observed that although maximum torque is independent of rotor resistance R_2, yet the exact location of T_m is dependent on it. *Greater the R_2, greater is the value of slip at which maximum torque occurs.*

15. OPERATING CHARACTERISTICS OF A 3-PHASE SQUIRREL-CAGE INDUCTION MOTOR

To start with, let us assume that the squirrel cage induction motor has been started without any load on it (*i.e.*, no electrical devices are connected to the generator, no metal is being cut by a machine tool, or the conveyor belt is completely disengaged). The motor will come to its no-load speed, which may be at a slip as low as 0.1 per cent. *Load is now placed on the motor* and the following reactions will be observed :

- *Speed decreases.*
- *Slip increases* (because the stator field continues to rotate at synchronous speed and therefore cuts the rotor bars more rapidly).
- As a consequence of increased cutting the *rotor induced e.m.f. increases.*
- *The current in the rotor bars increases.*
- *Torque developed by the rotor increases until it is sufficient to carry the load placed on the motor.*
- *The motor continues to rotate at that slip which will develop the required torque for that particular load.*

The motor may be loaded continuously till *pull-out (or break-down) torque* is developed, at which point the *motor will stop if more load is placed on it.* Pull-out occurs at a slip well beyond the

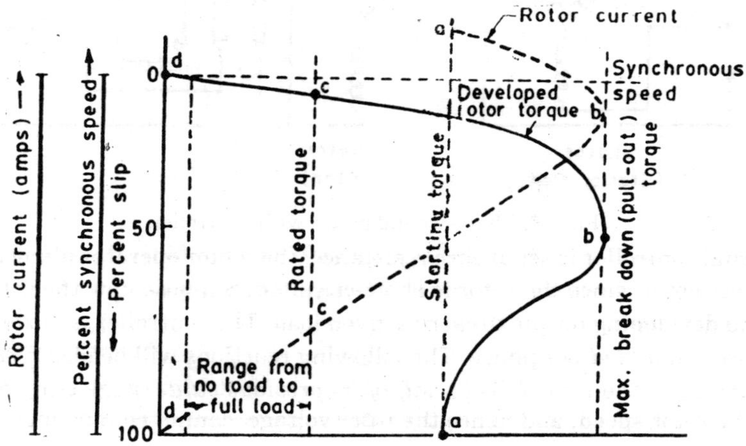

Fig. 14. Operating characteristics of a squirrel-cage induction motor.

normal range operation of the squirrel-cage induction motor. Even under normal operation, however, a suddenly applied load may require the temporary development of this maximum torque, after which the motor will speed up to its full-load value. Thus, in a manner similar to the shunt motor, the induction motor *inherently adjusts itself to the applied external torque.*

Fig. 14 shows typical operating characteristics of an induction motor.

Fig. 15 shows the effect of heavy loads on primary stator current and power factor.

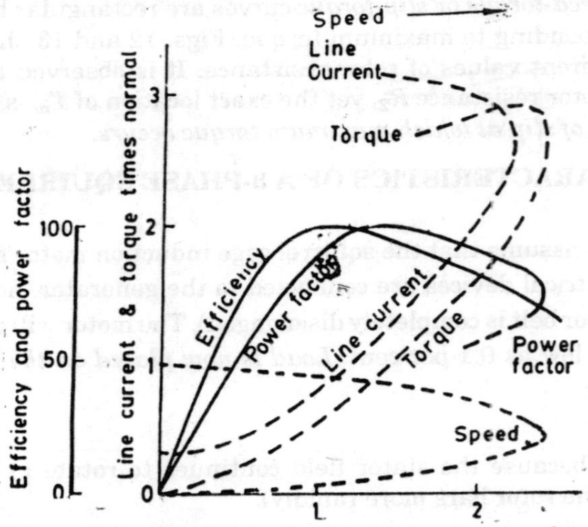

Fig. 15. Effect of heavy loads on primary stator current and power factor.

16. OPERATING CHARACTERISTICS OF A WOUND-ROTOR (SLIP RING) INDUCTION MOTOR

Fig. 16 shows a wound-rotor induction motor with controller (rheostat).

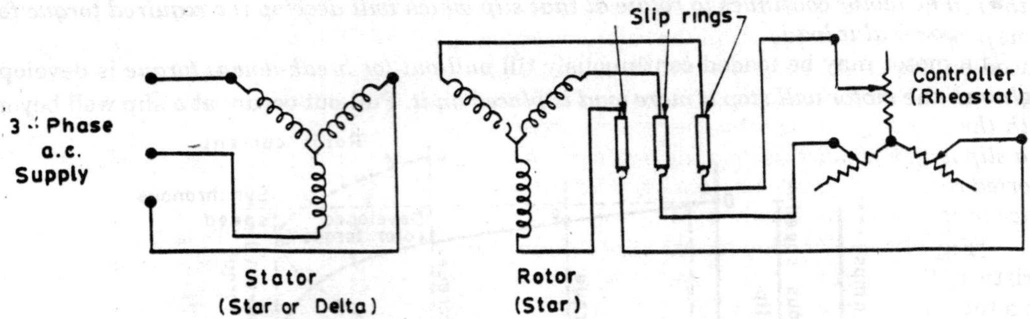

Fig. 16. Phase-wound motor with controller.

If the external controller is set at zero resistance, the motor operates almost exactly the same as the squirrel-cage motor, since the rotor is short-circuited. Suppose now the rotor is rotating at a constant speed and developing torque to carry a given load. The controller is moved so that the *rotor resistance per phase is doubled* per phase. The following reactions will be observed :

- The current in the rotor falls to *half of its previous value* (neglecting a transient effect), since the rotor speed, and hence the rotor voltage, cannot change instantaneously.
- The *developed torque decreases due to reduced rotor current.*

- The load on the motor has not changed, and with the external applied, or load, torque *now greater than the developed* torque, the *motor slows down.*
- The *slip increases* (increasing the rate at which the stator flux cuts the rotor conductors).
- The *rotor induced voltage increases.*
- The *rotor current now increases until it is sufficient to develop enough torque to again carry the load at a constant speed.*

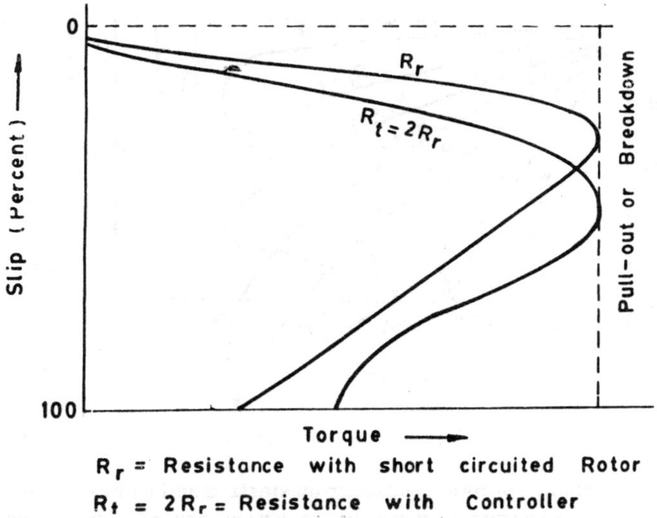

Fig. 17. Torque-slip characteristic of wound-rotor induction motor.

The induction motor now rotates at a lower speed, but with the same current as before, and develops the same torque as it did before the rotor circuit resistance was increased. The *frequency in the rotor is greater,* as is the rotor resistance. *Rotor resistance and reactance are both increased in same proportion,* however, and the *rotor power factor therefore remains the same.* If more load is now placed on the shaft, the rotor will develop the required torque, but always *at a lower speed i.e. a higher slip.* Eventually pull-out torque is attained. The *value of pull-out torque will be the same as with the short-circuited rotor, but it will occur at a slip which is twice as great as that with the shorted rotor, since the rotor circuit resistance is twice that of the rotor alone.*

Fig. 18 shows the torque-slip characteristic of the shorted rotor, together with that for a total resistance of, $R_t = 2R$.

It is worthnoting that the *torque developed at standstill is higher with increased rotor resistance.* This is explained by the fact that at standstill the rotor frequency is equal to the line frequency and rotor resistance is maximum. The addition of resistance in the rotor circuit obviously improves the power factor in that circuit while reducing the rotor current. Since standstill rotor resistance is much greater than

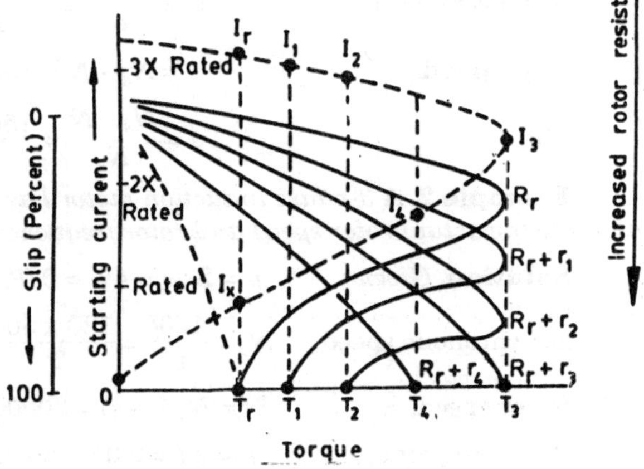

Fig. 18. Starting currents vs torque.

its resistance, the power factor increases more rapidly than the decrease of rotor current, and hence the developed torque is greater with added resistance.

Fig. 18 shows starting current vs torque.

Fig. 19 shows the effect of change of rotor resistance on starting and running characteristics.

In Figs. 18 and 19,

R_r = *basic rotor resistance*

r_1, r_2, r_3, r_4 = the resistance added to each phase of star-connected wound rotor respectively.

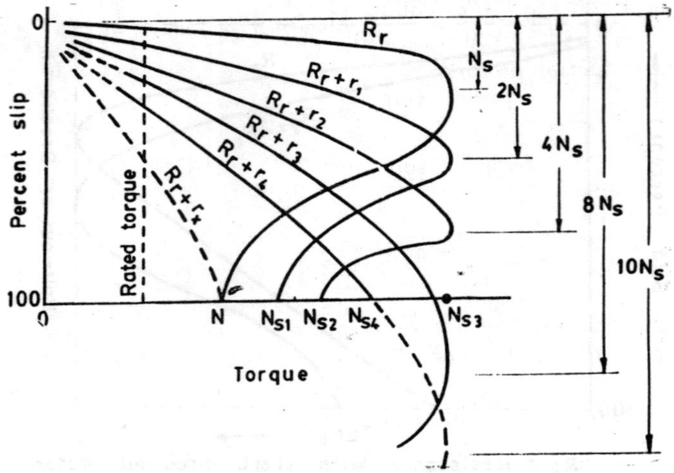

Fig. 19. Effect of change of rotor resistance on starting and running characteristics.

SLIP, STARTING TORQUE, MAXIMUM TORQUE

Example 1. *A 3-phase, 4-pole, 50-Hz induction motor is running at 1440 r.p.m. Determine the slip speed and slip.*

Solution. Number of poles, $p = 4$

Frequency, $f = 50$ Hz

Actual speed of rotor, $N = 1440$ r.p.m.

Slip speed = ?

Slip = ?

Synchronous speed, $N_s = \frac{120f}{p} = \frac{120 \times 50}{4} = 1500$ r.p.m.

Slip speed $= N_s - N = 1500 - 1440 = \mathbf{60}$ **r.p.m. (Ans.)**

Slip, $s = \frac{N_s - N}{N_s} = \frac{1500 - 1440}{1500} = \mathbf{0.04}$. **(Ans.)**

Example 2. *A 3-phase induction motor has 2 poles and is connected to 400 V, 50 Hz supply. Calculate the actual rotor speed and rotor frequency when the slip is 4%.* **(B.T.E. U.P., 1998)**

Solution. *Given* : $p = 2$; $s = 4\% = 0.04$; $f = 50$ Hz.

Synchronous speed, $N_s = \frac{120f}{p} = \frac{120 \times 50}{2} = 3000$ r.p.m.

Rotor speed, $\mathbf{N} = N_s(1 - s) = 3000(1 - 0.04) = \mathbf{2880}$ **r.p.m. (Ans.)**

Rotor frequency, $f_r = s.f = 0.04 \times 50 = \mathbf{2}$ **Hz. (Ans.)**

Example 3. *A 12-pole, 3-phase alternator is coupled to an engine running at 500 r.p.m. It supplies an induction motor which has a full speed of 1440 r.p.m.*

Find the percentage slip and number of poles of the motor. **(P.T.U., Jan., 2000)**

Solution. Poles of alternator, $p_a = 12$

Speed of the prime mover, $N_a = 500$ r.p.m.

Speed of induction motor, $N_m = 1440$ r.p.m.

Frequency of the supply generated by alternator,

$$f = \frac{N_a p_a}{120} = \frac{500 \times 12}{120} = 50 \text{ Hz}$$

No. of poles of the induction motor,

$$\mathbf{p_m} = \frac{120 f}{N_m} = \frac{120 \times 50}{1440} = 4.167 \text{ say } \mathbf{4.} \quad \textbf{(Ans.)}$$

$\therefore$ Synchronous speed, $N_s = \dfrac{120 f}{p_m} = \dfrac{120 \times 50}{4} = 1500$ r.p.m.

$\therefore$ Slip, $\mathbf{s} = \dfrac{N_s - N_m}{N_s} = \dfrac{1500 - 1440}{1500} = 0.04$ or **4%. (Ans.)**

Example 4. *A 3-phase alternator having 12-poles is driven at a speed of 500 r.p.m. It supplies power to an 8-pole, 3-phase induction motor. If the slip of the motor at full-load is 4%, calculate the full-load speed of the motor.*

Solution. Number of poles of the alternator = 12

Speed of alternator = 500 r.p.m.

Number of poles of induction motor = 8

Percentage slip = 4%

Full-load speed, N :

Supply frequency, $f = \dfrac{12 \times 500}{120} = 50$ Hz

Synchronous speed, $N_s = \dfrac{120 \times 50}{8} = 750$ r.p.m.

$\therefore$ % slip $= \dfrac{N_s - N}{N_s} \times 100$

$$4 = \frac{750 - N}{750} \times 100$$

$\therefore$ $N = 750 - \dfrac{4 \times 750}{100} = 720$ r.p.m.

Hence, **full-load speed, N = 720 r.p.m. (Ans.)**

☞**Example 5.** *A 3-phase, 6-pole, 50-Hz induction motor has a slip of 1% at no-load and 3% at full-load. Find :*

(i) Synchronous speed,
(ii) No-load speed,
(iii) Full-load speed,
(iv) Frequency of rotor current at standstill, and
(v) Frequency of rotor current at full-load.

Solution. Number of poles, $p = 6$

No-load slip, $s_0 = 1\%$

Full-load slip, $s_f = 3\%$

(*i*) **Synchronous speed,**

$$N_s = \frac{120f}{p} = \frac{120 \times 50}{6} = \mathbf{1000\ r.p.m.}\ \ \textbf{(Ans.)}$$

(*ii*) **No-load speed, N_0 :**

We know that, $s = \frac{N_s - N}{N_s}$ or $N = N_s(1 - s)$

or $\mathbf{N_0} = N_s(1 - s_0) = 1000\left(1 - \frac{1}{100}\right) = \mathbf{990\ r.p.m.}$ **(Ans.)**

(*iii*) **Full-load speed,**

$$N_f = N_s(1 - s_f) = 1000\left(1 - \frac{3}{100}\right) = \mathbf{970\ r.p.m.}\ \ \textbf{(Ans.)}$$

(*iv*) **Frequency of rotor current at standstill, f_r :**

At standstill, $s = 1$

$f_r = sf = 1 \times 50 =$ **50 Hz. (Ans.)**

(*v*) **Frequency of rotor current at full-load, f_r = ?**

$$f_r = s \times f = \frac{3}{100} \times 50 = \mathbf{1.5\ Hz.}\ \ \textbf{(Ans.)}$$

☞**Example 6.** *A 3-phase, 12-pole alternator is coupled to an engine running at 500 r.p.m. The alternator supplies an induction motor which has a full-load speed of 1455 r.p.m. Find the slip and number of poles of the motor.*

Solution. Number of poles of the alternator, $p_a = 12$

Speed of the engine, $N_e = 500$ r.p.m.

Full-load speed of the induction motor, $N_m = 1455$ r.p.m.

Slip, s = ?

Number of poles of the motor, p_m = ?

Supply frequency, $f = \frac{N_a p_a}{120} = \frac{500 \times 12}{120} = 50$ Hz.

When the supply frequency is 50-Hz, the synchronous speed can be 3000, 1500, 1000, 750 r.p.m. etc. Since the full-load speed is 1455 r.p.m. and the full-load slip is *always* less than 5%, the synchronous is 1500 r.p.m.

Slip, $s = \frac{N_s - N}{N_s} = \frac{1500 - 1455}{1500} =$ **0.03** or **3 per cent. (Ans.)**

Also, $N_s = \frac{120f}{p_m}$

$\therefore$ $p_m = \frac{120f}{N_s} = \frac{120 \times 50}{1500} = 4$ poles

Hence, **number of motor poles = 4. (Ans.)**

Example 7. *A 3-phase induction motor runs at almost 1000 rpm at no-load and 950 r.p.m. at full-load when supplied with power from a 50 Hz 3-phase line.*

(*i*) *How many poles has the motor ?*

(*ii*) *What is the percentage slip at full load ?*

(*iii*) *What is the corresponding frequency of rotor voltages ?*

(*iv*) *What is the corresponding speed of the rotor field with respect to rotor ?*

(*v*) *What is the corresponding speed of the rotor with respect to the stator ?*

(*vi*) *What is the corresponding speed of the rotor field with respect to the stator field ?*

(*vii*) *What is the rotor frequency at the slip of 10 per cent ?* **(AMIE, 1997)**

Solution. *Given* : $N_0 = 1000$ r.p.m., $N_f = 950$ r.p.m., $f = 50$ Hz

Since no-load speed of motor is almost 1000 r.p.m., hence synchronous speed near to 1000 r.p.m. is 1000 r.p.m.

(*i*) Poles on motor, $\mathbf{p} = \frac{120f}{N_s} = \frac{120 \times 50}{1000} = \mathbf{6.}$ **(Ans.)**

(*ii*) Percentage slip at full load, $\mathbf{s_f} = \frac{N_s - N_f}{N_s} \times 100 = \frac{1000 - 950}{1000} \times 100 = \mathbf{5\%}.$ **(Ans.)**

(*iii*) Frequency of rotor voltage, $f_r = s \times f = \frac{5}{100} \times 50 = \mathbf{2.5\ Hz.}$ **(Ans.)**

(*iv*) Speed of rotor field with respect to rotor $= \frac{120 \times f_r}{p} = \frac{120 \times 2.5}{6} = 50$ r.p.m. **(Ans.)**

(*v*) Speed of rotor with respect to stator = **950 r.p.m. (since stator is stationary). (Ans.)**

(*vi*) Rotor field and stator field are revolving at the same speed of 1000 r.p.m., therefore, *speed of rotor field w.r.t. stator field is* **zero. (Ans.)**

(*vii*) Rotor frequency at slip of 10%, $f_r = s' \times f = \frac{10}{100} \times 50 = \mathbf{5\ Hz.}$ **(Ans.)**

Example 8. *In a 6-pole, 3-phase, 50 Hz motor with star-connected rotor, the rotor resistance per phase is 0.3 Ω, the reactance at standstill is 1.5 Ω per phase, and an e.m.f. between the slip rings on open circuit is 175 V. Find :*

(*i*) *Slip at a speed of 950 r.p.m. ;*

(*ii*) *Rotor e.m.f. per phase ;*

(*iii*) *Rotor frequency and reactance at a speed of 950 r.p.m.* **(B.T.E. Pb., 1997)**

Solution. *Given* : $p = 6$; $f = 50$ Hz ; $R_2 = 0.3\ \Omega$, $X_2 = 1.5\ \Omega$; $N = 950$ r.p.m.

Synchronous speed, $N_s = \frac{120f}{p} = \frac{120 \times 50}{6} = 1000$ r.p.m.

(*i*) **Slip, s :**

$$\mathbf{s} = \frac{N_s - N}{N_s} = \frac{1000 - 950}{1000} = 0.05 \text{ or } \mathbf{5\%}. \textbf{ (Ans.)}$$

(*ii*) **Rotor e.m.f. per phase, E_2 :**

Rotor e.m.f. per phase at standstill $= \frac{E_2}{\sqrt{3}} = \frac{175}{\sqrt{3}} = 101$ V/phase

Rotor e.m.f. at 5% slip $= sE_2 = 0.05 \times 101 =$ **5.05 V/phase. (Ans.)**

(*iii*) **Rotor frequency and reactance at a speed of 950 r.p.m. :**

Rotor frequency, $\mathbf{f_r} = s \times f = 0.05 \times 50 =$ **2.5 Hz. (Ans.)**

Rotor reactance $= sX_2 = 0.05 \times 1.5 =$ **0.075 Ω/phase. (Ans.)**

Example 9. *A 50 Hz, 440 V, 3-phase, 4-pole induction motor develops half the rated torque at 1490 r.p.m. With the applied voltage magnitude remaining at the rated value, what should be its frequency if the motor has to develop the same torque at 1600 r.p.m. ? Neglect stator and rotor winding resistances, leakage reactances and iron losses.* **(GATE, 1995)**

Solution. *Given* : $f = 50$ Hz ; $p = 4$; $N = 1490$ r.p.m. ; New speed, $N_n = 1600$ r.p.m.

Synchronous speed, $N_s = \frac{120f}{p} = \frac{120 \times 50}{4} = 1500$ r.p.m.

Slip at a speed of 1490 r.p.m., $s = \frac{N_s - N}{N_s} = \frac{1500 - 1490}{1500} = 0.00667$

Since torque developed by an induction motor, $T \propto sV^2$ (Refer Art. 11)

Slip s for constant torque and constant applied voltage remains unchanged.

$\therefore$ New synchronous speed, $N_{sn} = \frac{N_n}{1-s} = \frac{1600}{1 - 0.00667} = 1610.7$ r.p.m.

New frequency, $f_n = \frac{p \times N_{sn}}{120} = \frac{4 \times 1610.7}{120} =$ **3.57 Hz. (Ans.)**

☞**Example 10.** *A 3-phase slip ring induction motor gives a reading of 60 V across slip rings on open circuit when at rest with normal stator voltage applied. The motor is star-connected and has impedance of (0.6 + j6) Ω per phase. Find the rotor current when the machine is : (i) at standstill with slip rings joined to a star-connected starter with a phase impedance of (5 + j4) Ω ; and (ii) running normally with a 4 per cent slip.*

Solution. Impedance of motor per phase = $(0.6 + j6)\ \Omega$

Phase impedance of star-connected starter = $(5 + j4)\ \Omega$

(*i*) **At standstill** :

Induced e.m.f. in the rotor winding per phase,

$$E_2 = \frac{60}{\sqrt{3}} = 34.64 \text{ V}$$

Resistance per phase in the rotor circuit,

$$R_2 = 0.6 + 5 = 5.6\ \Omega$$

Reactance per phase in the rotor circuit,

$$X_2 = 6 + 4 = 10\ \Omega$$

Impedance per phase in rotor circuit,

$$Z_2 = \sqrt{R_2^2 + X_2^2} = \sqrt{(5.6)^2 + (10)^2} = 11.46\ \Omega$$

Rotor current per phase, $I_2 = \frac{E_2}{Z_2} = \frac{34.64}{11.46} =$ **3.02 A. (Ans.)**

Power factor, $\cos\phi_2 = \frac{R_2}{Z_2} = \frac{5.6}{11.46} =$ **0.488 (lag). (Ans.)**

(*ii*) **Running at 4 per cent slip :**

Induced e.m.f. on the rotor winding per phase

$$= sE_2 = 0.04 \times 34.64 = 1.38 \text{ V}$$

Rotor impedance per phase,

$$Z_2 = \sqrt{R_2^2 + (sX_2)^2} = \sqrt{(0.6)^2 + (0.04 \times 6)^2}$$

$= 0.646\ \Omega$ [Here $R_2 = 0.6\ \Omega$]

Rotor current per phase,

$$I_2 = \frac{sE_2}{Z_2} = \frac{1.38}{0.646} = \textbf{2.136 A. (Ans.)}$$

Power factor, $\cos \phi_2 = \dfrac{R_2}{Z_2} = \dfrac{0.6}{0.646} = \mathbf{0.928\ (lag).}$ **(Ans.)**

☞**Example 11.** *A 3-phase induction motor has a star-connected rotor. The rotor e.m.f. (between slip rings) at standstill is 50 V. The rotor resistance and standstill reactance are 0.5 Ω and 3 Ω respectively. Find :*

(*i*) *Rotor current per phase at starting and the slip rings short-circuited.*

(*ii*) *Rotor current per phase at starting if a star connected rheostat of resistance 6 Ω per phase is connected across the slip rings.*

(*iii*) *Full-load rotor current and rotor power factor if the full-load slip is 4 per cent.*

(*iv*) *Rotor e.m.f. per phase under full-load condition.*

Solution. Rotor e.m.f. at standstill (between slip rings)

$$= 50 \text{ V}$$

∴ Rotor e.m.f./phase at standstill,

$$E_2 = \frac{50}{\sqrt{3}} = 28.87 \text{ V}$$

Rotor resistance/phase, $R_2 = 0.5\ \Omega$

Rotor reactance/phase, $X_2 = 3\ \Omega$

(*i*) At starting $s = 1$

We know that rotor current/phase (I_2) is given by,

$$I_2 = \frac{sE_2}{\sqrt{R_2^2 + (sX_2)^2}} \qquad ...(i)$$

∴ **Rotor current/phase at starting**

$$= \frac{E_2}{\sqrt{R_2^2 + X_2^2}} = \frac{28.87}{\sqrt{(0.5)^2 + (3)^2}} = \mathbf{9.49\ A.}\ \textbf{(Ans.)}$$

(*ii*) Total resistance in the rotor circuit

= rheostat resistance + rotor resistance = 6 + 0.5 = 6.5 Ω

∴ **Rotor current/phase with rheostat resistance,**

$$I_2 = \frac{28.87}{\sqrt{(6.5)^2 + (3)^2}} = \mathbf{4.03\ A.}\ \textbf{(Ans.)}$$

(*iii*) Full-load slip, $s_f = 4\% = 0.04$

Full-load rotor current (as per eqn. (*i*)),

$$I_2 = \frac{0.04 \times 28.87}{\sqrt{(0.5)^2 + (0.04 \times 3)^2}} = \mathbf{2.246\ A.}\ \textbf{(Ans.)}$$

Rotor power factor $= \dfrac{R_2}{\sqrt{R_2^2 + (s_f X_2)^2}} = \dfrac{0.5}{\sqrt{(0.5)^2 + (0.04 \times 3)^2}} = \mathbf{0.972.}$ **(Ans.)**

(*iv*) **Rotor e.m.f./phase under full-load condition**

$$= sE_2 = 0.04 \times 28.87 = \mathbf{1.155\ V.}\ \textbf{(Ans.)}$$

17. POWER STAGES IN AN INDUCTION MOTOR

- **Stator iron loss** (consisting of eddy and hysteresis losses) depends on the supply frequency and the flux density in the iron core. It is practically *constant*.

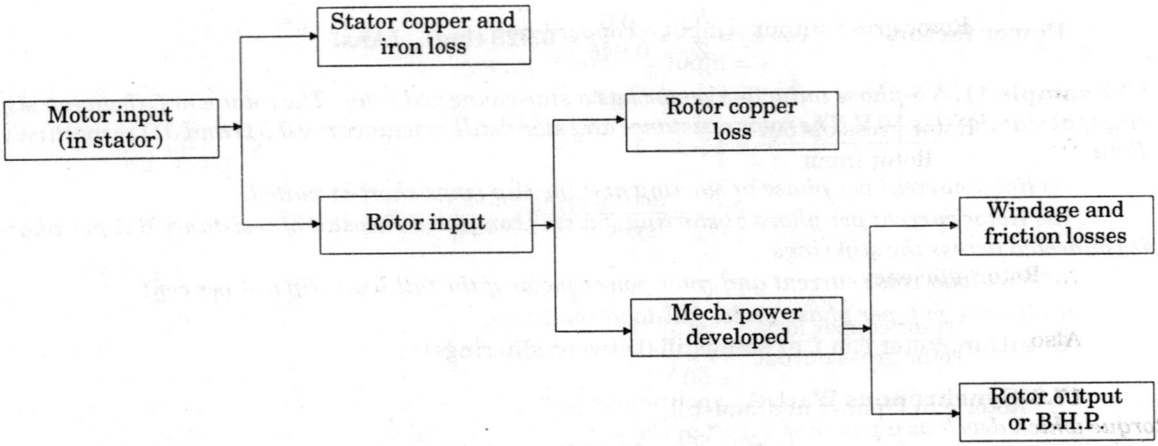

- Rotor iron loss is negligible because the frequency of rotor currents under normal running conditions is always small.
- Total rotor copper loss = $3I_2^2R_2$ (where I_2 and R_2 are current and resistance of the rotor respectively).

17.1. Rotor Output

Stator input = stator output + losses

Rotor input = stator output

[stator output is transferred entirely inductively to the rotor circuit]

Rotor gross output = rotor input – rotor copper losses.

This *rotor output is converted into mechanical* energy and gives rise to gross torque T_g. Out of this gross torque developed, some is lost due to windage and friction losses in the rotor and the rest appears as the *useful* or *shaft torque* T_{sh}. If T_g is in N-m and N is the actual speed of the rotor in r.p.m.

$$T_g = \frac{\text{rotor gross output in watts}}{(2\pi N/60)} \quad ...(19)$$

If there were no copper losses in the rotor, then rotor output will equal rotor input and the rotor will run at synchronous speed,

$$T_g = \frac{\text{rotor input}}{(2\pi N_s/60)} \quad ...(20)$$

From eqn. (19) and eqn. (20), we get

Rotor output $= T_g \times 2\pi N/60$

Rotor input $= T_g \times 2\pi N_s/60$ ...(21)

But rotor input = rotor output + rotor copper losses

$\therefore$ Rotor copper losses = rotor input – rotor output

$= T_g \times 2\pi/60\ (N_s - N)$ ...(22)

From eqn. (21) and eqn. (22), we get

$$\frac{\text{Rotor copper loss}}{\text{Rotor input}} = \frac{N_s - N}{N_s} = s$$

$\therefore$ Rotor copper loss = s × rotor input

= s × power across air-gap ...(23)

$$\text{Rotor gross output} = \text{input} - \text{copper loss}$$
$$= \text{input} - s \times \text{rotor input}$$
$$= (1 - s)\ \text{input} \qquad \text{...(24)}$$

$$\therefore \quad \frac{\text{Rotor gross output}}{\text{Rotor input}} = 1 - s = 1 - \frac{N_s - N}{N_s}$$

$$= \frac{N}{N_s} = \frac{\text{actual speed of rotor}}{\text{synchronous speed}}$$

$$\therefore \quad \text{Rotor efficiency} = \frac{N}{N_s}$$

Also, $$\frac{\text{rotor copper loss}}{\text{rotor gross output}} = \frac{s}{1-s}. \qquad \text{...(25)}$$

17.2. Synchronous Watt. A synchronous watt is a new unit of torque and is defined as the *torque which develops a power of 1 watt at the synchronous speed of motor*. When we say that motor is developing a torque of 500 synchronous watts, we mean that the rotor input is 500 watts and that the torque is such that power developed would be 500 watts provided the rotor is running synchronously and developing the same torque.

$$\text{Rotor input} = T_{sw} \times 2\pi N_s$$

or $$T_{sw} = \frac{\text{rotor input in watts}}{2\pi N_s}$$

Synchronous wattage of an induction motor equals the power transferred across the air-gap to the rotor.

17.3. Factors Determining Torque. When a 3-phase induction motor is operating on a certain load with slip s, then

Induced e.m.f. per phase in rotor = sE_2 volts (where E_2 = standstill rotor induced e.m.f.)

Rotor impedance per phase, $Z_2 = R_2 + jsX_2$

(where R_2 and X_2 are rotor resistance and standstill reactance per phase)

$$\text{Rotor current per phase} = \frac{sE_2}{Z_2} = \frac{sE_2}{\sqrt{R_2^2 + s^2X_2^2}}$$

$$\text{Rotor circuit power factor, } \cos\phi_2 = \frac{R_2}{\sqrt{R^2 + s^2X_2^2}}$$

Total rotor copper loss = number of phases × rotor induced e.m.f. per phase × rotor current per phase × rotor current power factor

$$= 3 \times sE_2 \times \frac{sE_2}{\sqrt{R_2^2 + s^2X_2^2}} \times \frac{R_2}{\sqrt{R_2^2 + s^2X_2^2}} = \frac{3R_2 s^2E_2^2}{R_2^2 + s^2X_2^2} \qquad \text{...(26)}$$

$$\text{Also, total rotor copper loss} = s \times \text{rotor input} = \frac{s \times T \times 2\pi N_s}{60}$$

$$\therefore \quad s \times \frac{T \times 2\pi N_s}{60} = \frac{3R_2 s^2E_2^2}{R_2^2 + s^2X_2^2}$$

or $$\text{Torque developed by the rotor, } T = \frac{3}{(2\pi N_s/60)} \cdot \frac{R_2 sE_2^2}{R_2^2 + s^2E_2^2} \qquad \text{...(27)}$$

$$= \frac{3}{\omega_s} \cdot \frac{R_2\, sE_2^2}{R_2^2 + s_2^2 E_2^2} \quad ...[27\ (a)]$$

or $$T \propto \frac{R_2\, sE_2^2}{R_2^2 + s^2 X_2^2} \propto \frac{R_2\, sV_1^2}{R_2^2 + s^2 X_2^2} \cdot (\because E_2 \propto V_1, \text{ the supply voltage})$$

ROTOR OUTPUT, LOSSES AND EFFICIENCY

Example 12. *The power input to a 3-phase induction motor is 50 kW. The stator-losses total 1.2 kW. If the motor is running with a slip of 3 per cent, find the total mechanical power developed.*

Solution. Stator power input = 50 kW

Stator losses = 1.2 kW

Slip = 3% (= 0.03)

Mechanical power developed :

Rotor input = stator output = stator input – stator losses

= 50 – 1.2 = 48.8 kW

Now *mechanical power developed*

= (1 – s) rotor input = (1 – 0.03) × 48.8 = **47.34 kW. (Ans.)**

☞**Example 13.** *A 6-pole, 50-Hz, 3-phase induction motor running on full-load with 3% slip develops a torque of 160 N-m at its pulley rim. The friction and windage losses are 210 W and the stator copper and iron losses equal 1640 W. Calculate :*

(i) Rotor output, *(ii) Rotor copper loss, and*

(iii) Efficiency at full-load.

Solution. Number of poles, $p = 6$

Supply frequency, $f = 50$ Hz

Slip, $s = 3\%$ or 0.03

Torque developed at full-load, $T_f = 160$ N-m

Friction and windage losses = 210 W

Stator copper and iron losses = 1640 W

Synchronous speed, $$N_s = \frac{120f}{p} = \frac{120 \times 50}{6} = 1000 \text{ r.p.m.}$$

Actual rotor speed, $N = (1 - s)\, N_s = (1 - 0.03) \times 1000 = 970$ r.p.m.

(i) **Rotor output**

$$= \frac{2\pi N T_f}{60} = \frac{2\pi \times 970 \times 160}{60} = \mathbf{16252.5\ W.}\ \textbf{(Ans.)}$$

(ii) **Rotor gross output**

= rotor output + friction and windage losses

= 16252.5 + 210 = 16462.5 W

Also, $$\frac{\text{rotor gross output}}{\text{rotor input}} = \frac{N}{N_s}$$

$\therefore$ $$\text{Power input} = \frac{\text{rotor gross output} \times N_s}{N}$$

$$= \frac{16462.5 \times 1000}{970} = 16971.6 \text{ W}$$

$\therefore$ **Rotor copper loss** = 16971.6 – 16462.5 = **509.1 W. (Ans.)**

(*iii*) **Stator output** = rotor input = 16971.6 W

∴ Stator input = stator output + stator copper and iron losses

= 16971.6 + 1640 = 18611.6 W

% Overall efficiency at full-load $= \frac{\text{rotor output}}{\text{stator input}}$

$= \frac{16252.5}{18611.6} \times 100 =$ **87.32%. (Ans.)**

Example 14. *The power input to the rotor of a 440 V, 50-Hz, 3-phase, 6-pole induction motor is 50 kW. It is observed that the rotor e.m.f. makes 120 complete cycles per minute. Calculate :*

(i) Slip, *(ii) Rotor speed,*

(iii) Rotor copper loss/phase, *(iv) Mechanical power developed, and*

(v) Rotor resistance/phase if rotor current is 50 A.

Solution. Supply voltage = 440 V

Supply frequency, f = 50-Hz

Number of poles, $p = 6$

Rotor input = 50 kW or 50000 W

Rotor frequency, $f_r = \frac{120}{60} = 2$ Hz

(*i*) **Slip,** $s = \frac{f_r}{f} = \frac{2}{50} = \mathbf{0.04}$ **or 4%. (Ans.)**

(*ii*) **Rotor speed,** $N = (1 - s)N_s = (1 - s) \times \frac{120f}{p}$

$= (1 - 0.04) \times \frac{120 \times 50}{6} =$ **960 r.p.m. (Ans.)**

(*iii*) Rotor copper loss $= s \times$ rotor input $= 0.04 \times 50000 = 2000$ W

Rotor copper loss/phase $= \frac{2000}{3} =$ **666.67 W. (Ans.)**

(*iv*) **Mechanical power developed**

$= (1 - s) \times$ rotor input $= (1 - 0.04) \times 50000$

= **48000 W or 48 kW. (Ans.)**

[or = rotor input – total copper loss]

(*v*) Let R_2 be the rotor resistance/phase

∴ $3I_2^2R_2 = 2000$

$$R_2 = \frac{2000}{3 \times 50 \times 50} = 0.267\ \Omega$$

Hence, **rotor resistance/phase = 0.267 Ω. (Ans.)**

Example 15. *A 50 H.P., 6-pole, 50 Hz, slip-ring induction motor runs at 960 r.p.m. on full-load with a rotor current of 40 A. Allowing 300 W for copper loss in the short-circuiting gear and 1200 W for mechanical losses, find the resistance R_2 per phase of the 3-phase rotor winding.*

(GATE, 1998)

Solution. P_{out} = 50 H.P. = 50 × 735.5 = 36775 W ; $p = 6$; f = 50 Hz ; N = 960 r.p.m. ; I_2 = 40 A ; copper loss in the short-circuiting gear = 300 W ; mechanical loss = 1200 W.

Resistance, R_2 :

Rotor output = motor output + short-circuiting gear loss + mechanical losses

= 36775 + 300 + 1200 = 38275 W

Synchronous speed, $N_s = \dfrac{120f}{p} = \dfrac{120 \times 50}{60} = 1000$ r.p.m.

Slip, $s = \dfrac{N_s - N}{N_s} = \dfrac{1000 - 960}{1000} = 0.04$

Rotor copper losses $= 3I_2^2 R_2 = \dfrac{\text{Rotor output}}{1 - s} \times s = \dfrac{38275}{(1 - 0.04)} \times 0.04 = 1594.8$ W

or Rotor resistance, $\mathbf{R_2} = \dfrac{1594.8}{3 \times 40^2} = \mathbf{0.332\ \Omega}$. **(Ans.)**

18. INDUCTION MOTOR AS TRANSFORMER

An induction motor is essentially a transformer with stator forming the primary and rotor forming (the short-circuited) rotating secondary (Fig. 20). This is so because the transfer of energy

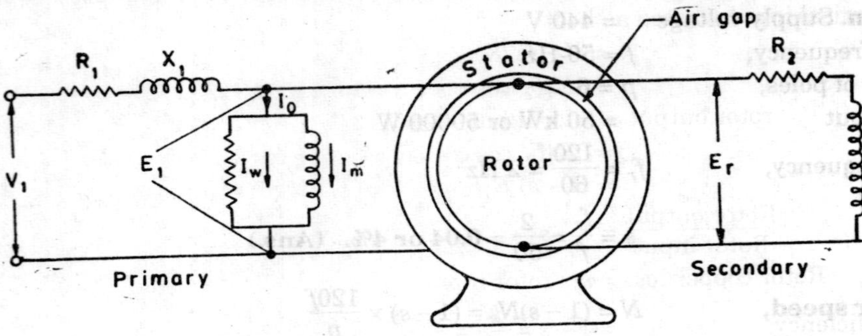

Fig. 20. Induction motor as transformer.

from stator to the rotor of an induction motor takes place entirely *inductively* with the help of flux mutually linking the two.

The vector diagram (Fig. 21) is similar to that of a transformer.

$$V_1 = E_1 + I_1 (R_1 + jX_1)$$

and

$$E_r = I_2 (R_2 + jsX_2)$$

However, the *important differences* between a transformer and an induction motor are :

1. In induction motor the magnetic leakage and leakage reactance of rotor and stator are *higher* than in a transformer.

2. The magnetic circuit of an induction motor has an air gap and this makes the per unit value of magnetising current *much higher* than that of a transformer.

3. Because of the distributed windings in an induction motor the *ratio of stator and rotor currents is not equal* to the ratio of turns per phase in the rotor and stator windings.

4. *The losses in an induction motor are higher and, therefore, the efficiency is lower than in a transformer.*

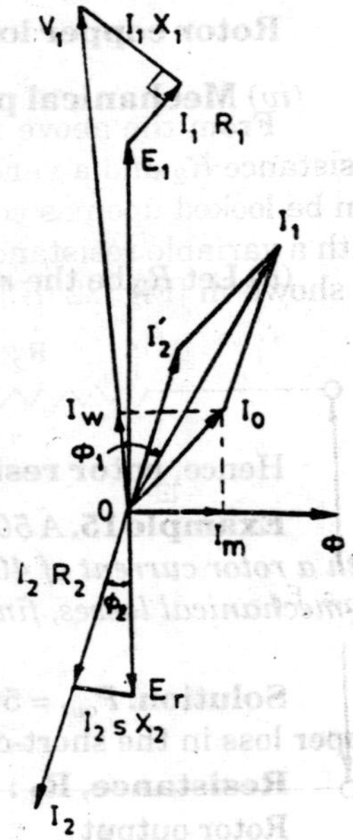

Fig. 21. Vector diagram of induction motor.

18.1. Rotor output. Primary current I_1 consists of two parts I_0 and I_2'. It is the latter which is transferred to the rotor, because I_0 is used in meeting the copper and iron losses in the stator itself. Out of the primary voltage V_1, some is absorbed in the primary itself ($= I_1Z_1$) and the remaining E_1 is transferred to the rotor. If the angle between E_2 and I_2' is ϕ, then

Rotor input/phase $= E_1I_1' \cos\phi$

Total rotor input $= 3E_1I_2' \cos\phi$

The electrical input to the rotor which is wasted in the form of the heat

$$= 3I_2E_r \cos\phi \text{ (or } 3I_2^2R_2)$$

Now $\quad I_2' = KI_2 \quad \text{or} \quad I_2 = \dfrac{I_2'}{K}$

$$E_r = sE_2 \text{ and } E_2 = KE_1$$

$\therefore \quad E_r = sKE_1$

$\therefore$ Electrical input wasted as heat

$$= 3 \times (I_2'/K) \times sKE_1 \cos\phi)$$

$$= 3E_1I_2' \cos\phi \times s = \text{rotor input} \times s$$

Now, $\quad$ rotor output = rotor input − losses

$$= 3E_1I_2' \cos\phi - 3E_1I_2' \cos\phi \times s$$

$$= (1-s)\, E_1I_2' \cos\phi = (1-s) \text{ rotor input}$$

$$\therefore \quad \frac{\text{Rotor output}}{\text{Rotor input}} = 1 - s$$

$\therefore$ Rotor copper loss $= s \times$ rotor input

$$\text{Rotor efficiency} = 1 - s = \frac{N}{N_s} = \frac{\text{actual speed}}{\text{synchronous speed}}$$

18.2. Equivalent Circuit of the Rotor. The rotor current I_2, when motor is loaded, is given by

$$I_2 = \frac{sE_2}{\sqrt{R_2^2 + (sX_2)^2}} = \frac{E_2}{\sqrt{(R_2/s)^2 + X_2^2}}$$

From the above relation it appears that the rotor circuit which actually consists of a fixed resistance R_2 and a variable reactance sX_2 (proportional to slip) connected across $E_r = sE_2$ [Fig. 22 (*i*)] can be looked upon as equivalent to a rotor circuit having a fixed reactance X_2 connected in series with a variable resistance R_2/s (inversely proportional to slip) and supplied with constant voltage E_2 as shown in [Fig. 22 (*ii*)].

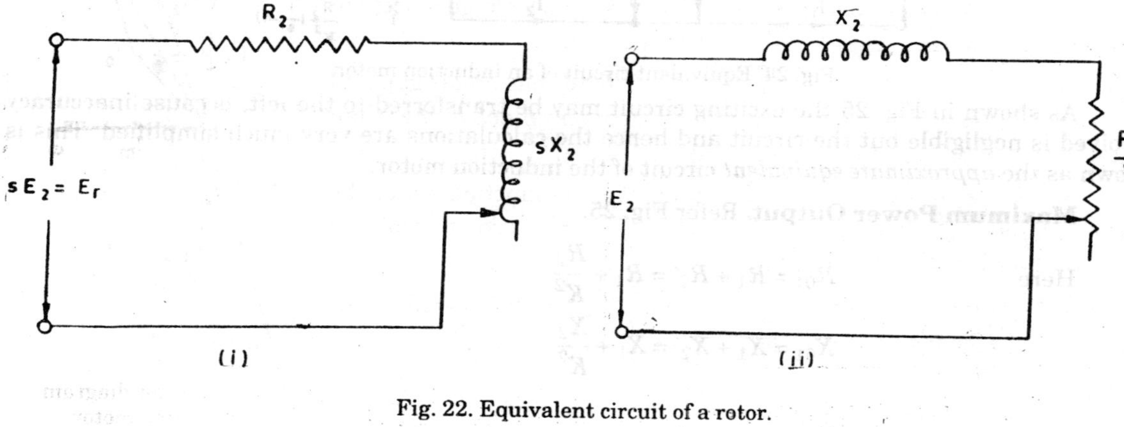

Fig. 22. Equivalent circuit of a rotor.

Also, the resistance R_2/s can be written as,

$$\frac{R_2}{s} = R_2 + R_2\left(\frac{1}{s} - 1\right).$$

It consists of two parts : (*i*) The part R_2 is the *rotor resistance* itself and *represents the rotor copper loss* (*ii*) the second part is $R_2\left(\frac{1}{s} - 1\right)$. This is known as *load resistance* R_L and is *the electrical equivalent of the mechanical load on the motor*. In other words the mechanical load on an induction motor can be represented by a *non-inductive resistance of the value* $R_2\left(\frac{1}{s} - 1\right)$.

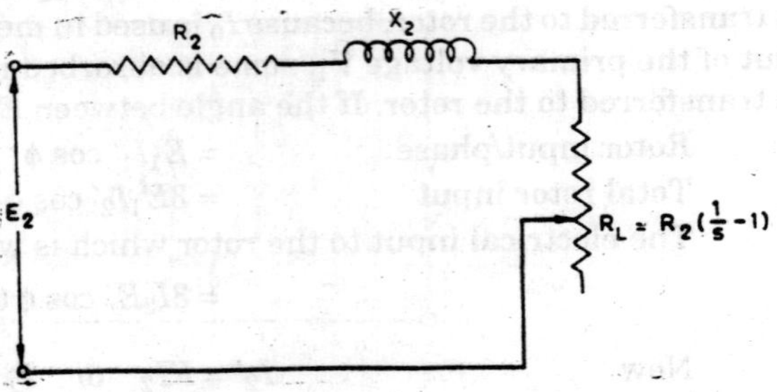

Fig. 23. Equivalent circuit of a rotor with load resistance R_L.

In Fig. 23 is shown the equivalent rotor circuit along with the load resistance R_L.

19. EQUIVALENT CIRCUIT OF AN INDUCTION MOTOR

The equivalent circuit for a polyphase induction motor is shown in Fig. 24, where

V_1 = applied voltage per phase
R_1 = stator resistance/phase
R_2 = rotor resistance/phase
X_1 = stator leakage reactance/phase
X_2 = rotor standstill leakage reactance/phase
K = turn-ratio of secondary to primary
R_0 = no-load resistance/phase, X_0 = no-load reactance/phase — Rotor being driven at synchronous speed.

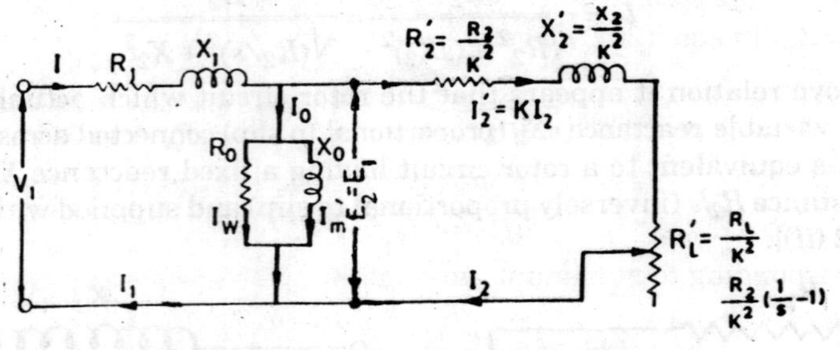

Fig. 24. Equivalent circuit of an induction motor.

As shown in Fig. 25 the exciting circuit may be transferred to the left, because inaccuracy involved is negligible but the circuit and hence the calculations are very much simplified. This is known as the *approximate equivalent* circuit of the induction motor.

Maximum Power Output. Refer Fig. 25.

Here
$$R_{01} = R_1 + R_2' = R_1 + \frac{R_2}{K^2}$$

$$X_{01} = X_1 + X_2' = X_1 + \frac{X_2}{K^2}$$

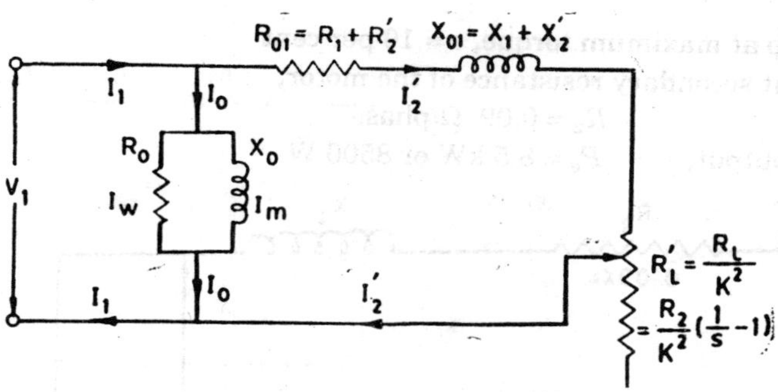

Fig. 25. Approximate equivalent circuit of an induction motor.

Load resistance, $$R_L' = R_L/K^2 = \frac{R_2}{K^2}\left(\frac{1}{s} - 1\right)$$

Gross mechanical power output,

$$P_g = 3I_2'^2 R_L'$$

But $$I_2' = \frac{V_1}{\sqrt{(R_{01} + R_L')^2 + (X_{01})^2}}$$

$\therefore$ $$P_g = \frac{3V_1^2 R_L'}{(R_{01} + R_L')^2 + (X_{01})^2} \quad ...(28)$$

The condition for maximum power output can be obtained by differentiating the above equation and by equating, the first derivative to zero. If it is done, it will be found that

$$R_L' = \sqrt{(R_{01})^2 + (X_{01})^2} = Z_{01}$$

Hence the gross mechanical power output of the motor is maximum when the equivalent *load resistance R_L' is equal to the standstill leakage impedance of the motor Z_{01}.*

Putting $R_L' = Z_{01}$ in eqn. (28), we get

$$P_{g\,(max)} = \frac{3V_1^2 Z_{01}}{(R_{01} + Z_{01})^2 + (X_{01})^2} = \frac{3V_1^2 Z_{01}}{R_{01}^2 + X_{01}^2 + Z_{01}^2 + 2R_{01}Z_{01}}$$

$$= \frac{3V_1^2}{2(R_{01} + Z_{01})} \quad ...(29)$$

The slip corresponding to maximum gross mechanical power output will be given as

$$Z_{01} = R_L' = \frac{R_2}{K^2}\left(\frac{1}{s} - 1\right)$$

or $$s = \frac{R_2/K^2}{R_2/K^2 + Z_{01}} \quad ...(30)$$

EQUIVALENT CIRCUIT OF AN INDUCTION MOTOR

Example 16. *In a 3-phase induction motor maximum torque occurs at a slip of 10 per cent. The equivalent secondary resistance of the motor is 0.09 Ω/phase. If the gross power output is 8.5 kW. Calculate :*

(i) Equivalent load resistance, *(ii) Equivalent load voltage, and*
(iii) Current at this slip.

Solution. Slip at maximum torque, s = 10 per cent

The equivalent secondary resistance of the motor,

$$R_2 = 0.09 \ \Omega/\text{phase}$$

Gross power output, P_g = 8.5 kW or 8500 W

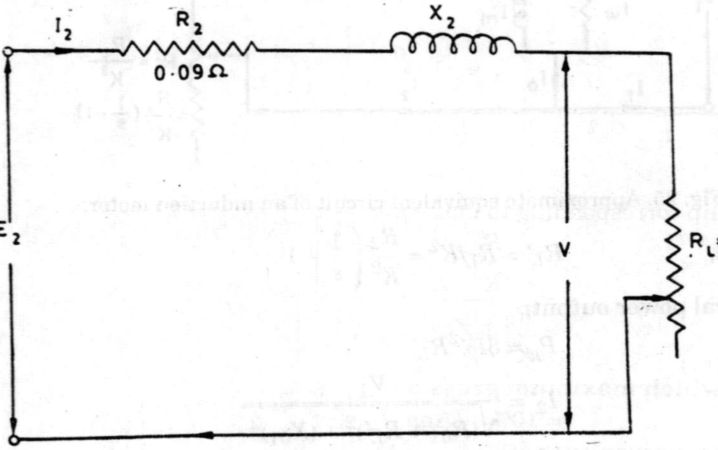

Fig. 26

(*i*) **Equivalent load resistance, R_L :**

We know that $R_L = R_2\left[\left(\frac{1}{s}\right) - 1\right] = 0.09\left(\frac{1}{0.1} - 1\right) = \mathbf{0.81\ \Omega.}$ **(Ans.)**

(*ii*) **Equivalent load voltage, V :**

As shown in the equivalent circuit of the rotor in Fig. 30, V is a fictitious voltage drop equivalent to that consumed in the load connected to the secondary *i.e.*, rotor, $V = I_2R_L$

But, gross power, $P_g = \dfrac{3V^2}{R_L}$

$$\therefore \quad V = \sqrt{\frac{P_g \times R_L}{3}} = \sqrt{\frac{8500 \times 0.81}{3}} = 47.9 \text{ V}$$

Hence, **equivalent load voltage = 47.9 V. (Ans.)**

(*ii*) **Equivalent load current :**

Equivalent load current $= \dfrac{V}{R_L} = \dfrac{47.9}{0.81} = \mathbf{59.1\ A.}$ **(Ans.)**

Example 17. *The stator impedance and equivalent rotor resistance of a 400 V, 3-phase star-connected induction motor are (0.05 + j0.2) Ω and (0.05 + j0.24) Ω respectively. Neglecting exciting current determine the maximum gross power and the slip at which it occurs.*

Solution. Supply voltage V_1 = 400 V

Stator impedance = (0.05 + j0.2) Ω

Equivalent rotor resistance = (0.05 + j0.24) Ω

The equivalent circuit is shown in Fig. 27.

$$R_{01} = R_1 + R_2' = 0.05 + 0.05 = 0.1\ \Omega$$

$$X_{01} = X_1 + X_2' = 0.2 + 0.24 = 0.44\ \Omega$$

$$\therefore \quad Z_{01} = \sqrt{R_{01}^2 + X_{01}^2} = \sqrt{(0.1)^2 + (0.44)^2} = 0.45\ \Omega$$

Slip = ?

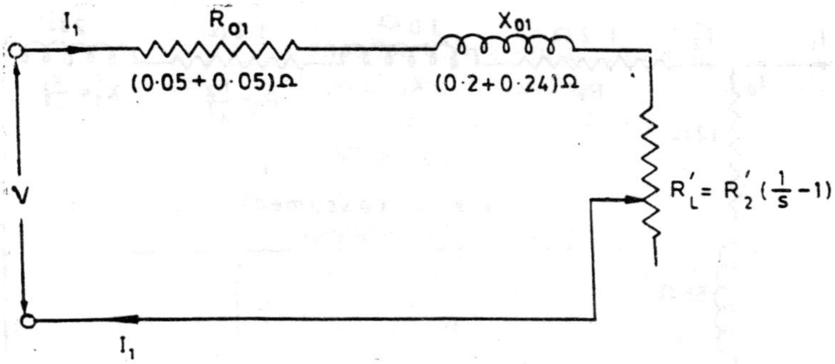

Fig. 27

We know that slip corresponding to maximum gross power output (Art. 19) is given by

$$s = \frac{\frac{R_2}{K^2}}{\frac{R_2}{K^2} + Z_{01}} = \frac{R_2'}{R_2' + Z_{01}} = \frac{0.05}{0.05 + 0.45} = 0.1 \text{ or } 10\%$$

Hence, **slip at which maximum gross power output occurs**
= 10%. (Ans.)

Maximum gross power output

$$= \frac{3V_1^2}{2(R_{01} + Z_{01})} = \frac{3\left(\frac{400}{\sqrt{3}}\right)^2}{2(0.1 + 0.45)}$$

= **145454 W** or **145.454 kW. (Ans.)**

☞**Example 18.** *The following data pertain to an induction motor : stator impedance = (1.2 + j3.0) Ω ; rotor standstill impedance = (1.2 + j2.0) Ω ; no-load shunt impedance = (12 + j55) Ω ; volts/phase = 240. Determine :*

(i) Stator current, *(ii) Equivalent rotor current,*
(iii) Input power factor, *(iv) Mechanical power developed, and*
(v) Efficiency of the motor at a slip of 5%.

Solution. Stator impedance = (1.2 + j3.0) Ω
Rotor standstill impedance = (1.2 + j2.0) Ω
No-load shunt impedance = (12 + j55) Ω
Volts/phase = 240 V

The equivalent circuit per phase of motor referred to stator is shown in Fig. 32. It is *assumed* that the given impedance figures are *phase values* and there is *unity turn-ratio* between stator and rotor.

Equivalent load resistance,

$$R_L' = R_2'\left(\frac{1}{s} - 1\right) = \frac{R_2}{K^2}\left(\frac{1}{s} - 1\right) = 1.2\left(\frac{1}{0.05} - 1\right) = 22.8\ \Omega$$

Effective impedance/phase $= (R_1 + R_2' + R_L') + j(X_1 + X_2') = (1.2 + 1.2 + 22.8) + j(3 + 2)$
$= (25.2 + j5)\ \Omega$

(*i*) **Stator current, I_1 :**

Stator load current, $$I_2' = \frac{240}{25.2 + j5} = \frac{240(25.2 - j5)}{(25.2 + j5)(25.2 - j5)}$$

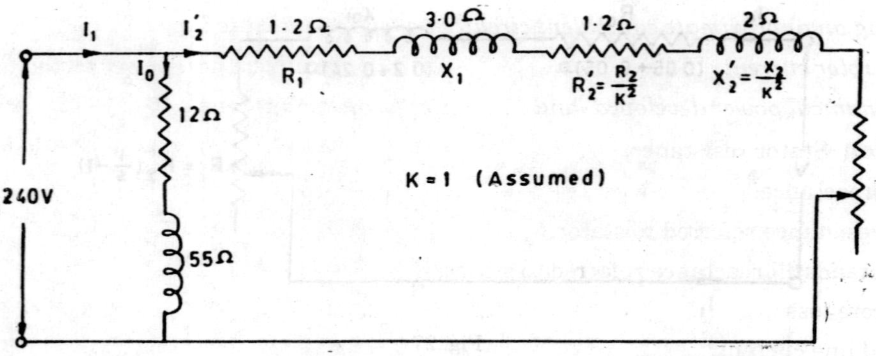

Fig. 28

$$= \frac{240(25.2 - j5)}{660} = (9.16 - j1.82) \text{ A}$$

Stator no-load current, $$I_0 = \frac{240}{(12 + j55)} = \frac{240}{12 + j55} \times \frac{(12 - j55)}{(12 - j55)}$$

$$= \frac{240(12 - j55)}{3169} = (0.91 - j4.16) \text{ A}$$

Total stator current, $$I_1 = I_2' + I_0 = (9.16 - j1.82) + (0.91 - j4.16)$$

$$= (10.07 - j5.98) \text{ A}$$

$\therefore$ $$I_1 = \sqrt{(10.07)^2 + (5.98)^2} = \mathbf{11.71\ A.}\ \ \textbf{(Ans.)}$$

(*ii*) **Equivalent rotor current,**

$$I_2' = \sqrt{(9.16)^2 + (1.82)^2} = \mathbf{9.34\ A.}\ \ \textbf{(Ans.)}$$

(*iii*) **Input power factor,**

$$\cos\phi = \frac{10.07}{11.71} = \mathbf{0.86.}\ \ \textbf{(Ans.)}$$

(*iv*) **Mechanical power developed,**

$$= 3(I_2')^2 R_L' = 3 \times (9.34)^2 \times 22.8$$

$$= \mathbf{5967\ W} \text{ or } \mathbf{5.967\ kW.}\ \ \textbf{(Ans.)}$$

(*v*) **Efficiency of the motor,**

$$\eta = \frac{\text{output}}{\text{input}} \times 100 = \frac{5967}{3 \times 240 \times 11.71 \times 0.86} \times 100$$

$$= \mathbf{82.29\%.}\ \ \textbf{(Ans.)}$$

☞**Example 19.** *The following data refer to a 10-pole, 400 V, 50-Hz, 3-phase induction rotor :* $R_1 = 1.75\ \Omega$, $X_1 = 5.5\ \Omega$, R_2' *(rotor resistance referred to stator)* $= 2.25\ \Omega$.

X_2' *(rotor standstill reactance referred to stator)* $= 6.6\ \Omega$.

When the motor is tested on no-load, it is observed that it takes 3.8 A (line current) and the total core loss is 310 W.

By using an approximate equivalent circuit at 4% slip, calculate :

(i) The rotor current, *(ii) Supply current and power factor,*

(iii) Mechanical power developed, and *(iv) Gross load torque.*

Solution. Stator resistance, $R_1 = 1.75\ \Omega$

Stator reactance, $X_1 = 5.5\ \Omega$

Rotor resistance referred to stator, $R_2' = 2.25\ \Omega$

Rotor standstill reactance referred to stator, $X_2' = 6.6\ \Omega$

Total core loss $= 310$ W

No-load line current, $I_0 = 3.8$ A

If $\cos \phi_0$ is the power factor at no-load, then

$$\sqrt{3}\ VI \cos \phi_0 = 310$$

$$\therefore \quad \cos \phi_0 = \frac{310}{\sqrt{3} \times 400 \times 3.8} = 0.1178$$

$$\phi_0 = 83.23°$$

$$\therefore \quad \sin \phi_0 = 0.993$$

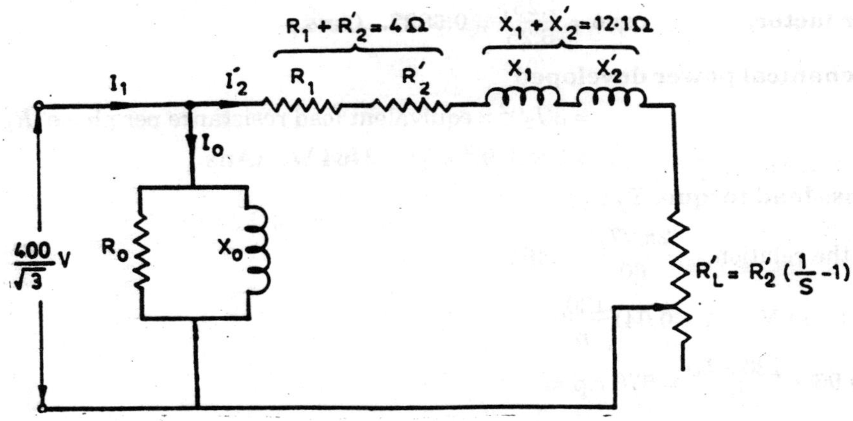

Fig. 29

So in the equivalent circuit,

$$R_0 = \frac{400}{\sqrt{3}\ I_0 \cos \phi_0} = \frac{400}{\sqrt{3} \times 3.8 \times 0.1178} = 515.9\ \Omega$$

$$X_0 = \frac{400}{\sqrt{3}\ I_0 \sin \phi_0} = \frac{400}{\sqrt{3} \times 3.8 \times 0.993} = 61.2\ \Omega$$

For a load slip of 4%

Equivalent load resistance,

$$R_L' = R_2' \left(\frac{1}{s} - 1 \right) = 2.25 \left(\frac{1}{0.04} - 1 \right) = 54\ \Omega$$

$$R_1 + R_2' = 1.75 + 2.25 = 4\ \Omega$$

$$X_1 + X_2' = 5.5 + 6.6 = 12.1\ \Omega$$

Phase voltage $= \frac{400}{\sqrt{3}}$ V = 231 V

(*i*) **Rotor current referred to stator,**

$$\mathbf{I_2}' = \frac{231 + j0}{(4 + 54) + j12.1} = \frac{231 + j0}{58 + j12.1} = \frac{231 + j0}{(58 + j12.1)} \times \left(\frac{58 - j12.1}{58 - j12.1}\right)$$

$$= \frac{13398 - j2795}{3364 + 146.4} = 3.82 - j0.976$$

$$\therefore \quad \mathbf{I_2}' = \sqrt{(3.82)^2 + (0.796)^2} = \mathbf{3.9\ A.}\ \ \mathbf{(Ans.)}$$

Also $I_0 = 3.8\,(\cos\phi_0 - j\sin\phi_0)$

$= 3.8\,(0.1178 - j0.993) = 0.447 - j3.77$

(*ii*) **Supply current and power factor :**

Supply (or motor) current,

$$\mathbf{I_1} = I_0 + I_2' = (0.447 - j3.77) + (3.82 - j0.796)$$

$$= 4.267 - j4.566$$

$$\therefore \quad \mathbf{I_1} = \sqrt{(4.267)^2 + (4.566)^2} = \mathbf{6.25\ A.}\ \ \mathbf{(Ans.)}$$

Power factor, $\cos\phi = \frac{4.267}{6.25} = \mathbf{0.6827.}$ **(Ans.)**

(*iii*) **Mechanical power developed**

$= 3(I_2')^2 \times$ equivalent load resistance per phase (R_L')

$= 3 \times (3.9)^2 \times 54 =$ **2464 W. (Ans.)**

(*iv*) **Gross load torque, $\mathbf{T_g}$:**

Using the relation, $\frac{2\pi N T_g}{60} = 2464$

where $N = (1 - s) N_s = (1 - 0.04)\frac{120f}{p}$

$= 0.96 \times \frac{120 \times 50}{10} = 576$ r.p.m.

$$\therefore \quad \frac{2\pi \times 576 \times T_g}{60} = 2464$$

or $$T_g = \frac{2464 \times 60}{2\pi \times 576} = 40.85 \text{ N-m}$$

Hence, **gross load torque = 40.85 N-m. (Ans.)**

20. STARTING OF INDUCTION MOTORS

Small induction motors (up to 2 kW) capacity may directly be switched on to the supply mains, but those of higher capacity must use some type of starting device, or starters as they are commonly called. *The function of these starters is to restrict the initial rush of current,* which, in the case of induction motors, is about 5 times the full-load current. This excessive current has two major upsetting effects, namely, a *large voltage drop in the distributing network* and *causing stoppage of machines which are already running on the supply mains.* Hence the Electrical Undertaking Authorities *forbid the users of large capacity induction motors to directly switch on their machines.*

The principle of all starting devices is to impress lower voltage on stator phases at the time of starting, or if the motor is slip ring or wound rotor, then to include external resistance in each rotor phase to keep the initial rotor current to a low value, this consequently means less current in the stator phases and therefore in the supply mains.

Direct-on-line starting of induction motors. This method means switching the motor directly on to the supply without using any device for reducing the starting current. The method is restricted to small motors up to about 2 kW. For these small motors, the *starting torque is about twice the full-load torque.* Hence the starting period lasts only a few seconds.

We know that,

Rotor input $= 2\pi N_s T = kT$

Also, rotor copper loss $= s \times$ rotor input

$$\therefore \quad 3I_2^2 R_2 = skT$$

$$\therefore \quad T \propto \frac{I_2^2}{s} \qquad \text{[if } R_2 \text{ is the same]}$$

Now $I_2 \propto I_1$

$$\therefore \quad T \propto \frac{I_1^2}{s} \quad \text{or} \quad T = \frac{KI_1^2}{s}$$

At starting $s = 1$

$$\therefore \quad T_{st} = KI_{st}^2 \quad \text{where } I_{st} = \text{starting current}$$

If I_f = normal full-load current

s_f = full-load slip

then
$$T_f = \frac{KI_f^2}{s_f}$$

$$\therefore \quad \frac{T_{st}}{T_f} = \left(\frac{I_{st}}{I_f}\right)^2 s_f \qquad \text{...[31 (a)]}$$

When the motor is direct-switched on to normal voltage, then starting current is the short-circuit current I_{sc}

$$\therefore \quad \frac{T_{st}}{T_f} = \left(\frac{I_{sc}}{I_f}\right)^2 s_f \qquad \text{...[31 (b)]}$$

Some of the starting devices for starting induction motors are discussed below :

Squirrel-Cage Motors :

(*i*) Stator rheostat starter. (*ii*) Auto-transformers (auto-starters).

(*iii*) Star-delta starter.

Slip Ring Motors :

(1) Rotor rheostat.

20.1. Squirrel-Cage Motors.

20.1.1. Stator rheostat starter. The connection diagram is shown in Fig. 30.

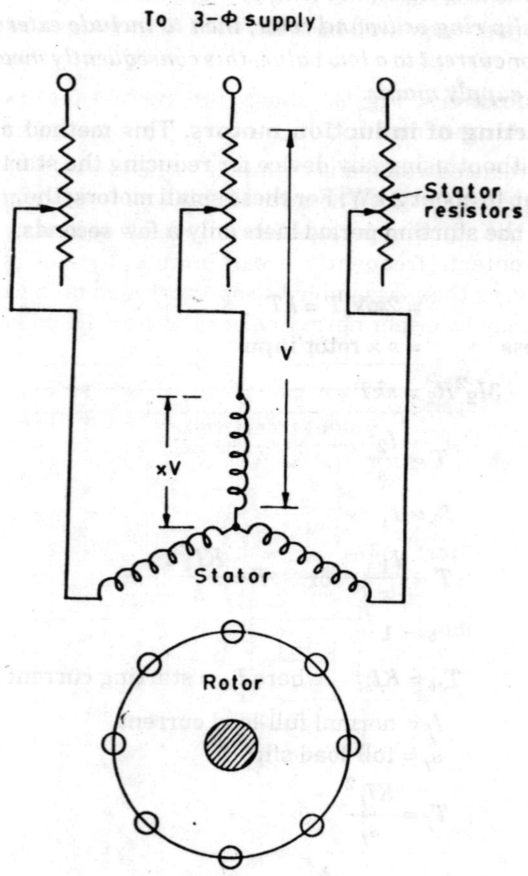

Fig. 30. Stator rheostat starter.

Reduced voltage is impressed on each stator phase due to the resistance R of the rheostat. Hence the initial current drawn from the supply mains will be *less* than if the machine were to be switched directly on to the supply mains.

If x = fraction of voltage (V) reduced by the stator resistors

Then $I_{st} = xI_{sc}$ **and** $T_{si} = x^2 T_{sc}$

$$\frac{T_{st}}{T_f} = \left(\frac{I_{st}}{I_f}\right)^2 s_f = \left(\frac{xI_{sc}}{I_f}\right)^2 s_f = x^2 \left(\frac{I_{sc}}{I_f}\right)^2 s_f \qquad ...(32)$$

- This method is suitable for starting of *small machines only*.

Advantages : 1. High power factor during start.

2. Smooth acceleration.

3. Less expensive than auto-transformer starter in lower output ratings.

4. Closed transition starting.

Disadvantages : 1. Heat is given off by the resistors.

2. Expensive resistors are required because starting duration usually exceeds 5 seconds.

3. Low torque efficiency.

20.1.2. Auto-transformers. Fig. 31 shows the connection diagram for auto-transformer starting of squirrel-cage induction motors.

In this method the reduced voltage is obtained by taking tappings at suitable points from a three-phase auto-transformer (Fig. 31). The auto-transformers are generally tapped at the 50, 60 and 80 per cent points, so that adjustment at these voltages may be made for proper starting torque requirements. Since the contacts frequently break large values of current arcing is sometimes quenched effectively by having them assembled to operate in an oil bath.

Auto-transformers may be either normally, or magnetically operated.

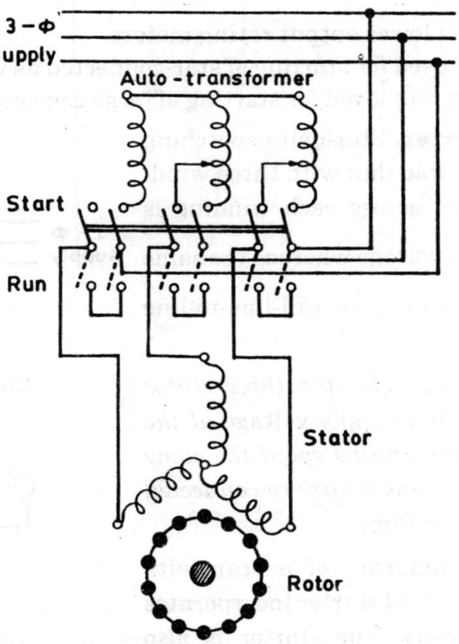

Fig. 31. Auto-transformer starter.

Relation between starting (T_{st}) and full-load (T_f) torques. Let the motor be started by an auto-transformer having transformation ratio K. If I_{sc} is the starting current when normal voltage is applied and applied voltage to stator winding at starting is KV then,

Motor input current, $I_{st} = KI_{sc}$

Supply current = Primary current of auto-transformer

$= K^2 I_{sc}$

$$\frac{T_{st}}{T_f} = \left(\frac{I_{st}}{I_f}\right)^2 \times s_f = K^2 \left(\frac{I_{sc}}{I_f}\right)^2 \times s_f \qquad ...(33)$$

It may be noted that this expression is similar to eqn. (32) except that x has been replaced by the transformation ratio K.

Advantages :

1. Voltage is reduced by transformation and not by dropping the voltage in resistors, and therefore, the current and power drawn from the supply mains are also reduced in comparison to resistor starting.
2. Availability of highest torque per ampere of supply current.
3. Adjustment of starting voltage by selection of proper tap on the auto-transformer.
4. The method is suitable for long starting periods.
5. Motor current larger than supply current.
6. Closed transition starting.

Disadvantages :

1. Low power factor.
2. Higher cost in case of lower output rating motors.

- This method can be used for starting of star-connected as well as delta-connected motors.
- This method is often employed for starting of large cage motors (rating exceeding 20 kW).

20.1.3. Star-Delta Starter. Star-delta switching method is based upon the principle that with three windings connected in star, voltage across each winding is $\frac{1}{\sqrt{3}}$ *i.e.*, 57.7% of the line to line voltage whereas the same windings connected in delta will have full-line-to-line voltage across each.

The star-delta starter *connects the three stator windings in star* across the rated supply voltage *at the starting instant. After the motor attains speed the same windings through a change over switch are re-connected in delta across the same supply voltage.*

The basic connection diagram of a star-delta starter is shown in Fig. 32. An actual starter incorporates *under-voltage* and *over-load coils*. The starter is also provided with a mechanical inter-locking device to prevent the handle from being put in the 'Run' position first. Such starters are employed for starting 3-phase squirrel-cage induction motors of rating between 4 kW and 15 kW.

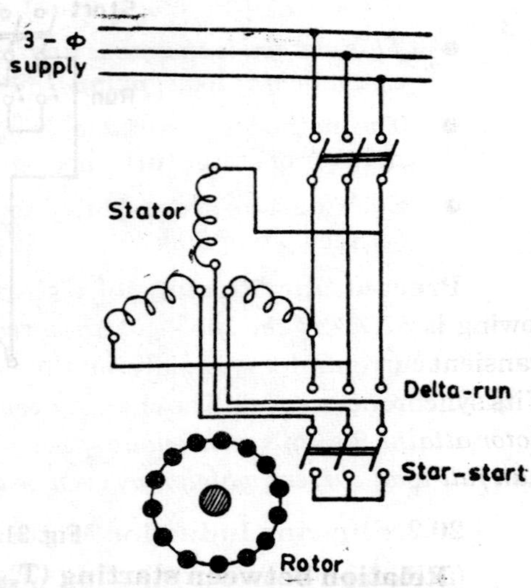

Fig. 32. Star-delta starter.

When *star-connected*, the applied voltage over each motor phase is reduced by a factor of $\frac{1}{\sqrt{3}}$ and hence the torque becomes $\frac{1}{3}$ of that which would have been developed if motor were directly connected in delta. The line current is reduced to $\frac{1}{3}$. Hence, during starting period when motor is star-connected, it takes $\frac{1}{3}$rd as much starting current and develops $\frac{1}{3}$rd as much torque as would have been developed were it directly connected in delta.

Relation between T_{st} and T_f :

$$I_{st} \text{ per phase} = \frac{1}{\sqrt{3}} I_{sc} \text{ per phase}$$

where I_{sc} is the current per phase which delta-connected motor would have taken if switched on to the supply directly ; however, line current at start is equal to $\frac{1}{3}$ of line I_{sc}.

Now, starting torque, $T_{st} \propto I_{st}^2 \quad (s = 1)$

Full-load torque, $T_f \propto I_f^2/s_f$

$$\therefore \quad \frac{T_{st}}{T_f} = \left(\frac{I_{st}}{I_f}\right)^2 s_f = \left(\frac{I_{sc}}{\sqrt{3} I_f}\right)^2 s_f = \frac{1}{3}\left(\frac{I_{sc}}{I_f}\right)^2 s_f$$

i.e.,

$$\frac{T_{st}}{T_f} = \frac{1}{3}\left(\frac{I_{sc}}{I_f}\right)^2 s_f \qquad \text{...(34)}$$

Here, I_{st} and I_{sc} represent phase values.

- This method reduces the starting line current to one-third but the starting torque is also reduced by the same amount.
- *This method is cheap but limited to applications where high starting torque is not necessary e.g., machine tools, pumps, motor-generator sets etc.*
- The method is unsuitable for motors for voltage exceeding 3000 V because of the excessive number of stator turns needed for delta connection.
- Such starters are employed for starting *3-phase squirrel cage induction motors of rating between 4 and 20 kW.*

Precaution with star-delta starting. When the motor is started in star the initial current flowing is *57.7% of the short-circuit current in delta together with a transient in each phase*. The transient currents decay rapidly but the steady state is not reached until the motor has attained 70% of its synchronous speed. The change-over from star to delta connection *should not be made until the motor attains 90% of synchronous speed, otherwise there will be a current surge considerably greater than full-load current which may even be greater than the standstill current with star-connection.*

20.2. Slip-ring Induction Motors—Starting of

(*i*) **Rotor rheostat.** The slip-ring induction motors are practically always started with full line voltage applied across the stator terminals. The value of starting current is adjusted by introducing a variable resistance in the rotor circuit. The controlling resistance is in the form of rheostat connected in star (Fig. 33), the resistance being gradually cut out of the rotor circuit as the motor gathers speed. *By increasing the rotor resistance, not only is the rotor* (and hence stator) *current reduced at starting* but *at the same time torque is also increased due to improvement in power factor.*

The rheostat is either of stud or contractor type and may be hand operated or automatic.

As discussed earlier, the introduction of additional external resistance in the rotor circuit enables slip-ring motor to develop a high starting torque with reasonably moderate starting current. Hence *such motors can be started under load.* When the motor runs under normal conditions the rings are short-circuited and brushes lifted from them.

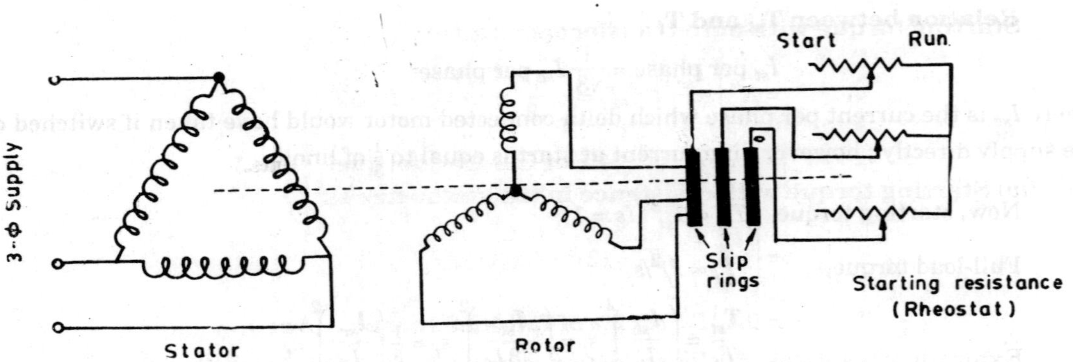

Fig. 33. Starting of slip-ring induction motor.

Example 20. *The short-circuit current of a small 3-phase induction motor is 3 times the full-load currents. Determine the starting torque as a percentage of full-load torque if full-load slip is 3 per cent.*

Solution. Short-circuit current, $I_{st} = 3I_f$

Full-load slip, $s_f = 3\%$ or 0.03

Starting torque, $T_{st} = T_f\left(\frac{I_{sc}}{I_f}\right)^2 s_f = T_f(3)^2 \times 0.03 = 0.27T_f$

Hence, **starting torque is 27% of full-load torque. (Ans.)**

Example 21. *A 3-phase, squirrel-cage induction motor has a short-circuit current equal to 4 times the full-load current. Find the starting torque as a percentage of full-load torque if the motor is started by :*

(i) direct switching to the supply ; *(ii) a star delta starter ;*
(iii) an auto-transformer ; and *(iv) a resistance in the stator circuit.*

The stator current in (iii) and (iv) is limited to 2 times the full-load current and the full-load slip is 3%.

Solution. Short-circuit current, $I_{sc} = 4I_f$

Full-load slip, $s_f = 3\%$ or 0.03

(i) **Starting torque with direct switch,**

$$T_{st} = \left(\frac{I_{sc}}{I_f}\right)^2 s_f T_f = (4)^2 \times 0.03 \times T_f$$

$= 0.48T_f$ *i.e.* **48% of full-load torque. (Ans.)**

(ii) **Starting torque with star-delta starter,**

$$T_{st} = \frac{1}{3}\left(\frac{I_{sc}}{I_f}\right)^2 s_f T_f = \frac{1}{3} \times (4)^2 \times 0.03 \times T_f$$

$= 0.16T_f$ *i.e.* **16% of full-load torque. (Ans.)**

(iii) **Supply line current**

$= K^2 I_{sc} = 2I_f$

or the transformation ratio,

$$K = \sqrt{\frac{2I_f}{I_{sc}}} = \sqrt{\frac{2I_f}{4I_f}} = \frac{1}{\sqrt{2}}$$

Starting torque with auto-transformer starter

$$= K^2 \left(\frac{I_{sc}}{I_f}\right)^2 s_f T_f = \left(\frac{1}{\sqrt{2}}\right)^2 \times (4)^2 \times 0.03 T_f$$

$= 0.24 T_f$ or **24% of full-load torque. (Ans.)**

(*iv*) **Starting torque with resistance in the stator circuit**

$$= \left(\frac{I_{sc}}{I_f}\right)^2 s_f T_f = (2)^2 \times 0.03 \times T_f$$

$= 0.12 T_f$ *i.e.* **12% of full-load torque. (Ans.)**

Example 22. *A 400 V, 50-Hz, induction motor, when started directly from the mains takes 4 times the full-load current and the torque produced is twice the full-load torque. Determine :*

(*i*) *The motor current, the line current and the starting torque when started by means of an auto- transformer of ratio 2.5 : 1.*

(*ii*) *The voltage to be applied and the motor current if the full-load torque is to be obtained at starting.*

Solution. Supply voltage $= 400$ V

Frequency $= 50$ Hz

Transformation ratio, $K = \frac{1}{2.5} = 0.4$

Let us *assume* that the winding is star-connected

Phase voltage, $E_{ph} = \frac{400}{\sqrt{3}} = 231$ V

Voltage across the motor terminal

$= KE_L = 0.4 \times 400 = 160$ V

The phase voltage when started by an auto-transformer

$= \frac{160}{\sqrt{3}} = 92.4$ V

As the current is proportional to the voltage per phase, therefore :

(*i*) **New starting phase current**

$= \frac{92.4}{231} \times 4I_f = \mathbf{1.6\ I_f}$. **(Ans.)**

Line current :

In case of a transformer, $\frac{E_2}{E_1} = \frac{N_2}{N_1} = \frac{I_1}{I_2}$

But $\frac{E_2}{E_1} = \frac{1}{2.5} = 0.4\ (= K)$ and $I_2 = 1.6 I_f$

$\therefore$ $\frac{I_1}{I_2} = K$ or $I_1 = KI_2 = K \times 1.6 I_f = 0.4 \times 1.6 I_f = 0.64\ I_f$

Hence, **line current = $\mathbf{0.64 I_f}$. (Ans.)**

Let T_f = full-load torque

$T \propto E^2$

$2T_f \propto (231)^2$ ...(*i*) [$\because$ $T = 2T_f$ (*given*)]

When the applied voltage is 92.4 V let the torque = T

$$T \propto (92.4)^2 \quad ...(ii)$$

$$\frac{T}{2T_f} = \left(\frac{92.4}{231}\right)^2 = (0.4)^2$$

$\therefore$ $$\mathbf{T = 0.32T_f.} \quad \textbf{(Ans.)}$$

(*ii*) Let the voltage per phase be V to get full-load starting torque,

or $$T_f \propto V^2 \quad ...(iii)$$

So from eqns. (*i*) and (*iii*), we get

$$\frac{T_f}{2T_f} = \frac{V^2}{(231)^2}$$

$\therefore$ $$V^2 = \frac{(231)^2}{2}$$

or $$\mathbf{V} = \frac{231}{\sqrt{2}} = \mathbf{163.3\ V.} \quad \textbf{(Ans.)}$$

$\therefore$ **Motor Voltage** $= 163.3 \times \sqrt{3} = \mathbf{282.8\ V.}$ **(Ans.)**

Motor current $= \dfrac{163.3}{231} \times 4I_f = \mathbf{2.827\ I_f.}$ **(Ans.)**

21. FACTORS GOVERNING PERFORMANCE OF INDUCTION MOTORS

The operating or performance characteristics of an induction motor are governed by the following factors/parameters :

1. Rotor resistance ;
2. Air gap length ;
3. Shapes of both stator and rotor slots and teeth.

1. Rotor resistance :

- The rotor resistance is the most important parameter and is determined primarily by the material used in construction and the area of cross-section of the end rings in case of a squirrel case induction motor. The commonly used materials are copper and aluminium.
- An increase in rotor resistance affects the performance of the motor as mentioned below :
  - With the *increase in starting resistance the starting torque increases until it attains the maximum value, after which it diminishes*.
  - An *increase in rotor resistance reduces the starting current and operating efficiency and causes increase in per cent regulation*. The maximum torque remains unaffected because it depends upon the rotor reactance only.

2. Air gap length :

- The air gap is one of the two main sources of the low power factor at which the induction motor operates.
- Too much reduction of air-gap may cause
  - *increase in noise, pulsation losses and magnetic pull* ;
  - *reduction in over-load capacity.*

3. Shapes of both stator and rotor slots and teeth :

- The shapes of rotor teeth and slots affect the reactance, which in turn, affects the starting current and maximum torque. The induction motors with deeper slots have more leakage reactance and so less value of starting current, maximum torque and slip corresponding to maximum torque.

- The slots in the rotor are *not always* made parallel to the shaft, but are given a slight twist or skew. The *motor noise and vibration, cagging and synchronous cups can be reduced or even eliminated by skewing either the stator or rotor slots* (in India, practice is to skew the rotor slots).

  Harmonic torques can be *reduced* by :

  —chording ;
  —integral slot winding ;
  —increasing air-gap length ;
  —skewing of rotor slots.

Note. Space harmonic fields are developed by the **windings, slotting, magnetic saturation, gap length irregularity**. These harmonic fields induce e.m.fs. and circulate harmonic currents in the rotor windings. These harmonic currents in the rotor interact with the harmonic fields and develop *harmonic torques, vibration* and *noise*.

22. EFFECTS OF OPERATING CONDITIONS

Effects of operating conditions are briefly described below :

1. Effect of unbalance supply voltage :

- An unbalance supply voltage produces a rotating field which will move at a non-uniform rate and in a non-uniform strength. It provides *unbalanced currents in the stator winding which produce unequal heating.*

2. Effect of loading :

- As the *load increases, speed falls and consequently torque increases, till a point is reached when the maximum torqué will be developed.*
- If the load is *increased further*, there is further *fall in the speed and the driving torque decreases and ultimately the rotor comes to standstill.*

3. Effects of variation in line frequency and line voltage :

- The variation in supply frequency and supply voltage not only affects the speed and torque of the motor but also the starting current, full-load current, starting torque, maximum torque and operating temperature of various parts of the machine and thus make their *operation unsatisfactory*.

4. Effects of break in one phase :

- When a polyphase induction motor is in operation and if its one phase breaks, may be due to any reason, the motor will continue to operate single phase, provided that load does not exceed 57.7 percent of the normal rating, with about the same temperature rise as when carrying rated load as a 3-phase induction motor. It will not, however, start single phase but if brought upto speed by some external means, such as putting on the belt, it will continue to operate satisfactorily on load below 57.7 percent of rated one.
- A break in one phase of the rotor of a wound-rotor induction motor may prevent it from starting but if it is brought upto speed before the circuit is opened, it will operate under reduced load but usually with considerable vibrations.

23. RATINGS OF 3-PHASE INDUCTION MOTORS

- Three-phase induction motors are rated in terms of the following :

  (*i*) Power output in kW ; (*ii*) Speed ;
  (*iii*) Voltage ; (*iv*) Frequency ;
  (*v*) Phase (single or three) ; (*vi*) Line current ;
  (*vii*) Temperature rise in specified time.

— If a motor is designed to be operated on more than one voltage or at more than one speed by reconnecting the winding, a connection diagram is also given on the name plate.

— Class of motors *A, B, C, D, E* or *F* is also to be mentioned on the name plate.

- *Standard types of squirrel cage motors are* :
  - — Class *A* motors—Low-impedance squirrel cage rotor motors.
  - — Class *B* motors—High-reactance squirrel cage rotor motors.
  - — Class *C* motors—High starting torque, low starting current-double squirrel cage motors..
  - — Class *D* motors—High resistance squirrel cage rotor motors.
  - — Class *E* motors.
  - — Class *F* motors.

24. SQUIRREL-CAGE MOTORS—ADVANTAGES, DISADVANTAGES AND APPLICATIONS

Advantages :

1. Cheaper in cost.
2. Simple and rugged in construction.
3. Maintenance cost is low.
4. Explosion proof.
5. Can be cooled better because of base end rings.
6. More pull-out torque and greater maximum power output.
7. Simple starting arrangement.
8. Nearly constant speed.
9. High overload capacity.
10. High power factor.

Disadvantages :

1. *Poor starting torque* due to low rotor resistance.
2. At the instant of switching on, *draws a large current* from the line.
3. Very *sensitive to change in supply voltage*.
4. *At light load the power factor is very low*.
5. Speed regulation *not possible*.

Applications. Suitable for industrial drives of small power where *speed control is not required* such as :

- Flour mills
- Printing machinery
- Other shaft drives of small power.

Note. When the squirrel cage motors are used for crane or hoist work where large starting torque is more important than high efficiency owing to the intermittent nature of load, the rotor resistance is increased by employing the end rings, and sometimes the bars as well, of high resistance metal such as German silver.

25. WOUND ROTOR (OR SLIP RING) INDUCTION MOTORS—ADVANTAGES, DISADVANTAGES AND APPLICATIONS

Advantages :

1. High starting torque.
2. High over-load capacity.

3. Nearly constant speed.
4. Low starting current (in comparison with squirrel cage motor).

Disadvantages :

1. Low power factor at light loads.
2. Lower efficiency and lower power factor (in comparison to squirrel cage motor).
3. Sensitivity to fluctuations in supply voltage.
4. Higher initial and maintenance costs.
5. Speed regulation is poor when operated with external resistances in the rotor circuit.

Applications. Suitable for most industrial drives of high power where high starting torque is required such as for driving :

- Line shafts
- Pumps
- Lifts
- Generators
- Winding machines
- Mills etc.

26. COMPARISON OF A SQUIRREL-CAGE AND A SLIP RING (OR PHASE WOUND) INDUCTION MOTORS

	Property	*Squirrel cage induction motor*	*Phase wound induction motor*
1.	**Starting**	The motor can be started with a star-delta starter and it eliminates the hazards of slip rings, brush gear etc.	The motor requires slip rings, brush gear, short-circuiting device and starting resistance etc.
2.	**Overhang**	Less	Large
3.	**Space factor in slots**	Better	Poor
4.	**Cooling**	The losses can be dissipated much efficiently since the end rings are bars and there is more space for providing an efficient fan.	Most of the space is taken up by the overhang and less space is left for the provision of a good cooling fan ; thus the cooling is not quite efficient.
5.	**Efficiency**	High efficiency (only for machines, which are not redesigned for high starting torque).	Low efficiency.
6.	**Starting torque**	For plain squirrel cage motor starting torque is poor which cannot be increased in an ordinary way.	Starting torque can be increased by the insertion of external resistance in each phase of the rotor.
7.	**Copper losses**	Since overhang is less and space factor is better the copper losses are *small*.	Due to large overhang and poor space factor, the copper losses are *more*.
8.	**Maintenance**	Less	More
9.	**Cost**	*Cheaper* (since there are less components and less labour required for winding).	Costlier

27. COMPARISON BETWEEN INDUCTION AND SYNCHRONOUS MOTORS

The comparison between induction and synchronous motors is given below :

S. No.	*Induction motor*	*Synchronous motor*
1.	It has got self-starting torque and no special means are required for starting.	It is inherently not self-starting and some external means are required for its starting.
2.	Does not require D.C. excitation.	Requires D.C. excitation.
3.	Speed can be controlled but to small extent.	Speed control not possible.
4.	Its speeds falls with the increase in load and is always less than synchronous speed.	Its average speed is constant and independent of load.
5.	It operates at only lagging power factor, which becomes very poor at light loads.	It can be operated under a wide range of power factor, both lagging and leading.
6.	Its torque is more sensitive to change in supply voltage.	Its torque is less sensitive to change in supply voltage.
7.	Breakdown torque is proportional to the square of the supply voltage.	Breakdown torque is proportional to the supply voltage.
8.	More simple and less costly comparatively.	More complicated and more costly comparatively.
9.	Employed for supplying mechanical load only.	Employed for supplying mechanical load as well as for power factor improvement.

HIGHLIGHTS

1. *An induction motor is simply an electric transformer whose magnetic circuit is separated by an air gap into two relatively movable portions, one carrying the primary and the other the secondary winding.*
2. *Synchronous speed,* $N_s = \frac{120f}{p}$, *where f is the frequency and p the number of poles.*
3. *Slip,* $s = \frac{N_s - N}{N_s}$, *where N is the motor speed (r.p.m).*
4. *Frequency of the rotor current,* $f_r = sf$, *where s is the slip and f is the supply frequency.*
5. *Torque,* $$T = \frac{ksR_2 E_2^2}{R_2^2 + s^2 X_2^2}$$
   Starting torque, $$T_{st} = \frac{kR_2 E_2^2}{R_2^2 + X_2^2}$$
   Condition for maximum torque, $s = \frac{R_2}{X_2}$
   Maximum torque, $$T_{max} = \frac{kE_2^2}{2X_2}$$
6. **Starting torque** *is proportional to the* **square of applied voltage.**
7. $$\frac{T_f}{T_m} = \frac{2s_f s_{mT}}{s_f^2 + s_{mT}^2}, \quad \text{where } s_f = \frac{\text{rotor resistance}}{\text{rotor standstill reactance}}.$$
   $$\frac{T_{st}}{T_{max}} = \frac{2s_{mT}}{s_{mT}^2 + 1}.$$

8. $$\frac{Rotor\ copper\ loss}{Rotor\ gross\ output} = \frac{s}{1-s}.$$

9. $$\frac{Rotor\ output}{Rotor\ input} = 1 - s.$$

OBJECTIVE TYPE QUESTIONS

(A) Choose the Correct Answer :

1. Which of the following components is usually fabricated out of silicon steel ?
(*a*) Bearings (*b*) Shaft
(*c*) Stator core (*d*) None of the above.
2. The frame of an induction motor is usually made of
(*a*) silicon steel (*b*) cast iron
(*c*) aluminium (*d*) bronze.
3. The shaft of an induction motor must be
(*a*) stiff (*b*) flexible
(*c*) hollow (*d*) any of the above.
4. The shaft of an induction motor is made of
(*a*) high speed steel (*b*) stainless steel
(*c*) carbon steel (*d*) cast iron.
5. In an induction motor, on no-load the slip is generally
(*a*) less than 1% (*b*) 1.5%
(*c*) 2% (*d*) 4%.
6. In medium sized induction motors, the slip is generally around
(*a*) 0.04% (*b*) 0.4%
(*c*) 4% (*d*) 14%.
7. In squirrel cage induction motors, the rotor slots are usually given slight skew in order to
(*a*) reduce windage losses (*b*) reduce eddy currents
(*c*) reduce accumulation of dirt and dust (*d*) reduce magnetic hum.
8. In case the air gap in an induction motor is increased
(*a*) the magnetising current of the rotor will decrease
(*b*) the power factor will decrease
(*c*) speed of motor will increase (*d*) the windage losses will increase.
9. Slip rings are usually made of
(*a*) copper (*b*) carbon
(*c*) phosphor bronze (*d*) aluminium.
10. A 3-phase 440 V, 50 Hz induction motor has 4% slip. The frequency of rotor e.m.f. will be
(*a*) 200 Hz (*b*) 50 Hz
(*c*) 2 Hz (*d*) 0.2 Hz.
11. If N_s is the synchronous speed and s the slip, then actual running speed of an induction motor will be
(*a*) N_s (*b*) $s \cdot N_s$
(*c*) $(1 - s)N_s$ (*d*) $(N_s - 1)s$.
12. The efficiency of an induction motor can be expected to be nearly
(*a*) 60 to 90% (*b*) 80 to 90%
(*c*) 95 to 98% (*d*) 99%.
13. The number of slip rings on a squirrel cage induction motor is usually
(*a*) two (*b*) three
(*c*) four (*d*) none.

14. The starting torque of a squirrel-cage induction motor is
(*a*) low
(*b*) negligible
(*c*) same as full-load torque
(*d*) slightly more than full-load torque.

15. A double squirrel-cage induction motor has
(*a*) two rotors moving in opposite direction
(*b*) two parallel windings in stator
(*c*) two parallel windings in rotor
(*d*) two series windings in stator.

16. The ratio of starting torque to normal torque in case of a star-delta starter will be
(*a*) 0.35
(*b*) 0.67
(*c*) 1.07
(*d*) 1.35.

17. Star-delta starting of motors is not possible in case of
(*a*) single phase motors
(*b*) variable speed motors
(*c*) low horse power motors
(*d*) high speed motors.

18. The terms 'cogging' is associated with
(*a*) three phase transformers
(*b*) compound generators
(*c*) D.C. series motors
(*d*) induction motors.

19. In case of the induction motors the torque is
(*a*) inversely proportional to ($\sqrt{\text{slip}}$)
(*b*) directly proportional to $(\text{slip})^2$
(*c*) inversely proportional to slip
(*d*) directly proportional to slip.

20. Stepless speed control of induction motor is possible by which of the following methods ?
(*a*) E.m.f. injection in rotor circuit
(*b*) Changing the number of poles
(*c*) Cascade operation
(*d*) None of the above.

21. Rotor rheostat control method of speed control is used for
(*a*) squirrel-cage induction motors only
(*b*) slip ring induction motors only
(*c*) both (*a*) and (*b*)
(*d*) none of the above.

22. In the circle diagram for induction motor, the diameter of the circle represents
(*a*) slip
(*b*) rotor current
(*c*) running torque
(*d*) line voltage.

23. For which motor the speed can be controlled from rotor side ?
(*a*) Squirrel-cage induction motor
(*b*) Slip-ring induction motor
(*c*) Both (*a*) and (*b*)
(*d*) None of the above.

24. If any two phases for an induction motor are interchanged
(*a*) the motor will run in reverse direction
(*b*) the motor will run at reduced speed
(*c*) the motor will not run
(*d*) the motor will burn.

25. An induction motor is
(*a*) self-starting with zero torque
(*b*) self-starting with high torque
(*c*) self-starting with low torque
(*d*) non-self starting.

26. The maximum torque in an induction motor depends on
(*a*) frequency
(*b*) rotor inductive reactance
(*c*) square of supply voltage
(*d*) all of the above.

27. In three-phase squirrel-cage induction motors
(*a*) rotor conductor ends are short circuited through slip rings
(*b*) rotor conductors are short-circuited through end rings
(*c*) rotor conductors are kept open
(*d*) rotor conductors are connected to insulation.

28. In a three-phase induction motor, the number of poles in the rotor winding is always
(*a*) zero
(*b*) more than the number of poles in stator
(*c*) less than number of poles in stator
(*d*) equal to number of poles in stator.

29. *DOL* starting of induction motors is usually restricted to
(*a*) low horsepower motors
(*b*) variable speed motors
(*c*) high horsepower motors
(*d*) high speed motors.

30. No-load current in an induction motor is nearly
(*a*) 0.1 per cent of full-load current (*b*) 1 to 5 per cent of full-load current
(*c*) 15 to 25 per cent of full-load current (*d*) 50 to 60 per cent of full-load current.

31. The speed of a squirrel-cage induction motor can be controlled by all of the following except
(*a*) changing supply frequency (*b*) changing number of poles
(*c*) changing winding resistance (*d*) reducing supply voltage.

32. The 'crawling' in an induction motor is caused by
(*a*) high loads (*b*) low voltage supply
(*c*) improper design of machine (*d*) harmonics developed in the motor.

33. The power factor of an induction motor under no-load conditions will be closer to
(*a*) 0.2 lagging (*b*) 0.2 leading
(*c*) 0.5 leading (*d*) unity.

34. The 'cogging' of an induction motor can be avoided by
(*a*) proper ventilation (*b*) using *DOL* starter
(*c*) auto-transformer starter
(*d*) having number of rotor slots more or less than the number of stator slots (not equal).

35. One phase of a three-phase partly loaded induction motor power supply got opened up while the motor was in motion. This is likely to result in
(*a*) stalling of the motor
(*b*) burning of one phase winding of the motor
(*c*) burning of all three phases of winding of the motor
(*d*) undisturbed running of motor.

36. If an induction motor with certain ratio of rotor to stator slots, runs at $\frac{1}{7}$ of the normal speed, the phenomenon will be termed as
(*a*) humming (*b*) hunting
(*c*) crawling (*d*) cogging.

37. It is advisable to avoid line starting of induction motor and use starter because
(*a*) it will run in reverse direction
(*b*) it will pick up very high speed and may go out of step
(*c*) motor takes five to seven times its full-load current
(*d*) starting torque is very high.

38. The speed characteristics of an induction motor closely resemble the speed-load characteristics of which of the following machines ?
(*a*) D.C. series motor (*b*) D.C. shunt motor
(*c*) Universal motor (*d*) None of the above.

39. In a slip-ring induction motor, resistance is connected in rotor phases
(*a*) to increase starting torque (*b*) to limit starting current
(*c*) to limit the starting current and increase the starting torque
(*d*) none of the above.

40. Which type of bearing is provided in small induction motors to support the rotor shaft ?
(*a*) Ball bearings (*b*) Cast iron bearings
(*c*) Bush bearings (*d*) None of the above.

41. A pump induction motor is switched on to a supply 30% lower than its rated voltage. The pump runs. What will eventually happen ? It will
(*a*) stall after sometime (*b*) stall immediately
(*c*) continue to run at lower speed without damage
(*d*) get heated and subsequently get damaged.

42. 5 H.P., 50-Hz, 3-phase, 440 V, induction motors are available for the following r.p.m. Which motor will be the costliest ?
(*a*) 730 r.p.m. (*b*) 960 r.p.m.
(*c*) 1440 r.p.m. (*d*) 2880 r.p.m.

43. A 3-phase slip ring motor has
(*a*) double cage rotor (*b*) wound rotor
(*c*) short-circuited rotor (*d*) any of the above.

44. The torque developed in an induction motor is nearly proportional to
(*a*) $\frac{1}{V}$ (*b*) V
(*c*) V^2 (*d*) $\frac{1}{V^2}$
where V = applied voltage.

45. Short-circuit test on an induction motor cannot be used to determine
(*a*) windage losses (*b*) copper losses
(*c*) transformation ratio (*d*) power scale of circle diagram.

46. In a three-phase induction motor
(*a*) iron losses in stator will be negligible as compared to that in rotor
(*b*) iron losses in motor will be negligible as compared to that in rotor
(*c*) iron losses in stator will be less than that in rotor
(*d*) iron losses in stator will be more than that in rotor.

47. In case of 3-phase induction motors, plugging means
(*a*) putting the motor directly on line without a starter
(*b*) locking of rotor due to harmonics
(*c*) starting the motor on load which is more than the rated load
(*d*) interchanging two supply phases for quick stopping.

48. Which is of the following data is required to draw the circle diagram for an induction motor ?
(*a*) Block rotor test only (*b*) No load test only
(*c*) Block rotor test, and no-load test
(*d*) Block rotor test, no-load test and stator resistance test.

49. In three-phase induction motors sometimes copper bars are placed deep in the rotor to
(*a*) improve starting torque (*b*) reduce copper losses
(*c*) improve efficiency (*d*) improve power factor.

50. In a three-phase induction motor
(*a*) power factor at starting is high as compared to that while running
(*b*) power factor at starting is low as compared to that while running
(*c*) power factor at starting is the same as that while running.

51. The circle diagram for an induction motor cannot be used to determine
(*a*) efficiency (*b*) power factor
(*c*) frequency (*d*) output.

52. The current taken by a three-phase induction motor during plugging will be
(*a*) normal full-load current (*b*) more than blocked rotor current
(*c*) equal to blocked rotor current (*d*) less than blocked rotor current.

53. Blocked rotor test on induction motors is used to find out
(*a*) leakage reactance (*b*) power factor on short circuit
(*c*) short-circuit current under rated voltage (*d*) all of the above.

54. Lubricant used for ball bearing is usually
(*a*) graphite (*b*) grease
(*c*) mineral oil (*d*) molasses.

55. An induction motor can run at synchronous speed when
(*a*) it is run on load
(*b*) it is run in reverse direction
(*c*) it is run on voltage higher than the rated voltage
(*d*) e.m.f. is injected in the rotor circuit.

56. Which of the following can be reduced by skewing in induction motor ?
(*a*) Noise
(*b*) Vibration
(*c*) Synchronous cusps
(*d*) All of the above.

57. Which motor is preferred for use in mines where explosive gases exist ?
(*a*) Air motor
(*b*) Induction motor
(*c*) D.C. shunt motor
(*d*) Synchronous motor.

58. The torque developed by a 3-phase induction motor least depends on
(*a*) rotor current
(*b*) rotor power factor
(*c*) rotor e.m.f.
(*d*) shaft diameter.

59. In an induction motor if air-gap is increased
(*a*) the power factor will be low
(*b*) windage losses will be more
(*c*) bearing friction will reduce
(*d*) copper loss will reduce.

60. In an induction motor
(*a*) $s = \frac{N - N_s}{N}$
(*b*) $s = \frac{N_s - N}{N_s}$
(*c*) $s = \frac{N_s - N}{N}$
(*d*) $s = \frac{N - N_s}{N_s}$.

61. When R_2 is the rotor resistance, X_2 the rotor reactance at supply frequency and s the slip, then the condition for maximum torque under running conditions will be
(*a*) $sR_2X_2 = 1$
(*b*) $sR_2 = X_2$
(*c*) $R_2 = sX_2$
(*d*) $R_2 = s^2X_2$.

62. How many different speeds can be obtained in cascade control of two induction motors ?
(*a*) 2
(*b*) 4
(*c*) 8
(*d*) 16.

63. In case of a double cage induction motor, the inner cage has
(*a*) high inductance and low resistance
(*b*) low inductance and high resistance
(*c*) low inductance and low resistance
(*d*) high inductance and high resistance.

64. Open circuit test is usually conducted on
(*a*) slip ring motors only
(*b*) wound rotor motors only
(*c*) either of (*a*) and (*b*) above
(*d*) none of the above.

65. When two motors of 12 and 8 poles respectively are connected to a 50-Hz supply, which of the following no-load speed cannot be obtained ?
(*a*) 300 r.p.m.
(*b*) 450 r.p.m.
(*c*) 500 r.p.m.
(*d*) 750 r.p.m.
(*e*) 1500 r.p.m.

66. The low power factor of induction motor is due to
(*a*) rotor leakage reactance
(*b*) stator reactance
(*c*) the reactive lagging magnetizing current necessary to generate the magnetic flux
(*d*) all of the above.

67. During back to back test
(*a*) both machines run as induction motors
(*b*) both machines run as induction generators
(*c*) one machine runs as induction motor and the other as induction generator
(*d*) one machine runs as induction motor and the other as synchronous generator.

68. Phase advancers are used with induction motors to
(*a*) reduce noise
(*b*) reduce vibrations
(*c*) reduce copper losses
(*d*) improve power factor.

69. **During dynamic braking of induction motor**
(*a*) stator connections are reversed **(*b*) motor is used as induction generator**
(*c*) stator winding is excited by direct current **(*d*) fictitious load is applied on motor.**

70. **The operation of repeated starting and stopping of an induction motor for exact positioning of the driven equipment is called**
(*a*) crawling **(*b*) balancing**
(*c*) plugging **(*d*) jogging.**

71. **In moist surroundings which frame of motor will be preferred ?**
(*a*) Open type **(*b*) Totally enclosed fan cooled**
(*c*) Drip proof **(*d*) Splash proof.**

72. **The principle of operation of a three-phase induction motor closely resembles to that of which of the following ?**
(*a*) Synchronous motor **(*b*) D.C. series motor**
(*c*) Transformer with short-circuited secondary **(*d*) Transformer with open-circuited secondary.**

73. **Under which of the following conditions a squirrel-cage induction motor would be preferred over a wound rotor induction motor ?**
(*a*) When external voltage is to be necessarily introduced into the rotor
(*b*) When wide range of speed control is desired
(*c*) When cost is the major consideration
(*d*) When higher starting torque is required.

74. **Which of the following statement is not valid when a squirrel-cage induction motor operates under no-load ?**
(*a*) Rotor induced e.m.f. is low **(*b*) Rotor current is low**
(*c*) Power factor is low **(*d*) Slip is low.**

ANSWERS

1. (*c*)	**2.** (*b*)	**3.** (*a*)	**4.** (*c*)	**5.** (*a*)	**6.** (*c*)	**7.** (*d*)
8. (*b*)	**9.** (*c*)	**10.** (*c*)	**11.** (*c*)	**12.** (*b*)	**13.** (*d*)	**14.** (*a*)
15. (*c*)	**16.** (*a*)	**17.** (*a*)	**18.** (*d*)	**19.** (*d*)	**20.** (*b*)	**21.** (*b*)
22. (*b*)	**23.** (*b*)	**24.** (*a*)	**25.** (*c*)	**26.** (*d*)	**27.** (*b*)	**28.** (*d*)
29. (*a*)	**30.** (*d*)	**31.** (*c*)	**32.** (*d*)	**33.** (*a*)	**34.** (*d*)	**35.** (*d*)
36. (*c*)	**37.** (*c*)	**38.** (*b*)	**39.** (*b*)	**40.** (*a*)	**41.** (*d*)	**42.** (*a*)
43. (*b*)	**44.** (*c*)	**45.** (*a*)	**46.** (*d*)	**47.** (*d*)	**48.** (*d*)	**49.** (*a*)
50. (*b*)	**51.** (*a*)	**52.** (*b*)	**53.** (*d*)	**54.** (*b*)	**55.** (*d*)	**56.** (*d*)
57. (*a*)	**58.** (*d*)	**59.** (*a*)	**60.** (*b*)	**61.** (*c*)	**62.** (*b*)	**63.** (*a*)
64. (*a*)	**65.** (*b*)	**66.** (*d*)	**67.** (*c*)	**68.** (*d*)	**69.** (*c*)	**70.** (*d*)
71. (*b*)	**72.** (*c*)	**73.** (*c*)	**74.** (*d*).			

(B) Say 'Yes' or 'No' :

1. **Three-phase induction motor has a low efficiency.**

2. **An induction motor is simply an electric transformer whose magnetic circuit is separated by an air gap into two relatively movable portions, one carrying the primary and the other the secondary winding.**

3. **Frames of electrical machines house the stator core.**

4. **The number of slots in the rotor should always be equal to the number of slots in the stator.**

5. **For large and heavy rotors journal bearings are used.**

6. **The slips rings are made of aluminium.**

7. **The difference between the synchronous speed and rotor speed is known as slip.**

8. **Starting torque,** $T_{st} = \dfrac{kR_2E_2^2}{R_2^2 + X_2^2}$.

9. Starting torque is inversely proportional to the square of the applied voltage.

10. Stator iron loss is practically constant.

11. $\dfrac{\text{Rotor copper loss}}{\text{Rotor gross output}} = \dfrac{1-s}{s}$.

12. Circle diagram of an induction motor can be drawn by using the data obtained from no-load test, short-circuit test and stator resistance test.

13. The synchronous speed of the nth order harmonic is $\frac{1}{n}$th of the synchronous speed of fundamental.

14. 'Cogging' of squirrel-cage motors can be easily overcome by making the number of rotor slots prime to the number of stator slots.

15. A double squirrel-cage motor has two independent squirrel-cage windings on the rotor, each having its own set of slots.

ANSWERS

1. No	**2.** Yes	**3.** Yes	**4.** No	**5.** Yes	**6.** No	**7.** Yes
8. Yes	**9.** No	**10.** Yes	**11.** No	**12.** Yes	**13.** Yes	**14.** Yes
15. Yes.						

THEORETICAL QUESTIONS

1. Explain the principle of operation of the polyphase induction motor.

2. Show that a rotating magnetic field can be produced by the use of 3-phase currents of equal magnitude.

3. What is meant by slip in an induction motor ? Develop an expression for the frequency of rotor currents in it.

4. Derive an expression for the torque of an induction motor and obtain the condition for maximum torque.

5. Show that in a 3-phase induction motor with negligible stator impedance maximum torque is developed at slip $s = \dfrac{R_2}{X_2}$, where R_2 and X_2 are rotor resistance and standstill reactance respectively.

6. State the effects of increasing rotor resistance on starting current, starting torque, maximum torque and full-load slip of an induction motor.

7. Show that in an induction motor the rotor input : power developed : rotor copper losses : : $1 : (1-s) : s$, where s is the fractional slip.

8. Draw equivalent circuit of a 3-phase induction motor.

9. Describe and explain how to perform a locked rotor test. What data can be obtained by the test ?

10. What are no-load and blocked rotor tests ? What sort of losses can be measured by these tests ?

11. How would you determine circle diagram of a 3-phase induction motor experimentally ?

12. Describe the constructional details of a 3-phase squirrel-cage and phase wound induction motors. Also discuss the applications of various types of starters used for starting these motors.

13. Why at all starters are necessary for starting the induction motors ? What are the various types of starters used for squirrel-cage motors ? Discuss them.

UNSOLVED EXAMPLES

1. A 3-phase, 4-pole, 50-Hz induction motor is running at 1455 r.p.m. Find the slip speed and slip. [**Ans.** 45 r.p.m., 3 per cent]

2. An induction motor having 8-poles runs at 50-Hz supply. If it operates at full-load at 720 r.p.m., calculate the slip. [**Ans.** 4 per cent]

3. A 6-pole alternator running at 1000 r.p.m. supplies an 8-pole induction motor. What is the actual speed of the motor if the slip is 2.5%. [**Ans.** 731.25 r.p.m.]

4. A slip ring motor runs at 290 r.p.m. on full-load when connected to 50-Hz supply. Calculate :

(*i*) Number of poles (*ii*) Absolute slip
(*iii*) Fractional slip (*iv*) Percentage slip
(*v*) Frequency of rotor currents. [**Ans.** 20, 10 r.p.m., 0.0333, 3.33%, 1.66 Hz]

5. A 12-pole, 3-phase alternator is coupled to an engine running at 500 r.p.m. It supplies a 3-phase induction motor having a full- load speed of 1440 r.p.m. Find the percentage slip and number of poles of motor. [**Ans.** 4 per cent, 4 poles]

6. A 4-pole induction motor is run at 1450 r.p.m. from a 50-Hz supply. Find the percentage slip and frequency of rotor current. [**Ans.** 3.33%, 1.67 Hz]

7. If the e.m.f. in the stator of an 8-pole induction motor has a frequency of 50-Hz and that in the rotor 1.5 Hz, at what speed is the motor running and what is the slip ? [**Ans.** 727.5 r.p.m., 3 per cent]

8. A three-phase slip ring induction motor gives a reading of 55 V across slip rings on open circuit when at rest with normal stator voltage applied. The rotor is star-connected and has impedance $(0.7 + j5)\ \Omega$ per phase. Find the rotor current when the machine is

(*i*) at standstill with the slip rings joined to a star-connected starter with a phase impedance of $(4 + j3)\ \Omega$; and
(*ii*) running normally with a 5 per cent slip.

[**Ans.** (*i*) 3.425 A at 0.506 p.f. (lag) ; (*ii*) 2.14 A at 0.942 p.f. (lag)

9. A 3-phase, 400 V, star-connected induction motor has a star-connected rotor with a stator to rotor turn ratio of 6.5. The rotor resistance and stand-still reactance per phase are 0.05 and 0.25 Ω respectively. What should be the value of external resistance per phase to be inserted in the rotor circuit to obtain maximum torque at starting and what will be the rotor starting current with this resistance ? [**Ans.** 0.2 Ω, 100 A (app.)]

10. If the motor has a rotor resistance of 0.02 Ω and a standstill reactance of 0.1 Ω what must be the value of the total resistance of a starter for the rotor circuit for maximum torque to be exerted at starting ? [**Ans.** 0.08 Ω]

11. The rotor of a 3-phase 440 V, 50 Hz, 4-pole induction motor has a resistance of 0.3 Ω per phase and an inductance of 0.008 H per phase. The ratio of stator to rotor turns is 2 : 1. The stator is delta connected and rotor is star-connected. Determine the voltage between slip rings when the motor is at standstill. Calculate the standstill rotor current and rotor current when the motor is operating with 3 per cent slip. [**Ans.** 381 V, 87 A, 22 A]

12. A 3300 V, 24-phase, 50-Hz, 3-phase star-connected induction motor has a slip-ring rotor resistance of 0.016 Ω and standstill reactance of 0.265 Ω per phase. Calculate :

(*i*) The speed at maximum torque.
(*ii*) Ratio of full-load torque to maximum torque if full-load torque is obtained at 247 r.p.m.

[**Ans.** 235 r.p.m., 0.382]

13. A 6-pole, 3-phase, 50 Hz induction motor runs on full-load with a slip of 4%. If the rotor standstill impedance per phase is $(0.01 + j0.05)\ \Omega$, calculate the available maximum torque in terms of full-load torque. Also determine the speed at which the maximum torque occurs. [**Ans.** $\frac{T_m}{T_f} = 2.6$, 800 r.p.m.]

14. A 6-pole, 3-phase, 50 Hz induction motor develops maximum torque of 200 Nm at a speed of 960 r.p.m. Determine the torque exerted by the motor at 5 per cent slip. The rotor resistance per phase is 0.5 Ω. [**Ans.** 195 Nm]

15. The rotor resistance and standstill reactance per phase of a 3-phase slip ring induction motor are 0.02 Ω and 0.1 Ω respectively. What should be the value of the external resistance per phase to be inserted in the rotor circuit to give maximum torque at starting ? [**Ans.** 0.08 Ω]

16. The star-connected rotor of an induction motor has a standstill impedance of $(0.4 + j4)\ \Omega$ per phase, and the rheostat impedance per phase is $(6 + j2)\ \Omega$. The motor has an induced e.m.f. of 80 V between slip rings at standstill when connected to its normal supply voltage. Find :

(*i*) The rotor current at standstill with the rheostat in the circuit.
(*ii*) When running short-circuited with slip of 0.03. [**Ans.** 5.27 A, 3.3 A]

17. A 4-pole, 50 Hz, 3-phase, induction motor develops a maximum torque of 115 N-m at 1365 r.p.m. The resistance of the star-connected rotor is 0.2 Ω/phase. Calculate the value of resistance that must be inserted in series with each rotor phase to produce a starting torque equal to half the maximum torque. [**Ans.** 0.4 Ω or 8.1 Ω]

18. A 4-pole, 50-Hz, 3-phase induction motor has rotor resistance and reactance of 0.03 Ω and 0.12 Ω per phase respectively. Determine :

(*i*) The value of speed at which maximum torque occurs.

(*ii*) The value of external rotor resistance per phase to be inserted to obtain 80% of maximum torque at starting. [**Ans.** 1125 r.p.m., 0.02418 Ω]

19. The resistance and standstill reactance of each phase of a 3-phase induction motor with star-connected rotor are 0.06 Ω and 0.4 Ω respectively. The full-load slip is 4%. Calculate the resistance per phase of a star-connected rheostat, which when connected to the rotor, will give a pull out torque at one half of the full-load speed. What is then the power factor ? [**Ans.** 0.148 Ω, 0.707 (lag)]

20. The rotor resistance and standstill reactance of a 3-phase induction motor are respectively 0.015 Ω and 0.09 Ω per phase. At normal voltage, the full-load slip is 3 per cent. Estimate the percentage reduction in stator voltage to develop full-load torque at half full-load speed. Also calculate the power factor. [**Ans.** 22.9%, 0.31 (app.)]

Rotor Output Losses, Efficiency

21. The power input to a 3-phase induction motor is 60 kW. The stator-losses total 1.5 kW. Find the total mechanical power developed if the motor is running with a slip of 4 per cent. [**Ans.** 56.16 kW]

22. The rotor e.m.f. of a 3-phase, 6-pole, 400 V, 50-Hz induction motor alternates at 3-Hz. Compute the speed and percentage slip of the motor. Find the rotor copper loss per phase if the full input to the rotor is 111.9 kW. [**Ans.** 940 r.p.m., 6%, 2.238 kW]

23. A 3-phase, 50-Hz, induction motor draws 50 kW from the mains. If the stator losses are 2 kW and the rotor e.m.f. is observed to make 100 complete oscillations/alterations per minute, determine :

(*i*) Rotor copper loss. (*ii*) Gross mechanical output. [**Ans.** 1.6 kW, 46.4 kW]

24. The power input to the rotor of a 3-phase, 50-Hz, 6-pole, slip ring induction motor is 40 kW and the motor runs at 960 r.p.m. The rotor resistance per phase is 0.25 Ω. Determine the value of the rotor current per phase. [**Ans.** 46.2 A]

25. The power input to the rotor of a 440 V, 50-Hz, 3-phase, 6-pole induction motor is 60 kW. It is observed that the rotor e.m.f. makes 90 complete cycles per minute. Calculate :

(*i*) Slip, (*ii*) Rotor speed,

(*iii*) Rotor copper loss per phase, (*iv*) Mechanical power developed, and

(*v*) Rotor resistance/phase if the rotor current is 60 A.

[**Ans.** 0.03, 970 r.p.m., 600 W, 58.2 kW, 0.167 Ω]

Equivalent Circuit of an Induction Motor

26. The maximum torque of a 3-phase induction motor occurs at a slip of 12 per cent. The motor has an equivalent secondary resistance of 0.08 Ω/phase. Calculate the equivalent load resistance R_L, the equivalent load voltage V_L and the current at this slip if the gross power output is 9 kW.

[**Ans.** 0.587 Ω/phase, 42 V, 71.6 A]

27. The stator impedance and equivalent rotor resistance of a 400 V, 3-phase star-connected induction motor are $(0.06 + j0.2)$ Ω and equivalent rotor impedance of $(0.06 + j0.22)$ Ω. Neglecting exciting current, find the maximum gross power and the slip at which it occurs. [**Ans.** 12%, 142.9 kW]

28. Estimate the stator current, equivalent rotor current, efficiency, output, and its power factor at a slip of 5% for a motor having the following data :

Stator impedance = $1.0 + j3.0$ Ω, rotor standstill impedance = $1.0 + j2.0$ Ω, no-load shunt impedance = $10 + j50$ Ω, volts/phase = 250 volts. [**Ans.** 14.34 A, 11.6 A, 83.6%, 7,670 W, 0.853]

Starting of Induction Motor

29. A small 3-phase induction motor has a short-circuit current equal to 4 times the full-load current. Determine the starting torque as a percentage of full-load torque if full-load slip is 2.5 per cent.

[**Ans.** 40% of full-load torque]

30. A 3-phase, squirrel-cage induction motor has a short-circuit current equal to 5 times the full-load current. Find the starting torque as a percentage of full-load torque if the motor is started by

(*i*) direct switching to the supply ; (*ii*) a star-delta starter ;
(*iii*) an auto-transformer ; and (*iv*) a resistance in the stator circuit.

The starting current in (*iii*) and (*iv*) is limited to 2.5 times the full-load current and the full-load slip is 4 per cent. [**Ans.** T_f, 33.3% T_f, 50%T_f, 25% T_f]

31. Determine a suitable auto-transformer ratio for starting an induction motor with a supply current not exceeding twice full-load current. Use the following data :

Short-circuit current ≐ 5 times the full-load current, full-load slip = 3%. Estimate the torque in terms of full-load torque. Ignore magnetising current. [**Ans.** 30%]

32. Find the percentage tapping required on an auto-transformer required for a cage motor to start the motor against $\frac{1}{4}$ of full-load torque. The short-circuit current at normal voltage is 4 times the full-load current and the full-load slip is 3 per cent. [**Ans.** K = 72.2%]

33. A squirrel-cage type induction motor when started by means of a star-delta starter takes 20 per cent of full-load line current and develops 40 per cent of full-load torque at starting. Calculate the starting torque and current in terms of full-load values, if an auto-transformer with 75 per cent tapping were employed.

[**Ans.** 3.375I_f, 67% full-load torque]

34. A 10 kW, 200 V, 3-phase induction motor has at full-load an efficiency of 87% and power factor of 0.85. At standstill the motor draws 5 times the full-load current and develops 1.5 times full-load torque. An auto-transformer is installed to reduce the starting current to give full-load torque at starting. Neglecting magnetising current, determine :

(*i*) The voltage to be applied at starting. (*ii*) The full-load line current.
(*iii*) The current drawn by the motor at start.
(*iv*) The current supplied to the primary of the autotransformer. [**Ans.** 163.3 V, 39 A, 159.2 A, 130 A]

35. A 3-phase squirrel-cage induction motor has a ratio of maximum torque to full-load torque as 2.5 : 1. Determine the ratio of actual starting torque to full-load torque for :

(*i*) direct starting ; (*ii*) star-delta starting ;
(*iii*) auto-transformer starting with tapping of 60%.

The rotor resistance and standstill reactance per phase are 0.4 Ω and 4 Ω respectively.

[**Ans.** 0.495, 0.165, 0.179]

7

Single Phase Motors

1. General aspects. 2. Types of single-phase motor. 3. Single-phase induction motors. 4. Split-phase motors. 5. Single-phase commutator motors. 6. Single-phase synchronous motors. 7. Stepper motors. 8. Servo motors. 9. Tachometer generators—Highlights—Objective type questions—Theoretical questions.

1. GENERAL ASPECTS

- The number of machines operating from single-phase supplies is greater than all other types taken in total. For the most part, however, they are only used in the smaller sizes, less than 5 kW and mostly in the fractional H.P. range. They operate at *lower power-factors* and are relatively *inefficient when compared with polyphase motors*. Though simplicity might be expected in view of the two-line supply, the *analysis is quite complicated*.
- *Single phase motors* perform a great variety of useful services in the home, the office, the factory, in business establishments, on the farm, and many other places where electricity is available. Since the requirements of the numerous applications differ so widely, the motor-manufacturing industry has developed several types of such machines, each type having operating characteristics that meet definite demands. For example, one type operates satisfactorily on direct current or any frequency up to 60 cycles ; another rotates at absolutely constant speed, regardless of load ; another develops considerable starting torque and still another, although not capable of developing much starting torque, is nevertheless extremely cheap to make and very rugged.

2. TYPES OF SINGLE-PHASE MOTOR

The single-phase motors may be of the following types :

1. **Single-phase Induction Motors :**

A. Split-phase motors

(*i*) Resistance-start motor (*ii*) Capacitor-start motor

(*iii*) Permanent-split (single-value) capacitor motor

(*iv*) Two-value capacitor motor.

B. Shaded-pole induction motor.

C. Reluctance-start induction motor.

D. Repulsion-start induction motor.

2. **Commutator-Type, Single-Phase Motors :**

A. Repulsion motor. B. Repulsion-induction motor.

C. A.C. series motor. D. Universal motor.

3. **Single-phase Synchronous Motors :**

A. Reluctance motor. B. Hysteresis motor.

C. Sub-synchronous motor.

3. SINGLE-PHASE INDUCTION MOTORS

3.1. Applications and Disadvantages

Applications :

- Single phase induction motors are in very wide use in industry especially in *fractional horse-power field*.

  They are extensively used for electric drive for *low power constant speed apparatus* such as *machine tools, domestic apparatus* and *agricultural machinery* in circumstances where a three-phase supply is *not* readily available.
- There is a large demand for single-phase induction motors in sizes ranging from a *fraction of horse-power upto about 5 H.P.*

Disadvantages :

Though these machines are useful for small outputs, they are not used for large powers as they suffer from many disadvantages and are never used in cases where three-phase machines can be adopted.

The main *disadvantages* of single-phase induction motors are :

1. Their output is only 50% of the three-phase motor, for a given frame size and temperature rise.
2. They have lower power factor.
3. Lower-efficiency.
4. These motors do not have inherent starting torque.
5. More expensive than three-phase motors of the same output.

3.2. Construction and Working

Construction :

- A single phase induction motor is similar to a 3-ϕ squirrel-cage induction motor in physical appearance. Its rotor is essentially the same as that used in 3-ϕ induction motors. Except for shaded pole motors, the stator is also very similar. There is a uniform air-gap between the stator and rotor but no electrical connection between them. It can be wound for any even number of poles, two, four and six being most common. Adjacent poles have opposite magnetic property and synchronous speed equation, $N_s = \dfrac{120f}{p}$ also applies.
- The stator windings differ in the following two aspects :
  - **Firstly** single phase motors are usually provided with concentric coils.
  - **Secondly,** these motors normally have two stator windings. In motors that operate with both windings energised, the winding with the *heaviest wire* is known as the **main winding** and the other is called the **auxiliary winding**. If the motor runs with auxiliary winding open, these windings are usually referred as *running* and *starting*.
  - In *most of motors the main winding is placed at the bottom of the slots and the starting winding on top but shifted 90° from the running winding*.

Working :

When the stator winding of a single phase induction motor is connected to single phase A.C. supply, a *magnetic field* is developed, *whose axis is always along the axis of stator coils*. The magnetic field produced by the stator coils is pulsating, varying sinusoidally with time. Currents are induced in the rotor conductors by transformer action, these currents being in such a direction as to oppose the stator m.m.f. Then the axis of the rotor m.m.f. wave coincides with that of the stator field, the *torque angle* is, therefore, *zero* and *no torque is developed on starting*. However, if the rotor is given a push by hand or by other means in either direction, it will pick-up the speed and continue to rotate in the same direction developing operating torque. Thus a single phase induction motor is not inherently self starting and requires some special means for starting.

The above mentioned behaviour of this type of motor can be explained by any one of the following theories :

1. Double revolving field theory
2. Cross-field theory.

The results given by both the theories are approximately same.

Double revolving field theory is described below :

The magnetic field produced by the stator coils is *pulsating*, varying sinusoidally with time. *Ferrari* pointed out that *such a field can be resolved into two equal fields but rotating in opposite directions with equal angular velocities. The maximum value of each component is equal to half the maximum of the pulsating field.*

If the initial time is such that the rotating vectors of the two component fields are along the *Y*-axis in the positive direction, the two component waves ϕ_1 and ϕ_2 coincide. The resultant of these

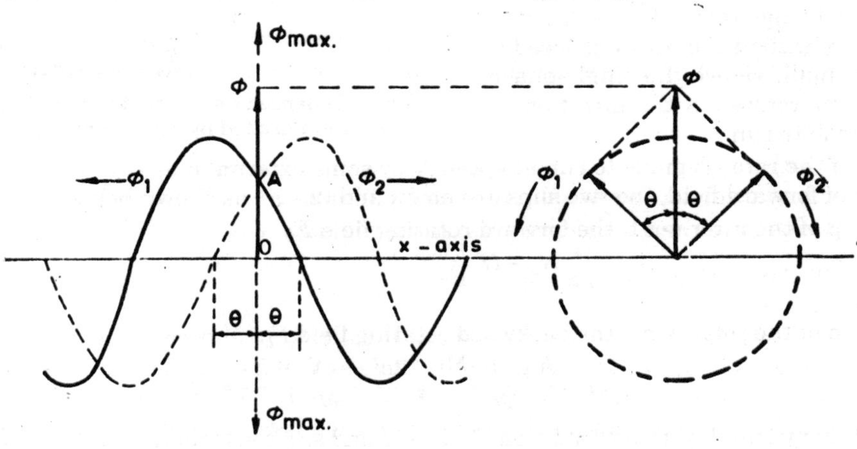

Fig. 1

two is ϕ_{max}. After a short interval of time the two vectors rotate, through an angle θ in their respective directions and the waves are shown to occupy the positions in Fig. 1. These waves intersect at *A* on the *Y*-axis and as the waves travel *A* moves along the *Y*-axis. Hence the resultant of these two component waves at any instant is equal to 2*OA*.

$$\phi_1 = OA = \phi_{1(max)} \cos (\omega t - \theta) \qquad ...(i)$$

$$\phi_2 = OA = \phi_{2(max)} \cos (\omega t + \theta) \qquad ...(ii)$$

and $$\phi_{1(max)} = \phi_{2(max)}$$

By expanding and adding (*i*) and (*ii*),

$$\phi_1 + \phi_2 = 2\phi_{1(max)} \cos \theta \cos \omega t$$

$$2OA = \phi_{(max)} \cos \theta \cos \omega t \qquad ...(iii)$$

which is the equation of the pulsating field and **proves Ferrari's statement.** Thus a single-phase induction motor is *not inherently self-starting*.

The existence of these two fluxes (*forward and backward*) rotating in opposite directions can be verified by supplying a fractional horsepower single-phase induction motor with rated voltage. The motor does not start, but if the shaft is turned by hand, say in clockwise direction, the rotor picks up speed. This means that the rotor conductors are rotating in the direction of that field which rotates in clockwise direction. When the motor is braked and stopped without switching off the supply, the rotor remains at rest. If now the shaft is turned by hand in anti-clockwise direction, the motor picks up speed in that direction. This means that the rotor conductors are now rotating in the direction of the other field.

This behaviour of the motor is *due to the presence of two opposing torques due to the two field.* When the rotor is at rest, (*i.e.* slip = 1) the two torques are equal but opposite in direction. Hence the net torque is zero and therefore the rotor remains at rest. Fig. 2. shows the torque variations due to the two fields. If the rotor is made to speed up in one direction, say in that in which T_1 increases, T_1 exceeds the opposing torque T_2 and the motor beings to accelerate. T_2 goes on diminishing until at the working speed it is negligibly small. Hence the single-phase induction motor rotates in the direction in which it is made to run.

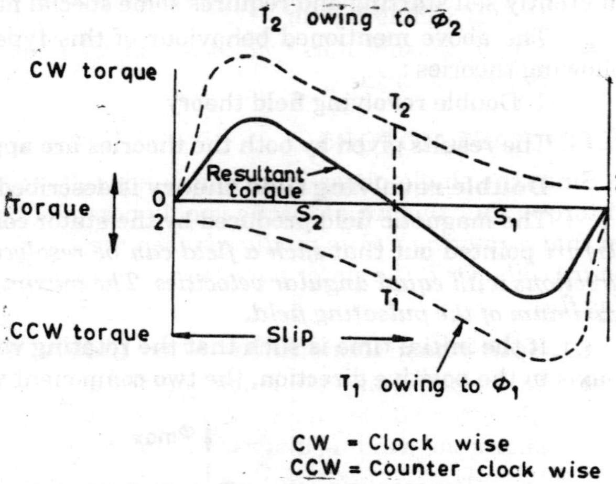

Fig. 2. Balanced torque at standstill in squirrel cage rotor excited by a single-phase winding.

Thus, if the rotor is made to run at speed N by some external means in any direction, say in the direction of forward field, the two slips are now s and $(2 - s)$, as shown below :

The slip of the rotor w.r.t. the forward rotating field F_f,

$$s_f = \frac{N_s - N}{N_s} = s \qquad ...(1)$$

The slip of the rotor w.r.t. the backward rotating field F_b,

$$s_b = \frac{N_s - (-N)}{N_s} = \frac{2N_s - (N_s - N)}{N_s} = (2 - s) \qquad ...(2)$$

- Under normal running condition $(2 - s) >> s$ and as a consequence the backward field rotor currents are much larger than at standstill and have a low power factor. The corresponding opposing rotor m.m.f., owing to stator impedance, causes the backward field to be greatly reduced in strength. On the other hand, the low-slip forward rotating field induces smaller currents of a high power factor in the rotor than at standstill. This leads to greatly strengthening of forward field. Thus weakening of backward field and strengthening of forward field depends upon the slip or speed of rotor and the difference increases with the decrease in slip w.r.t. the forward field or with the increase in rotor speed in forward direction.
- In a single phase induction motor, *the increase in rotor resistance increases the effectiveness of the backward field, which reduces the breakdown torque, lowers the efficiency and increases the slip corresponding to maximum torque.*
- The **performance characteristics** of a single phase induction motor are somewhat *inferior* to that of a 3-phase induction motor due to the presence of backward rotating field.
  — A single phase induction motor has a *lower breakdown torque at larger slip and increased power losses.*

- — Greater power input.
- — The *speed regulation* tends to be *poorer* than that for a polyphase motor.
- — The *power factor tends to be lower* (since the normal slip of a single phase induction motor under load conditions is rather greater than that of the corresponding 3-phase motor).

- In view of the above factors, a single phase induction motor has a *larger frame size* than that of 3-phase motor.
- Single phase motors *tend to be somewhat noiser than 3-phase motors which have no such pulsating torque.*

4. SPLIT-PHASE MOTORS

Since the single-phase induction is not self-starting, means must be provided to create an initial torque. But the initial torque is only possible if a rotating flux is created in the stator. It is known that *a rotating flux is produced when there is a difference of 90° between the currents of two stationary coils. Or if the stator possesses two fluxes having a large phase difference, the result is a rotating flux.*

4.1. Split-phase Resistance—Start Induction Motor

- In a split-phase induction motor the stator is provided with two parallel windings displaced 90 electrical degrees in space and somewhat less than 90° in time. Fig. 3 (*a*) shows the winding diagram of the *two windings* of a split-phase induction motor.

  The *starting winding* has fewer turns and is wound of smaller diameter copper than the running winding. The starting winding, therefore, *has a high resistance and low reactance.*

  The *running or main, winding* (heavier wire of more turns) has a *low resistance* and *high reactance.* Because of its lower impedance, the current in the running winding I_r is *higher* than the current in the starting winding, I_s.

  The phase relations of the lock-motor currents at the instant of starting are shown in Fig. 3 (*b*). The starting winding I_s lags the supply voltage by about 15°, while the greater running winding current lags the single-phase voltage by about 40°. Despite the fact that the current in the two space-quadrature windings are *not* equal, the quadrature components are practically equal.

  If the windings are displaced by 90° in space, and if their quadrature current components, which are displaced by 90° in time, are practically equal, an equivalent two-phase rotating field is produced at starting which develops sufficient starting torque to accelerate the rotor, in the direction of the rotating field produced by the currents.

- As the *motor speeds up,* the *torque developed increases.* Above 85 per cent of synchronous speed, the torque developed by the running winding (main winding) alone is actually greater than that developed by both windings, and it might be advantageous to *open the auxiliary circuit at this cross over point.* To allow for individual variations among motors and switches, however, the contacts are usually designed to open at 75 per cent of synchronous speed. This does not seriously affect the operation, because the running (or main) winding alone usually develops approximately 200 per cent of full-load torque at this speed.

- The *starting winding is not designed for continuous operation, and care should be exercised that it does not remain connected to the supply after it should have been disconnected by the switch.* This series switch is usually *centrifugally operated, and is rather inexpensive.*

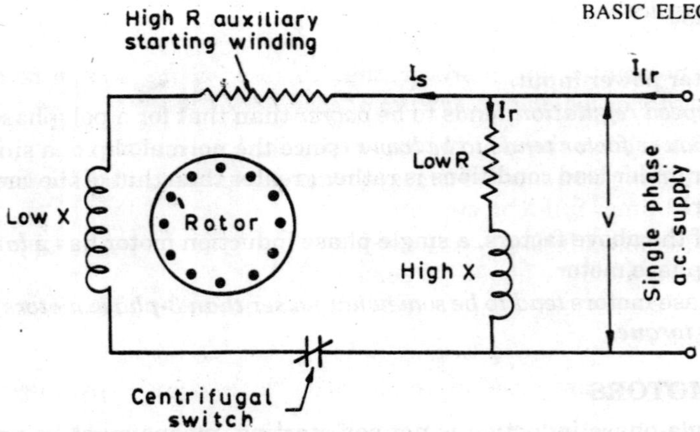

(*a*) Connection diagram.

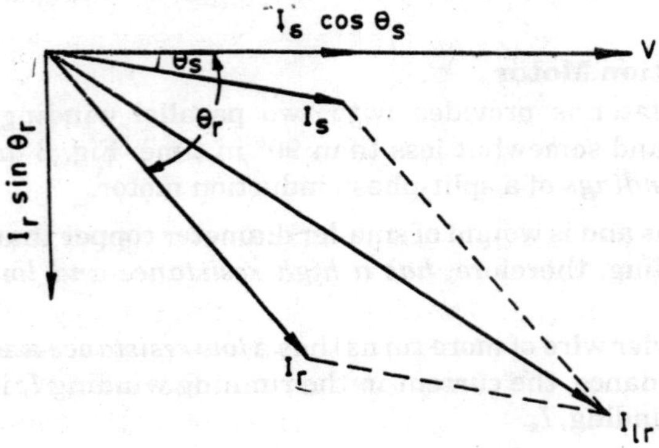

(*b*) Phase relations.

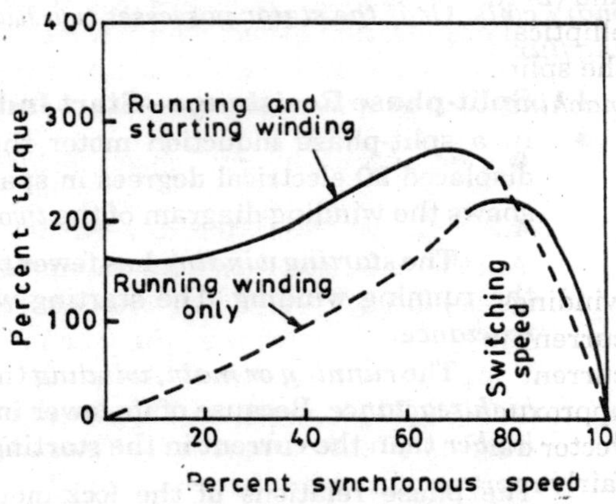

(*c*) Typical torque speed characteristic.

Fig. 3. Split phase resistance-start induction motor.

In case of a hermatically sealed motor, the switch is magnetically operated, and is opened in the de-energized condition.

- Split-phase induction motors may be *reversed by reversing the line connections of either the main or the auxiliary winding*. If however, reversal is attempted under normal running condition, nothing will happen.

  If it is necessary to *reverse the motor while it is rotating*, then some means must be incorporated to slow the motor down to the speed where the starting-switch contacts close, placing the starting winding across the supply lines. This may be done by *incorporating a timing device* which first disconnects the motor entirely from the line and then reverses one field at the proper time. A mechanical braking device which can be electrically operated may also be used.

- Speed control of split-phase windings is a *relatively difficult* matter since the synchronous speed of the rotating stator flux is determined by the frequency and number of poles developed in the running stator winding $\left(N_s = \dfrac{120f}{p}\right)$. By adding stator windings to change the number of poles, speed variation may be obtained. This, however, is a stepped speed change, as in polyphase induction motor, rather than a continuous variation. It must be pointed out, however, that *all speed changes must be accomplished in a range above that at which the centrifugal switch operates.*

Fig. 3 (*c*) shows the typical torque speed characteristics :

- The starting torque is 1.5 times to twice the full-load starting torque and starting current is 6 to 8 times full-load current.
- The speed regulation is very good.
- The percent slip is about 4-6 percent.
- Such a motor may operate with a power factor of 0.55—0.65 and efficiency of 60—65 percent.
- These motors are made in fractional kW $\left(\frac{1}{20} \text{ to } \frac{1}{4} \text{ kW}\right)$ ratings with speed ranging from 2875 to 700 r.p.m.

Shortcomings and uses. The major objections to the motor are (1) its *low starting torque* ; and (2) that, when heavily loaded, the slip exceeds 5 percent, reducing the e.m.f. and producing an elliptical or *pulsating torque which makes* the motor somewhat *annoyingly noisy*. For this reason, the split-phase motor is used in appliances to drive loads which are themselves noisy : *oil burners, machine tools, grinders, dish washers, washing machines, air blowers and air compressors.*

- *Because of low starting torque these motors are seldom employed in sizes larger than $\frac{1}{4}$ kW.*

4.2. Split-phase Capacitor—Start Induction Motor

Another method of splitting the single-phase supply into two phases to be applied to the stator windings is *placing a capacitor in series with the starting auxiliary winding.* In this manner, the current in the starting winding may be made to lead the line voltage. Since the running winding current lags the line voltage, the phase displacement between the two currents be made to approximately 90° on starting. The circuit of capacitor-start motor is shown in Fig. 4 (*a*), while the vector diagram of the currents and voltage is shown in Fig. 4 (*b*). The values of the angles shown are fairly representative, and are rounded off for convenience. *One of the factors upon which the starting torque depends is the sine of the angle between the currents in the two windings.* The value of series capacitor may therefore be reduced, while maintaining a phase-shift angle of about 90°.

- The increase in phase angle between starting and running winding currents is not the only difference between the split-phase and capacitor-start motors. *The split phase motor must keep the number of starting-winding turns low, so that the current may be nearly in phase with the line voltage.* This, *however, is unnecessary in a capacitor-start motor, since the capacitor can overcome the inductance of the winding while still providing the proper phase shift.* There are thus *more auxiliary starting turns* in the capacitor-start motor than in the comparable split-phase motor. This provides a greater number of ampere-turns, hence a *larger rotating flux,* and therefore a *further increase in the starting torque.*

Also it is seen that for the same magnitudes of field currents, the *current I_{lr} is less in capacitor-start motor,* because of the greater angle between the two field currents. In addition, the *starting power factor is also better*. For a given line current, the starting torque is thus much higher for a capacitor-start motor than for a split-phase induction motor. *The starting torque of capacitor-start motor is from 3 to 4.5 times the full-load torque, while that of split-phase resistance-start induction motor rarely exceeds twice the full-load torque.*

- These motors are manufactured in *ratings ranging from $\frac{1}{10}$ kW to $\frac{3}{4}$ kW, but larger sizes are also available.*

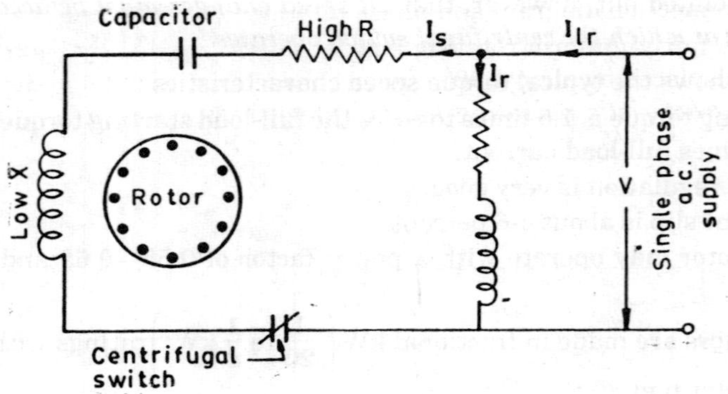

(*a*) Circuit diagram

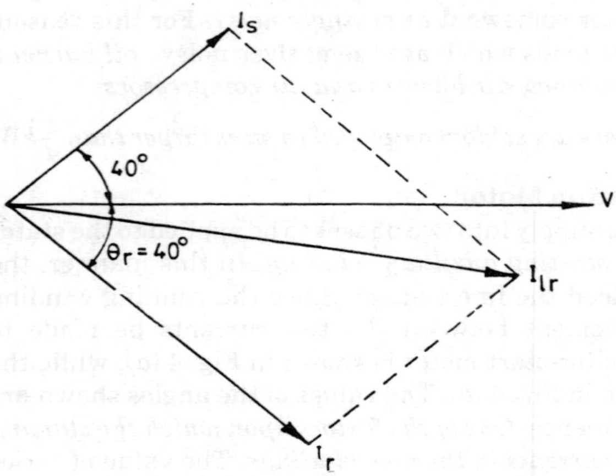

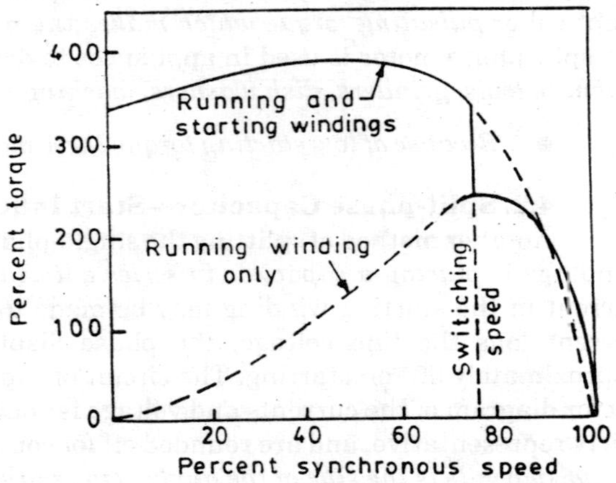

(*b*) Vector diagram (*c*) Typical torque-speed characteristic

Fig. 4. Capacitor-start induction motor.

- The capacitor-start motor may be reversed by changing the connections of one of the windings, but it is subject to the same limitations as the resistance-start induction motor.

Uses. By virtue of their higher starting torque, capacitor-start split-phase motors are used for *pumps, compressors, refrigeration units, air-conditioners,* and *large washing machines,* where a split-phase motor is required that will develop high starting torque under load.

4.3. Permanent-split Capacitor Induction Motor

- A permanent-split capacitor motor (also called single-value capacitor motor) is a single-phase induction motor which has the *same capacitor* in series with the starting (or auxiliary) winding for both starting and running. Because it runs continuously as a permanent split-phase motor, no centrifugal switch is required.
- The motor starts and runs by virtue of the quadrature phase-splitting produced by the two identical windings. As a result, *it does not possess the high running torque* produced by either the resistance-start or the capacitor-start motor.

Furthermore, the capacitor used is designed for continuous duty and is of an oil-filled type. The value of the capacitor is based on its optimum running rather than its starting characteristics. The result is that this motor *has a very poor starting torque,* about 50 to 100 per cent of rated torque.

Fig. 5 shows the connection diagram and phase relations of a permanent split-phase motor.

- Because of the fairly uniform rotating magnetic field created by equal windings whose currents are displaced by almost 90°, the torque is fairly uniform and the motor *does not exhibit the characteristic pulsating hum* developed by most single-phase motors when loaded.

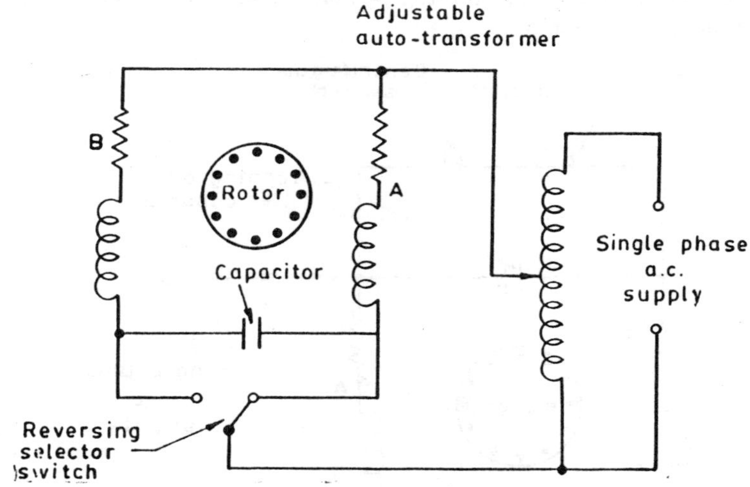

(*a*) Connection diagram.

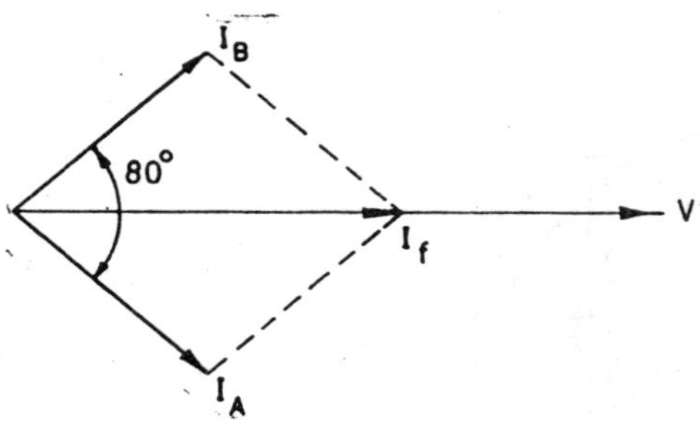

(*b*) Vector diagram or phase relations.

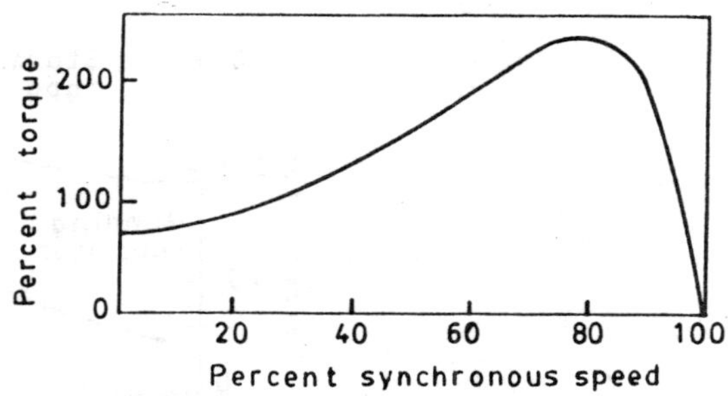

(*c*) Typical torque-speed characteristic.

Fig. 5. Permanent split-phase motor.

This motor possesses the following *merits* :

(*i*) Higher power factor at full-load

(*ii*) Lower full-load line current

(*iii*) Higher full-load efficiency

(*iv*) Increased pull-out torque.

- The permanent-split capacitor motor is *more expensive* than the equivalent split-phase or capacitor-start induction motor. This is primarily due to the fact that the auxiliary (or starting) winding is now also a running winding. It must therefore have a continuous duty rating and as such is heavier than if it were were short-time-rated.

Uses. Because of its *instant response as a reversing motor,* its *quieter operation,* and the *possibility of speed control,* the permanent-split capacitor motor is used for *exhaust and intake fans* and *blowers, unit heaters* and *office machines.*

4.4. Two-value Capacitor Induction Motor

- Refer Figs. 6 and 7. *The two-value capacitor motor combines the quiet operation and limited speed control advantages of a running permanent-split capacitor with high starting torque of the capacitor-start motor.*

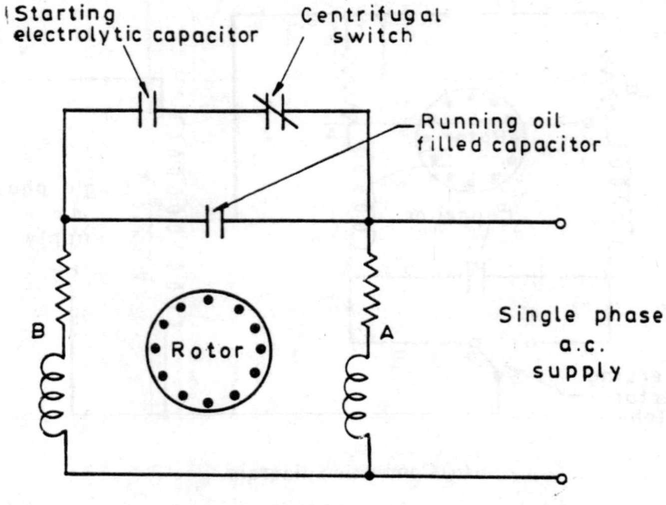

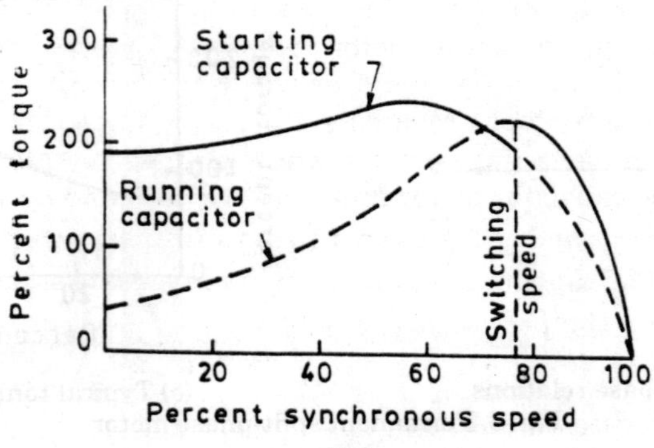

Fig. 6. Connection diagram for two-value capacitor motor and typical torque-speed characteristic.

- Two capacitors (Fig. 6) are employed during the starting period. One of these, an electrolytic starting capacitor, similar to that used for the intermittent duty of the capacitor-start split-phase motor, is of fairly high capacity (about 10 to 15 times the value of the running capacitor) and is cut out of the circuit by a centrifugal switch when slip reaches about 25 per cent.
- The *major advantage of the two value capacitor motor is its high starting torque, coupled with quiet operation and good running torque.* Still classed as a reversing motor, when

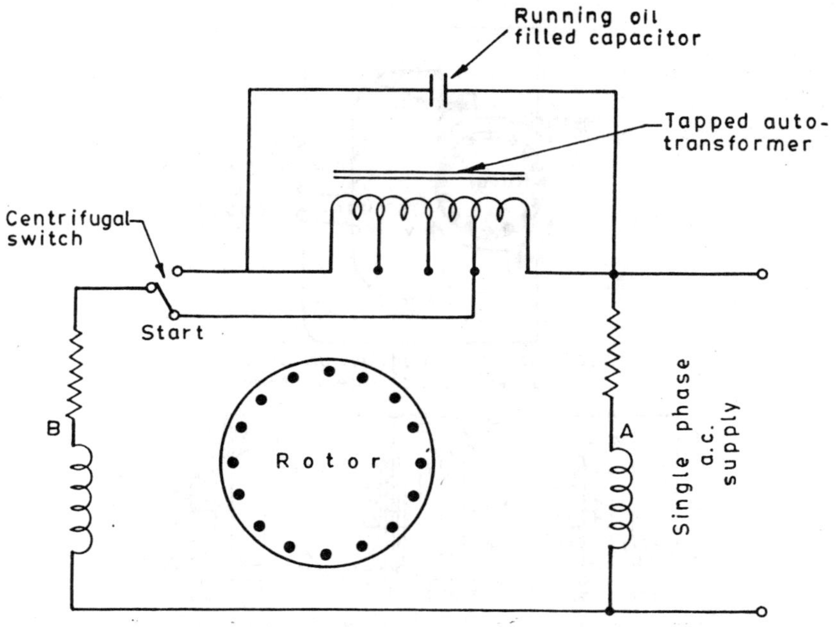

Fig. 7. Two-value capacitor motor with auto-transformer.

the line leads of one winding are reversed, it is reversed in the usual manner. When the speed drops below 25% slip during reversing, the centrifugal switch closes its starting contact, providing the maximum torque as the motor slows down and reverses. The contacts open again when the motor is up to 75% synchronous speed in the reverse direction. Frequent reversals, therefore will reduce the life of the centrifugal switch. For this reason, where frequent reversals are accomplished, permanent-split, single-value capacitor, using no centrifugal switch whatever, is preferable.

Uses. This two-value capacitor motor finds use in *smaller home air-conditioning units* which use this motor in its *compressor* and operates on a 15 amp. branch circuit. The lower starting current and lower running current (7.5 amp. maximum) at an improved power factor over the capacitor-start motor, are obtained through the *precise selection of starting and running capacitors* for the fixed compressor load.

4.5. Shaded-pole Induction Motor

A shaded-pole motor is one of the *simplest* and *cheapest* of manufactured motors. It is essentially an induction machine, since its squirrel-cage rotor receives power in much the same way as does the rotor of the polyphase induction motor. There is however, one extremely important difference between the two. Whereas the poly-phase induction motor creates a true revolving field, in the sense that it is constant in magnitude and rotates at synchronous speed *completely round the entire core*, the field of the shaded-pole motor is not constant in magnitude but *merely shifts from one side of the pole to the other.* Because the shaded-pole motor does not create a true revolving field, the *torque is not uniform but varies from instant to instant.*

Fig. 8 shows the *general construction and principle of shaded pole motor.*

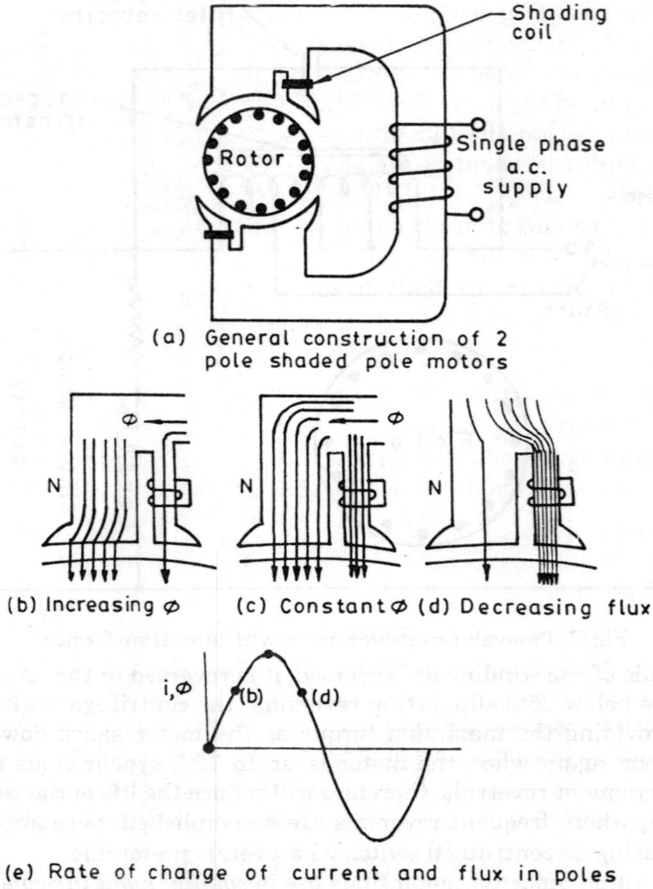

Fig. 8. General construction and principle of shaded pole motor.

Construction. Each of the laminated poles of the stator has a slot cut *across* the laminations about one-third the distance from one edge. Around the smaller of the two areas formed by this slot is placed a heavy *copper short-circuited coil,* called a *shading coil* ; the iron around which the shading coil is placed is called the *shaded* part of the pole, while the free portion of the pole is the *unshaded* part. The exciting coil surrounds the entire coil.

Principle of operation. When the exciting winding is connected to an A.C. source of supply, the *magnetic axis will shift from the unshaded part of the pole to the shaded part of the pole*. This shift in the magnetic axis is, in effect *equivalent to an actual physical motion of the pole* ; the result is that the squirrel-cage rotor will rotate in a direction from the unshaded part to the shaded part. The shifting of flux is explained below.

- Refer Fig. 8 (*b*). When the flux in the field poles tend to increase, a short-circuit current is induced in the shading coil, which by Lenz's law opposes the force and the flux producing it. Thus, as the flux increases in each field pole, there is a *concentration of flux in the main segment of each pole*, while the shaded segment opposes the main field flux.

- At point (*c*) shown in Fig. 8 (*e*), the rate of change of flux and of current is zero, and there is no voltage induced in the shaded coil. Consequently, the *flux is uniformly distributed across the poles* [Fig. 8 (*c*)].
- When the flux decreases, the current reverses in the shaded coil to maintain the flux in the same direction. The result is that the flux crowds in the *shaded segment of the pole* [Fig. 8 (*d*)]

  A typical torque-speed characteristic is shown in Fig. 9.
- Shaded pole motors are built up to about 40 W.

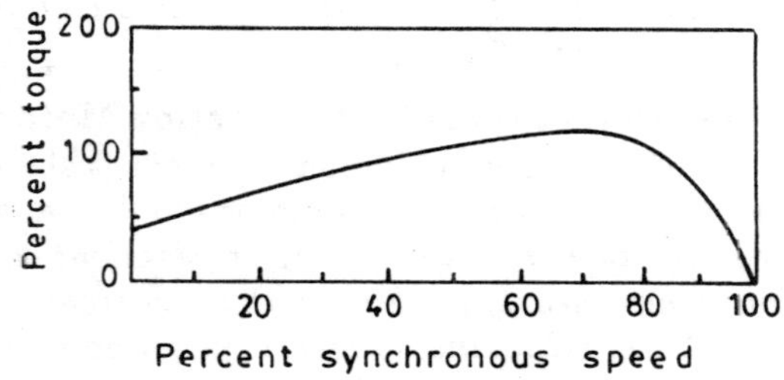

Fig. 9. Typical torque-speed characteristic of shaded pole motor.

Merits :

(*i*) Rugged construction (*ii*) Cheaper in cost

(*iii*) Small in size (*iv*) Requires little maintenance

(*v*) Its stalling locked-rotor current is only slightly higher than its normal rated current, so that it can remain stalled for short periods without harm.

Demerits :

(*i*) Very low starting torque (*ii*) Low efficiency

(*iii*) Low power factor.

Uses. Its low starting torque limits its application to *phono-motors* or *turntables, motion picture projectors, small fans and blowers, bending machines, rotating store-window display tables, and relatively light loads.*

Reversing shaded-pole motor. Fig. 10 shows a reversing shaded-pole motor with the switch in the clockwise rotation position. This places 1 and 3 actively in the circuit, being short-circuited by

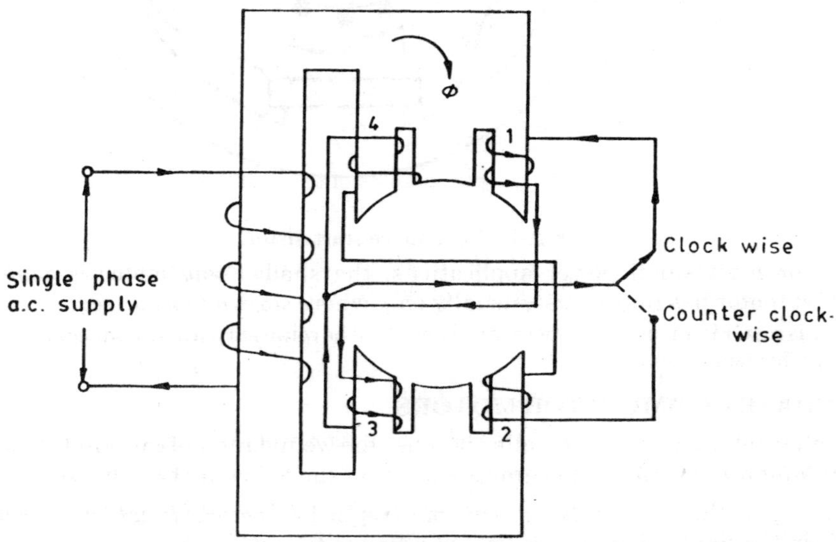

Fig. 10. Reversing shaded pole motor.

the reversing switch. Coils 2 and 4 remain in series with each other, but are on open circuit and so are inactive. The main-field flux is shown increasing vertically downward, and the arrows on the shading coils show the current in them. It is then seen that the *shading coil must be connected in such a manner as not to have their induced voltages in opposition, or there may not be any current in them, and hence no flux lag.*

4.6. Reluctance-start Induction Motor

A reluctance-start induction motor is shown in Fig. 11. Its characteristics are similar to that of shaded pole motor. In this motor too the magnetic field shifts across the pole, but the effect is obtained by the *non-uniform air gap of salient poles. Where there is a greater air gap, the flux in that portion is more nearly in phase with the current. There is a greater lag between flux and current where there is a lower reluctance or where the air gap is smaller.* Since both fluxes are produced by the same current, the flux across the larger air gap *leads* the flux across the smaller one. The two fluxes are obviously displaced in time, and so the *magnetic field shifts across the poles from larger air gap to the shorter gap.* Thus the *direction of rotation is firmly fixed by the construction, and the motor cannot be reversed at all.*

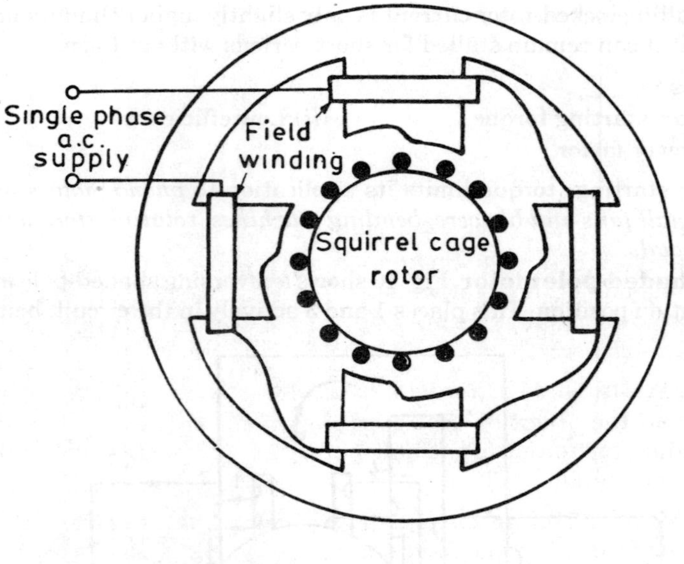

Fig. 11. Reluctance start motor.

Uses. For most small power applications, the shaded-pole motor is preferred, and the reluctance-start motor has limited use, usually only where *starting torque requirements are low.*

Note. This motor is an induction motor and should not be confused with *reluctance motor* which is actually a non-excited synchronous motor.

5. SINGLE-PHASE COMMUTATOR MOTORS

The commutator motors are so called because the wound rotor of this kind of motor is equipped with a *commutator and brushes.* This group consists of the following two classes :

1. Those operating on *'repulsion principle'* (repulsion motors) *in which energy is inductively transferred from the single-phase stator field winding to the rotor.*

2. Those operating on the *principle of the series motor* in which the energy is *conductively* carried both to the rotor armature and its series-connected single-phase stator field.

5.1. Repulsion Motor

A repulsion motor in its simplest form consists of a *field comprising a distributed winding*, housed in slots, in a smooth-cored stator and *an armature carrying a distributed winding connected to a commutator*. The stator winding, which produces the *main field*, is connected to the main supply. The armature or rotor winding is not connected electrically to the main circuit, but the brushes, which are set at an angle to the direction of the main flux, are short-circuited as shown in Fig. 12.

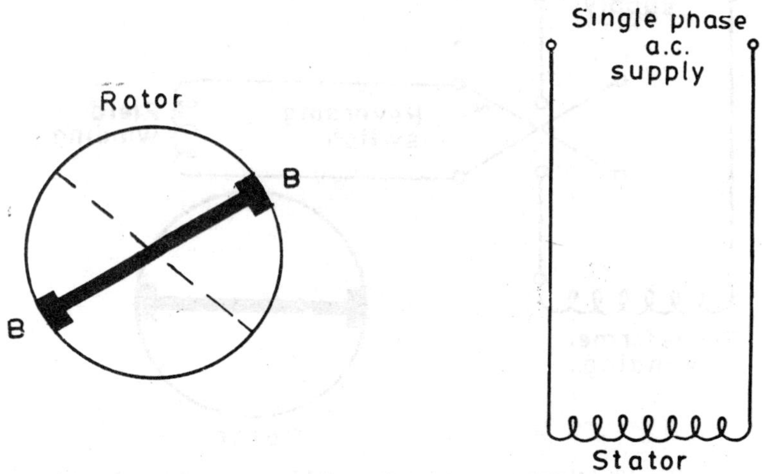

Fig. 12. Repulsion motor.

- If the brush axis *BB* is set *at right angles* to the direction of the stator flux, the e.m.f. induced in one-half of the rotor winding is *exactly balanced* by the e.m.f. induced in the other half, so that resultant e.m.f. is *zero* ; no current flows in the rotor winding and *no-torque is developed.*
- If the brush axis is placed *in line with the direction* of the stator flux, the e.m.f. in one half assists that in the other, so that *a maximum current flows.* Here again *no torque is developed,* since the torque of one half of the rotor conductors is exactly balanced by that due to the other half.
- If *the brush axis makes an angle with the stator flux as shown, a resultant torque is produced.* The torque is a *maximum theoretically, if this angle is 45°, but in practice the angle of inclination is about half this value.*

  It is clear from above that the *speed of repulsion motor depends upon the brush position. Speed control of* such a machine can be provided by *mounting the brushes on a rocker* which can be rotated by a lever handle mounted on the motor end-shield. If remote control is required, the lever handle may be manipulated by a simple system of rods and cranks.

  Alternatively, if the motor is to be totally enclosed, or remote control from a considerable distance is required, speed control may be obtained by the *use of an external series resistance with fixed brush gear.*
- The direction of rotation of a simple repulsion motor may be reversed by swinging the brushes into the position shown *dotted* in Fig. 12.

Atkinson repulsion motor. A modification of the simple repulsion motor is the Atkinson repulsion motor, in which the stator winding comprises two windings at right-angles to each other and connected in series, as shown in Fig. 13. One advantage obtained by this method is that the *direction of rotation can be reversed by reversing the connections to one of the stator windings.* Instead of moving the brush rocker, it is necessary only to throw the reversing switch, shown in Fig. 13.

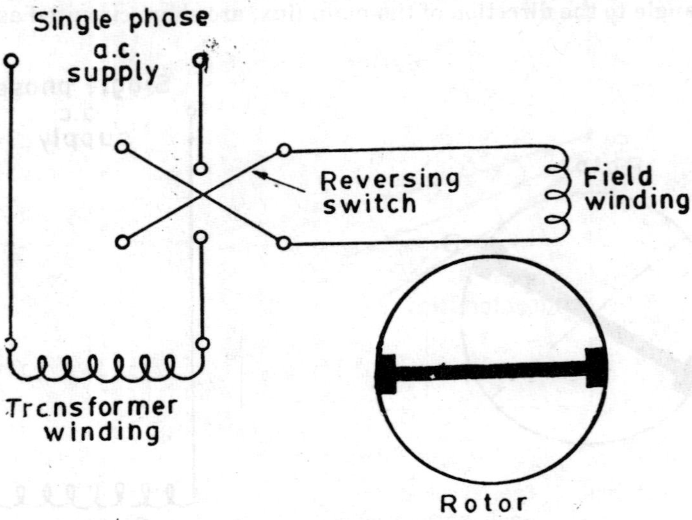

Fig. 13. Atkinson repulsion motor with reversing switch.

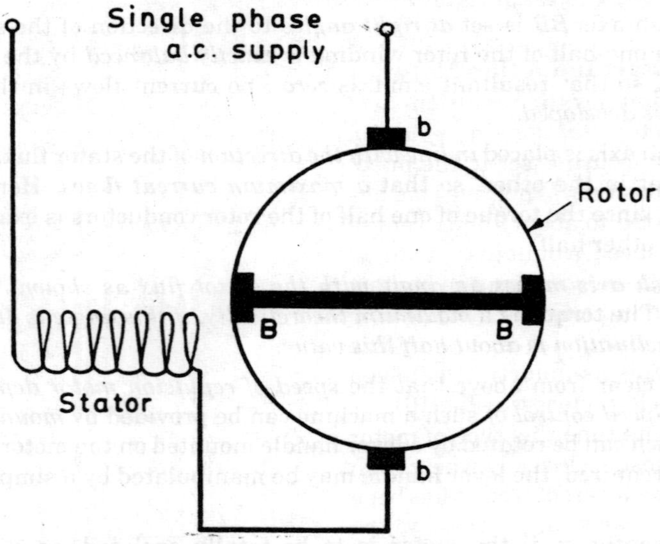

Fig. 14. Compensated repulsion motor.

It will be observed that, as the rotor is electrically connected to the stator, the compensated repulsion motor is not able to operate directly from a high-voltage supply, as was the case with the simple repulsion motor.

Fig. 15 shows the typical speed-torque characteristics of single-phase repulsion motor.

Compensated repulsion motor. The *power factor of the machine may be improved by compensation,* and this is done by providing *an additional set of brushes* as shown in Fig. 14. The

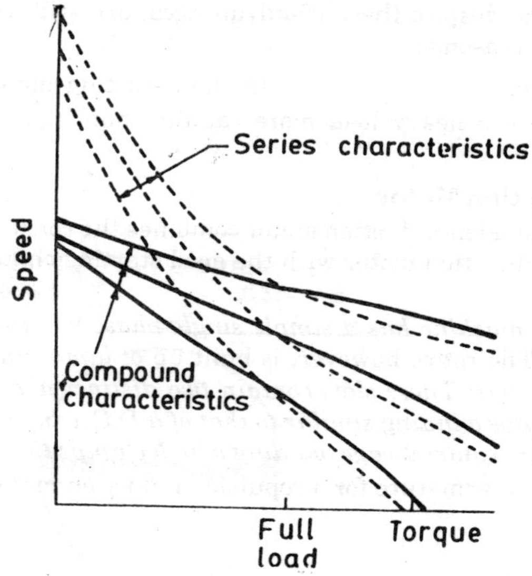

Fig. 15. Typical speed-torque characteristics of single-phase repulsion motors.

axis of one pair, *BB*, of brushes coincides with the stator winding, these brushes are *short-circuited.* The other pair, *bb*, of brushes is set at right-angles to the former and is connected in *series with the stator winding*.

5.2. Repulsion-start Induction Motor

- As its name implies the repulsion-start induction motor *starts as a repulsion motor* with its brushes set to the maximum torque position. When the load has been accelerated to about 75 per cent of synchronous speed, a built in centrifugal device *places a shorting ring* in contact with the commutating bars, *converting the armature to squirrel-cage rotor*. The motor then runs as induction motor on its induction characteristic (Fig. 16).
- Although at one time this type of motor was used almost exclusively where high starting torque was required, it has been replaced in nearly all cases by the capacitor motors because of the following reasons :

  (*i*) Requires more maintenance.

  (*ii*) More expensive.

  (*iii*) Makes quite a bit of noise on starting.

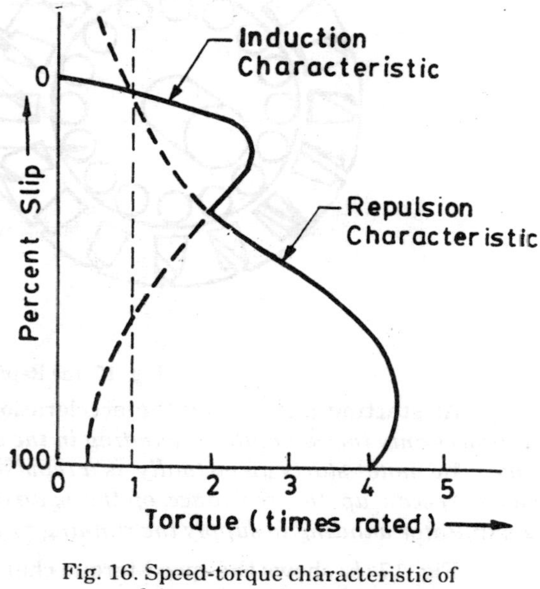

Fig. 16. Speed-torque characteristic of repulsion-start induction motor.

(*iv*) Causes radio interference when starting.

(*v*) Cannot be reversed easily.

Repulsion-start motors, despite these disadvantages, are still used in *integral-horsepower sizes* because of the following reasons :

(*i*) High starting torque. (*ii*) Low starting current.

(*iii*) Ability to accelerate a heavy load more rapidly than high-capacitance dual-capacitor motors.

5.3. Repulsion-induction Motor

- A single-phase repulsion-induction motor combines the constant-speed characteristics of the single-phase induction motor with the good starting characteristics of the repulsion motor.
- The *stator of this machine has a simple single-phase winding like that of single-phase induction motor.* The rotor, however, is built up of *laminations,* each of which has *two concentric sets of slots. These slots contain two distinct windings ; in the outer slots is wound a commutator winding similar to that of a D.C. armature, while in the inner slots is a cast aluminium squirrel-cage winding which clamps the laminations.*

Fig. 17 (*a*) illustrates an armature for a repulsion induction motor complete with a squirrel-cage winding.

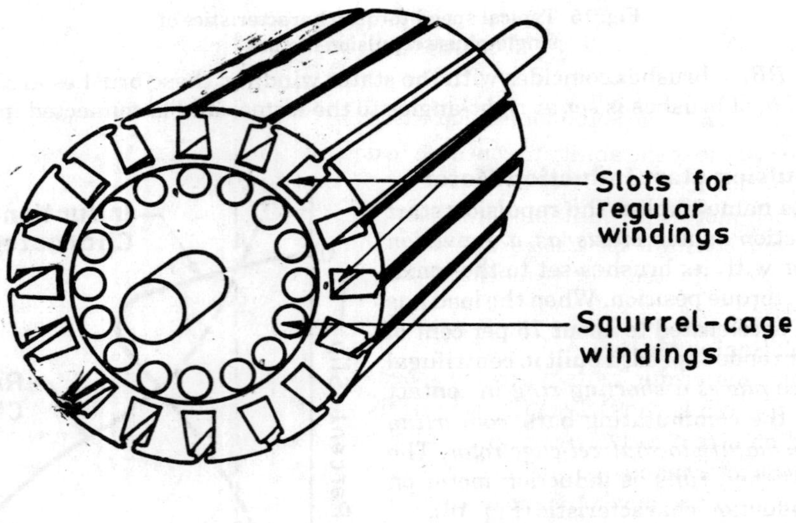

Fig. 17. (*a*) Repulsion induction motor.

At starting and during the acceleration period, the magnetic flux produced by the stator *embraces only the commutator winding in the outer slots owing to the high reactance of the squirrel-cage.* The *motor starts up virtually as a repulsion motor and develops a high starting torque. As the motor speeds up the reactance of the squirrel-cage decreases, so that this winding assists the commutator winding to supply the running torque.*

Fig. 17 (*b*) shows the speed-torque characteristic of repulsion-induction motor.

Merits. The repulsion-induction motor has the following merits :

(*i*) High starting torque. (*ii*) Fairly good speed regulation.

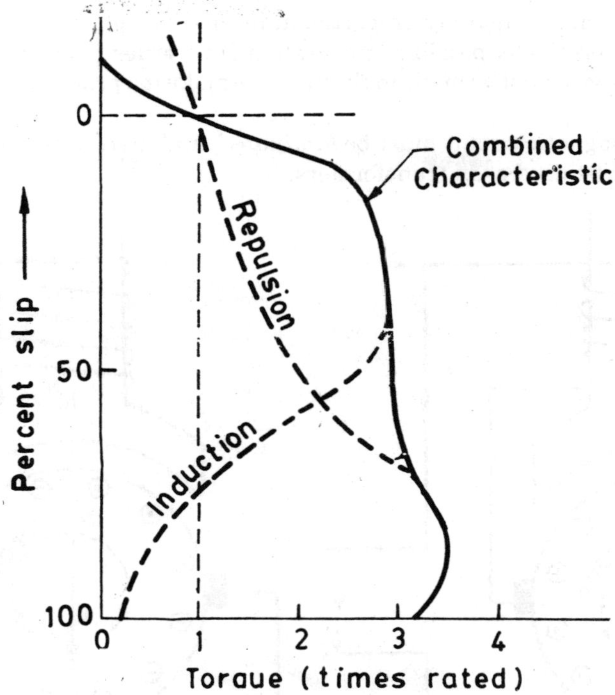

Fig. 17. (*b*) Speed-torque characteristic of repulsion-induction motor.

(*iii*) Major virtue is the ability to continue to develop torque under sudden, heavy applied loads without breaking down.

Uses. Such motors are suitable for all single-phase power applications *which require a high starting torque and constant speed when running* ; they also operate *at a very high power factor.* They are particularly well adapted to drive *machine tools,* lifts, hoists, *mixing machines, centrifugal pumps, fans and blowers.*

5.4. A.C. Series Motor

The series motor due to its desirable speed-torque characteristics *is almost exclusively used in railway service.* While the D.C. motor is entirely satisfactory for this class of work service and is generally used on street railway cars and trolley coaches, the fact that it is more convenient and more economical to transmit power and to transform voltages in A.C. systems than with direct currents has lead to the development of the A.C. series motor for use on some of the important steam-road electrifications.

Working principle. The working principle of an A.C. series motor is the same as that of the D.C. series motor. The armature and field are wound and interconnected in the same manner as the D.C. series motor.

When an alternating e.m.f. is applied to the terminals, since field and armature windings are connected in series, the field flux and armature current reverse simultaneously every half cycle, but the direction of the torque remains unchanged. The torque is pulsating, but its average value is equal to that which a D.C. motor will develop if it had the same r.m.s. value of flux and current. Motor connections, direction of torque, etc. for two successive half cycles are shown in Fig. 18. If the field and armature core are run at low saturation, the air-gap flux is approximately proportional to the current and the torque is approximately proportional to current squared.

Although it is theoretically possible to operate a D.C. series motor from an A.C. circuit, the following structural changes must be made in the motor to make it a practical and reasonable efficient machine :

- The entire magnetic circuit must be *laminated,* and materials with low iron-loss co-efficients should be used as in transformers.

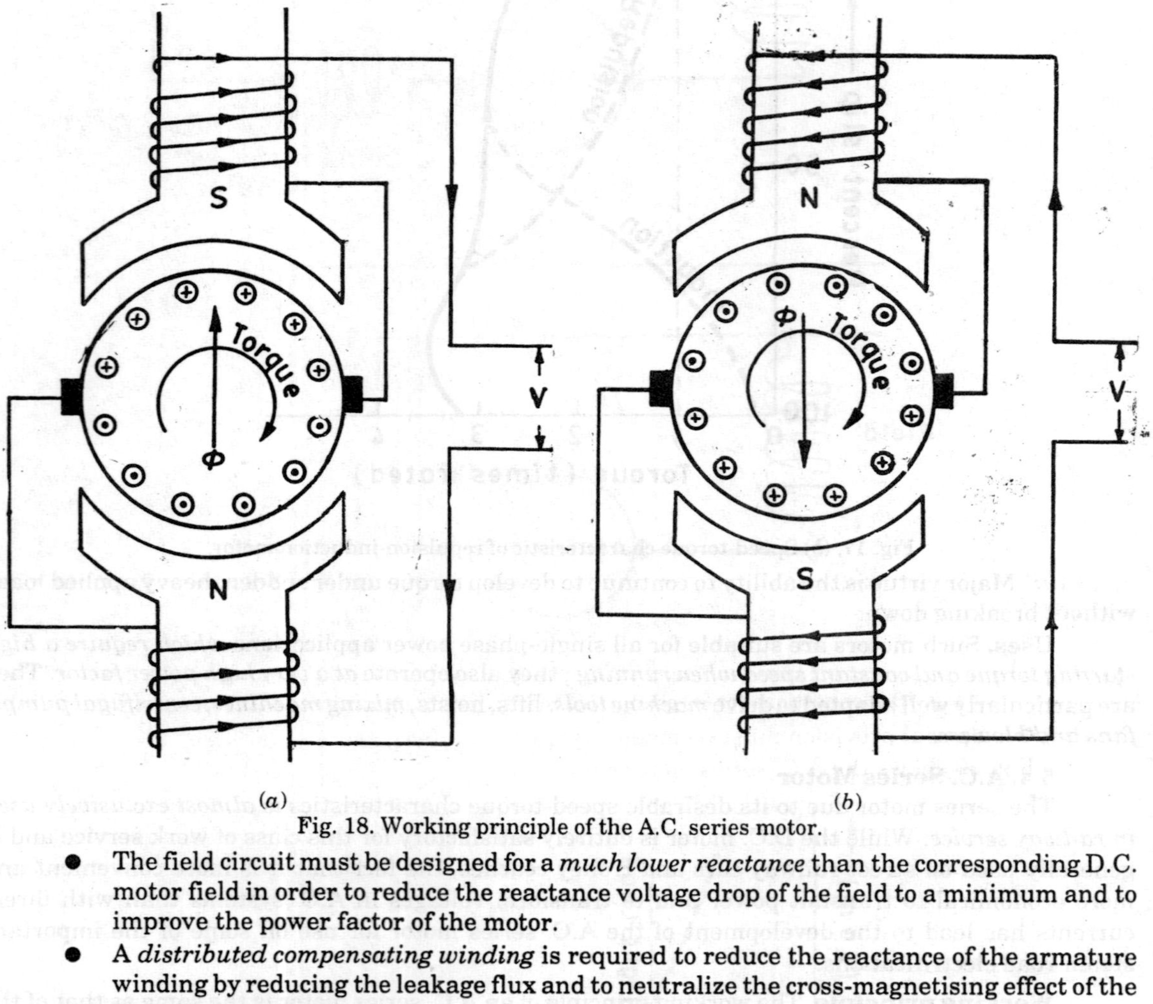

Fig. 18. Working principle of the A.C. series motor.

- The field circuit must be designed for a *much lower reactance* than the corresponding D.C. motor field in order to reduce the reactance voltage drop of the field to a minimum and to improve the power factor of the motor.
- A *distributed compensating winding* is required to reduce the reactance of the armature winding by reducing the leakage flux and to neutralize the cross-magnetising effect of the armature ampere turns.

The compensating winding may be connected in series with the series-field and armature windings, or it may be short-circuited upon itself and receive its excitation voltage by transformer action, since it is inductively coupled with the armature cross-field (Fig. 19). In the first case, the motor is said to be *conductively* compensated, while in the second it is *inductively* compensated. *Conductive compensation* is required on motors which are intended for operation in *D.C. as well as A.C. circuits*.

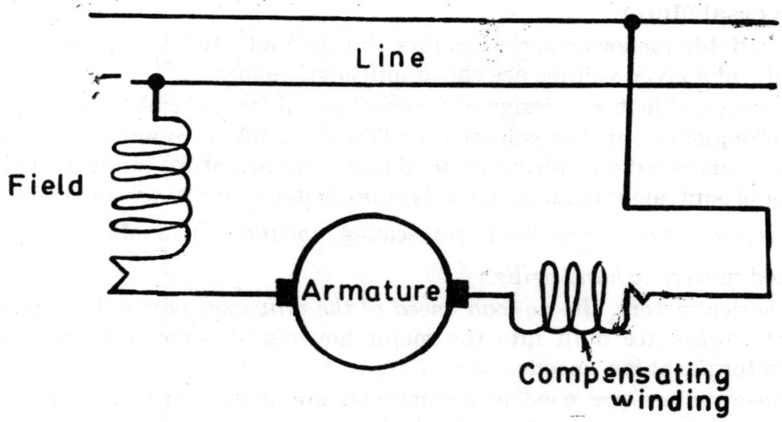

(a)

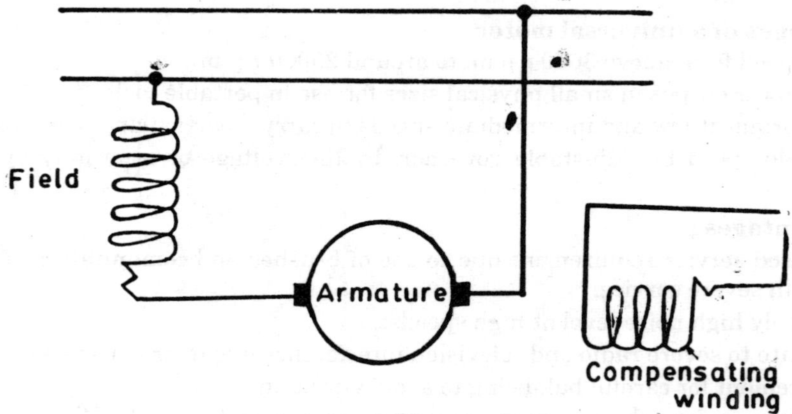

(b)

Fig. 19. Connections for (a) conductive compensation ; (b) inductive compensation.

- Special provision must be made to secure *satisfactory commutation*.

Fig. 20 shows the A.C. series motor characteristic.

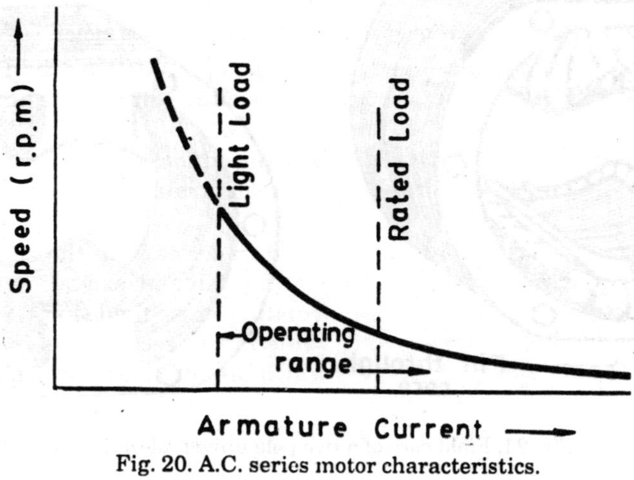

Fig. 20. A.C. series motor characteristics.

5.5. Universal Motor

- Fractional-horsepower series motors that are adapted for use on either D.C. or A.C. circuits of a given voltage are called *universal motors*.
- The universal motor is designed for commercial frequencies from 60 cycles down to D.C. (zero frequency), and for voltage from 250 V to 1.5 V. A commercial universal motor may have a somewhat weaker series field and more armature conductors than a D.C. series motor of equivalent horsepower. *It is manufactured in ratings up to 3 / 4 H.P., particularly for vacuum cleaners and industrial sewing machines.* In smaller sizes of $\frac{1}{4}$ H.P. or less, it is used in *electric hand drills*.

Like all series motors, the *no-load speed of the universal motor is universally high.* Quite frequently, *gears trains* are built into the motor housing of some universal motors to provide exceedingly high torque at low speeds.

When these motors are used in commercial appliances such as *electric shavers, sewing machines, office machines, and small hand hair dryers or vacuum cleaners,* they are always *directly loaded* with little danger of motor runaway.

Advantages of a universal motor :

1. High speed from above 3600 r.p.m. to around 25000 r.p.m.
2. High power output in small physical sizes for use in portable tools.
3. High torque at low and intermediate speeds to carry a particularly severe load.
4. Variable speed by adjustable governor, by line voltage or especially by modern pulse techniques.

Disadvantages :

1. Increased service requirement due to use of brushes and commutators. The life of these parts is limited in severe service.
2. Relatively high noise level at high speeds.
3. Moderate to severe radio and television interference due to brush sparking.
4. Requirement for careful balancing to avoid vibration.
5. Requirement for reduction gearing in most portable tools.

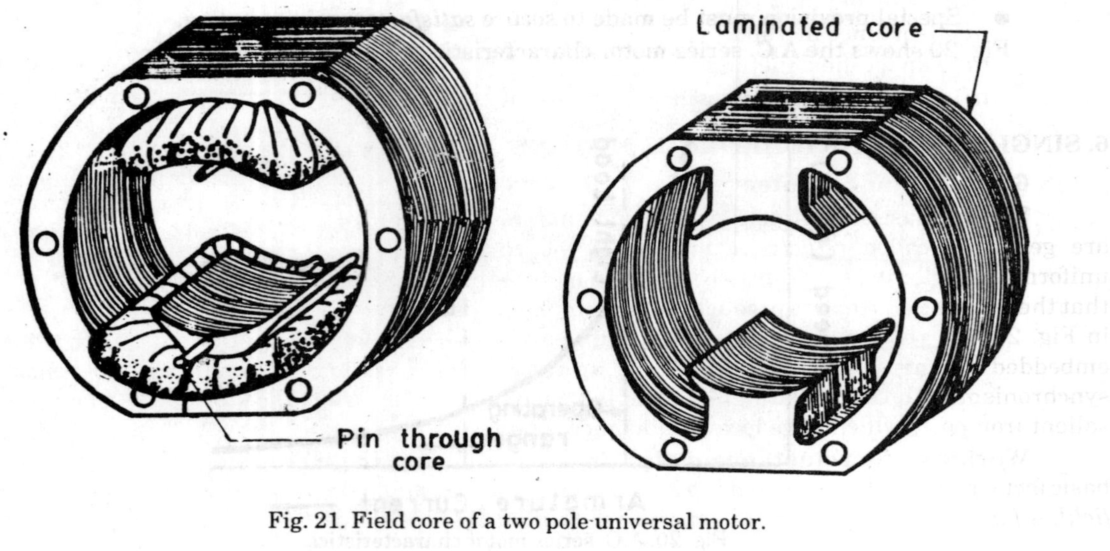

Fig. 21. Field core of a two pole-universal motor.

Universal motors are manufactured in two types :

1. Concentrated-pole, non-compensated type (low H.P. rating).
2. Distributed field compensated type (high H.P. rating).

Fig. 21 shows the laminated field structure of a typical concentrated field universal motor.

Operation of a Universal Motor. As explained in Art. 5.4, such motors develop unidirectional torque regardless of whether they operate on D.C. or A.C. supply. The production of unidirectional torque when the motor runs on A.C. supply can be easily understood from Fig. 22. The motor works on the same principle as a D.C. motor *i.e.*, the force between the main pole flux and the current carrying armature conductors. This is true regardless of whether current is alternating or direct.

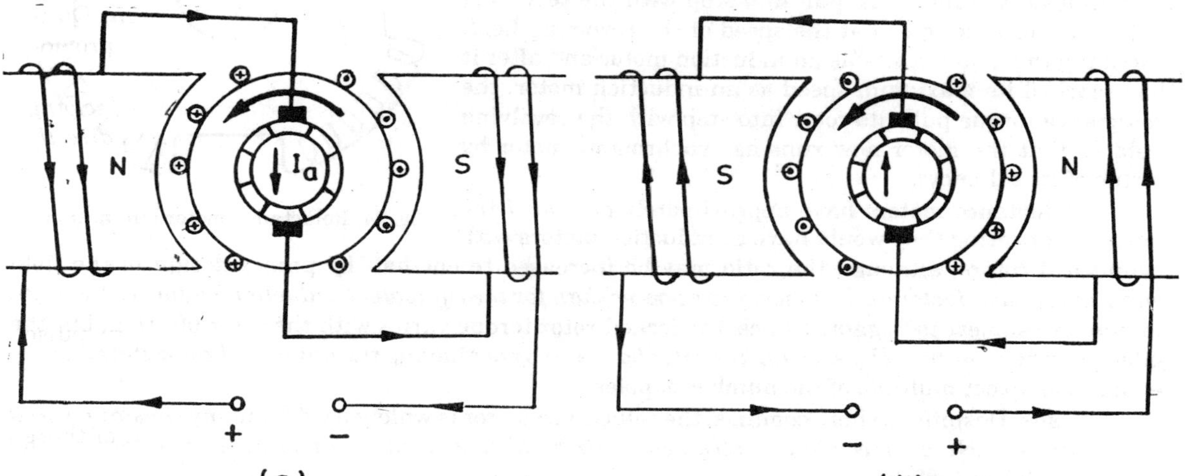

Fig. 22. Universal motor operation.

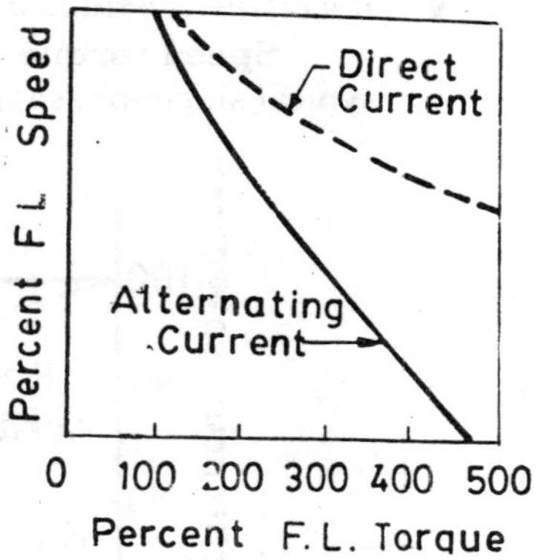

Fig. 23. Typical torque characteristics of a universal motor.

- Fig. 23 shows the typical torque characteristics of a universal motor both for D.C. and A.C. supply.
- The speed of a universal motor may be controlled by the following methods :
  (*i*) Reactance method.
  (*ii*) Tapped-field method.
  (*iii*) Centrifugal mechanism.

6. SINGLE-PHASE SYNCHRONOUS MOTORS

6.1. Reluctance Motor

Single-phase salient-pole synchronous-induction motors, are generally called *reluctance motors*. If the rotor of any uniformly distributed single-phase induction motor is altered so that the laminations tend to produce *salient rotor poles*, as shown in Fig. 24, the reluctance of the air-gap flux path will be greater where there are no conductors embedded in slots. Such a motor, coming up to speed as an induction motor, will be pulled into synchronism with the pulsating A.C. single-phase field by the reluctance torque developed at the salient iron poles which have lower-reluctance air gaps.

Working of a reluctance motor. In order to understand the working of such a motor the basic fact which must be kept in mind is that *when a piece of magnetic material is located in a magnetic field, a force acts on the material, tending to bring it into the densest portion of the field. The force* *tends to align the specimen of material in such a way that the reluctance of the magnetic path that passes through the material will be minimum.*

When supply is given to the stator winding, the revolving magnetic field will exert reluctance torque on the unsymmetrical rotor tending to align the salient pole axis of the rotor with the axis of the revolving magnetic field (because in this position, the reluctance of the magnetic path would be minimum). If the reluctance torque is sufficient to start the motor and its load, the rotor will pull into step with the revolving field and continue to run at the speed of the revolving field. (Actually the motor starts as an induction motor and after it has reached its maximum speed as an induction motor, the reluctance torque pulls its rotor into step with the revolving field so that the motor now runs as synchronous motor by virtue of its saliency).

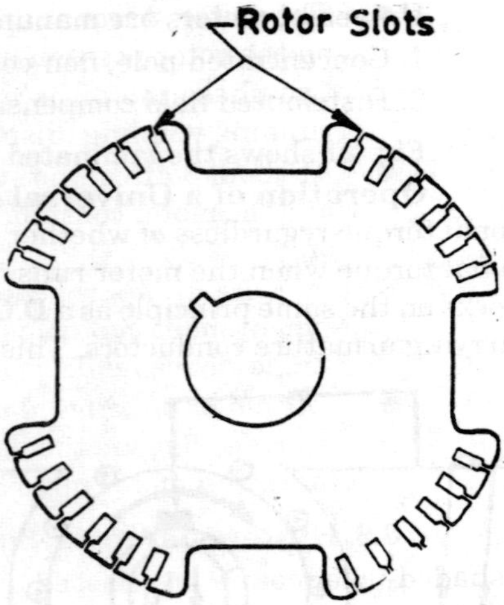

Fig. 24. Reluctance-motor lamination.

Reluctance motors have approximately *one-third* the horsepower rating they would have as induction motors with cylindrical rotors, although the ratio may be increased to one-half by proper design of the field windings. *Power factor and efficiency are poorer than for the equivalent induction motor*. Reluctance motors are subject to *'cogging'*, since, the locked-rotor torque varies with the rotor position, but the effect may be *minimized* by *skewing the rotor bars* and by *not* having the number of rotor slots exactly equal to an exact multiple of the number of poles.

Uses. Despite its short-comings, the reluctance motor is widely used for many *constant, speed* applications such as *recording instruments, time devices, control apparatus, regulators*, and *phonograph turntables*.

- *Reversing* is obtained as in any single-phase induction motor.

Speed-torque characteristics. Fig. 25 shows speed-torque characteristics of a typical single-phase reluctance motor.

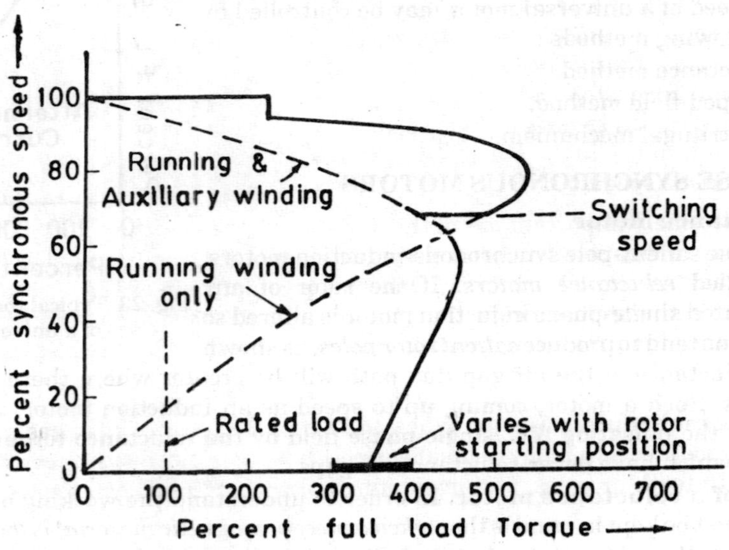

Fig. 25. Speed-torque characteristics of a single-phase reluctance motor.

- The motor starts at anywhere from 300 to 400 per cent of its full-load torque (depending on the rotor position of the unsymmetrical rotor with respect to the field windings) as a two-phase motor as a result of the magnetic rotating field created by a starting and running winding (displaced) 90° in both space and time.
- At about 3/4th of the synchronous speed, a centrifugal switch opens the starting winding, and the motor continues to develop a single-phase torque produced by its running winding only.

  As it approaches synchronous speed, the reluctance torque (developed as a synchronous motor) is sufficient to pull the rotor into synchronism with the pulsating single-phase field.
- *The motor operates at a constant speed up to a little over 200% of its full-load torque.* If it is loaded beyond the value of pull-out torque, it will continue to operate as a single-phase induction motor up to 500% of its rated output.

6.2. Hysteresis Motor. Single-phase cylindrical (non-salient-pole) synchronous-induction or shaded-pole motors are classed as *hysteresis motors.* A hysteresis motor has neither a salient-pole rotor nor direct excitation, but nevertheless it rotates at synchronous speed. This type of motor *runs into synchronism and runs on hysteresis torque.*

Hysteresis-type lamination, shown in Fig. 26, are usually made of *hardened, high retentivity steel* rather than commercial, low retentivity dynamo steel.

Working. As a result of a rotating magnetic field produced by phase splitting or a shaded-pole stator, *eddy currents are induced in the steel of the rotor which travel across the two bar paths of the rotor* as shown in Fig. 26. A high-retentivity steel produces a high hysteresis loss, and an appreciable amount of energy is consumed from the rotating field in reversing the current direction of the rotor. At the same time the *rotor magnetic field set up by the eddy currents causes the rotor to rotate. A high starting torque is produced as a result of the high resistance* (proportional to hysteresis). As the rotor approaches synchronous speed, the frequency of current reversal in the cross-bars decreases, and the rotor becomes *permanently magnetized* in one direction as a result of the high retentivity of the steel rotor. Consequently the motor continues to rotate at synchronous speed.

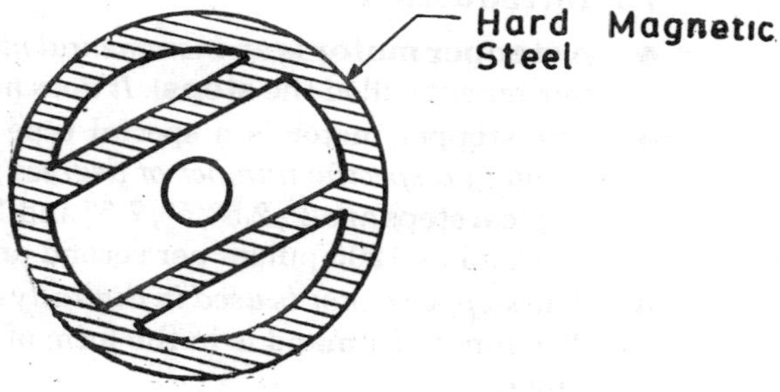

Fig. 26. Hysteresis rotor.

- An extremely important *use* of this type of motor is for the rotation of *gyroscope rotors* in inertial navigation and control systems. Here the requirement is for as near absolute accuracy as can be achieved. One major component of the instrument accuracy that contains the gyroscope is that the gyroscopic moment be *absolutely constant.* This constancy requires a synchronous motor that is driven by a regulated constant-frequency source.

6.3. Sub-synchronous Motor. When the motor has a rotor that has an overall cylindrical outline and yet is toothed as a many-pole salient-pole rotor, it is a **sub-synchronous motor.** A typical rotor may have 16 teeth or poles, and in conjunction with a 16-pole stator will normally rotate at synchronous 450 r.p.m. when operated on 60 Hz. If this motor were temporarily overloaded, it would drop out of synchronism. Then the speed drops down toward the maximum torque point, and the motor will *again lock into synchronism* at a sub-multiple speed of 225 r.p.m. Hence the name of sub-synchronous motor.

This type of motor starts and accelerates with hysteresis torque just as the hysteresis synchronous motor does. There is no equivalent of induction-motor torque as in the reluctance motors.

This type of motor in any given size will develop a higher starting torque but a lesser synchronous speed torque than a reluctance motor.

Fig. 27 shows a sub-synchronous rotor.

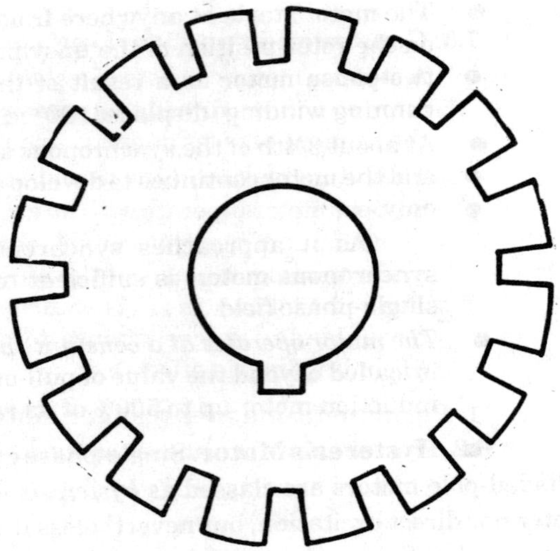

Fig. 27. Sub-synchronous rotor.

7. STEPPER MOTORS

7.1. Introduction

- A **stepper motor** is an *incremental motion machine* (*i.e.* the motor which turns in discrete movement called the **steps**). *It does not rotate continuously as a conventional motor does.*
- The stepper motor is a special type *of synchronous motor which is designed to rotate through a specific number of degrees for each electrical pulse received by its control unit.* Typical steps are 2°, 2.5°, 5°, 7.5° and 15° per pulse. These motors are built to follow signals as rapid as 1200 pulses per second and with equivalent power ratings upto several kW.
- The *stepper motor is used in digitally controlled position control system in open loop mode.* The input command is in the form of a train of pulses to turn a shaft through a specified angle.

7.2. Advantages and Applications

Advantages : The stepper motor (**a position control device**) entails the following **advantages** :

1. Compatibility with digital systems.
2. The angular displacement can be precisely controlled without any feedback arrangement.
3. No sensors are needed for position and speed sensing.
4. It can be readily interfaced with microprocessor (or computer based controller).

Applications : Stepper motors have a wide range of applications, mentioned below :

1. Paper feed motors in typewriters and printers.
2. Positioning of print heads.
3. Pens in XY-plotters.
4. Recording heads in computer disc drives.
5. Positioning of workables and tools in numerically controlled machining equipment.
6. Also employed to perform many other functions such as *metering, mixing, cutting, blending, stirring* etc. in several commercial, military and medical applications.

7.3. Construction and Working

- A stepper motor consists of a *slotted stator having multi-pole, multi-phase winding and a rotor structure carrying no winding*. They typically use three and four phase windings, the number of poles depends upon the required angular change per input pulse.
- The rotors may be of the *permanent magnet* or *variable reluctance type*.
- Stepper motors operate with an *external drive logic circuit*. When a train of pulse is applied to the input of the drive circuit, the circuit supplies currents to the stator windings of the motor to make the axis of the air-gap field around in coincidence with the input pulses. The rotor follows the axis of air-gap magnetic field by virtue of the permanent magnet torque and/or the reluctance torque, depending upon the pulse rate and load torque (including inertia effects).

1. **Permanent-magnet stepper motor :**

- Fig. 28 shows the phases or stacks of a 2-phase, 4-pole permanent-magnet stepper motor.

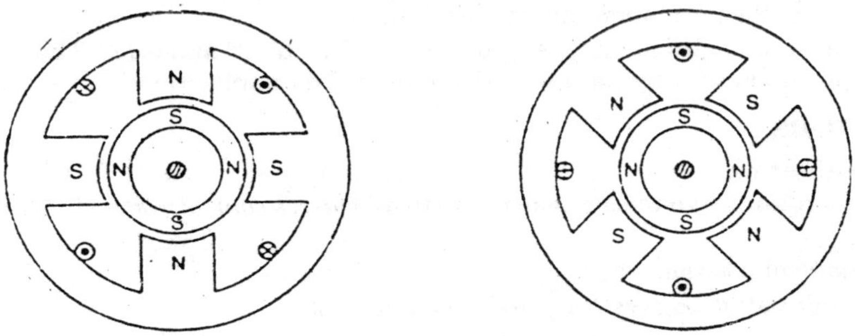

(i) Phase I (ii) Phase II

Fig. 28. Permanent-magnet stepper motor.

- The rotor is made of *ferrite or rare-earth material* which is *permanently magnetised*.
- The stator stack of phase II is staggered from that of phase 'I' by an angle of 90°.
- When the phase 'I' is excited, the rotor is aligned as shown in Fig. 28 (*i*). If now the phase 'II' is also excited, the effective stator poles shift anti-clockwise by 22.5° [Fig. 28 (*ii*)] causing the rotor to move accordingly. Now, keeping the phase 'II' still energised, if the phase 'I' is now de-energised, the rotor will move another step of 22.5°. The reversal of phase 'I' winding current will produce a further forward movement of 22.5°, and so on. It can be a easily observed/visualised as to how the direction of movement can be reversed.
- Each phase is provided with double coils to simplify the switching arrangement (which is electronically accomplished).

2. **Variable-reluctance stepper motor :**

- A variable-reluctance stepper motor has no permanent magnet on the rotor and the rotor employed is a ferro-magnetic multi-toothed one.
- *The large differences in magnetic reluctances that exist between the direct and quadrature axes develop the torque*. The stationary field developed by the direct current in some stator coils tends to develop a torque which causes the rotor to move to the position where the reluctance of the flux path is minimum.

Stepping angle, irrespective of the type of stepper motor is given as :

$$\alpha = \frac{360°}{\text{Number of phases} \times \text{number of poles}} = \frac{360}{np} \quad ...(3)$$

3. **Hybrid stepper motor :**

- This is *infact in permanent-magnet stepper motor with constructional features of toothed and stacked rotor adopted from the variable-reluctance motor.*
- The stator has only one set of winding-excited poles which interact with the two rotor stacks.
- The permanent magnet is placed *axially along the rotor in the form of an annular cylinder over the motor shaft.*
- The stacks at each end of the rotor are toothed. So all the teeth on the stack at one end of the rotor acquire the *same polarity* while the teeth of the stack at the other end of the rotor acquire the *opposite polarity*. The two sets of the teeth are displaced from each other by *one half of the tooth pitch* (also called pole pitch).
- The primary **advantage** of the hybrid motor is that if *stator excitation is removed, the rotor continues to remain locked into the same position, as before removal of excitation.* This is due to the reason that the rotor is prevented to move in either direction by torque because of the permanent magnet excitation.
- Typical step angles for stepper motors are 15°, 7.5°, 2° and 0.72°. The choice of the angle depends upon the angular resolution required for application.

8. SERVO-MOTORS

8.1. Introduction

The term *servo* or servo-mechanism refers to a *feedback control system* in which the controlled variable is :

— Mechanical position, or
— Time derivatives *e.g.* velocity and acceleration.

Following *characteristics* are usually required for a feedback control system :

(*i*) High accuracy ;
(*ii*) Remote operation ;
(*iii*) Fast-response ;
(*iv*) Unattended control.

Following are the *essentials* of a feedback control system :

1. **An error detecting device.** It determines when the regulated quantity is *different* from the reference quantity and sends out the error signal to the *amplifier*.

2. **An amplifier.** The amplifier receives the error signal and then supplies power to the error-correcting device, which in turn changes the regulated quantity so that it matches the reference input.

A *servo-motor* should entail the following *characteristics* :

1. *The output torque of the motor should be proportional to the voltage applied* (*i.e.* the control voltage which is developed by the amplifier in response to an error signal).

2. *The direction of the torque developed by the servo-motor should depend upon the instantaneous polarity of the control voltage.*

8.2. Types of Servo-motors

The servo-motors are of the following *two* types :

1. *D.C. servo-motors.*
2. *A.C. servo-motors.*

8.2.1. D.C. servo-motors

These motors are preferred for very high power systems since they operate more efficiently (as compared to A.C. servo-motors).

These motors may be of the following types :

— Series motors ;
— Split series motors ;
— Shunt control motors ;
— Permanent magnet (*fixed excitation*) shunt motor.

(*i*) **Series motors :**

- This motor has a high starting torque.
- It draws large current.
- The speed regulation is poor.
- Reversal can be obtained by reversing field voltage polarity with split series field winding.

(*ii*) **Split series motor :**

- The D.C. series motor with split field (small fractional kW) may be operated as a separately excited field-controlled motor (Fig. 29).

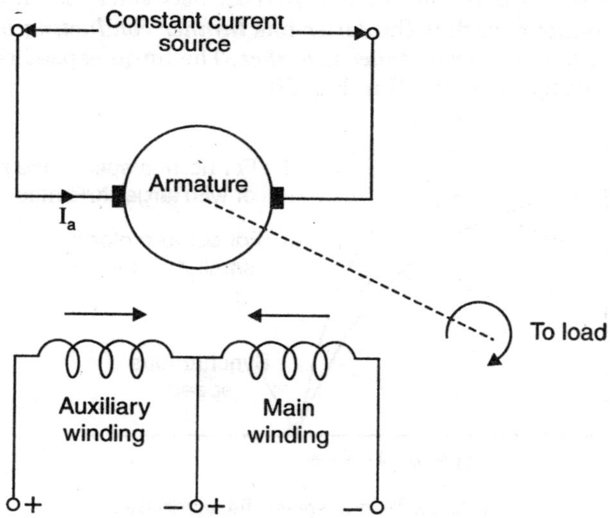

Fig. 29. From D.C. amplifier.

The armature may be supplied from a constant current source.

- A typical torque curve shows the following :
  — High stall torque ;
  — Rapid reduction in torque with increase in speed.

(*iii*) **Shunt control motor :**

- This type of motor has *two separate windings* : *Field winding placed on the stator and the armature winding placed on the rotor of the machine*. Both the windings are connected to a D.C. supply source.

- Whereas in a conventional D.C. shunt motor, the two windings are connected in parallel across the D.C. supply mains, but in a *servo-application* the windings are driven by *separate D.C. supplies*.

(*iv*) **Permanent magnet shunt motor :**

- It is a fixed excitation shunt motor where the field is actually supplied by a permanent magnet.
- Its performance is similar to that of armature controlled fixed field motor.

8.2.2. A.C. servo-motors

Applications :

- These motors are best suited for *low power applications*.
- Precision servo-motors are used in :
  — Instrument servos ;
  — Computers ;
  — Inertial guidance systems etc.
- The mechanical output power of A.C. servo-motor varies from 2 watts to a few hundred watts.
- An A.C. servo-motor is basically a two-phase induction motor except for certain special design features. The *main important difference* between a standard split-phase motor and an A.C. servo-motor is that the *latter* has *thinner conducting bars in the squirrel cage motor, so that the motor resistance is higher*. The torque-speed characteristics should be *linear* as shown by the curve II in Fig. 30.

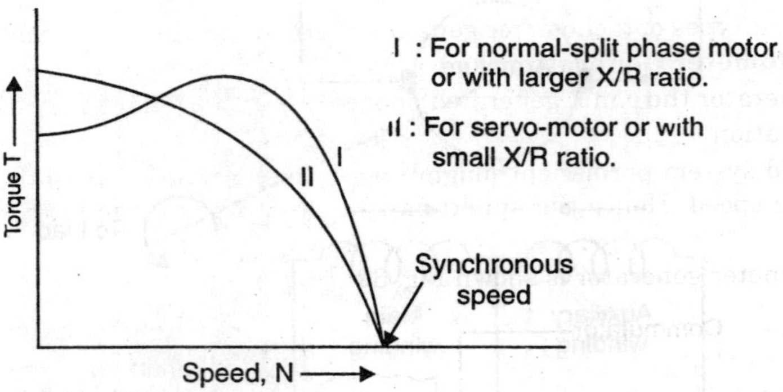

Fig. 30. Torque-speed characteristics.

Description of A.C. servo-motors :

1. **Drag-cup rotor servo-motor.** Refer Fig. 31.

- Drag-cup construction is used for *very low inertia applications*.
- In this type of motor the rotor construction is usually of squirrel cage or drag-cup type ; here only a *light cup rotates while the rotor core is stationary* (thus inertia is quite small).
- The servo-motors contains *two* windings namely, *main winding* (sometimes called fixed or reference winding) and *control winding*. The voltages applied to the windings are

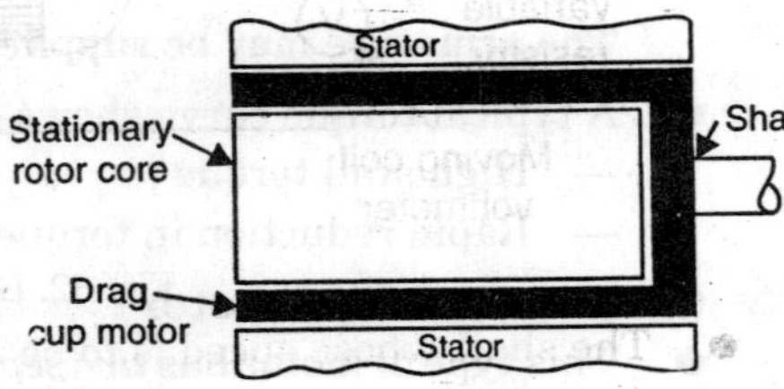

Fig. 31. Drag-cup rotor servo-motor.

at *right angles* to one another. Usually one winding is excited with a fixed voltage while the other one is excited by the control voltage (which is the output from servo-amplifier).

- *While in operation, the output torque of the motor is roughly proportional to the applied control voltage, and the direction of torque is determined by the polarity of the control voltage.*

2. **Shaded-pole type servo-motor :**

- This type of motor employs a *phase-sensitive relay to actuate those contacts which produce a short-circuit of the shaded-pole winding to produce rotation in the desired direction.*
- *The main short-coming of this motor is that it responds only when the amplifier error signal is of adequate magnitude to cause the relay to operate.*

9. TACHOMETER GENERATORS

Tachometers are the **devices employed for measuring the rotational speeds of motors and generators.**

The tachometers are **classified** as follows :

I. **Mechanical tachometers**

II. **Electrical tachometers**

1. D.C. tachometer generators
2. A.C. tachometer generators
3. Photo-electric tachometers (or Speed-meter)
4. Toothed motor variable reluctance tachometer etc. Electrical tachometers are preferred (as compared to mechanical tachometers) in view of the advantages they offer as **electrical transducers.**

Give important types of tachometer generators are briefly described below :

1. D.C. Tachometer Generator :

In a D.C. generator the e.m.f. generated depends upon the following two factors :

(*i*) Field excitation (*ii*) Speed

If for the field system permanent magnet pole pieces are used, then the generated voltage depends only on the speed. Hence the **speed can be computed by measuring the generated e.m.f.**

A D.C. tachometer generator is shown Fig. 32.

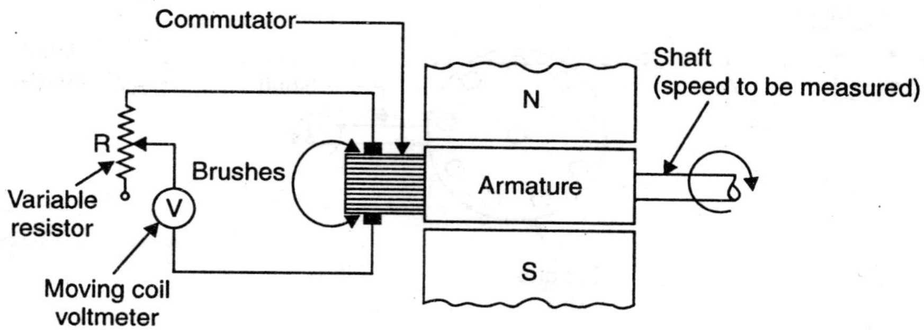

Fig. 32. D.C. tachometer generator.

- The shaft whose speed is to be measured is coupled to the armature.
- A moving coil voltmeter is connected across the brushes to measure the generated voltage. The variable resistance R is incorporated to limit the current through the voltmeter.

Since voltage is proportional to speed, the voltmeter may be calibrated in terms of speed (r.p.m.).

2. A.C. Tachometer Generator :

The inherent demerits associated with D.C. tachometer generator, due to the provision of commutor and brushes, are eliminated in A.C. tachometer generator.

An A.C. tachometer generator is shown in Fig. 33.

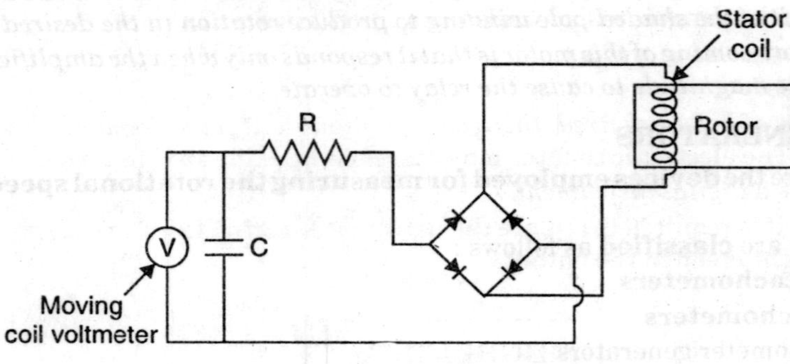

Fig. 33. A.C. tachometer generator.

- It consists of, like an alternator, a stationary armature (stator) and a rotating field system (rotor). Owing to the generation of e.m.f. in a stationary coil on a stator, commutation problems no longer exist.
- The alternating e.m.f. induced in the stationary coil is *rectified*, and the output D.C. voltage is measured with the help of a moving coil voltmeter (V).
- The *ripple content* of the rectified voltage is *smoothened* by the capacitor filter (C).
- As the speed depends on both the amplitude of the voltage and frequency, any one of them can be used as a measure of the speed. In an A.C. tachometer, it is the *induced voltage* that is considered as the required parameter.

3. Photo-electric Tachometer :

Fig. 34 shows a photo-electric tachometer.

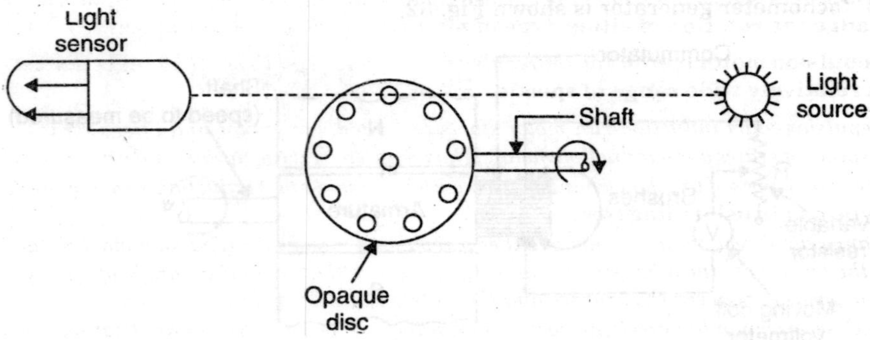

Fig. 34

- It consists of a *opaque disc* mounted on the shaft whose speed is to be measured. The disc has a number of equivalent holes around the periphery. On one side of the disc there is a source of light (L) while on the other side there is a light sensor (may be a photosensitive device or photo-tube) in line with it (light-source).

- On the rotation of the disc, holes and opaque portions of the disc come alternatory in between the light source and the light sensor. When a hole comes in between the two, light passes through the holes and falls on the light sensor, with the result that an *output pulse is generated.* But when the opaque portion of the disc comes in between, the light from the source is blocked and hence there is no pulse output. Thus *whenever a hole comes in line with the light source and sensor, a pulse is generated.* These pulses are counted/measured through an *electronic counter.*

The number of pulses generated depends upon the following factors :

(*i*) The number of holes in the disc ;

(*ii*) The shaft speed.

Since the number of holes are fixed, therefore, the number of pulses generated depends on the speed of the shaft only. The electronic counter may therefore be calibrated in terms of speed (r.p.m.)

Advantage. It is a *digital instrument.*

Disadvantage. It is required to replace the light source periodically, and if the grating period is small then errors might creep in the output.

HIGHLIGHTS

1. *There are two basic types of single-phase induction motors which start on the split-phase principle, namely resistance-start induction-run motor and the capacitor-start induction-run motor.*
2. *The resistance-start induction-run motor has a relatively poor starting torque because the phase angle between the running winding current and the starting winding current is only from 30 to 50 electrical degrees.*
3. *The capacitor-start induction-run motor has a relatively good starting torque because the phase angle between the running winding current and the starting current is practically 90 electrical degrees.*
4. *The capacitor-start, capacitor-run induction motor has relatively good starting torque and operates at a comparatively high power factor. The main winding and auxiliary windings are both energized at all times the motor is in operation.*
5. *A resistance-start induction-run motor or a capacitor-start induction-run motor may be reversed by interchanging the leads of the starting winding circuit. The direction of rotation of either type of motor can also be reversed by interchanging the leads of the running winding circuit. However,* **never attempt to change the rotation of either type of motor by reversing the line wires.**
6. *The* **repulsion motor** *operates on the repulsion principles at all times. It has* **excellent starting torque** *and* **a relatively wide range of speeds.**
7. *The repulsion-start induction-run motor starts as a repulsion motor and after it reaches 75 per cent of rated speed, operates as an induction motor. It has excellent starting torque and very good speed regulation.*
8. *There are two types of brush mechanism used with repulsion-start induction-run motors : The* **brush-lifting type** *and* **brush-riding** *type.*
9. *To change the direction of rotation of either a repulsion motor or a repulsion-start induction-run motor, move the brushes to the other side of stator field poles so that there is an angle of 15 electrical degrees in the new position between the brushes and the stator field pole centres.*
10. *The A.C. series motor has excellent starting torque and can be used in those applications requiring different speeds. It has characteristics comparable to those of the D.C. series motor.*
11. *The direction of rotation of an A.C. series motor can be reversed by interchanging the connections to the armature.*
12. *The shaded-pole induction motor is usually made in fractional horsepower sizes in excess of 1/10 horsepower. This type of motor has no centrifugal starting device or mechanism which requires maintenance. However, the* **starting torque** *for this type of motor is* **small.**

13. A *stepper motor* is an incremental motion machine. It does not rotate continuously as a conventional motor does.

14. The stepper motor is used in digitally controlled position control system in open loop mode.

15. Various types of stepper motors are :
(*i*) Permanent-magnet stepper motor ;
(*ii*) Variable-reluctance stepper motor ;
(*iii*) Hybrid stepper motor.

16. D.C. servo-motors are preferred for very high power systems.

17. A.C. servo-motors are best suited for low power applications.

OBJECTIVE TYPE QUESTIONS

(A) Choose the Correct Answer :

1. In a split-phase motor, the running winding should have
(*a*) high resistance and low inductance (*b*) low resistance and high inductance
(*c*) high resistance as well as high inductance (*d*) low resistance as well as low inductance.

2. If the capacitor of a single-phase motor is short-circuited
(*a*) the motor will not start (*b*) the motor will burn
(*c*) the motor will run in reverse direction
(*d*) the motor will run in the same direction at reduced r.p.m.

3. In capacitor-start single-phase motors
(*a*) current in the starting winding leads the voltage
(*b*) current in the starting winding lags the voltage
(*c*) current in the starting winding is in phase with voltage in running winding.
(*d*) none of the above.

4. In a capacitor-start and run motors the function of the running capacitor in series with the auxiliary winding is to
(*a*) improve power factor (*b*) increase overload capacity
(*c*) reduce fluctuations in torque (*d*) to improve torque.

5. In a capacitor-start motor, the phase displacement between starting and running winding can be nearly
(*a*) 10° (*b*) 30°
(*c*) 60° (*d*) 90°.

6. In a split-phase motor
(*a*) the starting winding is connected through a centrifugal switch
(*b*) the running winding is connected through a centrifugal switch
(*c*) both starting and running windings are connected through a centrifugal switch
(*d*) centrifugal switch is used to control supply voltage.

7. The torque developed by a single-phase motor at starting is
(*a*) more than the rated torque (*b*) rated torque
(*c*) less than the rated torque (*d*) zero.

8. Which of the following motors will give relatively high starting torque ?
(*a*) Capacitor-start motor (*b*) Capacitor run motor
(*c*) Split-phase motor (*d*) Shaded-pole motor.

9. Which type of capacitor is preferred for capacitor-start and run motor ?
(*a*) Electrolyte capacitor (*b*) Paper capacitor (oil filled)
(*c*) Ceramic capacitor (*d*) Air capacitor.

10. Which of the following motors will have relatively higher power factor ?
(*a*) Capacitor-run motor (*b*) Shaded-pole motor
(*c*) Capacitor-start motor (*d*) Split-phase motor.

11. In a shaded-pole motor, the shading coil usually consist of
(*a*) a single turn of heavy wire which is in parallel with running winding
(*b*) a single turn of heavy copper wire which is short-circuited and carries only induced current
(*c*) a multilayer fine gauge copper wire in parallel with running winding
(*d*) none of the above.

12. In a shaded-pole single-phase motor, the revolving field is produced by the use of
(*a*) inductor (*b*) capacitor
(*c*) resistor (*d*) shading coils.

13. A centrifugal switch is used to disconnect starting winding when motor has
(*a*) run for about 1 minute (*b*) run for about 5 minutes
(*c*) picked up about 50 to 70 per cent of rated speed
(*d*) picked up about 10 to 25 per cent of rated speed.

14. If a particular application needs high speed and high starting torque, then which of the following motors will be preferred ?
(*a*) Universal motor (*b*) Shaded-pole type motor
(*c*) Capacitor-start motor (*d*) Capacitor-start and run motor.

15. Which motor has the highest power to weight ratio ?
(*a*) Capacitor-run motor (*b*) Capacitor-start motor
(*c*) Universal motor (*d*) None of the above.

16. In a universal motor the width of the brush is nearly
(*a*) half the width of commutator segment (*b*) equal to width of the commutator segment
(*c*) equal to width of two commutator segment (*d*) equal to width of four commutator segment.

17. The range of efficiency for shaded-pole motors is
(*a*) 95% to 99% (*b*) 80% or 90%
(*c*) 50% to 75% (*d*) 5% to 35%.

18. In a capacitor-start single-phase motor, when capacitor is replaced by a resistance
(*a*) torque will increase (*b*) the motor will consume less power
(*c*) motor will run in reverse direction (*d*) motor will continue to run in same direction.

19. The power factor of a single-phase induction motor is usually
(*a*) lagging (*b*) always leading
(*c*) unity (*d*) unity to 0.8 leading.

20. A shaded-pole motor can be used for
(*a*) toys (*b*) hair dryers
(*c*) circulators (*d*) any of the above.

21. A hysteresis motor works on the principle of
(*a*) hysteresis loss (*b*) magnetisation of rotor
(*c*) eddy current loss (*d*) electromagnetic induction.

22. Which of the following motors will give the highest starting torque ?
(*a*) D.C. shunt motor (*b*) Schrage motor
(*c*) Repulsion-start and induction-run motor (*d*) Universal motor.

23. In repulsion motors by shifting the position of brushes
(*a*) only speed can be controlled (*b*) speed and direction of rotation can be controlled
(*c*) speed, direction of rotation and torque can be controlled
(*d*) none of the above.

24. Which of the following single-phase motors is reversible ?
(*a*) Capacitor-start split-phase motor (*b*) Resistance-start split-phase motor
(*c*) Reluctance motor (*d*) None of the above.

25. For which of the applications a reluctance motor is preferred ?
(*a*) Electric shavers (*b*) Refrigerators
(*c*) Signalling and timing devices (*d*) Lifts and hoists.

26. In repulsion motors
(*a*) speed remains constant (*b*) speed variation is within 5%
(*c*) speed varies with load (*d*) speed varies only with change in applied voltage.

27. In single-phase repulsion motors power factor is
(*a*) always leading (*b*) high at low speeds
(*c*) always unity (*d*) high at high speeds.

28. The motor used on small lathes is usually
(*a*) universal motor (*b*) d.c. shunt motor
(*c*) single-phase capacitor run motor (*d*) 3-phase synchronous motor.

29. Which of the following motors is preferred for tape recorders ?
(*a*) Shaded-pole motor (*b*) Hysteresis motor
(*c*) Two value capacitor motor (*d*) Universal motor.

30. A single-phase induction motor is
(*a*) inherently self-starting with high torque (*b*) inherently self-starting with low torque
(*c*) inherently non-self-starting with low torque (*d*) inherently non-self-starting with high torque.

31. A schrage motor can run on
(*a*) zero slip (*b*) negative slip
(*c*) positive slip (*d*) all of the above.

32. A universal motor can run on
(*a*) A.C. only (*b*) D.C. only
(*c*) either A.C. or D.C. (*d*) none of the above.

33. Which of the following single-phase motors is suitable for timing and control purposes ?
(*a*) Reluctance motor (*b*) Series motor
(*c*) Repulsion motor (*d*) Universal motor.

34. In case of a shaded-pole motor the direction of rotation of the motor is
(*a*) from main-pole to shaded-pole (*b*) from shaded-pole to main pole
(*c*) either of the above depending on voltage (*d*) either of the above depending on power factor.

35. In case of high speed universal motor which of the following needs more attention ?
(*a*) End play (*b*) Air gap
(*c*) Insulation in rotor (*d*) Balancing of rotor.

36. The wattage rating for a ceiling fan motor will be in the range
(*a*) 50 to 250 W (*b*) 250 to 500 W
(*c*) 50 to 150 W (*d*) 10 to 20 W.

37. The wattage of motor for driving domestic sewing machine will be around
(*a*) 100 to 150 W (*b*) 40 to 75 W
(*c*) 10 to 30 W (*d*) 5 to 10 W.

38. Which of the following single-phase motors has relatively poor starting torque ?
(*a*) Universal motor (*b*) Repulsion motor
(*c*) Capacitor motor (*d*) All single phase motors have zero starting torque.

39. Which type of load is offered by cranes and hoists ?
(*a*) gradually varying load (*b*) non-reversing, no-load start
(*c*) reversing, light start (*d*) reversing, heavy start.

40. The speed of a universal motor is generally reduced by using
(*a*) gear trains (*b*) V-belts
(*c*) brakes (*d*) chains.

41. Which of the following motors can be used for unity power factor ?
(*a*) Capacitor run motor (*b*) Shaded-pole motor
(*c*) Hysteresis motor (*d*) Schrage motor.

42. When a D.C. series motor is connected to A.C. supply, the power factor will be low because of
(*a*) high inductance of field and armature circuits
(*b*) induced current in rotor due to variations of flux
(*c*) fine copper wire winding (*d*) none of the above.

43. The direction of rotation of universal motor can be reversed by reversing the flow of current through
(*a*) armature winding (*b*) field winding
(*c*) either armature winding or field winding (*d*) none of the above.

44. In which single-phase motor, the rotor has no teeth or winding ?
(*a*) Split-phase motor (*b*) Reluctance motor
(*c*) Hysteresis motor (*d*) Universal motor.

45. Which motor is normally free from mechanical and magnetic vibrations ?
(*a*) Split-phase motor (*b*) Universal motor
(*c*) Hysteresis motor (*d*) Shaded-pole motor.

46. As hysteresis motors are free from mechanical and magnetic vibrations therefore these are considered as suitable for
(*a*) fans (*b*) blowers
(*c*) sound equipment (*d*) mixer grinders.

47. A reluctance motor
(*a*) is self-starting (*b*) is constant speed motor
(*c*) needs no D.C. excitation (*d*) all of the above.

48. In a hysteresis motor, the rotor must have
(*a*) retentivity (*b*) resistivity
(*c*) susceptibility (*d*) none of the above.

49. The rotor of a hysteresis motor is made of
(*a*) aluminium (*b*) cast iron
(*c*) chrome steel (*d*) copper.

50. The electric motor used in portable drills is
(*a*) capacitor-run motor (*b*) hysteresis motor
(*c*) universal motor (*d*) repulsion motor.

51. Which of the following applications always has some load whenever switched on ?
(*a*) Vacuum cleaners (*b*) Fan motors
(*c*) Pistol drills (*d*) All of the above.

52. The speed control of universal motor used for sewing machines is by
(*a*) friction (*b*) varying the resistance
(*c*) tapping the field (*d*) centrifugal mechanism.

ANSWERS

1. (*b*)	**2.** (*a*)	**3.** (*a*)	**4.** (*a*)	**5.** (*d*)	**6.** (*a*)
7. (*d*)	**8.** (*a*)	**9.** (*b*)	**10.** (*a*)	**11.** (*b*)	**12.** (*d*)
13. (*c*)	**14.** (*a*)	**15.** (*c*)	**16.** (*c*)	**17.** (*d*)	**18.** (*d*)
19. (*a*)	**20.** (*d*)	**21.** (*a*)	**22.** (*b*)	**23.** (*b*)	**24.** (*a*)
25. (*c*)	**26.** (*c*)	**27.** (*d*)	**28.** (*c*)	**29.** (*b*)	**30.** (*c*)
31. (*d*)	**32.** (*c*)	**33.** (*a*)	**34.** (*a*)	**35.** (*d*)	**36.** (*c*)
37. (*a*)	**38.** (*c*)	**39.** (*d*)	**40.** (*a*)	**41.** (*d*)	**42.** (*a*)
43. (*c*)	**44.** (*c*)	**45.** (*c*)	**46.** (*c*)	**47.** (*d*)	**48.** (*a*)
49. (*c*)	**50.** (*c*)	**51.** (*c*)	**52.** (*b*).		

(B) Say 'Yes' or 'No' :

1. In a capacitor-start motor, the phase displacement between starting and running winding can be nearly 90°.

2. In a split-phase motor the staring winding is connected through a centrifugal switch.
3. In a shaded-pole single-phase motor, the revolving field is produced by the use of capacitor.
4. A universal motor has the highest power to weight ratio.
5. The range of efficiency for shaded-pole motors is 80 to 90%.
6. The power factor of a single-phase induction motor is usually lagging.
7. A hysteresis motor works on the principle of eddy current loss.
8. Capacitor-start split-phase motor is reversible.
9. A reluctance motor is preferred for signalling and timing devices.
10. The wattage rating for a ceiling fan motor varies from 50 to 150 W.
11. A ...... motor is an incremental motion machine.
12. A stepper motor rotates continuously.
13. The stepper motor is used in digitally controlled position control system in open loop mode.
14. In stepper motors no sensors are needed for position and speed sensing.
15. A stepper motor cannot be readily interfaced with microprocessor.
16. In ...... stepper motor the rotor is made of ferite or rare earth material which is permanently magnetised.
17. In a ...... stepped motor the stator has only one set of winding-excited poles which interact with the two rotor stacks.
18. ...... servo-motors are preferred for very high power systems.
19. A.C. servo-motors are best suited for low power applications.
20. ...... construction is used for very low inertia applications.

ANSWERS

1. Yes	2. Yes	3. No	4. Yes	5. No	6. Yes
7. No	8. Yes	9. Yes	10. Yes	11. stepper	12. No
13. Yes	14. Yes	15. No	16. Permanent-magnet		17. hybrid
18. D.C.	19. Yes	20. Drag-cup.			

THEORETICAL QUESTIONS

1. Explain what is meant by the split-phase method of motor starting.
2. Explain why the starting torque of a capacitor-start induction run motor is better than that of a resistance-start induction-run motor.
3. How is the direction of rotation reversed for each of the following ?
   (*i*) Resistance-start induction-run motor, and
   (*ii*) Capacitor-start induction-run motor.
4. (*a*) Compare operating characteristics of a resistance-start induction-run motor with those of a capacitor-start induction-run motor.
   (*b*) List three applications for :
   (*i*) A resistance-start induction-run motor
   (*ii*) A capacitor-start induction-run motor.
5. Explain how a repulsion motor operates.
6. Explain how a repulsion-start induction-run motor operates.
7. How is the direction of rotation of each of the following motors reversed ?
   (*i*) The repulsion motor, and (*ii*) The repulsion-start induction-run motor.
8. What is the difference between A.C. and D.C. series motor ?
9. Why are small fractional horsepower A.C. series motors called universal motors ?
10. What is the difference between a conductively compensated series motor and an inductively compensated series motor ?
11. Describe the construction and operation of a shaded-pole motor.

12. Write a short note on a sub-synchronous motor and a hysteresis motor.
13. What is a stepper motor ? Enumerate its advantages and applications.
14. Explain briefly permanent-magnet stepper motor briefly. How does it differ from variable reluctance motor ?
15. What is a stepper motor ?
16. States the advantages of stepper motors.
17. Give the applications of stepper motors.
18. Explain briefly the construction and working of the following stepper motors :
    (*i*) Permanent-magnet stepper motor ; (*ii*) Variable-reluctance stepper motor ;
    (*iii*) Hybrid stepper motor.
19. What do you understand by the term servo or servo-mechanism ?
20. Explain briefly any two of the following D.C. servo-motors :
    (*i*) Series motor (*ii*) Split series motor
    (*iii*) Shunt control motor (*iv*) Permanent magnet shunt motor.
21. Give the description of drag-cup rotor servo-motor.
22. Write a short note on shaded-pole type servo-motor.

8

Direct Current (D.C.) Machines

1. Construction of D.C. Machines. 2. E.m.f. equation of a generator. 3. Types of D.C. generators—Separately excited generators—Self excited generators. 4. Power division in a D.C. generator. 5. Characteristics of D.C. generators—Separately excited generator—No-load saturation characteristic (or O.C.C.)—Internal and external characteristics (or Load characteristics)—Building up the voltage of self-excited shunt generator—Shunt generator characteristics—Series generator—Compound wound generator. 6. Applications of D.C. generators—Separately excited generators—Shunt generators—Series generators—Compound generators. 7. Direct current motor—General aspects—Principle of operation of D.C. motor—Back or counter e.m.f.—Comparison between motor and generator action—Torque developed in a motor—Mechanical power developed by motor armature—Types of D.C. motors—Speed of a D.C. motor—Speed regulation—Motor characteristics—Comparison of D.C. motor characteristics—Summary of characteristics and applications of D.C. motors—Starting of D.C. motors. 8. Speed control of D.C. motors—Factors controlling the speed—Field control method—Rheostatic control—Voltage control. 9. Losses and efficiency of D.C. machines—Introduction—Losses—Efficiency of D.C. machines—Highlights—Objective type questions—Theoretical questions—Unsolved examples.

1. CONSTRUCTION OF D.C. MACHINES

A D.C. machine consists of *two* main parts :

(*i*) **Stationary part.** It is designed mainly for *producing a magnetic flux*.

(*ii*) **Rotating part.** It is called the *armature*, where mechanical energy is converted into electrical (electrical generator), or conversely, electrical energy into mechanical (electric motor).

The stationary and rotating parts are separated from each other by an *air gap*.

- The *stationary part* of a D.C. machine consists of *main poles*, designed to create the magnetic flux, *commutating poles* interposed between the main poles and designed *to*

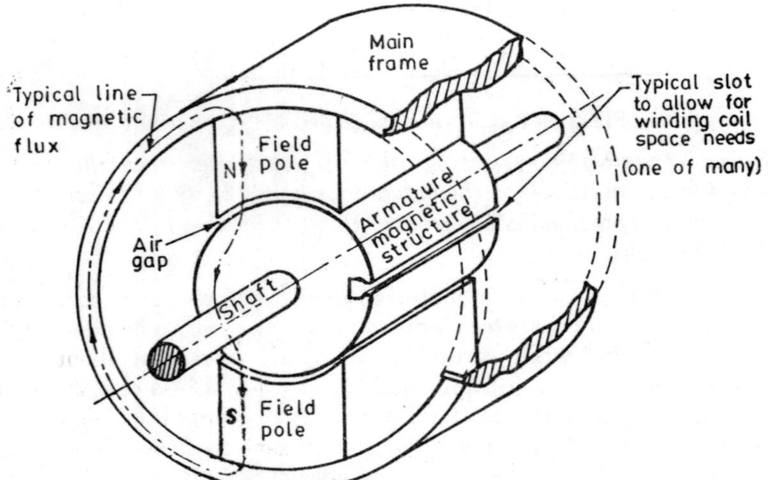

Fig. 1. Generator or motor magnetic structure.

ensure sparkless operation of the brushes at the commutator (in very small machines with a lack of space commutating poles are not used) ; and a *frame*/yoke.

- The *armature* is a cylindrical body rotating in the space between the poles and comprising a *slotted armature core*, a *winding* inserted in the armature core slots, a *commutator*, and *brush gear*.

Fig. 1 shows generator or motor magnetic structure.

Description of Parts of D.C. Machines :

1.1. Frame. Fig. 2 shows the sectional view of four pole D.C. machine.

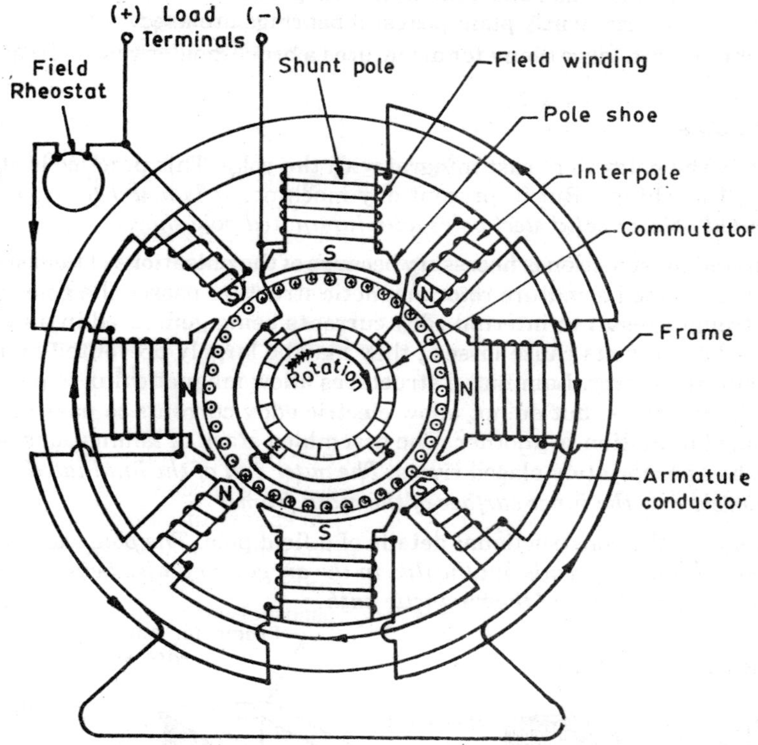

Fig. 2. Sectional view of a four pole D.C. machine.

- *The frame is the stationary part of a machine to which are fixed the main and commutating poles and by means of which the machine is bolted to its bed plate.*
- *The ring-shaped portion which serves as the path for the main and commutating pole fluxes is called the* '**yoke**'.

Cast iron used to be the material for the frame/yoke in *early machines* but now it has been *replaced by cast steel.* This is because cast iron is saturated by a flux density of about 0.8 Wb/m^2 while saturation with cast steel is at about 1.5 Wb/m^2. Thus the cross-section of a cast iron frame is about twice that of a cast steel frame for the same value of magnetic flux. Hence, if it is necessary to reduce the weight of machine, cast steel is used. Another disadvantage with the use of cast iron is that its mechanical and magnetic properties are uncertain due to the presence of blow holes in the casting. Lately, rolled steel yokes have been developed with the improvements in the welding techniques. The *advantages of fabricated yokes are that there are no pattern charges and the magnetic and mechanical properties of the frame are absolutely consistent.*

It may be advantageous to use cast iron for frames but for medium and large sizes usually rolled steel is used.

- If the armature diameter does not exceed 35 to 45 cm, then, in addition to the poles, end shields or frame-heads which carry the bearings are also attached to the frame. When the armature diameter exceeds 1 m, it is common practice to use pedestal-type bearings, mounted separately, on the machine bed plate outside the frame.
- The end shield bearings, and sometimes the pedestal bearings, are of ball or roller type. However, more frequently plain pedestal bearings are used.
- In machines with large diameter armatures a brush-holder yoke is frequently fixed to the frame.

1.2. Field Poles

- Formerly the poles were cast integral with the yoke. This practice is still being followed for small machines. But in present day machines *it is usual to use either a completely laminated pole, or solid steel poles with laminated pole shoes.*
- Laminated construction is necessary because of the pulsations of field strength that result when the notched armature rotor magnetic structure passes the pole shoe. Variations in field strength result in internal eddy currents being generated in a magnetic structure. These eddy currents cause losses ; they may be largely prevented by having laminated magnetic structures. Laminated structures allow magnetic flux to pass along the length of the laminations, but do not allow electric eddy currents to pass across the structure from one lamination to another. The assembled stack of laminations is held together as a unit by appropriately placed rivets. *The outer end of the laminated pole is curved to fit very closely into the inner surface of the main frame.*
- Fig. 3 shows the constructional details of a field pole. *The pole shoe acts as a support to the field coils and spreads out the flux in the air gap and also being of larger cross-section reduces the reluctance of the magnetic path.*

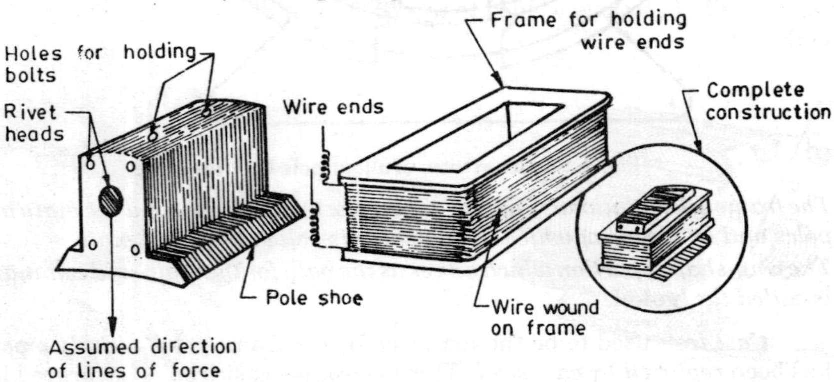

Fig. 3. Constructional details of a field pole.

- Different methods are used for attaching poles to the yoke. In case of *smaller sizes,* the back of the pole is drilled and tapped to receive pole bolts (*see* Fig. 4). In *larger sizes*, a circular or a rectangular pole bar is fitted to the pole. This pole bar is drilled and tapped and the pole bolts passing through laminations screw into the tapped bar (*see* Fig. 5).

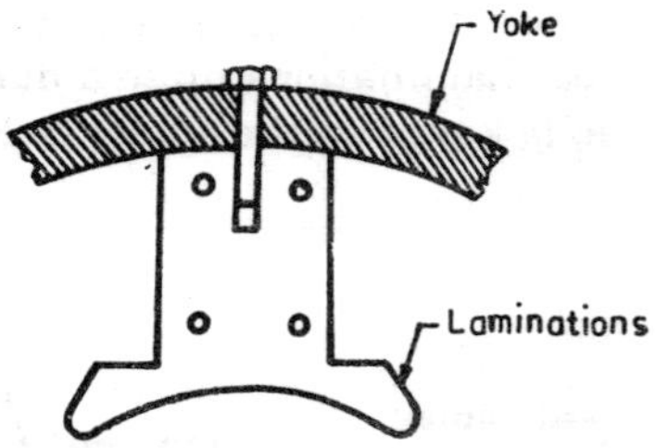

Fig. 4. Fixing pole to the yoke.

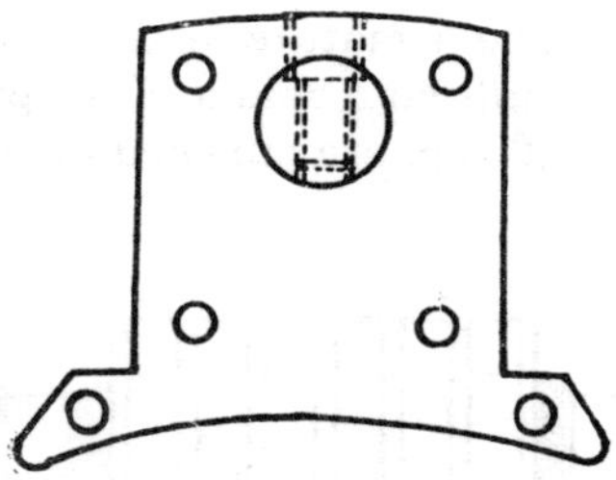

Fig. 5

1.3. Commutating Poles

- A commutating pole (also called *interpole*) is similar to a main pole and consists of core terminating in a pole shoe, which may have various shapes, and coil mounted on the core.
- The commutating poles *are arranged strictly midway between the main poles* and are bolted to the yoke.
- *Commutating poles are usually made of solid steel, but for machines operating on sharply varying loads they are made of sheet steel.*

1.4. Armature

- The armature consists of core and winding. *Iron being the magnetic material is used for armature core.* However, iron is also a good conductor of electricity. The rotation of solid iron core in the magnetic field results in eddy currents. The flow of eddy currents in the core leads to wastage of energy and creates the problem of heat dissipation. *To reduce the eddy currents the core is made of thin laminations.*
- *The armature of D.C. machines (see Fig. 6) is built up of thin laminations of low loss silicon steel.* The laminations are usually 0.4 to 0.5 mm thick and are insulated with varnish.
- The armature **laminations**, in *small machines,* are fitted directly on to the shaft and are clamped tightly between the flanges which also act as supports for the armature winding. One end flange rests against a shoulder on the shaft, the laminations are fitted and other end is pressed on the shaft and retained by a key.

The core (except in small size) is divided into number of packets by radial ventilation spacers. The spacers are usually I sections welded to thick steel laminations and arranged to pass centrally down each tooth.

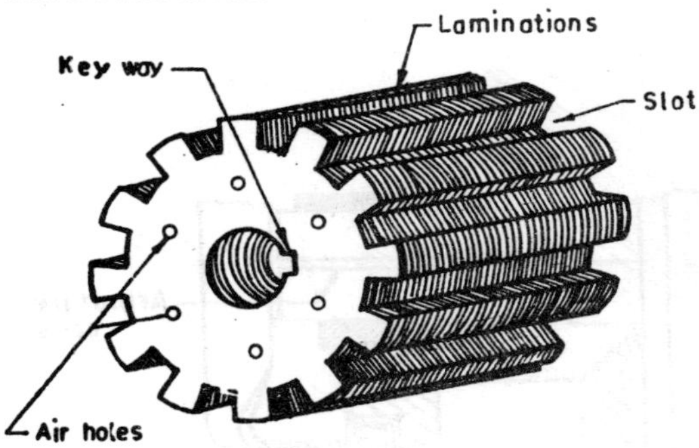

Fig. 6. Armature of a D.C. machine.

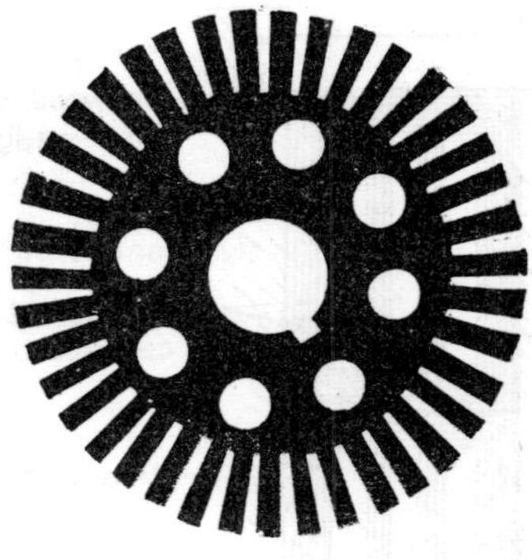

Fig. 7. Drum armature stamping with axial flow ventilation system.

— For *small machines* the laminations are punched in one piece (*see* Fig. 7). These laminations are built up directly on the shaft. With such an arrangement, it is necessary to provide *axial ventilation holes* so that air can pass into ventilating ducts.

— The armature laminations of *medium size machines* (having more than four poles) are built on a spider. The spider may be fabricated. Laminations up to a diameter of about 100 cm are punched in one piece and are directly keyed on the spider (*see* Fig. 8).

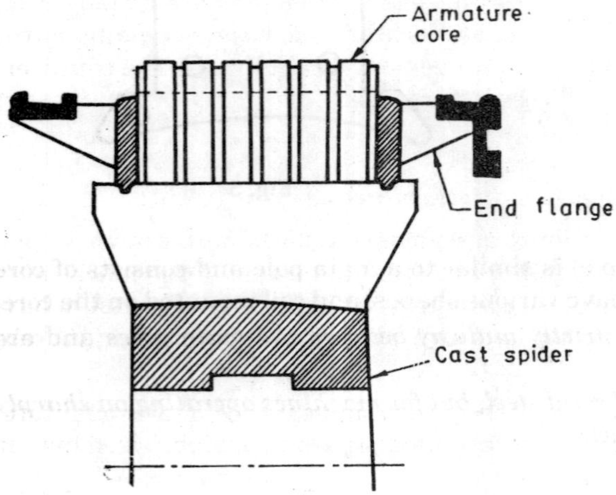

Fig. 8. Clamping of an armature core.

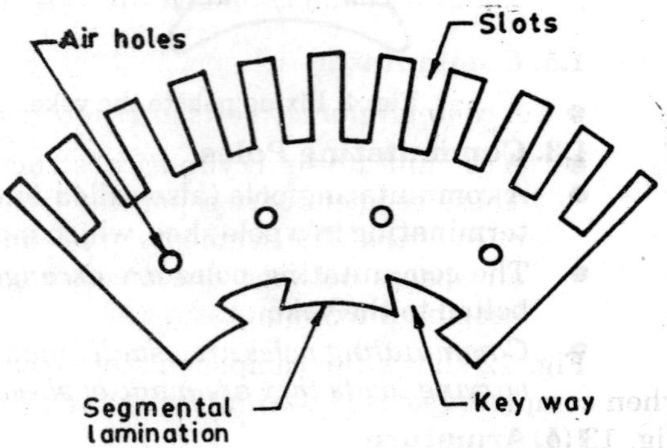

Fig. 9. Segmental stampings.

— In case of *large machines,* the laminations of such thin sections are difficult to handle because they tend to distort and become wavy when assembled together. Hence circular laminations instead of being cut in one piece are cut in a number of suitable sections or *segments* which form part of a complete ring (*see* Fig. 9). A complete circular limination is made up of four or six or even eight segmental laminations. Usually two keyways are notched in each segment and are dove-tailed or wedge shaped to make the laminations self-locking in position.

- The armature winding is housed in slots on the surface of the armature. The conductors of each coil are so spaced that when one side of the coil is under a north pole, the opposite is under a south pole.

Fig. 10 shows the arrangement of conductors and insulation in a slot.

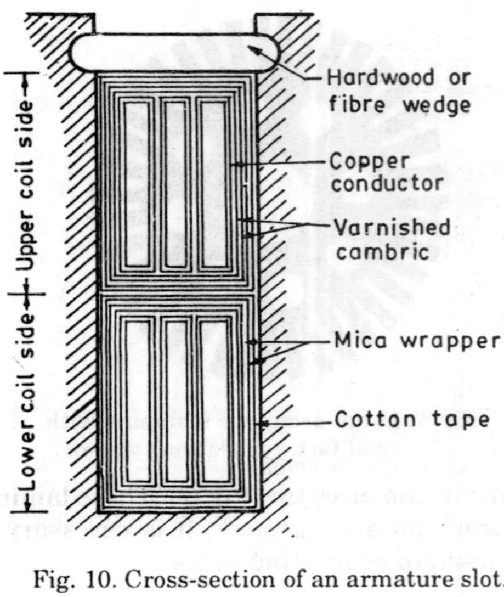

Fig. 10. Cross-section of an armature slot.

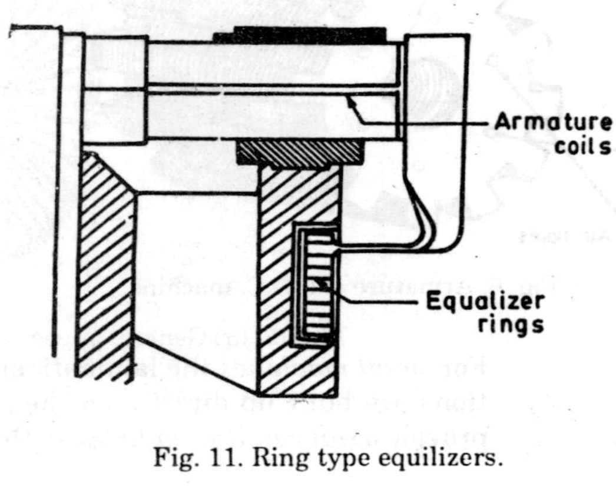

Fig. 11. Ring type equilizers.

- *In D.C. machines two layer winding with diamond shaped coils is used.* The coils are usually former wound. In *small machines,* the coils are held in position by band of steel wire, wound under tension along the core length. In *large machines*, it is useful to employ wedges of fibre or wood to hold coils in place in the slots. Wire bands are employed for holding the overhang. The *equilizer connections* are located under the overhang on the side of the commutator. Fig. 11 shows a typical arrangement for equilizers. The equilizers can be accommodated on the other end of the armature also.

1.5. Commutator

- *A commutator converts alternating voltage to a direct voltage.*
- A commutator is a cylindrical structure built up of segments made of hard drawn copper. These segments are separated from one another and from the frame of the machine by *mica strips.* The segments are connected to the winding through risers. The risers have air spaces between one another so that air is drawn across the commutator thereby keeping the commutator cool.

Fig. 12 shows the components of a commutator. The general appearance of a commutator when completed is as shown in Fig. 13 (*a*). The commutator and armature assembly is shown in Fig. 13 (*b*).

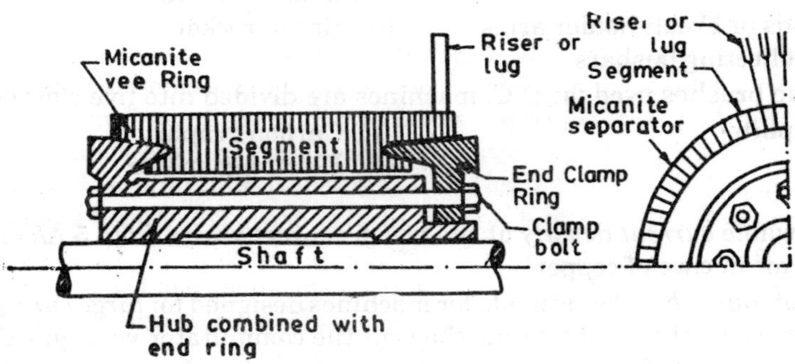

Fig. 12. Commutator components.

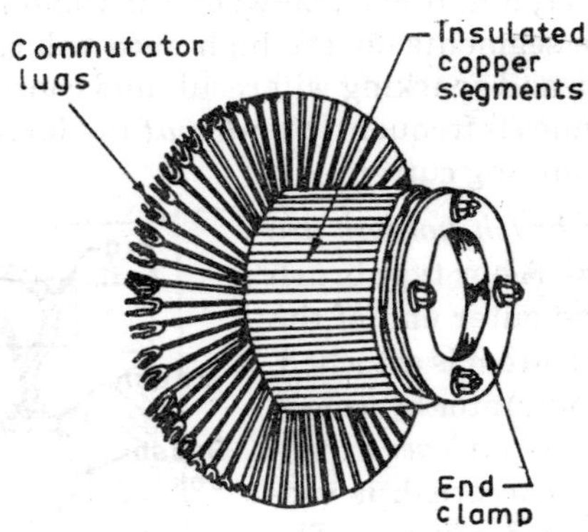

Fig. 13. (*a*) General appearance of a commutator after assembly.

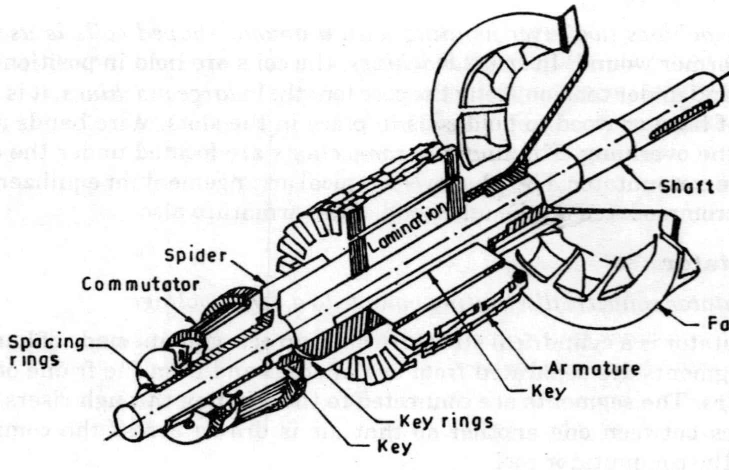

Fig. 13. (*b*) Commutator and armature assembly.

1.6. Brush Gear. To collect current from a rotating commutator or to feed current to it use is made of *brush-gear* which consists of :

(*i*) Brushes (*ii*) Brush holders
(*iii*) Brush studs or brush-holder arms (*iv*) Brush rocker
(*v*) Current-collecting busbars.

Brushes. The brushes used for D.C. machines are divided into five classes :

(*i*) Metal graphite (*ii*) Carbon graphite
(*iii*) Graphite (*iv*) Electro-graphite
(*v*) Copper.

- The allowable *current density* at the brush contact varies from 5 A/cm^2 in case of carbon to 23 A/cm^2 in case of copper.
- The use of *copper brushes* is made for machines designed for *large currents at low voltages.* Unless, very carefully lubricated, they cut the commutator very quickly and in any case, the wear is rapid. *Graphite and carbon graphite brushes are self-lubricating and, are, therefore, widely used.* Even with the softest brushes, however, there is a gradual wearing away of the commutator, and if mica between the commutator segments does not wear down so rapidly as the segments do, the high mica will cause the brushes to make poor contact with segments, and sparking will result and consequent damage to commutator. So to prevent this, the mica is frequently *'undercut'* to a level below the commutator surface by means of a narrow milling cutter.

Brush holders. *Box type brush holders* are used in all ordinary D.C. machines. A box type brush holder is shown in Fig. 14. At the outer end of the arm, a brush box, open at top and bottom is attached. The brush is pressed on to the commutator by a *clock spring.* The *pressure can be adjusted by a lever arrangement provided with the spring.* The brush is connected to a flexible conductor called *pig tail.* The flexible conductor may be attached to the brush by a screw or may be soldered.

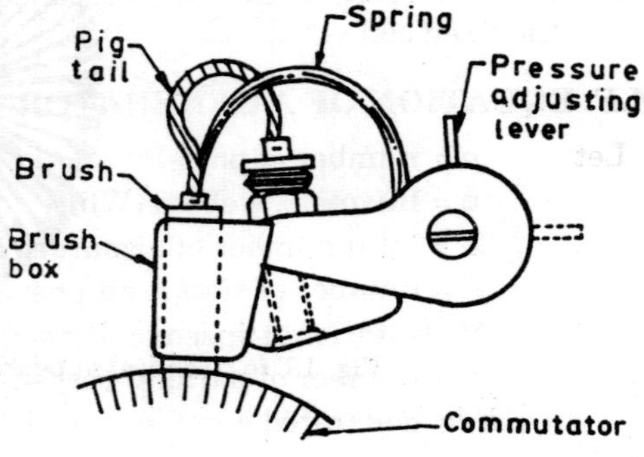

Fig. 14. Box type brush holder.

- The bush boxes are usually made of *bronze casting* or *sheet brass*. In low voltage D.C. machines where the commutation conditions are easy galvanised steel box may be used.
- Some manufacturers use individual brush holders while others use multiple holders, *i.e.*, a number of single boxes built up into one long assembly.

Brush rockers. Brush holders are fixed to brush rockers with bolts. The brush rocker is arranged concentrically round the commutator. *Cast iron is usually, used for brush rockers.*

1.7. Armature Shaft Bearings

- *With small machines roller bearings are used at both ends.*
- For *larger machines roller bearings are used for driving end and ball bearings are used for non-driving* (commutator) *end.*
- The bearings are housed in the end shields.
- For large machines pedestal bearings are used.

1.8. Armature Windings. The *armature winding* is very important element of a machine, as it directly takes part in the conversion of energy from one form into another. The requirements which a winding must meet are diverse and often of a conflicting nature. Among these requirements the following are of major importance.

- The winding must be designed with the *most advantageous utilisation of the material in respect to weight and efficiency*.
- The winding should *provide the necessary mechanical, thermal and electrical strength of the machine* to ensure the usual service life of 16-20 years.
- For D.C. machines proper current collection at the commutator (*i.e.*, absence of detrimental sparking) must be ensured.
- According to the degree of closure produced by winding, armature windings are of the following two types :

  1. Open coil winding 2. Closed coil winding.

  The closed armature windings are of two types :

  (*i*) Ring winding (*ii*) Drum winding

  In general there are two types of drum armature windings :

  (*i*) Lap winding (*ii*) Wave winding.

  "Lap winding" is *suitable for comparatively low voltage* but high current generators whereas **"wave of winding"** is used for high voltage, low current machines.

— In *'lap winding'* the finish of each coil is connected to the start of the next coil so that winding or commutator pitch is unity.

— In *'wave winding'* the finish of coil is connected to the start of another coil well away from the fixed coil.

2. E.M.F. EQUATION OF A GENERATOR

Let p = number of poles,

ϕ = flux/pole, webers (Wb),

Z = total number of armature conductors,

= number of slots × number of conductors/slot,

N = rotational speed of armature, r.p.m.,

a = number of parallel paths in armature, and

E_g = generated e.m.f. *per parallel path in armature*.

Average e.m.f. generated per conductor $= \frac{d\phi}{dt}$ volt.

Now, flux cut per conductor in one revolution, $d\phi = p\phi$ Wb.

Number of revolutions/second $= \frac{N}{60}$

$\therefore$ Time for one revolution, $dt = \frac{60}{N}$ seconds

Hence, according to Faraday's laws of electromagnetic induction,

E.m.f. generated per conductor $= \frac{p\phi N}{60}$ volts.

For a lap wound generator :

Number of parallel paths, $a = p$

Number of conductor (*in series*) in one path $= \frac{Z}{p}$

$\therefore$ E.m.f. generated *per path* $= \frac{\phi p N}{60} \times \frac{Z}{p} = \frac{\phi Z N}{60}$ volt.

For a wave wound generator :

Number of parallel paths, $a = p$

Number of conductor (in series) in one path $= \frac{Z}{2}$

$\therefore$ E.m.f. generator per path $= \frac{p\phi N}{60} \times \frac{Z}{2} = \frac{p\phi Z N}{120}$ volt.

In general, generated e.m.f.

$$E_g = \frac{\phi Z N}{60} \times \left(\frac{p}{a}\right) \text{volt} = \frac{p\phi Z N}{60a} \quad ...(i)$$

where $a = p$ ...... for lap winding

$= 2$ ...... for wave winding.

Example 1. *A six pole lap wound D.C. generator has 720 conductors, a flux of 40 m Wb per pole is driven at 400 r.p.m. Find the generated e.m.f.*

Solution. Number of poles, $p = 6$

Total number of conductors, $Z = 720$

Flux per pole, $\phi = 40$ m Wb $= 40 \times 10^{-3}$ Wb

Speed of rotation, $N = 400$ r.p.m.

Number of parallel paths, $a = p = 6$ [Since the generator is *lap wound.*]

Generated e.m.f. E_g :

Using the relation, $E_g = \frac{p\phi Z N}{60a} = \frac{6 \times 40 \times 10^{-3} \times 720 \times 400}{60 \times 6} = 192$ V.

Hence, **generated e.m.f. E_g = 192 V. (Ans.)**

Example 2. *A six-pole lap connected generator has a useful flux/pole of 0.045 Wb. If the no load voltage at 400 r.p.m. is 300 V, find the conductors on the armature periphery.*

Solution. Number of poles, $p = 6$

Useful flux/pole, $\phi = 0.045$ Wb

No load voltage, $E_g = 300$ V

Number of conductors, Z :

Number of parallel paths, $a = p = 6$ [Since the generator is *lap wound.*]

We know that, $E_g = \frac{p\phi Z N}{60a}$

$$300 = \frac{6 \times 0.045 \times Z \times 400}{60 \times 6}$$

$$\therefore \quad Z = \frac{300 \times 60 \times 6}{6 \times 0.045 \times 400} \quad i.e. \quad Z = 1000.$$

Hence, **total number of armature conductors = 1000. (Ans.)**

Example 3. *An 8-pole wave connected D.C. generator has 1000 armature conductors and flux/pole 0.035 Wb. At what speed must it be driven to generate 500 V?*

Solution. Number of poles, $p = 8$

Total number of armature conductor, $Z = 1000$

Flux/pole, $\phi = 0.035$ Wb

Generated voltage, $E_g = 500$ V

Number of parallel paths, $a = 2$ [Since the generator is wave wound.]

Speed of rotation, N :

Using the relation, $E_g = \frac{p\phi ZN}{60a}$

$$500 = \frac{8 \times 0.035 \times 1000 \times N}{60 \times 2}$$

$$\therefore \quad N = \frac{500 \times 60 \times 2}{8 \times 0.035 \times 1000} = 214.3 \text{ r.p.m.}$$

Hence, **speed of generator = 214.3 r.p.m. (Ans.)**

Example 4. *The armature of a 6-pole D.C. generator has a wave winding containing 650 conductors. Calculate the generated e.m.f. when the flux per pole is 0.055 Wb and the speed is 300 r.p.m.*

Calculate speed at which the armature must be driven to generate an e.m.f. of 550 V if the flux per pole is reduced to 0.05 Wb.

Solution. Number of poles, $p = 6$

Total number of conductors, $Z = 650$

Flux per pole, $\phi = 0.055$ Wb

Speed of rotation, $N = 300$ r.p.m.

E.m.f. generated, $E_g = ?$

Generated e.m.f. (2nd case) $= 550$ V

Flux per pole (2nd case) $= 0.05$ Wb

Speed of rotation, $N = ?$

Case I. E.m.f. generated, E_g :

Using the relation, $E_g = \frac{p\phi ZN}{60a}$

$$= \frac{6 \times 0.055 \times 650 \times 300}{60 \times 2} \quad [\because a = 2, \text{ as the generator is wave wound}]$$

$$= 536.25 \text{ V.}$$

Hence, **e.m.f. generated = 536.25 V. (Ans.)**

Case II. Speed of rotation, N :

$$E_g = \frac{p\phi ZN}{60a}$$

$$550 = \frac{6 \times 0.05 \times 650 \times N}{60 \times 2}$$

$$N = \frac{550 \times 60 \times 2}{6 \times 0.05 \times 650} = 338.46 \text{ r.p.m.}$$

Hence, **speed of rotation = 338.46 r.p.m. (Ans.)**

Example 5. *A six pole lap wound D.C. armature has 70 slots with 20 conductors/slot. The ratio of pole arc to pole pitch is 0.68. The diameter of bore of the pole shoe is 0.46 m. The length of pole shoe is 0.3 m. If the air gap flux density is 0.3 Wb/m² and the e.m.f. induced in the armature is 500 V, find the speed at which it runs.*

Solution. Number of poles, $p = 6$

Number of slots $= 70$

Conductors/slot $= 20$

$\therefore$ Total number of conductors, $Z = 70 \times 20 = 1400$

Ratio of pole arc to pole pitch $= 0.68$

Diameter of bore of the pole shoe $= 0.46$ m

Length of the pole shoe $= 0.3$ m

Air gap flux density, $B = 0.3 \text{ Wb/m}^2$

E.m.f. induced, $E_g = 500$ V

Speed of rotation, N :

$$\frac{\text{Pole arc}}{\text{Pole pitch}} = 0.68$$

$\therefore$ Pole arc $= 0.68 \times \text{pole pitch}$

$$= 0.68 \times \frac{\pi D}{p} = \frac{0.68 \times \pi \times 0.46}{6} = 0.1638 \text{ m}$$

Area of pole shoe, $A = \text{pole arc} \times \text{length of pole shoe}$

$$= 0.1638 \times 0.3 = 0.04914 \text{ m}^2$$

Now, $\phi = B \times A = 0.3 \times 0.04914 = 0.01474$ Wb

Using the relation, $E_g = \dfrac{p\phi ZN}{60a}$

$$500 = \frac{6 \times 0.01474 \times 1400 \times N}{60 \times 6} \quad [\because \ a = p = 6, \text{ generator being lap wound}]$$

$$\therefore \quad N = \frac{500 \times 60 \times 6}{6 \times 0.01474 \times 1400} = 1453.7 \text{ r.p.m.}$$

Hence, **speed of rotation = 1453.7 r.p.m. (Ans.)**

3. TYPES OF D.C. GENERATORS

- The power stations of modern design generate practically only three-phase alternating current. A large part of this power is used in the form of alternating current in industry, for *lighting* and *domestic needs*. When industrial needs make it necessary too or when it is of greater advantage to *use direct current* (for *chemical* and *metallurgical plants, electric traction*, etc.) *it is generally obtained by converting A.C. to D.C. with the help of converters of ionic or machine types*. In the latter case wide use is made of such installations as motor generator sets in which A.C. motor is coupled to a D.C. generator on a common shaft.
- As primary sources of power, D.C. generators are *mainly used in self-contained plants such as automobiles and air planes, for electric arc welding, train car lighting, in sub-marines, etc.*

3.1. Classification

According to *method of excitation* D.C. generators are classified as follows :

1. Separately excited generators, 2. Self-excited generators.

3.1.1. Separately excited generators

These are those generators whose field magnets are *energised* from an independent *external source* of D.C. current. Such a generator is shown in Fig. 15.

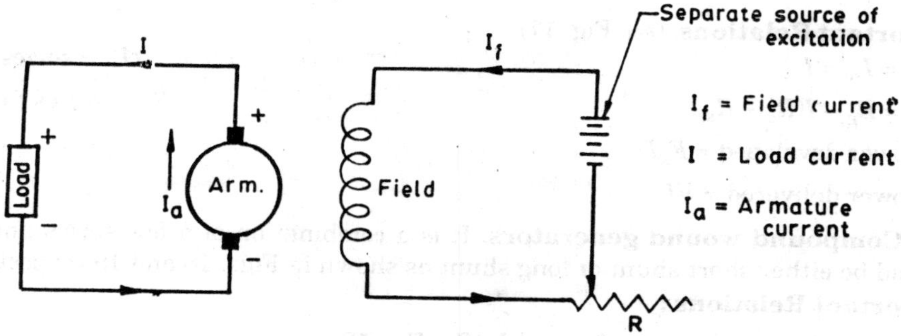

Fig. 15. Separately excited generator.

3.1.2. Self-excited generators

These are those generators whose field magnets are *energised by the current produced by the generators themselves.* Due to residual magnetism, there is always present some flux in the poles. When the armature is rotated, some e.m.f. and hence some induced current produced which is partly or fully passed through the field coils thereby *strengthening the pole flux.*

Self excited generators can be divided, in accordance with how the *field winding is connected into generators*, as follows :

(*i*) Shunt wound generators
(*ii*) Series wound generators
(*iii*) Compound wound generators :
(*a*) Short shunt
(*b*) Long shunt

(*i*) **Shunt wound generators :** Refer Fig. 16. In these generators the field windings are connected *across or in parallel with the armature conductors,* and have the full voltage of the generator across them.

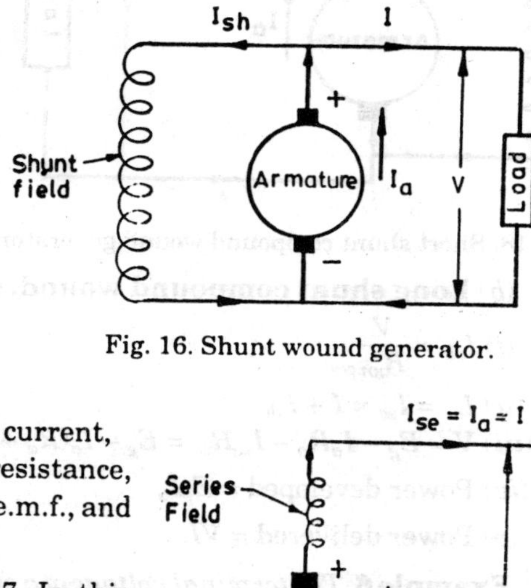

Fig. 16. Shunt wound generator.

Important relations : Refer Fig. 16.

(*i*) $I_{sh} = \dfrac{V}{R_{sh}}$

(*ii*) $I_a = I_{sh} + I$

(*iii*) $V = E_g - I_a R_a$

(*iv*) Power developed $= E_g I_a$

(*v*) Power delivered $= VI$

where I_{sh} = shunt field current, I_a = armature current,
I (or I_l) = load current, R_a = armature resistance,
R_{sh} = shunt field resistance, E_g = generated e.m.f., and
V = terminal voltage.

(*ii*) **Series wound generators :** Refer Fig. 17. In this case, the field windings are joined in series with armature conductors. *As they carry full load current, they consist of relatively few turns of thick wire or strip.* The use of such generators is limited to special purposes (as *boosters* etc.).

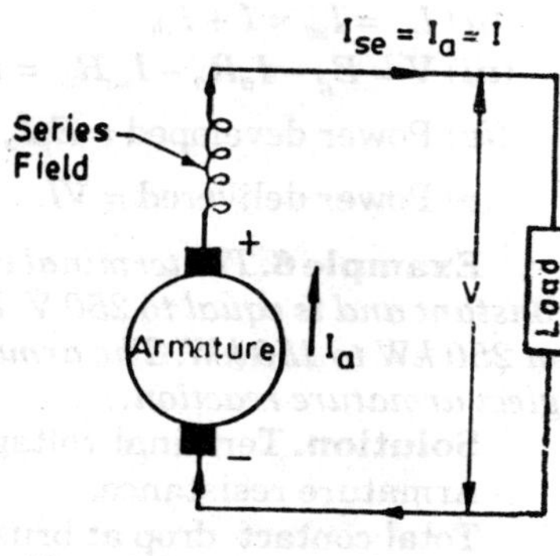

Fig. 17. Series wound generator.

Important Relations. (*see* Fig. 17) :

(*i*) $I_a = I_{se} = I$ (I_{se} = series field current)

(*ii*) $V = E_g - I(R_a + R_{se})$ (R_{se} = series field resistance)

(*iii*) Power developed = $E_g I$

(*iv*) Power delivered = VI.

(*iii*) **Compound wound generators.** It is a combination of a few series and a few shunt windings and be either short shunt or long shunt as shown in Figs. 18 and 19 respectively.

Important Relations :

(*a*) **Short shunt compound wound.** (*See* Fig. 18) :

(*i*) $I_{se} = I$

(*ii*) $I_{sh} = \dfrac{V + I_{se}R_{se}}{R_{sh}}$

(*iii*) $I_a = I + I_{sh}$

(*iv*) $V = E_g - I_aR_a - I_{se}R_{se}$

(*v*) Power developed = $E_g I_a$

(*vi*) Power delivered = VI.

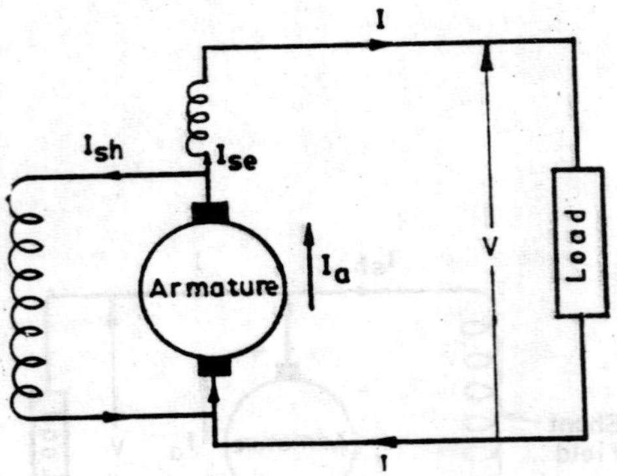

Fig. 18. Short shunt compound wound generator.

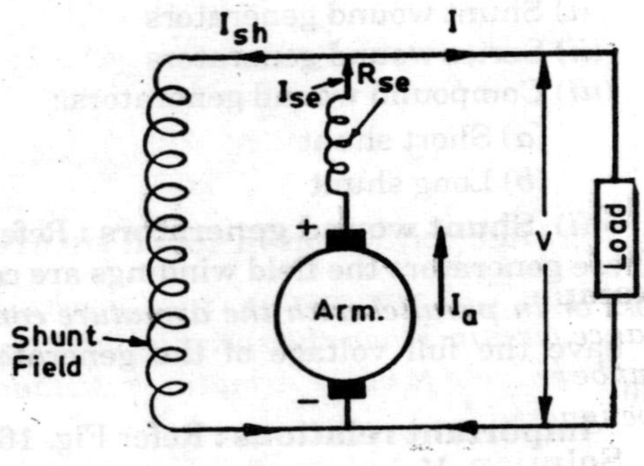

Fig. 19. Long shunt compound wound generator.

(*b*) **Long shunt compound wound.** (*See* Fig. 19) :

(*i*) $I_{sh} = \dfrac{V}{R_{sh}}$

(*ii*) $I_a = I_{se} = I + I_{sh}$

(*iii*) $V = E_g - I_aR_a - I_{se}R_{se} = E_g - I_a(R_a + R_{se})$

(*iv*) Power developed = $E_g I_a$

(*v*) Power delivered = VI.

Example 6. *The terminal voltage of a separately excited D.C. generator with constant excitation is constant and is equal to 250 V. Determine the percentage reduction in speed when the load changes from 250 kW to 150 kW. The armature resistance is 0.012 Ω and total contact drop at brushes = 2 V. Neglect armature reaction.*

Solution. Terminal voltage, V = 250 volts

Armature resistance, R_a = 0.012 Ω

Total contact drop at brushes = 2 V

Load change : From 250 kW to 150 kW

Percentage reduction in speed :

We know that generated e.m.f.,

$$E_g = \frac{p\phi ZN}{60a} \quad i.e. \quad E_g \propto \phi N$$

Since ϕ is constant, $\quad E_g \propto N$

— When the load is 250 kW, the armature current

$$I_{a1} = \frac{250 \times 1000}{250} = 1000 \text{ A}$$

and

$$E_{g1} = V + I_{a1} R_{a1} + \text{drop in brushes}$$
$$= 250 + 1000 \times 0.012 + 2 = 250 + 12 + 2 = 264 \text{ V}$$

— When the load is 150 kW, the armature current

$$I_{a2} = \frac{150 \times 1000}{250} = 600 \text{ A}$$

and

$$E_{g2} = V + I_{a2}R_a + \text{drop in brushes}$$
$$= 250 + 600 \times 0.012 + 2 = 250 + 7.2 + 2 = 259.2 \text{ V}$$

Now since $\quad E_g \propto N$

$$\therefore \quad \frac{E_{g2}}{E_{g1}} = \frac{N_2}{N_1}$$

i.e.

$$\frac{N_2}{N_1} = \frac{259.2}{264} \quad \text{or} \quad \frac{N_1}{N_2} = \frac{264}{259.2}$$

or

$$\frac{N_1 - N_2}{N_1} = \frac{264 - 259.2}{264} = 0.01818$$

$\therefore$ **%age reduction in speed = 0.01818 × 100 = 1.818%. (Ans.)**

☞**Example 7.** *A separately excited generator with constant excitation is connected to a constant resistance circuit. When the speed is 1200 r.p.m. it delivers 120 A at 500 V. At what speed will the current be reduced to 60 A ? Armature resistance = 0.1 Ω. Contact drop / brush = 1 V. Armature reaction may be ignored.*

Solution. Initial speed, $\quad N_1 = 1200$ r.p.m.

Initial load current, $\quad I_{a1} = 120$ A

Initial terminal voltage, $\quad V_1 = 500$ V

Final load current, $\quad I_{a2} = 60$ A

Armature resistance, $\quad R_a = 0.1\ \Omega$

Contact drop/brush $\quad = 1$ V

Speed N_2 :

Refer Fig. 20.

We know that e.m.f. generated (at load current of 120 A)

$$E_{g1} = V_1 + I_{a1}R_a + \text{drop in brushes}$$
$$= 500 + 120 \times 0.1 + 2 \times 1 = 514 \text{ V}$$

Load resistance (constant),

$$R = \frac{500}{120} = 4.167\ \Omega$$

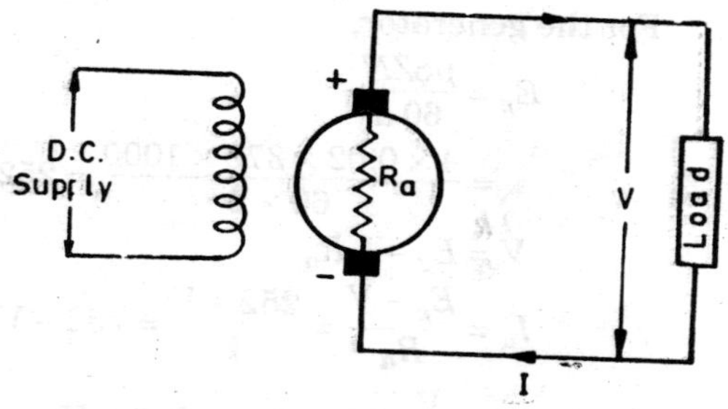

Fig. 20

Let E_{g2} be the generated e.m.f. when current delivered is 60 A.

$$V_2 = I_{a2}R = 60 \times 4.167 = 250 \text{ V}$$

$$E_{g2} = V_2 + I_{a2}R_a + 2 \times 1$$

$$= 250 + 60 \times 0.1 + 2 \times 1 = 258 \text{ V}$$

Now using the relation, $\frac{N_2}{N_1} = \frac{E_{g2}}{E_{g1}}$

$$\frac{N_2}{1200} = \frac{258}{514}$$

$$N_2 = \frac{1200 \times 258}{514} = 602.33 \text{ r.p.m.}$$

Hence, **speed at a load of 60 A = 602.33 r.p.m. (Ans.)**

Example 8. *A shunt generator supplied 500 A at 500 V. Calculate its generated e.m.f. if its armature and shunt field resistances are 0.02 Ω and 125 Ω respectively.*

Solution. Load current, $I = 500$ A

Terminal voltage, $V = 500$ volts

Armature resistance, $R_a = 0.2\ \Omega$

Shunt field resistance, $R_{sh} = 125\ \Omega$

Refer Fig. 21.

Shunt field current,

$$I_{sh} = \frac{V}{R_{sh}} = \frac{500}{125} = 4 \text{ A}$$

Armature current,

$$I_a = I + I_{sh} = 500 + 4 = 504 \text{ A}$$

Generated e.m.f.,

$$E_g = V + I_aR_a$$

$$= 500 + 504 \times 0.02 = 510.08 \text{ V}$$

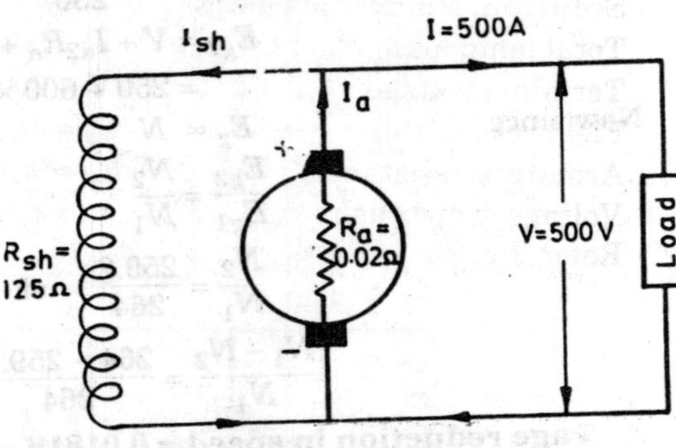

Fig. 21

Hence, **generated e.m.f. = 510.08 V. (Ans.)**

Example 9. *A 4-pole, D.C. shunt generator, with a shunt field resistance of 100 ohms and an armature resistance of 1 ohm, has 378 wave-connected conductors in its armature. The flux per pole is 0.02 Wb. If a load resistance of 10 ohms is connected across the armature terminals and the generator is driven at 1000 r.p.m., calculate power absorbed by load.*

Solution. *Given* : $p = 4$; $R_{sh} = 100\ \Omega$; $R_a = 1\ \Omega$, $Z = 378$; $a = 2$;
$\phi = 0.02$ Wb ; $R_L = 10\ \Omega$; $N = 1000$ r.p.m.

Power absorbed by the load :

The generator arrangement is shown in the Fig. 22.

For the generator,

$$E_g = \frac{p\phi ZN}{60\,a}$$

$$= \frac{4 \times 0.02 \times 378 \times 1000}{60 \times 2} = 252 \text{ V}$$

$$V = E_g - I_aR_a$$

$$\therefore \quad I_a = \frac{E_g - V}{R_a} = \frac{252 - V}{1} = 252 - V$$

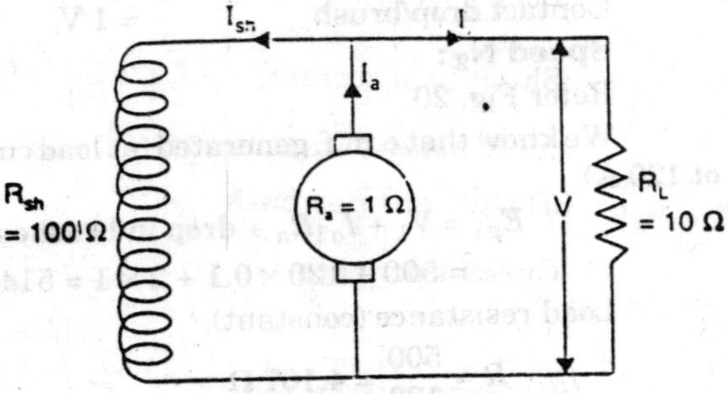

Fig. 22

Now $\quad I = \frac{V}{R_L} = I_a - I_{sh} = (252 - V) - \frac{V}{100}$

$$\therefore \quad \frac{V}{10} = (252 - V) - \frac{V}{100} \quad \text{or} \quad 252 - V = \frac{11\,V}{100}$$

$$\text{or} \quad 25200 - 100\,V = 11\,V \quad \text{or} \quad V = \frac{25200}{111} = 227 \text{ V}$$

$$I = \frac{227}{10} = 22.7 \text{ A}$$

$\therefore$ **Power absorbed by the load** $= I^2R_L$
$= 22.7^2 \times 10 = 5153$ W or **5.153 kW. (Ans.)**

Example 10. *A 4-pole lap wound shunt generator supplies to 50 lamps of 100 watts, 200 V each. The field and armature resistances are 50 Ω and 0.2 Ω respectively. Allowing a brush drop of 1 V each brush, calculate the following :*

(i) Armature current. *(ii) Current per path.*
(iii) Generated e.m.f. *(iv) Power output of D.C. armature.*

Solution. Number of poles, $p = 4$
Total lamp load, $P = 50 \times 100 = 5000$ W
Terminal voltage, $V = 200$ Volts
Field resistance, $R_{sh} = 50\ \Omega$
Armature resistance, $R_a = 0.2\ \Omega$
Voltage drop/brush $= 1$ V
Refer Fig. 23.

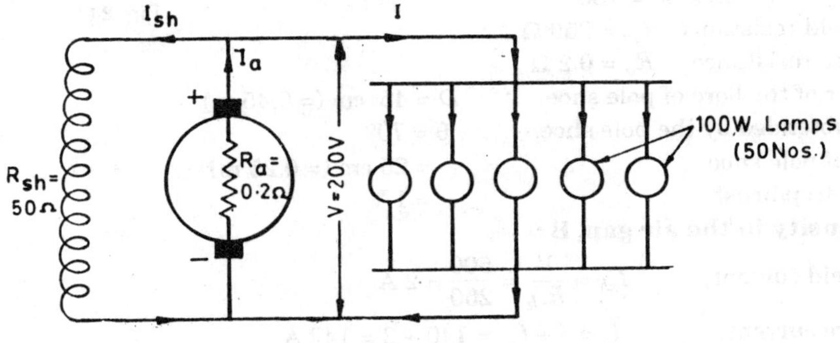

Fig. 23

(*i*) **Armature current, I_a :**

Load current, $I = \dfrac{\text{Power consumed}}{\text{Terminal voltage}} = \dfrac{P}{V} = \dfrac{5000}{200} = 25$ A

Shunt field current, $I_{sh} = \dfrac{V}{R_{sh}} = \dfrac{200}{50} = 4$ A

$\therefore$ Armature current, $I_a = I + I_{sh} = \mathbf{25 + 4 = 29\ A.}$ **(Ans.)**

(*ii*) **Current per path :**

Current per path $= \dfrac{I_a}{a} = \dfrac{29}{4}$ [$\because$ $a = p = 4$, generator being lap wound]

$= \mathbf{7.25\ A.}$ **(Ans.)**

(*iii*) **Generated e.m.f., E_g :**

$$E_g = V + I_aR_a + \text{brush drop} = 200 + 29 \times 0.2 + 2 \times 1 = 207.8 \text{ V}$$

Hence, **generated e.m.f. = 207.8 V. (Ans.)**

(*iv*) **Power output of D.C. armature :**

Power output of D.C. armature

$$= \frac{E_g I_a}{1000} = \frac{207.8 \times 29}{1000} \text{ kW} = \mathbf{6.026\ kW.}\ \ \textbf{(Ans.)}$$

☞**Example 11.** *A 4-pole, 500 V wave-wound shunt generator delivers a load current of 140 A. It has 65 slots with 12 conductors / slot and runs at 800 r.p.m. The shunt field and armature resistances are 250 Ω and 0.2 Ω respectively. The diameter of the bore of the pole shoe is 45 cm, the angle subtended by the pole shoe is 70° and it is 25 cm in length. Assuming contact drop / brush as 1 V, calculate the flux density in the air gap.*

Solution. Refer Fig. 24.

Number of poles, $p = 4$

Number of parallel paths, $a = 2$

[generator being wave wound]

Terminal voltage, $V = 500$ Volts

Load current, $I = 140$ A

Speed of rotation, $N = 800$ r.p.m.

Number of slots $= 65$

Number of conductors/slot = 12

∴ Total number of conductors,

$Z = 12 \times 65 = 780$

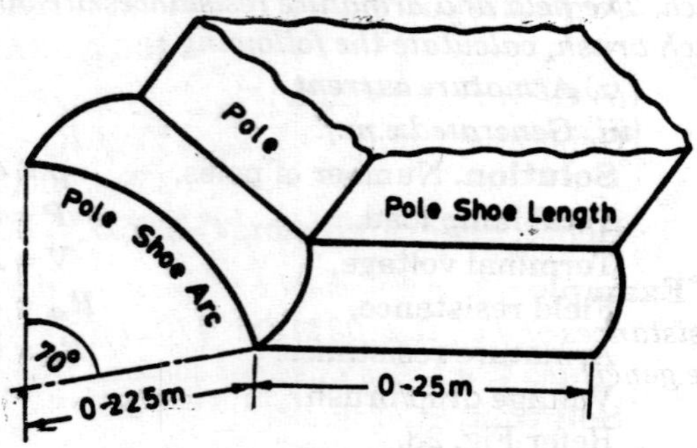

Fig. 24

Shunt field resistance, $R_{sh} = 250\ \Omega$

Armature resistance, $R_a = 0.2\ \Omega$

Diameter of the bore of pole shoe, $D = 45$ cm (= 0.45 m)

Angle subtended by the pole shoe, $\theta = 70°$

Length of pole shoe = 25 cm (= 0.25 m)

Contact drop/brush = 1 V

Flux density in the air gap, B :

Shunt field current, $I_{sh} = \dfrac{V}{R_{sh}} = \dfrac{500}{250} = 2$ A

Armature current, $I_a = I + I_{sh} = 140 + 2 = 142$ A

Generated e.m.f., $E_g = V + I_a R_a +$ brush drop

$= 500 + 142 \times 0.2 + 2 \times 1 = 530.4$ V

Also, $E_g = \dfrac{p\phi ZN}{60a}$

$$530.4 = \frac{4 \times \phi \times 780 \times 800}{60 \times 2}$$

or $$\phi = \frac{530.4 \times 60 \times 2}{4 \times 780 \times 800} = 0.0255 \text{ Wb}$$

Arc length of pole shoe $= \dfrac{\pi D\theta}{360} = \dfrac{\pi \times 0.45 \times 70}{360} = 0.275$ m

Area of pole shoe, A = Arc length × length of pole shoe

$= 0.275 \times 0.25 = 0.06875\ \text{m}^2$

Flux density in the air gap, $B = \dfrac{\phi}{A} = \dfrac{0.0255}{0.06875} = 0.371$ T (or Wb/m^2)

Hence, **flux density in the air gap = 0.371 T. (Ans.)**

Example 12. *A series generator delivers a current of 100 A at 250 V. Its armature and series field resistances are 0.1 Ω and 0.055 Ω respectively. Find :*

(*i*) *Armature current* (*ii*) *Generated e.m.f.*

Solution. *See* Fig. 25.

Load current, $I = 100$ A

Terminal voltage, $V = 250$ Volts

Armature resistance, $R_a = 0.1\ \Omega$

Series field resistance, $R_{se} = 0.055\ \Omega$

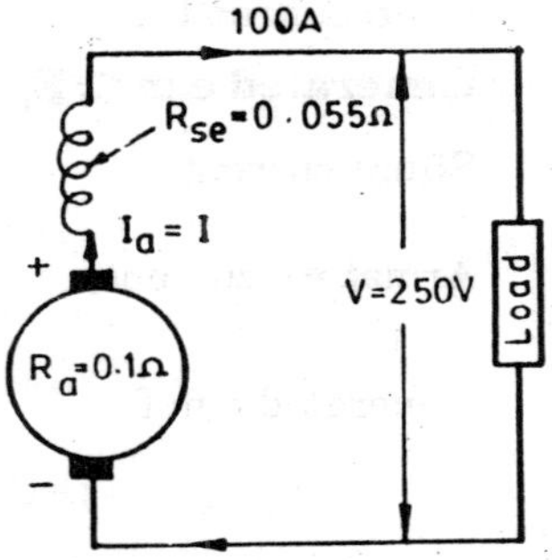

Fig. 25

(*i*) **Armature current,** I_a :

Armature current (I_a) = load current (I)

$\therefore$ I_a = **100 A. (Ans.)**

(*ii*) **Generated e.m.f.** :

Generated e.m.f. $E_g = V + I(R_a + R_{se})$

$= 250 + 100(0.1 + 0.055) = 265.5$ V

Hence, **generated e.m.f. = 265.5 V. (Ans.)**

☞**Example 13.** *A short shunt compound generator has armature, series field and shunt field resistances of 0.06 Ω, 0.03 Ω and 110 Ω respectively. It supplies 100 lamps rated at 250 V, 40 W. Find the generated e.m.f. Assume that contact drop / brush = 1 V.*

Solution. *See* Fig. 26.

Armature resistance, $R_a = 0.06\ \Omega$

Series field resistance, $R_{se} = 0.03\ \Omega$

Shunt field resistance, $R_{sh} = 110\ \Omega$

Terminal voltage, $V = 250$ Volts

Lamp load, $P = 100 \times 40 = 4000$ W

Contact drop/brush = 1 V

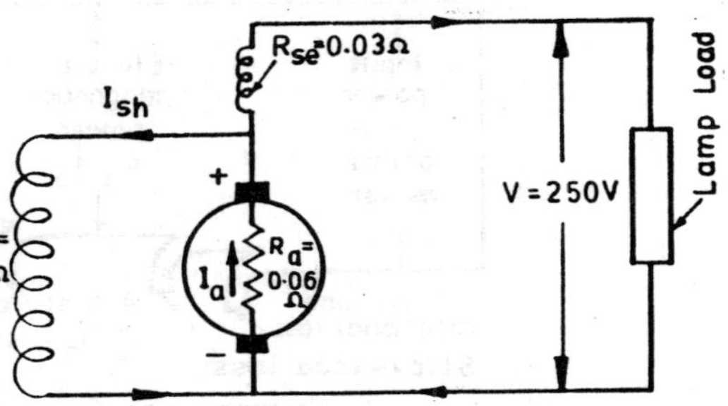

Fig. 26

Generated e.m.f., $\mathbf{E_g}$:

Load current, $I = \dfrac{P}{V} = \dfrac{4000}{250} = 16$ A

Voltage drop in series winding

$= IR_{se} = 16 \times 0.03 = 0.48$ V

Voltage across shunt field winding

$= V + IR_{se} = 250 + 0.48 = 250.48$ V

Shunt field current, $I_{sh} = \dfrac{250.48}{110} = 2.277$ A

Armature current, $I_a = I + I_{sh}$

$= 16 + 2.277 = 18.277$ A

Generated e.m.f., $\mathbf{E_g} = V + IR_{se} + I_aR_a$ + brush drop

$= 250 + 0.48 + 18.277 \times 0.06 + 2 \times 1$

= **253.58 V. (Ans.)**

Example 14. *A long shunt compound generator has an armature, series field and shunt field resistances of 0.04 Ω, 0.03 Ω and 200 Ω respectively. It supplies a load current of 180 A at 400 V. Calculate the generated e.m.f. Assume contact drop/brush = 1 V.*

Solution. Refer Fig. 27.

Armature resistance, $R_a = 0.04\ \Omega$

Series field resistance, $R_{se} = 0.03\ \Omega$

Shunt field resistance, $R_{sh} = 200\ \Omega$
Load current, $I = 180$ A
Terminal voltage, $V = 400$ Volts
Contact drop/brush $= 1$ V

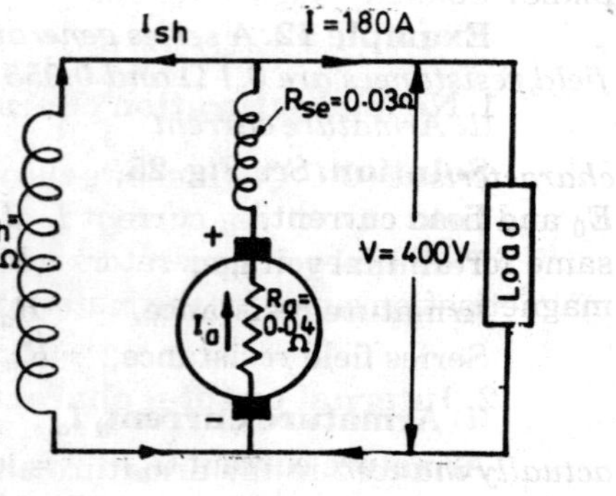

Fig. 27

Generated e.m.f., E_g :

Shunt current, $I_{sh} = \dfrac{V}{R_{sh}} = \dfrac{400}{200} = 2$ A

Armature current, $I_a = I + I_{sh}$
$= 180 + 2 = 182$ A

Generated e.m.f., $E_g = V + I_aR_a + I_aR_{se}$ + drop at brushes
$= 400 + 182 \times 0.04 + 182 \times 0.03 + 2 \times 1$
$= 400 + 7.28 + 5.46 + 2 = 414.74$ V

Hence, **generated e.m.f. = 414.74 V. (Ans.)**

4. POWER DIVISION IN A D.C. GENERATOR

Fig. 28 presents in graphical form power balances for D.C. generators with both shunt and series fields.

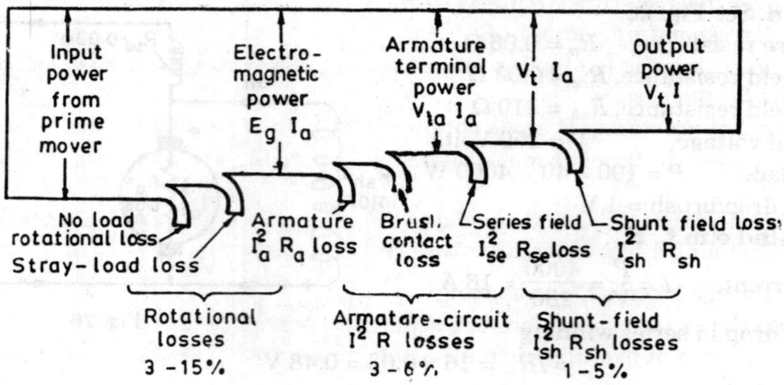

Fig. 28. Power division in a D.C. generator.

5. CHARACTERISTICS OF D.C. GENERATORS

The properties of generators are analysed with the aid of characteristics which give the relations between fundamental quantities determining the operation of a generator. These include the voltage across the generator terminals V, the field or exciting current I_f, the armature current I_a, and the speed of rotation N.

The three most important characteristics of D.C. generators are given below :

1. No load saturation characteristics $\left(\dfrac{E_0}{I_f}\right)$

2. Internal or total characteristics $\left(\dfrac{E}{I_a}\right)$

3. External characteristics $\left(\dfrac{V}{I}\right)$.

1. **No load saturation characteristic** $\left(\frac{E_0}{I_f}\right)$. It is also a known as *magnetic or open circuit characteristic* (O.C.C.). It shows the relationship between the no-load generated e.m.f. in armature, E_0 and field or exciting current I_f at a given fixed speed. The shape of the curve is practically the same for all types of generators whether they are separately excited or self-excited. It is just the magnetisation curve for the material of the electromagnets.

2. **Internal or total characteristic** $\left(\frac{E}{I_a}\right)$. It gives the relationship between the *e.m.f. E actually induced* in the armature after allowing for the demagnetising effect of armature reaction and the *armature current* I_a. This characteristic is of interest mainly to the designer.

3. **External characteristic** $\left(\frac{V}{I}\right)$

- This characteristic is also referred to as *performance characteristic* or sometimes *voltage-regulating curve*.
- It gives relation between the *terminal voltage* V and *load current I*.
- The curve lies below the internal characteristic because it takes into account the voltage drop over the armature circuit resistance. The *values of V are obtained by subtracting I_aR_a from corresponding values of E.*
- The characteristic is of great importance in judging the *suitability of a generator for a particular purpose.*

The external characteristic can be obtained by the following two ways :

(*i*) By making simultaneous measurements with a suitable voltmeter and an ammeter on a loaded generator.

(*ii*) Graphically from the O.C.C. provided the armature and field resistances are known and also if the demagnetising effect of the armature reaction is known.

5.1. Separately Excited Generator

- Fig. 29 shows the connections of a separately excited generator, a battery being indicated as the source of exciting current, although any other constant voltage source could be used.

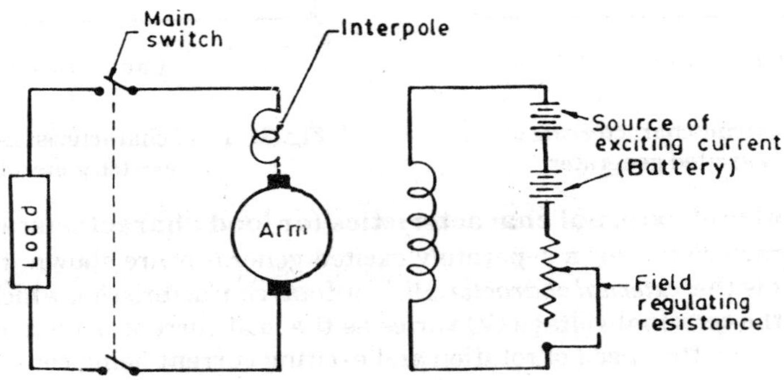

Fig. 29. Connection for a separately excited generator.

The field circuit is provided with a variable resistance and would normally contain a field switch and an ammeter, these being omitted from the diagram for simplicity. The armature is connected through 2-pole main switch to the bus bars, between which the load is connected.

5.1.1. No-load saturation characteristic (or O.C.C.) :

- If the generator is run at constant speed with the main switch open, and the terminal voltage is noted at various values of exciting or field current then the O.C.C. shown in Fig. 30 can be plotted. This is also referred to as the '*magnetisation curve*' since the same graph shows, to a suitably chosen scale, the amount of magnetic flux, there being a constant relationship (depending upon speed of rotation) between flux and induced voltage.
- It will be noticed that a small voltage is produced when the field current is zero, this being due to a small amount of permanent magnetism in the field poles. This is called *residual magnetism* and is usually sufficient to produce 2 or 3 per cent of normal terminal voltage, although in some special cases it is purposely increased to 10 per cent or more.
- The *first part of the curve is approximately straight* and *shows that the flux produced is proportional to the exciting current* ; but *after a certain point, saturation of the iron becomes perceptible as the curve departs from straight line form.*

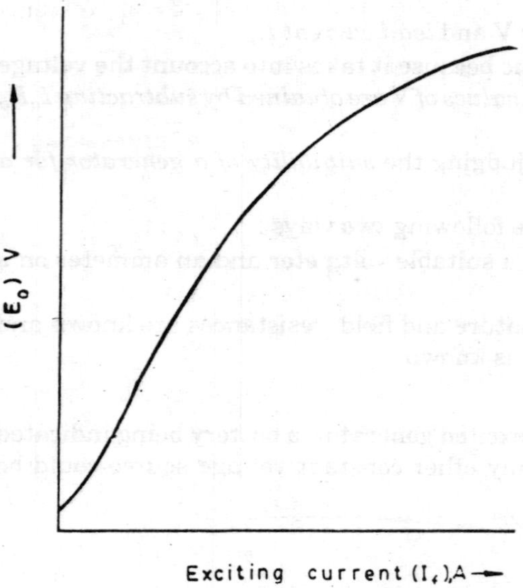

Fig. 30. Open-circuit characteristic of a separately excited generator.

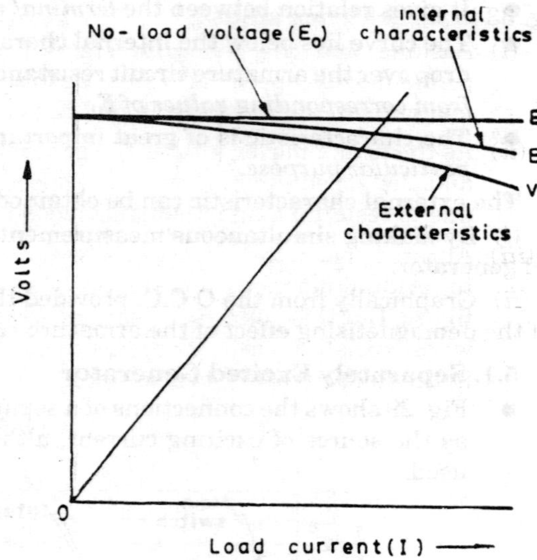

Fig. 31. Load characteristics of a separately excited generator.

5.1.2. Internal and external characteristics (or load characteristics) :

- Load characteristics for a separately excited generator are shown in Fig. 31. The most important is the '*external characteristic*' (or total characteristic), which indicates the way in which the terminal voltage (V) varies as the load current is increased from zero to its full load value, the speed of rotation and exciting current being constant.

The voltage drop (drop of volts) at any particular load current, *indicated by the vertical distance between the external characteristic and the no-load voltage* is brought about by two causes :

(*i*) *Armature reaction which has a demagnetising effect upon the field.*

(*ii*) *Resistance drop*, this being the product of the armature current and the total armature-circuit resistance, consisting of the armature resistance, interpole resistance and brush contact resistance.

- The *'internal characteristic'* is obtained by *calculating the resistance drop for a few values of current and adding this to the voltage shown by the external characteristic. The vertical distance between the internal characteristic and no load voltage then represents the effect of armature reaction alone.*
- When the resistance of load is R, then voltage across its terminals is $V = IR$, where I represents the current, so that if the values of V corresponding to various values of I are calculated, the values will all lie upon a straight line such as OL in Fig. 31. *The load current and terminal voltage corresponding to this resistance are given by the inter-section of the line OL with the external characteristics.*

Note. The great *advantage of separate excitation* over all other forms of excitation is that the *current is entirely independent of the load current in the armature.* It is however, rather inconvenient to have to depend upon a separate source of supply and, therefore, the *method is used only in special cases, where the generator has to operate over a wide range of terminal voltage.*

5.2. Building up the Voltage of Self-excited Shunt Generator

One of the simplest forms of 'self-excited' generator is the shunt-wound machine, the connection diagram (without load) of which is shown in Fig. 32. The manner in which a self-excited generator manages to excite its own field and build a D.C. voltage across its armature is described with reference to Fig. 33 in the following **steps** :

(*i*) Assume that the generator starts from rest, *i.e.*, prime-mover speed is zero. Despite a residual magnetism, the generated e.m.f. E, is zero.

(*ii*) As the prime-mover rotates the generator armature and the speed approaches rated speed, the voltage due to residual magnetism and speed increases.

(*iii*) At rated speed, the voltage across the armature due to residual magnetism is small, E_1, as shown in the figure. But this voltage is also across the field circuit whose resistance is R_f. Thus, the current which flows in the field circuit I_1, is also small.

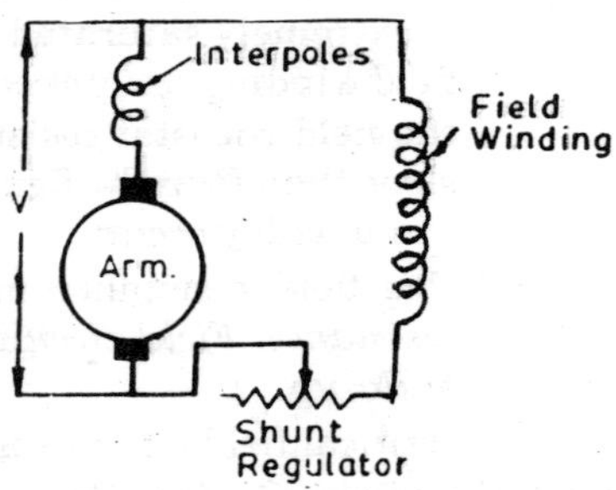

Fig. 32. Self-excited shunt generator.

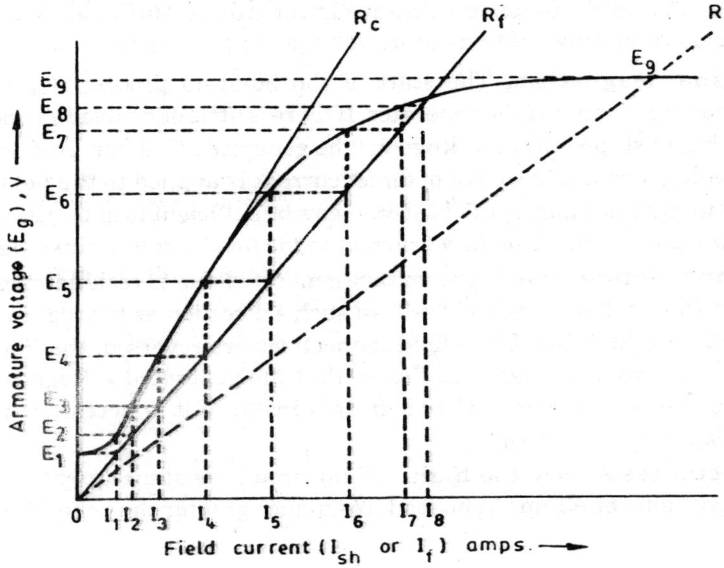

Fig. 33. Building up the voltage of a shunt generator.

(*iv*) When I_1 flows in the field circuit of the generator of Fig. 32, an increase in m.m.f. results (due to $I_f T_f$, T_f being field turns which aids the residual magnetism in increasing the induced voltage to E_2 as shown in Fig. 33.

(*v*) Voltage E_2 is now impressed across the field, causing a large current I_2 to flow in the field circuit. $I_2 T_f$ is an increased m.m.f., which produces generated voltage E_3.

(*vi*) E_3 yields I_3 in the field circuit, producing E_4. But E_4 causes I_4 to flow in the field producing E_5 ; and so on, up to E_8, the maximum value.

(*vii*) The *process continues until that point where the field resistance line crosses the magnetisation curve* in Fig. 33. Here the process stops. The induced voltage produced, when impressed across the field circuit, produces a current flow that in turn produces an induced voltage of the same magnitude, E_8, as shown in the figure.

Critical Resistance :

- In the above description a particular value of field resistance R_f was used for building up of self-excited shunt generator. If the field resistance were reduced by means of adjusting the field rheostat of Fig. 32 to a lower value say R_{f_1}. shown in Fig. 33, the build-up process would take place along field resistance line R_{f_1}, and build-up a somewhat higher value than E_8, *i.e.* the point where R_{f1} intersects the magnetisation curve, E_9. Since the curve is extremely saturated in the vicinity of E_9, reducing the field resistance (to its limiting field winding resistance) will not increase the voltage appreciably. Conversely, increasing the field rheostat resistance and the field circuit resistance (to a value having a higher slope than R_f in the figure) will cause a reduction of the maximum value to which build-up can possibly occur.
- The field resistance may be increased until the field circuit reaches a *critical field resistance. Field circuit resistance above the critical field resistance will fail to produce build-up.*

This critical field circuit resistance, R_c, is shown as tangent to the saturation curve passing through the origin, *O*, of the axes of the curve of Fig. 33. Thus a field circuit resistance higher than R_c will produce an armature voltage of E_1 approximately (and no more).

☞**Reasons for Failure of Self-excited Shunt Generator to Build-up Voltage.** The reasons why a self-excited generator may fail to build-up voltage are given below :

1. **No residual magnetism.** The start of the build-up process requires some residual magnetism in the magnetic circuit of the generator. If there is little or no residual magnetism, because of inactivity or jarring in shipment, no voltage will be generated that can produce field current. To overcome this difficulty, a separate source of direct current is applied to the field for a short period of time and then removed. The magnetic field should now be sufficient to allow the voltage to build-up. *The application of a separate source of direct current to the field is called 'flashing the field'*.

2. **Field connection reversed.** The voltage generated due to residual magnetism is applied to the field. Current should flow in the field coils in such a direction as to produce lines of flux in the same direction as the residual flux. If the field connections are reversed, the *lines of flux produced by the current flow will oppose the residual flux* so that the generated voltage will *decrease* rather than increase when the field circuit is closed. In this instance it is necessary to *reverse the field connections with respect to the armature.*

3. **Field circuit resistance too high.** A field circuit resistance greater than critical value will prevent an appreciable build-up. At no load, resistance greater than the critical may be caused by the following :

- **Open field circuit connection.** The effects of an open circuit are apparent. The field circuit resistance is *much greater than the critical value*; hence *generator will not build-up*.
- **Dirty commutator.** A dirty commutator does not permit good contact between the brushes and the commutator. This poor contact shows up as a high resistance to the flow of current in the field circuit and produces the same effect as a high field circuit resistance.

5.3. Shunt Generator Characteristics

In a shunt generator the field circuit is connected directly across the armature. Appliances, motors, light bulbs, and other electrical devices connected in *parallel* across the generator terminals represent a *load* on the generator. As more devices are connected in parallel, the load on the generator increases ; that is, the generator current increases. Because the generator current increases, the terminal voltage of the generator decreases. There are three factors that cause this decrease in voltage :

(*i*) Armature-circuit resistance (R_a),
(*ii*) Armature reaction, and
(*iii*) Reduction in field current.

(*i*) **Armature-circuit resistance.** The armature circuit of a generator, like every electrical circuit, contains resistance. This resistance includes the resistance of (*i*) *the copper conductors of the armature winding, (ii) the commutator, (iii) contact resistance between brushes and commutator,* and (*iv*) *the brushes themselves.* When no current flows through the armature, there is no *IR* drop in the armature and the voltage at the terminals is the same as the generated voltage. However, when there is current in the armature circuit, a voltage drop exists due to the armature resistance, and the terminal voltage is less than the generated voltage. The terminal voltage may be calculated from the following reaction :

$$V = E_g - I_a R_a$$

where V = voltage at terminals of generator,
E_g = generated or induced voltage,
I_a = total armature current, and
R_a = armature-circuit resistance.

(*ii*) **Armature reaction.** When current flows in the armature conductors a flux surrounds these conductors. The direction of this armature flux is such that it reduces the flux from the field poles, resulting in both a reduced generated voltage and terminal voltage.

(*iii*) **Reduction in the field current.** The field circuit is connected across the terminals of the generators. When the terminal voltage of the generator becomes smaller because of the armature-resistance volt drop and armature reaction, the voltage across the field circuit also becomes smaller and therefore field current will be less. A reduction in the magnitude of field current also reduces the flux from the field poles, which in turn reduces the generated voltage and also the terminal voltage.

External Characteristic

- See Figs. 34 and 35. The effect of the preceding three factors is shown in Fig. 35, which shows external (load-voltage) characteristic of a shunt generator.
- As shown in the circuit of Fig. 34, the readings of the voltage across the armature (and load), V are plotted

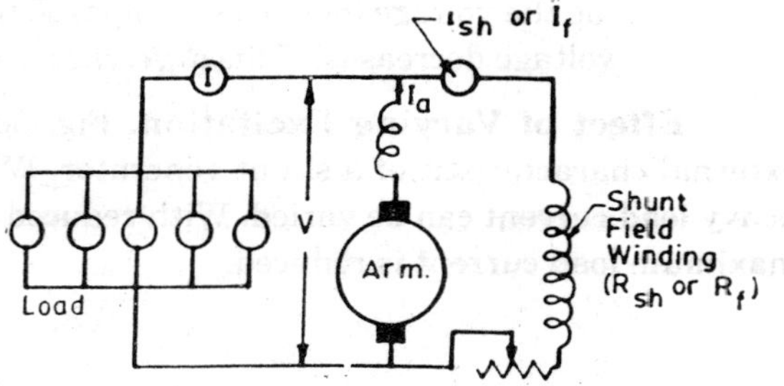

Fig. 34. Shunt generator under load.

as a function of load current, I. The voltage, V, is the same as E_g at no load (neglecting the I_aR_a and armature reaction drop produced by the field current). The effects of armature reaction, armature circuit voltage drop, and decrease in field current are all shown with progressive increase in load. Note that both the armature reaction and the I_aR_a drops are shown as dashed straight lines, representing theoretically linear voltage directly proportional to the increase in load current. The drop owing to decreased field current is a curved line, since it depends on the degree of saturation existing in the field at the value of load.

- Generally, the external load-voltage characteristic decreases with application of load only to *a small extent up to its rated load* (current) *value*. Thus, the shunt generator is considered as having a fairly constant output voltage with application of load, and in practice, is *rarely operated beyond the rated load current value continuously for any appreciable time.*

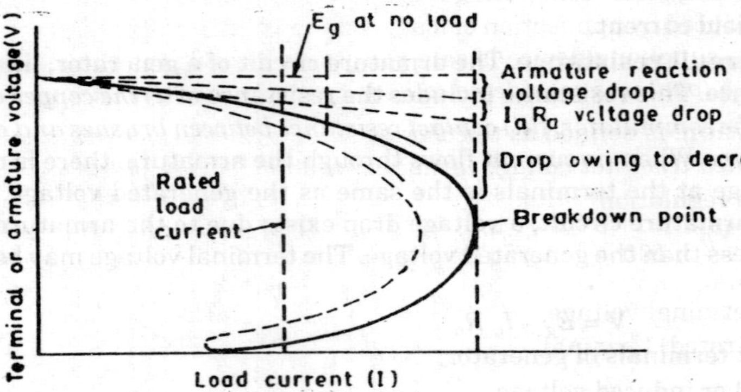

Fig. 35. Shunt generator load characteristics.

- As shown in Fig. 35 *further application of load causes the generator to reach a breakdown point beyond which further load causes it to 'unbuild' as it operates on the unsaturated portion of its magnetisation curve. This unbuilding process continues until the terminal voltage is zero, at which point the load current is of such magnitude that the internal armature circuit voltage drop equals the e.m.f. generated on the unsaturated or linear portion of its magnetisation curve.*
- It may be noted that if the *external load is decreased* (an increase of external load resistances), the *generator will tend to build up gradually along the dashed line* shown in Fig. 35. Note that for any value of load current, the terminal or armature voltage is less (as the voltage increases) compared to the solid lines which yield a higher voltage (as the voltage decreases). This *difference is due to hysteresis.*

Effect of Varying Excitation. Fig. 36 shows the effect of varying excitation upon the external characteristic of a shunt generator. With normal excitation the initial slope is small and heavy load current can be varied. With reduced excitation, the fall of voltage is more rapid and the maximum load current is reduced.

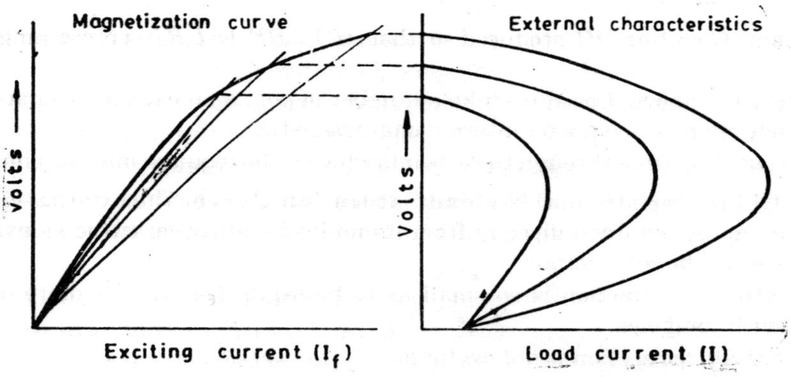

Fig. 36. Effect of varying excitation.

Voltage Regulation. The term *'voltage regulation'* is used to indicate the degree of change in armature voltage produced by application of load. If there is little change from no-load to full load, the generator or voltage-supplying device is said to possess good voltage regulation. If the *voltage changes appreciably with load, it is considered to have poor voltage regulation.*

'Voltage regulation' is defined as the *change in voltage from no-load to full load, expressed as a percentage of the rated terminal voltage* (armature voltage at full load).

i.e. Per cent voltage regulation

$$= \frac{V_{nl} - V_{fl}}{V_{fl}} \times 100 \qquad ...(1)$$

where V_{nl} = no load terminal voltage,
V_{fl} = full load (rated) terminal voltage.

Internal or Total Characteristic. To determine internal characteristic from external characteristic the following procedure is adopted [*see* Fig. 37].

Steps : 1. From the given data, draw the external characteristic (*I*).

2. Draw the shunt field resistance line *OL* and armature resistance line *OM*.

3. On the external characteristic take any point say *F*.

4. From point *F* draw vertical and horizontal lines intersecting *X* and *Y* axes respectively. Let these lines be *FC* and *FA* respectively.

5. Take point *D* on *X*-axis so that *CD* = *AB* representing the shunt field current, I_{sh}.

6. From point *D* draw vertical *DE* and produce it intersecting line *AF* produced at *H*.

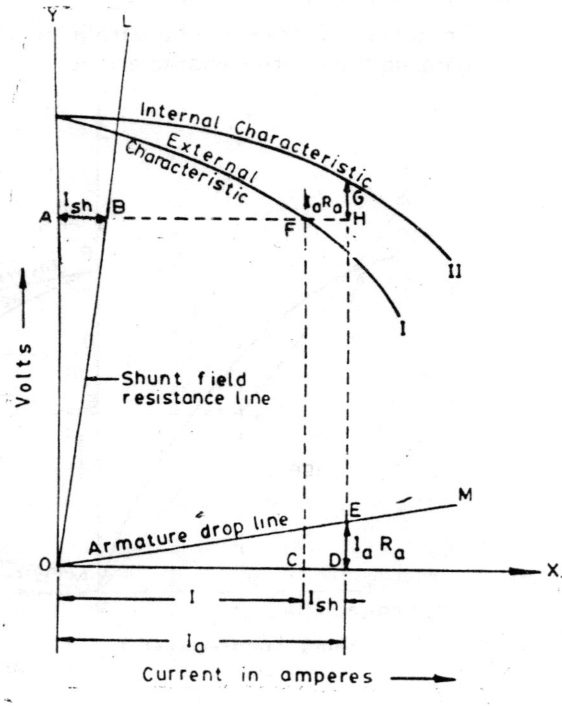

Fig. 37. Determination of internal characteristic from external characteristic.

7. Take point G on line DH produced so that $HG = DE$ $(= I_a R_a)$ representing the armature drop.
8. Following the above procedure take a number of points on external characteristic and find corresponding points lying on internal characteristic.
9. Draw a curve passing through these points which is the required internal characteristic (*II*).

External Characteristic and No-load Saturation Curve. The external characteristic of a shunt generator can be obtained directly from its no-load saturation curve as explained below. Following two cases will be considered :

(*A*) When armature reaction is so small as to be *negligible*. This is more or less true for generators fitted with compoles.

(*B*) When armature reaction is *not negligible*.

(A) Armature Reaction Negligible :

Steps 1. From given data draw O.C.C. [*see* Fig. 38].

2. Draw shunt field resistance line (say *OS*) meeting O.C.C. at any point (say *A*).
3. From point *A* draw horizontal line intersection *Y*-axis, say at point *B*.
   Hence, *OB* is the maximum no-load or open circuit voltage.
4. Take any point (say *L*) on the O.C.C. and draw an ordinate, say, *LMN* intersecting field resistance line at *M* and *X*-axis at *N*. Now *LN represents the generated e.m.f., MN represents the terminal voltage and LM represents voltage drop in armature.*
5. From points *L* and *M* draw horizontal lines cutting vertical axis say at point *D* and *E* respectively.
6. Draw armature resistance line *OC*.
7. From point *E*, draw line *EF* parallel to line *OC* cutting line *LD* produced at *F*. Hence, *F* is lying on the internal characteristic.

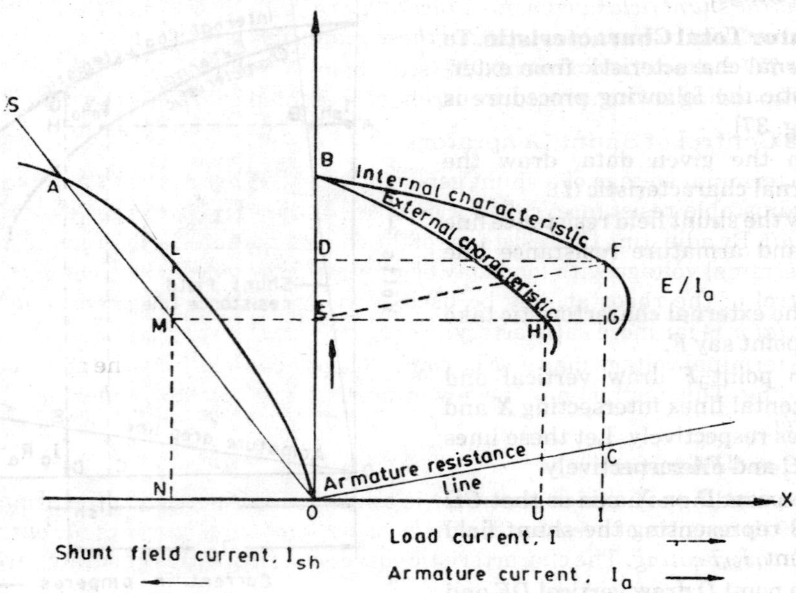

Fig. 38. Determination of external and internal characteristics from O.C.C.

Similarly other points can be obtained and internal characteristic may be drawn through these points :

8. From point F draw vertical line intersecting produced line ME at any point G and X-axis say, at point T. Since $FG = LM = CT$, hence point G lies on the *curve representing relation between armature current and terminal voltage.*

9. Take TU = shunt field current (I_{sh}) ON (scale being different). OU represents the load current corresponding to armature current represented by OT and terminal voltage OE.

10. From point U draw a vertical line intersecting line EG at H. Point H *lies on the external characteristic.*

Similarly other points may be obtained and curve may be drawn, which is the required *external characteristic.*

(B) Taking Armature Reaction into Account :

Here, in addition to considering the voltage drop in armature, voltage drop due to armature reaction is also taken into account.

Let I_aR_a = voltage drop in armature

I_{sh} = increase in shunt field current to counteract the demagnetising effect.

Now if a right-angled triangle say lmn is drawn as that ln = voltage drop in armature, and mn = shunt field current. The triangle lmn is called as the *drop reaction triangle.*

In order to draw external and internal characteristic repeat the process as in A with following modifications in steps 4 and 9 respectively.

4. Take any point L on the O.C.C. and draw line LM parallel to the line lm of triangle lmn and complete the triangle LMN. Now from the points L and M draw vertical lines cutting X-axis at points N' and M'. Now LN' *represents the generated e.m.f.* MM' represents the terminal voltage, LN *represents the voltage drop in armature due to armature resistance,* ON' *is the shunt field current to induce an e.m.f. represented by* LN' and $N'M'$ is the *increase in shunt field current to counteract the demagnetising effect.*

9. Take TU = shunt field current ON' (scale being different). OU represents the load current corresponding to the armature current represented by OT and terminal voltage MM'.

Voltage Control of Shunt Generators

- The terminal voltage of a shunt generator may be kept constant at all loads with the use of adjustable resistance, called a *field rheostat, connected in series with the shunt-field circuit.* By adjusting the resistance of the rheostat to suit the load on the machine, changes in terminal voltage with load may be prevented. When the load changes gradually, hand control of the rheostat may be used, although *automatic control* employing a *voltage regulator* is far more satisfactory.
- The terminal voltage may also be controlled *automatically* by the addition of a series-field winding. This method has the advantage of being *automatic cheap*, and *generally satisfactory.*

5.4. Series Generator

The field winding of a series generator is connected in series with the armature winding as shown in Fig. 39. It consists of a *few series of heavy wire, capable of carrying the output current of the machine without overheating.* The characteristic curves of a D.C. series generator are drawn as given below :

- The *saturation* curve (1) may be obtained in a manner exactly similar to that already described for the shunt machine. The series-connected generator is illustrated in Fig. 39.

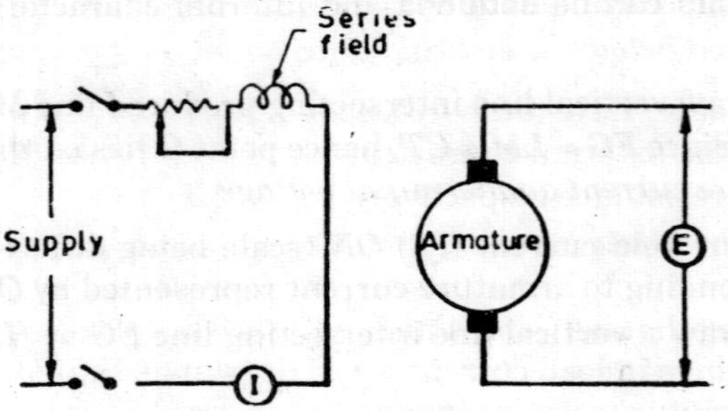

Fig. 39. Connection diagram for obtaining the saturation curve of a series generator.

The series connected generator illustrated in Fig. 40 must be capable of safely carrying the maximum current to be used, or about 125 per cent of rated load current. A plot of simultaneous readings of generated voltage and field current, taken at a rated speed, yields the magnetisation curve 1 of Fig. 41.

- **External Characteristic (curve 2).** To obtain the data for this curve, the machine is connected to the load as shown in Fig. 40, ammeter and voltmeter being inserted to read the load current I (= I_a) and the terminal voltage V respectively. The machine is run at constant (rated) speed, a series of simultaneous readings of voltage and current is taken while the load is varied from a minimum value to perhaps 125 per cent of rated load. When these readings are plotted, using V as co-ordinate and I_a (= I) as abscissa curve 2 (*see* Fig. 41) is obtained.

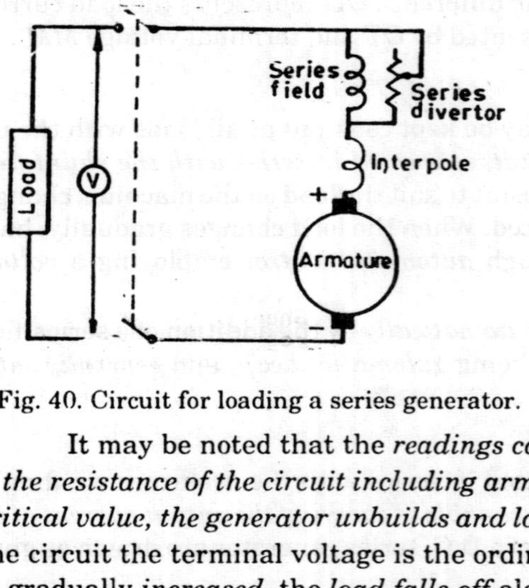

Fig. 40. Circuit for loading a series generator.

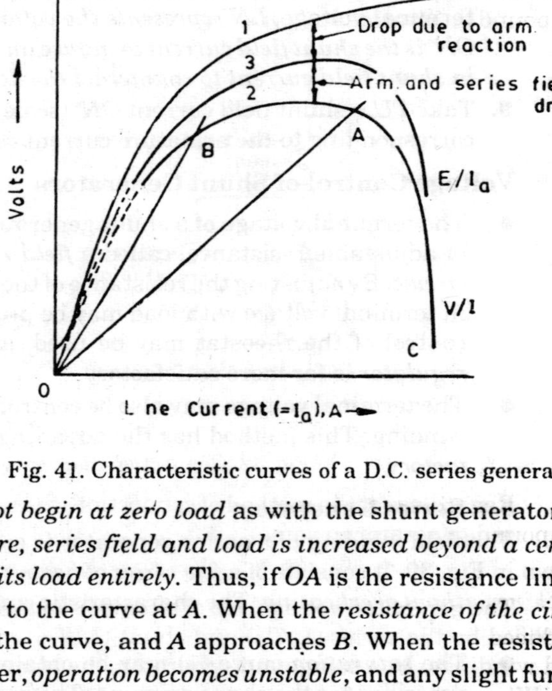

Fig. 41. Characteristic curves of a D.C. series generator.

It may be noted that the *readings cannot begin at zero load* as with the shunt generator, for if *the resistance of the circuit including armature, series field and load is increased beyond a certain critical value, the generator unbuilds and loses its load entirely.* Thus, if OA is the resistance line for the circuit the terminal voltage is the ordinate to the curve at A. When the *resistance of the circuit* is gradually *increased*, the *load falls off* along the curve, and A approaches B. When the resistance line finally becomes tangent to the curve, however, *operation becomes unstable*, and any slight further increase in the resistance causes the machine to unbuild its voltage and lose its load. The resistance that brings about this condition is called the *critical resistance*. Therefore, to *begin with, the resistance of the circuit must be reduced below the critical value before the generator delivers any load.*

- **Internal Characteristic (curve 3).** This curve is obtained by adding the resistance drop $I_a (R_a + R_{se})$ to the external characteristic curve ; R_a and R_{se} being armature resistance and series field resistance respectively.

  The difference between curves 1 and 3 is the *reduction in voltage caused by armature reaction.*
- It is worth noting that between A and C a considerable change in resistance brings about only a slight change in load current. Over this range the *voltage decreases rapidly,* owing to increasing armature reaction (particularly when the brushes are shifted forward), while the *current remains nearly constant*. Thus, between A and C the machine may be used to supply power to a *constant current variable-voltage circuit, such as series arc circuit.*
- Owing to initially *rising characteristic,* the series generator is often used as a *voltage booster* to give an increase of voltage practically proportional to the current.
- A series generator also finds application in electric traction where *'dynamic braking'* is employed. The connections of the series traction motors are changed by means of a controller so that they act as generators ; the power absorbed in braking the vehicle being dissipated in resistances, which are also used for starting purposes when the machines are reconnected as motors.

5.5. Compound Wound Generator

In case of a series generator the voltage regulation is very poor but the ability of the series field to produce additional useful magnetisation in response to increased load cannot be denied. This useful characteristic of the series field, combined with the relatively constant voltage characteristic of the shunt generator, led to the compound generator. Fig. 42 shows connection diagrams for a compound generator.

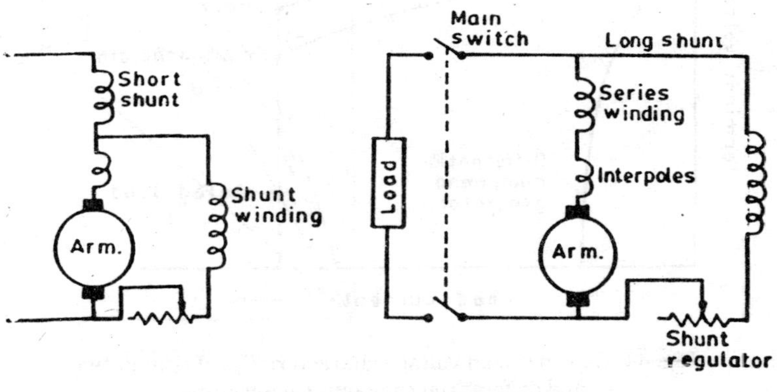

Fig. 42. Connection diagrams for a compound generator.

Regardless of the method of connection, the terminal voltage V of the short shunt or long shunt compound generator in same,

$$V = E_g - (I_{se}. R_{se} + I_a R_a).$$

The generated voltage, E_g, of a compound generator is the result of the combination of m.m.f.s produced by the series ($I_{se} T_{se}$) and shunt ($I_{sh} T_{sh}$) ampere-turns due to currents which flow in their field windings. In a compound generator, the *shunt field predominates and is much stronger of the two. When the series field m.m.f. aids the shunt field m.m.f.,* the generator is said to be *'cumulatively compounded* ' [*see* Fig. 43 (*a*)]. When the *series field m.m.f. opposes the shunt field m.m.f.*, the generator is said to be '*differentially compounded*' [*see* Fig. 43 (*b*)].

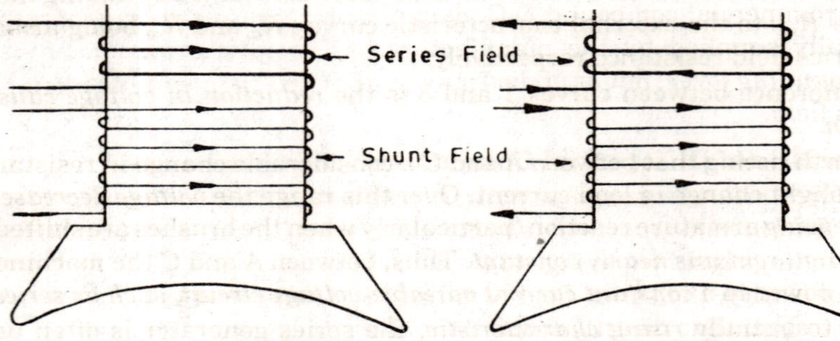

Fig. 43. Current directions in series and shunt-field coils of cumulative and differential-compound generators.

Characteristics of Cumulative Compound Generator

See Fig. 44. Depending on the relative additional aiding m.m.f., produced by the series field there are three types of load characteristics possible for the *cumulative* compound generator.

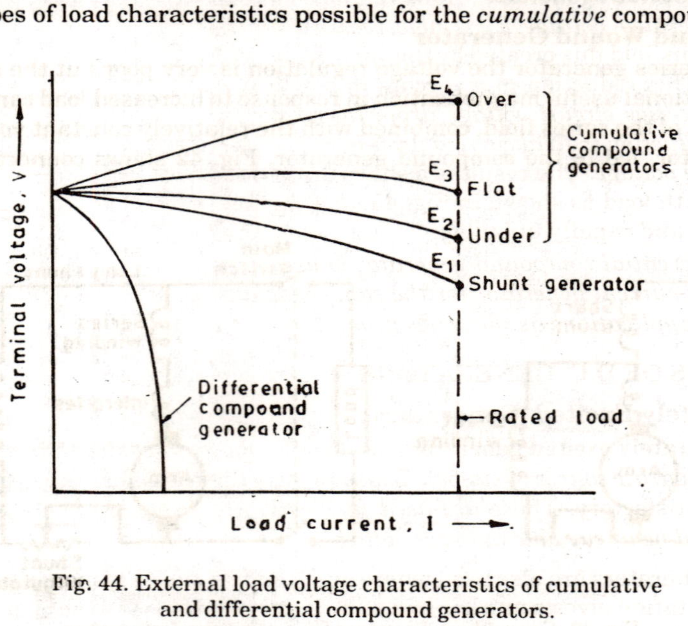

Fig. 44. External load voltage characteristics of cumulative and differential compound generators.

These types are called :

(*i*) Over-compound (*ii*) Flat-compound

(*iii*) Under-compound.

Over compound generator. An over-compound generator is one whose terminal voltage *rises with the application of load* so that its full-load voltage *exceeds* its no-load voltage (*negative regulation*).

Flat-compound generator. A flat compound generator has a load-voltage characteristic in which the no-load and full-load voltages are equal (*zero per cent regulation*).

Under-compound generator. An under-compound generator has a load characteristic in which the full load voltage is somewhat *less* than no-load voltage, but whose aiding series-field ampere-turns cause its characteristic to have *better regulation than an equivalent shunt generator*.

- Most commercial compound D.C. dynamos, whether used as generators or motors, are normally supplied by the manufacturer as *over-compound machines*. The *degree of compounding* (over, flat or under) may be adjusted by means of *divertor* which shunts the series field.

Characteristic of Differential Compound Generator

- The *differential compound generator is defined as that compounding produced when the series field m.m.f. opposes the shunt field m.m.f.* The difference in current direction of the two windings is shown in Fig. 43 (*b*), where for the sake of clarity, the series field winding is shown above (rather than directly around) the shunt field winding.
- Fig. 45 shows the load characteristic of differential compound generator.

When the differential compound generator is *without load* it builds up and self-excites its shunt field in much the same manner as the shunt generator. However, *when a load is applied, the generated voltage E_g is now reduced by the* reduction in the main field flux created by the *opposing m.m.f. of the series field*. This reduction in E_g occurs *in addition to the armature and series circuit voltage drop, the armature reaction, and the reduction in field current produced by reduction of the armature voltage*. The result is a sharp drop in the terminal voltage with load as shown in Fig. 45, and the field is below saturation and rapidly unbuilds.

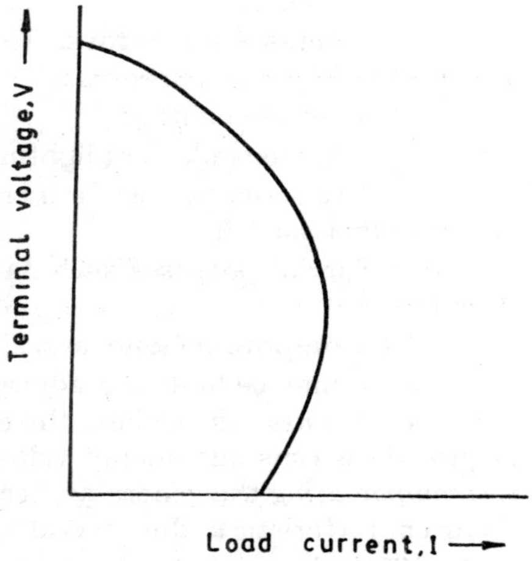

Fig. 45. Differential compound generator—load characteristic.

- *The differential compound generator is used as a constant-current generator for the same constant-current applications as the series generator.*

6. APPLICATIONS OF D.C. GENERATORS

6.1. Separately Excited Generators :

(*i*) The separately excited generators are usually *more expensive than self-excited generators* as they require *a separate source of supply*. Consequently they are generally used where self-excited are relatively unsatisfactory. These *are used in Ward Leonard systems of speed control, because self-excitation would be unsuitable at lower voltages*.

(*ii*) These generators are also used where quick and requisite response to control is important (since separate excitation gives a quicker and more precise response to the changes in the resistance of the field circuit).

6.2. Shunt Generators :

(*i*) These generators are used to advantage, in conjunction with automatic regulators, *as exciters for supplying the current required to excite the fields of A.C. generators*. The regulator controls the voltage of the exciter by cutting in and out some of the resistance of the shunt-field rheostat, thereby holding the voltage at whatever value is demanded by operating conditions. This is one of the most important applications of shunt generators.

(*ii*) Shunt generators are used *to charge batteries.* In this application the voltage should drop off slightly as the load increases, because the voltage of a lead battery is lower when battery is discharged than when battery is charged. When it is discharged, however, the battery can stand a large charging current than when it is charged. *Because of its drooping characteristic the shunt generator is admirably suited to battery charging service,* for, in a general way, the voltage curve of the generator has the same shape as the voltage curve of the battery itself. In both cases, as the load falls the voltage rises.

- Shunt generators can be operated in parallel without difficulty, and the wiring of parallel-operated shunt machines is quite a bit simpler than the corresponding wiring for compound machines. When a slight drop in voltage is not objectionable, as when a motor load is fed directly from the generator terminals, shunt machines may be used to advantage.

6.3. Series Generators. The field of application of series generator is limited. These are used for the following purposes :

(*i*) Series arc lighting.

(*ii*) Series incandescent lighting.

(*iii*) As a series booster for increasing the voltage across the feeder carrying current furnished by some other sources.

(*iv*) Special purposes such as supplying the field current for regenerative braking of D.C. locomotives.

6.4. Compound Generators. The compound generator is used for more than any other type.

(*i*) It may be built and adjusted automatically to supply an approximately *constant voltage at the point of use.,* throughout the entire range of load. This is *very great advantage.* It is possible to provide a constant supply voltage at the end of a long feeder by the simple expedient of *overcompounding* the generator, because the resistance drop in the line is compensated for by the rising characteristic of the generator.

When the point of utilisation is near the generator, a flat-compounded maehine may be used.

(*ii*) *Differentially compounded generator* finds an useful application *as an arc welding generator* where the generator is practically short circuited every time the electrode touches the metal plates to be welded.

(*iii*) Compound generators are used to supply power to :

- Railway circuits,
- Motors of electrified steam rail-roads,
- Industrial motors in many fields of industry,
- Incandescent lamps, and
- Elevator motors etc.

Example 15. *Draw armature drop line if armature resistance is 0.2 Ω*

Solution. The *procedure* of drawing an armature drop line is as follows (*see* Fig. 46) :

(*i*) Take any value of armature current and find corresponding voltage in the armature resistance. For example, take armature current of 200 A, voltage drop in the armature for this value of armature current is $200 \times 0.2 = 40$ V.

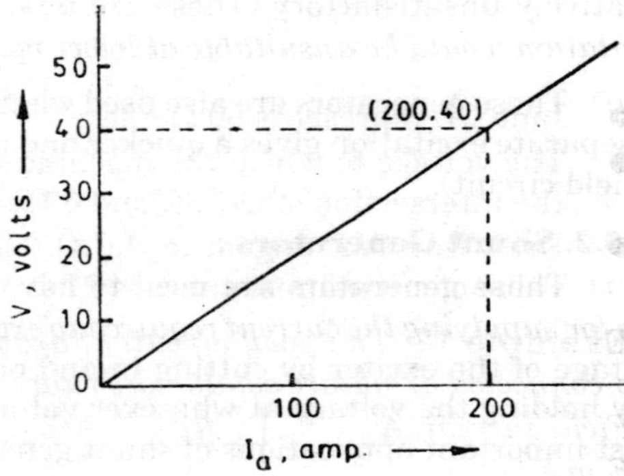

Fig. 46

So a point (200, 40) lying on this line is obtained.

(*ii*) Join the above point with origin *O*, the armature drop or ohmic drop line is obtained.

Example 16. *The open circuit characteristic (O.C.C.) of a separately excited generator, at 600 r.p.m., is as under :*

Field Current, A	:	*1.6*	*3.2*	*4.8*	*6.4*	*8.0*	*9.6*	*11.2*
E.m.f., V	:	*148*	*285*	*390*	*460*	*520*	*560*	*590*

Find :

(*i*) *The voltage to which the machine will excite as a shunt generator with a field circuit resistance of 60 ohm.*

(*ii*) *The critical resistance, at this speed.*

Solution.

- Plot O.C.C. (E_0/I_f) as shown in Fig. 47.

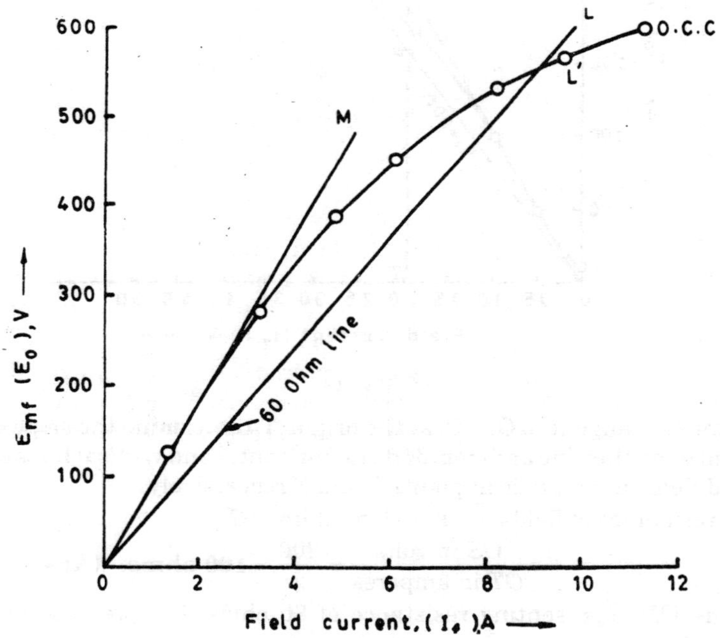

Fig. 47

- Line *OL* represents 60 Ω line.
- The voltage to which the machine will excite a shunt generator is given by point *L'* *i.e.* the intersection of O.C.C. and 60 Ω line. The machine will *excite* at 550 V.
- Draw line *OM* tangent to O.C.C. The *slope of this line represents critical resistance.* This value of critical resistance is 91 Ω.

Example 17. *The open circuit characteristic of a 4-pole, 250 V shunt generator having 610 lap-connected armature conductors running at 750 r.p.m. is as follows :*

Field current, A	:	*0*	*0.5*	*1.0*	*2.0*	*3.0*	*4.0*	*5.0*
E.m.f., V	:	*10*	*50*	*100*	*175*	*220*	*245*	*262*

Calculate :

(i) Field circuit critical resistance.

(ii) Critical speed for field circuit resistance of 80 Ω.

(iii) Residual flux per pole.

Solution. Number of poles of the generator, $p = 4$

Number of parallel paths, $a = p = 4$ [Generator being lap connected]

Number of armature conductors, $Z = 610$

- Plot O.C.C. from the given data as shown in Fig. 48.

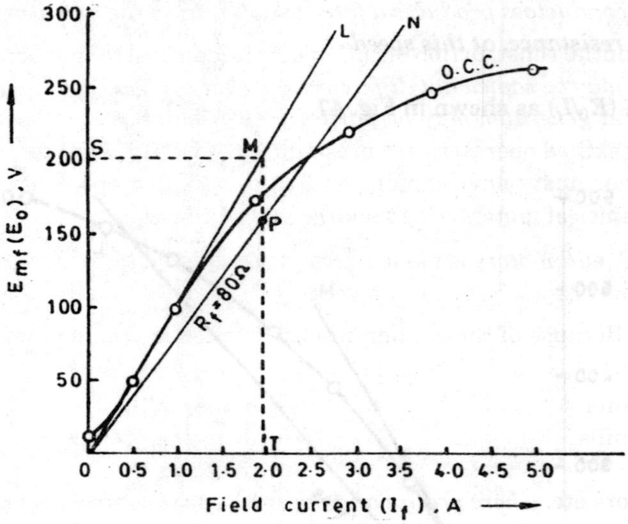

Fig. 48

- Draw line OL tangent to O.C.C. at the origin. To determine the slope of the line OL, take any point M on this line and from M draw horizontal and vertical lines cutting O.C. voltage axis and field current axis at points S and T respectively.

(*i*) Critical resistance of field circuit = slope of line OL

$$= \frac{OS \text{ in volts}}{OT \text{ in amperes}} = \frac{200}{2} = \textbf{100 ohms. (Ans.)}$$

(*ii*) Draw line ON representing resistance of 80 ohms. Let ordinate drawn from point M intersect line ON at P.

$$\textbf{Critical speed, } N_c = N \times \frac{TP}{TM} = 750 \times \frac{160}{200} = \textbf{600 r.p.m. (Ans.)}$$

(*iii*) **Residual flux per pole, φ :**

With no exciting current, the e.m.f. induced due to residual flux is 10 V at 750 r.p.m. (given).

Using the relation, $$E_g = \frac{p\phi ZN}{60a}$$

$$10 = \frac{4 \times \phi \times 610 \times 750}{60 \times 4}$$

$$\therefore \quad \phi = \frac{10 \times 60 \times 4}{4 \times 610 \times 750} = 0.00131 \text{ Wb.}$$

Hence, **residual flux/pole = 0.00131 Wb. (Ans.)**

7. DIRECT CURRENT MOTOR

7.1. General Aspects

The electric motor is a machine which converts electric energy into mechanical energy. It depends for its operation on the force which is known to exist on a conductor carrying current while situated in a magnetic field.

Construction. A D.C. motor is similar in construction to a D.C. generator. As a matter of fact any D.C. generator will run as a motor when its field and armature windings are connected to a source of direct current. The field winding produces the necessary magnetic field. *The flow of current through the armature conductors produces a force which rotates the armature.*

Though the essential construction of D.C. motor is identical to that of a generator, the *external appearance of a motor may be somewhat different from that of a generator.* This is mainly due to the fact that the *frame of a generator may be partially open* because it is located in relatively clean environment and only skilled operators are present in its vicinity. A motor, on the other hand, may be operating in a rather dusty environment and only unskilled operators may be working in its vicinity. Therefore, frames of motors are to a large extent closed.

The body of *D.C. mill motors* is made in two halves bolted together for easy access to the field windings and inter-poles.

Applications. Because of their inherent characteristics D.C. motors find extensive application in :

(*i*) Steel plants (*ii*) Paper mills
(*iii*) Textile mills (*iv*) Printing presses
(*v*) Cranes (*vi*) Winches
(*vii*) Excavators etc. *where precise and accurate speed control over a wide range is required.*

Advantages. The D.C. motors possess the following *advantages* :

(*i*) High starting torque.
(*ii*) Speed control over a wide range, both below and above the normal speed.
(*iii*) Accurate stepless speed control with constant torque.
(*iv*) Quick starting, stopping, reversing and accelerating.

Disadvantages. The *disadvantages* of D.C. motors are :

(*i*) High initial cost.
(*ii*) Increased operating and maintenance costs because of the commutators and brushgear.

7.2. Principle of Operation of D.C. Motor

The principle of motor action can be stated as follows :

"Whenever a current carrying conductor is placed in a magnetic field, it experiences a force whose direction is given by Fleming's left hand rule".

- Fig. 49 illustrates this principle.

 Fig. 49 (*a*) shows the field set up by the poles.

 Fig. 49 (*b*) shows the conductor field due to flow of current in the conductor.

Fig. 49 (*c*) shows the resultant field produced when the current carrying conductor wire of Fig. 49 (*b*) is inserted in the air gap of Fig. 49 (*a*) with the *axis of the conductor at right angles to the direction of the flux.*

On the *upper side* of the conductor in Fig. 49 (*c*) the magnetizing forces of the field and of the current in the conductor are *additive* while on the *lower side* these are *subtractive*. This explains why the resultant field is strengthened above and weakened below the conductor (wire).

The above experiment shows that the wire in Fig. 49 (*c*) has a force on it which tends to move it downward. Thus the *force acts in the direction of the weaker field*. When the current in the wire is reversed, the direction of the force is also reversed, as in Fig. 49 (*d*).

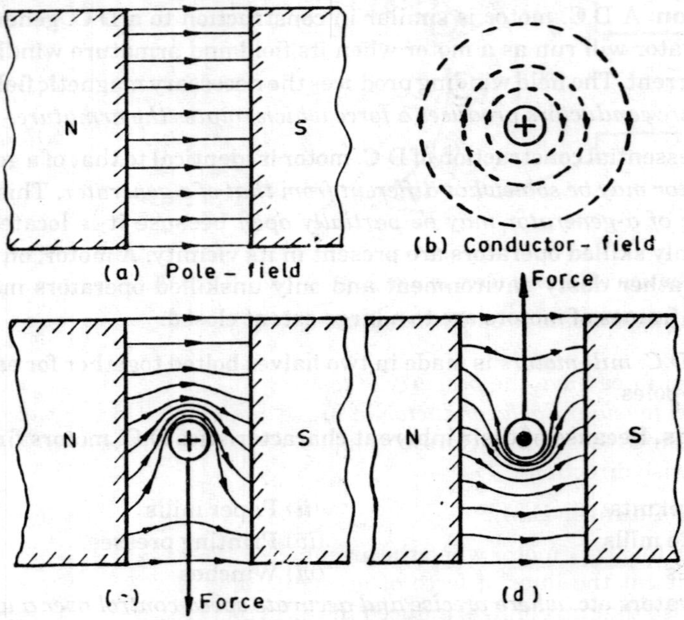

Fig. 49. The principle of motor action.

The force (F) developed in the conductor is given by the relation,

$$F = BIl \text{ newtons}$$

where B = flux density, T (Wb/m^2),

I = current in conductor, A, and

l = exposed length of conductor, m.

Now consider the magnetic field of a D.C. motor in which there is no current in the armature conductors ; the lines of force will be distributed as shown in Fig. 50.

Fig. 50. Distribution of lines of force in a motor due to magnetic field only.

If, now, the armature carries current, each of its conductors will produce a magnetic field which, when super-imposed on the main field, causes a distribution of magnetic lines as shown in Fig. 51.

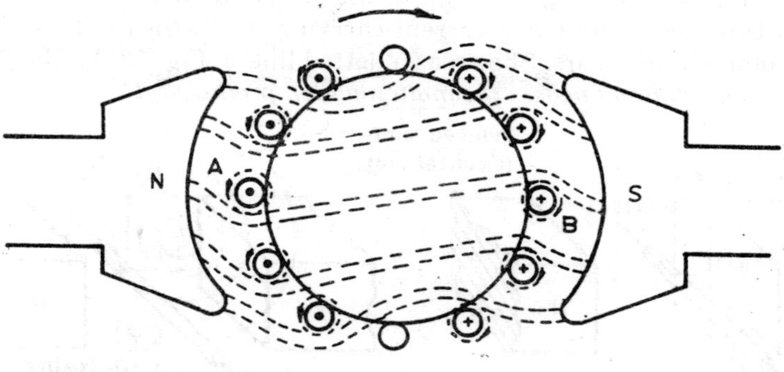

Fig. 51. Distribution of lines of force in a motor, on load, due to the armature and magnetic field.

The magnetic field is said to be distorted, since the lines of force no longer follow approximately straight paths.

These lines of force have the *property of tending to shorten* themselves, so that they may be regarded as *being in tension*. *Each conductor* in Fig. 51 *will experience a force like that exerted on a stone in a catapult*. Since these conductors are embedded in slots in the armature, the latter is caused to rotate in a clockwise direction.

7.3. Back or Counter E.M.F.

Refer Fig. 52. In a D.C. motor when the armature rotates, the conductors on it cut the lines of force of magnetic field in which they revolve, so that an e.m.f. is induced in the armature as in a generator. The *induced e.m.f. acts in opposition to the current in the machine* and, therefore, *to the applied voltage*, so that *it is customary to refer to this voltage as the 'back e.m.f.'* That this is so can be deduced by Lenz's law, which states that the *direction of an induced e.m.f. is such as to oppose the change causing it, which is, of course, the applied voltage.*

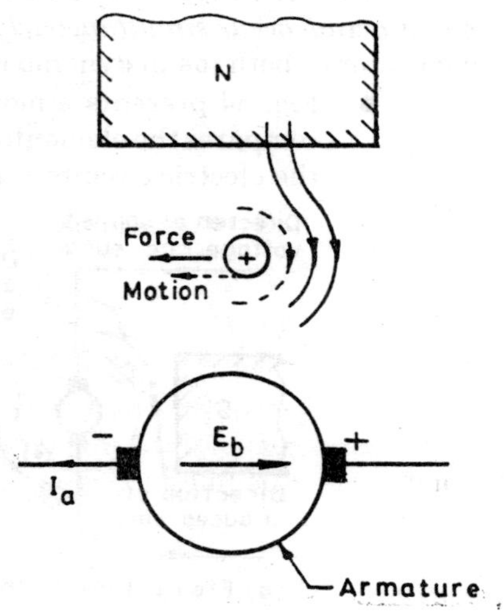

Fig. 52. Motoring operation.

The magnitude of the back or counter e.m.f. can be calculated by using formula for the induced e.m.f. in a generator, and it is important in the case of the motor, to appreciate that this is *proportional to the product of the flux and the speed*. Thus if E_b denotes the back e.m.f., ϕ the flux and N the speed, we may write,

$$E_b = k\,\phi\,N$$

where k is a number depending on *nature of armature winding*.

The value of back e.m.f. (E_b) is always less than the applied voltage, although difference is small when the machine is running under normal conditions. It is the difference between these two quantities which actually drives current through the resistance of the armature circuit. If this resistance is represented by R_a, the back e.m.f. by E_b and the applied voltage by V, then we have

$$V = E_b + I_a R_a$$

where I_a is the current in the armature circuit.

7.4. Comparison Between Motor and Generator Action

— Fig. 53 (*b*) shows a generator action, where a mechanical force moves a conductor in upward direction inducing an e.m.f. in the direction shown. When a current flows as a result of this e.m.f., there is a current-carrying conductor existing in a magnetic field ; hence motor action occurs. Shown as a dotted line in Fig. 53 (*b*), the *force developed as a result of motor action opposes the motion which produced it.*

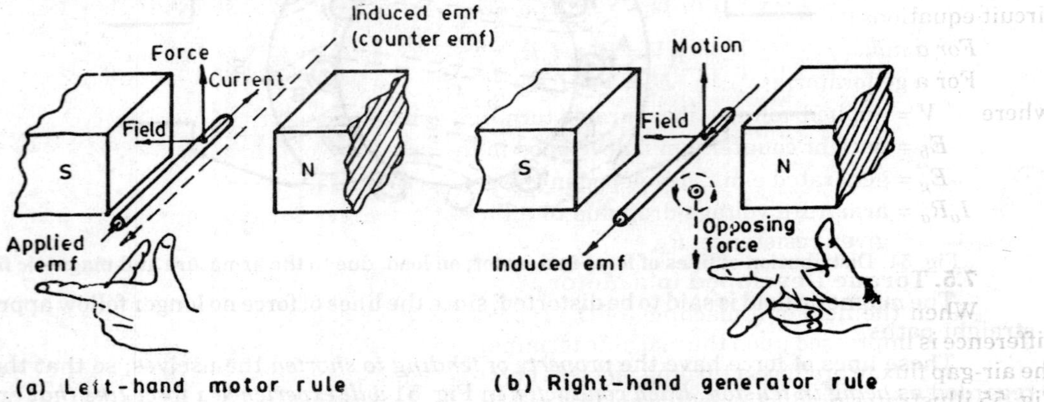

Fig. 53. Comparison of motor and generator action.

Thus it can be stated categorically that in rotating electric machines *generator action and motor action occur simultaneously*. Hence the same dynamo may be operated either as a motor or a generator or both (as in dynamo motor or synchronous converter).

- Fig. 54 presents a more graphic representation in terms of rotational elements, which compares the elementary motor and generator for the same direction of rotation and shows the electric circuits of each which is self-explanatory.

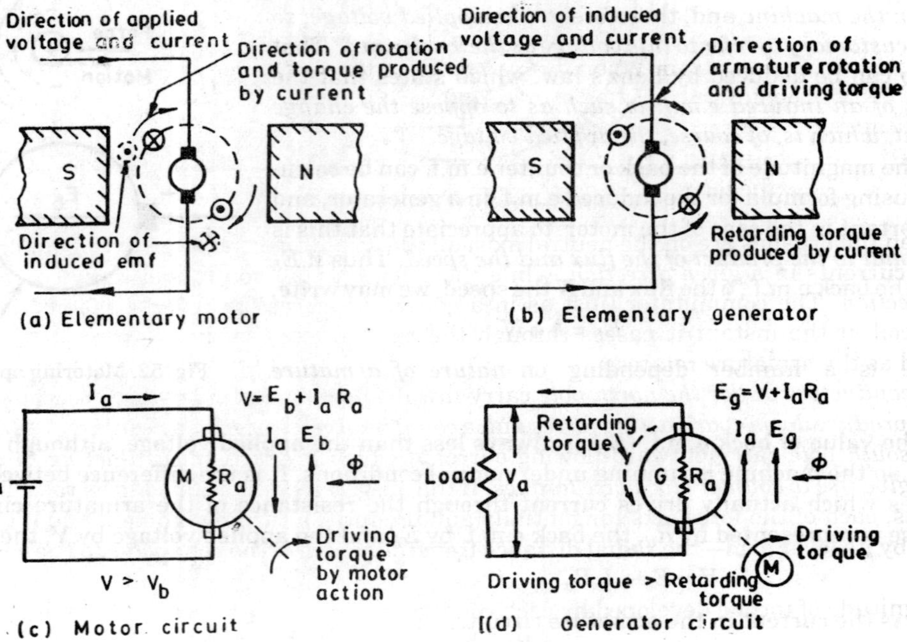

Fig. 54. Elementary motor action *versus* generator action.

It may be noted that when a dynamo is operating as a *motor*, the generated e.m.f. is *always less than the terminal voltage* (that produces motor action) and it *opposes the armature current.* On the other hand when a dynamo is operating as a *generator*, the *armature current is in the same direction as the generated e.m.f.*, and the *generated e.m.f.* E_g *exceeds the terminal voltage V applied across the load.* This distinction between generator and motor, in which the armature-generator voltage aids or opposes the armature current, respectively give rise to the following basic armature circuit equations :

For a motor, $V = E_b + I_aR_a$ ...(2)

For a generator, $E_g = V + I_aR_a$ ...(3)

where V = applied voltage (measurable terminal voltage) across the armature,

E_b = back or counter e.m.f. developed in the armature of the motor,

E_g = generated e.m.f. developed in the generator armature, and

I_aR_a = armature voltage drop due to a flow of armature current through an armature of a given resistance, R_a.

7.5. Torque Developed in a Motor

When the field of a machine (of the type described as generator) is excited and a potential difference is impressed upon the machine terminals, the current in the armature winding reacts with the air-gap flux to produce a turning moment or *torque* which tends to cause the armature to revolve. Fig. 55 illustrates production of torque in a motor.

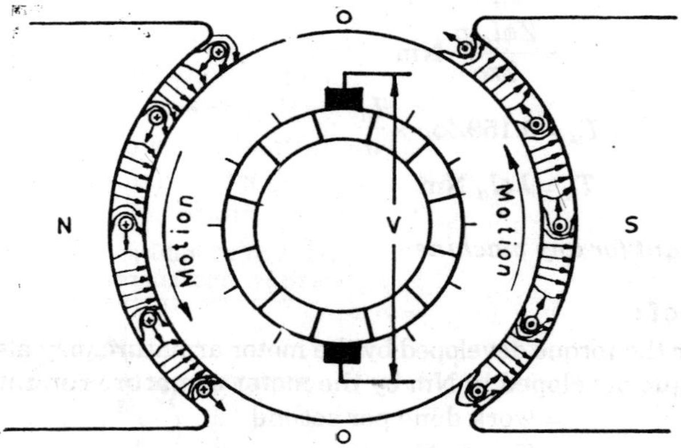

Fig. 55. Production of torque in a D.C. motor.

When the brushes are on the neutral axis, all the armature conductors lying under the north pole carry currents in a *given direction*, while those lying under south pole carry currents in the *reverse direction*. The commutator (just as in a generator) serves to reverse the current in each armature coil at the instant it passes through the neutral axis, so the above relation is always maintained as the armature rotates.

All *conductors under the north pole* carry inward-flowing currents which react with the air gap flux to *produce down-ward acting forces* and a *counter clockwise torque*. Similarly the *conductors under the south pole* carry outward-flowing currents which *produce upward-acting forces.* These *forces also give rise to counter clockwise torques.* If the air-gap flux is assumed to be radially directed at all points, each of the force acts tangentially and produces a turning moment equal to the force multiplied by its lever arm—the radial distance from the centre of the conductor to the centre of the shaft.

Magnitude of torque developed by each conductor

$$= BIlr \text{ Nm}$$

If the motor contains Z conductors, the total torque developed by the armature

$$T_a = BIlrZ \text{ Nm} \quad ...(4)$$

where B = gap density, T (Wb/m^2)
I = armature current in a conductor, A
l = active length of each conductor, m
r = average lever arm of a conductor or the average radius at which conductors are placed, m
Z = total number of armature conductors.

It is more convenient to express T_a in terms of armature current I_a, total flux per pole ϕ and number of poles p.

$$I = \frac{I_a}{a}$$

and

$$B = \frac{\phi}{A}$$

where a = number of parallel paths,
and A = the cross-sectional area of flux path at radius r.

$$A = \frac{2\pi rl}{p}$$

Then

$$T_a = \frac{Z\phi l I_a r}{2\pi rl} \times \frac{p}{a}$$

$$= \frac{Z\phi I_a p}{2\pi a} \text{ Nm}$$

i.e.

$$T_a = 0.159 Z\phi p \times \frac{I_a}{a} \quad ...(5)$$

or

$$T_a = k\phi I_a \text{ Nm} \quad ...(6)$$

where $k = \dfrac{Zp}{2\pi a}$ is a *constant for any machine.*

Alternative proof:

The expression for the torque developed by the motor armature may also be deduced as follows :
Let T_a be the torque developed in Nm by the motor armature running at N r.p.m.

Power developed = work done per second

$$= T_a \times 2\pi N \text{ watts} \quad ...(i)$$

Electrical equivalent of mechanical power developed by the armature also

$$= E_b I_a \text{ watts} \quad ...(ii)$$

Equating (*i*) and (*ii*), we get

$$T_a \times \frac{2\pi N}{60} = E_b I_a$$

or

$$T_a = \frac{E_b I_a}{2\pi\left(\dfrac{N}{60}\right)}$$

Also since

$$E_b = \frac{p\phi ZN}{60a}$$

$$\therefore \quad T_a \times 2\pi \frac{N}{60} = \frac{p\phi ZN}{60a} \, . \, I_a \quad \text{or} \quad T_a = \frac{Z\phi p}{2\pi} \, . \, \frac{I_a}{a} \text{ Nm}$$

i.e.

$$T_a = 0.159 \, Z\phi p \, . \, \frac{I_a}{a}.$$

Note. From the above equation for torque, we find that

$$T \propto \phi I_a$$

Then

(*i*) In the case of shunt motors, ϕ is practically constant,

hence $$T \propto I_a$$

(*ii*) In the case of series motors, ϕ is proportional to I_a before saturation (because field windings carry full armature current)

$$\therefore \quad T \propto I_a^2.$$

Shaft torque (T_{sh}). The torque developed by the armature is the *gross torque*. Whole of this torque is not available at the pulley, since certain percentage of torque developed by the armature is lost to overcome the iron and friction losses. *The torque which is available for useful work is known as* **shaft torque** T_{sh}. It is so called because it is *available at the shaft*. The horse power obtained by using shaft torque is called brake horse power (B.H.P.).

$$\text{B.H.P. (metric)} = \frac{T_{sh} \times 2\pi N}{735.5}$$

$$\therefore \quad T_{sh} = \frac{\text{B.H.P. (metric)} \times 735.5}{\frac{2\pi N}{60}} \qquad \ldots(7)$$

where N = speed of armature in r.p.m.

The difference $T_a - T_{sh}$ is known as **lost torque** (*i.e.* torque lost in iron and friction losses)

$$= 0.159 \times \frac{\text{iron and friction losses}}{\frac{N}{60}} \text{ Nm.}$$

7.6. Mechanical Power Developed by Motor Armature

Refer Fig. 56. The voltage V applied across the motor armature has to (*i*) overcome back e.m.f. E_b and (*ii*) supply the armature ohmic drop I_aR_a.

$$\therefore \quad V = E_b + I_aR_a$$

This is known as *voltage equation of a motor.*

Multiplying both sides by I_a, we get

$$VI_a = E_bI_a + I_a^2R_a$$

Here VI_a = electrical input to the armature,

E_bI_a = electrical equivalent of mechanical power P_m developed in the armature, and

$I_a^2R_a$ = copper loss in the armature.

The power available at the pulley for doing useful work is *somewhat less than the mechanical power developed by the armature.*

This is evident, since there are certain *mechanical losses* (such as bearing and windage friction and iron losses) that must be supplied by the driving power of the motor.

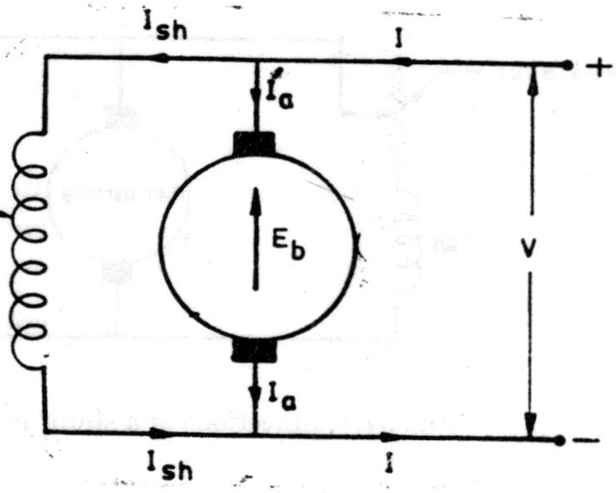

Fig. 56

Condition for maximum power. We know that, the mechanical power developed by the motor,

$$P_m = VI_a - I_a^2 R_a$$

Differentiating both sides with respect to I_a

$$\frac{d(P_m)}{dI_a} = V - 2I_aR_a = 0$$

$$\therefore \qquad I_aR_a = \frac{V}{2}$$

As $$V = E_b + I_aR_a$$

$$\therefore \qquad V = E_b + \frac{V}{2}$$

i.e. $$E_b = \frac{V}{2} \qquad ...(8)$$

Hence, *mechanical power developed by a motor is maximum when back e.m.f. is equal to half the applied voltage.* In practice, however, *this condition is not realized* because in that case current would be much beyond the normal current of the motor. Moreover, half the input would be wasted in the form of heat and taking other losses, such as mechanical and magnetic, into consideration, the efficiency of the motor will be below 50 per cent.

7.7. Types of D.C. Motors

There are three main types of motors characterised by the connection of the field winding in relation to the armature. These are :

1. **Shunt wound motor** or the shunt motor, in which the *field winding is connected in parallel with the armature.*

2. **Series motor,** in which the *armature and field windings are connected in series.*

3. **Compound motor,** which has two field windings, *one of which is connected in parallel with the armature and the other in series with it.*

Shunt wound motor. Fig. 57 shows the connections of a shunt motor. From these connections it may be observed at once that the *field current is constant*, since it is connected directly to the supply which is assumed to be at constant voltage. Hence the flux is approximately constant and, since also the *back e.m.f. is almost constant under normal conditions the speed is approximately constant*. This is not strictly true, but nevertheless, *it is usual for all practical purposes to regard the shunt motor as a constant speed machine*. It is, therefore, employed in practice for drives, the speeds of which are required to be *independent of the loads*. The speed can, of course, be varied when necessary and this is done by the inclusion of a variable resistor in series with the field winding, as shown in Fig. 57.

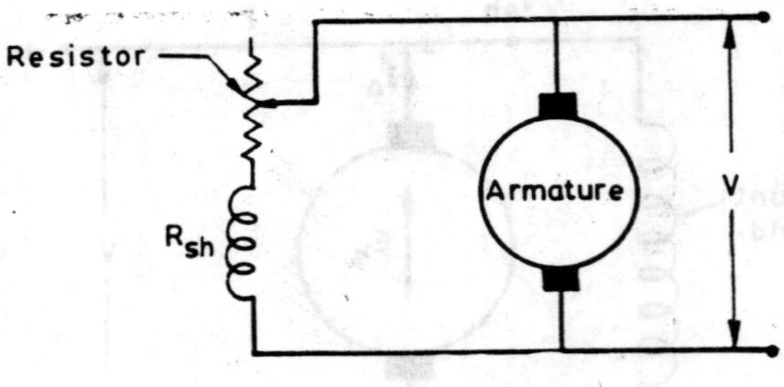

Fig. 57. Connections of a shunt motor.

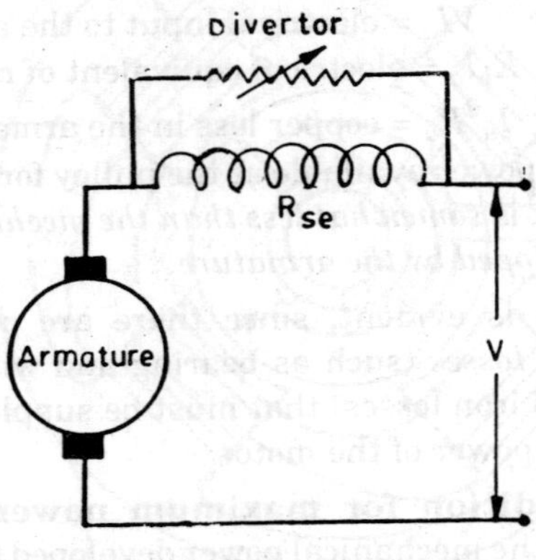

Fig. 58. Connections of a series motor.

Series motor. Fig. 58 shows the connections for the series motor. The current passing through the field winding is the same as that in the armature, since the armature current increases as the mechanical load on the shaft increases, so also does the field current. The resultant increase in magnetic flux causes a reduction in the speed, as can be observed from a consideration of the formula :

$$E_b = \frac{p\phi ZN}{60a} \quad \text{or} \quad E_b = k\phi N$$

where $k = \dfrac{pZ}{60a}$ being constant

or $$N = \frac{E_b}{k\phi}.$$

This is a useful property for many drives in which it is desirable that a heavy increase in the load should automatically bring about a compensating reduction in speed. As with the shunt motor, the speed may also be varied independently of the load by the inclusion of a variable resistor in the field circuit. In this case, however, it is connected in parallel with the series winding as shown in Fig. 58, and is called a *divertor compound motor*.

Refer Fig. 59. The compound motor has a shunt field winding in addition to the series winding so that the number of magnetic lines of force produced by each of its poles is the resultant of the flux produced by the shunt coil and that due to the series coil. The *flux so produced depends not only on the current and number of turns of each coil, but also on the winding direction of the shunt coil in relation to that of the series coil.* When the two fluxes *assist each other* the machine is a **cumulative compound motor,** while if they *oppose each other*, it is said to be a **differential compound motor.**

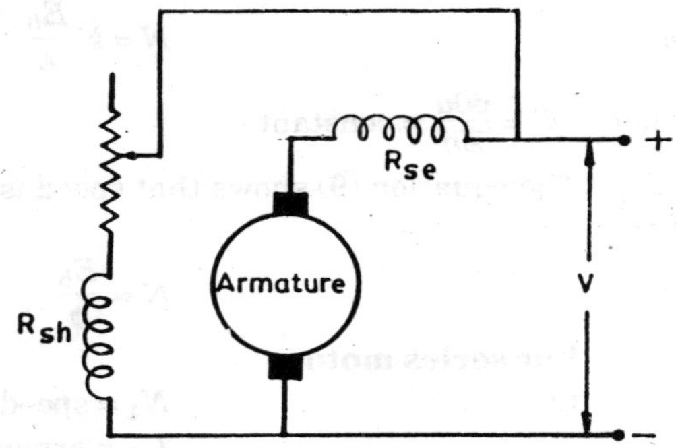

Fig. 59. Connections of a compound motor.

Fig. 60 shows the field windings and interpole connections of a *differential compound wound motor*. The shunt coil is made up of *many turns of fine wire,* whilst the series coil comprises relatively *few turns of thickwire*.

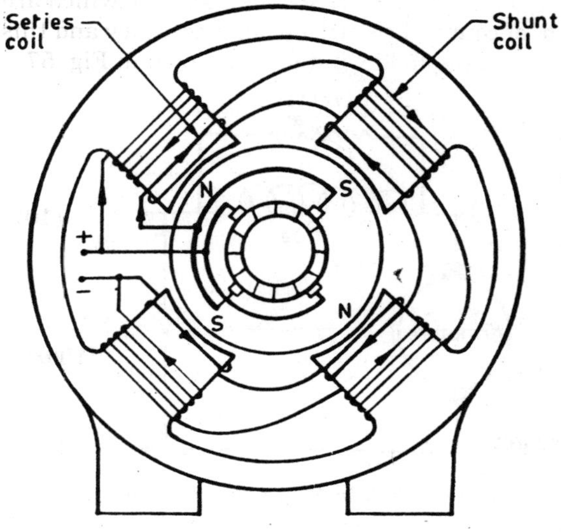

Fig. 60. Field windings of a differential compound motor.

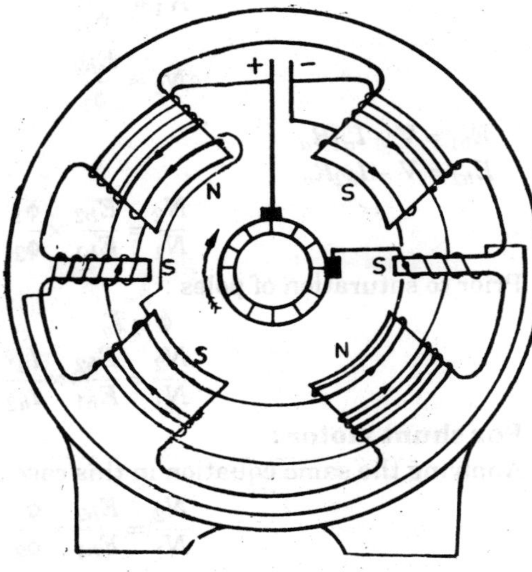

Fig. 61. Field windings and interpole connections of a cumulative compound wound motor.

Fig. 61 shows the field windings of a *cumulative compound motor*.

The flow of currents in the shunt and series coils is worth noting in Figs. 60 and 61.

7.8. Speed of a D.C. Motor

We know that the voltage equation of a motor is given by

$$V = E_b + I_a R_a$$

or $$E_b = V - I_a R_a$$

or $$\frac{p\phi ZN}{60a} = V - I_a R_a$$

$\therefore$ $$N = \frac{(V - I_a R_a)}{\phi} \times \frac{60a}{Zp}$$

But $$V - I_a R_a = E_b$$

$\therefore$ $$N = \frac{E_b}{\phi} \times \frac{60a}{Zp}$$

or $$N = k' \frac{E_b}{\phi} \quad ...(9)$$

where $k' = \frac{60a}{Zp}$ = constant.

The equation (9) shows that speed is *directly proportional to back e.m.f.* and *inversely to the flux*

or $$N \propto \frac{E_b}{\phi}.$$

For series motor :

Let N_1 = speed in the first case,

I_{a1} = armature current in the first case, and

ϕ_1 = flux in the first case.

N_2, I_{a2} and ϕ_2 = corresponding quantities in the second case.

Using the above relation $\left(i.e. \; N \propto \frac{E_b}{\phi}\right)$, we get

$$N_1 \propto \frac{E_{b1}}{\phi_1}$$

and $$N_2 \propto \frac{E_{b2}}{\phi_2}$$

where $E_{b1} = V - I_{a1} R_a$

and $E_{b2} = V - I_{a2} R_a$

$\therefore$ $$\frac{N_2}{N_1} = \frac{E_{b2}}{E_{b1}} \times \frac{\phi_1}{\phi_2} \quad ...(10)$$

Prior to saturation of poles :

$$\phi \propto I_a$$

$\therefore$ $$\frac{N_2}{N_1} = \frac{E_{b2}}{E_{b1}} \times \frac{I_{a1}}{I_{a2}} \quad ...[10(a)]$$

For shunt motor :

Applying the same equation in this case also, we get

$$\frac{N_2}{N_1} = \frac{E_{b2}}{E_{b1}} \times \frac{\phi_1}{\phi_2}$$

If $\phi_1 = \phi_2$

$$\frac{N_2}{N_1} = \frac{E_{b2}}{E_{b1}}. \quad ...(11)$$

7.9. Speed Regulation

The speed regulation of a D.C. motor is defined as follows :

"The change in speed when the load on the motor is reduced from rated value to zero, expressed as percent of the rated load speed."

$$\therefore \text{ Percent speed regulation} = \frac{\text{no load speed} - \text{full load speed}}{\text{full load speed}}.$$

SHUNT MOTORS

Example 18. *Determine the torque developed when a current of a 30 A passes through the armature of a motor with the following particulars : lap winding, 310 conductors, 4-pole, pole-shoes 16.2 cm long subtending an angle of 60° at the centre, bore radius 16.2 cm, flux density in air gap 0.7 tesla.*

Solution. Number of poles, $p = 4$

Number of parallel paths, $a = p = 4$ [Motor being lap-wound]

Number of armature conductors, $Z = 310$

Length of pole shoe $= 16.2$ cm

Bore radius $= 16.2$ cm

Flux density in air gap, $B = 0.7$ tesla

Armature current, $I_a = 30$ A

Torque developed, T :

We know that,

$$\text{Pole-shoe arc} = \pi D \times \frac{60}{360} = \pi(2 \times 16.2) \times \frac{60}{360} = 16.967 \text{ cm}$$

$$\therefore \text{ Pole area} = \text{pole-shoe length} \times \text{pole-shoe arc}$$

$$= (16.2 \times 10^{-2}) \times (16.967 \times 10^{-2})$$

$$= 16.2 \times 16.967 \times 10^{-4} \text{ m}^2 = 0.02748 \text{ m}^2$$

Also flux, $\phi = B \times \text{pole area}$

$$= 0.7 \times 0.02748 = 0.01924 \text{ Wb}$$

Using the relation,

$$T = 0.159 Z\phi p \left(\frac{I_a}{a}\right) = 0.159 \times 310 \times 0.01924 \times 4 \times \frac{30}{4}$$

$$= 28.45 \text{ Nm.}$$

Hence, **torque developed = 28.45 N-m. (Ans.)**

Example 19. *A 230 V D.C. shunt motor takes 32 A at full load. Find the back e.m.f. on full load if the resistances of motor armature and shunt field windings are 0.2 ohm and 115 ohms respectively.*

Solution. Supply voltage, $V = 230$ Volts

Full load current, $I = 32$ A

Armature resistance, $R_a = 0.2$ ohm

Shunt field windings' resistance, $R_{sh} = 115$ ohms

Back e.m.f., E_b :

$$\text{Shunt field current, } I_{sh} = \frac{230}{115} = 2 \text{ A}$$

$\therefore$ Armature current, $I_a = I - I_{sh} = 32 - 2 = 30$ A

Back e.m.f. on full load, $E_b = V - I_aR_a$

$= 230 - 30 \times 0.2 = 224$ V

Hence, **back e.m.f. on full load = 224 V. (Ans.)**

Example 20. *The power input to a 230 volts D.C. shunt motor is 8.477 kW. The field resistance is 230 Ω and armature resistance is 0.28 Ω. Find the input current, armature current and back e.m.f.*

(P.T.U., May, 2001)

Solution. *Given* : $P_{in} = 8.477$ kW ; $R_{sh} = 230\ \Omega$; $R_a = 0.28\ \Omega$

I : ; I_a : E_b :

Input current, $I = \dfrac{P_{in}}{V} = \dfrac{8.477 \times 1000}{230}$

$= \mathbf{36.86\ A.}$ **(Ans.)**

Shunt field current, $I_{sh} = \dfrac{V}{R_{sh}} = \dfrac{230}{230} = 1$ A

$\therefore$ *Armature current,* $I_a = I - I_{sh}$

$= 36.86 - 1$

$= \mathbf{35.86\ A.}$ **(Ans.)**

Back e.m.f. $E_b = V - I_aR_a$

$= 230 - 35.86 \times 0.28 = \mathbf{219.96\ V.}$ **(Ans.)**

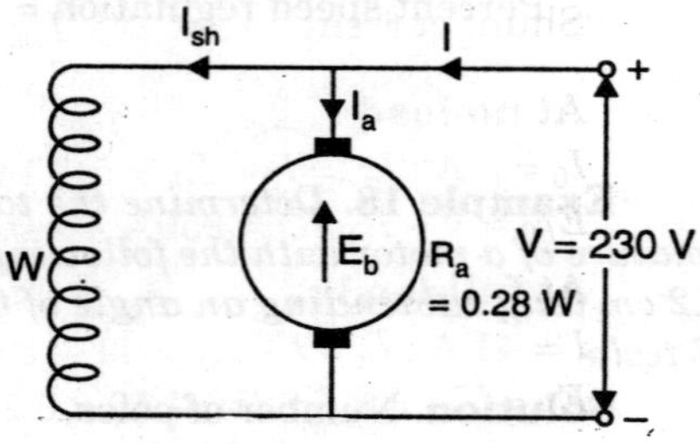

Fig. 62

Example 21. *A six-pole lap-connected 230 V shunt motor has 410 armature conductors. It takes 41 A on full load. The flux per pole is 0.05 weber. The armature and field resistances are 0.1 ohm and 230 ohms respectively. Contact drop per brush = 1 V.*

Determine the speed of motor at full load.

Solution. Number of poles, $p = 6$

Number of parallel paths, $a = 6$ [Motor being lap-connected]

Number of armature conductors, $Z = 410$

Full load current, $I = 41$ A

Flux per pole, $\phi = 0.05$ weber

Armature resistance, $R_a = 0.1$ ohm

Shunt field resistance, $R_{sh} = 230$ ohms

Contact drop/brush $= 1$ V

Speed of motor on full load, N :

Shunt field current, $I_{sh} = \dfrac{V}{R_{sh}} = \dfrac{230}{230} = 1$ A

Armature current, $I_a = I - I_{sh} = 41 - 1 = 40$ A

Back e.m.f. on full load, $E_b = V - I_aR_a$ – drop at brushes

$= 230 - 40 \times 0.1 - 2 \times 1 = 224$ V

We know that, $E_b = \dfrac{p\phi ZN}{60a}$

$$224 = \frac{6 \times 0.05 \times 410 \times N}{60 \times 6}$$

$$\therefore \quad N = \frac{224 \times 60 \times 6}{6 \times 0.05 \times 410} = 655.6 \text{ r.p.m.}$$

Hence, **speed of motor on full load = 655.6 r.p.m. (Ans.)**

Example 22. *A 250 volt d.c. shunt motor, on no load, runs at 1000 rpm and takes 5 A. The field and armature resistances are 250 ohms and 0.25 ohm respectively. Calculate the speed when the motor is loaded such that it takes 41 A if the armature reaction weakens the field by 3%.*

[P.T.U., May, 2001]

Solution. *Given* : $V = 250$ V ; $N_0 = 1000$ r.p.m. ; $I_0 = 5$ A ; $R_{sh} = 250\ \Omega$;

$$R_a = 0.25\ \Omega, I = 41\text{ A}, \phi = \left(1 - \frac{3}{100}\right)\phi_0 = 0.97\ \phi_0$$

Speed of the motor, N :

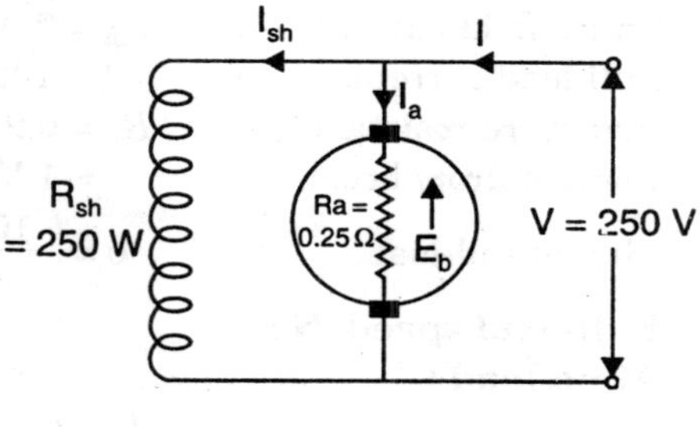

Fig. 63

Shunt current, $$I_{sh} = \frac{V}{R_{sh}} = \frac{250}{250} = 1\text{ A}$$

At no-load :

$I_0 = 5$ A ; $I_{as} = I_0 - I_{sh} = 5 - 1 = 4$ A

$E_{b0} = V - I_{a0} R_a = 250 - 4 \times 0.25 = 249$ V

At full-load :

$I = 41$ A ; $I_a = I - I_{sh} = 41 - 1 = 40$ A

$E_b = V - I_a R_a = 250 - 40 \times 0.25 = 240$ V

Using the relation, $$\frac{N}{N_0} = \frac{E_b}{E_{b0}} \times \frac{\phi_0}{\phi}$$

or $$\frac{N}{1000} = \frac{240}{249} \times \frac{\phi_0}{0.97\phi_0}$$

$\therefore$ $$\mathbf{N} = 1000 \times \frac{240}{249} \times \frac{1}{0.97} = \mathbf{993.67\ r.p.m.}\ \ \textbf{(Ans.)}$$

Example 23. *A 120 volt d.c. shunt motor has an armature resistance of 0.2 ohms and a field resistance of 60 ohms. The full-load line current is 60 A and full-load speed is 1800 r.p.m. If the brush contact drop is 3 V, find the speed of the motor at half-load.* **(P.T.U., June, 2000)**

Solution. *Given* : $V = 120$ V ; $R_a = 0.2\ \Omega$; $R_{sh} = 60\ \Omega$

$I_1 = 60$ A, $N_1 = 1800$ r.p.m. ;

Brush contact drop = 3 V

Speed of the motor at half-load, N_2 :

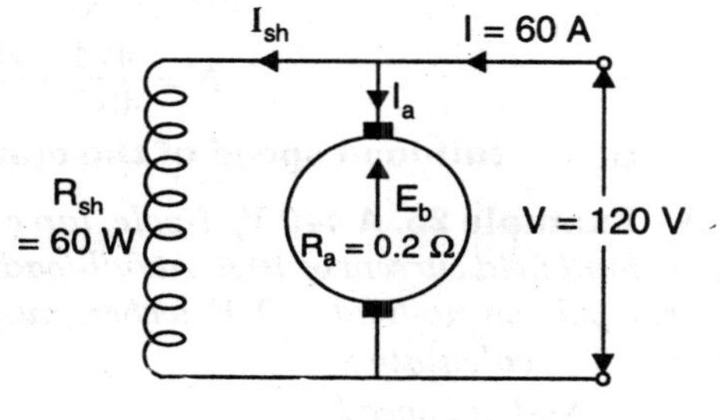

Fig. 64

$$I_{sh} = \frac{V}{R_{sh}} = \frac{120}{60} = 2\text{ A}$$

Full load armature current,

$$I_{a1} = I_1 - I_{sh} = 60 - 2 = 58\text{ A}$$

Back e.m.f. $E_{b1} = V - I_{a1} R_a$ – brush contact drop

$= 120 - 58 \times 0.2 - 3 = 105.4$ V

Armature current at half-load, $I_{a2} = \frac{I_{a1}}{2} = \frac{58}{2} = 29$ A

Back e.m.f. at half-load, $E_{b2} = V - I_{a2}R_a$ – brush contact drop

$= 120 - 29 \times 0.2 - 3 = 111.2$ V

Now, $$\frac{N_2}{N_1} = \frac{E_{b2}}{E_{b1}} \times \frac{\phi_1}{\phi_2} = \frac{E_{b2}}{E_{b1}}$$ [$\because$ $\phi_1 = \phi_2$ ...being shunt motor]

$\therefore$ $$\frac{N_2}{1800} = \frac{111.2}{105.4}$$

or $$\mathbf{N_2} = \frac{1800 \times 111.2}{105.4} = \mathbf{1899\ r.p.m.}\ \ \textbf{(Ans.)}$$

☞**Example 24.** *A 4-pole 500 V shunt motor takes 7 A on no-load, the no-load speed being 750 r.p.m. It has a shunt field current of 2 A. Calculate the full-load speed of the motor if it takes 122 A at full-load. Armature resistance = 0.2 ohm. Contact drop/brush = 1 V. Armature reaction weakens the field by 4% on full-load.*

Solution. Supply voltage, $V = 500$ Volts

No-load current, $I_0 = 7$ A

No-load speed, $N_0 = 750$ r.p.m.

Shunt field current, $I_{sh} = 2$ A

Full-load current, $I = 122$ A

Armature resistance, $R_a = 0.2$ ohm

Contact drop/ brush $= 1$ V

Flux at full-load, $\phi = \left(\frac{100-4}{100}\right)\phi_0 = 0.96\phi_0$

Full-load speed, N :

At no-load :

$$I_{a0} = I_0 - I_{sh} = 7 - 2 = 5 \text{ A}$$

$$E_{b0} = V - I_{a0}R_a - \text{contact drop at brushes}$$
$$= 500 - 5 \times 0.2 - 2 \times 1 = 497 \text{ V.}$$

At full-load :

$$I_a = I - I_{sh} = 122 - 2 = 120 \text{ A}$$

$$E_b = V - I_aR_a - \text{contact drop at brushes}$$
$$= 500 - 120 \times 0.2 - 2 \times 1 = 500 - 24 - 2 = 474 \text{ V}$$

We know that, $E_{b0} \propto \phi_0 N_0$

and $E_b \propto \phi N$

$$\therefore \quad \frac{E_b}{E_{b0}} = \frac{\phi}{\phi_0} \times \frac{N}{N_0}$$

$$\frac{474}{497} = \frac{0.96\phi_0}{\phi_0} \times \frac{N}{750}$$

$$\therefore \quad N = \frac{474 \times 750}{497 \times 0.96} = 745 \text{ r.p.m. (app.)}$$

Hence, **full-load speed of the motor = 745 r.p.m. (app.) (Ans.)**

☞**Example 25.** *A 440 V, 4-pole, lap connected shunt motor has a no-load input current of 15 A and a shunt field current of 10 A. At full-load it takes a current of 150 A. If armature resistance = 0.1 ohm, flux per pole on no-load = 0.05 weber, number of armature conductors = 750 and contact drop per brush = 1 V, calculate :*

(i) No-load speed *(ii) Full-load speed*

(iii) Speed regulation.

Armature reaction weakens the field by 1.5% on full-load.

Solution. Supply voltage, $V = 440$ Volts

Number of poles, $p = 4$

Number of parallel paths, $a = p = 4$ [Motor being lap-connected]

Armature resistance, $R_a = 0.1$ ohm

Flux per pole on no-load, $\phi_0 = 0.05$ weber

Number of armature conductors, $Z = 750$

Contact drop per brush, $= 1$ V

No-load input current, $I_0 = 15$ A
Shunt field current, $I_{sh} = 10$ A
Full-load current, $I = 150$ A

No-load speed, N_0 :
Full-load speed, N :
Speed regulation :

(*i*) **No-load speed :**

Armature current, $I_{a0} = I_0 - I_{sh} = 15 - 10 = 5$ A

$$E_{b0} = V - I_{a0}R_a - \text{brush contact drop}$$
$$= 440 - 5 \times 0.1 - 2 \times 1 = 437.5 \text{ V}$$

Using the relation : $$E_{b0} = \frac{p\phi_0 ZN_0}{60a}$$

or $$437.5 = \frac{4 \times 0.05 \times 750 \times N_0}{60 \times 4}$$

$\therefore$ $$N_0 = \frac{437.5 \times 60 \times 4}{4 \times 0.05 \times 750} = 700 \text{ r.p.m.}$$

Hence, **no-load speed = 700 r.p.m. (Ans.)**

(*ii*) **Full-load speed :**

Armature current, $I_a = I - I_{sh} = 150 - 10 = 140$ A

$$E_b = V - I_aR_a - \text{brush contact drop}$$
$$= 440 - 140 \times 0.1 - 2 \times 1 = 424 \text{ V}$$

$$\phi = (1 - 0.015)\,\phi_0$$ [Since armature reaction weakens the field by 1.5 per cent]

$$= 0.985\,\phi_0$$

Using the relation : $$\frac{N}{N_0} = \frac{E_b}{E_{b0}} \times \frac{\phi_0}{\phi}$$

or $$\frac{N}{700} = \frac{424}{437.5} \times \frac{\phi_0}{0.985\phi_0}$$

$\therefore$ $$N = 700 \times \frac{424}{437.5} \times \frac{\phi_0}{0.985\phi_0} = 688.73 \text{ r.p.m.}$$

Hence, **full-load speed = 688.73 r.p.m. (Ans.)**

(*iii*) **Percentage speed regulation**

$$= \frac{\text{no-load speed} - \text{full-load speed}}{\text{full-load speed}} \times 100$$
$$= \frac{700 - 688.73}{688.73} \times 100 = 1.637$$

Hence, **percentage speed regulation = 1.637%. (Ans.)**

Example 26. *A 250 V shunt motor takes a line current of 60 A and runs at 800 r.p.m. Its armature and field resistances are 0.2 ohm and 125 ohms respectively. Contact drop/brush = 1 V. Calculate :*

(*i*) *No-load speed if the no-load current is 6 A.*

(*ii*) *The percentage reduction in the flux per pole in order that the speed may be 1000 r.p.m. when the armature current is 40 A.*

Neglect the effects of armature reaction.

Solution. Supply voltage, $V = 250$ Volts

Load current, $I_1 = 60$ A

Load speed, $N_1 = 800$ r.p.m.

Armature resistance, $R_a = 0.2$ ohm

Shunt field resistance, $R_{sh} = 125$ ohms

Contact drop $= 2$ V

Armature current, $I_{a2} = 40$ A

Load speed, $N_2 = 1000$ r.p.m.

No-load current, $I_0 = 6$ A

No-load speed, N_0 :

Percentage reduction in flux/pole :

Shunt field current, $I_{sh} = \dfrac{250}{125} = 2$ A

At no-load :

$$I_{a0} = I_0 - I_{sh} = 6 - 2 = 4 \text{ A}$$

$$E_{b0} = V - I_{a0}R_a - \text{brush drop}$$

$$= 250 - 4 \times 0.2 - 2 \times 1 = 247.2 \text{ V}$$

At load ($I_1 = 60$ A) :

$$I_{a1} = I_1 - I_{sh} = 60 - 2 = 58 \text{ A}$$

$$E_{b1} = V - I_{a1}R_a - \text{brush drop}$$

$$= 250 - 58 \times 0.2 - 2 \times 1 = 236.4 \text{ V}$$

Using the relation, $\dfrac{N_1}{N_0} = \dfrac{E_{b1}}{E_{b0}} \times \dfrac{\phi_0}{\phi_1}$

Since the motor is shunt wound and armature reaction effect is negligible so *flux may be assumed to be constant* and therefore speed is directly proportional to the back e.m.f. developed.

$$\therefore \quad \frac{N_1}{N_0} = \frac{E_{b1}}{E_{b0}}$$

$$\frac{800}{N_0} = \frac{236.4}{247.2}$$

$$N_0 = \frac{800 \times 247.2}{236.4} = 836.5 \text{ r.p.m.}$$

Hence, **no-load speed = 836.5 r.p.m. (Ans.)**

At armature current, $I_{a2} = 40$ A

Let the flux be reduced to ϕ_2

$$E_{b2} = V - I_{a2}R_a - \text{brush drop}$$

$$= 250 - 40 \times 0.2 - 2 \times 1 = 240 \text{ V}$$

Using the relation, $\dfrac{N_2}{N_1} = \dfrac{E_{b2}}{E_{b1}} \times \dfrac{\phi_1}{\phi_2}$

or
$$\frac{1000}{800} = \frac{240}{236.4} \times \frac{\phi_1}{\phi_2}$$

$$\therefore \quad \frac{\phi_2}{\phi_1} = \frac{240 \times 800}{1000 \times 236.4} = 0.812$$

∴ Percentage reduction in flux

$$= \frac{\phi_1 - \phi_2}{\phi_1} \times 100 = \left(1 - \frac{\phi_2}{\phi_1}\right) \times 100 = (1 - 0.812) \times 100 = 18.8\%$$

Hence, **%age reduction in flux = 18.8%. (Ans.)**

☞**Example 27.** *A 220 V shunt motor takes 60 A when running at 800 r.p.m. It has an armature resistance of 0.1 ohm. Determine the speed and armature current if the magnetic flux is weakened by 20%. Contact drop per brush = 1 V.*

Total torque developed remains constant.

Solution. Supply voltage $= 220$ V

Load current, $= I_1 = I_{a1} = 60$ A [Neglecting shunt field current]

Speed at 60 *A*, $N_1 = 800$ r.p.m.

Armature resistance, $R_a = 0.1$ ohm

Contact drop/brush $= 1$ V

$\mathbf{N_2} = ?$, $\mathbf{I_{a2}} = ?$

$$\phi_2 = \left(1 - \frac{20}{100}\right)\phi_1 \quad \text{[Since magnetic flux is weakened by 20\%]}$$

$$= (1 - 0.20)\phi_1 = 0.8\phi_1$$

Since $T \propto \phi I_a$

and torque remains constant [Given]

$$\therefore \quad \phi_1 I_{a1} = \phi_2 I_{a2}$$

$$\phi_1 \times 60 = 0.8\phi_1 I_{a2}$$

$$\therefore \quad I_{a2} = \mathbf{75\ A.}$$

Back e.m.f., $E_{b1} = V - I_{a1}R_a -$ brush drop

$= 200 - 60 \times 0.1 - 2 \times 1 = 212$ V

Back e.m.f., $E_{b2} = V - I_{a2}R_a -$ brush drop

$= 220 - 75 \times 0.1 - 2 \times 1 = 210.5$ V

Using the relation, $\dfrac{N_2}{N_1} = \dfrac{E_{b2}}{E_{b1}} \times \dfrac{\phi_1}{\phi_2}$

$$\frac{N_2}{800} = \frac{210.5}{212} \times \frac{\phi_1}{0.8\phi_1}$$

$$\therefore \quad N_2 = \frac{800 \times 210.5}{212 \times 0.8} = \mathbf{992.9\ r.p.m.}$$

Hence, **Speed = 992.9 r.p.m. (Ans.)**

Armature current = 75 A. (Ans.)

☞**Example 28.** *A 440 V shunt motor takes 105 A (armature current) from the supply and runs at 1000 r.p.m. Its armature resistance is 0.15 ohm. If total torque developed is unchanged, calculate the speed and armature current if the magnetic field is reduced to 70% of the initial value.*

(Panjab University)

Solution. Supply voltage, $V = 440$ Volts

Armature current, $I_{a1} = 105$ A

Speed, $N_1 = 1000$ r.p.m.

Armature resistance, $R_a = 0.15$ ohm

$$\phi_2 = 0.7\ \phi_1.$$

(*i*) **Armature current, I_{a2} :**

As the torque developed is same

$$\therefore \quad \phi_1 I_{a1} = \phi_2 I_{a2}$$

$$I_{a2} = \frac{\phi_1}{\phi_2} \times I_{a1} = \frac{\phi_1}{0.7\phi_1} \times 105 = 150 \text{ A}$$

Hence, **armature current = 150 A. (Ans.)**

Now, $E_{b1} = V - I_{a1}R_a = 440 - 105 \times 0.15 = 424.25$

and $E_{b2} = V - I_{a2}R_a = 440 - 150 \times 0.15 = 417.5 \text{ V}.$

(*ii*) **Speed, N_2 :**

Using the relation,
$$\frac{N_2}{N_1} = \frac{E_{b2}}{E_{b1}} \times \frac{\phi_1}{\phi_2}$$

$$\frac{N_2}{1000} = \frac{417.5}{424.25} \times \frac{\phi_1}{0.7\phi_1}$$

$$\therefore \quad N_2 = 1000 \times \frac{417.5}{424.25} \times \frac{1}{0.7} = 1405 \text{ r.p.m.}$$

Hence, **speed of the motor when magnetic field is reduced to 70% of initial value = 1405 r.p.m. (Ans.)**

SERIES MOTORS

Example 29. *A 230 V series motor develops torque of 310 N-m at 800 r.p.m. The torque lost due to iron and friction loss is 10 N-m. If the efficiency is 85% determine the current taken by the motor at 800 r.p.m.*

Solution. Supply voltage $= 230$ V

Torque developed, $T_a = 310$ N-m

Torque lost due to iron and friction, $T_{lost} = 10$ N-m

Efficiency $= 85\%$

Speed of the motor, $N = 800$ r.p.m.

Current taken by motor at 800 r.p.m. :

$$T_{useful} = T_a - T_{lost} = 310 - 10 = 300 \text{ N-m}$$

Watts corresponding to useful torque

$$= \frac{T_{useful} \times 2\pi N}{60} = \frac{300 \times 2\pi \times 800}{60} = 25136 \text{ W}$$

Power in watts taken from mains $= \dfrac{25136}{0.85} = 29571.7$ W

Current taken by motor $= \dfrac{29571.7}{230} = 128.57$ A.

Example 30. *A 8-pole 240 V lap-wound, series motor has armature and series field resistances of 0.2 ohm and 0.02 ohm respectively. There are 660 armature conductors. If the flux/pole is 0.3 Wb and the total torque developed in the armature is 320 N-m, find the current taken by the motor and its speed.*

Solution. Number of poles, $p = 4$

Number of parallel paths, $a = p = 4$ [Motor being lap-wound.]

Armature resistance, $R_a = 0.2$ ohm

Series field resistance, $R_{se} = 0.02$ ohm

Number of armature conductors, $Z = 660$
Flux/pole, $\phi = 0.03$ Wb
Total torque developed, $I_a = 320$ N-m

(*i*) **Current taken by motor, I_a :**

$$E_b = \frac{p\phi ZN}{60a}$$

$$\therefore \quad \frac{E_b}{N} = \frac{p\phi Z}{60a} = \frac{4 \times 0.03 \times 660}{60 \times 4} = 0.33$$

Also, $$T_a \times \left(\frac{2\pi N}{60}\right) = E_b I_a$$

or $$T_a = \frac{E_b}{N} \cdot \frac{60 I_a}{2\pi} = 0.33 \times \frac{60 \times I_a}{2\pi}$$

$$\therefore \quad 320 = \frac{0.33 \times 60 \times I_a}{2\pi}$$

or $$I_a = \frac{320 \times 2\pi}{0.33 \times 60} = 101.56 \text{ A}$$

Hence, current taken by motor **= 101.56 A. (Ans.)**

(*ii*) **Speed of motor, N :**

$$E_b = V - I_a(R_a + R_{se})$$
$$= 240 - 101.56(0.2 + 0.02) = 217.66 \text{ V}$$

But $$\frac{E_b}{N} = 0.33$$ [Already calculated above]

$$\therefore \quad N = \frac{E_b}{0.33} = \frac{217.66}{0.33} = 659.6 \text{ r.p.m.}$$

☞**Example 31.** *A six-pole, lap-wound 400 V series motor has the following data : Number of armature conductors = 920, flux / pole = 0.045 Wb, total motor resistance = 0.6 ohm, iron and friction losses = 2 kW. If the current taken by the motor is 90 A, find :*

(*i*) *Total torque ;* (*ii*) *Useful torque at the shaft ;*
(*iii*) *Power output ;*
(*iv*) *Pull at the rim of a pulley of 40 cm diameter connected to the shaft.*

Solution. Number of poles, $p = 6$
Supply voltage, $V = 400$ Volts
Number of parallel paths, $a = p = 6$ [Motor being lap-wound]
Number of armature conductors, $Z = 920$
Flux/pole, $\phi = 0.045$ Wb
Motor resistance, $R_m = 0.6$ ohm
Iron and friction losses = 2 kW or 2000 W
Current taken by the motor, $I_a = 90$ A
Radius of the pulley, = 40/2 = 20 cm or 0.2 m
Using the relation, $E_b = V - I_a R_m = 400 - 90 \times 0.6$
$= 346$ V

Also, $$E_b = \frac{p\phi ZN}{60a}$$

$$346 = \frac{6 \times 0.045 \times 920 \times N}{60 \times 6}$$

$$\therefore \quad N = \frac{346 \times 60 \times 6}{6 \times 0.045 \times 920} = 501 \text{ r.p.m.}$$

(*i*) **Total torque, T_a :**

We know that, $T_a = 0.159 \times Z\phi p \times \left(\frac{I_a}{a}\right)$

$$= 0.159 \times 920 \times 0.045 \times 6 \times \frac{90}{6} = 592.4 \text{ Nm}$$

Hence, **Total torque = 592.04 N-m. (Ans.)**

(*ii*) **Useful torque, T_{useful} :**

$$T_{lost} \times \left(\frac{2\pi N}{60}\right) = \text{Iron and friction loss} = 2000 \text{ W}$$

$$\therefore \quad T_{lost} = \frac{2000 \times 60}{2\pi N} = \frac{2000 \times 60}{2\pi \times 501} = 38.11 \text{ Nm}$$

$$\therefore \quad T_{useful} = T_a - T_{lost} = 592.4 - 38.11$$

$$= 554.29 \text{ Nm}$$

Hence, **useful torque = 554.29 Nm. (Ans.)**

(*iii*) **Power output :**

Power output $= T_{useful}\left(\frac{2\pi N}{60}\right) = \frac{554.29 \times 2\pi \times 501}{60}$

$$= 29084.4 \text{ or } 29.08 \text{ kW}$$

Hence, **power output = 29.08 kW. (Ans.)**

(*iv*) If F is the pull at the rim of the pulley and r is the radius,

Torque at the shaft, $T_{useful} = F \times r$

i.e. $F \times 0.2 = 554.29$

$\therefore \quad F = 2771.45$ N

Hence, **pull at the rim of the pulley = 2771.45 N. (Ans.)**

Example 32. *A 240 V series motor takes 40 A when giving its rated output at 1500 r.p.m. Its resistance is 0.3 Ω. Find what resistance must be added to obtain rated torque (i) at starting and (ii) at 1000 r.p.m.*

Solution. *Given* : $V = 240$ volts ; $I (= I_a) = 40$ A, $N = 1500$ r.p.m., $R = 0.3\ \Omega$.

(*i*) **Resistance to be added to obtain rated torque at starting, R_{add} :**

Since the torque remains the same in both the cases, it is obvious that the current drawn by the motor remains constant at 40 A.

$$\therefore \quad 40 = \frac{240}{R_{add} + 0.3}$$

or $$R_{add} = \frac{240}{40} - 0.3 = 5.7\ \Omega.$$

(*ii*) **Resistance to be added to obtain rated torque at 1000 r.p.m., R_{add} :**

We know that, $\frac{N_2}{N_1} = \frac{E_{b2}}{E_{b1}}$

$$E_{b1} = 240 - 40 \times 0.3 = 228 \text{ V}$$

$$\therefore \quad \frac{1000}{1500} = \frac{E_{b2}}{228}$$

or $$E_{b2} = \frac{1000 \times 228}{1500} = 152 \text{ V}$$

Now $$E_{b2} = V - 40(R_{add} + 0.3) \quad \text{or} \quad 152 = 240 - 40(R_{add} + 0.3)$$

∴ $$\mathbf{R_{add}} = \frac{240 - 152}{40} - 0.3 = \mathbf{1.9\ \Omega.} \quad \textbf{(Ans.)}$$

Example 33. *A 200 V D.C. series motor runs at 700 r.p.m. when operating at its full-load current of 20 A. The motor resistance is 0.5 Ω and the magnetic circuit can be assumed unsaturated. What will be the speed if (i) the load torque is increased by 44% (ii) the motor current is 10 A.*

Solution. *Given* : $V = 200$ volts ; $N_1 = 700$ r.p.m., $I_1 = 20$ A, $R_m = (R_a + R_{sh}) = 0.5\ \Omega$

Speeds, (N_2, N_3) :

(*i*) **When the load torque is increased by 44% :**

$$T_2 = 1.44T_1$$

$$\phi_2 I_2 = 1.44\ \phi_1 I_1 \qquad (\because\ T \propto \phi I_a)$$

$$I_2^2 = 1.44\ I_1^2 \qquad (\because\ \phi \propto I)$$

or $$I_2 = I_1\sqrt{1.44} = 20 \times 1.2 = 24 \text{ A.}$$

Back e.m.f. $$E_{b1} = V - I_1(R_a + R_{se}) = 200 - 20 \times 0.5 = 190 \text{ V}$$

Back e.m.f. $$E_{b2} = V - I_2(R_a + R_{se}) = 200 - 24 \times 0.5 = 188 \text{ V}$$

Also $$\frac{N_2}{N_1} = \frac{E_{b2}}{E_{b1}} \times \frac{\phi_1}{\phi_2}$$

or $$\frac{N_2}{700} = \frac{188}{190} \times \frac{20}{24}$$

or $$\mathbf{N_2 = 577.2\ r.p.m.} \quad \textbf{(Ans.)}$$

(*ii*) **When the motor current is 10 A :**

Back e.m.f., $$E_{b3} = V - I_3(R_a + R_{se})$$
$$= 200 - 10 \times 0.5 = 195 \text{ V}$$

Also, $$\frac{N_3}{N_1} = \frac{E_{b3}}{E_{b1}} \times \frac{\phi_1}{\phi_3}$$

$$\frac{N_3}{700} = \frac{195}{190} \times \frac{20}{10}$$

or $$\mathbf{N_3 \simeq 1437\ r.p.m.} \quad \textbf{(Ans.)}$$

Example 34. *A 220 V D.C. series motor draws full-load line current of 38 A at the rated speed of 600 r.p.m. The motor has armature resistance of 0.4 Ω and the series field resistance is 0.2 Ω. The brush voltage drop irrespective of load is 3.0 volts, find :*

(*i*) *The speed of the motor when the load current drops to 19 A.*

(*ii*) *The speed on removal of load when the motor takes only 1 A from supply.*

(*iii*) *The internal horse power developed in each of the above cases. Neglect the effect of armature reaction and saturation.*

Solution. *Given* : $V = 220$ volts ; $I_1 = 38$ A, $N_1 = 600$ r.p.m., $R_a = 0.4\ \Omega$, $R_{se} = 0.2\ \Omega$, Brush drop = 3 V ; $I_2 = 19$ A, $I_0 = 1$ A.

(*i*) **Speed of the motor N_2 (at 19 A)**

On full-load :

Back e.m.f., $$E_{b1} = V - I_1(R_a + R_{se}) - \text{brush voltage drop}$$
$$= 220 - 38(0.4 + 0.2) - 3 = 194.2 \text{ V.}$$

On-load current of 19 A

Back e.m.f., $E_{b2} = 220 - 19(0.4 + 0.2) - 3 = 205.6$ V

No-load, back e.m.f., $E_{b0} = 220 - 1(0.4 + 0.2) - 3 = 216.4$ V

We know that $\dfrac{N_2}{N_1} = \dfrac{E_{b2}}{E_{b1}} \times \dfrac{\phi_1}{\phi_2}$

$$\frac{N_2}{600} = \frac{205.6}{194.2} \times \frac{I_1}{I_2} \quad \left[\text{Assuming unsaturated field } i.e. \ \frac{\phi_1}{\phi_2} = \frac{I_1}{I_2}\right]$$

$\therefore$ $$\mathbf{N_2} = 600 \times \frac{205.6}{194.2} \times \frac{38}{19} = \mathbf{1270 \text{ r.p.m.}} \quad \textbf{(Ans.)}$$

(*ii*) $\mathbf{N_0}$ **(at 1 A) :**

Now, $$\frac{N_0}{N_1} = \frac{E_{b0}}{E_{b1}} \times \frac{\phi_1}{\phi_0}$$

$$= \frac{E_{b0}}{E_{b1}} \times \frac{I_1}{I_0}$$

$\therefore$ $$N_0 = N_1 \times \frac{E_{b0}}{E_{b1}} \times \frac{I_1}{I_0} = 600 \times \frac{216.4}{194.2} \times \frac{38}{1} = \mathbf{25406 \text{ r.p.m.}} \quad \textbf{(Ans.)}$$

(*iii*) **Internal H.P. developed (in the above cases) :**

Internal H.P. developed at a load current 38 A

$$= \frac{194.2 \times 38}{735.5} = \mathbf{10.} \quad \textbf{(Ans.)}$$

Internal H.P. developed at a load current of 19 A

$$= \frac{205.6 \times 19}{735.5} = \mathbf{5.31.} \quad \textbf{(Ans.)}$$

Internal H.P. developed at no load current of 1 A

$$= \frac{216.4 \times 1}{735.5} = \mathbf{0.294.} \quad \textbf{(Ans.)}$$

Example 35. *A D.C. series motor draws a line current of 100 A from the mains while running at 1000 r.p.m. Its armature resistance is 0.15 Ω and the field resistance is 0.1 Ω. Assuming that the flux corresponding to a current of 25 A is 40% of that corresponding to 100 A, find the speed of the motor when it is drawing 25 A from 230 V supply.* **(GATE, 1996)**

Solution. *Given :* $V = 230$ volts ; $I_1 = 100$ A ; $N_1 = 1000$ r.p.m. ;

$R_a = 0.15\ \Omega$; $R_{se} = 0.1\ \Omega$, $I_2 = 25$ A ; $\phi_2 = 0.4\phi_1$

Speed, $\mathbf{N_2}$ **:**

While drawing line current of 100 A,

$$E_{b1} = V - I_1(R_a + R_{se})$$
$$= 230 - 100(0.15 + 0.1) = 205 \text{ V}$$

While drawing line current of 25 A,

$$E_{b2} = V - I_2(R_a + R_{se})$$
$$= 230 - 25(0.15 + 0.1) = 223.75 \text{ V}$$

We know that, $$\frac{N_2}{N_1} = \frac{E_{b2}}{E_{b1}} \times \frac{\phi_1}{\phi_2} \quad \text{or} \quad \frac{N_2}{1000} = \frac{223.75}{205} \times \frac{\phi_1}{0.4\phi_1}$$

$\therefore$ $$\mathbf{N_2} = 1000 \times \frac{223.75}{205} \times \frac{1}{0.4} = \mathbf{2729 \text{ r.p.m.}} \quad \textbf{(Ans.)}$$

7.10. Motor Characteristics

The properties of all motors and, in particular, D.C. motors are defined as a totality of the following characteristics :

(*i*) *Starting* ; (*ii*) *Operating and mechanical* ;
(*iii*) *Braking ; and* (*iv*) *Regulation.*

Starting characteristics. The starting characteristics *determine the operation of starting from the moment the motor begins running to the moment when steady-state operation is established* and *include* :

- The starting current I_{start} generally determined by the ratio $\frac{I_{start}}{I_{run}}$;
- The starting torque T_{start}, determined by the ratio $\frac{T_{start}}{T_{run}}$;
- The duration of starting t_{start} ;
- The economy of operation determined by the amount of energy consumed in starting ; and
- The cost and reliability of the starting equipment.

Operating characteristics. The operating characteristics are those that give the relation between *speed, torque* and *efficiency* as *functions* of the *useful power* or the *armature* current for V = constant and constant resistances in the armature and field circuit.

Mechanical characteristics. Of major importance for industrial drive mechanisms are the mechanical characteristics, which are the relation $N = f(T)$ (where N and T stand for speed and torque respectively) for conditions of constant voltage and resistances in the armature and field circuits. These also include the *braking characteristics.*

Regulation characteristics. These characteristics determine the *properties of motors when their speed is controlled.* These include :

- The regulation range determined by the ratio $\frac{N_{max}}{N_{min}}$;
- The efficiency of regulation from the point of view of the initial cost of the equipment and maintenance ;
- The nature of regulation—continuous or stepped ; and
- The simplicity of the control apparatus and methods.

The *D.C. motors possess versatile* and diverse regulation characteristics, and for this reason are indispensable in installations where wide-range control of speed is necessary.

The characteristic curves of a motor are those curves which show relation between the following quantities :

1. *Torque and armature current i.e.,* T_a/I_a characteristic. This is also known as **electrical characteristic.**

2. *Speed* and *armature current i.e.,* N/I_a characteristic.

3. *Speed* and *torque i.e.,* N/T_a characteristic. This is also known as **mechanical characteristic.** This can be obtained from (1) and (2) above.

Following relations are worth *keeping in mind* while discussing motor characteristics :

$$N \propto \frac{E_b}{\phi} \text{ and } T_a \propto \phi I_a.$$

7.10.1. Torque-current characteristics

Shunt motor :

- When running on *no-load,* a small armature current flows to supply the field and to *drive the machine against the friction and other losses in it.*
- As the load is applied to the motor, and is increased, the torque rises *almost proportionally to the increase in current*. This is not quite true, because the flux has been assumed to be constant, whereas it decreases slightly owing to armature reaction. The effect of this is to cause the top of the curve connecting torque and line current to *bend over* as shown in Fig. 65.
- The starting torque of a motor is determined by the starting resistance, which in turn, governs the initial current through the machine when the main switch is closed. At this moment the speed is zero, so that the back e.m.f. is zero and the starting current is given by $I = V/R$, where V is the supply voltage and R is the total resistance, which includes the armature and starting resistance.

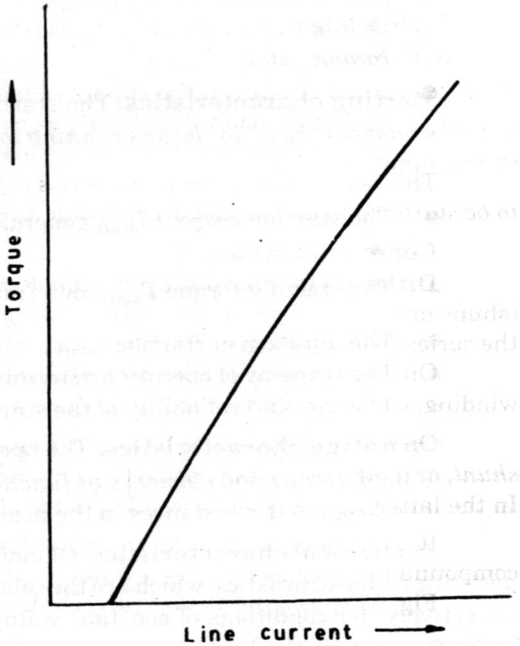

Fig. 65. Torque-current characteristic of a shunt motor.

If the starting current is *limited* by *heating considerations* to twice the full load current, then with normal supply voltage the starting torque of a shunt motor is twice the full load torque. If, however, the supply voltage is below normal, the flux is also less than twice full load torque. The importance of this will be appreciated when the starting torque of a series motor is compared with that of the shunt motor.

Series motor :

- In a series motor the torque ($T_a \propto \phi I_a$) increases much more than does the armature current. This is because the flux itself increases with the armature current, though, owing to the magnetic saturation, the two are not strictly proportional. Nevertheless, for all *but the heavy loads which tend to produce saturation of the field system*, it may be said that the *torque is approximately proportional to the square of the load current.*

Fig. 66 shows the relationship between torque and current. Here the current commences at the no-load value, rises *parabolically at first*, but *increases more slowly* as the *effects of armature reaction and magnetic saturation becomes appreciable*.

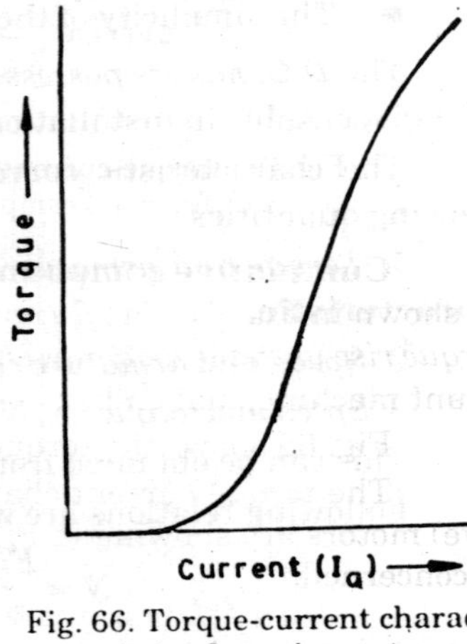

Fig. 66. Torque-current characteristic of a series motor.

- This property of a series motor, by virtue of which a *heavy current gives rise to a very high torque,* also influences its starting characteristics. In a case of a shunt motor, it has already been seen, that the current at the moment of starting may be as high as twice the full-load value ; if we allow for the armature of the magnetisation characteristic and for weakening effect of armature reaction and assume that the flux is increased to 1.5 times its full-load value, then it is obvious that the *starting torque of a series motor is three times the full-load torque.*

- Further more, if the supply voltage falls, the starting current may still be maintained at twice full-load value by cutting out some of the starting resistance, so that the high value of starting torque may still be maintained.

This type of motor (series motor) is *superior to shunt motor for drives in which machines have to be started and accelerated from rest when fully loaded,* as *is the case with traction equipment.*

Compound motors :

Differential compound motor. Refer Fig. 60. In this type of motor the two field windings (shunt and series) *oppose* each other. On *light loads,* such a machine *runs as a shunt motor,* since the series field winding, carrying only a small current, has relatively little effect.

On *heavy loads,* the series coils strengthen and since they are in opposition to the shunt winding, cause a *reduction in the flux* and a consequence *decrease in torque.*

On *heavy overloads* or *when starting up on load, the series winding may become as strong as shunt,* or *it may even predominate, in which the torque will be reduced to zero* or *may even be reversed.* In the latter case, the motor would tend to start-up in the *wrong direction.*

It is obvious that such characteristics may cause *dangerous results,* so that differential compound motors have only *very limited applications in practice.*

Fig. 67 shows the torque-current characteristic of a differential compound motor.

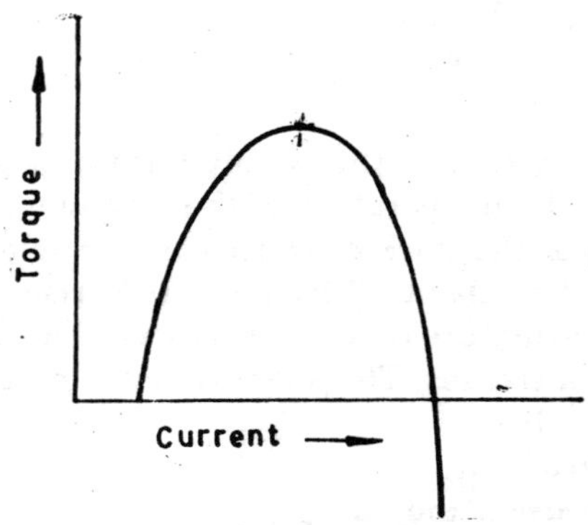

Fig. 67. Torque-current characteristic of a differential compound motor.

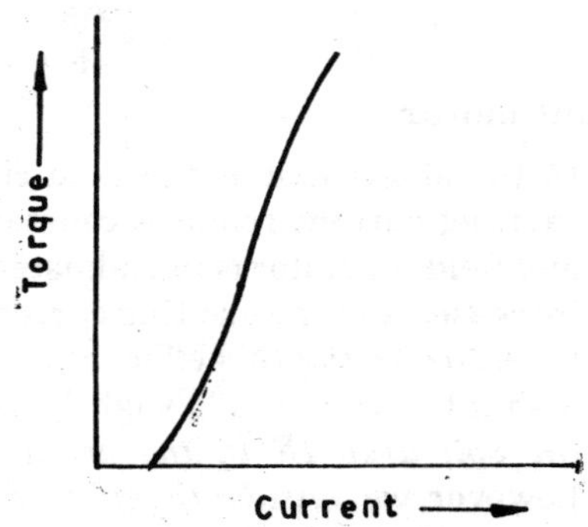

Fig. 68. Torque-current characteristic of a cumulative compound motor.

Cumulative compound motor. In this type of motor the two field windings *assist* each other as shown in Fig. 61. The *flux on no-load is that due to the shunt-winding, while on load the flux and torque rise with the load current.* The torque, therefore, increase more rapidly than in the case of the shunt machine, and on heavy loads it resembles the characteristic of the series motor.

Fig. 68 shows the torque-current characteristic of a cumulative compound motor.

The torque-current characteristics of shunt, series and compound (differential and cumulative) motors are showing in Fig. 69 from which their properties, may be compared, as far as torque is concerned.

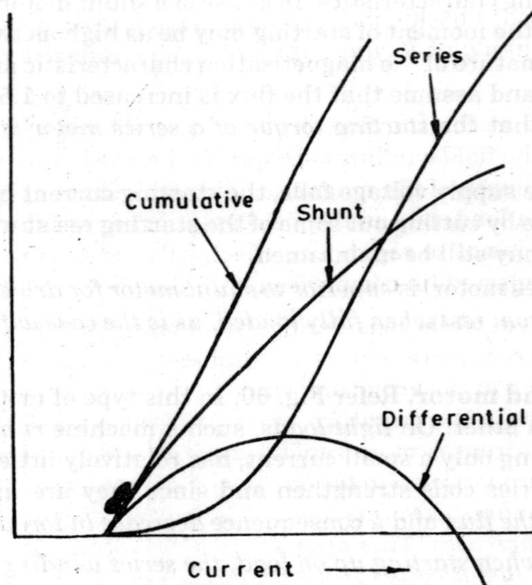

Fig. 69. Torque-current characteristics of shunt, series and compound motors.

7.10.2. Speed-current characteristics. The speed-current characteristics of various motors can be deduced from the following *motor equation,*

$$N \propto \frac{E_b}{\phi} \quad \text{or} \quad N \propto \frac{V - IR}{\phi}.$$

Shunt motor :

- In the shunt motor, the field circuit is connected to the supply terminals so that the exciting current remains constant as long the temperature of the machine is constant, and field regulator is not adjusted. Actually as the machine warms up, the field resistance increases and the exciting current decreases by about 4% for every 10°C rise in temperature. *Neglecting* this effect and also due to armature reaction, it is seen that the speed of a shunt machine falls slightly as the load increases. The *fall in speed is proportional to the volt drop IR in the armature circuit.* If, however, we *consider the effect of armature reaction, an increase of load causes a slight decrease in flux, unless the machine is fitted with compensating windings.* This weakening of the field *tends to raise the speed,* so that the actual fall in speed is less than that calculated by a consideration of the volt drop in the armature.

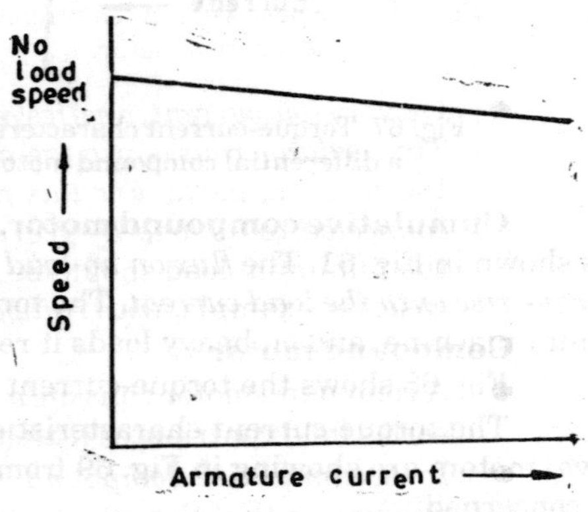

Fig. 70. Speed-current characteristic of a shunt motor.

On the whole, the shunt motor may be regarded as one in which the *speed is approximately constant,* falling slightly as the load increases (*see* Fig. 70).

- The *speed* of a shunt machine can be *increased by inserting resistance* in the field by means of a field regulator. This weakens the field and causes the motor to run faster in order to generate the necessary back e.m.f. of course, it is *impossible to reduce the speed by this method below that at which it runs with no field resistance in the circuit.*

Series motor :

- In case of a series motor the flux does not remain constant, or even approximately constant, because the field winding is in series with the load, so that as the load increases so also does the strength of the magnetic field. At first the flux increases approximately in proportion to the load, but as the field approaches saturation, owing to the heavier loads, the increase is not so rapid. The effects of temperature changes and of armature reaction may be neglected (in comparison with the above mentioned effect).
- It will be appreciated, while considering motor equation, that the back e.m.f. decreases as the armature current increases, as in shunt motor ; in the latter, however, the decrease is due to the volt drop in the armature, while in the series machine the loss in volts occurs in the field as well as in the armature, since they are in series. The back e.m.f. in a series motor, therefore, decreases more rapidly than in a corresponding shunt machine. *The speed,* however, is *proportional to the back e.m.f. divided by the flux ; the former decreases, while the latter increases with increasing load so that the speed decreases rapidly as the armature current increases.* This *property is a valuable feature in a drive of which the speed is required automatically to adjust to compensate for changes in load.*

The speed-current characteristic of a series motor are shown in Fig. 71 (*a*).

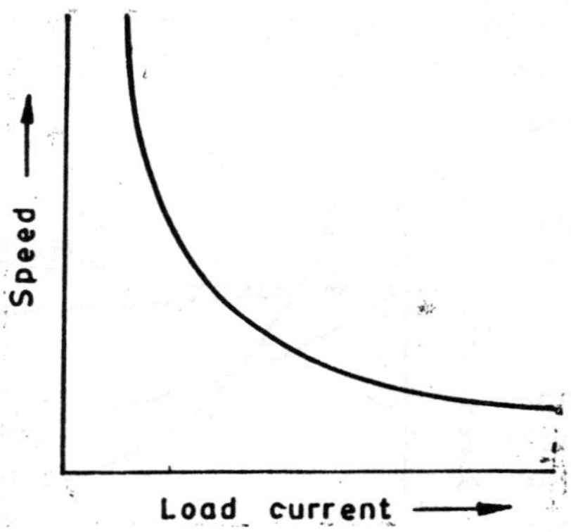

Fig. 71. (*a*) Speed-current characteristic of a series motor.

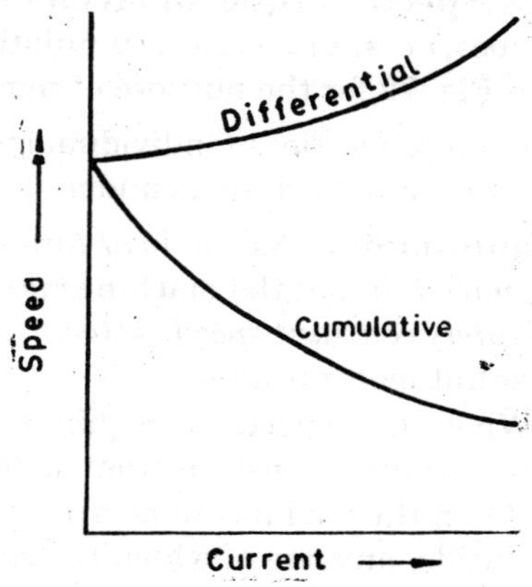

Fig. 71. (*b*) Speed-current characteristic of compound motors.

- On very *low current*, a *series motor runs at very high speeds*, or *tends to race*, as it is termed. This is *dangerous,* since the machine may be destroyed by the centrifugal forces set up in the rotating parts. *For this reason, when installing a series motor it must be positively connected to its load by gearing or by direct connection and never by belting.* Moreover, the minimum load should be great enough to keep the speed within safe limits, as is the case, for example, with *railway motors, hoists* and *rolling mills.*

Compound motor :

- A compound motor runs on '*no-load*' at a speed determined by its shunt winding since the series field contributes little to the total flux in this condition.
- In a *cumulative compound motor at 'no-load*', the series field strengthens the shunt winding so that the speed falls as in a series machine. Since the flux at any load is equal to the shunt and series fluxes, the speed is less than it would be if running on either field

alone. The speed-current characteristic of such a machine is shown in Fig. 71 (*b*). Such a characteristic has two important **advantages.** These are :

(*i*) *The machine has the compensating action of reducing its speed on heavy loads, as is the case with series machine.*

(*ii*) The maximum speed on no-load is limited by the shunt winding, since this produces a maximum flux even on no-load.

This type of motor, therefore, is suitable for driving machines which operate *on a cycle consisting of a power or working stroke followed by a return or idle stroke.* The series winding produces a fall in speed on the working stroke, while the shunt winding permits the return stroke to be completed at a high, but safe, speed. *A fly-wheel is also provided to act as a load equilizer in such a drive.*

In case of a *differential compound motor,* since the series winding opposes the shunt, the resultant flux decreases as the load increases ; thus the machine runs at a higher speed than it would do as a shunt motor. If the series windings were relatively weak, this reduction in flux might be just sufficient for the fall in speed, brought about by the volt drop in the machine. *Such a motor would have a useful application in driving loads at a constant speed.* If the series field were strong, however, an increase in load would result in a decrease in the magnetic flux and a rise in speed would take place as shown in Fig. 49, the heavier the load, the faster would the motor tend to run. This is the property which may have *dangerous consequences,* since a heavy overload would result in such a high speed that the motor would destroy itself.

7.10.3. Speed-Torque (or Mechanical) characteristics. The speed-torque characteristics of the four types, *i.e.* shunt, series, cumulative and differential of motors drawn on the same diagram are shown in Fig. 72 for the purpose of comparison.

The main properties of individual motors, from this diagram, may be summarised as under :

1. **Shunt motor.** As the *load torque increases* the *speed falls somewhat,* but the machine *may be regarded as an approximately constant speed motor.*

The shunt motor is *used* :

- When the speed is required to remain approximately constant from no-load to full-load.
- When the load has to be driven at a number of speeds, any one of which is required to remain approximately constant.

2. **Series motor.** As the *load torque increases the speed falls rapidly.* At low torque the speed becomes very high and machine tends to race.

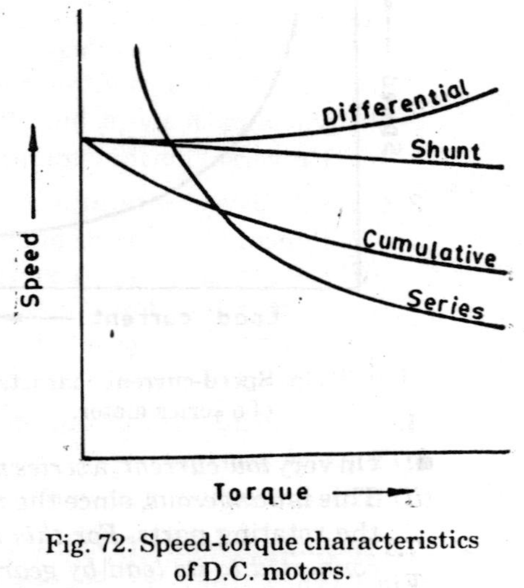

Fig. 72. Speed-torque characteristics of D.C. motors.

The series motors are used :

- When *large starting torque is required* (as in *traction motors*).
- When the load is subject to heavy fluctuations, and a reduced speed is desired to compensate for the high torque, provided that there is no possibility of the machine *'losing'* its load.

3. **Cumulative compound motor.** In this type of motor the *speed falls appreciably as the torque increases,* but on low torques the maximum speed is limited to a safe value. These motors are used :

- *When a large starting torque is required but when the load may fall so low that a series motor would race.*

- *When the load is of a fluctuating nature and a reduced speed is desirable on the heavy loads.*

In such a case a flywheel is usually fitted so that when speed is so reduced the kinetic energy stored in the flywheel at high speeds is given up to assist the motor in driving the heavy load.

- *When the supply voltage is subject to fluctuations* (as in traction systems).

4. **Differential compound motor.** The speed at low torque is limited by the shunt winding, as in the cumulative compound machine. At *high torques*, the speed may be arranged to remain constant or, with a stronger series field, the speed may rise with increasing load.

On *very heavy loads the machine may tend to race.*

Its use is usually restricted to applications *which require a constant speed.*

Industrial applications of D.C. Motors :

1. Shunt motors :

(*i*) Drills and milling machines
(*ii*) Line-shaft drives
(*iii*) Boring mills
(*iv*) Grinders and shapers
(*v*) Spinning and weaving machines
(*vi*) Wood working machines
(*vii*) Small printing presses
(*viii*) Light machine tools generally.

2. Series motors :

(*i*) Traction drives generally
(*ii*) Tram cars and railway cars
(*iii*) Cranes, derricks, hoists, elevators and winches
(*iv*) Fans and air compressors
(*v*) Vacuum cleaners, hair driers, sewing machines
(*vi*) Universal machines generally.

3. Cumulative compound motors :

(*i*) Punching, shearing and planing machines
(*ii*) Lifts, haulage gears and mine hoists
(*iii*) Pumps and power fans
(*iv*) Rolling mills, stamping presses and large printing presses
(*v*) Trolley buses.

4. Differential compound motors :

(*i*) Battery boosters
(*ii*) Experimental and research work.

7.11. Comparison of D.C. Motor Characteristics

Fig. 73 shows the comparison among shunt, series and compound operation for the *same motor* with its armature being acted upon by either or both fields. It is worth noting here that the outputs vary at the same armature current.

- Fig. 74 compares shunt, series and compound performance when the three are rated at the horse power (or kilowatts) and rpm (or radians/sec). In this type of plot all the characteristics cross at the design load point.
- Figs. 73 and 74 are drawn with the motors operating on their full rated voltage. *Different voltages will move the curves, but the relative shapes will remain similar.*

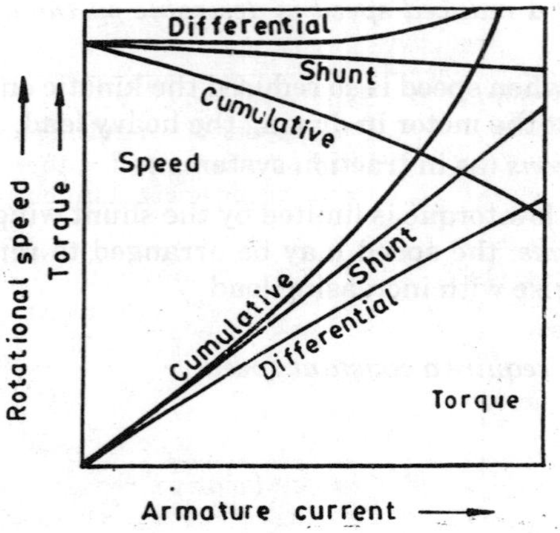

Fig. 73. Speed, torque relations for shunt and compound fields with the same armature current.

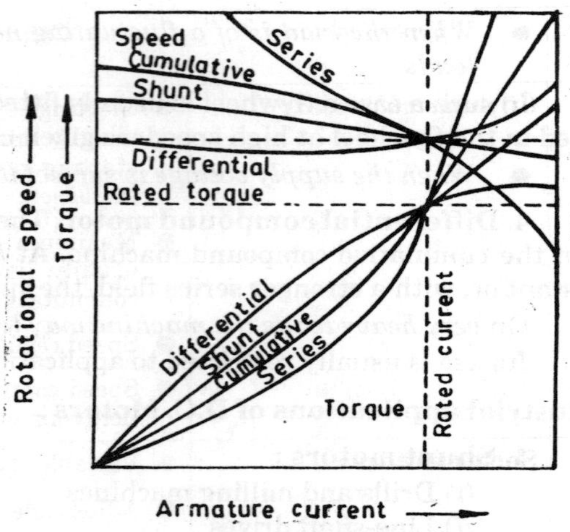

Fig. 74. Speed and torque relations for shunt, series and compound fields. Motors of same rated speed, torque and current.

The following may be deduced from Fig. 74.

(*i*) The curves diverge between no-load and full load. If the load torque were around mid-load, the shunt motor would draw the least current and the series the most.

(*ii*) At overloads the opposite is true. The series motor can develop an overload torque of around double the rated torque without drawing excessive current.

Fig. 73 shows that the speed characteristics for the same armature current are very much different. Therefore, the ability to take increasing torque loads that is the special contribution of a series field is accompanied by a much greater speed change.

7.12. Summary of Characteristics and Applications of D.C. Motors

The summary of characteristics and applications of D.C. motors is given in Table 1.

Table 1. Summary of Characteristics and Applications of D.C. motors

S No.	*Type of motor*	*Characteristics*	*Applications*
1.	**Separately excited D.C. motors :**	● Possible to *obtain very accurate speeds.* ● Most suitable for applications requiring *speed variation from very low value to high value.*	● Paper machines. ● Steel rolling units. ● Diesel electric propultion of ships.
2.	**Shunt motors :** (*i*) *Constant speed :*	● Starting torque-medium, usually limited to 250% by a starting resistor but may be increased. ● Maximum operating torque usually limited to about 200% by commutation. ● Speed control : — increase upto 200% speed by field control. — decrease by armature voltage control.	● Employed for constant-speed applications ; may be used for adjustable speed not greater than 2 : 1 range. Field of applications includes : — Lathes ; — Centrifugal pumps ; — Fans and blowers — Machine tools ; — Wood working machines ; — Reciprocating pumps ; — Spinning and weaving machines ; — Printing presses, etc.

S No.	Type of motor	Characteristics	Applications
	(ii) Adjustable speed :	● Starting torque-medium, usually limited to 250% by a starting resistor but may be increased. ● Maximum operating torque-usually limited to 200% by commutation. ● Speed regulation – 10 to 15%. ● Speed control : 6 : 1 range by field control.	● Same as above, For applications requiring adjustable speeds control, either constant torque or constant output.
3.	**Series motors :**	● Variable speed. ● Adjustable varying speed. ● *Starting torque* very high upto 500%. ● *Maximum momentary operating* torque upto 400%. ● *Speed regulation* : Widely variable, very high at no-lead. ● *Speed control* : By series resistance.	● Suitable for drives requiring high starting torque and where adjustable, varying speed is satisfactory. Fields of application include : — Cranes ; — Hoists ; — Trolley cars ; — Conveyors ; — Electric locomotives etc. **Loads must be positively connected, not belted.* *To prevent overspeed, lightest load should not be much less than 15 to 20% of full-load torque.
4.	**Compound motors :**		
	(i) Cumulative compound wound :	● Variable speed. ● Adjustable varying speed. ● *Starting torque* high upto 450% depending upon the degree of compounding. ● *Maximum momentary operating torque* higher than shunt, upto 350%. ● *Speed regulation* : Varying, depending upon degree of compounding, upto 25-30%. ● *Speed control* : Usually not used but may be upto 125% by field control.	● Suitable for drives requiring high starting torque and only fairly constant speed, pulsuating loads with flywheel action. Fields of application includes : — Shears ; — Punches ; — Elevators ; — Conveyors ; — Rolling mills ; — Heavy planes etc.
	(ii) Differentially compound wound :	● Torque and speed *almost constant.* ● Tendency towards speed instability with a possibility of motor running away and strong possibility of motor starting in wrong direction.	● Employed for experimental and research work.

7.13. Starting of D.C. Motors

7.13.1. Need for starters. A motor at *rest* has no back or counter e.m.f. At staring therefore, the *armature current is limited only by the resistance of the armature circuit.* The *armature resistance is very low*, however, and if full voltage were impressed upon the motor terminals at stand still, the resulting armature current would be *many times full-load* value—*usually sufficient to damage the machine.* For this reason, *additional resistance* is introduced into the armature circuit at starting. As the motor gains speed, its back e.m.f. builds up and the starting resistance is cut out.

Note. Very small D.C. motors, either shunt, series or compound wound, have sufficient armature resistance so that they may be started directly from the line without the use of a staring resistance and without injury to the motor.

Fig. 75 shows the connections of a starting resistance in three types of D.C. motors :

(*a*) A series motor ; (*b*) A shunt motor ; and

(*c*) A compound motor.

- In the case of *series motor* [Fig. 75 (*a*)], the armature, field and starting resistance are all in series.
- In the case of *shunt motor* [Fig. 75 (*b*)], it will be seen that the *top end of shunt field is connected to the first contact* on the starting resistance. This is to *ensure* that the *field winding receives the full supply at the moment of switching on.* If the fields were connected to the *last stud* of the starting resistance, then on starting, the field would receive only a *proportion of the supply voltage, the field current would be correspondingly weak and the torque might be too small to start the motor against the friction of the moving parts.*

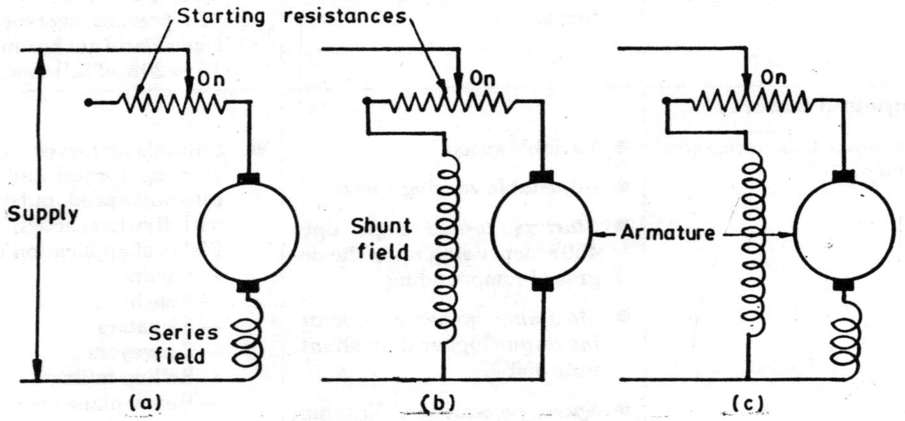

Fig. 75. Circuits incorporating starting resistances.

- The connections for the *compound motor* are seen from [Fig. 75 (*c*)] to be a combination of those of the series and the shunt connections.

7.13.2. Starters for shunt and compound motors

- The starters of D.C. motors are generally manufactured in convenient sizes and styles for use as auxiliaries with D.C. shunt and compound motors. Their *primary function is to limit the current in the armature circuit during the starting accelerating period.*
- The motor starters are always rated on the basis of output power and voltage of the motors with which they are to be used.

- There are two standard types of motor starters for shunt and compound motors. These are :

  (i) Three-point type ; and (ii) Four-point type.

Three-point starters are not completely satisfactory when used with motors whose speeds must be controlled by inserting resistance in the shunt field circuit. However, when applications require *little or no speed control,* either may be employed.

Three-point starter. Refer Fig. 76. The starter has three terminals *L, F* and *A*. The line terminal *L* must be connected to either side, positive or negative of the D.C. source on the main switch ; the field terminal *F* is connected to one field terminal on the motor ; the armature terminal *A* must be connected to either one of the motor armature terminals. The final connection must then be made from the second line terminal on the main switch to a junction of the remaining two armature and field terminals of the motor. If it is desired that the speed of the motor is controlled, a field rheostat is added as shown in Fig. 76 (*a*).

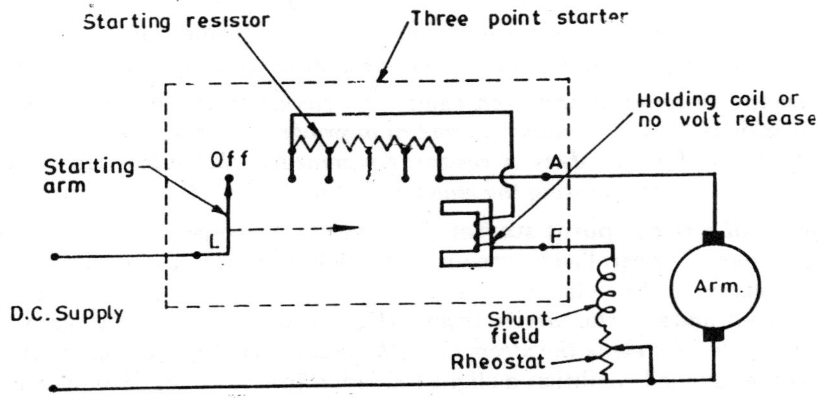

(*a*)

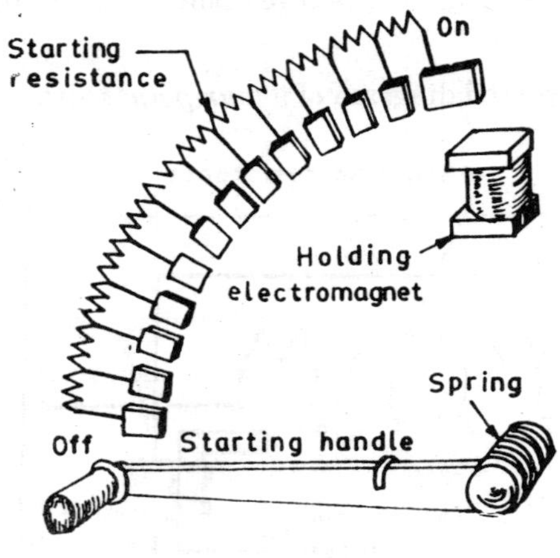

(*b*)

Fig. 76. Three-point starter connected to a shunt motor.

When the motor is at rest, the starter arm [represented by an arrow in Fig. 76 (*a*)] is held in the OFF position by a strong spiral spring.

Starting of motor :

- In order to *start the motor*, one hand is held on the handle of the open main switch while the starter arm is moved to the first stud [Figs. 76 (*a*) and (*b*)] with the other hand ; then the main switch is closed. If all the wiring is correct and the armature is free to turn, the motor will start.
- After the armature has accelerated sufficiently on the first stud, the starter arm is slowly moved to studs 2, 3, 4 etc. until the arm rests firmly against the iron poles of the *holding-coil* electromagnet. The entire starting process should take from 5 to 10 seconds. In the final position, the electromagnetic pull exerted by the holding coil will be greater than the force exerted by the spiral spring. *Should there be a power failure or should the field circuit be opened accidentally, the starter arm will fall back to its OFF position.* This function of starter is particularly important because :

  (*i*) if the power fails and starter arm is not restored to the OFF position, the motor might be damaged should the power, come on again ; and

  (*ii*) if the shunt field circuit were opened accidentally and the starter arm did not return to the OFF position, the motor speed might become dangerously high.
- Often the motors are protected against *overloads* by thermal *overload relays* in which *bimetallic is heated by motor current at approximately the same rate at which the motor is itself heating up. Above a certain temperature, this relay trips and opens the line contractor thereby isolating the motor from the supply.*

Drawback of a three-point starter. The use of a three-point starter presents a problem. The speed of the motor is controlled by means of the field rheostat. To increase the speed of motor necessitates the setting of the field rheostat to a higher resistance value. The current through the shunt field is reduced, and so is the current through the coil of the holding electromagnet. The *reduced current through the coil weakens the strength of the magnet and makes susceptible to line-voltage variations.* In the weakened condition a slight reduction in line voltage would further weaken the holding magnet, releasing the arm of the starter and thus *disconnecting the motor from the line.* Unscheduled stoppages of the motor make the three-point starter quite unpopular.

Four-point starter :

- Fig. 77 shows a simplified diagram of a *four-point starter*.

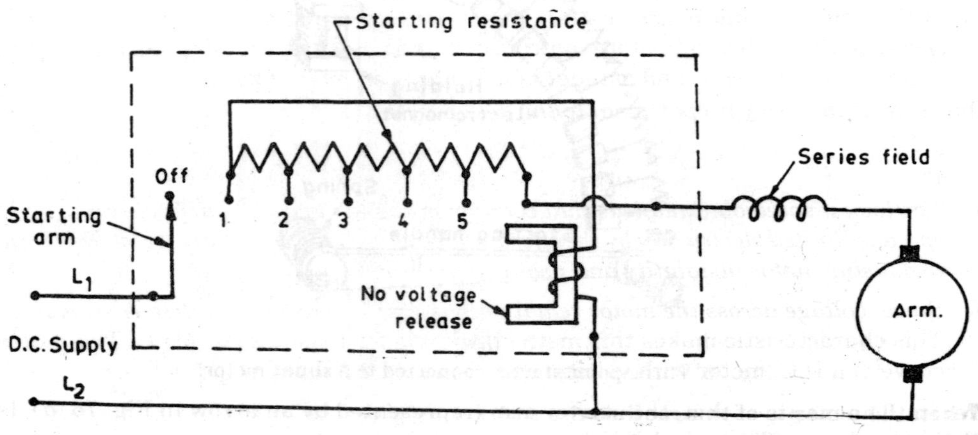

Fig. 77. Four-point starter.

- In this starter the drawback/disadvantage of the three-point starter is eliminated. In addition to the same three-points that were used with the three-point starter, the other side of the line, L_2 is the *fourth point* brought to the starter. The coil of the holding magnet is connected across the line when the arm is moved from the 'off' position. The holding magnet and starting resistors function as in the three-point starter. *The possibility of accidentally opening the field circuit is quite remote ; hence the greater acceptance of the four-point starter over the three-point starter.*

— The four-point starter *provides the motor with no voltage protection.* Should the power fail, the motor must be disconnected from the line. If not, full line voltage will be applied to the armature without the benefit of starting resistors when power is restored. The holding magnet, being connected across the line, releases the arm *when the voltage drops below a specific value, thus protecting the motor when the power is restored.*

8. SPEED CONTROL OF D.C. MOTORS

8.1. Factors Controlling the Speed

D.C. machines are generally much more adaptable to adjustable speed service. *The ready availability of D.C. motors to adjustment of their operating speed over wide ranges and by a variety of methods is one of the important reasons for the strong competitive position of D.C. machinery in modern industrial applications.*

The speed of a D.C. motor can be expressed by the following relationship.

$$N \propto \frac{V - I_a R_a}{\phi}.$$

Therefore, the speed of D.C. motor can be regulated by changing ϕ, R or V, or in other words, by,

1. Field control
2. Rheostatic control
3. Voltage control.

Note. Thyristor control of D.C. motor is out of scope of this book.

8.2. Field Control Method

- **Field control** is the most common method and forms one of the *outstanding advantages of shunt motors*. The method is, of course, also applicable to compound motors. Adjustment of field current and hence the flux and speed by adjustment of the shunt field circuit resistance or with a solid-state control when the field is *separately excited* is accomplished simply, *inexpensively*, and *without much change in motor losses*.

The *speed is inversely proportional to the field current*

i.e.
$$N \propto \frac{1}{I_f} \propto \frac{1}{\phi}.$$

- The *lowest speed obtainable is that corresponding to maximum field current* ; the *highest speed is limited electrically by the effects of armature reaction under weak- field conditions in causing motor instability and poor commutation.*
- *Since, voltage across the motor remains constant, it continues to deliver constant output.* This characteristic makes this method suitable for fixed output loads. The performance curve of a D.C. motor with voltage and field control is shown in Fig. 78.

Merits. The merits of this method are :

1. Good working efficiency.
2. Compact controlling equipment.

3. Capability of minute speed control.

4. The speed is not effected by load, and speed control can be performed effectively even at light loads.

5. Relatively inexpensive and simple to accomplish, both manually and automatically.

6. Within limits, field control does not affect speed regulation in the cases of shunt, compound, and series motors.

7. Provides relatively smooth and stepless control of speed.

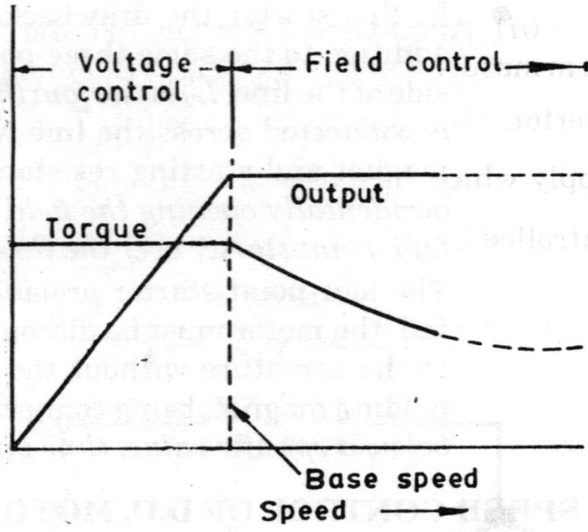

Fig. 78

Demerits. The demerits of field control as a method of speed control are :

1. Inability to obtain speeds below the basic speed.

2. Instability at high speeds because of armature reaction.

3. Commutation difficulties and possible commutator damage at high speeds.

Shunt Motors :

- The flux of a D.C. shunt motor can be changed by changing shunt field current (I_{sh}) with the help of a shunt field rheostat as shown in Fig. 79. Since the field current is very small, the power wasted in the controlling resistance is very small.

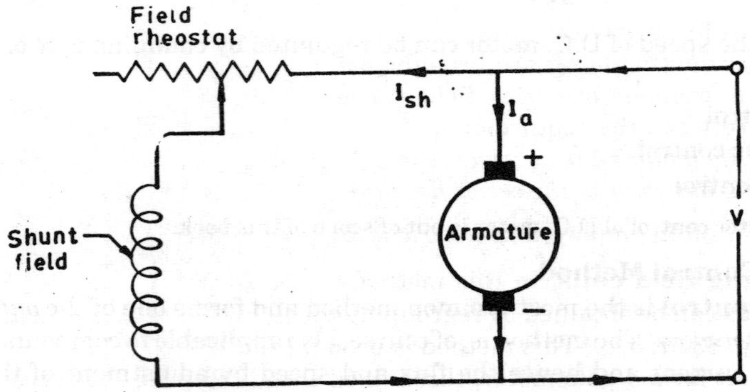

Fig. 79. Field rheostatic control for a D.C. shunt motor.

- In *non-interpolar machines* the speed can be increased by this method in the ratio **2 : 1**. In machines fitted *with interpoles* a ratio of maximum to minimum speeds of **6 : 1** is fairly common.

Series Motors. In a series motor, *variations of flux* can be brought about in any one of the following ways :

(*i*) Field divertors (*ii*) Armature divertor

(*iii*) Tapped field control (*iv*) Paralleling field coils.

(*i*) **Field divertors.** A variable resistance, known as field divertor (Fig. 80) *shunts* the series windings. Any desired amount of current can be passed through the divertor by adjusting its resistance. Hence, the *flux can be decreased and consequently the speed of the motor increased*.

(*ii*) **Armature divertor.** In order to *get speeds lower than the normal speed* a divertor across the armature can be used (Fig. 81). For a *given constant load torque,* if I_a is reduced due to armature divertor, then ϕ must increase ($\because\ T_a \propto I_a$). This results in an increase in current taken from the supply which increases the flux and a fall in speed $\left(\because\ N \propto \frac{1}{\phi}\right)$. The variations in speed can be controlled by varying the divertor resistance.

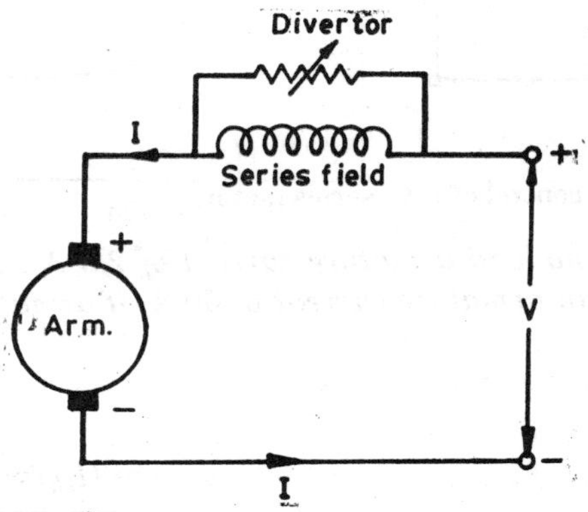

Fig. 80. Field divertor method of speed control for D.C. series motor.

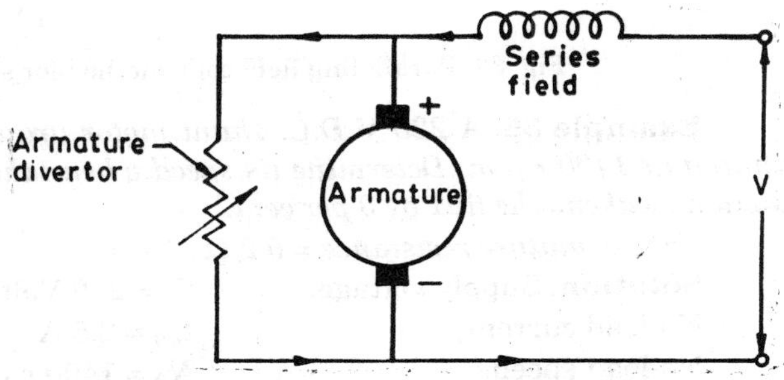

Fig. 81. Armature divertor method of speed control of D.C. series motor.

(*iii*) **Tapped field control.** In this method a number of tappings from the field winding are brought outside, as shown in Fig. 82. A number of series field turns can be short-circuited according to the requirement. *When all field turns are in circuit, the motor runs at lowest speed and speed increases with cutting out some of the series field turns.*

This method is often employed in *electric traction.*

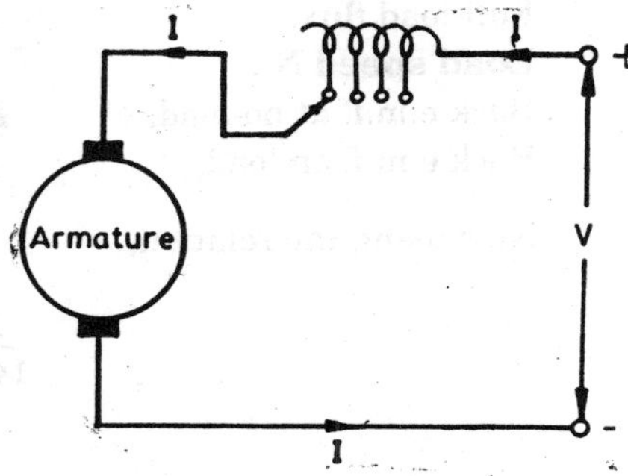

Fig. 82. Tapped field control for D.C. series motor.

(*iv*) **Paralleling field coils.** In this method of speed control several speeds can be obtained by regrouping the field coils as shown in Fig. 83 (*a*, *b*, *c*). This method is used for *fan motors.* It is seen that for a 4-pole motor, three fixed speeds can be obtained.

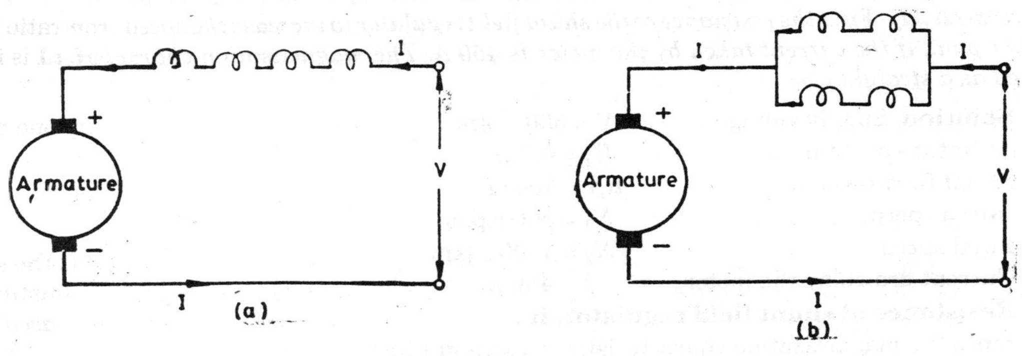

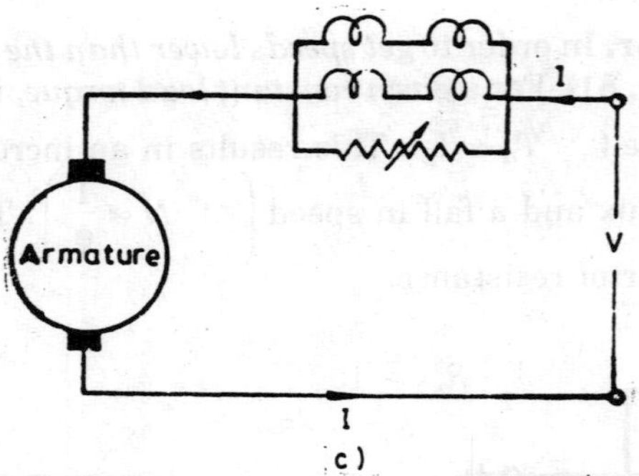

Fig. 83. Paralleling field coils method for speed control of D.C. series motor.

Example 36. *A 220 V D.C. shunt motor draws a no-load armature current of 2.5 A when running at 1400 r.p.m. Determine its speed when taking an armature current of 60 A, if armature reaction weakens the flux by 3 per cent.*

Take armature resistance = 0.2 Ω.

Solution. Supply voltage, $V = 220$ Volts

No-load current, $I_{a0} = 2.5$ A

No-load speed, $N_0 = 1400$ r.p.m.

Armature resistance, $R_a = 0.2\ \Omega$

Armature current, $I_a = 60$ A

Full-load flux, $\phi = 0.97\ \phi_0$

Load speed N :

Back e.m.f. at no-load, $E_{b0} = V - I_{a0}R_a = 200 - 2.5 \times 0.2 = 219.5$ V

Back e.m.f. on load, $E_b = V - I_aR_a = 220 - 60 \times 0.2 = 208$ V.

Now using the relation, $$\frac{N}{N_0} = \frac{E_b}{E_{b0}} \times \frac{\phi_0}{\phi}$$

$$\frac{N}{1400} = \frac{208}{219.5} \times \frac{\phi_0}{0.97\phi_0}$$

$$\therefore \quad N = \frac{1400 \times 208}{219.5 \times 0.97} = 1367.7 \text{ r.p.m.}$$

Hence, **load speed = 1367.7 r.p.m. (Ans.)**

Example 37. *The armature and shunt field resistances of a 500 V shunt motor are 0.2 Ω and 100 Ω respectively. Find the resistance of the shunt field regulator to increase the speed from 800 r.p.m. to 1000 r.p.m., if the current taken by the motor is 450 A. The magnetisation characteristic may be assumed as a straight line.*

Solution. Supply voltage, $V = 500$ Volts

Armature resistance, $R_a = 0.2\ \Omega$

Shunt field resistance, $R_{sh} = 100\ \Omega$

Initial speed, $N_1 = 800$ r.p.m.

Final speed, $N_2 = 1000$ r.p.m.

Current drawn by the motor, $I = 450$ A

Resistance of shunt field regulator, R :

Since the magnetization characteristic is a straight line,

$$\phi \propto I_{sh}$$

Also, $\frac{N_2}{N_1} = \frac{E_{b2}}{E_{b1}} \times \frac{\phi_1}{\phi_2}$

$\therefore$ $\frac{N_2}{N_1} = \frac{E_{b2}}{E_{b1}} \times \frac{I_{sh1}}{I_{sh2}}$.

At 800 r.p.m.

$$I_{sh1} = \frac{V}{R_{sh}} = \frac{500}{100} = 5 \text{ A}$$

Armature current, $I_{a1} = I - I_{sh1} = 450 - 5 = 445$ A

Back e.m.f. $E_{b1} = V - I_{a1}R_a = 500 - 445 \times 0.2 = 411$ V

At 1000 r.p.m.

Armature current, $I_{a2} = I - I_{sh2} = 450 - I_{sh2}$

Back e.m.f., $E_{b2} = V - I_{a2}R_a = 500 - (450 - I_{sh2}) \times 0.2 = 410 + 0.2I_{sh2}$

Now $\frac{1000}{800} = \frac{(410 + 0.2I_{sh2})}{411} \times \frac{5}{I_{sh2}}$

$$410 + 0.2I_{sh2} = \frac{1000}{800} \times \frac{411}{5} I_{sh2} = 102.75I_{sh2}$$

or $102.55I_{sh2} = 410$

$\therefore$ $I_{sh2} = 4$ A app.

Total resistance in the field circuit

$$= \frac{V}{I_{sh2}} = \frac{500}{4} = 125\ \Omega$$

Resistance of the shunt field $= 100\ \Omega$

Resistance of the shunt field regulator,

$$R = 125 - 100 = 25\ \Omega$$

Hence, **resistance of the shunt field regulator = 25 Ω. (Ans.)**

☞**Example 38.** *A 500 V series motor has an armature resistance of 0.4 ohm and series field resistance of 0.3 ohm. It takes a current of 100 A at a speed of 600 r.p.m. Find the speed of the motor if a divertor of resistance 0.6 ohm is connected across the field, the load torque being kept constant. Neglect armature reaction and assume that flux is proportional to the current.*

Solution. Supply voltage, $V = 500$ Volts

Armature resistance, $R_a = 0.4$ ohm

Series field resistance, $R_{se} = 0.3$ ohm

Divertor resistance, $R_{\text{div.}} = 0.6$ ohm

$I_{a1} = 100$ A, $N_1 = 600$ r.p.m.

Speed, N_2 :

Back e.m.f., $E_{b1} = V - I_{a1}(R_a + R_{se})$

$= 500 - 100(0.4 + 0.3) = 430$ V.

Let I_{a2} be the current taken and ϕ_2 be the flux produced when a divertor is connected across the series field (Fig. 84)

Since the torque remains constant,

$\therefore$ $\phi_1 I_{a1} = \phi_2 I_{a2}$ ...(*i*)

But $\phi \propto$ current through series field

$\therefore$ $\phi_1 \propto I_{a1}$

Current through the series field when a divertor is connected

$$= I_{a2} \times \frac{R_{div.}}{R_{div.} + R_{se}}$$

$$= I_{a2} \times \frac{0.6}{0.6 + 0.3} = 0.667 I_{a2}$$

Flux in this case $\phi_2 \propto 0.667 I_{a2}$

Substituting in (*i*), we get

$$I_{a1}^2 = 0.667 I_{a2}^2$$

$$I_{a2}^2 = \frac{I_{a1}^2}{0.667}$$

$$\therefore \quad I_{a2} = \frac{I_{a1}}{\sqrt{0.667}} = \frac{100}{\sqrt{0.667}} = 122.44 \text{ A}$$

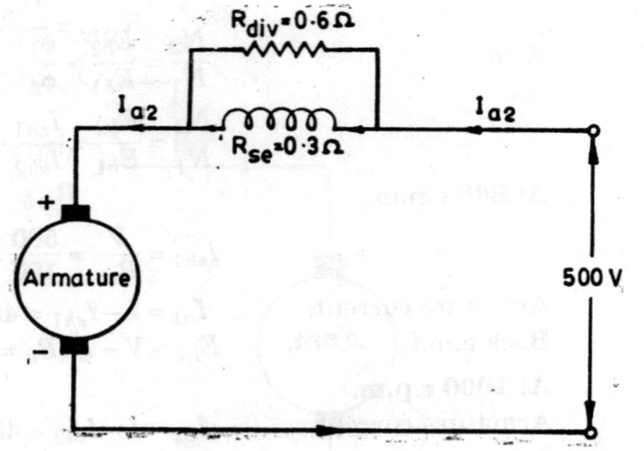

Fig. 84

Series field current, $\quad I_{se} = 0.667 I_{a2}$

$$= 0.667 \times 122.44 = 81.64 \text{ A}$$

Back e.m.f., $\quad E_{b2} = V - I_{a2}R_a - I_{se}R_{se}$

$$= 500 - 122.44 \times 0.4 - 81.64 \times 0.3$$

$$= 500 - 48.97 - 24.49 = 426.54 \text{ V}$$

Using the relation, $\quad \frac{N_2}{N_1} = \frac{E_{b2}}{E_{b1}} \times \frac{\phi_1}{\phi_2}$

$$\frac{N_2}{600} = \frac{426.54}{430} \times \frac{100}{81.64}$$

$$\therefore \quad N_2 = 729 \text{ r.p.m.}$$

Hence, **speed of the motor = 729 r.p.m. (Ans.)**

☞**Example 39.** *A D.C. series motor drives a load the torque of which varies are square of the speed. The motor takes a current of 20 A when the speed is 800 r.p.m. Calculate the speed and current when the motor field winding is shunted by a divertor of the same resistance as that of the field winding. Neglect all motor losses and assume that the magnetic circuit is unsaturated.*

Solution. $\quad I_{a1} = 20$ A, $N_1 = 800$ r.p.m.

$\mathbf{I_{a2} = ?,\ N_2 = ?}$

When the field winding is shunted by a divertor of equal resistance, then current through either is half the armature current. If I_{a2} is the new armature current, then $\frac{I_{a2}}{2}$ passes through the winding (Fig. 85).

$$\therefore \quad \phi_2 = \frac{I_{a2}}{2}$$

Now $\quad T_1 \propto \phi_1 I_{a1} \propto N_1^2$ (Given)

and $\quad T_2 \propto \phi_2 I_{a2} \propto N_2^2$ (Given)

From (*i*) and (*ii*), we get

$$\left(\frac{N_2}{N_1}\right)^2 = \frac{\phi_2 I_{a2}}{\phi_1 I_{a1}} \quad ...(i)$$

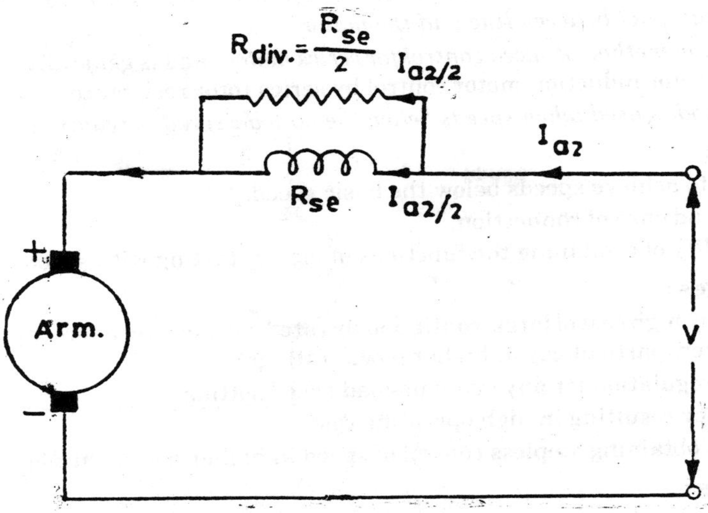

Fig. 85

Because all losses are negligible, hence the armature and series field resistances are negligible. This means that back e.m.f. in both cases is the same as the applied voltage.

$$\therefore \quad \frac{N_2}{N_1} = \frac{E_{b2}}{E_{b1}} \times \frac{\phi_1}{\phi_2} \text{ becomes}$$

$$\frac{N_2}{N_1} = \frac{\phi_1}{\phi_2} \quad \text{...(ii)}$$

$[\because \; E_{b2} = E_{b1} = \text{applied voltage}]$

Putting this value in (*i*) above, we get

$$\left(\frac{\phi_1}{\phi_2}\right)^2 = \frac{\phi_2 I_{a2}}{\phi_1 I_{a1}}$$

or
$$\frac{I_{a2}}{I_{a1}} = \left(\frac{\phi_1}{\phi_2}\right)^3$$

$$\frac{I_{a2}}{20} = \left(\frac{20}{I_{a2}/2}\right)^3 \quad [\because \; \phi_2 \propto I_a/2]$$

$$\therefore \quad I^4_{a2} = 20 \times 40^3$$

or
$$I_{a2} = 33.63 \text{ A}$$

Hence, **current = 33.63 A. (Ans.)**

From (*ii*), we get
$$\frac{N_2}{800} = \frac{20}{(33.63)/2}$$

$$N_2 = 800 \times \frac{40}{33.63} = 951.52 \text{ r.p.m.}$$

Hence, **speed of the motor = 951.53 r.p.m. (Ans.)**

8.3. Rheostatic Control

- This method consists of obtaining reduced speeds by the insertion of external series resistance in the armature circuit. It can be used with series, shunt and compound motors ;

for the last two types, the *series resistor must be connected between the shunt field and the armature, not between line and the motor.*

- It is *common method of speed control for series motors* and is generally analogous in action to wound-rotor induction-motor control by series rotor resistance.
- *This method is used when speeds below the no-load speed is required.*

Advantages :

1. The ability to achieve speeds below the basic speed.
2. Simplicity and ease of connection.
3. The possibility of combining the functions of motor starting with speed control.

Disadvantages :

1. The relatively high cost of large, continuously rated, variable resistors capable of dissipating large amounts of power (particularly at higher power ratings).
2. Poor speed regulation for any given no-load speed setting.
3. Low efficiency resulting in high operating cost.
4. Difficulty in obtaining stepless control of speed in higher power ratings.

Shunt motors :

- In armature or rheostatic control method of speed the voltage across the armature (which in normally constant) is varied by inserting a variable rheostat or resistance, called *controller resistance,* in *series* with the armature circuit. As the controller resistance is increased, the potential difference across the armature is decreased thereby decreasing the armature speed. For a load of constant torque, speed is approximately proportional to the potential difference across the armature

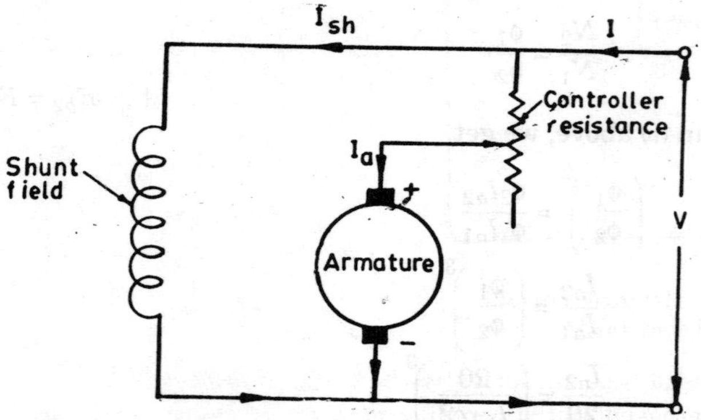

Fig. 86. Armature resistance control for D.C. shunt motor.

From the speed/armature current characteristic it is seen that *greater the resistance in armature, greater is the fall in speed.*

There is a particular load current for which the *speed would be zero. This is the maximum current and is known as 'stalling current'.*

- This method is *very wasteful, expensive and unsuitable for rapidly changing load,* because for a given value of R_t the speed will change with load. A more stable operation can be obtained by using a *divertor across the armature* (Fig. 88) in addition to armature control resistance. Now, the changes in armature current due to changes in the load torque *will not be so effective* in changing the potential difference across the armature and hence the speed of the armature.

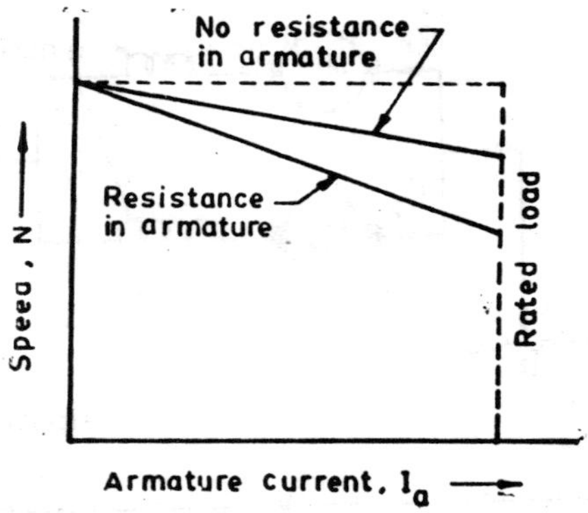

Fig. 87. Speed-current characteristic of D.C. shunt motor.

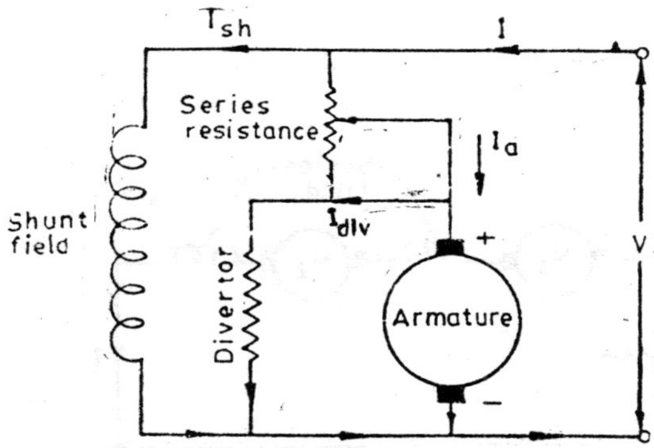

Fig. 88. Use of divertor across the armature for speed control of D.C. shunt motor.

Series motors. Armature resistance control is the most common method employed for D.C. series motors (Figs. 89 and 90).

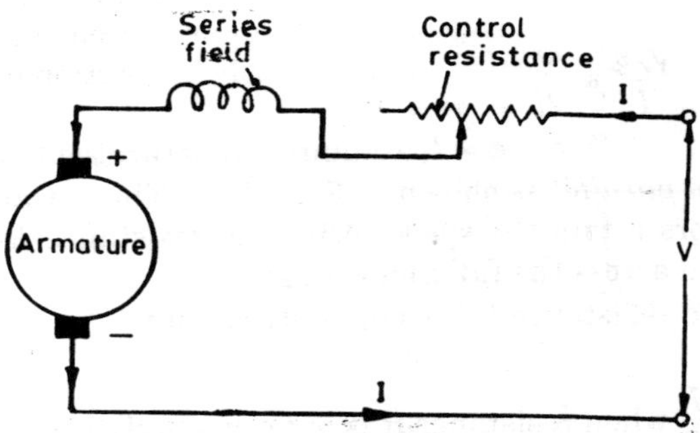

Fig. 89. Armature resistance control for D.C. series motor.

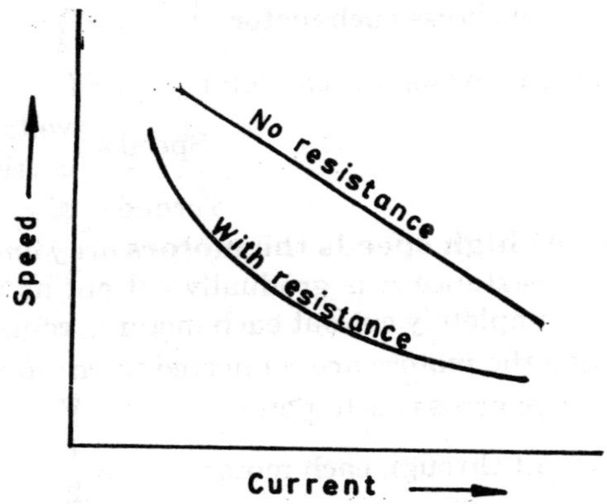

Fig. 90

By increasing the resistance in series with the armature the voltage applied across the armature terminals can be decreased. With the reduced voltage across the armature, the speed is reduced.

Since full motor current passes through the resistance, the loss of power is considerable.

Although terminal-voltage control by means of a variable voltage supply would effectively control the speed of a D.C. series motor, the high cost of the control equipment is seldom warranted.

Series-parallel control :

- This system is widely used in *electric traction*. Here two or more similar mechanically-coupled motors are employed.
- **At low speeds** the *motors are joined in series* as shown in Fig. 91 (*a*). The additional resistance is gradually cut out by controller as the motors attain the speed, and finally the resistance is totally removed, then *each motor has half of line voltage*. In this arrangement, for *any given value of armature current, each motor will run at half of its normal speed*. As there is no external resistance in the circuit, therefore *there is no waste of energy and so motors operate at an efficiency nearly equal to that obtainable with full line voltage across the terminals of each motor.*

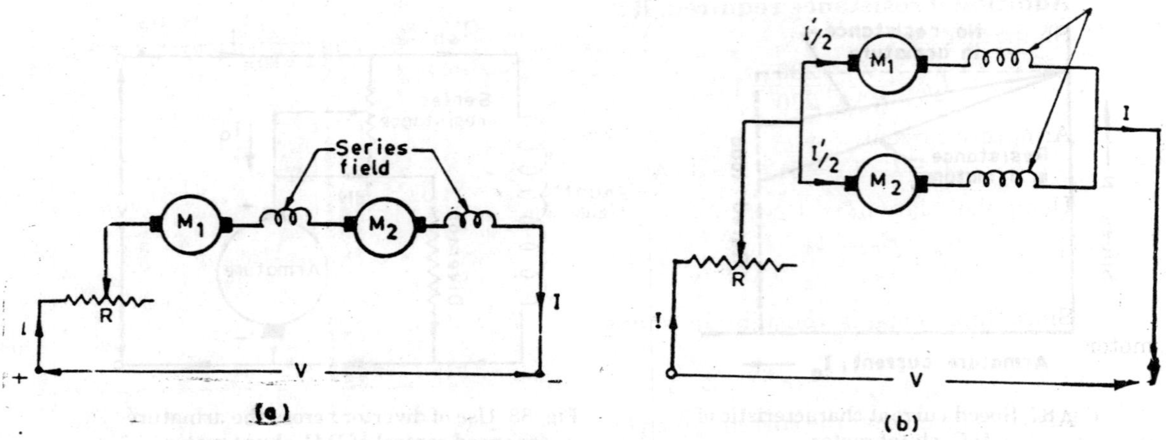

Fig. 91. Series-parallel control.

When the motors are connected in *series* and resistance R is *completely cut out* :

Voltage across each motor $= \dfrac{V}{2}$

Current through each motor $= I$

$$\text{Speed} \propto \frac{\text{voltage}}{\text{current}} \propto \frac{V/2}{I} \propto \frac{V}{2I}$$

$$\text{Torque} \propto \phi I \propto I^2 \qquad \text{[Since } \phi \propto I\text{, assuming unsaturated field]}$$

- **At high speeds the motors** *are joined in parallel* as shown in Fig. 91 (*b*). The variable resistance R is gradually cut out as motors attain the speed. After the resistance R is completely cut out each motor is connected across the full line voltage.

When the motors are connected in *parallel* and resistance R is completely cut out :

Voltage across each motor $= V$

Current through each motor $= \dfrac{I}{2}$ $\left[= \dfrac{I'}{2} \text{ when resistance } R \text{ is } \textit{not} \text{ completely cut out} \right]$

$$\text{Speed} \propto \frac{\text{voltage}}{\text{current}} \propto \frac{V}{I/2} \propto \frac{2V}{I}$$

Also, $\text{torque} \propto \phi\, I \propto I^2$ $[\because \phi \propto I]$

$$\therefore \quad T \propto \left(\frac{I}{2}\right)^2 \propto \frac{I^2}{4}$$

The *torque is* $\frac{1}{4}$ *times that produced by motors when in series.*

Example 40. *The armature and shunt field resistances of a 230 V shunt motor are 0.1 ohm and 230 ohms respectively. It takes a current of 61 A at 1000 r.p.m. If the current taken remains unaltered find the resistance to be included in series with the armature circuit to reduce the speed to 750 r.p.m.*

Solution. Supply voltage, $V = 230$ Volts
Armature resistance, $R_a = 0.1$ ohm
Shunt field resistance, $R_{sh} = 230$ ohms
Load current, $I = 61$ A
Speed, $N_1 = 1000$ r.p.m.
Speed, $N_2 = 750$ r.p.m.

Additional resistance required, R :

Shunt field current,

$$I_{sh} = \frac{V}{R_{sh}} = \frac{230}{230} = 1 \text{ A}$$

Armature current,

$$I_1 = I - I_{sh} = 61 - 1 = 60 \text{ A}$$

Using the relation,

$$\frac{N_2}{N_1} = \frac{E_{b2}}{E_{b1}} \times \frac{\phi_1}{\phi_2} \quad ...(i)$$

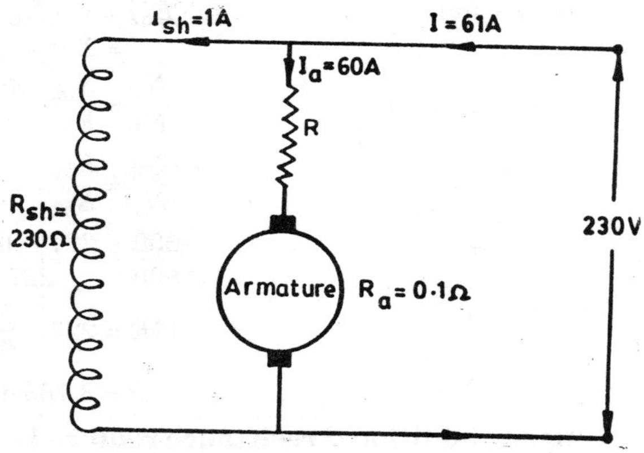

Fig. 92

Since flux remains constant in shunt motor

$$\phi_1 = \phi_2$$

Also it is given that the current taken by the motor remains constant

$$\therefore \quad I = I_1 = I_2 = 61 \text{ A}$$

and

$$I_{a1} = I_{a2} = 60 \text{ A}$$

Thus eqn. (*i*) reduces to

$$\frac{N_2}{N_1} = \frac{E_{b2}}{E_{b1}} \quad ...(ii)$$

$$E_{b1} = V - I_{a1}R_a = 230 - 60 \times 0.1 = 224 \text{ V}$$

$$E_{b2} = V - I_{a2}(R_a + R)$$

$$= 230 - 60(0.1 + R) = 230 - 6 - 60R$$

$$= 224 - 60\,R$$

Putting this value in (*ii*), we get

$$\frac{750}{1000} = \frac{224 - 60R}{224}$$

or

$$60R = 224 - \frac{750}{1000} \times 224$$

$$\therefore \quad R = 0.933 \text{ ohm}$$

Hence, **additional resistance required = 0.933 ohm. (Ans.)**

Example 41. *The armature resistance of a 230 V D.C. shunt motor is 0.2 ohm. It takes 15 A at rated voltage and runs at 800 r.p.m. Calculate the value of additional resistance required in the armature circuit to reduce the speed to 600 r.p.m. when the load torque is independent of speed. Ignore the field current.*

Solution. Supply voltage, $V = 230$ Volts

Armature resistance, $R_a = 0.2$ ohm

Armature current, $I_1 = I_{a1} = 15$ A

Speed, $N_1 = 800$ r.p.m.

Speed, $N_2 = 600$ r.p.m.

Additional resistance required, R :

Back e.m.f., $E_{b1} = V - I_{a1}R_a = 230 - 15 \times 0.2 = 227$ V

Since as per given data load torque is independent of speed and flux is constant

$$I_{a1} = I_{a2} = 15 \text{ A}$$

$$\phi_1 = \phi_2$$

Back e.m.f., $E_{b2} = V - I_{a2}(R_a + R) = 230 - 15(0.2 + R)$

$= 230 - 3 - 15R = 227 - 15R$

Using the relation, $\frac{N_2}{N_1} = \frac{E_{b2}}{E_{b1}} \times \frac{\phi_1}{\phi_2}$

$\frac{N_2}{N_1} = \frac{E_{b2}}{E_{b1}}$ $[\because \phi_1 = \phi_2]$

$\therefore$ $\frac{600}{800} = \frac{227 - 15R}{227}$

$15R = 227 - \frac{600}{800} \times 227$

$\therefore$ $R = 3.783$ ohms

Hence, **additional resistance required = 3.783 ohms. (Ans.)**

8.4. Voltage Control

When the speed is controlled by regulating the motor terminal voltage while maintaining constant field current, it is called **voltage control.**

With voltage control, the change in speed is almost proportional to the change in voltage as shown in Fig. 78. The *output varies directly with speed and the torque remains constant.* Since the voltage has to be regulated without affecting the field, the application of voltage control is limited to *separately excited motors* (Fig. 93) *only.*

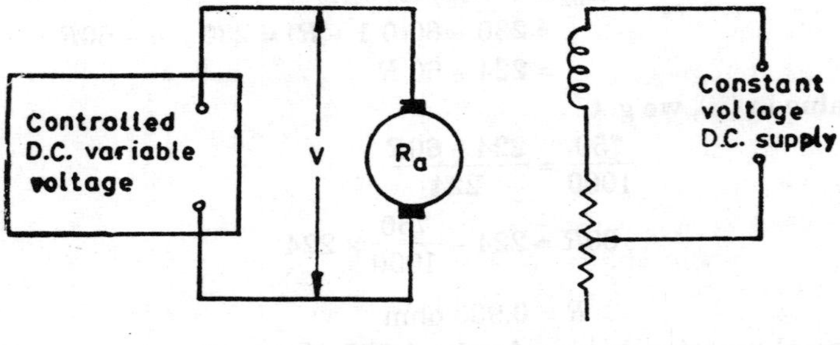

Fig. 93. Voltage control method.

- For D.C. motors of *fractional* and *relatively low power* rating, the variable D.C. voltage source may be a D.C. vacuum tube, a gas or thyraton tube, or a semi-conductor (silicon controlled rectifier) amplifier, operating from a three-phase or single phase A.C. supply.
- *Motors of moderate rating up to* 75 kW may be controlled by this method using Rototrol or Regulex or magnetic amplifiers as the adjustable D.C. voltage source.
- *Large D.C. motors* are controlled in this manner by means of rotary amplifiers such as the amplidyne or the Ward-Leonard control system.

Advantages :

1. Speed control over a wide range is possible.
2. This method eliminates the need for series armature starting resistance.
3. Uniform acceleration can be obtained.
4. Speed regulation is good.

Disadvantages :

1. Arrangement is costly as two extra machines are required.
2. The overall efficiency of the system is low, especially at light loads.

Applications :

Inspite of the high capital cost, this method finds wide applications in :

(*i*) Steel mills for reversing the rolling mills.

(*ii*) Seamless tube mills and shears.

(*iii*) High and medium speed elevators in tall buildings, mine hoists, paper machine drives and electric shovels.

Ward-Leonard System. This method of control not only gives a wide range of operating speeds, but reduces to the very minimum the wastage of energy that may take place at staring and stopping.

Fig. 94 shows the schematic arrangement of Ward-Leonard method.

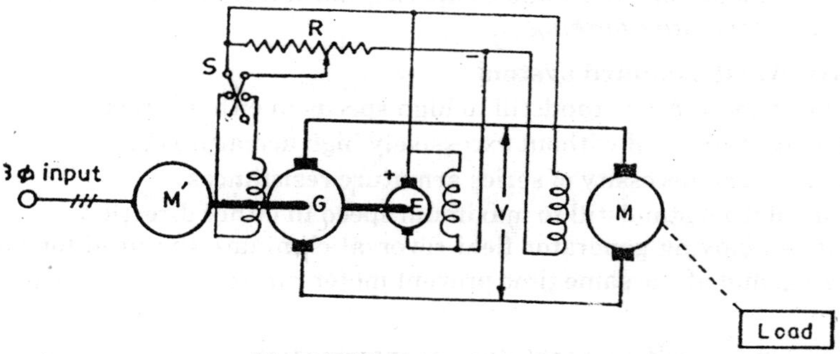

Fig. 94. Ward-Leonard method.

M = main motor whose speed is to be controlled

G = separately excited generator which feeds the armature of the motor *M*

E = an exciter (a small shunt generator) which provides field excitation to the generator *G* and motor *M*

M′ = driving motor—a constant speed motor which drives *G* and *E*.

[If the system is to work on A.C. supply, the driving motor *M′* is a 3-phase induction motor. If the system is to work on D.C. supply, the motor *M′* is a shunt motor. In the latter case the exciter *E* is not necessary because the excitation for the generator *G* and motor *M* can be obtained from D.C. mains. A diesel engine can also be used in place of motor *M′*].

R = a potentiometer rheostat

S = a double throw switch.

The working of this system is as follows :

- The motor *M′* drives the generator *G* and excitor *E* at constant speed. The voltage fed to motor *M* can be controlled by varying the setting of *R*. A change in voltage applied to motor *M* changes its speed. The speed can be adjusted to any value from zero to maximum in either direction by means of a rheostat *R* and switch *S*.
- When the sliding contact of *R* is at extreme right, the motor is running at full speed in one direction. To decrease the speed the sliding contact is moved to the left. When the sliding contact is at the extreme left position, the speed of motor *M* is zero. In order to reverse the speed of the motor, the sliding contact is shifted to the extreme left, the switch *S* is reversed and the sliding contact shifted to right again.
- A modification of the Ward-Leonard system is known as *Ward-Leonard Ilgner system*, which uses a small motor-generator set with the *addition of a flywheel whose function is to reduce fluctuations in the power demand from the supply circuit*. When the main motor *M* becomes suddenly overloaded, the driving motor *M′* slows down, thus allowing the

inertia of the flywheel to supply a part of the overload. However, when the load is suddenly thrown off the main motor M, then M' speeds up thereby again storing energy in the flywheel. When Ilgner system is driven by means of an A.C. motor (whether induction or synchronous) another refinement in the form of a 'slip regulator' can be usefully employed thus giving an additional control.

- One important feature of the Ward-leonard system is its *regenerative action.* When a locomotive, fitted with this system, is descending a slope, it speeds up due to the action of gravity. The speed of motor M increases until its back e.m.f. exceeds the applied voltage. Motor M then runs as generator and feeds the machine G which now works as a generator and feeds electrical energy back into the trolley wire. This results in salvaging of considerable amount of energy and a superior and smooth braking action. Such an action is known as *regenerative braking*.

Advantages of Ward-Leonard system :

1. A wide range of speed from standstill to high speeds in either direction.
2. Rapid and instant reversal without excessively high armature currents.
3. Starting without the necessity of series armature resistances.
4. Stepless control from standstill to maximum speed in either direction.
5. Larger units employing generator field reversal eliminate the need for heavy armature conductors for reversing, and at the same time prevent motor runway since the motor field is always excited.
6. The method lends itself to adaptation of intermediate electronic, semi-conductor, and magnetic amplifiers to provide stages of amplification for an extremely large motor. Thus the power in the control circuit may be extremely small.
7. Extremely good speed regulation at any speed.

Disadvantages :

1. High initial cost.
2. Since the efficiency, neglecting the exciter efficiency, is essentially the product of the individual efficiencies of the two larger machines, the efficiency of this method is *not as high as rheostat speed control or the field control method.*

Example 42. *A series motor drives a fan for which the torque varies as square of the speed. Its resistance between terminals is 1.2 ohm. On 220 V, it runs at 350 r.p.m. and takes 30 A. The speed is to be raised to 450 r.p.m. by increasing the voltage. Find the voltage.*

Assume that flux varies directly as current.

Solution. Resistance between terminals = 1.2 ohm

$$I_{a1} = 30 \text{ A}, N_1 = 350 \text{ r.p.m.}, \; N_2 = 450 \text{ r.p.m.}$$

Since $\phi \propto I_a$

$\therefore$ $$T \propto \phi I_a \propto I_a^2 \quad ...(i)$$

Also, $$T \propto N^2 \quad ...\text{[Given]} \quad ...(ii)$$

From (i) and (ii), we get $I_a^2 \propto N^2$ or $I_a \propto N$

or $$\frac{I_{a2}}{I_{a1}} = \frac{N_2}{N_1} = \frac{450}{350}$$

$\therefore$ $$I_{a2} = I_{a1} \times \frac{450}{350} = 30 \times \frac{450}{350} = 38.57 \text{ A}$$

Back e.m.f., $$E_{b1} = 220 - 30 \times 1.2 = 184 \text{ V}$$

Back e.m.f., $$E_{b2} = V - 38.57 \times 1.2 = V - 46.28$$

$$\frac{\phi_1}{\phi_2} = \frac{I_{a1}}{I_{a2}} = \frac{30}{38.57}$$

Now using the relation, $\frac{N_2}{N_1} = \frac{E_{b2}}{E_{b1}} \times \frac{\phi_1}{\phi_2}$

$$\frac{450}{350} = \frac{V - 46.28}{184} \times \frac{30}{38.57}$$

$$\therefore \quad V - 46.28 = \frac{450}{350} \times 184 \times \frac{38.57}{30} = 304.15$$

$$\therefore \quad \mathbf{V = 304.15\ Volts.\ (Ans.)}$$

9. LOSSES AND EFFICIENCY OF D.C. MACHINES

9.1. Introduction

The output of a generator or motor is always less than the input because some of the energy supplied is lost as heat. These losses raise the temperatures of the machine parts above that of surrounding air until such temperatures are reached that the heat losses are radiated as fast as they are generated. Certain of the losses depend upon the load. The temperature rise therefore depends upon the load also, and the maximum allowable temperature rise determines the maximum permissible load that the machine may carry. The limit of output occurs at the load for which the temperature rise becomes high enough to endanger the insulation of the windings.

Thus the consideration of machine losses is important for the following three reasons :

1. Losses appreciably influence the operating cost of the machine.
2. Losses determine the heating of the machine and hence the rating or power output that can be obtained without undue deterioration of the insulation.
3. The voltage drops or current components associated with supplying the losses must be properly accounted for in a machine representation.

Machine efficiency is given by

$$\text{Efficiency} = \frac{\text{output}}{\text{input}} \quad ...(12)$$

which can also be expressed as

$$\text{Efficiency} = \frac{\text{input} - \text{losses}}{\text{input}} = 1 - \frac{\text{losses}}{\text{input}} \quad ...(13)$$

$$\text{Efficiency} = \frac{\text{output}}{\text{output} + \text{losses}} \quad ...(14)$$

9.2. Losses

Losses in a D.C. machine can be classified as follows :

1. **Electrical losses** (*copper loss*) :

(*i*) Armature $= I_a^2 R_a$ (*ii*) Series field $= I_{se}^2 R_{se}$

(*iii*) Shunt field $= VI_{sh}$ (*iv*) Commutating field $= I_c^2 R_c$.

2. **Rotational losses** (*stray-power loss*) :

(*i*) Core loss :

(*a*) Hysteresis (*b*) Eddy currents.

(*ii*) Mechanical (or friction) loss :

(*a*) Bearings (*b*) Brushes

(*c*) Windage.

- The *core losses and friction losses* are supplied from the mechanical power developed by the machine. They are put into a single group called *mechanical losses* or more generally, *stray-power losses*. When a generator or motor runs at a fixed speed and generates a given voltage, the *stray-power loss is constant regardless* of the electrical output or input of the machine for *speed and flux density are the only factors that influence the stray-power loss*.
- The *electrical losses* are supplied from the electrical power generated by or delivered to the machine, as the case may be. Of these, the shunt field loss is somewhat, though not entirely independent of the load, while the remaining electrical losses are nearly proportional to the square of the load current. Fig. 95 shows the combined power-flow diagram for motor or generator action which is self-explanatory.

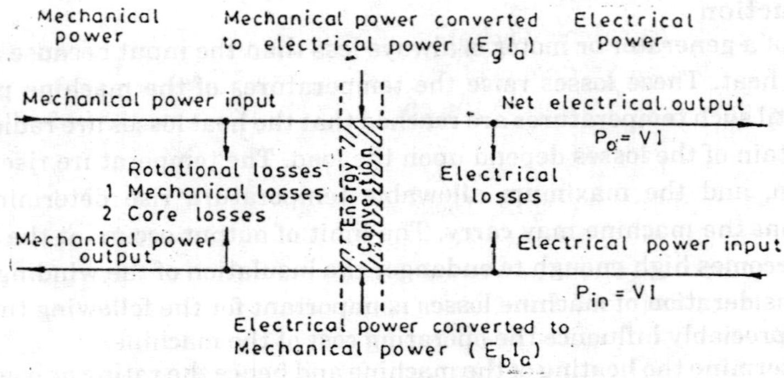

Fig. 95. Combined power-flow diagram for motor or generator action.

9.2.1. Copper (or electrical) losses. When an electric current of I ampere flows in a resistance of R ohms, heat energy is lost at the rate of I^2R joules/sec, and the power loss is I^2R watts. Generators and motors have one or more field circuits and an armature circuit in which such losses occur. All resistance losses of kind are classed as *copper loss*.

(*i*) **Armature copper loss** = $I_a^2 R_a$. This is about 30% to 40% of total full-load losses.

(*ii*) **Field copper loss.** This loss equals $I_{sh}^2 R_{sh}$ for the shunt field winding and $I_{se}^2 R_{se}$ for the series field winding. This loss constitutes about 30% of the total full-load losses.

(*iii*) **Brush contact loss.** This loss is due to the resistance of brush contacts. The voltage drop at the brush is almost independent of I_a. For carbon brushes the voltage drop is around 1 volt per brush. The power loss due to brush contact resistance is $2e_b I_a$ where e_b is the voltage drop at one brush.

(*iv*) **Loss in commutating pole winding.** This loss equals I_a^2 × resistance of commutating pole winding.

(*v*) **Loss in compensating winding.** This loss equals I_a^2 × resistance of compensating winding.

9.2.2. Iron losses. *Iron losses are a function of both flux and speed.*

Hysteresis loss. The hysteresis loss P_h is a *measure of the electric energy required to overcome the retentivity of the iron in the magnetic flux path,* using watts as unit,

$$P_h = K_h B^x f V \qquad ...(15)$$

where V = volume of iron in dynamo subject to change of flux

K_h = constant for the grade of iron employed

B = flux density raised to the Steinmetz exponent. With modern values of dynamo x is no longer 1.6 but closer to 2.0. This is not to imply that for a given volume, V, of iron the loss has increased, because K_h has been reduced considerably

f = frequency (Hz) of reversal of flux.

Eddy current losses. These losses occur not only in the dynamo iron but in all conductive materials with the flux path of the rotating or varying magnetic field of the dynamo. The eddy current loss P_e, in watts is

$$P_s = K_e t^2 B^2 f^2 V \quad ...(16)$$

where K_e = an eddy current constant for the grade of iron employed

t = thickness of the laminations of the pole core and armature

B = flux density

f = frequency (Hz) of reversal of flux

V = volume of iron subject to change of flux.

For a D.C. dynamo the frequency, f, reversal of flux varies with speed. Thus the hysteresis loss varies directly with speed, whereas the eddy current loss varies as the square of speed. Both hysteresis loss and eddy current loss vary approximately as the square of the flux density. For this reason core losses are considered a function of both flux and speed.

9.2.3. Mechanical (or friction) losses. When a machine is running, there are various frictional forces to be overcome, each of which requires a continuous expenditure of energy and results in heating the rubbed parts. There is friction loss in the machine bearings, at the surface of the commutator due to the rubbing of the brushes, and in the armature core due to its fanning action. *These losses depend upon the speed* but are independent of the load on the machine. They are difficult to estimate by direct calculation but may be found by measurement.

9.3. Efficiency of D.C. Machines :

Generator :

1. *Mechanical efficiency* (η_m)

$$\eta_m = \frac{\text{Electrical power developed by armature}}{\text{Total mechanical power input}}$$

$$= \frac{E_g I_a}{\text{B.H.P. of prime mover} \times 735.5} \quad ...(17)$$

2. *Electrical efficiency* (η_e)

$$\eta_e = \frac{\text{Useful electrical power output}}{\text{Electrical power developed}} = \frac{VI}{E_g I_a}. \quad ...(18)$$

3. *Overall or commercial efficiency* ($\eta_{og} = \eta_m \times \eta_e$)

$$\eta_{og} = \frac{\text{Useful electrical power output}}{\text{Total mechanical power input}}$$

$$= \frac{VI}{\text{B.H.P. of prime mover} \times 735.5} \quad ...(19)$$

The overall efficiency of generator can also be expressed as follows :

$$\eta_{og} = \frac{\text{Useful power output}}{\text{Useful power output + total losses}}$$

$$= \frac{VI}{VI + \text{total losses}} \quad ...(20)$$

where E_g = generated e.m.f. V = terminal voltage

I = load current, I_a = armature current.

For good generators the value of overall or commercial efficiency may be *as high as 95%*.

Motor :

1. *Electrical efficiency* (η_e)

$$\eta_e = \frac{\text{Mechanical power developed}}{\text{Total electrical power input}} = \frac{E_b I_a}{VI}. \quad ...(21)$$

2. *Mechanical efficiency* (η_m)

$$\eta_m = \frac{\text{Useful mechanical power output}}{\text{Mechanical power developed}}$$

$$= \frac{\text{B.H.P. of motor} \times 735.5}{E_b I_a} \quad ...(22)$$

3. *Overall or commercial efficiency* ($\eta_{om} = \eta_m \times \eta_e$)

$$\eta_{om} = \frac{\text{Useful mechanical power output}}{\text{Total electrical power input}}$$

$$= \frac{\text{B.H.P. of motor} \times 735.5}{VI} \quad ...(23)$$

η_{om} can also be expressed as follows :

Overall efficiency of motor

$$\eta_{om} = \frac{\text{Useful power output}}{\text{Total power input}} = \frac{\text{Total power input} - \text{total losses}}{\text{Total power input}}$$

$$= \frac{VI - \text{total losses}}{VI} \quad ...(24)$$

where E_b = back e.m.f. V = supply voltage
I = load current I_a = armature current.

9.3.1. Condition for maximum efficiency. Condition for maximum efficiency for a D.C. generator or D.C. motor is same. For a D.C. generator the condition for maximum efficiency is derived as follows :

Generator power output $= VI$.

If flux and speed are constant all losses except armature copper loss are constant.

$$\text{Losses} = \text{armature copper loss} + \text{constant loss}$$

$$= (I + I_{sh})^2 R_a + P_c = I^2 R_a + P_c$$

[Neglecting I_{sh} in comparison with load current I]

where I_{sh} is the shunt field current and P_c denotes constant losses which include iron loss, field winding loss and mechanical loss.

$$\text{Efficiency,} \ \eta = \frac{\text{Output}}{\text{Input}} = \frac{\text{Output}}{\text{Output} + \text{losses}}$$

$$= \frac{VI}{VI + I^2 R_a + P_c} \quad ...(25)$$

η is maximum when

$$\frac{d\eta}{dI} = 0 = \frac{(VI + I^2 R_a + P_c)V - VI(V + 2IR_a)}{(VI + I^2 R_a + P_c)^2}$$

or

$$I^2 R_a = P_c \quad ...(26)$$

Hence, *efficiency will be maximum when variable losses are equal to constant losses*

Further (from eqn. 26)

$$I = \sqrt{\frac{P_c}{R_a}} \quad ...(27)$$

Thus the efficiency increases with increase in load current, reaches a maximum value when load current equals the value given by eqn. (27) and then starts decreasing.

Efficiency curve. The efficiency of a machine is different at different values of power output. *As the output increases,* the *efficiency increases till it reaches a maximum value.* As the *output is further increased, the efficiency starts decreasing.* A graph of efficiency *vs.* output is called efficiency curve. A typical efficiency curve is shown in Fig. 96. *The machines are so designed as to give maximum efficiency at or near the rated output of the machine.* Since the generators operate at a constant terminal voltage V, the efficiency curve of a generator can be drawn between efficiency and load current *I*.

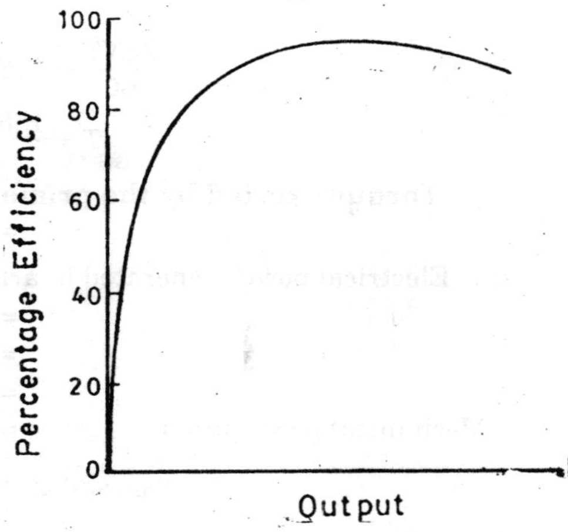

Fig. 96. Efficiency curve.

☞**Example 43.** *The armature and shunt field resistances of 110 V shunt generator running at 800 r.p.m. are 0.12 Ω, 55 Ω respectively. It supplies 220 lamps each rated at 60 W, 110 V. The friction, windage and core loss of the machine is equal to 410 W. Calculate :*

(*i*) *Armature copper loss on full-load and shunt field copper loss.*

(*ii*) *Torque in Nm exerted by the prime-mover.*

(*iii*) *Mechanical, electrical and overall efficiencies.*

Solution. Terminal voltage, $V = 110$ V

Speed of the generator, $N = 800$ r.p.m.

Armature resistance, $R_a = 0.12\ \Omega$

Shunt field resistance, $R_{sh} = 55\ \Omega$

Friction, windage and core loss = 410 W

Full-load current, $I = \dfrac{220 \times 60}{110} = 120$ A

Shunt field current, $I_{sh} = \dfrac{V}{R_{sh}} = \dfrac{110}{55} = 2$ A

$I_a = I + I_{sh} = 120 + 2 = 122$ A

(*i*) Armature copper loss $= I_a^2 R_a$

$= 122^2 \times 0.12 = 1786$ W

∴ **Armature copper loss = 1786 W. (Ans.)**

Shunt field copper loss $= VI_{sh} = 110 \times 2 =$ **220 W. (Ans.)**

[Shunt field copper loss $= I_{sh}^2 R_{sh} = 2^2 \times 55 = 220$ W]

(*ii*) Total copper loss = 1786 + 220 = 2006 W

Core, friction and windage loss = 410 W

∴ Total loss = 2006 + 410 = 2416 W

Output = 220 × 60 = 13200 W

Input = output + losses = 13200 + 2416 = 15616 W

If *T* is the torque in Nm exerted by the prime mover

$$\frac{2\pi NT}{60} = \text{input in watts}$$

$\therefore \qquad \frac{2\pi NT}{60} = 15616$

i.e., $\qquad T = \frac{15616 \times 60}{2\pi N} = \frac{15616 \times 60}{2\pi \times 800} = 186.4$ Nm

$\therefore$ **Torque exerted by the prime mover**

$= \mathbf{186.4}$ **Nm. (Ans.)**

(*iii*) Electrical power generated in armature

$= E_b I_a$

= mechanical input – stray losses

= 15616 – 410 = 15206 W

Mechanical power input = 15616 W.

$$\therefore \qquad \eta_{mechanical} = \frac{\text{Power generated in armature}}{\text{Mechanical power input}} \times 100$$

$$= \frac{15206}{15616} \times 100 = \mathbf{97.4\%.} \quad \textbf{(Ans.)}$$

Electrical power generated = 220 × 60 = 13200 W

$$\text{Electrical efficiency, } \eta_{electrical} = \frac{\text{Electrical power output}}{\text{Total power developed in armature}} \times 100$$

$$= \frac{13200}{15206} \times 100 = \mathbf{86.8\%.} \quad \textbf{(Ans.)}$$

$$\eta_{overall} = \frac{\text{Electrical power output}}{\text{Mechanical power input}} \times 100$$

$$= \frac{13200}{15616} \times 100 = \mathbf{84.53\%.} \quad \textbf{(Ans.)}$$

Example 44. *Find the efficiency at full-load and at half-load for a 1200 kW, 600 V shunt generator using the following data :*

Armature resistance = 0.005 Ω, shunt field resistance = 60 Ω, brush contact drop = 1 V per brush, mechanical and iron losses at rated load = 20 kW, stray load loss = 1.2% of output.

Solution. Terminal voltage, $V = 600$ V

Full-load output = 1200 kW

Armature resistance, $R_a = 0.005\ \Omega$

Shunt field resistance, $R_{sh} = 60\ \Omega$

Brush contact drop = 1 V per brush

Mechanical losses at rated load = 20 kW

Stray load loss = 1.2% of output

Efficiency at full-load and at half-load :

(*i*) **At full-load**

Full load current, $I = \frac{\text{Full-load output}}{\text{Voltage}} = \frac{1200 \times 1000}{600} = 2000$ A

Shunt field current, $I_{sh} = \frac{V}{R_{sh}} = \frac{600}{60} = 10$ A

$\therefore$ Armature current, $I_a = I + I_{sh} = 2000 + 10 = 2010$ A

Armature copper loss $= I_a^2 R_a = 2010^2 \times 0.005 = 20200$ W

Shunt field loss $= VI_{sh} = 600 \times 10 = 6000$ W

Brush contact loss = 2 × 2010 = 4020 W

Mechanical and iron losses = 20 × 1000 = 20000 W

Stray load losses $= 1.2\%$ of full-load output

$$= \frac{1.2}{100} \times 1200 \times 1000 = 14400 \text{ W}$$

Efficiency $= \dfrac{\text{Output}}{\text{Output + Losses}}$

$$= \frac{1200 \times 1000}{1200 \times 1000 + 20200 + 6000 + 4020 + 20000 + 14400}$$

$$= \frac{12 \times 10^5}{1200000 + 64620} = \frac{12 \times 10^5}{1264620} = 0.949 \text{ or } 94.9\%$$

Hence, **efficiency at full-load = 94.9%. (Ans.)**

At half-load :

Output $= \dfrac{1200 \times 1000}{2} = 600000$ W

Load current $= \dfrac{2000}{2} = 1000$ A

Armature current, $I_a = 1000 + 10 = 1010$ A

Armature copper loss, $I_a{}^2R_a = 1010^2 \times 0.005 = 5100$ W

Shunt field loss $= VI_{sh} = 600 \times 10 = 6000$ W

Brush contact loss $= 2 \times 1010 = 2020$ W

Mechanical and iron losses $= 20000$ W

Stray load losses $= 1.2\%$ of $600000 = 7200$ W

Efficiency $= \dfrac{\text{Output}}{\text{Output + Losses}}$

$$= \frac{600000}{600000 + 5100 + 6000 + 2020 + 20000 + 7200}$$

$$= \frac{600000}{640320} = 0.937 \text{ or } 93.7\%.$$

Hence, **efficiency at half-load = 93.7%. (Ans.)**

Example 45. *The following data pertains to a short-shunt compound generator :*

Armature resistance = 0.04 Ω, series field resistance = 0.025 Ω, shunt field resistance = 45 Ω, contact drop / brush = 1 V, magnetic losses = 2.5 kW, mechanical losses = 1 kW.

Find the efficiency of the generator when it delivers 400 A at 440 V.

Solution. Terminal voltage, $V = 440$ V

Load current, $I = 400$ A

Armature resistance, $R_a = 0.04\ \Omega$

Series field resistance, $R_{se} = 0.025\ \Omega$

Shunt field resistance, $R_{sh} = 45\ \Omega$

Copper drop/brush $= 1$ V

Magnetic losses $= 2.5$ kW $= 2500$ W

Mechanical losses $= 1$ kW $= 1000$ W

Refer Fig. 97

Output of the generator $= VI = 440 \times 400 = 176000$ W

Voltage across shunt field winding $= V + IR_{se} = 440 + 400 \times 0.025 = 450$ V

Shunt field current, $I_{sh} = \dfrac{V}{R_{sh}} = \dfrac{450}{45} = 10$ A

Armature current, $I_a = I + I_{sh} = 400 + 10 = 410$ A

Armature copper loss $= I_a{}^2R_a = 410^2 \times 0.04 = 6724$ W

Brush contact loss = voltage drop/pair of brush × I_a

$= 2 \times 410 = 820$ W

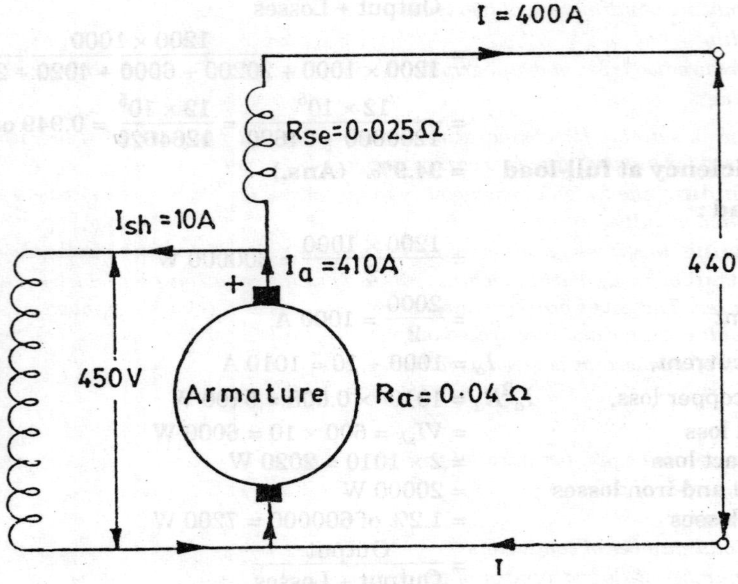

Fig. 97

Series field copper loss $= I^2R_{se} = 400^2 \times 0.025 = 4000$ W

Shunt field copper loss $= VI_{sh} = 450 \times 10 = 4500$ W

Magnetic losses = 2500 W [Given]

Mechanical losses = 1000 W [Given]

Total losses = 6724 + 820 + 4000 + 4500 + 2500 + 1000 = 19544 W

Efficiency $= \dfrac{\text{output}}{\text{output + losses}} = \dfrac{176000}{176000 + 19544}$

$= \dfrac{176000}{195544} = 0.90$ or 90%

Hence, **efficiency of the generator = 90%. (Ans.)**

HIGHLIGHTS

1. **Basic type of D.C. machine** *is that of commutator type. This is actually an alternating current (A.C.) machine, but furnished with a special device, a commutator, which under certain conditions converts alternating current into direct current.*
2. *An electrical* **generator** *is a machine which converts mechanical energy (or power) into electrical energy (or power). It works on the following principle :*

   "Whenever a conductor cuts magnetic flux, dynamically induced e.m.f. is produced in it according to Faraday's Laws of Electromagnetic induction".
3. *A D.C. machine consists of two* **main** *parts :*
   — **Stationary Part :** *designed mainly for producing magnetic flux.*
   — **Rotating Part :** *called armature, where mechanical energy is converted into electrical (electric generator) or, conversely, electrical energy into mechanical (electric motor).*

The various parts of a D.C. machine are enumerated below :
Frame, Field poles, Commutating poles, Armature, Commutator, Brush gear, Armature shaft bearings, Armature windings.

4. *According to the degree of closure produced by winding, armature windings are of following two types :*
1. Open coil winding *2. Closed coil winding*
The closed armature windings are of two types :
(i) Ring winding *(ii) Drum winding*
In general there are two types of drum armature windings :
(i) Lap winding *(ii) Wave winding.*

5. **'Lap winding'** *is suitable for comparatively low voltage but high current generators whereas* **'wave winding'** *is used for high voltage, low current machines.*

6. *In the* **'lap winding'** *the finish of each coil is connected to the start of the next coil so that winding or commulator pitch is unity.*
In **'wave winding'** *the finish of coil is connected to the start of another coil well away from the first coil.*

7. *The* **rated output** *of an electrical machine is its duty in the conditions for which it has been designed by the manufacturer. The rated duty is characterised by values given on the machine's rating plate, and termed rated value, as, for instance : rated output, rated voltage, rated current, rated speed, etc.*

8. *E.m.f. equation of a generator is given as follows :*

$$E_g = \frac{p\phi ZN}{60a}$$

where E_g = *generated e.m.f. per parallel path in armature.*
p = *number of poles*
ϕ = *flux/pole, Wb*
Z = *total number of conductors*
= *number of slots × number of conductors/slot*
N = *rotational speed of armature, r.p.m.*
a = *number of parallel paths in armature*

[*For a* **wave wound** *generator :* $a = 2$
For a **lap wound** *generator :* $a = p$]

9. *According to* **method of excitation** *D.C. generators are classified as follows :*
1. Separately excited generators *2. Self excited generators.*
Self excited generators can be divided in accordance with how the field winding is connected into generators as follows :
(i) Shunt wound generators *(ii) Series wound generators*
(iii) Compound wound generators :
(a) Short shunt
(b) Long shunt.

10. *The electric motor is a machine which converts electrical energy into mechanical energy.*

11. *Voltage equation of a motor is* $V = E_b + I_a R_a$.

12. *Condition for maximum power developed in armature is* $E_b = \frac{V}{2}$.

13. *Armature torque of a motor is given by* $T_a = 0.159\, Z\phi p \times \frac{I_a}{a}$ *Nm.*

14. *Shaft or useful torque is given by* $T_{sh} = \frac{B.H.P.\ (metric) \times 735.5}{2\pi N/60}$ *Nm*

Torque lost in iron and friction losses $= 0.159 \times \frac{\text{iron and friction losses}}{N/60}$ *Nm.*

15. *Speed of a D.C. motor is given by* $N = \frac{E_b}{\phi}\left(\frac{60a}{Zp}\right)$ *r.p.m.*

16. *Speed regulation of a motor is given by*

$$\text{Percentage speed regulation} = \frac{\text{no-load speed} - \text{full-load speed}}{\text{full-load speed}}.$$

17. *In a D.C. motor reversal is accomplished by changing the polarity of either the armature or the field, but not by changing both.*

OBJECTIVE TYPE QUESTIONS

D.C. GENERATOR

Choose the Correct Answer :

1. Laminations of core are generally made of
   (*a*) cast iron (*b*) carbon
   (*c*) silicon steel (*d*) stainless steel.
2. Which of the following could be approximately the thickness of laminations of a D.C. machine ?
   (*a*) 0.005 mm (*b*) 0.05 mm
   (*c*) 0.5 mm (*d*) 5 mm.
3. The armature of D.C. generator is laminated to
   (*a*) reduce the bulk (*b*) provide passage for cooling air
   (*c*) insulate the core (*d*) reduce eddy current loss.
4. The resistance of armature winding depends on
   (*a*) length of conductor (*b*) cross-sectional area of the conductor
   (*c*) number of conductors (*d*) all of the above.
5. The field coils of D.C. generator are usually made of
   (*a*) mica (*b*) copper
   (*c*) cast iron (*d*) carbon.
6. The commutator segments are connected to the armature conductors by means of
   (*a*) copper lugs (*b*) resistance wires
   (*c*) insulation pads (*d*) brazing.
7. In a commutator
   (*a*) copper is harder than mica (*b*) mica and copper are equally hard
   (*c*) mica is harder than copper (*d*) none of the above.
8. In D.C. generators the pole shoes are fastened to the pole core by
   (*a*) rivets (*b*) counter sunk screws
   (*c*) brazing (*d*) welding.
9. According to Fleming's right-hand rule for finding the direction of induced e.m.f., when middle finger points in the direction of induced e.m.f., forefinger will point in the direction of
   (*a*) motion of conductor (*b*) lines of force
   (*c*) either of the above (*d*) none of the above.
10. Fleming's right-hand rule regarding direction of induced e.m.f., correlates
    (*a*) magnetic flux, direction of current flow and resultant force
    (*b*) magnetic flux, direction of motion and the direction of e.m.f. induced
    (*c*) magnetic field strength, induced voltage and current
    (*d*) magnetic flux, direction of force and direction of motion of conductor.
11. While applying Fleming's right-hand rule to find the direction of induced e.m.f., the thumb points towards
    (*a*) direction of induced e.m.f.
    (*b*) direction of flux

(*c*) direction of motion of the conductor if forefinger points in the direction of generated e.m.f.
(*d*) direction of motion of conductor, if forefinger points along the lines of flux.

12. The bearings used to support the rotor shafts are generally
(*a*) ball bearings (*b*) bush bearings
(*c*) magnetic bearings (*d*) needle bearings.

13. In D.C. generators, the cause of rapid brush wear may be
(*a*) severe sparking (*b*) rough commutator surface
(*c*) imperfect contact (*d*) any of the above.

14. In lap winding, the number of brushes is always
(*a*) double the number of poles (*b*) same as the number of poles
(*c*) half the number of poles (*d*) two.

15. For a D.C. generator when the number of poles and the number of armature conductors are fixed, then which winding will give the higher e.m.f. ?
(*a*) Lap winding (*b*) Wave winding
(*c*) Either of (*a*) and (*b*) above (*d*) Depends on other features of design.

16. A 4 pole D.C. generator rotates at 1500 r.p.m. How many times in a second will the e.m.f. induced in each armature conductor change its direction ?
(*a*) 100 (*b*) 50
(*c*) 400 (*d*) 200.

17. If B is the flux density, l the length of conductor and v the velocity of conductor, then induced e.m.f. is given by
(*a*) Blv (*b*) Blv^2
(*c*) Bl^2v (*d*) Bl^2v^2.

18. In case of a wave wound generator, the average pitch
(*a*) must be even (*b*) must be odd
(*c*) may be odd or even (*d*) is generally fractional.

19. In case of a 4 pole D.C. generator provided with a two layer lap winding with sixteen coils, the pole pitch will be
(*a*) 4 (*b*) 8
(*c*) 16 (*d*) 32.

20. The material for commutator brushes is generally
(*a*) mica (*b*) copper
(*c*) cast iron (*d*) carbon.

21. The insulating material used between the commutator segments is normally
(*a*) graphite (*b*) paper
(*c*) mica (*d*) insulating varnish.

22. In a D.C. generator, if p be the number of poles and N be the r.p.m. of rotor, then the frequency of magnetic reversals will be
(*a*) $\frac{Np}{2}$ (*b*) $\frac{Np}{60}$
(*c*) $\frac{Np}{120}$ (*d*) $\frac{Np}{3000}$.

23. For generating large currents on D.C. generators which winding is generally preferred ?
(*a*) Progressive wave winding (*b*) Lap winding
(*c*) Retrogressive wave winding (*d*) Current depends on design.

24. The purpose of providing dummy coils in a generator is
(*a*) to enhance flux density (*b*) to amplify voltage
(*c*) to provide mechanical balance for the rotor (*d*) to reduce eddy currents.

D.C. MOTOR

25. No-load speed of which of the following motor will be highest ?
(*a*) Shunt motor (*b*) Series motor
(*c*) Cumulative compound motor (*d*) Differentially compound motor.

26. The direction of rotation of a D.C. series motor can be changed by
(*a*) interchanging supply terminals (*b*) interchanging field terminals
(*c*) either of (*a*) and (*b*) above (*d*) none of the above.

27. Which of the following applications requires high starting torque ?
(*a*) Lathe machine (*b*) Centrifugal pump
(*c*) Locomotive (*d*) Air blower.

28. In case a D.C. motor is connected across A.C. supplies
(*a*) the motor will burn (*b*) the motor will run at rated speed
(*c*) the motor will run at corresponding synchronous speed
(*d*) the motor will run at slow speed.

29. If a D.C. motor is to be selected for conveyors, which motor would be preferred ?
(*a*) Series motor (*b*) Shunt motor
(*c*) Differentially compound motor (*d*) Cumulative compound motor.

30. Which D.C. motor will be preferred for machine tools ?
(*a*) Series motor (*b*) Shunt motor
(*c*) Cumulative compound motor (*d*) Differentially compound motor.

31. Which D.C. motor will be preferred for constant speed line shafting ?
(*a*) Cumulatively compound motor (*b*) Differentially compound motor
(*c*) Shunt motor (*d*) Series motor.

32. Differentially compound D.C. motors can find applications requiring
(*a*) high starting torque (*b*) low starting torque
(*c*) variable speed (*d*) frequent on-off cycles.

33. Which D.C. motor is preferred for elevators ?
(*a*) Shunt motor (*b*) Series motor
(*c*) Differentially compound motor (*d*) Cumulatively compound motor.

34. According to Fleming's left-hand rule, when the forefinger points in the direction of the field or flux, the middle finger will point in the direction of
(*a*) current in the conductor (*b*) movement of conductor
(*c*) resultant force on conductor (*d*) none of the above.

35. If the field of a D.C. shunt motor gets opened while motor is running
(*a*) the speed of motor will be reduced (*b*) the armature current will reduce
(*c*) the motor will attain dangerously high speed
(*d*) the motor will continue to run at constant speed.

36. Starters are used with D.C. motors because
(*a*) these motors have high starting torque (*b*) these motors are not self-starting
(*c*) back e.m.f. of these motors is zero initially
(*d*) to restrict armature current as there is no back e.m.f. while staring.

37. Which of the following D.C. motors has self-load properties ?
(*a*) Series motor (*b*) Shunt motor
(*c*) Cumulatively compound motor (*d*) Differentially compound motor.

38. In D.C. shunt motors as load is reduced
(*a*) the speed will increase abruptly
(*b*) the speed will increase in proportion to reduction in load
(*c*) the speed will remain almost constant
(*d*) the speed will reduce.

39. What will happen if the back e.m.f. of a D.C. motor vanishes
(*a*) the motor will stop
(*b*) the motor will continue to run
(*c*) the armature may burn
(*d*) the motor will run noisy.

40. In case of D.C. shunt motors the speed is dependent on back e.m.f. only because
(*a*) back e.m.f. is equal to armature drop
(*b*) armature drop is negligible
(*c*) flux is proportional to armature current
(*d*) flux is practically constant in D.C. shunt motors.

41. In a D.C. shunt motor, under the conditions of maximum power, the current in the armature will be
(*a*) almost negligible
(*b*) rated full-load current
(*c*) less than full-load current
(*d*) more than full-load current.

42. Which D.C. motor will have least percentage increase of input current, for the same percentage increase in the torque ?
(*a*) Shunt motor
(*b*) Series motor
(*c*) Cumulatively compound motor
(*d*) Separately excited motor.

43. These days D.C. motors are widely used in
(*a*) pumping sets
(*b*) air compressors
(*c*) electric traction
(*d*) machine shops.

44. By looking at which part of the motor, it can be easily confirmed that a particular motor is D.C. motor ?
(*a*) frame
(*b*) shaft
(*c*) commutator
(*d*) stator.

45. In which of the following applications D.C. series motor is invariably used ?
(*a*) Starter for a car
(*b*) Drive for a water pump
(*c*) Fan motor
(*d*) Motor operation in A.C. or D.C.

46. In D.C. machines fractional pitch winding is used
(*a*) to improve cooling
(*b*) to reduce copper losses
(*c*) to increase the generated e.m.f.
(*d*) to reduce the sparking.

47. A three point starter is considered as suitable for
(*a*) shunt motors
(*b*) shunt as well as compound motors
(*c*) shunt, compound and series motors
(*d*) all D.C. motors.

48. Small D.C. motors up to 5 H.P. usually have
(*a*) 2 poles
(*b*) 4 poles
(*c*) 6 poles
(*d*) 8 poles.

49. In case the conditions for maximum power for a D.C. motor are established, the efficiency of the motor will be
(*a*) 100%
(*b*) around 90%
(*c*) anywhere between 75% and 90%
(*d*) less than 50%.

50. A shearing machine has cyclic load consisting of intermittent light and heavy loads. Which of the following D.C. motors will be suitable for this purpose ?
(*a*) Series motor
(*b*) Shunt motor
(*c*) Cumulative compound motor
(*d*) Differentially compound motor.

51. The ratio of starting torque to full-load torque is least in case of
(*a*) series motor
(*b*) shunt motor
(*c*) compound motors
(*d*) none of the above.

52. In D.C. motor which of the following can sustain the maximum temperature rise ?
(*a*) Slip rings
(*b*) Commutator
(*c*) Field winding
(*d*) Armature winding.

53. **Which of the following law/rule can be used to determine the direction of rotation of D.C. motor ?**
(*a*) Lenz's law (*b*) Faraday's law
(*c*) Coloumb's law (*d*) Fleming's left-hand rule.

54. **Which of the following loads normally needs staring torque more than the rated torque ?**
(*a*) Blowers (*b*) Conveyors
(*c*) Air compressors (*d*) Centrifugal pumps.

55. **The starting resistance of a D.C. motor is generally**
(*a*) low (*b*) around 500 Ω
(*c*) 1000 Ω (*d*) infinitely large.

56. **In case of conductively compensated D.C. series motors, the compensating winding is provided**
(*a*) as separately wound unit (*b*) in parallel with armature winding
(*c*) in series with armature winding (*d*) in parallel with field winding.

57. **Sparking at the commutator of a D.C. motor may result in**
(*a*) damage to commutator segments (*b*) damage to commutator insulation
(*c*) increased power consumption (*d*) all of the above.

58. **Which of the following motors is preferred for operation in highly explosive atmosphere ?**
(*a*) Series motor (*b*) Shunt motor
(*c*) Air motor (*d*) Battery operated motor.

59. **If the supply voltage for a D.C. motor is increased, which of the following will decrease ?**
(*a*) Staring torque (*b*) Operating speed
(*c*) Full-load current (*d*) All of the above.

60. **When the speed of a D.C. motor increases**
(*a*) back e.m.f. increases and current drawn decreases
(*b*) back e.m.f. as well as current drawn both increase
(*c*) back e.m.f. as well as current drawn both decrease
(*d*) back e.m.f. decreases and current drawn increases.

61. **As compared to an induction motor, the air gap in a D.C. motor is**
(*a*) less than 50% (*b*) between 50% and 90%
(*c*) same (*d*) more.

62. **Field winding of a D.C. series motor is usually provided with thick wire**
(*a*) to provide large flux (*b*) to reduce the use of insulating materials
(*c*) as it carries large load current (*d*) in order to reduce eddy current.

63. **Which one of the following is not the function of pole shoes in a D.C. machine ?**
(*a*) To reduce eddy current loss (*b*) To support the field coils
(*c*) To spread out flux for better uniformity (*d*) To reduce the reluctance of the magnetic path.

64. **The mechanical power developed by a shunt motor will be maximum when the ratio of back e.m.f. to applied voltage is**
(*a*) 4.0 (*b*) 2.0
(*c*) 1.0 (*d*) 0.5.

65. **The condition for maximum power in case of D.C. motor is**
(*a*) back e.m.f. = 2 × supply voltage (*b*) back e.m.f. = $\frac{1}{2}$ × supply voltage
(*c*) supply voltage = $\frac{1}{2}$ × back e.m.f. (*d*) supply voltage = back e.m.f.

66. **For which of the following applications a D.C. motor is preferred over an A.C. motor ?**
(*a*) Low speed operation (*b*) High speed operation
(*c*) Variable speed operation (*d*) Fixed speed operation.

67. **What will happen in case 220 V D.C. series motor is connected to 220 V A.C. supply?**
(*a*) The armature winding of motor will burn
(*b*) The motor will vibrate violently

(*c*) The motor will run with less efficiency and more sparking
(*d*) The motor will not run.

68. In D.C. machines the residual magnetism is of the order of
(*a*) 2 to 3 per cent (*b*) 10 to 15 per cent
(*c*) 20 to 25 per cent (*d*) 50 to 75 per cent.

69. If T_a be the torque and I_a the armature current for a D.C. motor, then which of the following relation is valid before saturation ?
(*a*) $T_a \propto I_a$ (*b*) $T_a \propto \frac{I}{I_a}$
(*c*) $T_a \propto I_a^2$ (*d*) $T_a \propto \frac{I}{(I_a)^2}$.

70. Which D.C. motor is generally preferred for cranes and hoists ?
(*a*) Series motor (*b*) Shunt motor
(*c*) Cumulatively compound motor (*d*) Differentially compound motor.

ANSWERS

1. (*c*)	**2.** (*c*)	**3.** (*d*)	**4.** (*d*)	**5.** (*b*)	**6.** (*a*)	**7.** (*c*)
8. (*b*)	**9.** (*b*)	**10.** (*b*)	**11.** (*d*)	**12.** (*a*)	**13.** (*d*)	**14.** (*b*)
15. (*b*)	**16.** (*a*)	**17.** (*a*)	**18.** (*c*)	**19.** (*b*)	**20.** (*d*)	**21.** (*c*)
22. (*c*)	**23.** (*b*)	**24.** (*c*)	**25.** (*b*)	**26.** (*b*)	**27.** (*c*)	**28.** (*a*)
29. (*a*)	**30.** (*b*)	**31.** (*c*)	**32.** (*b*)	**33.** (*d*)	**34.** (*a*)	**35.** (*c*)
36. (*d*)	**37.** (*d*)	**38.** (*c*)	**39.** (*c*)	**40.** (*d*)	**41.** (*d*)	**42.** (*b*)
43. (*c*)	**44.** (*c*)	**45.** (*a*)	**46.** (*d*)	**47.** (*b*)	**48.** (*a*)	**49.** (*d*)
50. (*c*)	**51.** (*b*)	**52.** (*c*)	**53.** (*d*)	**54.** (*b*)	**55.** (*a*)	**56.** (*c*)
57. (*d*)	**58.** (*c*)	**59.** (*c*)	**60.** (*a*)	**61.** (*d*)	**62.** (*c*)	**63.** (*a*)
64. (*d*)	**65.** (*b*)	**66.** (*c*)	**67.** (*c*)	**68.** (*a*)	**69.** (*c*)	**70.** (*a*).

THEORETICAL QUESTIONS

D.C. GENERATOR

1. What is a basic type of D.C. machine ?
2. Explain the principle on which a generator works.
3. Explain the construction and working of an elementary generator.
4. Describe briefly various parts of a D.C. machine.
5. Why is a commutator and brush arrangement necessary for operation of a D.C. machine ?
6. Name the main parts of a D.C. machine and state the materials of which each part is made.
7. Enumerate all the parts of a D.C. machine and indicate their functions.
8. Give the materials and functions of the following parts of a D.C. machine :
   (*i*) Field poles (*ii*) Yoke
   (*iii*) Commutator (*iv*) Commutating poles
   (*v*) Armature.
9. Give the constructional features and working principle of a D.C. generator. Draw a cross-sectional view of a 4 pole D.C. generator and label there on all parts. Also sketch the shapes of :
   (*i*) Commutator segment with riser, and
   (*ii*) Armature stamping.
10. Give the advantages and uses of lap and wave windings.
11. Explain the following terms as applied to a D.C. armature winding.
    (*i*) Front pitch (*ii*) Back pitch
    (*iii*) Pole pitch (*iv*) Commutator pitch.

12. Draw up the winding table for a 4 pole, wave connected armature having 30 coil sides and give a developed diagram of the winding showing the polarity and position of the brushes, the main poles and the direction of motion of the armature for a D.C. motor.
13. Draw the developed winding diagram of lap winding for 6 poles, 18 slots with two coil sides/slot, double layer showing therein position of poles, direction of motion, direction of generated e.m.f. and position of brushes.
14. Derive the e.m.f. equation of a D.C. generator.
15. Deduce an expression for the voltage induced in a D.C. generator.
16. Write a short note on equiliser rings.
17. How are ratings of electrical machines expressed ?
18. Draw the magnetic circuit of a 4 pole D.C. machine.
19. Explain different methods of excitation of D.C. generators with suitable diagrams.
20. How are D.C. generators classified ?
21. What is the difference between a separately excited and a self-excited generator ?
22. Sketch the following types of D.C. generators :
    (*i*) Shunt (*ii*) Series
    (*iii*) Compound.
    State with reason(s) where each is used ?
23. What is the difference between the short-shunt and long shunt compound generators ?
24. With the help of a neat diagram show power division in a D.C. generator.
25. Answer the following questions briefly :
    (*i*) State the principle on which generators operate.
    (*ii*) Name the materials of which the following parts of a D.C. machine are made and why ?
    (*a*) Frame (*b*) Armature
    (*c*) Bushes.
    (*iii*) Why are field coils provided in a D.C. generator ?
    (*iv*) What is the function of an armature in a D.C. generator ?
    (*v*) Of what material are the laminations of an armature made up of ?
    (*vi*) What is the function of a commutator ?
    (*vii*) What purpose is served by brushes in a D.C. machine ?

D.C. MOTOR

26. What is the working principle of a D.C. motor ?
27. "In every D.C. generator motor action occurs and in every D.C. motor a generator action occurs". Explain.
28. What is torque ? What is the source of the torque force in a D.C. motor ?
29. What is back e.m.f. or counter voltage ?
30. What is the effective voltage across a D.C. motor armature ?
31. What is the meaning of armature power ?
32. Is electrical power related to mechanical power ?
33. Explain the function of commutator in a D.C. motor. In what respects is the commutation process in D.C. motors different from that in generators ?
34. What is mean by speed regulation ?
35. What is meant by the term 'shunt motor' ?
36. What is meant by the term 'series motor' ?
37. What is the dominant speed characteristic of a shunt motor ?
38. Explain speed-current and torque-current characteristics of a series motor ?
39. What is meant by a compound motor ?
40. What conditions require the use of a compound motor ?
41. How is D.C. motor reversed ?

42. Why is a starter necessary for a motor ? Give the diagram and explain the working of a three-point starter for a shunt motor.

43. What are the functions of 'no-volt release' and 'overload release' in a starter ? Discuss the operation of these two features in a shunt motor starter.

44. Give the reasons for the following :

(*i*) A series motor should not be connected to a load through a belt.

(*ii*) A series motor develops a high starting torque.

(*iii*) A differential compound motor is very rarely used.

(*iv*) A shunt motor runs at almost constant speed irrespective of load current.

45. Derive an expression for the torque of a D.C. motor.

46. Differentiate between the generator action and motor action of a D.C. machine.

47. How may the direction of rotation of a D.C. shunt motor be reversed ? What is the effect of reversing the line terminals ?

48. Develop the general expression for the speed of a motor in terms of supply voltage, armature resistance and flux per pole.

49. Explain the 'armature reaction', in D.C. shunt motors, indicating also few remedies to its adverse effects.

50. Explain in detail how the shunt motor behaves as a constant speed motor and the series motor as a variable speed motor.

51. Draw the characteristic curves of D.C. shunt and series motors. Use these curves to explain the applications for which these motors are used.

52. Discuss the applications of shunt, series and compound motors.

53. State the types of loads for which shunt, series and compound D.C. motors are suitable. Give reasons in support of your answer.

54. Why is a starter necessary for D.C. motor ? Explain the working of a three-point starter with the help of a neat diagram.

55. What is the difference between three-point and four-point starters ? What are the additional features incorporated in a four-point starter.

56. Explain what happens, giving reasons thereof, in the following cases :

(*i*) In a D.C. shunt motor armature terminals are reversed.

(*ii*) In a D.C. shunt motor field terminals are reversed.

(*iii*) The line terminals to a D.C. motor interchanged.

(*iv*) A D.C. shunt motor is directly connected to the supply mains.

(*v*) A D.C. motor is stopped by throwing back the starting arm.

57. Explain with neat diagrams construction and working of a series motor starter.

58. How does D.C. motor automatically adjust input to match the mechanical load on the motor ?

UNSOLVED EXAMPLES

D.C. GENERATOR

1. An eight-pole lap connected armature runs at 750 r.p.m. It has 800 conductors and flux/pole of 60 mWb. Calculate the e.m.f. generated. [**Ans.** 600 volts]

2. An 8-pole lap connected generator has a useful flux/pole of 0.05 Wb. If the no load voltage at 400 r.p.m. is 300 V find the conductors on the armature periphery. [**Ans.** 900]

3. An eight-pole wave connected armature has 300 conductors and runs at 400 r.p.m. Determine the useful flux/pole if the electromotive force generated on open circuit is 500 V. [**Ans.** 6.25×10^{-2} Wb]

4. A 6-pole wave connected D.C. generator has 52 slots and each slot has 20 conductors. Find the speed of the generator. The generated e.m.f. is 240 V and the flux per pole is 5 mWb. [**Ans.** 923 r.p.m.]

5. An 8-pole wave connected D.C. generator has 900 armature conductors and a flux/pole of 0.04 Wb. At what speed it must be driven to generate 500 V ? [**Ans.** 208.3 r.p.m.]

6. The lap wound armature of a 4-pole generator has 51 slots. Each slot contains 20 conductors. What will be the e.m.f. generated in machine when driven at 1500 r.p.m. if useful flux per pole is 0.01 Wb ? [**Ans.** 255 V]

7. An 8-pole lap wound generator armature has 960 conductors, a flux of 40 mWb and a speed of 400 r.p.m. Calculate the e.m.f. generated on open circuit. If the same armature is wave wound, at what speed must it be driven to generate 400 volts ? [**Ans.** 256 V, 156.25 r.p.m.]

8. An 8-pole D.C. shunt generator with 788 wave connected conductors and running at 500 r.p.m. supplies a load of 12.5 Ω resistance at a terminal voltage of 250 V. The armature resistance is 0.24 Ω and the field resistance is 250 Ω. Find :

(*i*) Armature current (*ii*) Induced e.m.f.

(*iii*) Flux/pole. [**Ans.** 21 A, 255.04 V, 9.83 mWb]

9. A 4-pole lap wound D.C. shunt generator has a useful flux/pole 0.07 Wb. The armature winding consists of 220 turns, each turn having a resistance of 0.004 Ω. Calculate the terminal voltage when running at 900 r.p.m. if the armature current is 50 A. [**Ans.** 459.25 V]

10. A shunt generator has a generated voltage on open circuit of 127 V. When the machine is on load, terminal voltage is 120 V. Find the load current if the field resistance be 15 Ω and the armature resistance 0.02 Ω. Neglect armature reaction. [**Ans.** 342 A]

11. A short shunt compound generator has armature, series field and shunt field resistances of 0.08 Ω, 0.04 Ω and 100 Ω respectively. It supplies 125 lamps each rated at 250 V, 40 W. Find the generated e.m.f. and the armature current. Drop under each brush may be taken as 1 V. [**Ans.** 22.508 A, 254.6 V]

D.C. MOTOR

12. A 250 V D.C. shunt motor takes 41 A at full-load. Find the back e.m.f. on full-load if the resistances of motor armature and shunt field windings are 0.1 ohm and 250 ohms respectively. [**Ans.** 246 V]

13. A 230 V motor has an armature circuit resistance of 0.6 ohm. If the full-load armature current is 30 A and the no-load armature current is 4 A, find the change in back e.m.f. from no-load to full-load. [**Ans.** 15.6 V]

14. A six-pole lap-connected 250 V shunt motor has 396 armature conductors. It takes 30 A on full-load. The flux per pole is 0.04 weber. The armature and field resistances are 0.1 ohm and 200 ohms respectively. Contact drop per brush = 1 V.

Determine the speed on full-load. [**Ans.** 928.5 r.p.m.]

15. A four-pole, 440 V shunt motor takes 8 A on no-load, the no-load speed being 750 r.p.m. It has a shunt field current of 1.5 A. Calculate the full-load speed of the motor if it takes 100 A at full-load. Armature resistance = 0.25 ohm. Contact drop per brush = 1 V. Armature reaction weakens the field by 5 per cent on full-load. [**Ans.** 748 r.p.m.]

16. A four-pole D.C. shunt motor working on 250 V takes a current of 2 A when running at 1,000 r.p.m. What will be its back e.m.f., speed and the percentage speed drop if the motor takes 51 A at a certain load ? Armature resistance = 0.2 ohm and shunt field resistance = 250 ohms. [**Ans.** 240 V, 963 r.p.m., 3.84%]

17. A four-pole lap wound 500 V shunt motor takes a full-load armature current of 50 A, the full-load speed being 800 r.p.m. There are 440 armature conductors and armature resistance is 0.3 ohm, contact drop per brush = 1 V.

Find the useful flux per pole. [**Ans.** 0.08233 weber]

18. A 500 V, 6-pole lap connected shunt motor has a no-load input current of 16 A and a shunt field current of 10 A. At full-load it takes a current of 160 A. If armature resistance = 0.1 ohm, flux/pole at no-load = 0.06 weber, number of armature conductors = 744 and contact drop/brush = 1 V, calculate :

(*i*) No-load speed (*ii*) Full-load speed

(*iii*) Speed regulation.

Armature reaction weakens the field by 2 per cent. [**Ans.** 669 r.p.m., 662.7 r.p.m., 0.9504%]

19. 250 V shunt motor takes a line current of 50 A and runs at 600 r.p.m. Its armature and field resistances are 0.4 ohm and 125 ohms respectively. Neglecting the effects of armature reaction and allowing 2 V brush drop, calculate :

(*i*) No-load speed if the no-load line current is 5 A

(*ii*) Percentage reduction in the flux per pole in order that the speed may be 800 r.p.m. when the armature current is 40 A. **[Ans.** 647 r.p.m., 23.7%]

20. A 250 V shunt motor on no-load runs at 1000 r.p.m. and takes 5 A. The total armature and shunt field resistances are respectively 0.2 ohm and 250 ohms. Calculate the speed when loaded and taking a current of 50 A, if armature reaction weakens the field by 3%. **[Ans.** 994 r.p.m.]

21. A D.C. series motor of resistance 1 ohm between terminals runs at 800 r.p.m. at 200 V with a current of 15 A. Find the speed at which it will run when connected in series with a 5 ohms resistance and taking the same current. **[Ans.** 476 r.p.m.]

22. A 250 V shunt motor takes 70 A when running at 1000 r.p.m. It has an armature resistance of 0.1 ohm. Determine the speed and armature current if the magnetic flux is weakened by 30 per cent. Contact drop per brush = 1 V.
Total torque developed remains constant. **[Ans.** 1411 r.p.m., 100 A]

23. A 500 V shunt motor takes a no-load current of 10 A at 500 V. It takes 100 A at 500 V on full-load. Armature and shunt field resistances are 0.2 ohm and 200 ohms respectively. Contact drop/brush = 1 V. If the no-load speed is 800 r.p.m. find the full-load speed.
Armature reaction weakens the field by 3 per cent on full-load. **[Ans.** 795 r.p.m.]

24. A 250 V shunt motor on no-load runs at 1000 r.p.m. and takes 5 A. The total armature and shunt field resistances are respectively 0.2 ohm and 250 ohms. Calculate the speed when loaded and taking a current of 50 A, if armature reaction weakens the field by 3 per cent. **[Ans.** 994 r.p.m.]

25. A 250 V shunt motor takes a current of 52 A and runs at 1000 r.p.m. at full-load. Armature and shunt field resistances are 0.3 ohm and 125 ohms respectively. Determine :

(*i*) Speed at full-load torque

(*ii*) Speed at double full-load torque

(*iii*) Stalling torque in terms of the full-load torque.

Assume that the flux remains constant. **[Ans.** 681 r.p.m., 298 r.p.m., 2.78 times]

26. A 4-pole, 250 V lap connected shunt motor has armature and field resistances of 0.05 ohm and 100 ohms respectively. There are 492 armature conductors and the useful flux per pole is 0.03 weber. It takes a current of 7.5 A on no-load and a current of 120 A on full-load. Determine :

(*i*) No-load speed (*ii*) Full-load speed

(*iii*) Speed regulation.

Neglect drop/brush = 1 V. Armature reaction may be neglected. **[Ans.** 1007 r.p.m., 984 r.p.m., 2.337%]

27. A 500 V shunt motor has an armature resistance of 0.2 ohm. It takes an armature current of 60 A at 750 r.p.m. Find the speed and armature current if the flux is increased to 150% of its initial value. Assume the torque to be constant throughout. Contact drop/brush = 1 V. **[Ans.** 504.2 r.p.m., 40 A]

28. Determine the value of torque in Nm established by the armature of a 4-pole motor having 774 conductors, two paths in parallel, 24 m Wb flux per pole when the total armature current is 50 amperes. **[Ans.** 295.3 Nm]

29. Calculate torque on an armature carrying a total of 25000 ampere-conductors, the diameter of the core being 0.9144 m, the length 30.5 cm, the polar arc 70% and the flux density in the air gap 0.6 tesla. **[Ans.** 1464 Nm]

30. A 220 V series motor in which the total armature and field resistance is 0.1 ohm is working with unsaturated field, taking 100 A and running at 800 r.p.m. Calculate at what speed the motor will run when developing half the torque ? **[Ans.** 1147 r.p.m.]

31. A 4-pole wave wound shunt motor, running at 1000 r.p.m, takes 100 A at 250 V. There are 252 armature conductors. Useful flux per pole = 0.04 Wb. The shunt field current of motor is 5 A. Determine :

(*i*) Total torque developed

(*ii*) Actual torque if the friction and iron loss is 2 kW. **[Ans.** 304.7 Nm, 285.61 Nm]

32. A 4-pole 250 V D.C. shunt motor has lap connected 960 conductors. The flux per pole is 20 mWb. Determine the torque developed by the armature and the useful torque in Nm when current drawn by the motor is 32 A. The armature resistance is 0.1 ohm and shunt field resistance is 125 ohms. The rotational losses of the machine amount to 825 watts. **[Ans.** 91.6 Nm ; 81.4 Nm]

33. A shunt motor operating on 230 V takes an armature current of 6 A at no-load and runs at 1200 r.p.m. The armature resistance is 0.25 ohm. Determine the speed and electromagnetic torque when the armature takes 36 A with the same flux. **[Ans.** 1161 r.p.m, 65.44 Nm**]**

34. A 44.76 kW, 250 V, 4-pole lap-connected D.C. shunt motor has 30 slots with 12 conductors/slot. The armature and shunt field resistances are 0.06 ohm and 100 ohms respectively. The flux/pole is 0.03 Wb. If the full-load efficiency is 88%, find at full-load :

(*i*) The speed ;

(*ii*) Useful torque at shaft. **[Ans.** 1323 r.p.m., 323.3 Nm**]**

35. A 4-pole, lap-wound, 250 V shunt motor has the following particulars : No. of armature conductors = 720 ; armature resistance = 0.311 ohm, useful flux/pole = 0.04 Wb. If the total torque developed by the motor is 147.15 Nm, find the armature current taken and the speed. **[Ans.** 32.12 A ; 500 r.p.m.**]**

36. The input to a 220 V D.C. shunt motor is 11 kW. The other particulars of the motor are : no-load current = 5 A ; no-load speed = 1150 r.p.m. ; armature resistance = 0.5 ohm ; shunt field resistance = 110 ohms. Calculate :

(*i*) The torque developed (*ii*) The efficiency

(*iii*) The speed at this load. **[Ans.** 87.1 Nm ; 79.6% ; 1031 r.p.m.**]**

9

Measuring Instruments

1. INTRODUCTION AND CLASSIFICATION

The instruments used for all electrical measurements are called *measuring instruments.* They include *ammeters, voltmeters, wattmeters,* energy meters etc. The various electrical instruments may broadly be divided into two groups :

1. **Absolute instruments.** Absolute instruments are those instruments which indicate the quantity to be measured in terms of the *constants of the instrument* (dimensions, turns etc.) and in order to find out the quantity in the practical units it is necessary to *multiply such deflections with an instrument constant.* No previous calibration or comparison is necessary in this case. The most common absolute instrument is *tangent* galvanometer which gives the measured current in terms of tangent of the deflected angle, the radius and the number of turns of the galvanometer. Such instruments are rarely used (the use being merely confined within laboratories as standardizing instruments).

2. **Secondary instruments.** Secondary instruments are those in which the value of electrical quantity to be measured can be determined from the deflection of instrument only when they have been *pre-calibrated* by comparison with an absolute instrument. The deflection of the instrument gives directly the quantity to be measured. These instruments are most generally used in everyday work.

Secondary instruments may also be *classified* as follows :

1. **Indicating instruments.** Indicating instruments are those which indicate the instantaneous value of the electrical quantity being measured at the time at which it is being measured. Their indications are given by pointers moving over calibrated chief.

Example. *Ammeters, voltmeters* and *wattmeters.*

2. **Recording instruments.** Recording instruments are those which give a *continuous record* of the variations of an electrical quantity over a selected period of time. The pointer in these types of instruments is an infed pen which leaves a trace on a paper put over a moving drum.

3. **Integrating instruments.** Integrating instruments are those which measure the total quantity of electricity delivered in a particular time.

Example. *Ampere-hour and watt-hour meters.*

Electrical measuring instruments may also be *classified* as follows :

1. **According to the quantity being measured :**

Ammeters. ...... for measuring the magnitude of current.

Voltmeters. ...... for measuring voltages.

Ohmmeters and resistance bridges. ...... for measuring resistances.

Wattmeters. ...... for power measurements.

Watt-hour meters. ...... for energy measurements.

Frequency meters. ...... for frequency measurements.

Power factor meters. ...... for power-factor measurements.

2. **According to the kind of current :**

Instruments are classified into *D.C.*, *A.C.* and *A.C./D.C. instruments.*

3. **According to accuracy limits :**

4. **According to the principle of operation :**

Instruments are grouped into :

— Moving coil
— Moving iron
— Electrodynamic
— Induction
— Hot-wire
— Thermo-electric
— Rectifier types.

5. **According to the type of indication :**

Instruments may be :

— Indicating type
— Recording type.

6. **According to application :**

— Switch board
— Portable.

2. ELECTRICAL PRINCIPLES OF OPERATION

All electrical measuring instruments depend for their action on any of many physical effects of electric current or potential. The following are the effects generally used in the manufacture :

(*i*) **Magnetic effect.** ...... Voltmeters, ammeters, wattmeters, power factor meters etc.

(*ii*) **Thermal effect.** ...... Ammeters, voltmeters, maximum demand meters etc.

(*iii*) **Chemical effect.** ...... D.C. ampere hour meters (integrating meters).

(*iv*) **Electrostatic effect.** ...... Voltmeters which can indirectly be used as ammeters and wattmeters.

(*v*) **Electro-magnetic induction effect.** ...... Voltmeters, ammeters, wattmeters and integrating meters used in A.C. only.

3. ELECTRICAL INDICATING INSTRUMENTS

Almost invariably an indicating instrument is fitted with a pointer which indicates on a scale the value of the quantity being measured. The moving system of such an instrument is usually carried by a *spindle of hardened steel*, having its ends tapered and highly polished to form pivots which rest in hollow-ground bearings, usually of saphine, set in steel screws. In some instruments, the moving system is attached to *thin ribbons of spring material* such as beryllium-copper alloy, held taut by tension springs mounted on the frame of movement. This arrangement *eliminates pivot friction* and the instrument *is less susceptible to damage by shock or vibration*.

3.1. Essential Features. Indicating instruments possess **three essential features**.

1. **Deflecting device.** ...... whereby a mechanical force is produced by the electric current, voltage or power.

2. **Controlling device.** ...... whereby the value of deflection is dependent upon the magnitude of the quantity being measured.

3. **Damping device.** ...... to prevent oscillation of the moving system and enable the latter to reach its final position quickly.

3.2. Deflecting Device. A deflecting device produces a deflecting torque which is caused by anyone of the previously mentioned effects (*i.e.*, thermal effect, chemical effect, electrostatic effect etc.) ; with the help of this deflecting torque the needle or the pointer moves from zero position to the final position. The arrangement of the deflecting device with each type of instrument will be discussed individually.

3.3. Controlling Devices.

There are two types of controlling devices :

(*i*) Spring control

(*ii*) Gravity control.

(*i*) **Spring control.** Fig. 1 shows a commonly used spring control arrangement. It utilises two spiral hair springs, 1 and 2, the inner ends of which are attached to the spindle *S*. The outer end of spring 2 is fixed while that of 1 is attached to a lever, the adjustment of which gives zero

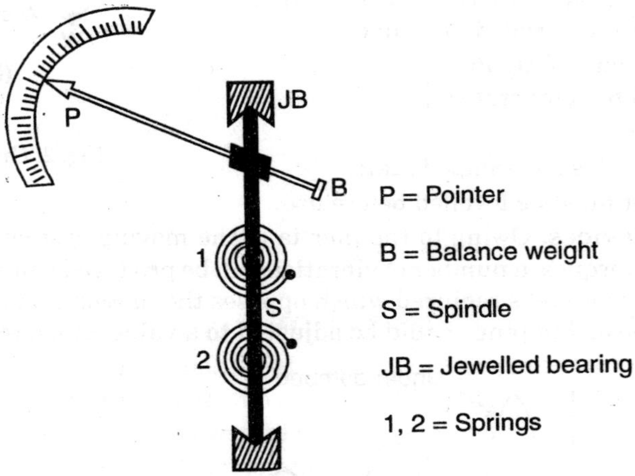

Fig. 1. Spring control.

adjustments. The two springs 1 and 2 are wound in *opposite directions* so that when the moving system is deflected, one spring winds up while the other unwinds, and the controlling torque is due to the *combined torsions* of the springs.

Since the torsional torque of a spiral spring is proportional to the angle of twist, the controlling torque (T_c) is directly proportional to the angular deflection of the pointer (θ).

i.e., $T_c \propto \theta$.

The spring material should have the following **properties** :

(*i*) It should be non-magnetic.

(*ii*) It must be of low temperature co-efficient.

(*iii*) It should have low specific resistance.

(*iv*) It should not be subjected to fatigue.

(*ii*) **Gravity control.** With gravity control, weights L and M are attached to the spindle S [Fig. 2 (*a*)], the function of L being to *balance* the weight of the pointer P. Weight M therefore provides the controlling torque. When the pointer is at zero, M hangs vertically downwards. When P is deflected through angle θ, the controlling torque is equal to (weight of $M \times$ distance d) and is therefore proportional to the *sine* of the angular deflection [Fig. 2. (*b*)],

i.e., $T_c \propto \sin\theta$.

The degree of control is adjusted by screwing the weight up or down the carrying system.

It may be seen from Fig. 2 (*b*) that as θ approaches 90°, the distance 1-2 increases by a relatively small amount for a given change in the angle that when θ is just increasing from its zero value. Hence *gravity-controlled instruments have scales which are not uniform but are cramped or crowded at their lower ends.*

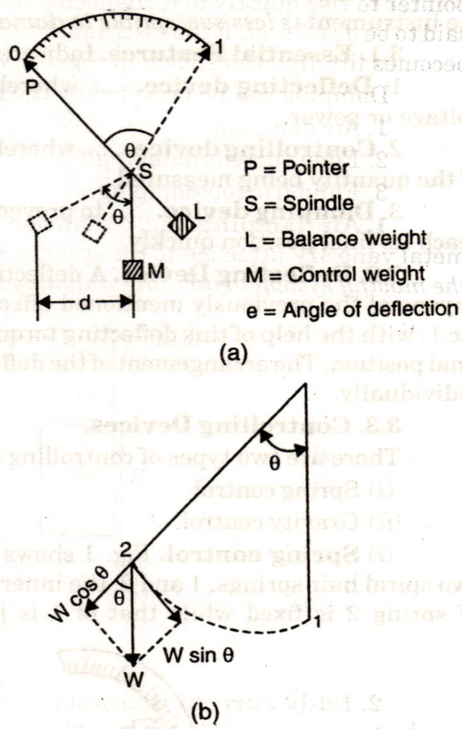

Fig. 2. Gravity control.

Advantages :

1. The gravity controlled instrument is cheaper than corresponding spring-controlled instrument.
2. It is not subjected to fatigue.
3. It is unaffected by temperature.

Disadvantages :

1. Gravity control gives a cramped scale.
2. The instrument must be levelled before use.

3.4. Damping Devices. Owing to the inertia of the moving system, when subjected to the deflecting and restoring torques, a number of vibration will be produced before coming finally to rest. To-avoid this, a *damping torque* is required which opposes the motion and ceases when the pointer comes to rest. The degree of damping should be adjusted to a value which is sufficient to enable the

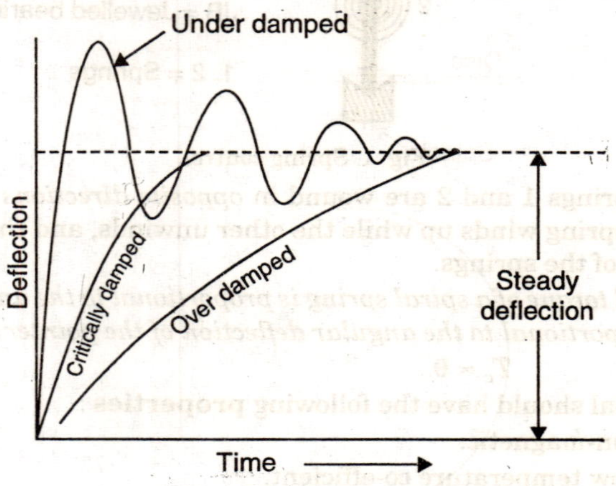

Fig. 3. Damping curves.

pointer to rise quickly to its deflected position without overshooting. In that case, the instrument is said to be *dead-beat.* If the instrument is over-damped the movement is very slow (and the instrument becomes lithargic) as shown in Fig. 3.

Damping can be provided by the following methods :

1. Air damping.
2. Eddy current damping.
3. Fluid friction damping.

1. Air damping. Fig. 4 shows an arrangement for obtaining air damping. It consists of a thin metal vane *MV* attached to the spindle *S* ; the vane moves in a sector-shaped box *B*. *Any tendency of the moving system to oscillate is damped by the action of the air on the vane.*

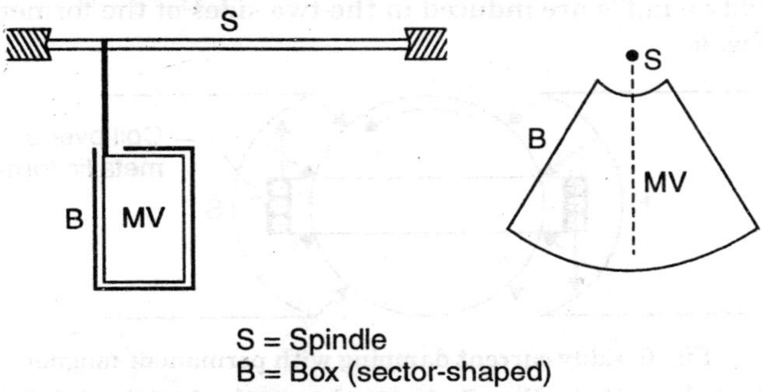

Fig. 4. Air damping.

2. Eddy current damping. This method of damping is based on the *principle* that when a conducting non-magnetic material is moved in a magnetic field an e.m.f. is induced in it which causes

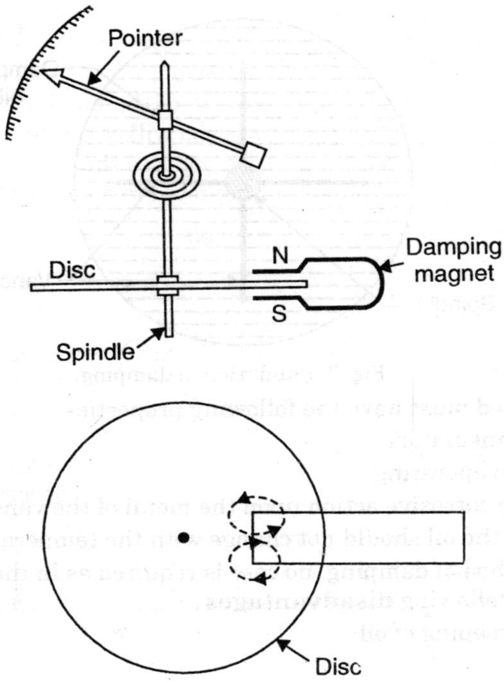

Fig. 5. Eddy current damping.

currents called the *eddy currents*. Due to these eddy currents a force exists between them and the field. Due to Lenz's law this force is always in opposition to the force causing rotation of the conducting material, thus, it provides the necessary damping.

- One form of eddy-current damping is shown in Fig. 5. Here a copper or aluminium disc, curried by a spindle, can move between the poles of a permanent magnet. If the disc moves clockwise, the e.m.f.'s induced in the disc circulate eddy currents as shown dotted. It follows from Lenz's law that these currents exert a force *opposing* the motion producing them, namely the clockwise movement of the disc.
- Another form of providing damping is used in the *moving coil instruments* using permanent magnet. The moving coil is mounted over a metallic former. When the coil is deflected eddy e.m.f.'s are induced in the two sides of the former, causing eddy forces as shown in Fig. 6.

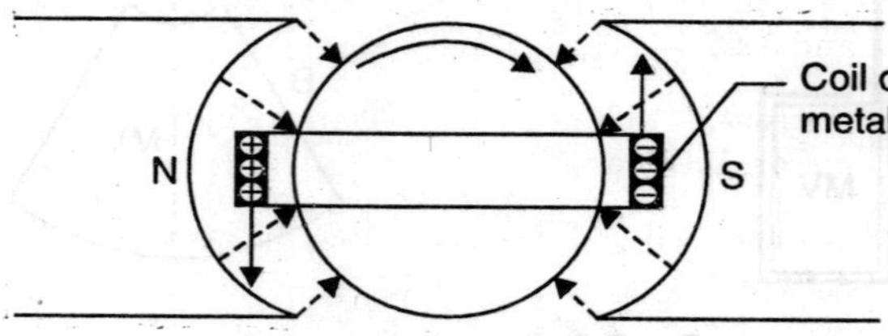

Fig. 6. Eddy current damping with permanent magnet.

3. Fluid friction damping. Fig. 7 shows the method of fluid friction damping. Here light vanes are attached to the spindle of the moving system. The vanes are dipped into a pot of damping oil and are completely submerged by the oil. The motion of the moving system is always opposed by the friction of the damping oil on the vanes. The damping force thus created always increases with the increase in velocity of vanes. There is no damping force when the vanes are stationary.

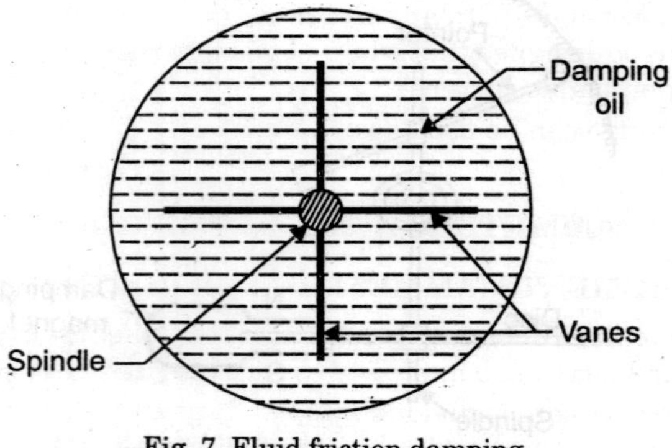

Fig. 7. Fluid friction damping.

The damping oil used must have the following properties :

(*i*) Must be a good insulator.

(*ii*) Should be non-evaporating.

(*iii*) Should not have corrosive action upon the metal of the vane.

(*iv*) The viscosity of the oil should not change with the temperature.

Though in this method of damping, no case is required as in the air friction damping but it is not much used due to the following **disadvantages** :

(*i*) Objectionable creeping of oil.

(*ii*) Using the instrument always in the vertical position and its obvious unsuitability for use in portable instruments.

The principle types of electrical indicating instruments, together with the methods of control and damping, are summarized below :

S. No.	*Type of instrument*	*Suitable for measuring*	*Method of control*	*Method of damping*
1.	Moving-iron	Current and voltage, D.C. and A.C.	Hair springs	Air
2.	Permanent magnet moving coil	Current and voltage, D.C. only	Hair springs	Eddy current
3.	Thermocouple	Current and voltage, D.C. and A.C.	As for moving coil	As for moving coil
4.	Electro-dynamic or dynamometer	Current, voltage and power, D.C. and A.C.	Hair springs	Air
5.	Electrostatic	Voltage only, D.C. and A.C.	Hair springs	Air or eddy current
6.	Rectifier	Current and voltage, A.C. only	As for moving coil	As for moving coil

Note. Apart from the electrostatic type of voltmeter, all voltmeters are in effect milliammeters connected in series with non-reactive resistor having a high resistance.

Difference between an ammeter and a voltmeter

An ammeter and a voltmeter work on the same principle. The ammeter has a *low resistance* so that when it is connected in **series** with any circuit, it does not change the current. The voltmeter has a *high resistance* and it is so designed that when connected in **parallel** to the circuit for measuring voltages it does not take appreciable current.

An ammeter of low range can be used as a voltmeter by connecting an external resistance in series with it.

AMMETERS AND VOLTMETERS

4. MOVING-IRON INSTRUMENTS (AMMETERS AND VOLTMETERS)

Moving-iron instruments are commonly used in laboratories and switch boards at commercial frequencies because they are *very cheap and can be manufactured with required accuracy*.

Moving-iron instruments can be divided into two types :

1. Attraction type ...... in which a sheet of soft iron is *attracted towards a solenoid.*

2. Repulsion type ...... in which two parallel rods or strips of soft iron, magnetised inside a solenoid, are regarded as *repelling each other.*

4.1. Attraction Type. Fig. 8 shows the sectional front and an end elevation of the attracted-iron type instrument. It consists of a solenoid (or coil) C and oval shaped soft-iron disc D in such a way that it can move in or out of the solenoid. To this iron a pointer P is attached so that it may deflect along with the moving iron over a graduated scale. The soft-iron disc is made of sheet metal specially shaped to give a scale as nearly uniform as possible.

When the current to be measured (or a definite fraction of the current to be measured or proportional to the voltage to be measured) is passed through the solenoid, a magnetic field is set up inside the solenoid, which in turn magnetises the iron. Thus the soft-iron disc is attracted into the

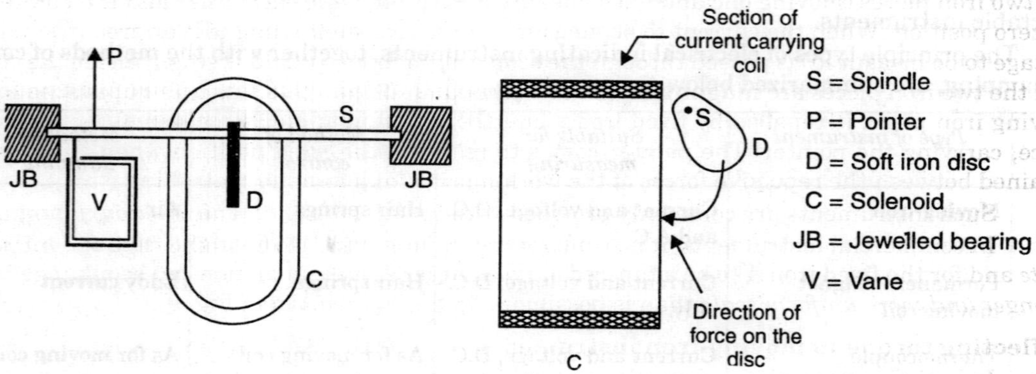

Fig. 8. Attraction-type moving-iron instrument.

solenoid/coil, causing the spindle and the pointer to rotate. *Damping* is provided by vane *V* attached to the spindle and moving in an *air chamber*, and *control* is by *hair spring*.

4.2. Repulsion Type. Repulsion-type moving-iron instrument is shown in Fig. 9. Here there are two irons, one fixed (*A*) and the other mounted on a short arm fixed (*B*) to the instrument spindle. The two irons lie in the magnetic field due to a solenoid/coil *C*. When there is no current in the coil

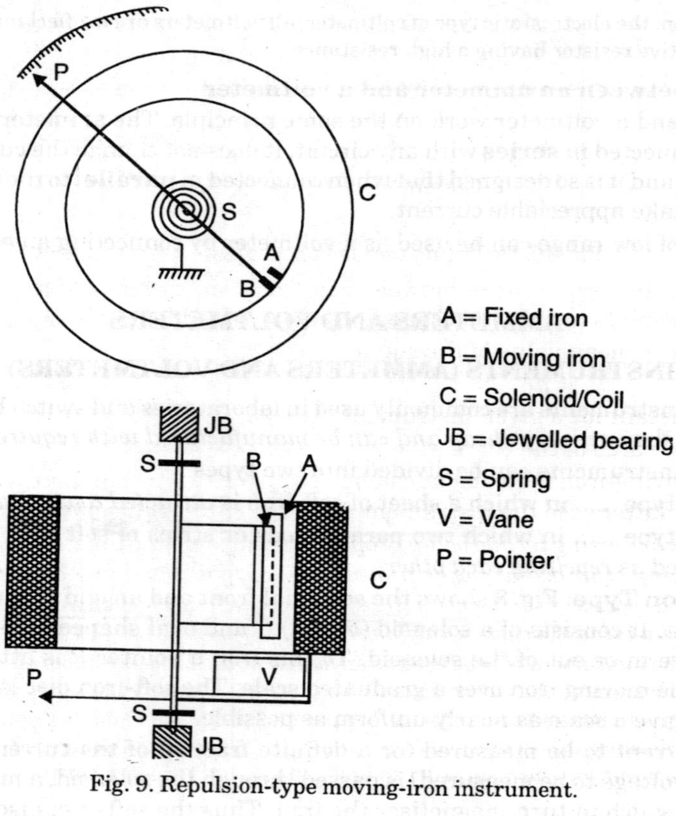

Fig. 9. Repulsion-type moving-iron instrument.

the two iron pieces (moving one and fixed one) are almost touching each other and the pointer rests on zero position. When the current to be measured (or a definite fraction of it or proportional to the voltage to be measured) is passed through the solenoid, a magnetic field is set up inside the solenoid and the two iron pieces are **magentised in the same direction.** This sets up a **repulsive force** so moving iron piece, is repelled by fixed iron piece, thereby results in the motion of the moving iron piece, carrying the pointer. The pointer comes to rest in a deflected position when equilibrium is attained between the repulsive forces of the working elements and the controlling force.

Such instruments are commonly provided with *spring control* and *air friction damping.*

In commercial instruments, it is usual for the moving-iron B to be in the form of a *thin curved plate* and for the fixed iron A to be a tapered curved sheet. This construction can be arranged to give *a longer and more uniform scale* than is possible with the rods shown in Fig. 9.

Deflecting torque in moving-iron instruments :

In both the attraction and repulsion type moving-iron instruments it is found that for a given position of the moving system, the value of the *deflecting torque is proportional to the square of the current,* so long as the *iron is working below saturation.* Hence, if the current waveform is as shown in Fig. 10, the variation of the deflecting torque is represented by the dotted wave. If the supply

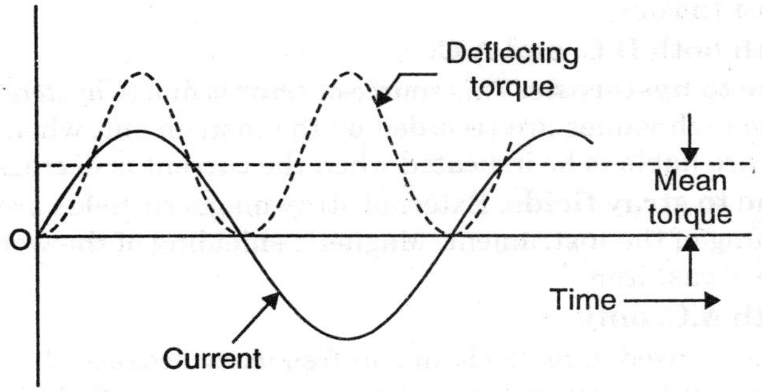

Fig. 10. Deflecting torque in a moving-iron instrument.

frequency is 50 Hz, the torque varies between zero and a maximum 100 times a second, so that the moving system (due to its inertia) takes up a position corresponding to the mean torque, where

$$\text{mean torque} \propto \text{mean value of the square of the current}$$
$$= kI^2$$

where k = a constant for a given instrument

and I = r.m.s. value of the current.

Thus the moving-iron instrument can be used to *measure both direct current and alternating current,* and in the latter case the instrument gives the r.m.s. value of the current.

Owing to the deflecting torque being proportional to the square of the current, the scale divisions are not uniform, being cramped at the beginning and open at the upper end of the scale.

Note. For both types of instruments (attraction-type and repulsion type) the necessary magnetic field is produced by the *ampere-turns of a current-carrying coil.*

— In case the instrument is to be used as an *ammeter,* the coil has *comparatively few turns of thick wire so* that the ammeter has *low resistance* because it is connected in *series* with the circuit.

— In case it is to be used as a *voltmeter,* the coil has *high impedance* so as to draw as small a current as possible since it is connected in *parallel with the circuit.* As current through the coil is small it has large *number of turns* in order to produce sufficient ampere-turns.

4.3. Advantages and Disadvantages of Moving-Iron Instruments

Advantages :

1. Can be used both in D.C. as well as in A.C. circuits.
2. Robust and simple in construction.
3. Possess high operating torque.
4. Can withstand overload momentarily.
5. Since the stationary parts and the moving parts of the instrument are simple so they are cheapest.
6. Suitable for low frequency and high power circuits.
7. Capable of giving an accuracy within limits of both precision and industrial grades.

Disadvantages :

1. Scales not uniform.
2. For low voltage range the power consumption is higher.
3. The errors are caused due to hysteresis in the iron of the operating system and due to stray magnetic field.
4. In case of A.C. measurements, change in frequency causes serious error.
5. With the increase in temperature the stiffness of the spring decreases.

4.4. Sources of Errors

A. Errors with both D.C. and A.C.

(*i*) **Errors due to hysteresis.** This source of error is due to *hysteresis* in the soft iron moving part, due to which too high values are recorded by the instrument, when the current is increasing and too low readings are liable to be indicated when the current is decreasing.

(*ii*) **Errors due to stray fields.** External stray magnetic fields are liable to affect adversely the accurate functioning of the instrument. Magnetic shielding of the working parts is obtained by using a covering case of cast iron.

B. Errors with A.C. only

Errors may be caused due to change in frequency because change in frequency produces (*i*) *change in impedance of the coil* and (*ii*) *change in magnitude of eddy currents.* The error due to the former is negligible in ammeters, as the coil current is determined by the external circuit and the error due to the latter can normally be made small.

5. MOVING-COIL INSTRUMENTS

The moving-coil instruments are of the following two types :

1. **Permanent-magnet type**......can be used for D.C. only.
2. **Dynamometer type**......can be used both for A.C. and D.C.

5.1. Permanent-magnet Moving-Coil Type (PMMC) Instruments

A permanent-magnet moving coil-type instrument works on the principle that *"when a current-carrying conductor is placed in a magnetic field, it is acted upon by a force which tends to move it to one side and out of the field"*.

Construction

- The instrument consists of a permanent magnet *M* and a rectangular coil *C* which consists of insulated copper wire wound on light aluminium frame fitted with polished steel pivots resting in jewel bearings. The magnet is made of Alnico and has soft-iron pole-pieces *PP* which are bored out cylindrically.
- The rectangular coil *C* is free to move in air gaps between the soft-iron pole pieces and a soft-iron cylinder *A* (central core), supported by a brass plate (not shown).

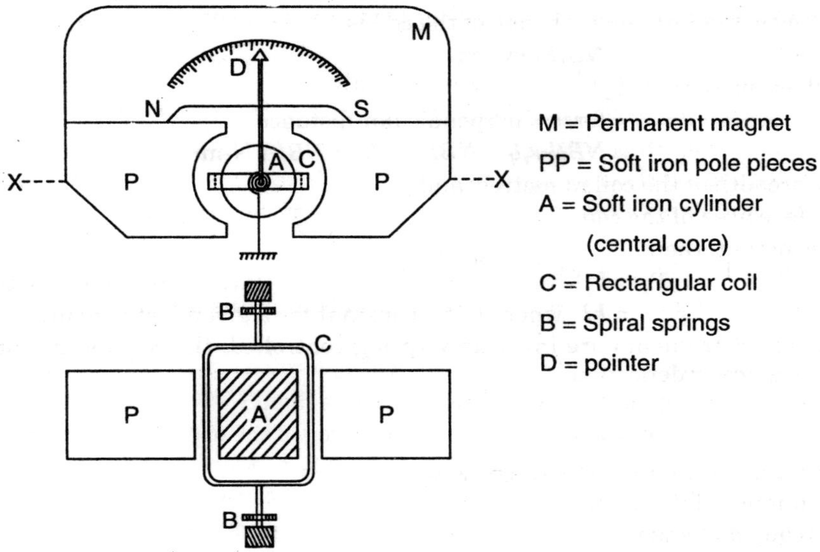

Fig. 11. Permanent-magnet moving-coil instrument.

The *functions of the central core A* are :

(*i*) To intensify the magnetic field by reducing the length of air gap across which the magnetic flux has to pass.

(*ii*) To give a radial magnetic flux of uniform density, thereby enabling the scale to be uniformly divided.

— The movement of the coil is controlled by two phosphor bronze hair springs *BB* (one above and one below), which additionally serve the purpose of leading the current in and out of the coil. The two springs are spiralled in *opposite directions* for neutralizing the effects of changes in temperature.

— The aluminium frame not only provides support for the coil but also *provides damping by eddy currents induced in it.*

Deflecting torque. Refer Fig. 12. When current is passed through the coil, forces are set up on its both sides which produce *deflection torque.* If I amperes is the current passing through the coil, the magnitude of the force (F) experienced by each of its sides is given by

$$F = BIl \text{ newton}$$

where B = flux density in Wb/m^2, and

l = length or depth of coil in metres.

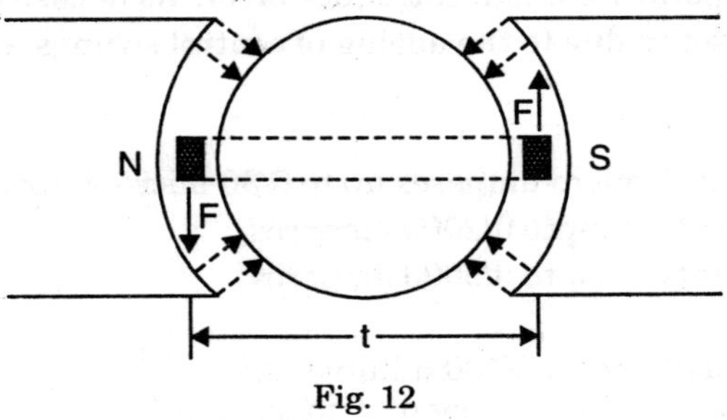

Fig. 12

For N turns, the force on each side of the coil is

$$= NBIl \text{ newton}$$

$\therefore$ Deflecting torque (T_d)

$$= \text{force} \times \text{perpendicular distance}$$

$$= NBIl \times b = NBI\,(l \times b) = NBIA \text{ N-m}$$

where b = breadth of the coil in metres, and

A = face area of the coil.

If B is constant, then

$$T_d \propto I \quad (i.e., \text{ current passing through the coil})$$

$$= kI \quad \text{where } k \text{ is a constant for a given instrument.}$$

Since such instruments are invariably spring controlled, the controlling torque (T_c) of the spiral springs $\propto$ angular deflection

i.e., $T_c \propto \theta$

or $T_c = c\,\theta$

where c = a constant for given springs, and

θ = angular deflection.

For a steady deflection,

Controlling torque (T_c) = deflecting torque (T_d)

Hence $c\theta = kI$

$$\therefore \quad \theta = \frac{k}{c} I$$

i.e., the *deflection is proportional to the current* and the *scale is therefore uniformly divided.*

Advantages and Disadvantages. The moving-coil permanent-magnet type instruments have the following advantages and disadvantages :

Advantages :

(*i*) Low power consumption.
(*ii*) Their scales are uniform.
(*iii*) No hysteresis loss.
(*iv*) High torque/weight ratio.
(*v*) They have very effective and efficient eddy-current damping.
(*vi*) Range can be extended with shunts or multipliers.
(*vii*) No effect of stray magnetic field as intense polarised or unidirectional field is employed.

Disadvantages :

(*i*) Somewhat costlier as compared to moving-iron instruments.
(*ii*) Cannot be used for A.C. measurements.
(*iii*) Friction and temperature might introduce errors as in case of other instruments.
(*iv*) Some errors are set in due to the ageing of control springs and the permanent magnets.

Ranges :

D.C. Ammeters

(*i*) *Without shunt*......0/5 micro-amperes up to 0/30 micro-amperes.
(*ii*) *With internal shunts*......upto 0/2000 amperes.
(*iii*) *With external shunts*......upto 0/5000 amperes.

D.C. Voltmeters

(*i*) *Without series resistance*......0/100 milli-volts.
(*ii*) *With series resistance*......upto 20000 or 30000 volts.

— Moving-coil permanent-magnet instruments can be used as :

(*i*) Ammeters......by using a low resistance shunt.

(*ii*) Voltmeters......by using a high series resistance.

(*iii*) Flux-meters......by eliminating the control springs.

(*iv*) Basllistic galvanometers......by making control springs of large moment of inertia.

Extension of Range. The following devices may be used for extending the range of instruments : (*i*) Shunts (*ii*) Multipliers (*iii*) Current transformers and (*iv*) Potential transformers. Use of ammeter shunts and voltage multipliers is discussed below.

1. Ammeter shunts. An ammeter shunt is merely a *low resistance* that is placed in parallel with the coil circuit of the instrument in order to *measure fairly large currents.* The greater part of the current in the main circuit is then diverted around the coil through the shunt. The connection diagram for a shunt and milli-ammeter for measuring large currents is shown in Fig. 13.

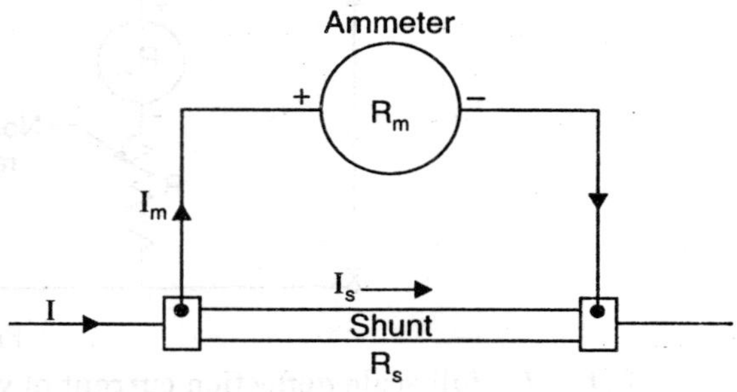

Fig. 13. Use of a shunt for measuring large currents.

The shunt is provided with four terminals, the milliammeter being connected across the potential terminals. If the instrument were connected across the current terminals, there might be considerable error due to the contact resistance at these terminals being appreciable compared with the resistance of the shunt.

Let I = current of the circuit to be measured,

I_m = current passing through the ammeter,

I_s = current passing through the shunt,

R_s = resistance of the shunt, and

R_m = resistance of the ammeter (plus its leads to the potential terminals).

Thus $$I = I_m + I_s \quad ...(i)$$

Also, since the voltage drop across the shunt and the instrument in same,

$$\therefore \quad I_s R_s = I_m R_m$$

or $$R_s = \frac{I_m}{I_s} R_m$$

Substituting $I_s = I - I_m$ from equation (*i*), we get

$$R_s = \frac{I_m}{(I - I_m)} \cdot R_m$$

or $$R_s = \frac{R_m}{\left(\frac{I}{I_m} - 1\right)} \quad ...(ii)$$

The ratio of $\frac{I}{I_m}$ is known as '*multiplying power*' of the shunt. Denoting, the ratio by N, we have shunt resistance,

$$R_s = \frac{R_m}{N - 1} \quad ...(iii)$$

The shunts are made of a material such as *manganin* (copper, manganese and nickel), which has a *negligible temperature coefficient of resistance.* The material, is employed in the form of thin strips, the ends of which are soldered to two large copper blocks. Each copper block carries two terminals-one current terminal and other potential terminal. The strips which form the shunt are spaced from each other to promote a good circulation of air and thus efficient cooling.

Note. A *'swamping'* resistor *r*, of material having negligible temperature co-efficient of resistance, is connected in series with the instrument (moving coil). The latter is wound with copper wire and the function of *r* is to *reduce* the error due to variation of resistance of the instrument with variation of temperature.

2. Voltmeter multipliers. The range of the instrument, when used as a voltmeter can be extended or multiplied by using a high non-inductive series resistance *R* connected in series with it as shown in Fig. 14.

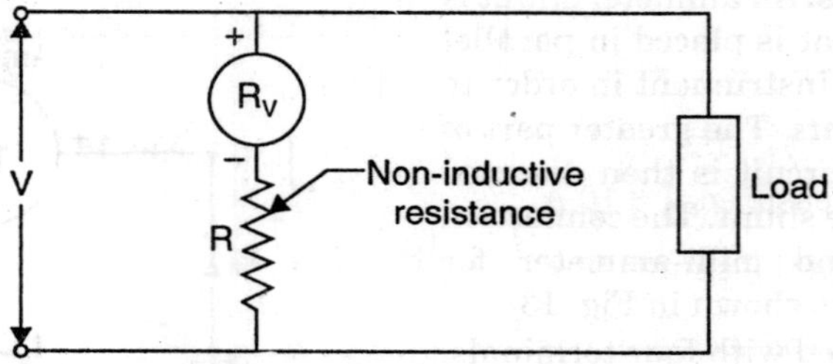

Fig. 14

Let I = full scale deflection current of voltmeter,

V = voltage of the circuit to be measured,

R_v = resistance of the voltmeter, and

R = external series resistance.

Now, voltage across supply leads

= voltage drop across the voltmeter + voltage drop across external resistance

$$\therefore \quad V = IR_v + IR$$

or
$$R = \frac{V - IR_v}{I} = \frac{V}{I} - R_v$$

The voltage multipliers to be used for D.C. measurements should satisfy the following requirements :

(*i*) The resistance should not change with time of usage.

(*ii*) The temperature co-efficient of resistance must be very low.

Note. The frequency error introduced by the inductance of the instrument coil can be compensated by shunting *R* by a capacitor.

Example 1. *A milliammeter of 2.5 ohms resistance reads upto 100 milli amperes. What resistance is necessary to enable it to be used as :*

(*i*) *A voltmeter reading upto 10 V.*

(*ii*) *An ammeter reading upto 10 A.*

Draw the connection diagram in each case.

Solution. Resistance of the milli-ammeter, $R_m = 2.5\ \Omega$

Maximum current of the milli-ammeter, I_m = 100 mA = 0.1 A.

(*i*) Voltage to be measured, V = 10 volts

Resistance to be connected in series.

$$\mathbf{R} = \frac{V}{I_m} - R_m = \frac{10}{0.1} - 2.5 = \mathbf{97.5\ \Omega} \quad \textbf{(Ans.)}$$

Connection diagram is shown in Fig. 15.

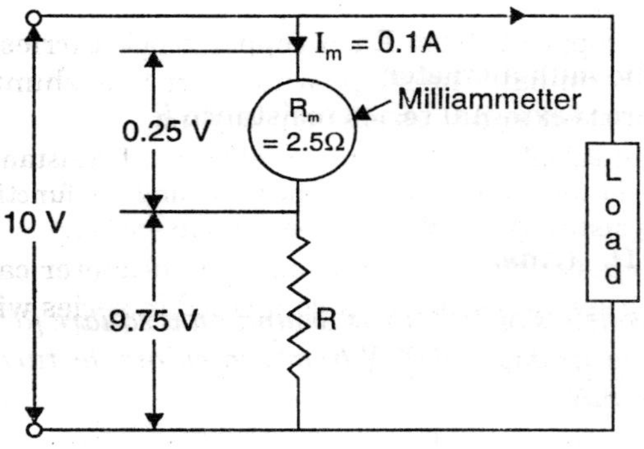

Fig. 15

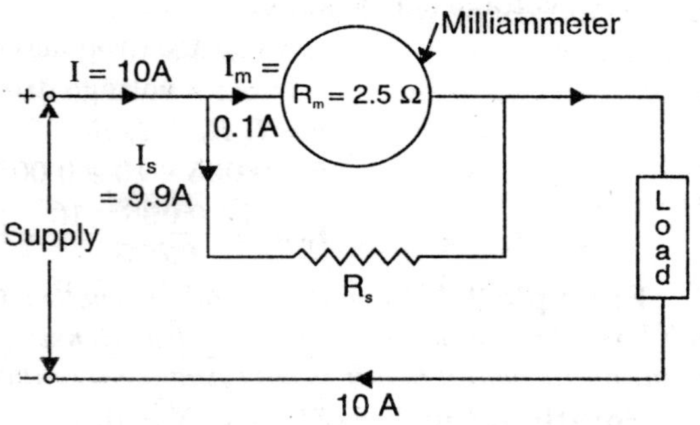

Fig. 16

(*ii*) Current to be measured, $I = 10$ A

Multiplying power of the shunt, $N = \frac{I}{I_m} = \frac{10}{0.1} = 100$

∴ Resistance to be connected in parallel,

$$R_s = \frac{R_m}{N-1} = \frac{2.5}{100-1} = \mathbf{0.02525\ \Omega.}\ \ \textbf{(Ans.)}$$

Connection diagram is shown in Fig. 16.

Example 2. *A moving-coil milli-ammeter having a resistance of 10 ohms gives full scale deflection when a current of 5 mA is passed through it. Explain how this instrument can be used for measurement of :*

(*i*) *Current upto 1 A.*

(*ii*) *Voltage upto 5 V.*

Solution. Resistance of the milli-ammeter, $R_m = 10\ \Omega$

Full scale deflection current, $I_m = 5$ mA = 0.005 A.

(*i*) **To measure current upto 1 A :**

Resistance of the shunt, R_s : (Refer Fig. 17)

Since voltage drop across the milli-ammeter and the shunt are equal

$$\therefore \quad I_m R_m = I_s R_s = (I - I_m) R_s$$

$$\therefore \quad \mathbf{R_s} = \frac{I_m R_m}{(I - I_m)} = \frac{0.005 \times 10}{(1 - 0.005)} = \mathbf{0.05025\ \Omega.}\ \ \textbf{(Ans.)}$$

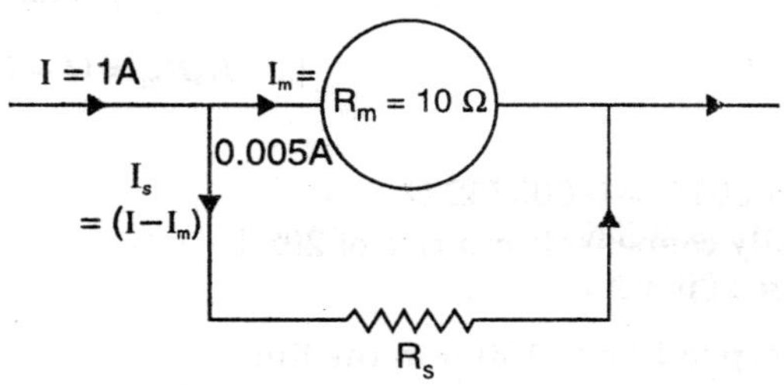

Fig. 17

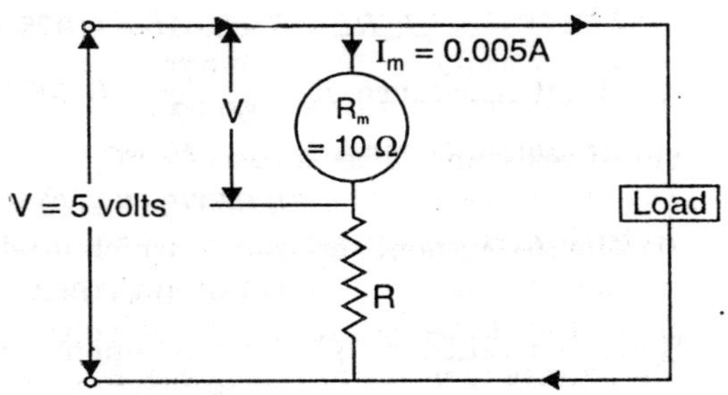

Fig. 18

(*ii*) **To measure voltage upto 5 V :**

The value of external series resistance, R : (Refer Fig. 18)

Now, voltage across supply loads

= voltage drop across the milli-ammeter

+ voltage drop across external series resistance R

$\therefore \quad V = I_m R_m + I_m R$

or $\quad 5 = 0.005 \times 10 + 0.005\, R$

or $\quad \mathbf{R} = \dfrac{5 - 0.005 \times 10}{0.005} = \mathbf{990\ \Omega}$. **(Ans.)**

Example 3. *If the moving coil of a voltmeter consists of 100 turns wound on a square former which has a length of 30 mm and the flux density in the air gap is 0.09 Wb / m², calculate the turning moment on the coil when it is carrying a current of 10 mA.*

Solution. Number of turns, $N = 100$

Length of each side, $l = 30\text{ mm} = 0.03\text{ m}$

Flux density, $B = 0.09\text{ Wb/m}^2$

Current through the coil, $I = 10\text{ mA} = 0.01\text{ A}$

We know that the force on each side of the coil,

$$F = NBIl \text{ newton}$$

$\therefore$ Turning moment (*i.e.*, deflecting torque),

$$T = F \times \text{breadth} = F \times l = NBIl^2 \text{ N-m}$$

$$= 100 \times 0.09 \times 0.01 \times (0.03)^2 \text{ N-m}$$

$$= \mathbf{8.1 \times 10^{-5}\ N\text{-}m}. \quad \textbf{(Ans.)}$$

Example 4. *A moving-coil instrument has a resistance of 5 Ω between terminals and full-scale deflection is obtained with a current of 0.015 A. This instrument is to be used with a manganin shunt to measure 100 A full scale. Calculate the error caused by a 20°C rise in temperature.*

(i) When the internal resistance of 5 Ω is due to copper only.

(ii) When a 4 Ω manganin swamping resistor is used in series with a copper resistor of l Ω.

The temperature-resistance co-efficients are :

Copper : $\alpha_c = 0.4\%$ *per °C, Manganin :* $\alpha_m = 0.015\%$ *per °C.*

Solution. Resistance of the instrument, $R_m = 5\ \Omega$

Current through the instrument, $I_m = 0.015\text{ A}$

Current to be measured, $I = 100\text{ A}$

Current through the shunt,

$$I_s = I - I_m = 100 - 0.015 = 99.985\text{ A}$$

Voltage across the shunt

$$I_m R_m = 5 \times 0.015 = 0.075\text{ V}$$

$\therefore$ Shunt resistance, $R_s = \dfrac{0.075}{99.985} = 0.00075\ \Omega \qquad [\because\ I_m R_m = (I - I_m) R_s]$

Shunt resistance after a rise of 20°C

$$= 0.00075(1 + 20 \times 0.00015) = 0.000752\ \Omega.$$

(*i*) The instrument resistance (which is wholly copper) after a rise of 20°C

$$= 5(1 + 20 \times 0.004) = 5.4\ \Omega$$

Hence, current through the instrument corresponding to 100 A in the line

$$= \frac{0.000752}{(5.4 + 0.000752)} \times 100 = 0.01392\text{ A}$$

Reading of the instrument $= 0.01392 \times 100/0.015 = 92.8$ A

$\therefore$ **Percentage error** $= 100 - 92.8 = \mathbf{7.2}$. **(Ans.)**

(*ii*) The instrument resistance after a rise of 20°C

$$= 1(1 + 20 \times 0.004) + 4(1 + 20 \times 0.00015)$$
$$= 1.08 + 4.012 + 5.092\ \Omega$$

Instrument current with a line current of 100 A

$$= \frac{0.000752}{(5.092 + 0.000752)} \times 100 = 0.01476 \text{ A}$$

$\therefore$ Instrument reading $= 0.01476 \times \dfrac{100}{0.015} = 98.4$ A

Percentage error = 100 – 98.4 = **1.6. (Ans.)**

Example 5. *The coil of a 250 V moving iron voltmeter has a resistance of 500 Ω and inductance of 1 H. The current taken by the instrument when placed on 250 V, D.C. supply is 0.05 A. Determine the percentage error when the instrument is placed on 250 V, A.C. supply at 100 Hz.*

Solution. Total ohmic resistance, $R = \dfrac{250}{0.05} = 500\ \Omega$

(Original calibration of the instrument is with direct current)

Reactance of the coil, $X_L = 2\pi fL = 2\pi \times 100 \times 1 = 628\ \Omega$

Coil impedance, $Z = \sqrt{R^2 + X_L{}^2} = \sqrt{(5000)^2 + (628)^2} = 5039\ \Omega$

$\therefore$ Voltage reading on A.C. $= \dfrac{250 \times 5000}{5039} = 248$ V

$\therefore$ Error = 248 – 250 = – 2 V

Percentage error $= \dfrac{2}{250} \times 100 =$ **0.8. (Ans.)**

Example 6. *A 15-volt moving iron voltmeter has a resistance of 300 Ω and an inductance of 0.12 H. Assuming that this instrument reads correctly on D.C., what will be its readings on A.C. at 15 volts when frequency is (i) 25 Hz and (ii) 100 Hz ?*

Solution. On *D.C.*, *only ohmic resistance* is involved and the voltmeter reads correctly. But on *A.C.*, it is the *impedance* of the instrument which has to be taken into account.

(*i*) **When frequency is 25 Hz :**

Impedance at 25 Hz,

$$Z = \sqrt{R^2 + X_L{}^2}\ \sqrt{R^2 + (2\pi fL)^2}$$
$$= \sqrt{(300)^2 + (2\pi \times 25 \times 0.12)^2} = 300.6\ \Omega$$

$\therefore$ **Voltmeter reading** $= 15 \times \dfrac{300}{300.6} =$ **14.97 V. (Ans.)**

(*ii*) **When frequency is 100 Hz :**

Impedance at 100 Hz,

$$Z = \sqrt{(300)^2 + (2\pi \times 100 \times 0.12)^2} = 309.3\ \Omega$$

$\therefore$ **Voltmeter reading** $= 15 \times \dfrac{300}{309.3} =$ **14.55 V. (Ans.)**

Incidently, it may be noted that as the frequency is increased, the impedance of the voltmeter is also increased. Hence, the current is decreased and, therefore, the voltmeter readings are lower.

5.2. Electrodynamic or Dynamometer Instruments

In an electrodynamic instrument the *operating field is produced by another fixed coil* and *not* by *permanent magnet.* This instrument can be used as an ammeter or as voltmeter but is *generally used as a wattmeter.*

Refer Fig. 19 (*a*), (*b*).

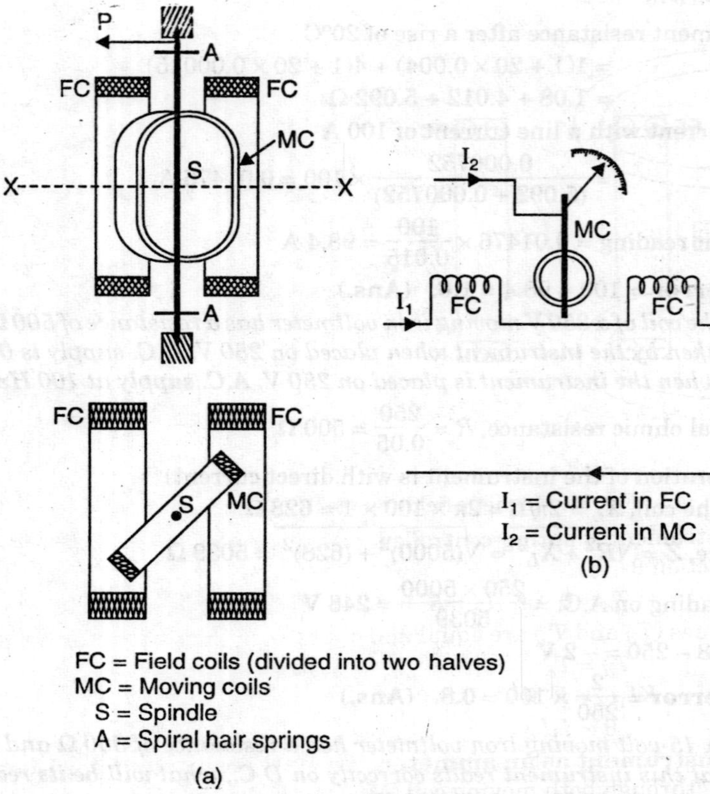

Fig. 19. Electrodynamic or dynamometer instrument.

These instruments essentially consist of fine wire *moving coil* placed in the magnetic field produced by *another fixed coil* when carrying currents. The coils are usually *air cored* to *avoid hysteresis, eddy currents and other errors* when the instrument is used on *A.C.* The *fixed coil FC* is divided into *two halves* placed close together and *parallel* to each other in order to provide a *fairly uniform field* within the range of the movement of the moving coil.

The upper diagram in Fig. 19 (*a*) shows a sectional elevation through fixed coil *FC* and the lower diagram represents a sectional plan on *XX*. The moving coil *MC* is carried by a spindle *S* and the controlling torque is exerted by spiral hair springs *A*, which may also serve to lead the current into and out of *MC*.

Deflecting torque. The deflecting torque is due to *interaction of the magnetic fields produced by currents in the fixed and moving coils.*

- Fig. 20 (*a*) shows the magnetic field due to current flowing through *FC* (I_1) in the direction indicated by the dots and cross.
- Fig. 20 (*b*) shows the magnetic field due to current (I_2) in *MC*.
- Fig. 20 (*c*) shows the combined effect of the above magnetic fields. By combining these magnetic fields it will be seen that when currents (I_1 and I_2) flow simultaneously through *FC* and *MC*, the resultant magnetic field is *distorted* and effect is to exert a clockwise torque on *MC*.

Since *MC* is carrying current (I_2) at right angles to the magnetic field produced by *FC*, deflecting torque, $T_d \propto I_1 \times I_2$

or $T_d = KI_1I_2$ where *K* is a constant.

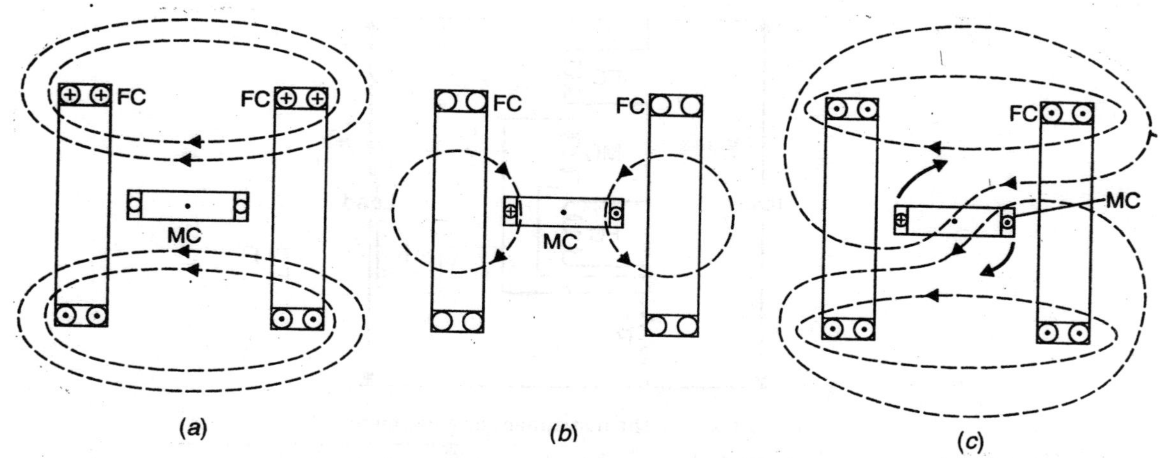

Fig. 20. Magnetic fields due to fixed and moving coils.

Since the instrument is spring-controlled, the restoring of control torque (T_c) is proportional to the angular deflection θ.

∴ $T_c \propto \theta$ or $T_c = K'\theta$

The two torques (T_d and T_c) are equal and opposite in the final deflected position.

∴ $T_d = T_c$

or $KI_1I_2 = K'\theta$

or $\theta \propto I_1I_2$

Use of the instrument as an ammeter. When the instrument is used as an ammeter then same current passes through both moving coil (*MC*) and fixed coils (*FC*) as shown in Fig. 21. In this case, $I_1 = I_2 = I$, hence $\theta \propto I^2$ or $I \propto \sqrt{\theta}$. The connections of Fig. 21. are used when *small currents* are to be measured.

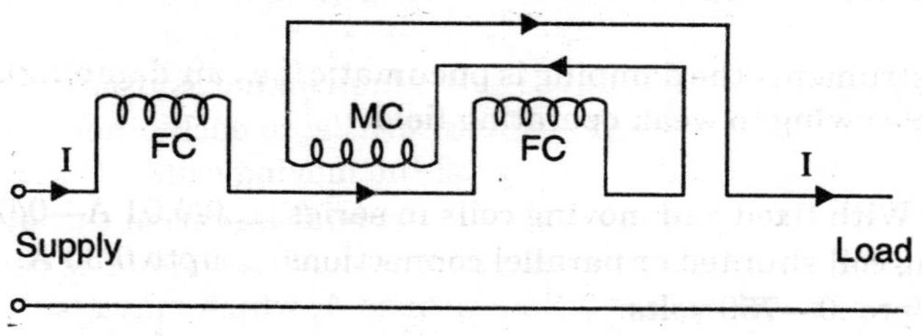

Fig. 21. Measurement of small currents.

In the case of *heavy currents,* a shunt is used to limit current through the moving coil as shown in Fig. 22.

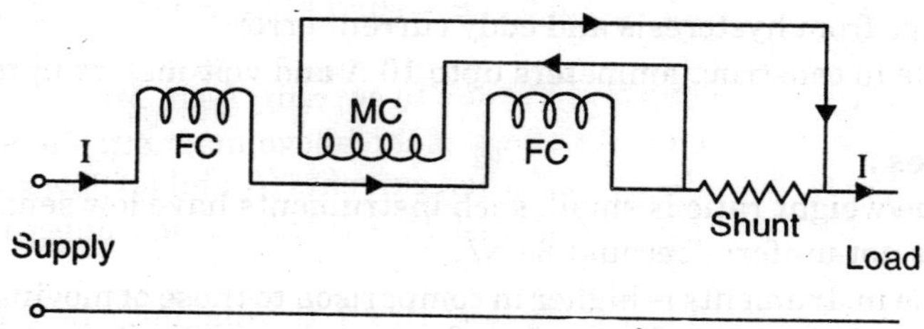

Fig. 22. Measurement of heavy currents.

Use of the instrument as voltmeter. When the instrument is used as a voltmeter, the fixed and moving coils are used in series along with a high resistance as shown in Fig. 23.

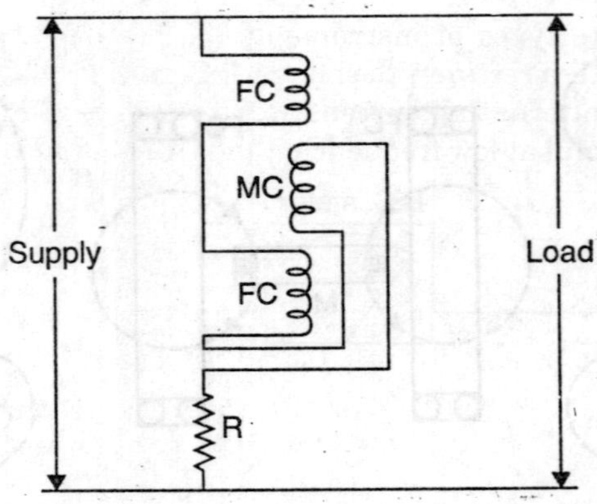

Fig. 23. Use of the instrument as a voltmeter.

Here again $I_1 = I_2 = I$,

where $I = \frac{V}{R}$......in D.C. circuits

and $I = \frac{V}{Z}$......in A.C. circuits

$\therefore$ $\theta \propto V \times V$ or $\theta \times V^2$

or $V = \sqrt{\theta}$

Thus whether the instrument is used as an ammeter or voltmeter its scale is *uneven* through the whole of its range and is cramped or crowded near the zero in particular.

Note. When the dynamometer instrument is used to measure an alternating current or voltage, the moving coil-due to its inertia-takes up a position where the average deflecting torque over one cycle is balanced by the restoring torque of the spiral springs. For that position, the deflecting torque is proportional to the mean value of the square of current or voltage, and the instrument scale can therefore be *calibrated to read the r.m.s. value.*

- In these instruments the damping is pneumatic (*i.e.*, air damping). Eddy current damping is admissible owing to weak operating field.

Ranges :

Ammeters. (*i*) With fixed and moving coils in series......0/0.01 A—0/0.05 A

(*ii*) With moving coil shunted or parallel connections......upto 0/30 A.

Voltmeters. Upto .0—750 volts.

Advantages and Disadvantages

Advantages :

(*i*) Can be used on both D.C. as well as A.C. systems.

(*ii*) They are free from hysteresis and eddy current errors.

(*iii*) It is possible to construct ammeters upto 10 A and volt-meters upto 600 V with precision grade accuracy.

Disadvantages :

(*i*) Since torque/weight ratio is small, such instruments have low sensitivity.

(*ii*) The scale is not uniform because $\theta \propto \sqrt{I}$.

(*iii*) Cost of these instruments is higher in comparison to those of moving iron instruments. So, these are only used as voltmeters and ammeters for precision measurements.

(*iv*) Higher friction losses.

6. RECTIFIER INSTRUMENTS

These are not separate types of instruments, but rather a means of using a D'Arsonval movement, in conjunction with a *rectifier,* to change A.C. to D.C. Thus, a direct current movement can be adopted for use with alternating current as shown in Fig. 24. Rectifier type meters, using copper oxide rectifiers, are useful at low frequencies and will give good indication upto about 20 kHz.

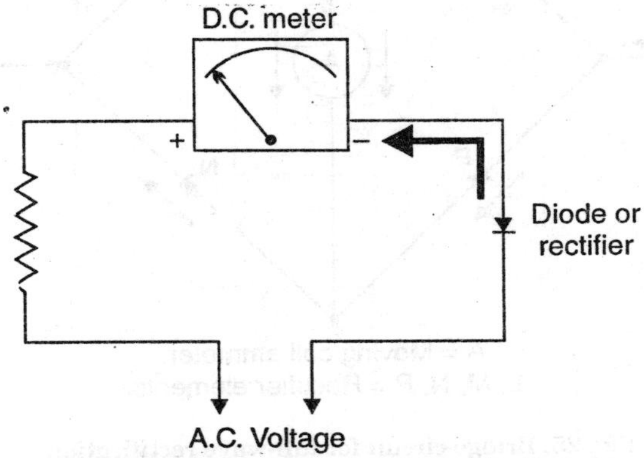

Fig. 24. A D.C. moving-coil meter can be used to measure A.C. voltage by putting a diode or rectifier in the meter circuit.

- Rectifier instruments can operate well into the R.F. (radio-frequency) range with the proper use of silicon or germanium rectifiers.
- *Measurements of current and voltage at several hundred megahertz are possible.*
- Another advantage is that this type of *meter is more sensitive* than any other type of A.C. meter.

Characteristics of Rectifier Instruments :

The *characteristics* of rectifier instruments are given below :

1. They provide an economical and practical means of A.C. measurements in radio and communication circuits at audio frequencies (1000 to 10000 Hz) and in other A.C. circuits where small available power makes it necessary to use sensitive low-energy instruments.

2. A rectifier instrument is, in general, approximately *fifty times* more sensitive than either an electrodynamometer or a moving-iron instrument.

3. In most instances rectifier instruments have essentially *linear scales.*

4. They commonly possess sensitivity of the order of 1000 to 2000 ohms per volt and more.

5. The power consumed by a rectifier instrument is generally several times as high as its permanent magnet moving coil (PMMC) mechanism because of the resistance of its rectifier.

6. Rectifier instruments are manufactured as microammeters and milliammeters in ratings from 100 μ A to 15 mA for full scale deflection. The higher rating of 15 mA is more or less governed by the special size of the rectifier in comparison with the mechanism with which it is to be used.

7. Shunting of rectifier instruments is not practical because of the change in resistance of the rectifier with both temperature and the amount of current.

Rectifier Ammeters. The *rectifier ammeters usually* consist of four rectifier elements *L, M, N* and *P* arranged in the form of a bridge (as shown in Fig. 25) and *A* represents a moving-coil ammeter. The apex of the black triangle (shown in the bridge) indicates the direction in which the resistance is low.

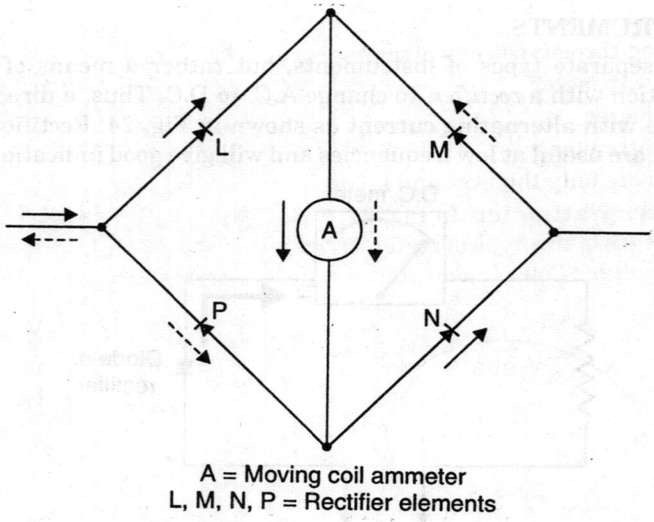

Fig. 25. Bridge circuit for full-wave rectification.

As shown in Fig. 25, during the half-cycles current flows through elements L and N from left to right as shown by the *full arrows.* During other half-cycles, the current flows through M and N, as shown by the *dotted arrows.* The waveform of the current flowing through A is therefore as shown in Fig. 26. The deflection of moving-coil ammeter A, thus depends upon the *average value of the current* ; and the scale of A can be calibrated to read the r.m.s. value of the current on the assumption that the waveform of the latter is sinusoidal with a form factor of 1.11.

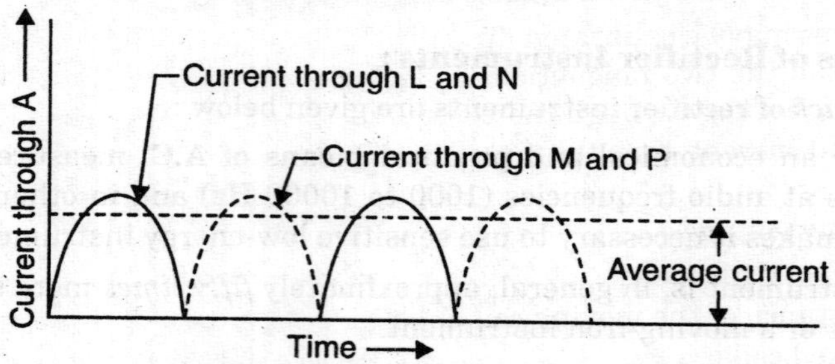

Fig. 26. Waveform of current through moving-coil ammeter.

Rectifier Voltmeters. In a *rectifier voltmeter*, A is a milliammeter and the bridge circuit of Fig. 25 is *connected in series with a suitable resistor*.

Advantages of rectifier instruments :

1. The primary advantage of the rectifier *voltmeter* is that it is *far more sensitive* as compared to other types of voltmeter suitable for measuring A.C. voltages.

2. Metal rectifiers can be incorporated in universal instruments, such as ammeter, thereby enabling a moving-coil milliammeter to be used in combination with shunt and series resistances to measure various ranges of *D.C.* and *D.C. voltage,* and in combination with a *bridge rectifier and suitable resistors* to measure various ranges of *A.C.* and *A.C. voltage.*

7. WATTMETERS

A wattmeter is a combination of an ammeter and a voltmeter and, therefore consists of two coils known as *current coil* and *pressure coil*. The operating torque is produced due to interaction of fluxes on account of currents in current and pressure coils.

There are following three types of wattmeters :

1. Dynamometer wattmeter
2. Induction wattmeter
3. Electrostatic wattmeter.

We shall discuss here only the first and second type.

7.1. Dynamometer Wattmeter. In Fig. 27, the dynamometer is connected as a wattmeter. This is one of the advantages of this type of meter. If the coils are connected so that a value of current *proportional to the load voltage* flows in one, and a value of current *proportional to the load current*

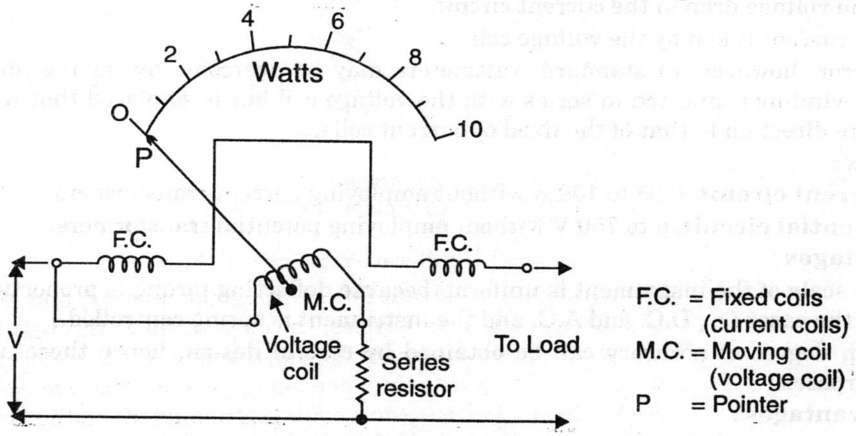

Fig. 27. Connection of dynamometer for measuring power.

flows in the other, the meter may be calibrated directly in watts. This is true because the indication depends upon the product of the two magnetic fields. The *strength of the magnetic fields depends upon the values of currents flowing* through the coils. If one current is proportional to load voltage and other current is the load current, then the meter can be calibrated in terms of watts or true power consumed by the load.

Let v = supply voltage,
i = load current, and
R = resistance of the moving coil circuit.

Current through fixed coils, $i_f = i$.

Current through the moving coil, $i_m = \dfrac{v}{R}$

Deflecting torque, $T_d \propto i_f \times i_m \propto \dfrac{iv}{R}$.

— For a *D.C. circuit the deflecting torque is thus proportional to the power.*

— For any circuit with *fluctuating torque, the instantaneous torque is proportional instantaneous power.* In this case due to inertia of moving parts the deflection will be proportional to the average torque *i.e.*, the deflection will be proportional to the *average power.* For sinusoidal alternating quantities the average power is $VI \cos \phi$, where

V = r.m.s. value of voltage,
I = r.m.s. value of current, and
ϕ = phase angle between V and I.

Hence an electrodynamic instrument, when connected as shown in Fig. 27, indicates the power, irrespective of the fact it is conected in an A.C. or D.C. circuit.

- Scales of such wattmeters are more or less uniform because the deflection is proportional to the average power and for spring control, controlling torque is proportional to the deflection, hence $\theta \propto$ power. Damping is *pneumatic*.

Errors :

- The error may creep in due to the inductance of the moving or voltage coil. However, the high non-inductive resistance connected in series with coil swamps, to a great extent, the phasing effect of the voltage coil inductance.
- There may be error in the indicated power due to the following :

(*i*) Some voltage drop in the current circuit.

(*ii*) The current taken by the voltage coil.

This error, however, in standard wattmeters may be overcome by having an additional compensating winding connected in series with the voltage coil but is so placed that it produces a field in opposite direction to that of the fixed or current coils.

Ranges :

(*i*) **Current circuit.** 0.25 to 100 A without employing current transformers.

(*ii*) **Potential circuit.** 5 to 750 V without employing potential transformers.

Advantages :

(*i*) The scale of the instrument is uniform (because deflecting torque is proportional to true power in both the cases *i.e.*, D.C. and A.C. and the instrument is spring controlled.)

(*ii*) High degree of accuracy can be obtained by careful design, hence these are used for calibration purposes.

Disadvantages :

(*i*) The error due to the inductance of pressure coil at low power factor is very serious (unless special features are incorporated to reduce its effect).

(*ii*) Stray field may effect the reading of the instrument. To reduce it, magnetic shielding is provided by enclosing the instrument in an iron case.

7.2. Induction Wattmeters. Induction wattmeters can be used on A.C. circuit only (in contrast with dynamometer wattmeters can be used both on D.C. and A.C. circuits) and are useful only when the frequency and supply voltage are constant.

The *operation* of all induction instruments depends on the *production of torque due to reaction between a flux* ϕ_1 (whose magnitude depends on the current or voltage to be measured) *and eddy currents induced in a metal disc or drum by another flux* ϕ_2 (whose magnitude also depends on the current or voltage to be measured). Since the magnitude of eddy currents also depends on the flux producing them, the *instantaneous value of the deflecting torque is proportional to the square of the current or voltage under measurement and the value of mean deflecting torque is proportional to the mean square of the current or voltage.*

Fig. 28 shows an induction wattmeter. It has two laminated electromagnets one is excited by the current in the main circuit—its exciting winding being joined in series with the circuit, hence it is also called *series magnet.*

The other electromagnet is excited by current which is *proportional to the voltage of the circuit*. Its exciting coil is joined in parallel with the circuit, hence this magnet is sometimes referred to as *shunt magnet*.

A thin aluminium disc is mounted in such a way that it cuts the fluxes of both the magnets. Hence two eddy currents are produced in the disc. *The deflection torque is produced due to the interaction of these eddy currents and the inducing fluxes.* Two or three copper rings are fitted on the central limb of the shunt magnet and can be so adjusted as *to make the resultant flux in the shunt magnet lag behind the applied voltage by a suitable angle.*

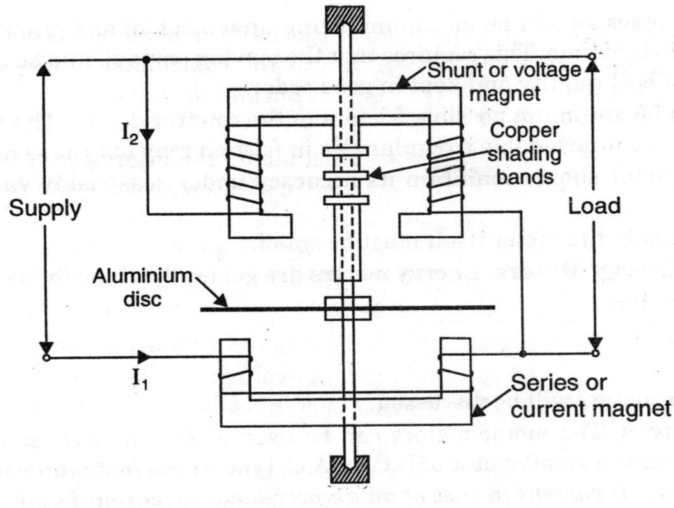

Fig. 28. Induction wattmeter.

This instrument is spring controlled, the spring being fitted to the spindle of the moving system which also carries a pointer. The scale is uniformly even and extends over 300°.

Such wattmeters can handle current upto 100 A. For handling greater currents they are used in conjunction with *current transformers*.

Advantages :

(*i*) Fairly long scale (extending over 300°).

(*ii*) Free from the effects of stray fields.

(*iii*) Good damping.

(*iv*) Practically free from frequency errors.

Disadvantages :

Sometimes subjected to serious temperature errors.

8. INTEGRATING METERS (ENERGY METERS)

Integrating or energy meters are *used to measure the quantity of electric energy supplied to a circuit in a given time.* They give *no direct indication of power i.e.*, as to the rate at which energy is being suppl‥‛ because their registrations are independent of the rate at which at given quantity of electric energy ‥being consumed.

The main difference between an *energy meter* and *a wattmeter* is that the former is fitted with some type of *registration mechanism* whereby all the instantaneous readings of power are summed over a definite period of time whereas the latter indicates the value at a *particular instant* where it is read.

8.1. Essential Characteristics of Energy Meters. The essential characteristics of energy meters are given below :

1. They must be simple in design and must not contain any parts which may rapidly deteriorate.

2. The readings may be given directly by the dials and must avoid any multiplying factors.

3. The casing of the meter should be dust, water and insect proof.

4. Permanency of calibration is a prime requisite and to attain it, the friction at the pivots etc., and retarding torque of the magnetic brakes must remain constant. The magnets should be so placed that they are not affected in their strength by the magnetic field of the current coil.

5. The friction losses should be minimum (being unavoidable) and must remain practically constant over long periods of time. This requires that the moving parts should be light, and the jewels and pivots should be of best quality and kept in good order.

6. There should be minimum possible friction in the counter device. The torque of the meter should be high so that the unavoidable irregularities in friction may not cause inaccuracies.

7. The energy meter should maintain its accuracy under reasonably varying conditions of voltage and load.

8. The energy loss in the meter itself must be small.

8.2. Types of Energy Meters. Energy meters are generally of the following three types :

1. Electrolytic meters
2. Motor meters
3. Clock meters.

Here only motor meters will be discussed.

8.3. Motor Meters. The motor meters can be used in *D.C. as well as in A.C. circuits.* In principle the motor meter is a small motor of D.C. or A.C. type whose *instantaneous speed of rotation is proportional to the circuit current in case of an ampere-hour meter and to the power of the circuit in case of a watt-hour meter.*

The following are the essential parts of the motor meters :

1. **An operating torque system.** It produces a torque and causes the moving system to rotate continuously.

2. **A braking device.** It is usually a permanent magnet, known as *brake magnet.* This brake magnet induces current in some part of the moving system which in turn produces the braking torque. Thus the braking torque is *proportional to the induced currents* whereas the induced currents are *proportional to the speed of the moving system* (and hence the *braking torque is proportional to the speed of the moving system* (*disc*). When the braking torque is equal to the driving torque the moving system attains a steady speed.

3. **Revolution registering device.** This device is obtained by having a worm cut on the spindle of the instrument. The worm engages with a pinion and thus drives the train of wheels and registers ampere-hours and watt-hours directly.

Types of motor meters. The various types of motor meters are :

(*i*) Mercury motor meters
(*ii*) Commutator motor meters
(*iii*) Induction motor/energy meters.

— *Mercury motor meters* and *commutator meters* are used on *D.C. circuits*
— *Induction meters* are used on *A.C. circuit.*

8.4. Motor-driven Meter—Watt-hour Meter. The motor-driven meter shown in Fig. 29 can be used on direct or alternating current. It contains a small motor and an aluminium retarding disc. *The field winding is connected in series with the load, and the field strength is proportional to the load current.*

The armature is connected *across* the source, and the *current in the armature is proportional to the source or line voltage.* The torque produced in the armature is *proportional to the power* consumed by the load.

The armature shaft drives a series of counters that are calibrated in watt-hours. The power that is used can be read directly from the dials.

The aluminium disc attached to the armature is used to control the armature speed. The disc turns in a magnetic field produced by the permanent magnets, and the retarding force increases as the rotation increases and stops when the disc stops. The retarding force is produced by the

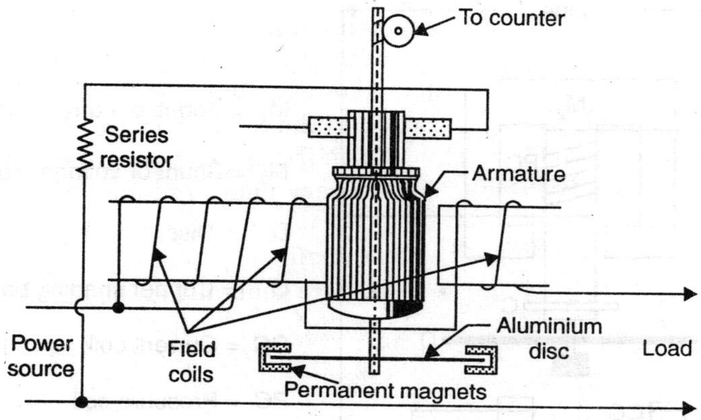

Fig. 29. Motor-driven meter (watt-hour meter) designed to operate on direct or alternating current.

aluminium conductor cutting through the lines of force of the permanent magnets. This is a form of magnetic damping.

The meter must overcome the friction of the bearings and indicators at *very light loads.* A portion of the field is produced by the armature current (coil in series with the armature winding). This coil is (called as compensating coil) wound to aid the field and is adjusted to the point where it just overcomes the meter friction.

8.5. Induction Type Watt-hour Meter. This is the most commonly used meter on A.C. circuits for measurement of energy.

Advantages :

(*i*) Simple in operation
(*ii*) High torque/weight ratio
(*iii*) Cheap in cost
(*iv*) Correct registration even at very low power factor
(*v*) Unaffected by temperature variations
(*vi*) More accurate than commutator type energy meter on light loads (owing to absence of a commutator with its accompanying friction).

Induction Type Single Phase Energy Meters. These are, by far, the most common form of A.C. meters met with in every-day *domestic* and *industrial installations.* These meters measure electric energy in kWh.

The principle of these meters is practically the same as that of induction watt meters. Instead of the control spring and the pointer of the watt-meter, the watt-hour meter, (energy meter) employs a *brake magnet and a counter attached to the spindle.* Just like other watt-hour meters, the eddy currents induced in the aluminium disc by the brake magnet due to the revolution of the disc, are utilised to control the continuously rotating disc.

Construction. The construction of a typical meter of this type is shown in Fig. 30. The brake-magnet and recording wheel-train being omitted for clearances. It consists of the following :

(*i*) Series magnet M_1
(*ii*) Shunt magnet M_2
(*iii*) Brake magnet
(*iv*) A rotating disc.

The *series electro-magnet* M_1 consists of a number of U-shaped iron laminations assembled together to form a core, wound with a *few turns of a heavy gauge wire.* This wound coil is known as

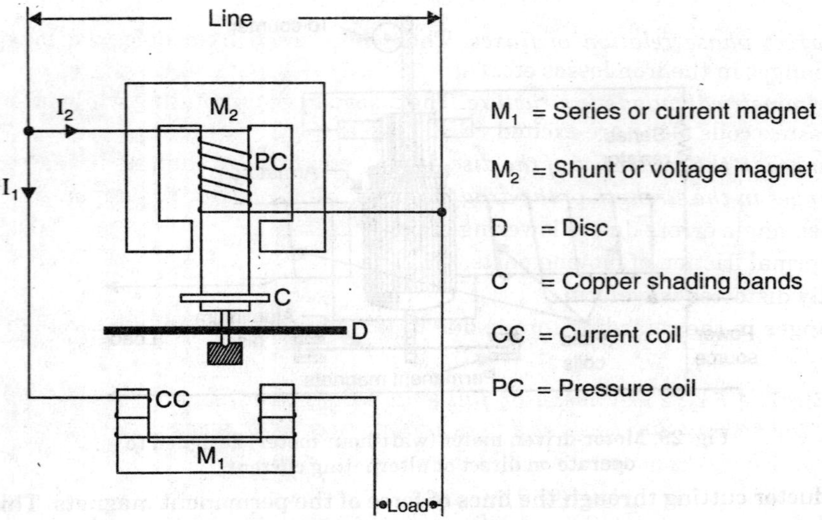

Fig. 30. Induction type single phase energy meter.

current coil and is connected in one of the lines and in series with the load to be metered. The series electromagnet is energized and sets up a magnetic field cutting through the rotating disc, when load current flows through the current coil C.C. The rotating disc is an aluminum disc mounted on a vertical spindle and supported on a sapphire cup contained in a bottom screw. The bottom pivot, which is usually removable, is of hardened steel, and the end, which is hemispherical in shape, rests in the sapphire cup. The top pivot (not shown) merely serves to maintain the spindle in a vertical position under working condition and does not support any weight or exert any appreciable thrust in any direction.

The *shunt magnet* M_2 consists of a number of *M* shaped iron laminations assembled together to form a core. A core having *large number of turns of fine wire* is fitted on the middle limb of the shunt magnet, this coil is known as *pressure coil P.C.* and is connected across the supply mains.

The *brake magnet* consists of *C* shaped piece of alloy steel bent round to form a complete magnetic circuit, with the exception of a narrow gap between the poles. This magnet is mounted so that the disc revolves in the air gap between the polar extremities. The movement of the rotating disc through the magnetic field crossing the air gap *sets up eddy currents in the disc which react with the field and exerts a braking effect.* The speed of the rotating disc may be adjusted by changing the position of the brake magnet or by diverting some of the flux there from.

Working. The shunt electromagnet produces a magnetic field which is of pulsating character ; it cuts through the rotating disc and induces eddy currents there in, but normally *does not in itself produce any driving force.* Similarly series electromagnet induces eddy currents in the rotating disc, but *does not in itself produce any driving force.* In order *to obtain driving force in this type of meter, phase displacement of 90° between the magnetic field set up by shunt electromagnet and applied voltage V is achieved by adjustment of copper shading band C* (also known as power factor compensator or compensating loop). *The reaction between these magnetic fields and eddy currents set up a driving torque in the disc.*

Sources of Errors. The various sources of errors in an induction-type energy meter are given below :

(*i*) *Incorrect magnitude of the fluxes.* These may arise from abnormal voltages and load currents.

(*ii*) *Incorrect phase relation of fluxes.* These may arise from defective lagging, abnormal frequencies, changes in the iron losses etc.

(*iii*) *Unsymmetrical magnetic structure.* The disc may go on rotating while no current is being drawn but pressure coils alone are excited.

(*iv*) *Changes in the resistance of the disc.* It may occur due to changes in temperature.

(*v*) *Changes in the strength of the drag magnets.* It may be due to temperature or ageing.

(*vi*) Phase-angle errors due to lowering of power factor.

(*vii*) Abnormal friction of moving parts.

(*viii*) Badly distorted waveform.

(*ix*) Changes in the retarding torque due to the disc moving through the field of the current coils.

Example 7. *A 5 A, 230 V meter on full load unity power factor test makes 60 revolutions in 360 seconds. If the normal disc speed is 520 revolutions per kWh, what is the percentage error ?*

Solution. Energy consumed in 360 seconds

$$= \frac{VI \cos \phi \times t}{3600 \times 1000} \text{ kWh}$$

$$= \frac{230 \times 5 \times 1 \times 360}{3600 \times 1000} = 0.115 \text{ kWh}$$

where, t is in seconds.

Energy recorded by the meter

$$= \frac{60}{520} = 0.11538 \text{ kWh}$$

$\therefore$ % **age error** $= \frac{0.11538 - 0.115}{0.115} \times 100 = \mathbf{0.33\%}$ **(fast). (Ans.)**

Example 8. *The constant of a 230 V, 50 Hz, single phase energy meter is 185 revolutions per kWh. The meter takes 190 seconds for 10 revolutions while supplying a non-inductive load of 4.5 A at normal voltage. What is the percentage error of the instrument ?*

Solution. Energy consumed in 190 seconds

$$= \frac{VI \cos \phi}{1000} \times t = \frac{230 \times 4.5 \times 1}{1000} \times \frac{190}{3600} = 0.0546 \text{ kWh}$$

[cos ϕ = 1, since load supplied is non-inductive]

Energy registered by the meter $= \frac{10}{185} = 0.054$ kWh

$\therefore$ % age error $= \frac{0.054 - 0.0546}{(0.054)} = \mathbf{0.06\%}$ **(slow). (Ans.)**

Example 9. *The name plate of a meter reads "1 kWh = 15000 revoltions". In a check up, the meter completed 150 revolutions during 45 seconds. Calculate the power in the circuit.*

Solution. Power metered in 150 revolutions

$$= 1 \times 150/15000 = 0.01 \text{ kWh}$$

If P kilowatt is the power in the circuit, then energy consumed in 45 seconds

$$= \frac{P \times 45}{3600} \text{ kWh} = 0.0125 \, P \text{ kWh}$$

Equating the two amounts of energy, we have

$$0.0125 \, P = 0.01$$

$$P = 0.8 \text{ kW} = \mathbf{800 \text{ W}}. \quad \textbf{(Ans.)}$$

Example 10. *A 230 V ampere-hour type meter is connected to a 230 V D.C. supply. If the meter completes 225 revolutions in 10 minutes when carrying 14 A, calculate : (i) The kWh registered by the meter, and (ii) The percentage error of the meter above or below the original calibration.*

The timing constant of the meter is 40 A-s/revolution.

Solution. During 225 revolutions the meter would register 40 × 225 A-*s* or coloumbs. since time taken is 10 minutes or 600 seconds it corresponds to a current of $\frac{40 \times 225}{600} = 15$ A.

(*i*) Energy recorded by the meter $= \frac{VIt}{1000}$ kWh

where *t* is in hour

$$= \frac{230 \times 15}{1000} \times \frac{10}{60} = 0.575 \text{ kWh}.$$

(*ii*) Actual energy consumed $= \frac{230 \times 14}{1000} \times \frac{10}{60} = 0.5367$ kWh

Percentage error $= \frac{0.575 - 0.5367}{0.575} \times 100 =$ **6.66%. (Ans.)**

9. MEASUREMENT OF RESISTANCE

9.1. Voltmeter-Ammeter Method. Fig. 31 (*a*) and (*b*) shows two methods of connections for measurement of resistance.

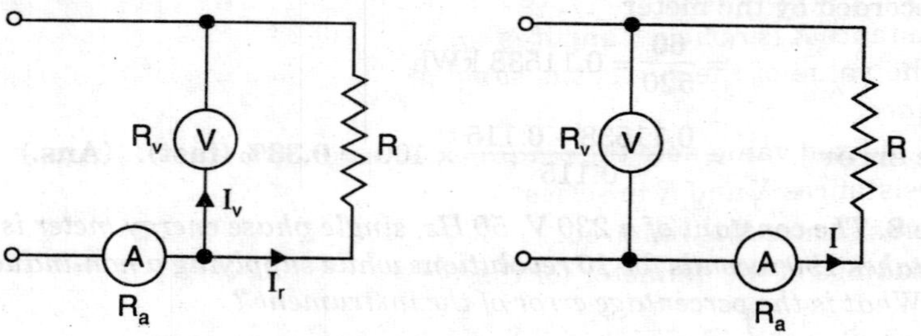

Fig. 31. Methods for measuring resistance.

Let R_a = resistance of the ammeter (which is very small),
R_v = resistance of the voltmeter (which is a large value),
R = resistance to be measured,
V = reading of the voltmeter, and
I = reading of the ammeter.

Fig. 31 (*a*) shows the connections for measurement of *low resistances*.

Here $I = I_v + I_r$

$$= \frac{V}{R_v} + \frac{V}{R}$$

$$\therefore \quad \frac{V}{R} = I - \frac{V}{R_v}$$

or $$R = V \Big/ \left(1 - \frac{V}{R_v}\right) \quad ...(1)$$

If R_v is *very high*, $$R = \frac{V}{I} \quad ...(2)$$

Fig. 31 (*b*) shows the connections for measurement of high and low resistances.

Here
$$V = IR_a + IR$$
$$\therefore \quad IR = V - IR_a$$
or
$$R = \frac{V}{I} - R_a \quad ...(3)$$
If the value of R_a *is very small*
$$R = \frac{V}{I} \quad ...(4)$$

9.2. Substitution Method. The connections for measurement of resistance by substitution are shown in Fig. 32.

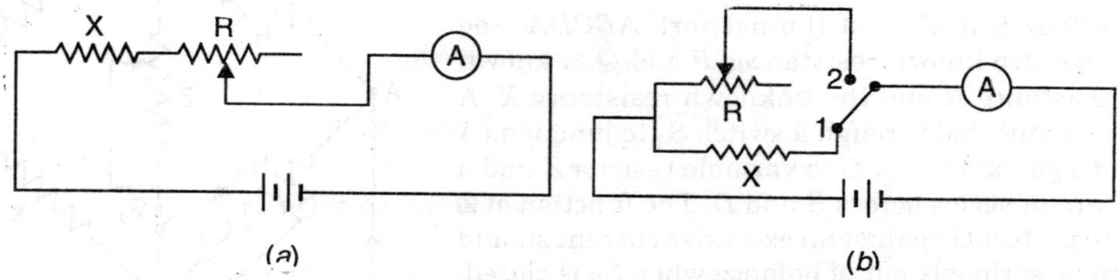

Fig. 32. Measurement of resistance by substitution.

Let R be a variable resistance which can be changed in small steps, say of 0.1 Ω.

(*i*) Refer Fig. 32 (*a*). First resistance X is put in the circuit and the value of current noted.

Then resistance X is *removed* and it is *substituted* by a known variable resistance R which is varied so that the value of the current is same in both the cases. *This value of R is equal to the unknown resistance.*

(*ii*) If R is of fixed value note the *readings* of the ammeter for the following cases :

(*a*) For resistances X and R in series.

(*b*) When resistance X is removed.

Let the readings of the ammeter for these cases be I_1 and I_2

$$\therefore \quad I_1 = \frac{V}{R + X} \quad ...(5)$$
and
$$I_2 = \frac{V}{R} \quad ...(6)$$
$$\frac{I_2}{I_1} = \frac{R + X}{R} = 1 + \frac{X}{R} \quad ...(7)$$
or
$$\frac{X}{R} = \frac{I_2}{I_1} - 1$$
or
$$X = R\left(\frac{I_2 - I_1}{I_1}\right) \quad ...(8)$$

(*iii*) Refer Fig. 32 (*b*). The two-way switch first makes contact with 1 and then with 2 and let these readings be I_1 and I_2.

$$I_1 = \frac{V}{X} \text{ (when } X \text{ is in the circuit)} \quad ...(9)$$
$$I_2 = \frac{V}{R} \text{ (when } R \text{ is in the circuit)} \quad ...(10)$$

$\therefore \quad I_2/I_1 = X/R$

so that $$X = R \times \frac{I_2}{I_1} \quad ...(11)$$

9.3. Measurement of Resistance by the Wheatstone Bridge. In Wheatstone bridge method of measuring resistances, a resistor of *unknown* resistance is balanced against resistors of known resistances. Though based on comparison of resistances, this is *one of the most accurate methods of measuring resistances as it is independent of the calibration of the indicating instrument and relies upon the null-point method.*

The four branches of the network *ABCDA* (see Fig. 33) have two known resistances *P* and *Q*, a known variable resistance *R* and the unknown resistance *X*. A battery *E* is connected through a switch S_1 to junctions *A* and *C*; and a galvanometer *G*, a variable resistor *Z* and a switch S_2 are in series across *B* and *D*. The function of *Z* is merely to protect *G* against an excessive current should the system be seriously out of balance when S_2 is closed.

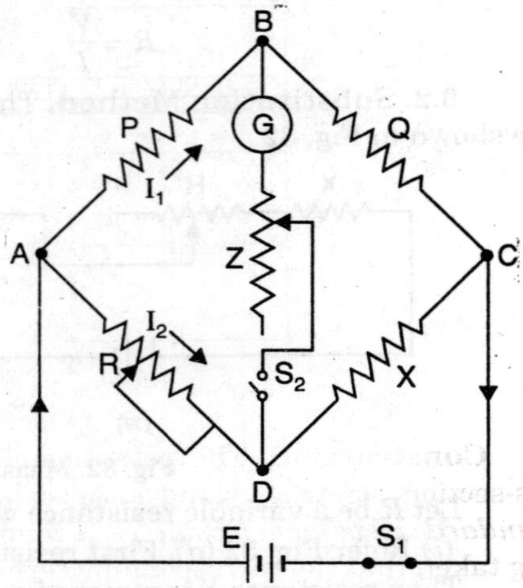

Fig. 33. Wheatstone bridge.

After closing S_1 and S_2, *R* is adjusted until there is no deflection on *G* with the resistance of *Z* reduced to zero. Junctions *B* and *D* are then at the *same potential*, so that p.d. between *A* and *B* is the same as that between *A* and *D*, and the p.d. between *B* and *C* is the same as that between *C* and *D*.

Let I_1 and I_2 be the currents through *P* and *R* respectively when the bridge is balanced. From Kirchhoff's first law it follows that since there is no current through *G*, the currents through *Q* and *X* are also I_1 and I_2 respectively.

But potential difference (p.d.) across $P = PI_1$

Potential differences across $R = RI_2$

$$\therefore \quad PI_1 = RI_2 \quad ...(12)$$

Also potential difference across $Q = QI_1$

Potential difference across $X = XI_2$

$$QI_1 = XI_2 \quad ...(13)$$

Dividing (13) by (12), we get

$$\frac{Q}{P} = \frac{X}{R}$$

$$\therefore \quad X = R \times \frac{Q}{P} \quad ...(14)$$

which is the relation of Wheatstone bridge. *P* and *Q* are usually called the *ratio arms* and *X* as the *balance* or *rheostat arm*. The battery and the galvanometer can be interchanged without affecting the relation.

Note. The resistances *P* and *Q* may take the form of the resistance of a slide-wire, in which case *R* may be a fixed value and the balance is obtained by moving a sliding contact along the wire. If the wire is homogeneous and of uniform section, the ratio of *P* to *Q* is the same as the ratio of the lengths of wire in the respective arms.

10. THE POTENTIOMETER

For the accurate measurement of potential difference, current and resistance the *potentiometer is one of the most useful instruments.*

Its principle of action is that *an unknown e.m.f. or p.d. is measured by balancing it, wholly or in part, against a known difference of potential.*

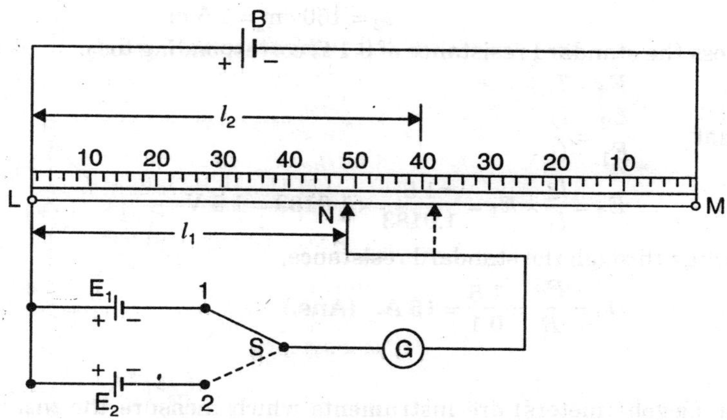

Fig. 34. A simple potentiometer.

Construction. The potentiometer, in its simplest form, consists of wire *LM* of uniform cross-section, stretched alongside a scale and connected across an accumulator *B* of ample capacity. *A standard cell of known e.m.f.* E_1 is connected between *L* and terminal 1 of a two-way switch *S*, care being taken that the corresponding terminals of *B* and E_1 are connected to *L*.

Working

— Slider *N* is pressed momentarily against wire *LM* and its position adjusted until the galvanometer deflection is zero when *N* is making contact with *LM*. Let l_1 be the corresponding distance between *L* and *N*. The fall of potential over length l_1 of the wire is then the same as the e.m.f. E_1.

— Then move the switch *S* to 2, thereby replacing the standard cell by another cell, the e.m.f. E_2 of *which is to be measured.* Adjust the slider *N* again to give zero deflection on *G*. If l_2 be the new distance between *L* and *N*, then

$$\frac{E_2}{E_1} = \frac{l_2}{l_1}$$

or $$E_2 = \frac{l_2}{l_1} \times E_1 \qquad ...(15)$$

Applications of potentiometer. The following are the applications/uses of potentiometers :

1. Measurement of small e.m.fs. (upto 2 volts).
2. Comparison of e.m.fs. of two cells.
3. Measurement of high e.m.fs. (say 250 volts).
4. Measurement of resistance.
5. Measurement of current.
6. Calibration of ammeter.
7. Calibration of voltmeter.

Example 11. *Using a Weston cadium cell of 1.0183 V and a standard resistance of 0.1 Ω a potentiometer was adjusted so that 1.0183 m was equivalent to the e.m.f. of the cell ; when a certain direct current was flowing through the standard resistance, the voltage across it correspond to 150 cm. What was the value of current ?*

Solution. E.m.f. of the standard cell, $E_1 = 1.0183$ V

Potentiometer adjustment, $l_1 = 1.0183$ m

and $l_2 = 150$ cm $= 1.5$ m

Voltage across the standard resistance of 0.1 Ω corresponding to l_2,

$$E_2 = ?$$

We know that $\frac{E_2}{E_1} = \frac{l_2}{l_1}$

$$\therefore \quad E_2 = \frac{l_2}{l_1} \times E_1 = \frac{1.5}{1.0183} \times 1.0183 = 1.5 \text{ V}$$

Current flowing through the standard resistance,

$$I_2 = \frac{E_2}{R} = \frac{1.5}{0.1} = 15 \text{ A.} \quad \textbf{(Ans.)}$$

11. MEGGER

Meggers (or megohmmeters) are instruments which measure the *insulation resistance* of *electric circuits relative to earth and one another.*

A megger consists of an *e.m.f. source* and a *voltmeter*. The scale of the voltmeter is calibrated in ohms (kilo-ohms or megohms, as the case may be). In measurements the e.m.f. of the self-contained source must be equal to that of the source used in calibration.

Fig. 35 shows diagrammatically a megger whose readings are independent of the speed of the self-contained generator. The moving system incorporates two coils 1 (current coil) and 2 (pressure coil) mounted on the same shaft and *placed in the field of a permanent magnet* (not shown) 90° apart. The generator energizes the two coils over separate wires. Connected in series with one coil is a fixed resistance R_1 (or several different resistances in order to extend the range of the instrument). The unknown resistance R_x is connected in series with the other coil. *The currents in the coils interact with the magnetic field and produce opposing torques.*

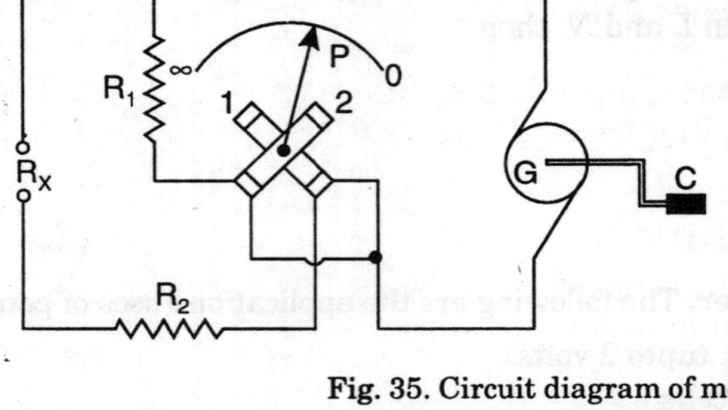

G = Generator
C = Crank
1, 2 = Coils
P = Pointer
1 = Current coil
2 = Pressure/Voltage Coil
R_x = Unknown resistance
R_1 = Fixed resistance
R_2 = Safety resistance

Fig. 35. Circuit diagram of megger.

The deflection of the moving system depends on the ratio of the currents in the coils and is independent of the applied voltage. The unknown resistance is read *directly* from the scale of the instrument. (The accuracy of measurement is unaffected by variations in the speed of the generator between 60 and 180 r.p.m.).

12. INSTRUMENT TRANSFORMERS

It is not practicable to connect instruments and meters directly to the lines in high voltage circuits. Instead instrument transformers are used. The following are the *two basic advantages* inherent in this method :

(*i*) Standard rated instruments may be used.

(*ii*) Operating personnel coming in contact with the instruments are not subjected to high voltage and current of the lines, and so there is less danger to them. Even with a low-voltage system, instrument transformers are used for measuring large currents, so that heavy leads to the instrument panel and to the ammeter and other current terminals are avoided.

The principle of the instrument transformer is fundamentally the same as that of the power transformer. The instrument transformers are classified as follows :

1. Potential transformers.
2. Current transformers.

12.1. Potential Transformers (P.T.)

- A potential transformer is a *step down transformer* used along with a low range voltmeter for measuring a high voltage. The primary is connected across the high voltage supply and the secondary to the voltmeter or potential coil of the wattmeter. Since the voltmeter (or potential coil) impedance is very high, the secondary current is very small and the potential transformer behaves as an ordinary two winding transformer operating on no-load. Fig. 36 shows a potential transformer used to measure the voltage of a circuit. It may be noted that the secondary is grounded. This is done so that if the insulation breaks down, the high voltage does not endanger personnel who may be reading the meters.

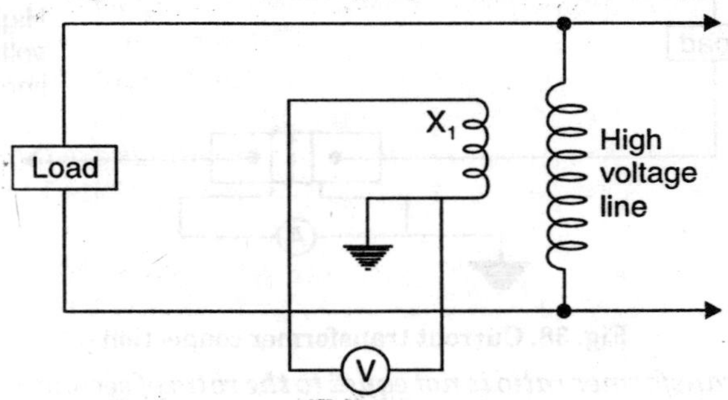

Fig. 36. Potential transformer connections.

- These transformers are *made with high quality iron core operating at very low flux densities so that the magnetising current may be very small.* Careful design ensures minimum variation of voltage ratio with load and minimum phase shift between input and output voltages. Potential transformers secondary are commonly designed for an output of 110 V.

12.2. Current Transformers (C.T.)

- Just as a shunt extends the range of a D.C. ammeter, so does the current transformer perform the same function in A.C. circuits. Thus a *high magnitude alternating current can be easily measured by a combination of a current transformer and a low range ammeter.*
- The primary of a current transformer (C.T.) consists of a few turns of thick cross-section connected in *series* with the high current line. Very often the primary is just one turn formed by taking the line conductor through the secondary winding (Fig. 37). The secondary winding consists of a large number of turns of fine wire designed for either 5A or 1 A rating. Thus a current transformer is *step-up* transformer. The current transformer has the secondary effectively short-circuited through the low impedance of the ammeter. Fig. 38 shows the current transformer connections.

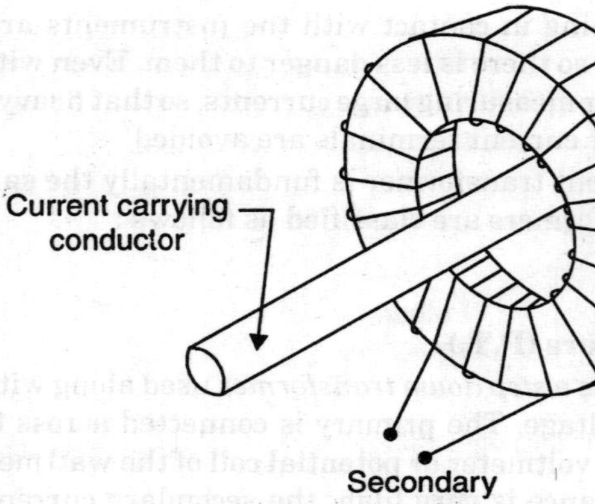

Fig. 37. Line conductor acting as primary.

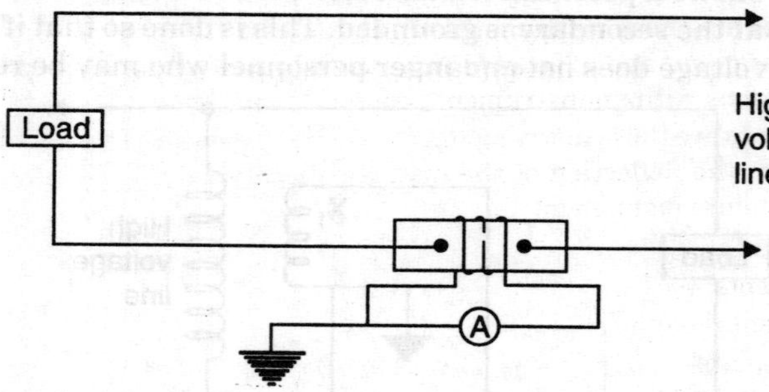

Fig. 38. Current transformer connections.

- The *current transformer ratio is not equal to the ratio of secondary to primary turns, mainly because of the effect of the magnetising current.* The primary current can be thought of as the sum of two currents, the first to balance secondary current so that primary and secondary m.m.fs. may balance and the second being the no-load current I_0. The component I_0 besides being responsible for a slight error in the current ratio, is also responsible for a phase angle error. The transformer *must be carefully designed to minimise the ratio and phase angle error.*

It may be noted that current transformer *must never be operated on open-circuit* for the following two reasons :

(*i*) There will be no secondary m.m.f. and since the primary current (and m.m.f.) is fixed, the core flux will increase enormously. This will cause large eddy current and hysteresis losses and the *resulting high temperature may damage the insulation or even the core.*

(*ii*) A very high voltage will be induced in the multi-turn secondary and this high voltage may be dangerous both to life and to the insulation.

Fig. 39 shows the wiring diagrams for potential and current instrument transformers.

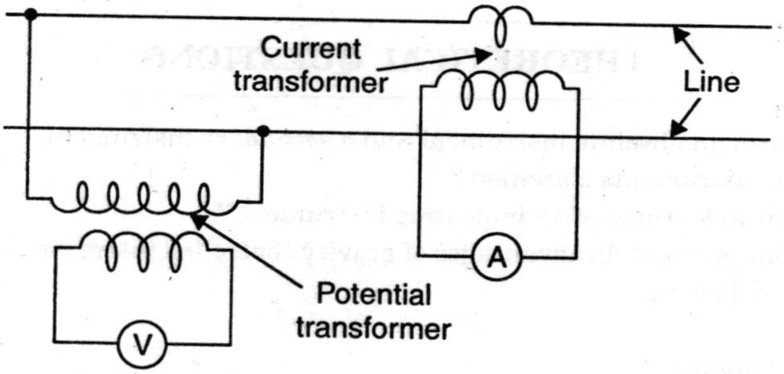

Fig. 39. Wiring diagrams for potential and current instrument transformers.

HIGHLIGHTS

1. **Absolute instruments** are those instruments which indicate the quantity to be measured in terms of the *constants of the instrument* and in order to find out the quantity in the practical units it is necessary to multiply such deflections with an instrument constant.
2. **Secondary instruments** are those instruments in which the value of electrical quantity to be measured can be determined from the deflection of the instrument only when they have been precalibrated by comparison with an absolute instrument.

   These are classified as :

   (*i*) Indicating instruments
   (*ii*) Recording instruments
   (*iii*) Integrating instruments.
3. Essential features of indicating instruments are :

   (*i*) Deflecting device
   (*ii*) Controlling device
   (*iii*) Damping device.
4. Moving iron instruments are of the following two types :

   (*i*) Attraction type
   (*ii*) Repulsion type.
5. Moving-coil instruments are of the following two types :

   (*i*) Permanent magnet type
   (*ii*) Dynamometer type.
6. Rectifier instruments can operate well into the R.F. range with the proper use of silicon or germanium rectifiers. They are more sensitive than any other type of A.C. meter.
7. A wattmeter consists of two coils known as current coil and pressure coil.
8. Integrating (or energy) meters are used to measure the quantity of electric energy supplied to a circuit in a given time. These are generally of the following three types :

   (*i*) Electrolytic meters
   (*ii*) Motor meters
   (*iii*) Clock meters.
9. The resistance can be measured by the following methods :

   (*i*) Voltmeter-ammeter method
   (*ii*) Substitution method
   (*iii*) By using wheatstone bridge
   (*iv*) By using a potentiometer.
10. Meggars are instruments which measure the insulation resistance of electric circuits relative to earth and one another.

THEORETICAL QUESTIONS

1. Differentiate between an absolute instrument and a secondary instrument.
2. How are secondary instruments classified ?
3. What are the essential features of an indicating instrument ?
4. What are the advantages and disadvantages of gravity controlled instruments ?
5. Explain briefly the following :
   (*i*) Air damping
   (*ii*) Fluid friction damping
   (*iii*) Eddy current damping.
6. Explain briefly, with the help of sketches, the construction and working of the following moving iron instruments :
   (*i*) Attraction type
   (*ii*) Repulsion type.
7. What are the advantages and disadvantages of moving-iron instruments ?
8. How are moving coil instruments classified ?
9. Give the construction and working of a permanent magnet moving-coil type instrument. Also enumerate its advantages and disadvantages.
10. Explain the construction and working of a dynamometer type instrument. How it can be used as an ammeter and a voltmeter.
11. Give the characteristics of rectifier instruments.
12. Explain briefly the following :
    (*i*) Rectifier ammeters
    (*ii*) Rectifier voltmeters.
13. What is a wattmeter ?
14. Enumerate types of wattmeter.
15. Discuss the construction working of a dynamometer wattmeter with the help of a neat diagram.
16. Draw a neat sketch of an induction wattmeter and explain its working. Also state its advantages and disadvantages.
17. What is an integrating or energy meter ?
18. What are the essential characteristics of energy meters ?
19. Give the construction and working of a motor-driven meter-watt hour meter.
20. With the help of a neat sketch explain the construction and working of an induction type single phase energy meter. Also discuss the sources of errors prevalent in this type of energy meter.
21. What are the various methods by which a resistance can be measured ?
22. Explain any two of the following methods of measuring resistance :
    (*i*) Voltmeter-ammeter method
    (*ii*) Substitution method
    (*iii*) Measurement of resistance by the Wheatstone bridge.
23. Explain briefly the following :
    (*i*) The potentiometer
    (*ii*) Meggar.
24. What are instrument transformers ?
25. How are instrument transformers classified ?
26. Explain briefly, with near sketches the following :
    (*i*) Potential transformers
    (*ii*) Current transformers.

UNSOLVED EXAMPLES

1. A moving-coil instrument gives full-scale deflection with 15 mA and has a resistance of 5 Ω. Calculate the resistance to be connected :
   (*i*) in parallel to enable the instrument to read upto 1 A.
   (*ii*) in series to enable it to read upto 100 V. [**Ans.** (*i*) 0.076 Ω (*ii*) 6661.7 Ω]
2. A moving-coil instrument has a resistance of 10 ohms and gives a full-scale deflection when carrying 50 mA. Show how it can be adopted to measure voltage upto 750 volts and current upto 100 amperes. [**Ans.** R = 14990 Ω, R_s = 0.005 Ω]
3. A 20 V moving-iron voltmeter reads correctly when put on D.C. and the instrument has a resistance of 600 ohms and inductance of 0.15 H. Find out the reading on 20 V A.C. mains (*i*) at 250 Hz (*ii*) 50 HZ. [**Ans** (*i*) 18.6 V, (*ii*) 19.87 V]
4. A 15 V moving-iron voltmeter has a resistance of 500 Ω and inductance of 0.12 H. Assuming that this instrument reads correctly on D.C. what will be its reading on A.C. at 15 V when the frequency is (*i*) 25 Hz and (*ii*) 100 Hz ? [**Ans.** (*i*) 14.99 V, (*ii*) 14.83 V]
5. The total resistance of a moving-iron voltmeter is 1000 Ω and coil has an inductance of 0.765 H. The instrument is calibrated with a full-scale deflection of 50 V D.C. Calculate the percentage error when the instrument is used on (*i*) 25 Hz supply, (*ii*) 50 Hz supply, the applied voltage being 50 V in each case. [**Ans.** (*i*) 0.72%, (*ii*) 36%]
6. In a moving-coil instrument, the moving coil has 40 turns and is of square shape with a mean length of 40 mm along each side. The coil hangs in a uniform radial field of 0.08 Wb/m^2. Find the turning moment of the coil when it is carrying a current of 10 mA. [**Ans.** 512×10^{-7} N-m]
7. A moving-coil instrument has a resistance of 5 Ω between terminals and full scale deflection is obtained with a current of 0.015 A. The instrument is to be used with a manganin shunt to measure 100 A full scale. Calculate the error caused by a 10°C rise in temperature. (*i*) when the internal resistance of 5 Ω is due to copper only, (*ii*) when a 4 Ω manganin swamping resistor is used in series with a copper resistor of 1 Ω.
   Take : α_{copper} = 0.004/°C and $\alpha_{manganin}$ = 0.00015/°C. [**Ans.** (*i*) 3.7%, (*ii*) 0.8%]
8. A 230 V, 50 Hz, single phase energy meter has a constant of 200 revolutions per kWh. While supplying a non-inductive load of 4.4 A at normal voltage the meter takes 3 minutes for 10 revolutions. Calculate the percentage error of the instrument. [**Ans.** 1.186% (slow)]
9. The disc of an energy meter makes 600 revolutions per unit of energy. When a 1000 W load is connected, the disc rotates at 10.2 r.p.m. If the load is on for 12 hours, how many units are recorded as error ? [**Ans.** 0.24 kWh more]
10. The name plate of a meter reads "1 kWh = 15000 revolutions". In a check up, the meter completed 150 revolutions during 50 seconds. Calculate the power in the circuit. [**Ans.** 720 W]

10

Power Supply System

1. Sources of Energy. 2. Power Plants—Steam power plant—Hydro-electric power plant—Nuclear power plant—Simple gas turbine plant—Diesel engine power plant. 3. Transmission and Distribution—General aspects—Systems of transmission—Line supports—Conductor material—Line insulators—Distribution systems—Electric supply system (A-C)—Overhead and underground systems—Kelvin's law—Corona—Skin effect—Proximity effect—Classification of transmission lines—Regulation and efficiency of a transmission line—Classification of cables—Surge impedance or neutral impedance—π and T-circuit of long transmission line. 4. Power Factor Improvement—Highlights—Objective type questions—Theoretical questions.

1. SOURCES OF ENERGY

The various sources of energy are :

1. Fuels
   - Solids—Coal, coke, anthracite etc.
   - Liquids—Petroleum and its derivatives.
   - Gases—Natural gas, blast furnace gas etc.
2. Energy stored in water
3. Nuclear energy
4. Wind power
5. Solar energy
6. Tidal power
7. Geothermal energy
8. Thermoelectric power.

2. POWER PLANTS

2.1. Steam Power Plant

2.1.1. Layout

Refer Fig. 1. The layout of a modern steam power plant comprises of the following four circuits :

1. Coal and ash circuit
2. Ash and gas circuit
3. Feed water and steam flow circuit
4. Cooling water circuit.

1. Coal and ash Circuit. Coal arrives at the storage yard and after necessary landing, passes on to the furnaces through the *fuel feeding device*. Ash resulting from combustion of coal collects at the back of the boiler and is removed to the ash storage yard through *ash handling equipment*.

2. Ash and gas Circuit. Ash is taken in from atmosphere through the action of a forced or induced draught fan and passes on to the furnace through the *air preheater*, where it has been heated by the heat of flue gases which pass to the chimney via the preheater. The flue gases after passing around boiler tubes and superheater tubes in the furnace pass, through a *dust* catching device or precipitator, then through the economiser, and finally through the air preheater before being exhausted to the atmosphere.

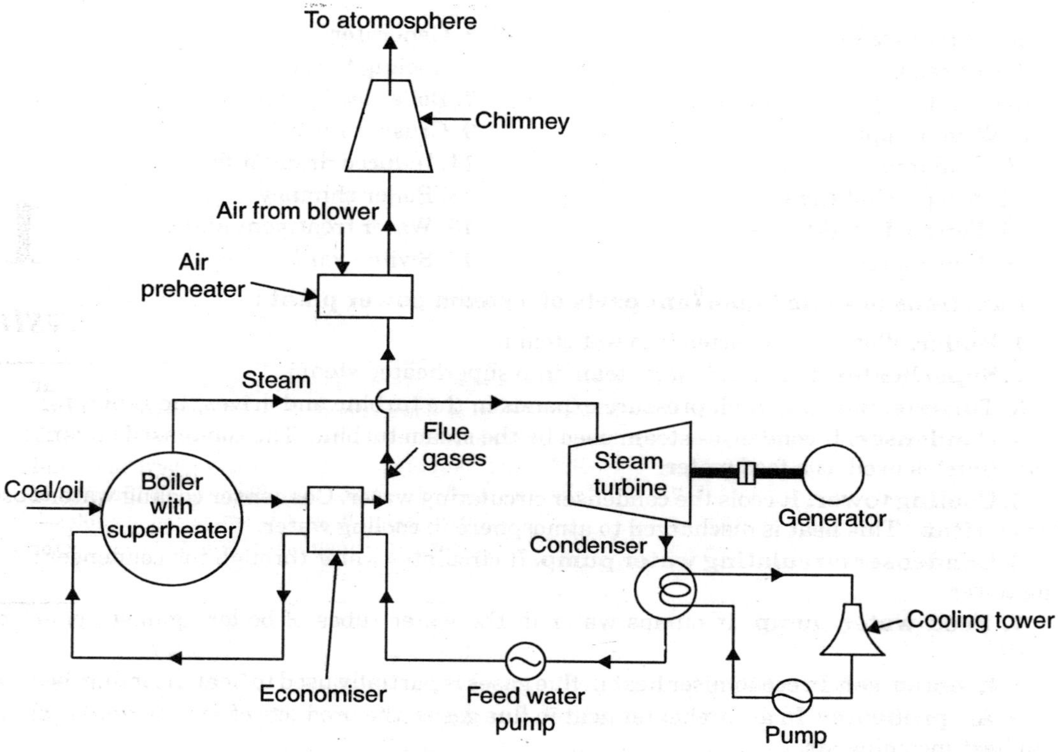

Fig. 1. Layout of a steam power plant.

3. Feed water and steam flow circuit. In the water and steam circuit condensate leaving the condenser is first heated in a closed feed water heater through extracted steam from the lowest pressure extraction point of the turbine. It then passes through the *deaerator* and a few more water heaters before going into the boiler through *economiser*.

In the boiler drum and tubes, water circulates due to the difference between the density of water in the lower temperature and the higher temperature sections of the boiler. Wet steam from the drum is further heated up in the superheater for being supplied to the prime mover. After expanding in high pressure turbine steam in taken to the reheat boiler (not shown) and brought to its original dryness or superheated before being passed on to the low pressure turbine. From there it is exhausted through the condenser into the hot well. The condensate is heated in the feed heaters (not shown) using the steam trapped (bled-steam) from different points of turbine.

A part of steam and water is lost while passing through different components and this is compensated by supplying additional feed water. This feed water should be purified before hand, to avoid the scaling of the tubes of the boiler.

4. Cooling water circuit. The cooling water supply to the condenser helps in maintaining a low pressure in it. The water may be taken from a natural source such as river, lake or sea or the same water may be cooled and circulated once again. In the latter case the cooling arrangement is made through spray pond or cooling lower.

2.1.2. Components of a modern steam power plant

A modern steam power plant comprises of the following components :

1. Boiler
   - (*i*) Superheater
   - (*ii*) Reheater
   - (*iii*) Economiser
   - (*iv*) Air-heater

2. Steam turbine
3. Generator
4. Condenser
5. Cooling towers
6. Circulating water pump
7. Boiler feed pump
8. Wagon tippler
9. Crusher house
10. Coal mill
11. Induced draught fans
12. Ash precipitators
13. Boiler chimney
14. Forced draught fans
15. Water treatment plant
16. Control room
17. Switch yard.

Functions of some important parts of a steam power plant :

1. Boiler. Water is converted into wet steam.

2. Superheater. It converts wet steam into superheated steam.

3. Turbine. Steam at high pressure expands in the turbine and drives the generator.

4. Condenser. It condenses steam used by the steam turbine. The condensed steam (known as *condensate*) is used as a feed water.

5. Cooling tower. It cools the condenser circulating water. Condenser cooling water absorbs heat from steam. This heat is discharged to atmosphere in cooling water.

6. Condenser circulating water pump. It circulates water through the condenser and the cooling tower.

7. Feed water pump. It pumps water in the water tubes of boiler against boiler steam pressure.

8. Economiser. In economiser heat in flue gases is partially used to heat incoming feed water.

9. Air preheater. In air preheater heat in flue gases (the products of combustion) is partially used to heat incoming air.

2.2. Hydro-electric Power Plant

2.2.1. Classification

Hydro-electric power stations may be classified as follows :

A. **According to availability of head**

1. High head power plants
2. Medium head power plants
3. Low head power plants.

B. **According to the nature of load**

1. Base load plants
2. Peak load plants.

C. **According to the quantity of water available**

1. Run-of-river plant without pondage
2. Run-of-river plant with pondage
3. Storage type plants
4. Pump storage plants
5. Mini and micro-hydel plants.

A. According to Availability of Head :

The following figures give a rough idea of the heads under which the various types of plants work :

(*i*) High head power plants ...... 100 m and above

(*ii*) Medium head power plants ...... 30 to 500 m

(*iii*) Low head power plants ...... 25 to 80 m.

Note. It may be noted that figures given above overlap each other. Therefore it is difficult to classify the plants directly on the basis of head alone. The basis, therefore, technically adopted is the *specific speed* of the turbine used for a particular plant.

1. High head power plants :

Water is usually stored up in lakes on high mountains during the rainy season or during the season when the snow melts. The rate of flow should be such that water can last throughout the year.

Fig. 2 shows high head power plant layout. Surplus water discharged by the spillway cannot endanger the stability of the main dam by erosion because they are separated. The tunnel through the mountain has a surge chamber excavated near the exit. Flow is controlled by head gates at the tunnel intake, butterfly valves at the top of the penstocks, and gate valves at the turbines. This type of site might also be suitable for an underground station.

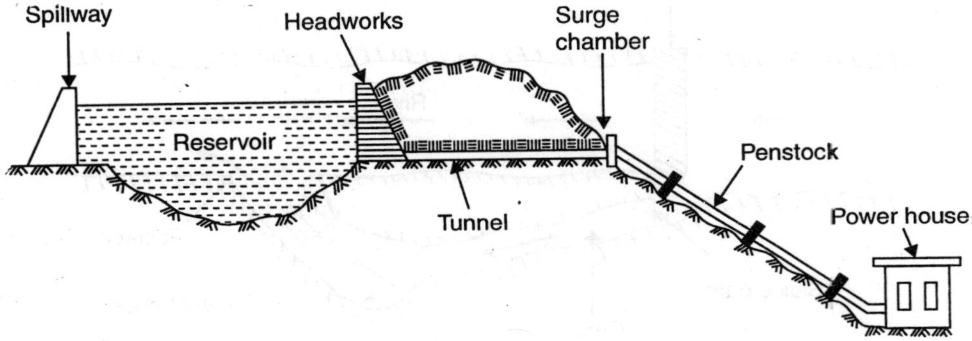

Fig. 2. High head power plant layout. The main dam, spillway, and power house stand at widely separated locations. Water flows from the reservoir through a tunnel and penstock to the turbines.

The Pelton *wheel* is the common prime-mover used in high head power plants.

2. Medium head power plants :

Refer Fig. 3. This type of plant commonly uses *Francis turbines*. The forebay provided at the beginning of the penstock serves as water reservoir. In such plants, the water is generally carried in

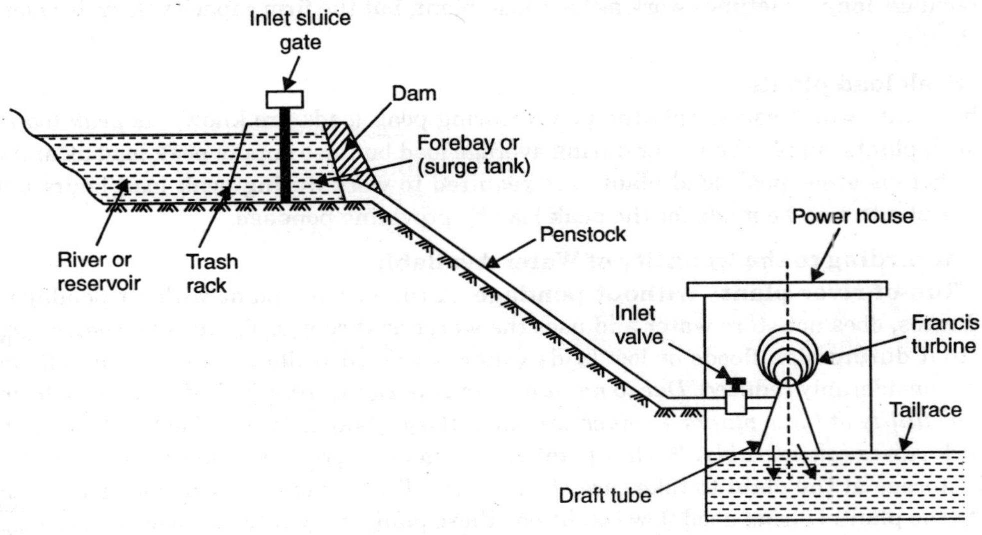

Fig. 3. Medium head power plant layout.

open canals from main reservoir to the forebay and then to the powerhouse through the penstock. The forebay itself works as a surge tank in this plant.

3. Low head power plants :

Refer Fig. 4. These plants usually consist of a dam across a river. A sideway stream diverges from the river at the dam. Over this stream the power house is constructed. Later this channel joins the river further downstream. This type of plant uses vertical shaft Francis turbine or Kaplan turbine.

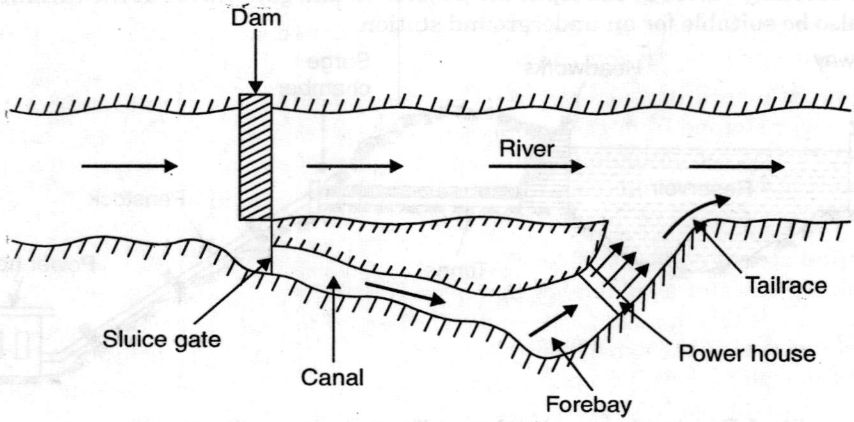

Fig. 4. Low head power plant layout.

B. According to the Nature of Load :

1. Base load plants :

The plants which cater for the base load of the system are called *base load plants*. These plants are required to *supply a constant power when connected to the grid*. Thus they *run without stop* and are often remote-controlled with which least staff is *required for such plants.* Run-of-river plants without pondage may sometimes work as base load plant, but the firm capacity in such cases will be very much less.

2. Peak load plants :

The plants which can supply the power during peak loads are known as *peak load plants.* Some of such plants supply the power during average load but also supply peak load as and when it is there, whereas other peak load plants are required to work during peak load hours only. The run-of-river plants may be made for the peak load by providing pondage.

C. According to the Quantity of Water Available :

1. Run-of-river plants without pondage. A run-of-river plant without pondage, as the name indicates, does not store water and uses the water as it comes. There is *no control on flow of water* so that during high floods or low loads water is wasted while during low run-off the plant capacity is considerably reduced. *Due to non-uniformity of supply and lack of assistance from a firm capacity the utility of these plants is much less* than those of other types. The head on which these plants work *varies considerably*. Such a plant can be made a great deal more useful by providing sufficient storage at the plant to take care of the hourly fluctuations in load. This lends some *firm capacity* to the plant. During good flow conditions these plants may cater to base load of the system, when flow reduces they may supply the peak demands. *Head water elevation for plant fluctuates with* *the flow conditions*. These plants without storage may sometimes be made to supply the base load, but the firm capacity depends on the minimum flow of river. The run-of-river plant may be made for load service with pondage, though storage is usually seasonal.

2. Run-of-river plant with pondage. *Pondage* usually refers to the collection of water behind a dam at the plant and increases the stream capacity for a short period, say a week. *Storage* mean collection of water in upstream reservoirs and this increases the capacity of the stream over an extended period of several months. Storage plants may work satisfactorily as base load and peak load plants.

This type of plant, as compared to that without pondage, is *more reliable* and its generating capacity *is less dependent on the flow rates of water available*.

3. Storage type plants. A storage type plant is one with a reservoir of sufficiently large size to permit carry-over storage from the wet season to the dry season, and thus to supply firm flow substantially more than the minimum natural flow. This plant can be used as base load plant as well as peak load plant as water is available with control as required. The majority of hydro-electric plants are of this type.

4. Pumped storage plants. Refer Fig. 5. Pumped storage plants are employed at the places where the quantity of water available for power generation is *inadequate*. Here the water passing through the turbines is stored in *'tail race pond'*. During low load periods this water is pumped back to the head reservoir using the extra energy available. This water can be again used for generating power during peak load periods. Pumping of water may be done seasonally or daily depending upon the conditions of the site and the nature of the load on the plant.

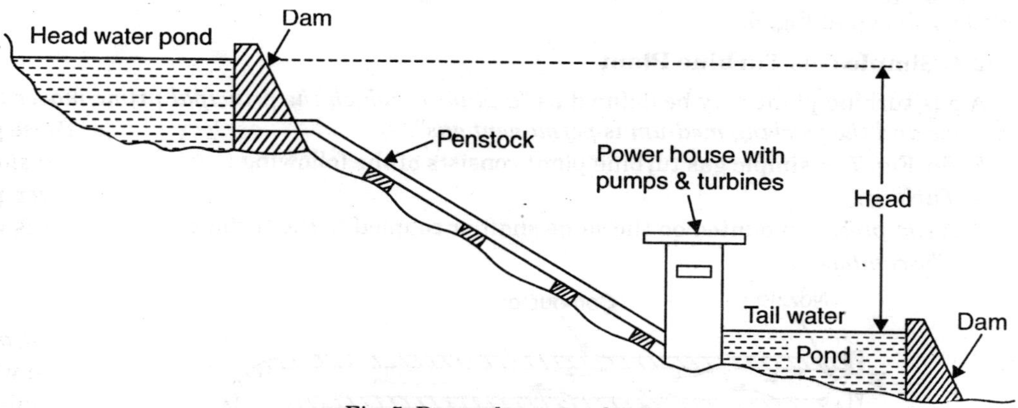

Fig. 5. Pumped storage plant.

Such plants are *usually interconnected* with steam or diesel engine plants so that off peak capacity of interconnecting stations is used in pumping water and the same is used during peak load periods. Of course, the energy available from the quantity of water pumped by the plant is *less* than the energy input during pumping operation. Again while using pumped water the *power available is reduced* on account of losses occuring in prime-movers.

2.3. Nuclear Power Plant

Fig. 6 shows schematically a nuclear power plant.

The main components of a nuclear power plant are :

1. Nuclear reactor
2. Heat exchanger (steam generator)
3. Steam turbine
4. Condenser
5. Electric generator.

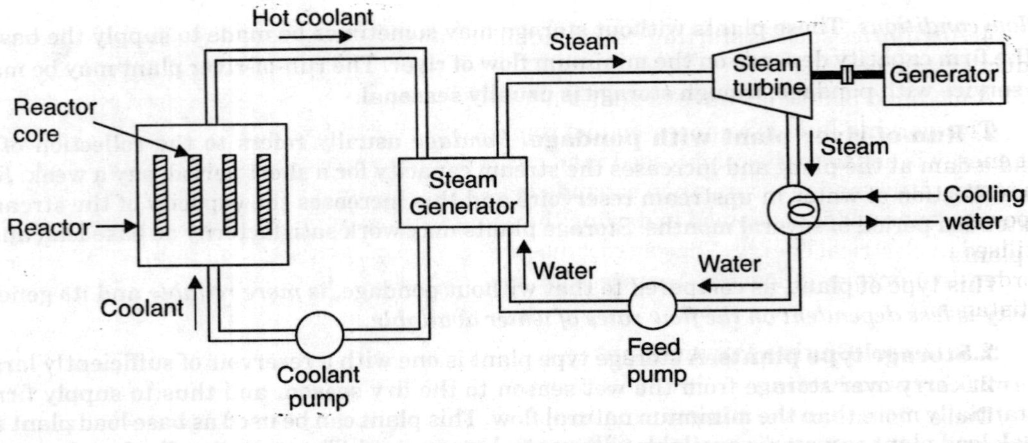

Fig. 6. Nuclear power plant.

In a nuclear power plant the reactor performs the same function as that of the furnace of steam power plant (*i.e.*, produces heat). The heat liberated in the reactor as a result of the nuclear fission of the fuel is taken up by the coolant circulating through the reactor core. Hot coolant leaves the reactor at the top and then flows through the tubes of steam generator and passes on its heat to the feed water. The steam so produced expands in the steam turbine, producing work, and thereafter is condensed in the condenser. The steam turbine in turn runs an electric generator thereby producing electrical energy. In order to maintain the flow of coolant, condensate and feed water pumps are provided as shown in Fig. 6.

2.4. Simple Gas Turbine Plant

A gas turbine plant may be defined as *"a plant in which the principal prime-mover is of the turbine type and the working medium is permanent gas"*.

Refer Fig. 7. A simple gas turbine plant consists of the following :

1. *Turbine.*
2. *A compressor* mounted on the same shaft or coupled to the turbine.
3. *The combustor*.

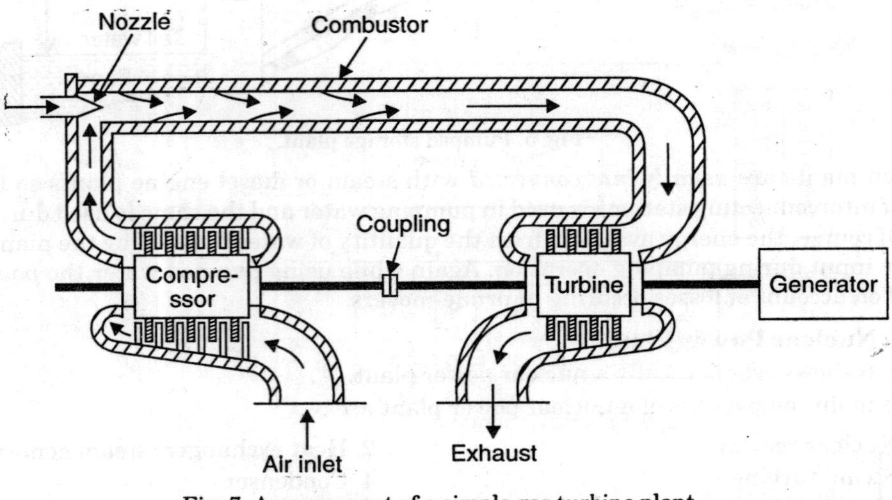

Fig. 7. Arrangement of a simple gas turbine plant.

4. *Auxiliaries* such as starting device, auxiliary lubrication pump, fuel system, oil system and the duct system etc.

A modified plant may have in addition to above an *intercooler, regenerator, a reheater* etc.

The working fluid is compressed in a compressor which is generally rotary, multistage type. Heat energy is added to the compressed fluid in the combustion chamber. This high energy fluid, at high temperature and pressure, then expands in the turbine unit thereby generating slower. Part of the power generated is consumed in driving the generating compressor and accessories and the rest is utilised in electrical energy. The gas turbines work on open cycle, semi-closed cycle or closed cycle. In order to improve efficiency, compression and expansion of working fluid is carried out in multistages.

2.5. Diesel Engine Power Plant

2.5.1. Introduction

Diesel engine power plants are installed *where supply of coal and water is not available in sufficient quantity* or *where power is to be generated in small quantity* or *where standby sets are required for continuity of supply such as in hospitals, telephone exchanges, radio stations and cinemas*. These plants in the range of 2 to 50 MW capacity are used as *central stations* for supply authorities and works and they are universally adapted to supplement hydro-electric or thermal stations where stand-by generating plants are essential for starting from cold and under emergency conditions.

In several countries, the demand for diesel power plants is increased for electric power generation because of difficulties experienced in construction of new hydraulic plants and enlargement of old hydro-plants. A long term planning is required for the development of thermo and hydro-plants which cannot keep the pace many times with the increased demand by the people and industries.

The diesel units used for electric generation are *more reliable* and *long-lived piece of equipment* compared with other types of plants.

2.5.2. Essential components of a diesel power plant

Refer Fig. 8. The essential components of a diesel power plant are listed and discussed below :

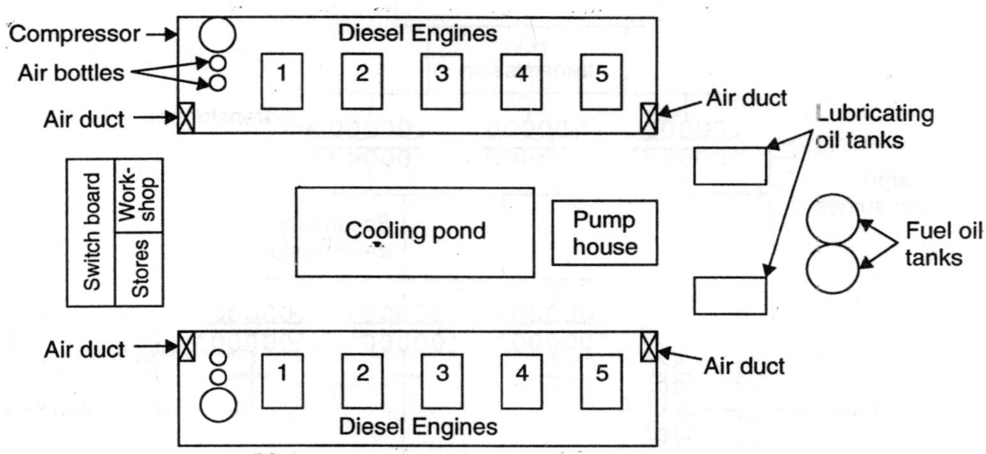

Fig. 8. Schematic arrangement of a diesel power plant.

1. Engine
2. Air intake system
3. Exhaust system
4. Fuel system

5. Cooling system
6. Lubrication system
7. Engine starting system
8. Governing system.

3. TRANSMISSION AND DISTRIBUTION

3.1. General Aspects

Electrical energy is generated in large hydroelectric, thermal and nuclear power stations. These stations are mostly situated away from the load centres. Therefore an extensive power supply network is necessary between the generating plants and consumers' loads.

By *transmission and distribution of electric power is meant its conveyance from the central station where it is generated to the places where it is demanded by the consumers* (like pumping stations, residential and commercial buildings, mills, factories etc.)

- The maximum voltage in advanced countries is 33 kV while that in India is 11 kV.
- The amount of power that has to be transmitted through transmission lines is very large and if this power is transmitted at 11 kV (or 33 kV) the line current and the power loss would be very large. Therefore this *voltage is stepped up to a higher value by using step-up transformers* located in *sub-stations.*
- The transmission voltages in India are 400 kV, 220 kV and 132 kV.
- The high voltage transmission lines transmit electrical power from the generating stations to main receiving end sub-stations. At these stations the voltage is stepped down to a lower value of 66 kV or 33 kV.

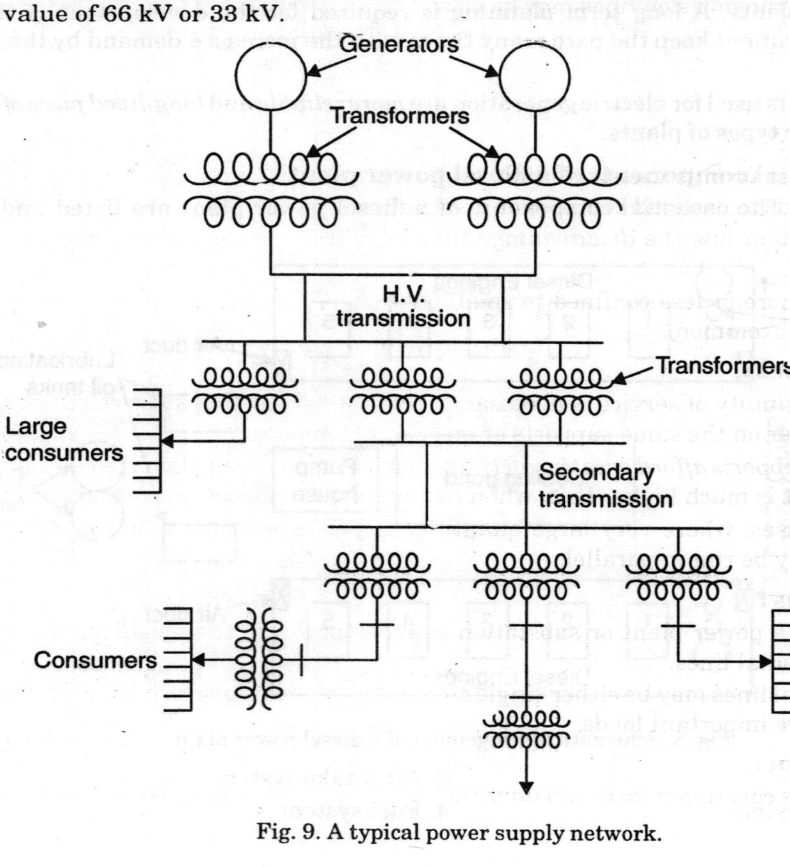

Fig. 9. A typical power supply network.

- The secondary transmission system forms the link between the main receiving end sub-stations and secondary sub-stations. At the secondary sub-stations the voltage is stepped down to 33 kV or 11 kV and the power is fed into the primary distribution system.
- The 33 kV or 11 kV distribution lines (usually known as *feeders*) emanate from the secondary sub-stations and terminate in distribution sub-stations. The distribution sub-stations consist of step-down transformers and are located at convenient places in the area in which the power is to be supplied. Sometimes these distribution sub-stations consist of pole mounted transformers located on the road side. These transformers step down the voltage to 400 V.
- The 400 V distribution lines (usually known as *distributors*) are laid along the roads and service connections to consumers are tapped off from the distributors.
- All *transmission and distribution systems are 3-phase systems.*
- The transmission lines and feeders are 3-phase 3 wire circuits.
- The distributors are 3-phase 4 wire circuits because a neutral wire is necessary to supply the single phase loads of domestic and commercial consumers.

The transmission network is commonly known as '**grid**'.

Fig. 9 shows a typical power supply network.

3.2. Systems of Transmission

For transmission of electrical power three-phase circuits are generally used because of economical reasons. Transmission lines may be classified as follows :

1. Single line
2. Parallel lines
3. Radial lines
4. Ring system
5. Network.

1. **Single line :**

- The simplest form is the single line, such as obtained from a power plant supplying its entire output to one load centre over a single-circuit line.
- Such a system has the *disadvantage* that in case of damage to the line the service is *interrupted.*
- Its use is more or less confined to small power systems and is therefore becoming more and more uncommon.

2. **Parallel lines :**

- Where continuity of service is necessary, it is best to use at least two circuits in parallel, placed either on the same supports or on separate supports.
- *Separate supports afford greater safety* against both lines being damaged at the same time, but the cost is much higher than when two circuits are placed on one support.
- In some cases, where very large quantities of power must be handled, more than two circuits may be run in parallel.

3. **Radial lines :**

- Invariably a power plant or substation supplies power to the neighbouring territory by means of radial lines.
- These radial lines may be either single circuit for the less important loads or double circuit for the more important loads.

4. **Ring system :**

- For *systems covering a large territory* the ring system of transmission is very important.

- With this system the main high-voltage power line makes a closed ring, taps being taken off at any advantageous point of the ring, thus supplying a large territory.
- In case of damage to any section of the ring, that section may be disconnected for repairs, and power will be supplied from both ends of the rings, thereby maintaining continuity of service.

5. Network :

- A network often constitutes several ring systems with sections of single, parallel, or radial lines.

3.3. Line Supports

Electrical power may be transmitted by overhead or underground conductors. Underground transmission, with the exception of a few notable cases, is limited to voltage less than 45 kV.

The supports for the *overhead transmission lines* may be of any one of the following classes :

1. Poles 2. Towers

1. **Poles.** Poles may be made of materials like *wood, steel, cement concrete* etc. Usually R.C.C. poles are used. Figs. 10 and 11 show the wooden poles and R.C.C. poles respectively.

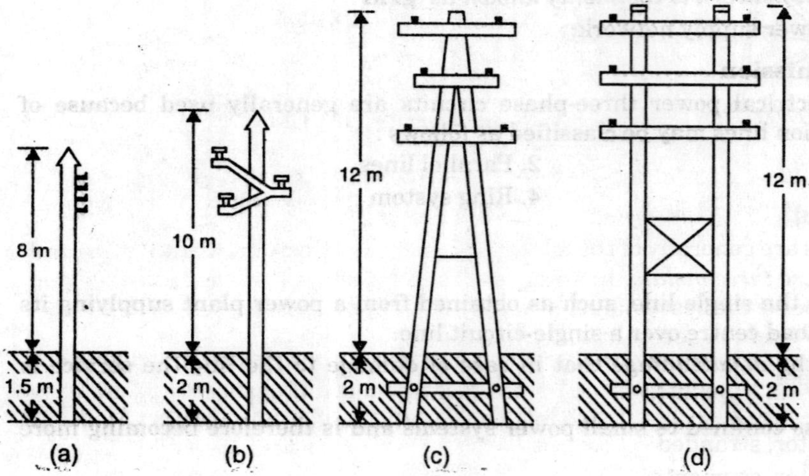

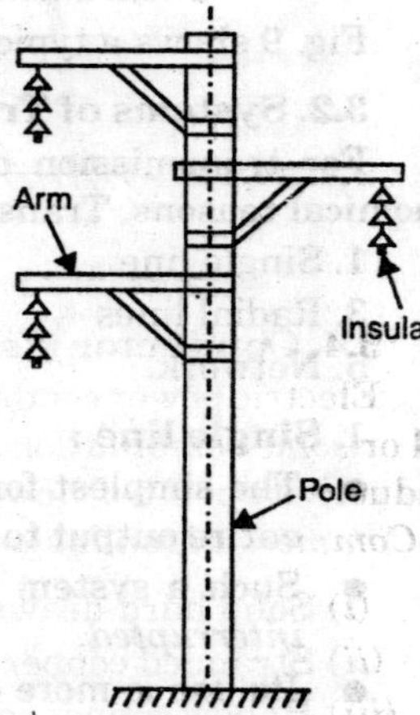

Fig. 10. Wooden poles : (*a*) 400 V, 4-phase, 4-wire distributor, (*b*) 11 kV, 3-phase feeder, (*c*) 'A' pole, double circuit, 33 kV feeder, (*d*) 'H' pole double circuit 33 kV feeder.

Fig. 11. R.C.C. pole.

2. **Towers.** Towers made of steel are mechanically sturdy and durable and are commonly used. These towers (steel) have been shown in Fig. 12 and Fig. 13. These towers are generally four legged, each leg anchored properly.

The line supports discussed above should have the following *characteristics* :

(*i*) High mechanical strength.
(*ii*) Cheap in cost.
(*iii*) Light in weight.
(*iv*) Longer life.
(*v*) Good looking.
(*vi*) Easy accessible for painting and erection of line conductors.
(*vii*) Low maintenance cost.

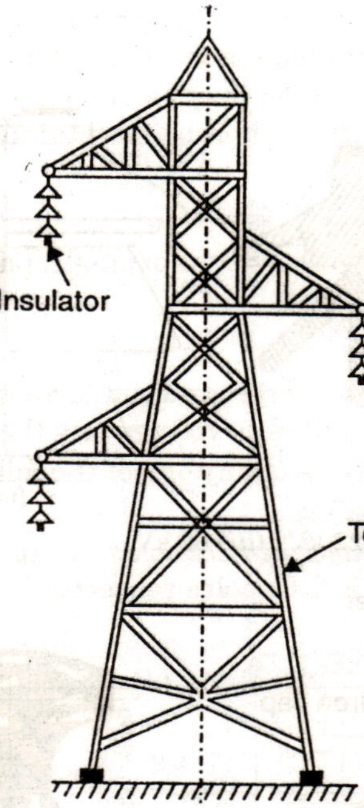

Fig. 12

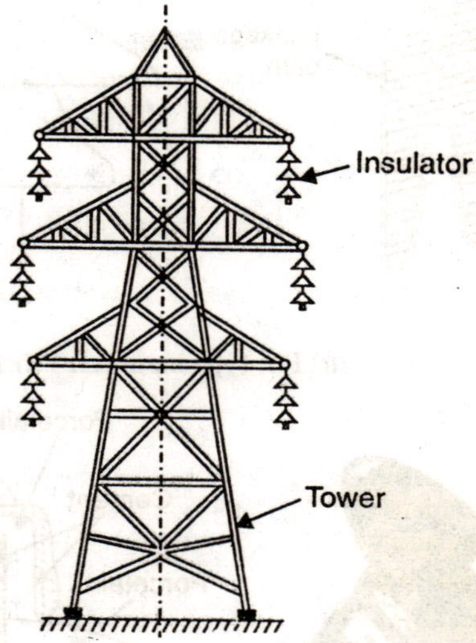

Fig. 13

3.4. Conductor Material

Electric power conductors are generally of the following materials : **Copper, aluminium** and **steel** or *some combination of these three metals.* In some cases special alloys have been used. From a conductivity point of view, copper is the best conductor material, aluminium being second, and steel last. *Commercial conductors* are made in several forms, namely :

(*i*) Solid hard-drawn copper
(*ii*) Stranded copper
(*iii*) Hollow copper conductor, stranded
(*iv*) Hollow copper conductor, segmental
(*v*) Copperweld copper
(*vi*) Aluminium stranded, steel cored.

3.5. Line Insulators

The insulators of a transmission line are its most important item, since the operation of a line cannot be any better than the insulators that support the conductors. Transmission line insulators must possess *good mechanical strength and good insulating qualities under all conditions of weather and temperature and must not deteriorate fast.* Insulators are made of **glass, porcelain,** and **patented compounds.** Glass is the cheapest material and when properly made will produce satisfactory insulators for low-voltage work, such as telephone and telegraph, and under favourable conditions may be used upto 25 kV. Though there are a number of patented compounds on the market, these do not seem to offer much competition with porcelain, since *porcelain has very good electrical characteristics as well as high mechanical strength.*

Transmission line insulators may be classified as follows :

1. Pin type 2. Suspension type 3. Strain type.

Fig. 14 shows the various types of insulators.

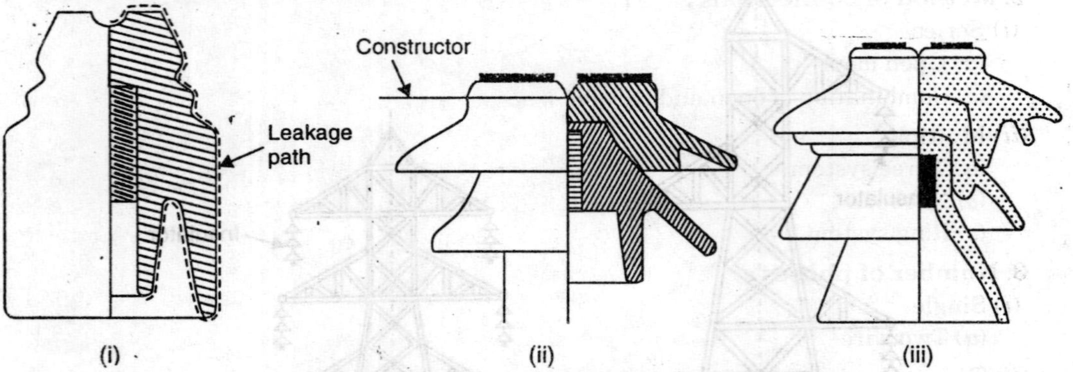

(*a*) Pin type insulator : (*i*) 11 kV ; (*ii*) 22 kV ; (*iii*) 66 kV

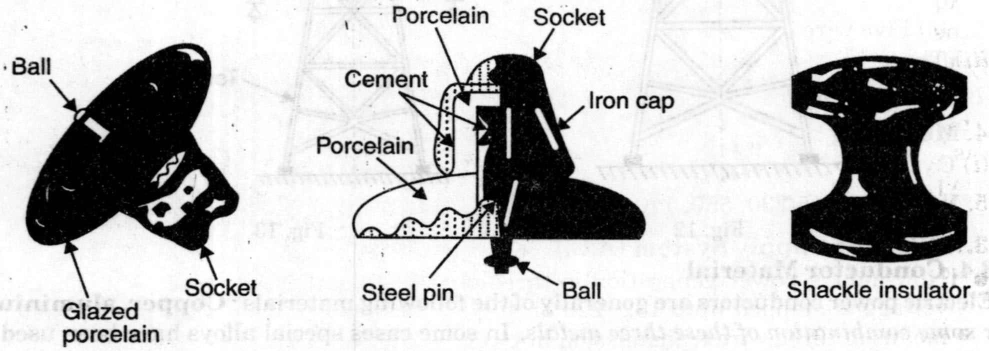

(*b*) Suspension type insulators

Fig. 14. Various types of insulators.

- For *low voltages, pin-type insulators* made of glass are generally used. Pin-type insulators made of porcelain are designed for voltages upto about 90 kV but are seldom used on lines above 66 kV.
- For voltages above 66 kV it is generally desirable to use *suspension insulators.*
- *Strain insulators* may be of pin or suspension type. Upto about 30 kV pin-type insulators are satisfactory, but for higher voltages the suspension type is generally used. *Strain insulators are used on dead-end towers at bends or corners of transmission lines, or when making very long spans*. Extra heavy suspension units are made for such service, but often standard units may be used. On ordinary straight line dead-end towers a single string is often sufficient but for severe service two more strings may be connected in parallel.

3.6. Distribution Systems

Oftenly it is impossible to draw a line between the distribution and transmission systems of a large power network. In general, *distribution systems comprise that part of the network of a power system which distributes power for local use*.

Distribution systems may be classified as follows :

1. Nature of current :

(*i*) Direct current

(*a*) Two wire (*b*) Three wire

(*ii*) Alternating current.

2. Method of connections :

(*i*) Series

(*a*) Open loop (*b*) Parallel loop

(*c*) Combination of open and parallel loops

(*ii*) Multiple

(*a*) Three system (*b*) Feeder and main

(*c*) Network (*d*) Loop system

(*e*) Ring system.

3. Number of phases :

(*i*) Single

(*a*) Two wire (*b*) Three wire

(*ii*) Two

(*a*) Three wire (*b*) Four wire

(*c*) Five wire

(*iii*) Three

(*a*) Three wire (*b*) Four wire

4. Mounting :

(*i*) Overhead (*ii*) Underground

5. **Voltage** : 115/230, 550, 1100, 2200, 6600, 11000, 12000, 13200, 32000 V.

3.7. Electric Supply System (A.C.)

- Now-a-days power generation is three-phase, 50-Hz, usually at 11 kV. The generated voltage is stepped upto 132, 220 or 400 kV by means of *step-up transformers.* Then by means of three wire transmission lines electric power is carried to different places where it is received in sub-stations situated near cities. Here the voltage is stepped down to 66 or 33 kV and further carried through three-wire transmission lines to various sub-stations in the cities where the voltage is further stepped down to 11, 6.6 or 3.3 kV. These voltages are further stepped down to 415 V to make available power to small consumers through 415/240 V, three-phase, four-wire distributors.
- Small consumers receive power at 415/240 V where they can use both three-phase and single-phase supply. Very small domestic consumers are often provided with single-phase supply at 240 V. Distribution of power at 415/240 V is called *secondary or low voltage distribution system.* Electric power to large consumers is, however, often supplied direct at 11 kV, 6.6 kV or 3.3 kV. Such consumers install their own sub-station to step-down voltage to 415/240 volts.

3.7.1. Three-phase four wire distribution system

- Electric power in India is supplied to the consumers by the State Electricity Boards. The following are the different systems by which power is distributed to the consumer :

1. Single-phase A.C. supply using a 2-wire system ;
2. Three-phase A.C. supply using a 3-wire system ;
3. Supply of three-phase and neutral using a 4-wire system.

D.C. supply may also be available in certain areas, on either a two wire system at 220 V, or a three wire system with 440 V between the two outer conductors and 220 V between the outer conductors and the centre wire. However, D.C. supply is no longer commonly used in India and, therefore, we rarely come across D.C. installations.

- The standard voltages at which the supply authorities deliver power to the consumer are generally as follows :
  Single phase : 240 V, 50 Hz, 2 wire
  Three phase : 415 V, 50 Hz, 4 wire.
  In India and in many European countries the frequency of the supply is 50 Hz. In the USA the power supply frequency is 60 Hz.
- Standardisation of supply voltage and frequency is extremely important to both electricity supply authorities as well as the manufacturers of electrical equipment. In the 3-phase, 4-wire power distribution system, power is supplied from the sub-station through 4-wires. Three of these wires are called *live or phase or line wires*. The fourth wire is usually at *zero voltage and is called neutral wire. The neutral wire is earthed at the sub-station*.
- The voltage between any two of the line wires in a 3-phase system is $\sqrt{3}$ times the voltage between any phase wire and the neutral.
- Electric loads of the consumers are connected in such a way that all the three phases are equally loaded. If the individual phases of the 3-phase systems are equally loaded the current through the neutral wire will be zero. In case of 3-phase motors, ovens etc. the three phases are equally loaded by their very design. In such cases the neutral wire may be omitted. But in cases where a large number of single phase loads have to be supplied, loads on the three phases are balanced by connecting various single phase consumers or groups of consumers to different phases of the 3- phase supply.
- Large consumers, particularly those with heavy motor loads, are provided with a 3-phase, 4-wire supply. Motors and heavy heating loads are connected to 415 V, 3-phase supply and are called *three phase loads*. The load is balanced over the three phases by equally distributing the lighting load and small power loads over the three phases. Consumers with load requirement more than 250 kVA are provided with supply at high voltage with a substation installed in the consumer's premises where voltage is stepped down to 415/240 V. Fig. 15 shows typical 3-phase 4-wire distribution from a sub-station to the consumer.

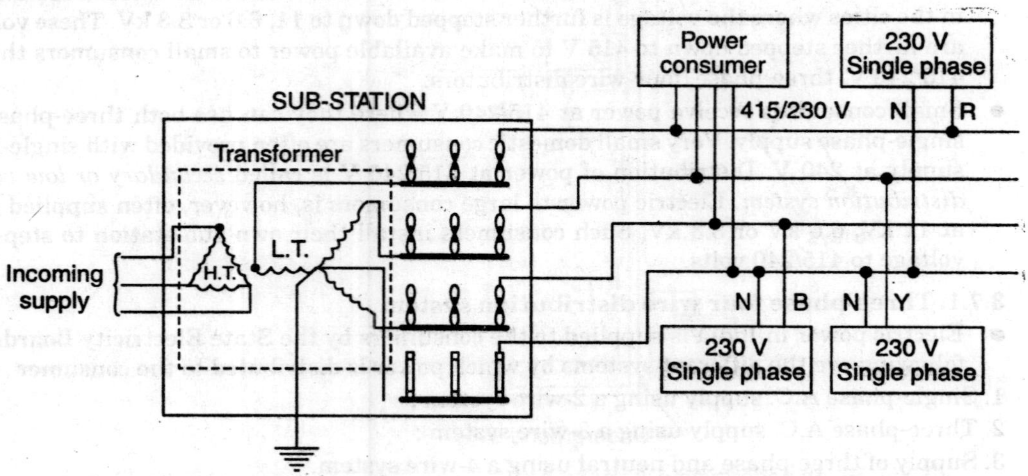

Fig. 15. Three phase 4-wire distribution.

3.7.2. Voltage tolerances

- Current carried by the electrical power distribution network varies at different times of the day. This leads to varying voltage drops in the supply cables. It is thus impracticable to give each consumer exact nominal supply voltage *i.e.*, 415/240 V at his terminals. Supply authorities are, therefore, permitted certain tolerances. Under the Indian Electricity Rules, the *voltage fluctuation may not vary by more than 5% above or below the declared nominal voltage and the frequency must be within ± 1% of the declared frequency of 50 Hz.* Thus the terminal voltage must be within the range.

  228 V to 252 V for a nominal voltage of 240 V ;

  394.25 V to 435.75 V for a nominal voltage of 415 V.

3.7.3. Service connections

- The supplier's distribution system brings power to the consumer through overhead lines or by means of underground cables to a place just outside the consumer's premises.
- The *line bringing electric power from supplier's low voltage distributor upto the energy meter installed at the consumer's premises is called the* **service connection.**
- Service connection may be by means of **underground cables** or by means of **overhead conductors or cables.**

Fig. 16 shows an *underground service connection* taken from an overhead distributor.

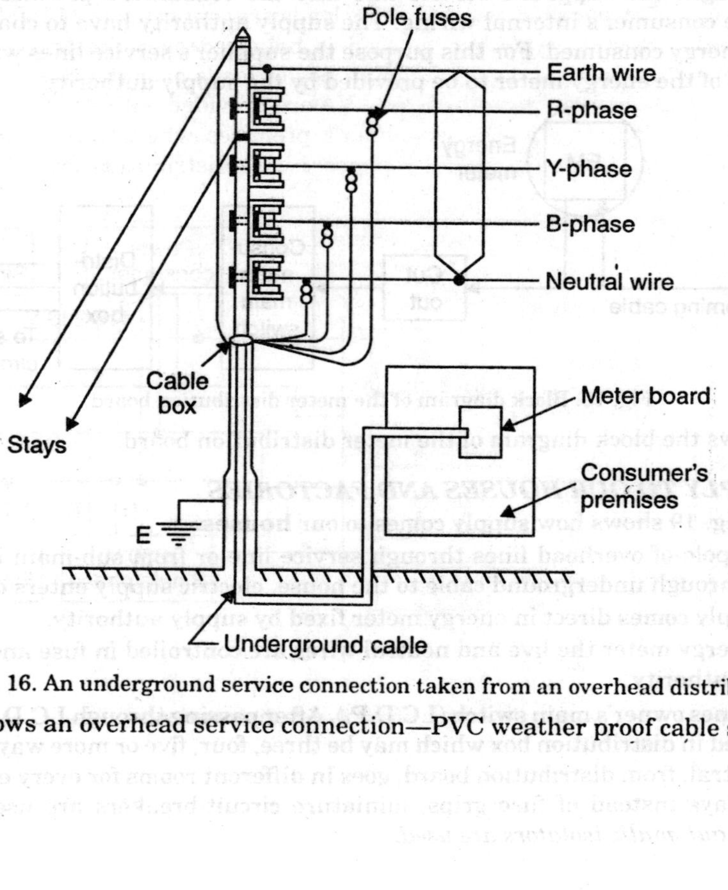

Fig. 16. An underground service connection taken from an overhead distributor.

Fig. 17 shows an overhead service connection—PVC weather proof cable service line.

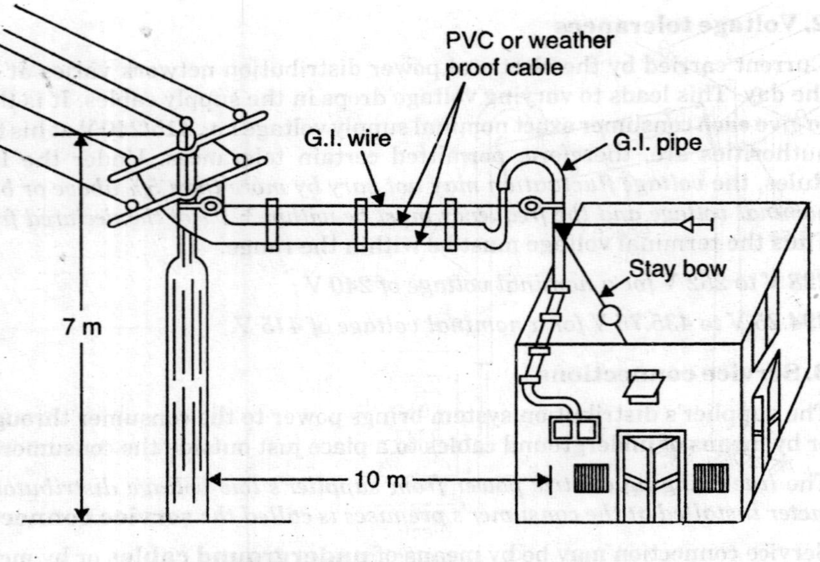

Fig. 17. PVC weather proof cable service-line.

3.7.4. Service mains

Having brought the supplier's service line into the consumer's premises it is now to be connected with the consumer's internal wiring. The supply authority have to charge the consumer for the electrical energy consumed. For this purpose the supplier's service lines will be connected to the input terminal of the energy meter to be provided by the supply authority.

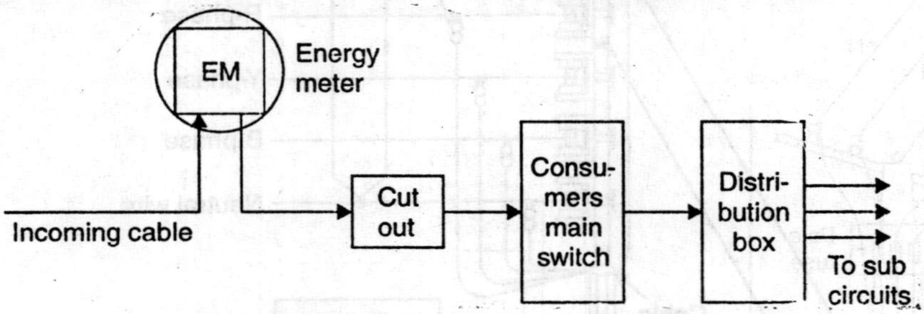

Fig. 18. Block diagram of the meter distribution board.

Fig. 18 shows the block diagram of the meter distribution board.

ELECTRIC SUPPLY TO OUR HOUSES AND FACTORIES

Houses : Fig. 19 shows how supply comes to our **houses.**

- From a pole of overhead lines through service line or from sub-main in the street (L.T. Pillar) through underground cable to the house, electric supply enters our house.
- The supply comes direct in energy meter fixed by supply authority.
- After energy meter the live and neutral wires are controlled in fuse and link supplied by supply authority.
- Then comes owner's main switch (I.C.D.P.). After passing through I.C.D.P. wires are again controlled in distribution box which may be three, four, five or more ways. Separate phase and neutral, from distribution board, goes in different rooms for every eight to ten points. Now-a-days instead of fuse grips, miniature circuit-breakers are used and instead of I.C.D.P. *automatic isolators are used.*

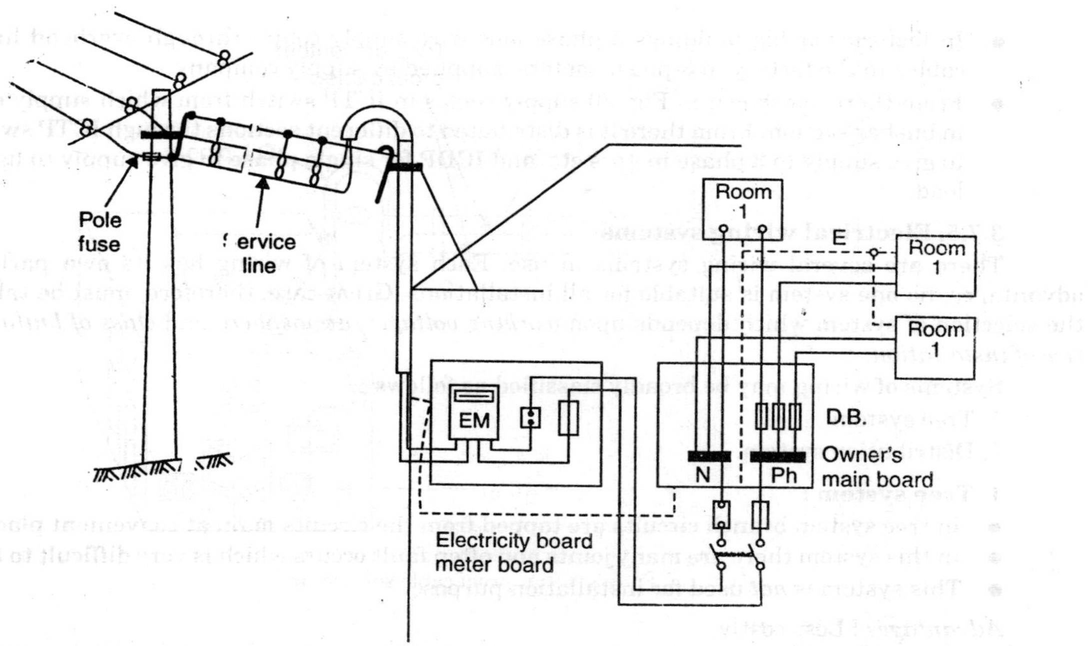

Fig. 19. Electric supply to houses.

Factories : Fig. 20 showing how supply comes to **factories.**

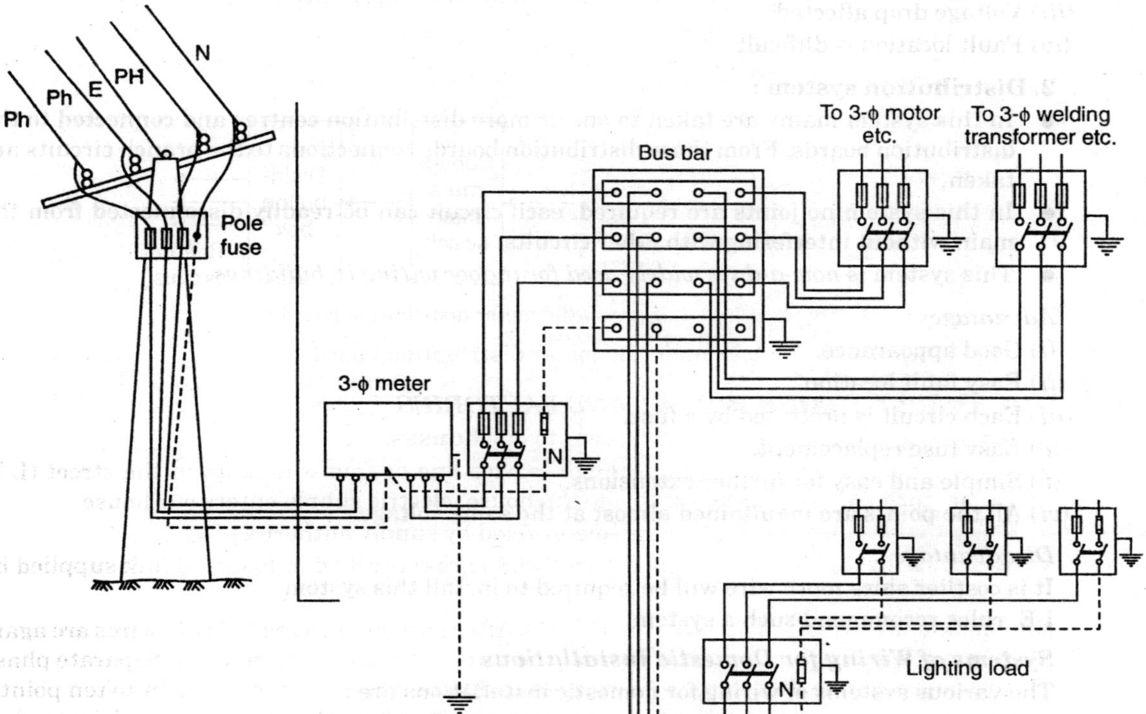

Fig. 20. Factory wiring.

- In factories or big buildings 3 phase four wire supply comes through overhead lines or cables to the factory in 3-phase meters, supplied by supply company.
- From there, as shown in Fig. 20 supply comes in ICTP switch from which supply enters in busbar section. From there it is distributed to different sections through ICTP switches to give supply to 3 phase motors etc. and ICDP for single phase to give supply to lighting load.

3.7.5. Electrical wiring systems

There are several wiring systems in use. Each system of wiring has its own particular advantages, no one system is suitable for all installations. Great care, therefore, must be taken in the selection of system which depends upon *working voltages, atmosphere* and *class of building or size of installation.*

Systems of wiring may be broadly classified as follows :

1. Tree system
2. Distribution system.

1. Tree system :

- In tree system branch circuits are tapped from the circuits main at convenient places.
- In this system there are many joints and often fault occurs which is very difficult to locate.
- This system is *not* used for installation purpose.

Advantages : Less costly.

Disadvantages :

(*i*) Poor appearance.
(*ii*) The fuses are scattered.
(*iii*) Voltage drop affected.
(*iv*) Fault location is difficult.

2. Distribution system :

- In this system mains are taken to one or more distribution centres and connected to the distribution boards. From those distribution boards connections to the branch circuits are taken.
- In this system no joints are required, each circuit can be readily disconnected from the main without interfering with other circuits.
- This system is *now-a-days widely used for indoor wiring in buildings*.

Advantages :

(*i*) Good appearance.
(*ii*) Easy fault location.
(*iii*) Each circuit is protected by a fuse.
(*iv*) Easy fuse replacement.
(*v*) Simple and easy for further extensions.
(*vi*) All the points are maintained almost at the same voltage.

Disadvantage :

It is costlier since more wire will be required to install this system.

I.E. rules recommend such a system.

Systems of Wiring for Domestic Installations

The various systems of wiring for domestic installations are :

1. Cleat wiring 2. C.T.S. (T.R.S.) wiring

3. Wooden casing and capping wiring
4. Lead-sheathed wiring
5. Conduit wiring
   — Surface conduct ;
   — Underground or concealed conduct.

1. Cleat wiring :

- This type of wiring is *not used practically for permanent wiring* and is *only suitable for temporary wiring purposes* such as for marriages with *advantages of saving in labour and overall cost*.
- In this type of wiring, insulated conductors are supported on porcelain or wooden cleats.
- The cleats are very easy to erect and fixed at a distance of 4.5 cm to 15 cm apart. V.I.R. (Volcanised India Rubber) or P.V.C. (Polyvinyl chloride) wires are normally used in this system, though T.R.S. (Tough rubber sheathed system) wires can also be installed in special cleats for certain types of installation.

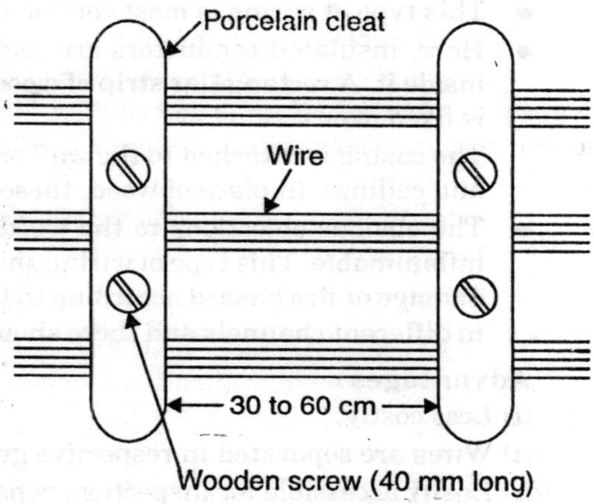

Fig. 21. Wires laid in three groove cleats.

Fig. 21 shows wires laid in three groove cleats.

Advantages :

(*i*) It is easy to retrieve the materials used.
(*ii*) Rare fire accidents.
(*iii*) Cheaper in cost comparatively.
(*iv*) Reduced risk of short-circuit (since the conductors are insulated from each other).
(*v*) Inspections are easy.
(*vi*) Alterations can be made easily.

Disadvantages :

(*i*) Appearance not attractive.
(*ii*) Regular cleaning is needed.
(*iii*) Due to the exposure of the conductors, there is a risk of mechanical injury.

2. C.T.S. T.R.S. (Batten Wiring) :

- In such a wiring, the wires used are sheathed in tough rubber of P.V.C. wires and they are *clipped on wooden button with clips*. The button is fixed on the wall or ceiling.
- This wiring is *suitable for damp climate*, but *cannot* withstand much heat and so is not suitable for places of very hot weather and there is also danger of mechanical damage and fire hazard. C.T.S. wires are not suitable for outdoor use. Therefore, they should not be exposed to direct sunlight and where there are corrosive acid fumes. At such places now P.V.C. wire can be easily used.

Clips used are of the following two types :

1. Link clips
2. Joint link clips.

Advantages :

(*i*) Less risk of short circuit.
(*ii*) Least risk of mechanical injury.
(*iii*) Cheaper than wooden casing and capping wiring.

Disadvantages :

(*i*) Skilled workmen are required to do the wiring.

(*ii*) Need to be protected from sunlight, rain etc.

(*iii*) Sharp bends should be avoided.

3. Wooden casing and capping wiring :

- This type of wiring is most commonly used for indoor and domestic installations.
- Here, insulated conductors are laid inside rectangular teak wood boxes having grooves inside it. A rectangular strip of wood called *capping*, having same width as that of *casing*, is fixed over it.

  The casing is attached to the wall or ceiling (and they can be painted matching with walls and ceiling). In place of wood, these days PVC is used.
- The obvious objections to the wooden casing and capping system are that wood is very inflammable. This type of wiring should not be used where there is a danger of mechanical damage or fire hazard according to I.E. Rules. Positive and negative wires should be taken in different channels and there should be no crossing of + ve and – ve wires in the channel.

Advantages :

(*i*) Less costly.

(*ii*) Wires are separated in respective grooves of the casing.

(*iii*) Easily accessible for inspection, repairs and alterations.

(*iv*) Suitable for low voltage installations.

Disadvantages :

(*i*) Wood may get damaged by white ants etc. (if wood is not properly seasoned)

(*ii*) There is a risk of damage due to fire.

(*iii*) For making proper casings and cappings, skilled carpenters are required.

4. Lead-sheathed wiring :

- In lead-sheathed system the conductor, either twin core or three core, each being separately insulated and covered with the common lead sheath, is used. The lead-sheathed wires are easily fixed by means of metal clips on wooden batten and form a good surface system.
- Whenever the wires are broken for connections to switch or light point, a junction box must be used. The junction boxes are usually of metal and provided with some means of earthing so that continuity of the sheath is obtained.
- Being very costly, these wires are not used now-a-days.

5. Conduit wiring. Refer Fig. 22.

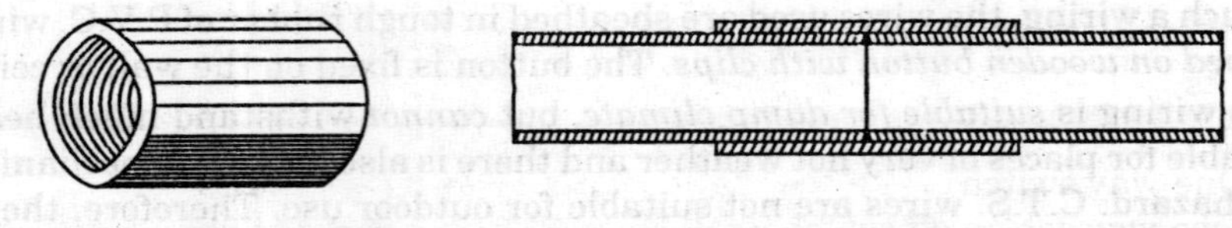

Fig. 22. Rigid conduit coupling (Screwed).

- In conduit wiring V.I.R. wires, or P.V.C. wires are carried through *steel or iron pipes giving good protection from mechanical injury or fire risks.*
- This system is the **best and most desirable system of wiring for workshops and public buildings**. It gives very good appearance when concealed.

- In this wiring the pipes are cut with hacksaw and are threaded with die set for bend, tee junction box etc. and are then fixed on the walls on wooden gutties or plugs with saddles. Then wires are drawn with the help of *fish wire* (steel wire).
- *Now-a-days P.V.C. conduit pipes are also available which do not require threading. Jointing is done with a special made solution*. These are flexible and can be bent easily. For concealed wiring, the pipes are directly buried in the wall and roofs and then wires are drawn through them.

Advantages :

(*i*) No risk of fire.
(*ii*) Protection against mechanical injury.
(*iii*) The lead and return wires are carried in the same tube/pipe.
(*iv*) Earthing and electrical continuity is easily assured.
(*v*) It is shock proof if earthing and bonding is properly done.
(*vi*) Replacement/alteration of defective wiring is easy.
(*vii*) The whole system is water proof.

Disadvantages :

(*i*) As compared to other systems, it is costlier.
(*ii*) In order to avoid drop in voltage due to induction, bunching of wires needs to be done.
(*iii*) This type of wiring requires skilled workers.
(*iv*) There is risk of short circuit under wet conditions (due to condensation of water in the tubes/pipes).
(*v*) Its erection is not so easy and requires time.

Comparison of wiring in domestic use :

S. No.	*Aspect*	*Cleat wiring*	*Wooden casing capping wiring*	*C.T.S. Wiring*	*Metal sheathed wiring*	*Conduct wiring*
1.	*Cost*	Low	Medium	Medium	Medium	High
2.	*Life*	Short	Fair	Long	Long	Very long
3.	*Possibility of fire*	Nil	Good	Poor	Poor	Nil
4.	*Mechanical protection*	Nil	Fair	Fair	Good	Very good
5.	*Protection from dampness*	Nil	Slight	Good	Good	Poor
6.	*Type of personnal required*	Semi-skilled	Highly skilled	Skilled	Skilled	Highly skilled

Wiring Accessories :

The **wiring accessory** is *any device associated with the wiring and electrical appliance of an installation*, such as a fuse, plug, switch etc.

In a complete electrical wiring system the following wiring accessories are used :

1. Cut-outs
2. Switches
   (*i*) One-way switch
   (*ii*) Two-way switch
   (*iii*) Two-way centre off switch
   (*iv*) Double pole switch
   (*v*) Push button switch
   (*vi*) Table lamp switch
   (*vii*) Bed switch.
3. Meter
4. Distribution board
5. Socket 5A
6. Socket 15A
7. Plug top

8. Holder
   (*i*) Bracket holder (*ii*) Batten holder
   (*iii*) Pendant holder
9. Ceiling rose
10. Junction boxes
11. Fuses
   (*i*) Round type fuse unit (*ii*) Rewirable or kit-kat type fuses
   (*iii*) Carbridge type fuses (*iv*) High rupturing capacity (HRC).

Conventional symbols :

WIRING

General Wiring

Wiring on Wall surface 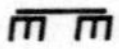

Wiring below wall surface (concealed wiring) 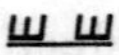

Surface Conduit wiring 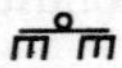

Concealed conduit wiring 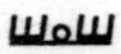

Wiring going upward 

Wiring going downward 

Wiring passing vertically through a room. 

SWITCHES :

Single pole switch 

Two pole switch 

Three pole switch 

Single pole pull switch 

Multiposition switch 

Interlocking switch and socket outlet 15 ampere 

Two way switch 

Intermediate switch 

Period limiting switch 

Time switch 

Pendent switch 

Push button or bell push 

Luminous push button 

Restricted Access push button 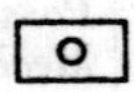

SOCKET OUTLETS :

Socket outlet 5 amperes 

Socket outlet 15 amperes 

Socket outlet 5 amperes with switch 

Socket outlet 15 amperes with switch 

Interlocking switch and socket outlet 5 ampere 

Water tight light fitting 

Specifications of wires :

- For ordinary wiring the wires are of single solid conductors or multistrands.
- Where flexibility is required, flexible cords are used.
- Following types of wires are commonly used :
  1. Volcanised India Rubber (V.I.R.) and tough rubber sheathed (T.R.S.)
  2. — Polyvinyl chloride (P.V.C.) insulated
     — Polyvinyl chloride (P.V.C.) insulated polyvinyl chloride (P.V.C.) sheathed
  3. Weather proof wires
  4. Flexible wires.

3.7.6. D.C. distribution

In a D.C. system the electrical energy may be fed and distributed either by two wire or by three wire system.

Two-wire D.C. distribution systems :

Fig. 23 shows a two-wire D.C. distribution system from generating station or substation to the consumer's terminals.

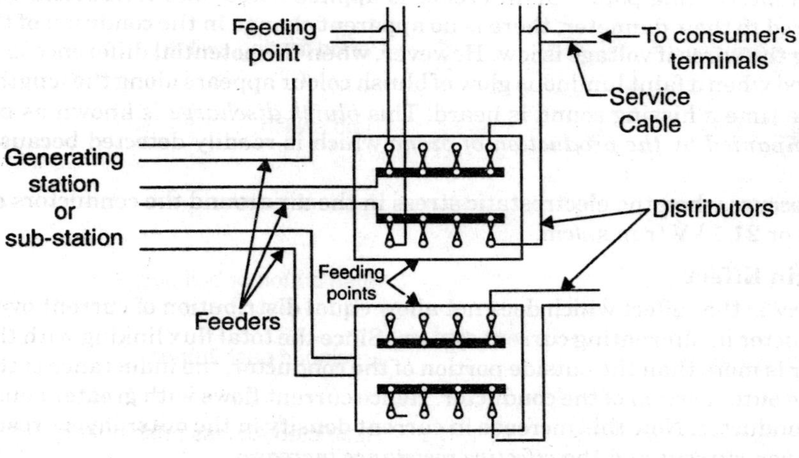

Fig. 23. Two-wire D.C. system.

- This system consists of two wires, one outgoing (known as *positive wire*) and another returning one (*negative wire*).
- Untapped feeders run into bus bars in suitable feeding points in the distribution area.
- Distribution cables are connected to the bus bars through fuses or links.
- Each separate consumer is fed from the distributor by a service cable tapped on to the distributor at the nearest convenient point.
- The electrical appliances and motors etc. are connected in parallel between the two wires.

Three-wire D.C. distribution systems :

- This system consists of *two outers* and an *earthed middle wire* known as *neutral wire*.
- The lamps or consumer apparatus are connected between the neutral and one of the outers.
- The voltage between the outers being twice that of consumer's terminals *increases the transmission efficiency and reduces the copper cost*.

3.8. Overhead and Underground Systems

Electric power may be transmitted by two methods :

1. *Overhead system*

2. *Underground system.* Its main advantage is that it is less prone to natural hazards like rain, wind, lightning and that it does not interfere with other amenities. However it is *more costly* as compared to an overhead system.

Because of cost consideration the transmission systems in India, are generally overhead. For distribution also the *use of underground cables is limited to densely populated areas.*

3.9. Kelvin's Law

The statement of Kelvin's law is as follows :

"If variable part of annual cost on account of interest and depreciation on the capital outlay is equal to the annual cost of electrical energy wasted in the conductors, the total annual cost will be minimum and the corresponding size of the conductor will be most economical."

3.10. Corona

When an alternating potential difference is applied across two conductors whose spacing is large as compared to their diameter, there is no apparent change in the condition of the atmospheric air surrounding the wires if voltage is low. However, when the potential difference is increased, then a point is reached when a faint luminous glow of bluish colour appears along the lengths of conductors and at the same time a hissing sound is heard. This *bluish discharge* is known as **corona.** *Corona is always accompanied by the production of ozone* which is readily detected because of its characteristic odour.

Corona occurs when the electrostatic stress in the air around the conductors exceeds 30 kVA (maximum)/cm or 21.1 kV (r.m.s.)/cm.

3.11. Skin Effect

Skin effect is that effect which does not allow equal distribution of current over the cross-section of the conductor in alternating current system. Since the total flux linking with the inner portion of the conductor is more than the outside portion of the conductor, the inductance of the inner portion is more than the outer portion of the conductor. Hence current flows with greater density at the outer surface of the conductor. Now this increase in current density in the outer layers results in a greater I^2R loss for a given current and the *effective resistance increases.*

At low frequency the skin effect is very small, but at high frequency it is very large.

3.12. Proximity Effect

Proximity effect is also an effect of non-uniform current distribution in a conductor due to the effect of linkage of flux of *another current carrying conductor placed nearby.* More flux will link with the *nearer half portion of the conductor and thus an increase in the resistance of the conductor results.*

3.13. Classification of Transmission Lines

The transmission lines are classified according to distance of transmission as follows :

1. **Short transmission line.** Length less than 60 km and operating voltage less than 20 kV.

2. **Medium transmission line.** Length between 60 and 150 km and line voltage between 20 kV and 100 kV.

3. **Long transmission line.** Length above 150 km and line voltage above 100 kV.

3.14. Regulation and Efficiency of a Transmission Line

Regulation. Regulation is defined as *the change in voltage at the receiving end when the full load is thrown off, the sending end voltage remaining constant.*

Mathematically, percentage regulation $= \frac{V_S - V_R}{V_S} \times 100$

where V_S = voltage at sending end, and
V_R = voltage at receiving end.

Efficiency. Efficiency of a transmission line is defined as the *ratio of power received to the power sent out.*

Mathematically, percent $\eta_T = \frac{V_R I_R \cos \phi_R}{V_S I_S \cos \phi_S} \times 100$

where V_R, I_R and $\cos \phi_R$ are the receiving end voltage, current and power factor while V_S, I_S and $\cos \phi_S$ are the sending end voltage, current and power factor.

3.15. Classification of Cables

1. Low voltage (L.T.) cables ... operating voltage upto 1 kV
2. High voltage (H.T.) cables ... operating voltage upto 1.1. kV
3. Super tension (S.T.) cables ... operating voltage upto 33 kV
4. Extra high tension (E.H.T.) cables ... operating voltage upto 66 kV
5. Extra super voltage power cables ... operating voltage beyond 132 kV.

Insulation resistance of a single core sheathed cable :

Insulation resistance of a single core sheathed cable is given by :

$$R = \frac{\rho}{2\pi l} \log_e \frac{r_2}{r_1}$$

where R = insulation resistance,
ρ = resistivity of dielectric,
r_1 = conductor radius, and
r_2 = sheath radius.

The insulation resistance of a cable can be measured by the following two methods :
1. Galvanometer method
2. By meggar.

3.16. Surge Impedance or Neutral Impedance

It is defined as $Z_c = \sqrt{Y/Z}$

where Y = shunt admittance, and
Z = series impedance.

For lines having negligible resistance and no shunt leakage,

$$Z_c = \sqrt{L/C}$$

The value of surge impedance :

(*i*) *For overhead line* 400 to 600 Ω

(*ii*) *For an underground cable* 4 to 60 Ω.

3.17. π and T-circuit of Long Transmission Line

For experimental model representation of long lines in the steady state at rated frequency and sometimes for computation of the receiving end and sending end quantities, it is convenient to use a *lumped constant system*. That is why equivalent π and *T*-circuits are developed. Those are called equivalent π and equivalent *T*-circuits.

4. POWER FACTOR IMPROVEMENT

The power factors of different loads are as follows :

Type of load	Power factor (range)
Heating and lighting	0.95 to unity
Motor loads	0.5 to 0.9
Single-phase motors	as low as 0.4
Electric welding units	0.2 to 0.3

The power factor is given by :

$$\cos \phi = \frac{\text{kW}}{\text{kVA}} \quad \text{or} \quad \text{kVA} = \frac{\text{kW}}{\cos \phi}$$

In case of *single-phase supply*,

$$\text{kVA} = \frac{VI}{1000} \quad \text{or} \quad I = \frac{1000 \text{ kVA}}{V}$$

$$\therefore \quad I \propto \text{kVA}$$

In case of *three phase supply*,

$$\text{kVA} = \frac{\sqrt{3}\, E_L I_L}{1000} \quad \text{or} \quad I_L = \frac{1000 \text{ kVA}}{\sqrt{3}\, E_L}$$

$$\therefore \quad I_L \propto \text{kVA}$$

In each case, the *kVA is directly proportional to current.*

The higher current due to poor power factor affects the system and lead to the following undesirable results :

1. Rating of alternators and transformers are proportional to their output current hence inversely proportional to power factor, therefore large *generators and transformers are required to deliver same load but at a low power factor*.

2. When a load having a low lagging power is switched on, there is a large voltage drop in the supply voltage because of the increased voltage drop in the supply lines and transformers. This drop in voltage adversely affects the starting torque of motors and necessitates *expensive voltage stabilizing equipment* for keeping the consumer's voltage fluctuations within the statutory limits.

3. The cross-sectional area of the bus-bar, and the contact surface of switch gear is required to be *enlarged* for the same power to be delivered but at a low power factor.

4. To transmit same power at low power factor, more current will have to be carried by the transmission line or the distributor or cable. If the current density in the line is to be kept constant *the size of the conductor will have to be increased.* Thus *more copper is required to deliver the same load but at a low power factor.*

5. Copper losses are proportional to the square of the current hence inversely proportional to the square of the power factor *i.e.* more copper losses at low power factor and hence *poor efficiency*.

Thus we find that the capital cost of transformers, alternators, distributors, transmission lines etc. *increase with the low power factor*.

Causes of Low Power Factor :

(*i*) All A.C. motors (except overexcited synchronous motors and certain types of commutators motors) and transformers operate at lagging power factor.

(*ii*) Due to typical characteristic of the arc, arc lamps operate at low power factor.

(*iii*) When there is increase in supply voltage, which usually occurs during low load periods (such as lunch hours, night hours etc.) the magnetizing current of inductive reactances increases and power factor of the electrical plant as a whole decreases.

(*iv*) Arc and induction furnaces etc. operate at a very low lagging power factor.

(*v*) Due to improper maintenance and repairs of motors the power factor at which motors operate fall.

Methods of Improving Power Factor. The power factor may improve by using the following methods :

1. Use of high power factor motors.
2. Use of induction motors with phase advancers.
3. Use of static capacitors.
4. Use of capacitance boosters.
5. Use of synchronous condensers.

Example 1. *A 40-MVA, 11 kV 3-phase alternator supplies full-load at a lagging power factor of 0.6. Find the percentage increase in earning capacity if the power factor is increased to 0.9.*

Solution. Rating of the alternator = 40 MVA

Supply voltage = 11 kV

Initial power factor = 0.6 (lagging)

Final power factor after increase = 0.9 (lagging)

Percentage increase in earning capacity :

The earning capacity is proportional to the power (in MW or kW) supplied by the alternator.

MW supplied at 0.6 lagging $= 40 \times 0.6 = 24$

MW supplied at 0.9 lagging $= 40 \times 0.9 = 36$

Increase in MW $= 36 - 24 = 12$

The increase in earning capacity is proportional to 12.

$\therefore$ **Percentage increase in earning capacity** $= \frac{12}{24} \times 100 =$ **50%. (Ans.)**

Example 2. *A 3-phase, 40 kW, 440 V, 50 Hz induction motor operates on full-load with an efficiency of 90 per cent and at a power factor of 0.8 lagging. Calculate the total kVA rating of capacitors required to raise the full-load power factor to 0.9 lagging. What will be capacitance per phase if the capacitances are :*

(*i*) *Delta connected* (*ii*) *Star-connected.*

Solution. Motor output = 40 kW

Supply voltage = 440 V

Efficiency of motor = 90%

Motor power input, $P = \frac{\text{output}}{\text{efficiency}} = \frac{40}{0.9} = 44.44$ kW

Power factor 0.8 (lagging)

$$\cos \phi_1 = 0.8$$

$$\therefore \quad \phi_1 = \cos^{-1} 0.8 = 36.9°$$

$$\tan \phi_1 = 0.75$$

$$\text{Motor kVAR}_1 = P \tan \phi_1 = 44.44 \times 0.75 = 33.33$$

Power factor 0.9 (lagging)

Motor power input = 44.44 kW [As before]

[Power is same as before since the capacitors are *loss free i.e.*, they *do not absorb any power*]

$$\cos \phi_2 = 0.9$$
$$\therefore \quad \phi_2 = 25.8°$$
$$\tan \phi_2 = 0.484$$
$$\therefore \quad \text{Motor kVAR}_2 = P \tan \phi_2 = 44.44 \times 0.484 = 21.5.$$

The difference in values of kVAR is ***due to the capacitors which supply leading kVAR to partially neutralize the lagging kVAR of the machine.***

$\therefore$ **Leading kVAR supplied by capacitors is**

$$= \text{kVAR}_1 - \text{kVAR}_2 = 33.33 - 21.5$$
$$= 11.83 \text{ ...... MQ or NT.}$$

Since capacitors are loss-free, their kVAR is same as kVA

$\therefore$ **Total kVA rating of capacitors = 11.83 (Ans.)**

$$\text{kVAR/capacitor} = \frac{11.83}{3} = 3.943$$
$$\therefore \quad \text{VAR/capacitor} = 3943.$$

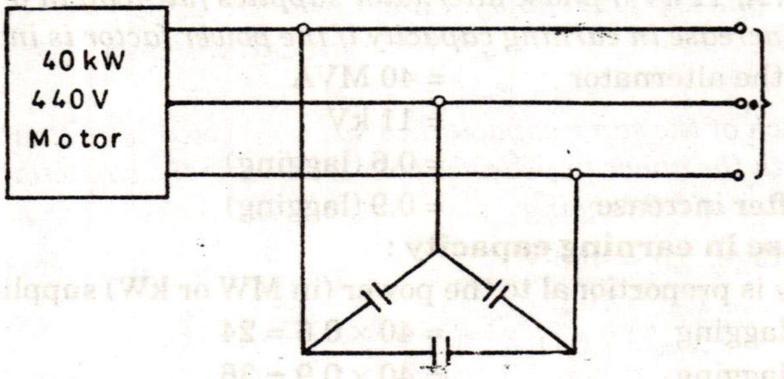

Fig. 24

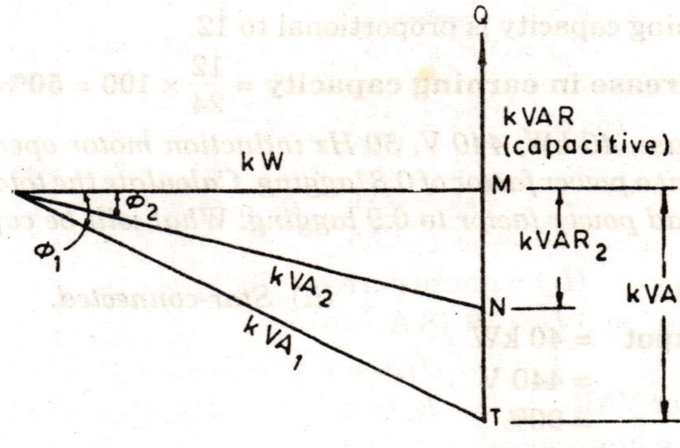

Fig. 25

(i) **Delta connection.** In delta-connection, voltage across each capacitor is 440 V.

$$\text{Current drawn by each capacitor, } I_C = \frac{3943}{440} = 8.96 \text{ A}$$

Now,
$$I_C = \frac{V}{X_C} = \frac{V}{\frac{1}{2\pi fC}} = 2\pi fCV$$

$$\therefore \quad C = \frac{I_C}{2\pi fV} = \frac{6.96}{2\pi \times 50 \times 440} = 50.35 \times 10^{-6} \text{ F} = 50.35 \text{ μF}$$

Hence, capacitance of each capacitor = 50.35 μF. (Ans.)

(*ii*) **Star-connection.** In star-connection, voltage across each capacitor

$$= \frac{440}{\sqrt{3}} = 254 \text{ V}$$

Current drawn by each capacitor,

$$I_C = \frac{3943}{254} = 15.52 \text{ A}$$

Now, $$I_C = \frac{V}{X_C} = \frac{V}{\frac{1}{2\pi fC}} = 2\pi fCV$$

$$\therefore \quad C = \frac{I_C}{2\pi fV} = \frac{15.52}{2\pi \times 50 \times 440} = 112.3 \times 10^{-6} \text{ F} = 112.3 \ \mu\text{F}$$

Hence, **capacitance of each capacitor = 112.3 μF. (Ans.)**

Example 3. *Three impedance coils each having a resistance of 10 Ω and a reactance of 7.5 Ω are connected in star to a 400 V, 3-phase 50 Hz supply. Calculate :*

(*i*) *Power factor* (*ii*) *Line current, and*

(*iii*) *Power supplied.*

If three capacitors, each of the same capacitance, are connected in delta to the same supply so as to from parallel circuit with the above impedance coils, calculate the capacitance of each capacitor to obtain resultant power factor of 0.9 lagging.

Solution. Supply voltage, $E_L = 400$ V [Line-to-line]

Phase resistance, $R_{ph} = 10 \ \Omega$

Phase reactance, $X_{ph} = 7.5 \ \Omega$

Phase voltage, $$E_{ph} = \frac{400}{\sqrt{3}} = 231 \text{ V}$$

Phase impedance, $$Z_{ph} = \sqrt{R_{ph}^2 + X_{ph}^2} = \sqrt{10^2 + (7.5)^2} = 12.5 \ \Omega$$

(*i*) **Power factor,** $$\cos \phi_1 = \frac{R_{ph}}{Z_{ph}} = \frac{10}{12.5} = \mathbf{0.8.} \ \textbf{(Ans.)}$$

Phase current, $$I_{ph} = \frac{E_{ph}}{Z_{ph}} = \frac{231}{12.5} = \mathbf{18.48 \ A.} \ \textbf{(Ans.)}$$

(*ii*) Line current, (I_L) = phase current, (I_{ph})

$\therefore$ I_L = **18.48 A. (Ans.)**

(*iii*) **Power supplied,** $P = \sqrt{3} E_L I_L \cos \phi = \sqrt{3} \times 400 \times 18.48 \times 0.8 =$ **10243 W. (Ans.)**

Motor $\text{VAR}_1 = \sqrt{3} \times E_L I_L \sin \phi_1 = \sqrt{3} \times 400 \times 18.48 \times 0.6 =$ **7682. (Ans.)**

When capacitors are connected :

Power factor, $\cos \phi_2 = 0.9$

$\therefore$ $\phi_2 = 25.8°$

and $\tan \phi_2 = 0.4834$

Since capacitors themselves do not absorb any power, power remains the same *i.e.*, 10243 W even when capacitors are connected. The only thing that changes is the VAR.

Now $\text{VAR}_2 = P \tan \phi_2 = 10243 \times 0.4834 = 4951$

Leading VAR supplied by three capacitors is

$= \text{VAR}_1 - \text{VAR}_2 = 7682 - 4951 = 2731$ = MQ or NT (Fig. 27)

$$\text{VAR/capacitor} = \frac{2731}{3} = 910.3$$

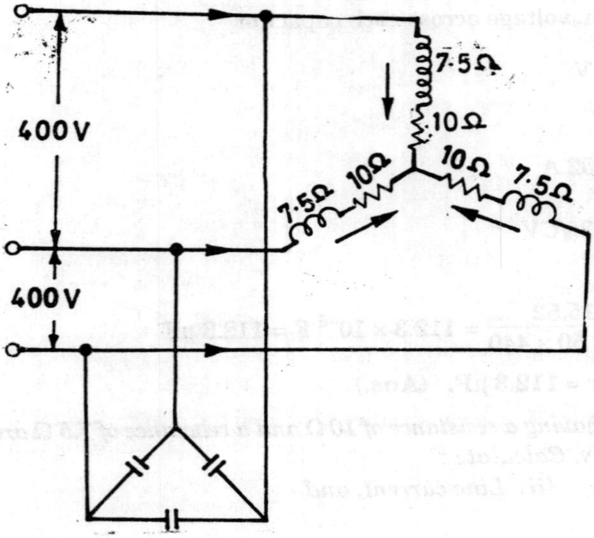

Fig. 26

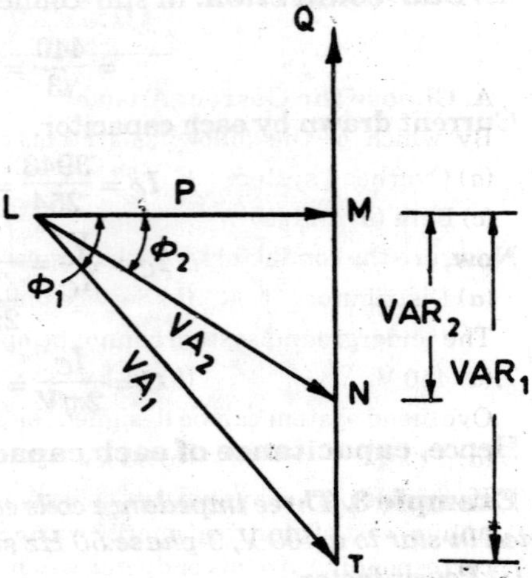

Fig. 27

For delta connection voltage across each capacitor is 400 V

$$\therefore \qquad I_C = \frac{910.3}{400} = 2.275 \text{ A}$$

Also
$$I_C = \frac{V}{\dfrac{1}{2\pi fC}} = V \times 2\pi fC$$

$$\therefore \qquad C = \frac{I_C}{V \times 2\pi f} = \frac{2.275}{400 \times 2 \times 3.142 \times 50} = 18.10 \times 10^{-6} \text{ F or } 18.10 \text{ μF}$$

Hence, **capacitance of each capacitor = 18.10 μF. (Ans.)**

HIGHLIGHTS

1. **The various types of power plants are :**
Steam power point, Hydro-electric power plant, Nuclear power plant, Simple gas power plant, Diesel engine power plant.
2. **Transmission lines may be classified as follows :**
Single line, Parallel lines, Radial lines, Ring system, Network.
3. **In general, distribution systems comprise that part of the network of a power system which distributes power for local use.**
4. **Systems of wiring may be broadly classified as follows :**
(*i*) Tree system **(*ii*) Distribution system.**
5. **The various systems of wiring for domestic installations are :**
(*i*) Cleat wiring **(*ii*) C.T.S. (T.R.S.) wiring**
(*iii*) Wooden casing and capping wiring **(*iv*) Lead-sheathed wiring**
(*v*) Conduit wiring.

OBJECTIVE TYPE QUESTIONS

A. Choose the Correct Answer :

1. By which of the following systems electric power may be transmitted ?
(*a*) Overhead system (*b*) Underground system
(*c*) Both (*a*) and (*b*) (*d*) None of the above.
2. ...... are the conductors, which connect the consumer's terminals to the distribution
(*a*) Distributors (*b*) Service mains (*c*) Feeders (*d*) None of the above.
3. The underground system cannot be operated above
(*a*) 440 V (*b*) 11 kV (*c*) 33 kV (*d*) 66 kV.
4. Overhead system can be designed for operation upto
(*a*) 11 kV (*b*) 33 kV (*c*) 66 kV (*d*) 400 kV.
5. If variable part of annual cost on account of interest and depreciation on the capital outlay is equal to the annual cost of electrical energy wasted in the conductors, the total annual cost will be minimum and the corresponding size of conductor will be most economical. This statement is known as
(*a*) Kelvin's law (*b*) Ohm's law (*c*) Kirchhoff's law (*d*) Faraday's law
(*e*) none of the above.
6. The wooden poles well impregnated with creosite oil or any preservative compound have life
(*a*) from 2 to 5 years (*b*) 10 to 15 years (*c*) 25 to 30 years (*d*) 60 to 70 years.
7. Which of the following materials is not used for transmission and distribution of electrical power ?
(*a*) Copper (*b*) Aluminium (*c*) Steel (*d*) Tungsten.
8. Galvanised steel wire is generally used as
(*a*) stay wire (*b*) earth wire (*c*) structural components (*d*) all of the above.
9. The usual spans with R.C.C. poles are
(*a*) 40—50 metres (*b*) 60—100 metres (*c*) 80—100 metres (*d*) 300—500 metres.
10. The corona is considerably affected by which of the following ?
(*a*) Size of the conductor (*b*) Shape of the conductor
(*c*) Surface condition of the conductor (*d*) All of the above.
11. Which of the following are the constants of the transmission lines ?
(*a*) Resistánce (*b*) Inductance (*c*) Capacitance (*d*) All of the above.
12. %age regulation of a transmission line is given by
(*a*) $\dfrac{V_R - V_S}{V_R^2} \times 100$ (*b*) $\dfrac{V_S - V_R}{V_R} \times 100$ (*c*) $\dfrac{V_S - V_R}{V_S} \times 100$ (*d*) $\dfrac{V_S - V_R}{V_R^2} \times 100$
where V_S and V_R are the voltages at the sending end and receiving end respectively.
13. The phenomenon of rise in voltage at the receiving end of the open-circuited or lightly loaded line is called the
(*a*) Seeback effect (*b*) Ferranti effect (*c*) Raman effect (*d*) none of the above.
14. The square root of the ratio of line impedance and shunt admittance is called the
(*a*) surge impedance of the line (*b*) conductance of the line
(*c*) regulation of the line (*d*) none of the above.
15. Which of the following is the *demerit* of a 'constant voltage transmission system' ?
(*a*) Increase of short-circuit current of the system
(*b*) Availability of steady voltage at all loads at the line terminals
(*c*) Possibility of better protection for the line due to possible use of higher terminal reactances
(*d*) Improvement of power factor at times of moderate and heavy loads
(*e*) Possibility of carrying increased power for a given conductor size in case of long-distance heavy power transmission.

16. Low voltage cables are meant for use up to
(*a*) 1.1 kV (*b*) 3.3 kV (*c*) 6.6 kV (*d*) 11 kV.

17. The operating voltage of high voltage cables is upto
(*a*) 1.1 kV (*b*) 3.3 kV (*c*) 6.6 kV (*d*) 11 kV.

18. The operating voltage of supertension cables is upto
(*a*) 3.3 kV (*b*) 6.6 kV (*c*) 11 kV (*d*) 33 kV.

19. The operating voltage of extra high tension cables is upto
(*a*) 6.6. kV (*b*) 11 kV (*c*) 33 kV (*d*) 66 kV
(*e*) 132 kV.

20. Which of the following methods is used for laying of underground cables ?
(*a*) Direct laying (*b*) Draw-in-system (*c*) Solid system (*d*) All of the above.

21. Which of the following is the source of heat generation in the cables ?
(*a*) Dielectric losses in cable insulation (*b*) I^2R losses in the conductor
(*c*) Losses in the metallic sheathings and armourings
(*d*) All of the above.

22. Due to which of the following reasons the cables should not be operated too hot ?
(*a*) The oil may loose its viscosity and it may start drawing off from higher levels
(*b*) Expansion of the oil may cause the sheath to burst
(*c*) Unequal expansion may create voids in the insulation which will lead to ionization
(*d*) The thermal instability may rise due to the rapid increase of dielectric losses with temperature
(*e*) All of the above.

23. Which of the following D.C. distribution system is the simplest and lowest in first cost ?
(*a*) Radial system (*b*) Ring system
(*c*) Inter-connected system (*d*) None of the above.

24. A booster is a
(*a*) series wound generator (*b*) shunt wound generator
(*c*) synchronous generator (*d*) none of the above.

25. Besides a method of trial and error, which of the following methods is employed for solution of network problems in interconnected system ?
(*a*) Circulating current method (*b*) Thevenin's theorem
(*c*) Superposition of currents (*d*) Direct application of Kirchhoff's laws
(*e*) All of the above.

26. Which of the following faults is most likely to occur in cables ?
(*a*) Cross or short-circuit fault (*b*) Open circuit fault
(*c*) Breakdown of cable insulation (*d*) All of the above.

27. The cause of damage to the lead sheath of a cable is
(*a*) crystallisation of the lead through vibration
(*b*) chemical action on the lead when buried in the earth
(*c*) mechanical damage (*d*) all of the above.

28. The voltage of the single phase supply to residential consumers is
(*a*) 110 V (*b*) 210 V (*c*) 230 V (*d*) 400 V.

29. Most of the high voltage transmission lines in India are
(*a*) underground (*b*) overhead (*c*) either of the above (*d*) none of the above.

30. The distributors for residential areas are
(*a*) single phase (*b*) three-phase three wire
(*c*) three-phase four wire (*d*) none of the above.

31. The conductors of the overhead lines are
(*a*) solid (*b*) stranded
(*c*) both solid and stranded (*d*) none of the above.

32. High voltage transmission lines use
(*a*) suspension insulators (*b*) pin insulators
(*c*) both (*a*) and (*b*) (*d*) none of the above.

33. Multicore cables generally use
(*a*) square conductors (*b*) circular conductors
(*c*) rectangular conductors (*d*) sector-shaped conductors
(*e*) none of the above.

34. Distribution lines in India generally use
(*a*) wooden poles (*b*) R.C.C. poles (*c*) steel towers (*d*) none of the above.

35. The material commonly used for insulation in high voltage cables is
(*a*) lead (*b*) paper (*c*) rubber (*d*) none of the above.

36. The loads on distributors systems are generally
(*a*) balanced (*b*) unbalanced (*c*) either of the above (*d*) none of the above.

37. The power factor of industrial loads is generally
(*a*) unity (*b*) lagging (*c*) leading (*d*) zero.

38. Overhead lines generally use
(*a*) copper conductors (*b*) all aluminium conductors
(*c*) A.C.S.R. conductors (*d*) none of these.

39. In transmission lines the cross-arms are made of
(*a*) copper (*b*) wood (*c*) R.C.C. (*d*) steel.

40. The material generally used for armour of high voltage cables is
(*a*) aluminium (*b*) steel (*c*) brass (*d*) copper.

41. Transmission line insulators are made of
(*a*) glass (*b*) porcelain (*c*) iron (*d*) P.V.C.

42. The material commonly used for sheaths of underground cables is
(*a*) lead (*b*) rubber (*c*) copper (*d*) iron.

43. The minimum clearance between the ground and a 220 kV line is about
(*a*) 4.3 m (*b*) 5.5 m (*c*) 7.0 m (*d*) 10.5 m.

44. The spacing between phase conductors of a 220 kV line is approximately equal to
(*a*) 2 m (*b*) 3.5 m (*c*) 6 m (*d*) 8.5 m.

45. Large industrial consumers are supplied electrical energy at
(*a*) 400 V (*b*) 11 kV (*c*) 66 kV (*d*) 400 kV.

ANSWERS

A. Choose the Correct Answer :

1. (*c*)	**2.** (*b*)	**3.** (*d*)	**4.** (*d*)	**5.** (*a*)	**6.** (*c*)
7. (*d*)	**8.** (*d*)	**9.** (*c*)	**10.** (*d*)	**11.** (*d*)	**12.** (*b*)
13. (*b*)	**14.** (*a*)	**15.** (*a*)	**16.** (*c*)	**17.** (*d*)	**18.** (*d*)
19. (*d*)	**20.** (*d*)	**21.** (*d*)	**22.** (*e*)	**23.** (*a*)	**24.** (*a*)
25. (*e*)	**26.** (*d*)	**27.** (*d*)	**28.** (*c*)	**29.** (*b*)	**30.** (*c*)
31. (*b*)	**32.** (*a*)	**33.** (*d*)	**34.** (*b*)	**35.** (*b*)	**36.** (*b*)
37. (*b*)	**38.** (*c*)	**39.** (*d*)	**40.** (*b*)	**41.** (*b*)	**42.** (*a*)
43. (*c*)	**44.** (*c*)	**45.** (*c*)			

B. Fill in the Blanks or Say 'Yes' or 'No' :

1. ...... system is more safer than ...... system.
2. Underground system is less expensive. (Yes/No)
3. Underground system cannot be operated above 66 kV. (Yes/No)
4. The chances of accidents in underground system are very high as compared to overhead system. (Yes/No)
5. Maintenance cost of underground system is very ...... in comparison to that of overhead system.
6. In underground system there is no interference to communication circuits. (Yes/No)
7. Mostly the high voltage transmission is carried out by overhead system due to ......
8. ...... system is more flexible than ...... system.
9. All conductors used for overhead lines are preferably ...... in order to increase the flexibility.
10. Hard drawn copper conductor is the best conductor owing to its high ...... and great tensile strength for all type of transmission.
11. The conductivity of aluminium is ...... percent that of copper.
12. ...... conductor consists of a core of galvanised steed strand surrounded by a number of aluminium strands.
13. Galvanised iron and steel conductors have been used for extremely long spans. (Yes/No)
14. Copper clad steel conductors are well suited for very long spans. (Yes/No)
15. The difference in level between the points of supports and the lowest point is known as ......
16. Sag is ..... proportional to the square of the span length.
17. ...... chart is helpful in knowing the sag and tension at any temperature.
18. The material employed for ...... should not be porous and there should be no effect of change in temperature.
19. ...... is extensively used material (for insulators) and is produced by firing at a controlled temperature a mixture of kaolin, felspar and quartz.
20. Glass insulator can be used upto ...... kV under ordinary atmospheric conditions and well up to ...... kV in dry atmosphere.
21. The suspension insulator hangs from the cross arm, as opposed to the ...... insulator which sits on the top of it.
22. ...... type insulators give more fllexibility to the line and mechanical stresses are reduced in this arrangement.
23. Egg or stay insulators are used in guy cables. (Yes/No)
24. The corona decreases with the ...... in diameter of conductor.
25. A ...... conductor gives rise to more corona than a solid conductor.
26. If the spacing between the conductors is made very ...... as compared with their diameters, there may not be any corona effect.
27. Line voltage largely affects the corona. (Yes/No)
28. At ...... voltage, there is no corona effect.
29. Due to corona, the probability of flash over is ......
30. The transmission lines are basically electrical circuits having ...... constants.
31. At ...... frequencies the skin effect is very small.
32. The proximity effect results in uniform distribution of current in the cross-section of a conductor. (Yes/No)
33. Any two conductors separated by an insulating medium constitute a ......
34. The transposition of conductors increases the disturbances to the nearby communication circuits. (Yes/No)
35. The regulation of a transmission line, for any given load power factor, can be graphically determined by ...... regulation diagram.

36. The phenomenon of rise in voltage at the receiving end of the open-circuited or lightly loaded line is called the ...... effect.

37. The load of unity power factor that can be delivered by the line of negligible resistance is called ...... impedance loading.

38. The operating voltage of extra super voltage power cables is beyond 132 kV. (Yes/No)

39. The insulation resistance of a cable can be measured by ...... method and by ......

40. The energy losses occuring in the dielectric of cables are due to leakage and so called dielectric ......

41. The dielectric losses are ...... to square of the voltage.

42. The sheath loss in open-circuited sheath is about 10% of the total loss. (Yes/No)

43. When two or more generating stations are connected together, the system is known as ...... system.

44. When the two ends of a distributor fed at equal voltages brought together, then such distributor is known as ......

45. A distributor arranged to form a closed circuit and fed at one or more than one point is called the ...... distributor.

46. A ...... is a series wound generator, which is inserted into a circuit to add or inject a certain voltage so that the excessive voltage drop in the feeder is compensated.

47. Blavier's test is used to locate the ground fault of a single cable. (Yes/No)

48. Loop tests employ the principle of ...... bridge.

49. The maximum generation voltage in advanced countries is ...... kV while that in India is ...... kV.

ANSWERS

B. Fill in the Blanks/Say 'Yes' or 'No' :

1. Underground, overhead	**2.** No	**3.** Yes
4. No	**5.** low	**6.** Yes
7. low cost	**8.** Overhead, underground	**9.** stranded
10. electrical conductivity	**11.** 60	**12.** A.C.S.R.
13. Yes	**14.** Yes	**15.** sag
16. directly	**17.** Stringing	**18.** insulators
19. Porcelain	**20.** 25, 50	**21.** pin
22. Suspension	**23.** Yes	**24.** increase
25. stranded	**26.** large	**27.** Yes
28. low	**29.** reduced	**30.** distributed
31. low	**32.** No	**33.** capacitor
34. No	**35.** Kapp's	**36.** Ferranti
37. surge	**38.** Yes	**39.** galvanometer, meggar
40. hysteresis	**41.** proportional	**42.** No
43. interconnected	**44.** ring main	**45.** ring
46. booster	**47.** Yes	**48.** wheatstone
49. 33, 11.		

THEORETICAL QUESTIONS

1. Enumerate various sources of energy.

2. Explain briefly any two of the following power plants :

(*i*) Steam power plant
(*ii*) Hydro-electric power plant
(*iii*) Diesel engine power plant
(*iv*) Nuclear power plant.

3. **Give typical power supply network.**

4. **How are transmission lines classified ?**

5. **List the important characteristics which the line supports should possess ?**

6. **Enumerate and explain briefly the various types of transmission line insulators.**

7. **How are distribution systems classified ?**

8. **Explain briefly three-phase four-wire distribution system.**

9. **How are systems of wiring classified ?**

10. **Explain briefly the following systems of wiring :**

(*i*) Tree system **(*ii*) Distribution system.**

11. **Explain briefly any two systems of wiring for domestic installations :**

(*i*) Cleat wiring **(*ii*) C.T.S. (T.R.S.) wiring**

(*iii*) Wooden casing and capping wiring **(*iv*) Conduit wiring.**

12. **Explain briefly, with the help of a sketch, two-wire D.C. distribution system.**

13. **What are the causes of low power factor ?**

14. **Enumerate various methods of improving power factor.**

ADDITIONAL QUESTIONS

(A) QUESTIONS WITH SHORT ANSWERS

(B) QUESTIONS WITH VERY SHORT ANSWERS

A. QUESTIONS WITH SHORT ANSWERS

I. Basic Concepts

Q. 1. *What do you mean by Electric Current ?*

Ans. *"Electric current" is the flow of electrons.* If an electromotive force is applied across a bar of material having free electrons, the electrons in the bar are attracted to the positive terminal of the source and repelled at the negative terminal. Thus flow of electrons occurs and this phenomenon is defined as Electric Current.

Q. 2. *What is electromotive force ?*

Ans. When a group of negatively charged atoms is separated from a group of positively charged atoms, a difference of potential occurs between the groups. The property or state by virtue of which a potential difference occurs between the two points of device is termed its electromotive force.

Q. 3. *What are the sources of electromotive force ?*

Ans. The sources of electromotive force are as follows :

(1) Electromagnetic induction.

(2) Contact of dissimilar substances.

(3) Thermoelectric action.

(4) Chemical action.

(5) Friction between dissimilar substances.

(6) Photo-electric effect.

Q. 4. *What do you mean by 'Resistance' ?*

Ans. Resistance is the property of the device which opposes the flow of electrons in an electric circuit. The resistance can be expressed mathematically as follows :

$$R = \rho \frac{l}{a} \text{ ohm}$$

where ρ is the resistivity of the device in ohm metre.

l is the length of the device in metre

a is the cross-section of the device in metre^2.

Q. 5. *What is Ohm's law ?*

Ans. When a simple conductor carries a steady current, the potential difference between any two points on that conductor is directly proportional to the current and to the resistance of the current path between the points. This can be expressed mathematically as follows :

$$V = IR$$

where V is the steady voltage in volt

I is the steady current in ampere, and

R is the constant resistance in ohm.

[**Note.** The simple conductor indicates that there are no e.m.f.s within the conductor.]

Q. 6. *What do you mean by direct and alternating current ?*

Ans. Ideally a direct current is defined as the unidirectional current of constant magnitude developed in an electric circuit having a constant source of e.m.f. and resistance. The current which is unidirectional but is not constant is also termed as direct current. Of course, these variations of current must be small.

An alternating current is defined as the current developed in a constant resistance electric circuit whose source of e.m.f. is regularly reversing the polarities of its terminals. Thus alternately positive and negative values of currents are produced uniformly. Therefore, the average algebraic value of an ideal alternating current over a complete reversal is zero.

Q. 7. *What is power ?*

Ans. Electric energy is derived from some form of energy. For example chemical energy is converted to electric energy through battery. Again this electric energy is converted into heat in the resistance of electric circuit. The rate of this energy dissipation from the time point of view is termed as power. The power can be expressed mathematically as follows :

$$P = ei = (iR)\, i = i^2R \text{ watts}$$

where e is the voltage across a resistor in volt.

R is the resistance of the resistor in ohms.

i is the amount of current in ampere in the electric circuit.

Q. 8. *What is the field intensity at a point due to a charged sphere ?*

Ans. The field intensity at a radial distance R is $\dfrac{Q}{4\pi\varepsilon R^2}$

where Q is the total charge on the sphere in coulombs.

R is the distance of the point from the centre of the sphere in metres.

ε is the absolute permittivity of the medium.

Q. 9. *What is the field intensity in a charged hollow spherical conductor ?*

Ans. The field intensity at any point inside a charged hollow spherical conductor is zero as it can be proved that there can be no lines of force inside a closed charged sphere.

Q. 10. *What do you mean by potential ?*

Ans. The potential at any point in the vicinity of a system of charged bodies may be defined as the work done in bringing a unit positive charge from infinity to the point considered. Of course, the distribution of the charges on the bodies remains unaffected during this time. The potential at a point can be expressed mathematically as follows :

$$V = \frac{Q}{4\pi\varepsilon d} \text{ volts}$$

where Q is the charge in coulombs.

d is the distance of the point from the charge Q in metres.

ε is the absolute permittivity of the medium.

Q. 11. *What is the relation between the field intensity and potential ?*

Ans. The field intensity at any point is equal in magnitude to the potential gradient at that point. This can be expressed mathematically as follows :

If E be the field intensity of a particular point, the work done in moving the unit charge is small distance dx is $E \,.\, dx$. The difference of potential covered in this distance dx is dv and therefore from the definition of potential

$$dv = -E \,.\, dx \quad \text{or} \quad E = -\frac{dv}{dx} \text{ volts/metre.}$$

Q. 12. *What is equipotential surface ?*

Ans. Equipotential surface is a surface such that all points on it are at same potential. Hence the potential gradient for such a surface is zero. The field strength, being equal to negative of the potential gradient, is found to be zero along equipotential surface.

Therefore, the lines of force always cross an equipotential surface normally. Different spherical surfaces around a charged sphere are the example of equipotential surfaces.

Q. 13. *What is capacitance ?*

Ans. The capacitance of a system is the ratio of its charge to potential. This system is termed as capacitor and it is an arrangement of two conductors placed comparatively closed together so that a strong electric field exists between them.

The capacitance can be expressed mathematically as follows :

$$C = \frac{Q}{V} \text{ farads}$$

where Q is the charge in coulombs.

V is the potential difference between conductors in volts.

Q. 14. *What is the value of capacitance between two parallel plates ?*

Ans. The value of the capacitance between two parallel plates is $\frac{\varepsilon A}{d}$ farads, considering the electric field over the plate area is uniform. [A is the plate area in metre^2, d is the distance between the plates and ε is the absolute permittivity of the medium].

Q. 15. *What is the value of capacitance between two concentric spheres ?*

Ans. The value of capacitance between two concentric spheres is $\frac{4\pi\varepsilon}{\frac{1}{r_1} - \frac{1}{r_2}}$ farad, where r_1 and r_2 are the radii of the inner and outer spheres respectively in metres and ε is the absolute permittivity of the medium.

Q. 16. *What is the capacitance between co-axial cylinders ?*

Ans. The value of capacitance between two co-axial cylinder is $\frac{2\pi\varepsilon}{\log_e (b/a)}$ farad, where a and b are the radii of the inner and outer cylinder respectively in metres and ε is the absolute permittivity of the medium.

Q. 17. *What is the capacitance between two parallel plates having mediums of different permittivities ?*

Ans. If a space between two parallel plates consists of three mediums having absolute permittivity and thickness ε_1, d_1 ; ε_2, d_2 and ε_3, d_3 respectively, then capacitance between parallel plates will be

$$\frac{A}{\frac{d_1}{\varepsilon_1} + \frac{d_2}{\varepsilon_2} + \frac{d_3}{\varepsilon_3}} \text{ farad}$$

where A is the cross-section of the plates in square metre.

Q. 18. *What is the value of capacitance when a system capacitances are connected in series ?*

Ans. If a number of capacitors having capacitances $C_1, C_2, C_3, ..., C_n$ are connected in series, then the resultant capacitance 'C' can be expressed mathematically as follows :

$$\frac{1}{C} = \frac{1}{C_1} + \frac{1}{C_2} + \frac{1}{C_3} + ... + \frac{1}{C_n}.$$

Q. 19. *What is the value of capacitance when a system of capacitances are connected in parallel ?*

Ans. If a number of capacitors having capacitances $C_1, C_2, C_3, ..., C_n$ are connected in parallel, then the resultant capacitance 'C' can be expressed mathematically as follows :

$$C = C_1 + C_2 + C_3 + ... + C_n.$$

Q. 20. *What do you mean by the term "Energy stored in an Electric Field" ?*

Ans. When two conductors are charged to have a certain potential difference, an electric field will exist between them. This field represents a quantity of stored energy. Fundamentally, a potential.

Q. 21. *How is the conventional average value of a sine wave determined ?*

Ans. Consider the +ve half cycle and divide it into n number of equal parts as shown in Fig. 1. Let $i_1, i_2, ..., i_n$ be the mid-ordinates.

Average value = mean of mid-ordinates

$$= \frac{i_1 + i_2 + i_3 + ... + i_n}{n}.$$

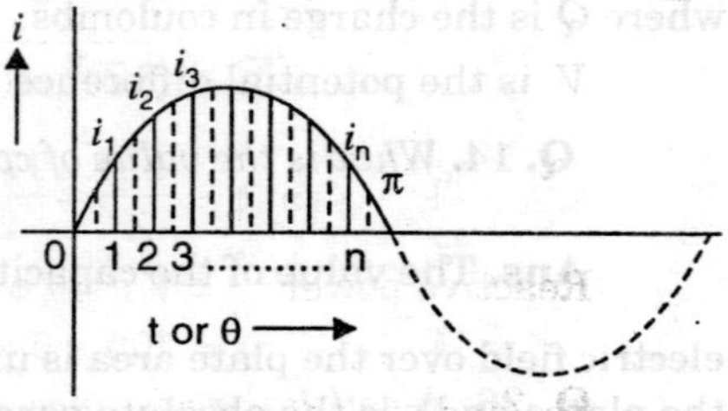

Fig. 1

Q. 22. *How can the effective value of an A.C. wave be determined ?*

Ans. To determine the effective value of an A.C. wave, consider half cycle of the wave and divide it into n equal parts. Let $i_1, i_2, i_3, ..., i_n$ be the mid-ordinates of n-equal parts. Then

Effective value,

$$I_{eff.} = \sqrt{\frac{i_1^2 + i_2^2 + i_3^2 + ... + i_n^2}{n}}$$

$$= \sqrt{\frac{\text{Area under the squared wave of an alternation}}{\text{No. of equal divisions}}}$$

= root mean square *(rms) value.*

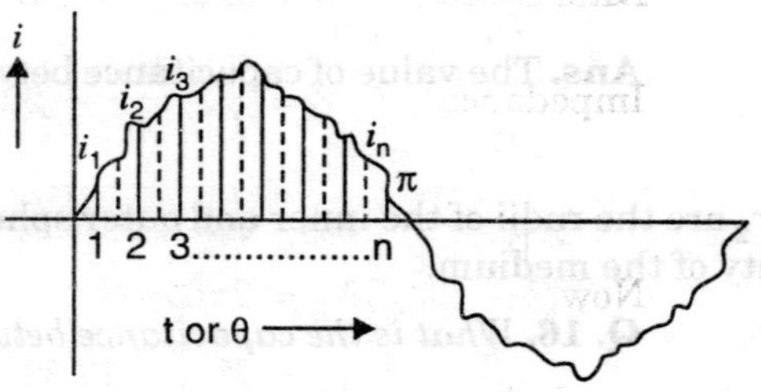

Fig. 2

Q. 23. *A coil has resistance of 10 ohms and draws a current of 5 A when connected across a 100 volt, 50 Hz source. Determine the power factor of the circuit.*

Ans. Impedance of the coil, $Z = \frac{V}{I} = \frac{100}{5} = 20\ \Omega$

Resistance of the coil, $R = 10\ \Omega$

Power factor of the circuit, $\cos\phi = \frac{R}{Z} = \frac{10}{20} =$ **0.5 lag. (Ans.)**

Q. 24. *Do wave shapes other than the sine wave have effective values ? Explain.*

Ans.

- Yes, all *the wave shapes other than sine wave have effective value since in each half cycle of the wave, work is being done.*
- In fact, the *effective value of current (or voltage) is that value of current (or voltage) which represents a steady current which when flows through a resistor of known resistance for a given time produces the same amount of heat as produced by alternating current when flows through the same resistance for the same time.*

Q. 25. *A coil has resistance of 10 ohms and draws a current of 5 A when connected across a 100 volt, 50 Hz source. Determine the reactive power of the circuit.*

Ans. *Given* : $R = 10\ \Omega\ ;\ I = 5\text{ A}\ ;\ V = 100\text{ V}\ ;\ f = 50\text{ Hz}$

$$Z = \frac{V}{I} = \frac{100}{5} = 20\ \Omega$$

$$X_L = \sqrt{(Z)^2 - (R)^2} = \sqrt{(20)^2 - (10)^2} = 17.32\ \Omega$$

$$\sin\phi = \frac{X_L}{Z} = \frac{17.32}{20} = 0.866$$

Reactive power $= VI \sin\phi$

$= 100 \times 5 \times 0.866 =$ **433 VAR. (Ans.)**

Q. 26. *A coil has resistance of 10 ohms and draws a current of 5 A when connected across a 100 volt, 50 Hz source. Determine the inductance of the coil.*

Ans. *Given* : $R = 10\ \Omega\ ;\ I = 5\text{ A}\ ;\ V = 100\text{ V}\ ;\ f = 50\text{ Hz}$

Impedance, $Z = \frac{V}{I} = \frac{100}{5} = 20\ \Omega$

$$X_L = \sqrt{(Z)^2 - (R)^2} = \sqrt{(20)^2 - (10)^2} = 17.32\ \Omega$$

Now, $X_L = 2\pi f L$

$\therefore$ Inductance, $L = \frac{X_L}{2\pi f} = \frac{17.32}{2\pi \times 50} =$ **55.13 mH. (Ans.)**

D.C. Machines

Q. 1. *What are the two main types of direct current machines and what is the major difference between them ?*

Ans. The two main types of direct current machines are homopolar and heteropolar. In a homopolar type a single conductor is continuously moved at a constant speed through a constant magnetic field. This results a constant d.c. voltage induced on the conductor. Hence no commutator is needed to develop d.c. voltage. In a heteropolar type the active magnetic field system being a number of poles arranged alternately, north pole and south pole, the voltage induced in each conductor alternates with frequency. Hence a commutator is needed to develop d.c. voltage. Moreover, in heteropolar d.c. machine any desired number of conductors can be grouped in series. But for homopolar d.c. machines any desired number of conductors can be connected in series only by providing slip rings, the number of which will be twice the number of conductors.

Q. 2. *What are the essential parts of a d.c. machine ?*

Ans. The essential parts of the D.C. machines are as follows :

(1) Yoke	(2) Pole cores
(3) Field winding	(4) Armature core
(5) Armature winding	(6) Commutator
(7) Brush.	

Q. 3. *What is lamination in d.c. machine ?*

Ans. The armature cores and pole cores of d.c. machines are built up of a number of plates which are lightly insulated from each other by varnish, paper or some other suitable material to reduce the eddy currents to a reasonable proportions. These are called *laminations.*

The plane of the laminations is always perpendicular to the direction of generated voltage, in other words, parallel to the direction of flux and to the direction of motion. For a steel of good permeability, the thickness of lamination is about 0.025 inch.

Q. 4. *What is the necessity of skewing the slots of d.c. machine ?*

Ans. When the tooth approaches a pole, the tension along the lines of magnetic force that pass between the pole face and the teeth tends to pull the tooth in one direction, and similarly in the other direction when the tooth leaves the pole. Therefore, the tooth begins to vibrate in the manner of a short reed clamped at one end. A severe condition may occur when the natural period of vibration of the tooth coincides with the frequency of magnetic blows it withstands at the time of entering and leaving the pole face. Hence, the slots are skewed to prevent this action.

Q. 5. *Why is shunt d.c. generator superior to series d.c. generator from the point of view of self-excitation ?*

Ans. The series d.c. generator cannot be self-excited until the load circuit is completed. Whereas in d.c. shunt generator, there is no necessity of completing load circuit during self-excitation.

Again, in series d.c. generator, the resistance of the exciting circuit is the complete load circuit resistance of the generator. Hence, more attention is to be taken so that the resistance cannot exceed the critical resistance.

In series machine, the induced voltage is completely depending upon the load conditions, *i.e.*, it builds up practically zero voltage at no load and normal voltage at full load, but the excitation of the shunt d.c. generator is practically free from this load condition.

Q. 6. *What is armature reaction in a d.c. machine ?*

Ans. Under load condition, the current in the armature conductors produces an independent excitation which changes both the magnitude and distribution of the flux developed by the field winding. This action of the armature is called armature reaction.

Q. 7. *Is it possible to use a d.c. Shunt generator as a constant potential d.c. source ?*

Ans. Shunt generator can be used as a constant potential d.c. source by adjusting resistance in field circuit and thus the drop in voltage between no load and full load can be made minimum.

Q. 8. *What does the name plate of d.c. generator generally indicates ?*

Ans. The name plate of a d.c. generator generally indicates :

(*a*) output in kW	(*b*) output voltage
(*c*) speed	(*d*) permissible temperature rise
(*e*) rated current.	

Q. 9. *Why is series generator not generally used ?*

Ans. The external characteristic of a series generator shows that during the initial portion of the curve, the terminal voltage increases with the increases in load. This is an unstable condition for this reason, a series generator has very limited applications.

Q. 10. *Upon what factor the armature resistance (R_a) of a d.c. machine depends ?*

Ans. The armature resistance of a d.c. machine depends upon

(*a*) number, size and connection of armature coils

(*b*) contact resistance between the carbon brush and the copper commutator.

Q. 11. *How does flashover in commutator occur ?*

Ans. When a sudden heavy overload occurs in a d.c. machine, there results a violent field distortion due to cross-magnetizing effect of armature reaction and an arc may form between plus and minus brushes. This is termed as commutator flashover. Since the cross magnetizing field is approximately proportional to the current rise, the transient overload increases the cross field and the self-induced voltage during commutation will be developed in those coils which lie in the regions of high flux density. Moreover, more suddenly the overload occurs, the greater will be this self-induced voltage as it depends on the rate of change of flux. Due to this high e.m.f., the commutator flashover occurs.

Q. 12. *What is dummy coil ?*

Ans. The number of coil sides should be such in the wave winding that both front pitch and back pitch should be odd integer. Due to this restriction upon the number of coil sides in wave windings, there may be necessity of dummy coils, which are electrically inactive and provide mechanical balance for armature. These type of dummy coils are generally used when the requirements of wave winding are not met by the standard armature punchings available in armature winding shops.

Q. 13. *What are equalizing rings ?*

Ans. In a multipolar lap wound armature there is every possibility of flow of circulating currents since the values of the electromotive force in the various circuits may not be identical. To prevent circulating currents flowing through the brushes, some steps are taken to provide other paths for flow of the same. Certain equipotential points on the windings are connected by equalizing ring. These connections are made either from the corresponding commutator segments or from the back of the coils. In general, 10 to 20 equalizing rings are utilised and the number of coils connected to each ring are equal to the number of pole pairs.

Q. 14. *What do you mean by coil pitch ?*

Ans. Coil pitch is the distance between two sides of the individual coils. It is measured in terms of slots.

Q. 15. *What do you mean by front pitch and back pitch ?*

Ans. The back pitch is the distance between the first and second active sides of the same coil section as shown in Fig. 3 (*a*) and (*b*) by Y_1 both in lap and wave windings.

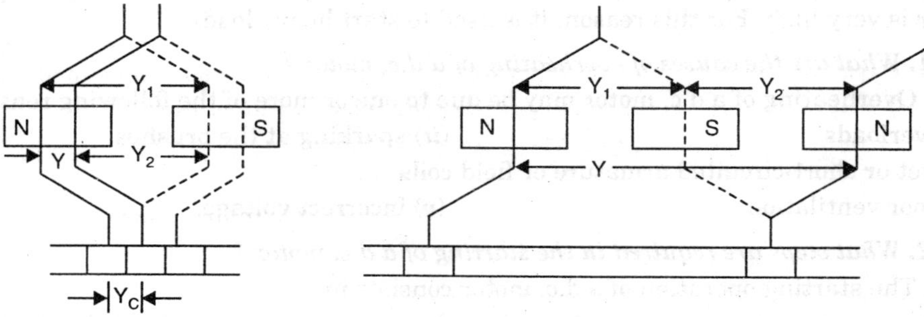

(*a*) (*b*)

Fig. 3

The front pitch is the distance between the second active side of a coil section and the first active side of the section following it as shown in Fig. 3 (*a*) and (*b*) by Y_2 both in lap and wave windings.

Q. 16. *What is the significance of back e.m.f. ?*

Ans. When voltage V is applied to the motor, it starts rotating and an e.m.f. is induced in the armature conductors. This is known as back e.m.f. E_b and opposes the applied voltage (By lenz's law). The applied voltage has to force current through the armature against this back e.m.f. E_b. The electric work done in overcoming this opposition is converted into mechanical energy developed by in the armature. But for the production of back e.m.f., energy conversion would not have been possible.

Q. 17. *What is the condition for developing maximum power in a d.c. motor ?*

Ans. The condition for developing maximum mechanical power is

$$E_b = V/2$$

where V = applied voltage ; E_b = Back e.m.f.

However, such a condition is not sought in practice because in that case current in the motor would be much more than normal current and at the same time efficiency would be low (less than 50%).

Q. 18. *What do you mean by the term " starter" ?*

Ans. At the instant of starting, the back, e.m.f. of d.c. motor is zero as the armature is not revolving. Hence some external resistance must be inserted with the low resistance of armature winding to avoid excessive armature current. The device by dint of which the resistances are inserted is called starter.

Q. 19. *Why is the speed of a shunt motor practically constant ?*

Ans. The speed of a d.c. motor is given by :

$$N \propto \frac{V - I_a R_a}{\phi}$$

In a shunt motor, V, R_a and ϕ are practically constant. At no load, the armature current is very small and $I_a R_a$ drop is negligible as compared to V. At full-load, $I_a R_a$ drop is ordinarily about 5% of V. Consequently, the full-load speed of the motor is about 95% of the non-load value. This fall in speed is reduced slightly due to armature reaction. In some cases, armature reaction is sufficient to cause the speed to remain nearly constant. For these reasons, a shunt motor is considered a constant speed motor, though the speed decreases slightly with the increase in load.

Q. 20. *Why is a d.c. series motor used to start heavy loads ?*

Ans. In a d.c. motor, $T_\alpha \phi I_a$. For a d.c. series motor, at first, $\phi \alpha I_a$ so that $T_a \propto I_a^2$. If current increases two-fold, the torque will increase four-load. Consequently, the starting torque of a d.c. series motor is very high. For this reason, it is used to start heavy loads.

Q. 21. *What are the causes of overheating of a d.c. motor ?*

Ans. Overheating of a d.c. motor may be due to one or more of the following reasons :

(*i*) overloads (*ii*) sparking at the brushes

(*iii*) wet or short-circuited armature or field coils

(*iv*) poor ventilation (*v*) incorrect voltage.

Q. 22. *What steps are required in the starting of a d.c. motor ?*

Ans. The starting operation of a d.c. motor consists in

(*i*) the insertion of external resistance into the armature circuit to limit the starting current taken by the motor.

(*ii*) the removal of this resistance in steps as the motor accelerates.

(*iii*) the insertion and removal of resistance may be done manually or automatically.

Q. 23. *What is the necessity of a starter in a d.c. motor ?*

Ans. When we apply full voltage (V) to a d.c. motor, the starting current in the armature is very high because of the absence of back e.m.f. The high starting current may result in :

(*i*) burning out the armature

(*ii*) damaging the commutator and brushes due to heavy sparking.

(*iii*) overloading the line to which the machine is connected.

In order to overcome above ill effects during starting of the d.c. motor, a starter is used with it.

Q. 24. *Can a d.c. series motor be operated on a.c. supply.*

Ans. If current direction through a series motor is reversed, the polarities of the field poles and the premises conductors are both reversed, resulting in a pushing effort between them in the same direction. Because of this, a series motor will operate on a.c. supply, its direction of rotation being the same for both half cycles. The lower the frequency of a.c., the greater the torque that can be produced. If the frequency is high, the inductive reactance of the coil limits the current that can flow and the torque of the motor.

Q. 25. *Calculate the voltage induced in the armature winding of a 4-pole, lap wound, D.C. machine having 728 active conductors and running at 1800 r.p.m. The flux per pole is 30 mWb.*

Ans. *Given* : $p = 4$; $a = p = 4$ (lap wound) ; $Z = 728$; $N = 1800$ r.p.m. ; $\phi = 30 \times 10^{-3}$ Wb.

$$E_g = \frac{P\phi ZN}{60a} = \frac{4 \times 30 \times 10^{-3} \times 728 \times 1800}{60 \times 4} = \mathbf{655.2\ V.} \quad \textbf{(Ans.)}$$

Q. 26. *A 4-pole lap wound armature has 144 slots with two coil sides per slot, each coil having two turns. If the flux per pole is 20 milliwebers and the armature rotates at 720 r.p.m., what is the induced voltage ?*

Ans. *Given* : $p = 4$; $a = p = 4$ (lap wound) ; No. of slots = 144 ;

No. of conductors/slot = $2 \times 2 = 4$; $\phi = 20 \times 10^{-3}$ Wb ; $N = 720$ r.p.m.

Total No. of conductor, $Z = 4 \times 144 = 576$

$$E_g = \frac{p\phi ZN}{60a} = \frac{4 \times 20 \times 10^{-3} \times 576 \times 720}{60 \times 4} = \mathbf{138.24\ V.} \quad \textbf{(Ans.)}$$

Q. 27. *Calculate the voltage induced in the armature winding of a 4-pole, wave wound, D.C. machine having 728 active conductors and running at 1800 r.p.m. The flux per pole is 35 mWb.*

Ans. *Given* : $p = 4$; $Z = 728$; $\phi = 35 \times 10^{-3}$ Wb ; $N = 1800$ r.p.m. ; $a = 2$

$$E_g = \frac{p\phi ZN}{60a}$$

$$\therefore \quad E_g = \frac{4 \times 35 \times 10^{-3} \times 728 \times 1800}{60 \times 2} = \mathbf{1528.8\ V.} \quad \textbf{(Ans.)}$$

II. Transformers

Q. 1. *What is the function of a transformer ?*

Ans. The function of a transformer is to change an alternating voltage from one value to another.

Q. 2. *What do you mean by step-up and step-down transformer ?*

Ans. When the transformer receives energy at a given voltage and delivers the same energy at a higher voltage, then the transformer is termed as step up transformer. On the other hand, when it delivers the same energy at a lower voltage, then the transformer is termed as step down transformer.

Q. 3. *What is the necessity of step-up and step-down transformer in transmission system ?*

Ans. The step-up transformer is necessary for transmitting a given amount of energy at a high voltage since more will be the voltage less will be the current and correspondingly the line loss. Moreover, it is not feasible to generate high voltage from the alternator itself.

On the other hand, step-down transformer is necessary for reducing the high transmission voltage to safe and usable values.

Q. 4. *What are the different types of transformers from the point of view of construction ?*

Ans. From the construction, point of view the transformers can be classified as follows :

(*a*) Core type.

(*b*) Shell type.

Q. 5. *What is the main difference between core and shell type construction ?*

Ans. The main difference between core and shell type construction is that in the core type the impression is created that the coils have been wound around the core, on the other hand, in the shell type the core has been built around the coils.

Q. 6. *What is the material of the core ?*

Ans. The material of the core is high grade silicon steel.

Q. 7. *Why is the silicon steel used ?*

Ans. The silicon steel is used due to the fact that it reduces losses and prevents ageing.

Q. 8. *Why does a transformer have an iron core ? How does the leakage flux occur in a transformer ?*

Ans.

- To *provide low reluctance path for the magnetic flux to setup in the core, it is made of iron.* Because of iron, core provides low reluctance path and a sufficient quantity of magnetic flux is set-up in the core which induces mutually induced e.m.f. and self induced e.m.f. in the secondary and primary respectively.
- The flux produced by the primary or secondary windings which is setup in air *linking with its own turns and not linking with the other is* called **leakage flux.**

Q. 9. *List four applications of transformers.*

Ans. Four important applications of transformer are :

(*a*) It can raise or lower the voltage or current in an a.c. circuit.

(*b*) It can increase or decrease the value of a capacitor, an inductor or resistance in an a.c. circuit. It can thus act as an impedance transferring device.

(*c*) It can isolate two circuits electrically.

(*d*) It can be used to prevent d.c. from passing from one circuit to the other.

Q. 10. *What will happen if the primary of a transformer is connected to d.c. supply ?*

Ans. If the primary of a transformer is connected to a d.c. supply, the primary will draw a steady current and hence produce constant flux. Consequently, no back e.m.f. will be produced. The primary winding will draw excessive current due to low resistance of the primary. The result is that the primary will over heat and burn out or the fuses will blow care must be taken not to connect the primary of a transformer across the d.c. supply.

Q. 11. *Why do we use iron-core in a transformer ?*

Ans. Transformer action demands only the existence of alternating flux linking the two windings. No doubt such action will be obtained if an air core is used but it will be obtained much more effectively if an iron core is used. It is because the flux is then substantially confined to a definite path having a much higher permeability than air.

Q. 12. *Why is the efficiency of transformer generally high ?*

Ans. The efficiency of transformer is high because it is a static equipment having no moving parts. There is no mechanical friction, windage and other mechanical losses.

Electrical losses including iron losses due to eddy currents and hysteresis are small due to use of superior quantity laminated cold rolled grain oriented silicon iron sheets. Copper losses are less due to use of sufficient cross-section of copper for conductor for windings.

Efficiency of a well designed transformer is the order of 99%.

Q. 13. *What are the advantages of open and short-circuit tests on a transformer ?*

Ans. The efficiency of a transformer is always determined by open-circuit and short-circuit tests due to the following reasons.

(*a*) The power required to carry out these tests is very small.

(*b*) These tests give the core loss and copper losses separately.

Q. 14. *Why are iron losses constant at all loads in a Transformer ?*

Ans. Since the induced primary ampere-turns and secondary ampere-turns always neutralise one another, the flux in the core on load is the same as the flux on no load. Hence, the iron losses are constant and are independent of load.

Q. 15. *What is the conservator tank ? Why is it needed ?*

Ans. The conservator tank is fitted over the transformer tank. It is needed for taking up the cyclic expansion and contraction of oil with the change of temperature without allowing the transformer tank oil to come in contact with the ambient air. Thus the transformer oil will have no chance of absorbing moisture from air.

Q. 16. *On what factors does the hysteresis loss depend ?*

Ans. Mathematically, Hysteresis loss P_h is given by the relation :

$$P_h = K_h f V B_{max}^{1.6}$$

where K_h = constant, its value depends upon type of iron ;

V = volume of iron core ;

f = frequency of reversal of magnetic field ;

B_{max} = maximum value of flux density in the core.

Q. 17. *When the load current of a transformer increases, how does the input current increase to meet the new conditions ?*

Ans. When the load current of a transformer increases, the secondary ampere-turns ($N_2 I_2$) set up more magnetic field in opposite direction to the main magnetic field in the transformer core. In other words, the secondary ampere-turns weaken the magnetic field in the core. Momentarily, self induced e.m.f. E_1 decreases due to which primary draws extra current from the mains. This extra current drawn from the mains increase the primary ampere-turns ($N_1 I_1'$) and compensate the extra secondary ampere-turns.

$$N_1 I_1' = N_2 I_2$$

Thus, in a transformer when the load current increases, the transformer draws extra current from the mains.

Q. 18. *It is desired to have a 4.13 mWb maximum core flux in a transformer operating at 110 volt and 50 Hz. Determine the required number of turns in the primary.*

Ans. *Given* : $\phi_{max} = 4.13 \times 10^{-3}$ Wb ; $E_1 = 110$ V ; $f = 50$ Hz

We know that, $E_1 = 4.44 f \phi_{max} N_1$

$$\therefore \text{ Primary turns, } N_1 = \frac{E_1}{4.44 f \phi_{max}} = \frac{110}{4.44 \times 50 \times 4.13 \times 10^{-3}} = \mathbf{120.} \quad \textbf{(Ans.)}$$

Q. 19. *Is the voltage regulation better or worse at 0.6 power factor lagging than 0.8 power factor lagging for the same kVA output ? Explain.*

Ans. The voltage regulation will be worse at 0.6 p.f. lagging than at 0.8 p.f. lagging. It can be explained by considering the mathematical expression ;

$$_0V_2 (= E_2) = V_2 + I_2R_2 \cos \phi_2 + I_2X_2 \sin \phi_2$$

$$\frac{_0V_2 - V_2}{_0V_2} = \frac{I_2R_2 \cos \phi_2 + I_2X_2 \sin \phi_2}{_0V_2}$$

When the value of $\cos \phi_2$ decreases, the value of $\sin \phi_2$ increases. The overall voltage drop increases since $X_2 > R_2$ and the voltage regulation is worse.

Q. 20. *A 50 Hz transformer takes 75 watts input power at 1.5 A and 120 V. If primary winding resistance is 0.4 ohm, find the core loss.*

Ans. Considering that the transformer is operating at no-load,

Copper losses $= I^2 R = (1.5)^2 \times 0.4 = 0.9$ W

Core losses = Power input at no-load-Cu loss

$= 75 - 0.9 =$ **74.1 W. (Ans.)**

Q. 21. *What is the effect of leakage flux in a transformer ? What is meant by Magnetising current ?*

Ans.

- Leakage flux produces self inductances $L_1 (= N_1 \phi l_1/I_1)$ and $L_2 (= N_2 \phi l_2/I_2)$ in the respective windings *i.e.*, primary and secondary. These inductances, in turn, produce inductive reactances $X_1 (= 2\pi f L_1)$ and $X_2 (= 2\pi f L_2)$. Under loaded conditions, when current flows through these windings, *some voltage is dropped due to these reactances.*

The current drawn from the mains which is used to set-up flux in the transformer core is called *magnetising current.*

Q. 22. *On what factors the eddy-current loss depend ? Why can this loss be assumed to be constant ?*

Ans.

- Eddy current losses are given by the relation,

$$P_e = K_e V f^2 t^2 B_{max}^2$$

Hence, the factors on which eddy current losses depend are :

K_e = a constant whose value depends upon the type of magnetic material ;
V = volume of magnetic material ;
f = frequency of reversal of magnetic field ;
t = thickness of the lamination ;
B_{max} = maximum value of flux density.

- *These losses are independent of the load,* therefore, these are considered as *constant losses.* All the above factors remain the same from no-load to full-load expect frequency of reversal of field (f). In most of the machines this factor is also constant.

Q. 23. *A 50 Hz transformer takes 75 watts input power at 1.5 A and 120 V. If primary winding resistance is 0.4 ohms, find the no-load power factor.*

Ans. *Given :* $f = 50$ Hz, $P_{in} = 75$ W ; $I_0 = 1.5$ A ; $V = 120$ W ; $R_1 = 0.4\ \Omega$

Primary copper loss $= I_0^{\ 2} R_1 = (1.5)^2 \times 0.4 = 0.9$ W

No-load iron loss $= P_{in} - I_0^2 R_1 = 75 - 0.9 = 74.1$ W

$$V_1 I_0 \cos \phi_0 = 74.1 \text{ W}$$

No-load power factor, $\cos \phi_0 = \dfrac{74.1}{120 \times 1.5} = $ **0.412 lag. (Ans.)**

Q. 24. *Where is the auto-transformer in general used ?*

Ans. The auto-transformer is in general used in cases where the voltage ratio is not less than 2 : 1 approximately and electrical isolation of the primary and the secondary circuit is not essential for example application for starters, high voltage transmission line.

III. Induction Motors

Q. 1. *What is an induction motor ?*

Ans. The induction motor is a motor where no current is conducted but the current results from the induced voltage. That is why, this is termed as an induction motor.

Q. 2. *What are the principal types of induction motors ?*

Ans. The principal types of induction motors are as follows :

(*a*) Squirrel cage.

(*b*) Wound rotor machine.

Q. 3. *What is the difference between squirrel cage and wound rotor induction motor ?*

Ans. The difference between squirrel cage and wound rotor induction motor is as follows :

The difference between the squirrel cage and induction motor is in the construction of rotor. A series of rotor conductors are evenly spaced around the periphery of rotor of squirrel cage with their ends short circuited by conductors in the form of end rings. But in case of wound rotor machine, the winding in the motor will be just like that of stator and the electrical connections are being made between the rotor winding and slip rings mounted on the shaft.

Q. 4. *Is it possible to have a three phase stator and two phase rotor in an induction motor ?*

Ans. Yes, it is possible to have a three phase stator and two phase rotor in an induction motor, but it should be wound rotor machine.

Q. 5. *Are the number of poles same for both stator and rotor windings of induction motor ?*

Ans. Yes, the number of poles are same for both stator and rotor windings of induction motors irrespective of their types *i.e.*, whether they are squirrel cage or wound rotor.

Q. 6. *How can the direction of the stator revolving field be reversed.*

Ans. Reversing two of the motor terminals connected to the mains, cause the phase reversal. This results in reversal of direction of rotation of the stator field.

Q. 7. *At what speed does the revolving field rotate ? On which factors does the speed depend ?*

Ans. Field rotates at synchronous speed given by

$$n_s = \frac{120f}{P}$$

Factors determining the speed

(*i*) Frequency of currents

(*ii*) Number of poles for which the stator is wound.

Q. 8. *Explain why the rotor of a squirrel cage induction motor rotates in the direction of the air-gap magnetic field ?*

Ans. Rotating field induces currents in its own pattern in the short circuited rotor conductors. In turn the interaction of rotor currents with the air-gap field produces a torque in the direction of field rotation (as per Lenz's law). As a result rotor rotates in the direction of the field. Of course it cannot reach the field speed, otherwise the induced current and so the torque would vanish.

Q. 9. *Why an IM is not operated under conditions of maximum load ?*

Ans. Maximum power output condition corresponds to matching of the load resistance in the circuit model to the circuit impedance as seen by the load resistance. This condition corresponds to much larger than the normal operating slip and hence larger current than the full-load value. As copper loss is proportional to square of motor current, the motor efficiency under condition of maximum power output is well below 50% which is unacceptable for a power device. Acceptable efficiencies are about 85% or higher.

The above statement holds for all electromechanical power devices.

Q. 10. *What is the effect of air-gap of an induction motor on break down torque and why ?*

Ans. The effect of air-gap of an induction motor on breakdown torque is that the increase in air-gap decreases the breakdown torque of an induction motor. Since the leakage reactance of an induction motor increases with the increase of the length of air-gap and the breakdown torque of an induction motor decreases with the increase in leakage reactance, the breakdown torque of an induction motor decreases with the increase in air-gap.

Q. 11. *Define Cogging ?*

Ans. *Cogging* : A squirrel cage rotor may exhibit a peculiar behaviour in starting for certain relationship between the number of poles and the stator and rotor slots. With the number of stator slots S_1 equal to or an integral multiple of rotor slots S_2, the variation of reluctance as a function of space will be quite pronounced resulting in strong alignment forces at the instant of starting. These forces may create an aligning torque stronger than the accelerating torque with consequent failure of the motor to start. This phenomenon is known as Cogging. Such combination of stator and rotor slots must, therefore, be avoided in machine design.

Q. 12. *The induction motor can be called a generalized transformer. Explain, why ?*

Ans.

- An induction motor and a transformer both operates on the same principle *i.e., induction phenomenon.*
- When A.C. supply is given to the primary winding of a transformer, an alternating flux is set-up in the core which links with the secondary winding and an e.m.f. in induced in it.

Thus, electrical power is transformed from primary to secondary through magnetic field.

In case of induction motor, when A.C. supply is given to the stator (primary) winding, an alternating (revolving) field is set up in the core which links with the rotor conductors or winding (secondary) and an e.m.f. is induced in it. Because of winding design torque is developed and rotor rotates. Thus, electrical power is transformed from stator to rotor and converted into mechanical power.

Since, the basic phenomenon is the same, an induction motor is called a *generalized transformer.*

Q. 13. *The speed of an induction motor cannot be equal to synchronous speed. Explain. Why ?*

Ans. The rotor of an induction motor can not pick the speed as that of speed of revolving field (synchronous speed), otherwise, there *will not be relative speed between the revolving field and rotor conductors* and the basic induction phenomenon may not occur to develop torque. Hence, an induction motor can never run at synchronous speed.

Q. 14. *A 3-phase induction motor develops starting torque but a single-phase induction motor develops no starting torque, why ?*

Ans.

- In case of 3-phase induction motor, when 3-phase supply is given to a 3-phase wound stator of an induction motor, a *resultant field is set-up which rotates w.r.t. stator at synchronous speed. This rotating field develops starting torque.*
- In case of 1-phase induction motor, when 1-phase supply is given to a 1-phase wound stator of an induction motor, an *alternating field is set up.* This alternating field may be represented by two fields each having half the magnitude revolving in opposite direction at constant synchronous speed. These revolving fields *develop equal and opposite torques due to which no starting torque is developed.*

Q. 15. *The reactance of the rotor of the induction motor varies greatly between starting and running conditions. Explain, why ?*

Ans. Rotor reactance of an induction motor is given by the relation :

$$X_2 = 2\pi f_r L_2 = s \times 2\pi f L_2 \qquad (\text{since } f_r = sf)$$

At start slip is 1 and the rotor reactance is maximum called *standstill rotor reactance,*

$$X_{2s} = 2\pi f L_2$$

$$X_2 = sX_{2s}.$$

As the slip varies greatly between starting and running conditions, therefore, rotor reactance also varies greatly between starting and running conditions.

Q. 16. *Justify your choice of electric motor in the following applications :*

(i) Refrigerator (ii) Mixie (iii) Wall-clock (iv) Ceiling fan (v) Sewing machine (vi) Lathe (vii) Rolling mill (viii) Electric traction (ix) Lift (x) Vacuum cleaner.

Ans.

(i) **Refrigerator.** *Single-phase, capacitor start capacitor run induction motor.*

This motor provides heavy starting torque. Under running condition, one of the capacitors is disconnected which provides heavy starting torque and the other remains in circuit to run the machine at unit p.f.

(ii) **Mixie.** *Single-phase series motor.*

It can operate at high speed to crush the material and can pick heavy loads at start.

(iii) **Wall clock.** *Single phase synchronous motor.*

In rotates only at constant (synchronous) speed to give accurate time.

(iv) **Ceiling fan.** *Single-phase capacitor run induction motor.*

It operates at almost unity p.f. Its speed is almost constant at a particular supply voltage. Its speed can be changed by changing the voltage applied across its terminals.

(v) **Sewing machine.** *Single-phase A.C. series motor.*

It can operate at a very high speed and stitching can be done very quickly.

(vi) **Lathe.** *Single-phase capacitor run induction motor or 3-phase squirrel cage induction motor.*

These motors operate almost at constant speed at all loads.

(*vii*) **Rolling mills.** *3-phase slip-ring induction motors.*

These motors have the ability to pick heavy loads. In the rolling mills, when steel to be rolled is employed between the rollers, this acts as a heavy load. This load can only be picked by 3-phase slip-ring induction motors.

(*viii*) **Electric traction.** *D.C. series motors.*

The heavy load of the train can be picked up at the start by D.C. series motors very efficiently. Moreover, the speed of these motors (or train) can be regulated very easily by using various electric circuits.

(*ix*) **Lifts.** *3-phase slip-ring induction motors.*

These motors have the ability to lift the heavy loads at start by inserting external resistance in the rotor circuit.

(*x*) **Vacuum cleaners.** *A.C. series motors.*

To create vacuum, high speed motors are required. AC series motors are best suited as they can be operated at very high speeds.

Q. 17. *State the difference between "Cage rotor" and "Phase wound rotor" construction.*

S. No.	Cage rotor construction	Phase wound rotor construction
1.	Bare bar conductors are used.	Insulated winding is used.
2.	No insulation is provided between rotor conductors and rotor core.	Insulation is provided between rotor winding and rotor core.
3.	Rotor conductors are short-circuited by providing short-circuiting rings at both the sides of the rotor.	Rotor winding is connected is star and the remaining three ends are connected to each slip-ring separately.
4.	No slip ring is provided.	Three slip rings are provided.
5.	Rotor conductors are usually aluminium die cast.	Rotor conductors (winding) are generally of copper material.
6.	Rotor resistance is fixed.	Rotor circuit resistance can be changed by inserting external resistance.

IV. Synchronous Generator or Alternator

Q. 1. *What is the basic difference between salient and non-salient poles ?*

Ans. The basic difference between salient and non-salient poles are as follows :

The salient pole machine consists of laminated salient or projecting poles in the field. Whereas non-salient pole machine consists of cylindrical rotor having field windings placed in slots around the surface of the rotor.

Q. 2. *What are the disadvantages of salient pole machine ?*

Ans. The disadvantages of salient pole machine are as follows :

(*a*) The salient field poles cause excessive windage loss when generator is driven at high speed and thus produce excessive noise.

(*b*) At high speed, if salient field poles are to be used, the field structure is to be made strong enough so that the poles can withstand the greater stresses to which they will be subjected at high speeds.

Q. 3. *Why is the hole provided through the laminations of the generator ?*

Ans. The hole is provided through the laminations of the generator due to the fact that cooling of the armature is made possible by forcing cooling medium through the hole.

Q. 4. *Write the major parts of synchronous generator.*

Ans. The major parts of synchronous generator can be described as follows :

(*a*) Armature core.

(*b*) Field core.

(*c*) Armature winding and insulation.

(*d*) Field winding and insulation.

(*e*) Cooling arrangement.

Q. 5. *What is the ratio between effective resistance and d.c. resistance of the armature ?*

Ans. The ratio between effective resistance and d.c. resistance of armature of synchronous generator varies between 1.25 to 1.75.

Q. 6. *How do you determine the effective resistance of the synchronous generator ?*

Ans. The effective resistance of synchronous generator can be determined by applying an external alternating voltage to the armature winding and measuring power and current input to one phase. Then the effective resistance will be the ratio of the input power and square of the current which is flowing.

Q. 7. *What is the error in determining the effective resistance by the above method ?*

Ans. The error in determining the effective resistance by the above method can be described as follows :

Since the input power is utilised both for copper loss and iron loss, the effective resistance, determined by this method, will be found high.

Q. 8. *How do you calculate the synchronous reactance for providing data in synchronous impedance method of finding out regulation ?*

Ans. With the help of open circuit characteristic and short circuit characteristic of the synchronous generator, the synchronous reactance can be calculated. From these characteristic curves the synchronous impedance for a particular field current can be determined by dividing the open circuit voltage by corresponding short circuit current. Of course, this synchronous impedance is approximately equal to synchronous reactance if armature resistance is neglected.

Q. 9. *Discuss, why the rotor windings of alternator are generally star connected ?*

Ans.

- The stator winding of an alternator is generally connected in star so *as to obtain neutral which is properly earthed at the generating station.*
- At the sub-stations, the neutral point of the star connected winding of transformer is also earthed. Thus neutral completes its path through the earth.

Q. 10. *Discuss, why the winding of an alternator is usually short pitched ?*

Ans. The winding of an alternator is usually short-pitched due to following reasons :

- The copper used for the short pitched winding is slightly less than the full-pitch winding as the overhang is reduced.
- The wave shape of the induced e.m.f. is a coil and hence across the phase windings can be improved by proper short pitching the coils.

Q. 11. *Discuss, why the phase spread of three-phase winding is 60° and not 120° ?*

Ans. Three-phase winding is placed under each pole and each pole occupies 180° elect. Hence each phase is spread by 60° $\left(i.e., \frac{180°}{3} = 60°\right)$.

Q. 12. *What do you mean by synchronous reactance of a synchronous generator ?*

Ans. The synchronous reactance of a synchronous generator is the reactance of the generator which includes both the leakage reactance and fictitious reactance which replaces the effect of armature reaction at a definite frequency.

Q. 13. *How does the synchronous reactance of synchronous generator vary ?*

Ans. The synchronous reactance of synchronous generator varies with the degree of saturation of the magnetic circuit. If the degree of saturation increases, the synchronous reactance decreases.

Q. 14. *What are the factors on which the regulation of a synchronous generator depends?*

Ans. The factors on which the regulation of synchronous generator depends can be expressed as follows :

(*a*) Armature reaction

(*b*) Armature reactance

(*c*) Armature effective resistance

(*d*) Change in the pole leakage with change in load.

Q. 15. *What is the armature reaction of a synchronous generator ?*

Ans. The armature reaction of synchronous generator can be defined as follows :

When the synchronous generator is loaded the m.m.f. of the armature current modifies the flux produced by the field winding. This effect of the armature m.m.f. is termed as armature reaction.

Q. 16. *What are the factors on which the armature reaction synchronous generator depends ?*

Ans. The factors on which the armature reaction of synchronous generator depends are as follows :

(*a*) The arrangement of the armature winding.

(*b*) The current it carries.

(*c*) The reluctance of the magnetic circuit.

(*d*) Power factor of the load.

Q. 17. *What is the error in determining the voltage regulation by synchronous impedance method ?*

Ans. The error in determining the voltage regulation by synchronous impedance method is as follows :

The regulation obtained by this method is found higher since the synchronous reactance calculated by this method is in the straight part of its magnetiating curve whereas the value of the synchronous reactance varies with the degree of saturation. Hence, in the actual practice, the degree of saturation will be higher than the straight part of the magnetization curve.

Q. 18. *Why are alternator rated in kVA ?*

Ans. The power $\sqrt{3}\, V_L\, I_L \cos\phi$ delivered by the alternator, for the same value of current, depends upon p.f. (cos ϕ) of the load. But the alternator conductors are calculated for a definite current and the insulation and magnetic system are designed for a definite voltage independent of p.f. of the load. For this reason, apparent power measured in kVA is regarded as the rated power of the alternator.

Q. 19. *How is the efficiency of an alternator effected by load power factor ?*

Ans. The efficiency of an alternator depends not only on kVA output but also on p.f. on the load. For a given load efficiency is maximum at unity p.f. and decreases as the p.f. falls.

Q. 20. *How does a 3-phase alternator differ from a d.c. generator ?*

Ans. Like a d.c. generator, an alternator has also an armature and a field winding. However, there are following two important differences between the two :

(*a*) The armature of an alternator is an a.c. winding *i.e.*, it is connected to produce 3-phase a.c. supply while the field winding is connected to d.c. supply. Therefore, no commutator is required in an alternator which makes its construction quite simpler than a d.c. generator.

(*b*) In a d.c. generator, the armature winding rotates and the field system is stationary. But practically an alternator how armature winding on the stationary part called stator and field winding on the rotating part called rotor.

V. Single-Phase Motors

Q. 1. *Where do we require single-phase induction motors ?*

Ans. Single-phase induction motors are required where :

(*i*) Three-phase supply is not available.

(*ii*) Efficiency is of lesser importance.

(*iii*) Equipment is portable.

(*iv*) Rating is less than one H.P.

Q. 2. *What is split phase type motor ?*

Ans. The split phase type of single phase induction motor is that type of motor where resistance and inductance are connected in series with one winding or resistance is connected with one winding and inductance is connected with the other winding. Of course, this added resistance or inductance is cut out from the circuit after the starting of single phase induction motor with the help of a centrifugally operated switch.

Q. 3. *What is capacitor start single phase induction motor ?*

Ans. The capacitor start single phase induction motor is that type of split phase motor where main winding is connected direct to the source and auxiliary winding is connected with a capacitor to the source. When the motor speeds up the auxiliary winding is cut out from the circuit with the help of a centrifugally operated switch.

Q. 4. *What is the basic principle of utilising auxiliary winding for starting single phase induction motor ?*

Ans. The basic principle of utilising auxiliary winding for starting single phase induction motor is as follows :

An auxiliary winding in the stator is to be placed in the space quadrature with the main winding.

The two stator windings are to be fed from the supply in parallel. Furthermore, a phase displacement in the currents of both the winding is to be made by the difference between their constants or by inserting resistance inductance, capacitance with one of the windings. The rotor is generally a squirrel-cage type. Now, with this arrangement, a rotating field is developed for the start of the single phase induction motor.

Q. 5. *How is starting winding disconnected when single-phase induction motor runs upto speed ?*

Ans. The purpose of starting winding is that it provides a rotating flux in conjunction with the main winding at the time of starting. A centrifugal switch (which is speed-sensitive) is connected in series with the starting winding. It disconnects the starting winding when the motor reaches about 75% of synchronous speed.

Q. 6. *What is the function of centrifugal starting switch in a single-phase induction motor ?*

Ans. The centrifugal switch is connected in series with the starting winding. The primary function of the starting winding is to produce rotating flux in conjunction with main winding at the start. When the motor has started and reach nearly 75% of synchronous speed. It produces its own rotating field from the cross-field effect. The starting winding now has no function to perform and is removed from the circuit by a centrifugally operated switch.

Q. 7. *Why is power factor of a single-phase induction motor low ?*

Ans. The low power factor of a single-phase induction motor is mainly due to the large magnetising current which ranges between 60% to 70% of full-load current. Consequently, even at no-load, their motors reach temperatures close to the full-load temperature.

Q. 8. *How do you control the speed of the single phase capacitor motor ?*

Ans. The speed of the single phase capacitor motor can be controlled by changing the impressed voltage. Of course, the speed-torque curve of the load will be of proper shape with respect to that of the motor for giving stable operating conditions.

Q. 9. *What are the applications of shaded-pole motors ?*

Ans. Shaded-pole motors produce very weak rotating magnetic field. Consequently, the starting torque of such motors is very small. They are suitable only for driving small devices *e.g.*, small fans, electric clocks, relays, etc.

Q. 10. *Why should the auxiliary winding in a capacitor start motor be disconnected after the motor has picked up speed ?*

Ans. In this case, the capacitor used is a dry-type electrolytic capacitor. This type of capacitor is designed for a *definite duty cycle and not for continuous use.* A faulty switch may keep the auxiliary winding energised for long time and thereby shorten its life span.

Q. 11. *A single-phase induction motor has poorer performance as compared to a three-phase induction motor. Why ?*

Ans. The *flux distribution in single-phase induction motor is poor in comparison to three-phase induction motors.* Therefore, the performance characteristics of 1-phase induction motors are poorer than 3-phase induction motors.

B. QUESTIONS WITH VERY SHORT ANSWERS

Q. 1. *A particular application employing a D.C. generator requires high voltage and low current. Should be machine be lap wound or wave wound ?*

Ans. Wave wound.

Q. 2. *Iron loss in a D.C. machine increases with increasing load. What is the mode of excitation ?*

Ans. Series excitation.

Q. 3. *Why is that the V/F ratio is kept constant while controlling the speed of a 3-phase induction motor by varying the supply frequency ?*

Ans. To keep the flux constant.

Q. 4. *A full pitch coil in an alternator has a span of 18 slots. What should be the span if the third harmonic component of the e.m.f. is to be eliminated ?*

Ans. A span of 12 slots.

Q. 5. *A 3-phase induction motor is to be started first using an auto-transformers with 80% tapping and next direct on-line. What will be the ratio of the starting torques ?*

Ans. $\frac{T_{auto}}{T_{d.o.l.}} = 0.64.$

Q. 6. *Why does synchronous impedance method give a poorer voltage regulation ?*

Ans. It does not account for saturation.

Q. 7. *A synchronous motor is to be used as a synchronous condenser. Under what condition should it operate ?*

Ans. It should be over-excited.

Q. 8. *Can a repulsion-induction motor ever run at super-synchronous speed ?*

Ans. Yes, at no-load.

Q. 9. *Does the power factor of a single phase series motor increase or decrease with increasing load torque ?*

Ans. Power factor decreases with increase in load torque.

Q. 10. *What is the dimension of* $\sqrt{LC}$ *?*

Ans. Reciprocal of velocity *viz.*, LT^{-1}.

Q. 11. *Is maximum torque of a 3-phase induction motor dependent on the rotor resistance ?*

Ans. Independent of rotor resistance.

Q. 12. *What is the relationship of development torque of a 3-phase induction motor with the supply voltage ?*

Ans. Directly proportional.

Q. 13. *In a 3-phase synchronous motor field m.m.f. lags the air gap flux, then what is the position of air gap flux vis-a-vis armature m.m.f. ?*

Ans. It also lags.

Q. 14. *By which test armature reaction m.m.f. and leakage reactance of a synchronous machine are determined ?*

Ans. O.C. and S.C. tests.

Q. 15. *What happens to the speed and p.f. of a single-phase series motor when the load increases ?*

Ans. Speed decreases and p.f. increases.

Q. 16. *Two single-phase transformers A and B of identical voltage and turns ratio operate in parallel. If the impedances of the two transformers are equal while resistance/reactance ratio of A is higher than that B, what would be the magnitudes of* I_A *and* I_B *and their phase positions ?*

Ans. Magnitudes of I_A and I_B are equal but phase angle of I_A is less than that of I_B.

Q. 17. *A 2-winding transformer is used to step down the voltage from 220 V to 110 V. If an auto-transformer is used for the same purpose, what would be the ratio of copper weights in them ?*

Ans. $\frac{\text{Weight of copper of autotransformer}}{\text{Weight of copper of two-winding transformer}} = 0.5.$

Q. 18. *While controlling the speed of a D.C. shunt motor what should be done to achieve a constant torque drive ?*

Ans. Supply voltage is to be controlled.

Q. 19. *How plugging of a D.C. motor is done ?*

Ans. By cutting the field supply.

Q. 20. *Alternators have no commutators, why ?*

Ans. They have to supply electrical energy with an alternating voltage.

Q. 21. *An alternator can be used as a synchronous condenser. Under what conditions it should operate ?*

Ans. Over excited.

Q. 22. *Which test gives the copper-loss of a transformer ?*

Ans. S.C. test.

Q. 23. *What is dimension of L/R ?*

Ans. T.

Q. 24. *Stator of an induction motor has full pitch coil span of 18 slots. What must be short pitch angle in degree if the 5th harmonics components of the e.m.f. is to be eliminated ?*

Ans. 36°.

Q. 25. *A 3-phase, 50 Hz induction motor runs at 960 r.p.m. on no-load and 940 r.p.m. at full-load. Find full-load slip.*

Ans. 6%.

Q. 26. *What is critical speed in a D.C. shunt generator ?*

Ans. Speed for which shunt field resistance represents critical resistance.

Q. 27. *Which method we must adopt to control the speed of a D.C. shunt motor above the base speed ?*

Ans. Field control method.

Q. 28. *What are the advantages of two wattmeters method over single wattmeter for 3-phase power measurement ?*

Ans. Can be used for balanced as well as unbalanced system.

Q. 29. *Name the motor being used in the ceiling fan.*

Ans. Single phase induction motor with split phase.

Q. 30. *Which is the cheapest method of starting a 3-phase induction motor ?*

Ans. D.O.L. starting.

Q. 31. *When a D.C. motor produces maximum output power ?*

Ans. When back e.m.f. is equal to half of the applied voltage.

Q. 32. *What type of load should be connected to a transformer for getting maximum voltage regulation ?*

Ans. Load required to be connected for maximum voltage regulation is inductive and for maximum regulation $\phi = \tan^{-1}(X_1/R_1)$.

Q. 33. *What is the condition for producing maximum torque in a 3-phase induction motor ?*

Ans. $R_2 = sX_2$.

Q. 34. *What measures can be taken for minimising the effect of crawling ?*

Ans. Effect of crawling can be minimised by chording, skewing and integral slot winding.

Q. 35. *Which parameter of the equivalent circuit should change the diameter of the circle diagram and how it effects the pull-out torque ?*

Ans. Total standstill leakage reactance per phase referred to stator. Pull-out torque is reduced by increased value of X.

Q. 36. *What parameter of load influences the armature reaction of an alternator ?*

Ans. Power factor of the load.

Q. 37. *What is the use of Potier's triangle ?*

Ans. To find voltage regulation.

Q. 38. *How the power factor of a synchronous motor is changed keeping the shaft load undisturbed ?*

Ans. By changing the excitation.

Q. 39. *What are the advantages of synchronous motor ?*

Ans. It runs at constant synchronous speed and is capable of being operated under a wide range of power factors both lagging and leading.

Q. 40. *How the direction of rotation of a capacitor start 1-phase induction motor is changed ?*

Ans. By changing the connection of one of the windings.

Q. 41. *Does the armature m.m.f. of an alternator delivering a capacitive load magnetise or demagnetise the main field ?*

Ans. Will demagnetise the field.

Q. 42. *The percentage leakage impedance of a 1-phase 2000 V/100 V, 5 kVA transformer is 2.5. What voltage should be applied to the HV side for carrying out SC test at rated current ?*

Ans. 125 V.

Q. 43. *A particular application employing a D.C. generator requires high voltage and low current. Should the machine be lap wound or wave wound ?*

Ans. Wave wound.

Q. 44. *A full pitch coil in an alternator has a span of 18 slots. What should be the span if the third harmonic component of the e.m.f. is to be eliminated ?*

Ans. 12 slots.

Q. 45. *Can a repulsion-induction motor ever run at super-synchronous speed ?*

Ans. Yes, at no-load.

Q. 46. *What is dimension of* $\sqrt{LC}$ *?*

Ans. LT^{-1}.

Q. 47. *On switching a D.C. motor it is found to rotate in direction opposite to that for which it is designed. Identify the motor.*

Ans. D.C. series motors.

Q. 48. *A synchronous motor runs at* N_s *r.p.m. at full-load. What will be its speed if the load is reduced to half the full-load ?*

Ans. N_s.

Q. 49. *A 3-phase induction motor is to be started first using an auto-transformer with 80% tapping and next direct-on-line. What will be the ratio of the starting torques ?*

Ans. $\dfrac{T_{auto}}{T_{d.o.l.}} = 0.64.$

Q. 50. *Three identical impedances having the same $\frac{X}{R}$ ratio are first connected in star and then in delta to a 400 V, 50 Hz supply. What is the ratio of active powers drawn ?*

Ans. $\frac{1}{\sqrt{3}}$.

Q. 51. *What is the power drawn by a 3-phase balanced load ?*

Ans. $\sqrt{3}\ V_L I_L \cos\phi$.

Q. 52. *A dummy coil is in the armature winding of a D.C. machine. Is the machine lap wound or wave wound ?*

Ans. Lap wound.

Q. 53. *The air gap reluctance of a resistance motor is uniform. Is the statement true or false ?*

Ans. False.

Q. 54. *Is the interpole AT required for a compensated D.C. machine more or less than that of an uncompensated one ?*

Ans. Less.

Q. 55. *A 3-phase induction motor is to be started first using an auto transformer with 80% tapping and next direct-on-line. What will be the ratio of the starting torques ?*

Ans. 0.64.

Q. 56. *Why does synchronous impedance method give a poorer voltage regulation ?*

Ans. Saturation conditions are lower.

Q. 57. *A synchronous motor is to be used as a synchronous condenser. Under what condition should it operate ?*

Ans. Over excited.

Q. 58. *Can a repulsion-induction motor ever run at super synchronous speed ?*

Ans. Yes.

Q. 59. *A full pitch coil in an alternator has a span of 18 slots. What should be the span if the third harmonic component of the e.m.f. is to be eliminated ?*

Ans. 12 slots.

Q. 60. *If the applied voltage to the primary of a transformer is reduced to 50 per cent of its initial value, at the same time the number of turns in the secondary are doubled. What will be the secondary voltage ?*

Ans. Same.

Q. 61. *Derive the relation between electrical and mechanical angle in case of rotating machine.*

Ans. Mechanical degrees in rotation = 360°

Electrical degrees = No. of pairs of poles times 360°

$$\therefore \quad \text{Mechanical degrees} = \frac{\text{electrical degrees}}{\text{pair of poles}} \quad i.e., \ Q_{mech} = \frac{Q_{elect}}{(p/2)}.$$

Q. 62. *A 6-pole D.C. generator has a rated speed of 600 r.p.m. Calculate the frequency of generated e.m.f. in armature conductors and at its brushes.*

Ans.

$$N = \frac{120f}{p}$$

$$\therefore \quad f = \frac{pN}{120} = \frac{6 \times 600}{120} = 30 \text{ Hz}.$$

Q. 63. *Why does an induction motor always operate at lagging power factor ?*

Ans. Induction motor has large magnetising current at very low power factor, hence even at load it will always work at lagging power factor.

Q. 64. *Can the excitation level give the idea of operating power factor of synchronous motor ?*

Ans. Excitation decides operating power factor of synchronous motor, for example, over excited motor has leading power factor.

Q. 65. *Explain the advantages of using closed slots for the rotor of small rating squirrel cage induction motor.*

Ans. Close side slot gives low air gap reluctance and hence has reduced value of magnetising current. This provides high value of leakage reactance, due to which starting current is limited.

Q. 66. *Transformers, belonging to phase group DY1 and YY0 cannot be operated in parallel, state why.*

Ans. There exists a phase angle difference between the two secondary voltages, hence cannot be connected in parallel.

Q. 67. *Comment on the statement, 'triplex order harmonics current can flow in a delta connected winding but are absent in the line current'.*

Ans. Triplex harmonic currents are cophased, hence they can flow in delta local circuit, their resultant in line becomes zero.

Q. 68. *A D.C. shunt motor may run away at heavy loads. Comment.*

Ans. At heavy loads the internal drops become very large and hence terminal voltages are reduced to a very low value, finally resulting into run away.

Q. 69. *Why, the pole changing method of speed control, can be used in squirrel cage induction motor ?*

Ans. By providing different sets of windings in the stator of squirrel cage induction motor and selecting a particular winding at a time the speed of the motor can be controlled.

Q. 70. *An induction generator is running at a slip of 0.03 and is feeding power to the infinite bus. What will happen if the input is suddenly removed ?*

Ans. The induction generator will start functioning as induction motor by taking power from the infinite bus.

LABORATORY PRACTICALS

(EXPERIMENTS 1 TO 16)

LABORATORY PRACTICALS

(EXPERIMENTS 1 TO 16)

EXPERIMENT NO. 1

Objective : *To find voltage, current relationship and power factor of a given R-L circuit.*

Material and Equipment Required :

1. Single-phase A.C. load (or lamps and choke coil) and supply.
2. Variac or 1-phase auto-transformer.
3. Moving-iron voltmeter of range 0 to 250 V .... 1 No.
4. Moving-iron ammeter of range 0 to 5 A .... 1 No.
5. Dynamometer type wattmeter 250 V, 5 A .... 1 No.
6. Power factor meter .... 1 No.
7. Double pole, single throw (DPST) switch .... 1 No.
8. Connecting leads.

Theory :

- The current flowing (I) through an A.C. circuit is given by that relation :

$$I = \frac{V}{Z}$$

where, V = A.C. supply voltage, (in *volts*) and

Z = Impedance of the circuit (in *ohms*).

- The power factor (cos φ) of the A.C. circuit is given as :

$$\cos\phi = \frac{P}{VI}$$

where, P = Power of the given circuit (in *watts*),

V = voltage applied (in *volts*), and

I = current flowing in the circuit (in *amperes*).

Circuit/Connection Diagram :

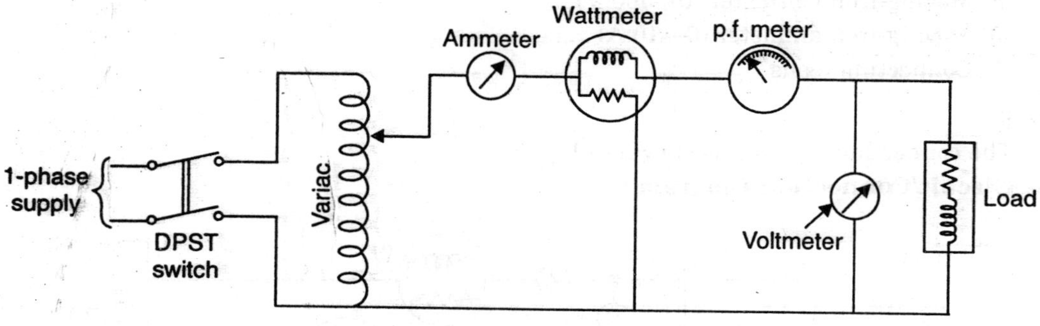

Fig. 1

Procedure :

1. Connect Variac, ammeter, wattmeter, p.f. meter, voltmeter, load (lamps in series with choke coil), through double pole single throw switch, to single phase supply mains as shown in Fig. 1.
2. Bring the variac at the lowest point and switch on DPST switch.

3. Through the variac, increase the voltage in steps and take the readings of ammeter, wattmeter, p.f. meter and voltmeter.

Observations :

S. No.	*Voltmeter reading (V) in volts*	*Ammeter reading (I) in amps.*	*Wattmeter reading (W) in watts*	*P.F. meter reading*	*Calculate* $p.f. = \frac{W}{VI}$
1.					
2.					
3.					
4.					

Conclusion :

1. The current (*I*) increases proportionately to the applied voltage V.
2. For a given load, the power factor (p.f.) of the circuit is same.
3. Calculated p.f. value matches measured one.

EXPERIMENT NO. 2

Objective : *To measure power and power factor of 3-phase balanced circuit by two-wattmeter method.*

Material and Equipment required :

1. 3-phase A.C. supply.
2. 3-phase balanced load, say an induction motor.
3. Triple pole iron clad (TPIC) switch ...... 1 No.
4. Dynamometer type wattmeters (500 V, 10 A), ... 2 No.
5. Moving-iron voltmeter (0—500 V) ...... 1 No.
6. Moving-iron ammeter (0—10 A) ...... 1 No.
7. Connecting leads.

Theory :

The connection diagram is shown in Fig. 2.

Circuit/Connection Diagram :

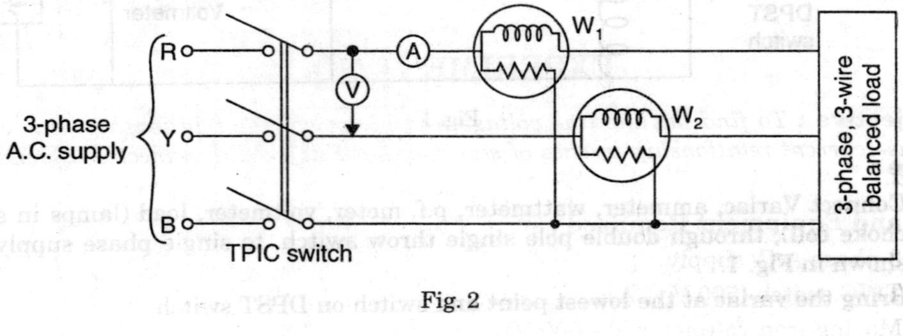

Fig. 2

The total power (P) = sum of the two wattmeters' readings W_1 and W_2 *i.e.*, $P = W_1 + W_2$

Power factor of the load, $\cos \phi = \cos \left[\tan^{-1} \left\{ \sqrt{3} \left(\frac{W_1 - W_2}{W_1 + W_2} \right) \right\} \right]$ ...(*i*)

Also, power factor, $\cos \phi = \frac{W_1 + W_2}{\sqrt{3}\, VI}$ ...(*ii*)

Procedure :

1. Make the connections as shown in the connection diagram (Fig. 2).
2. Switch-ON the supply through TPIC switch.
3. Switch-ON the load.
4. Note the readings of wattmeters W_1 and W_2, ammeter A and voltmeter V with different loads (in case of an induction motor the load can be varied by changing the mechanical load on it.)

   It may be noted that while observing the readings of the two wattmeters, if any one of them gives down scale reading (it happens when the load power factor is below 0.5), switch off the supply and reverse the connections of current coil (CC) or potential coil (PC) of that wattmeter and while doing calculation consider this reading as *negative*.
5. Note down the readings with their proper signs in the observation table :

Observations :

S. No.	*Ammeter reading (A) in amps.*	*Voltmeter reading (V) in volts*	*Wattmeter reading in watts*		*Total power, $P = W_1 + W_2$*	*Power factor (cos φ)*	
			W_1	W_2		$\cos \left[\tan^{-1} \sqrt{3} \left(\frac{W_1 - W_2}{W_1 + W_2} \right) \right]$	$\frac{W_1 + W_2}{\sqrt{3}\, VI}$
1.							
2.							
3.							
4.							

Conclusions :

1. When the induction motor is taken as a load and is operated at no-load (or light load), one of the wattmeters gives down scale reading *i.e.*, p.f. is less than 0.5.
2. Power factor calculated by the two relations given in the theory is the same.

EXPERIMENT NO. 3

Objective : *To find out the line voltage and phase voltage relationship, line current and phase current relationship in case of star-connected and delta-connected, 3-phase balanced load.*

Material and Equipment Required :

1. 3-phase A.C. supply.
2. TPIC switch (500 V, 1 A).. 1 No.
3. Moving-iron voltmeter (0—500 V) .... 2 No.

4. Moving-iron ammeter (0—10 A) .... 2 No.
5. 3-phase balanced load (say a 3-phase, 440 V, 50 Hz, 7.5 kW induction motor).
6. Board containing six open terminals .... 1 No.
7. Connecting leads.

Theory :

In *star-connections* :

Line voltage (E_L) = $\sqrt{3}$ phase voltage (E_{ph}).
Line current (I_L) = phase current (I_{ph}).
In *delta-connections* :
Line voltage (E_L) = phase voltage (E_{ph}).
Line current (I_L) = $\sqrt{3}$ phase current (I_{ph}).

Circuit/Connection Diagram :

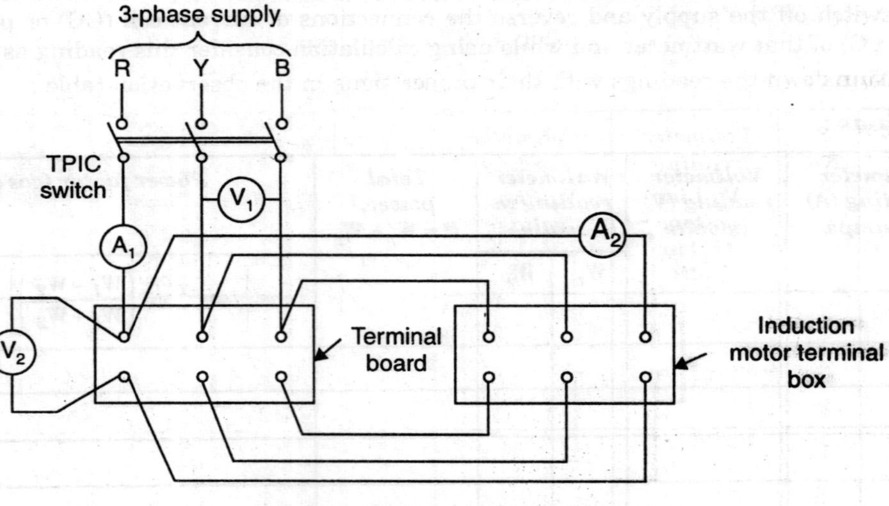

Fig. 3. Star-connections.

Procedure :

1. Make all the connections in the terminal box as shown in Fig. 3.
2. Start the induction motor and put on load and note down the readings of both the voltmeters and ammeters for different loads on the motor.
3. Now make-off the TPIC switch and change the connections in terminal board, as illustrated in Fig. 4, for delta-connections.
4. Start the motor and put on load and note down the readings of both the voltmeters and ammeters for different loads on the motor.
5. Dismantle the circuit.

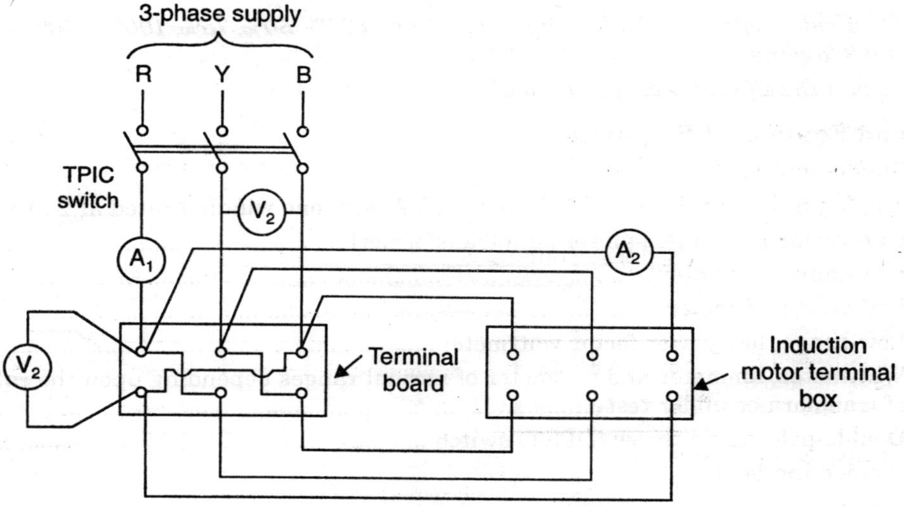

Fig. 4. Delta-connections.

Observations :

Connections	S. No.	Voltmeter reading V_1, Line voltage (V_L) in volts	Voltmeter reading V_2, Phase voltage (V_{ph}), in volts	Ammeter reading A_1, Line current (I_L) in amps.	Ammeter reading A_2, Phase current (I_{ph}) in amps.	$\frac{V_L}{V_{ph}}$	$\frac{I_L}{I_{ph}}$
	1.						
Star	2.						
	3.						
	1.						
Delta	2.						
	3.						

Conclusion :

- *In star connections.* Line voltage (V_L) comes out be $\sqrt{3}$ times phase voltage (V_{ph}) and line current (I_L) comes out to be equal to phase current (I_{ph}).
- *In delta connections.* Line voltage (V_L) comes out to be equal to phase voltage (V_{ph}) and line current (I_L) comes out to be $\sqrt{3}$ times phase current (I_{ph}).

EXPERIMENT NO. 4

Objective : *Open-Circuit and Short-Circuit Tests on a Single-Phase Transformer.*

Objectives :

(*i*) *To calculate the complete parameters of the equivalent circuit of transformer.*
(*ii*) *To determine iron and copper losses.*

(*iii*) *To calculate efficiency and voltage regulation at 25%, 50%, 75%, 100%, 150% load at a p.f. of 0.8 lagging.*

(*iv*) *To plot the efficiency curve v/s load.*

Material and Equipment Required :

1. Single-phase A.C. supply.
2. Single-phase transformer (preferably 1 kVA with one winding rated at 230 V)...... 1 No.
3. One-variac [*i.e.*, single-phase auto-transformer] .... 1 No.
4. Low-range voltmeter .... 1 No.
5. Low-range ammeter .... 1 No.
6. Low-range, low-power factor wattmeter .... 1 No.
7. Wattmeter, ammeter and voltmeter of normal ranges depending upon the rating of transformer under test .... one each.
8. Double-pole single-throw (DPST) switch .... 1 No.
9. Connecting leads.

Theory and Circuit/Connection diagram : Refer Fig. 5

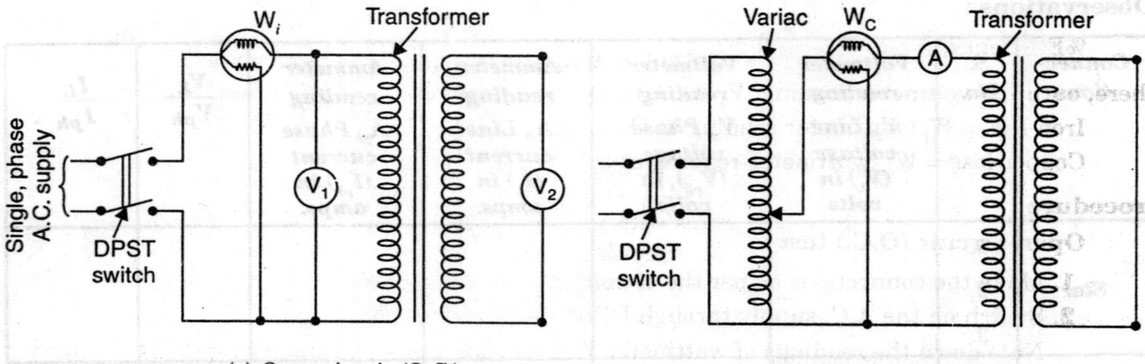

(a) Open circuit (O.C.) test. (b) Short circuit (S.C.) test

Fig. 5

- *Open-circuit test.* This test is performed to determine the *transformation ratio* (*K*) and the *iron loss*. It also determines no-load current I_0 which is helpful in finding the no-load parameters R_0 and X_0 of the transformer.

The *transformation ratio,*

$$K = \frac{V_2 \text{ (i.e., secondary voltage on open-circuit)}}{V_1 \text{ (i.e., voltage applied to primary)}}$$

The *iron loss* = wattmeter reading W_i (at open circuit). The wattmeter reading represents practically the core loss (*copper loss being negligibly small*) under no-load conditions and this loss is *same for all loads.*

Fig. 6 shows the no-load vector diagram

$$W_i = V_1 I_0 \cos \phi_0$$

(I_0 is only 2 to 10 % of the rated full-load current)

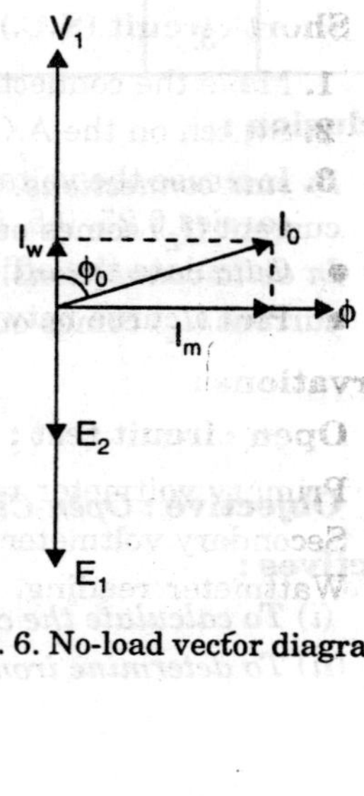

Fig. 6. No-load vector diagram.

$\therefore$ No-load power factor, $\cos\phi_0 = \dfrac{W_i}{V_1 I_0}$

Active or *working* or *iron loss component*, $I_w = I_0 \cos\phi_0 = \dfrac{W_i}{V_1}$

Magnetising component, $I_m = I_0 \sin\phi_0$ or $\sqrt{I_0^2 - I_w^2}$

No-load parameters are :

No-load exciting resistance, $R_0 = \dfrac{V_1}{I_w}\left(= \dfrac{V_1^2}{W_i}\right)$

No-load exciting reactance, $X_0 = \dfrac{V_1}{I_m}\left(= \dfrac{V_1}{\sqrt{I_0^2 - I_w^2}}\right)$

The iron losses measured by this test are used for calculating the efficiency of the transformer.

- **Short-circuit test.** This test is performed to determine the *full load copper loss*.

Here copper loss = wattmeter reading W_c (at short-circuit). (The wattmeter reading represents full-load copper losses, *the iron-loss being negligibly small.*)

Transformer efficiency (η) at any load is given by the relation,

%Efficiency, $$\eta = \frac{\text{Output}}{\text{Output + iron loss + copper loss at given load}} \times 100$$

where, output = voltmeter reading, V_1 (O.C. test) × ammeter reading A (S.C. test) × p.f.

Iron loss = W_i (wattmeter reading at O.C. test)

Copper loss = W_c (wattmeter reading at S.C. test).

Procedure :

Open-circuit (O.C.) test :

1. Make the connections as per the circuit shown in Fig. 5 (*a*).
2. Switch on the A.C. supply through DPST switch at *rated voltage*.
   Note down the readings of wattmeter W_i. (for iron losses), voltmeter readings V_1 and V_2 (for transformation ratio).
3. Switch off the supply.

Short-circuit (S.C.) test :

1. Make the connections as per the circuit shown in Fig. 5 (*b*).
2. Switch on the A.C. supply through DPST switch to variac.
3. Increase the voltage fed to the primary of the main transformer in steps so that ammeter A carries 0.25, 0.5, 0.75, and 1.5 times the rated current.

- Calculate the efficiency of the transformer at different loads (as mentioned above).
- Plot a curve between load current and efficiency of a transformer.

Observations :

Open-circuit test :

Primary voltmeter reading, V_1 = ...

Secondary voltmeter reading, V_2 = ...

Wattmeter reading, W_i = ...

Transformation ratio, $K = \frac{V_2}{V_1}$

Calculations : $\cos \phi_0 = \frac{W_i}{V_1 I_0} = \frac{\text{Wattmeter reading}}{(\text{Voltmeter} \times \text{ammeter}) \text{ reading}}$

Iron loss $= W_i$

$\therefore$ $R_0 = \frac{W_i}{I_0^2}$ or $\frac{V_0}{I_0 \cos \phi_0}$,

and $X_0 = \frac{V_0}{I_0 \sin \phi_0}$.

S. No.	*Voltmeter readings* V_1 *(in volts)*	*Ammeter readings* I_0 *(in amps.)*	*Wattmeter readings* W_i *(in watts)*
1.			
2.			
3.			
4.			
5.			

Short-circuit test :

Rated primary current, $I = \frac{\text{Rated kVA} \times 1000}{\text{Rated primary voltage } (V_1)} = \ldots..$

Power factor, $\cos \phi = 0.8$ lagging.

Calculations :

Total resistance as referred to primary sides $R_{01} = \frac{W_{sc}}{I_{sc}^2}$

Total impedance referred to primary side, $Z_{01} = \frac{V_{sc}}{I_{sc}}$

Total reactance referred to primary side, $X_{01} = \sqrt{Z_{01}^2 - R_{01}^2}$

S. No.	*Voltmeter reading* $(V_{sc}$ *in volts)*	*Ammeter reading* (I_{sc})*, in amps.*	*Wattmeter reading* (W_c)*, in watts*	% $\eta = \frac{V_1 I_{sc} \cos \phi}{V_1 I_{sc} \cos \phi + W_i + W_c} \times 100$ $= \frac{V_i I_{sc} \times 0.8}{V_1 I_{sc} \times 0.8 + W_i + W_c} \times 100$
1.				
2.				
3.				
4.				
5.				

Conclusions :

- Fig. 7 shows a curve between the load current and efficiency of the transformer. It may be observed that the efficiency of the transformer is *maximum when* $W_c = W_i$.
- With the parameters of equivalent circuit known, draw the equivalent circuit of the transformer referred to primary.

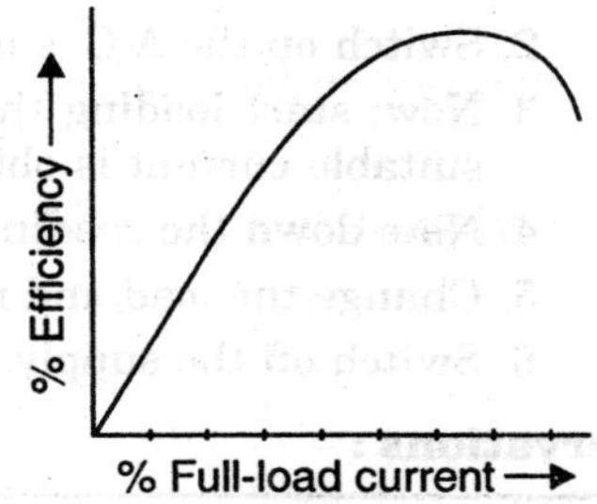

Fig. 7

EXPERIMENT NO. 5

Measurement of Efficiency of a Single-phase Transformer by Load Test.

Objectives : (*i*) *To determine the efficiency.*

(*ii*) *To determine voltage regulation.*

Material and equipment required :

1. Transformer under test -- 1 No.
2. Auto-transformer -- 1 No.
3. Moving-iron voltmeter -- 2 No.
4. Moving-iron ammeter --- 2 No.
5. Dynamometer type wattmeter --- 2 No.
6. Connecting leads.

Theory :

When the secondary winding of a transformer is completed through a load and the voltage V_1 is supplied to the primary winding, the transformer is said to operate under load. The voltage applied across the primary remains practically constant (or should be kept-constant so long as the test is conducted).

Circuit/Connection diagram : Refer Fig. 8.

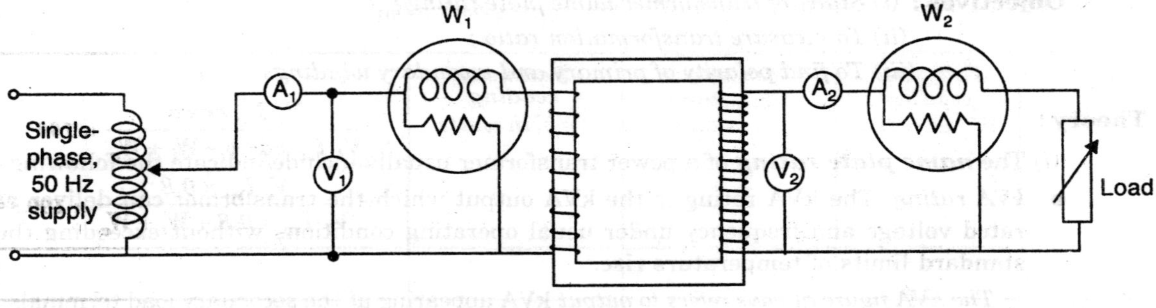

Fig. 8

Procedure :

1. Make the connections as shown in Fig. 8, keeping the auto-transformer in zero position and all switches of load in off-position.

2. Switch on the A.C. supply and then vary the voltage upto rated voltage.
3. Now, start loading the transformer by putting on some of the load switches so that a suitable current is obtained.
4. Note down the readings of the instruments.
5. Change the load and note down different readings.
6. Switch off the supply.

Observations :

S. No.	*Primary side*			*Secondary side* *open circuit voltage* = ${}_0V_2$			*Efficiency,* $\eta = \frac{W_2}{W_1}$
	Voltage V_1	*Current* I_1	*Power* W_1	*Voltage* V_2	*Current* I_2	*Power* W_2	
1.							
2.							
3.							
4.							
5.							

Calculations : (*i*) $\% \ \eta = \dfrac{\text{Output power}}{\text{Input power}} \times 100 = \dfrac{W_2}{W_1} \times 100$

(*ii*) % Voltage regulation $= \dfrac{{}_0V_2 - V_2}{V_2} \times 100.$

EXPERIMENT NO. 6

Study of Transformer Name Plate Rating, Determination of Transformation Ratio and Polarity.

Objectives : (*i*) *Study of transformer name plate rating ;*
(*ii*) *To measure transformation ratio ;*
(*iii*) *To find polarity of primary and secondary windings.*

Theory :

(*i*) The ***name plate rating*** of a power transformer usually include/indicate the following :

- ***kVA rating.*** The kVA rating is the kVA output which the transformer can deliver at rated voltage and frequency under usual operating conditions without exceeding the standard limits of temperature rise.

  The kVA figure always refers to output kVA appearing at the secondary load terminals ; the input kVA, of course, is slightly higher because of internal losses (core and winding losses).

- *Voltage ratio* = Rated primary voltage/secondary voltage or V_1/V_2. *Rated secondary voltage* is the voltage that appears across the secondary terminals *when rated current flows.*

Rated primary voltage is equal to the rated secondary voltage multiplied by the turn-ratio between primary and secondary *i.e.*, V_1 (rated) = V_2 (rated) × $\frac{N_1}{N_2}$.

- *Frequency.*
- *Number of phases / Type (1-phase or 3-phase).*
- *Equivalent impedance (%).*

(*ii*) ***Transformation ratio.***

For an ideal transformer $E_1 = V_1$ and $E_2 = V_2$

(where $E_1 = 4.44 f \phi N_1$ and $E_2 = 4.44 f \phi N_2$)

$$\therefore \quad \text{Transformation ratio } (K) = \frac{E_2}{E_1} = \frac{V_2}{V_1} = \frac{N_2}{N_1}.$$

(*iii*) ***Polarity test.*** A polarity test is carried out to find out the ***terminals having the same instantaneous polarity*** assuming that the terminals are not marked.

Circuit/Connection diagram : Refer Fig. 9.

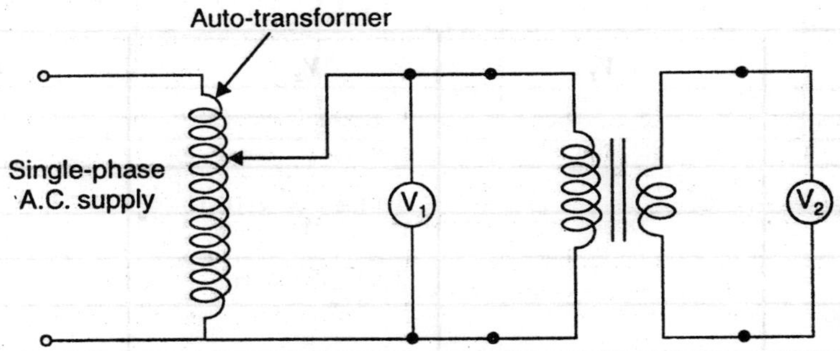

Fig. 9. Transformation ratio test.

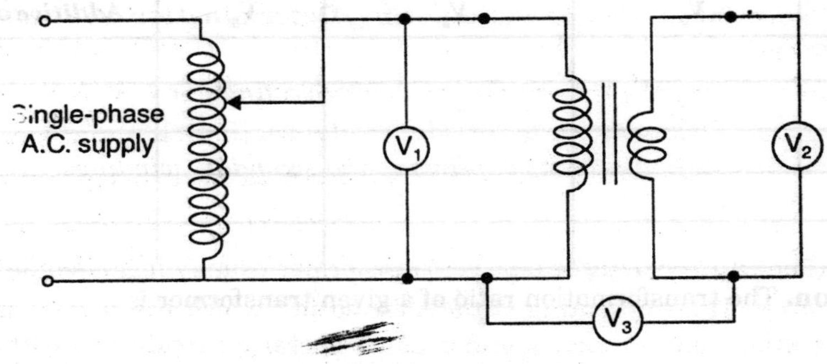

Fig. 10. Polarity test.

Procedure :

Transformation ratio test :

1. Make the connections as shown in Fig. 9. Set the auto-transformer to zero position.
2. Switch-ON the supply and adjust the transformer till a suitable (below rated) voltage is obtained.

3. Note down the voltages V_1 (across the primary) and V_2 (across the secondary).
4. Take a number of readings by varying the auto-transformer and note them in the observation table.
5. Switch off the supply.

Polarity test :

1. Make the connections as shown in Fig. 10 and set the auto-transformer to zero position.
2. Switch-ON the supply.
3. Note down the voltages V_1, V_2 and V_3. If V_3 is more than V_1 the polarity is additive, otherwise subtractive.
4. Take a number of readings by varying the varying the auto-transformer and note down in the observation table.
5. Switch-off the supply.

Observations :

Transformation ratio test :

S. No.	V_1	V_2	$K = \frac{V_2}{V_1}$
1.			
2.			
3.			
4.			
5.			

Polarity test :

S. No.	V_1	V_2	V_3	*Additive or subtractive*
1.				
2.				
3.				
4.				
5.				

Conclusion. The transformation ratio of a given transformer is ... .

EXPERIMENT NO. 7

Objective : *To connect three identical single phase transformers for three phase power transformation through following connections :*

(*a*) *Star-delta* (*b*) *Star-star*

(*c*) *Delta-star* (*d*) *Delta-delta*

Material and Equipment Required :

1. Identical single phase transformers .... 3 No.
2. Moving-iron type voltmeters .... 4 No.
3. TPIC swith .... 1 No.
4. Connecting leads.

Theory :

Virtually all power distribution is by poly-phase system of voltages. Three-phase transformations may be made with the use of properly connected single-phase transformers. These connections are in extensive commercial use. The most frequently used connections are the following :

(*i*) Primary *Y*—secondary *Y*. (Fig. 12)
(*ii*) Primary Δ—secondary Δ. (Fig. 14)
(*iii*) Primary Δ—secondary *Y*, *or vice versa*. (Fig. 11 and 12)
(*iv*) Primary and secondary open Δ.
(*v*) Primary *T* secondary *T* (Scott connection).

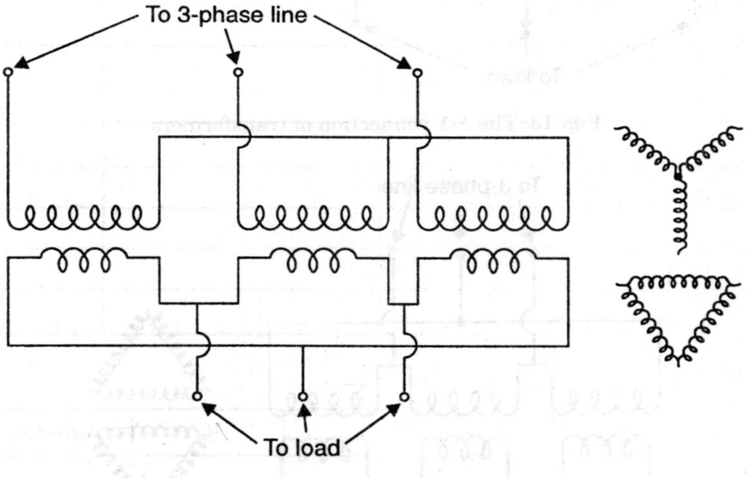

Fig. 11. The *Y*-Δ connection.

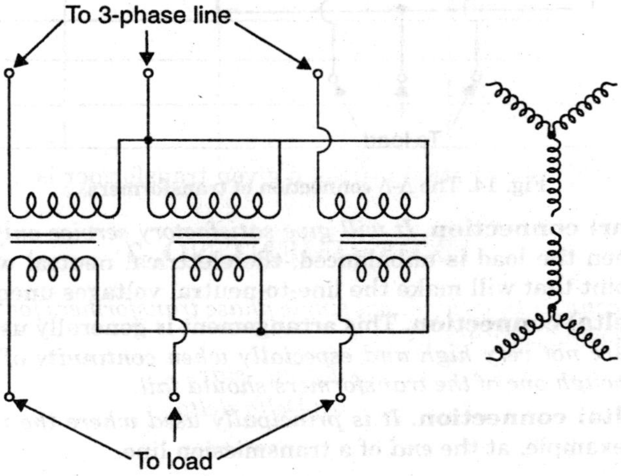

Fig. 12. The *Y*-*Y* connection of transformers.

Thus the most common connections are Y-Y, Δ-Δ, Y-Δ, Δ-Y, open delta or V-V and Scott connection or T-T connection.

Refer Figs. 11 to 14.

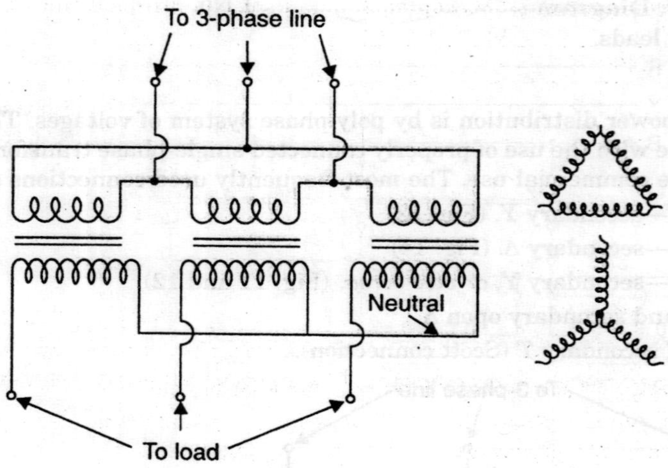

Fig. 13. The Δ-Y connection of transformers.

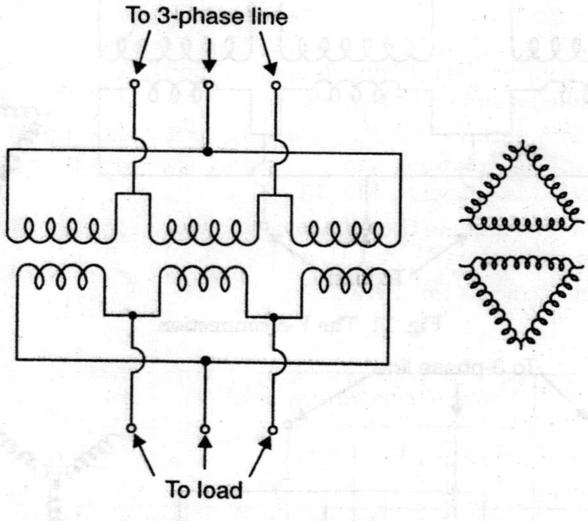

Fig. 14. The Δ-Δ connection of transformers.

- **Y-Y (star-star) connection.** *It will give satisfactory service only if three-phase load is balanced.* When the load is unbalanced, the electrical neutral will shift from its exact centre to a point that will make the line-to-neutral voltages unequal.
- **Δ-Δ (delta-delta) connection.** This arrangement is generally used in systems in *which the voltages are not very high and especially when continuity of service must be maintained even though one of the transformers should fail.*
- **Y-Δ (star-delta) connection.** *It is principally used where the voltage is to be stepped down,* as for example, at the *end* of a transmission line.

- **Δ-Y (delta-star) connection.** This type of connection is employed where it is necessary to *step-up the voltage,* as for example, at the *beginning* of a high-tension transmission system.

Circuit/Connection Diagram :

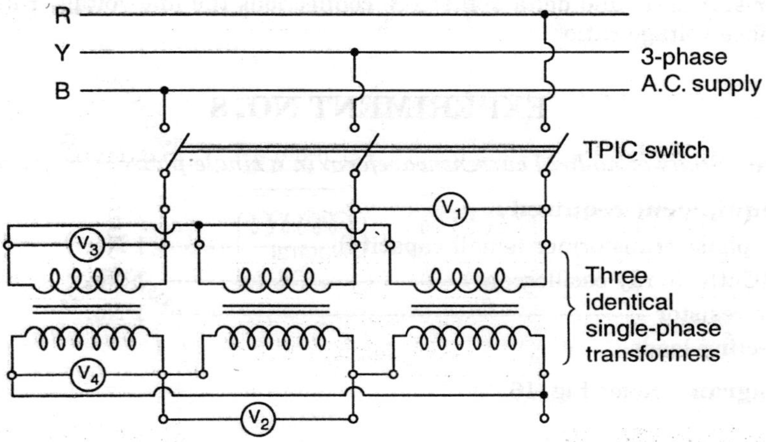

Fig. 15

Procedure :

1. Connect three identical single phase transformers in star on primary side and in delta on secondary side as shown in Fig. 15.
2. Connect the voltmeters V_1 and V_2 across *line terminals* on primary and secondary sides respectively and voltmeters V_3 and V_4 across *phase terminals* on primary and secondary sides respectively, as shown in Fig. 15.
3. Connect the transformers to the 3-phase A.C. supply through TPIC switch and note down the readings of voltmeters V_1, V_2, V_3 and V_4.
4. Repeat the same process for delta-star and delta-delta connections.

Observations :

Connections	***Voltmeters readings in volts***				***Line voltage ratio V_2/V_1***	***Phase voltage ratio V_4/V_3***	***Relationship between V_2/V_1 and V_4/V_3***
	V_1	V_2	V_3	V_4			
Star-delta (*Y*-Δ) connections							$\frac{V_2}{V_1}=\frac{1}{\sqrt{3}}\frac{V_4}{V_3}$
Star-star (*Y*-*Y*) connections							$\frac{V_2}{V_1}=\frac{V_4}{V_3}$
Delta-star (Δ-*Y*) connections							$\frac{V_2}{V_1}=\sqrt{3}\frac{V_4}{V_3}$
Delta-delta (Δ-Δ) connections							$\frac{V_2}{V_1}=\frac{V_4}{V_3}$

Conclusion :

- In star-delta (Y-Δ) connections the line voltage ratio is $\frac{1}{\sqrt{3}}$ time phse voltage ratio whereas in delta-star connections the line voltage ratio is $\sqrt{3}$ times phase voltage ratio.
- In star-star (Y-Y) and delta-delta (Δ-Δ) connections the line voltage ratio is the same as the phase voltage ratio.

EXPERIMENT NO. 8

Objective : *Study of no-load current waveform in a single-phase transformer using CRO.*

Material and equipment required :

1. Single-phase transformer (small capacity) -------------------- 1 No.
2. CRO (Cathode ray oscilloscope) ------------------------------- 1 No.
3. Purely resistor --- 1 No.
4. Connecting leads.

Connection diagram : Refer Fig. 16

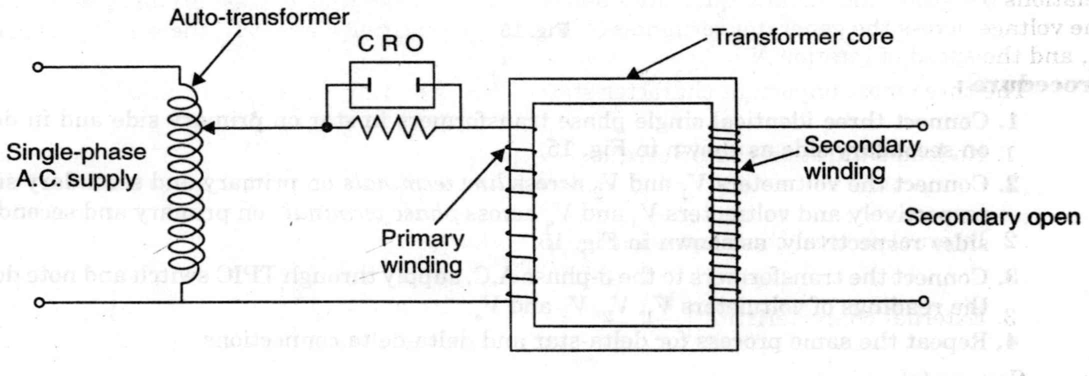

Fig. 16

Theory :

The no-load current flowing in the transformer lags the induced e.m.f. by an angle *slightly less than 90°. The saturation non-linearity results in a family of odd harmonics components* in the no-load or exciting current. Consequently, *the no-load current has a waveform which is non-sinusoidal in nature ; this can be displayed on a CRO.*

Procedure :

1. Connect the primary of the transformer to the single-phase supply with a rheostat or purely resistor as shown in the Fig. 10. Set up the auto-transformer to zero position.
2. Switch-ON the supply, and vary the voltage by auto-transformer.
3. Observe the waveform on CRO.

Conclusion :

- The current wave is non-sinusoidal and peaky while odd symmetry is maintained.
- The current and flux variations occur simultaneously.

EXPERIMENT NO. 9

No-load Saturation Characteristics (O.C.C.) of a D.C. Shunt Generator.

Objectives : (*i*) *To determine the effects of field current* (I_f) *and speed on the e.m.f. induced in the armature of the generator under no-load condition.*

(*ii*) *To determine the critical field resistance.*

(*iii*) *To determine the maximum voltage developed by the generator on particular speed.*

Material and Equipment required :

1. Shunt generator -- 1 No.
2. Moving-coil ammeter -- 1 No.
3. Moving-coil voltmeter -- 1 No.
4. Rheostat -- 1 No.
5. Tachometer (digital) -- 1 No.
6. Connecting leads.

Theory :

The properties of generators are analysed with the aid of characteristics which give the relations between fundamental quantities determining the operation of a generator. These include the voltage across the generator terminals V, the field or exciting current I_f, the armature current I_a, and the speed of rotation N.

The three most important characteristics of D.C. generators are given below :

1. No load saturation characteristics $\left(\dfrac{E_0}{I_f}\right)$

2. Internal or total characteristics $\left(\dfrac{E}{I_a}\right)$

3. External characteristics $\left(\dfrac{V}{I}\right)$.

Separately excited generators :

- Fig. 17 shows the connections of a separately excited generator, a battery being indicated as the source of exciting current, although any other constant voltage source could be used.

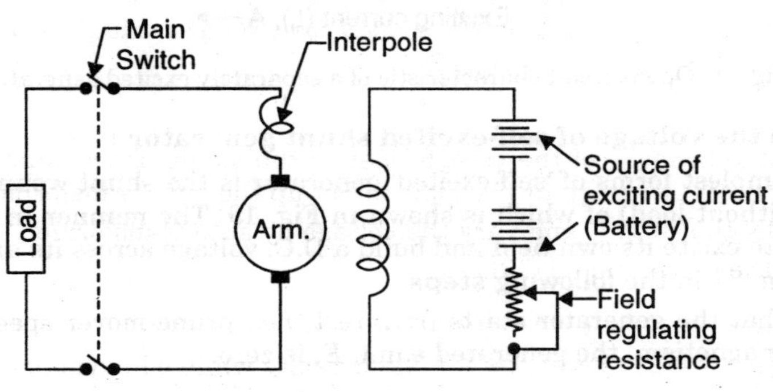

Fig. 17. Connection for a separately excited generator.

The field circuit is provided with a variable resistance and would normally contain a field switch and an ammeter, these being omitted from the diagram for simplicity. The armature is connected through 2-pole main switch to the bus bars, between which the load is connected.

No-load saturation characteristics (or O.C.C.) :

- If the generator is run at constant speed with the main switch open, and the terminal voltage is noted at various values of exciting or field current then the O.C.C. shown in Fig. 18 can be plotted. This is also referred to as the '*magnetisation curve*' since the same graph shows, to a suitably chosen scale, the amount of magnetic flux, there being a constant relationship (depending upon speed of rotation) between flux and induced voltage.
- It will be noticed that a small voltage is produced when the field current is zero, this being due to a small amount of permanent magnetism in the field poles. This is called *residual magnetism* and is usually sufficient to produce 2 or 3 per cent of normal terminal voltage, although in some special cases it is purposely increased to 10 per cent or more.
- The *first part of the curve is approximately straight* and *shows that the flux produced is proportional to the exciting current* ; but *after a certain point, saturation of the iron becomes perceptible as the curve departs from straight line form.*

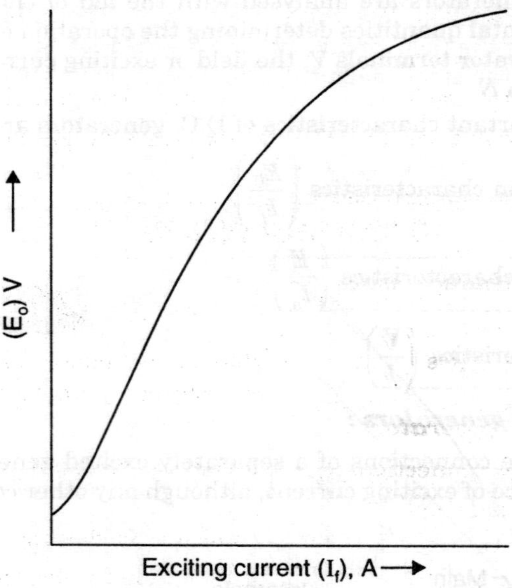

Fig. 18. Open-circuit characteristic of a separately excited generator.

Building up the voltage of self-excited shunt generator :

One of the simplest forms of 'self-excited' generator is the shunt-wound machine, the connection diagram (without load) of which is shown in Fig. 19. The manner in which a self-excited generator manages to excite its own field and build a D.C. voltage across its armature is described with reference to Fig. 31 in the following **steps** :

(*i*) Assume that the generator starts from rest, *i.e.*, prime-mover speed is zero. Despite a residual magnetism, the generated e.m.f. *E*, is zero.

(*ii*) As the prime-mover rotates the generator armature and the speed approaches rated speed, the voltage due to residual magnetism and speed increases.

(*iii*) At rated speed, the voltage across the armature due to residual magnetism is small, E_1, as shown in the figure. But this voltage is also across the field circuit whose resistance is R_f. Thus, the current which flows in the field circuit I_1, is also small.

(*iv*) When I_1 flows in the field circuit of the generator of Fig. 19, an increases in m.m.f. results (due to $I_f T_f$, T_f being field turns, which aids the residual magnetism in increasing the induced voltage to E_2 as shown in Fig. 20).

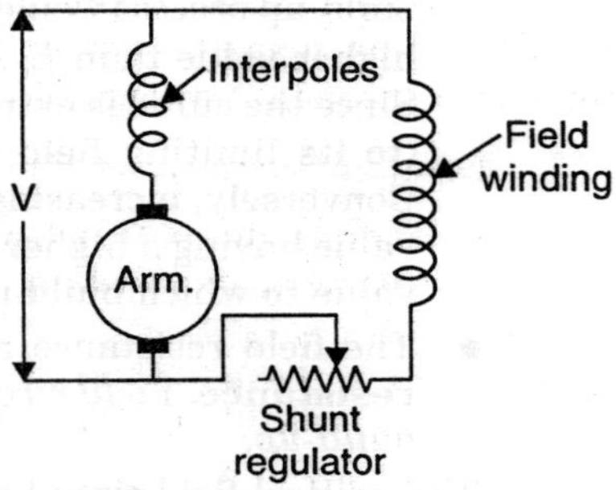

Fig. 19. Self-excited shunt generator.

(*v*) Voltage E_2 is now impressed across the field, causing a large current I_2 to flow in the field circuit. $I_2 T_f$ is an increased m.m.f., which produces generated voltage E_3.

(*vi*) E_3 yields I_3 in the field circuit, producing E_4. But E_4 causes I_4 to flow in the field producing E_5 ; and so on, up to E_8, the maximum value.

(*vii*) The *process continues until that point where the field resistance line crosses the magnetisation curve* in Fig. 20. Here the process stops. The induced voltage produced, when impressed across the field circuit, produces a current flow that in turn produces an induced voltage of the same magnitude, E_8, as shown in the figure.

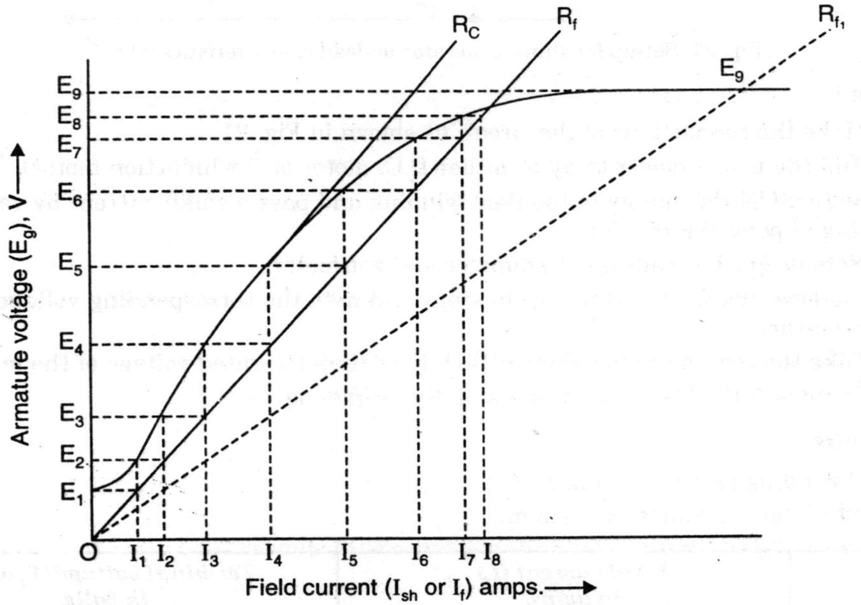

Fig. 20. Building up the voltage of a shunt generator.

Critical resistance :

- In the above description a particular value of field resistance R_f was used for building up of self-exited shunt generator. If the field resistance were reduced by means of adjusting the field reheostat of Fig. 19 to a lower value say R_{f_1}, shown in Fig. 20, the build-up process would take place along field resistance line R_{f_1}, and build-up a somewhat higher value than E_8, *i.e.*, the point where R_{f1} intersects the magnetisation curve, E_9. Since the curve is extremely saturated in the vicinity of E_9, reducing the field resistance (to its limiting field winding resistance) will not increase the voltage appreciably. Conversely, increasing the field rheostat resistance and the field circuit resistance (to a value having a higher slope than R_f in the figure) will cause a reduction of the maximum value to which build-up can possibly occur.

- The field resistance may be increased until the field circuit reaches a **critical field resistance.** *Field circuit resistance above the critical field resistance will fail to produce build-up.*

This critical field circuit resistance, R_c, is shown as *tangent to the saturation curve* passing through the origin, *O*, of the axes of the curve of Fig. 20. Thus a field circuit resistance higher than R_c will produce an armature voltage of E_1 approximately (and no more).

Circuit/Connection diagram :

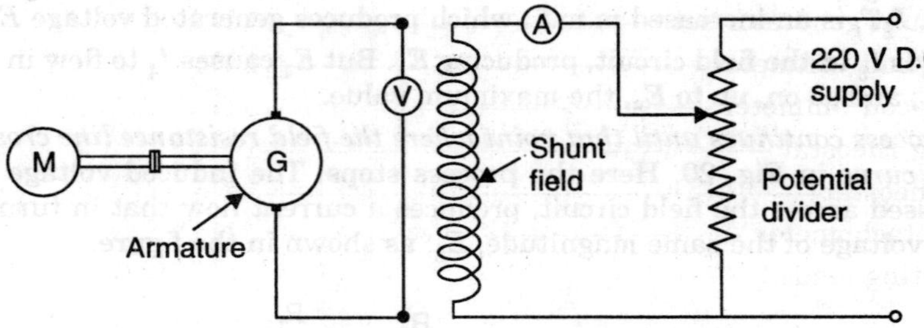

Fig. 21. Set-up for shunt generator no-load characteristics (O.C.C.).

Procedure :

1. Make the connections of the circuit as shown in Fig. 21.
2. Run the prime-mover (may be either D.C. motor or 3-ϕ induction motor).
3. Switch-ON the supply to the field winding and pass a small current by varying the setting of potential divider.
4. Note down the readings of ammeter and voltmeter.
5. Increase the field current (I_f) in steps and note the corresponding voltage(s) across the armature.
6. Take the reading upto values little higher than the rated voltage of the machine.
7. Switch-off the D.C. supply and stop the prime-mover.

Observations :

Field winding resistance = ... ohms

Speed of the machine(s) = ... r.p.m.

S. No.	*Field current (I_f) in amps.*	*Terminal voltage (V_t or E_o) in volts.*
1.		
2.		
3.		
4.		
5.		

- Plot a graph between (V_t or E_0) and I_f, which represents the no-load characteristics (O.C.C.) of the generator.
- Draw the field resistance line on the same graph, the point at which resistance line cuts the O.C.C. curve gives the maximum voltage built up by the generator.
- Draw a tangent line to the curve, the slope of this tangent will be the critical resistance of the generator.

EXPERIMENT NO. 10

Load Characteristics of D.C. Shunt Generator.

Objective : *To obtain and plot the external characteristics of a D.C. shunt generator and to deduce the internal characteristics from the external characteristics.*

Material and equipment required :

1. D.C. shunt generator set -- 1 No.
2. Moving-coil voltmeter -- 1 No.
3. Moving-coil ammeter -- 1 No.
4. Loading rheostat -- 1 No.
5. Field rheostat --- 1 No.
6. Digital tachometer -- 1 No.
7. Connecting leads.

Theory :

In a shunt generator, the field circuit is connected directly across the armature. Appliances, motors, light bulbs, and other electrical devices connected in *parallel* across the generator terminals represent a *load* on the generator. As more devices are connected in parallel, the load on the generator increases ; that is, the generator current increases. Because the generator current increases, the terminal voltage of the generator decreases. There are three factors that cause this decrease in voltage :

(*i*) Armature-circuit resistance (R_a),
(*ii*) Armature reaction, and
(*iii*) Reduction in field current.

(*i*) **Armature-circuit resistance.** The armature circuit of a generator, like every electrical circuit, contains resistance. This resistance includes the resistance of (*i*) *the copper conductors of the armature winding, (ii) the commutator, (iii) contact resistance between brushes and commutator*, and (*iv*) *the brushes themselves.* When no current flows through the armature, there is no *IR* drop in the armature and the voltage at the terminals is the same as the generated voltage. However, when there is current in the armature circuit, a voltage drop exists due to the armature resistance, and the terminal voltage is less than the generated voltage. The terminal voltage may be calculated from the following reaction :

$$V = E_g - I_a R_a$$

where V = voltage at terminals of generator,
E_g = generated or induced voltage,
I_a = total armature current, and
R_a = armature-circuit resistance.

(*ii*) **Armature reaction.** When current flows in the armature conductors a flux surrounds these conductors. The direction of this armature flux is such that it reduces the flux from the field poles, resulting in both a reduced generated voltage and terminal voltage.

(*iii*) **Reduction in the field current.** The field circuit is connected across the terminals of the generators. When the terminal voltage of the generator becomes smaller because of the armature-resistance volt drop and armature reaction, the voltage across the field circuit also becomes smaller and therefore field current will be less. A reduction in the magnitude of field current also reduces the flux from the field poles, which in turn reduces the generated voltage and also the terminal voltage.

External Characteristics :

- See Figs. 22 and 23. The effect of the preceding three factors is shown in Fig. 23, which shows external (load-voltage) characteristic of a shunt generator.

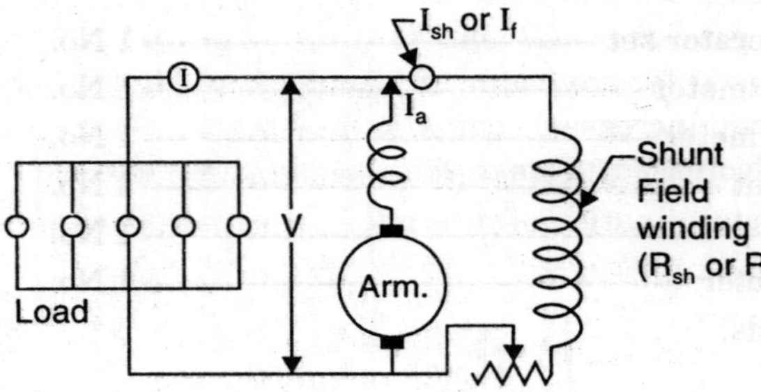

Fig. 22. Shunt generator under load.

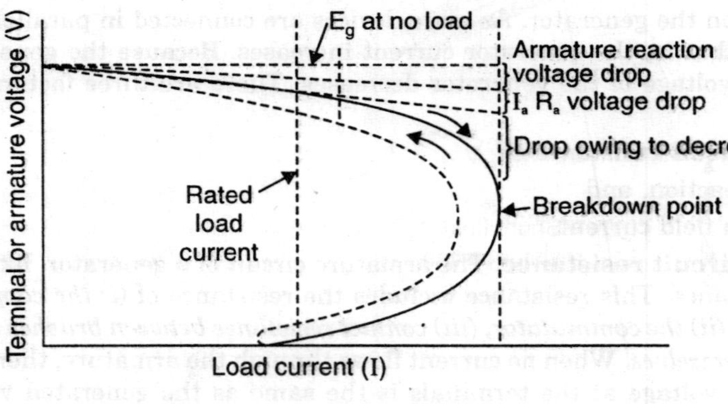

Fig. 23. Shunt generator load characteristics.

- As shown in the circuit of Fig. 22, the readings of the voltage across the armature (and load), V are plotted as a function of load current, I. The voltage, V, is the same as E_g at no load (neglecting the I_aR_a and armature reaction drop produced by the field current). The effects of armature reaction, armature circuit voltage drop, and decrease in field current are all shown with progressive increase in load. Note that both the armature reaction and the I_aR_a drops are shown as dashed straight lines, representing theoretically linear voltage directly proportional to the increase in load current. The drop owing to decreased field current is a curved line, since it depends on the degree of saturation existing in the field at the value of load.

- Generally, the external load-voltage characteristic decreases with application of load only to *a small extent up to its rated load* (current) *value*. Thus, the shunt generator is considered as having a fairly constant output voltage with application of load, and in practice, is *rarely operated beyond the rated load current value continuously for any appreciable time.*
- As shown in Fig. 23 *further application of load causes the generator to reach a breakdown point beyond which further load causes it to 'unbuild' as it operates on the unsaturated portion of its magnetisation curve. This unbuilding process continues until the terminal voltage is zero, at which point the load current is of such magnitude that the internal armature circuit voltage drop equals the e.m.f. generated on the unsaturated or linear portion of its magnetisation curve.*
- It may be noted that if the *external load is decreased* (an increase of external load resistances), the *generator will tend to build up gradually along the dashed line* shown in Fig. 23. Note that for any value of load current, the terminal or armature voltage is less (as the voltage increases) compared to the solid lines which yield a higher voltage (as the voltage decreases). This *difference is due to hysteresis.*

Fig. 24 shows the determination of internal characteristics from external characteristics.

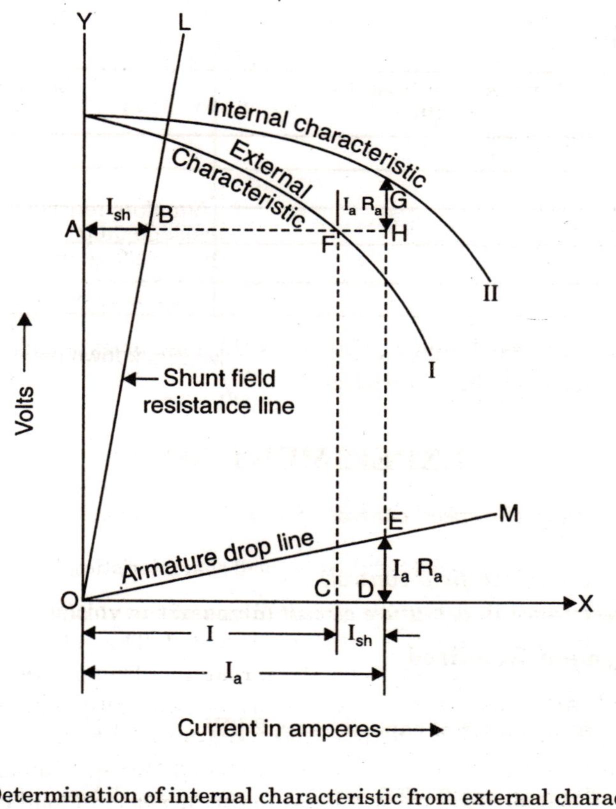

Fig. 24. Determination of internal characteristic from external characteristic.

Circuit/Connection diagram :

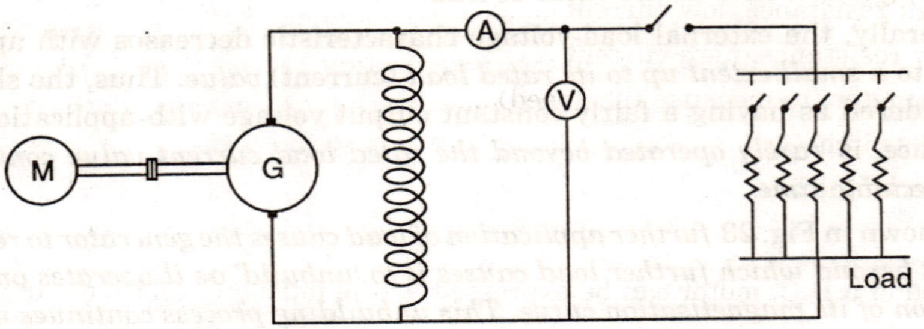

Fig. 25. Set-up for shunt generator load characteristics.

Procedure :

1. Make the connections as shown in Fig. 25.
2. Run the prime-mover, and increase the load current by operating the switches in the load resistance in five steps.
3. Note down the readings of the load current and the termianl voltage. Speed should be maintained constant for each set of reading.
4. Switch-OFF the supply to stop the prime-mover.

Observations :

Speed = ... r.p.m.

S. No.	*Terminal voltage (V) in volts*	*Load current (I) in amps.*
1.		
2.		
3.		
4.		
5.		

- Plot a graph between *V* and *I* ; this gives the load or external characteristics.
- Deduce the internal characteristics.

EXPERIMENT NO. 11

Objective : *To study the speed control of a D.C. shunt motor and draw the speed variation with respect to :*

(a) change of field current (field control) ;

(b) change of resistance in armature circuit (armature control).

Material and Equipment Required :

1. 250 V D.C. supply.
2. 250 V D.C. shunt motor of capacity say, 2 kW .... 1 No.
3. Ammeters :
   - PMMC of range, say 0—2.5 A ; .... 1 No.
   - PMMC of range, say 0—10 A .... 1 No.

4. Rheostats :
   - High resistance, low current .... 1 No.
   - Low resistance, high current .... 1 No.
5. Tachometer (to measure the speed) .... 1 No.
6. DPIC switch .... 1 No.
7. Connecting leads.

Theory :

The speed of a D.C. motor can be expressed by the following relationship :

$$N \propto \frac{V - I_a R_a}{\phi} \quad \text{or} \quad N = K\,\frac{(V - I_a R_a)}{I_{sh}}$$

Hence the speed of D.C. shunt motor can be changed by either *changing the shunt field current* I_{sh} by inserting an external resistance in the field circuit $\left(i.e.,\ I_{sh} = \dfrac{V}{R_{sh} + R}\right)$ or *changing the back e.m.f.* E_b $(i.e.,\ V - I_a\,(R_a + R_{as}))$ by inserting an external resistance in the armature circuit.

Circuit/Connection Diagram :

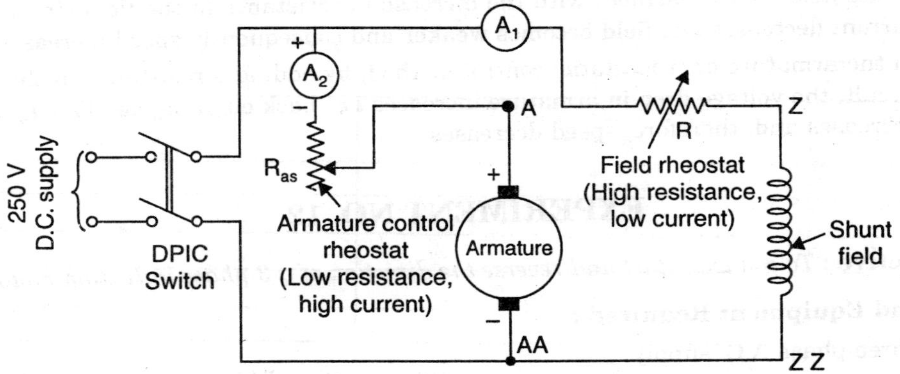

Fig. 26. Circuit diagram.

Procedure :

1. Make the connections as shown in the Fig. 7.
2. Bring the rheostats in zero positions.
3. Switch on the D.C. supply through DPIC switch.

Field control :

4. Note down the reading of ammeter A_1 and measure speed of the motor by tachometer.
5. Change the *field rheostat* and setting in steps and note down the readings of ammeter A_1 and measure and record the speed of the motor at *each step.*
6. Bring the field rheostat reading back to zero.

Armature control :

7. Increase the resistance in steps in the armature circuit through *armature control rheostat.*

8. Note down the readings of ammeter A_2 and measure and record the speed of the motor *at each step*.

9. Bring the armature control rheostat back to zero and switch off the supply through DPIC switch.

Observations :

S. No.	*Field control*		*Armature*	
	Ammeter A_1 readings in amps., I_{sh}	*Speed of the motor in r.p.m.*	*Ammeter A_2 readings in amps., I_a*	*Speed of the motor in r.p.m.*
1.				
2.				
3.				
4.				
5.				

Conclusion :

1. In the *field control method*, with the increase in resistance in the field circuit, the field current decreases *i.e.*, field becomes weaker and consequently speed increases.
2. In the armature or (rheostatic) control method, by adding a resistance in the armature circuit, the voltage drop in armature increases *i.e.*, back e.m.f. E_b *i.e.*, $[V - I_a (R_a + R_{as})]$ decreases and, therefore, speed decreases.

EXPERIMENT NO. 12

Objective : *To connect, start and reverse the direction of a 3-phase induction motor.*

Material and Equipment Required :

1. Three-phase A.C. supply.
2. Three-phase induction motor ...... 1 No.
3. Star-delta starter ...... 1 No.
4. TPIC switch ...... 1 No.
5. Tools such as screw driver, plier, test-pin etc.
6. Connecting leads.

Theory :

Motor is connected to the mains through a star-delta starter and a TPIC switch. *The direction of rotation can be reversed by interchanging any two terminals at the TPIC switch.*

Circuit/Connection Diagram :

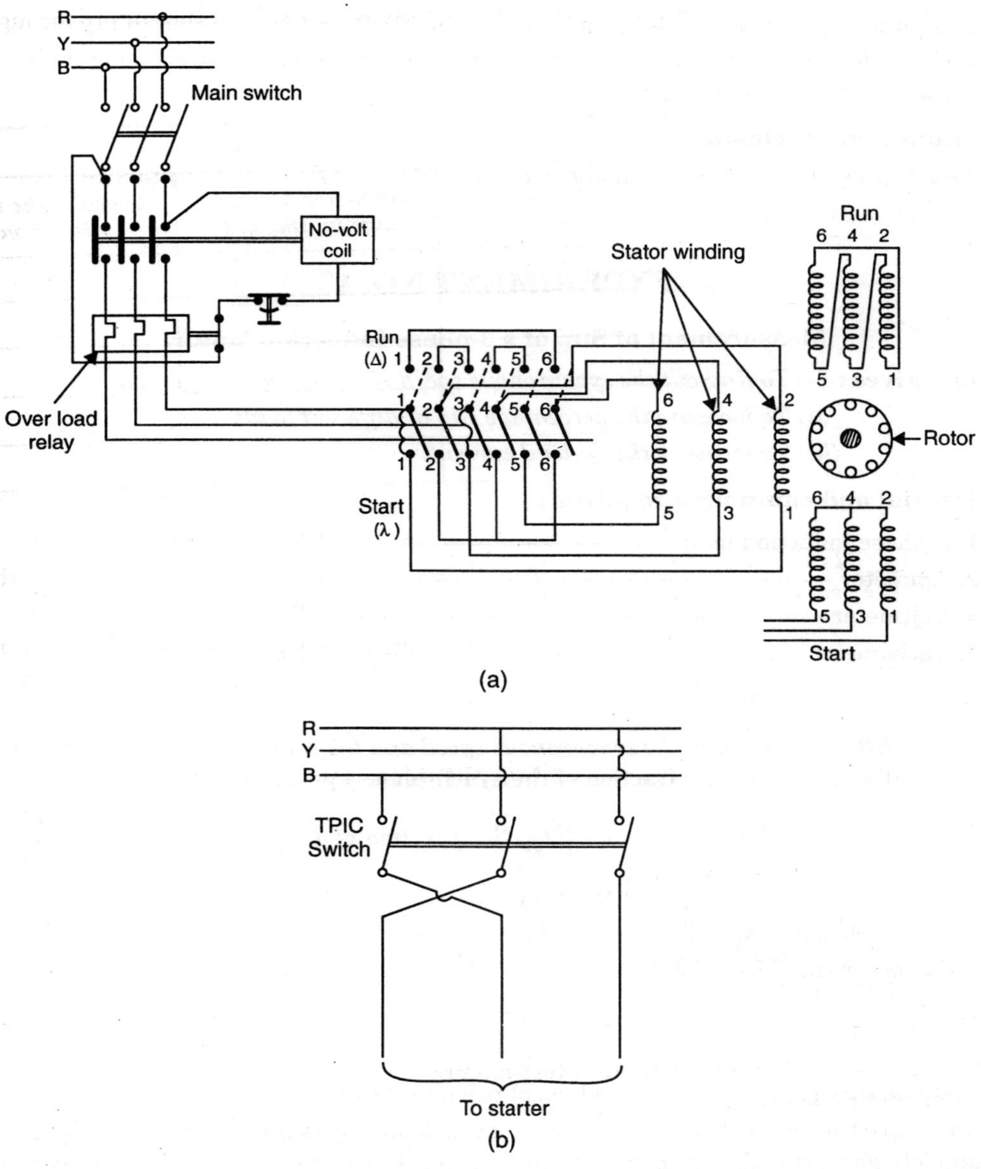

Fig. 27

Procedure :

1. **Make the connections to the motor through TPIC switch and star-delta starter as shown in Fig. 27 (*a*).**
2. **Close the TPIC switch and start the motor by taking the lever of the starter to the start (star) position and then with a jerk to the run (or delta) position.**
3. ***Observe the direction of rotation of the motor.***

4. Stop the motor by pushing the stop button and remove the supply to the motor by opening the TPIC switch.
5. Interchange the two leads of the motor to the TPIC switch as shown in Fig. 27 (*b*).
6. Close the TPIC switch and start the motor again. Observe the direction of the motor.
7. Switch-off the motor and supply.

Observations and conclusion :

The *direction of rotation of the motor in the second case is found to be* **opposite** *to that in the first case.*

EXPERIMENT NO. 13

Measurement of Slip of a 3-phase Induction Motor.

Objectives : (*i*) *To find out the synchronous speed N_s of the induction motor.*
(*ii*) *To measure the percentage slip at different loads.*
(*iii*) *To measure the no-load current.*

Material and equipment required :

1. 3-phase induction motor -- 1 No.
2. Ammeter -- 1 No.
3. Voltmeter -- 1 No.
4. Tachometer -- 1 No.

Theory :

- *The difference between the synchronous speed and the rotor speed is known as* **'slip'.** It is usually expressed as a fraction of the synchronous speed. Thus slip s is

$$s = \frac{N_s - N}{N_s} \quad ...(i)$$

or

$$N = N_s(1 - s) \quad ...(ii)$$

where N_s = synchronous speed (r.p.m.), and
N = motor speed (r.p.m.).

$$N_s = \frac{120f}{p} \quad ...(iii)$$

where f and p are frequency of supply and no. of poles respectively.

In practice the value of slip is very small. At no-load, slip is around 1% or so and at full-load it is around 3%. For large efficient machines the slip at full-load may be around 1% only. The induction motor, is therefore, a motor with substantially constant speed and fills the same role as *D.C. shunt motor*.

- When the rotor is stationary (standstill) its speed is zero and s = 1. The rotor cannot run at synchronous speed because then there will be no rotor e.m.f. and no rotor current and torque. If the rotor is to run at synchronous speed an external torque is necessary. *If the rotor is driven such that $N > N_s$, the slip becomes negative, the rotor torque opposes the external driving torque and the* **machine acts as induction generator.**